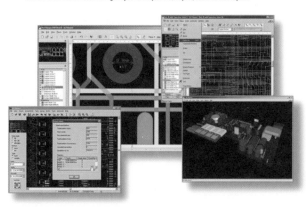

Essentials of Circuit Analysis

Robert L. Boylestad

PEARSON
Prentice
Hall

Upper Saddle River, New Jersey
Columbus, Ohio

Library of Congress Cataloging-in-Publication Data

Boylestad, Robert L.
    Essentials of circuit analysis / Robert L. Boylestad.
        p. cm.
    ISBN 0-13-061655-9
        1. Electronic circuits. 2. Electronic circuit design. 3. Electric circuit analysis. I. Title.

TK7867.B645 2004
621.3815—dc21

2003043364

Editor in Chief: Stephen Helba
Acquisitions Editor: Dennis Williams
Development Editor: Kate Linsner
Production Editor: Rex Davidson
Text Designer: Rebecca M. Bobb
Cover Art: Painting by Sigmund Årseth, artist and teacher, Valdres, Norway
Design Coordinator: Karrie Converse-Jones
Cover Designer: Ali Mohrman
Production Manager: Patricia A. Tonneman
Marketing Manager: Ben Leonard

This book was set in Times Roman by Carlisle Communications, Ltd., and was printed and bound by Courier
Kendallville, Inc. The cover was printed by Phoenix Color Corp.

Cadence and the Cadence logo, OrCAD, PSpice, and OrCAD Capture are registered trademarks of Cadence
Design Systems, Inc.

Electronics Workbench and Multisim are registered trademarks of Electronics Workbench, Inc.

Pearson Education Ltd.
Pearson Education Singapore Pte. Ltd.
Pearson Education Canada, Ltd.
Pearson Education—Japan

Pearson Education Australia Pty. Limited
Pearson Education North Asia Ltd.
Pearson Educación de Mexico, S.A. de C.V.
Pearson Education Malaysia Pte. Ltd.

10 9 8 7 6 5 4 3 2 1
ISBN: 0-13-061655-9

Current and past users of *Introductory Circuit Analysis* (ICA), now in its tenth edition, have a right to wonder why I have chosen to write another text on essentially the same subject matter. The reason lies in my belief that there has been a growing need in recent years for a text dedicated primarily to those concepts that a graduate of this program of study must retain in order to be successful in the industrial community. In other words, a text is needed that has an increased measure of detail in specific important areas to ensure a clear, correct understanding of the most important laws and concepts, with less concern for special cases and material of less importance. The goal has been to create a text with a more practical orientation to better prepare the student for the laboratory and real-world experience. Admittedly, in comparing the tables of contents of the two texts, one can immediately see that there is a close correspondence in content (although numerous sections are moved and a number of chapters are combined). This correspondence, however, should not suggest that this text is merely a cut-and-paste revision of the ICA text. In fact, as I worked through the copyedited pages of this new text, I realized how little of the original ICA presentation remains. Except for the practical examples and the computer coverage, the books are very different. If you examine any section of particular interest, you will find that the pace, level of presentation, and content of *Essentials of Circuit Analysis* are all different from those of the ICA text.

From the very beginning, I decided that any cut-and-paste approach simply would not work. I examined each topic, decided what was really important, and wrote the corresponding sections in almost exactly the same way that I might teach the subject in the classroom. I believe that the differences between ICA and this text are evident immediately in just leafing through the pages. *Essentials of Circuit Analysis* truly has an exciting new appearance that invites examination and further investigation.

This text includes the actual construction of numerous networks to help define how a circuit diagram is drawn. Meters are also included throughout the text to maintain a close link with real-world experience. Methods of analysis are simplified by removing concern about special cases. Controlled sources are *not* included, while subjects such as Bode plots are included but are only touched upon rather than covered in depth. Theorems such as the substitution theorem, Millman's theorem, and the reciprocity theorem are eliminated in favor of giving more coverage to more important concepts. As a developmental aid, a list of objectives is introduced at the beginning of each chapter, and a chapter summary list and an equation list are added to the end of each chapter. The format of the text is designed to ensure that students are aware of the concepts they should take away from the course. All of the artwork utilizes color and shading to clarify the analysis being described, and many new photos are added. Each problem section is written to complement the content and level of coverage presented, progressing from the simple to the complex within each section, with emphasis on developing a student's confidence before moving on to the more complex problems.

FEATURES

Important changes in content begin in Chapter 1 and progress throughout the entire text. Chapter 1 includes expanded coverage of the proper use of calculators. As confident as students might appear in their use of the calculator, they continue to generate impossible results because they do not know the correct operating sequences. Section 1.12 describes in detail the specifications provided with a computer. This general information about a computer system is provided so that users of a computer can understand its capabilities and its limitations.

Throughout the introductory chapters, new sections are added and older ones are deleted. The ampere-hour rating of a battery has its own section (Section 2.6), and a section is added (Section 2.7) to cover factors that affect the life of a battery. The coverage of voltage and current (Sections 2.3 and 2.4, respectively) are reversed to emphasize that it is the voltage that initiates the flow of charge in an electrical system, not vice versa. The use of the metric system to calculate body resistance is not included in this text in order to provide more coverage of the circular mil approach. A number of derivations are eliminated in favor of spending more time helping the student understand the conclusions and apply the resulting equations. Film resistors, rather than carbon composition resistors, are used throughout the text to reflect recent trends. Instrumentation appears as often as possible to prepare the student for both real-world and laboratory experiences. Analogies are also used whenever possible to clarify important concepts.

To ensure that students understand the process of creating network diagrams, the first few diagrams are derived from a drawing of the wired network. Ammeters and voltmeters are also applied to a drawing of the actual wired network to ensure a proper understanding of the use of these basic instruments. Too often, students simply apply the meters to a line drawing, and confusion and misunderstanding result.

In many ways, I feel that the chapters on series and parallel networks (Chapters 5 and 6, respectively) are two of the most important chapters because they expose the student to a number of the fundamental laws of electric circuits. These laws are carried throughout the remaining chapters of the text, and thus expanded coverage of each concept is provided through analogies, wired circuits, examples—whatever it takes to give the student a solid understanding of the principles involved. Chapter 6 on parallel networks is covered from a purely resistance viewpoint rather than emphasizing conductance levels. The use of conductance levels is covered in a later chapter, but early descriptions using the resistance approach are more in line with what I would normally cover in an actual lecture on the subject. In the same light, the chapter on series-parallel circuits (Chapter 7) also receives special attention because it provides a test of the concepts just learned in the previous two chapters. Single- and double-subscript notation is covered in detail but is relegated to a later section to be sure that the introductory sections are not clouded by this special notation.

Every effort is made to ensure that the methods of analysis (Chapter 8) initially appear friendly and less complex. The number of steps to apply each method is reduced, and the details associated with special cases are in a later section in Chapter 8. Each step in the application of each method is covered in detail, and the example problems are in line with the needs of the student. The concepts of supermesh and supernodes are placed in a later section to avoid unnecessary complexity in the early stages of development. An approximate method developed by this author is also introduced to provide an alternative to the more complex supermesh and supernode approaches. Although Δ-Y and Y-Δ conversions are covered, the derivations associated with each are not, permitting extended coverage of the application of each concluding equation.

In the theorem chapter (Chapter 9), the examples associated with the superposition theorem and Thévenin's theorem are carefully chosen to match the needs of the student, and thus these examples are not overly complex. The topic of maximum power transfer is treated in a simple manner, and topics of lesser importance have reduced coverage. Experimental techniques are discussed in detail to prepare the student for the laboratory experience. Norton's theorem is included because it provides an excellent opportunity to test the student's knowledge of current sources and the impact of a short circuit in a network.

The chapters on capacitors and inductors (Chapters 10 and 11, respectively) use a less mathematical approach. These chapters have extended coverage of the fundamental behavior of each component, the transient response, calculator usage, reading the nameplate data on the unit, and associated instrumentation. The general equation for the transient behavior of each component with initial conditions is developed at an early stage to provide a general equation for the discussions and examples to follow. Current and past users of the ICA text will find that the chapter on magnetic circuits is not included here. Many of the important topics covered in that chapter are now scattered throughout this text, but in less depth.

The content of the first chapter on the ac response (Chapter 12) is very similar to that of the ICA text, but you will find extended coverage of instrumentation and the use of the calculator. Further, in deference to industry usage, the subscript "rms" rather than the subscript "eff" is now used throughout. The following ac chapter (Chapter 13) represents a major change in pedagogy, with a beginning section devoted to adding and subtracting sinusoidal waveforms rather than a somewhat complex discussion of derivatives. The result is a more natural flow from the previous chapter. In addition, the student avoids a complex topic so early in the introduction to ac waveforms. Phasors are then introduced and applied to the basic elements. The entire discussion of the response of the capacitor and inductor to an ac signal is written using a less mathematical approach, with general statements that should be helpful to a student whenever working with capacitors or inductors. Although there is no power chapter in this text, the important concepts concerning power are introduced as needed in the ac section. Furthermore, the frequency response of the basic elements is treated at two levels—the ideal and the practical—with more general comments and less mathematics.

The basic chapter on series and parallel ac networks (Chapter 14) is written with more detail and with less dependence on letting the mathematics generate the solution. The approach to the frequency response of each configuration includes enhanced artwork and an expanded commentary rather than leaning on a purely mathematical derivation of the results. Parallel ac networks are treated in the same manner as appearing in the dc section, in order to build student confidence through repetition and to demonstrate the similarities that exist between solutions in the ac and dc domains. Phase measurements are covered in detail in Chapter 14 rather than interspersed throughout the ac chapters.

The content of the chapter on series-parallel ac networks (Chapter 15) is very similar to that in the ICA text, except that the power triangle is added here. The power triangle is an important component of any background in this field and deserves to be covered in some detail. The detail, however, is not overwhelming, with a few straightforward examples to demonstrate the important facets of the subject. The chapter on ac methods of analysis and theorems (Chapter 16) is a close match with the corresponding dc chapters except for the use of phasors and impedances in the analysis rather than just fixed sources and resistors. A major difference between this text and the ICA text is that this one does not contain sections that apply the methods and theorems to networks with controlled sources.

Resonance and filters are covered in a single chapter (Chapter 17) primarily because the majority of the material on Bode plots is removed. Most of the important conclusions associated with each subject are present, but derivations and special cases are avoided. The detailed discussion of decibels is included in its entirety in response to current ICA users who have found the material unique in its coverage.

Transformers and three-phase systems are combined into one chapter (Chapter 18), with content limited solely to the ideal iron-core transformer. Most of the important concepts associated with each topic appear in the chapter to establish a level of understanding that should prepare a student for his/her initial exposure to both areas in an industrial environment.

ANCILLARIES

The laboratory manual to support the text is very similar to that provided with the ICA text, with minor changes. The CD-ROM enclosed with this textbook contains 100 figures from the text rendered in Multisim® as well as the Enhanced Textbook Edition of Multisim 2001®, which is all that is required to open and work with all of the circuit files. However, anyone who wishes to order the full version of Multisim may do so through Prentice Hall by visiting **http://www.prenhall.com**, phoning 800-282-0693, or sending a fax request to 800-835-5327. Technical questions relating specifically to Multisim should be referred to Electronics Workbench directly by writing to **support @electronicsworkbench.com.**

In addition to the laboratory manual and the CD-ROM, a full package of ancillaries accompanies this text:

Instructor's Solutions Manual
Laboratory Solutions Manual
Prentice Hall TestGen, an electronic test bank
PowerPoint® Transparencies
Companion Website: **http://www.prenhall.com/boylestad**
Electronics Supersite: **http://www.prenhall.com/electronics**

ACKNOWLEDGMENTS

When I started this project, I had a vision of how the text should be written, how the material should be presented, and how Prentice Hall could help in the process. After a few false starts, the material developed quite rapidly and the pieces started to fall together. To be honest, I had initially hoped that more of the ICA material could be conserved, but in time it became clear that this was to be a major revision. Throughout the process, the support from Prentice Hall was wonderful. Rex Davidson, a good friend and someone I have worked with for many years, has wonderful control over all the elements that bring a text together. Kate Linsner, the development editor, had to suffer through all my frustration but managed somehow to keep everything moving and on an even keel. Maggie Diehl, the copy editor, managed to wade through a manuscript with numerous inserts and changes and raise its content to a level beyond one that simply bears some resemblance to the English language.

Since accuracy is extremely important to me, I want to thank Professor Franz Monssen for taking on the exhausting process of double-checking all the solutions in the text. In addition, I thank Professor Ron Kolody for his detailed review of the manuscript, resulting in some very important questions about content and direction. On the practical side I have to extend my deepest thanks to Mr. Jerry Sitbon for providing important comments and suggestions about the content and direction of a number of important areas of investigation. Ms. Kelly Barber is responsible for taking my handwritten solutions and putting them in the beautiful form you will find in the solutions manual. A number of the beautiful photographs are the result of the special talents of photographer Michael Gallitelli. Often forgotten is the marketing and sales process, and the extensive efforts of such people as Ben Leonard, Marketing Manager; Adam Kloza, Senior Marketing Coordinator; and David Gesell, Director of Marketing, Prentice Hall Career and Technology.

Finally, I thank the following reviewers of the manuscript for their helpful comments: Don Abernathy, DeVry University; Gerald W. Cockrell, Indiana State University; M. David Luneau, Jr., University of Arkansas at Little Rock; Larry Masten, Texas Tech University; Jack Noonan, DeVry University; James M. Rhodes, Blue Ridge Community College; Ross E. Sacco, College of Eastern Utah; Cecil Stuerke, DeVry University; and Paul T. Svatik, Owens Community College.

The art studio was amazing in its ability to take a rough sketch and generate some of the beautiful pieces you will find throughout the text. Then there is my editor, Dennis Williams, who was always there when important financial and schedule decisions had to be made. In all, it was an exhausting but rewarding effort, and one that came together because of the help that was available. All who provided the support I needed to put the text together have my deepest thanks and appreciation.

BRIEF CONTENTS

DEDICATION

In Loving Memory of Bestemor
Johanna O. Olufsen

CONTENTS

Introduction

OBJECTIVES

- Become aware of how quickly the electrical-electronics industry has developed in recent years.
- Realize the importance of both knowing the unit of measurement for a quantity and applying a unit of measurement to any result.
- Become familiar with the SI system of units which is used throughout this text.
- Understand the need for powers of ten in the mathematical calculations required in this field, and be able to perform all the basic algebraic operations with numbers using the power-of-ten notation.
- Be able to switch between fixed-point, scientific, and engineering notation for numbers as the need arises.
- Learn the conversion procedures to change a number from one power of ten to another and to change from one unit of measurement to another.
- Be able to recognize the meaning of the various symbols used in the text.
- Understand how to use a calculator correctly, and be aware of the mistakes that are often made when using a calculator.
- Develop an understanding of the computer field and realize how computers can be used as an effective analytical tool.
- Become aware of the meaning of the typical nameplate data provided with a computer in the retail environment.

1.1 THE ELECTRICAL-ELECTRONICS INDUSTRY

The growing sensitivity to NASDAQ on Wall Street is clear evidence that the electrical-electronics industry is one that will have a sweeping impact on future development in a wide range of areas that affect our life style, general health, and capabilities. Even the arts, initially so determined not to utilize technological methods, are embracing some of the new, innovative techniques that permit exploration into areas they never thought possible. The Windows approach to computer simulation has made computer systems much friendlier to the average person, resulting in an expanding market which further stimulates growth in the field. The computer in the home is now as common as the telephone or television.

Every facet of our lives seems touched by developments that appear to surface at an ever-increasing rate. For the layperson, the most obvious improvement of recent years has been the reduced size of electrical-electronics systems. Televisions are now small enough to be hand-held and have a battery capability that allows them to be more portable. Computers with significant memory capacity are now smaller than this textbook (and certainly lighter!). The size of radios is limited simply by our ability to read the numbers on the face of the dial.

Hearing aids are no longer visible, and pacemakers are significantly smaller and more reliable. All the reduction in size is due primarily to a marvelous development of the last few decades—the **integrated circuit (IC).** First developed in the late 1950s, the IC has now reached a point where cutting 0.18 micrometer lines is commonplace. The integrated circuit shown in Fig. 1.1 is the Intel® Pentium® 4 processor, which has 42 million transistors in an area measuring only 0.34 square inch. Intel Corporation recently presented a technical paper describing 0.02 micrometer (20 nanometer) transistors, developed in its silicon research laboratory. These small, ultrafast transistors will permit placing nearly 1 billion transistors on a sliver of silicon no larger than a fingernail, leaving us to wonder about the limits of such development.

It is natural to wonder what the limits to growth may be when we consider the changes over the last few decades. Rather than following a steady growth curve that would be somewhat predictable, the industry is subject to surges that revolve around significant developments in the field. Present indications are that the level of miniaturization will continue, but at a more moderate pace. Interest has turned toward increasing the quality and yield levels (percentage of good integrated circuits in the production process).

History reveals that there have been peaks and valleys in industry growth but that revenues continue to rise at a steady rate and funds set aside for research and development continue to command an increasing share of the budget. The field changes at a rate that requires constant retraining of employees from the entry to the director level. Many companies have instituted their own training programs and have encouraged local universities to develop programs to ensure that the latest concepts and procedures are brought to the attention of their employees. A period of relaxation could be disastrous to a company dealing in competitive products.

No matter what the pressures on an individual in this field may be to keep up with the latest technology, there is one saving grace that becomes immediately obvious: Once a concept or procedure is clearly and correctly understood, it will bear fruit throughout the career of the individual at any level of the industry. For example, once a fundamental equation such as Ohm's law (Chapter 4) is understood, it will not be *replaced* by another equation as more advanced theory is considered. It is a relationship of fundamental quantities that can have application in the most advanced setting. In addition, once a procedure or method of analysis is understood, it can usually be applied to a wide (if not infinite) variety of problems, making it unnecessary to learn a different technique for each slight variation in the system. The content of this text is such that every morsel of information will have application in more advanced courses. It will not be replaced by a different set of equations and procedures unless required by the specific area of application. Even then, the new procedures will usually be an expanded application of concepts already presented in the text.

It is paramount, therefore, that the material presented in this introductory course be clearly and precisely understood. It is the foundation for the material to follow and will be applied throughout your working days in this growing and exciting field.

1.2 A BRIEF HISTORY

In the sciences, once a hypothesis is proven and accepted, it becomes one of the building blocks of that area of study, permitting additional investigation and development. Naturally, the more pieces of a puzzle available, the more obvious the avenue toward a possible solution. In fact, history demonstrates that a single development may provide the key that will result in a mushroom effect that brings the science to a new plateau of understanding and impact.

If the opportunity presents itself, read one of the many publications reviewing the history of this field. Space requirements are such that only a brief review can be provided here. There are many more contributors than could be listed, and their efforts have often provided important keys to the solution of some very important concepts.

As noted earlier, there were periods characterized by what appeared to be an explosion of interest and development in particular areas. As you will see from the discussion of the late 1700s and the early 1800s, inventions, discoveries, and theories came fast and furiously. Each new concept has broadened the possible areas of application until it becomes almost impossible to trace developments without picking a particular area of interest and following it through. In the review, as you read about the development of the

FIG. 1.1

Computer chip.
(Courtesy of Corbis
Stock Market.)

radio, television, and computer, keep in mind that similar progressive steps were occurring in the areas of the telegraph, the telephone, power generation, the phonograph, appliances, and so on.

There is a tendency when reading about the great scientists, inventors, and innovators to believe that their contribution was a totally individual effort. In many instances, this was not the case. In fact, many of the great contributors were friends or associates who provided support and encouragement in their efforts to investigate various theories. At the very least, they were aware of one another's efforts to the degree possible in the days when a letter was often the best form of communication. In particular, note the closeness of the dates during periods of rapid development. One contributor seemed to spur on the efforts of the others or possibly provided the key needed to continue with the area of interest.

In the early stages, the contributors were not electrical, electronic, or computer engineers as we know them today. In most cases, they were physicists, chemists, mathematicians, or even philosophers. In addition, they were not from one or two communities of the Old World. The home country of many of the major contributors introduced in the paragraphs to follow is provided to show that almost every established community had some impact on the development of the fundamental laws of electrical circuits.

As you proceed through the remaining chapters of the text, you will find that a number of the units of measurement bear the name of major contributors in those areas—*volt* after Count Alessandro Volta, *ampere* after André Ampère, *ohm* after Georg Ohm, and so forth—fitting recognition for their important contributions to the birth of a major field of study.

Time charts indicating a limited number of major developments are provided in Fig. 1.2, primarily to identify specific periods of rapid development and to reveal how far we have come in the last few decades. In essence, the current state of the art is a result of efforts that began in earnest some 250 years ago, with progress in the last 100 years almost exponential.

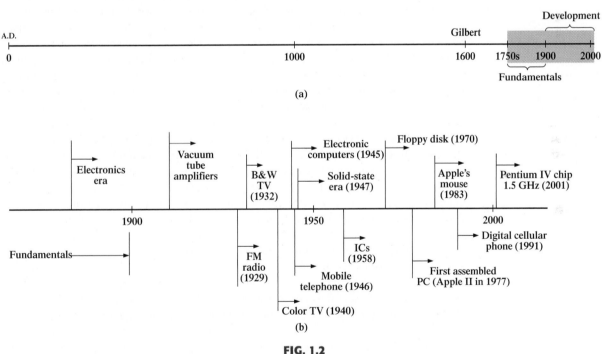

FIG. 1.2

Time charts: (a) long-range; (b) expanded.

As you read through the following brief review, try to sense the growing interest in the field and the enthusiasm and excitement that must have accompanied each new revelation. Although you may find some of the terms used in the review new and essentially meaningless, the remaining chapters will explain them thoroughly.

The Beginning

The phenomenon of **static electricity** has been toyed with since antiquity. The Greeks called the fossil resin substance so often used to demonstrate the effects of static electricity *elektron,* but no extensive study was made of the subject until **William Gilbert** researched the event in 1600. In the years to follow, there was a continuing investigation of electrostatic charge by many individuals such as **Otto von Guericke,** who developed the first machine to generate large amounts of charge, and **Stephen Gray,** who was able to transmit electrical charge over long distances on silk threads. **Charles DuFay** demonstrated that charges either attract or repel each other, leading him to believe that there were two types of charge—a theory we subscribe to today with our defined positive and negative charges.

Many believe that the true beginnings of the electrical era lie with the efforts of **Pieter van Musschenbroek** and **Benjamin Franklin.** In 1745, van Musschenbroek introduced the **Leyden jar** for the storage of electrical charge (the first capacitor) and demonstrated electrical shock (and therefore the power of this new form of energy). Franklin used the Leyden jar some seven years later to establish that lightning is simply an electrical discharge, and he expanded on a number of other important theories, including the definition of the two types of charge as *positive* and *negative.* From this point on, new discoveries and theories seemed to occur at an increasing rate as the number of individuals performing research in the area grew.

In 1784, **Charles Coulomb** demonstrated in Paris that the force between charges is inversely related to the square of the distance between the charges. In 1791, **Luigi Galvani,** professor of anatomy at the University of Bologna, Italy, performed experiments on the effects of electricity on animal nerves and muscles. The first **voltaic cell,** with its ability to produce electricity through the chemical action of a metal dissolving in an acid, was developed by another Italian, **Alessandro Volta,** in 1799.

The fever pitch continued into the early 1800s with **Hans Christian Oersted,** a Swedish professor of physics, announcing in 1820 a relationship between magnetism and electricity that serves as the foundation for the theory of **electromagnetism** as we know it today. In the same year, a French physicist, **André Ampère,** demonstrated that there are magnetic effects around every current-carrying conductor and that current-carrying conductors can attract and repel each other just like magnets. In the period from 1826 to 1827, a German physicist, **Georg Ohm,** introduced an important relationship between potential, current, and resistance which we now refer to as *Ohm's law.* In 1831, an English physicist, **Michael Faraday,** demonstrated his theory of *electromagnetic induction,* whereby a changing current in one coil can induce a changing current in another coil, even though the two coils are not directly connected. Professor Faraday also did extensive work on a storage device he called the *condenser,* which we refer to today as a *capacitor.* He introduced the idea of adding a dielectric between the plates of a capacitor to increase the storage capacity (Chapter 10). **James Clerk Maxwell,** a Scottish professor of natural philosophy, performed extensive mathematical analyses to develop what are currently called *Maxwell's equations,* which support the efforts of Faraday linking electric and magnetic effects. Maxwell also developed the *electromagnetic theory of light* in 1862, which, among other things, revealed that electromagnetic waves travel through air at the velocity of light (186,000 miles per second or 3×10^8 meters per second). In 1888, a German physicist, **Heinrich Rudolph Hertz,** through experimentation with lower-frequency electromagnetic waves (microwaves), substantiated Maxwell's predictions and equations. In the mid-1800s, Professor **Gustav Robert Kirchhoff** introduced a series of laws of voltages and currents that find application at every level and area of this field (Chapters 5 and 6). In 1895, another German physicist, **Wilhelm Röntgen,** discovered electromagnetic waves of high frequency, commonly called *X rays* today.

By the end of the 1800s, a significant number of the fundamental equations, laws, and relationships had been established, and various fields of study, including electronics, power generation, and calculating equipment, started to develop in earnest.

The Age of Electronics

Radio   The true beginning of the electronics era is open to debate and is sometimes attributed to efforts by early scientists in applying potentials across evacuated glass envelopes. However, many trace the beginning to **Thomas Edison,** who added a metallic electrode to the vacuum of the tube and discovered that a current was established between the metal electrode and the filament when a positive voltage was applied to the metal electrode. The phenomenon, demonstrated in 1883, was referred to as the **Edison effect.** In the period to follow, the transmission of radio waves and the development of the radio received widespread attention. In 1887, **Heinrich Hertz,** in his efforts to verify Maxwell's equations, transmitted radio waves for the first time in his laboratory. In 1896, an Italian scientist, **Guglielmo Marconi** (often called the father of the radio), demonstrated that telegraph signals could be sent through the air over long distances (2.5 kilometers) using a grounded antenna. In the same year, **Aleksandr Popov** sent what might have been the first radio message some 300 yards. The message was the name "Heinrich Hertz" in respect for Hertz's earlier contributions. In 1901, Marconi established radio communication across the Atlantic.

In 1904, **John Ambrose Fleming** expanded on the efforts of Edison to develop the first diode, commonly called **Fleming's valve**—actually the first of the *electronic devices*. The device had a profound impact on the design of detectors in the receiving section of radios. In 1906, **Lee De Forest** added a third element to the vacuum structure and created the first amplifier, the triode. Shortly thereafter, in 1912, **Edwin Armstrong** built the first regenerative circuit to improve receiver capabilities and then used the same contribution to develop the first nonmechanical oscillator. By 1915, radio signals were being transmitted across the United States, and in 1918, Armstrong applied for a patent for the superheterodyne circuit employed in virtually every television and radio to permit amplification at one frequency rather than at the full range of incoming signals. The major components of the modern-day radio were now in place, and sales in radios grew from a few million dollars in the early 1920s to over $1 billion by the 1930s. The 1930s were truly the golden years of radio, with a wide range of productions for the listening audience.

Television   The 1930s were also the true beginnings of the television era, although development on the picture tube began in earlier years with **Paul Nipkow** and his *electrical telescope* in 1884 and **John Baird** and his long list of successes, including the transmission of television pictures over telephone lines in 1927 and over radio waves in 1928, and simultaneous transmission of pictures and sound in 1930. In 1932, NBC installed the first commercial television antenna on top of the Empire State Building in New York City, and RCA began regular broadcasting in 1939. The war slowed development and sales, but in the mid-1940s the number of sets grew from a few thousand to a few million. Color television became popular in the early 1960s.

Computers   The earliest computer system can be traced back to **Blaise Pascal** in 1642 with his mechanical machine for adding and subtracting numbers. In 1673 **Gottfried Wilhelm von Leibniz** used the *Leibniz wheel* to add multiplication and division to the range of operations, and in 1823 **Charles Babbage** developed the **difference engine** to add the mathematical operations of sine, cosine, logs, and several others. In the years to follow, improvements were made, but the system remained primarily mechanical until the 1930s when electromechanical systems using components such as relays were introduced. It was not until the 1940s that totally electronic systems became the new wave. It is interesting to note that even though IBM was formed in 1924, it did not enter the computer industry until 1937. An entirely electronic system known as **ENIAC** was dedicated at the University of Pennsylvania in 1946. It contained 18,000 tubes and weighed 30 tons but was several times faster than most electromechanical systems. Although other vacuum tube systems were built, it was not until the birth of the solid-state era that computer systems experienced a major change in size, speed, and capability.

The Solid-State Era

In 1947, physicists **William Shockley, John Bardeen,** and **Walter H. Brattain** of Bell Telephone Laboratories demonstrated the point-contact **transistor** (Fig. 1.3), an amplifier constructed entirely of solid-state materials with no requirement for a vacuum, glass envelope, or heater voltage for the filament. It was called a *point-contact* transistor because amplification or transistor action occurred when two pointed metal contacts were pressed onto the surface of the semiconductor material. The contacts, which are supported by a wedge-shaped piece of insulating material, are placed extremely close together so that they are separated by only a few thousandths of an inch. The contacts are made of gold, and the semiconductor is germanium. The semiconductor rests on a metal base.

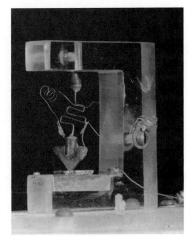

FIG. 1.3
The first transistor.
(Used with permission of Lucent Technologies Inc./Bell Labs.)

Although reluctant at first due to the vast amount of material available on the design, analysis, and synthesis of tube networks, the industry eventually accepted the new transistor technology as the wave of the future. In 1958 the first **integrated circuit (IC)** was developed at Texas Instruments, and in 1961 the first commercial integrated circuit was manufactured by the Fairchild Corporation.

It is impossible to review properly the entire history of the electrical-electronics field in a few pages. The effort here, both through the discussion and the time graphs of Fig. 1.2, was to reveal the amazing progress of this field in the last 50 years. The growth appears to be truly exponential since the early 1900s, raising the interesting question, Where do we go from here? The time chart suggests that the next few decades will probably contain many important innovative contributions that may cause an even faster growth curve than we are now experiencing.

1.3   UNITS OF MEASUREMENT

In any technical field it is naturally important to understand the basic concepts and the impact they will have on certain parameters. However, the application of these rules and laws will be successful only if the mathematical operations involved are applied correctly. In particular, **it is vital that the importance of applying the proper unit of measurement to a quantity is understood and appreciated.** Students often generate a numerical solution but decide not to apply a unit of measurement to the result because they are somewhat unsure of which unit should be applied. Consider, for example, the following very fundamental physics equation:

$$v = \frac{d}{t}$$
$$\begin{aligned} v &= \text{velocity} \\ d &= \text{distance} \\ t &= \text{time} \end{aligned} \qquad \textbf{(1.1)}$$

Assume, for the moment, that the following data are obtained for a moving object:

$$d = 4000 \text{ ft}$$
$$t = 1 \text{ min}$$

and v is desired in miles per hour. Often, without a second thought or consideration, the numerical values are simply substituted into the equation, with the result here that

$$v = \frac{d}{t} = \frac{4000 \text{ ft}}{1 \text{ min}} = \cancel{4000 \text{ mi/h}}$$

As indicated above, the solution is totally incorrect. If the result is desired in *miles per hour,* the unit of measurement for distance must be *miles,* and that for time, *hours.* In a moment, when the problem is analyzed properly, the extent of the error will demonstrate the importance of ensuring that

the numerical value substituted into an equation must have the unit of measurement specified by the equation.

The next question is normally, How do I convert the distance and time to the proper unit of measurement? A method will be presented in a later section of this chapter, but for now it is given that

$$1 \text{ mi} = 5280 \text{ ft}$$
$$4000 \text{ ft} = 0.7576 \text{ mi}$$
$$1 \text{ min} = \frac{1}{60} \text{ h} = 0.0167 \text{ h}$$

Substituting into Eq. (1.1), we have

$$v = \frac{d}{t} = \frac{0.7576 \text{ mi}}{0.0167 \text{ h}} = \textbf{45.37 mph}$$

which is significantly different from the result obtained before.

To complicate the matter further, suppose the distance is given in kilometers, as is now the case on many road signs. First, we must realize that the prefix *kilo* stands for a multiplier of 1000 (to be introduced in Section 1.5), and then we must find the conversion factor between kilometers and miles. If this conversion factor is not readily available, we must be able to make the conversion between units using the conversion factors between meters and feet or inches, as described in Section 1.6.

Before substituting numerical values into an equation, try to mentally establish a reasonable range of solutions for comparison purposes. For instance, if a car travels 4000 ft in 1 min, does it seem reasonable that the speed would be 4000 mph? Obviously not! This self-checking procedure is particularly important in this day of the hand-held calculator, when ridiculous results may be accepted simply because they appear on the digital display of the instrument.

Finally,

if a unit of measurement is applicable to a result or piece of data, then it must be applied to the numerical value.

To state that $v = 45.37$ without including the unit of measurement *mph* is meaningless.

Equation (1.1) is not a difficult one. A simple algebraic manipulation will result in the solution for any one of the three variables. However, in light of the number of questions arising from this equation, you may wonder if the difficulty associated with an equation will increase at the same rate as the number of terms in the equation. In the broad sense, this will not be the case. There is, of course, more room for a mathematical error with a more complex equation, but once the proper system of units is chosen and each term is properly found in that system, there should be very little added difficulty associated with an equation requiring an increased number of mathematical calculations.

In review, before substituting numerical values into an equation, be absolutely sure of the following:

1. *Each quantity has the proper unit of measurement as defined by the equation.*
2. *The proper magnitude of each quantity as determined by the defining equation is substituted.*
3. *Each quantity is in the same system of units (or as defined by the equation).*
4. *The magnitude of the result is of a reasonable nature when compared to the level of the substituted quantities.*
5. *The proper unit of measurement is applied to the result.*

1.4   SYSTEMS OF UNITS

In the past, the *systems of units* most commonly used were the English and metric, as outlined in Table 1.1. Note that while the English system is based on a single standard, the metric is subdivided into two interrelated standards: the **MKS** and the **CGS**. Fundamental quantities of these systems are compared in Table 1.1 along with their abbreviations. The MKS and CGS systems draw their names from the units of measurement used with each system; the MKS system uses **Meters, Kilograms,** and **Seconds,** while the CGS system uses **Centimeters, Grams,** and **Seconds.**

Understandably, the use of more than one system of units in a world that finds itself continually shrinking in size, due to advanced technical developments in communications and transportation, would introduce unnecessary complications to the basic understanding of any technical data. The need for a standard set of units to be adopted by all nations has become increasingly obvious. The International Bureau of Weights and Measures located at Sèvres, France, has been the host for the General Conference of Weights and Measures, attended by representatives from all nations of the world. In 1960, the General Conference adopted a system called Le Système International d'Unités (International System of Units), which has the international abbreviation **SI.** Since then, it has been adopted by the Institute of Electrical and Electronic Engineers, Inc. (IEEE) in 1965 and by the United States of America Standards Institute in 1967 as a standard for all scientific and engineering literature.

For comparison, the SI units of measurement and their abbreviations appear in Table 1.1. These abbreviations are those usually applied to each unit of measurement, and they were carefully chosen to be the most effective. Therefore, it is important that they be used whenever applicable to ensure universal understanding. Note the similarities of the SI system to the MKS system. This text will employ, whenever possible and practical, all of the major units and abbreviations of the SI system in an effort to support the need for a universal system.

TABLE 1.1

Comparison of the English and metric systems of units.

English	Metric		
	MKS	**CGS**	**SI**
Length: Yard (yd) (0.914 m)	Meter (m) (39.37 in.) (100 cm)	Centimeter (cm) (2.54 cm = 1 in.)	**Meter (m)**
Mass: Slug (14.6 kg)	Kilogram (kg) (1000 g)	Gram (g)	**Kilogram (kg)**
Force: Pound (lb) (4.45 N)	Newton (N) (100,000 dynes)	Dyne	**Newton (N)**
Temperature: Fahrenheit (°F) $\left(= \dfrac{9}{5}°C + 32\right)$	Celsius or Centigrade (°C) $\left(= \dfrac{5}{9}(°F - 32)\right)$	Centigrade (°C)	**Kelvin (K)** K = 273.15 + °C
Energy: Foot-pound (ft-lb) (1.356 joules)	Newton-meter (N·m) or joule (J) (0.7376 ft-lb)	Dyne-centimeter or erg (1 joule = 10^7 ergs)	**Joule (J)**
Time: Second (s)	Second (s)	Second (s)	**Second (s)**

Those readers requiring additional information on the SI system should contact the information office of the American Society for Engineering Education (ASEE).[*]

Fig. 1.4 should help you develop some feeling for the relative magnitudes of the units of measurement of each system of units. Note in the figure the relatively small magnitude of the units of measurement for the CGS system.

A standard exists for each unit of measurement of each system. The standards of some units are quite interesting.

The **meter** was originally defined in 1790 to be 1/10,000,000 the distance between the equator and either pole at sea level, a length preserved on a platinum-iridium bar at the International Bureau of Weights and Measures at Sèvres, France.

The meter is now defined with reference to the speed of light in a vacuum, which is 299,792,458 m/s.

The kilogram is defined as a mass equal to 1000 times the mass of 1 cubic centimeter of pure water at 4°C.

This standard is preserved in the form of a platinum-iridium cylinder in Sèvres.

The **second** was originally defined as 1/86,400 of the mean solar day. However, since Earth's rotation is slowing down by almost 1 second every 10 years,

the second was redefined in 1967 as 9,192,631,770 periods of the electromagnetic radiation emitted by a particular transition of a cesium atom.

---

[*]American Society for Engineering Education (ASEE), 1818 N Street, N.W., Suite 600, Washington, DC, 20036, (202) 331–3500; Website http://www.asee.org

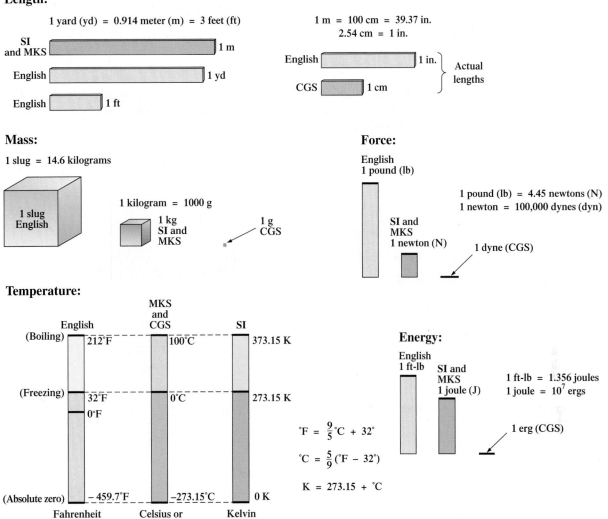

FIG. 1.4
Comparison of the various systems of units.

1.5   POWERS OF TEN

It should be apparent from the relative magnitude of the various units of measurement that very large and very small numbers will frequently be encountered in the sciences. To ease the difficulty of mathematical operations with numbers of such varying size, *powers of ten* are usually employed. This notation takes full advantage of the mathematical properties of powers of ten. The notation used to represent numbers that are integer powers of ten is as follows:

$$
\begin{array}{ll}
1 = 10^0 & 1/10 = \quad 0.1 = 10^{-1} \\
10 = 10^1 & 1/100 = \quad 0.01 = 10^{-2} \\
100 = 10^2 & 1/1000 = \quad 0.001 = 10^{-3} \\
1000 = 10^3 & 1/10{,}000 = 0.0001 = 10^{-4}
\end{array}
$$

In particular, note that $10^0 = 1$, and, in fact, any quantity to the zero power is 1 ($x^0 = 1$, $1000^0 = 1$, and so on). Also, note that the numbers in the list that are **greater than 1 have positive powers of ten,** and numbers in the list that are **less than 1 have negative powers of ten.**

   A quick method of determining the proper power of ten is to place a caret mark to the right of the numeral 1 wherever it may occur; then count from this point to the number of places to the right or left before arriving at the decimal point. Moving to the right indicates a positive power of ten, whereas moving to the left indicates a negative power. For example,

$$10,000.0 = 10,0\,0\,0. = 10^{+4}$$

$$\underset{1\ 2\ 3\ 4}{}$$

$$0.00001 = 0.0\,0\,0\,0\,1 = 10^{-5}$$

$$\underset{5\ 4\ \ 3\ 2\ 1}{}$$

Some important mathematical equations and relationships pertaining to powers of ten are listed below, along with a few examples. In each case, n and m can be any positive or negative real number.

$$\frac{1}{10^{n}} = 10^{-n} \qquad \frac{1}{10^{-n}} = 10^{n} \tag{1.2}$$

Equation (1.2) clearly reveals that shifting a power of ten from the denominator to the numerator, or the reverse, requires simply changing the sign of the power.

EXAMPLE 1.1

a. $\dfrac{1}{1000} = \dfrac{1}{10^{+3}} = \mathbf{10^{-3}}$

b. $\dfrac{1}{0.00001} = \dfrac{1}{10^{-5}} = \mathbf{10^{+5}}$

The product of powers of ten:

$$(10^{n})(10^{m}) = 10^{(n+m)} \tag{1.3}$$

EXAMPLE 1.2

a. $(1000)(10,000) = (10^{3})(10^{4}) = 10^{(3+4)} = \mathbf{10^{7}}$
b. $(0.00001)(100) = (10^{-5})(10^{2}) = 10^{(-5+2)} = \mathbf{10^{-3}}$

The division of powers of ten:

$$\frac{10^{n}}{10^{m}} = 10^{(n-m)} \tag{1.4}$$

EXAMPLE 1.3

a. $\dfrac{100,000}{100} = \dfrac{10^{5}}{10^{2}} = 10^{(5-2)} = \mathbf{10^{3}}$

b. $\dfrac{1000}{0.0001} = \dfrac{10^{3}}{10^{-4}} = 10^{(3-(-4))} = 10^{(3+4)} = \mathbf{10^{7}}$

Note the use of parentheses in Example 1.3(b) to ensure that the proper sign is established between operators.

The power of powers of ten:

$$(10^{n})^{m} = 10^{(nm)} \tag{1.5}$$

EXAMPLE 1.4

a. $(100)^4 = (10^2)^4 = 10^{(2)(4)} = \mathbf{10^8}$
b. $(1000)^{-2} = (10^3)^{-2} = 10^{(3)(-2)} = \mathbf{10^{-6}}$
c. $(0.01)^{-3} = (10^{-2})^{-3} = 10^{(-2)(-3)} = \mathbf{10^6}$

Basic Arithmetic Operations

Let us now examine the use of powers of ten to perform some basic arithmetic operations using numbers that are not just powers of ten. The number 5000 can be written as $5 \times 1000 = 5 \times 10^3$, and the number 0.0004 can be written as $4 \times 0.0001 = 4 \times 10^{-4}$. Of course, 10^5 can also be written as 1×10^5 if it clarifies the operation to be performed.

Addition and Subtraction To perform addition or subtraction using powers of ten, the power of ten *must be the same for each term;* that is,

$$A \times 10^n \pm B \times 10^n = (A \pm B) \times 10^n \qquad \textbf{(1.6)}$$

Equation (1.6) covers all possibilities, but students often prefer to remember a verbal description of how to perform the operation.
Equation (1.6) states

when adding or subtracting numbers in a powers-of-ten format, be sure that the power of ten is the same for each number. Then separate the multipliers, perform the required operation, and apply the same power of ten to the result.

EXAMPLE 1.5

a. $6300 + 75{,}000 = (6.3)(1000) + (75)(1000)$
$= 6.3 \times 10^3 + 75 \times 10^3$
$= (6.3 + 75) \times 10^3$
$= \mathbf{81.3 \times 10^3}$

b. $0.00096 - 0.000086 = (96)(0.00001) - (8.6)(0.00001)$
$= 96 \times 10^{-5} - 8.6 \times 10^{-5}$
$= (96 - 8.6) \times 10^{-5}$
$= \mathbf{87.4 \times 10^{-5}}$

Multiplication In general,

$$(A \times 10^n)(B \times 10^m) = (A)(B) \times 10^{n+m} \qquad \textbf{(1.7)}$$

revealing that the *operations with the powers of ten can be separated from the operation with the multipliers.*
Equation (1.7) states

when multiplying numbers in the powers-of-ten format, first find the product of the multipliers and then determine the power of ten for the result by adding the power-of-ten exponents.

EXAMPLE 1.6

a. $(0.0002)(0.000007) = [(2)(0.0001)][(7)(0.000001)]$
$= (2 \times 10^{-4})(7 \times 10^{-6})$
$= (2)(7) \times (10^{-4})(10^{-6})$
$= \mathbf{14 \times 10^{-10}}$

b. $(340{,}000)(0.00061) = (3.4 \times 10^5)(61 \times 10^{-5})$
$= (3.4)(61) \times (10^5)(10^{-5})$
$= 207.4 \times 10^0$
$= \mathbf{207.4}$

Division In general,

$$\boxed{\frac{A \times 10^n}{B \times 10^m} = \frac{A}{B} \times 10^{n-m}}$$ **(1.8)**

revealing again that the *operations with the powers of ten can be separated from the same operation with the multipliers.*

Equation (1.8) states

when dividing numbers in the powers-of-ten format, first find the result of dividing the multipliers. Then determine the associated power for the result by subtracting the power of ten of the denominator from the power of ten of the numerator.

EXAMPLE 1.7

a. $\dfrac{0.00047}{0.002} = \dfrac{47 \times 10^{-5}}{2 \times 10^{-3}} = \left(\dfrac{47}{2}\right) \times \left(\dfrac{10^{-5}}{10^{-3}}\right)$
$= \mathbf{23.5 \times 10^{-2}}$

b. $\dfrac{690{,}000}{0.00000013} = \dfrac{69 \times 10^4}{13 \times 10^{-8}} = \left(\dfrac{69}{13}\right) \times \left(\dfrac{10^4}{10^{-8}}\right)$
$= \mathbf{5.31 \times 10^{12}}$

Powers In general,

$$\boxed{(A \times 10^n)^m = A^m \times 10^{nm}}$$ **(1.9)**

which again permits the separation of the *operation with the powers of ten from the multipliers.*

Equation (1.9) states

when finding the power of a number in the power-of-ten format, first separate the multiplier from the power of ten and determine each separately. Determine the power-of-ten component by multiplying the power of ten by the power to be determined.

EXAMPLE 1.8

a. $(0.00003)^3 = (3 \times 10^{-5})^3 = (3)^3 \times (10^{-5})^3$
$= \mathbf{27 \times 10^{-15}}$

b. $(90{,}800{,}000)^2 = (9.08 \times 10^7)^2 = (9.08)^2 \times (10^7)^2$
$= \mathbf{82.446 \times 10^{14}}$

In particular, remember that the following operations are not the same. One is the product of two numbers in the powers-of-ten format, while the other is a number in the powers-of-ten format taken to a power. As noted below, the results of each are quite different:

$$(10^3)(10^3) \neq (10^3)^3$$
$$(10^3)(10^3) = 10^6 = 1{,}000{,}000$$
$$(10^3)^3 = (10^3)(10^3)(10^3) = 10^9 = 1{,}000{,}000{,}000$$

Fixed-Point, Floating-Point, Scientific, and Engineering Notation

When you are using a computer or a calculator, numbers will generally appear in one of four ways. If powers of ten are not employed, numbers are written in the **fixed-point** or **floating-point notation.**

The fixed-point format requires that the decimal point appear in the same place each time. In the floating-point format, the decimal point will appear in a location defined by the number to be displayed.

Most computers and calculators permit a choice of fixed- or floating-point notation. In the fixed format, the user can choose the level of accuracy for the output as tenths place, hundredths place, thousandths place, and so on. Every output will then fix the decimal point to one location, such as the following examples using thousandths place accuracy:

$$\frac{1}{3} = 0.333 \qquad \frac{1}{16} = 0.063 \qquad \frac{2300}{2} = 1150.000$$

If left in the floating-point format, the results will appear as follows for the above operations:

$$\frac{1}{3} = 0.333333333333 \qquad \frac{1}{16} = 0.0625 \qquad \frac{2300}{2} = 1150$$

Powers of ten will creep into the fixed- or floating-point notation if the number is too small or too large to be displayed properly.

Scientific (also called *standard*) **notation** and **engineering notation** make use of powers of ten, with restrictions on the mantissa (multiplier) or scale factor (power of the power of ten).

Scientific notation requires that the decimal point appear directly after the first digit greater than or equal to 1 but less than 10.

A power of ten will then appear with the number (usually following the power notation E), even if it has to be to the zero power. A few examples:

$$\frac{1}{3} = 3.33333333333E-1 \qquad \frac{1}{16} = 6.25E-2 \qquad \frac{2300}{2} = 1.15E3$$

Within scientific notation, the fixed- or floating-point format can be chosen. In the above examples, floating was employed. If fixed is chosen and set at the thousandths point accuracy, the following will result for the above operations:

$$\frac{1}{3} = 3.333E-1 \qquad \frac{1}{16} = 6.250E-2 \qquad \frac{2300}{2} = 1.150E3$$

The last format to be introduced is **engineering notation,** which specifies that

all powers of ten must be multiples of 3, and the mantissa must be greater than or equal to 1 but less than 1000.

This restriction on the powers of ten is due to the fact that specific powers of ten have been assigned prefixes that will be introduced in the next few paragraphs. Using scientific notation in the floating-point mode will result in the following for the above operations:

$$\frac{1}{3} = 333.333333333E-3 \qquad \frac{1}{16} = 62.5E-3 \qquad \frac{2300}{2} = 1.15E3$$

Using engineering notation with three-place accuracy will result in the following:

$$\frac{1}{3} = 333.333E-3 \qquad \frac{1}{16} = 62.500E-3 \qquad \frac{2300}{2} = 1.150E3$$

Prefixes

Specific powers of ten in engineering notation have been assigned prefixes and symbols, as appearing in Table 1.2. They permit easy recognition of the power of ten and an improved channel of communication between technologists.

TABLE 1.2

Multiplication Factors	SI Prefix	SI Symbol
$1\ 000\ 000\ 000\ 000 = 10^{12}$	tera	T
$1\ 000\ 000\ 000 = 10^{9}$	giga	G
$1\ 000\ 000 = 10^{6}$	mega	M
$1\ 000 = 10^{3}$	kilo	k
$0.001 = 10^{-3}$	milli	m
$0.000\ 001 = 10^{-6}$	micro	μ
$0.000\ 000\ 001 = 10^{-9}$	nano	n
$0.000\ 000\ 000\ 001 = 10^{-12}$	pico	p

EXAMPLE 1.9

a. $1,000,000$ ohms $= 1 \times 10^{6}$ ohms
$$= \textbf{1 megohm} = \textbf{1 M}\Omega$$

b. $100,000$ meters $= 100 \times 10^{3}$ meters
$$= \textbf{100 kilometers} = \textbf{100 km}$$

c. 0.0001 second $= 0.1 \times 10^{-3}$ second
$$= \textbf{0.1 millisecond} = \textbf{0.1 ms}$$

d. 0.000001 farad $= 1 \times 10^{-6}$ farad
$$= \textbf{1 microfarad} = \textbf{1 }\mu\textbf{F}$$

Here are a few examples with numbers that are not strictly powers of ten.

EXAMPLE 1.10

a. $41,200$ m is equivalent to 41.2×10^{3} m $= 41.2$ kilometers $= \textbf{41.2 km.}$
b. 0.00956 J is equivalent to 9.56×10^{-3} J $= 9.56$ millijoules $= \textbf{9.56 mJ.}$
c. 0.000768 s is equivalent to 768×10^{-6} s $= 768$ microseconds $= \textbf{768 }\mu\textbf{s.}$

d. $\dfrac{8400 \text{ m}}{0.06} = \dfrac{8.4 \times 10^{3} \text{ m}}{6 \times 10^{-2}} = \left(\dfrac{8.4}{6}\right) \times \left(\dfrac{10^{3}}{10^{-2}}\right)$ m

$$= 1.4 \times 10^{5} \text{ m} = 140 \times 10^{3} \text{ m} = 140 \text{ kilometers} = \textbf{140 km}$$

e. $(0.0003)^{4}$ s $= (3 \times 10^{-4})^{4}$ s $= 81 \times 10^{-16}$ s
$$= 0.0081 \times 10^{-12} \text{ s} = 0.008 \text{ picosecond} = \textbf{0.0081 ps}$$

1.6 CONVERSION BETWEEN LEVELS OF POWERS OF TEN

It is often necessary to convert from one power of ten to another. For instance, if a meter measures kilohertz (kHz—a unit of measurement for the frequency of an ac waveform), it may be necessary to find the corresponding level in megahertz (MHz); or if time is measured in milliseconds (ms), it may be necessary to find the corresponding time in microseconds (μs) for a graphical plot. The process is not a difficult one if we simply keep in mind that an increase or a decrease in the power of ten must be associated with the opposite effect on the multiplying factor. The procedure is best described by a few examples.

EXAMPLE 1.11

a. Convert 20 kHz to megahertz.

Solution: In the power-of-ten format:

$$20 \text{ kHz} = 20 \times 10^{3} \text{ Hz}$$

$$\underbrace{\overbrace{20 \times 10^3 \text{ Hz} \Rightarrow \underline{\quad} \times 10^6 \text{ Hz}}^{\text{Increase by 3}}}_{\text{Decrease by 3}}$$

Since the power of ten will be *increased* by a factor of *three*, the multiplying factor must be *decreased* by moving the decimal point *three* places to the left, as shown below:

$$\underbrace{020.}_{3} = 0.02$$

and
$$20 \times 10^3 \text{ Hz} = 0.02 \times 10^6 \text{ Hz} = \textbf{0.02 MHz}$$

b. Convert 0.01 ms to microseconds.

Solution: In the power-of-ten format:

$$0.01 \text{ ms} = 0.01 \times 10^{-3} \text{ s}$$

and
$$\underbrace{\overbrace{0.01 \times 10^{-3} \text{ s} = \underline{\quad} \times 10^{-6} \text{ s}}^{\text{Reduce by 3}}}_{\text{Increase by 3}}$$

Since the power of ten will be *reduced* by a factor of three, the multiplying factor must be *increased* by moving the decimal point three places to the right, as follows:

$$\underbrace{0.010}_{3} = 10$$

and
$$0.01 \times 10^{-3} \text{ s} = 10 \times 10^{-6} \text{ s} = \textbf{10 } \boldsymbol{\mu}\textbf{s}$$

There is a tendency when comparing -3 to -6 to think that the power of ten has increased, but keep in mind when making your judgment about increasing or decreasing the magnitude of the multiplier that 10^{-6} is a great deal smaller than 10^{-3}.

c. Convert 0.002 km to millimeters.

Solution:

$$\underbrace{\overbrace{0.002 \times 10^3 \text{ m} \Rightarrow \underline{\quad} \times 10^{-3} \text{ m}}^{\text{Reduce by 6}}}_{\text{Increase by 6}}$$

In this example we have to be very careful because the difference between $+3$ and -3 is a factor of 6, requiring that the multiplying factor be modified as follows:

$$\underbrace{0.002000}_{6} = 200$$

and
$$0.002 \times 10^3 \text{ m} = 2000 \times 10^{-3} \text{ m} = \textbf{2000 mm}$$

1.7   CONVERSION WITHIN AND BETWEEN SYSTEMS OF UNITS

The conversion within and between systems of units is a process that cannot be avoided in the study of any technical field. It is an operation, however, that is performed incorrectly so often that this section was included to provide one approach that, if applied properly, will lead to the correct result.

There is more than one method of performing the conversion process. In fact, some people prefer to determine mentally whether the conversion factor is multiplied or divided. This approach is acceptable for some elementary conversions, but it is risky with more complex operations.

The procedure to be described here is best introduced by examining a relatively simple problem such as converting inches to meters. Specifically, let us convert 48 in. (4 ft) to meters.

If we multiply the 48 in. by a factor of **1**, the magnitude of the quantity remains the same:

$$\boxed{48 \text{ in.} = 48 \text{ in.}(\mathbf{1})} \qquad (1.10)$$

Let us now look at the conversion factor, which is the following for this example:

$$1 \text{ m} = 39.37 \text{ in.}$$

Dividing both sides of the conversion factor by 39.37 in. will result in the following format:

$$\frac{1 \text{ m}}{39.37 \text{ in.}} = (\mathbf{1})$$

Note that the end result is that the ratio 1 m/39.37 in. equals 1, as it should since they are equal quantities. If we now substitute this factor (**1**) into Eq. (1.10), we obtain

$$48 \text{ in.}(\mathbf{1}) = 48 \text{ in.}\left(\frac{1 \text{ m}}{39.37 \text{ in.}}\right)$$

which results in the cancellation of inches as a unit of measure and leaves meters as the unit of measure. In addition, since the 39.37 is in the denominator, it must be divided into the 48 to complete the operation:

$$\frac{48}{39.37} \text{ m} = \mathbf{1.219 \text{ m}}$$

Let us now review the method, which has the following steps:

1. *Set up the conversion factor to form a numerical value of (1) with the unit of measurement to be removed from the original quantity in the denominator.*
2. *Perform the required mathematics to obtain the proper magnitude for the remaining unit of measurement.*

EXAMPLE 1.12

a. Convert 6.8 min to seconds.

Solution: The conversion factor is

$$1 \text{ min} = 60 \text{ s}$$

Since the minute is to be removed as the unit of measurement, it must appear in the denominator of the (**1**) factor, as follows:

Step 1:
$$\left(\frac{60 \text{ s}}{1 \text{ min}}\right) = (\mathbf{1})$$

Step 2:
$$6.8 \text{ min}(\mathbf{1}) = 6.8 \text{ min}\left(\frac{60 \text{ s}}{1 \text{ min}}\right) = (6.8)(60) \text{ s}$$
$$= \mathbf{408 \text{ s}}$$

b. Convert 0.24 m to centimeters.

Solution: The conversion factor is

$$1 \text{ m} = 100 \text{ cm}$$

Since the meter is to be removed as the unit of measurement, it must appear in the denominator of the (**1**) factor as follows:

Step 1:
$$\left(\frac{100 \text{ cm}}{1 \text{ m}}\right) = \mathbf{1}$$

Step 2:
$$0.24 \text{ m}(\mathbf{1}) = 0.24 \text{ m}\left(\frac{100 \text{ cm}}{1 \text{ m}}\right) = (0.24)(100) \text{ cm}$$
$$= \mathbf{24 \text{ cm}}$$

The products (**1**)(**1**) and (**1**)(**1**)(**1**) are still **1**. Using this fact, we can perform a series of conversions in the same operation.

EXAMPLE 1.13

a. Determine the number of minutes in half a day.

Solution: Working our way through from days to hours to minutes, always ensuring that the unit of measurement to be removed is in the denominator, will result in the following sequence:

$$0.5 \, \cancel{day} \left(\frac{24 \, \cancel{h}}{1 \, \cancel{day}} \right) \left(\frac{60 \, min}{1 \, \cancel{h}} \right) = (0.5)(24)(60) \, min$$

$$= \mathbf{720 \, min}$$

b. Convert 2.2 yards to meters.

Solution: Working our way through from yards to feet to inches to meters will result in the following:

$$2.2 \, \cancel{yards} \left(\frac{3 \, \cancel{ft}}{1 \, \cancel{yard}} \right) \left(\frac{12 \, \cancel{in.}}{1 \, \cancel{ft}} \right) \left(\frac{1 \, m}{39.37 \, \cancel{in.}} \right) = \frac{(2.2)(3)(12)}{39.37} \, m$$

$$= \mathbf{2.012 \, m}$$

The following examples are variations of the above in practical situations.

EXAMPLE 1.14

a. In Europe and Canada and many other locations throughout the world, the speed limit is posted in kilometers per hour. How fast in miles per hour is 100 km/h?

Solution:

$$\left(\frac{100 \, km}{h} \right) (1)(1)(1)(1)$$

$$= \left(\frac{100 \, \cancel{km}}{h} \right) \left(\frac{1000 \, \cancel{m}}{1 \, \cancel{km}} \right) \left(\frac{39.37 \, \cancel{in.}}{1 \, \cancel{m}} \right) \left(\frac{1 \, \cancel{ft}}{12 \, \cancel{in.}} \right) \left(\frac{1 \, mi}{5280 \, \cancel{ft}} \right)$$

$$= \frac{(100)(1000)(39.37)}{(12)(5280)} \, \frac{mi}{h}$$

$$= \mathbf{62.14 \, mph}$$

Many travelers use 0.6 as a conversion factor to simplify the math involved; that is,

$$(100 \, km/h)(0.6) \cong 60 \, mi/h$$

and

$$(60 \, km/h)(0.6) \cong 36 \, mi/h$$

b. Determine the speed in miles per hour of a competitor who can run a 4 min mile.

Solution: Inverting the factor 4 min/1 mi to 1 mi/4 min, we can proceed as follows:

$$\left(\frac{1 \, mi}{4 \, \cancel{min}} \right) \left(\frac{60 \, \cancel{min}}{h} \right) = \frac{60}{4} \, mi/h = \mathbf{15 \, mph}$$

1.8   SYMBOLS

Throughout the text, various symbols will be employed that you may not have had occasion to use. Some are defined in Table 1.3, and others will be defined in the text as the need arises.

1.9   CONVERSION TABLES

Conversion tables such as those appearing in Appendix A can be very useful when time does not permit the application of methods described in this chapter. However, even though such tables appear easy to use, frequent errors occur because the operations appearing at the head

TABLE 1.3

Symbol	Meaning
$\neq$	Not equal to $\quad 6.12 \neq 6.13$
$>$	Greater than $\quad 4.78 > 4.20$
$\gg$	Much greater than $\quad 840 \gg 16$
$<$	Less than $\quad 430 < 540$
$\ll$	Much less than $\quad 0.002 \ll 46$
$\geq$	Greater than or equal to $\quad x \geq y$ is satisfied for $y = 3$ and $x > 3$ or $x = 3$
$\leq$	Less than or equal to $\quad x \leq y$ is satisfied for $y = 3$ and $x < 3$ or $x = 3$
$\cong$	Approximately equal to $\quad 3.14159 \cong 3.14$
Σ	Sum of $\quad \Sigma (4 + 6 + 8) = 18$
$| \ |$	Absolute magnitude of $\quad |a| = 4$, where $a = -4$ or $+4$
$\therefore$	Therefore $\quad x = \sqrt{4} \quad \therefore x = \pm 2$
$\equiv$	By definition $\quad$ Establishes a relationship between two or more quantities

of the table are not performed properly. In any case, when using such tables, try to establish mentally some order of magnitude for the quantity to be determined compared to the magnitude of the quantity in its original set of units. This simple operation should prevent several impossible results that may occur if the conversion operation is improperly applied.

For example, consider the following from such a conversion table:

To convert from	To	Multiply by
Miles	Meters	1.609×10^3

A conversion of 2.5 mi to meters would require that we multiply 2.5 by the conversion factor; that is,

$$2.5 \text{ mi}(1.609 \times 10^3) = 4.0225 \times 10^3 \text{ m}$$

A conversion from 4000 m to miles would require a division process:

$$\frac{4000 \text{ m}}{1.609 \times 10^3} = 2486.02 \times 10^{-3} = 2.486 \text{ mi}$$

In each of the above, there should have been little difficulty realizing that 2.5 mi would convert to a few thousand meters and 4000 m would be only a few miles. As indicated above, this kind of anticipatory thinking will eliminate the possibility of ridiculous conversion results.

1.10 CALCULATORS

In most texts, the calculator is not discussed in detail. Instead, students are left with the general exercise of choosing an appropriate calculator and learning to use it properly on their own. However, it is now clear that there must be some discussion about the use of the calculator to eliminate some of the impossible results obtained (and often strongly defended by the user—because the calculator says so) through a correct understanding of the process by which a calculator performs the various tasks. Time and space do not permit a detailed explanation of all the possible operations, but it is assumed that the following discussion will enlighten the user to the fact that it is important to understand the manner in which a calculator proceeds with a calculation and not to expect the unit to accept data in any form and always generate the correct answer.

When choosing a calculator (scientific for our use), be absolutely sure that it has the ability to operate on complex numbers (polar and rectangular) which will be described in detail in Chapter 12. For now simply look up the terms in the index of the operator's manual, and be sure that the terms appear and that the basic operations with them are discussed. Next, be aware that some calculators perform the operations with a minimum number of steps while others can require a downright lengthy or complex series of steps. Speak to your instructor if unsure about your purchase. For this text, the TI-86 of Fig. 1.5 was chosen because of its treatment of complex numbers.

FIG. 1.5

TI-86 calculator.
(Courtesy of Texas Instruments, Inc.)

Initial Settings

Format and accuracy are the first two settings that must be made on any scientific calculator. For most calculators the choices of formats are *Normal, Scientific,* and *Engineering.* For the TI-86 calculator, pressing the 2nd function (yellow) key followed by the MORE key will provide a list of options for the initial settings of the calculator. For calculators without a MORE choice, consult the operator's manual for the manner in which the format and accuracy level are set.

Examples of each are shown below:

Normal: $1/3 = 0.33$
Scientific: $1/3 = 3.33E-1$
Engineering: $1/3 = 333.33E-3$

Note that the Normal format simply places the decimal point in the most logical location. The Scientific ensures that the number preceding the decimal point is a single digit followed by the required power of ten. The Engineering format will always ensure that the power of ten is a multiple of 3 (whether it be positive, negative, or zero).

In the above examples the accuracy was hundredths place. To set this accuracy for the TI-86 calculator, return to the MODE selection and choose 2 to represent two-place accuracy or hundredths place.

Initially, you will probably be most comfortable with the Normal mode with hundredths place accuracy. However, as you begin to analyze networks, you may find the Engineering mode more appropriate since you will be working with component levels and results that have powers of ten that have been assigned abbreviations and names. On the other hand, the Scientific mode may the best choice for a particular analysis. In any event, take the time now to become familiar with the differences between the various modes and how to set them on your calculator.

Order of Operations

Although being able to set the format and accuracy is important, these features are not the source of the impossible results that often arise because of improper use of the calculator. Improper results occur primarily because users fail to realize that no matter how simple or complex an equation, the calculator will perform the required operations in a specific order.

For instance, the operation

$$\frac{8}{3 + 1}$$

is often entered as

$$\boxed{8}\ \boxed{\div}\ \boxed{3}\ \boxed{+}\ \boxed{1}\ = \frac{8}{3} + 1 = 2.67 + 1 = 3.67$$

which is totally incorrect (2 is the answer).

The user must be aware that the calculator ***will not*** perform the addition first and then the division. In fact, addition and subtraction are the last operations to be performed in any equation. It is therefore very important that you carefully study and thoroughly understand the next few paragraphs in order to use the calculator properly.

1. *The first operations to be performed by a calculator can be set using parentheses ( ). It does not matter which operations are within the parentheses. The parentheses simply dictate that this part of the equation is to be determined first. There is no limit to the number of parentheses in each equation—all operations within parentheses will be performed first. For instance, for the example above, if parentheses are added as shown below, the addition will be performed first and the correct answer obtained:*

$$\frac{8}{(3 + 1)} = \boxed{8}\ \boxed{\div}\ \boxed{(}\ \boxed{3}\ \boxed{+}\ \boxed{1}\ \boxed{)} = \frac{8}{4} = 2$$

2. *Next, powers and roots are performed, such as x^2, $\sqrt{x}$, and so on.*
3. *Negation (applying a negative sign to a quantity) and single-key operations such as sin, $\tan^{-1}$, and so on, are performed.*
4. *Multiplication and division are then performed.*
5. *Addition and subtraction are performed last.*

It may take a few moments and some repetition to remember the order, but at least you are now aware that there is an order to the operations and are aware that ignoring them can result in meaningless results.

EXAMPLE 1.15   Determine

$$\sqrt{\frac{9}{3}}$$

Solution: The following calculator operations will result in an **incorrect answer** of 1 because the square-root operation will be performed before the division.

$$\boxed{\sqrt{}}\ \boxed{9}\ \boxed{\div}\ \boxed{3} = \frac{\sqrt{9}}{3} = \frac{3}{3} = 1$$

However, recognizing that we must first divide 9 by 3, we can use parentheses as follows to define this operation as the first to be performed, and the correct answer will be obtained:

$$\boxed{\sqrt{}}\ \boxed{(}\ \boxed{9}\ \boxed{\div}\ \boxed{3}\ \boxed{)} = \sqrt{\left(\frac{9}{3}\right)} = \sqrt{3} = \mathbf{1.67}$$

EXAMPLE 1.16   Find

$$\frac{3 + 9}{4}$$

Solution: If the problem is entered as it appears, the **incorrect answer** of 5.25 will result.

$$\boxed{3}\ \boxed{+}\ \boxed{9}\ \boxed{\div}\ \boxed{4} = 3 + \frac{9}{4} = 5.25$$

Using brackets to ensure that the addition takes place before the division will result in the correct answer as shown below:

$$\boxed{(}\ \boxed{3}\ \boxed{+}\ \boxed{9}\ \boxed{)}\ \boxed{\div}\ \boxed{4} = \frac{(3 + 9)}{4} = \frac{12}{4} = \mathbf{3}$$

EXAMPLE 1.17   Determine

$$\frac{1}{4} + \frac{1}{6} + \frac{2}{3}$$

Solution: Since the division will occur first, the correct result will be obtained by simply performing the operations as indicated. That is,

$$\boxed{1}\ \boxed{\div}\ \boxed{4}\ \boxed{+}\ \boxed{1}\ \boxed{\div}\ \boxed{6}\ \boxed{+}\ \boxed{2}\ \boxed{\div}\ \boxed{3} = \frac{1}{4} + \frac{1}{6} + \frac{2}{3} = \mathbf{1.08}$$

Powers of Ten

The $\boxed{\text{EE}}$ key is used to set the power of ten of a number. For the examples that follow, the ENG mode was set to provide results that were a power of three. Setting up the number $2200 = 2.2 \times 10^3$ requires the following key pad selections:

$$\boxed{2}\ \boxed{\cdot}\ \boxed{2}\ \boxed{\text{EE}}\ \boxed{3}\ \boxed{\text{ENTER}} = \mathbf{2.200E3}$$

Setting up the number 8.2×10^{-6} requires the negative sign $(-)$ **from the numerical key pad**. **Do not** use the negative sign from the mathematical listing of $\div$, $\times$, $-$, and $+$. That is,

$$\boxed{8}\ \boxed{\cdot}\ \boxed{2}\ \boxed{EE}\ \boxed{-}\ \boxed{6}\ \boxed{ENTER} = \mathbf{8.200E-6}$$

EXAMPLE 1.18   Perform the addition $6.3 \times 10^3 + 75 \times 10^3$ and compare your answer with the longhand solution of Example 1.5(a).

Solution:

$$\boxed{6}\ \boxed{\cdot}\ \boxed{3}\ \boxed{EE}\ \boxed{3}\ \boxed{+}\ \boxed{7}\ \boxed{5}\ \boxed{EE}\ \boxed{3}\ \boxed{ENTER} = \mathbf{81.300E3}$$

which confirms the results of Example 1.5(a).

EXAMPLE 1.19   Perform the division $(69 \times 10^4)/(13 \times 10^{-8})$ and compare your answer with the longhand solution of Example 1.7(b).

Solution:

$$\boxed{6}\ \boxed{9}\ \boxed{EE}\ \boxed{4}\ \boxed{\div}\ \boxed{1}\ \boxed{3}\ \boxed{EE}\ \boxed{-}\ \boxed{8}\ \boxed{ENTER} = \mathbf{5.308E12}$$

which confirms the results of Example 1.7(b).

EXAMPLE 1.20   Using the provided format of each number, perform the following calculation in one series of key pad entries:

$$\frac{(0.004)(6 \times 10^{-4})}{(2 \times 10^{-3})^2} = ?$$

Solution:

$$\boxed{(}\ \boxed{(}\ \boxed{0}\ \boxed{\cdot}\ \boxed{0}\ \boxed{0}\ \boxed{4}\ \boxed{)}\ \boxed{\times}\ \boxed{(}\ \boxed{6}\ \boxed{EE}\ \boxed{-}\ \boxed{4}\ \boxed{)}\ \boxed{)}\ \boxed{\div}\ \boxed{2}$$

$$\boxed{EE}\ \boxed{-}\ \boxed{3}\ \boxed{x^2}\ \boxed{ENTER} = \mathbf{600.000E-3} = \mathbf{0.6}$$

Brackets were used to ensure that the calculations were performed in the correct order.

1.11   COMPUTER ANALYSIS

The use of computers in the educational process has grown exponentially in the past decade. Very few texts at this introductory level fail to include some discussion of current popular computer techniques. In fact, the very accreditation of a technology program may be a function of the depth to which computer methods are incorporated in the program.

There is no question that a basic knowledge of computer methods is something that the graduating student should carry away from a two-year or four-year program. Industry is now expecting students to have a basic knowledge of computer jargon and some hands-on experience.

For some students, the thought of having to become proficient in the use of a computer may result in an insecure, uncomfortable feeling. Be assured, however, that through the proper learning experience and exposure, the computer can become a very "friendly," useful, and supportive tool in the development and application of your technical skills in a professional environment.

For the new student of computers, two general directions can be taken to develop the necessary computer skills: the study of computer languages or the use of software packages.

Languages

There are several languages that provide a direct line of communication with the computer and the operations it can perform. A **language** is a set of symbols, letters, words, or statements that the user can enter into the computer. The computer system will "understand" these entries and will perform them in the order established by a series of commands called a **program.** The program tells the computer what to do on a sequential, line-by-line basis in the same order a student would perform the calculations in longhand. The computer can respond only to the commands entered by the user. This requires that the programmer understand fully the sequence of operations and calculations required to obtain a particular solution. In other words, the computer can only respond to the user's input—it does not have some mysterious way of providing solutions unless told how to obtain those solutions. A lengthy analysis can result in a program having hundreds or thousands of lines. Once written, the program must be checked carefully to ensure that the results have meaning and are valid for an expected range of input variables. Writing a program can, therefore, be a long, tedious process, but keep in mind that once the program is tested and true, it can be stored in memory for future use. The user can be assured that any future results obtained have a high degree of accuracy but require a minimum expenditure of energy and time. Some of the popular languages applied in the electrical-electronics field today include C++, QBASIC, Pascal, and FORTRAN. Each has its own set of commands and statements to communicate with the computer, but each can be used to perform the same type of analysis.

Software Packages

The second approach to computer analysis—**software packages**—avoids the need to know a particular language; in fact, the user may not be aware of which language was used to write the programs within the package. All that is required is a knowledge of how to input the network parameters, define the operations to be performed, and extract the results; the package will do the rest. The individual steps toward a solution are beyond the needs of the user—all the user needs is an idea of how to get the network parameters into the computer and how to extract the results. Herein lie two of the concerns of the author with packaged programs: obtaining a solution without the faintest idea of either how the solution was obtained or whether the results are valid or way off base. It is imperative that you realize that the computer should be used as a tool to assist the user—it must not be allowed to control the scope and potential of the user! Therefore, as we progress through the chapters of the text, be sure that you clearly understand the concepts before turning to the computer for support and efficiency.

Each software package has a **menu,** which defines the range of application of the package. Once the software is entered into the computer, the system will perform all the functions appearing in the menu, as it was preprogrammed to do. Be aware, however, that if a particular type of analysis is requested that is not on the menu, the software package cannot provide the desired results. The package is limited solely to those maneuvers developed by the team of programmers who developed the software package. In such situations the user must turn to another software package or write a program using one of the languages listed above.

In broad terms, if a software package is available to perform a particular analysis, then that package should be used rather than developing new routines. Most popular software packages are the result of many hours of effort by teams of programmers with years of experience. However, if the results are not in the desired format, or if the software package does not provide all the desired results, then the user's innovative talents should be put to use to develop a software package. As noted above, any program the user writes that passes the tests of range and accuracy can be considered a software package of his or her authorship for future use.

Three software packages are used throughout this text: Cadence's OrCAD's PSpice 9.2, Electronics Workbench Multisim 2001, and MathSoft's Mathcad 2000, all of which are shown in Fig. 1.6. Although both PSpice and Electronics Workbench are designed to analyze electric circuits, there are sufficient differences between the two to warrant covering each approach. The growing use of some form of mathematical support in the educational and industrial environments supports the introduction and use of Mathcad throughout the text. However, you are not required to obtain all three programs in order to proceed with the content of this text. The primary reason for including these programs is simply to introduce each and demonstrate how each can support the learning process. In most cases, sufficient detail has been provided to actually use the software package to perform the examples provided, although it would certainly be helpful to have someone to turn to if questions arise. In addition, the literature supporting all three packages has improved dramatically in recent years and should be available through your bookstore or a major publisher. Appendix B lists all the system requirements, including how to get in touch with each company.

1.12 COMPUTER (PC) SPECIFICATIONS

This section will provide a surface description of the data usually provided with a computer system such as the Dell Dimension 8200 depicted in Fig. 1.7. A detailed description would take many more pages than permitted in this text. The topics chosen cover specific data about a computer system that users of the system should remember so that they can talk intelligently to other individuals and also so that they can properly review potential software programs to determine whether they will function well in the system.

It is important to realize from the start that there is no one simple way of defining which computer will be the best for each user. Buying a computer is like buying a car—each will have its own design and its own operating characteristics. For instance, although two cars may have the same horsepower, how is it achieved in each car? Is a turbo-booster required? What is the associated torque? What is the gear ratio at different gear levels? What is the associated gas mileage? The questions can seem endless. The following information is presented simply to provide some understanding of each piece of data provided so that more intelligent decisions can be made about which computer is best for which person and which application.

(a)

(b)

Processor

The term *processor* is short for **central processing unit,** or **CPU.** It is a microprocessor in the form of an integrated circuit (IC) that contains millions of electronic components. In every respect,

the CPU is the heart (some prefer the term "brains") of the computer and will define the maximum speed of operation for the computer.

It will also determine the size of the data string for each byte of information, the manner in which data will be operated on, the amount of memory that can be accessed, and so on. However, it is important to realize that the speed provided on the data sheet for the CPU is not sufficient to determine which computer will perform particular tasks the fastest. CPUs of the same speed but different manufacturers will not perform the same tasks at the same rate due to the way the operations are performed, the size of the **cache** memory (internal

(c)

FIG. 1.6
Software packages: (a) Cadence's OrCAD PSpice release 9.2; (b) Electronics Workbench Multisim 2001; (c) MathSoft's Mathcad 2000.

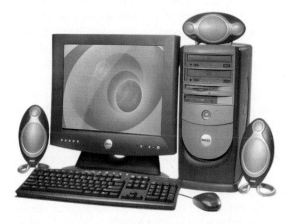

Dell™ Dimension™ 8200

Intel® Pentium 4 Processor at 2.40 GHz with 533 MHz system bus and 512K L2 cache
256MB RDRAM
40GB Ultra ATA/100 hard drive (7200 rpm)
17" E772 monitor
64 MB DDR NVIDIA® GeForce™ Mx graphics card with TV out
16x max. DVD-ROM drive
SB Live! 1024 V digital sound card
Harmon Kardon HK-695 surround sound speakers with subwoofer
56K PCI telephony modem
10/100 PCI fast Ethernet network card
Microsoft® Windows® XP Home Edition; Microsoft Works Suite 2002

FIG. 1.7
Dell™ Dimension™ 8200 desktop computer system and specifications.
(Courtesy of Dell.)

CPU memory), the bus speed, data and address access size, and so on. In total, therefore, it is important to realize that

the CPU speed alone will not reveal how fast the computer system will perform particular tasks.

It is possible that one computer model using a CPU with a clock speed of 2 GHz will perform particular tasks more slowly than another computer model with a speed of 1.86 Hz, given all the other important factors that are associated with the CPU. However,

in general, the CPU speed will provide a relative idea how fast a computer will perform particular tasks.

That is, there is no question that a 2 GHz system is faster than most 1.6 GHz systems and that a 2.8 GHz system is faster than a 2.2 GHz system. Today, most new PCs have CPU speeds from 1.6 GHz to 3.06 GHz, and faster models are released every month.

We often use terms such as MHz and GHz without giving a moment's thought as to what we are really dealing with. A CPU with a clock speed of 2 GHz is capable of performing not 2000 or 200,000 or even 2,000,000, but 2000 million or 2,000,000,000 operations (steps) in 1 second. Now that is an incredible achievement. Just think how many you can handle in 1 second.

In total, the CPU will determine how fast a set of tasks can be performed, that is, how long it will take to perform a series of calculations, transfer data, store and retrieve information, and so on. A system that is too slow, with insufficient memory, might require that you wait for the computer to complete the tasks required to display the information. In other words, after finishing a few pages of a file, you may have to wait for a line to appear after you have finished typing the line.

When you are purchasing software, the required CPU speed is an important factor. Windows 2000 requires a CPU of at least 133 MHz, whereas Windows XP requires a minimum speed of 233 MHz. In general,

the processor speed must always equal or exceed the specified speed of the software package's specifications.

Memory

The next piece of information normally provided is the amount of memory installed. Although information can be saved and retrieved from the hard drive, floppy disks, or CDs, the term *memory* in the advertising literature generally refers to the random access memory (RAM) only. Most well-written software packages will include the amount of memory required to operate the program and the amount of free hard disk space needed to save the program.

In general, random access memory (RAM) is a temporary (also called "volatile") read/write storage area in which data and programs can be entered and accessed as needed.

When a computer is first turned on, the data in RAM storage are a mix of random information of no real value. However, the CPU will immediately direct the computer system to retrieve information from the hard drive and place it in RAM memory so that the computer can execute the task at hand. As a task is performed by the computer, information will be flowing into and out of RAM memory at a very high rate of speed. When the computer is shut down, the information stored in RAM is lost and worthless until the system is turned on again.

As with the CPU, there are a wide variety of RAM modules with different labels. Three of the most popular include **SDRAM (synchonous dynamic random access memory—also called DIMM), RDRAM (RAMBUS dynamic random access memory), and DDR (double data rate SDRAM).** SDRAMs are typically available from 64 MB to 512 MB (the B stands for "byte" of information), and RDRAMs and DDRs from 128 MB to 1 GB.

A second piece of data provided with RAMs is their **operating speed.** The faster the speed, the more quickly information can be transferred to or from RAM. Typical speeds for SDRAMs extend from 100 MHz to 133 MHz, whereas RAMBUS units range from 600 MHz to 800 MHz and DDRs from 200 MHz to 266 MHz.

The memory requirement is one that often goes hand in hand with the CPU speed. For instance, Windows 2000 requires a 133 MHz CPU as mentioned above, but it also requires at least 32 MB to 64 MB of memory. For Windows XP, the minimum amount of memory is 128 MB. For any software package,

the computer memory must be sufficient to store both the operating system and the application programs.

Hard Drive

The next information typically provided consists of the hard drive specifications—a READ/WRITE storage mechanism. This mechanism is used by the computer to store information such as the operating system of the computer, system documentation for the computer, programs such as Windows, user data, and so on, with the ability to have additional information stored on the drive at any time. When the computer is first turned on, the operating system will be transferred from the hard drive to RAM memory for use by the computer. The hard drive is so named because the information sits on a hard disk that typically rotates at 3000 to 10,000 rpm (over 166 revolutions per second—pretty fast!), with 9 ms access speed (the average time required to retrieve data). In general, therefore,

the hard drive is a storage device whose effectiveness is determined by a number of factors, including its storage capacity, rotational speed, and access speed.

For today's computers, a range of 20 GB to 80 GB is typical. Most operate between 5400 rpm and 7200 rpm, with access speeds that range from 6 ms to 10 ms.

The hard drive is the "library" for the computer. It contains the Windows file, special operating files, programs such as Mathcad and Electronics Workbench, stored data, Word files, and so on. All software packages specify a minimum free space for the storage of the program. You must be continually aware of the remaining disk space when considering a program or new project of any size.

Floppy/Diskette Drive

For years the floppy disk was the storage mechanism for data, files, and so on. Floppies came in 8", 5¼", and 3½" sizes. The 8" and 5¼" are now considered obsolete, and the 3½" will follow in the near future. Since so much information has been placed on these disks over the years, most new computers still provide a floppy disk drive, although it may not appear in the prominent list of specifications. Most of the current floppies have 1.44 MB storage capacities which 10 years ago was considerable. The main purpose of a floppy disk today is to initially install an operating system and to transport small data or program files.

CD Drives

There are a variety of **compact disk (CD) 5¼" drives** on the market today. A basic **CD** or **CD-ROM** is a **read only** device, and data cannot be stored or modified on it. If the disk is labeled **CD-R,** the data on it can be read and information can be stored, but the recordable feature can be applied only once. A **CD-RW** has the ability to read and write more than once (it is rewritable). For CD-R or CD-RW drives, special disks must be used.

The term **burner** is applied to CD drives that have the record capability. A standard CD has a storage capacity of 650 MB, with CD-RWs at 550 MB or 650 MB. Take a second to appreciate that a 650 MB CD has over 450 times the storage of a 1.44 MB floppy—no wonder floppies are rapidly becoming obsolete! Yes, it is true that writable CDs may be slightly more expensive today, but as with everything else the price will drop with time and demand. The speed on specification sheets is often written as $1\times$, $10\times$, $40\times$, and so on, rather than as kBps or MBps. The $\times$ should be read as "times" a standard, the standard (or $1\times$) speed being 150 kBps. Therefore, if you see $48\times$, you know that the speed is 48 times 150 kBps, or 7.2 MBps. For a CD-RW burner the specifications may appear as "$16\times/10\times/32\times$," revealing that the drive will write on CD-R disks at 16× or 2.4 MBps and on CD-RW disks at 10× or 1.5 MBps, and it will read at 32×, or 4.8 MBps.

If a disk drive is labeled **DVD**, it can be used to play **digital video disks**, whereas **DVD-RW** drives have a read/write capability. DVDs and DVD-RWs have storage capacities that range from 4.7 GB to 17 GB. For DVDs the standard or 1× speed is 1.38 MBps, with 6× being 8.28 MBps. If a DVD-RW is labeled "18×/12×/36×/4×," the first three are CD multipliers for CD-R write, CD-RW write, and CD Read. However, the last is a multiplier of the DVD standard, so the read/write speed is 5.5 MBps.

In summary, it is important to realize that a CD drive cannot play a DVD disk, but a DVD drive can play anything—a characteristic known as **downward compatible**.

Display Adapter Card

The **display adapter card,** also known as the *graphics card, video adapter card*, and combinations thereof, is the link between the data generated by the computer and the way the data are displayed on the monitor. All display adapters must be compatible with the established standard, referred to as the **video graphics adapter (VGA) which has 640 × 480 dpi (dots per inch) and 16 colors, with at least 256 kB of memory (also specified as 320 × 200 dpi with 256 colors).** The standard for the **extended graphics adapter (XGA) is 1024 × 768 dpi with 16 colors and 512 kB of memory.** If the letters **DVI** (digital visual interface) appear in the description, the adapter has a special connector that supports digital signals. The adapter card used will define the type of monitor that can be used, the number of colors, and the overall resolution.

Monitor

The monitors used today are either the **CRT (cathode-ray tube)** or the **FPD (flat-panel display).** Most are of the 15" to 19" variety, with the dimension measured diagonally between the corners of the display. Naturally, in order for you to take full advantage of the monitor, it must equal or exceed the specifications of the display adapter just described. It is important to realize that cathode-ray tube displays are typically analog devices. The signals from the computer must be converted from digital to analog to operate the display monitor. Flat-panel displays, on the other hand, are digital devices but often include an analog interface so that they are compatible with older display adapters. The chosen monitor will determine the depth of color, the brightness, and the general quality of the image.

Modem

A modem is a device used to connect your computer to other informational systems using a telephone line, cable system, or dedicated line. Without a modem, your computer is an isolated unit that cannot communicate with the outside world. The modem determines the maximum rate at which information is transmitted to your computer. Higher speeds are luxuries that most frequent modem users would not be willing to forgo. The telephone connection uses a **standard modem** that can be built in or added on to the computer; it normally operates at a speed of 48 kBps to 56.6 kBps. For the cable TV connection, a **cable modem** is used that can operate at speeds up to 1 MBps. Separately installed high-speed lines, referred to as *DSL (digital subscriber line),* typically have speeds that range from 64 kBps to 500 kBps; but higher speeds from 500 kBps to 2.5 MBps are available if you are willing to pay the price. In any case, be sure that the modem employed can handle the provider signal and speed.

The above listing covers most of the important parameters that you should be aware of when purchasing a computer or checking whether a software package can be installed without overtaxing the computer system. Manufacturers also use other abbreviations which have not been discussed above, but a salesperson should be able to expand on those with little difficulty. The material has been provided to provide a basis of comparison for your investigation and to point out the major areas of concern.

CHAPTER SUMMARY

UNITS OF MEASUREMENT

- One must always be aware of the unit of measurement applied to a particular quantity.
- A unit of measurement must be applied to any result to give it full meaning.
- The SI system is the system of units applied to the electrical-electronics industry.

POWERS OF TEN

- Powers of ten are employed to make it easier to work with the very small and very large numbers that appear in this industry.
- To perform addition or subtraction using powers of ten, the power of ten for each term must be the same.
- Multiplication is performed by adding (algebraically) the powers of ten.
- Division is performed by subtracting the power of ten of the denominator from the power of ten of the numerator.
- The power of a number in power-of-ten format is found by multiplying the power of ten of the number by the power to be determined.

NOTATION

- Fixed-point notation requires that the decimal point appear in the same place for any number.
- Floating-point notation simply means that the decimal point will be placed where it makes the most sense for the given number.
- Scientific notation requires that the decimal point appear directly after the first number that is greater than or equal to 1 but less than 10.
- Engineering notation requires that the power of ten associated with the number be a multiple of (or divisible by) three and that the mantissa be greater than or equal to 1 but less than 1000.

CONVERSIONS

- When converting from one power of ten to another, remember that if the power increases, the multiplier decreases, and vice versa. The decimal point is moved the same number of places as the change in power of ten.
- When converting from one unit of measurement to another, be sure that the units of each term are placed such that the result has the units you are looking for.

CALCULATORS

- Always check the result of a calculator to be sure that the result makes sense. It is possible that you entered the numbers incorrectly or made an invalid maneuver with the calculator.
- Take the time to read the manual that comes with your calculator to be sure that you are setting the mode correctly and that you are aware of all its special features.

COMPUTERS

- Whenever a software program is available that can perform the operations you require, it is usually wise to use the program rather than try to write your own program. The commercial program is the result of many hours of labor that would be difficult to match on an individual basis. Of course, if the software package does not perform the operations you need, you should turn to a programming language.
- Knowing the meaning of the terms that define a computer will help you determine which system is best for your needs and will help you talk intelligently with others in your field.

Important Equations

$$\frac{1}{10^n} = 10^{-n} \qquad \frac{1}{10^{-n}} = 10^n$$

$$(10^n)(10^m) = 10^{(n+m)}$$

$$\frac{10^n}{10^m} = 10^{(n-m)}$$

$$(10^n)^m = 10^{nm}$$

PROBLEMS

Note: More difficult problems are denoted by an asterisk (*) throughout the text.

SECTION 1.2   A Brief History

1. Visit your local library (at school or home) and describe the extent to which it provides literature and computer support for the technologies—in particular, electricity, electronics, electromagnetics, and computers.

2. Choose an area of particular interest in this field and write a very brief report on the history of the subject.

3. Choose an individual of particular importance in this field and write a very brief review of his or her life and important contributions.

SECTION 1.3   Units of Measurement

4. What is the velocity of a rocket in mph if it travels 20,000 ft in 10 s?

5. In a recent Tour de France time trial, Lance Armstrong traveled 31 miles in a time trial in 1 hour and 4 minutes. What was his average speed in mph?

*6. A pitcher has the ability to throw a baseball at 95 mph.
   a. How fast is the speed in ft/s?
   b. How long does the hitter have to make a decision about swinging at the ball if the plate and the mound are separated by 60 feet?
   c. If the batter wanted a full second to make a decision, what would the speed in mph have to be?

SECTION 1.4   Systems of Units

7. Are there any relative advantages associated with the metric system compared to the English system with respect to length, mass, force, and temperature? If so, explain.

8. Which of the four systems of units appearing in Table 1.1 has the smallest units for length, mass, and force? When would this system be used most effectively?

*9. Which system of Table 1.1 is closest in definition to the SI system? How are the two systems different? Why do you think the units of measurement for the SI system were chosen as listed in Table 1.1? Give the best reasons you can without referencing additional literature.

10. What is room temperature (68°F) in the MKS, CGS, and SI systems?

11. How many foot-pounds of energy are associated with 1000 J?

12. How many centimeters are there in 1/2 yd?

SECTION 1.5   Powers of Ten

13. Express the following numbers as powers of ten:
   a. 10,000   b. 1,000,000   c. 1000
   d. 0.001   e. 1   f. 0.1

14. Using only those powers of ten listed in Table 1.2, express the following numbers in what seems to you the most logical form for future calculations:
   a. 15,000   b. 0.030   c. 2,400,000
   d. 150,000   e. 0.00040200   f. 0.0000000002

15. Perform the following operations and express your answer as a power of ten:
   a. $4200 + 48{,}000$ b. $9 \times 10^4 + 3.6 \times 10^5$
   c. $0.5 \times 10^{-3} - 6 \times 10^{-5}$ d. $1.2 \times 10^3 + 50{,}000 \times 10^{-3} - 0.006 \times 10^6$

16. Perform the following operations and express your answer as a power of ten:
   a. (100) (1000)
   b. (0.01) (1000)
   c. $(10^3) (10^6)$
   d. (100) (0.00001)
   e. $(10^{-6}) (10,000,000)$
   f. $(10,000) (10^{-8}) (10^{28})$

17. Perform the following operations.
   a. (50,000) (0.0003)
   b. 2200×0.002
   c. (0.000082) (2,800,000)
   d. $(30 \times 10^{-4}) (0.004) (7 \times 10^8)$

18. Perform the following operations.
   a. $\dfrac{100}{10,000}$
   b. $\dfrac{0.010}{1000}$
   c. $\dfrac{10,000}{0.001}$

   d. $\dfrac{0.0000001}{100}$
   e. $\dfrac{10^{38}}{0.000100}$
   f. $\dfrac{(100)^{1/2}}{0.01}$

19. Perform the following operations.
   a. $\dfrac{2000}{0.00008}$
   b. $\dfrac{0.004}{60,000}$
   c. $\dfrac{0.000220}{0.00005}$
   d. $\dfrac{78 \times 10^{18}}{4 \times 10^{-6}}$

20. Perform the following operations.
   a. $(100)^3$
   b. $(0.0001)^{1/2}$
   c. $(10,000)^8$
   d. $(0.00000010)^9$

21. Perform the following operations.
   a. $(400)^2$
   b. $(0.006)^3$
   c. $(0.004) (6 \times 10^2)^2$
   d. $((2 \times 10^{-3}) (0.8 \times 10^4) (0.003 \times 10^5))^3$

22. Perform the following operations and express your answer in scientific notation:
   a. $(-0.001)^2$
   b. $\dfrac{(100)(10^{-4})}{1000}$

   c. $\dfrac{(0.001)^2(100)}{10,000}$
   d. $\dfrac{(10^3)(10,000)}{1 \times 10^{-4}}$

   e. $\dfrac{(0.0001)^3(100)}{1 \times 10^6}$
   *f. $\dfrac{[(100)(0.01)]^{-3}}{[(100)^2][0.001]}$

23. Perform the following operations and express your answer in engineering notation:
   a. $\dfrac{(300)^2(100)}{3 \times 10^4}$
   b. $[(40,000)^2] [(20)^{-3}]$

   c. $\dfrac{(60,000)^2}{(0.02)^2}$
   d. $\dfrac{(0.000027)^{1/3}}{200,000}$

   e. $\dfrac{[(4000)^2][300]}{2 \times 10^{-4}}$
   f. $[(0.000016)^{1/2}] [(100,000)^5] [0.02]$

   *g. $\dfrac{[(0.003)^3][(0.00007)^2][(800)^2]}{[(100)(0.0009)]^{1/2}}$ (a challenge)

SECTION 1.6   Conversion between Levels of Powers of Ten

24. Fill in the blanks of the following conversions:
   a. $6 \times 10^3 = \underline{\quad} \times 10^6$
   b. $4 \times 10^{-3} = \underline{\quad} \times 10^{-6}$
   c. $50 \times 10^5 = \underline{\quad} \times 10^3 = \underline{\quad} \times 10^6$
   $\qquad\qquad\quad = \underline{\quad} \times 10^9$
   d. $30 \times 10^{-8} = \underline{\quad} \times 10^{-3} = \underline{\quad} \times 10^{-6}$
   $\qquad\qquad\qquad = \underline{\quad} \times 10^{-9}$

25. Perform the following conversions:
   a. 0.05 s to milliseconds
   b. 2000 μs to milliseconds
   c. 0.04 ms to microseconds
   d. 8400 ps to microseconds
   e. 4×10^{-3} km to millimeters
   f. 260×10^3 mm to kilometers

26. Perform the following conversions:
    a. 1.5 min to seconds
    b. 0.04 h to seconds
    c. 0.05 s to microseconds
    d. 0.16 m to millimeters
    e. 0.00000012 s to nanoseconds
    f. 3,620,000 s to days
    g. 1000 mm to meters

27. Perform the following conversions:
    a. 0.1 μF (microfarad) to picofarads
    b. 0.5 km to meters
    c. 80 mm to centimeters
    d. 60 cm to kilometers
    e. 3.2 h to milliseconds
    f. 0.016 mm to micrometers
    g. 60 sq cm (cm^2) to square meters (m^2)

28. Perform the following conversions:
    a. 100 in. to meters
    b. 4 ft to meters
    c. 6 lb to newtons
    d. 60,000 dyn to pounds
    e. 150,000 cm to feet
    f. 0.002 mi to meters (5280 ft = 1 mi)
    g. 7800 m to yards

29. What is a mile in feet, yards, meters, and kilometers?

30. Calculate the speed of light in miles per hour using the speed defined in Section 1.4.

31. How long in seconds will it take a car traveling at 60 mph to travel the length of a football field (100 yd)?

32. Convert 6 mph to meters per second.

33. If an athlete can row at a rate of 50 yd/min, how many days would it take to cross the Atlantic ($\cong$ 3000 mi)?

34. How long would it take a runner to complete a 10 km race if a pace of 6.5 min/mi were maintained?

35. Quarters are about 1 in. in diameter. How many would be required to stretch from one end of a football field to the other (100 yd)?

36. Compare the total time in hours required to drive across the United States ($\cong$ 3000 mi) at an average speed of 60 mph versus an average speed of 70 mph. Is the time saved for such a long trip worth the added risk of the higher speed?

*37. Find the distance in meters that a mass traveling at 600 cm/s will cover in 0.016 h.

*38. Each spring there is a race up 86 floors of the 102 story Empire State Building in New York City. If you were able to climb 2 steps/second, how long would it take you to reach the 86th floor if each floor is 14 ft high and each step is about 9 in.?

*39. The record for the race in Problem 38 is 10 minutes, 47 seconds. What was the racer's speed in min/mi for the race?

*40. If the race of Problem 38 were a horizontal distance, how long would it take a runner who can run 5 min miles to cover the distance? Compare this with the record speed of Problem 39. Gravity is certainly a factor to be reckoned with!

SECTION 1.9 Conversion Tables

41. Using Appendix A, determine the number of
    a. Btu in 5 J of energy.
    b. cubic meters in 24 oz of a liquid.
    c. seconds in 1.4 days.
    d. pints in 1 m^3 of a liquid.

SECTION 1.10 Calculators

Perform the following operations using a calculator:

42. $6(4 + 8) =$

43. $\dfrac{(20 + 32)}{4} =$

44. $\sqrt{5^2 + 12^2} =$

45. $\cos 50° =$

46. $\tan^{-1} \dfrac{3}{4} =$

47. $\sqrt{\dfrac{400}{6^2 + 10}} =$

48. $\dfrac{8.2 \times 10^{-3}}{0.04 \times 10^{3}} =$

*49. $\dfrac{(0.06 \times 10^{5})(20 \times 10^{3})}{(0.01)^{2}} =$

*50. $\dfrac{4 \times 10^{4}}{2 \times 10^{-3} + 400 \times 10^{-5}} + \dfrac{1}{2 \times 10^{-6}} =$

SECTION 1.11 Computer Analysis

51. Investigate the availability of computer courses and computer time in your curriculum. Which languages are commonly used, and which software packages are popular?

52. Develop a list of three popular computer languages, including a few characteristics of each. Why do you think some languages are better for the analysis of electric circuits than others?

GLOSSARY

Burner   A CD drive capable of recording data on a CD.

Cache   A continually updated memory location that stores data which the computer has recently used to speed up the overall performance of the system.

Cathode-ray tube (CRT)   A glass enclosure with a relatively flat face (screen) and a vacuum inside that will display the light generated from the bombardment of the screen by electrons.

Central processing unit (CPU)   The main IC (often called the "brains" or "heart") of a computer that interprets the input instructions, performs calculations, and controls the flow of information throughout the system.

CGS system   The system of units employing the *C*entimeter, *G*ram, and *S*econd as its fundamental units of measure.

Compact disk (CD) drive   An electromechanical package that controls the retrieving and recording of information on a compact disk (CD).

Compact disk-recordable (CD-R)   A CD disk permitting only one writing of information on its surface.

Compact disk-rewritable (CD-RW)   A CD disk permitting more than one writing of information on its surface.

Difference engine   One of the first mechanical calculators.

Digital video disk (DVD)   A high-density CD that can carry the equivalent of 13 typical CDs; it was initially developed primarily to store entire movies on a single CD.

Display adapter card   A circuit board that plugs into an expansion slot of the computer to translate the provided video signal to one that can be understood by the monitor.

DRAM (dynamic RAM)   A computer chip constituting the main memory of the computer.

DVD-RW drive   A DVD drive that permits more than one writing of information on the surface of the DVD disk.

Edison effect   Establishing a flow of charge between two elements in an evacuated tube.

Electromagnetism   The relationship between magnetic and electrical effects.

Engineering notation   A method of notation that specifies that all powers of ten used to define a number be multiples of 3 with a mantissa greater than or equal to 1 but less than 1000.

ENIAC   The first totally electronic computer.

Extended graphics adapter (XGA)   A circuit board with enhanced memory characteristics to permit an improved video display.

Fixed-point notation   Notation using a decimal point in a particular location to define the magnitude of a number.

Flat-panel display (FPD)   A monitor that employs LCDs (liquid-crystal displays) to permit a thinner monitor that takes up less space and uses less electricity.

Fleming's valve   The first of the electronic devices, the diode.

Floating-point notation   Notation that allows the magnitude of a number to define where the decimal point should be placed.

Integrated circuit (IC)   A subminiature structure containing a vast number of electronic devices designed to perform a particular set of functions.

Joule (J)   A unit of measurement for energy in the SI or MKS system. Equal to 0.7378 foot-pound in the English system and 10^{7} ergs in the CGS system.

Kelvin (K) A unit of measurement for temperature in the SI system. Equal to 273.15 + °C in the MKS and CGS systems.

Kilogram (kg) A unit of measure for mass in the SI and MKS systems. Equal to 1000 grams in the CGS system.

Language A communication link between user and computer to define the operations to be performed and the results to be displayed or printed.

Leyden jar One of the first charge-storage devices.

Menu A computer-generated list of choices for the user to determine the next operation to be performed.

Meter (m) A unit of measure for length in the SI and MKS systems. Equal to 1.094 yards in the English system and 100 centimeters in the CGS system.

MKS system The system of units employing the *Meter*, *Kilogram*, and *Second* as its fundamental units of measure.

Modem A device that permits communication with the computer using telephone lines, TV cables, or dedicated lines.

Newton (N) A unit of measurement for force in the SI and MKS systems. Equal to 100,000 dynes in the CGS system.

Pound (lb) A unit of measurement for force in the English system. Equal to 4.45 newtons in the SI or MKS system.

Program A sequential list of commands, instructions, and so on, to perform a specified task using a computer.

PSpice A software package designed to analyze various dc, ac, and transient electrical and electronic systems.

Random access memory (RAM) A temporary storage location inside the computer that releases the stored information when the computer is turned off.

RDRAM (RAMBUS DRAM) A memory chip that works faster to increase the clock speed of a computer to reduce the time it takes to complete various tasks.

Scientific notation A method for describing very large and very small numbers through the use of powers of ten, which requires that the multiplier be a number between 1 and 10.

SDRAM (synchronous dynamic random access memory) A memory chip that can operate in lock-step with the clock signals of the memory bus (hence the term *synchronous*).

Second (s) A unit of measurement for time in the SI, MKS, English, and CGS systems.

SI system The system of units adopted by the IEEE in 1965 and the USASI in 1967 as the International System of Units (*Système International d'Unités*).

Slug A unit of measure for mass in the English system. Equal to 14.6 kilograms in the SI or MKS system.

Software package A computer program designed to perform specific analysis and design operations or generate results in a particular format.

Static electricity Stationary charge in a state of equilibrium.

Transistor The first semiconductor amplifier.

Voltaic cell A storage device that converts chemical to electrical energy.

Current and Voltage

OBJECTIVES

- Become aware of the basic atomic structure of copper and understand why copper is used so extensively in this field.
- Understand how the terminal voltage of a battery is established and what role it plays in a basic electric circuit.
- Become aware of the properties of current in an electric circuit and learn how current is calculated.
- Become familiar with the properties of different types of voltage sources such as batteries, generators, and laboratory power supplies.
- Be able to use the ampere-hour rating of a battery to calculate its expected life at a particular current level.
- Become aware of the factors that affect the life of a battery.
- Understand the differences among conductors, insulators, and semiconductors.
- Be able to use an ammeter and a voltmeter correctly to measure a circuit's current and voltage, respectively.
- Develop a clear understanding of the working mechanism of a flashlight and a battery charger.

2.1 INTRODUCTION

Now that the foundation for the study of electricity/electronics has been established, the concepts of voltage and current can be investigated. In today's world it would be absolutely impossible not to come in contact with the term **voltage** at some point in our everyday lives. We have all replaced batteries in our flashlights, answering machines, calculators, automobiles, and so on, that had specific voltage ratings. The fact that most outlets in the home are 120 volts is also quite well known. Although **current** may be a less familiar term, we become aware of its importance when we place too many appliances on the same outlet—the circuit breaker opens due to the excessive current that results. It is fairly common knowledge that current is something that moves through the wires and causes sparks and possibly fire if there is a "short circuit"; it will heat up the coils of an electric heater or the range of an electric stove; it generates light when passing through the filament of a bulb; it causes twists and kinks in the wire of an electric iron over time, and so on. All in all, the terms *voltage* and *current* are part of the vocabulary of most individuals.

In this chapter, the basic impact of current and voltage and the properties of each will be introduced and discussed in some detail. Hopefully, any mysteries surrounding the general characteristics of each will be eliminated, and you will gain a clear understanding of the impact of each on an electric/electronics circuit.

2.2 ATOMS AND THEIR STRUCTURE

A basic understanding of the fundamental concepts of current and voltage requires a degree of familiarity with the atom and its structure. The simplest of all atoms is the hydrogen atom, made up of two basic particles, the **proton** and the **electron,** in the relative positions shown in Fig. 2.1(a). The **nucleus** of the hydrogen atom is the proton, a positively charged particle.

The orbiting electron carries a negative charge equal in magnitude to the positive charge of the proton.

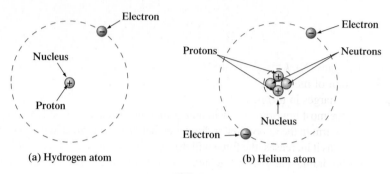

(a) Hydrogen atom   (b) Helium atom

FIG. 2.1

The hydrogen and helium atoms.

In all other elements, the nucleus also contains **neutrons,** which are slightly heavier than protons and have **no electrical charge.** The helium atom, for example, has two neutrons in addition to two electrons and two protons, as shown in Fig. 2.1(b). In general,

the atomic structure of any stable atom has an equal number of electrons and protons.

Different atoms have various numbers of electrons in concentric orbits called *shells* around the nucleus. The first shell, which is closest to the nucleus, can contain only two electrons. If an atom has three electrons, the extra electron must be placed in the next shell. The number of electrons in each succeeding shell is determined by $2n^2$ where n is the shell number. Each shell is then broken down into subshells where the number of electrons is limited to 2, 6, 10, and 14 in that order as you move away from the nucleus.

Copper is the most commonly used metal in the electrical/electronics industry. An examination of its atomic structure will reveal why it has such widespread application. As shown in Fig. 2.2, it has 29 electrons in orbits around the nucleus, with the 29th electron appearing all by itself in the 4th shell. Note that the number of electrons in each shell and subshell is as defined above. There are two important things to note in Fig. 2.2. First, the 4th shell, which can have a total of $2n^2 = 2(4)^2 = 32$ electrons, has only one electron. **The**

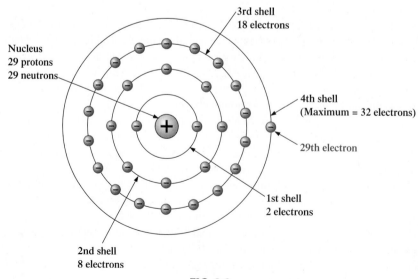

FIG. 2.2

The atomic structure of copper.

outermost shell is incomplete and, in fact, is far from complete because it has only 1 electron. Atoms with complete shells (that is, a number of electrons equal to $2n^2$) are usually quite stable. Those atoms with a small percentage of the defined number for the outermost shell are normally considered somewhat unstable and volatile. Second, the 29th electron is the **farthest electron from the nucleus.** Opposite charges are attracted to each other, but the farther apart they are, the less the attraction. In fact, the force of attraction between the nucleus and the 29th electron of copper can be determined by **Coulomb's law** developed by Charles Augustin Coulomb (Fig. 2.3) in the late 18th century:

$$F = k\frac{Q_1Q_2}{r^2} \qquad \text{(newtons, N)} \qquad \textbf{(2.1)}$$

where F is in newtons (N), k = a constant = 9.0×10^9 N·m²/C², Q_1 and Q_2 are the charges in coulombs (a unit of measure to be introduced in the next section), and r is the distance between the two charges in meters.

At this point, the most important thing to note is that the distance between the charges appears as a squared term in the denominator. First, the fact that this term is in the denominator clearly reveals that as it increases, the force will decrease. However, since it is a squared term, the force will drop dramatically with distance. For instance, if the distance is doubled, the force will drop to 1/4 because $(2)^2 = 4$. If the distance is increased by a factor of 4, it will drop by 1/16, and so on. The result, therefore, is that the force of attraction between the 29th electron and the nucleus is significantly less than that between an electron in the first shell and the nucleus. The result is that the 29th electron is **loosely bound to the atomic structure** and with a little bit of pressure from outside sources could be encouraged to leave the parent atom.

If this 29th electron gains sufficient energy from the surrounding medium to leave the parent atom, it is called a **free electron.** In 1 cubic inch of copper at room temperature, there are approximately 1.4×10^{24} free electrons. Expanded, that is 1,400,000,000,000,000,000,000,000 free electrons in a 1 in. square cube. The point is that we are dealing with enormous numbers of electrons when we talk about the number of free electrons in a copper wire—not just a few that you could leisurely count. Further, the numbers involved are clear evidence of the need to become proficient in the use of powers of ten to represent numbers and use them in mathematical calculations.

Other metals that exhibit the same properties as copper, but to a different degree, are silver, gold, and aluminum, and some rarer metals such as tungsten. Additional comments on the characteristics of conductors will be found in the sections to follow.

2.3   VOLTAGE

If we separate the 29th electron in Fig. 2.2 from the rest of the atomic structure of copper by a dashed line as shown in Fig. 2.4(a), we create regions that have a net positive and negative charge as shown in Fig. 2.4(b) and (c). For the region inside the dashed boundary, the number of protons in the nucleus exceeds the number of orbiting electrons by 1, so the net charge is positive as shown in both figures. This positive region created by separating the free electron from the basic atomic structure is called a **positive ion.** If the free electron then leaves the vicinity of the parent atom as shown in Fig. 2.4(d), regions of positive and negative charge have been established.

This separation of charge to establish regions of positive and negative charge is the action that occurs in every battery. Through chemical action a heavy concentration of positive charge (positive ions) is established at the positive terminal, with an equally heavy concentration of negative charge (electrons) at the negative terminal.

In general,

every source of voltage is established by simply creating a separation of positive and negative charges.

It is that simple: If you want to create a voltage level of any magnitude, simply establish a region of positive and negative charge. The more the required voltage, the greater the quantity of positive and negative charge.

In Fig. 2.5(a), for example, a region of positive charge has been established by a packaged number of positive ions, and a region of negative charge by a similar number of electrons, both

Courtesy of the Smithsonian Institution. Photo No. 52,597

FIG. 2.3
Charles Augustin Coulomb.

French (Angoulême, Paris)
(1736–1806) Scientist and Inventor
Military Engineer, West Indies

Attended the engineering school at Mezieres, the first such school of its kind. Formulated *Coulomb's law,* which defines the force between two electrical charges and is, in fact, one of the principal forces in atomic reactions. Performed extensive research on the friction encountered in machinery and windmills and the elasticity of metal and silk fibers.

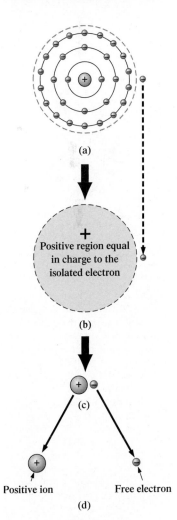

(a)

+
Positive region equal
in charge to the
isolated electron

(b)

(c)

Positive ion   Free electron

(d)

FIG. 2.4
Defining the positive ion.

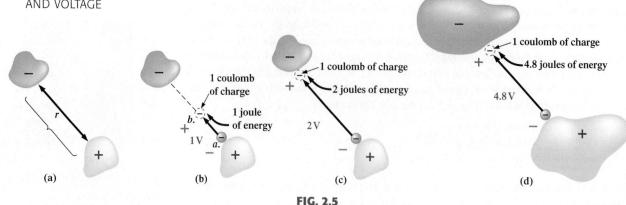

FIG. 2.5

Defining the voltage between two points.

separated by a distance *r*. Since it would be inconsequential to talk about the voltage established by the separation of a single electron, a package of electrons called a **coulomb (C)** of charge was defined as follows:

One coulomb of charge is the total charge associated with 6.242 × 10¹⁸ electrons.

A coulomb of positive charge would have the same magnitude but opposite polarity.

In Fig. 2.5(b), if we take a coulomb of negative charge near the surface of the positive charge and move it toward the negative charge, energy must be expended to overcome the repulsive forces of the larger negative charge and the attractive forces of the positive charge. In the process of moving the charge from point *a* to point *b* in Fig 2.5(b):

if a total of 1 joule (J) of energy is used to move the negative charge of 1 coulomb (C), there is a difference of 1 volt (V) between the two points.

The defining equation is

$$V = \frac{W}{Q}$$
$$\begin{aligned} V &= \text{volts (V)} \\ W &= \text{joules (J)} \\ Q &= \text{coulombs (C)} \end{aligned}$$

(2.2)

Take particular note that the charge is measured in coulombs, the energy in joules, and the voltage in volts. The unit of measurement, **volt,** was chosen to honor the efforts of Alessandro Volta, who first demonstrated that a voltage could be established through chemical action (Fig. 2.6).

If the charge is now moved all the way to the surface of the larger negative charge as shown in Fig. 2.5(c), using 2 joules of energy for the whole trip, there are 2 volts between the two charged bodies. If the package of positive and negative charge is larger, as shown in Fig. 2.5(d), more energy will have to be expended to overcome the larger repulsive forces of the large negative charge and attractive forces of the large positive charge. As shown in Fig. 2.5(d), 4.8 joules of energy were expended, resulting in a voltage of 4.8 V between the two points. We can therefore conclude that it would take 12 joules of energy to move 1 coulomb of negative charge from the positive terminal to the negative terminal of a 12 V car battery.

Through algebraic manipulations, we can define an equation to determine the energy required to move charge through a difference in voltage:

$$W = QV \qquad \text{(joules, J)}$$

(2.3)

Finally, if we want to know how much charge was involved:

$$Q = \frac{W}{V} \qquad \text{(coulombs, C)}$$

(2.4)

FIG. 2.6

Count Alessandro Volta.

Italian (Como, Pavia) **(1745–1827)**
Physicist
Professor of Physics, Pavia, Italy

Began electrical experiments at the age of 18 working with other European investigators. Major contribution was the development of an electrical energy source from chemical action in 1800. For the first time, electrical energy was available on a continuous basis and could be used for practical purposes. Developed the first *condenser* known today as the *capacitor.* Was invited to Paris to demonstrate the *voltaic cell* to Napoleon. The International Electrical Congress meeting in Paris in 1881 honored his efforts by choosing the *volt* as the unit of measure for electromotive force.

EXAMPLE 2.1   Find the voltage between two points if 60 J of energy are required to move a charge of 20 C between the two points.

Solution: Eq. (2.2): $V = \dfrac{W}{Q} = \dfrac{60 \text{ J}}{20 \text{ C}} = \textbf{3 V}$

EXAMPLE 2.2   Determine the energy expended moving a charge of 50 μC between two points if the voltage between the points is 6 V.

Solution: Eq. (2.3) $W = QV = (50 \times 10^{-6} \text{ C})(6 \text{ V}) = 300 \times 10^{-6} \text{ J} = \textbf{300 } \boldsymbol{\mu}\textbf{J}$

There are a variety of ways to separate charge to establish the desired voltage. The most common is the **chemical action** used in car batteries, flashlight batteries, and, in fact, all portable batteries. Other sources use **mechanical methods** such as car generators and steam power plants or **alternative sources** such as solar cells and windmills. In total, however, the sole purpose of the system is to create a separation of charge. In the future, therefore, when you see a positive and a negative terminal on any type of battery, you can think of it as a point where a large concentration of charge has gathered to create a voltage between the two points. More important is to recognize that a voltage exists between two points—for a battery between the positive and negative terminals. Hooking up just the positive or the negative terminal of a battery and not the other would be meaningless.

Both terminals must be connected to define the applied voltage.

As we moved the 1 coulomb of charge in Fig. 2.5(b), the energy expended would depend on where we were in the crossing. The **position** of the charge is therefore a factor in determining the voltage level at each point in the crossing. Since the **potential energy** associated with a body is defined by its position, the term **potential** is often applied to define voltage levels. For example, the **difference in potential** is 4 V between the two points; or the **potential difference** between a point and ground is 12 V; and so on.

2.4   CURRENT

The question, "Which came first—the chicken or the egg?" can be applied here also because the layperson has a tendency to use the terms *current* and *voltage* interchangeably as if both were sources of energy. It is time to set things straight:

The applied voltage is the starting mechanism—the current is simply a reaction to the applied voltage.

In Fig. 2.7(a), a copper wire sits isolated on a laboratory bench. If we cut the wire with an imaginary perpendicular plane, producing the circular cross section shown in Fig. 2.7(b), we would be amazed to find that there are free electrons crossing the surface in both directions.

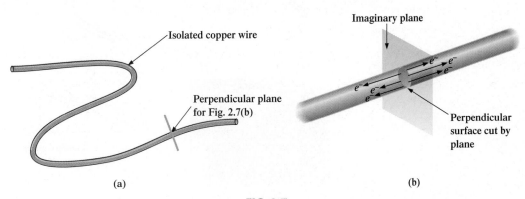

FIG. 2.7

There is motion of free carriers in an isolated piece of copper wire, but the flow of charge fails to have a particular direction.

Those free electrons generated at room temperature are in constant motion in random directions. However, at any instant of time the number of electrons crossing the imaginary plane in one direction is exactly equal to that crossing in the opposite direction, so the **net flow in any one direction is zero.** Even though the wire seems dead to the world sitting by itself on the bench, internally, it is quite active. The same would be true for any other good conductor.

Now, to make this electron flow do work for us, we need to give it a direction and be able to control its magnitude. This is accomplished by simply applying a voltage across the wire to force the electrons to move toward the positive terminal of the battery, as shown in Fig. 2.8. The instant the wire is placed across the terminals, the free electrons in the wire will drift toward the positive terminal. The positive ions in the copper wire will simply oscillate in a mean fixed position. As the electrons pass through the wire, the negative terminal of the battery will act as a supply of additional electrons to keep the process moving. The electrons arriving at the positive terminal will be absorbed, and through the chemical action of the battery, additional electrons will be deposited at the negative terminal to make up for those that left.

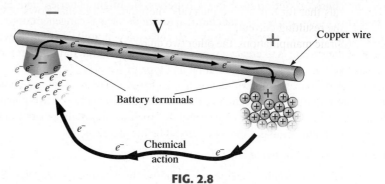

FIG. 2.8

Motion of negatively charged electrons in a copper wire when placed across battery terminals with a difference in potential of V volts.

To take the process a step further, consider the configuration of Fig. 2.9, where a copper wire has been used to connect a light bulb to a battery to create the simplest of electric circuits. The instant the final connection is made, the free electrons of negative charge will drift toward the positive terminal, while the positive ions left behind in the copper wire will simply oscillate in a mean fixed position. The flow of charge (the electrons) through the bulb will heat up the filament of the bulb through friction to the point that it will glow red-hot and emit the desired light.

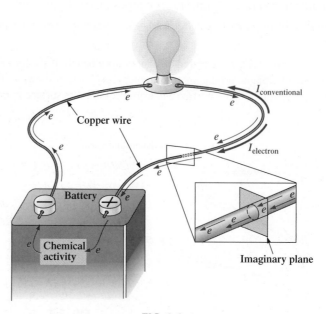

FIG. 2.9

Basic electric circuit.

In total, therefore, the applied voltage has established a flow of electrons in a particular direction. In fact, by definition,

if 6.242 × 10¹⁸ electrons (1 coulomb) pass through the imaginary plane of Fig. 2.7 in 1 second, the flow of charge, or current, is said to be 1 ampere (A).

The unit of current measurement, **ampere,** was chosen to honor the efforts of André Ampère in the study of electricity in motion (Fig. 2.10).

Using the coulomb as the unit of charge, the current in amperes can be determined using the following equation:

$$I = \frac{Q}{t} \qquad \begin{array}{l} I = \text{amperes (A)} \\ Q = \text{coulombs (C)} \\ t = \text{time (s)} \end{array} \qquad \textbf{(2.5)}$$

The capital letter *I* was chosen from the French word for current: *intensité*. The SI abbreviation for each quantity in Eq. (2.5) is provided to the right of the equation. The equation clearly reveals that for equal time intervals, the more charge that flows through the wire, the larger the resulting current.

Through algebraic manipulations, the other two quantities can be determined as follows:

$$Q = It \qquad \text{(coulombs, C)} \qquad \textbf{(2.6)}$$

and

$$t = \frac{Q}{I} \qquad \text{(seconds, s)} \qquad \textbf{(2.7)}$$

EXAMPLE 2.3 The charge flowing through the imaginary surface of Fig. 2.9 is 0.16 C every 64 ms. Determine the current in amperes.

Solution: Eq.(2.5): $I = \dfrac{Q}{t} = \dfrac{0.16 \text{ C}}{64 \times 10^{-3} \text{ s}} = \dfrac{160 \times 10^{-3} \text{ C}}{64 \times 10^{-3} \text{ s}} = \textbf{2.50 A}$

EXAMPLE 2.4 Determine how long it will take 4×10^{16} electrons to pass through the imaginary surface of Fig 2.9 if the current is 5 mA.

Solution: Determine the charge in coulombs:

$$4 \times 10^{16} \text{ electrons} \left(\frac{1 \text{ C}}{6.242 \times 10^{18} \text{ electrons}} \right) = 0.641 \times 10^{-2} \text{ C} = 6.41 \text{ mC}$$

Eq. (2.7): $t = \dfrac{Q}{I} = \dfrac{6.41 \times 10^{-3} \text{ C}}{5 \times 10^{-3} \text{ A}} = \textbf{1.282 s}$

In summary, therefore,

the applied voltage (or potential difference) in an electrical/electronics system is the "pressure" to set the system in motion, and the current is the reaction to that pressure.

A mechanical analogy often used to explain the above is the simple garden hose. In the absence of any pressure, the water sits quietly in the hose with no general direction, just as electrons do not have a net direction in the absence of an applied voltage. However, release the spigot, and the applied pressure will force the water to flow through the hose. Similarly, apply a voltage to the circuit, and a flow of charge or current will result.

A second glance at Fig. 2.9 will reveal that two directions of charge flow have been indicated. One is called *conventional flow,* and the other is called *electron flow.* This text will deal only with conventional flow for a variety of reasons, including the fact that it is the most widely used at educational institutions and in industry, it is employed in the design of

Courtesy of the Smithsonian Institution. Photo No. 76,524

FIG. 2.10
André Marie Ampère.

French (Lyon, Paris) **(1775–1836)**
Mathematician and Physicist
Professor of Mathematics,
École Polytechnique in Paris

On September 18, 1820, introduced a new field of study, *electrodynamics,* devoted to the effect of electricity in motion, including the interaction between currents in adjoining conductors and the interplay of the surrounding magnetic fields. Constructed the first *solenoid* and demonstrated how it could behave like a magnet (the first *electromagnet*). Suggested the name *galvanometer* for an instrument designed to measure current levels.

all electronic device symbols, and it is the popular choice for all major computer software packages. The flow controversy is a result of an assumption made at the time electricity was discovered that the positive charge was the moving particle in metallic conductors. Be assured that the choice of conventional flow will not create great difficulty and confusion in the chapters to follow. Once the direction of I is established, the issue is dropped and the analysis can continue without confusion.

Safety Considerations

It is important to realize that even small levels of current through the human body can cause serious, dangerous side effects. Experimental results reveal that the human body begins to react to currents of only a few milliamperes. Although most individuals can withstand currents up to perhaps 10 mA for very short periods of time without serious side effects, any current over 10 mA should be considered dangerous. In fact, currents of 50 mA can cause severe shock, and currents of over 100 mA can be fatal. In most cases the skin resistance of the body when dry is sufficiently high to limit the current through the body to relatively safe levels for voltage levels typically found in the home. However, be aware that when the skin is wet due to perspiration, bathing, and so on, or when the skin barrier is broken due to an injury, the skin resistance drops dramatically, and current levels could rise to dangerous levels for the same voltage shock. In general, therefore, simply remember that *water and electricity don't mix*. Granted, there are safety devices in the home today [such as the ground fault current interrupt (GFCI) breaker to be introduced in Chapter 4] that are designed specifically for use in wet areas such as the bathroom and kitchen, but accidents happen. Treat electricity with respect—not fear.

2.5   VOLTAGE SOURCES

The terminology **dc** which is used throughout the chapters to follow is an abbreviation for **direct current,** which encompasses all systems where there is a unidirectional (one direction) flow of charge. This section will review dc voltage supplies that apply a fixed voltage to electrical/electronics systems.

The graphic symbol for all dc voltage sources is provided in Fig. 2.11. Note that the relative length of the bars at each end define the polarity of the supply: The long bar represents the positive side, and the short bar the negative. Note also the use of the letter E to denote *voltage source*. It comes from the fact that

an electromotive force (emf) is a force that establishes the flow of charge (or current) in a system due to the application of a difference in potential.

In general, dc voltage sources can be divided into three basic types: (1) batteries (chemical action or solar energy), (2) generators (electromechanical), and (3) power supplies (rectification—a conversion process to be described in your electronics courses). Each will be described under a separate heading.

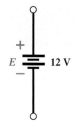

FIG. 2.11
Standard symbol for a dc voltage source.

Batteries

General Information   For the layperson, the battery is the most common of the dc sources. By definition, a battery (derived from the expression "battery of cells") consists of a combination of two or more similar **cells,** a cell being the fundamental source of electrical energy developed through the conversion of chemical or solar energy. All cells can be divided into the **primary** or **secondary** types. The secondary is rechargeable, whereas the primary is not. That is, the chemical reaction of the secondary cell can be reversed to restore its capacity. The two most common rechargeable batteries are the lead-acid unit (used primarily in automobiles) and the nickel-metal hydride (NiMH) battery (used in calculators, tools, photoflash units, shavers, and so on). The obvious advantage of the rechargeable unit is the reduced costs associated with not having to continually replace discharged primary cells.

All the cells appearing in this chapter except the **solar cell,** which absorbs energy from incident light in the form of photons, establish a potential difference at the expense of chemical energy. In addition, each has a positive and a negative *electrode* and an **electrolyte** to complete the circuit between electrodes within the battery. The electrolyte is the contact element and the source of ions for conduction between the terminals.

Primary Cells The popular alkaline primary battery employs a powdered zinc anode (+); a potassium (alkali metal) hydroxide electrolyte; and a manganese dioxide, carbon cathode (−) as shown in Fig. 2.12(a). In Fig. 2.12(b), note that for the cylindrical types (AAA, AA, C, and D) the voltage is the same for each, but **the ampere-hour (Ah) rating increases significantly with size.** The ampere-hour rating is an indication of the level of current that the battery can provide for a specified period of time (to be discussed in detail in Section 2.6). In particular, note that for the large, lantern-type battery, the voltage is only 4 times that of the AAA battery, but the ampere-hour rating of 52 Ah is almost 42 times that of the AAA battery.

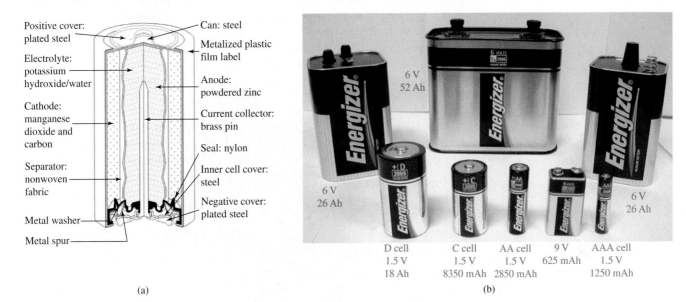

(a)

(b)

FIG. 2.12

Alkaline primary cell: (a) Cutaway of cylindrical Energizer® cell; (b) various types of Eveready Energizer® primary cells.
(Courtesy of Eveready Battery Company, Inc.)

Another type of popular primary cell is the lithium battery, shown in Fig. 2.13. Again, note that the voltage is the same for each, but the size increases substantially with ampere-hour rating and the rated drain current.

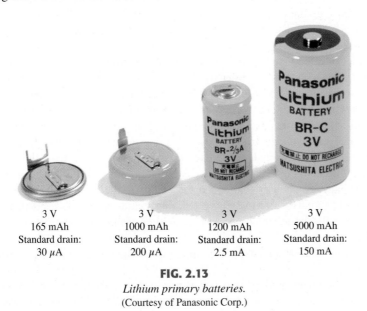

FIG. 2.13

Lithium primary batteries.
(Courtesy of Panasonic Corp.)

In general, therefore,

for batteries of the same type, the size is dictated primarily by the standard drain current or ampere-hour rating, not by the terminal voltage rating.

Lead-Acid Secondary Cell For the secondary lead-acid unit appearing in Fig. 2.14, the electrolyte is sulfuric acid, and the electrodes are spongy lead (Pb) and lead peroxide (PbO_2). When a load is applied to the battery terminals, there is a transfer of electrons from the spongy lead electrode to the lead peroxide electrode through the load. This transfer of electrons will continue until the battery is completely discharged. The discharge time is determined by how diluted the acid has become and how heavy the coating of lead sulfate is on each plate. The state of discharge of a lead storage cell can be determined by measuring the **specific gravity** of the electrolyte with a hydrometer. The specific gravity of a substance is defined to be the ratio of the weight of a given volume of the substance to the weight of an equal volume of water at 4°C. For fully charged batteries, the specific gravity should be somewhere between 1.28 and 1.30. When the specific gravity drops to about 1.1, the battery should be recharged.

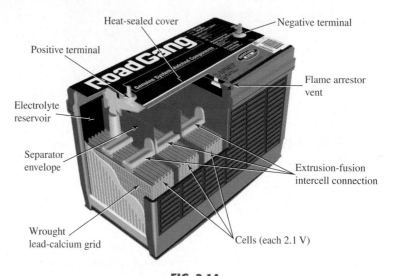

FIG. 2.14

Maintenance-free 12 V (actually 12.6 V) lead-acid battery.
(Courtesy of Delco-Remy, a division of General Motors Corp.)

Since the lead storage cell is a secondary cell, it can be recharged at any point during the discharge phase simply by applying an external **dc current source** across the cell that will pass current through the cell in a direction opposite to that in which the cell supplied current to the load. This will remove the lead sulfate from the plates and restore the concentration of sulfuric acid.

The output of a lead storage cell over most of the discharge phase is about 2.1 V. In the commercial lead storage batteries used in the automobile, 12.6 V can be produced by six cells in series, as shown in Fig. 2.14. In general, lead-acid storage batteries are used in situations where a high current is required for relatively short periods of time. At one time all lead-acid batteries were vented. Gases created during the discharge cycle could escape, and the vent plugs provided access to replace the water or electrolyte and to check the acid level with a hydrometer. The use of a grid made from a wrought lead–calcium alloy strip, rather than the lead-antimony cast grid commonly used, has resulted in maintenance-free batteries such as that appearing in Fig. 2.14. The lead-antimony structure was susceptible to corrosion, overcharge, gasing, water usage, and self-discharge. Improved design with the lead-calcium grid has either eliminated or substantially reduced most of these problems.

It would seem with all the years of technology surrounding batteries that smaller, more powerful units would now be available. However, when it comes to the electric car, which is slowly gaining interest and popularity throughout the world, the lead-acid battery is still the primary source of power. A "station car," manufactured in Norway and used on a test basis in San Francisco for typical commuter runs, has a total weight of 1650 pounds, with 550 pounds (a third of its weight) for the lead-acid rechargeable batteries. Although the station car will travel at speeds of 55 mph, its range is limited to 65 miles on a charge. It would appear that long-distance travel with significantly reduced weight factors for the batteries will depend on a new, innovative approach to battery design.

Nickel–Metal Hydride Secondary Cells The rechargeable battery has been receiving enormous interest and development in recent years. For applications such as flashlights, shavers, portable televisions, power drills, and so on, rechargeable batteries such as the nickel–metal hydride (NiMH) batteries of Fig. 2.15 are the secondary batteries of choice. These batteries are so well made that they can survive over 1000 charge/discharge cycles over a period of time and can last for years.

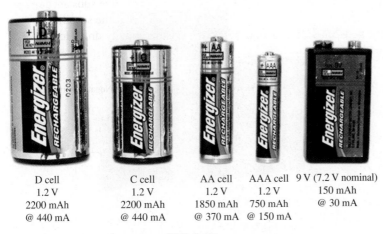

D cell	C cell	AA cell	AAA cell	9 V (7.2 V nominal)
1.2 V	1.2 V	1.2 V	1.2 V	150 mAh
2200 mAh	2200 mAh	1850 mAh	750 mAh	@ 30 mA
@ 440 mA	@ 440 mA	@ 370 mA	@ 150 mA	

FIG. 2.15

Nickel–metal hydride (NiMH) rechargeable batteries.

(Courtesy of Eveready Battery Company, Inc.)

It is important to recognize that if an appliance calls for a rechargeable battery such as a NiMH battery, a primary cell should not be used. The appliance may have an internal charging network that would be dysfunctional with a primary cell. In addition, be aware, as shown in Fig. 2.15, that NiMH batteries are about 1.2 V per cell, while the common primary cells are typically 1.5 V per cell.

There is some ambiguity about how often a secondary cell should be recharged. For the vast majority of situations, the battery can be used until there is some indication that the energy level is low, such as a dimming light from a flashlight, less power from a drill, or a blinking light if one is provided with the equipment. Keep in mind that secondary cells do have some "memory." If they are recharged continuously after being used for a short period of time, they may begin to believe they are short-term units and actually fail to hold the charge for the rated period of time. In any event, always try to avoid a "hard" discharge, which results when every bit of energy is drained from a cell. Too many hard-discharge cycles will reduce the cycle life of the battery. Finally, be aware that the charging mechanism for nickel-cadmium cells is quite different from that for lead-acid batteries. The nickel-cadmium battery is charged by a constant current source, with the terminal voltage staying pretty steady through the entire charging cycle. The lead-acid battery is charged by a constant voltage source, permitting the current to vary as determined by the state of the battery. The capacity of the NiMH battery increases almost linearly throughout most of the charging cycle. One may find that Ni-Cad batteries are relatively warm when charging. The lower the capacity level of the battery when charging, the higher the temperature of the cell. As the battery approaches rated capacity, the temperature of the cell approaches room temperature.

Other types of rechargeable batteries include the nickel-cadmium (Ni-Cad) and nickel hydrogen (Ni-H) batteries. In reality, however, the NiMH battery is a hybrid of the nickel-cadmium and nickel-hydrogen cells, combining the positive characteristics of each to create a product with a high power level in a small package that has a long life. Another type of rechargeable battery is the lithium-ion variety of Fig. 2.16, used in the IBM lap-top computer.

10.8 V, 10.8 Ah

Charge time:

    System operational: 6 h max.

    Power off: 2.5 h max.

FIG. 2.16

IBM ThinkPad T-20 lithium-ion rechargeable battery.

(Courtesy of IBM.)

Solar Cell The SX 20 and SX 30 solar modules (a combination of connected cells) of Fig. 2.17 provide 20 W and 30 W of electrical power, respectively. The size and the orientation of such units are important because the maximum available wattage on an average bright, sunlit day is 100 mW/cm². Since conversion efficiencies are currently only at 10% to 14%, the maximum available power per square centimeter from most commercial units is between 10 mW and 14 mW. For a square meter, however, the return would be 100 W to 140 W. The units appearing in Fig. 2.17 are typically used for remote telemetry, isolated instrumentation systems, security sensors, signal sources, and land-based navigation aids. A more detailed description of the solar cell will appear in your electronics courses, but for now it is important to realize that a fairly steady source of electrical dc power can be obtained from the sun.

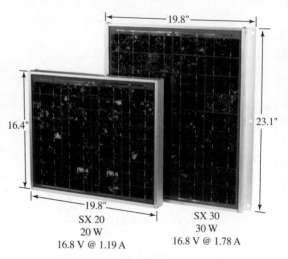

19.8"

16.4"

23.1"

19.8"

SX 20
20 W
16.8 V @ 1.19 A

SX 30
30 W
16.8 V @ 1.78 A

FIG. 2.17
Photovoltaic solar module.
(Photo courtesy BP Solar.)

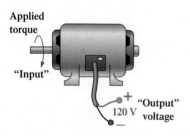

Applied
torque

"Input"

+
120 V
−

"Output"
voltage

FIG. 2.18
dc generator.

Generators

The **dc generator** is quite different, both in construction (Fig. 2.18) and in mode of operation, from the battery. When the shaft of the generator is rotating at the nameplate speed due to the applied torque of some external source of mechanical power, a voltage of rated value will appear across the external terminals. The terminal voltage and power-handling capabilities of the dc generator are typically higher than those of most batteries, and its lifetime is determined only by its construction. Commercially used dc generators are typically of the 120 V or 240 V variety. As pointed out earlier in this section, for the purposes of this text, no distinction will be made between the symbols for a battery and a generator.

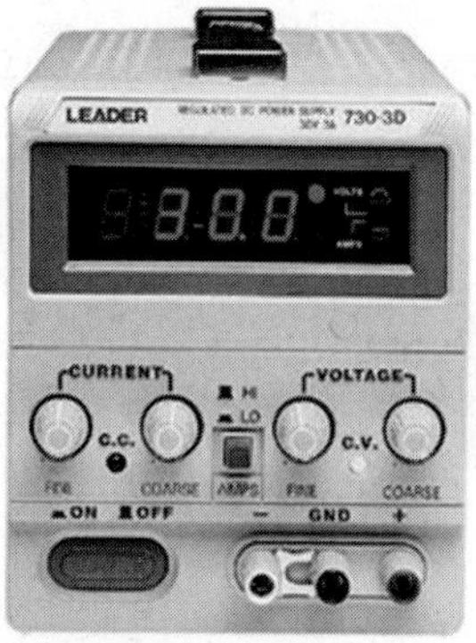

Power Supplies

The dc supply encountered most frequently in the laboratory employs the **rectification** and *filtering* processes as its means toward obtaining a steady dc voltage. Both processes will be covered in detail in your basic electronics courses. In total, a time-varying voltage (such as ac voltage available from a home outlet) is converted to one of a fixed magnitude. A dc laboratory supply of this type appears in Fig. 2.19.

FIG. 2.19
dc laboratory supply (30 V, 3 A).
(Image compliments of Leader
Instruments Corporation.)

Most dc laboratory supplies have a regulated, adjustable voltage output with three available terminals, as indicated in Figs. 2.19 and 2.20(a). The symbol for ground or zero potential (the reference) is also shown in Fig. 2.20(a). If 10 V above ground potential are required, then the connections are made as shown in Fig. 2.20(b). If 15 V below ground potential are required, then the connections are made as shown in Fig. 2.20(c). If connections are as shown in Fig. 2.20(d), we say we have a "floating" voltage of 5 V since the reference level is not included. Seldom is the configuration of Fig. 2.20(d) employed since it fails to protect the operator by providing a direct low-resistance path to ground and to

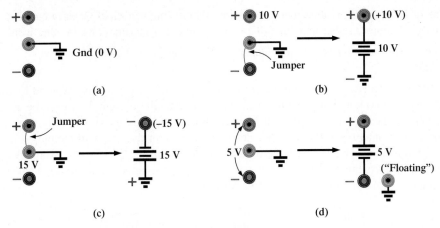

FIG. 2.20

dc laboratory supply: (a) available terminals; (b) positive voltage with respect to (w.r.t.) ground; (c) negative voltage w.r.t. ground; (d) floating supply.

establish a common ground for the system. In any case, **the positive and negative terminals must be part of any circuit configuration.**

2.6   AMPERE-HOUR RATING

The most important piece of data for any battery (other than its voltage rating) is its **ampere-hour (Ah) rating.** You have probably noted in the photographs of batteries in this chapter that both the voltage and the ampere-hour rating have been provided for each battery.

The ampere-hour (Ah) rating provides an indication of how long a battery of fixed voltage will be able to supply a particular current.

A battery with an ampere-hour rating of 100 will theoretically provide a current of 1 A for 100 hours, 10 A for 10 hours, or 100 A for 1 hour. Quite obviously, the greater the current, the shorter the time. An equation for determining the length of time a battery will supply a particular current is the following:

$$\text{Life (hours)} = \frac{\text{ampere-hour (Ah) rating}}{\text{amperes drawn (A)}} \tag{2.8}$$

EXAMPLE 2.5   How long will a 9 V transistor battery with an ampere-hour rating of 520 mAh provide a current of 20 mA?

Solution: Eq.(2.8): $\text{Life} = \dfrac{520 \text{ mAh}}{20 \text{ mA}} = \dfrac{520}{20} \text{h} = \textbf{26 h}$

EXAMPLE 2.6   How long can a 1.5 V flashlight battery provide a current of 250 mA to light the bulb if the ampere-hour rating is 16 Ah?

Solution: Eq.(2.8): $\text{Life} = \dfrac{16 \text{ Ah}}{250 \text{ mA}} = \dfrac{16}{250 \times 10^{-3}} \text{h} = \textbf{64 h}$

2.7   BATTERY LIFE FACTORS

The previous section made it clear that the life of a battery is directly related to the magnitude of the current drawn from the supply. However, there are factors that will affect the given ampere-hour rating of a battery, so we may find that a battery with an ampere-hour rating of 100 can supply a current of 10 A for 10 hours but can supply a current of 100 A for only 20 minutes rather than the full 1 hour calculated using Eq. (2.8). In other words,

the capacity of a battery (in ampere-hours) will change with change in current demand.

This is not to say that Equation (2.8) is totally invalid. It can always be used to gain some insight into how long a battery can supply a particular current. However, be aware that there are factors that will affect the ampere-hour rating. Just as with most systems, including the human body, the more we demand, the shorter the time that the output level can be maintained. This is clearly verified by the curves of Fig. 2.21 for the Eveready Energizer D cell. As the constant current drain increased, the ampere-hour rating decreased from about 18 Ah at 25 mA to around 12 Ah at 300 mA.

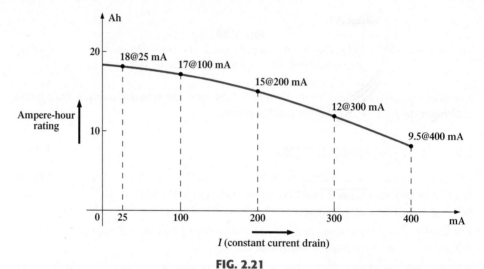

FIG. 2.21
Ampere-hour rating (capacity) versus drain current for an Energizer® D cell.

Another factor that will affect the ampere-hour rating is the temperature of the unit and of the surrounding medium. In Fig. 2.22, the capacity of the same battery plotted in Fig. 2.21 is clearly showing a peak value near the common room temperature of 68°F. At very cold temperatures and very warm temperatures, the capacity dropped. Clearly, the ampere-hour rating will be provided at or near room temperature to give it a maximum value, but be aware that it will drop off with increase or decrease in temperature. Most of us have experienced the fact that the battery in a car, radio, two-way radio, flashlight, or whatever seems to have less power in really cold weather. However, most of us have probably thought the battery capacity would increase with higher temperatures—apparently not the case. In general, therefore,

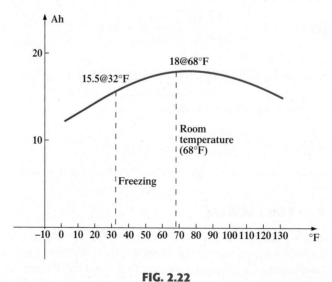

FIG. 2.22
Ampere-hour rating (capacity) versus temperature for an Energizer® D cell.

the ampere-hour rating of a battery will decrease from the room-temperature level with very cold and very warm temperatures.

CONDUCTORS AND
INSULATORS   **47**

Another interesting factor that affects the performance of a battery is how long it is asked to supply a particular voltage at a continuous drain current. Note the curves of Fig. 2.23, where the terminal voltage dropped at each level of drain current as the time period increased. The lower the current drain, the longer it could supply the desired current. At 100 mA, it was limited to about 100 hours near the rated voltage, but at 25 mA, it did not drop below 1.2 V until about 500 hours had passed. That is an increase in time of 5 : 1, and that is significant. The result is that

the terminal voltage of a battery will eventually drop (at any level of current drain) if the time period of continuous discharge is too long.

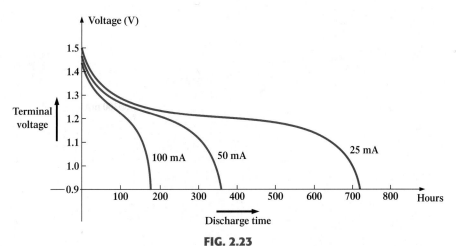

FIG. 2.23
Terminal voltage versus discharge time for specific drain currents for an Energizer® D cell.

2.8   CONDUCTORS AND INSULATORS

Different wires placed across the same two battery terminals will allow different amounts of charge to flow between the terminals. Many factors, such as the density, mobility, and stability characteristics of a material, account for these variations in charge flow. In general, however,

conductors are those materials that permit a generous flow of electrons with very little external force (voltage) applied.

In addition,

good conductors typically have only one electron in the valence (most distant from the nucleus) ring.

Since **copper** is used most frequently, it serves as the standard of comparison for the relative conductivity in Table 2.1. Note that aluminum, which has seen some commercial use,

TABLE 2.1
Relative conductivity of various materials.

Metal	Relative Conductivity (%)
Silver	105
Copper	**100**
Gold	70.5
Aluminum	61
Tungsten	31.2
Nickel	22.1
Iron	14
Constantan	3.52
Nichrome	1.73
Calorite	1.44

has only 61% of the conductivity level of copper, but keep in mind that this must be weighed against the cost and weight factors.

Insulators are those materials that have very few free electrons and require a large applied potential (voltage) to establish a measurable current level.

A common use of insulating material is for covering current-carrying wire, which, if uninsulated, could cause dangerous side effects. Power linemen wear rubber gloves and stand on rubber mats as safety measures when working on high-voltage transmission lines. A number of different types of insulators and their applications appear in Fig. 2.24.

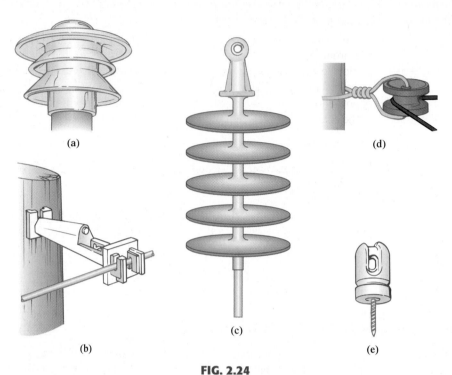

(a)

(b)

(c)

(d)

(e)

FIG. 2.24

Various types of insulators and their applications: (a) Corning Glass Works Pyrex™ power line insulator; (b) Fi-Shock extender insulator; (c) Lapp power line insulator; (d) Fi-Shock corner insulator; (e) Fi-Shock screw-in post insulator.

It must be pointed out, however, that even the best insulator will break down (permit charge to flow through it) if a sufficiently large potential is applied across it. The breakdown strengths of some common insulators are listed in Table 2.2. According to this table, for insulators with the same geometric shape, it would require 270/30 = 9 times as much potential to pass current through rubber compared to air and approximately 67 times as much voltage to pass current through mica as through air.

TABLE 2.2

Breakdown strength of some common insulators.

Material	Average Breakdown Strength (kV/cm)
Air	30
Porcelain	70
Oils	140
Bakelite®	150
Rubber	270
Paper (paraffin-coated)	500
Teflon®	600
Glass	900
Mica	2000

2.9 SEMICONDUCTORS

Semiconductors are a specific group of elements that exhibit characteristics between those of insulators and those of conductors.

The prefix *semi,* included in the terminology, has the dictionary definition of *half, partial,* or *between,* as defined by its use. The entire electronics industry is dependent on this class of materials since the electronic devices and integrated circuits (ICs) are constructed of semiconductor materials. Although *silicon* (Si) is the most extensively employed material, *germanium* (Ge) and *gallium arsenide* (GaAs) are also used in many important devices.

Semiconductor materials typically have four electrons in the outermost valence ring.

Semiconductors are further characterized as being photoconductive and having a negative temperature coefficient. Photoconductivity is a phenomenon where the photons (small packages of energy) from incident light can increase the carrier density in the material and thereby the charge flow level. A negative temperature coefficient reveals that the resistance (a characteristic to be described in detail in the next chapter) will decrease with an increase in temperature (opposite to that of most conductors). A great deal more will be said about semiconductors in the chapters to follow and in your basic electronics courses.

2.10 AMMETERS AND VOLTMETERS

It is important to be able to measure the current and voltage levels of an operating electrical system to check its operation, isolate malfunctions, and investigate effects impossible to predict on paper. As the names imply, **ammeters** are used to measure current levels, and **voltmeters,** the potential difference between two points. If the current levels are usually of the order of milliamperes, the instrument will typically be referred to as a *milliammeter,* and if the current levels are in the microampere range, as a *microammeter.* Similar statements can be made for voltage levels. Throughout the industry, voltage levels are measured more frequently than current levels, primarily because measurement of the former does not require that the network connections be disturbed.

The potential difference between two points can be measured by simply connecting the leads of the meter *across the two points,* as indicated in Fig. 2.25. An up-scale reading is obtained by placing the positive lead of the meter to the point of higher potential of the network and the common or negative lead to the point of lower potential. The reverse connection will result in a negative reading or a below-zero indication.

Ammeters are connected as shown in Fig. 2.26. Since ammeters measure the rate of flow of charge, the meter must be placed in the network such that the charge will flow through the

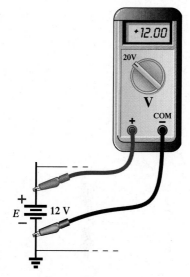

FIG. 2.25
Voltmeter connection for an up-scale (+) reading.

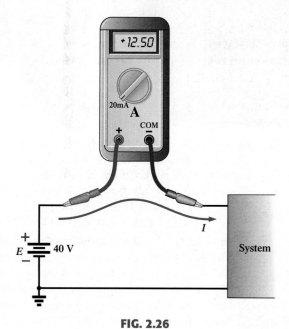

FIG. 2.26
Ammeter connection for an up-scale (+) reading.

meter. The only way this can be accomplished is to open the path in which the current is to be measured and place the meter between the two resulting terminals. For the configuration of Fig. 2.26, the voltage source lead (+) must be disconnected from the system and the ammeter inserted as shown. An up-scale reading will be obtained if the polarities on the terminals of the ammeter are such that the current of the system enters the positive terminal.

The introduction of any meter into an electrical/electronic system raises a concern about whether the meter will affect the behavior of the system. This question and others will be examined in Chapters 5 and 6 after additional terms and concepts have been introduced. For the moment, let it be said that since voltmeters and ammeters do not have internal components, they will affect the network when introduced for measurement purposes. The design of each, however, is such that the impact is minimized.

There are instruments designed to measure just current or just voltage levels. However, the most common laboratory meters include the *volt-ohm-milliammeter* (VOM) and the *digital multimeter* (DMM) of Figs. 2.27 and 2.28, respectively. Both instruments will measure voltage and current and a third quantity, resistance, to be introduced in the next chapter. The VOM uses an analog scale, which requires interpreting the position of a pointer on a continuous scale, while the DMM provides a display of numbers with decimal-point accuracy determined by the chosen scale. Comments on the characteristics and use of various meters will be made throughout the text. However, the major study of meters will be left for the laboratory sessions.

FIG. 2.27
Volt-ohm-milliammeter (VOM) analog meter.
(Courtesy of Simpson Electric Co.)

FIG. 2.28
Digital multimeter (DMM).
(Courtesy of Fluke Corporation.
Reproduced with permission.)

2.11 APPLICATIONS

Throughout the text, Applications sections such as this one have been included to permit a further investigation of terms, quantities, or systems introduced in the chapter. The primary purpose of these Applications is to establish a link between the theoretical concepts of the text and the real, practical world. Although the majority of components that appear in a system may not have been introduced (and, in fact, some components will not be examined until more advanced studies), the topics were chosen very carefully and should be quite interesting to a new student of the subject matter. Sufficient comment is included to provide a surface understanding of the role of each part of the system, with the understanding that the details will come at a later date. Since exercises on the subject matter of the Applications do not appear at the end of the chapter, the content is designed not to challenge the student but rather to stimulate his or her interest and answer some basic questions such as how the system looks inside, what role specific elements play in the system, and, of course, how the system works. In essence, therefore, each Applications section provides an opportunity to begin to establish a practical background beyond simply the content of the chapter. Do not be concerned if you do not understand every detail of each application. Understanding will come with time and experience. For now, take what you can from the examples and then proceed with the material.

Flashlight

Although the flashlight employs one of the simplest of electrical circuits, a few fundamentals about its operation do carry over to more sophisticated systems. First, and quite obviously, it is a dc system with a lifetime totally dependent on the state of the batteries and bulb. Unless it is the rechargeable type, each time you use it, you take some of the life out of it. For many hours the brightness will not diminish noticeably. Then, however, as it reaches the end of its ampere-hour capacity, the light will become dimmer at an increasingly rapid rate (almost exponentially). For the standard two-battery flashlight appearing in Fig. 2.29(a) with its electrical schematic in Fig. 2.29(b), each 1.5 V battery has an ampere-hour rating of about 18 as indicated in Fig. 2.12. The single-contact miniature flange-base bulb is rated at 2.5 V and 300 mA with good brightness and a lifetime of about 30 hours. Thirty hours may not seem like a long lifetime, but you have to consider how long you usually use a flashlight on each occasion. If we assume a 300 mA drain from the battery for the bulb when in use, the lifetime of the battery, by Eq. (2.8), is about 60 hours. Comparing the 60 hour lifetime of the battery to the 30 hour life expectancy of the bulb suggests that we normally have to replace bulbs more frequently than batteries.

However, most of us have experienced the opposite effect: We can change batteries two or three times before we need to replace the bulb. This is simply one example of the fact that one cannot be guided solely by the specifications of each component of an electrical design. The operating conditions, terminal characteristics, and details about the actual response of the system for short and long periods of time must be taken into account. As mentioned earlier, the battery loses some of its power each time it is used. Although the terminal voltage may not change much at first, its ability to provide the same level of current will drop with each usage. Further, batteries will slowly discharge due to "leakage currents" even if the switch is not on. The air surrounding the battery is not "clean" in the sense that

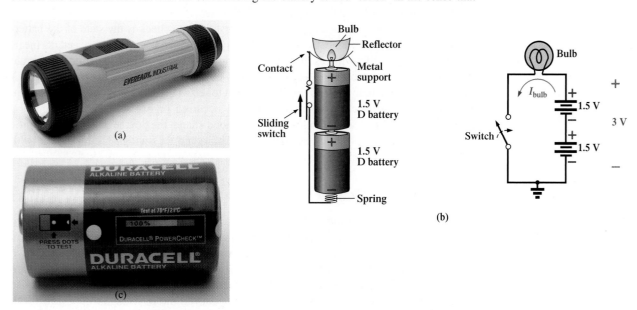

FIG. 2.29

(a) Eveready® D cell flashlight; (b) electrical schematic of flashlight of part (a); (c) Duracell® Powercheck™ D cell battery.

moisture and other elements in the air can provide a conduction path for leakage currents through the air, through the surface of the battery itself, or through other nearby surfaces, and the battery will eventually discharge. How often have we left a flashlight with new batteries in a car for a long period of time only to find the light very dim or the batteries dead when we need the flashlight the most? An additional problem is acid leaks that appear as brown stains or corrosion on the casing of the battery. These leaks will also affect the life of the battery. Further, when the flashlight is turned on, there is an initial surge in current that will drain the battery more than continuous use for a period of time. In other words, continually turning the flashlight on and off will have a very detrimental effect on its life. We must also realize that the 30 hour rating of the bulb is for continuous use, that is, 300 mA flowing through the bulb for a continuous 30 hours. Certainly, the filament in the bulb and the bulb itself will get hotter with time, and this heat has a detrimental effect on the filament wire. When the flashlight is turned on and off, it gives the bulb a chance to cool down and regain its normal characteristics, thereby avoiding any real damage. Therefore, with normal use we can expect the bulb to last longer than the 30 hours specified for continuous use.

Even though the bulb is rated for 2.5 V operation, it would appear that the two batteries would result in an applied voltage of 3 V which suggests poor operating conditions. However, a bulb rated at 2.5 V can easily handle 2.5 V to 3 V. In addition, as was pointed out in this chapter, the terminal voltage will drop with the current demand and usage. Under normal operating conditions, a 1.5 V battery is considered to be in good condition if the loaded terminal voltage is 1.3 V to 1.5 V. When it drops to the range from 1 V to 1.1 V, it is weak, and when it drops to the range from 0.8 V to 0.9 V, it has lost its effectiveness. The levels can be related directly to the test band now appearing on Duracell® batteries, such as on the one shown in Fig. 2.29(c). In the test band on this battery, the upper voltage area (green) is near 1.5 V (labeled 100%); the lighter area to the right, from about 1.3 V down to 1 V; and the replace area (red) on the far right, below 1 V.

Be aware that the total supplied voltage of 3 V will be obtained only if the batteries are connected as shown in Fig. 2.29(b). Accidentally placing the two positive terminals together will result in a total voltage of 0 V, and the bulb will not light at all. *For the vast majority of systems with more than one battery, the positive terminal of one battery will always be connected to the negative terminal of another. For all low-voltage batteries, the end with the nipple is the positive terminal, and the end with the flat end is the negative terminal. In addition, the flat or negative end of a battery is always connected to the battery casing with the helical coil to keep the batteries in place. The positive end of the battery is always connected to a flat spring connection or the element to be operated.* If you look carefully at the bulb, you will find that the nipple connected to the positive end of the battery is insulated from the jacket around the base of the bulb. The jacket is the second terminal of the battery used to complete the circuit through the on/off switch.

If a flashlight fails to operate properly, the first thing to check is the state of the batteries. It is best to replace both batteries at once. A system with one good battery and one nearing the end of its life will result in pressure on the good battery to supply the current demand, and, in fact, the bad battery will actually be a drain on the good battery. Next check the condition of the bulb by checking the filament to see whether it has opened at some point because a long-term, continuous current level occurred or because the flashlight was dropped. If the battery and bulb seem to be in good shape, the next area of concern is the contacts between the positive terminal and the bulb and the switch. Cleaning both with emery cloth will often eliminate this problem.

12 V Car Battery Charger

Battery chargers are a common household piece of equipment used to charge everything from small flashlight batteries to heavy-duty, marine, lead-acid batteries. Since all are plugged into a 120 V ac outlet such as found in the home, the basic construction of each is quite similar. In every charging system a *transformer* (Chapter 18) must be included to cut the ac voltage to a level appropriate for the dc level to be established. A *diode* (also called *rectifier*) arrangement must be included to convert the ac voltage which varies with time to a fixed dc level such as described in this chapter. Diodes and/or rectifiers will be discussed in detail in your first electronics course. Some dc chargers will also include a *regulator* to provide an improved dc level (one that varies less with time or load). Since the car battery charger is one of the most common, it will be described in the next few paragraphs.

The outside appearance and the internal construction of a Sears 6/2 AMP Manual Battery Charger are provided in Fig. 2.30. Note in Fig. 2.30(b) that the transformer (as in most chargers) takes up most of the internal space. The additional air space and the holes in the casing are there to ensure an outlet for the heat that will develop due to the resulting current levels.

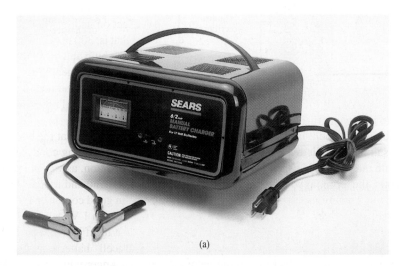

(a)

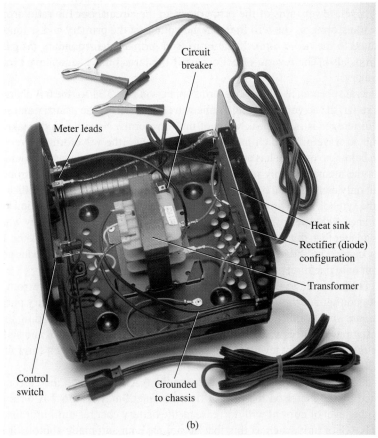

(b)

FIG. 2.30
Battery charger: (a) external appearance; (b) internal construction.

The schematic of Fig. 2.31 includes all the basic components of the charger. Note first that the 120 V from the outlet are applied directly across the primary of the transformer. The charging rate of 6 A or 2 A is determined by the switch, which simply controls how many windings of the primary will be in the circuit for the chosen charging rate. If the battery is charging at the 2 A level, the full primary will be in the circuit, and the ratio of

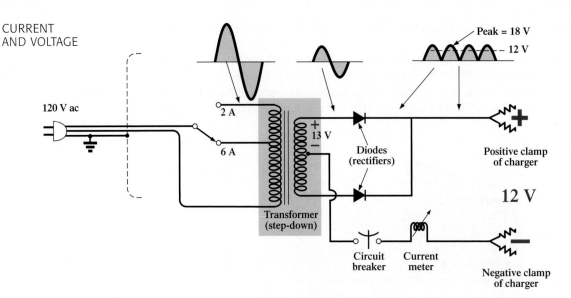

FIG. 2.31

Electrical schematic for the battery charger of Fig. 2.30.

the turns in the primary to the turns in the secondary will be a maximum. If it is charging at the 6 A level, fewer turns of the primary are in the circuit, and the ratio drops. When you study transformers, you will find that the voltage at the primary and secondary is directly related to the *turns ratio*. If the ratio from primary to secondary drops, then the voltage drops also. The reverse effect occurs if the turns on the secondary exceed those on the primary.

The general appearance of the waveforms appears in Fig. 2.31 for the 6 A charging level. Note that so far, the ac voltage has the same wave shape across the primary and secondary. The only difference is in the peak value of the waveforms. Now the diodes take over and convert the ac waveform which has zero average value (the waveform above equals the waveform below) to one that has an average value (all above the axis) as shown in the same figure. For the moment simply recognize that diodes are semiconductor electronic devices that permit only conventional current to flow through them in the direction indicated by the arrow in the symbol. Even though the waveform resulting from the diode action has a pulsing appearance with a peak value of about 18 V, it will charge the 12 V battery whenever its voltage is greater than that of the battery, as shown by the shaded area. Below the 12 V level the battery cannot discharge back into the charging network because the diodes permit current flow in only one direction.

In particular, note in Fig. 2.30(b) the large plate that carries the current from the rectifier (diode) configuration to the positive terminal of the battery. Its primary purpose is to provide a *heat sink* (a place for the heat to be distributed to the surrounding air) for the diode configuration. Otherwise, the diodes would eventually melt down and self-destruct due to the resulting current levels. Each component of Fig 2.31 has been carefully labeled in Fig. 2.30(b) for reference.

When current is first applied to a battery at the 6 A charge rate, the current demand as indicated by the meter on the face of the instrument may rise to 7 A or almost 8 A. However, the level of current will decrease as the battery charges until it drops to a level of 2 A or 3 A. For units such as this that do not have an automatic shutoff, it is important to disconnect the charger when the current drops to the fully charged level; otherwise, the battery will become overcharged and may be damaged. A battery that is at its 50% level can take as long as 10 hours to charge, so don't expect it to be a 10 minute operation. In addition, if a battery is in very bad shape with a lower than normal voltage, the initial charging current may be too high for the design. To protect against such situations, the circuit breaker will open and stop the charging process. Because of the high current levels, it is important that the directions provided with the charger be carefully read and applied.

A wide variety of systems in the home and office receive their dc operating voltage from an ac/dc conversion system plugged right into a 120 V ac outlet. Lap-top computers, answering machines/phones, radios, clocks, cellular phones, CD players, and so on, all receive their dc power from a packaged system such as appearing in Fig. 2.32. The conversion from ac to dc occurs within the unit which is plugged directly into the outlet. The dc voltage is available at the end of the long wire which is designed to be plugged into the operating unit. As small as the unit may be, it contains basically the same components as appearing in the battery charger of Fig. 2.30.

In Fig. 2.33 you can see the transformer used to cut the voltage down to appropriate levels (again the largest component of the system). Note that two diodes establish a dc level, and a capacitive filter (Chapter 10) is added to smooth out the dc as shown. The system can be relatively small because the operating current levels are quite small, permitting the use of thin wires to construct the transformer and limit its size. The lower currents also reduce the concerns about heating effects, permitting a small housing structure. The unit of Fig. 2.33, rated at 9 V at 200 mA, is commonly used to provide power to answering machines/phones. Further smoothing of the dc voltage will be accomplished by a regulator built into the receiving unit. The regulator is normally a small IC chip placed in the receiving unit to separate the heat that it generates from the heat generated by the transformer, thereby reducing the net heat at the outlet close to the wall. In addition, its placement in the receiving unit will reduce the possibility of picking up noise and oscillations along the long wire from the conversion unit to the operating unit, and it will ensure that the full rated voltage is available at the unit itself, not a lesser value due to losses along the line.

FIG. 2.32
Answering machine/phone 9 V dc supply.

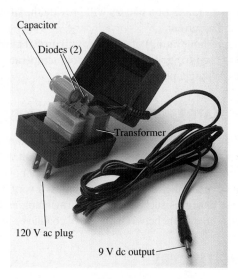

FIG. 2.33
Internal construction of the 9 V dc supply of Fig. 2.32.

2.12 COMPUTER ANALYSIS

In some texts, the procedure for choosing a dc voltage source and placing it on the schematic using computer methods is introduced at this point. This approach, however, results in a need for the student to turn back to this chapter for the procedure when the first complete network is installed and examined. Therefore, the procedure will be introduced in Chapter 4 when the first complete network is examined, thereby localizing the material and removing the need to reread this chapter and the chapter to follow.

CHAPTER SUMMARY
ATOMIC STRUCTURE

- The orbiting electron carries a negative charge equal in magnitude to the positive charge of the proton.
- The atomic structure of any stable atom has an equal number of electrons and protons.

- Copper has 29 electrons, with the 29th electron the only electron in the outermost shell, making it relatively easy to separate from the parent atom.

VOLTAGE

- A source of voltage or difference in potential can be established by simply creating a separation between a cluster of positive and negative charges.
- The basic unit for charge is the coulomb which has a charge equal to that of 6.242×10^{18} electrons.
- If 1 joule of energy is required to move a charge of 1 coulomb between two points, the difference in potential or voltage between the two points is 1 volt.
- Voltage is a two-point phenomenon; that is, both ends of a supply must be connected to establish a voltage or difference in potential between two points.

CURRENT

- Current is a reaction to the "pressure" established by an applied voltage.
- The more charge that passes through a wire per unit time, the greater the current.
- If 1 coulomb of charge passes through a wire in 1 second, the current is 1 ampere.

VOLTAGE SOURCES

- A primary voltage cell cannot be recharged, whereas a secondary type can be recharged.
- The ampere-hour (Ah) rating of a battery will provide the life of the battery in hours if divided by the drain current.
- The capacity (ampere-hour rating) of a battery will drop if current demand increases, if the battery is exposed to very high or very cold temperatures, or if the discharge current exists for too long a period.

CONDUCTORS, INSULATORS, AND SEMICONDUCTORS

- Conductors are materials such as copper, gold, and aluminum that will permit a generous flow of charge with very little external pressure (voltage) applied.
- Insulators are materials such as rubber, Bakelite™, and a host of other ceramics that do not generate a large number of free carriers unless a very high voltage is applied.
- Semiconductors are materials such as silicon, germanium, and gallium arsenide that have characteristics somewhere between those of a conductor and those of an insulator.

AMMETERS AND VOLTMETERS

- Ammeters are always connected such that the current to be measured will pass right through the meter.
- Voltmeters are connected across the two points where the difference in potential is to be determined. Both ends must be connected to make a reading.

Important Equations

$$F = k\frac{Q_1 Q_2}{r^2} \qquad k = 9 \times 10^9 \text{ N·m}^2/\text{C}^2$$

$$V = \frac{W}{Q}$$

$$I = \frac{Q}{t}$$

$$\text{Life (hours)} = \frac{\text{ampere-hour (Ah) rating}}{\text{amperes drawn (A)}}$$

PROBLEMS

SECTION 2.2   Atoms and Their Structure

1. The numbers of orbiting electrons in aluminum and silver are 13 and 47, respectively. Draw the electronic configuration for each, and discuss briefly why each is a good conductor.

2. Find the force of attraction in newtons between the charges Q_1 and Q_2 in Fig. 2.34 when
   a. $r = 1$ m   **b.** $r = 3$ m   **c.** $r = 10$ m
   d. Did the force drop off quickly with an increase in distance?

*3. Find the force of repulsion in newtons between Q_1 and Q_2 in Fig. 2.35 when
   a. $r = 1$ mi   **b.** $r = 0.01$ m   **c.** $r = 1/16$ in.

*4. Plot the force of attraction (in newtons) versus separation (in meters) for two charges of 2 C and 8 C. Set r to 0.5 m and 1 m, followed by 1 m intervals to 10 m. Comment on the shape of the curve. Is it linear or nonlinear? What does it tell you about the force of attraction between charges as they are separated? What does it tell you about any function plotted against a squared term in the denominator?

5. Determine the distance between two charges of 20 μC if the force between the two charges is 3.6×10^4 N.

*6. Two charged bodies, Q_1 and Q_2, when separated by a distance of 2 m, experience a force of repulsion equal to 1.8 N.
   a. What will the force of repulsion be when they are 10 m apart?
   b. If the ratio $Q_1/Q_2 = 1/2$, find Q_1 and Q_2 ($r = 10$ m).

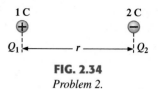

FIG. 2.34
Problem 2.

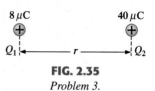

FIG. 2.35
Problem 3.

SECTION 2.3   Voltage

7. What is the voltage between two points if 120 mJ of energy are required to move 4.8 C between the two points?

8. If the potential difference between two points is 42 V, how much work is required to bring 6 C from one point to the other?

9. Find the charge Q that requires 96 J of energy to be moved through a potential difference of 16 V.

10. How much charge passes through a radio battery of 9 V if the energy expended is 72 J?

SECTION 2.4   Current

11. Find the current in amperes if 650 C of charge pass through a wire in 50 s.

12. If 465 C of charge pass through a wire in 2.5 min, find the current in amperes.

13. If a current of 40 A exists for 1 min, how many coulombs of charge have passed through the wire?

14. How many coulombs of charge pass through a lamp in 2 min if the current is constant at 750 mA?

15. If the current in a conductor is constant at 2 mA, how much time is required for 6 mC to pass through the conductor?

16. If $21.847 \times 10^{+18}$ electrons pass through a wire in 12 s, find the current.

17. How many electrons pass through a conductor in 1 min and 15 s if the current is 4 mA?

18. Will a fuse rated at 1 A "blow" if 86 C pass through it in 1.2 min?

*19. If $0.84 \times 10^{+16}$ electrons pass through a wire in 60 ms, find the current.

*20. Which would you prefer?
   a. A penny for every electron that passes through a wire in 0.01 μs at a current of 2 mA, or
   b. A dollar for every electron that passes through a wire in 1.5 ns if the current is 100 μA.

*21. If a conductor with a current of 200 mA passing through it converts 40 J of electrical energy into heat in 30 s, what is the potential drop across the conductor?

*22. Charge is flowing through a conductor at the rate of 420 C/min. If 742 J of electrical energy are converted to heat in 30 s, what is the potential drop across the conductor?

*23. The potential difference between two points in an electric circuit is 24 V. If 0.4 J of energy were dissipated in a period of 5 ms, what would the current be between the two points?

SECTION 2.6   Ampere-Hour Rating

24. What current will a battery with an Ah rating of 200 theoretically provide for 40 h?

25. What is the Ah rating of a battery that can provide 0.8 A for 75 h?

26. For how many hours will a battery with an Ah rating of 32 theoretically provide a current of 1.28 A?

27. A standard 12 V car battery has an ampere-hour rating of 40 Ah, whereas a heavy-duty battery has a rating of 60 Ah. How would you compare the energy levels of each and the available current for starting purposes?

*28. A portable television using a 12 V, 3 Ah rechargeable battery can operate for a period of about 5.5 h. What is the average current drawn during this period? What is the energy expended by the battery in joules?

SECTION 2.8 Conductors and Insulators

29. Discuss two properties of the atomic structure of copper that make it a good conductor.

30. Explain the terms *insulator* and *breakdown strength*.

31. List three uses of insulators not mentioned in Section 2.8.

SECTION 2.9 Semiconductors

32. What is a semiconductor? How does it compare with a conductor and an insulator?

33. Consult a semiconductor electronics text and note the extensive use of germanium and silicon semiconductor materials. Review the characteristics of each material.

SECTION 2.10 Ammeters and Voltmeters

34. What are the significant differences in the way ammeters and voltmeters are connected?

35. If an ammeter reads 2.5 A for a period of 4 min, determine the charge that has passed through the meter.

36. Between two points in an electric circuit, a voltmeter reads 12.5 V for a period of 20 s. If the current measured by an ammeter is 10 mA, determine the energy expended and the charge that flowed between the two points.

GLOSSARY

Ammeter An instrument designed to read the current through elements in series with the meter.

Ampere (A) The SI unit of measurement applied to the flow of charge through a conductor.

Ampere-hour (Ah) rating The rating applied to a source of energy that will reveal how long a particular level of current can be drawn from that source.

Cell A fundamental source of electrical energy developed through the conversion of chemical or solar energy.

Conductors Materials that permit a generous flow of electrons with very little voltage applied.

Copper A material possessing physical properties that make it particularly useful as a conductor of electricity.

Coulomb (C) The fundamental SI unit of measure for charge. It is equal to the charge carried by 6.242×10^{18} electrons.

Coulomb's law An equation defining the force of attraction or repulsion between two charges.

Current The flow of charge resulting from the application of a difference in potential between two points in an electrical system.

dc current source A source that will provide a fixed current level even though the load to which it is applied may cause its terminal voltage to change.

dc generator A source of dc voltage available through the turning of the shaft of the device by some external means.

Direct current (dc) Current having a single direction (unidirectional) and a fixed magnitude over time.

Electrolytes The contact element and the source of ions between the electrodes of the battery.

Electron The particle with negative polarity that orbits the nucleus of an atom.

Free electron An electron unassociated with any particular atom, relatively free to move through a crystal lattice structure under the influence of external forces.

Insulators Materials in which a very high voltage must be applied to produce any measurable current flow.

Neutron The particle having no electrical charge, found in the nucleus of the atom.

Nucleus The structural center of an atom that contains both protons and neutrons.

Positive ion An atom having a net positive charge due to the loss of one of its negatively charged electrons.

Potential difference The algebraic difference in potential (or voltage) between two points in an electrical system.

Potential energy The energy that a mass possesses by virtue of its position.

Primary cell Sources of voltage that cannot be recharged.

Proton The particle of positive polarity found in the nucleus of an atom.

Rectification The process by which an ac signal is converted to one that has an average dc level.

Secondary cell Sources of voltage that can be recharged.

Semiconductor A material having a conductance value between that of an insulator and that of a conductor. Of significant importance in the manufacture of semiconductor electronic devices.

Solar cell Sources of voltage available through the conversion of light energy (photons) into electrical energy.

Specific gravity The ratio of the weight of a given volume of a substance to the weight of an equal volume of water at 4°C.

Volt (V) The unit of measurement applied to the difference in potential between two points. If 1 joule of energy is required to move 1 coulomb of charge between two points, the difference in potential is said to be 1 volt.

Voltage The term applied to the difference in potential between two points as established by a separation of opposite charges.

Voltmeter An instrument designed to read the voltage across an element or between any two points in a network.

Resistance

OBJECTIVES

- Become familiar with the parameters that determine the resistance of an element and learn how to calculate resistance.
- Understand the definition of *circular mil (CM)* and learn how the CM relates to square mils.
- Become familiar with wire tables, including how to use them.
- Be aware of the effect of temperature on conductors, insulators, and semiconductors, and be able to calculate resistance at any temperature.
- Become familiar with the characteristics of a superconductor and be aware of how superconductors may affect future development in the industry.
- Become familiar with the variety of fixed and variable resistors available today and learn how to determine resistance using the standard color code.
- Understand the relationship between conductance and resistance, including how one is calculated from the other.
- Become aware of how to use an ohmmeter to measure resistance and perform a variety of other important functions.
- Understand the basic operation of an electric baseboard heater and a strain gauge.
- Spend a few minutes using the computer software package Mathcad to appreciate how it can be used to perform mathematical calculations.

3.1 INTRODUCTION

In the previous chapter we found that placing a voltage across a wire or simple circuit would result in a flow of charge or current through the wire or circuit. The question remains, however, as to what determines the level of current that will result due to the application of a particular voltage. Why is the current heavier in some circuits than in others? The answer lies in the fact that there is an opposition to the flow of charge in the system that depends on the components of the circuit. This opposition to the flow of charge through an electrical circuit, called **resistance,** has the units of **ohms** and uses the Greek letter omega Ω as its symbol. The graphic symbol for resistance, which resembles the cutting edge of a saw, is provided in Fig. 3.1.

This opposition, due primarily to collisions and friction between the free electrons and other electrons, ions, and atoms in the path of motion, will convert the supplied electrical energy into **heat** that will raise the temperature of the electrical component and surrounding medium. The heat you feel from an electrical heater is simply due to passing current through a high-resistance material.

Each material with its unique atomic structure will react differently to pressures to establish current through its core. Conductors that permit a generous flow of charge with little external pressure will have low resistance levels, while insulators will have high resistance characteristics.

FIG. 3.1
Resistance symbol and notation.

3.2 RESISTANCE

The resistance of any material is due primarily to four factors:

1. *Material*
2. *Length*
3. *Cross-sectional area*
4. *Temperature of the material*

As noted in Section 3.1, the atomic structure will certainly determine how easily a free electron will pass through a material. It would also seem natural that the longer the path through which the free electron must pass, the greater the resistance factor. Free electrons pass more easily through conductors with larger cross-sectional areas. In addition, the higher the temperature of the conductive materials, the greater the internal vibration and motion of the components that make up the atomic structure of the wire, and the more difficult it is for the free electrons to find a path through the material.

The first three elements are related by the following basic equation for resistance:

$$R = \rho \frac{l}{A}$$

$\rho = \text{CM·}\Omega/\text{ft at } T = 20°C$
$l = \text{feet}$
$A = \text{area in circular mils (CM)}$

(3.1)

with each component of the equation defined by Fig. 3.2.

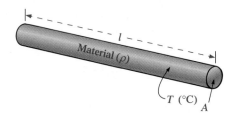

FIG. 3.2
Factors affecting the resistance of a conductor.

The material is identified by a factor called the **resistivity,** which uses the Greek letter rho as its symbol and is measured in CM·Ω /ft. Its value at a temperature of 20°C (room temperature = 68°F) is provided in Table 3.1 for a variety of common materials. Since the larger the resistivity, the greater the resistance to setting up a flow of charge, it appears as a multiplying factor in Eq. (3.1); that is, it appears in the numerator of the equation. It is important to realize at this point that since the resistivity is provided at a particular temperature, **Equation (3.1) is applicable only at room temperature.** The effect of higher and lower temperatures will be considered in Section 3.4.

Next, since the longer the material, the greater the resistance, the length appears as a multiplying factor in the numerator of Eq. (3.1). It is measured in feet for the given value of resistivity. Finally, the larger the area, the smaller the resistance (an inverse relationship), so the area appears in the denominator of the equation. The circular mil unit of measurement will be discussed later in this section.

The effect of all three factors is pictured in Fig. 3.3 for a copper conductor.

TABLE 3.1
Resistivity (ρ) of various materials.

Material	ρ @ 20°C
Silver	9.9
Copper	**10.37**
Gold	14.7
Aluminum	17.0
Tungsten	33.0
Nickel	47.0
Iron	74.0
Constantan	295.0
Nichrome	600.0
Calorite	720.0
Carbon	21,000.0

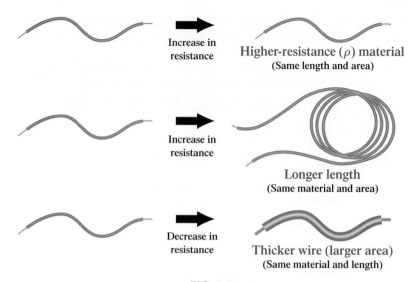

FIG. 3.3

Factors that will affect the resistance of a conductor.

Circular Mils (CM)

In Eq. (3.1), the area is measured in a quantity called **circular mils,** abbreviated **CM.** It is the quantity used in most commercial wire tables, and thus it needs to be carefully defined. The *mil* is a unit of measurement for length and is related to the inch by

$$1 \text{ mil} = \frac{1}{1000} \text{ in.}$$

or

$$1000 \text{ mils} = 1 \text{ in.}$$

In general, therefore, the mil is a very small unit of measurement for length. There are 1000 mils in an inch, or 1 mil is only 1/1000 of an inch. It is a length that is not visible with the naked eye although it can be measured with special instrumentation. The phrase *milling* used in steel factories is derived from the fact that a few mils of material are often removed by heavy machinery such as a lathe, and the thickness of steel is usually measured in mils.

By definition,

a wire with a diameter of 1 mil has an area of 1 CM.

as shown in Fig. 3.4.

An interesting result of such a definition is that the area of a circular wire in circular mils can be defined by the following rather simple equation:

$$\boxed{A_{\text{CM}} = (d_{\text{mils}})^2} \quad \text{(circular mils, CM)} \tag{3.2}$$

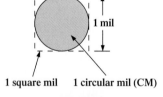

1 square mil   1 circular mil (CM)

FIG. 3.4

Defining the circular mil (CM).

Verification of this equation appears in Fig. 3.5 which shows that a wire with a diameter of 2 mils has a total area of 4 CM and a wire with a diameter of 3 mils has a total area of 9 CM.

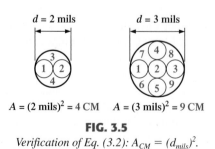

$A = (2 \text{ mils})^2 = 4 \text{ CM}$ $A = (3 \text{ mils})^2 = 9 \text{ CM}$

FIG. 3.5

Verification of Eq. (3.2): $A_{CM} = (d_{mils})^2$.

In the future, if you need to know the area of a wire in circular mils and the diameter is given in inches, first convert the diameter to mils by simply writing the diameter in decimal form and moving the decimal point three places to the right. For example,

$$\frac{1}{8} \text{ in.} = 0.\underset{\substack{\smile \\ 3 \text{ places}}}{125} \text{in.} = 125 \text{ mils}$$

Then the area is determined by

$$A_{\text{CM}} = (d_{\text{mils}})^2 = (125 \text{ mils})^2 = \textbf{15,625 CM}$$

Sometimes when you are working with conductors that are not circular, you will need to convert square mils to circular mils, and vice versa. Applying the basic equation for the area of a circle and substituting a diameter of 1 mil will result in

$$A = \frac{\pi d^2}{4} = \frac{\pi}{4} (1 \text{ mil})^2 \overset{\substack{\text{by definition} \\ \downarrow}}{=} \frac{\pi}{4} \text{ sq mils} \equiv 1 \text{ CM}$$

from which we can conclude the following:

$$\boxed{1 \text{ CM} = \frac{\pi}{4} \text{ sq mils}} \qquad \textbf{(3.3)}$$

EXAMPLE 3.1 What is the resistance of a 100 ft length of copper wire with a diameter of 0.020 in. at 20°C?

Solution:

$$\rho = 10.37 \, \frac{\text{CM}\cdot\Omega}{\text{ft}} \qquad 0.020 \text{ in.} = 20 \text{ mils}$$

$$A_{\text{CM}} = (d_{\text{mils}})^2 = (20 \text{ mils})^2 = 400 \text{ CM}$$

$$R = \rho \frac{l}{A} = \frac{(10.37 \text{ CM}\cdot\Omega/\text{ft})(100 \text{ ft})}{400 \text{ CM}}$$

$$R = \textbf{2.59 } \boldsymbol{\Omega}$$

EXAMPLE 3.2 An undetermined number of feet of wire have been used from the carton of Fig. 3.6. Find the length of the remaining copper wire if it has a diameter of 1/16 in. and a resistance of 0.5 Ω.

Solution:

$$\rho = 10.37 \text{ CM}\cdot\Omega/\text{ft} \qquad \frac{1}{16} \text{ in.} = 0.0625 \text{ in.} = 62.5 \text{ mils}$$

$$A_{\text{CM}} = (d_{\text{mils}})^2 = (62.5 \text{ mils})^2 = 3906.25 \text{ CM}$$

$$R = \rho \frac{l}{A} \Rightarrow l = \frac{RA}{\rho} = \frac{(0.5 \, \Omega)(3906.25 \text{ CM})}{10.37 \, \dfrac{\text{CM}\cdot\Omega}{\text{ft}}} = \frac{1953.125}{10.37}$$

$$l = \textbf{188.34 ft}$$

FIG. 3.6
Example 3.2.

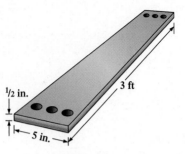

FIG. 3.7
Example 3.3.

½ in.

3 ft

5 in.

EXAMPLE 3.3 What is the resistance of a copper bus-bar, as used in the power distribution panel of a high-rise office building, with the dimensions indicated in Fig. 3.7?

Solution:

$$A_{CM}\begin{cases} 5.0\text{ in.} = 5000\text{ mils} \\[4pt] \dfrac{1}{2}\text{ in.} = 500\text{ mils} \\[4pt] A = (5000\text{ mils})(500\text{ mils}) = 2.5 \times 10^6\text{ sq mils} \\[4pt] \quad = 2.5 \times 10^6\text{ sq mils}\left(\dfrac{4/\pi\text{ CM}}{1\text{ sq mil}}\right) \\[4pt] A = 3.183 \times 10^6\text{ CM} \end{cases}$$

$$R = \rho\frac{l}{A} = \frac{(10.37\text{ CM·}\Omega/\text{ft})(3\text{ ft})}{3.183 \times 10^6\text{ CM}} = \frac{31.11}{3.183 \times 10^6}$$

$$R = \mathbf{9.774 \times 10^{-6}\ \Omega}$$

(quite small, $0.000009774\ \Omega \cong 0\ \Omega$)

We will find in the chapters to follow that the less the resistance of a conductor, the lower the losses in conduction from the source to the load. Similarly, since resistivity is a major factor in determining the resistance of a conductor, the lower the resistivity, the lower the resistance for the same size conductor. Table 3.1 would suggest, therefore, that silver, copper, gold, and aluminum would be the best conductors and the most common. In general, there are other factors, however, such as **malleability** (ability of a material to be shaped), **ductility** (ability of a material to be drawn into long, thin wires), temperature sensitivity, resistance to abuse, and, of course, cost, that must all be weighed when choosing a conductor for a particular application.

In general, copper is the most widely used material because it is quite malleable, ductile, and available; has good thermal characteristics; and is less expensive than silver or gold. It is certainly not cheap, however. Wiring is removed quickly from buildings to be torn down, for example, to extract the copper. At one time aluminum was introduced for general wiring because it is cheaper than copper, but its thermal characteristics created some difficulties. It was found that the heating due to current flow and the cooling that occurred when the circuit was turned off resulted in expansion and contraction of the aluminum wire to the point where connections could eventually work themselves loose and dangerous side effects could result. Aluminum is still used today, however, in areas such as integrated circuit manufacturing and in situations where the connections can be made secure. Silver and gold are, of course, much more expensive than copper or aluminum, but there are places where the cost is justified. Silver has excellent plating characteristics for surface preparations, and gold is used quite extensively in integrated circuits. Tungsten has a resistivity three times that of copper, but there are occasions when its physical characteristics (durability, hardness) are the overriding considerations.

3.3 WIRE TABLES

The wire table was designed primarily to standardize the size of wire produced by manufacturers. As a result, the manufacturer has a larger market and the consumer knows that standard wire sizes will always be available. The table was designed to assist the user in every way possible; it usually includes data such as the cross-sectional area in circular mils, diameter in mils, ohms per 1000 feet at 20°C, and weight per 1000 feet.

The American Wire Gage (AWG) sizes are given in Table 3.2 for solid round copper wire. A column indicating the maximum allowable current in amperes, as determined by the National Fire Protection Association, has also been included.

The chosen sizes have an interesting relationship: For every drop in 3 gage numbers, the area is doubled; and for every drop in 10 gage numbers, the area increases by a factor of 10.

Examining Eq. (3.1), we note also that *doubling the area cuts the resistance in half, and increasing the area by a factor of 10 decreases the resistance of 1/10 the original*, everything else kept constant.

The actual sizes of some of the gage wires listed in Table 3.2 are shown in Fig. 3.8 with a few of their areas of application. A few examples using Table 3.2 follow.

TABLE 3.2
American Wire Gage (AWG) sizes.

	AWG #	Area (CM)	Ω/1000 ft at 20°C	Maximum Allowable Current for RHW Insulation (A)[*]
(4/0)	0000	211,600	0.0490	230
(3/0)	000	167,810	0.0618	200
(2/0)	00	133,080	0.0780	175
(1/0)	0	105,530	0.0983	150
	1	83,694	0.1240	130
	2	66,373	0.1563	115
	3	52,634	0.1970	100
	4	41,742	0.2485	85
	5	33,102	0.3133	—
	6	26,250	0.3951	65
	7	20,816	0.4982	—
	8	16,509	0.6282	50
	9	13,094	0.7921	—
	10	10,381	0.9989	30
	11	8,234.0	1.260	—
	12	6,529.9	1.588	20
	13	5,178.4	2.003	—
	14	4,106.8	2.525	15
	15	3,256.7	3.184	
	16	2,582.9	4.016	
	17	2,048.2	5.064	
	18	1,624.3	6.385	
	19	1,288.1	8.051	
	20	1,021.5	10.15	
	21	810.10	12.80	
	22	642.40	16.14	
	23	509.45	20.36	
	24	404.01	25.67	
	25	320.40	32.37	
	26	254.10	40.81	
	27	201.50	51.47	
	28	159.79	64.90	
	29	126.72	81.83	
	30	100.50	103.2	
	31	79.70	130.1	
	32	63.21	164.1	
	33	50.13	206.9	
	34	39.75	260.9	
	35	31.52	329.0	
	36	25.00	414.8	
	37	19.83	523.1	
	38	15.72	659.6	
	39	12.47	831.8	
	40	9.89	1049.0	

[*]Not more than three conductors in raceway, cable, or direct burial.

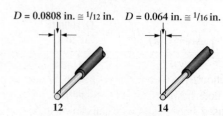

$D = 0.365$ in. $\cong$ 1/3 in.

Stranded for increased flexibility

00

Power distribution

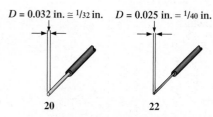

$D = 0.0808$ in. $\cong$ 1/12 in. $D = 0.064$ in. $\cong$ 1/16 in.

12   14

Lighting, outlets, general home use

$D = 0.032$ in. $\cong$ 1/32 in. $D = 0.025$ in. $= $ 1/40 in.

20   22

Radio, television

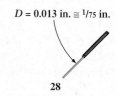

$D = 0.013$ in. $\cong$ 1/75 in.

28

Telephone, instruments

FIG. 3.8

Popular wire sizes and some of their areas of application.

EXAMPLE 3.4   Find the resistance of 650 ft of #8 copper wire ($T = 20°C$).

Solution: For #8 copper wire (solid), Ω/1000 ft at 20°C = 0.6282 Ω, and

$$650 \,\text{ft}\left(\frac{0.6282 \;\Omega}{1000 \,\text{ft}}\right) = \mathbf{0.408 \;\Omega}$$

EXAMPLE 3.5   What is the diameter, in inches, of a #12 copper wire?

Solution: For #12 copper wire (solid), $A = 6529.9$ CM, and

$$d_{\text{mils}} = \sqrt{A_{\text{CM}}} = \sqrt{6529.9 \text{ CM}} \cong 80.81 \text{ mils}$$

$$d = \mathbf{0.0808 \text{ in.}} \text{ (or close to } 1/12 \text{ in.)}$$

EXAMPLE 3.6   For the system of Fig. 3.9, the total resistance of *each* power line cannot exceed 0.025 Ω, and the maximum current to be drawn by the load is 95 A. What gage wire should be used?

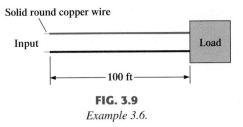

FIG. 3.9

Example 3.6.

Solution:

$$R = \rho\frac{l}{A} \Rightarrow A = \rho\frac{l}{R} = \frac{(10.37 \text{ CM}\cdot\Omega/\text{ft})(100 \text{ ft})}{0.025 \text{ }\Omega} = \textbf{41,480 CM}$$

Using the wire table, we choose the wire with the next largest area, which is #4, to satisfy the resistance requirement. We note, however, that 95 A must flow through the line. This specification requires that **#3 wire be used** since the #4 wire can carry a maximum current of only 85 A.

3.4   TEMPERATURE EFFECTS

Temperature has a significant effect on the resistance of conductors, semiconductors, and insulators.

Conductors

Conductors have a generous number of free electrons, and any introduction of thermal energy will have little impact on the total number of free carriers. In fact, the thermal energy will only increase the intensity of the random motion of the particles within the material and make it increasingly difficult for a general drift of electrons in any one direction to be established. The result is that

for good conductors, an increase in temperature will result in an increase in the resistance level. Consequently, conductors have a positive temperature coefficient.

The plot of Fig. 3.10(a) has a positive temperature coefficient.

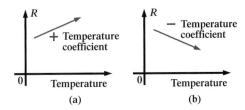

FIG. 3.10

Demonstrating the effect of a positive and a negative temperature coefficient on the resistance of a conductor.

Semiconductors

In semiconductors an increase in temperature will impart a measure of thermal energy to the system that will result in an increase in the number of free carriers in the material for conduction. The result is that

for semiconductor materials, an increase in temperature will result in a decrease in the resistance level. Consequently, semiconductors have negative temperature coefficients.

The plot of Fig. 3.10(b) has a negative temperature coefficient.

Insulators

As with semiconductors, an increase in temperature will result in a decrease in the resistance of an insulator. The result is a negative temperature coefficient.

Inferred Absolute Temperature

Figure 3.11 reveals that for copper (and most other metallic conductors), the resistance increases almost linearly (in a straight-line relationship) with an increase in temperature. Since temperature can have such a pronounced effect on the resistance of a conductor, it is important that we have some method of determining the resistance at any temperature within operating limits. An equation for this purpose can be obtained by approximating the curve of Fig. 3.11 by the straight dashed line that intersects the temperature scale at $-234.5°C$. Although the actual curve extends to **absolute zero** $(-273.15°C$, or 0 K), the straight-line approximation is quite accurate for the normal operating temperature range. At two different temperatures, T_1 and T_2, the resistance of copper is R_1 and R_2, as indicated on the curve. Using a property of similar triangles, we may develop a mathematical relationship between these values of resistances at different temperatures. Let x equal the distance from $-234.5°C$ to T_1 and y the distance from $-234.5°C$ to T_2, as shown in Fig. 3.11. From similar triangles,

$$\frac{x}{R_1} = \frac{y}{R_2}$$

or

$$\frac{234.5 + T_1}{R_1} = \frac{234.5 + T_2}{R_2} \tag{3.4}$$

The temperature of $-234.5°C$ is called the **inferred absolute temperature** of copper. For different conducting materials, the intersection of the straight-line approximation will occur at different temperatures. A few typical values are listed in Table 3.3.

TABLE 3.3

Inferred absolute temperatures (T_i).

Material	°C
Silver	−243
Copper	**−234.5**
Gold	−274
Aluminum	−236
Tungsten	−204
Nickel	−147
Iron	−162
Nichrome	−2,250
Constantan	−125,000

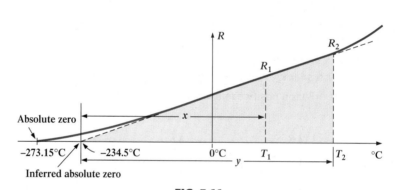

FIG. 3.11

Effect of temperature on the resistance of copper.

The minus sign does not appear with the inferred absolute temperature on either side of Eq. (3.4) because x and y are the *distances* from $-234.5°C$ to T_1 and T_2, respectively, and therefore are simply magnitudes. For T_1 and T_2 less than zero, x and y are less than $-234.5°C$, and the distances are the differences between the inferred absolute temperature and the temperature of interest.

Equation (3.4) can easily be adapted to any material by inserting the proper inferred absolute temperature. It may therefore be written as follows:

$$\frac{|T_i| + T_1}{R_1} = \frac{|T_i| + T_2}{R_2} \tag{3.5}$$

where $|T_i|$ indicates that the inferred absolute temperature of the material involved is inserted as a positive value in the equation. In general, therefore, associate the sign only with T_1 and T_2.

EXAMPLE 3.7   If the resistance of a copper wire is 50 Ω at 20°C, what is its resistance at 100°C (the boiling point of water)?

Solution: Eq. (3.4):

$$\frac{234.5°C + 20°C}{50\ \Omega} = \frac{234.5°C + 100°C}{R_2}$$

$$R_2 = \frac{(50\ \Omega)(334.5°C)}{254.5°C} = \mathbf{65.72\ \Omega}$$

EXAMPLE 3.8   If the resistance of a copper wire at freezing (0°C) is 30 Ω, what is its resistance at −40°C?

Solution: Eq. (3.4):

$$\frac{234.5°C + 0}{30\ \Omega} = \frac{234.5°C - 40°C}{R_2}$$

$$R_2 = \frac{(30\ \Omega)(194.5°C)}{234.5°C} = \mathbf{24.88\ \Omega}$$

EXAMPLE 3.9   If the resistance of an aluminum wire at room temperature (20°C) is 100 mΩ (measured by a milliohmmeter), at what temperature will its resistance increase to 120 mΩ?

Solution: Eq. (3.5):

$$\frac{236°C + 20°C}{100\ \text{m}\Omega} = \frac{236°C + T_2}{120\ \text{m}\Omega}$$

and

$$T_2 = 120\ \text{m}\Omega \left(\frac{256°C}{100\ \text{m}\Omega} \right) - 236°C$$

$$T_2 = \mathbf{71.2°C}$$

Temperature Coefficient of Resistance

There is a second popular equation for calculating the resistance of a conductor at different temperatures. Defining

$$\alpha_{20} = \frac{1}{|T_i| + 20°C} \qquad (\Omega/°C/\Omega) \tag{3.6}$$

as the **temperature coefficient of resistance** at a temperature of 20°C, and R_{20} as the resistance of the sample at 20°C, the resistance R_1 at a temperature T_1 is determined by

$$R_1 = R_{20}[1 + \alpha_{20}(T_1 - 20°C)] \tag{3.7}$$

The values of α_{20} for different materials have been evaluated, and a few are listed in Table 3.4.

Equation (3.7) can be written in the following form:

$$\alpha_{20} = \frac{\left(\dfrac{R_1 - R_{20}}{T_1 - 20°C} \right)}{R_{20}} = \frac{\dfrac{\Delta R}{\Delta T}}{R_{20}}$$

from which the units of $\Omega/°C/\Omega$ for α_{20} are defined.

Since $\Delta R/\Delta T$ is the slope of the curve of Fig. 3.11, we can conclude that

the higher the temperature coefficient of resistance for a material, the more sensitive the resistance level to changes in temperature.

TABLE 3.4

Temperature coefficient of resistance for various conductors at 20°C.

Material	Temperature Coefficient (α_{20})
Silver	0.0038
Copper	0.00393
Gold	0.0034
Aluminum	0.00391
Tungsten	0.005
Nickel	0.006
Iron	0.0055
Constantan	0.000008
Nichrome	0.00044

Referring to Table 3.3, we find that copper is more sensitive to temperature variations than is silver, gold, or aluminum, although the differences are quite small. The slope defined by α_{20} for constantan is so small that the curve is almost horizontal.

Since R_{20} of Eq. (3.7) is the resistance of the conductor at 20°C and $(T_1 - 20°C)$ is the change in temperature from 20°C, Equation (3.7) can be written in the following form:

$$R = \rho \frac{l}{A} [1 + \alpha_{20} \Delta T] \qquad \textbf{(3.8)}$$

providing an equation for resistance in terms of all the controlling parameters.

PPM/°C

For resistors, as for conductors, resistance changes with a change in temperature. The specification is normally provided in parts per million per degree Celsius (**PPM/°C**), providing an immediate indication of the sensitivity level of the resistor to temperature. For resistors, a 5000 PPM level is considered high, whereas 20 PPM is quite low. A 1000 PPM/°C characteristic reveals that a 1° change in temperature will result in a change in resistance equal to 1000 PPM, or $1000/1,000,000 = 1/1000$ of its nameplate value—not a significant change for most applications. However, a 10° change would result in a change equal to 1/100 (1%) of its nameplate value, which is becoming significant. The concern, therefore, lies not only with the PPM level but with the range of expected temperature variation.

In equation form, the change in resistance is given by

$$\Delta R = \frac{R_{\text{nominal}}}{10^6} (\text{PPM})(\Delta T) \qquad \textbf{(3.9)}$$

where R_{nominal} is the nameplate value of the resistor at room temperature and ΔT is the change in temperature from the reference level of 20°C.

EXAMPLE 3.10   For a 1 kΩ carbon composition resistor with a PPM of 2500, determine the resistance at 60°C.

Solution:

$$\Delta R = \frac{1000 \; \Omega}{10^6} (2500)(60°C - 20°C)$$

$$= 100 \; \Omega$$

and

$$R = R_{\text{nominal}} + \Delta R = 1000 \; \Omega + 100 \; \Omega$$

$$= \textbf{1100} \; \boldsymbol{\Omega}$$

3.5   SUPERCONDUCTORS

There is no question that the field of electricity/electronics has been one of the most exciting of the 20th century, and this excitement continues into the 21st century. Even though new developments appear almost weekly from extensive research and development activities, every once in a while there is some very special step forward that has the whole field at the edge of its seat waiting to see what might develop in the near future. Such a level of excitement and interest surrounds the research drive to develop a room-temperature **superconductor**—an advance that will rival the introduction of semiconductor devices such as the transistor (to replace tubes), wireless communication, or the electric light. The implications of such a development are so far-reaching that it is difficult to forecast the vast impact it will have on the entire field.

The intensity of the research effort throughout the world today to develop a room-temperature superconductor is described by some researchers as "unbelievable, contagious, exciting, and demanding" but an adventure in which they treasure the opportunity to be involved. Progress in the field since 1986 suggests that the use of superconductivity in commercial applications will grow quite rapidly in the next few decades. It is indeed an exciting era full of growing anticipation! Why this interest in superconductors? What are they all about? In a nutshell,

superconductors are conductors of electric charge that, for all practical purposes, have zero resistance.

In a conventional conductor, electrons travel at average speeds in the neighborhood of 1000 mi/s (they can cross the United States in about 3 seconds), even though Einstein's theory of relativity suggests that the maximum speed of information transmission is the speed of light, or 186,000 mi/s. The relatively slow speed of conventional conduction is due to collisions of atoms in the material, repulsive forces between electrons (like charges repel), thermal agitation that results in indirect paths due to the increased motion of the neighboring atoms, impurities in the conductor, and so on. In the superconductive state, there is a pairing of electrons, denoted by the **Cooper effect,** in which electrons travel in pairs and help each other maintain a significantly higher velocity through the medium. In some ways this is like "drafting" by competitive cyclists or runners. There is an oscillation of energy between partners or even "new" partners (as the need arises) to ensure passage through the conductor at the highest possible velocity with the least total expenditure of energy.

Even though the concept of superconductivity first surfaced in 1911, it was not until 1986 that the possibility of superconductivity at room temperature became a renewed goal of the research community. For some 74 years superconductivity could be established only at temperatures colder than 23 K. (Kelvin temperature is universally accepted as the unit of measurement for temperature for superconductive effects. Recall that K = 273.15° + °C, so a temperature of 23 K is −250°C, or −418°F.) In 1986, however, physicists Alex Muller and George Bednorz of the IBM Zurich Research Center found a ceramic material, lanthanum barium copper oxide, that exhibited superconductivity at 30 K. Although it would not appear to be a significant step forward, it introduced a new direction to the research effort and spurred others to improve on the new standard. In October 1987 both scientists received the Nobel prize for their contribution to an important area of development.

In just a few short months, Professors Paul Chu of the University of Houston and Man Kven Wu of the University of Alabama raised the temperature to 95 K using a superconductor of yttrium barium copper oxide. The result was a level of excitement in the scientific community that brought research in the area to a new level of effort and investment. The major impact of such a discovery was that liquid nitrogen (boiling point of 77 K) could now be used to bring the material down to the required temperature rather than liquid helium, which boils at 4 K. The result is a tremendous saving in the cooling expense since liquid helium is at least ten times more expensive than liquid nitrogen. Pursuing the same direction, some success has been achieved at 125 K and 162 K using a thallium compound (unfortunately, however, thallium is a very poisonous substance).

Figure 3.12(a) clearly reveals that there was little change in the temperature for superconductors until the discovery of 1986. The curve then takes a sharp curve upward, suggesting that room-temperature superconductors may become available in a few short years. However, unless there is a significant breakthrough in the near future, this goal no longer seems feasible. The effort continues and is receiving an increasing level of financing and worldwide attention. Now, increasing numbers of corporations are trying to capitalize on the success already attained, as will be discussed later in this section.

The fact that ceramics have provided the recent breakthrough in superconductivity is probably a surprise when you consider that they are also an important class of insulators. However, the ceramics that exhibit the characteristics of superconductivity are compounds that include copper, oxygen, and rare earth elements such as yttrium, lanthanum, and thallium. There are also indicators that the current compounds may be limited to a maximum temperature of 200 K (about 100 K short of room temperature), leaving the door wide open to innovative approaches to compound selection. The temperature at which a superconductor reverts back to the characteristics of a conventional conductor is called the *critical temperature,* denoted by T_c. Note in Fig. 3.12(b) that the resistivity level changes abruptly at T_c. The sharpness of the transition region is a function of the purity of the sample. Long listings of critical temperatures for a variety of tested compounds can be found in reference materi-

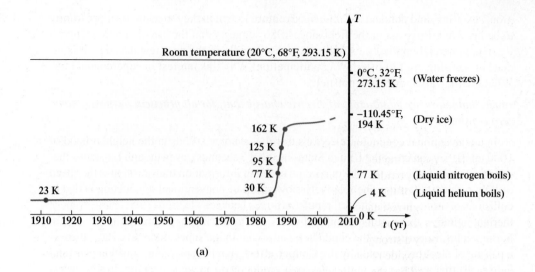

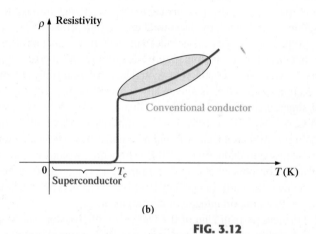

FIG. 3.12

Superconductors: (a) rising temperatures; (b) defining the critical temperature T_c.

als providing tables of a wide variety to support research in physics, chemistry, geology, and related fields. Two such publications include the CRC (The Chemical Rubber Co.) *Handbook of Tables for Applied Engineering Science* and the CRC *Handbook of Chemistry and Physics.*

Even though ceramic compounds have established higher transition temperatures, there is concern about their brittleness and current density limitations. In the area of integrated circuit manufacturing, current density levels must equal or exceed 1 MA/cm², or 1 million amperes through a cross-sectional area about one-half the size of a dime. Recently IBM attained a level of 4 MA/cm² at 77 K, permitting the use of superconductors in the design of some new-generation, high-speed computers.

Although room-temperature success has not been attained, there are numerous applications for some of the superconductors developed thus far. It is simply a matter of balancing the additional cost against the results obtained or deciding whether any results at all can be obtained without the use of this zero-resistance state. Some research efforts require high-energy accelerators or strong magnets attainable only with superconductive materials. Superconductivity is currently applied in the design of 300 mi/h Meglev trains (trains that ride on a cushion of air established by opposite magnetic poles), in powerful motors and generators, in nuclear magnetic resonance imaging systems to obtain cross-sectional images of the brain (and other parts of the body), in the design of computers with operating speeds four times that of conventional systems, and in improved power distribution systems.

The range of future uses for superconductors is a function of how much success physicists have in raising the operating temperature and how well they can utilize the successes obtained thus far. However, it would appear that it is only a matter of time (the eternal optimist) before magnetically levitated trains increase in number, improved medical diagnostic equipment is available, computers operate at much higher speeds, high-efficiency power and storage systems are available, and transmission systems operate at very high efficiency levels due to this area of developing interest. Only time will reveal the impact that this new direction will have on the quality of life.

3.6   TYPES OF RESISTORS

Fixed Resistors

Resistors are made in many forms, but all belong to either of two groups: fixed or variable. Today the most common of the low-wattage, fixed-type resistors is the film resistor of Fig. 3.13. It is constructed by depositing a thin layer of resistive material (typically carbon, metal, or metal oxide) on a ceramic rod. The desired resistance is then obtained by cutting away some of the resistive material in a helical manner to establish a long, continuous band of high-resistance material from one end of the resistor to the other. In general, carbon-film resistors have a beige body and a lower wattage rating. The metal-film resistor is typically a stronger color, such as brick red or dark green, with higher wattage ratings. The metal-oxide resistor is usually a softer pastel color, such as the powder blue of Fig. 3.13(b), and has the highest wattage raging of the three.

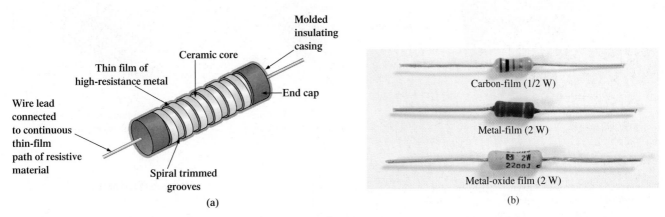

(a)

(b)

FIG. 3.13

Film-resistors: (a) construction; (b) types.

When you search through most electronics catalogs or visit a local electronics dealer such as Radio Shack to purchase resistors, you will find that the most common resistor is the film resistor. In years past, the carbon composition resistor of Fig. 3.14 was the most common, but fewer and fewer companies are manufacturing this variety, with its range of applications reduced to special applications such as where very high temperatures and inductive effects (Chapter 11) can be a problem. Its resistance is determined by the carbon composition material molded directly to each end of the resistor. The high resistivity characteristics of carbon ($\rho = 21{,}000$ CM·Ω/ft) provide a high-resistance path for the current through the element.

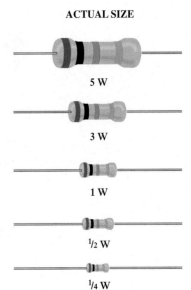

ACTUAL SIZE

5 W

3 W

1 W

$^{1}/_{2}$ W

$^{1}/_{4}$ W

FIG. 3.15

Fixed metal-oxide resistors of different wattage ratings.

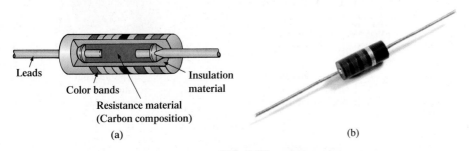

(a)

(b)

FIG. 3.14

Fixed composition resistors: (a) construction; (b) appearance.

For a particular style and manufacturer, the size of a resistor will increase with the power or wattage rating.

The concept of power will be covered in detail in Chapter 4, but for the moment recognize that increased power ratings are normally associated with the ability to handle higher current and temperature levels. Figure 3.15 depicts the actual size of thin-film, metal-oxide resistors in the 1/4 W to 5 W rating range. All the resistors of Fig. 3.15 are 1 MΩ, revealing that

the size of a resistor does not define its resistance level.

A variety of other fixed resistors are depicted in Fig. 3.16. The wire-wound resistors of Fig. 3.16(a) are formed by winding a high-resistance wire around a ceramic core. The entire structure is then baked in a ceramic cement to provide a protective covering. Wire-wound resistors are typically used for larger power applications, although they are also available with very small wattage ratings and very high accuracy.

Figures 3.16(c) and (g) are special types of wire-wound resistors with a low percent tolerance. Note, in particular, the high power ratings for the wire-wound resistors for their relatively small size. Figures 3.16(b), (d), and (f) are power film resistors that use a thicker layer of film material than used in the variety of Fig. 3.13. The chip resistors of Fig. 3.16(f)

100 Ω, 25 W

2 kΩ, 8 W

470 Ω, 35 W
Thick-film power resistor

(b)

1 kΩ, 25 W
Aluminum-housed, chassis-mount
resistor—precision wire-wound

(c)

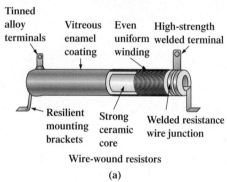

Tinned alloy terminals   Vitreous enamel coating   Even uniform winding   High-strength welded terminal

Resilient mounting brackets   Strong ceramic core   Welded resistance wire junction

Wire-wound resistors

(a)

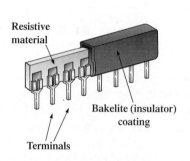

Resistive material

Bakelite (insulator) coating

Terminals

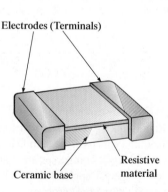

Electrodes (Terminals)

Ceramic base   Resistive material

100 M Ω, 0.75 W
Precision power film resistor

(d)

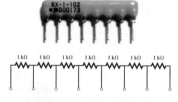

1 kΩ   1 kΩ   1 kΩ   1 kΩ   1 kΩ   1 kΩ   1 kΩ

1 kΩ bussed (all connected
on one side) single
in-line resistor network

(e)

22 kΩ, 1 W
thick-film chip
resistors with gold
electrodes

(f)

25 kΩ, 5 W
Silicon-coated, wire-wound resistor

(g)

FIG. 3.16
Various types of fixed resistors.

are used where space is a priority, such as on the surface of circuit board. Units of this type can be less than 1/4" in length or width, with thickness as small as 1/30", yet they can still handle 0.5 W of power with resistance levels as high as 1000 MΩ—clear evidence that size does not determine the resistance level. The fixed resistor of Fig. 3.16(e) has terminals applied to a layer of resistor material, with the resistance between the terminals a function of the dimensions of the resistive material and the placement of the terminal pads.

Variable Resistors

Variable resistors, as the name implies, have a terminal resistance that can be varied by turning a dial, knob, screw, or whatever seems appropriate for the application. They can have two or three terminals, but most have three terminals. If the two- or three-terminal device is used as a variable resistor, it is usually referred to as a **rheostat.** If the three-terminal device is used for controlling potential levels, it is then commonly called a **potentiometer.** Even though a three-terminal device can be used as a rheostat or a potentiometer (depending on how it is connected), it is typically called a *potentiometer* when listed in trade magazines or requested for a particular application.

The symbol for a three-terminal potentiometer appears in Fig. 3.17(a). When used as a variable resistor (or rheostat), it can be hooked up in one of two ways, as shown in Fig. 3.17(b) and (c). In Fig. 3.17(b), points *a* and *b* are hooked up to the circuit, and the remaining terminal is left hanging. The resistance introduced is determined by that portion of the resistive element between points *a* and *b*. In Fig. 3.17(c), the resistance is again between points *a* and *b,* but now the remaining resistance is "shorted-out" (effect removed) by the connection from *b* to *c*. The universally accepted symbol for a rheostat appears in Fig. 3.17(d).

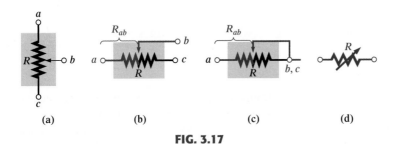

| (a) | (b) | (c) | (d) |

FIG. 3.17

Potentiometer: (a) symbol; (b) and (c) rheostat connections; (d) rheostat symbol.

Most potentiometers have three terminals in the relative positions shown in Fig. 3.18. The knob, dial, or screw in the center of the housing controls the motion of a contact that can move along the resistive element connected between the outer two terminals. The contact is connected to the center terminal, establishing a resistance from movable contact to each outer terminal.

The resistance between the outside terminals a and c of Fig. 3.19(a) (and Fig. 3.18) is always fixed at the full rated value of the potentiometer, regardless of the position of the wiper arm b.

In other words, the resistance between terminals *a* and *c* of Fig. 3.19(a) for a 1 MΩ potentiometer will always be 1 MΩ, no matter how we turn the control element and move the contact. In Fig. 3.19(a), the center contact is not part of the network configuration.

The resistance between the wiper arm and either outside terminal can be varied from a minimum of 0 Ω to a maximum value equal to the full rated value of the potentiometer.

In Fig. 3.19(b), the wiper arm has been placed 1/4 of the way down from point *a* to point *c*. The resulting resistance between points *a* and *b* will therefore be 1/4 of the total, or 250 kΩ (for a 1 MΩ potentiometer), and the resistance between *b* and *c* will be 3/4 of the total, or 750 kΩ.

The sum of the resistances between the wiper arm and each outside terminal will equal the full rated resistance of the potentiometer.

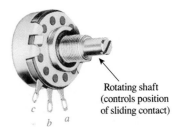

(a) External view

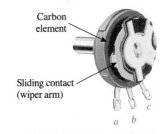

(b) Internal view

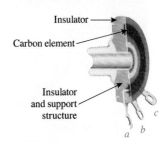

(c) Carbon element

FIG. 3.18

Molded composition–type potentiometer.

(Courtesy of Allen-Bradley Co.)

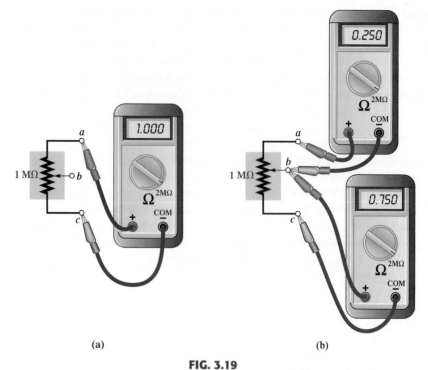

(a)   **(b)**

FIG. 3.19

Resistance components of a potentiometer: (a) between outside terminals;
(b) between wiper arm and each outside terminal.

This was demonstrated by Fig. 3.19(b), where 250 kΩ + 750 kΩ = 1 MΩ. Specifically:

$$R_{ac} = R_{ab} + R_{bc}$$

(3.10)

Therefore, as the resistance from the wiper arm to one outside contact increases, the resistance between the wiper arm and the other outside terminal must decrease accordingly. For example, if R_{ab} of a 1 kΩ potentiometer is 200 Ω, then the resistance R_{bc} must be 800 Ω. If R_{ab} is further decreased to 50 Ω, then R_{bc} must increase to 950 Ω, and so on.

The molded carbon composition potentiometer is typically applied in networks with smaller power demands, and it ranges in size from 20 Ω to 22 MΩ (maximum values). A miniature trimmer (less than 1/4" in diameter) appears in Fig. 3.20(a), and a variety of potentiometers that use a cermet resistive material appear in Fig. 3.20(b). The contact point of the three point wire-wound resistor of Fig. 3.20(c) can be moved to set the resistance between the three terminals.

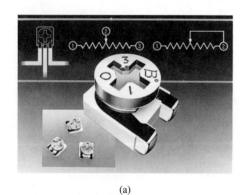

(a)

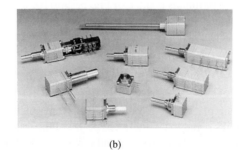

(b)

(c)

FIG. 3.20

Variable resistors: (a) 4 mm (≈5/32") trimmer (courtesy of Bourns, Inc.); (b) conductive plastic and cermet elements
(courtesy of Honeywell Clarostat); (c) three-point wire-wound resistor.

When the device is used as a potentiometer, the connections are as shown in Fig. 3.21. It can be used to control the level of V_{ab}, V_{bc}, or both, depending on the application. Additional discussion of the potentiometer in a loaded situation can be found in the chapters that follow.

3.7 COLOR CODING AND STANDARD RESISTOR VALUES

A wide variety of resistors, fixed or variable, are large enough to have their resistance in ohms printed on the casing. Some, however, are too small to have numbers printed on them, so a system of **color coding** is used. For the thin-film resistor, four, five, or six bands may be used. Initially, however, you must become accustomed to the four-band scheme. Later in this section the purpose of the fifth and sixth bands will be described.

For the four-band scheme the bands are **always read from the end that has a band closest to it,** as shown in Fig. 3.22. The bands are numbered as shown for reference in the discussion to follow.

The first two bands represent the first and second digits, respectively.

They are the actual first two numbers that define the numerical value of the resistor.

The third band determines the power-of-ten multiplier for the first two digits (actually the number of zeros that follow the second digit for resistors greater than 10 Ω).

The fourth band is the manufacturer's tolerance, which is an indication of the precision by which the resistor was made.

If the fourth band is omitted, the tolerance is assumed to be $\pm 20\%$.

The number corresponding to each color is defined by Fig. 3.23. The fourth band will be either $\pm 5\%$ or $\pm 10\%$ as defined by gold and silver, respectively. To remember which color goes with which percent, simply remember that $\pm 5\%$ resistors cost more and gold is more valuable than silver.

Remembering which color goes with each digit takes a bit of practice. In general, the colors start with the very dark shades and move toward the lighter shades. The best way to memorize is to simply repeat to yourself over and over again that red is 2, yellow is 4, and so on. In time you will be able to associate one with the other pretty quickly. Remembering a phrase, whatever it may be, is time-consuming, and to repeat the phrase for every color is nonsense. Simply practice with a friend or a fellow student, and you will learn most of the colors in short order. It seems that students have the most trouble with green, blue, and violet which are in the middle of the pack. However, simply remember that blue and violet are similar tones at the end of the pack, and green is a combination of yellow (just before green) and brown. Practice is the key, however, and proficiency will come with time.

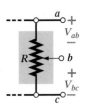

FIG. 3.21
Potentiometer control of voltage levels.

1 2 3 4

FIG. 3.22
Color coding for fixed resistors.

Number		Color
0		Black
1		Brown
2		Red
3		Orange
4		Yellow
5		Green
6		Blue
7		Violet
8		Gray
9		White
$\pm 5\%$ (0.1 multiplier if 3rd band)		Gold
$\pm 10\%$ (0.01 multiplier if 3rd band)		Silver

FIG. 3.23
Color coding.

EXAMPLE 3.11 Find the value of the resistor of Fig. 3.24.

Solution: Reading from the band closest to the left edge, we find that the first two colors of brown and red represent the numbers 1 and 2, respectively. The third band is orange, representing the number 3 for the power of the multiplier as follows:

$$12 \times 10^3 \ \Omega$$

resulting in a value of 12 kΩ. As indicated above, if 12 kΩ is written as 12,000 Ω, the third band reveals the number of zeros that follow the first two digits.

Now for the fourth band of gold, representing a tolerance of $\pm 5\%$: In order to find the range into which the manufacturer has guaranteed the resistor will fall, first convert the 5% to a decimal number by moving the decimal point two places to the left:

$$5\% \Rightarrow 0.05$$

Then multiply the resistor value by this decimal number:

$$0.05(12 \text{ k}\Omega) = 600 \ \Omega$$

Finally, add the resulting number to the resistor value to determine the maximum value, and subtract the number to find the minimum value. That is,

$$\text{Maximum} = 12,000 \ \Omega + 600 \ \Omega = 12.6 \text{ k}\Omega$$
$$\text{Minimum} = 12,000 \ \Omega - 600 \ \Omega = 11.4 \text{ k}\Omega$$
$$\textbf{Range} = \textbf{11.4 k}\boldsymbol{\Omega} \textbf{ to 12.6 k}\boldsymbol{\Omega}$$

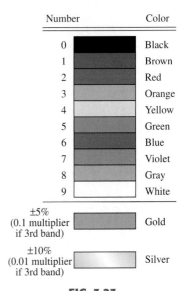

FIG. 3.24
Example 3.11.

The result is that the manufacturer has guaranteed with the 5% gold band that the resistor will fall in the range just determined. In other words, the manufacturer is not guaranteeing that the resistor will be exactly 12 kΩ but rather that it will fall in a range as defined above.

Using the above procedure, the smallest resistor that could be labeled with the color code would be 10 Ω. However,

the range can be extended to include resistors from 0.1 Ω to 10 Ω by simply using gold as a multiplier color (third band) to represent 0.1 and using silver to represent 0.01.

This is demonstrated in the next example.

FIG. 3.25
Example 3.12.

EXAMPLE 3.12   Find the value of the resistor of Fig. 3.25.

Solution: The first two colors are gray and red, representing the numbers 8 and 2. The third color is gold, representing a multiplier of 0.1. Using the multiplier, we obtain a resistance of

$$(0.1)(82 \ \Omega) = 8.2 \ \Omega$$

The fourth band is silver, representing a tolerance of ±10%. Converting to a decimal number and multiplying through will yield

$$10\% = 0.10 \quad \text{and} \quad (0.1)(8.2 \ \Omega) = 0.82 \ \Omega$$

$$\text{Maximum} = 8.2 \ \Omega + 0.82 \ \Omega = 9.02 \ \Omega$$
$$\text{Minimum} = 8.2 \ \Omega - 0.82 \ \Omega = 7.38 \ \Omega$$

so that $\quad\quad\quad\quad\quad$ **Range = 7.38 Ω to 9.02 Ω**

Although it will take some time to learn the numbers associated with each color, it is certainly encouraging to become aware that

the same color scheme to represent numbers is used for all the important elements of electric circuits.

Later on, when we encounter capacitors and inductors, we will find that the numerical value associated with each color is the same. Therefore, once learned, the scheme has repeated areas of application.

Some manufacturers prefer to use a **5-band color code.** In such cases, as shown in the top portion of Fig. 3.26, three digits are provided before the multiplier. The fifth band remains the tolerance indicator. If the manufacturer decides to include the temperature coefficient, a sixth band will appear as shown in the lower portion of Fig. 3.26, with the color indicating the ppm level.

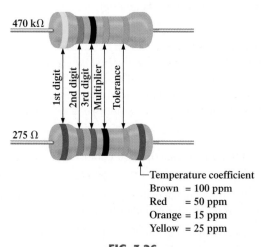

FIG. 3.26
Five-band color coding for fixed resistors.

For four, five, or six bands, if the tolerance is less than 5%, the following colors are used to reflect the % tolerances:

brown = ±1%, red = ±2%, green = ±0.5%, blue = ±0.25%, and violet = ±0.1%.

You might expect that resistors would be available for a full range of values such as 10 Ω, 20 Ω, 30 Ω, 40 Ω, 50 Ω, and so on. However, this is not the case with some typical commercial values, such as 27 Ω, 56 Ω, and 68 Ω. This may seem somewhat strange and out of place. There is a reason for the chosen values, which is best demonstrated by examining the list of standard values of commercially available resistors in Table 3.5. The values in boldface blue are available with 5%, 10%, and 20% tolerances, making them the most common of the commercial variety. The values in boldface black are typically available with 5% and 10% tolerances, and those in normal print are available only in the 5% variety.

TABLE 3.5
Standard values of commercially available resistors.

Ohms (Ω)					Kilohms (kΩ)		Megohms (MΩ)	
0.10	1.0	10	100	1000	10	100	1.0	10.0
0.11	1.1	11	110	1100	11	110	1.1	11.0
0.12	**1.2**	**12**	**120**	**1200**	**12**	**120**	**1.2**	**12.0**
0.13	1.3	13	130	1300	13	130	1.3	13.0
0.15	1.5	15	150	1500	15	150	1.5	15.0
0.16	1.6	16	160	1600	16	160	1.6	16.0
0.18	**1.8**	**18**	**180**	**1800**	**18**	**180**	**1.8**	**18.0**
0.20	2.0	20	200	2000	20	200	2.0	20.0
0.22	2.2	22	220	2200	22	220	2.2	22.0
0.24	2.4	24	240	2400	24	240	2.4	
0.27	**2.7**	**27**	**270**	**2700**	**27**	**270**	**2.7**	
0.30	3.0	30	300	3000	30	300	3.0	
0.33	3.3	33	330	3300	33	330	3.3	
0.36	3.6	36	360	3600	36	360	3.6	
0.39	**3.9**	**39**	**390**	**3900**	**39**	**390**	**3.9**	
0.43	4.3	43	430	4300	43	430	4.3	
0.47	4.7	47	470	4700	47	470	4.7	
0.51	5.1	51	510	5100	51	510	5.1	
0.56	**5.6**	**56**	**560**	**5600**	**56**	**560**	**5.6**	
0.62	6.2	62	620	6200	62	620	6.2	
0.68	**6.8**	**68**	**680**	**6800**	**68**	**680**	**6.8**	
0.75	7.5	75	750	7500	75	750	7.5	
0.82	**8.2**	**82**	**820**	**8200**	**82**	**820**	**8.2**	
0.91	9.1	91	910	9100	91	910	9.1	

FIG. 3.27
Werner von Siemens.

German (Lenthe, Berlin)
(1816–92) Electrical Engineer
Telegraph Manufacturer,
Siemens & Halske AG

Developed an *electroplating process* during a brief stay in prison for acting as a second in a duel between fellow officers of the Prussian army. Inspired by the electronic telegraph invented by Sir Charles Wheatstone in 1817, he improved on the design and proceeded to lay cable with the help of his brother Carl across the Mediterranean and from Europe to India. His inventions included the first *self-excited generator,* which depended on the *residual* magnetism of its electromagnet rather than an inefficient permanent magnet. In 1888 he was raised to the rank of nobility with the addition of *von* to his name. The current firm of Siemens AG has manufacturing outlets in some 35 countries with sales offices in some 125 countries.

3.8   CONDUCTANCE

By finding the reciprocal of the resistance of a material, we have a measure of how well the material will conduct electricity. The quantity is called **conductance,** has the symbol G, and is measured in *siemens* (S) (note Fig. 3.27). In equation form, conductance is

$$G = \frac{1}{R} \quad \text{(siemens, S)} \tag{3.11}$$

A resistance of 1 MΩ is equivalent to a conductance of 10^{-6} S, and a resistance of 10 Ω is equivalent to a conductance of 10^{-1} S. The larger the conductance, therefore, the less the resistance and the greater the conductivity.

In equation form, conductance is determined by

$$G = \frac{A}{\rho l} \quad \text{(S)} \tag{3.12}$$

indicating that increasing the area or decreasing either the length or the resistivity will increase the conductance.

EXAMPLE 3.13 What is the relative increase or decrease in conductivity of a conductor if the area is reduced by 30% and the length is increased by 40%? The resistivity is fixed.

Solution: Eq. (3.12):

$$G_i = \frac{A_i}{\rho_i l_i}$$

with the subscript i for the initial value. Using the subscript n for the new value:

$$G_n = \frac{A_n}{\rho_n l_n} = \frac{0.70 A_i}{\rho_i (1.4 l_i)} = \frac{0.70}{1.4} \frac{A_i}{\rho_i l_i} = \frac{0.70}{1.4 G_i}$$

and

$$G_n = \mathbf{0.5 G_i}$$

3.9 OHMMETERS

The **ohmmeter** is an instrument used to perform the following tasks and several other useful functions:

1. *Measure the resistance of individual or combined elements.*
2. *Detect open-circuit (high-resistance) and short-circuit (low-resistance) situations.*
3. *Check the continuity of network connections and identify wires of a multilead cable.*
4. *Test some semiconductor (electronic) devices.*

For most applications, the ohmmeters used most frequently are the ohmmeter section of a VOM or DMM. The details of the internal circuitry and the method of using the meter will be left primarily for a laboratory exercise. In general, however, the resistance of a resistor can be measured by simply connecting the two leads of the meter across the resistor, as shown in Fig. 3.28. There is no need to be concerned about which lead goes on which end; the result will be the same in either case since resistors offer the same resistance to the flow of charge (current) in either direction. If the VOM is employed, a switch must be set to the proper resistance range, and a nonlinear scale (usually the top scale of the meter) must be properly read to obtain the resistance value. The DMM also requires choosing the best scale setting for the resistance to be measured, but the result appears as a numerical display, with the proper placement of the decimal point as determined by the chosen scale. When measuring the resistance of a single resistor, it is usually best to remove the resistor from the network before making the measurement. If this is difficult or impossible, at least one end of the resistor must not be connected to the network, or the reading may include the effects of the other elements of the system.

If the two leads of the meter are touching in the ohmmeter mode, the resulting resistance is zero. A connection can be checked as shown in Fig. 3.29 by simply hooking up the meter to either side of the connection. If the resistance is zero, the connection is secure. If it is other than zero, the connection could be weak; and if it is infinite, there is no connection at all.

If one wire of a harness is known, a second can be found as shown in Fig. 3.30. Simply connect the end of the known lead to the end of any other lead. When the ohmmeter indicates zero ohms (or very low resistance), the second lead has been identified. The above procedure can also be used to determine the first known lead by simply connecting the meter to any wire at one end and then touching all the leads at the other end until a zero ohm indication is obtained.

Preliminary measurements of the condition of some electronic devices such as the diode and the transistor can be made using the ohmmeter. The meter can also be used to identify the terminals of such devices.

One important note about the use of any ohmmeter:

Never hook up an ohmmeter to a live circuit!

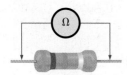

FIG. 3.28

Measuring the resistance of a single element.

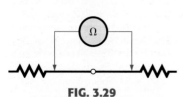

FIG. 3.29

Checking the continuity of a connection.

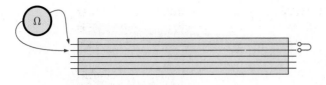

FIG. 3.30

Identifying the leads of a multilead cable.

The reading will be meaningless and you may damage the instrument. The ohmmeter section of any meter is designed to pass a small sensing current through the resistance to be measured. A large external current could damage the movement and would certainly throw off the calibration of the instrument. In addition:

Never store a VOM or a DMM in the resistance mode.

The two leads of the meter could touch and the small sensing current could drain the internal battery. VOMs should be stored with the selector switch on the highest voltage range, and the selector switch of DMMs should be in the off position.

3.10   APPLICATIONS

The following are examples of how resistance can be used to perform the task of heating and the task of measuring the stress or strain on a supporting member of a structure. In general, resistance is a component of every electrical or electronic application.

Electric Baseboard Heating Element

One of the most common applications of resistance is in household fixtures such as toasters and baseboard heating where the heat generated by current passing through a resistive element is employed to perform a useful function.

Recently, as we remodeled our house, the local electrician informed us that we were limited to 16 ft of electric baseboard on a single circuit. That naturally had me wondering about the wattage per foot, the resulting current level, and whether the 16 ft limitation was a national standard. Reading the label on the 2 ft section appearing in Fig. 3.31(a), I found VOLTS AC 240/208, WATTS 750/575 [the power rating will be described in Chapter 4] AMPS 3.2/2.8. Since my panel is rated 208 V (as are those in most residential homes), the wattage rating per foot is 575 W/2 or 287.5 W at a current of 2.8 A. The total wattage for the 16 ft is therefore 16×287.5 W or 4600 W. In Chapter 4 you will find that the power to a resistive load is related to the current and applied voltage by the equation $P = VI$. The total resulting current can then be determined using this equation in the following manner: $I = P/V = 4600$ W/208 V $= 22.12$ A. The result was that we needed a circuit breaker larger than 22.12 A; otherwise, the circuit breaker would trip every time we turned the heat on. In my case the electrician used a 30 A breaker to meet the National Fire Code requirement that does not permit exceeding 80% of the rated current for a conductor or breaker. In most panels a 30 A breaker takes two slots of your panel, whereas the more common 20 A breaker takes only one slot. If you have a moment, take a look in your own panel and note the rating of the breakers used for various circuits of your home.

Going back to Table 3.2, we find that the #12 wire commonly used for most circuits in the home has a maximum rating of 20 A and would not be suitable for the electric baseboard. Since #11 is usually not commercially available, a #10 wire with a maximum rating of 30 A was used. You might wonder why the current drawn from the supply is 22.12 A while that required for one unit was only 2.8 A. This difference is due to the parallel combination of sections of the heating elements, a configuration that will be described in Chapter 6. It is now clear why they specify a 16 ft limitation on a single circuit. Additional elements would raise the current to a level that would exceed the code level for #10 wire and would approach the maximum rating of the circuit breaker.

Figure 3.31(b) shows a photo of the interior construction of the heating element. The red feed wire on the right is connected to the core of the heating element, and the black wire at the other end passes through a protective heater element and back to the terminal box of the

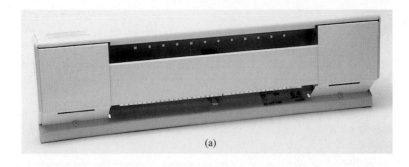

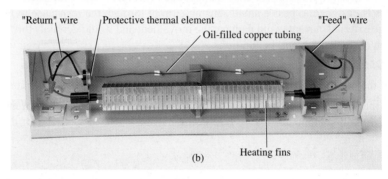

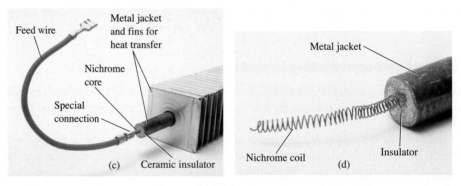

FIG. 3.31

Electric baseboard: (a) 2 ft section; (b) interior; (c) heating element; (d) nichrome coil.

unit (the place where the exterior wires are brought in and connected). If you look carefully at the end of the heating unit as shown in Fig. 3.31(c), you will find that the heating wire that runs through the core of the heater is not connected directly to the round jacket holding the fins in place. A ceramic material (insulator) separates the heating wire from the fins to remove any possibility of conduction between the current passing through the bare heating element and the outer fin structure. Ceramic materials are used because they are excellent conductors of heat and because they have a high retentivity for heat so the surrounding area will remain heated for a period of time even after the current has been turned off. As shown in Fig. 3.31(d), the heating wire that runs through the metal jacket is normally a nichrome composite (because pure nichrome is quite brittle) wound in the shape of a coil to compensate for expansion and contraction with heating and also to permit a longer heating element in standard-length baseboard. For interest sake we opened up the core and found that the nichrome wire in the core of a 2 ft baseboard was actually 7 ft long, or a 3.5 : 1 ratio. The thinness of the wire was particularly noteworthy, measuring out at about 8 mils in diameter, not much thicker than a hair. Recall from this chapter that the longer the conductor and the thinner the wire, the greater the resistance. We took a section of the nichrome wire and tried to heat it with a reasonable level of current and the application of a hair dryer. The change in resistance was almost unnoticeable. In other words, all our effort to increase the resistance with the basic elements available to us in the lab was fruitless. This was an excellent demonstration of the meaning of the temperature coefficient of resistance in Table 3.4. Since the coefficient is so small for nichrome, the resistance does not measurably change unless the change in temperature is truly significant. The curve of Fig. 3.11 would therefore be close to horizontal for nichrome. For baseboard heaters this is an excellent characteristic because the heat developed, and the power dissipated, will not vary with time as the conductor heats up with time. The flow of heat from the unit will remain fairly constant.

The feed and return cannot be soldered to the nichrome heater wire for two reasons. First, you cannot solder nichrome wires to each other or to other types of wire. Second, if you could, there could be a problem because the heat of the unit could rise above 880°F at the point where the wires are connected, and the solder could melt and the connection could be broken. Nichrome must be spot welded or crimped onto the copper wires of the unit. Using Eq. (3.1) and the 8 mil measured diameter, and assuming pure nichrome for the moment, the resistance of the 7 ft length is

$$R = \frac{\rho l}{A}$$

$$= \frac{(600)(7')}{(8\text{ mils})^2} = \frac{4200}{64}$$

$$R = \mathbf{65.6\ \Omega}$$

In the next chapter a power equation will be introduced in detail relating power, current, and resistance in the following manner: $P = I^2 R$. Using the above data and solving for the resistance, we obtain

$$R = \frac{P}{I^2}$$

$$= \frac{575\text{ W}}{(2.8\text{ A})^2}$$

$$R = \mathbf{73.34\ \Omega}$$

which is very close to the value calculated above from the geometric shape since we cannot be absolutely sure about the resistivity value for the composite.

During normal operation the wire heats up and passes that heat on to the fins, which in turn heat the room via the air flowing through them. The flow of air through the unit is enhanced by the fact that hot air rises, so when the heated air leaves the top of the unit, it draws cold area from the bottom to contribute to the convection effect. Closing off the top or bottom of the unit would effectively eliminate the convection effect, and the room would not heat up. A condition could occur in which the inside of the heater became too hot, causing the metal casing also to get too hot. This concern is the primary reason for the thermal protective element introduced above and appearing in Fig. 3.31(b). The long, thin copper tubing in Fig. 3.31 is actually filled with an oil-type fluid that will expand when heated. If too hot, it will expand, depress a switch in the housing, and turn off the heater by cutting off the current to the heater wire.

Strain Gauges

Any change in the shape of a structure can be detected using strain gauges whose resistance will change with applied stress or flex. An example of a strain gauge is shown in Fig. 3.32. Strain gauges are semiconductor devices whose terminal resistance will change in a nonlinear (not a straight-line) fashion through a fairly wide range in values when they are stressed by compression or extension. Since the stress gauge does emit a signal, a signal processor must also be part of the system to translate the change in resistance to some meaningful output. One simple example of the use of resistive strain gauges is to monitor earthquake activity. When the gauge is placed across an area of suspected earthquake activity, the slightest separation in the earth will change the terminal resistance, and the processor will display a result sensitive to the amount of separation. Another example is in alarm systems where the slightest change in the shape of a supporting beam when someone walks overhead will result in a change in terminal resistance, and an alarm will sound. Other examples include placing strain gauges on bridges to maintain an awareness of their rigidity and on very large generators to check whether various moving components are beginning to separate because of a wearing of the bearings or spacers. The small mouse control within the keyboard of a portable computer can be a series of stress gauges that reveal the direction of stress applied to the controlling element on the keyboard. Movement in one direction can extend or compress a resistance gauge which can monitor and control the motion of the mouse on the screen.

Terminals   Resistive material

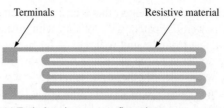

(a) Typical strain gauge configuration.

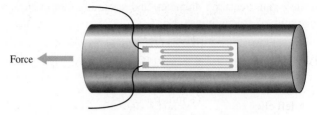

Force ←

(b) The strain gauge is bonded to the surface to be measured along the line
of force. When the surface lengthens, the strain gauge stretches.

FIG. 3.32

Resistive strain gauge.

3.11 MATHCAD

Throughout the text a mathematical software package called Mathcad will be used to introduce a variety of operations that a math software package can perform. There is no
need to obtain a copy of the software package to continue with the material covered in
this text. The coverage is at a very introductory level simply to introduce the scope and
power of the package. All the exercises appearing at the end of each chapter can be done
without Mathcad.

Once the package is installed, all operations begin with the basic screen of Fig. 3.33. The
operations must be performed in the sequence appearing in Fig. 3.34, that is, from left to
right and then from top to bottom. For example, if an equation on the second line is to operate on a specific variable, the variable must be defined to the left of or above the equation.

To perform any mathematical calculation, simply click on the screen at any convenient point to establish a crosshair on the display (the location of the first entry). Then type

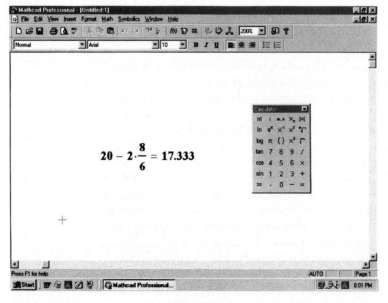

FIG. 3.33

Using Mathcad to perform a basic mathematical operation.

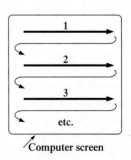

Computer screen

FIG. 3.34

*Defining the order of mathematical
operations for Mathcad.*

in the mathematical operation such as **20 − 2 · 8/6** as shown in Fig. 3.33; the instant the equal sign is selected, the result, **17.333,** will appear as shown in Fig. 3.33. The multiplication is obtained using the asterisk (*) appearing at the top of the number 8 key (under the SHIFT CONTROL key). The division is set by the / key at the bottom right of the keyboard. The equal sign can be selected from the top right corner of the keyboard. Another option is to apply the sequence **View-Toolbars-Calculator** to obtain the **Calculator** of Fig. 3.33. Then use the calculator to enter the entire expression and obtain the result using the left clicker of the mouse.

As an example in which variables must be defined, the resistance of a 200 ft length of copper wire with a diameter of 0.01 in. will be determined. First, as shown in Fig. 3.35, the variables for resistivity, length, and diameter must be defined. This is accomplished by first calling for the **Greek** palette through **View-Toolbars-Greek** and selecting the Greek letter rho (ρ) followed by a combined **Shift-colon (Shift:)** operation. A colon and an equal sign will appear, after which **10.37** is entered. For all the calculations to follow, the value of ρ has been defined. A left click on the screen will then remove the rectangular enclosure and place the variable and its value in memory. Proceed in the same way to define the length l and the diameter d. Next the diameter in millimeters is defined by multiplying the diameter in inches by 1000, and the area is defined by the diameter in millimeters squared. Note that m had to be defined to the left of the expression for the area, and the variable d was defined in the line above. The power of 2 was obtained by first selecting the superscript symbol (^) at the top of the number 6 on the keyboard and then entering the number 2 in the Mathcad bracket. Or you can simply type the letter m and choose x^2 from the **Calculator** palette. In fact, all the operations of multiplication, division, and so on, required to determine the resistance R can be lifted from the **Calculator** palette.

On the next line of Fig. 3.35, the values of m and A were calculated by simply typing in m followed by the keyboard equal sign. Finally, the equation for the resistance R is defined in terms of the variables, and the result is obtained. The true value of developing in the above sequence is the fact that you can place the program in memory and, when the need arises, call it up and change a variable or two—the result will appear immediately. There is no need to reenter all the definitions—just change the numerical value.

In the chapters to follow, Mathcad will appear at every opportunity to demonstrate its ability to perform calculations in a quick, effective manner. You will probably want to learn more about this time-saving and accuracy-checking option.

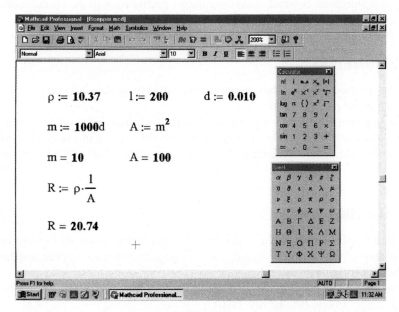

FIG. 3.35
Using Mathcad to calculate the resistance of a copper conductor.

CHAPTER SUMMARY
RESISTANCE

- The four factors that determine the resistance of a body are its material, length, cross-sectional area, and temperature.
- The chosen material defines the resistivity to be used in the resistance equation.
- One circular mil (CM) is equal to the area defined by a diameter of 1 mil where 1 mil is equal to 1/1000 inch.
- In general, for resistors, increased power levels are normally associated with an increase in size.
- The size of a resistor does not define its resistance level.
- Conductance is the direct inverse of resistance: The higher the level of resistance, the lower the level of conductance.

WIRE TABLES

- As the wire gage number increases, the area decreases. For every drop in three gage numbers, the area is doubled; and for every drop in 10 gage numbers, the area increases by a factor of 10.
- The resistance per 1000 ft of wire in a wire table is at room temperature only.

TEMPERATURE EFFECTS

- Conductors have a positive temperature coefficient; that is, the resistance will increase with an increase in temperature.
- The inferred absolute temperature is the temperature at which a plot of resistance versus temperature would appear to cross the zero-resistance line if the plot were approximated by a straight line.
- At absolute zero temperature of $-273.15°C$, the resistance of copper is zero ohms.
- The higher the temperature coefficient of resistance for a material, the more sensitive the resistance level is to changes in temperature.

SUPERCONDUCTORS

- Superconductors are conductors of electric charge that, for all practical purposes, have zero resistance.
- The temperature at which a superconductor reverts back to the characteristics of a conventional conductor is called the *critical temperature.*

POTENTIOMETERS

- The resistance between the outermost terminals is fixed and unaffected by turning the shaft of the potentiometer.
- As you turn the shaft of a potentiometer, the resistance between the center terminal and one of the outside terminals will either decrease or increase. The resistance between the center terminal and the other outside terminal will have the opposite response.

COLOR CODING

- For a four-band color scheme, the first two bands represent the first and second digits, respectively. The third band determines the power-of-ten multiplier, and the fourth band the manufacturer's tolerance.
- For resistors in the range of 0.1 Ω to 10 Ω, the third band will be a multiplier, with gold representing 0.1 and silver 0.01.
- The numbers associated with the colors are applicable to the wide variety of components you will encounter in this field.

- Ohmmeters are never used on a live circuit.
- Ohmmeters are connected directly across the resistance to be measured.
- When measuring the resistance of a resistor in a circuit, be sure to disconnect one lead of the resistor from the circuit. Otherwise, the resistance of some neighboring resistors may affect the reading.

Important Equations

$$R = \rho \frac{l}{A} \qquad \rho = 10.37 \text{ (for copper)}$$

$$1 \text{ mil} = \frac{1}{1000} \text{ in.}$$

$$A_{CM} = (d_{mils})^2$$

$$1 \text{ CM} = \frac{\pi}{4} \text{ sq mils}$$

$$\frac{|T_i| + T_1}{R_1} = \frac{|T_i| + T_2}{R_2} \qquad T_i = -234.5°C \text{ (for copper)}$$

$$\alpha_{20} = \frac{1}{|T_i| + 20°C}$$

$$R_1 = R_{20}[1 + \alpha_{20}(T_1 - 20°C)]$$

$$\Delta R = \frac{R_{nominal}}{10^6}(\text{PPM})(\Delta T)$$

$$G = \frac{1}{R}$$

PROBLEMS

SECTION 3.2   Resistance

1. Convert the following to mils:
   a. 0.5 in.   b. 0.02 in.   c. 1/4 in.
   d. 1 in.   e. 0.02 ft   f. 0.1 cm

2. Calculate the area in circular mils (CM) of wires having the following diameters:
   a. 0.050 in.   b. 0.016 in.   c. 0.30 in.
   d. 1 cm   e. 0.02 ft   f. 0.0042 m

3. The area in circular mils is
   a. 1600 CM   b. 900 CM   c. 40,000 CM
   d. 625 CM   e. 6.25 CM   f. 120 CM

   What is the diameter of each wire in inches?

4. What is the resistance of a copper wire 200 ft long and 0.01 in. in diameter ($T = 20°C$)?

5. Find the resistance of a silver wire 50 yd long and 4 mils in diameter ($T = 20°C$).

6. a. What is the area in circular mils of an aluminum conductor that is 80 ft long with a resistance of 2.5 Ω?
   b. What is its diameter in inches?

7. A 2.2 Ω resistor is to be made of nichrome wire. If the available wire is 1/32 in. in diameter, how much wire is required?

8. a. What is the area in circular mils of a copper wire that has a resistance of 2.5 Ω and is as long as a football field (100 yd) ($T = 20°C$)?
   b. Without working out the numerical solution, determine whether the area of an aluminum wire will be smaller or larger than that of the copper wire. Explain.
   c. Repeat (b) for a silver wire.

9. In Fig. 3.36, three conductors of different materials are presented.
   a. Without working out the numerical solution, which do you think has the most resistance? Explain.
   b. Find the resistance of each section and compare your answer with the result of (a) ($T = 20°C$).

10. A wire 1000 ft long has a resistance of 0.5 kΩ and an area of 94 CM. Of what material is the wire made ($T = 20°C$)?

*11. a. What is the resistance of a copper bus-bar for a high-rise building with the dimensions shown ($T = 20°C$) in Fig. 3.37?
   b. Repeat (a) for aluminum and compare the results.

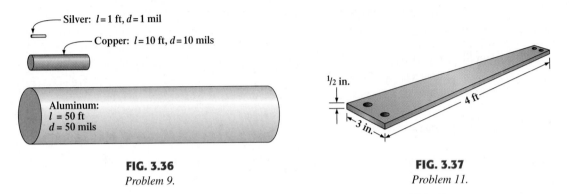

Silver: $l = 1$ ft, $d = 1$ mil

Copper: $l = 10$ ft, $d = 10$ mils

Aluminum:
$l = 50$ ft
$d = 50$ mils

½ in.

3 in.

4 ft

FIG. 3.36
Problem 9.

FIG. 3.37
Problem 11.

12. Determine the increase in resistance of a copper conductor if the area is reduced by a factor of 4 and the length is doubled. The original resistance was 0.2 Ω. The temperature remains fixed.

*13. What is the new resistance level of a copper wire if the length is changed from 200 ft to 100 yd, the area is changed from 40,000 CM to 0.04 in.^{2}, and the original resistance was 800 mΩ?

SECTION 3.3 Wire Tables

14. a. Using Table 3.2, find the resistance of 450 ft of #11 and #14 AWG wires.
   b. Compare the resistances of the two wires.
   c. Compare the areas of the two wires.

15. a. Using Table 3.2, find the resistance of 1800 ft of #8 and #18 AWG wires.
   b. Compare the resistances of the two wires.
   c. Compare the areas of the two wires.

16. a. For the system of Fig. 3.38, the resistance of each line cannot exceed 0.006 Ω, and the maximum current drawn by the load is 110 A. What gage wire should be used?
   b. Repeat (a) for a maximum resistance of 0.003 Ω, $d = 30$ ft, and a maximum current of 110 A.

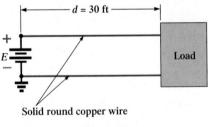

$d = 30$ ft

E

Load

Solid round copper wire

FIG. 3.38
Problem 16.

*17. a. From Table 3.2, determine the maximum permissible current density (A/CM) for an AWG #0000 wire.
   b. Convert the result of (a) to A/in.^{2}.
   c. Using the result of (b), determine the cross-sectional area required to carry a current of 5000 A.

SECTION 3.4 Temperature Effects

18. The resistance of a copper wire is 2 Ω at room temperature (20°C). What is its resistance at 100°C (the boiling point of water)?

19. The resistance of an aluminum bus-bar is 0.02 Ω at 0°C. What is its resistance at 100°C?

20. The resistance of a copper wire is 4 Ω at 70°F. What is its resistance at 32°F?

21. The resistance of a copper wire is 0.76 Ω at 30°C. What is its resistance at −40°C?

22. If the resistance of a silver wire is 0.04 Ω at −30°C, what is its resistance at 0°C?

23. **a.** The resistance of a copper wire is 1 Ω at 4°C. At what temperature (°C) will it be 1.1 Ω?
    b. At what temperature will it be 0.2 Ω?

*24. **a.** If the resistance of a 1000 ft length of copper wire is 10 Ω at room temperature (20°C), what will its resistance be at 50 K (Kelvin units) using Eq. (3.5)?
    b. Repeat part (a) for a temperature of 38.65 K. Comment on the results obtained by reviewing the curve of Fig. 3.11.
    c. What is the temperature of absolute zero in Fahrenheit units?

25. **a.** Verify the value of α_{20} for copper in Table 3.4 by substituting the inferred absolute temperature into Eq. (3.6).
    b. Using Eq. (3.7) find the temperature at which the resistance of a copper conductor will increase to 1 Ω from a level of 0.8 Ω at 20°C.

26. Using Eq. (3.7), find the resistance of a copper wire at 16°C if its resistance at 20°C is 0.4 Ω.

*27. Determine the resistance of a 1000 ft coil of #12 copper wire sitting in the desert at a temperature of 115°F.

28. A 22 Ω wire-wound resistor is rated at +200 PPM for a temperature range of −10°C to +75°C. Determine its resistance at 65°C.

SECTION 3.5   Superconductors

29. Visit your local library and find a table listing the critical temperatures for a variety of materials. List at least five materials with the critical temperatures that are not mentioned in this text. Choose a few materials that have relatively high critical temperatures.

30. Find at least one article on the application of superconductivity in the commercial sector, and write a short summary, including all interesting facts and figures.

31. Using the required 1 MA/cm^2 density level for IC manufacturing, determine what the resulting current would be through a #12 house wire. Compare the result obtained with the allowable limit of Table 3.2.

SECTION 3.6   Types of Resistors

32. **a.** What is the approximate increase in size from a 1 W to a 2 W metal-film resistor?
    b. What is the approximate increase in size from a 1/2 W to a 2 W metal-film resistor?
    c. In general, can we conclude that for the same type of resistor, an increase in wattage rating requires an increase in size (volume)? Is it almost a linear relationship? That is, does twice the wattage require an increase in size of 2 : 1?

33. If the resistance between the outside terminals of a linear potentiometer is 10 kΩ, what is its resistance between the wiper (movable) arm and an outside terminal if the resistance between the wiper arm and the other outside terminal is 2.8 kΩ?

34. If the wiper arm of a linear potentiometer is one-quarter the way around the contact surface, what is the resistance between the wiper arm and each terminal if the total resistance is 25 kΩ?

*35. Show the connections required to establish 4 kΩ between the wiper arm and one outside terminal of a 10 kΩ potentiometer while having only zero ohms between the other outside terminal and the wiper arm.

SECTION 3.7   Color Coding and Standard Resistor Values

36. Find the range in which a resistor having the following color bands must exist to satisfy the manufacturer's tolerance:

	1st band	2nd band	3rd band	4th band
a.	green	blue	yellow	gold
b.	red	red	brown	silver
c.	brown	black	red	—

37. Find the color code for the following 10% resistors:
    a. 220 Ω **b.** 3300 Ω
    c. 6.8 kΩ **d.** 2 MΩ

38. Is there an overlap in coverage between 20% resistors? That is, determine the tolerance range for a 10 Ω 20% resistor and a 15 Ω 20% resistor, and note whether their tolerance ranges overlap.

39. Repeat Problem 38 for 10% resistors of the same value.

SECTION 3.8 Conductance

40. Find the conductance of each of the following resistances:
  a. $100\ \Omega$ **b.** $4\ k\Omega$ **c.** $2.2\ M\Omega$

Compare the three results.

41. Find the conductance of 1000 ft of #18 AWG wire made of
  a. copper   **b.** aluminum

*42. The conductance of a wire is 100 S. If the area of the wire is increased by 2/3 and the length is reduced by the same amount, find the new conductance of the wire if the temperature remains fixed.

SECTION 3.9 Ohmmeters

43. How would you check the status of a fuse with an ohmmeter?

44. How would you determine the on and off states of a switch using an ohmmeter?

45. How would you use an ohmmeter to check the status of a light bulb?

SECTION 3.11 Mathcad

46. Verify the results of Example 3.3 using Mathcad.

47. Verify the results of Example 3.9 using Mathcad.

GLOSSARY

Absolute zero   The temperature at which all molecular motion ceases; $-273.15°C$.
Circular mil (CM)   The cross-sectional area of a wire having a diameter of 1 mil.
Color coding   A technique employing bands of color to indicate the resistance levels and tolerance of resistors.
Conductance (G)   An indication of the relative ease with which current can be established in a material. It is measured in siemens (S).
Cooper effect   The "pairing" of electrons as they travel through a medium.
Ductility   The property of a material that allows it to be drawn into long, thin wires.
Inferred absolute temperature   The temperature through which a straight-line approximation for the actual resistance-versus-temperature curve will intersect the temperature axis.
Malleability   The property of a material that allows it to be worked into many different shapes.
Negative temperature coefficient of resistance   The value revealing that the resistance of a material will decrease with an increase in temperature.
Ohm (Ω)   The unit of measurement applied to resistance.
Ohmmeter   An instrument for measuring resistance levels.
Positive temperature coefficient of resistance   The value revealing that the resistance of a material will increase with an increase in temperature.
Potentiometer   A three-terminal device through which potential levels can be varied in a linear or nonlinear manner.
PPM/°C   Temperature sensitivity of a resistor in parts per million per degree Celsius.
Resistance (R)   A measure of the opposition to the flow of charge through a material.
Resistivity (ρ)   A constant of proportionality between the resistance of a material and its physical dimensions.
Rheostat   An element whose terminal resistance can be varied in a linear or a nonlinear manner.
Superconductor   Conductors of electric charge that have, for all practical purposes, zero ohms.

Ohm's Law, Power, and Energy

4.1 INTRODUCTION

Now that the three important elements of an electric circuit have been introduced, this chapter will reveal how they are interrelated. The most important equation in the study of electric circuits will be introduced, and various other equations that allow us to find power and energy levels will be discussed in detail. It is the first chapter where we tie things together and develop a feeling for the way an electric circuit behaves and what affects its response. For the first time, the data provided on the labels of household appliances and the manner in which your electric bill is calculated will have some meaning. It is indeed a chapter that should open your eyes to a wide array of past experiences with electrical systems.

4.2 OHM'S LAW

As mentioned above, the first equation to be described is without question one of the most important to be learned in this field. It is not particularly difficult mathematically, but it is very powerful because it can be applied to any network in any time frame. That is, it is applicable to dc circuits, ac circuits, digital and microwave circuits, and, in fact, any type of applied signal. In addition, it can be applied over a period of time or for instantaneous responses. The equation can be derived directly from the following basic equation for all physical systems:

$$\text{Effect} = \frac{\text{cause}}{\text{opposition}} \qquad \textbf{(4.1)}$$

Every conversion of energy from one form to another can be related to this equation. In electric circuits, the *effect* we are trying to establish is the flow of charge, or *current*. The *potential difference,* or voltage, between two points is the *cause* ("pressure"), and the opposition is the *resistance* encountered.

An excellent analogy for the simplest of electrical circuits is the water in a hose connected to a pressure valve, as discussed in Chapter 2. Think of the electrons in the copper wire as the water in the hose, the pressure valve as the applied voltage, and the size of the hose as the factor that determines the resistance. If the pressure valve is closed, the water simply sits in the hose without a general direction, much like the oscillating electrons in a conductor without an applied voltage. When we open the pressure valve, water will flow through the hose much like the electrons in a copper wire when the voltage is applied. In other words, the absence of the "pressure" in one case and the voltage in the other will simply result in a system without direction or reaction. The rate at which the water will flow in the hose is a function of the size of the hose. A hose with a very small diameter will limit the rate at which water can flow through the hose, just as a copper wire with a small diameter will have a high resistance and will limit the current.

In summary, therefore, the absence of an applied "pressure" such as voltage in an electric circuit will result in no reaction in the system and no current in the electric circuit. Current is a reaction to the applied voltage and not the factor that gets the system in motion. To continue with the analogy, the more the pressure at the spigot, the greater the rate of water flow through the hose, just as applying a higher voltage to the same circuit will result in a higher current.

Substituting the terms introduced above into Eq. (4.1) results in

$$\text{Current} = \frac{\text{potential difference}}{\text{resistance}}$$

and

$$I = \frac{E}{R} \qquad (\text{amperes, A}) \qquad \textbf{(4.2)}$$

Equation (4.2) is known as **Ohm's law** in honor of Georg Simon Ohm (Fig. 4.1). The law clearly reveals that for a fixed resistance, the greater the voltage (or pressure) across a resistor, the more the current, and the more the resistance for the same voltage, the less the current. In other words, the current is proportional to the applied voltage and inversely proportional to the resistance.

By simple mathematical manipulations, the voltage and resistance can be found in terms of the other two quantities:

$$E = IR \qquad (\text{volts, V}) \qquad \textbf{(4.3)}$$

and

$$R = \frac{E}{I} \qquad (\text{ohms, }\Omega) \qquad \textbf{(4.4)}$$

All the parameters of Eq. (4.2) appear in the simple electrical circuit of Fig. 4.2. A resistor has been connected directly across a battery to establish a current through the resistor and supply. Note that

the symbol E is applied to all sources of voltage

and

the symbol V is applied to all voltage drops across components of the network.

Both are measured in volts and can be applied interchangeably in Eqs. (4.2) through (4.4).

Since the battery of Fig. 4.2 is connected directly across the resistor, the voltage V_R across the resistor must be equal to that of the supply. Applying Ohm's law:

$$I = \frac{V_R}{R} = \frac{E}{R}$$

FIG. 4.1
Georg Simon Ohm.

German (Erlangen, Cologne)
(1789–1854)
Physicist and Mathematician
Professor of Physics,
University of Cologne

In 1827, developed one of the most important laws of electric circuits: *Ohm's law.* When the law was first introduced, the supporting documentation was considered lacking and foolish, causing him to lose his teaching position and search for a living doing odd jobs and some tutoring. It took some 22 years for his work to be recognized as a major contribution to the field. He was then awarded a chair at the University of Munich and received the Copley Medal of the Royal Society in 1841. His research also extended into the areas of molecular physics, acoustics, and telegraphic communication.

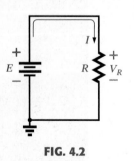

FIG. 4.2
Basic circuit.

Note in Fig. 4.2 that the voltage source "pressures" current (conventional current) in a direction that passes from the negative to the positive terminal of the battery. **This will always be the case for single-source networks.** (The effect of more than one source in the same network will be investigated in a later chapter.) Note also that the current enters the positive terminal and leaves the negative terminal for the load resistor R.

For any resistor, in any network, the direction of current through a resistor will define the polarity of the voltage drop across the resistor

as shown in Fig. 4.3 for two directions of current. Polarities as established by current direction will become increasingly important in the analyses to follow.

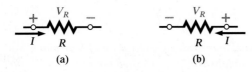

FIG. 4.3

Defining polarities.

EXAMPLE 4.1 Determine the current resulting from the application of a 9 V battery across a network with a resistance of 2.2 Ω.

Solution: Eq. (4.2):

$$I = \frac{V_R}{R} = \frac{E}{R} = \frac{9 \text{ V}}{2.2 \text{ } \Omega} = \textbf{4.09 A}$$

EXAMPLE 4.2 Calculate the resistance of a 60 W bulb if a current of 500 mA results from an applied voltage of 120 V.

Solution: Eq. (4.4):

$$R = \frac{V_R}{I} = \frac{E}{I} = \frac{120 \text{ V}}{500 \times 10^{-3} \text{ A}} = \textbf{240 } \boldsymbol{\Omega}$$

EXAMPLE 4.3 Calculate the current through the 2 kΩ resistor of Fig. 4.4 if the voltage drop across it is 16 V.

Solution:

$$I = \frac{V}{R} = \frac{16 \text{ V}}{2 \times 10^3 \text{ } \Omega} = \textbf{8 mA}$$

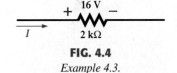

FIG. 4.4

Example 4.3.

EXAMPLE 4.4 Calculate the voltage that must be applied across the soldering iron of Fig. 4.5 to establish a current of 1.5 A through the iron if its internal resistance is 80 Ω.

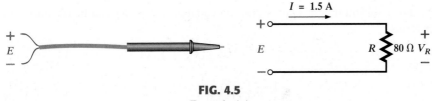

FIG. 4.5

Example 4.4.

Solution:

$$E = V_R = IR = (1.5 \text{ A})(80 \text{ } \Omega) = \textbf{120 V}$$

In a number of the examples in this chapter, such as Example 4.4 above, the voltage applied is actually that obtained from an ac outlet in the home, office, or laboratory. This approach was used to provide an opportunity for the student to relate to real-world situations as soon as possible and to demonstrate that a number of the equations derived in this chapter are applicable to ac networks also. Chapter 12 will provide a direct relationship between ac and dc voltages that permits the mathematical substitutions used in this chapter. In other words, don't be concerned about the fact that some of the voltages and currents appearing in the examples of this chapter are actually ac voltages, because the equations for dc networks have exactly the same format, and all the solutions will be correct.

4.3 POWER

In general,

the term power is applied to provide an indication of how much work (energy conversion) can be accomplished in a specified amount of time; that is, power is a rate of doing work.

For instance, a large motor has more power than a smaller motor because it has the ability to convert more electrical energy into mechanical energy in the same period of time. Since energy is measured in joules (J) and time in seconds (s), power is measured in joules/second (J/s). The electrical unit of measurement for power is the watt (W) defined by

$$1 \text{ watt (W)} = 1 \text{ joule/second (J/s)} \tag{4.5}$$

In equation form, power is determined by

$$P = \frac{W}{t} \qquad \text{(watts, W, or joules/second, J/s)} \tag{4.6}$$

with the **energy** (W) measured in joules and the time t in seconds.

The unit of measurement, the watt, is derived from the surname of James Watt (Fig. 4.6), who was instrumental in establishing the standards for power measurements. He introduced the **horsepower** (hp) as a measure of the average power of a strong dray horse over a full working day. It is approximately 50% more than can be expected from the average horse. The horsepower and watt are related in the following manner:

$$1 \text{ horsepower} \cong 746 \text{ watts}$$

The power delivered to, or absorbed by, an electrical device or system can be found in terms of the current and voltage by first substituting Eq. (2.5) into Eq. (4.6):

$$P = \frac{W}{t} = \frac{QV}{t} = V\frac{Q}{t}$$

But

$$I = \frac{Q}{t}$$

so that

$$\boxed{P = VI} \qquad \text{(watts, W)} \tag{4.7}$$

By direct substitution of Ohm's law, the equation for power can be obtained in two other forms:

$$P = VI = V\left(\frac{V}{R}\right)$$

and

$$\boxed{P = \frac{V^2}{R}} \qquad \text{(watts, W)} \tag{4.8}$$

or

$$P = VI = (IR)I$$

and

$$\boxed{P = I^2 R} \qquad \text{(watts, W)} \tag{4.9}$$

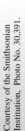

FIG. 4.6
James Watt.

Scottish (Greenock, Birmingham)
(1736–1819)
Instrument Maker and Inventor
Elected Fellow of the Royal
Society of London in 1785

In 1757, at the age of 21, used his innovative talents to design mathematical instruments such as the *quadrant, compass,* and various *scales.* In 1765, introduced the use of a separate *condenser* to increase the efficiency of steam engines. In the years to follow he received a number of important patents on improved engine design, including a rotary motion for the steam engine (versus the reciprocating action) and a double-action engine, in which the piston pulled as well as pushed in its cyclic motion. Introduced the term **horsepower** as the average power of a strong dray (small cart) horse over a full working day.

The result is that the power absorbed by the resistor of Fig. 4.7 can be found directly, depending on the information available. In other words, if the current and resistance are known, it pays to use Eq. (4.9) directly, and if V and I are known, use of Eq. (4.7) is appropriate. It saves having to apply Ohm's law before determining the power.

The power supplied by a battery can be determined by simply inserting the supply voltage into Eq. (4.7) to produce

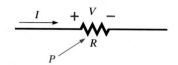

$$\boxed{P = EI} \qquad \text{(watts, W)} \qquad\qquad \textbf{(4.10)}$$

FIG. 4.7
Defining the power to a resistive element.

The importance of Eq. (4.10) cannot be overstated. It clearly states the following:

The power associated with any supply is not simply a function of the supply voltage. It is determined by the product of the supply voltage and its maximum current rating.

The simplest example is the car battery—a battery that is large, difficult to handle, and relatively heavy. It is only 12 V, a voltage level that could be supplied by a battery slightly larger than the small 9 V portable radio battery. However, to provide the **power** necessary to start a car, the battery must be able to supply the high surge current required at starting—a component that requires size and mass. In total, therefore, it is not the voltage or current rating of a supply that determines its power capabilities; it is the product of the two.

Throughout the text, the abbreviation for energy (W) can be distinguished from that for the watt (W) by the fact that the one for energy is in italics while the one for watt is in roman. In fact, all variables in the dc section appear in italics while the units appear in roman.

EXAMPLE 4.5 Find the power delivered to the dc motor of Fig. 4.8.

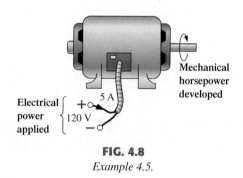

FIG. 4.8
Example 4.5.

Solution:

$$P = EI = (120 \text{ V})(5 \text{ A}) = 600 \text{ W} = \textbf{0.6 kW}$$

EXAMPLE 4.6 What is the power dissipated by a 5 Ω resistor if the current is 4 A?

Solution:

$$P = I^2R = (4 \text{ A})^2(5 \text{ Ω}) = \textbf{80 W}$$

EXAMPLE 4.7 The *I-V* characteristics of a light bulb are provided in Fig. 4.9. Note the nonlinearity of the curve, indicating a wide range in resistance of the bulb with applied voltage. If the rated voltage is 120 V, find the wattage rating of the bulb. Also calculate the resistance of the bulb under rated conditions.

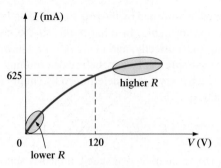

FIG. 4.9

The nonlinear I-V characteristics of a 75 W light bulb (Example 4.7).

Solution: At 120 V,

$$I = 0.625 \text{ A}$$

and

$$P = VI = (120 \text{ V})(0.625 \text{ A}) = \textbf{75 W}$$

At 120 V,

$$R = \frac{V}{I} = \frac{120 \text{ V}}{0.625 \text{ A}} = \textbf{192 } \boldsymbol{\Omega}$$

Sometimes the power is given and the current or voltage must be determined. Through algebraic manipulations, an equation for each variable is derived as follows:

$$P = I^2 R \Rightarrow I^2 = \frac{P}{R}$$

and

$$\boxed{I = \sqrt{\frac{P}{R}}} \qquad \text{(amperes, A)} \qquad \textbf{(4.11)}$$

$$P = \frac{V^2}{R} \Rightarrow V^2 = PR$$

and

$$\boxed{V = \sqrt{PR}} \qquad \text{(volts, V)} \qquad \textbf{(4.12)}$$

EXAMPLE 4.8 Determine the current through a 5 kΩ resistor when the power dissipated by the element is 20 mW.

Solution: Eq. (4.11):

$$I = \sqrt{\frac{P}{R}} = \sqrt{\frac{20 \times 10^{-3} \text{ W}}{5 \times 10^3 \text{ }\Omega}} = \sqrt{4 \times 10^{-6}} = 2 \times 10^{-3} \text{ A}$$

$$= \textbf{2 mA}$$

4.4 ENERGY

For power, which is the rate of doing work, to produce an energy conversion of any form, it must be *used over a period of time.* For example, a motor may have the horsepower to run a heavy load, but unless the motor is *used* over a period of time, there will be no energy conversion. In addition, the longer the motor is used to drive the load, the greater will be the energy expended.

The **energy** (W) lost or gained by any system is therefore determined by

$$\boxed{W = Pt} \qquad \text{(wattseconds, Ws, or joules)} \qquad \textbf{(4.13)}$$

Since power is measured in watts (or joules per second) and time in seconds, the unit of energy is the *wattsecond* or *joule* (note Fig. 4.10), as indicated above. The wattsecond, however, is too small a quantity for most practical purposes, so the *watthour* (Wh) and the *kilowatthour* (kWh) were defined, as follows:

$$\boxed{\text{Energy (Wh)} = \text{power (W)} \times \text{time (h)}} \qquad \textbf{(4.14)}$$

$$\boxed{\text{Energy (kWh)} = \frac{\text{power (W)} \times \text{time (h)}}{1000}} \qquad \textbf{(4.15)}$$

Note that the energy in kilowatthours is simply the energy in watthours divided by 1000. To develop some sense for the kilowatthour energy level, consider that **1 kWh is the energy dissipated by a 100 W bulb in 10 h.**

The **kilowatthour meter** is an instrument for measuring the energy supplied to the residential or commercial user of electricity. It is normally connected directly to the lines at a point just prior to entering the power distribution panel of the building. A typical set of dials is shown in Fig. 4.11, along with a photograph of an analog kilowatthour meter. As indicated, each power of ten below a dial is in kilowatthours. The more rapidly the aluminum disc rotates, the greater the energy demand. The dials are connected through a set of gears to the rotation of this disc. A solid-state digital meter with an extended range of capabilities also appears in Fig. 4.11.

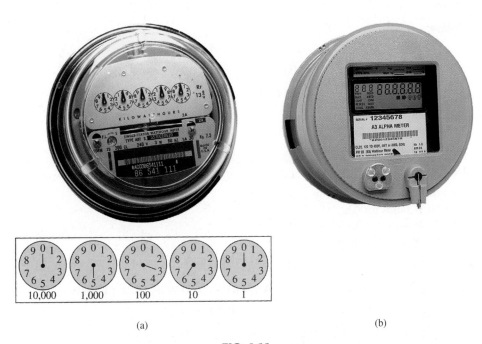

(a) (b)

FIG. 4.11
(a) Analog (electromechanical) kilowatthour meter;
(b) solid-state digital (electronic) kilowatthour meter.
(Photos provided courtesy of Elster Electricity, LCC.)

FIG. 4.10
James Prescott Joule.
British (Salford, Manchester)
(1818–89)
Physicist
Honorary Doctorates from the Universities of Dublin and Oxford

Contributed to the important fundamental *law of conservation of energy* by establishing that various forms of energy, whether electrical, mechanical, or heat, are in the same family and can be exchanged from one form to another. In 1841 introduced *Joule's law*, which stated that the heat developed by electric current in a wire is proportional to the product of the current squared and the resistance of the wire (I^2R). He further determined that the heat emitted was equivalent to the power absorbed, and therefore heat is a form of energy.

EXAMPLE 4.9 For the dial positions of Fig. 4.11, calculate the electricity bill if the previous reading was 4650 kWh and the average cost in your area is 9¢ per kilowatthour.

Solution:

$$5360 \text{ kWh} - 4650 \text{ kWh} = 710 \text{ kWh used}$$

$$710 \text{ kWh} \left(\frac{9¢}{\text{kWh}} \right) = \textbf{\$63.90}$$

EXAMPLE 4.10 How much energy (in kilowatthours) is required to light a 60 W bulb continuously for 1 year (365 days)?

Solution:

$$W = \frac{Pt}{1000} = \frac{(60 \text{ W})(24 \text{ h/day})(365 \text{ days})}{1000} = \frac{525,600 \text{ Wh}}{1000}$$

$$= \textbf{525.60 kWh}$$

EXAMPLE 4.11 How long can a 205 W television set be on before using more than 4 kWh of energy?

Solution:

$$W = \frac{Pt}{1000} \Rightarrow t \text{ (hours)} = \frac{(W)(1000)}{P} = \frac{(4 \text{ kWh})(1000)}{205 \text{ W}} = \textbf{19.51 h}$$

EXAMPLE 4.12 What is the cost of using a 5 hp motor for 2 h if the rate is 9¢ per kilowatthour?

Solution:

$$W \text{ (kilowatthours)} = \frac{Pt}{1000} = \frac{(5 \text{ hp} \times 746 \text{ W/hp})(2 \text{ h})}{1000} = 7.46 \text{ kWh}$$

$$\text{Cost} = (7.46 \text{ kWh})(9\text{¢/kWh}) = \textbf{67.14¢}$$

EXAMPLE 4.13 What is the total cost of using all of the following at 9¢ per kilowatthour?

A 1200 W toaster for 30 min
Six 50 W bulbs for 4 h
A 400 W washing machine for 45 min
A 4800 W electric clothes dryer for 20 min

Solution:

$$W = \frac{(1200 \text{ W})(\frac{1}{2} \text{ h}) + (6)(50 \text{ W})(4 \text{ h}) + (400 \text{ W})(\frac{3}{4} \text{ h}) + (4800 \text{ W})(\frac{1}{3} \text{ h})}{1000}$$

$$= \frac{600 \text{ Wh} + 1200 \text{ Wh} + 300 \text{ Wh} + 1600 \text{ Wh}}{1000} = \frac{3700 \text{ Wh}}{1000}$$

$$W = 3.7 \text{ kWh}$$

$$\text{Cost} = (3.7 \text{ kWh})(9\text{¢/kWh}) = \textbf{33.3¢}$$

The chart in Fig. 4.12 shows the national average cost per kilowatthour compared to the kilowatthours used per customer. Note that the cost today is just above the level of 1926, but the average customer uses more than 20 times as much electrical energy in a year. Keep in mind that the chart of Fig. 4.12 is the average cost across the nation. Some states have average rates close to 5¢ per kilowatthour, whereas others approach 12¢ per kilowatthour.

Table 4.1 lists some common household appliances with their typical wattage ratings. You might find it interesting to calculate the cost of operating some of these appliances over a period of time using the chart in Fig. 4.12 to find the cost per kilowatthour.

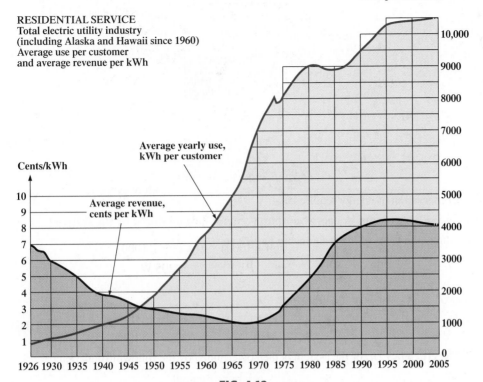

kWh per customer

RESIDENTIAL SERVICE
Total electric utility industry
(including Alaska and Hawaii since 1960)
Average use per customer
and average revenue per kWh

FIG. 4.12

Cost per kWh and average kWh per customer versus time.
(Courtesy of Edison Electric Institute.)

TABLE 4.1

Typical wattage ratings of some common household items.

Appliance	Wattage Rating	Appliance	Wattage Rating
Air conditioner	860	Lap-top computer:	
Blow dryer	1,300	Sleep	< 1 W (Typically 0.3 W to 0.5 W)
Cassette			
player/recorder	5	Normal	10–20 W
Cellular phone:		High	25–35 W
Standby	≅ 35 mW	Microwave oven	1,200
Talk	≅ 4.3 W	Pager	1–2 mW
Clock	2	Phonograph	75
Clothes dryer		Projector	1,200
(electric)	4,800	Radio	70
Coffee maker	900	Range (self-cleaning)	12,200
Dishwasher	1,200	Refrigerator	
Fan:		(automatic defrost)	1,800
Portable	90	Shaver	15
Window	200	Stereo equipment	110
Heater	1,322	Sun lamp	280
Heating equipment:		Toaster	1,200
Furnace fan	320	Trash compactor	400
Oil-burner motor	230	TV (color)	200
Iron, dry or steam	1,100	Videocassette recorder	110
		Washing machine	500
		Water heater	4,500

4.5 EFFICIENCY

A flowchart for the energy levels associated with any system that converts energy from one form to another is provided in Fig. 4.13. Take particular note of the fact that the output energy level must always be less than the applied energy due to losses and storage within the system. The best one can hope for is that W_o and W_i are relatively close in magnitude.

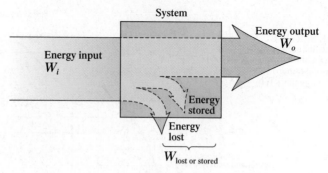

FIG. 4.13

Energy flow through a system.

Conservation of energy requires that

Energy input = energy output + energy lost or stored by the system

Dividing both sides of the relationship by t gives

$$\frac{W_{\text{in}}}{t} = \frac{W_{\text{out}}}{t} + \frac{W_{\text{lost or stored by the system}}}{t}$$

Since $P = W/t$, we have the following:

$$\boxed{P_i = P_o + P_{\text{lost or stored}}} \qquad \text{(W)} \qquad\qquad \textbf{(4.16)}$$

The **efficiency** (η) of the system is then determined by the following equation:

$$\text{Efficiency} = \frac{\text{power output}}{\text{power input}}$$

and

$$\boxed{\eta = \frac{P_o}{P_i}} \qquad \text{(decimal number)} \qquad\qquad \textbf{(4.17)}$$

where η (the lowercase Greek letter eta) is a decimal number. Expressed as a percentage,

$$\boxed{\eta\% = \frac{P_o}{P_i} \times 100\%} \qquad \text{(percent)} \qquad\qquad \textbf{(4.18)}$$

In terms of the input and output energy, the efficiency in percent is given by

$$\boxed{\eta\% = \frac{W_o}{W_i} \times 100\%} \qquad \text{(percent)} \qquad\qquad \textbf{(4.19)}$$

The maximum possible efficiency is 100%, which occurs when $P_o = P_i$, or when the power lost or stored in the system is zero. Obviously, the greater the internal losses of the system in generating the necessary output power or energy, the lower the net efficiency.

EXAMPLE 4.14   A 2 hp motor operates at an efficiency of 75%. What is the power input in watts? If the applied voltage is 220 V, what is the input current?

Solution:

$$\eta\% = \frac{P_o}{P_i} \times 100\%$$

$$0.75 = \frac{(2 \text{ hp})(746 \text{ W/hp})}{P_i}$$

and

$$P_i = \frac{1492 \text{ W}}{0.75} = \textbf{1989.33 W}$$

$$P_i = EI \quad \text{or} \quad I = \frac{P_i}{E} = \frac{1989.33 \text{ W}}{220 \text{ V}} = \textbf{9.04 A}$$

EXAMPLE 4.15   What is the output in horsepower of a motor with an efficiency of 80% and an input current of 8 A at 120 V?

Solution:

$$\eta\% = \frac{P_o}{P_i} \times 100\%$$

$$0.80 = \frac{P_o}{(120\ \text{V})(8\ \text{A})}$$

and

$$P_o = (0.80)(120\ \text{V})(8\ \text{A}) = 768\ \text{W}$$

with

$$768\ \cancel{\text{W}}\left(\frac{1\ \text{hp}}{746\ \cancel{\text{W}}}\right) = \textbf{1.029 hp}$$

EXAMPLE 4.16   If $\eta = 0.85$, determine the output energy level if the applied energy is 50 J.

Solution:

$$\eta = \frac{W_o}{W_i} \Rightarrow W_o = \eta W_i = (0.85)(50\ \text{J}) = \textbf{42.5 J}$$

The very basic components of a generating (voltage) system are depicted in Fig. 4.14. The source of mechanical power is a structure such as a paddlewheel that is turned by water rushing over the dam. The gear train will ensure that the rotating member of the generator is turning at rated speed. The output voltage must then be fed through a transmission system to the load. For each component of the system, an input and output power have been indicated. The efficiency of each system is given by

$$\eta_1 = \frac{P_{o_1}}{P_{i_1}} \qquad \eta_2 = \frac{P_{o_2}}{P_{i_2}} \qquad \eta_3 = \frac{P_{o_3}}{P_{i_3}}$$

If we form the product of these three efficiencies,

$$\eta_1 \cdot \eta_2 \cdot \eta_3 = \frac{\cancel{P_{o_1}}}{P_{i_1}} \cdot \frac{\cancel{P_{o_2}}}{\cancel{P_{i_2}}} \cdot \frac{P_{o_3}}{\cancel{P_{i_3}}} = \frac{P_{o_3}}{P_{i_1}}$$

and substitute the fact that $P_{i_2} = P_{o_1}$ and $P_{i_3} = P_{o_2}$, we find that the quantities indicated above will cancel, resulting in P_{o_3}/P_{i_1}, which is a measure of the efficiency of the entire system.

In general, for the representative cascaded system of Fig. 4.15,

$$\boxed{\eta_{\text{total}} = \eta_1 \cdot \eta_2 \cdot \eta_3 \cdots \eta_n} \qquad\qquad \textbf{(4.20)}$$

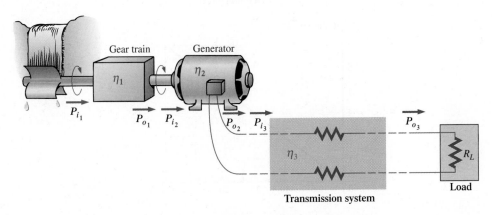

FIG. 4.14

Basic components of a generating system.

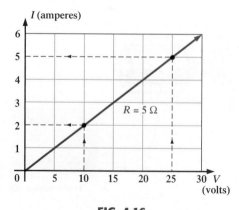

FIG. 4.15

Cascaded system.

EXAMPLE 4.17   Find the overall efficiency of the system of Fig. 4.14 if $\eta_1 = 90\%$, $\eta_2 = 85\%$, and $\eta_3 = 95\%$.

Solution:

$$\eta_T = \eta_1 \cdot \eta_2 \cdot \eta_3 = (0.90)(0.85)(0.95) = 0.727, \text{ or } \mathbf{72.7\%}$$

EXAMPLE 4.18   If the efficiency η_1 drops to 40%, find the new overall efficiency and compare the result with that obtained in Example 4.17.

Solution:

$$\eta_T = \eta_1 \cdot \eta_2 \cdot \eta_3 = (0.40)(0.85)(0.95) = 0.323, \text{ or } \mathbf{32.3\%}$$

Certainly 32.3% is noticeably less than 72.7%. The total efficiency of a cascaded system is therefore determined primarily by the lowest efficiency (weakest link) and is less than (or equal to if the remaining efficiencies are 100%) the least efficient link of the system.

4.6   PLOTTING OHM'S LAW

Graphs, characteristics, plots, and the like, play an important role in every technical field as a mode through which the broad picture of the behavior or response of a system can be conveniently displayed. It is therefore critical to develop the skills necessary both to read data and to plot them in such a manner that they can be interpreted easily.

For most sets of characteristics of electronic devices, the current is represented by the vertical axis (ordinate), and the voltage by the horizontal axis (abscissa), as shown in Fig. 4.16. First note that the vertical axis is in amperes and the horizontal axis is in volts. For some plots, I may be in milliamperes (mA), microamperes (μA), or whatever is appropriate for the range of interest. The same is true for the levels of voltage on the horizontal axis. Note also that the chosen parameters require that the spacing between numerical

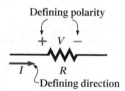

FIG. 4.16

Plotting Ohm's law.

values of the vertical axis be different from that of the horizontal axis. The linear (straight-line) graph reveals that the resistance is not changing with current or voltage level; rather, it is a fixed quantity throughout. The current direction and the voltage polarity appearing at the top of Fig. 4.16 are the defined direction and polarity for the provided plot. If the current direction is opposite to the defined direction, the region below the horizontal axis is the region of interest for the current *I*. If the voltage polarity is opposite to that defined, the region to the left of the current axis is the region of interest. For the standard fixed resistor, the first quadrant, or region, of Fig. 4.16 is the only region of interest. However, you will encounter many devices in your electronics courses that will use the other quadrants of a graph.

Once a graph such as Fig. 4.16 is developed, the current or voltage at any level can be found from the other quantity by simply using the resulting plot. For instance, at *V* = 25 V, if a vertical line is drawn on Fig. 4.16 to the curve as shown, the resulting current can be found by drawing a horizontal line over to the current axis, where a result of 5 A is obtained. Similarly, at *V* = 10 V, a vertical line to the plot and a horizontal line to the current axis will result in a current of 2 A, as determined by Ohm's law.

If the resistance of a plot is unknown; it can be determined at any point on the plot since a straight line indicates a fixed resistance. At any point on the plot, find the resulting current and voltage, and simply substitute into the following equation:

$$R_{dc} = \frac{V}{I}$$

(4.21)

To test Eq. (4.21), consider a point on the plot where *V* = 20 V and *I* = 4 A. The resulting resistance is R_{dc} = 20 V/*I* = 20 V/4 A = 5 Ω. For comparison purposes, a 1 Ω and a 10 Ω resistor were plotted on the graph of Fig. 4.17. Note that the less the resistance, the steeper the slope (closer to the vertical axis) of the curve.

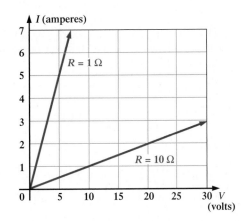

FIG. 4.17

Demonstrating on an I-V plot that the less the resistance, the steeper the slope.

If we write Ohm's law in the following manner and relate it to the basic straight-line equation

$$I = \frac{1}{R} \cdot E + 0$$

$$\downarrow \quad \downarrow \quad \downarrow \quad \downarrow$$

$$y = m \cdot x + b$$

we find that the slope is equal to 1 divided by the resistance value, as indicated by the following:

$$m = \text{slope} = \frac{\Delta y}{\Delta x} = \frac{\Delta I}{\Delta V} = \frac{1}{R}$$

(4.22)

where Δ signifies a small, finite change in the variable.

Equation (4.22) clearly reveals that the greater the resistance, the less the slope. If written in the following form, Equation (4.22) can be used to determine the resistance from the linear curve:

$$R = \frac{\Delta V}{\Delta I} \quad \text{(ohms)} \qquad \textbf{(4.23)}$$

The equation states that by choosing a particular ΔV (or ΔI), you can obtain the corresponding ΔI (or ΔV, respectively) from the graph, as shown in Fig. 4.18, and then determine the resistance. It the plot is a straight line, Equation (4.23) will provide the same result no matter where the equation is applied. However, if the plot curves at all, the resistance will change.

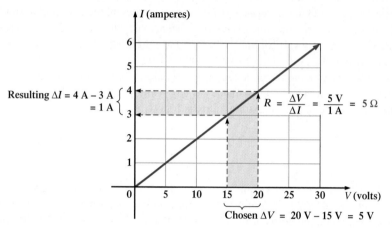

FIG. 4.18

Applying Eq. (4.23).

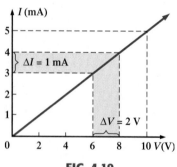

FIG. 4.19

Example 4.19.

EXAMPLE 4.19 Determine the resistance associated with the curve of Fig. 4.19 using Eqs. (4.21) and (4.23), and compare results.

Solution: At $V = 6$ V, $I = 3$ mA, and

$$R_{dc} = \frac{V}{I} = \frac{6\text{ V}}{3\text{ mA}} = \textbf{2 k}\boldsymbol{\Omega}$$

For the interval between 6 V and 8 V,

$$R = \frac{\Delta V}{\Delta I} = \frac{2\text{ V}}{1\text{ mA}} = \textbf{2 k}\boldsymbol{\Omega}$$

The results are equivalent.

Before leaving the subject, let us first investigate the characteristics of a very important semiconductor device called the **diode,** which will be examined in detail in basic electronics courses. This device will ideally act as a low-resistance path to current in one direction and a high-resistance path to current in the reverse direction, much like a switch that will pass current in only one direction. A typical set of characteristics appears in Fig. 4.20. Without any mathematical calculations, the closeness of the characteristic to the voltage axis for negative values of applied voltage indicates that this is the low-conductance (high resistance, switch opened) region. Note that this region extends to approximately 0.7 V positive. However, for values of applied voltage greater than 0.7 V, the vertical rise in the characteristics indicates a high-conductivity (low resistance, switch closed) region. Application of Ohm's law will now verify the above conclusions.

At $V_D = +1$ V,

$$R_{\text{diode}} = \frac{V_D}{I_D} = \frac{1\text{ V}}{50\text{ mA}} = \frac{1\text{ V}}{50 \times 10^{-3}\text{ A}}$$
$$= 20\ \Omega \qquad \text{(a relatively low value for most applications)}$$

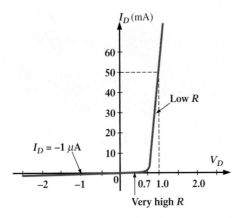

FIG. 4.20

Semiconductor diode characteristics.

At $V_D = -1$ V,

$$R_{\text{diode}} = \frac{V_D}{I_D} = \frac{1\text{ V}}{1\ \mu\text{A}}$$

$$= 1\text{ M}\Omega \qquad (\text{which is often represented by an open-circuit equivalent})$$

4.7 CIRCUIT BREAKERS, GFCIs, AND FUSES

The incoming power to any large industrial plant, heavy equipment, simple circuit in the home, or meters used in the laboratory must be limited to ensure that the current through the lines is not above the rated value. Otherwise, the conductors or the electrical or electronic equipment may be damaged, or dangerous side effects such as fire or smoke may result. To limit the current level, **fuses** or **circuit breakers** are installed where the power enters the installation, such as in the panel in the basement of most homes at the point where the outside feeder lines enter the dwelling. The fuses of Fig. 4.21 have an internal metallic conductor through which the current will pass; a fuse will begin to melt if the current through the system exceeds the rated value printed on the casing. Of course, if the fuse melts through, the current path is broken and the load in its path is protected.

(a)

(b)

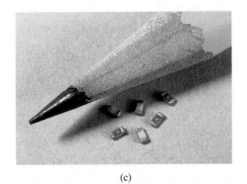

(c)

FIG. 4.21

Fuses: (a) CC-TRON® (0–10 A); (b) Semitron (0–600 A);
(c) subminiature surface-mount chip fuses.
(Courtesy Cooper Bussmann 2002.)

In homes built in recent years, fuses have been replaced by circuit breakers such as those appearing in Fig. 4.22. When the current exceeds rated conditions, an electromagnet in the device will have sufficient strength to draw the connecting metallic link in the breaker out of the circuit and open the current path. When conditions have been corrected, the breaker can be reset and used again.

FIG. 4.22
Circuit breakers.
(Reprinted with permission of Tyco Electronics Corporation, Potter & Brumfield Division.)

FIG. 4.23
*Ground fault current interrupter
(GFCI): 125 V ac, 60 Hz,
15 A outlet.*
(Reprinted with permission of the
Leviton Manufacturing Company.
Leviton SmartLock™ GFCI.)

The most recent National Electrical Code requires that outlets in the bathroom and other sensitive areas be of the ground fault current interrupt (GFCI) variety; GFCIs are designed to trip more quickly than the standard circuit breaker. The commercial unit of Fig. 4.23 trips in 5 ms. It has been determined that 6 mA is the maximum level that most individuals can be exposed to for a short period of time without the risk of serious injury. A current higher than 11 mA can cause involuntary muscle contractions that could prevent a person from letting go of the conductor and possibly cause him or her to enter a state of shock. Higher currents lasting more than a second can cause the heart to go into fibrillation and possibly cause death in a few minutes. The GFCI is able to react as quickly as it does by sensing the difference between the input and output currents to the outlet; the currents should be the same if everything is working properly. An errant path such as through an individual establishes a difference in the two current levels and causes the breaker to trip and disconnect the power source.

4.8 APPLICATION

Household Wiring

A number of facets of household wiring can be discussed without examining the manner in which they are physically connected. In the chapters to follow, additional coverage will be provided to ensure that you develop a solid fundamental understanding of the overall household wiring system. At the very least you will establish a background that will permit you to answer questions that you should be able to answer as a student of this field.

The one specification that defines the overall system is the maximum current that can be drawn from the power lines since the voltage is fixed at 120 V or 240 V (sometimes 208 V). For most older homes with a heating system other than electric, a 100 A service is the norm. Today, with all the electronic systems becoming commonplace in the home, many people are opting for the 200 A service even if they don't have electric heat. A 100 A service specifies that the maximum current that can be drawn through the power lines into your home is 100 A. Using the line-to-line rated voltage and the full-service current (and assuming all resistive-type loads), we can determine the maximum power that can be delivered using the basic power equation:

$$P = EI = (240 \text{ V})(100 \text{ A}) = 24,000 \text{ W} = 24 \text{ kW}$$

This rating reveals that the total rating of all the units turned on in the home cannot exceed 24 kW at any one time. If it did, we could expect the main breaker at the top of the power panel to open. Initially, 24 kW may seem like quite a large rating, but when you consider that a self-cleaning electric oven may draw 12.2 kW, a dryer 4.8 kW, a water heater 4.5 kW, and a dishwasher 1.2 kW, we are already at 22.7 kW (if all the units are operating at peak demand), and we haven't turned the lights or TV on yet. Obviously, the use of an electric oven alone may strongly suggest considering a 200 A service. However, be aware that seldom are all the burners of a stove used at once, and the oven incorporates a thermostat to control the temperature so that it is not on all the time. The same is true for the water heater and dishwasher, so the chances of all the units in a home demanding full service at the same time is very slim. Certainly, a typical home with electric heat that may draw 16 kW just for heating in cold weather must consider a 200 A service. You must also understand that there is some leeway in maximum ratings for safety purposes. In other words, a system designed for a maximum load of 100 A can accept a slightly higher current for short periods of time without significant damage. For the long term, however, it should not be exceeded.

Changing the service to 200 A is not simply a matter of changing the panel in the basement—a new, heavier line must be run from the road to the house. In some areas feeder cables are aluminum because of the reduced cost and weight. In other areas, aluminum is not permitted because of its temperature sensitivity (expansion and contraction), and copper must be used. In any event, when aluminum is used, the contractor must be absolutely sure that the connections at both ends are very secure. **The National Electric Code specifies that 100 A service must use a #4 AWG copper conductor or #2 aluminum conductor. For 200 A service, a 2/0 copper wire or a 4/0 aluminum conductor must be used as shown in Fig. 4.24(a).** A 100 A or 200 A service must have two lines and a service neutral as shown in Fig. 4.24(b). Note in Fig. 4.24(b) that the lines are coated and insulated from each other, and the service neutral is spread around the inside of the wire coating. At the terminal point, all the strands of the service neutral are gathered together and securely attached to the panel. It is fairly obvious that the cables of Fig. 4.24(a) are stranded for added flexibility.

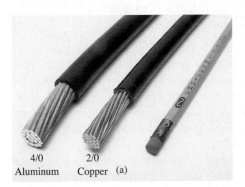
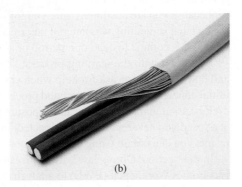

4/0   2/0
Aluminum   Copper   (a)   (b)

FIG. 4.24

200 A service conductors: (a) 4/0 aluminum and 2/0 copper; (b) three-wire 4/0 aluminum service.

Within the system the incoming power is broken down into a number of circuits with lower current ratings utilizing 15 A, 20 A, 30 A, and 40 A protective breakers. Since the load on each breaker should not exceed 80% of its rating, in a 15 A breaker the maximum current should be limited to 80% of 15 A, or 12 A, with 16 A for a 20 A breaker, 24 A for a 30 A breaker, and 32 A for a 40 A breaker. The result is that a home with 200 A service can theoretically have a maximum of 12 circuits (200 A/16 A = 12.5) utilizing the 16 A maximum current ratings associated with 20 A breakers. However, if aware of the loads on each circuit, the electrician can install as many circuits as he feels appropriate. The code further specifies that a #14 wire shall not carry a current in excess of 15 A, a #12 in excess of 20 A, and a #10 in excess of 30 A. Thus, #12 wire is now the most common for general home wiring to ensure that it can handle any excursions beyond 15 A on the 20 A breaker (the most common breaker size). The #14 wire is often used in conjunction with the #12 wire in areas where it is known that the current levels are limited. The #10 wire is typically used for high-demand appliances such as dryers and ovens. The circuits themselves are usually broken down into those that provide lighting, outlets, and so on. Some circuits (such as ovens and dryers) require a higher voltage of 240 V, obtained by using two power lines and the neutral. The higher voltage reduces the current requirement for the same power rating, with the net result that the appliance can usually be smaller. For example, the size of an air conditioner with the same cooling ability is measurably smaller when designed for a 240 V line than when designed for 120 V. Most 240 V lines, however, demand a current level that requires 30 A or 40 A breakers and special outlets to ensure that appliances rated at 120 V are not connected to the same outlet. If time permits, check the panel in your home and take note of the number of circuits—in particular the rating of each breaker and the number of 240 V lines indicated by breakers requiring two slots of the panel. Total the current ratings of all the breakers in your panel, and explain, using the above information, why the total exceeds your feed level.

For safety sake, grounding is a very important part of the electrical system in your home. The National Electric Code requires that the neutral wire of the above system be grounded to an earth-driven rod, a metallic water piping system of 10 ft or more, or a buried metal plate. That ground is then passed on through the electrical circuits of the home for further protection. In a later chapter the details of the connections and grounding methods will be introduced.

4.9 COMPUTER ANALYSIS

Now that a complete circuit has been introduced and examined in detail, we can begin the application of computer methods. As mentioned in Chapter 1, three software packages will be introduced to demonstrate the options available with each and the differences that exist. All have a broad range of support in the educational and industrial communities. **The student version of PSpice (OrCAD Release 9.2 from Cadence Design Systems) will receive the most attention, followed by Electronics Workbench from Multisim.** Each approach has its own characteristics with procedures that must be followed exactly; otherwise, error messages will appear. Do not assume that you can "force" the system to respond the way you would prefer—every step is well defined, and one error on the input side can result in results of a meaningless nature. At times you may believe that the system is in error because you are absolutely sure you followed every step correctly. In such cases, accept the fact that something was entered incorrectly, and review all your work very carefully. All it takes is a comma instead of a period or a decimal point to generate incorrect results.

Be patient with the learning process; keep notes of specific maneuvers that you learn; and don't be afraid to ask for help when you need it. For each approach there is always the initial concern about how to start and proceed through the first phases of the analysis. However, be assured that with time and exposure you will work through the required maneuvers at a speed you never would have expected. In time you will be absolutely delighted with the results you can obtain using computer methods.

In this section, Ohm's law will be investigated using the software packages PSpice and Electronics Workbench (EWB) to analyze the circuit in Fig. 4.25. Both require that the circuit first be "drawn" on the computer screen and then analyzed (simulated) to obtain the desired results. As mentioned above, the analysis program is fixed in stone and cannot be changed by the user. The proficient user is one who can draw the most out of a computer software package.

Although the author feels that there is sufficient material in the text to carry a new student of the material through the programs provided, be aware that this is not a computer text. Rather, it is one whose primary purpose is simply to introduce the different approaches and how they can be applied effectively. Today, some excellent texts and manuals are available that cover the material in a great deal more detail and perhaps at a slower pace. In fact, the quality of the available literature has improved dramatically in recent years.

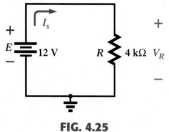

FIG. 4.25

Circuit to be analyzed using PSpice and Electronics Workbench.

PSpice

Readers who were familiar with older versions of PSpice such as version 8 will find that the major changes in this latest 9.2 version are primarily in the front end and the simulation process. After executing a few programs, you will find that most of the procedures you learned from older versions will be applicable here also—at least the sequential process has a number of strong similarities.

Once 9.2 OrCAD Lite has been installed, the first required procedure is to open a **Folder** in the **C:** drive for storage of the circuit files that will result from the analysis. Be aware, however, that

once the folder has been defined, it does not have to be defined for each new project unless you choose to do so. If you are satisfied with one location (folder) for all your projects, this is a one-time operation that does not have to be repeated with each network.

To establish the **Folder,** simply right-click the mouse on **Start** to obtain a listing that includes **Explore.** Select **Explore,** and then use the sequence **File-New Folder** to obtain a new folder on the screen waiting for a name. Type in **PSpice** (the author's choice) followed by a left click of the mouse to install. Then exit (using the **X** at the top right of the screen) the **Exploring-Start Menu,** and the first step is complete—you're on your way. The folder **PSpice** has been established for all the projects you plan to work on in this text.

Our first project can now be initiated by double-clicking on the **Orcad Lite Edition** icon on the screen, or you can use the sequence **Start-Programs-Orcad Family Release 9.2 Lite Edition.** The resulting screen has only a few active keys on the top toolbar. The first at the top left is the **Create new document** key (or you can use the sequence **File-New Project**). Selecting the key will result in a **New Project** dialog box in which the **Name** of the project must be entered. For our purposes we will choose **Ohmslaw** as shown in the heading of Fig. 4.26 and select **Analog or Mixed A/D** (to be used for all the analyses of this text). Note at the bottom of the dialog box that the **Location** appears as **C:\PSpice** as set above. Click **OK,** and another dialog box will appear titled **Create PSpice Project.** Select **Create a blank project** (again, for all the analyses to be performed in this text). Click **OK,** and a third toolbar will appear at the top of the screen with some of the keys enabled. A **Project Manager Window** will appear with **Ohmslaw** as its heading. The new project listing will appear with an icon and an associated + sign in a small square. Clicking on the + sign will take the listing a step further to **SCHEMATIC1.** Click + again, and **PAGE1** will appear; clicking on a − sign will reverse the process. Double-clicking on **PAGE1** will create a working window titled **SCHEMATIC1: PAGE1,** revealing that a project can have more than one schematic file and more than one associated page. The width and height of the window can be adjusted by grabbing an edge to obtain a double-headed arrow and dragging the border to the desired location. Either window on the screen can be moved by clicking on the top heading to make it dark blue and then dragging it to any location.

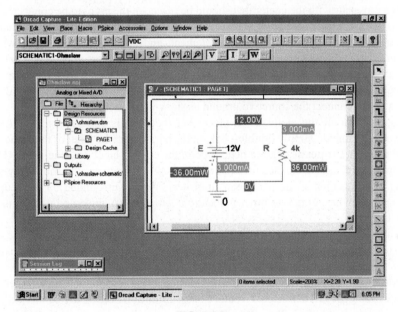

FIG. 4.26

Using PSpice to determine the voltage, current, and power levels for the circuit of Fig. 4.25.

Now we are ready to build the simple circuit of Fig. 4.25. Select the **Place a part** key (the second key from the top of the toolbar on the right) to obtain the **Place Part** dialog box. Since this is the first circuit to be constructed, we must ensure that the parts appear in the list of active libraries. Select **Add Library-Browse File,** and select **analog.olb, eval.olb,** and **source.olb.** When each appears under the **File name** heading, select **Open.** All three files will be required to build the networks appearing in this text. However, it is important to realize that

once the library files have been selected, they will appear in the active listing for each new project without having to add them each time—a step, such as the Folder *step above, that does not have to be repeated with each similar project.*

Click **OK,** and we can now place components on the screen. For the dc voltage source, first select the **Place a part** key and then select **SOURCE** in the library listing. Under **Part List,** a list of available sources will appear; select **VDC** for this project. Once **VDC** has been selected, its symbol, label, and value will appear on the picture window at the bottom right of the dialog box. Click **OK,** and the **VDC** source will follow the cursor across the screen. Move it to a convenient location, left-click the mouse, and it will be set in place as shown in Fig. 4.26. Since only one source is required, a right click of the mouse will result in a list of options, in which **End Mode** appears at the top. Choosing this option will end the procedure, leaving the source in a red dashed box. The fact that it is red indicates that it is an active mode and can be operated on. One more left click of the mouse, and the source will be in place and the red active status removed.

One of the most important steps in the procedure is to ensure that a 0 V ground potential is defined for the network so that voltages at any point in the network have a reference point. *The result is a requirement that every network must have a ground defined.* For our purposes, the **0/SOURCE** option will be our choice when the **GND** key is selected. It will ensure that one side of the source is defined as 0 V. Finally, we need to add a resistor to the network by selecting the **Place a part** key again and then selecting the **ANALOG** library. Scrolling the options, note that **R** will appear and should be selected. Click **OK,** and the resistor will appear next to the cursor on the screen. Move it to the desired location and click it in place. Then right-click the mouse and **End Mode,** and the resistor has been entered into the schematic's memory. Unfortunately, the resistor ended up in the horizontal position, and the circuit of Fig. 4.25 has the resistor in the vertical position. No problem: Simply select the resistor again to make it red, and right-click the mouse. A listing will appear in which **Rotate** is an option. It will turn the resistor 90° in the counterclockwise direction. It can also be rotated 90° by simultaneously selecting **Ctrl-R.**

All the required elements are on the screen, but they need to be connected. This is accomplished by selecting the **Place a wire** key that looks like a step in the right toolbar. The result is a crosshair with the center that should be placed at the point to be connected. Place the crosshair at the top of the voltage source, and left-click it once to connect it to that point. Then draw a line to the end of the next element, and click the mouse again when the crosshair is at the correct point. A red line will result with a square at each end to confirm that the connection has been made. Then move the crosshair to the other elements, and build the circuit. Once everything is connected, a right click will provide the **End Mode** option. Don't forget to connect the source to ground as shown in Fig. 4.26.

Now we have all the elements in place, but their labels and values are wrong. To change any parameter, simply double-click on the parameter (the label or the value) to obtain the **Display Properties** dialog box. Type in the correct label or value, click **OK,** and the quantity is changed on the screen. The labels and values can be moved by simply clicking on the center of the parameter until it is closely surrounded by the four small squares and then dragging it to the new location. Another left click, and it is deposited in its new location.

Finally, we can initiate the analysis process, called **Simulation,** by selecting the **Create a new simulation profile** key near the top left of the display—it resembles a data page with a star in the top left corner. **A New Simulation** dialog box will result that first asks for the **Name** of the simulation. **Bias Point** is selected for a dc solution, and **none** is left in the **Inherit From** request. Then select **Create,** and a **Simulation Setting** dialog box will appear in which **Analysis-Analysis Type-Bias Point** is sequentially selected. Click **OK,** and select the **Run** key (which looks like an isolated blue arrowhead) or choose **PSpice-Run** from the menu bar. An **Output Window** will result that appears to be somewhat inactive. It will not be used in the current analysis, so close (**X**) the window, and the circuit of Fig. 4.26 will appear with the voltage, current, and power levels of the network. The voltage, current, or power levels can be removed (or replaced) from the display by simply selecting the **V, I,** or **W** in the third toolbar from the top. Individual values can be removed by simply selecting the value and pressing the **Delete** key or the scissors key in the top menu bar. Resulting values can be moved by simply left-clicking the value and dragging it to the desired location.

Note in Fig. 4.26 that the current is 3 mA (as expected) at each point in the network, and the power delivered by the source and dissipated by the resistor is the same, or 36 mW. There are also 12 V across the resistor as required by the configuration.

There is no question that the description above was long for such a trivial circuit. However, keep in mind that we needed to introduce many new facets of using PSpice that will not be touched on again in the future. By the time you finish analyzing your third or fourth network, the above procedure will appear routine and will move rather quickly. You will once again be looking for new challenges.

For comparison purposes, Electronics Workbench (EWB) will be used to analyze the circuit in Fig. 4.25. Although there are differences between PSpice and EWB, such as initiating the process, constructing the networks, making the measurements, and setting up the simulation procedure, there are sufficient similarities between the two approaches to make it easier to learn one if you are already familiar with the other. The similarities will be obvious only if you make an attempt to learn both. One of the major differences between the two is the option to use actual instruments in EWB to make the measurements—a positive trait in preparation for the laboratory experience. However, in EWB you may not find the extensive list of options available with PSpice. In general, however, both software packages are well prepared to take us through the types of analyses to be encountered in this text.

When the **Multisim 2001** icon is selected from the opening window, a screen will appear with the heading **Multisim-Circuit 1** (see Fig. 4.27). A menu bar appears across the top of the screen, with one toolbar below the menu bar and one to each side of the screen. The toolbars appearing can be controlled by the sequence **View-Toolbars** followed by a selection of which toolbars you want to appear. For the analysis of this text, all the toolbars were selected. For the placement of components, **View-Show Grid** was selected so that a grid would appear on the screen. As you place an element, it will automatically be placed in a relationship specific to the grid structure.

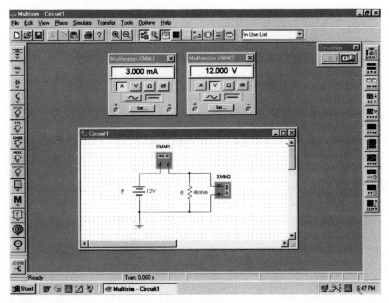

FIG. 4.27

Using Electronics Workbench to determine the voltage and current levels for the circuit of Fig. 4.25.

Now to build the circuit of Fig. 4.25. First take the cursor and place it on the battery symbol at the top of the component toolbar at the left of the screen. One left click of the mouse, and a list of sources will appear. Place the cursor on any one of the sources, and text will appear on the screen defining the type of source. Placing the cursor on the third key pad down will result in **DC VOLTAGE SOURCE.** Left-click again, and the battery symbol will appear on the screen next to the location of the cursor. Move the cursor to the desired location, and with a single left click of the mouse the battery symbol can be set in place. The operation is complete. If you want to delete the source, simply click on the symbol again with a left click of the mouse to create four small squares around the source. These squares indicate that the source is in the active mode and can be operated on. If you want to delete it, simply click on the **Delete** key or select the scissor key pad on the top toolbar. If you want to modify the source, perform a right click of the mouse *outside* the four small squares, and you get one list. Perform the right click *within* the four squares, and you have a different set of options. At any time, if you want to remove the active state, simply perform a left click anywhere on the screen. If you want to move the source, simply click on the source symbol to create the four squares, but do not release the clicker. Hold it down and drag the source to the preferred location. When the source is in place, release the clicker, and one more click will remove the active state. To remove the **SOURCES** toolbar, simply click on the **X** in the top right corner of the toolbar.

The next key down from the source key that looks like a resistor controls the display of the **Basic** passive components of a network. Click once on the symbol, and two columns of components will appear. *From now on, whenever possible, the word* click *will imply a left click of the mouse. The need for a right click will continue to be spelled out.*

For the circuit of Fig. 4.25 we need a resistor. When you place the cursor over the left resistor, the text **RESISTOR** will appear. When you place it over the right resistor, the text **RESISTOR_VIRTUAL** will appear. For all the analyses in this text using EWB, the virtual resistor will be used. The term **RESISTOR** is used for all resistors of a standard commercial value—the values typically made commercially. The term **VIRTUAL** is applied to any component in which you, the user, can define the value you want. Click once on the virtual resistor, and it will appear on the screen next to the cursor in the horizontal position. In Fig. 4.25 it is in the vertical position, so a rotation must be made. This can be done by clicking on the resistor to obtain the active state and then performing a right click of the mouse within the four squares. A number of options appear, including deleting (**Cut**) the component, copy, change position, and color. Since we want to rotate 90° clockwise, we select that option, and the resistor will automatically be rotated 90°. Now, as with the battery, to place the resistor in position, simply click on the resistor symbol to create the four small squares, and then, holding the left clicker down, drag the resistor to the desired position. When the resistor is in place, release the clicker, and click again to remove the four squares—the resistor is in place.

Finally, we need a ground for all networks. Going back to the **SOURCES** parts bin, we find that a ground is the first option at the top of the toolbar. Select the **GROUND** on the left, and place it on the screen below the voltage source as shown in Fig. 4.27. Now, before connecting the components together, move the labels and the value of each component to the relative positions shown in Fig. 4.27. This is accomplished by simply clicking on the label or value to create a small set of squares around the element and then dragging the element to the desired location. Release the clicker, and then click again to set the element in place. To change the label or value, simply double-click on the label (such as **V1**), and a **Battery** dialog box will appear. Select **Label** and enter E as the **Reference ID.** Then, before leaving the dialog box, go to **Value** and change the value if necessary. It is very important to realize that you cannot type in the units where the **V** now appears. The prefix is controlled by the scroll keys at the left of the unit of measure. For practice, try the scroll keys, and you will find that you can go from **pV** to **TV.** For now leave it as simply **V.** Click **OK,** and both have been changed on the screen. The same process can be applied to the resistive element to obtain the label and value appearing in Fig. 4.27.

Next, we should tell the system which results should be generated and how they should be displayed. For this example we will use a multimeter to measure both the current and the voltage of the circuit. The **Multimeter** is the first option in the list of instruments appearing in the toolbar to the right of the screen. When selected, it will appear on the screen and be placed anywhere using the same procedure defined for the components above. The voltmeter was turned clockwise using the procedure described above for the elements. Double-click on either meter symbol, and a **Multimeter** dialog box will appear in which the function of the meter must be defined. Since the meter **XMM1** will be used as an ammeter, the letter **A** will be selected and the horizontal line to indicate dc level. There is no need to select **Set** for the default values since they have been chosen for the broad range of applications. The dialog meters can be moved to any location by simply clicking on their heading bar to make it dark blue and then dragging the meter to the preferred position. For the voltmeter, **V** and the horizontal bar were selected as shown in Fig. 4.27.

Finally, the elements need to be connected. This is accomplished by simply bringing the cursor to one end of an element, say, the top of the voltage source, with the result that a small dot and a crosshair will appear at the top end of the element. Click the mouse once, follow the path you want, and place the crosshair over the positive terminal of the ammeter. Then click again and the wire will appear in place.

At this point you should be aware that the software package has its preferences about how it wants the elements to be connected. That is, you may try to draw it one way, but the computer generation may be a different path. In time you will be aware of those preferences and will be able to set up the network to your liking. Now continue making the connections appearing in Fig. 4.27, moving elements or adjusting lines as necessary. Be sure that the small dot appears at any point where you want a connection. Its absence suggests that the connection has not been made and the software program has not accepted the entry.

Now we are ready to run the program and view the solution. The analysis can be initiated in a number of ways. One option is to select **Simulate** from the top toolbar, followed by **RUN/STOP.** Another is to select the **Simulate** key in the design bar grouping in the top toolbar. It appears as a sharp, jagged, green plot on a black background. The last option, and the one we will use the most, requires an **OFF/ON, 0/1** switch on the screen. It is obtained through **VIEW-Show Simulate Switch** and will appear as shown in the top right corner of Fig. 4.27. Using this last option, the analysis (called **Simulation**) is initiated by placing the cursor on the 1 of the switch and left-clicking the mouse. The analysis will be performed, and the current and voltage will appear on the meter as shown in Fig. 4.27. Note that both are providing the expected results.

Now for one of the most important things to learn about applying EWB:

Always stop or end the simulation (clicking on 0 or choosing OFF) before making any changes in the network. When the simulation is initiated, it will stay in that mode until turned off.

There was obviously a great deal of material to learn in this first exercise using Electronics Workbench. Be assured, however, that as we continue with more examples, you will find the procedure quite straightforward and actually enjoyable to apply.

CHAPTER SUMMARY

OHM'S LAW

- The applied voltage is the "pressure" to establish a current level that is controlled by the resistance of the circuit.
- Ohm's law can be applied to any circuit, with any type of source applied, for any time frame.

POWER

- Power is a measure of how much work can be done over a particular period of time; that is, power is a rate of doing work.
- Power is measured in watts (W), with 1 watt equal to a rate of doing work equal to 1 joule per second (J/s).
- The power delivered by a source or to a unit is determined by the product of the voltage and current; it is not simply the magnitude of either quantity.

ENERGY

- The energy delivered or expended in any electrical system is determined by the product of the power involved and the time of operation. The longer the system is used, the greater will be the amount of energy converted.
- The consumer pays for the amount of energy used and not the sum total of the power levels of the appliances in the home. The largest home or factory will not use any electrical energy if the appliances are never turned on.

EFFICIENCY

- Efficiency is a measure of how much energy is applied to the job at hand compared to the energy level required from the source. It can never be more than 100% and will usually be less due to losses in the conversion process.
- For cascaded systems the system with the lowest efficiency rating will have the most impact on the total efficiency of the entire system.

PLOTTING OHM'S LAW

- Always be aware of the defined polarities and directions when plotting quantities on a graph. They will determine whether a plot point should be in the positive or the negative region of the graph.
- The resistance at any point can be determined by simply dividing the voltage at that point by the current level.
- For a plot where the current is the vertical axis and the voltage the horizontal axis, the steeper the curve, the less the resistance.

Important Equations

$$I = \frac{E}{R} \qquad E = IR \qquad R = \frac{E}{I}$$

1 watt (W) = 1 joule/second (J/s)

1 horsepower = 746 W

$$P = VI \qquad P = \frac{V^2}{R}$$

$$P = I^2R \qquad P = EI$$

$$I = \sqrt{\frac{P}{R}} \qquad V = \sqrt{PR}$$

$$W = Pt$$

$$\text{Energy (kWh)} = \frac{\text{power (W)} \times \text{time (h)}}{1000}$$

$$\eta\% = \frac{P_o}{P_i} \times 100\% \qquad \eta\% = \frac{W_o}{W_i} \times 100\%$$

$$\eta_{\text{total}} = \eta_1 \cdot \eta_2 \cdot \eta_3 \cdots \eta_n$$

PROBLEMS

SECTION 4.2 Ohm's Law

1. Find the current through a standard 4.7 Ω resistor if the applied voltage is 9 V.

2. Calculate the voltage across a standard 22 kΩ resistor if the current through the resistor is 5 mA.

3. How much resistance is required to limit the current to 1.5 mA if the potential drop across the resistor is 6 V?

4. At starting, what is the current drain on a 12 V car battery if the resistance of the starting motor is 0.06 Ω?

5. If the current through a 0.02 MΩ resistor is 3.6 μA, what is the voltage drop across the resistor?

6. If a voltmeter has an internal resistance of 11 MΩ, find the current through the meter when it reads 22 V.

7. If a refrigerator draws 2.2 A at 120 V, what is its resistance?

8. If a clock has an internal resistance of 7.5 kΩ, find the current through the clock if it is plugged into a 120 V outlet.

9. A washing machine is rated at 4.8 A at 120 V. What is its internal resistance?

10. If a soldering iron draws 14 A at 120 V, what is its resistance?

11. The input current to a transistor is 20 μA. If the applied (input) voltage is 24 mV, determine the input resistance of the transistor.

12. The internal resistance of a dc generator is 0.5 Ω. Determine the loss in terminal voltage across this internal resistance if the current is 15 A.

*13. **a.** If an electric heater draws 9.5 A when connected to a 120 V supply, what is the internal resistance of the heater?
   b. How much energy is converted in 1 h?

14. If 420 J of energy are absorbed by a resistor in 4 min, what is the power to the resistor?

15. The power to a device is 40 joules per second (J/s). How long will it take to deliver 640 J?

16. a. How many joules of energy does a 2 W nightlight dissipate in 8 h?
  b. How many kilowatthours does it dissipate?

17. A resistor of 22 Ω has charge flowing through it at the rate of 300 coulombs per minute (C/min). How much power is dissipated?

18. How long must a steady current of 2 A exist in a resistor that has 3 V across it to dissipate 12 J of energy?

19. What is the power delivered by a 12 V battery if charge flows at the rate of 480 C/min?

20. The current through a 4 Ω resistor is 7 mA. What is the power delivered to the resistor?

21. The voltage drop across a 3 Ω resistor is 9 mV. What is the power input to the resistor?

22. If the power input to a 4 Ω resistor is 64 W, what is the current through the resistor?

23. A 1/2 W resistor has a resistance of 1000 Ω. What is the maximum current that it can safely handle?

24. A 2.2 kΩ resistor in a stereo system dissipates 42 mW of power. What is the voltage across the resistor?

25. A dc battery can deliver 40 mA at 9 V. What is the power rating?

26. What are the "hot" resistance level and current rating of a 120 V, 100 W bulb?

27. What are the internal resistance and voltage rating of a 450 W automatic washer that draws 3.75 A?

28. A calculator with an internal 3 V battery draws 0.4 mW when fully functional.
  a. What is the current demand from the supply?
  b. If the calculator is rated to operate 500 h on the same battery, what is the ampere-hour rating of the battery?

29. A 20 kΩ resistor has a rating of 100 W. What are the maximum current and the maximum voltage that can be applied to the resistor?

*__30.__ **a.** Plot power versus current for a 100 Ω resistor. Use a power scale from 0 to 1 W and a current scale from 0 to 100 mA with divisions of 0.1 W and 10 mA, respectively.
  b. Is the curve linear or nonlinear?
  c. Using the resulting plot, determine the current at a power level of 500 mW.

*__31.__ A small, portable black-and-white television draws 0.455 A at 9 V.
  a. What is the power rating of the television?
  b. What is the internal resistance of the television?
  c. What is the energy converted in 6 h of typical battery life?

*__32.__ **a.** If a home is supplied with a 120 V, 100 A service, find the maximum power capability.
  b. Can the homeowner safely operate the following loads at the same time?
      5 hp motor
      3000 W clothes dryer
      2400 W electric range
      1000 W steam iron

SECTION 4.4 **Energy**

33. A 10 Ω resistor is connected across a 12 V battery.
  a. How many joules of energy will it dissipate in 1 min?
  b. If the resistor is left connected for 2 min instead of 1 min, will the energy used increase? Will the power dissipation level increase?

34. How much energy in kilowatthours is required to keep a 230 W oil-burner motor running 12 h a week for 5 months during the winter season? (Use 30 days = 1 month.)

35. How long can a 1500 W heater be on before using more than 10 kWh of energy?

36. How much does it cost to use a 30 W radio for 3 h at 9¢ per kilowatthour?

37. a. In 10 h an electrical system converts 500 kWh of electrical energy into heat. What is the power level of the system?
  b. If the applied voltage is 208 V, what is the current drawn from the supply?

38. a. At 9¢ per kilowatthour, how long can you watch a 250 W color television for $1?
  b. For $1, how long can you use a 4.8 kW dryer?
  c. Compare the results of parts (a) and (b), and comment on the effect of the wattage level on the relative cost of using an appliance.

39. What is the total cost of using the following at 9¢ per kilowatthour?

860 W air conditioner for 24 h

4800 W clothes dryer for 30 min

400 W washing machine for 1 h

1200 W dishwasher for 45 min

*40. What is the total cost of using the following at 9¢ per kilowatthour?

110 W stereo set for 4 h

1200 W projector for 20 min

60 W tape recorder for 1.5 h

150 W color television set for 3 h 45 min

SECTION 4.5 Efficiency

41. What is the efficiency of a motor that has an output of 0.5 hp with an input of 450 W?

42. The motor of a power saw is rated 70% efficient. If 1.8 hp are required to cut a particular piece of lumber, what is the current drawn from a 120 V supply?

43. What is the efficiency of a dryer motor that delivers 1 hp when the input current and voltage are 4 A and 220 V, respectively?

44. A stereo system draws 2.4 A at 120 V. The audio output power is 50 W.

a. How much power is lost in the form of heat in the system?

b. What is the efficiency of the system?

45. If an electric motor having an efficiency of 87% and operating off a 220 V line delivers 3.6 hp, what input current does the motor draw?

46. A motor is rated to deliver 2 hp.

a. If it runs on 110 V and is 90% efficient, how many watts does it draw from the power line?

b. What is the input current?

c. What is the input current if the motor is only 70% efficient?

47. An electric motor used in an elevator system has an efficiency of 90%. If the input voltage is 220 V, what is the input current when the motor is delivering 15 hp?

48. A 2 hp motor drives a sanding belt. If the efficiency of the motor is 87% and that of the sanding belt 75% due to slippage, what is the overall efficiency of the system?

49. If two systems in cascade each have an efficiency of 80% and the input energy is 60 J, what is the output energy?

50. The overall efficiency of two systems in cascade is 72%. If the efficiency of one is 0.9, what is the efficiency in percent of the other?

*51. If the total input and output power of two systems in cascade are 400 W and 128 W, respectively, what is the efficiency of each system if one has twice the efficiency of the other?

52. a. What is the total efficiency of three systems in cascade with efficiencies of 98%, 87%, and 21%?

b. If the system with the least efficiency (21%) were removed and replaced by one with an efficiency of 90%, what would be the percentage increase in total efficiency?

53. a. Perform the following conversions:

1 Wh to joules

1 kWh to joules

b. Based on the results of part (a), discuss when it is more appropriate to use one unit versus the other.

SECTION 4.6 Plotting Ohm's Law

54. a. Plot the curve of I (vertical axis) versus V (horizontal axis) for a 120 Ω resistor. Use a horizontal scale of 0 to 100 V and a vertical scale of 0 to 1 A.

b. Using the graph of part (a), find the current at a voltage of 20 V and 50 V.

55. a. Plot the I-V curve for a 5 Ω and a 20 Ω resistor on the same graph. Use a horizontal scale of 0 to 40 V and a vertical scale of 0 to 2 A.

b. Which is the steeper curve? Can you offer any general conclusions based on results?

c. If the horizontal and vertical scales were interchanged, which would be the steeper curve?

56. a. Plot the I-V characteristics of a 1 Ω, 100 Ω, and 1000 Ω resistor on the same graph. Use a horizontal axis of 0 to 100 V and a vertical axis of 0 to 100 A.

b. Comment on the steepness of a curve with increasing levels of resistance.

*57. Sketch the internal resistance characteristics of a device that has an internal resistance of 20 Ω from 0 to 10 V, an internal resistance of 4 Ω from 10 V to 15 V, and an internal resistance of 1 Ω for any voltage greater than 15 V. Use a horizontal scale that extends from 0 to 20 V and a vertical scale that permits plotting the current for all values of voltage from 0 to 20 V.

SECTION 4.9 Computer Analysis

58. Using PSpice or EWB, repeat the analysis of the circuit of Fig. 4.25 with $E = 400$ mV and $R = 0.04$ MΩ.

59. Using PSpice or EWB, repeat the analysis of the circuit of Fig. 4.25, but reverse the polarity of the battery and use $E = 0.02$ V and $R = 240$ Ω.

Programming Language

60. Write a program to calculate the cost of using five different appliances for varying lengths of time if the cost is 9¢ per kilowatthour.

61. Request I, R, and t and determine V, P, and W. Print out the results with the proper units.

GLOSSARY

Circuit breaker   A two-terminal device designed to ensure that current levels do not exceed safe levels. If "tripped," it can be reset with a switch or a reset button.

Diode   A semiconductor device whose behavior is much like that of a simple switch; that is, it will pass current ideally in only one direction when operating within specified limits.

Efficiency (η)   A ratio of output to input power that provides immediate information about the energy-converting characteristics of a system.

Energy (W)   A quantity whose change in state is determined by the product of the rate of conversion (P) and the period involved (t). It is measured in joules (J) or wattseconds (Ws).

Fuse   A two-terminal device whose sole purpose is to ensure that current levels in a circuit do not exceed safe levels.

Horsepower (hp)   Equivalent to 746 watts in the electrical system.

Kilowatthour meter   An instrument for measuring kilowatthours of energy supplied to a residential or commercial user of electricity.

Ohm's law   An equation that establishes a relationship among the current, voltage, and resistance of an electrical system.

Power   An indication of how much work can be done in a specified amount of time; a *rate* of doing work. It is measured in joules/second (J/s) or watts (W).

Series dc Circuits

OBJECTIVES

- Become familiar with the characteristics of a series circuit and learn how to solve for the quantities of interest.
- Develop a clear understanding of Kirchhoff's voltage law and its impact on the analysis of electric circuits.
- Be able to properly apply an ohmmeter, voltmeter, ammeter, and power meter to measure the quantities of importance in a series circuit.
- Appreciate how the applied voltage divides in a series circuit and be able to apply the voltage divider rule effectively.
- Become aware of how the voltage regulation of a supply affects its terminal voltage.
- Understand the loading effects of applying meters to a circuit.
- Develop a sense of how the series configuration can be used effectively in everyday applications.
- Become aware of how computer software can be used to solve for all the quantities of interest in a series dc circuit.

5.1 INTRODUCTION

The characteristics and appearance of a number of circuit elements were introduced in previous chapters. We are now ready to connect the elements together so that each performs a particular function in an electrical system.

Fundamentally, there are three distinct ways in which elements can be connected together. The first to be described is the **series connection** which will be covered in detail in this chapter. The second is the **parallel connection** to be covered in detail in the next chapter. The last grouping is simply **any connection in which the elements are not in series or parallel.** Proficiency in identifying a type of connection will come only with exposure and experience. It is particularly helpful, however, to fall back on the basic definitions of each to help define the case at hand. This will become evident in the practice exercises provided.

5.2 SERIES RESISTORS

Before the series connection is described, first recognize that every fixed resistor has only two terminals to connect in a configuration—it is therefore referred to as a **two-terminal device.** In Fig. 5.1, one terminal of resistor R_2 is connected to resistor R_1 on one side, and the remaining terminal is connected to resistor R_3 on the other side, resulting in one, and only one, connection between adjoining resistors. When connected in this manner, the resistors have established a series connection. If three elements were connected to the same point, as shown in Fig. 5.2, there would not be a series connection between resistors R_1 and R_2.

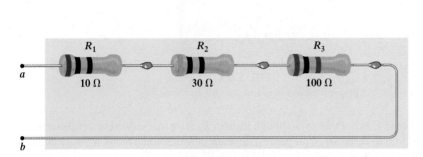

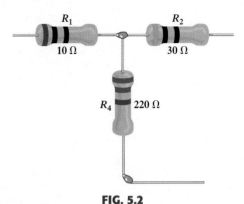

FIG. 5.1

Series connection of resistors.

FIG. 5.2

Configuration in which none of the resistors are in series.

On a schematic, the series connection of resistors of Fig. 5.1 would appear as shown in Fig. 5.3 using the resistor symbol introduced in Chapter 3.

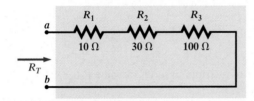

FIG. 5.3

Schematic representation of Fig. 5.1.

The effect of connecting resistors in series is to increase the resistance between points a and b of Fig. 5.1. In fact, the total resistance is determined by finding the sum of the resistance levels of the three resistors; that is, $R_T = R_1 + R_2 + R_3$. In general, therefore,

the total resistance of a series configuration is the sum of the resistance levels.

In equation form for any number (N) of resistors,

$$R_T = R_1 + R_2 + R_3 + R_4 + \cdots + R_N \qquad \textbf{(5.1)}$$

A result of Eq. (5.1) is that

the more resistors we add in series, the greater the resistance, no matter what their value.

Further,

the largest resistor in a series combination will have the most impact on the total resistance.

For the configuration of Fig. 5.3, the total resistance would be

$$R_T = R_1 + R_2 + R_3$$
$$= 10\ \Omega + 30\ \Omega + 100\ \Omega$$

and $\qquad\qquad R_T = \textbf{140}\ \boldsymbol{\Omega}$

EXAMPLE 5.1 Determine the total resistance of the series connection of Fig. 5.4. Note that all the resistors appearing in this network are standard values.

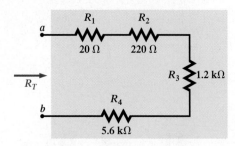

FIG. 5.4

Series connection of resistors for Example 5.1.

Solution: Note in Fig. 5.4 that even though resistor R_3 is on the vertical and resistor R_4 returns at the bottom to terminal b, all the resistors are in series since there are only two terminals at each connection point.

Applying Eq. (5.1):

$$R_T = R_1 + R_2 + R_3 + R_4$$

$$R_T = 20\ \Omega + 220\ \Omega + 1.2\ k\Omega + 5.6\ k\Omega$$

and

$$R_T = 7040\ \Omega = \mathbf{7.04\ k\Omega}$$

For the special case where resistors are the same value, Equation (5.1) can be modified as follows:

$$\boxed{R_T = NR} \tag{5.2}$$

where N is the number of resistors in series of value R.

EXAMPLE 5.2 Find the total resistance of the series resistors of Fig. 5.5. Again, recognize 3.3 kΩ as a standard value.

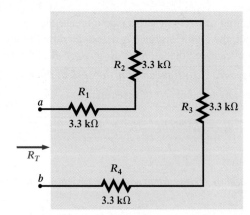

FIG. 5.5

Series connection of four resistors of the same value (Example 5.2).

Solution: Again, don't be concerned about the change in configuration. Neighboring resistors are connected only at one point, satisfying the definition of series elements.

Eq. (5.2):

$$R_T = NR$$

$$= (4)(3.3\ k\Omega) = \mathbf{13.2\ k\Omega}$$

It is important to realize that since the parameters of Eq. (5.1) can be put in any order,
the total resistance of resistors in series is unaffected by the order in which they are connected.

The result is that the total resistance of Fig. 5.6(a) and (b) are both the same. Again, note that all the resistors are standard values.

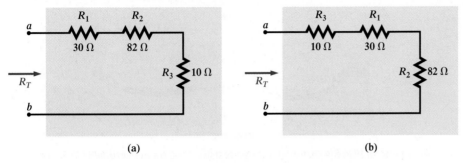

(a)   (b)

FIG. 5.6

Two series combinations of the same elements with the same total resistance.

EXAMPLE 5.3 Determine the total resistance for the series resistors (standard values) of Fig. 5.7.

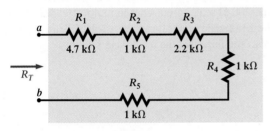 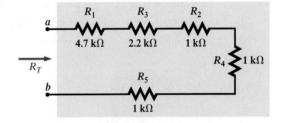

FIG. 5.7   **FIG. 5.8**

Series combination of resistors for Example 5.3.   *Series circuit of Fig. 5.7 redrawn to permit the use of Eq. (5.2): $R_T = NR$.*

Solution: First, the order of the resistors is changed as shown in Fig. 5.8 to permit the use of Eq. (5.2). The total resistance is then

$$R_T = R_1 + R_3 + NR_2$$
$$= 4.7 \text{ k}\Omega + 2.2 \text{ k}\Omega + (3)(1 \text{ k}\Omega) = \textbf{9.9 k}\boldsymbol{\Omega}$$

Analogies

Throughout the text, analogies will be used to help explain some of the important fundamental relationships in electrical circuits. An analogy is simply a combination of elements of a different type that are helpful in explaining a particular concept, relationship, or equation.

Two analogies that work well for the series combination of elements are connecting different lengths of rope together to make the rope longer, as shown in Fig. 5.9(a). Adjoining pieces of rope are connected at only one point, satisfying the definition of series elements. Connecting a third rope to the common point, as shown in Fig. 5.9(b), would mean that none of the sections of rope are in series any longer.

Another analogy is the connecting of hoses together to form a longer hose, as shown in Fig. 5.9(c). Again, there is only one connection point between adjoining sections, resulting in a series connection.

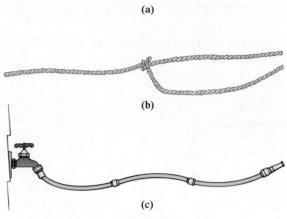

FIG. 5.9

Series circuit analogies: (a) series connection of ropes; (b) three ropes not in series; (c) three hoses in series.

Instrumentation

The total resistance of any configuration can be measured by simply connecting an ohmmeter across the access terminals as shown in Fig. 5.10 for the circuit of Fig. 5.1. **Since there is no polarity associated with resistance,** either lead can be connected to point *a*, with the other lead connected to point *b*. Choose a scale that will exceed the total resistance of the circuit, and remember when you read the response on the meter, if a kilohm scale was selected, the result will be in kilohms. For Fig. 5.10, the 200 Ω scale of our chosen multimeter was used because the total resistance is 140 Ω. If the 2 kΩ scale of our meter were selected, the digital display would read 0.140, and the user must recognize that the result is in kilohms.

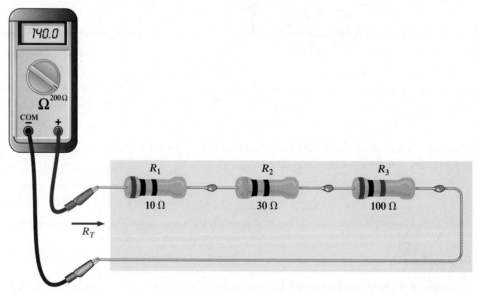

FIG. 5.10

Using an ohmmeter to measure the total resistance of a series circuit.

In the next section, another method for determining the total resistance of a circuit will be introduced using Ohm's law.

5.3 SERIES CIRCUITS

If we now take an 8.4 V dc supply and connect it in series with the series resistors of Fig. 5.1, we have the **series circuit** of Fig. 5.11(a).

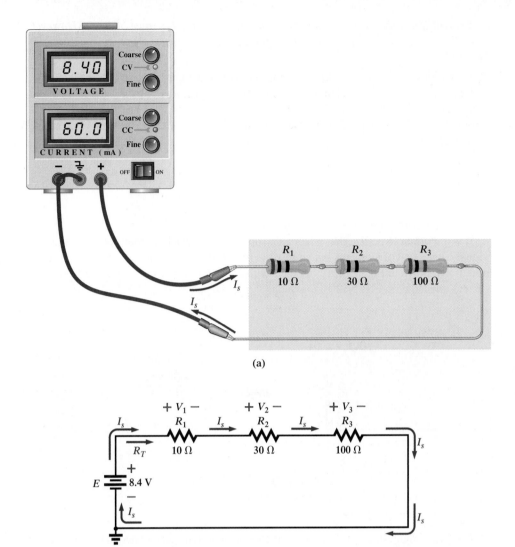

FIG. 5.11

*Applying a laboratory dc supply set at 8.4 V to the series circuit of Fig. 5.1: (a) actual circuit;
(b) schematic representation showing that the current is the same at every point.*

*A circuit is any combination of elements that will result in a flow of charge, or current,
in the configuration.*

The schematic representation of Fig. 5.11(a) appears as Fig. 5.11(b). Note the connection
from the negative terminal to ground in both figures to ground the configuration.

First recognize that the **dc supply is also a two-terminal device** with two points to be connected. If we simply ensure that there is only one connection made at each end of the supply
to the series combination of resistors, we can be sure that we have established a series circuit.

The manner in which the supply is connected will determine the direction of the resulting conventional current. For series dc circuits:

*the direction of conventional current in a series dc circuit is from the negative to the
positive terminal of the supply, that is, leaving the positive terminal and returning to
the negative terminal, as shown in Fig. 5.11.*

The next statement is probably one of the most important to remember when it comes to
analyzing series circuits and defining elements that are in series:

The current is the same at every point in a series circuit.

For the circuit of Fig. 5.11, the above statement dictates that the current is the same through
the three resistors and the voltage source. In addition, if you are ever concerned about whether
two elements are in series, simply check whether the current is the same through each element.

In any configuration, if two elements are in series, the current must be same. However, if the current is the same for two adjoining elements, the elements may or may not be in series.

The need for this constraint in the last sentence will be demonstrated in a later chapter.

Now that we have a complete circuit and current has been established, the level of current and the voltage across each resistor should be determined. This is accomplished by returning to Ohm's law and replacing the resistance in the equation by the total resistance of the circuit. That is,

$$I_s = \frac{E}{R_T} \qquad (5.3)$$

with the subscript s used to indicate source current.

It is important to realize that when a dc supply is connected, it does not "see" the individual connection of elements but simply the total resistance "seen" at the connection terminals, as shown in Fig. 5.12(a). In other words, it reduces the entire configuration to one such as appearing in Fig. 5.12(b) to which Ohm's law can easily be applied.

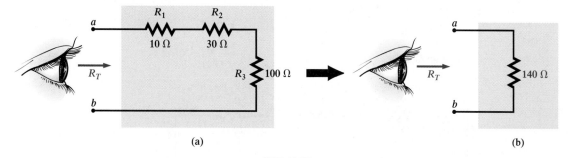

(a)

(b)

FIG. 5.12

Resistance "seen" at the terminals of a series circuit.

For the configuration of Fig. 5.11, using the total resistance calculated in the last section, the resulting current is

$$I_s = \frac{E}{R_T} = \frac{8.4 \text{ V}}{140 \, \Omega} = 0.06 \text{ A} = \textbf{60 mA}$$

as also indicated in Fig. 5.11. Note that the current at every point or corner of the network is the same. Furthermore, note that the current is also indicated on the current display of the power supply.

Now that we have the current level, we can calculate the voltage across each resistor. First recognize that

the polarity of the voltage across a resistor is determined by the direction of the current.

Current entering a resistor creates a drop in voltage with the polarity indicated in Fig. 5.13(a). Reverse the direction of the current, and the polarity will reverse as shown in Fig. 5.13(b). Change the orientation of the resistor, and the same rules apply as shown in Fig. 5.13(c). Applying the above to the circuit of Fig. 5.11(b) will result in the indicated polarities.

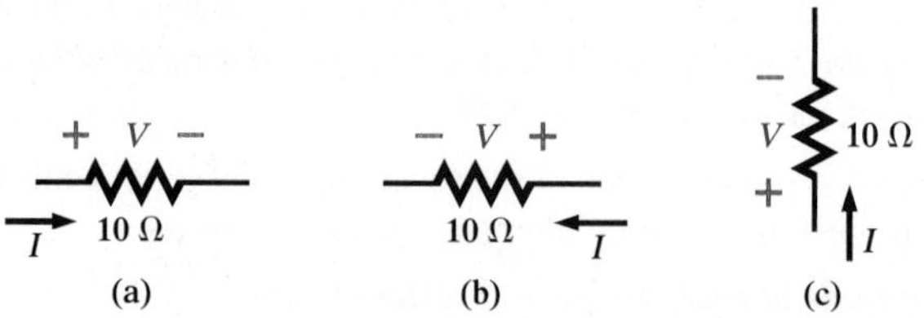

+ V −
⟶ I 10 Ω

(a)

− V +
10 Ω ⟵ I

(b)

− V 10 Ω
+ ↑ I

(c)

FIG. 5.13

Inserting the polarities across a resistor as determined by the direction of the current.

The magnitude of the voltage drop across each resistor can then be found by simply applying Ohm's law using only the resistance of each resistor. That is,

$$\boxed{\begin{array}{l} V_1 = I_1R_1 \\ V_2 = I_2R_2 \\ V_3 = I_3R_3 \end{array}} \qquad (5.4)$$

which for Fig. 5.11 results in

$$V_1 = I_1R_1 = I_sR_1 = (60 \text{ mA})(10 \text{ } \Omega) = \mathbf{0.6 \text{ V}}$$

$$V_2 = I_2R_2 = I_sR_2 = (60 \text{ mA})(30 \text{ } \Omega) = \mathbf{1.8 \text{ V}}$$

$$V_3 = I_3R_3 = I_sR_3 = (60 \text{ mA})(100 \text{ } \Omega) = \mathbf{6.0 \text{ V}}$$

Note that in all the numerical calculations appearing in the text thus far, a unit of measurement has been applied to each calculated quantity. Always remember that a quantity without a unit of measurement is often meaningless.

EXAMPLE 5.4 For the series circuit of Fig. 5.14:

a. Find the total resistance R_T.
b. Calculate the resulting source current I_s.
c. Determine the voltage across each resistor.

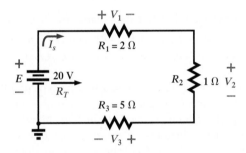

FIG. 5.14
Series circuit to be investigated in Example 5.4.

Solutions:

a. $R_T = R_1 + R_2 + R_3$
   $\quad = 2 \text{ } \Omega + 1 \text{ } \Omega + 5 \text{ } \Omega$
   $R_T = \mathbf{8 \text{ } \Omega}$

b. $I_s = \dfrac{E}{R_T} = \dfrac{20 \text{ V}}{8 \text{ } \Omega} = \mathbf{2.5 \text{ A}}$

c. $V_1 = I_1R_1 = I_sR_1 = (2.5 \text{ A})(2 \text{ } \Omega) = \mathbf{5 \text{ V}}$
   $\quad V_2 = I_2R_2 = I_sR_2 = (2.5 \text{ A})(1 \text{ } \Omega) = \mathbf{2.5 \text{ V}}$
   $\quad V_3 = I_3R_3 = I_sR_3 = (2.5 \text{ A})(5 \text{ } \Omega) = \mathbf{12.5 \text{ V}}$

EXAMPLE 5.5 For the series circuit of Fig. 5.15:

a. Find the total resistance R_T.
b. Determine the source current I_s and indicate its direction on the circuit.
c. Find the voltage across resistor R_2 and indicated its polarity on the circuit.

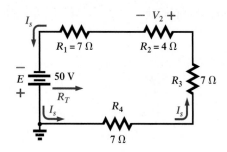

FIG. 5.15

Series circuit to be analyzed in Example 5.5.

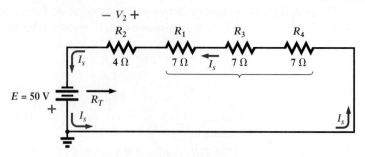

FIG. 5.16

Circuit of Fig. 5.15 redrawn to permit the use of Eq. (5.2).

Solutions:

a. The elements of the circuit are rearranged as shown in Fig. 5.16.

$$R_T = R_2 + NR$$

$$= 4\ \Omega + (3)(7\ \Omega)$$

$$= 4\ \Omega + 21\ \Omega$$

$$R_T = \mathbf{25\ \Omega}$$

b. Note that because of the manner in which the dc supply was connected, the current now has a counterclockwise direction as shown in Fig. 5.16.

$$I_s = \frac{E}{R_T} = \frac{50\ \text{V}}{25\ \Omega} = \mathbf{2\ A}$$

c. The direction of the current will define the polarity for V_2 appearing in Fig. 5.16.

$$V_2 = I_2 R_2 = I_s R_2 = (2\ \text{A})(4\ \Omega) = \mathbf{8\ V}$$

Examples 5.4 and 5.5 are straightforward, substitution-type problems that are relatively easy to solve with some practice. Example 5.6, however, is another type of problem that re-quires both a firm grasp of the fundamental laws and equations and an ability to identify which quantity should be determined first. The best preparation for this type of exercise is simply to work through as many problems of this kind as possible.

EXAMPLE 5.6   Given R_T and I_3, calculate R_T and E for the circuit of Fig. 5.17.

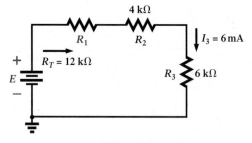

FIG. 5.17

Series circuit to be analyzed in Example 5.6.

Solution: Since we are given the total resistance, it seems natural to first write the equa-tion for the total resistance and then insert what we know.

$$R_T = R_1 + R_2 + R_3$$

We find that there is only one unknown which can be determined with some simple math-ematical manipulations. That is,

$$12\ \text{k}\Omega = R_1 + 4\ \text{k}\Omega + 6\ \text{k}\Omega = R_1 + 10\ \text{k}\Omega$$

and $12\ \text{k}\Omega - 10\ \text{k}\Omega = R_1$

so that $R_1 = \mathbf{2\ k\Omega}$

The dc voltage can be determined directly from Ohm's law.

$$E = I_s R_T = I_3 R_T = (6 \text{ mA})(12 \text{ k}\Omega) = \mathbf{72 \text{ V}}$$

Analogies

The analogies used earlier to define the series connection are also excellent for the current of a series circuit. For instance, for the series-connected ropes in Fig. 5.18(a), the stress on each rope **is the same** as they try to hold the heavy weight. In Fig. 5.18(b), the flow of water **is the same** through each section of hose as the water is carried to its destination.

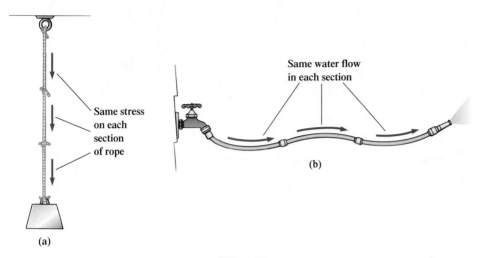

FIG. 5.18

Analogies for the current in a series circuit: (a) series connection of ropes with a load;
(b) series connection of hoses.

Instrumentation

It is important to realize the following:

The insertion of any meter in a circuit will affect the circuit.

It is therefore important to use meters that will minimize the impact on the response of the circuit. The loading effects of meters will be discussed in detail in a later section of this chapter. For now, we will assume that the meters are ideal and do not affect the networks to which they are applied so that we can concentrate on their proper usage.

Further, it is particularly helpful in the laboratory to realize that

the voltages of a circuit can be measured without disturbing (breaking the connections in) the circuit.

In Fig. 5.19, all the voltages of the circuit of Fig. 5.11 are being measured by voltmeters that were connected without disturbing the original configuration. Note that all the voltmeters are placed **across** the resistive elements. In addition, note that the positive (normally red) lead of the voltmeter is connected to the point of higher potential (positive sign), with the negative (normally black) lead of the voltmeter connected to the point of lower potential (negative sign) for V_1 and V_2. The result is a positive reading on the display. If the leads were reversed, the magnitude would remain the same, but a negative sign would appear as shown for V_3.

Take special note that the 20 V scale of our meter was used to measure the -6 V level, while the 2 V scale of our meter was used to measure the 0.6 V and 1.8 V levels. The maximum value of the chosen scale must always exceed the maximum value to be measured. In general,

when using a voltmeter, start with a scale that will ensure that the reading is less than the maximum value of the scale. Then work your way down in scales until the reading with the highest level of precision is obtained.

If we now turn our attention to the current of the circuit, we find that

using an ammeter to measure the current of a circuit requires that the circuit be broken at some point and the meter inserted in series with the branch in which the current is to be determined.

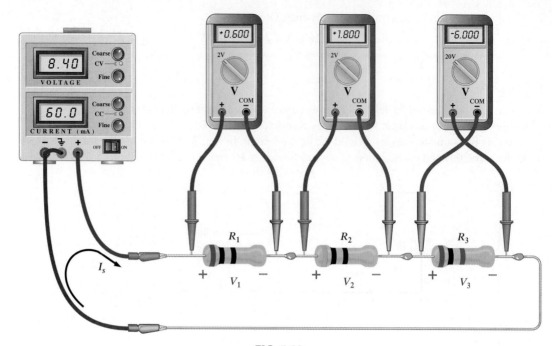

FIG. 5.19
Using voltmeters to measure the voltages across the resistors of Fig. 5.11.

For instance, in order to measure the current leaving the positive terminal of the supply, the connection to the positive terminal must be removed to create an open circuit between the supply and resistor R_1. The ammeter would then be inserted between these two points to form a bridge between the supply and the first resistor, as shown in Fig. 5.20. The ammeter is now in series with the supply and the other elements of the circuit. If each meter is to provide a positive reading, the connection must be made such that conventional current enters the positive terminal of the meter and leaves the negative terminal. This was done for three of the ammeters, with the ammeter to the right of R_3 connected in the reverse manner. The result is a negative sign for the current. However, also note that the current has the correct magnitude. Since the current is 60 mA, the 200 mA scale of our meter was employed for each meter.

As expected, the current at each point in the series circuit is the same using our ideal ammeters.

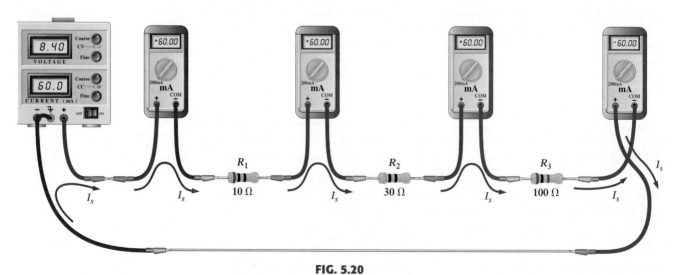

FIG. 5.20
Measuring the current throughout the series circuit of Fig. 5.11.

5.4   POWER DISTRIBUTION IN A SERIES CIRCUIT

In any electrical system, the power applied will equal the power dissipated or absorbed. For any series circuit, such as that of Fig. 5.21,

the power applied by the dc supply must equal that dissipated by the resistive elements.

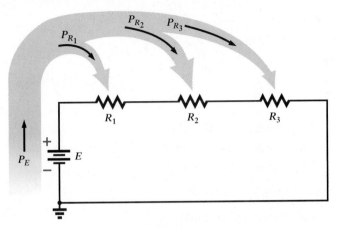

FIG. 5.21
Power distribution in a series circuit.

In equation form,

$$P_E = P_{R_1} + P_{R_2} + P_{R_3} \tag{5.5}$$

The power delivered by the supply can be determined using

$$P_E = EI_s \qquad \text{(watts, W)} \tag{5.6}$$

The power dissipated by the resistive elements can be determined by any of the following forms (shown for resistor R_1 only):

$$P_1 = V_1 I_1 = I_1^2 R_1 = \frac{V_1^2}{R_1} \qquad \text{(watts, W)} \tag{5.7}$$

Since the current is the same through series elements, you will find in the examples to follow that

in a series configuration, maximum power is delivered to the largest resistor.

EXAMPLE 5.7   For the series circuit of Fig. 5.22 (all standard values):

a. Determine the total resistance R_T.
b. Calculate the current I_s.

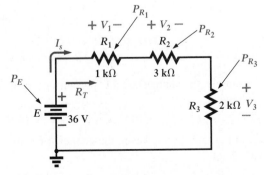

FIG. 5.22
Series circuit to be investigated in Example 5.7.

c. Determine the voltage across each resistor.
d. Find the power supplied by the battery.
e. Determine the power dissipated by each resistor.
f. Comment on whether the total power supplied equals the total power dissipated.

Solutions:

a. $R_T = R_1 + R_2 + R_3$

$\quad = 1\ k\Omega + 3\ k\Omega + 2\ k\Omega$

$R_T = \textbf{6 k}\boldsymbol{\Omega}$

b. $I_s = \dfrac{E}{R_T} = \dfrac{36\ V}{6\ k\Omega} = \textbf{6 mA}$

c. $V_1 = I_1 R_1 = I_s R_1 = (6\ mA)(1\ k\Omega) = \textbf{6 V}$

$\quad V_2 = I_2 R_2 = I_s R_2 = (6\ mA)(3\ k\Omega) = \textbf{18 V}$

$\quad V_3 = I_3 R_3 = I_s R_3 = (6\ mA)(2\ k\Omega) = \textbf{12 V}$

d. $P_E = EI_s = (36\ V)(6\ mA) = \textbf{216 mW}$

e. $P_1 = V_1 I_1 = (6\ V)(6\ mA) = \textbf{36 mW}$

$\quad P_2 = I_2^2 R_2 = (6\ mA)^2(3\ k\Omega) = \textbf{108 mW}$

$\quad P_3 = \dfrac{V_3^2}{R_3} = \dfrac{(12\ V)^2}{2\ k\Omega} = \textbf{72 mW}$

f. $P_E = P_{R_1} + P_{R_2} + P_{R_3}$

$\quad 216\ mW = 36\ mW + 108\ mW + 72\ mW = \textbf{216 mW}$ (checks)

5.5 VOLTAGE SOURCES IN SERIES

Voltage sources can be connected in series, as shown in Fig. 5.23, to increase or decrease the total voltage applied to a system. The net voltage is determined by simply summing the sources with the same polarity and subtracting the total of the sources with the opposite polarity. The net polarity is the polarity of the larger sum.

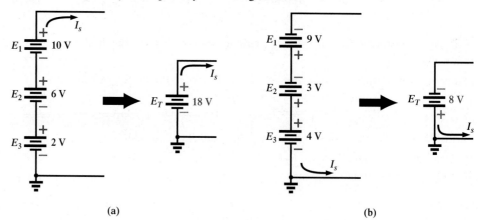

FIG. 5.23

Reducing series dc voltage sources to a single source.

In Fig. 5.23(a), for example, the sources are all "pressuring" current to follow a clockwise path, so the net voltage is

$$E_T = E_1 + E_2 + E_3 = 10\ V + 6\ V + 2\ V = \textbf{18 V}$$

as shown in the figure. In Fig. 5.23(b), however, the greater "pressure" is in the clockwise direction with a net voltage of

$$E_T = E_1 + E_2 - E_3 = 9\ V + 3\ V - 4\ V = \textbf{8 V}$$

and with the polarity shown in the figure.

The connection of batteries in series to obtain a higher voltage is common in much of to-day's portable electronic equipment. For example, in Fig. 5.24(a), four 1.5 V AAA batteries have been connected in series to obtain a source voltage of 6 V. Although the voltage has increased, keep in mind that the maximum current for each AAA battery and for the 6 V supply is still the same. However, the power available has increased by a factor of 4 due to the increase in terminal voltage. Note also, as mentioned in Chapter 2, that the negative end of each battery is connected to the spring, and the positive end to the solid contact. In addition, note how the connection is made between batteries using the horizontal connecting tabs.

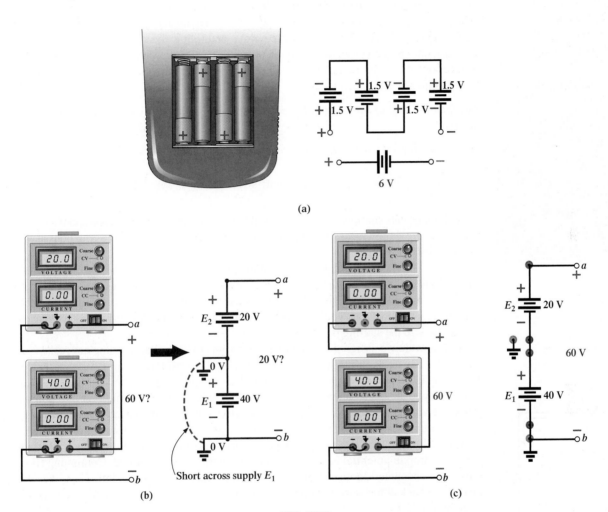

FIG. 5.24

Series connection of dc supplies: (a) four 1.5 V batteries in series to establish a terminal voltage of 6 V; (b) incorrect connections for two series dc supplies; (c) correct connection of two series supplies to establish 60 V at the output terminals.

In general, supplies with only two terminals (+ and −) can be connected as shown for the batteries. A problem arises, however, if the supply has an optional or fixed internal ground connection. In Fig. 5.24(b), two laboratory supplies have been connected in series with both grounds connected. The result is a shorting out of the lower source E_1 (which may damage the supply if the protective fuse does not activate quickly enough) because both grounds are at zero potential. In such cases the supply E_2 must be left ungrounded (floating), as shown in Fig. 5.24(c), to provide the 60 V terminal voltage. If the laboratory supplies have an internal connection from the negative terminal to ground as a protective feature for the users, a series connection of supplies cannot be made. Be aware of this fact, because some educational institutions will add an internal ground to the supplies as a protective feature even though the panel still displays the ground connection as an optional feature.

Courtesy of the Smithsonian
Institution, Photo No. 58,283.

FIG. 5.25

Gustav Robert Kirchhoff.

German (Königsberg, Berlin)
(1824–87)
Physicist
Professor of Physics, University of
Heildelberg

Although a contributor to a number of areas in the physics domain, he is best known for his work in the electrical area with his definition of the relationships between the currents and voltages of a network in 1847. Did extensive research with German chemist Robert Bunsen (developed the *Bunsen burner*), resulting in the discovery of the important elements of *cesium* and *rubidium*.

5.6 KIRCHHOFF'S VOLTAGE LAW

The law to be described in this section is one of the most important in this field. It has application not only to dc circuits but also to any type of signal—whether it be ac, digital, and so on. This law is far-reaching and can be very helpful in working out solutions to networks that sometimes leave us lost for a direction of investigation.

The law, called **Kirchhoff's voltage law (KVL),** was developed by Gustav Kirchhoff (Fig. 5.25) in the mid-1800s, so it has been around for quite a while and, in fact, will never be outdated or replaced—it is a cornerstone of the entire field.

The application of the law requires that we define a closed path of investigation, permitting us to start at one point in the network, travel through the network, and find our way back to the original starting point. The path does not have to be circular, square, or any other defined shape; it must simply provide a way to leave a point and get back to it without leaving the network. In Fig. 5.26, if we leave point a and follow the current, we will end up at point b. Continuing, we can pass through points c and d and eventually return through the voltage source to point a, our starting point. The path $abcda$ is therefore a closed path, or **closed loop.** The law specifies that

the algebraic sum of the potential rises and drops around a closed path (or closed loop) is zero.

In symbolic form it can be written as

$$\Sigma_C\, V = 0 \qquad \text{(Kirchhoff's voltage law in symbolic form)} \qquad \textbf{(5.8)}$$

where Σ represents summation, $\circlearrowright$ the closed loop, and V the potential drops and rises. The term *algebraic* simply means paying attention to the signs that result in the equations as we add and subtract terms.

The first question that often arises is, Which way should I go around the closed path? Should I always follow the direction of the current? To simplify matters, this text will always try to move in a clockwise direction. By selecting a direction, you eliminate the need to think about which way would be more appropriate. Any direction will work as long as you get back to the starting point.

Now, how do we assign positive or negative signs to the voltages of the network? Going from a negative sign to a positive sign is usually considered a positive experience. Therefore, if we pass through a voltage with this sequence of polarities, it is given a plus sign; the opposite is given a negative sign. In Fig. 5.26, as we go from point d to point a across the voltage source, we move from a negative potential (the negative sign) to a positive potential (the positive sign), so a positive sign is given to the source voltage E. As we proceed from point a to point b, we encounter a positive sign followed by a negative sign, so a drop in potential has occurred and a negative sign is applied. Continuing from b to c, we encounter another drop in potential, so another negative sign is applied. We then arrive back at the starting point (d), and the resulting sum is set equal to zero as defined by Eq. (5.8).

Writing out the sequence with the voltages and the signs will result in the following:

$$+E - V_1 - V_2 = 0$$

which can be rewritten as $\qquad E = V_1 + V_2$

The result is particularly interesting because it tells us that

the applied voltage of a series dc circuit will equal the sum of the voltage drops of the circuit.

Kirchhoff's voltage law can also be written in the following form:

$$\Sigma_C\, V_{\text{rises}} = \Sigma_C\, V_{\text{drops}} \qquad \textbf{(5.9)}$$

revealing that

the sum of the voltage rises around a closed path will always equal the sum of the voltage drops.

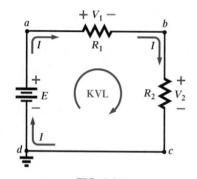

FIG. 5.26

Applying Kirchhoff's voltage law to a series dc circuit.

To demonstrate that the direction that you take around the loop will have no effect on the results, let's take the counterclockwise path and compare results. The resulting sequence will appear as

$$-E + V_2 + V_1 = 0$$

yielding the same result of

$$E = V_1 + V_2$$

EXAMPLE 5.8 Use Kirchhoff's voltage law to determine the unknown voltage of Fig. 5.27.

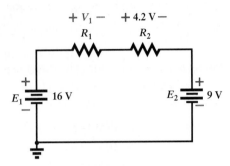

FIG. 5.27
Series circuit to be examined in Example 5.8.

Solution: When applying Kirchhoff's voltage law, be sure to concentrate on the polarities of the voltage rise or drop rather than on the type of element. In other words, do not treat a voltage drop across a resistive element differently from a voltage rise (or drop) across a source. If the polarity dictates that a drop has occurred, that is the important fact, not whether it is a resistive element or source.

Application of Kirchhoff's voltage law to the circuit of Fig. 5.27 in the clockwise direction will result in

$$+E_1 - V_1 - V_2 - E_2 = 0$$

and

$$V_1 = E_1 - V_2 - E_2$$
$$= 16\,V - 4.2\,V - 9\,V$$

so

$$V_1 = \textbf{2.8 V}$$

The result clearly indicates that there is no need to know the values of the resistors or the current to determine the unknown voltage. Sufficient information was carried by the other voltage levels to permit a determination of the unknown.

EXAMPLE 5.9 Determine the unknown voltage for the circuit of Fig. 5.28.

Solution: In this case, the unknown voltage is not across a single resistive element but between two arbitrary points in the circuit. Simply apply Kirchhoff's voltage law around a path, including the source or resistor R_3. For the clockwise path, including the source, the resulting equation is the following:

$$+E - V_1 - V_x = 0$$

and

$$V_x = E - V_1 = 32\,V - 12\,V = \textbf{20 V}$$

For the clockwise path, including resistor R_3, the following will result:

$$+V_x - V_2 - V_3 = 0$$

and

$$V_x = V_2 + V_3$$
$$= 6\,V + 14\,V$$

with

$$V_x = \textbf{20 V}$$

providing exactly the same solution.

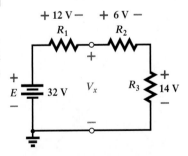

FIG. 5.28
Series dc circuit to be analyzed in Example 5.9.

The next example will demonstrate that it is unnecessary to know what elements are inside a container when applying Kirchhoff's voltage law. They could all be voltage sources or a mix of sources and resistors. It doesn't matter—simply pay strict attention to the polarities encountered.

There is no requirement that the followed path have charge flow or current. In the next example the current is zero everywhere, but Kirchhoff's voltage law can still be applied to determine the voltage between the points of interest. Also, there will be situations where the actual polarity will not be provided. In such cases, simply assume a polarity. If the answer is negative, the magnitude of the result is correct, but the polarity should be reversed.

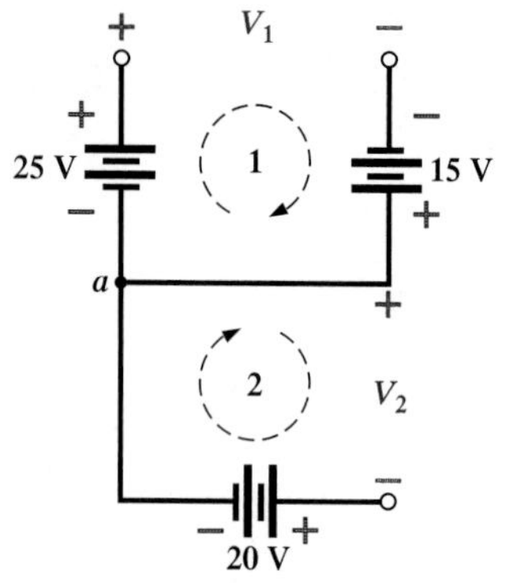

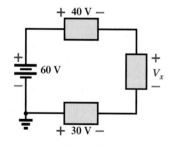

FIG. 5.29

Combination of voltage sources to be examined in Example 5.10.

EXAMPLE 5.10 Using Kirchhoff's voltage law, determine voltages V_1 and V_2 for the network of Fig. 5.29.

Solution: For path 1, starting at point a in a clockwise direction,
$$+25\,V - V_1 + 15\,V = 0$$
and
$$V_1 = \textbf{40 V}$$
For path 2, starting at point a in a clockwise direction,
$$-V_2 - 20\,V = 0$$
and
$$V_2 = \textbf{-20 V}$$

The minus sign in the solution simply indicates that the actual polarities are different from those assumed.

The description will now continue with a few examples. It would be wise to try to find the unknown quantities without looking at the solutions. It will help define where you may be having trouble.

The next example will emphasize the fact that when you are applying Kirchhoff's voltage law, it is the polarities of the voltage rise or drop that are the important parameters, not the type of element involved.

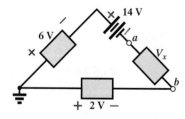

FIG. 5.30

Series configuration to be examined in Example 5.11.

EXAMPLE 5.11 Using Kirchhoff's voltage law, determine the unknown voltage for the circuit of Fig. 5.30.

Solution: Note that in this circuit, there are various polarities across the unknown elements since they can contain any mixture of components. Applying Kirchhoff's voltage law in the clockwise direction will result in
$$+60\,V - 40\,V - V_x + 30\,V = 0$$
and
$$V_x = 60\,V + 30\,V - 40\,V = 90\,V - 40\,V$$
with
$$V_x = \textbf{50 V}$$

EXAMPLE 5.12 Determine the voltage V_x for the circuit of Fig. 5.31. Note that the polarity of V_x was not provided.

Solution: For cases where the polarity is not included, simply make an assumption about the polarity, and apply Kirchhoff's voltage law as before. If the result has a positive sign, the assumed polarity was correct. If the result has a minus sign, the **magnitude is correct,** but the assumed polarity must be reversed. In this case, if we assume point a to be positive and point b to be negative, an application of Kirchhoff's voltage law in the clockwise direction will result in
$$-6\,V - 14\,V - V_x + 2\,V = 0$$
and
$$V_x = -20\,V + 2\,V$$
so that
$$V_x = \textbf{-18 V}$$

Since the result is negative, we know that point a should be negative and point b should be positive, but the magnitude of 18 V is correct.

FIG. 5.31

Applying Kirchhoff's voltage law to a circuit in which the polarities have not been provided for one of the voltages (Example 5.12).

EXAMPLE 5.13   For the series circuit of Fig. 5.32:

a. Determine V_2 using Kirchhoff's voltage law.
b. Determine current I_2.
c. Find R_1 and R_3.

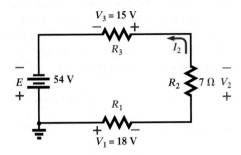

FIG. 5.32
Series configuration to be examined in Example 5.13.

Solutions:

a. Applying Kirchhoff's voltage law in the clockwise direction starting at the negative terminal of the supply will result in

$$-E + V_3 + V_2 + V_1 = 0$$

and $$E = V_1 + V_2 + V_3 \quad \text{(as expected)}$$

so that $$V_2 = E - V_1 - V_3 = 54\text{ V} - 18\text{ V} - 15\text{ V}$$

and $$V_2 = \textbf{21 V}$$

b. $I_2 = \dfrac{V_2}{R_2} = \dfrac{21\text{ V}}{7\ \Omega}$

$I_2 = \textbf{3 A}$

c. $R_1 = \dfrac{V_1}{I_1} = \dfrac{18\text{ V}}{3\text{ A}} = \textbf{6 }\boldsymbol{\Omega}$

with $R_3 = \dfrac{V_3}{I_3} = \dfrac{15\text{ V}}{3\text{ A}} = \textbf{5 }\boldsymbol{\Omega}$

5.7   VOLTAGE DIVISION IN A SERIES CIRCUIT

The previous section clearly demonstrated that the sum of the voltages across the resistors
of a series circuit will always equal the applied voltage. It cannot be more or less than that
value. The next question is, How will a resistor's value affect the voltage across the resistor? It turns out that

the voltage across series resistive elements will divide as the magnitude of the
resistance levels.

In other words,

in a series resistive circuit, the larger the resistance, the more of the applied voltage it
will capture.

In addition,

the ratio of the voltages across series resistors will be the same as the ratio of their
resistance levels.

All of the above statements can best be described by a few simple examples. In Fig. 5.33,
all the voltages across the resistive elements are provided. The largest resistor of 6 Ω cap-
tures the bulk of the applied voltage, while the smallest resistor, R_3, has the least. In addition, note that since the resistance level of R_1 is six times that of R_3, the voltage across R_1 is

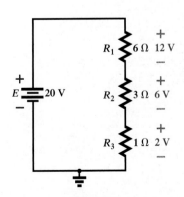

FIG. 5.33
Revealing how the voltage will divide
across series resistive elements.

six times that of R_3. The fact that the resistance level of R_2 is three times that of R_1 results in three times the voltage across R_2. Finally, since R_1 is twice R_2, the voltage across R_1 is twice that of R_2. In general, therefore, the voltage across series resistors will have the same ratio as their resistance levels.

It is particularly interesting to note that if the resistance levels of all the resistors of Fig. 5.33 are increased by the same amount, as shown in Fig. 5.34, the voltage levels will all remain the same. In other words, even though the resistance levels were increased by a factor of 1 million, the voltage ratios remained the same. Clearly, therefore, it is the ratio of resistor values that counts when it comes to voltage division, not the magnitude of the resistors. The current level of the network will be severely affected by this change in resistance level, but the voltage levels remain unaffected.

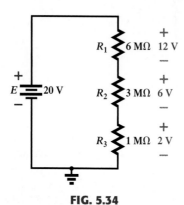

FIG. 5.34

The ratio of the resistive values determines the voltage division of a series dc circuit.

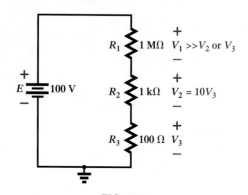

FIG. 5.35

The largest of the series resistive elements will capture the major share of the applied voltage.

Based on the above, it should now be clear that when you first encounter a circuit such as that of Fig. 5.35, you will expect that the voltage across the 1 MΩ resistor will be much greater than that across the 1 kΩ or the 100 Ω resistors. In addition, based on a statement above, the voltage across the 1 kΩ resistor will be 10 times as great as that across the 100 Ω resistor since the resistance level is 10 times as much. Certainly, you would expect that very little voltage will be left for the 100 Ω resistor. Note that the current was never mentioned in the above analysis. The distribution of the applied voltage is determined solely by the ratio of the resistance levels. Of course, the magnitude of the resistors will determine the resulting current level.

To continue with the above, since 1 MΩ is 1000 times larger than 1 kΩ, voltage V_1 will be 1000 times larger than V_2. In addition, voltage V_2 will be 10 times larger than V_3. Finally, the voltage across the largest resistor of 1 MΩ will be $(10)(1000) = 10,000$ times larger than V_3.

Now for some details. The total resistance is

$$R_T = R_1 + R_2 + R_3$$
$$= 1 \text{ M}\Omega + 1 \text{ k}\Omega + 100 \ \Omega$$
$$R_T = \textbf{1,001,100 } \boldsymbol{\Omega}$$

The current is

$$I_s = \frac{E}{R_T} = \frac{100 \text{ V}}{1,001,100 \ \Omega} \cong 99.89 \ \mu A \quad (\text{about } 100 \ \mu A)$$

with

$$V_1 = I_1 R_1 = I_s R_1 = (99.89 \ \mu A)(1 \text{ M}\Omega) = \textbf{99.89 V} \quad (\text{almost the full } 100 \text{ V})$$
$$V_2 = I_2 R_2 = I_s R_2 = (99.89 \ \mu A)(1 \text{ k}\Omega) = \textbf{99.89 mV} \quad (\text{about } 100 \text{ mV})$$
$$V_3 = I_3 R_3 = I_s R_3 = (99.89 \ \mu A)(100 \ \Omega) = \textbf{9.989 mV} \quad (\text{about } 10 \text{ mV})$$

Clearly, as illustrated above, the major part of the applied voltage is across the 1 MΩ resistor. The current is in the microampere due primarily to the large 1 MΩ resistor. Voltage V_2 is about 0.1 V compared to almost 100 V for V_1. The voltage across R_3 is only about 10 mV, or 0.010 V.

For the future, therefore, before making any detailed, lengthy calculations, examine the resistance levels of the series resistors to develop some idea as to how the applied voltage will be divided through the circuit. It will reveal, with a minimum amount of effort, what you should expect when performing the calculations (a checking mechanism). It also allows you to speak intelligently about the response of the circuit without having to resort to any calculations at all.

Voltage Divider Rule (VDR)

The **voltage divider rule (VDR)** permits the determination of the voltage across a series resistor without first having to determine the current of the circuit. The rule itself can be derived by analyzing the simple series circuit of Fig. 5.36.

First, determine the total resistance as follows:

$$R_T = R_1 + R_2$$

Then

$$I_s = I_1 = I_2 = \frac{E}{R_T}$$

Apply Ohm's law to each resistor:

$$V_1 = I_1 R_1 = \left(\frac{E}{R_T}\right) R_1 = R_1 \frac{E}{R_T}$$

$$V_2 = I_2 R_2 = \left(\frac{E}{R_T}\right) R_2 = R_2 \frac{E}{R_T}$$

The resulting format for V_1 and V_2 is

$$\boxed{V_x = R_x \frac{E}{R_T}} \qquad \text{(voltage divider rule)} \qquad \textbf{(5.10)}$$

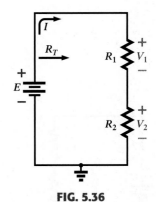

FIG. 5.36

Developing the voltage divider rule.

where V_x is the voltage across the resistor R_x, E is the impressed voltage across the series elements, and R_T is the total resistance of the series circuit.

In words, the voltage divider rule states that

the voltage across a resistor in a series circuit is equal to the value of that resistor times the total applied voltage divided by the total resistance of the series configuration.

Although Equation (5.10) was derived using a series circuit of only two elements, it can be used for series circuits with any number of series resistors.

EXAMPLE 5.14   For the series circuit of Fig. 5.37:

a. Without making any calculations, how much larger would you expect the voltage across R_2 to be compared to that across R_1?
b. Find the voltage V_1 using only the voltage divider rule.
c. Using the conclusion of part (a), determine the voltage across R_2.
d. Use the voltage divider rule to determine the voltage across R_2, and compare your answer to your conclusion in part (c).
e. How does the sum of V_1 and V_2 compare to the applied voltage?

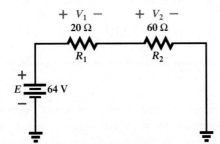

FIG. 5.37

Series circuit to be examined using the voltage divider rule in Example 5.14.

Solutions:

a. Since resistor R_2 is three times R_1, it is expected that $V_2 = 3V_1$.

b. $V_1 = R_1 \dfrac{E}{R_T} = 20 \, \Omega \left(\dfrac{64 \text{ V}}{20 \, \Omega + 60 \, \Omega} \right) = 20 \, \Omega \left(\dfrac{64 \text{ V}}{80 \, \Omega} \right) = \mathbf{16 \text{ V}}$

c. $V_2 = 3V_1 = 3(16 \text{ V}) = \mathbf{48 \text{ V}}$

d. $V_2 = R_2 \dfrac{E}{R_T} = (60 \, \Omega) \left(\dfrac{64 \text{ V}}{80 \, \Omega} \right) = \mathbf{48 \text{ V}}$

The results are an exact match.

e. $E = V_1 + V_2$

$64 \text{ V} = 16 \text{ V} + 48 \text{ V} = 64 \text{ V}$ (checks)

EXAMPLE 5.15 Using the voltage divider rule, determine voltages V_1 and V_3 for the series circuit of Fig. 5.38.

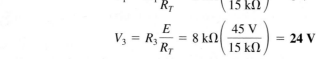

Solution:

$$R_T = R_1 + R_2 + R_3$$
$$= 2 \text{ k}\Omega + 5 \text{ k}\Omega + 8 \text{ k}\Omega$$
$$R_T = 15 \text{ k}\Omega$$

$$V_1 = R_1 \frac{E}{R_T} = 2 \text{ k}\Omega \left(\frac{45 \text{ V}}{15 \text{ k}\Omega} \right) = \mathbf{6 \text{ V}}$$

and

$$V_3 = R_3 \frac{E}{R_T} = 8 \text{ k}\Omega \left(\frac{45 \text{ V}}{15 \text{ k}\Omega} \right) = \mathbf{24 \text{ V}}$$

FIG. 5.38

Series circuit to be investigated in Examples 5.15 and 5.16.

The voltage divider rule can be extended to the voltage across two or more series elements if the resistance in the numerator of Eq. (5.10) is expanded to include the total resistance of the series resistors across which the voltage is to be found (R'). That is,

$$\boxed{V' = R' \frac{E}{R_T}}$$

(5.11)

EXAMPLE 5.16 Determine the voltage (denoted V') across the series combination of resistors R_1 and R_2 in Fig. 5.38.

Solution: Since the voltage desired is across both R_1 and R_2, the sum of R_1 and R_2 will be substituted as R' in Eq. (5.11). The result is

$$R' = R_1 + R_2 = 2 \text{ k}\Omega + 5 \text{ k}\Omega = 7 \text{ k}\Omega$$

and

$$V' = R' \frac{E}{R_T} = 7 \text{ k}\Omega \left(\frac{45 \text{ V}}{15 \text{ k}\Omega} \right) = \mathbf{21 \text{ V}}$$

In the next example you are presented with a problem of the other kind: Given the voltage division, you must determine the required resistor values. In most cases, problems of this kind simply require that you be able to use the basic equations introduced thus far in the text.

EXAMPLE 5.17   Given the voltmeter reading of Fig. 5.39, find voltage V_3.

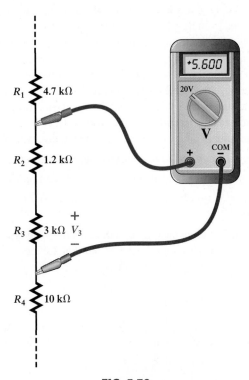

FIG. 5.39
Voltage divider action for Example 5.17.

Solution: Even though the rest of the network is not shown and the current level has not been determined, the voltage divider rule can be applied by using the voltmeter reading as the full voltage across the series combination of resistors. That is,

$$V_3 = \frac{R_3(V_{\text{meter}})}{R_3 + R_2} = \frac{3 \text{ k}\Omega(5.6 \text{ V})}{3 \text{ k}\Omega + 1.2 \text{ k}\Omega}$$

$$V_3 = \textbf{4 V}$$

EXAMPLE 5.18   Design the voltage divider circuit of Fig. 5.40 such that the voltage across R_1 will be four times the voltage across R_2; that is, $V_{R_1} = 4V_{R_2}$.

Solution: The total resistance is defined by

$$R_T = R_1 + R_2$$

However, if $$V_{R_1} = 4V_{R_2}$$

then $$R_1 = 4R_2$$

so that $$R_T = R_1 + R_2 = 4R_2 + R_2 = 5R_2$$

Applying Ohm's law, we can determine the total resistance of the circuit:

$$R_T = \frac{E}{I_s} = \frac{20 \text{ V}}{4 \text{ mA}} = 5 \text{ k}\Omega$$

so $$R_T = 5R_2 = 5 \text{ k}\Omega$$

and $$R_2 = \frac{5 \text{ k}\Omega}{5} = \textbf{1 k}\Omega$$

Then $$R_1 = 4R_2 = 4(1 \text{ k}\Omega) = \textbf{4 k}\Omega$$

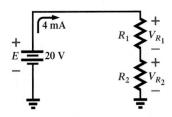

FIG. 5.40
Designing a voltage divider circuit (Example 5.18).

5.8 VOLTAGE REGULATION AND THE INTERNAL RESISTANCE OF VOLTAGE SOURCES

When you pick up a dc supply of any kind, you initially assume that it will provide the desired voltage for any resistive load you may hook up to the supply. In other words, if the battery is labeled 1.5 V or the supply is set at 20 V, you might assume that they will provide that voltage no matter what load we may apply. Unfortunately, this is not the case. For instance, if we apply a 1 kΩ resistor to a dc laboratory supply, it is fairly easy to set the voltage across the resistor to 20 V. However, if we remove the 1 kΩ resistor and replace it with a 100 Ω resistor and don't touch the controls on the supply at all, we find that the voltage has dropped to 19.14 V. Change the load to a 68 Ω resistor, and the terminal voltage drops to 18.72 V. Reality has set in: The load applied will affect the terminal voltage of the supply. In fact, this example points out that

a network should always be connected to a supply before the level of supply voltage is set.

The reason the terminal voltage drops with changes in load (current demand) is that

every practical (real-world) supply has an internal resistance in series with the idealized voltage source

as shown in Fig. 5.41(b). The resistance level will depend on the type of supply, but it will always be present. Every year new supplies come out that are less sensitive to the load applied, but even so, some sensitivity still remains.

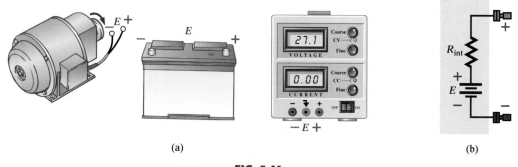

(a) (b)

FIG. 5.41

(a) Sources of dc voltage; (b) equivalent circuit.

The supply of Fig. 5.42 will help us explain the action that occurred above as we changed the load resistor. Due to the internal resistance of the supply, the ideal internal supply must be set to 20.1 V in Fig. 5.42(a) if 20 V are to appear across the 1 kΩ resistor. The internal resistance will capture 0.1 V of the applied voltage. The current in the circuit is determined by simply looking at the load and using Ohm's law; that is, $I_L = V_L/R_L = 20 \text{ V}/1 \text{ k}\Omega = 20 \text{ mA}$, which is a relatively low current.

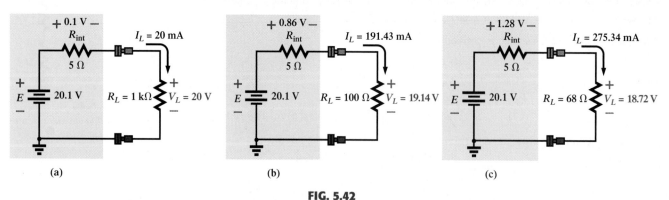

(a) (b) (c)

FIG. 5.42

Demonstrating the effect of changing a load on the terminal voltage of a supply.

In Fig. 5.42(b), all the settings of the supply are left untouched, but the 1 kΩ load is replaced by a 100 Ω resistor. The resulting current is now $I_L = E/R_T = 20.1 \text{ V}/105 \Omega = 191.43 \text{ mA}$, and the output voltage is $V_L = I_L R = (191.43 \text{ mA})(100 \Omega) = 19.14 \text{ V}$, a drop of 0.86 V. In Fig. 5.42(c), a 68 Ω load is applied, and the current increases substantially to 275.34 mA with a terminal voltage of only 18.72 V, which is a drop of 1.28 V from the expected level. Quite obviously, therefore, as the current drawn from the supply increases, the terminal voltage will continue to drop.

If we plot the terminal voltage versus current demand from 0 A to 275.34 mA, we obtain the plot of Fig. 5.43. Interestingly enough, it turns out to be a straight line that will continue to drop with an increase in current demand. Note, in particular, that the curve begins at a current level of 0 A. Under no-load conditions, where the output terminals of the supply are not connected to any load, the current will be 0 A due to the absence of a complete circuit. The output voltage will be the internal ideal supply level of 20.1 V.

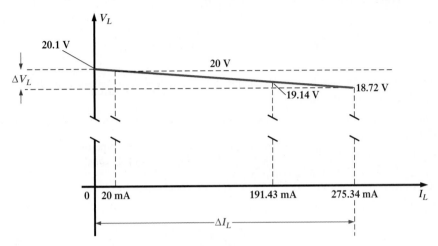

FIG. 5.43
Plotting V_L versus I_L for the supply of Fig. 5.42.

The slope of the line is defined by the internal resistance of the supply. That is,

$$R_{int} = \frac{\Delta V_L}{\Delta I_L} \qquad \text{(ohms, } \Omega\text{)} \qquad \textbf{(5.12)}$$

which for the plot of Fig. 5.43 results in

$$R_{int} = \frac{\Delta V_L}{\Delta I_L} = \frac{20.1 \text{ V} - 18.72 \text{ V}}{275.34 \text{ mA} - 0 \text{ mA}} = \frac{1.38 \text{ V}}{275.34 \text{ mA}} = \textbf{5 } \boldsymbol{\Omega}$$

For supplies of any kind, the plot of particular importance is the output voltage versus current drawn from the supply, as appearing in Fig. 5.44(a). Note that it is its maximum value under no-load (NL) conditions as defined by Fig. 5.44(b) and the description above. Full-load (FL) conditions are defined by the maximum current the supply can provide on a continuous basis, as shown in Fig. 5.44(c).

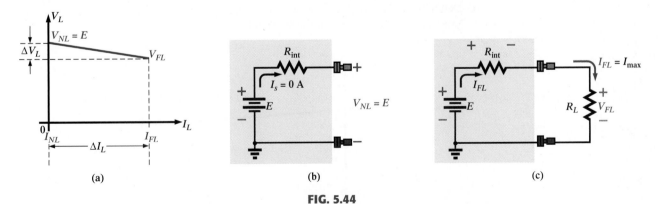

FIG. 5.44
Defining the properties of importance for a power supply.

As a basis for comparison, an ideal power supply and its response curve are provided in Fig. 5.45. Note the absence of the internal resistance and the fact that the plot is a horizontal line (no variation at all with load demand)—an impossible response curve. When we compare the curve of Fig. 5.45 with that of Fig. 5.44(a), however, we now realize that the **steeper the slope,**

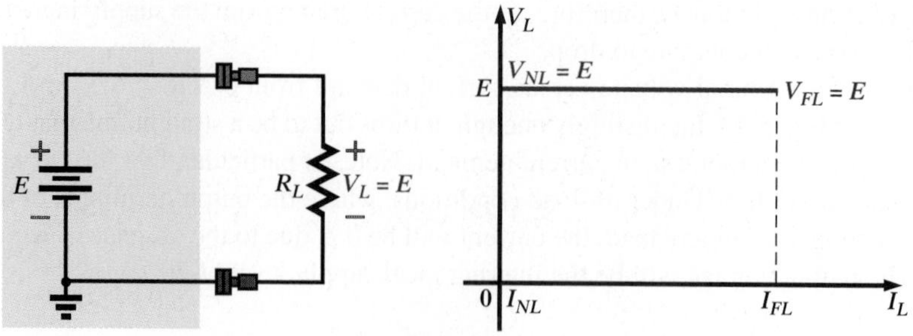

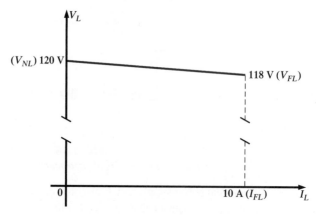

FIG. 5.45

Ideal supply and its terminal characteristics.

the more sensitive the supply is to the change in load and therefore the **less desirable** it is for many laboratory procedures. In fact,

the larger the internal resistance, the steeper the drop in voltage with an increase in load demand (current).

In order for us to develop some sense for the expected response of a supply, a defining quantity called **voltage regulation** (abbreviated *VR*; often called *load regulation* on specification sheets) was established. The basic equation in terms of the quantities of Fig. 5.44(a) is the following:

$$VR = \frac{V_{NL} - V_{FL}}{V_{FL}} \times 100\% \qquad (5.13)$$

The examples to follow will make it clear that

the smaller the voltage or load regulation of a supply, the less the terminal voltage will change with increasing levels of current demand.

For the supply above with a no-load voltage of 20.1 V and a full-load voltage of 18.72 V, at 275.34 mA the voltage regulation would be

$$VR = \frac{V_{NL} - V_{FL}}{V_{FL}} \times 100\% = \frac{20.1 \text{ V} - 18.72 \text{ V}}{18.72 \text{ V}} \times 100\% \cong \mathbf{7.37\%}$$

which is quite high, revealing that we have a very sensitive supply. Most modern commercial supplies have regulation factors less than 1%, with 0.01% being very typical.

EXAMPLE 5.19

a. Given the characteristics of Fig. 5.46, determine the voltage regulation of the supply.
b. Determine the internal resistance of the supply.
c. Sketch the equivalent circuit for the supply.

FIG. 5.46

Terminal characteristics for the supply of Example 5.19.

Solutions:

a. $VR = \dfrac{V_{NL} - V_{FL}}{V_{FL}} \times 100\%$

$\qquad = \dfrac{120\ \text{V} - 118\ \text{V}}{118\ \text{V}} \times 100\% = \dfrac{2}{118} \times 100\%$

$\qquad VR \cong \mathbf{1.7\%}$

b. $R_{int} = \dfrac{\Delta V_L}{\Delta I_L} = \dfrac{120\ \text{V} - 118\ \text{V}}{10\ \text{A} - 0\ \text{A}} = \dfrac{2\ \text{V}}{10\ \text{A}} = \mathbf{0.2\ \Omega}$

c. See Fig. 5.47.

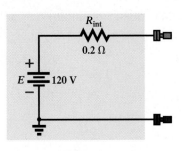

FIG. 5.47
dc supply with the terminal characteristics of Fig. 5.46.

EXAMPLE 5.20 Given a 60 V supply with a voltage regulation of 2%:

a. Determine the terminal voltage of the supply under full-load conditions.
b. If the half-load current is 5 A, determine the internal resistance of the supply.
c. Sketch the curve of the terminal voltage versus load demand and the equivalent circuit for the supply.

Solutions:

a. $\qquad VR = \dfrac{V_{NL} - V_{FL}}{V_{FL}} \times 100\%$

$\qquad 2\% = \dfrac{60\ \text{V} - V_{FL}}{V_{FL}} \times 100\%$

$\qquad \dfrac{2\%}{100\%} = \dfrac{60\ \text{V} - V_{FL}}{V_{FL}}$

$\qquad 0.02 V_{FL} = 60\ \text{V} - V_{FL}$

$\qquad 1.02 V_{FL} = 60\ \text{V}$

$\qquad V_{FL} = \dfrac{60\ \text{V}}{1.02} = \mathbf{58.82\ V}$

b. $I_{FL} = 10\ \text{A}$

$\qquad R_{int} = \dfrac{\Delta V_L}{\Delta I_L} = \dfrac{60\ \text{V} - 58.82\ \text{V}}{10\ \text{A} - 0\ \text{A}} = \dfrac{1.18\ \text{V}}{10\ \text{A}} \cong \mathbf{0.12\ \Omega}$

c. See Fig. 5.48.

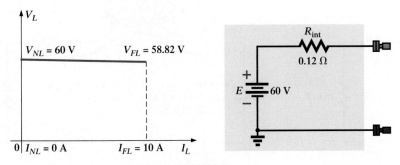

FIG. 5.48
Characteristics and equivalent circuit for the supply of Example 5.20.

5.9 LOADING EFFECTS OF INSTRUMENTS

In the previous section we became aware that power supplies are not the ideal instruments we may have thought they were. The applied load can have an effect on the terminal voltage. Fortunately, since today's supplies have such small load regulation factors, the change in terminal voltage with load can usually be ignored for most applications. If we now turn our attention to the various meters we use in the lab, we again find that they are not totally ideal:

Whenever you apply a meter to a circuit, you change the circuit and the response of the system. Fortunately, however, for most applications, considering the meters to be ideal is a valid approximation as long as certain factors are considered.

For instance,

any ammeter connected in a series circuit will introduce resistance to the series combination that will affect the current and voltages of the configuration.

The resistance between the terminals of an ammeter is determined by the chosen scale of the ammeter. In general,

for ammeters, the higher the maximum value of the current for a particular scale, the smaller will be the internal resistance.

For example, it is not uncommon for the resistance between the terminals of an ammeter to be 250 Ω for a 2 mA scale but only 1.5 Ω for the 2 A scale, as shown in Fig. 5.49(a) and (b). If analyzing a circuit in detail, you can include the internal resistance as shown in Fig. 5.49 as a resistor between the two terminals of the meter.

At first reading, such resistance levels at low currents give the impression that ammeters are far from ideal, that they should be used only to obtain a general idea of the current and should not be expected to provide a true reading. Fortunately, however, when you are reading currents below the 2 mA range, the resistors in series with the ammeter are typically in the kilohm range. For example, in Fig. 5.50(a), using an ideal ammeter, the current displayed is 0.6 mA as determined from $I_s = E/R_T = 12$ V/20 kΩ = 0.6 mA. If we now insert a meter with an internal resistance of 250 Ω as shown in Fig. 5.50(b), the additional resistance in the circuit will drop the current to 0.593 mA as determined from $I_s = E/R_T = 12$ V/20.25 kΩ = 0.593 mA. Now, certainly the current has dropped from the ideal level, but the difference in results is only about 1%—nothing major, and the measurement can be used for most purposes. If the series resistors were in the same range as the 250 Ω resistors, we would have a different problem, and we would have to look at the results very carefully.

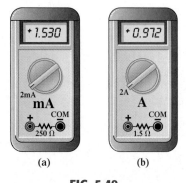

FIG. 5.49

Including the effects of the internal resistance of an ammeter: (a) 2 mA scale; (b) 2 A scale.

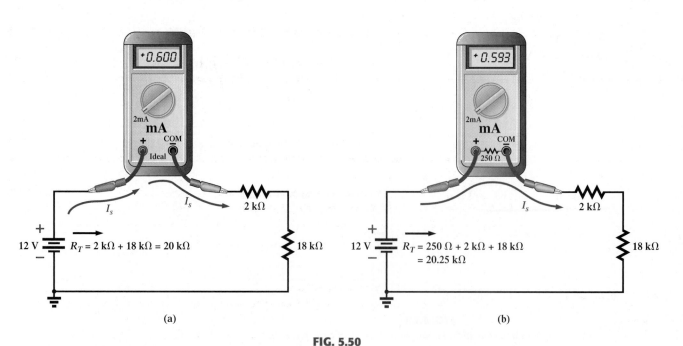

FIG. 5.50

Applying an ammeter, set on the 2 mA scale, to a circuit with resistors in the kilohm range: (a) ideal; (b) practical.

For interest sake, let us go back to Fig. 5.20 and determine the actual current if each meter on the 2 A scale has an internal resistance of 1.5 Ω. The fact that there are four meters will result in an additional resistance of $(4)(1.5\ \Omega) = 6\ \Omega$ in the circuit, and the current will be $I_s = E/R_T = 8.4\ \text{V}/146\ \Omega \cong 58$ mA, rather than the 60 mA under ideal conditions. This value is still close enough to be considered a helpful reading. However, keep in mind that if we were measuring the current in the circuit, we would use only one ammeter, and the current would be $I_s = E/R_T = 8.4\ \text{V}/141.5\ \Omega \cong 59$ mA, which can certainly be approximated as 60 mA.

In general, therefore, be aware that this internal resistance must be factored in, but for the reasons just described, most readings can be used as an excellent first approximation to the actual current.

It should be added that because of this **insertion problem** with ammeters, and because of the very important fact that the **circuit must be disturbed** to measure a current, ammeters are not used as much as you might initially expect. Rather than break a circuit to insert a meter, the voltage across a resistor is often measured and then the current calculated using Ohm's law. This eliminates the need to worry about the level of the meter resistance and having to disturb the circuit. Another option is to use the clamp-type ammeters introduced in Chapter 3, removing the concerns about insertion loss and disturbing the circuit. Of course, for many practical applications (such as on power supplies), it is convenient to have an ammeter permanently installed so that the current can quickly be read from the display. In such cases, however, the design is such as to compensate for the insertion loss.

In summary, therefore, keep in mind that the insertion of an ammeter will add resistance to the branch and will affect the current and voltage levels. However, in most cases the effect is minimal, and the reading will provide a good first approximation to the actual level.

The loading effect of voltmeters will be discussed in detail in the next chapter because loading is not a series effect. In general, however, the results will be similar in many ways to those of the ammeter, but the major difference is that the circuit does not have to be disturbed to apply the meter.

5.10 PROTOBOARDS (BREADBOARDS)

At some point in the design of any electrical/electronic system, a prototype must be built and tested. One of the most effective ways to build a testing model is to use the **protoboard** (in the past most commonly called a **breadboard**) of Fig. 5.51. It permits a direct connection of the power supply and provides a convenient method to hold and connect the components. There isn't a great deal to learn about the protoboard, but it is important to point out some of its characteristics, including the way the elements are typically connected.

The red terminal V_a is connected directly to the positive terminal of the dc power supply, with the black lead V_b connected to the negative terminal and the green terminal used for the ground connection. Under the hole pattern there are **continuous horizontal copper**

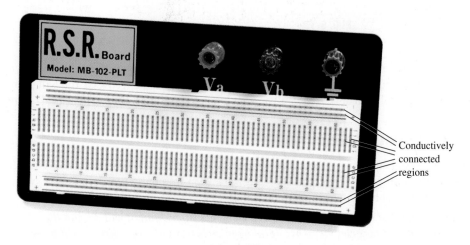

Conductively
connected
regions

FIG. 5.51

Protoboard with areas of conductivity defined using two different approaches.

strips under the top and bottom rows, as shown by the copper bands in Fig. 5.51. **In the center region, the conductive strips are vertical but do not extend beyond the deep notch running the horizontal length of the board.** That's all there is to it, although it will take some practice to make the most effective use of the conductive patterns.

As examples, the network of Fig. 5.11 is connected on the protoboard in the photo of Fig. 5.52(a) using two different approaches. After the dc power supply has been hooked up, a lead is brought down from the positive red terminal to the top conductive strip marked "+." Keep in mind that now the entire strip is connected to the positive terminal of the supply. The negative terminal is connected to the bottom strip marked with a minus sign (−), so that 8.4 V can be read at any point between the top positive strip and the bottom negative strip. A ground connection to the negative terminal of the battery was made at the site of the three terminals. For the user's convenience, kits are available in which the length of the wires is color coded. Otherwise, a spool of 24 gage wire is cut to length and the ends are stripped. In general, feel free to use the extra length—everything doesn't have to be at right angles. For most protoboards, 1/4 through 1 W resistors will insert nicely in the board. For clarity 1/2 W resistors were used in Fig. 5.52. The voltage across any component can be easily read by simply inserting additional leads as shown in the figure (yellow leads) for the voltage V_3 of each configuration (the yellow wires) and attaching the meter. For any network there are a variety of ways in which the components can be wired. Note in the configuration on the right that the horizontal break through the center of the board was used to isolate the two terminals of each resistor. Even though there are no set standards, it is important that the arrangement is such that it can **easily be understood by someone else.**

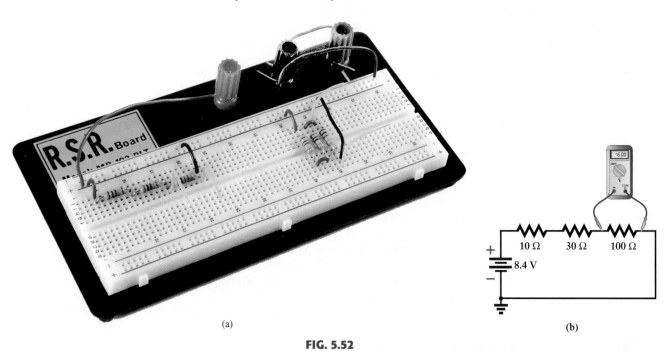

(a)

(b)

FIG. 5.52

Network of Fig. 5.11 on a protoboard with yellow leads added to measure voltage V_3 with a voltmeter.

Additional setups using the protoboard will appear in the chapters to follow so that you can become accustomed to the manner in which it is used most effectively. You will probably see the protoboard quite frequently in your laboratory sessions or in an industrial setting.

5.11 APPLICATIONS

Before looking at a few applications, we need to consider a few general characteristics of the series configuration that you should always keep in mind when designing a system. First, and probably the most important, is that

if one element of a series combination of elements should fail, it will disrupt the response of all the series elements. If an open circuit occurs, the current will be zero. If a short circuit results, the voltage will increase across the other elements, and the current will increase in magnitude.

Second, and just a thought you should always keep in mind, is that

for the same source voltage, the more elements you place in series, the less the current and the less the voltage across all the elements of the series combination.

Last, and a result discussed in detail in this chapter, is that

the current is the same for each element of a series combination, but the voltage across each element is a function of its terminal resistance.

There are other characteristics of importance that you will become aware of as you investigate possible areas of application, but the above is a list of the most important characteristics.

Series Control

One common use of the series configuration is in setting up a system that will ensure that everything is in place before full power is applied. In Fig. 5.53, various sensing mechanisms can be tied to series switches, preventing power to the load until all the switches are in the closed or on position. For instance, as shown in the figure, one component may test the environment for dangers such as gases, high temperatures, and so on. The next component may be sensitive to the properties of the system to be energized to be sure everything is in running order. Security could be another factor in the series sequence, and finally a timing mechanism may be present to ensure limited hours of operation or to restrict operating periods. The list is endless, but the fact remains that "all systems must be go" before power will reach the operating system.

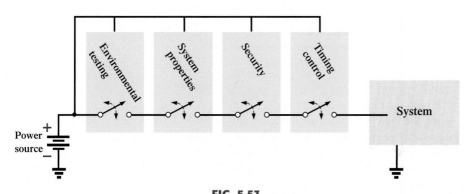

FIG. 5.53
Series control over an operating system.

Holiday Lights

In recent years the small blinking holiday lights with as many as 50 to 100 bulbs on a string have become very popular [see Fig. 5.54(a)]. Although holiday lights can be connected in series or parallel (to be described in the next chapter), the smaller blinking light sets are normally connected in series. It is relatively easy to determine if the lights are connected in series. If one wire enters and leaves the bulb casing, they are in series. If two wires enter and leave, they are probably in parallel. **Normally, when bulbs are connected in series, if one should burn out (the filament breaks and the circuit opens), all the bulbs will go out since the current path has been interrupted.** However, the bulbs of Fig. 5.54(a) are specially designed, as shown in Fig. 5.54(b), to permit current to continue to flow to the other bulbs when the filament burns out. At the base of each bulb there is a fuse link wrapped around the two posts holding the filament. The fuse link of a soft conducting metal appears to be touching the two vertical posts, but in fact a coating on the posts or fuse link prevents conduction from one to the other under normal operating conditions. If a filament should burn out and create an open circuit between the posts, the current through the bulb and other bulbs would be interrupted if it were not for the fuse link. At the instant a bulb opens up, current through the circuit is zero, and the full 120 V from the outlet will appear across the bad bulb.

This high voltage from post to post of a single bulb is of sufficient potential difference to establish current through the insulating coatings and spot-weld the fuse link to the two posts. The circuit is again complete, and all the bulbs will light except the one with the activated fuse link. Keep in mind, however, that each time a bulb burns out, there will be more voltage across the other bulbs of the circuit, making them burn brighter. Eventually, if too many bulbs burn out, the voltage will reach a point where the other bulbs will burn out in rapid succession. The result is that one must replace burned-out bulbs at the earliest opportunity.

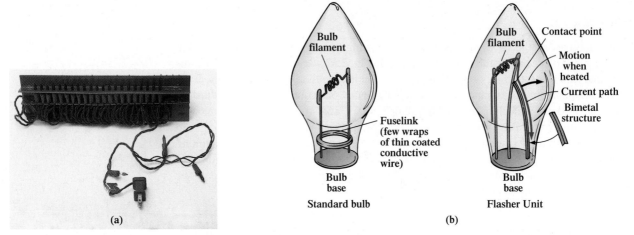

FIG. 5.54

Holiday lights: (a) 50-unit set; (b) bulb construction.

The bulbs of Fig. 5.54(b) are rated 2.5 V at 0.2 A or 200 mA. Since there are 50 bulbs in series, the total voltage across the bulbs will be 50×2.5 V or 125 V which matches the voltage available at the typical home outlet. Since the bulbs are in series, the current through each bulb will be 200 mA. The power rating of each bulb is therefore $P = VI = (2.5 \text{ V})(0.2 \text{ A}) = 0.5$ W with a total wattage demand of 50×0.5 W $= 25$ W.

A schematic representation for the set of Fig. 5.54(a) is provided in Fig. 5.55(a). Note that only one flasher unit is required. Since the bulbs are in series, when the flasher unit interrupts the current flow, it will turn off all the bulbs. As shown in Fig. 5.54(b), the flasher unit incorporates a bimetal thermal switch that will open when heated to a preset level by the current. As soon as it opens, it will begin to cool down and close again so that current can return to the bulbs. It will then heat up again, open up, and repeat the entire process. The result is an on-and-off action that creates the flashing pattern we are so familiar with. Naturally, in a colder climate (for example, outside in the snow and ice), it will initially take longer to heat up, so the flashing pattern will be reduced at first; but in time as the bulbs warm up, the frequency will increase.

The manufacturer specifies that no more than six sets should be connected together. The first question that then arises is how sets can be connected together, end to end, without reducing the voltage across each bulb and making all the lights dimmer. If you look closely at the wiring, you will find that since the bulbs are connected in series, there is one wire to each bulb with additional wires from plug to plug. Why would they need two additional wires if the bulbs are connected in series? The answer lies in the fact that when each set is connected together, they will actually be in parallel (to be discussed in the next chapter) by a unique wiring arrangement shown in Fig. 5.55(b) and redrawn in Fig. 5.55(c) to clearly show the parallel arrangement. Note that the top line is the hot line to all the connected sets, and the bottom line is the return, neutral, or ground line for all the sets. Inside the plug of Fig. 5.55(d), the hot line and return are connected to each set, with the connections to the metal spades of the plug as shown in Fig. 5.55(b). We will find in the next chapter that the current drawn from the wall outlet for parallel loads is the sum of the cur-

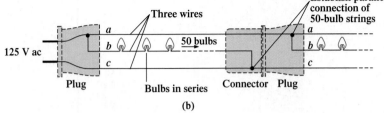

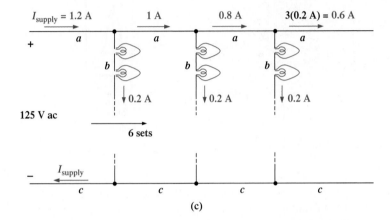

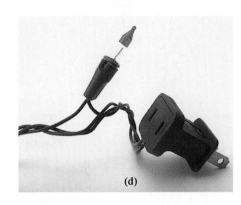

FIG. 5.55

(a) Single-set wiring diagram; (b) special wiring arrangement;
(c) redrawn schematic; (d) special plug and flasher unit.

rent to each branch. The result, as shown in Fig. 5.55(c), is that the current drawn from the supply is 6×200 mA $= 1.2$ A, and the total wattage for all six sets is the product of the applied voltage and the source current or $(120\ V)(1.2\ A) = 144$ W with 144 W/6 = 24 W per set.

Microwave Oven

Series circuits can be very effective in the design of safety equipment. Although we all recognize the usefulness of the microwave oven, it can be quite dangerous if the door is not closed or sealed properly. It is not enough to test the closure at only one point around the door because the door may be bent or distorted from continual use, and leakage could result at some point distant from the test point. One common safety arrangement appears in Fig. 5.56. Note that magnetic switches are located all around the door, with the magnet in the door itself and the magnetic door switch in the main frame. Magnetic switches are simply switches where the magnet draws a magnetic conducting bar between two contacts to complete the circuit—somewhat revealed by the symbol for the device in the circuit diagram of Fig. 5.56. Since the magnetic switches are all in series, they must all be closed to complete the circuit and turn on the power unit. If the door is sufficiently out of shape to prevent a single magnet from getting close enough to the switching mechanism, the circuit will not be complete, and the power cannot be turned on. Within the control unit of the power supply, either the series circuit completes a circuit for operation or a sensing current is established and monitored that controls the system operation.

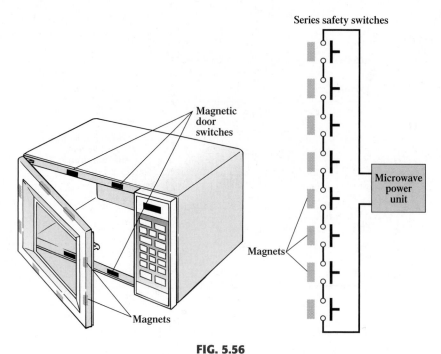

FIG. 5.56

Series safety switches in a microwave oven.

5.12 COMPUTER ANALYSIS

PSpice

In Section 4.9, the basic procedures for setting up the folder and running a program were presented. Because of the detail provided in that section, you should review it before proceeding with this example. Because this is only the second example using PSpice, some detail will be provided, but not at the level of Section 4.9.

The circuit to be investigated appears in Fig. 5.57. Since the **PSpice** folder was established in Section 4.9, there is no need to repeat the process here—the folder is immediately available. Double-clicking on the **Orcad Lite Edition** icon will generate the **Orcad Capture-Lite Edition** window. A new project is then initiated by selecting the **Create document** key at the top left of the screen (it looks like a page with a star in the upper-left corner). The result is the **New Project** dialog box in which **SeriesDC** is inserted as the **Name.** The **Analog or Mixed A/D** is already selected, and **C:\PSpice** appears as the **Location**—only the **Name** had to be entered! Click **OK,** and the **Create PSpice Project** dialog box will appear. Select **Create a blank project,** click **OK,** and the working windows will appear. Grabbing the left edge of the **SCHEMATIC1:PAGE1** window will allow you to move it to the right so that you can see both screens. Clicking the + sign in the **Project Manager** window will allow you to set the sequence down to **PAGE1.** If you prefer to change the name of the **SCHEMATIC1,** just select it and right-click on the mouse. A listing will appear in which **Rename** is an option; selecting it will result in a **Rename Schematic** dialog box in which **SeriesDC** can be entered. In Fig. 5.58, it was left as **SCHEMATIC1.**

Now this next step is important! If the toolbar on the right edge does not appear, be sure to double-click on **PAGE1** in the **Project Manager** window, or select the **Schematic Window.** When the heading of the **Schematic Window** is dark blue, the toolbar will appear. To start building the circuit, select the **Place a part** key (the second one down) to obtain the **Place Part** dialog box. Note that now the **SOURCE** library is already in place in the **Library** list from the efforts of Chapter 4; it does not have to be reinstalled. Selecting **SOURCE** will result in the list of sources under **Part List,** and **VDC** can be selected. Click **OK,** and the cursor can put it in place with a single left click. Right-click and select **End Mode** to end the process since the network has only one source. One more left click and the source is in place.

Now the **Place a part** key is selected again, followed by **ANALOG** library to find the resistor **R.** Once **R** is selected, an **OK** will place it next to the cursor on the screen. This

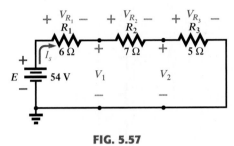

FIG. 5.57
Series dc network to be analyzed using PSpice.

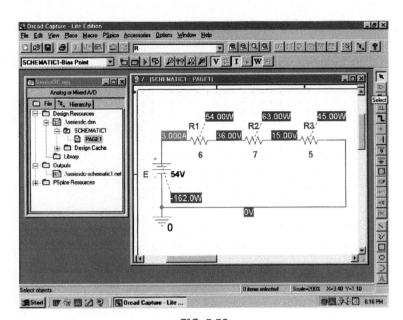

FIG. 5.58
Applying PSpice to a series dc circuit.

time, since three resistors need to be placed, there is no need to go to **End Mode** between depositing each. Simply click one in place, then the next, and finally the third. Then right-click to end the process with **End Mode.** Finally, a **GND** must be added by selecting the appropriate key and selecting **0/SOURCE** in the **Place Ground** dialog box. Click **OK,** and place the ground as shown in Fig. 5.58.

The elements must now be connected using the **Place a wire** key to obtain the crosshair on the screen. Start at the top of the voltage source with a left click and draw the wire, left-clicking it at every 90° turn. When a wire is connected from one element to another, move on to the next connection to be made—there is no need to go to **End Mode** between connections. Now the labels and values must be set by double-clicking on each parameter to obtain a **Display Properties** dialog box. Since the dialog box appears with the quantity of interest in a blue background, simply type in the desired label or value, followed by **OK.** The network is now complete and ready to be analyzed.

Before simulation, select the **V, I,** and **W** in the toolbar at the top of the window to ensure that the voltages, currents, and power are displayed on the screen. To simulate, select the **New Simulation Profile** key (which appears as a data sheet on the second toolbar down, with a star in the top-left corner) to obtain the **New Simulation** dialog box. Enter **Bias Point** for a dc solution under **Name,** and hit the **Create** key. A **Simulation Settings-Bias Point** dialog box will appear in which **Analysis** is selected and **Bias Point** is found under the **Analysis type** heading. Click **OK,** and then select the **Run PSpice** key (the blue arrow) to initiate the simulation. Exit the resulting screen, and the display of Fig. 5.58 will result.

The current is clearly 3 A for the circuit, with 15 V across R_3 and 36 V from a point between R_1 and R_2 to ground. The voltage across R_2 is 36 V $-$ 15 V $=$ 21 V, and the voltage across R_1 is 54 V $-$ 36 V $=$ 18 V. The power supplied or dissipated by each element is also listed. There is no question that the results of Fig. 5.58 include a very nice display of voltage, current, and power levels.

Electronics Workbench (EWB)

Since this is only the second circuit to be constructed using EWB, a detailed list of steps will be included as a review. Essentially, however, the entire circuit of Fig. 5.59 can be "drawn" using simply the construction information introduced in Chapter 4.

After you have selected the **Multisim 2001** icon, a **Multisim-Circuit 1** window will appear ready to accept the circuit elements. Select the **Sources** key at the top of the left toolbar, and a **Sources** parts bin will appear with 30 options. Selecting the top option will place the **GROUND** on the screen of Fig. 5.59, and selecting the third option down will result in **DC_VOLTAGE_SOURCE.** The resistors are obtained by choosing the second

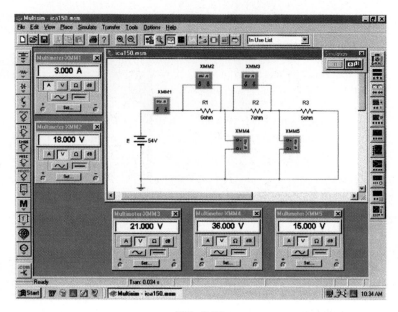

FIG. 5.59

Applying EWB to a series dc circuit.

key down on the left toolbar called the **Basic** key. The result is 25 options in which **RESISTOR_ VIRTUAL** is selected. We must return to the **RESISTOR_VIRTUAL** key to place each resistor on the screen. However, each new resistor is numbered in sequence, although they are all given the default value of 1 kΩ. Remember from the discussion of Chapter 4 that you should add the meters before connecting the elements together because the meters take space and must be properly oriented. The current will be determined by the **XMM1** ammeter and the voltages by **XMM2** through **XMM5.** Of particular importance, note that

in EWB the meters are connected in exactly the same way they would be placed in an active circuit in the laboratory. Ammeters are in series with the branch in which the current is to be determined, and voltmeters are connected between the two points of interest (across resistors). In addition, for positive readings, ammeters are connected so that conventional current enters the positive terminal, and voltmeters are connected so that the point of higher potential is connected to the positive terminal.

The meter settings are made by double-clicking on the meter symbol on the schematic. In each case, **V** or **I** had to be chosen, but the horizontal line for dc analysis is the same for each. Again, the **Set** key can be selected to see what it controls, but the default values of meter input resistance levels will be fine for all the analyses described in this text. Leave the meters on the screen so that the various voltages and the current level will be displayed after the simulation.

Recall from Chapter 4 that elements can be moved by simply clicking on each schematic symbol and dragging it to the desired location. The same is true for labels and values. Labels and values are set by double-clicking on the label or value and entering your preference. Click **OK,** and they will appear changed on the schematic. There is no need to first select a special key to connect the elements. Simply bring the cursor to the starting point to generate the small circle and crosshair. Click on the starting point, and follow the desired path to the next connection path. When the crosshair is in the correct location, click again and the line will appear. All connecting lines can make 90° turns. However, you cannot follow a diagonal path from one point to another. To remove any element, label, or line, simply click on the quantity to obtain the four-square active status, and select the **Delete** key or the scissors key on the top menu bar.

Before simulating, be sure that the **Simulate Switch** is visible by selecting **View-Show Simulate Switch.** Then select the **1** option on the switch, and the analysis will begin. The results appearing in Fig. 5.59 verify those obtained using PSpice and the longhand solution.

CHAPTER SUMMARY

SERIES RESISTORS

- The total resistance of resistors in series is the sum of the resistance values.
- The largest resistor in a series combination will have the most impact on the total resistance and the resulting current.
- The total resistance is unaffected by the order in which series resistors are connected.
- When an ohmmeter is used to measure resistance, there is no polarity associated with the measurement.

SERIES CIRCUIT

- The current is the same at every point in a series circuit.
- The polarity of the voltage across a resistor is determined by the direction of the current.
- The power applied to a dc circuit must equal that dissipated by the resistive elements.
- In a series circuit, maximum power is delivered to the largest resistor.
- In a series circuit, the applied voltage is equal to the sum of the voltage drops across the resistive elements.
- The algebraic sum of the potential rises and drops around a closed path is zero.

VOLTAGE DIVIDER RULE

- The voltage across series resistive elements divides as the magnitude of the values.
- The ratio of the voltages across series resistors will be the same as the ratio of their resistance levels.
- The voltage across a resistor in a series circuit is equal to the value of that resistor times the total applied voltage divided by the total resistance of the series configuration.

VOLTAGE REGULATION

- A network should always be connected to a supply before the level of the supply voltage is set.
- Every practical supply has some internal resistance.
- The larger the internal resistance, the steeper the drop in voltage with increase in current.
- The smaller the voltage regulation, the less sensitive the supply to a change in load.

AMMETERS

- The insertion of an ammeter into a circuit will affect the current and voltages of the configuration.
- The higher the maximum current for a particular scale, the less the internal resistance of the ammeter.
- In most cases, the insertion losses of an ammeter can be ignored because those scales with the higher resistance levels are normally applied in circuits with larger resistor values.

Important Equations

$$R_T = R_1 + R_2 + R_3 + \cdots + R_N$$

$$R_T = NR$$

$$I_s = \frac{E}{R_T} \qquad V_1 = I_1 R_1 \qquad V_2 = I_2 R_2 \qquad V_3 = I_3 R_3 \ldots$$

$$P_E = EI_s \qquad P_E = P_{R_1} + P_{R_2} + P_{R_3} + \cdots$$

$$P_1 = V_1 I_1 = I_1^2 R_1 = \frac{V_1^2}{R_1}$$

$$\Sigma_C V = 0$$

$$V_x = R_x \frac{E}{R_T} \quad \text{(voltage divider rule)}$$

$$R_{\text{int}} = \frac{\Delta V_L}{\Delta I_L}$$

$$VR = \frac{V_{NL} - V_{FL}}{V_{FL}} \times 100\%$$

PROBLEMS

SECTION 5.2   Series Resistors

1. For each configuration of Fig. 5.60, find the individual (not combinations of) elements (voltage sources and/or resistors) that are in series. If necessary, use the fact that elements in series have the same current. Do not consider elements that are not in series. Simply list those that satisfy the conditions for a series relationship. We will learn more about other combinations later.

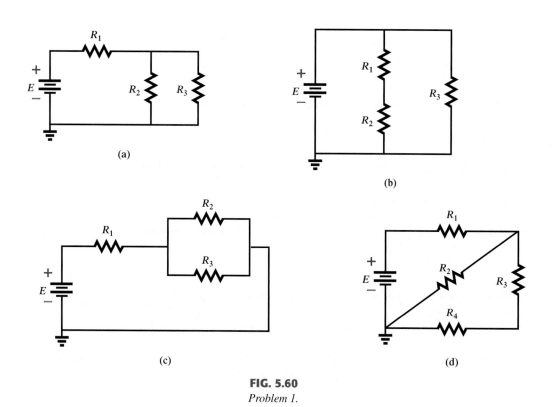

FIG. 5.60
Problem 1.

2. Find the total resistance for each configuration of Fig. 5.61. Note that only standard resistor values were employed.

3. For each circuit board of Fig. 5.62, find the total resistance between connection tabs 1 and 2.

4. For the circuit of Fig. 5.63, composed of standard values:
   a. Which resistor will have the most impact on the total resistance?
   b. On an approximate basis, which resistors can be ignored when determining the total resistance?
   c. Find the total resistance, and comment on your results for parts (a) and (b).

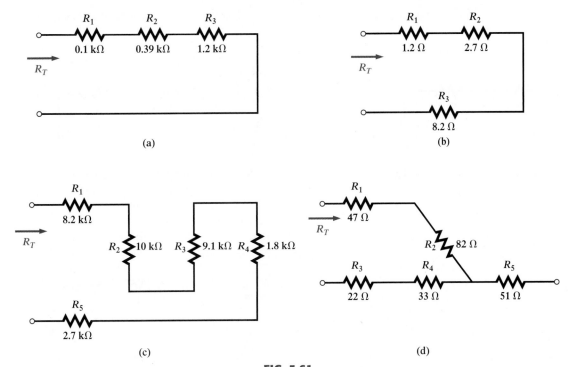

FIG. 5.61
Problem 2.

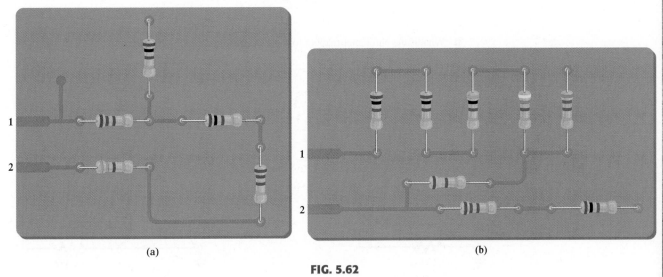

FIG. 5.62
Problem 3.

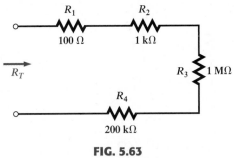

FIG. 5.63
Problem 4.

5. For each configuration of Fig. 5.64, find the unknown resistors using the ohmmeter reading.

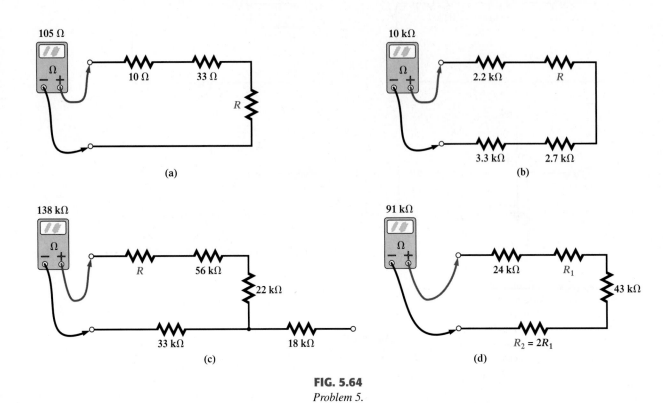

FIG. 5.64
Problem 5.

6. What is the ohmmeter reading for each configuration of Fig. 5.65?

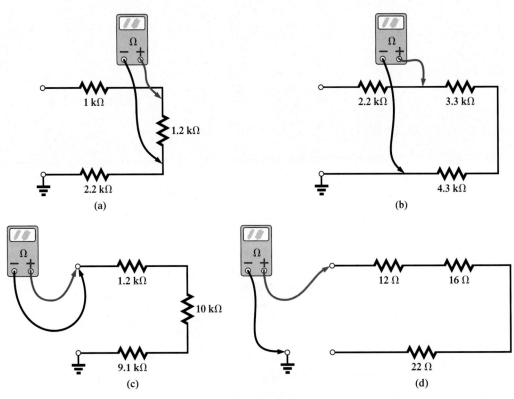

FIG. 5.65
Problem 6.

7. For the series configuration of Fig. 5.66, constructed of standard values:
   a. Find the total resistance.
   b. Calculate the current.
   c. Find the voltage across each resistive element.

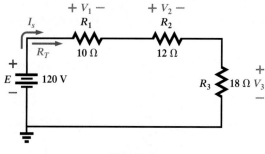

FIG. 5.66
Problem 7.

8. For the series configuration of Fig. 5.67, constructed using standard value resistors:
   a. Without making a single calculation, which resistive element will have the most voltage across it? Which will have the least?
   b. Which resistor will have the most impact on the total resistance and the resulting current? Find the total resistance and the current.
   c. Find the voltage across each element and review your response to part (a).

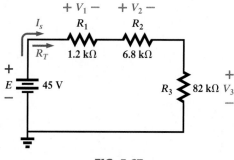

FIG. 5.67
Problem 8.

9. Find the applied voltage necessary to develop the current specified in each circuit of Fig. 5.68.

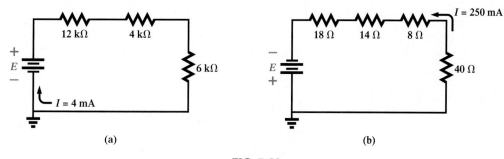

FIG. 5.68
Problem 9.

10. For each network of Fig. 5.69, constructed of standard values, determine:
    a. The current I.
    b. The source voltage E.
    c. The unknown resistance.
    d. The voltage across each element.

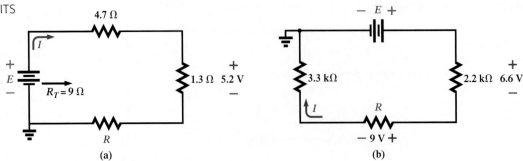

FIG. 5.69
Problem 10.

11. For each configuration of Fig. 5.70, what are the readings of the ammeter and the voltmeter?

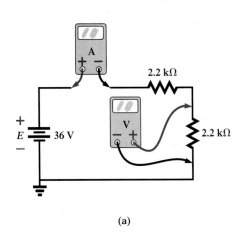

(a)

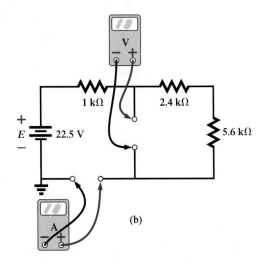

(b)

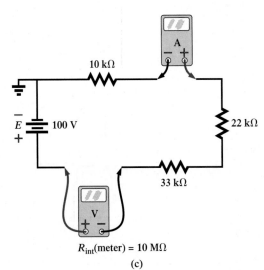

R_{int}(meter) = 10 MΩ

(c)

FIG. 5.70
Problem 11.

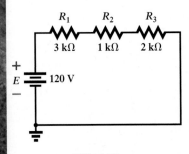

FIG. 5.71
Problem 12.

SECTION 5.4   Power Distribution in a Series Circuit

12. For the circuit of Fig. 5.71, constructed of standard value resistors:
    a. Find the total resistance, current, and voltage across each element.
    b. Find the power delivered to each resistor.
    c. Calculate the total power delivered to all the resistors.
    d. Find the power delivered by the source.

e. How does the power delivered by the source compare to that delivered to all the resistors?

f. Which resistor received the most power? Why?

g. What happened to all the power delivered to the resistors?

h. If the resistors are available with wattage ratings of 1/2 W, 1 W, 2 W, and 5 W, what minimum wattage rating can be used for each resistor?

13. Repeat Problem 12 for the circuit of Fig. 5.72.

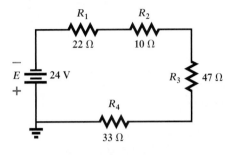

FIG. 5.72
Problem 13.

14. Find the unknown quantities for the circuits of Fig. 5.73 using the information provided.

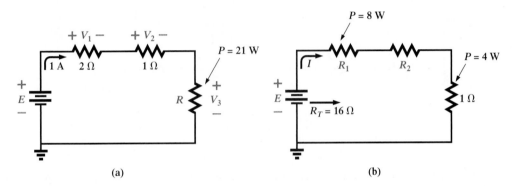

(a)   (b)

FIG. 5.73
Problem 14.

15. Eight holiday lights are connected in series as shown in Fig. 5.74.

a. If the set is connected to a 120 V source, what is the current through the bulbs if each bulb has an internal resistance of 28⅛ Ω?

b. Determine the power delivered to each bulb.

c. Calculate the voltage drop across each bulb.

d. If one bulb burns out (that is, the filament opens), what is the effect on the remaining bulbs? Why?

16. For the conditions specified in Fig. 5.75, determine the unknown resistance.

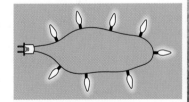

FIG. 5.74
Problem 15.

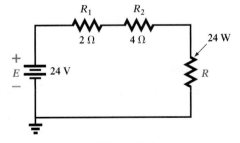

FIG. 5.75
Problem 16.

SECTION 5.5   Voltage Sources in Series

17. Combine the series voltage sources of Fig. 5.76 into a single voltage source between points *a* and *b*.

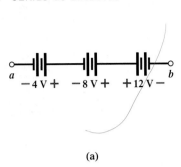

(a)

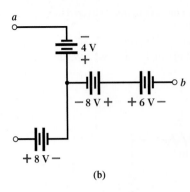

(b)

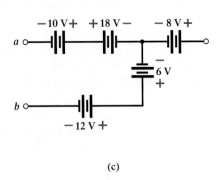

(c)

FIG. 5.76
Problem 17.

18. Determine the current *I* and its direction for each network of Fig. 5.77. Before solving for *I*, re-draw each network with a single voltage source.

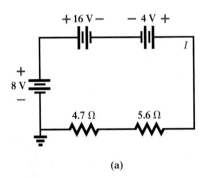

(a)

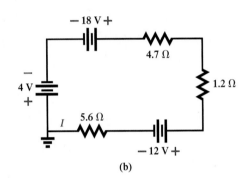

(b)

FIG. 5.77
Problem 18.

19. Find the unknown voltage source and resistor for the networks of Fig. 5.78. First combine the series voltage sources into a single source. Indicate the direction of the resulting current.

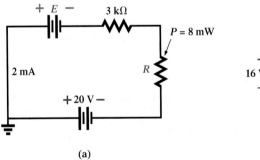

(a)

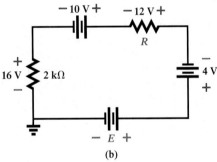

(b)

FIG. 5.78
Problem 19.

SECTION 5.6 Kirchhoff's Voltage Law

20. Using Kirchhoff's voltage law, find the unknown voltages for the circuits of Fig. 5.79.

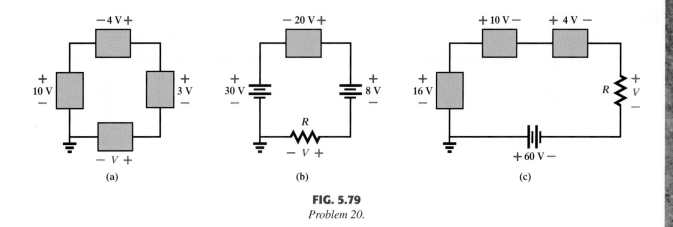

FIG. 5.79
Problem 20.

21. Using Kirchhoff's voltage law, determine the unknown voltages for the configurations of Fig. 5.80.

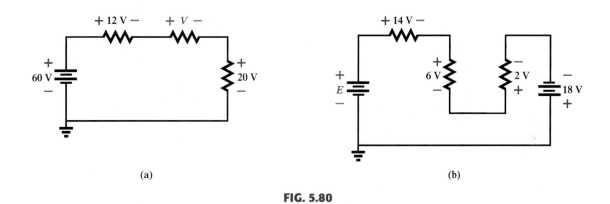

FIG. 5.80
Problem 21.

22. Using Kirchhoff's voltage law, determine the unknown voltages for the series circuits of Fig. 5.81.

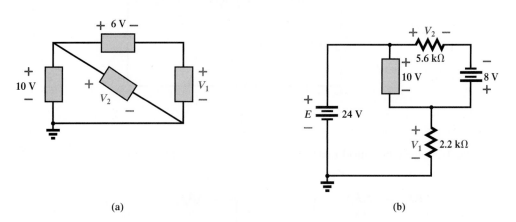

FIG. 5.81
Problem 22.

23. Using Kirchhoff's voltage law, find the unknown voltages for the configurations of Fig. 5.82.

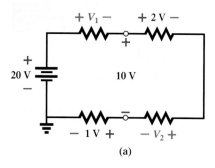
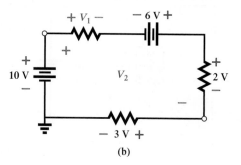

(a) (b)

FIG. 5.82
Problem 23.

SECTION 5.7 Voltage Division in a Series Circuit

24. Determine the values of the unknown resistors of Fig. 5.83 using the provided voltage levels.

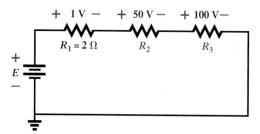

FIG. 5.83
Problem 24.

25. For the configuration of Fig. 5.84, with standard resistor values:
a. By inspection, which resistor will receive the largest share of the applied voltage? Why?
b. How much larger will voltage V_3 be compared to V_2 and V_1?
c. Find the voltage across the largest resistor using the voltage divider rule.
d. Find the voltage across the series combination of resistors R_2 and R_3.

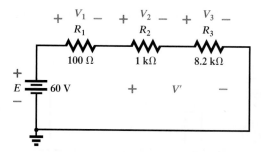

FIG. 5.84
Problem 25.

26. Using the voltage divider rule, find the indicated voltages of Fig. 5.85.

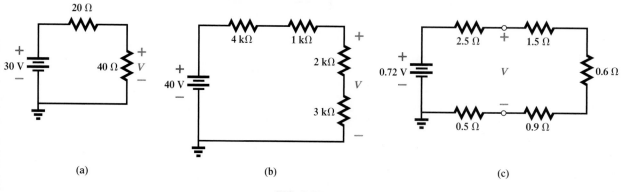

(a) (b) (c)

FIG. 5.85
Problem 26.

27. Using the voltage divider rule or Kirchhoff's voltage law, determine the unknown voltages for the configurations of Fig. 5.86.

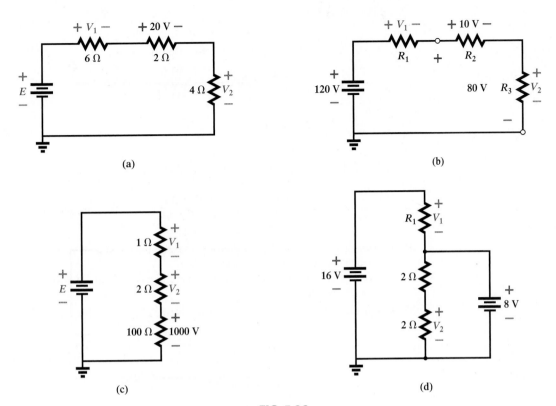

(a)

(b)

(c)

(d)

FIG. 5.86
Problem 27.

28. Using the information provided, find the unknown quantities appearing in Fig. 5.87.

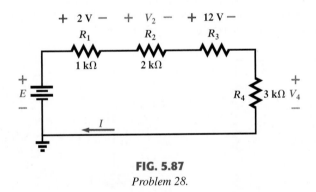

FIG. 5.87
Problem 28.

*29. Using the voltage divider rule, find the unknown resistance for the configurations of Fig. 5.88.

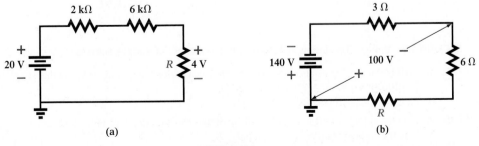

(a)

(b)

FIG. 5.88
Problem 29.

30. Referring to Fig. 5.89:
  a. Determine V_2.
  b. Calculate V_3.
  c. Determine R_3.

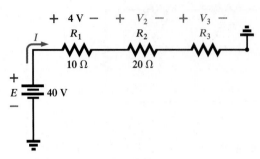

FIG. 5.89
Problem 30.

31. a. Design a voltage divider circuit that will permit the use of an 8 V, 50 mA bulb in an automobile with a 12 V electrical system.
  b. What is the minimum wattage rating of the chosen resistor if 1/4 W, 1/2 W, and 1 W resistors are available?

32. Design the voltage divider of Fig. 5.90 such that $V_{R_1} = 1/5 V_{R_2}$. That is, find R_1 and R_2.

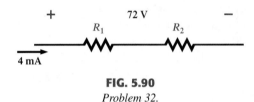

FIG. 5.90
Problem 32.

33. Find the voltage across each resistor of Fig. 5.91 if $R_1 = 2R_3$ and $R_2 = 7R_3$.

*34. a. Design the circuit of Fig. 5.92 such that $V_{R_2} = 3V_{R_1}$ and $V_{R_3} = 4V_{R_2}$.
  b. If the current is reduced to 10 μA, what are the new values of R_1, R_2, and R_3? How do they compare to the results of part (a)?

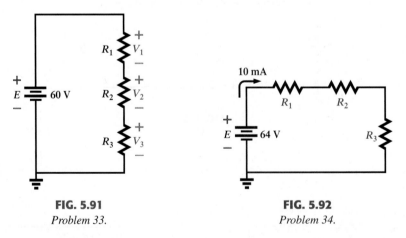

FIG. 5.91
Problem 33.

FIG. 5.92
Problem 34.

SECTION 5.8 Voltage Regulation and the Internal Resistance of Voltage Sources

35. a. Find the internal resistance of a battery that has a no-load output of 60 V and that supplies a full-load current of 2 A to a load of 28 Ω.
  b. Find the voltage regulation of the supply.

36. a. Find the voltage to the load (full-load conditions) for the supply of Fig. 5.93.
  b. Find the voltage regulation of the supply.
  c. How much power is supplied by the source and lost to the internal resistance under full-load conditions?

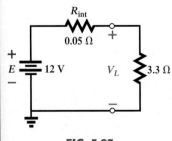

FIG. 5.93
Problem 36.

37. a. Determine the current through the circuit of Fig. 5.94.
   b. If an ammeter with an internal resistance of 250 Ω is inserted into the circuit of Fig. 5.94, what effect will it have on the current level?
   c. Is the difference in current level a major concern for most applications?

SECTION 5.12   Computer Analysis

38. Use the computer to verify the results of Example 5.4.

39. Use the computer to verify the results of Example 5.5.

40. Use the computer to verify the results of Example 5.15.

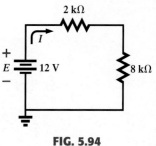

FIG. 5.94
Problem 37.

GLOSSARY

Circuit   A combination of a number of elements joined at terminal points providing at least one closed path through which charge can flow.

Closed loop   Any continuous connection of branches that allows tracing of a path that leaves a point in one direction and returns to that same point from another direction without leaving the circuit.

Internal resistance   The inherent resistance found internal to any source of energy.

Kirchhoff's voltage law (KVL)   The algebraic sum of the potential rises and drops around a closed loop (or path) is zero.

Protoboard (breadboard)   A flat board with a set pattern of conductively connected holes designed to accept 24-gage wire and components with leads of about the same diameter.

Series circuit   A circuit configuration in which the elements have only one point in common and each terminal is not connected to a third, current-carrying element.

Two-terminal device   Any element or component with two external terminals for connection to a network configuration.

Voltage divider rule (VDR)   A method by which a voltage in a series circuit can be determined without first calculating the current in the circuit.

Voltage regulation (VR)   A value, given as a percent, that provides an indication of the change in terminal voltage of a supply with a change in load demand.

Parallel dc Circuits

OBJECTIVES

- Become familiar with the characteristics of a parallel circuit.
- Be able to find the total resistance of any parallel combination of resistive elements in a quick and efficient manner.
- Understand how to use Ohm's law to determine the various unknowns of a parallel circuit.
- Become aware of the power flow in a parallel circuit.
- Develop a clear understanding of Kirchhoff's current law and its impact on the analysis of electric circuits.
- Be able to properly apply an ohmmeter, voltmeter, and ammeter to measure the quantities of importance in a parallel circuit.
- Appreciate how current divides in a parallel circuit, and be able to apply the current divider rule effectively.
- Understand the impact of connecting voltage sources in parallel.
- Become aware of the loading effects of voltmeters on a circuit.
- Develop a sense for how the parallel configuration can be used effectively in everyday applications.
- Begin to understand how computer software can be used effectively in the analysis of parallel dc circuits.

6.1 INTRODUCTION

The series combination of elements was covered in detail in the previous chapter. We will now move on to another combination of elements of equal importance: the **parallel connection.** Be assured that if you now feel comfortable with the series configuration, the parallel combination will be just as straightforward to learn. There are a few fundamental rules, and once you have learned them, it will be quite easy to identify the combination and apply the appropriate laws. In many ways this chapter will parallel that of the previous chapter because there is subject matter for each that is similar in nature.

6.2 PARALLEL RESISTORS

The term *parallel* is used so often to describe a physical arrangement between two elements that most individuals are aware of its general characteristics. For instance, the two lines of Fig. 6.1(a) are in parallel. At every point between the two lines, the distance is the same. If we now simply connect the ends as shown in Fig. 6.1(b), we have the basic structure of a parallel connection of elements.

In general,

two elements, branches, or circuits are in parallel if they have two, and only two, points in common.

For instance, in Fig. 6.2(a), the two resistors are in parallel because they are connected at points a and b. If both ends were **not** connected as shown, the resistors would not be in parallel. In Fig. 6.2(b), resistors R_1 and R_2 are in parallel because they again have points a and b in common. R_1 is not in parallel with R_3 because they are connected at only one point (b). Further, R_1 and R_3 are not in series because a third connection appears at point b. The same can be said for resistors R_2 and R_3. In Fig. 6.2(c), resistors R_1 and R_2 are in series because they have only one point in common that is not connected elsewhere in the network. Resistors R_1 and R_3 are not in parallel because they have only point a in common. In addition, they are not in series because of the third connection to point a. The same can be said for resistors R_2 and R_3. In a broader context it can be said that the series combination of resistors R_1 and R_2 is in parallel with resistor R_3 (more will be said about this option in Chapter 7). Furthermore, even though the discussion above was only for resistors, it can be applied to any two-terminal elements such as voltage sources and meters.

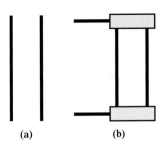

FIG. 6.1
(a) Parallel lines; (b) parallel connection.

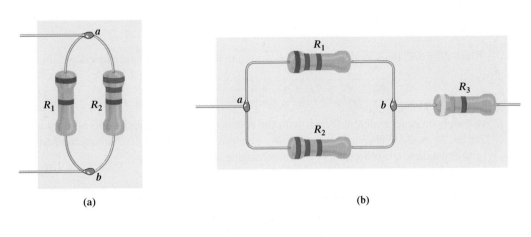

(a)

(b)

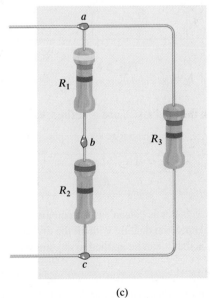

(c)

FIG. 6.2
(a) Parallel resistors; (b) R_1 and R_2 are in parallel; (c) R_3 is in parallel with the series combination of R_1 and R_2.

On schematics the parallel combination can appear in a number of ways, as shown in Fig. 6.3. In each case, the three resistors are in parallel. They all have points a and b in common.

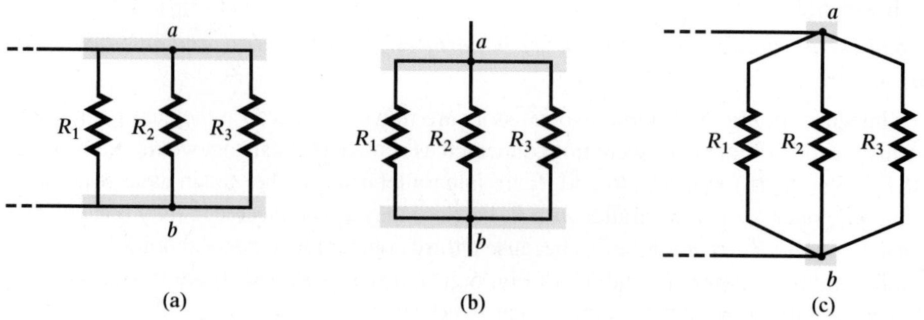

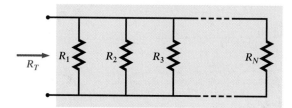

FIG. 6.3
Schematic representations of three parallel resistors.

For resistors in parallel as shown in Fig. 6.4, the total resistance is determined from the following equation:

$$\frac{1}{R_T} = \frac{1}{R_1} + \frac{1}{R_2} + \frac{1}{R_3} + \cdots + \frac{1}{R_N}$$ **(6.1)**

Since $G = 1/R$, the equation can also be written in terms of conductance levels as follows:

$$G_T = G_1 + G_2 + G_3 + \cdots + G_N \quad \text{(siemens, S)}$$ **(6.2)**

which is an exact match in format with the equation for the total resistance of resistors in series: $R_T = R_1 + R_2 + R_3 + \cdots + R_N$. The result of this duality is that you can go from one equation to the other simply by interchanging R and G.

FIG. 6.4
Parallel combination of resistors.

In general, however, when the total resistance is desired, the following format is applied:

$$R_T = \frac{1}{\dfrac{1}{R_1} + \dfrac{1}{R_2} + \dfrac{1}{R_3} + \cdots + \dfrac{1}{R_N}}$$ **(6.3)**

Quite obviously, Equation (6.3) is not as "clean" as the equation for the total resistance of series resistors. Care must be taken when dealing with all the divisions into 1. The great feature about the equation, however, is that it can be applied to any number of resistors in parallel.

EXAMPLE 6.1 Find the total resistance of the parallel resistors of Fig. 6.5.

Solution: Applying Eq. (6.3):

$$R_T = \frac{1}{\dfrac{1}{R_1} + \dfrac{1}{R_2}} = \frac{1}{\dfrac{1}{3\ \Omega} + \dfrac{1}{6\ \Omega}}$$

$$= \frac{1}{0.333\ \text{S} + 0.167\ \text{S}} = \frac{1}{0.5\ \text{S}} = 2\ \Omega$$

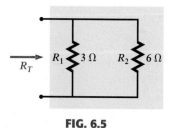

FIG. 6.5
Parallel resistors for Example 6.1.

EXAMPLE 6.2 Find the total resistance of the parallel resistors of Fig. 6.6.

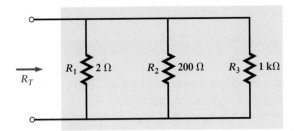

FIG. 6.6
Parallel resistors for Example 6.2.

Solution: Applying Eq. (6.3):

$$R_T = \cfrac{1}{\cfrac{1}{R_1} + \cfrac{1}{R_2} + \cfrac{1}{R_3}} = \cfrac{1}{\cfrac{1}{2\ \Omega} + \cfrac{1}{200\ \Omega} + \cfrac{1}{1\ k\Omega}}$$

$$= \cfrac{1}{0.5\ S + 0.005\ S + 0.001\ S} = \cfrac{1}{0.506\ S} = \mathbf{1.976\ \Omega}$$

EXAMPLE 6.3 Find the total resistance of the configuration of Fig. 6.7.

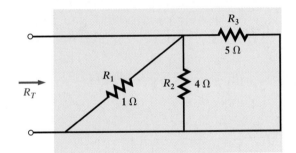

FIG. 6.7
Network to be investigated in Example 6.3.

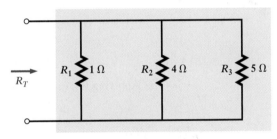

FIG. 6.8
Network of Fig. 6.7 redrawn.

Solution: First the network is redrawn as shown in Fig. 6.8 to clearly demonstrate that all the resistors are in parallel.

Applying Eq. (6.3):

$$R_T = \cfrac{1}{\cfrac{1}{R_1} + \cfrac{1}{R_2} + \cfrac{1}{R_3}} = \cfrac{1}{\cfrac{1}{1\ \Omega} + \cfrac{1}{4\ \Omega} + \cfrac{1}{5\ \Omega}}$$

$$= \cfrac{1}{1\ S + 0.25\ S + 0.2\ S} = \cfrac{1}{1.45\ S} \cong \mathbf{0.69\ \Omega}$$

If you review the examples above, you will find that the total resistance is less than the smallest parallel resistor. That is, in Example 6.1, 2 Ω is less than 3 Ω or 6 Ω. In Example 6.2, 1.976 Ω is less than 2 Ω, 100 Ω, or 1 kΩ; and in Example 6.3, 0.69 Ω is less than 1 Ω, 4 Ω, or 5 Ω. In general, therefore,

the total resistance of parallel resistors is always less than the value of the smallest resistor.

This is particularly important when you want a quick estimate of the total resistance of a parallel combination. Simply find the smallest value, and you know that the total resistance will be less than that value. It is also a great check on your calculations. In addition, you will find that

if the smallest resistor of a parallel combination is much smaller than the other parallel resistors, the total resistance will be very close to the smallest resistor value.

This fact is pretty obvious in Example 6.2 where the total resistance of 1.976 Ω is very close to the smallest resistor of 2 Ω.

Another interesting characteristic of parallel resistors will now be demonstrated in Example 6.4.

EXAMPLE 6.4

a. What is the effect of adding another resistor of 100 Ω in parallel with the parallel resistors of Example 6.1 as shown in Fig. 6.9?
b. What is the effect of adding a parallel 1 Ω resistor to the configuration of Fig. 6.9?

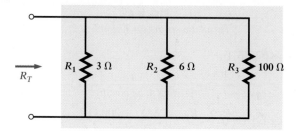

FIG. 6.9
Adding a parallel 100 Ω resistor to the network of Fig. 6.5.

Solutions:
a. Applying Eq. (6.3):

$$R_T = \frac{1}{\dfrac{1}{R_1} + \dfrac{1}{R_2} + \dfrac{1}{R_3}} = \frac{1}{\dfrac{1}{3\ \Omega} + \dfrac{1}{6\ \Omega} + \dfrac{1}{100\ \Omega}}$$

$$= \frac{1}{0.333\ \text{S} + 0.167\ \text{S} + 0.010\ \text{S}} = \frac{1}{0.510\ \text{S}} = \mathbf{1.961\ \Omega}$$

b. Applying Eq. (6.3):

$$R_T = \frac{1}{\dfrac{1}{R_1} + \dfrac{1}{R_2} + \dfrac{1}{R_3} + \dfrac{1}{R_4}} = \frac{1}{\dfrac{1}{3\ \Omega} + \dfrac{1}{6\ \Omega} + \dfrac{1}{100\ \Omega} + \dfrac{1}{1\ \Omega}}$$

$$= \frac{1}{0.333\ \text{S} + 0.167\ \text{S} + 0.010\ \text{S} + 1\ \text{S}} = \frac{1}{1.51\ \text{S}} = \mathbf{0.662\ \Omega}$$

In part (a) of Example 6.4, the total resistance dropped from 2 Ω to 1.961 Ω. In part (b), it dropped to 0.662 Ω. The results clearly reveal that

the total resistance of parallel resistors will always drop as new resistors are added in parallel, irrespective of their value.

Recall that this is the opposite of series resistors, where additional resistors of any value increase the total resistance.

For equal resistors in parallel, the equation for the total resistance becomes significantly easier to apply. For *N* equal resistors in parallel, Equation (6.3) becomes

$$R_T = \frac{1}{\dfrac{1}{R} + \dfrac{1}{R} + \dfrac{1}{R} + \cdots + \dfrac{1}{R_N}}$$

$$= \frac{1}{N\left(\dfrac{1}{R}\right)} = \frac{1}{\dfrac{N}{R}}$$

and
$$\boxed{R_T = \frac{R}{N}}$$
(6.4)

In other words,

the total resistance of N parallel resistors of equal value is the resistance of one resistor divided by the number (N) of parallel resistors.

EXAMPLE 6.5 Find the total resistance of the parallel resistors of Fig. 6.10.

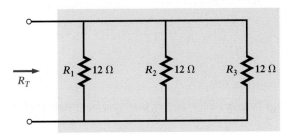

FIG. 6.10

Three equal parallel resistors to be investigated in Example 6.5.

Solution: Applying Eq. (6.4):

$$R_T = \frac{R}{N} = \frac{12\ \Omega}{3} = \textbf{4}\ \boldsymbol{\Omega}$$

EXAMPLE 6.6 Find the total resistance for the configuration of Fig. 6.11.

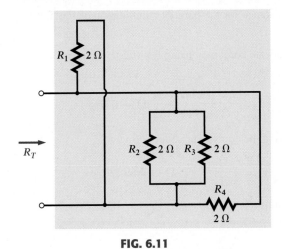

FIG. 6.11

Parallel configuration for Example 6.6.

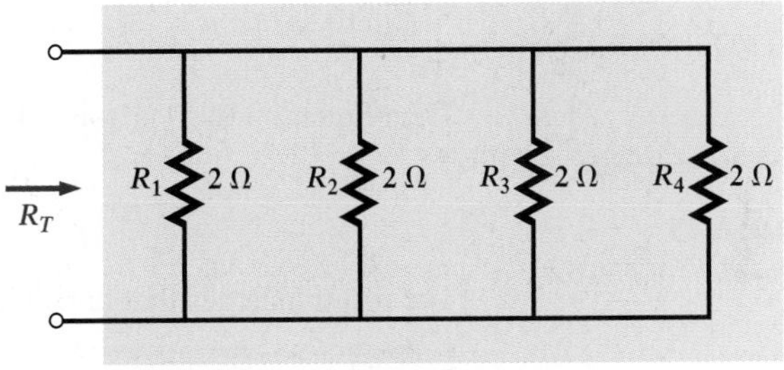

FIG. 6.12

Network of Fig. 6.11 redrawn.

Solution: Redrawing the network will result in the parallel network of Fig. 6.12.
Applying Eq. (6.4):

$$R_T = \frac{R}{N} = \frac{2\ \Omega}{4} = \textbf{0.5}\ \boldsymbol{\Omega}$$

Special Case: Two Parallel Resistors

In the vast majority of cases, only two or three parallel resistors will have to be combined. With this in mind, an equation has been derived for two parallel resistors that is easy to apply and removes the need to continually worry about dividing into 1 and possibly misplacing a decimal point. For three parallel resistors, the equation to be derived here can be applied twice, or Equation (6.3) can be used.

For two parallel resistors, the total resistance is determined by Eq. (6.1):

$$\frac{1}{R_T} = \frac{1}{R_1} + \frac{1}{R_2}$$

Multiplying the top and bottom of each term of the right side of the equation by the other resistor will result in

$$\frac{1}{R_T} = \left(\frac{R_2}{R_2}\right)\frac{1}{R_1} + \left(\frac{R_1}{R_1}\right)\frac{1}{R_2} = \frac{R_2}{R_1R_2} + \frac{R_1}{R_1R_2}$$

$$\frac{1}{R_T} = \frac{R_2 + R_1}{R_1R_2}$$

and

$$\boxed{R_T = \frac{R_1R_2}{R_1 + R_2}}$$

(6.5)

In words, the equation states that

the total resistance of two parallel resistors is simply the product of their values divided by their sum.

EXAMPLE 6.7 Repeat Example 6.1 using Eq. (6.5).

Solution: Eq. (6.5):

$$R_T = \frac{R_1R_2}{R_1 + R_2} = \frac{(3\ \Omega)(6\ \Omega)}{3\ \Omega + 6\ \Omega} = \frac{18}{9}\Omega = \mathbf{2\ \Omega}$$

which matches the earlier solution.

EXAMPLE 6.8 Determine the total resistance for the parallel combination of Fig. 6.8 using two applications of Eq. (6.5).

Solution: First the 1 Ω and 4 Ω resistors are combined using Eq. (6.5), resulting in the reduced network of Fig. 6.13.

Eq. (6.4):

$$R'_T = \frac{R_1R_2}{R_1 + R_2} = \frac{(1\ \Omega)(4\ \Omega)}{1\ \Omega + 4\ \Omega} = \frac{4}{5}\Omega = 0.8\ \Omega$$

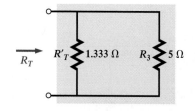

FIG. 6.13
Reduced equivalent of Fig. 6.8.

Then Equation (6.5) is applied again using the equivalent value:

$$R_T = \frac{R'_T R_3}{R'_T + R_3} = \frac{(0.8\ \Omega)(5\ \Omega)}{0.8\ \Omega + 5\ \Omega} = \frac{4}{5.8}\Omega = \mathbf{0.69\ \Omega}$$

The result matches that obtained in Example 6.3.

Recall that series elements can be interchanged without affecting the magnitude of the total resistance. In parallel networks,

parallel resistors can be interchanged without affecting the total resistance.

The next example will demonstrate this fact and reveal how redrawing a network can often define which operations or equations should be applied.

EXAMPLE 6.9 Determine the total resistance of the parallel elements of Fig. 6.14.

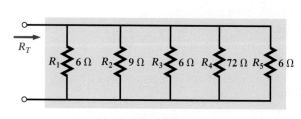

FIG. 6.14
Parallel network for Example 6.9.

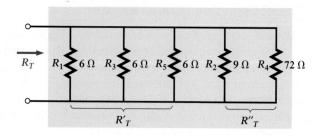

FIG. 6.15
Redrawn network of Fig. 6.14 (Example 6.9).

Solution: The network is redrawn in Fig. 6.15.

Eq. (6.4):
$$R_T' = \frac{R}{N} = \frac{6\,\Omega}{3} = 2\,\Omega$$

Eq. (6.5):
$$R_T'' = \frac{R_2 R_4}{R_2 + R_4} = \frac{(9\,\Omega)(72\,\Omega)}{9\,\Omega + 72\,\Omega} = \frac{648}{81}\,\Omega = 8\,\Omega$$

Eq. (6.5):
$$R_T = \frac{R_T' R_T''}{R_T' + R_T''} = \frac{(2\,\Omega)(8\,\Omega)}{2\,\Omega + 8\,\Omega} = \frac{16}{10}\,\Omega = \mathbf{1.6\,\Omega}$$

The preceding examples involve direct substitution; that is, once the proper equation has been defined, it is only a matter of plugging in the numbers and performing the required algebraic manipulations. The next two examples have a design orientation, in which specific network parameters are defined and the circuit elements must be determined.

EXAMPLE 6.10 Determine the value of R_2 in Fig. 6.16 to establish a total resistance of 9 kΩ.

Solution:

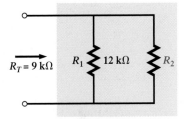

FIG. 6.16
Parallel network for Example 6.10.

$$R_T = \frac{R_1 R_2}{R_1 + R_2}$$

$$R_T(R_1 + R_2) = R_1 R_2$$

$$R_T R_1 + R_T R_2 = R_1 R_2$$

$$R_T R_1 = R_1 R_2 - R_T R_2$$

$$R_T R_1 = (R_1 - R_T)R_2$$

and
$$R_2 = \frac{R_T R_1}{R_1 - R_T}$$

Substituting values:

$$R_2 = \frac{(9\,k\Omega)(12\,k\Omega)}{12\,k\Omega - 9\,k\Omega} = \frac{108}{3}\,k\Omega = \mathbf{36\,k\Omega}$$

EXAMPLE 6.11 Determine the values of R_1, R_2, and R_3 in Fig. 6.17 if $R_2 = 2R_1$, $R_3 = 2R_2$, and the total resistance is 16 kΩ.

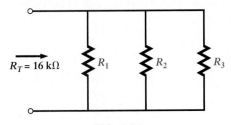

FIG. 6.17
Parallel network for Example 6.11.

Solution: Eq. (6.1):

$$\frac{1}{R_T} = \frac{1}{R_1} + \frac{1}{R_2} + \frac{1}{R_3}$$

However,
$$R_2 = 2R_1 \quad \text{and} \quad R_3 = 2R_2 = 2(2R_1) = 4R_1$$

so that
$$\frac{1}{16\,k\Omega} = \frac{1}{R_1} + \frac{1}{2R_1} + \frac{1}{4R_1}$$

and
$$\frac{1}{16\,k\Omega} = \frac{1}{R_1} + \frac{1}{2}\left(\frac{1}{R_1}\right) + \frac{1}{4}\left(\frac{1}{R_1}\right)$$

or

$$\frac{1}{16 \text{ k}\Omega} = 1.75 \left(\frac{1}{R_1} \right)$$

resulting in $R_1 = 1.75(16 \text{ k}\Omega) = \textbf{28 k}\boldsymbol{\Omega}$

so that $R_2 = 2R_1 = 2(28 \text{ k}\Omega) = \textbf{56 k}\boldsymbol{\Omega}$

and $R_3 = 2R_2 = 2(56 \text{ k}\Omega) = \textbf{112 k}\boldsymbol{\Omega}$

Analogies

Analogies were effectively used to introduce the concept of series elements. They can also be used for parallel elements, as shown in Fig. 6.18. In Fig. 6.18(a), parallel-connected cables are used to hold up the roadway. The top of each cable is connected to the supporting beams, while the bottom of each cable is connected to the roadway. In Fig. 6.18(b), a ladder set on its side shows that its rungs are in parallel because they are connected by the running sides of the ladder.

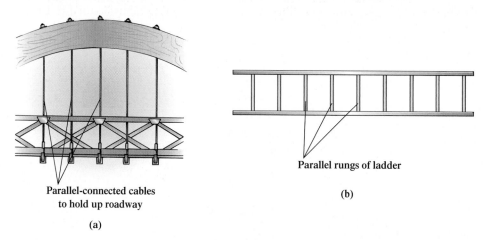

Parallel rungs of ladder

(b)

Parallel-connected cables
to hold up roadway

(a)

FIG. 6.18

Analogies for the parallel connection of elements.

Instrumentation

As shown in Fig. 6.19, the total resistance of a parallel combination of resistive elements can be found by simply applying an ohmmeter. There is no polarity to resistance, so either lead of the ohmmeter can be connected to either side of the network. Although there are no

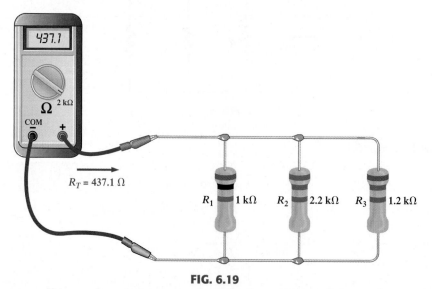

$R_T = 437.1 \ \Omega$

R_1 1 kΩ R_2 2.2 kΩ R_3 1.2 kΩ

FIG. 6.19

Using an ohmmeter to measure the total resistance of a parallel network.

supplies in Fig. 6.19, always keep in mind that ohmmeters can never be applied to a "live" circuit. It is not enough to simply set the supply to 0 V or to turn it off. It may still load down (change the network configuration of) the circuit and change the reading. It is best to remove the supply and apply the ohmmeter to the two resulting terminals. Since all the resistors are in the kilohm range, the 20 kΩ scale was chosen first. We then moved down to the 2 kΩ scale for increased precision. Moving down to the 200 Ω scale resulted in an "OL" indication since we were below the measured resistance value.

6.3 PARALLEL CIRCUITS

A **parallel circuit** can now be established by connecting a supply across a set of parallel resistors as shown in Fig. 6.20(a). Note that the negative terminal of the supply is again connected to the ground terminal of the supply to ground the circuit. The supply was set at 12 V, resulting in a current of 16 mA through the supply. The schematic representation of the network appears in Fig. 6.20(b). Note again that the negative terminal of the supply is connected directly to ground.

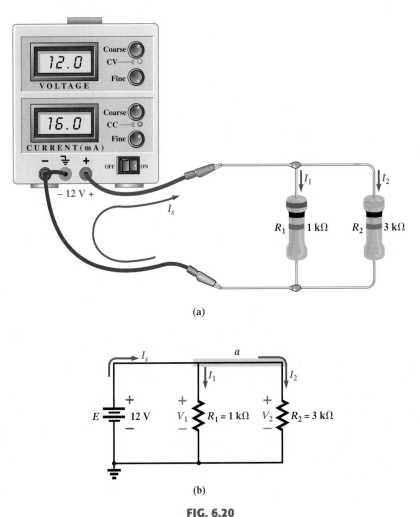

(a)

(b)

FIG. 6.20

Parallel circuit: (a) actual; (b) schematic.

The positive terminal of the supply is directly connected to the top of each resistor, while the negative terminal is connected to the bottom of each resistor. Therefore, it should be quite clear that the applied voltage is the same across each resistor. In general,

the voltage is always the same across parallel elements.

For the future, therefore, remember that

if two elements are in parallel, the voltage across them must be the same. However, if the voltage across two neighboring elements is the same, the two elements may or may not be in parallel.

The reason for this qualifying comment in the above statement will be discussed in detail in Chapter 7.

For the voltages of the circuit of Fig. 6.20, the result is that

$$V_1 = V_2 = E \qquad (6.6)$$

Once the supply has been connected, a source current will be established through the supply that will pass through the parallel resistors. The current that will result is a direct function of the total resistance of the parallel circuit. The smaller the total resistance, the more the current, as occurred for series circuits also.

Recall from series circuits that the source does not "see" the parallel combination of elements. It reacts only to the total resistance of the circuit, as shown in Fig. 6.21. The source current can then be determined using Ohm's law:

$$I_s = \frac{E}{R_T} \qquad (6.7)$$

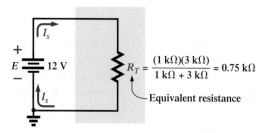

$$R_T = \frac{(1\ \text{k}\Omega)(3\ \text{k}\Omega)}{1\ \text{k}\Omega + 3\ \text{k}\Omega} = 0.75\ \text{k}\Omega$$

Equivalent resistance

FIG. 6.21

Replacing the parallel resistors of Fig. 6.20 with the equivalent total resistance.

Since the voltage is the same across parallel elements, the current through each resistor can also be determined using Ohm's law. That is,

$$I_1 = \frac{V_1}{R_1} = \frac{E}{R_1} \quad \text{and} \quad I_2 = \frac{V_2}{R_2} = \frac{E}{R_2} \qquad (6.8)$$

The direction for the currents is dictated by the polarity of the voltage across the resistors. Recall that for a resistor, current enters the positive side of a potential drop and leaves the negative. The result, as shown in Fig. 6.20(b), is that the source current enters point *a*, and currents I_1 and I_2 leave the same point. An excellent analogy for describing the flow of charge through the network of Fig. 6.20 is the flow of water through the parallel pipes of Fig. 6.22. The larger pipe with less "resistance" to the flow of water will have a larger flow of water through it. The thinner pipe with its increased "resistance" level will have less water through it. In any case, the total water entering the pipes at the top Q_T must equal that leaving at the bottom, with $Q_T = Q_1 + Q_2$.

The relationship between the source current and the parallel resistor currents can be derived by simply taking the equation for the total resistance in Eq. (6.1) form:

$$\frac{1}{R_T} = \frac{1}{R_1} + \frac{1}{R_2}$$

Multiplying both sides by the applied voltage:

$$E\left(\frac{1}{R_T}\right) = E\left(\frac{1}{R_1} + \frac{1}{R_2}\right)$$

resulting in

$$\frac{E}{R_T} = \frac{E}{R_1} + \frac{E}{R_2}$$

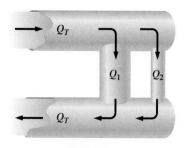

FIG. 6.22

Mechanical analogy for Fig. 6.20.

Then note that $E/R_1 = I_1$ and $E/R_2 = I_2$ to obtain

$$\boxed{I_s = I_1 + I_2}$$ **(6.9)**

The result reveals a very important property of parallel circuits:

For single-source parallel networks, the source current (I_s) is always equal to the sum of the individual branch currents.

The duality that exists between series and parallel circuits continues to pop up as we proceed through the basic equations for electric circuits. This is fortunate because it provides a way of remembering the characteristics of one using the results of another. For instance, in Fig. 6.23(a), we have a parallel circuit where it is clear that $I_T = I_1 + I_2$. By simply replacing the currents of the equation in Fig. 6.23(a) by a voltage level, as shown in Fig. 6.23(b), we have Kirchhoff's voltage law for a series circuit: $E = V_1 + V_2$. In other words,

for a parallel circuit, the source current equals the sum of the branch currents, while for a series circuit, the applied voltage equals the sum of the voltage drops.

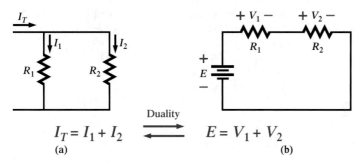

FIG. 6.23
Demonstrating the duality that exists between series and parallel circuits.

EXAMPLE 6.12   For the parallel network of Fig. 6.24:

a. Find the total resistance.
b. Calculate the source current.
c. Determine the current through each parallel branch.
d. Show that Equation (6.9) is satisfied.

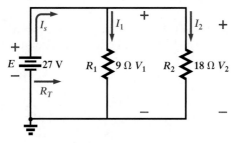

FIG. 6.24
Parallel circuit for Example 6.12.

Solutions:

a. Using Eq. (6.5):

$$R_T = \frac{R_1 R_2}{R_1 + R_2} = \frac{(9 \ \Omega)(18 \ \Omega)}{9 \ \Omega + 18 \ \Omega} = \frac{162}{27} \ \Omega = \textbf{6 } \boldsymbol{\Omega}$$

b. Applying Ohm's law:

$$I_s = \frac{E}{R_T} = \frac{27 \ \text{V}}{6 \ \Omega} = \textbf{4.5 A}$$

c. Applying Ohm's law:

$$I_1 = \frac{V_1}{R_1} = \frac{E}{R_1} = \frac{27 \text{ V}}{9 \text{ }\Omega} = \textbf{3 A}$$

$$I_2 = \frac{V_2}{R_2} = \frac{E}{R_2} = \frac{27 \text{ V}}{18 \text{ }\Omega} = \textbf{1.5 A}$$

d. Substituting values from parts (b) and (c):

$$I_s = \textbf{4.5 A} = I_1 + I_2 = 3 \text{ A} + 1.5 \text{ A} = \textbf{4.5 A} \quad \text{(checks)}$$

EXAMPLE 6.13 For the parallel network of Fig. 6.25:

a. Find the total resistance.
b. Calculate the source current.
c. Determine the current through each branch.

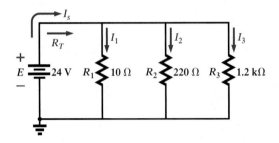

FIG. 6.25
Parallel network for Example 6.13.

Solutions:

a. Applying Eq. (6.3):

$$R_T = \frac{1}{\dfrac{1}{R_1} + \dfrac{1}{R_2} + \dfrac{1}{R_3}} = \frac{1}{\dfrac{1}{10 \text{ }\Omega} + \dfrac{1}{220 \text{ }\Omega} + \dfrac{1}{1.2 \text{ k}\Omega}}$$

$$= \frac{1}{100 \times 10^{-3} + 4.545 \times 10^{-3} + 0.833 \times 10^{-3}} = \frac{1}{105.38 \times 10^{-3}}$$

$$R_T = \textbf{9.489 }\Omega$$

Note that the total resistance is less than the smallest parallel resistor, and the magnitude is very close to the smallest resistor because the other resistors are larger by a factor greater than 10 : 1.

b. Using Ohm's law:

$$I_s = \frac{E}{R_T} = \frac{24 \text{ V}}{9.489 \text{ }\Omega} = \textbf{2.529 A}$$

c. Applying Ohm's law:

$$I_1 = \frac{V_1}{R_1} = \frac{E}{R_1} = \frac{24 \text{ V}}{10 \text{ }\Omega} = \textbf{2.4 A}$$

$$I_2 = \frac{V_2}{R_2} = \frac{E}{R_2} = \frac{24 \text{ V}}{220 \text{ }\Omega} = \textbf{0.109 A}$$

$$I_3 = \frac{V_3}{R_3} = \frac{E}{R_3} = \frac{24 \text{ V}}{1.2 \text{ k}\Omega} = \textbf{0.020 A}$$

A careful examination of the results of Example 6.13 will reveal that the larger the parallel resistor, the smaller the branch current. In general, therefore, *for parallel resistors, the greatest current will exist in the branch with the least resistance.*

A more powerful statement is that

current always seeks the path of least resistance.

EXAMPLE 6.14 Given the information provided in Fig. 6.26:

a. Determine R_3.
b. Find the applied voltage E.
c. Find the source current I_s.
d. Find I_2.

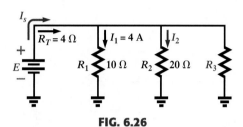

FIG. 6.26
Parallel network for Example 6.14.

Solutions:

a. Applying Eq. (6.1):

$$\frac{1}{R_T} = \frac{1}{R_1} + \frac{1}{R_2} + \frac{1}{R_3}$$

Substituting:

$$\frac{1}{4\ \Omega} = \frac{1}{10\ \Omega} + \frac{1}{20\ \Omega} + \frac{1}{R_3}$$

so that

$$0.25\ \text{S} = 0.1\ \text{S} + 0.05\ \text{S} + \frac{1}{R_3}$$

and

$$0.25\ \text{S} = 0.15\ \text{S} + \frac{1}{R_3}$$

with

$$\frac{1}{R_3} = 0.1\ \text{S}$$

and

$$R_3 = \frac{1}{0.1\ \text{S}} = \mathbf{10\ \Omega}$$

b. Using Ohm's law:

$$E = V_1 = I_1 R_1 = (4\ \text{A})(10\ \Omega) = \mathbf{40\ V}$$

c.

$$I_s = \frac{E}{R_T} = \frac{40\ \text{V}}{4\ \Omega} = \mathbf{10\ A}$$

d. Applying Ohm's law:

$$I_2 = \frac{V_2}{R_2} = \frac{E}{R_2} = \frac{40\ \text{V}}{20\ \Omega} = \mathbf{2\ A}$$

Instrumentation

In Fig. 6.27, voltmeters have been connected to verify that the voltage across parallel elements is the same. Note that the positive or red lead of each voltmeter is connected to the high (positive) side of the voltage across each resistor to obtain a positive reading. The 20 V scale was used because the applied voltage exceeded the range of the 2 V scale.

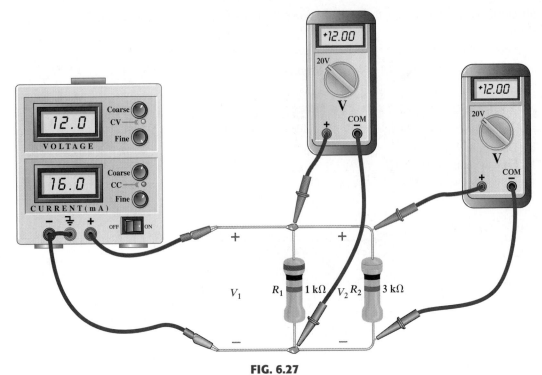

FIG. 6.27
Measuring the voltages of a parallel dc network.

In Fig. 6.28, an ammeter has been hooked up to measure the source current. First, the connection to the supply had to be broken at the positive terminal and the meter inserted as shown. Be sure to use the ammeter terminals on your meter for such measurements. The red or positive lead of the meter is connected such that the source current will enter that lead and leave the negative or black lead to ensure a positive reading. The 200 mA scale was used because the source current exceeded the maximum value of the 2 mA scale. For the moment we will assume that the internal resistance of the meter can be ignored. Since the internal resistance of an ammeter on the 200 mA scale is typically only a few ohms, compared to the parallel resistors in the kilohm range, it is an excellent assumption.

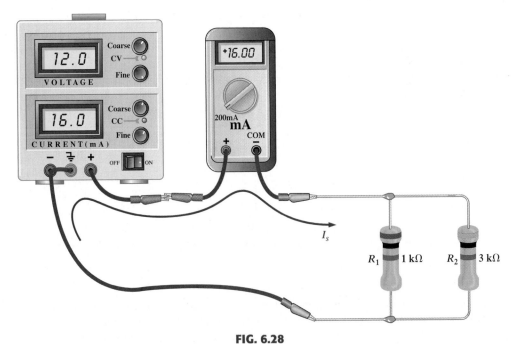

FIG. 6.28
Measuring the source current of a parallel network.

A more difficult measurement is for the current through resistor R_1. This measurement often gives trouble in the laboratory session. First, as shown in Fig. 6.29(a), resistor R_1 must be disconnected from the upper connection point to establish an open circuit. The ammeter is then inserted between the resulting terminals such that the current will enter the positive or red terminal, as shown in Fig. 6.29(b). Always remember: When using an ammeter, first establish an open circuit in the branch in which the current is to be measured, and then insert the meter.

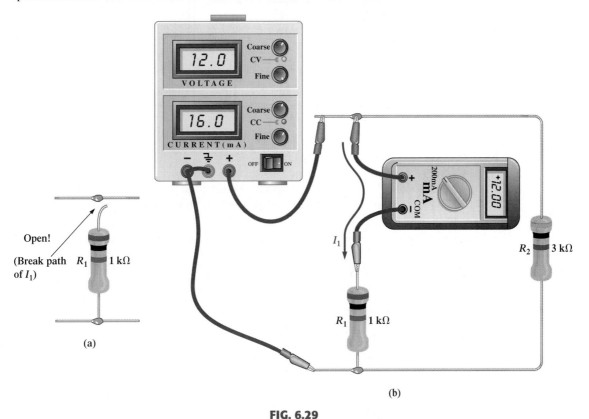

FIG. 6.29
Measuring the current through resistor R_1.

The easiest measurement is for the current through resistor R_2. Simply break the connection to R_2 above or below the resistor, and insert the ammeter with the current entering the positive or red lead to obtain a positive reading.

6.4 POWER DISTRIBUTION IN A PARALLEL CIRCUIT

Recall from the discussion of series circuits that the power applied to a series resistive circuit was equal to the power dissipated by the resistive elements. The same is true for parallel resistive networks. In fact,

for any network composed of resistive elements, the power applied by the battery will equal that dissipated by the resistive elements.

For the parallel circuit of Fig. 6.30:

$$P_E = P_{R_1} + P_{R_2} + P_{R_3}$$
(6.10)

which is exactly the same as obtained for the series combination.

The power delivered by the source is the same:

$$P_E = EI_s \qquad \text{(watts, W)}$$
(6.11)

as is the equation for the power to each resistor (shown for R_1 only):

$$P_1 = V_1 I_1 = I_1^2 R_1 = \frac{V_1^2}{R_1} \qquad \text{(watts, W)}$$
(6.12)

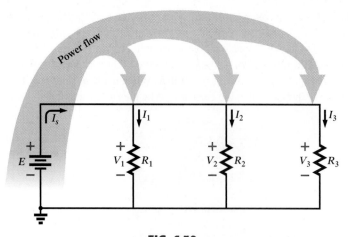

FIG. 6.30
Power flow in a dc parallel network.

In the equation $P = V^2/R$, the voltage across each resistor in a parallel circuit will be the same. The only factor that will change is the resistance in the denominator of the equation. The result is that

in a parallel resistive network, the larger the resistor, the less the power absorbed.

EXAMPLE 6.15 For the parallel network of Fig. 6.31 (all standard values):

a. Determine the total resistance R_T.
b. Find the source current and the current through each resistor.
c. Calculate the power delivered by the source.
d. Determine the power absorbed by each parallel resistor.
e. Verify Eq. (6.10).

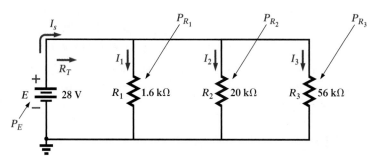

FIG. 6.31
Parallel network for Example 6.15.

Solutions:

a. Without making a single calculation, it should now be apparent from previous examples that the total resistance is less than 1.6 kΩ and very close to this value because of the magnitude of the other resistance levels.

$$R_T = \frac{1}{\dfrac{1}{R_1} + \dfrac{1}{R_2} + \dfrac{1}{R_3}} = \frac{1}{\dfrac{1}{1.6 \text{ k}\Omega} + \dfrac{1}{20 \text{ k}\Omega} + \dfrac{1}{56 \text{ k}\Omega}}$$

$$= \frac{1}{625 \times 10^{-6} + 50 \times 10^{-6} + 17.867 \times 10^{-6}} = \frac{1}{692.867 \times 10^{-6}}$$

and $R_T = \mathbf{1.443 \text{ k}\Omega}$

b. Applying Ohm's law:

$$I_s = \frac{E}{R_T} = \frac{28 \text{ V}}{1.443 \text{ k}\Omega} = \mathbf{19.4 \text{ mA}}$$

Recalling that current always seeks the path of least resistance immediately tells us that the current through the 1.6 kΩ resistor will be the largest and the current through the 56 kΩ resistor the least.

Applying Ohm's law again:

$$I_1 = \frac{V_1}{R_1} = \frac{E}{R_1} = \frac{28 \text{ V}}{1.6 \text{ k}\Omega} = \textbf{17.5 mA}$$

$$I_2 = \frac{V_2}{R_2} = \frac{E}{R_2} = \frac{28 \text{ V}}{20 \text{ k}\Omega} = \textbf{1.4 mA}$$

$$I_3 = \frac{V_3}{R_3} = \frac{E}{R_3} = \frac{28 \text{ V}}{56 \text{ k}\Omega} = \textbf{0.5 mA}$$

c. Applying Eq. (6.11):

$$P_E = EI_s = (28 \text{ V})(19.4 \text{ mA}) = \textbf{543.2 mW}$$

d. Applying each form of the power equation:

$$P_1 = V_1 I_1 = EI_1 = (28 \text{ V})(17.5 \text{ mA}) = \textbf{490 mW}$$

$$P_2 = I_2^2 R_2 = (1.4 \text{ mA})^2 (20 \text{ k}\Omega) = \textbf{39.2 mW}$$

$$P_3 = \frac{V_3^2}{R_3} = \frac{E^2}{R_3} = \frac{(28 \text{ V})^2}{56 \text{ k}\Omega} = \textbf{14 mW}$$

A review of the results clearly substantiates the fact that the larger the resistor, the less the power absorbed.

e. $$P_E = P_{R_1} + P_{R_2} + P_{R_3}$$

$$\textbf{543.2 mW} = 490 \text{ mW} + 39.2 \text{ mW} + 14 \text{ mW} = \textbf{543.2 mW} \quad \text{(checks)}$$

6.5 KIRCHHOFF'S CURRENT LAW

In the previous chapter, Kirchhoff's voltage law was introduced, providing a very important relationship between the voltages of a closed path. Professor Gustav Kirchhoff is also credited with developing the following equally important relationship between the currents of a network, called **Kirchhoff's current law (KCL):**

The algebraic sum of the currents entering and leaving a junction (or region) of a network is zero.

The law can also be stated in the following way:

The sum of the currents entering a junction (or region) of a network must equal the sum of the currents leaving the same junction (or region).

In equation form the above statement can be written as follows:

$$\boxed{\Sigma I_i = \Sigma I_o} \tag{6.13}$$

with I_i representing the current entering, or "in," and I_o representing the current leaving, or "out."

In Fig. 6.32, for example, the shaded area can enclose an entire system or a complex network, or it can simply provide a connection point (junction) for the displayed currents. In each case, the current entering must equal that leaving, as required by Eq. (6.13):

$$\Sigma I_i = \Sigma I_o$$

$$I_1 + I_4 = I_2 + I_3$$

$$4 \text{ A} + 8 \text{ A} = 2 \text{ A} + 10 \text{ A}$$

$$\textbf{12 A} = \textbf{12 A} \quad \text{(checks)}$$

The most common application of the law will be at a junction of two or more current paths, as shown in Fig. 6.33(a). For some students it is difficult initially to determine whether a current is entering or leaving a junction. One approach that may help is to picture yourself as

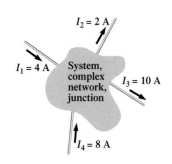

FIG. 6.32
Introducing Kirchhoff's current law.

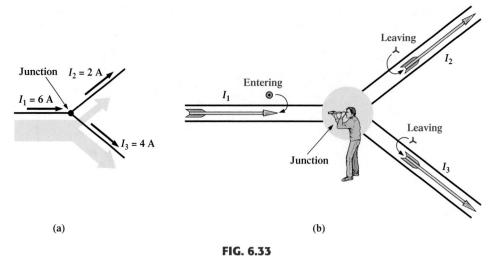

FIG. 6.33

*(a) Demonstrating Kirchhoff's current law; (b) determining whether a current
is entering or leaving a junction.*

standing at the junction and treating the currents as arrows, as shown in Fig. 6.33(b). For I_1,
the arrow is definitely coming in our direction. Looking at I_2 and I_3, we simply see the tail of
the arrow as it leaves the area. Applying Kirchhoff's current law at the junction of Fig. 6.33(a)
or (b) will result in

$$\Sigma I_i = \Sigma I_o$$

$$I_1 = I_2 + I_3$$

$$6 \text{ A} = 2 \text{ A} + 4 \text{ A}$$

$$6 \text{ A} = 6 \text{ A} \quad \text{(checks)}$$

Perhaps the water analogy of Fig. 6.34 will be helpful in understanding the application
of Eq. (6.13). As we stand on the small bridge at the point where the brook divides, we can
see the water coming at us as Q_1 and the water leaving us as Q_2 and Q_3. The result is $Q_1 =
Q_2 + Q_3$.

FIG. 6.34

Water analogy for the junction of Fig. 6.33(a).

In the next few examples, unknown currents can be determined by applying Kirchhoff's current law. Simply remember to place all current levels entering the junction to
the left of the equals sign and the sum of all currents leaving the junction to the right of
the equals sign.

In technology the term **node** is commonly used to refer to a junction of two or more
branches. Therefore, this term will be used frequently in the analyses to follow.

EXAMPLE 6.16 Determine currents I_3 and I_4 of Fig. 6.35 using Kirchhoff's current law.

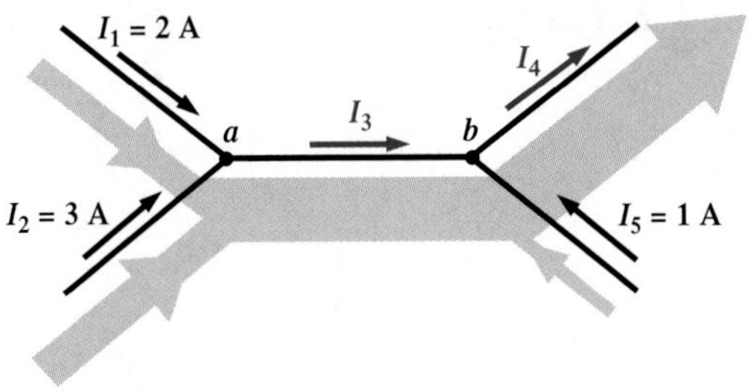

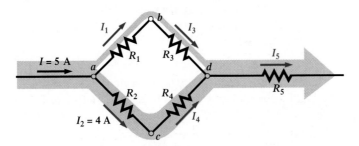

FIG. 6.35

Two-node configuration for Example 6.16.

Solution: There are two junctions or nodes in Fig. 6.35. Node *a* has only one unknown, while node *b* has two unknowns. Since a single equation can be used to solve for only one unknown, we must apply Kirchhoff's current law to node *a* first.

At node *a*:

$$\Sigma I_i = \Sigma I_o$$

$$I_1 + I_2 = I_3$$

$$2\,A + 3\,A = I_3 = \mathbf{5\,A}$$

At node *b*, using the result just obtained:

$$\Sigma I_i = \Sigma I_o$$

$$I_3 + I_5 = I_4$$

$$5\,A + 1\,A = I_4 = \mathbf{6\,A}$$

Note that in Fig. 6.35, the width of the blue shaded regions matches the magnitude of the current in that region.

EXAMPLE 6.17   Determine currents I_1, I_3, I_4, and I_5 for the network of Fig. 6.36.

FIG. 6.36

Four-node configuration for Example 6.17.

Solution: In this configuration four nodes are defined. Nodes *a* and *c* have only one unknown current at the junction, so Kirchhoff's current law can be applied at either junction.

At node *a*:

$$\Sigma I_i = \Sigma I_o$$

$$I = I_1 + I_2$$

$$5\,A = I_1 + 4\,A$$

and

$$I_1 = 5\,A - 4\,A = \mathbf{1\,A}$$

At node *c*:

$$\Sigma I_i = \Sigma I_o$$

$$I_2 = I_4$$

and

$$I_4 = I_2 = \mathbf{4\,A}$$

Using the above results at the other junctions will result in the following.
   At node b:

$$\Sigma I_i = \Sigma I_o$$
$$I_1 = I_3$$

and
$$I_3 = I_1 = \mathbf{1\,A}$$

   At node d:

$$\Sigma I_i = \Sigma I_o$$
$$I_3 + I_4 = I_5$$
$$1\,A + 4\,A = I_5 = \mathbf{5\,A}$$

   If we enclose the entire network, we find that the current entering from the far left is $I = 5$ A, while the current leaving from the far right is $I_5 = 5$ A. The two must be equal since the net current entering any system must equal the net current leaving.

EXAMPLE 6.18   Determine currents I_3 and I_5 of Fig. 6.37 through applications of Kirchhoff's current law.

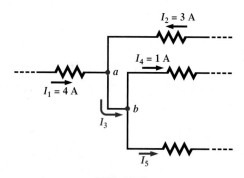

FIG. 6.37
Network for Example 6.18.

Solution: Note first that since node b has two unknown quantities (I_3 and I_5), and node a has only one, Kirchhoff's current law must first be applied to node a. The result will then be applied to node b.
   At node a:

$$\Sigma I_i = \Sigma I_o$$
$$I_1 + I_2 = I_3$$
$$4\,A + 3\,A = I_3 = \mathbf{7\,A}$$

   At node b:

$$\Sigma I_i = \Sigma I_o$$
$$I_3 = I_4 + I_5$$
$$7\,A = 1\,A + I_5$$

and
$$I_5 = 7\,A - 1\,A = \mathbf{6\,A}$$

EXAMPLE 6.19   For the parallel dc network of Fig. 6.38:

a. Determine the source current I_s.
b. Find the source voltage E.
c. Determine R_3.
d. Calculate R_T.

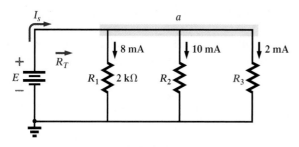

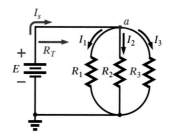

FIG. 6.38

Parallel network for Example 6.19.

FIG. 6.39

Redrawn network of Fig. 6.38.

Solutions:

a. First apply Eq. (6.13) at node a. Although node a in Fig. 6.38 may not initially appear as a single junction, it can be redrawn as shown in Fig. 6.39, where it is clearly a common point for all the branches.

The result is

$$\Sigma I_i = \Sigma I_o$$

$$I_s = I_1 + I_2 + I_3$$

Substituting values:　　$I_s = 8 \text{ mA} + 10 \text{ mA} + 2 \text{ mA} = \textbf{20 mA}$

Note in this solution that there is no need to know the resistor values or the voltage applied. The solution was determined solely by the current levels.

b. Applying Ohm's law:

$$E = V_1 = I_1 R_1 = (8 \text{ mA})(2 \text{ k}\Omega) = \textbf{16 V}$$

c. Applying Ohm's law in a different form:

$$R_3 = \frac{V_3}{I_3} = \frac{E}{I_3} = \frac{16 \text{ V}}{2 \text{ mA}} = \textbf{8 k}\Omega$$

d. Applying Ohm's law again:

$$R_T = \frac{E}{I_s} = \frac{16 \text{ V}}{20 \text{ mA}} = \textbf{0.8 k}\Omega$$

The application of Kirchhoff's current law is not limited to networks where all the internal connections are known or visible. For instance, all the currents of the integrated circuit of Fig. 6.40 are known except I_1. By treating the entire system (which could contain over a million elements) as a single node, we can apply Kirchhoff's current law as shown in Example 6.20.

Before looking at Example 6.20 in detail, note that the direction of the unknown current I_1 is not provided in Fig. 6.40. On many occasions this will be the case. With so many currents entering or leaving the system, it is difficult to know by inspection which direction should be assigned to I_1. **In such cases, simply make an assumption about the direction and then check out the result. If the result is negative, the wrong direction was assumed. If the result is positive, the correct direction was assumed. In either case, the magnitude of the current will be correct.**

EXAMPLE 6.20　Determine I_1 for the integrated circuit of Fig. 6.40.

Solution:　Assuming that the current I_1 enters the chip will result in the following when Kirchhoff's current law is applied:

$$\Sigma I_i = \Sigma I_o$$

$$I_1 + 10 \text{ mA} + 4 \text{ mA} + 8 \text{ mA} = 5 \text{ mA} + 4 \text{ mA} + 2 \text{ mA} + 6 \text{ mA}$$

$$I_1 + 22 \text{ mA} = 17 \text{ mA}$$

$$I_1 = 17 \text{ mA} - 22 \text{ mA} = \textbf{-5 mA}$$

We find that the direction for I_1 is **leaving** the IC, although the magnitude of 5 mA is correct.

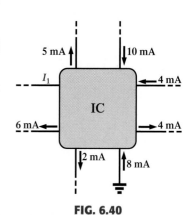

FIG. 6.40

Integrated circuit for Example 6.20.

As we leave this important section, be aware that Kirchhoff's current law will be applied in one form or another throughout the text. **Kirchhoff's laws are unquestionably two of the most important in this field because they are applicable to the most complex configurations in existence today.** They will not be replaced by a more important law or dropped for a more sophisticated approach.

6.6   CURRENT DIVIDER RULE

For series circuits we have the powerful voltage divider rule for finding the voltage across a resistor in a series circuit. We now introduce the equally powerful **current divider rule (CDR)** for finding the current through a resistor in a parallel circuit.

In Section 6.4 it was pointed out that current will always seek the path of least resistance. In Fig. 6.41, for example, the current of 9 A is faced with splitting between the three parallel resistors. Based on the previous sections, it should now be clear without a single calculation that the majority of the current will pass through the smallest resistor of 10 Ω, and the least current will pass through the 1 kΩ resistor. In fact, the current through the 100 Ω resistor will also exceed that through the 1 kΩ resistor. We can take it one step further by recognizing that the resistance of the 100 Ω resistor is 10 times that of the 10 Ω resistor. The result will be a current through the 10 Ω resistor that is 10 times that of the 100 Ω resistor. Similarly, the current through the 100 Ω resistor will be 10 times that through the 1 kΩ resistor.

In general, therefore,

the current entering parallel resistive elements will split as the inverse of their resistive values.

Further,

the current through equal parallel resistors will be the same.

The result of this last statement is that if the 1 kΩ resistor of Fig. 6.41 were changed to 100 Ω, the current through each 100 Ω resistor would be the same **no matter how many other elements were in parallel.**

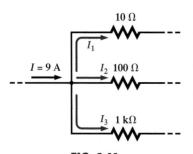

FIG. 6.41
Discussing the manner in which the current will split between three parallel branches of different resistive value.

EXAMPLE 6.21

a. Determine currents I_1 and I_3 for the network of Fig. 6.42.
b. Find the source current I_s.

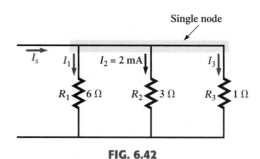

FIG. 6.42
Parallel network for Example 6.21.

Solutions:

a. Since R_1 is twice R_2, the current I_1 must be one-half I_2, and

$$I_1 = \frac{I_2}{2} = \frac{2 \text{ mA}}{2} = \textbf{1 mA}$$

Since R_2 is three times R_3, the current I_3 must be three times I_2, and

$$I_3 = 3I_2 = 3(2 \text{ mA}) = \textbf{6 mA}$$

b. Applying Kirchhoff's current law:

$$\Sigma I_i = \Sigma I_o$$

$$I_s = I_1 + I_2 + I_3$$

$$I_s = 1 \text{ mA} + 2 \text{ mA} + 6 \text{ mA} = \mathbf{9 \text{ mA}}$$

Although the above discussions and examples permitted a determination of the relative magnitude of a current based on a known level, they do not provide the magnitude of a current through a branch of a parallel network if only the total entering current is known. The result is a need for the current divider rule which will be derived using the parallel configuration of Fig. 6.43(a). The current I_T (using the subscript T to indicate the total entering current) will split between the N parallel resistors and then gather itself together again at the bottom of the configuration. In Fig. 6.43(b), the parallel combination of resistors has been replaced by a single resistor equal to the total resistance of the parallel combination as determined in the previous sections.

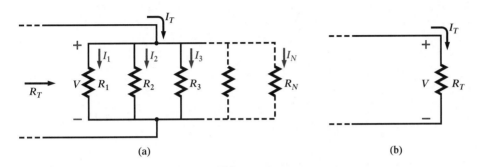

(a)   (b)

FIG. 6.43

Deriving the current divider rule: (a) parallel network of N parallel resistors; (b) reduced equivalent of part (a).

The current I_T can then be determined using Ohm's law:

$$I_T = \frac{V}{R_T}$$

Since the voltage V is the same across parallel elements, the following is true:

$$V = I_1 R_1 = I_2 R_2 = I_3 R_3 = \cdots = I_x R_x$$

where the product $I_x R_x$ refers to any combination in the series.

Substituting for V in the above equation for I_T, we have

$$I_T = \frac{I_x R_x}{R_T}$$

Solving for I_x, the final result is the **current divider rule:**

$$\boxed{I_x = \frac{R_T}{R_x} I_T} \qquad\qquad \textbf{(6.14)}$$

In words, it states that

the current through any branch of a parallel resistive network is equal to the total resistance of the parallel network divided by the resistor of interest and multiplied by the total current entering the parallel configuration.

Since R_T and I_T are constants, for a particular configuration the larger the value of R_x (in the denominator), the smaller the value of I_x for that branch, confirming the fact that current always seeks the path of least resistance.

EXAMPLE 6.22 For the parallel network of Fig. 6.44, determine current I_1 using Eq. (6.14).

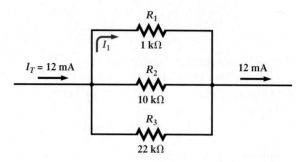

FIG. 6.44

Using the current divider rule to calculate current I_1 in Example 6.22.

Solution: Eq. (6.3):

$$R_T = \frac{1}{\dfrac{1}{R_1} + \dfrac{1}{R_2} + \dfrac{1}{R_3}}$$

$$= \frac{1}{\dfrac{1}{1\ k\Omega} + \dfrac{1}{10\ k\Omega} + \dfrac{1}{22\ k\Omega}} = \frac{1}{1 \times 10^{-3} + 100 \times 10^{-6} + 45.46 \times 10^{-6}}$$

$$= \frac{1}{1.145 \times 10^{-3}} = \textbf{873.01 } \boldsymbol{\Omega}$$

Eq. (6.14): $\quad I_1 = \dfrac{R_T}{R_1} I_T$

$$= \frac{(873.01\ \Omega)}{1\ k\Omega}(12\ \text{mA}) = (0.873)(12\ \text{mA}) = \textbf{10.48 mA}$$

and the smallest parallel resistor receives the majority of the current.

At this point, it should also be pointed out that

for a parallel network, the current through the smallest resistor will be very close to the total entering current if the other parallel elements of the configuration are much larger in magnitude.

Note that in Example 6.22 the current through R_1 is very close to the total current because R_1 is 10 times less than the next smallest resistor.

Special Case: Two Parallel Resistors

For the case of two parallel resistors as shown in Fig. 6.45, the total resistance is determined by

$$R_T = \frac{R_1 R_2}{R_1 + R_2}$$

Substituting R_T into Eq. (6.14) for current I_1 will result in

$$I_1 = \frac{R_T}{R_1} I_T = \frac{\left(\dfrac{R_1 R_2}{R_1 + R_2}\right)}{R_1} I_T$$

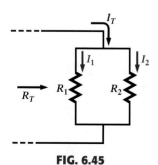

FIG. 6.45

Deriving the current divider rule for the special case of only two parallel resistors.

and

$$I_1 = \left(\frac{R_2}{R_1 + R_2} \right) I_T$$

(6.15a)

Similarly, for I_2,

$$I_2 = \left(\frac{R_1}{R_1 + R_2} \right) I_T$$

(6.15b)

In words, Equation (6.15) states that

for two parallel resistors, the current through one is equal to the other resistor times the total entering current divided by the sum of the two resistors.

Since the combination of two parallel resistors is probably the most common parallel configuration, the simplicity of the format for Eq. (6.15) suggests that it is worth memorizing. Take particular note, however, that the denominator of the equation is simply the **sum,** not the total resistance, of the combination.

EXAMPLE 6.23 Determine current I_2 for the network of Fig. 6.46 using the current divider rule.

Solution: Using Eq. (6.15b):

$$I_2 = \left(\frac{R_1}{R_1 + R_2} \right) I_T$$

$$= \left(\frac{4 \text{ k}\Omega}{4 \text{ k}\Omega + 8 \text{ k}\Omega} \right) 6 \text{ A} = (0.333)(6 \text{ A}) = \mathbf{2\ A}$$

Using Eq. (6.14):

$$I_2 = \frac{R_T}{R_2} I_T$$

with

$$R_T = 4 \text{ k}\Omega \parallel 8 \text{ k}\Omega = \frac{(4 \text{ k}\Omega)(8 \text{ k}\Omega)}{4 \text{ k}\Omega + 8 \text{ k}\Omega} = 2.667 \text{ k}\Omega$$

and

$$I_2 = \left(\frac{2.667 \text{ k}\Omega}{8 \text{ k}\Omega} \right) 6 \text{ A} = (0.333)(6 \text{ A}) = \mathbf{2\ A}$$

matching the above solution.

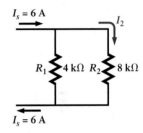

FIG. 6.46

Using the current divider rule to determine current I_2 in Example 6.23.

It would appear that the solution with Eq. [6.14(a)] is more direct in Example 6.23. However, keep in mind that Equation (6.14) is applicable to any parallel configuration, removing the necessity to remember two equations.

Now we present a design-type problem.

EXAMPLE 6.24 Determine resistor R_1 in Fig. 6.47 to implement the division of current shown.

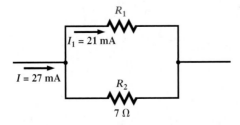

FIG. 6.47

A design-type problem for two parallel resistors (Example 6.24).

Solution: There are essentially two approaches to this type of problem. One involves the direct substitution of known values into the current divider rule equation followed by a mathematical analysis. The other is the sequential application of the basic laws of electric circuits. First we will use the latter approach.

Applying Kirchhoff's current law:

$$\Sigma I_i = \Sigma I_o$$

$$I = I_1 + I_2$$

$$27 \text{ mA} = 21 \text{ mA} + I_2$$

and

$$I_2 = 27 \text{ mA} - 21 \text{ mA} = 6 \text{ mA}$$

The voltage V_2:

$$V_2 = I_2 R_2 = (6 \text{ mA})(7 \text{ } \Omega) = 42 \text{ mV}$$

so that

$$V_1 = V_2 = 42 \text{ mV}$$

Finally,

$$R_1 = \frac{V_1}{I_1} = \frac{42 \text{ mV}}{21 \text{ mA}} = \mathbf{2 \text{ } \Omega}$$

Now for the other approach using the current divider rule:

$$I_1 = \frac{R_2}{R_1 + R_2} I_T$$

$$21 \text{ mA} = \left(\frac{7 \text{ } \Omega}{R_1 + 7 \text{ } \Omega} \right) 27 \text{ mA}$$

$$(R_1 + 7 \text{ } \Omega)(21 \text{ mA}) = (7 \text{ } \Omega)(27 \text{ mA})$$

$$(21 \text{ mA})R_1 + 147 \text{ mV} = 189 \text{ mV}$$

$$(21 \text{ mA})R_1 = 189 \text{ mV} - 147 \text{ mV} = 42 \text{ mV}$$

and

$$R_1 = \frac{42 \text{ mV}}{21 \text{ mA}} = \mathbf{2 \text{ } \Omega}$$

In summary, therefore, keep in mind that current always seeks the path of least resistance, and the ratio of the resistor values is the inverse of the resulting current levels, as shown in Fig. 6.48. The thickness of the blue bands in Fig. 6.48 reflects the relative magnitude of the current in each branch.

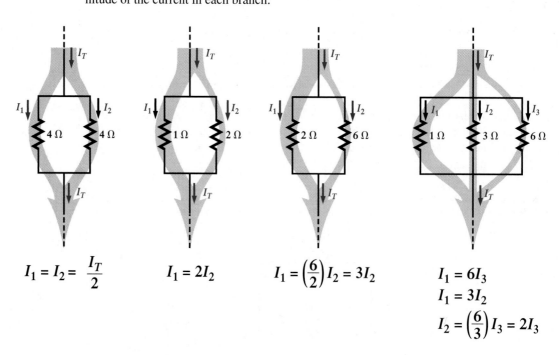

$$I_1 = I_2 = \frac{I_T}{2} \qquad I_1 = 2I_2 \qquad I_1 = \left(\frac{6}{2}\right) I_2 = 3I_2 \qquad \begin{aligned} I_1 &= 6I_3 \\ I_1 &= 3I_2 \\ I_2 &= \left(\frac{6}{3}\right) I_3 = 2I_3 \end{aligned}$$

FIG. 6.48

Demonstrating how current will divide through equal and unequal parallel resistors.

6.7 VOLTAGE SOURCES IN PARALLEL

Because the voltage is the same across parallel elements,

voltage sources can be placed in parallel only if they have the same voltage.

The primary reason for placing two or more batteries or supplies in parallel would be to increase the current rating above that of a single supply. For example, in Fig. 6.49, two ideal batteries of 12 V have been placed in parallel. The total source current using Kirchhoff's current law is now the sum of the rated currents of each supply. The resulting power available will be twice that of a single supply if the rated supply current of each is the same. That is,

with $\qquad I_1 = I_2 = I$

then $\qquad P_T = E(I_1 + I_2) = E(I + I) = E(2I) = 2(EI) = 2P_{\text{(one supply)}}$

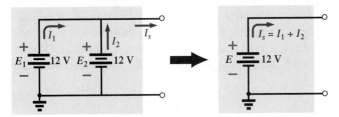

FIG. 6.49

Demonstrating the effect of placing two ideal supplies of the same voltage in parallel.

If for some reason two batteries of different voltages were placed in parallel, both would be left ineffective or damaged because the battery with the larger voltage would rapidly discharge through the battery with the smaller terminal voltage. For example, consider two lead-acid car batteries of different terminal voltages placed in parallel as shown in Fig. 6.50. It makes no sense to talk about placing an ideal 12 V battery in parallel with a 6 V battery, because Kirchhoff's voltage law would be violated. However, we can examine the effects if we include the internal resistance levels as shown in Fig. 6.50.

The only current-limiting resistors in the network are the internal resistances, resulting in a very high discharge current for the battery with the larger supply voltage. The resulting current for the case of Fig. 6.50 would be

$$I = \frac{E_1 - E_2}{R_{\text{int}_1} + R_{\text{int}_2}} = \frac{12\ \text{V} - 6\ \text{V}}{0.03\ \Omega + 0.02\ \Omega} = \frac{6\ \text{V}}{0.05\ \Omega} = 120\ \text{A}$$

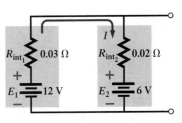

FIG. 6.50

Examining the impact of placing two lead-acid batteries of different terminal voltages in parallel.

This value far exceeds the rated drain current of the 12 V battery, resulting in rapid discharge of E_1 and a destructive impact on the smaller supply due to the excessive currents. This type of situation did arise on occasion when some cars still had 6 V batteries. It was typical of a layman to think, "If I have a 6 V battery, a 12 V battery will work twice as well"—not the case!

In general,

it is always recommended that when you are replacing batteries in series or parallel, all the batteries be replaced.

A fresh battery placed in parallel with an older battery will probably have a higher terminal voltage and will immediately start discharging through the older battery. In addition, the available current will be less for the older battery, resulting in a higher-than-rated current drain from the newer battery when a load is applied.

6.8 VOLTMETER LOADING EFFECTS

In the previous chapters we found that ammeters are not ideal instruments. When you insert an ammeter, you actually introduce an additional resistance in series with the branch in which you are measuring the current. Although in most cases it is not a serious problem, it can have a troubling effect on your readings, so it is important to be aware of its presence.

Voltmeters also have an internal resistance that will appear between the two terminals of interest when a measurement is being made. Unlike the ammeter, therefore, which places an additional resistance in series with the branch of interest, a voltmeter places an additional resistance **across** the element, as shown in Fig. 6.51. Since it appears in parallel with the element of interest, **the ideal level for the internal resistance of a voltmeter would be infinite**

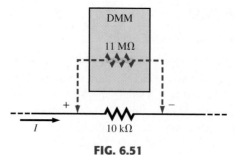

FIG. 6.51

Voltmeter loading.

ohms, just as zero ohms would be ideal for an ammeter. Unfortunately, the internal resistance of any voltmeter is not infinite and changes from one type of meter to another.

Most digital meters have a fixed internal resistance level in the megohm range that remains the same **for all its scales.** For example, the meter of Fig. 6.51 has the typical level of 11 MΩ for its internal resistance, no matter which voltage scale is used. When the meter is placed across the 10 kΩ resistor, the total resistance of the combination is

$$R_T = 10 \text{ k}\Omega \parallel 11 \text{ M}\Omega = \frac{(10^4 \ \Omega)(11 \times 10^6 \ \Omega)}{10^4 \ \Omega + (11 \times 10^6 \ )} = 9.99 \text{ k}\Omega$$

and the behavior of the network will not be seriously affected. The result, therefore, is that *most digital voltmeters can be used in circuits with resistances up to the high-kilohm range without concern for the effect of the internal resistance on the reading.*

However, if the resistances are in the megohm range, then some investigation into the effect of the internal resistance should be initiated.

An analog VOM is a different matter, however, because the internal resistance levels are much lower and the internal resistance levels are a function of the scale used. If a VOM on the 2.5 V scale were placed across the 10 kΩ resistor of Fig. 6.51, the internal resistance might be 50 kΩ, resulting in a combined resistance of

$$R_T = 10 \text{ k}\Omega \parallel 50 \text{ k}\Omega = \frac{(10^4 \ \Omega)(50 \times 10^3 \ \Omega)}{10^4 \ \Omega + (50 \times 10^3 \ \Omega)} = 8.33 \text{ k}\Omega$$

and the behavior of the network would be affected because the 10 kΩ resistor would appear as an 8.33 kΩ resistor.

To determine the resistance R_m of any scale of a VOM, simply multiply the **maximum voltage** of the chosen scale by the **ohm/volt (Ω/V) rating** normally appearing at the bottom of the face of the meter. That is,

$$R_m \text{ (VOM)} = \text{(scale)}(\Omega/\text{V rating})$$

For a typical Ω/V rating of 20,000, the 2.5 V scale would have an internal resistance of

$$(2.5 \text{ V})(20,000 \ \Omega/\text{V}) = \textbf{50 k}\Omega$$

whereas for the 100 V scale the internal resistance of the VOM would be

$$(100 \text{ V})(20,000 \ \Omega/\text{V}) = \textbf{2 M}\Omega$$

and for the 250 V scale,

$$(250 \text{ V})(20,000 \ \Omega/\text{V}) = \textbf{5 M}\Omega$$

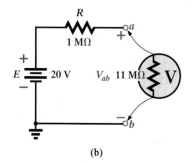

(a) Measuring an open-circuit voltage with a voltmeter; (b) determining the effect of using a digital voltmeter with an internal resistance of 11 MΩ on measuring an open-circuit voltage (Example 6.25).

FIG. 6.52

EXAMPLE 6.25 For the relatively simple circuit of Fig. 6.52(a):

a. What is the open-circuit voltage V_{ab}?
b. What will a DMM indicate if it has an internal resistance of 11 MΩ? Compare your answer to that of part (a).
c. Repeat part (b) for a VOM with an Ω/V rating of 20,000 on the 100 V scale.

Solutions:

a. Due to the open circuit, the current is zero and the voltage drop across the 1 MΩ resistor would be zero volts. The result is that the entire source voltage appears between points *a* and *b*, and

$$V_{ab} = \textbf{20 V}$$

b. When the meter is connected as shown in Fig. 6.52(b), a complete circuit has been established, and current can pass through the circuit. The voltmeter reading can be determined using the voltage divider rule as follows:

$$V_{ab} = \frac{(11\text{ M}\Omega)(20\text{ V})}{11\text{ M}\Omega + 1\text{ M}\Omega} = \mathbf{18.33\text{ V}}$$

and the reading is affected somewhat.

c. For the VOM, the internal resistance of the meter is

$$R_m = (100\text{ V})\,(20{,}000\text{ }\Omega/\text{V}) = 2\text{ M}\Omega$$

and

$$V_{ab} = \frac{(2\text{ M}\Omega)(20\text{ V})}{2\text{ M}\Omega + 1\text{ M}\Omega} = \mathbf{13.33\text{ V}}$$

which is considerably below the desired level of 20 V.

In general, therefore, as mentioned earlier, if you encounter circuits with resistances in the megohm range, be aware that most voltmeters will have internal resistance levels that may have to be considered.

6.9 PROTOBOARDS (BREADBOARDS)

In Section 5.10, the protoboard was introduced with the connections for a simple series circuit. To continue the development, the network of Fig. 6.19 was set up on the board of Fig. 6.53(a) using two different techniques. The possibilities are endless, but these two solutions use a fairly straightforward approach.

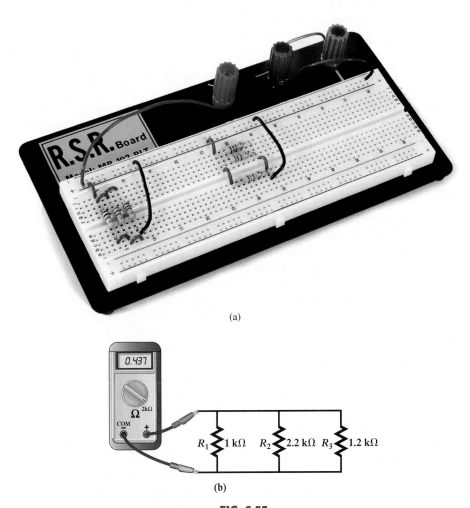

(a)

(b)

FIG. 6.53

Using a protoboard to set up the circuit of Fig. 6.19.

First note that the supply lines and ground are established across the length of the board using the horizontal conduction zones at the top and bottom of the board through the connections to the terminals. The network to the left on the board was used to set up the circuit in much the same manner as it appears in the schematic of Fig. 6.53(b). This approach required that the resistors be connected between two vertical conducting strips. If placed perfectly vertical in a single conducting strip, the resistors would have shorted themselves out. Often times, setting the network up in a manner that best copies the original can make it easier to check and make measurements. The network to the right in part (a) used the vertical conducting strips to connect the resistors together at each end. Since there wasn't enough room for all three, a connection had to be added from the upper vertical set to the lower set. The resistors are in order R_1, R_2, and R_3 from the top-down. For both configurations, the ohmmeter can simply be connected to the positive lead of the supply terminal and the negative or ground terminal.

Take a moment to review the connections and think of other possibilities. Improvements can often be made, but it can be an enjoyable challenge to find the most effective setup with the least number of connecting wires.

6.10 APPLICATIONS

One of the most important advantages of the parallel configuration is that

if one branch of the configuration should fail (open circuit), the remaining branches will still have full operating power.

In a home, the parallel connection is used throughout to ensure that if one circuit has a problem and opens the circuit breaker, the remaining circuits will still have the full 120 V. The same is true in automobiles, computer systems, industrial plants, and wherever it would be disastrous for one circuit to control the total power distribution.

Another important advantage is that

branches can be added at any time without affecting the behavior of those already in place.

In other words, unlike the series connection where an additional component will reduce the current level and perhaps affect the response of some of the components in place, an additional parallel branch will not affect the current level in the other branches. Of course, the current demand from the supply will increase as determined by Kirchhoff's current law, but you must simply be aware of the limitations of the supply.

The following are some of the most common applications of the parallel configuration.

Car System

As you begin to examine the electrical system of an automobile, the most important thing to understand is that the entire electrical system of a car is run as a *dc system*. Although the generator will produce a varying ac signal, rectification will convert it to one having an average dc level for charging the battery. In particular, note the filter capacitor in the alternator branch of Fig. 6.54 to smooth out the rectified ac waveform and to provide an improved dc supply. The charged battery must therefore provide the required direct current for the entire electrical system of the car. Thus, the power demand on the battery at any instant is the product of the terminal voltage and the current drain of the total load of every operating system of the car. This certainly places an enormous burden on the battery and its internal chemical reaction and warrants all the battery care we can provide.

Since the electrical system of a car is essentially a parallel system, the total current drain on the battery is the sum of the currents to all the parallel branches of the car connected directly to the battery. In Fig. 6.54, a few branches of the wiring diagram for a car have been sketched to provide some background information on basic wiring, current levels, and fuse configurations. Every automobile has fuse links and fuses, and some also have circuit breakers, to protect the various components of the car and to ensure that a dangerous fire situation does not develop. Except for a few branches that may have series elements, the operating voltage for most components of a car is the terminal voltage of the battery which we will designate as 12 V even though it will typically vary between 12 V and the charging level of 14.6 V. In other words, each component is connected to the battery at one end and to the ground or chassis of the car at the other end.

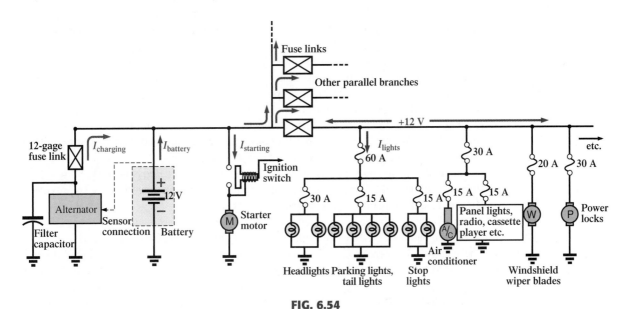

FIG. 6.54

Expanded view of an automobile's electrical system.

Referring to Fig. 6.54, we find that the alternator or charging branch of the system is connected directly across the battery to provide the charging current as indicated. Once the car is started, the rotor of the alternator will turn, generating an ac varying voltage which will then pass through a rectifier network and filter to provide the dc charging voltage for the battery. Charging will occur only when the sensor, connected directly to the battery, signals that the terminal voltage of the battery is too low. Just to the right of the battery the starter branch was included to demonstrate that there is no fusing action between the battery and starter when the ignition switch is activated. The lack of fusing action is provided because enormous starting currents (hundreds of amperes) will flow through the starter to start a car that may not have been used for days and/or that may have been sitting in a cold climate—and high friction occurs between components until the oil starts flowing. The starting level can vary so much that it would be difficult to find the right fuse level, and frequent high currents may damage the fuse link and cause a failure at expected levels of current. When the ignition switch is activated, the starting relay will complete the circuit between the battery and starter, and hopefully the car will start. If a car should fail to start, the first point of attack should be to check the connections at the battery, starting relay, and starter to be sure that they are not providing an unexpected open circuit due to vibration, corrosion, or moisture.

Once the car has started, the starting relay will open and the battery can turn its attention to the operating components of the car. Although the diagram of Fig. 6.54 does not display the switching mechanism, the entire electrical network of the car, except for the important external lights, is usually disengaged so that the full strength of the battery can be dedicated to the starting process. The lights are included for situations where turning the lights off, even for short periods of time, could create a dangerous situation. If the car is in a safe environment, it is best to leave the lights off at starting to save the battery an additional 30 A of drain. If the lights are on at starting, a dimming of the lights can be expected due to the starter drain which may exceed 500 A. Today, batteries are typically rated in cranking (starting) current rather than ampere-hours. Batteries rated with cold cranking ampere ratings between 700 A and 1000 A are typical today.

Separating the alternator from the battery and the battery from the numerous networks of the car are fuse links such as shown in Fig. 6.55(a). They are actually wires of specific gage designed to open at fairly high current levels of 100 A or more. They are included to protect against those situations where there is an unexpected current drawn from the many circuits it

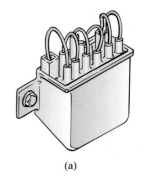

(a)

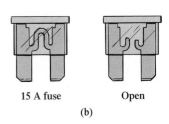

15 A fuse   Open

(b)

FIG. 6.55

Car fuses: (a) fuse link; (b) plug-in.

is connected to. That heavy drain can, of course, be from a short circuit in one of the branches, but in such cases the fuse in that branch will probably release. The fuse link is an additional protection for the line if the total current drawn by the parallel-connected branches begins to exceed safe levels. The fuses following the fuse link have the appearance shown in Fig. 6.55(b), where a gap between the legs of the fuse indicates a blown fuse. As shown in Fig. 6.54, the 60 A fuse (often called a *power distribution fuse*) for the lights is a second-tier fuse sensitive to the total drain from the three light circuits. Finally, the third fuse level is for the individual units of a car such as the lights, air conditioner, and power locks. In each case, the fuse rating exceeds the normal load (current level) of the operating component, but the level of each fuse does give some indication of the demand to be expected under normal operating conditions. For instance, the headlights will typically draw more than 10 A, the tail lights more than 5 A, the air conditioner about 10 A (when the clutch engages), and the power windows 10 A to 20 A depending on how many are operated at once.

Some details for only one section of the total car network are provided in Fig. 6.54. In the same figure, additional parallel paths with their respective fuses have been provided to further reveal the parallel arrangement of all the circuits.

In all vehicles made in the United States and in some vehicles made in European countries, the return path to the battery through the ground connection is actually through the chassis of the car. That is, there is only one wire to each electrical load, with the other end simply grounded to the chassis. The return to the battery (chassis to negative terminal) is therefore a heavy-gage wire matching that connected to the positive terminal. In some European cars constructed of a mixture of materials such as metal, plastic, and rubber, the return path through the metallic chassis is lost, and two wires must be connected to each electrical load of the car.

House Wiring

In Chapter 4, the basic power levels of importance were discussed for various services to the home. We are now ready to take the next step and examine the actual connection of elements in the home.

First, it is important to realize that except for some very special circumstances, the basic wiring is done in a parallel configuration. Each parallel branch, however, can have a combination of parallel and series elements. Every full branch of the circuit receives the full 120 V or 208 V, with the current determined by the applied load. Figure 6.56(a) provides the detailed wiring of a single circuit having a light bulb and two outlets. Figure 6.56(b) shows the schematic representation. First note that although each load is in parallel with the supply, switches are always connected in series with the load. The power will get to the lamp only when the switch is closed and the full 120 V appears across the bulb. The connection point for the two outlets is in the ceiling box holding the light bulb. Since a switch is not present, both outlets are always "hot" unless the circuit breaker in the main panel is opened. It is important that you understand this because you may be tempted to change the light fixture by simply turning off the wall switch. True, if you're very careful, you can work with one line at a time (being sure that you don't touch the other line at any time), but it is standard procedure to throw the circuit breaker on the panel whenever working on a circuit. Note in Fig. 6.56(a) that the *feed* wire (black) into the fixture from the panel is connected to the switch and both outlets at one point. It is not connected directly to the light fixture because that would put it on all the time. Power to the light fixture is made available through the switch. The continuous connection to the outlets from the panel ensures that the outlets are "hot" whenever the circuit breaker in the panel is on. Note also how the *return* wire (white) is connected directly to the light switch and outlets to provide a return for each component. There is no need for the white wire to go through the switch since an applied voltage is a two-point connection and the black wire is controlled by the switch.

Proper grounding of the system in total and of the individual loads is one of the most important facets in the installation of any system. There is a tendency at times to be satisfied that the system is working and to pay less attention to proper grounding technique. Always keep in mind that a properly grounded system has a direct path to ground if an undesirable situation should develop. The absence of a direct ground will cause the system to determine its own path to ground, and you could be that path if you happened to touch the wrong wire, metal box, metal pipe, and so on. In Fig. 6.56(a), the connections for the ground wires have been in-

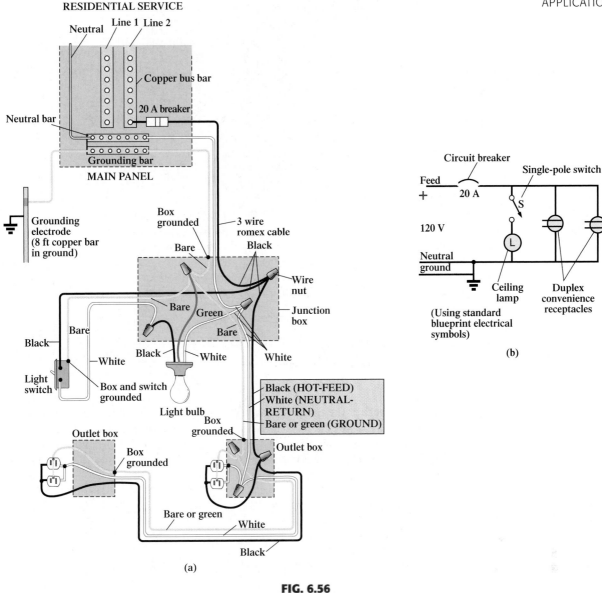

FIG. 6.56

Single phase of house wiring: (a) physical details; (b) schematic representation.

cluded. For the romex (plastic-coated wire) used in Fig. 6.56(a), the ground wire is provided as a bare copper wire. Note that it is connected to the panel which in turn is directly connected to the grounded 8 ft copper rod. In addition, note that the ground connection is carried through the entire circuit, including the switch, light fixture, and outlets. It is one continuous connection. If the outlet box, switch box, and housing for the light fixture are made of a conductive material such as metal, the ground will be connected to each. If each is plastic, there is no need for the ground connection. However, the switch, both outlets, and the fixture itself are connected to ground. For the switch and outlets there is usually a green screw for the ground wire which is connected to the entire framework of the switch or outlet as shown in Fig. 6.57, including the ground connection of the outlet. For both the switch and the outlet, even the screw or screws used to hold the outside plate in place are grounded since they are screwed into the metal housing of the switch or outlet. When screwed into a metal box, the ground connection can be made by the screws that hold the switch or outlet in the box as shown in Fig. 6.57. In any event, always pay strict attention to the grounding process whenever installing any electrical equipment. It is a facet of electrical installation that is often treated too lightly.

On the practical side, whenever hooking up a wire to a screw-type terminal, always wrap the wire around the screw in the clockwise manner so that when you tighten the screw, it will grab the wire and turn it in the same direction. An expanded view of a typical house-wiring arrangement will appear in Chapter 14.

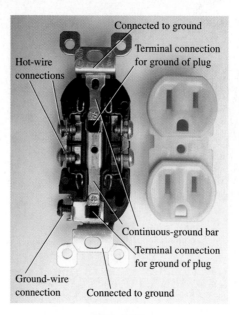

Hot-wire
connections

Connected to ground

Terminal connection
for ground of plug

Continuous-ground bar

Terminal connection
for ground of plug

Ground-wire
connection

Connected to ground

FIG. 6.57

Continuous ground connection in a duplex outlet.

6.11   COMPUTER ANALYSIS

PSpice

Parallel dc Network   The computer analysis coverage for parallel dc circuits will be very similar to that for series dc circuits. However, in this case the voltage will be the same across all the parallel elements, and the current through each branch will change with the resistance value. The parallel network to be analyzed will have a wide range of resistor values to demonstrate the effect on the resulting current. The following is a list of abbreviations for any parameter of a network when using PSpice:

$$\mathbf{f} = 10^{-15}$$
$$\mathbf{p} = 10^{-12}$$
$$\mathbf{n} = 10^{-9}$$
$$\boldsymbol{\mu} = 10^{-6}$$
$$\mathbf{m} = 10^{-3}$$
$$\mathbf{k} = 10^{+3}$$
$$\mathbf{MEG} = 10^{+6}$$
$$\mathbf{G} = 10^{+9}$$
$$\mathbf{T} = 10^{+12}$$

In particular, note that **m** (or **M**) is used for "milli," and **MEG** for "megohms." Also, PSpice does not distinguish between upper- and lowercase units, but certain parameters typically use either the upper- or lowercase abbreviation as shown above.

Since the details of setting up a network and going through the simulation process were covered in detail in Sections 4.9 and 5.12 for dc circuits, the coverage here will be limited solely to the various steps required. These steps should make it obvious that after some exposure, getting to the point where you can "draw" the circuit and then run the simulation is pretty quick and direct.

After selecting the **Create document** key (the top left of the screen), the following sequence will bring up the **Schematic** window: **ParallelDC-OK-Create a blank project-OK-PAGE1** (if necessary).

The voltage source and resistors are introduced as described in detail in earlier sections, but now the resistors need to be turned 90°. You can accomplish this by a right click of the mouse before setting a resistor in place. The resulting long list of options includes **Rotate,** which if selected will turn the resistor counterclockwise 90°. It can also be rotated by simultaneously selecting **Ctrl-R.** The resistor can then be placed in position by a left click of the mouse. An additional benefit of this maneuver is that the remaining resistors to be placed will already be in the vertical position. The values selected for the voltage source and resistors appear in Fig. 6.58.

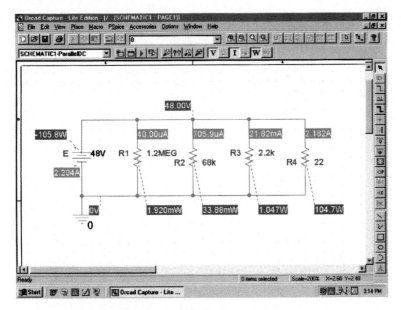

FIG. 6.58

Applying PSpice to a parallel network.

Once the network is complete, the simulation and the results of Fig. 6.58 can be obtained through the following sequence: **Select New Simulation Profile** key-**Bias Point-Create-Analysis-Bias Point-OK-Run PSpice** key-**Exit(X).**

The result is Fig. 6.58 which clearly reveals that the voltage is the same across all the parallel elements and the current increases significantly with decrease in resistance. The range in resistor values suggests, by inspection, that the total resistance will be just less than the smallest resistance of 22 Ω. Using Ohm's law and the source current of 2.204 A results in a total resistance of $R_T = E/I_s = 48$ V/2.204 A = 21.78 Ω, confirming the above conclusion.

Electronics Workbench (EWB)

Parallel dc Network　For comparison purposes the parallel network of Fig. 6.58 will now be analyzed using EWB. The source and ground are selected and placed as shown in Fig. 6.59 using the procedure defined in previous chapters. For the resistors, the

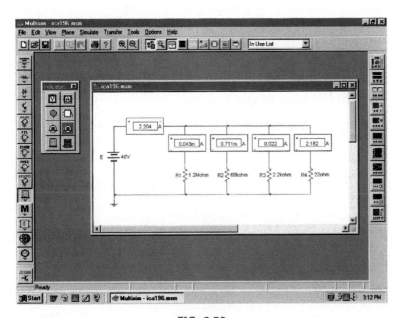

FIG. 6.59

Using the indicators of EWB to display the currents of a parallel dc network.

VIRTUAL_RESISTOR is chosen, but it must be rotated 90° to match the configuration of Fig. 6.58. You can accomplish this by first clicking on the resistor symbol to place it in the active state. (Be sure that the resulting small black squares surround the symbol, label, and value; otherwise, you may have activated only the label or value.) Then right-click the mouse. The **90° Clockwise** can then be selected, and the resistor will be turned automatically. Unfortunately, there is no continuum here, so the next resistor will have to be turned using the same procedure. The values of each resistor are set by double-clicking on the resistor symbol to obtain the **Virtual Resistor** dialog box. Remember that the unit of measurement is controlled by the scrolls at the right of the unit of measurement. For EWB, unlike PSpice, megohm uses capital **M** and milliohm uses lowercase **m.**

This time, rather than using meters to make the measurements, we will use indicators. The **Indicators** key pad is the tenth down on the left toolbar. It has the appearance of an LCD display with the number 8. Once it has been selected, eight possible indicators will appear. For this example, the **A** indicator, representing an ammeter, will be used since we are interested only in the current levels. When **A** has been selected, a **Component Browser** will appear with four choices under the **Component Name List;** each option refers to a position for the ammeter. The **H** means "horizontal" as shown in the picture window when the dialog box is first opened. The **HR** means "horizontal," but with the polarity reversed. The **V** is for a vertical configuration with the positive sign at the top, and the **VR** is the vertical position with the positive sign at the bottom. Simply select the one you want followed by an **OK,** and your choice will appear in that position on the screen. Click it into position, and you can return for the next indicator. Once all the elements are in place and their values set, simulation can be initiated with the sequence **Simulate-Run.** The results shown in Fig. 6.59 will appear.

Note that all the results appear with the indicator boxes. All are positive results because the ammeters were all entered with a configuration that would result in conventional current entering the positive current. Also note that as was true for inserting the meters, the indicators are placed in series with the branch in which the current is to be measured.

CHAPTER SUMMARY
PARALLEL RESISTORS

- Two elements, branches, or circuits are in parallel if they have two, and only two, points in common.
- The total resistance of parallel resistors is always less than the value of the smallest resistor.
- The total resistance of parallel resistors will always drop as new resistors are added in parallel, irrespective of their value.
- The total resistance of parallel resistors is unaffected by the order in which resistors are connected.
- The total resistance of two parallel resistors is their product divided by their sum.

PARALLEL CIRCUITS

- The voltage is always the same across parallel elements.
- In a parallel circuit, the source current must equal the sum of the currents through all its branches.
- Current always seeks the path of least resistance.
- The power supplied by the source must equal that dissipated by the resistive elements.
- Voltage sources can be placed in parallel only if they have the same voltage rating.

KIRCHHOFF'S CURRENT LAW

- Kirchhoff's current law states that the sum of the currents entering a junction (or region) must equal the sum of the currents leaving.
- The current entering parallel resistive elements will split as the inverse of their resistive values.

VOLTMETERS

- Voltmeters have an internal resistance that is placed in parallel with the element or branch when a measurement is being made. The larger the internal resistance compared to the circuit elements, the more accurate the reading.
- For DMMs the internal resistance is the same for each scale, while for VOMs it is equal to the Ω/V rating times the maximum voltage of the scale being used.

Important Equations

$$\frac{1}{R_T} = \frac{1}{R_1} + \frac{1}{R_2} + \frac{1}{R_3} + \cdots + \frac{1}{R_N}$$

$$G_T = G_1 + G_2 + G_3 + \cdots + G_N$$

$$R_T = \frac{R}{N}$$

$$R_T = \frac{R_1 R_2}{R_1 + R_2}$$

$$V_1 = V_2 = E$$

$$I_s = \frac{E}{R_T} \qquad I_1 = \frac{V_1}{R_1} = \frac{E}{R_1} \qquad I_2 = \frac{V_2}{R_2} = \frac{E}{R_2}$$

$$I_s = I_1 + I_2 + I_3 + \cdots$$

$$P_E = EI_s = P_{R_1} + P_{R_2} + P_{R_3}$$

$$P_1 = V_1 I_1 = I_1^2 R_1 = \frac{V_1^2}{R_1}$$

$$\Sigma I_i = \Sigma I_o$$

$$I_x = \frac{R_T}{R_x} I_T$$

$$I_1 = \left(\frac{R_2}{R_1 + R_2}\right) I_T \qquad I_2 = \left(\frac{R_1}{R_1 + R_2}\right) I_T$$

$$R_m(\text{VOM}) = (\text{scale})(\Omega/\text{V rating})$$

PROBLEMS

SECTION 6.2   Parallel Resistors

1. For each configuration of Fig. 6.60, find the elements (individual elements, not combinations of elements)—voltage sources and/or resistors—that are in parallel. If necessary, use the fact that elements in parallel have the same voltage.

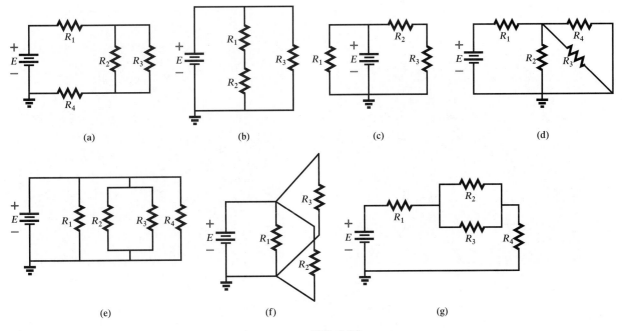

FIG. 6.60
Problem 1.

2. For the network of Fig. 6.61:
  a. Find the elements (voltage sources and/or resistors) that are in parallel.
  b. Find the elements (voltage sources and/or resistors) that are in series.

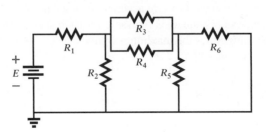

FIG. 6.61
Problem 2.

3. Find the total resistance for each configuration of Fig. 6.62. Note that only standard value resistors were employed.

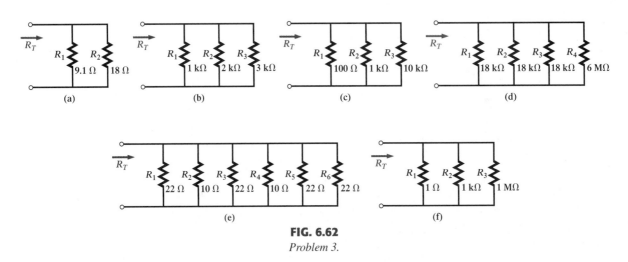

FIG. 6.62
Problem 3.

4. For each circuit board of Fig. 6.63, find the total resistance between connection tabs 1 and 2.

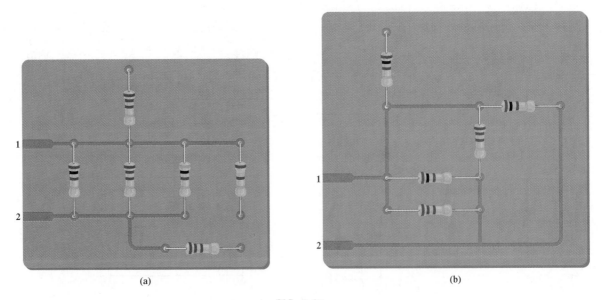

FIG. 6.63
Problem 4.

5. The total resistance of each of the configurations of Fig. 6.64 is specified. Find the unknown resistance.

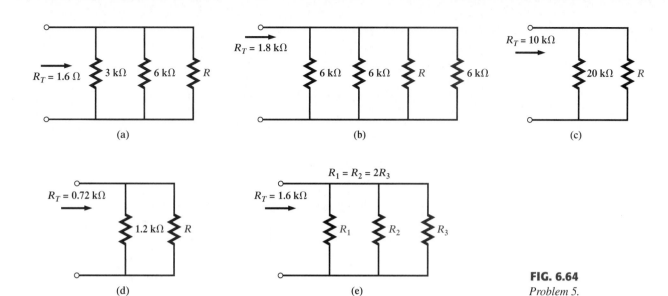

(a)　　　(b)　　　(c)

(d)　　　(e)

FIG. 6.64
Problem 5.

6. For the parallel network of Fig. 6.65, composed of standard values:
   a. Which resistor will have the most impact on the total resistance?
   b. Without making a single calculation, what is an approximate value for the total resistance?
   c. Calculate the total resistance and comment on your response to part (b).
   d. On an approximate basis, which resistors can be ignored when determining the total resistance?
   e. If we add another parallel resistor of any value to the network, what will the impact be on the total resistance?

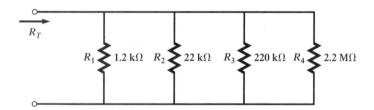

FIG. 6.65
Problem 6.

7. What is the ohmmeter reading for each configuration of Fig. 6.66?

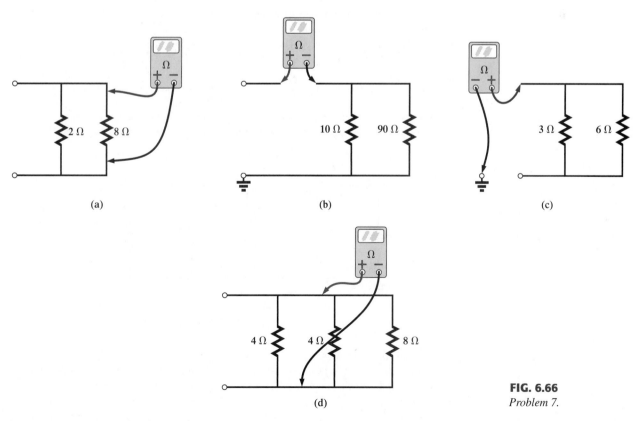

(a)　　　(b)　　　(c)

(d)

FIG. 6.66
Problem 7.

205

SECTION 6.3 Parallel Circuits

8. For the parallel network of Fig. 6.67:
   a. Find the total resistance.
   b. What is the voltage across each branch?
   c. Determine the source current and the current through each branch.
   d. Verify that the source current equals the sum of the branch currents.

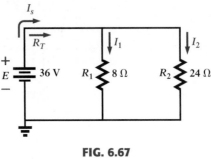

FIG. 6.67
Problem 8.

9. Repeat the analysis of Problem 8 for the network of Fig. 6.68.

10. Repeat the analysis of Problem 8 for the network of Fig. 6.69, constructed of standard value resistors.

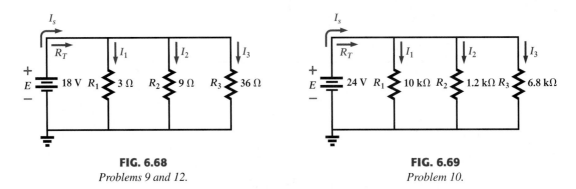

FIG. 6.68
Problems 9 and 12.

FIG. 6.69
Problem 10.

11. For the parallel network of Fig. 6.70:
   a. Without making a single calculation, make a guess as to the total resistance.
   b. Calculate the total resistance and compare it to your guess in part (a).
   c. Without making a single calculation, which branch will have the most current? Which will have the least?
   d. Calculate the current through each branch, and compare your results to the assumptions of part (c).
   e. Find the source current and test whether it equals the sum of the branch currents.
   f. How does the magnitude of the source current compare to that of the branch currents?

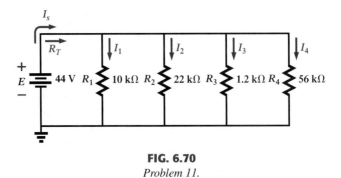

FIG. 6.70
Problem 11.

12. For the network of Fig. 6.68:
   a. Redraw the network and insert ammeters to measure the source current and the current through each branch.

b. Connect a voltmeter to measure the source voltage and the voltage across resistor R_3. Is there any difference in the connections? Why?

13. What is the response of the voltmeter and ammeters connected in Fig. 6.71?

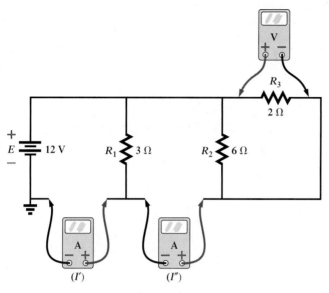

FIG. 6.71
Problem 13.

14. Given the information provided in Fig. 6.72, find the unknown quantities: E, R_1, and I_3.

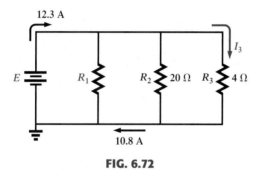

FIG. 6.72
Problem 14.

SECTION 6.4   Power Distribution in a Parallel Network

15. For the configuration of Fig. 6.73:
   a. Find the total resistance and the current through each branch.
   b. Find the power delivered to each resistor.
   c. Calculate the power delivered by the source.
   d. Compare the power delivered by the source to the sum of the powers delivered to the resistors.
   e. Which resistor received the most power? Why?

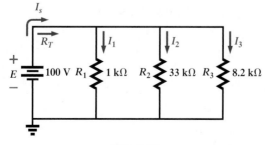

FIG. 6.73
Problem 15.

16. Eight holiday lights are connected in parallel as shown in Fig. 6.74.
    a. If the set is connected to a 120 V source, what is the current through each bulb if each bulb has an internal resistance of 1.8 kΩ?
    b. Determine the total resistance of the network.
    c. Find the current drain from the supply.
    d. What is the power delivered to each bulb?
    e. Using the results of part (d), what is the power delivered by the source?
    f. If one bulb burns out (that is, the filament opens up), what is the effect on the remaining bulbs? What is the effect on the source current? Why?

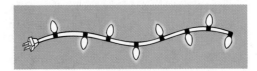

FIG. 6.74
Problem 16.

17. Determine the power delivered by the dc battery of Fig. 6.75.

18. A portion of a residential service to a home is depicted in Fig. 6.76.
    a. Determine the current through each parallel branch of the system.
    b. Calculate the current drawn from the 120 V source. Will the 20 A breaker trip?
    c. What is the total resistance of the network?
    d. Determine the power delivered by the source. How does it compare to the sum of the wattage ratings appearing in Fig. 6.76?

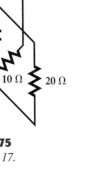

FIG. 6.75
Problem 17.

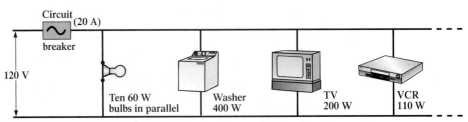

FIG. 6.76
Problem 18.

SECTION 6.5 Kirchhoff's Current Law

19. Using Kirchhoff's current law, determine the unknown currents for the parallel network of Fig. 6.77.

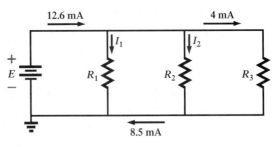

FIG. 6.77
Problem 19.

20. Using Kirchhoff's current law, find the unknown currents for the complex configurations of Fig. 6.78.

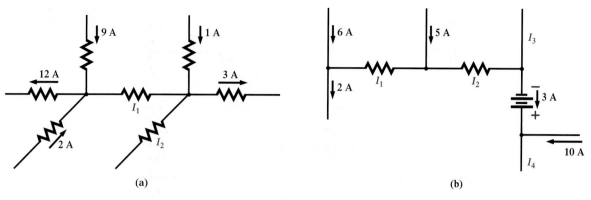

(a)   (b)

FIG. 6.78
Problem 20.

21. Using Kirchhoff's current law, determine the unknown currents for the networks of Fig. 6.79.

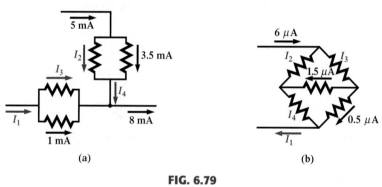

(a)   (b)

FIG. 6.79
Problem 21.

22. Using the information provided in Fig. 6.80, find the branch resistors R_1 and R_3, the total resistance R_T, and the voltage source E.

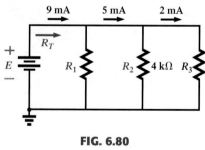

FIG. 6.80
Problem 22.

23. Find the unknown quantities for the networks of Fig. 6.81 using the information provided.

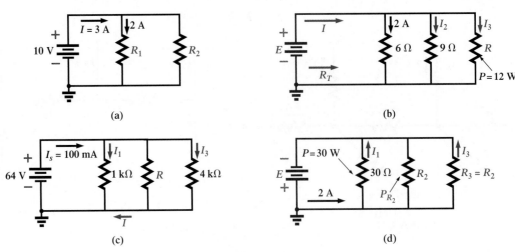

(a)

(b)

(c)

(d)

FIG. 6.81
Problem 23.

SECTION 6.6   Current Divider Rule

24. Based solely on the resistor values, determine all the currents for the configuration of Fig. 6.82. Do not use Ohm's law.

25. Determine the currents for the configurations of Fig. 6.83.

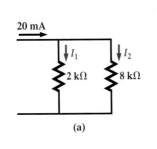

FIG. 6.82
Problem 24.

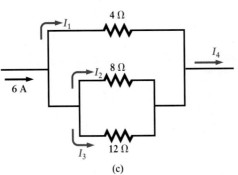

(a)

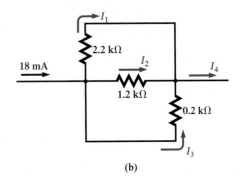

(b)

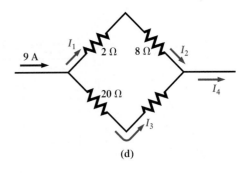

(c)

(d)

FIG. 6.83
Problem 25.

26. Parts (a) through (e) of this problem should be done by inspection—that is, mentally. The intent is to obtain an approximate solution without a lengthy series of calculations. For the network of Fig. 6.84:
   a. What is the approximate value of I_1 considering the magnitude of the parallel elements?
   b. What is the ratio I_1/I_2? Using the result of part (a), what is an approximate value of I_2?
   c. What is the ratio I_1/I_3? Using the result, what is an approximate value of I_3?
   d. What is the ratio I_1/I_4? Using the result, what is an approximate value of I_4?

FIG. 6.84
Problem 26.

e. What is the effect of the parallel 100 kΩ resistor on the above calculations? How much smaller will the current I_4 be than the current I_1?

f. Calculate the current through the 1 Ω resistor using the current divider rule. How does it compare to the result of part (a)?

g. Calculate the current through the 10 Ω resistor. How does it compare to the result of part (b)?

h. Calculate the current through the 1 kΩ resistor. How does it compare to the result of part (c)?

i. Calculate the current through the 100 kΩ resistor. How does it compare to the solutions to part (e)?

27. Find the unknown quantities for the networks of Fig. 6.85 using the information provided.

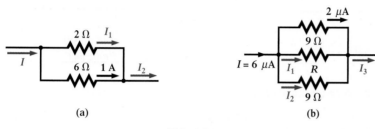

(a)   (b)

FIG. 6.85
Problem 27.

28. Find resistor R for the network of Fig. 6.86 that will ensure that $I_1 = 3I_2$.

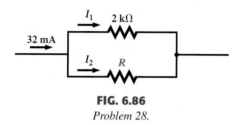

FIG. 6.86
Problem 28.

29. Design the network of Fig. 6.87 such that $I_2 = 2I_1$ and $I_3 = 2I_2$.

SECTION 6.7 Voltage Sources in Parallel

30. Assuming identical supplies in Fig. 6.88:

a. Find the indicated currents.

b. Find the power delivered by each source.

c. Find the total power delivered by both sources, and compare it to the power delivered to the load R_L.

d. If only source current were available, what would the current drain be to supply the same power to the load? How does the current level compare to the calculated level of part (a)?

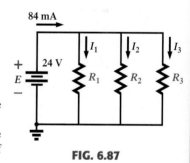

FIG. 6.87
Problem 29.

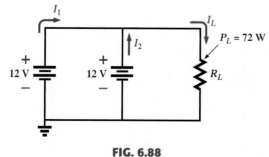

FIG. 6.88
Problem 30.

31. Assuming identical supplies, determine currents I_1, I_2, and I_3 for the configuration of Fig. 6.89.

32. Assuming identical supplies, determine the current I and resistance R for the parallel network of Fig. 6.90.

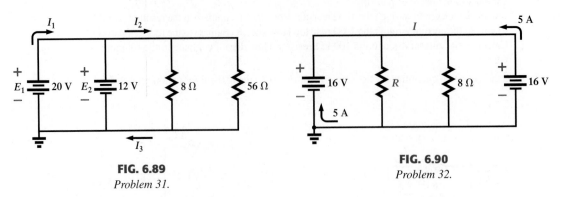

FIG. 6.89
Problem 31.

FIG. 6.90
Problem 32.

SECTION 6.8 Voltmeter Loading Effects

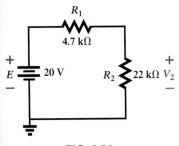

FIG. 6.91
Problem 33.

33. For the simple series configuration of Fig. 6.91:
 a. Determine voltage V_2.
 b. Determine the reading of a DMM having an internal resistance of 11 MΩ when used to measure V_2.
 c. Repeat part (b) with a VOM having an Ω/V rating of 20,000 using the 20 V scale. Compare the results of parts (b) and (c). Explain any differences.
 d. Repeat parts (a) through (d) with $R_1 = 100$ kΩ and $R_2 = 200$ kΩ.
 e. Based on the above, what general conclusions can you make about the use of a DMM or a VOM in the voltmeter mode?

34. Given the configuration of Fig. 6.92:
 a. What is the voltage between points a and b?
 b. What will the reading of a DMM be when placed across terminals a and b if the internal resistance of the meter is 11 MΩ?
 c. Repeat part (b) if a VOM having an Ω/V rating of 20,000 using the 200 V scale is used. What is the reading using the 20 V scale? Is there a difference? Why?

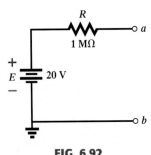

FIG. 6.92
Problem 34.

35. Based on the measurements of Fig. 6.93, determine whether the network is operating properly. If not, try to determine why.

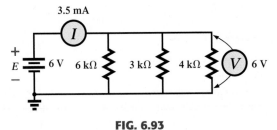

FIG. 6.93
Problem 35.

36. Referring to Fig. 6.94, find the voltage V_{ab} without the meter in place. When the meter is applied to the active network, it reads 8.8 V. If the measured value does not equal the theoretical value, which element or elements may have been connected incorrectly?

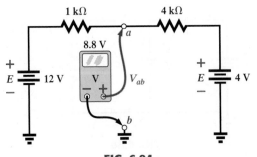

FIG. 6.94
Problem 36.

37. Using PSpice or EWB, verify the results of Example 6.13.

38. Using PSpice or EWB, determine the solution to Problem 8, and compare your answer to the longhand solution.

39. Using PSpice or EWB, determine the solution to Problem 10, and compare your answer to the longhand solution.

GLOSSARY

Current divider rule (CDR)   A method by which the current through parallel elements can be determined without first finding the voltage across those parallel elements.

Kirchhoff's current law (KCL)   The algebraic sum of the currents entering and leaving a node is zero.

Node   A junction of two or more branches.

Ohm/volt (Ω/V) rating   A rating used to determine both the current sensitivity of the movement and the internal resistance of the meter.

Parallel circuit   A circuit configuration in which the elements have two points in common.

Series-Parallel Circuits

OBJECTIVES

- Become aware of the characteristics of a series-parallel circuit.
- Understand how to apply the reduce and return approach.
- Recognize and be able to apply the special notation for voltage sources and understand the use of double- and single-subscript notation.
- Become familiar with the ladder network and its methods of analysis.
- Understand the conditions under which the voltage divider supply can provide a number of terminal voltages.
- Be able to use a potentiometer to control the voltage across any given load.
- Develop some self-confidence in the analysis of circuits with shorts and open circuits.
- Become aware of how PSpice can be applied to solve complex series-parallel circuits.

7.1 INTRODUCTION

The previous two chapters were dedicated to ensuring that the fundamentals of series and parallel circuits were clearly defined. In some ways, these chapters might be the most important ones in the text, because they form a foundation for all the material to follow. The remaining network configurations cannot be defined by a strict list of conditions because of the variety of configurations that exists. In broad terms we can look upon the remaining possibilities as either **series-parallel** or **complex.**

A series-parallel configuration is one that is formed by a combination of series and parallel elements.

A complex configuration is one in which none of the elements are in series or parallel.

In this chapter, we will examine the series-parallel combination using the basic laws introduced for series and parallel circuits. There are no new laws or rules to learn—simply an approach that permits the analysis of such structures. In the next chapter, we will consider complex networks using methods of analysis that will allow us to analyze any type of network.

The possibilities for series-parallel configurations are endless—yes, in fact, infinite. The result is a need to examine each network as a separate entity and define the approach that will provide the best path to determining the unknown quantities. In time, you will find similarities between configurations that will make it easier to define the best route to a solution, but this will occur only with exposure, practice, and patience. As mentioned above, however, the best preparation for the analysis of series-parallel networks is a firm understanding of the concepts introduced for series and parallel networks. All the rules and laws to be applied in this chapter have already been introduced in the previous two chapters.

7.2 SERIES-PARALLEL NETWORKS

The network of Fig. 7.1 is a series-parallel network. At first exposure, you must be very careful to determine which elements are in series and which are in parallel. For instance, resistors R_1 and R_2 are **not** in series due to resistor R_3 connected to the common point (b) between R_1 and R_2. Resistors R_2 and R_4 are **not** in parallel because they are not connected at both ends. They are separated at one end by resistor R_3. The need to be absolutely sure of your definitions from the last two chapters now becomes obvious. In fact, it may be a good idea to refer to those rules as we progress through this chapter.

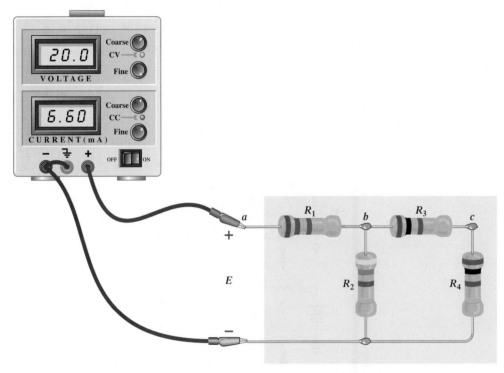

FIG. 7.1
Series-parallel dc network.

If we look carefully enough at Fig. 7.1, we do find that the two resistors R_3 and R_4 are in series because they share only point c, and no other element is connected to that point. Further, the voltage source E and resistor R_1 are in series because they share point a, with no other elements connected to the same point. In the entire configuration there are no two elements in parallel.

The problem now arises as to how we analyze such configurations. The approach is one that requires us to first identify elements that can be combined. Since there are no parallel elements, we must turn to the possibilities with series elements. The voltage source and the series resistor cannot be combined because they are different types of elements. However, resistors R_3 and R_4 can be combined to form a single resistor. The total resistance of the two will be their sum as defined by series circuits. The resulting resistance will then be in parallel with resistor R_2, and they can be combined using the laws for parallel elements. The process has begun: We are slowly reducing the network to one that will be represented by a single resistor equal to the total resistance "seen" by the source.

The source current can now be determined using Ohm's law, and we can "work back" through the network to find all the other currents and voltages. The ability to define the first step in the analysis can sometimes be difficult. However, combinations can be made only by using the rules for series or parallel elements, so naturally the first step may simply be to define which elements are in series or parallel. You must then define how to find such things as the total resistance and the source current and proceed with the analysis. In general, the following steps will provide some guidance for the wide variety of possible combinations that you might encounter.

General Approach:

1. Take a moment to study the problem "in total" and make a brief mental sketch of the overall approach you plan to use. The result may be time- and energy-saving shortcuts.

2. Next examine each region of the network independently before tying them together in series-parallel combinations. This will usually simplify the network and possibly reveal a direct approach toward obtaining one or more desired unknowns. It also eliminates many of the errors that might result due to the lack of a systematic approach.

3. Redraw the network as often as possible with the reduced branches and undisturbed unknown quantities to maintain clarity and provide the reduced networks for the trip back to unknown quantities from the source.

4. When you have a solution, check that it is reasonable by considering the magnitudes of the energy source and the elements in the network. If it does not seem reasonable, either solve the circuit using another approach, or check over your work very carefully.

7.3 REDUCE AND RETURN APPROACH

The schematic representation of the network of Fig. 7.1 appears as Fig. 7.2(a). For this discussion, let us assume that voltage V_4 is desired. As described in Section 7.2, the first step is to combine the series resistors R_3 and R_4 to form an equivalent resistor R' as shown

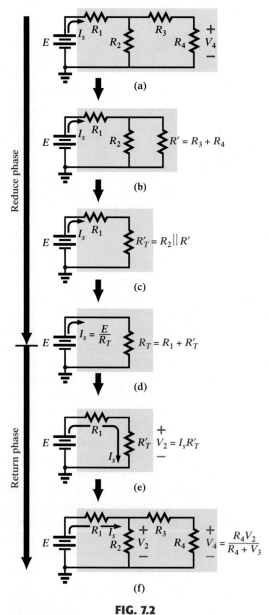

FIG. 7.2

Introducing the reduce and return approach.

in Fig. 7.2(b). Resistors R_2 and R' are then in parallel and can be combined to establish an equivalent resistor R'_T as shown in Fig. 7.2(c). Resistors R_1 and R'_T are then in series and can be combined to establish the total resistance of the network as shown in Fig. 7.2(d). The **reduction phase** of the analysis is now complete. The network cannot be put in a simpler form.

We can now proceed with the **return phase** whereby we work our way back to the desired voltage V_4. Due to the resulting series configuration, the source current is also the current through R_1 and R'_T. The voltage across R'_T (and therefore across R_2) can be determined using Ohm's law as shown in Fig. 7.2(e). Finally, the desired voltage V_4 can be determined by an application of the voltage divider rule as shown in Fig. 7.2(f).

The **reduce and return approach** has now been introduced. It is simply a process whereby the network is reduced to its simplest form across the source, and the source current is determined. There is then a return phase in which you use the resulting source current to work back to the desired unknown. For most single-source series-parallel networks, the above approach provides a viable option toward the solution. In some cases, shortcuts can be applied that will save some time and energy. Now for a few examples.

EXAMPLE 7.1   Find current I_3 for the series-parallel network of Fig. 7.3.

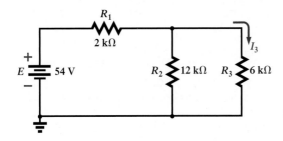

FIG. 7.3
Series-parallel network for Example 7.1.

Solution: Checking for series and parallel elements, we find that resistors R_2 and R_3 are in parallel. Their total resistance is

$$R' = R_2 \parallel R_3 = \frac{R_2 R_3}{R_2 + R_3} = \frac{(12 \text{ k}\Omega)(6 \text{ k}\Omega)}{12 \text{ k}\Omega + 6 \text{ k}\Omega} = 4 \text{ k}\Omega$$

Replacing the parallel combination with a single equivalent resistance will result in the configuration of Fig. 7.4. Resistors R_1 and R' are then in series, resulting in a total resistance of

$$R_T = R_1 + R' = 2 \text{ k}\Omega + 4 \text{ k}\Omega = 6 \text{ k}\Omega$$

The source current is then determined using Ohm's law:

$$I_s = \frac{E}{R_T} = \frac{54 \text{ V}}{6 \text{ k}\Omega} = 9 \text{ mA}$$

In Fig. 7.4, since R_1 and R' are in series, they will have the same current I_s. The result is

$$I_1 = I_s = 9 \text{ mA}$$

Returning to Fig. 7.3, we find that I_1 is the total current entering the parallel combination of R_2 and R_3. Applying the current divider rule will result in the desired current:

$$I_3 = \left(\frac{R_2}{R_2 + R_3} \right) I_1 = \left(\frac{12 \text{ k}\Omega}{12 \text{ k}\Omega + 6 \text{ k}\Omega} \right) 9 \text{ mA} = \mathbf{6 \text{ mA}}$$

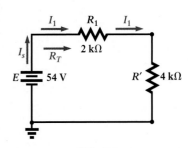

FIG. 7.4
Substituting the parallel equivalent resistance for resistors R_2 and R_3 of Fig. 7.3.

Note in the solution for Example 7.1 that all of the equations employed were introduced in the last two chapters—nothing new was introduced except how to approach the problem and use the equations properly.

EXAMPLE 7.2 For the network of Fig. 7.5:

a. Determine current I_4 and voltage V_2.
b. Insert the meters to measure current I_4 and voltage V_2.

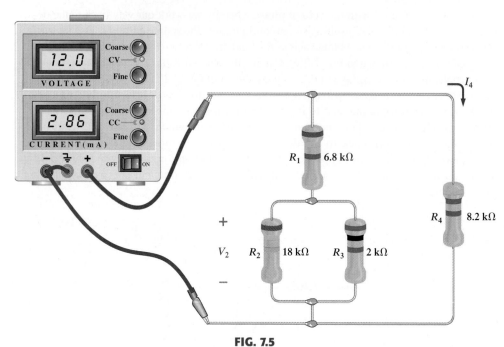

FIG. 7.5
Series-parallel network for Example 7.2.

Solutions:

a. Checking out the network, we find that there are no two resistors in series and the only parallel combination is resistors R_2 and R_3. Combining the two parallel resistors will result in a total resistance of

$$R' = R_2 \parallel R_3 = \frac{R_2 R_3}{R_2 + R_3} = \frac{(18 \text{ k}\Omega)(2 \text{ k}\Omega)}{18 \text{ k}\Omega + 2 \text{ k}\Omega} = 1.8 \text{ k}\Omega$$

Redrawing the network with resistance R' inserted will result in the configuration of Fig. 7.6.

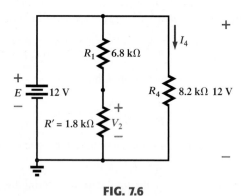

FIG. 7.6
Schematic representation of the network of Fig. 7.5 after substituting the equivalent resistance R' for the parallel combination of R_2 and R_3.

You may now be tempted to combine the series resistors R_1 and R' and redraw the network once more. However, a careful examination of Fig. 7.6 reveals that since the two resistive branches are in parallel, the voltage is the same across each branch. That is, the voltage across the series combination of R_1 and R' is 12 V and that across resistor R_4 is 12 V. The result is that I_4 can be determined directly using Ohm's law as follows:

$$I_4 = \frac{V_4}{R_4} = \frac{E}{R_4} = \frac{12\ V}{8.2\ k\Omega} = \textbf{1.463 mA}$$

In fact, for the same reason, I_4 could have been determined directly from Fig. 7.5. Because the total voltage across the series combination of R_1 and R'_T is 12 V, the voltage divider rule can be applied to determine voltage V_2 as follows:

$$V_2 = \left(\frac{R'}{R' + R_1} \right) E = \left(\frac{1.8\ k\Omega}{1.8\ k\Omega\ +\ 6.8\ k\Omega} \right) 12\ V = \textbf{2.512 V}$$

b. The meters have been properly inserted in Fig. 7.7. Note that the voltmeter is across both resistors since the voltage across parallel elements is the same. In addition, note that the ammeter is in series with resistor R_4, forcing the current through the meter to be the same as that through the series resistor.

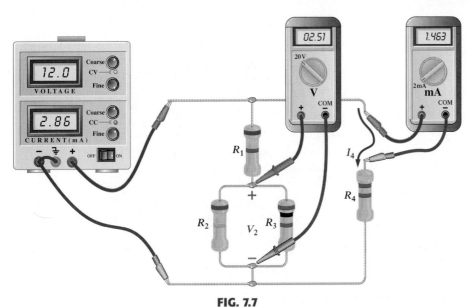

FIG. 7.7
Inserting an ammeter and a voltmeter to measure I_4 and V_2, respectively.

Clearly, Example 7.2 revealed how a careful study of a network can eliminate unnecessary steps toward the desired solution. It is often worth the extra time to sit back and carefully examine a network before diving in with every equation that might seem appropriate.

EXAMPLE 7.3   Find the indicated currents and voltages for the network of Fig. 7.8.

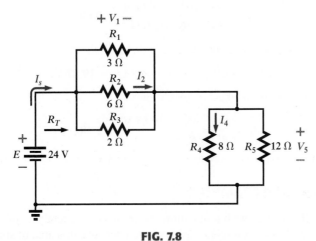

FIG. 7.8
Series-parallel network for Example 7.3.

Solution: In this case we find that there are no series combinations but two obvious parallel configurations. Since R_T is required in order to determine the source current, the two parallel combinations will first be reduced to their simplest forms. That is,

$$R' = R_1 \| R_2 \| R_3 = \cfrac{1}{\cfrac{1}{R_1} + \cfrac{1}{R_2} + \cfrac{1}{R_3}} = \cfrac{1}{\cfrac{1}{3\ \Omega} + \cfrac{1}{6\ \Omega} + \cfrac{1}{2\ \Omega}}$$

$$= \frac{1}{0.333\ \text{S} + 0.167\ \text{S} + 0.5\ \text{S}} = \frac{1}{1.0\ \text{S}} = 1\ \Omega$$

and
$$R'' = R_4 \| R_5 = \frac{R_4 R_5}{R_4 + R_5} = \frac{(8\ \Omega)(12\ \Omega)}{8\ \Omega + 12\ \Omega} = 4.8\ \Omega$$

resulting in the configuration of Fig. 7.9.

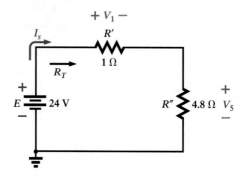

FIG. 7.9

Network of Fig. 7.8 after substituting the equivalent resistors R′ and R″.

The total resistance is then

$$R_T = R' + R'' = 1\ \Omega + 4.8\ \Omega = 5.8\ \Omega$$

and the source current is

$$I_s = \frac{E}{R_T} = \frac{24\ \text{V}}{5.8\ \Omega} = \textbf{4.14 A}$$

Since voltage V_1 is the same for Fig. 7.8 as it is for Fig. 7.9, we can determine V_1 using Ohm's law and the combined resistance of Fig. 7.9:

$$V_1 = I_s\, R' = (4.14\ \text{A})(1\ \Omega) = \textbf{4.14 V}$$

You might be tempted to use the current divider rule to determine I_2, but a more direct route is to simply realize that voltage V_1 is the same across resistor R_2, and I_2 can be determined using Ohm's law:

$$I_2 = \frac{V_2}{R_2} = \frac{V_1}{R_2} = \frac{4.14\ \text{V}}{6\ \Omega} = \textbf{0.69 A}$$

Voltage V_5 can also be determined using Fig. 7.9 as follows:

$$V_5 = I_s\, R'' = (4.14\ \text{A})(4.8\ \Omega) = \textbf{19.87 V}$$

Current I_4 can be determined using Ohm's law:

$$I_4 = \frac{V_4}{R_4} = \frac{V_5}{R_4} = \frac{19.87\ \text{V}}{8\ \Omega} = \textbf{2.484 A}$$

7.4 DESCRIPTIVE EXAMPLES

This section will continue with a few more examples of the analysis of series-parallel networks. The additional examples were included because this area of study is one that requires exposure and practice to develop the skills necessary to approach series-parallel combinations with confidence and success. The more difficult networks can provide a challenge that can be enjoyed in much the same way as a checkers or chess match. The basic rules for the pieces of either game are relatively simple and direct, but their placement on the board can develop a whole new set of challenges—just as the placement of a single resistor can change a straightforward network to a fairly complex one.

Although the solutions are provided, it would wise to try to solve each before looking at the solution. This will provide an opportunity to discover where you are having difficulty: Is it in getting started? Determining whether elements are in series or in parallel? Finding the most direct path to a solution?

EXAMPLE 7.4   Find the indicated currents and voltages for the network of Fig. 7.10.

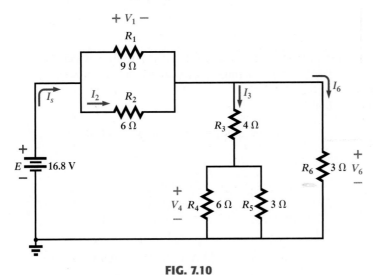

FIG. 7.10

Series-parallel network for Example 7.4.

Solution: A careful inspection of the entire network reveals that there are no two elements in series. There are, however, two parallel combinations that can be worked on immediately. Before proceeding, however, we should be sure that there is no shortcut to the desired currents and voltages. There is nothing obvious, so the reduce and return approach will be employed to combine some elements and redraw the network.

Combining parallel elements:

$$R' = R_1 \| R_2 = \frac{R_1 R_2}{R_1 + R_2} = \frac{(9\ \Omega)(6\ \Omega)}{9\ \Omega + 6\ \Omega} = 3.6\ \Omega$$

$$R'' = R_4 \| R_5 = \frac{R_4 R_5}{R_4 + R_5} = \frac{(6\ \Omega)(3\ \Omega)}{6\ \Omega + 3\ \Omega} = 2\ \Omega$$

Redrawing the network will result in Fig. 7.11 where we now find that R_3 and R'' are in series.

Combining series elements:

$$R''' = R_3 + R'' = 4\ \Omega + 2\ \Omega = 6\ \Omega$$

Redrawing the network again will result in Fig. 7.12 where we find that resistors R'' and R_6 are now in parallel.

Combining parallel elements:

$$R'_T = R''' \| R_6 = \frac{(R''')(R_6)}{R''' + R_6} = \frac{(6\ \Omega)(3\ \Omega)}{6\ \Omega + 3\ \Omega} = 2\ \Omega$$

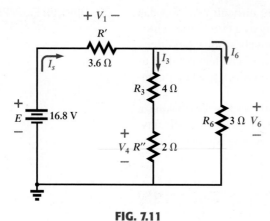

FIG. 7.11

Network of Fig. 7.10 after substituting R' and R".

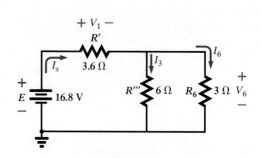

FIG. 7.12

*Network of Fig. 7.11 redrawn
after substituting R"'.*

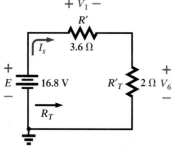

FIG. 7.13

*Network of Fig. 7.12 after
substituting R'_T.*

Redrawing the network once more will result in Fig. 7.13 from which the source current I_s and voltages V_1 and V_6 can be determined; they are the only unknowns conserved from the original network of Fig. 7.10.

The total resistance:

$$R_T = R' + R'_T = 3.6\ \Omega + 2\ \Omega = 5.6\ \Omega$$

The source current:

$$I_s = \frac{E}{R_T} = \frac{16.8\ \text{V}}{5.6\ \Omega} = \textbf{3 A}$$

and voltage V_1 and V_6:

$$V_1 = I_s R' = (3\ \text{A})(3.6\ \Omega) = \textbf{10.8 V}$$
$$V_6 = I_s R'_T = (3\ \text{A})(2\ \Omega) = \textbf{6 V}$$

Since we have voltage V_6, current I_6 can be found using Ohm's law:

$$I_6 = \frac{V_6}{R_6} = \frac{6\ \text{V}}{3\ \Omega} = \textbf{2 A}$$

and voltage V_1 leads directly to I_2 from Fig. 7.10:

$$I_2 = \frac{V_2}{R_2} = \frac{V_1}{R_2} = \frac{10.8\ \text{V}}{6\ \Omega} = \textbf{1.8 A}$$

Using Fig. 7.12, we can determine I_3 using Ohm's law:

$$I_3 = \frac{V_6}{R'''} = \frac{6\ \text{V}}{6\ \Omega} = \textbf{1 A}$$

Since V_4 also appears in Fig. 7.11, we can apply Ohm's law as follows:

$$V_4 = I_3 R'' = (1\ \text{A})(2\ \Omega) = \textbf{2 V}$$

The solution for Example 7.4 required that the original network be redrawn a number of times. Redrawing the network may take time, but it will ensure that you are aware of what voltages and currents are conserved as you move from one network to the other. For instance, in Example 7.4, voltages V_1 and V_6 and the source current I_s were never lost. They were the same for each configuration, so their level could be determined at any point. In time, you will probably feel more at ease about not redrawing the network so many times, but in the beginning it is probably a good idea.

EXAMPLE 7.5 For the network of Fig. 7.14:

a. Find voltages V_1, V_3, and V_x.
b. Determine currents I_1, I_3, and I_s.
c. Find the source current for supply E_2.

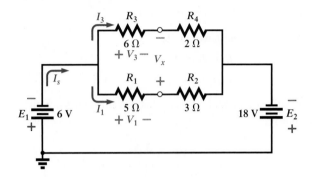

FIG. 7.14

Series-parallel network for Example 7.5.

Solutions:

a. For this network, we find three series combinations but no parallel combinations (of single elements). Yes, the series combination of R_1 and R_2 is in parallel with the series combination of R_3 and R_4. However, initially our interest is in defining the smallest series or parallel combination of elements. One of the series combinations is the two voltage sources. Since they are in opposition, the magnitude will be the difference and the polarity that of the larger, as shown in the redrawn network of Fig. 7.15.

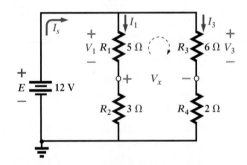

FIG. 7.15

Network of Fig. 7.14 redrawn after combining series voltage sources.

In this case, the redrawing of the network was quite revealing. Note that the resultant supply voltage of 12 V is directly across the two parallel branches of series elements. That is, we know that the voltage across the series combination of R_1 and R_2 is 12 V, and we also realize that the voltage across the series combination of R_3 and R_4 is 12 V. Therefore, we can use the voltage divider rule to find V_1 and V_3 as follows:

$$V_1 = \frac{R_1 E}{R_1 + R_2} = \frac{(5\ \Omega)(12\ \text{V})}{5\ \Omega + 3\ \Omega} = \textbf{7.5 V}$$

$$V_3 = \frac{R_3 E}{R_3 + R_4} = \frac{(6\ \Omega)(12\ \text{V})}{6\ \Omega + 2\ \Omega} = \textbf{9 V}$$

Because we already know V_1 and V_3, voltage V_x must be found by applying Kirchhoff's voltage law around the indicated loop of Fig. 7.15. Starting at point a in the clockwise direction will result in

$$+V_1 - V_3 + V_x = 0$$

and $\qquad\qquad V_x = V_3 - V_1 = 9\ \text{V} - 7.5\ \text{V} = \textbf{1.5 V}$

If we knew V_2 and V_4, we could have used the bottom loop to find V_x.

b. Applying Ohm's law to Fig. 7.15, we find

$$I_2 = I_1 = \frac{V_1}{R_1} = \frac{7.5\text{ V}}{5\ \Omega} = \textbf{1.5 A}$$

and

$$I_4 = I_3 = \frac{V_3}{R_3} = \frac{9\text{ V}}{6\ \Omega} = \textbf{1.5 A}$$

Through Kirchhoff's current law and Fig. 7.15:

$$I_s = I_1 + I_3$$

and

$$I_s = 1.5\text{ A} + 1.5\text{ A} = \textbf{3 A}$$

c. The current through source E_2 is the same as that through E_1, with the direction shown in Fig. 7.14. That is,

$$I_{E_2} = \textbf{3 A}$$

Example 7.5 was certainly a case where redrawing the network was very helpful. In fact, we never had to combine resistors R_1 and R_2, or R_3 and R_4, to find the desired unknowns.

EXAMPLE 7.6   Determine the unknown currents and voltages for the series-parallel network of Fig. 7.16.

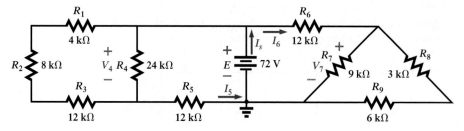

FIG. 7.16
Series-parallel network for Example 7.6.

Solution: A careful examination of Fig. 7.16 reveals that R_1, R_2, and R_3 are in series and can be combined without affecting any of the desired unknowns. In addition, R_8 and R_9 are in series and can also be combined without affecting any of the unknown quantities. In the entire configuration there are no other series or parallel elements. Performing the indicated combinations and redrawing the network will result in the configuration of Fig. 7.17.

$$R' = R_1 + R_2 + R_3$$
$$= 4\text{ k}\Omega + 8\text{ k}\Omega + 12\text{ k}\Omega = 24\text{ k}\Omega$$
$$R'' = R_8 + R_9$$
$$= 3\text{ k}\Omega + 6\text{ k}\Omega = 9\text{ k}\Omega$$

We can now combine resistors R' and R_4, and resistors R_7 and R'', and still hold on to all our unknown quantities as shown in Fig. 7.18.

$$R_x = R' \parallel R_4 = \frac{R'}{N} = \frac{24\text{ k}\Omega}{2} = 12\text{ k}\Omega$$

$$R_y = R_7 \parallel R'' = \frac{R_7}{N} = \frac{9\text{ k}\Omega}{2} = 4.5\text{ k}\Omega$$

Using Ohm's law, current I_5 is

$$I_5 = \frac{E}{R_x + R_5} = \frac{72\text{ V}}{12\text{ k}\Omega + 12\text{ k}\Omega} = \frac{72\text{ V}}{24\text{ k}\Omega} = \textbf{3 mA}$$

and current I_6 is

NOTATION **225**

$$I_6 = \frac{E}{R_6 + R_y} = \frac{72 \text{ V}}{12 \text{ k}\Omega + 4.5 \text{ k}\Omega} = \frac{72 \text{ V}}{16.5 \text{ k}\Omega} \cong \mathbf{4.36 \text{ mA}}$$

The source current is then determined using Kirchhoff's current law:

$$I_s = I_5 + I_6 = 3 \text{ mA} + 4.36 \text{ mA} = \mathbf{7.36 \text{ mA}}$$

Using Fig. 7.17, we can find V_4 and V_7 using Ohm's law:

$$V_4 = I_5 R_x = (3 \text{ mA})(12 \text{ k}\Omega) = \mathbf{36 \text{ V}}$$

and
$$V_7 = I_6 R_y = (4.36 \text{ mA})(4.5 \text{ k}\Omega) = \mathbf{19.62 \text{ V}}$$

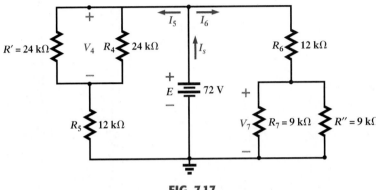
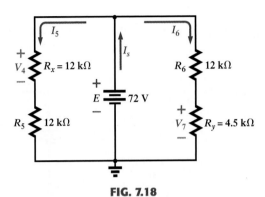

FIG. 7.17

Network of Fig. 7.16 redrawn using equivalent resistors R' and R''.

FIG. 7.18

Network of Fig. 7.17 redrawn with equivalent resistors R_x and R_y.

It is interesting to note that since the voltage source is in the middle of the configuration, the network of Fig. 7.16 can be redrawn as shown in Fig. 7.19. The networks to either side are unaffected by the maneuver. Of course, now the source current for the original circuit must be determined by the sum of the source currents from each section; that is, $I_s = I_5 + I_6$.

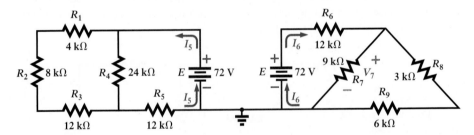

FIG. 7.19

Network of Fig. 7.16 redrawn to show the impact of placing the source in the center of the network.

7.5 NOTATION

We have reached a point where a word or two should be said about special notation before continuing with the examples. The notation to be introduced is the standard in the industry, and you must understand it in order to read complex schematics properly.

Voltage Sources and Ground

Except for a few special cases, electrical and electronic systems are grounded to establish a reference point and to ensure safety. In all the circuits analyzed thus far, the ground symbol of Fig. 7.20 has been employed. Although it has been used to ground each circuit, no mention has been made of the fact that its defined potential level is 0 V. The result is that

the potential at every ground symbol of a network, no matter how many ground symbols there are, is 0 V.

FIG. 7.20

Ground potential.

Furthermore,

the ground symbols of a network can be connected together or left separate without affecting the behavior of the network.

The result is that the circuit of Fig. 7.21 can be drawn in three ways as shown. The use of Fig. 7.21(b) provides a cleaner appearance because we do not have to draw the line between the two points. A third option appears in Fig. 7.21(c) to reinforce the fact that all grounds are at 0 V no matter what their location on a schematic.

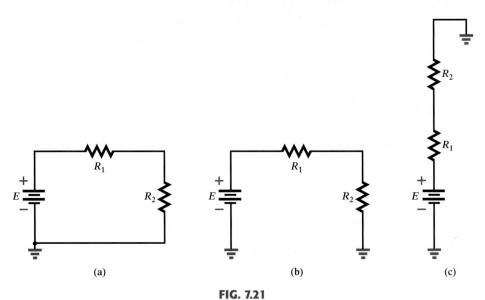

FIG. 7.21

Three ways to sketch the same series circuit.

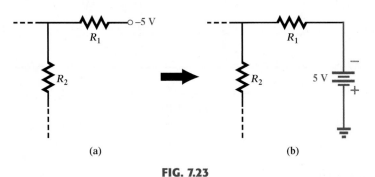

On large schematics where space is at a premium and clarity is important, voltage sources may be represented as shown in Figs. 7.22(b) and 7.23(b). In addition, potential levels may be indicated on a network as shown in Fig. 7.24, to permit a rapid check of whether the circuit is performing correctly. The potential level is from that point to ground, so a voltmeter can be placed from that point to ground to see if a match occurs. If not, the system needs to be carefully examined for a fault.

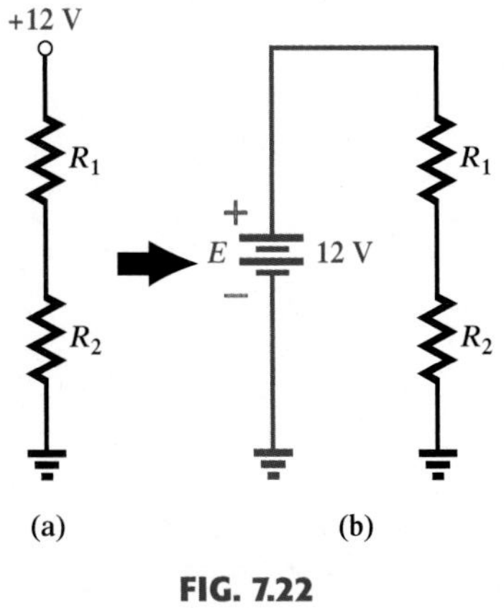

FIG. 7.22

Replacing the special notation for a dc voltage source with the standard symbol.

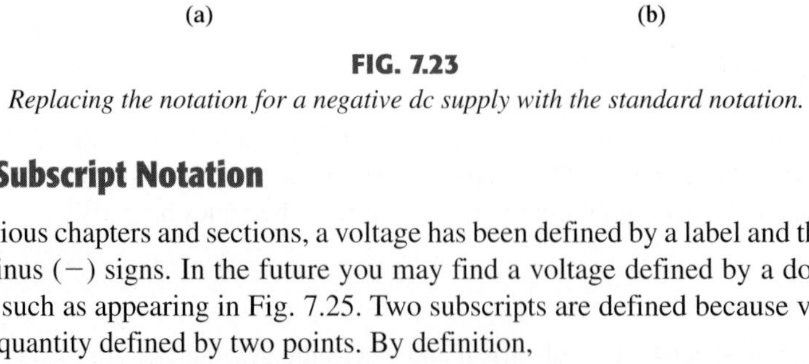

FIG. 7.23

Replacing the notation for a negative dc supply with the standard notation.

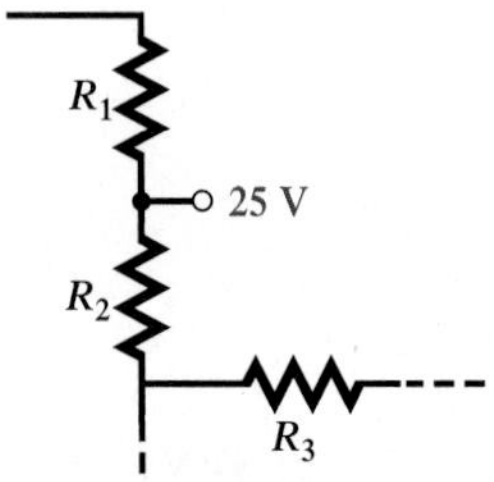

Double-Subscript Notation

In the previous chapters and sections, a voltage has been defined by a label and the plus (+) and the minus (−) signs. In the future you may find a voltage defined by a double set of subscripts such as appearing in Fig. 7.25. Two subscripts are defined because voltages are an **across** quantity defined by two points. By definition,

for all double-subscript voltages, the first subscript is assumed to be at the point of higher potential. If this is not the case, a negative sign must be associated with the magnitude of the voltage.

In Fig. 7.25(a), since point a is at a higher potential, voltage V_{ab} is +4 V. In Fig. 7.25(b), where point b is at the higher potential, $V_{ab} = -4$ V.

FIG. 7.24

The expected voltage level at a particular point in a network of the system is functioning properly.

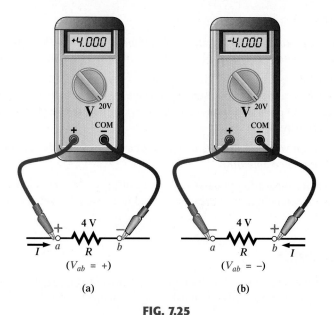

FIG. 7.25
Defining the sign for double-subscript notation.

In words,

the label V_{ab} defines the voltage at point a with respect to (w.r.t.) point b.

All meters internally assume that the red lead, or the lead connected to the positive terminal of the meter, will be connected to the point of higher potential. In Fig. 7.25(a), since this is the case, the reading has a positive sign (or no sign at all). In Fig. 7.25(b), a negative sign is the result since the red lead is connected to the point of lower potential. In either case, the magnitude of the reading is correct. The resulting sign simply confirms or reveals the polarity across the resistor.

Single-Subscript Notation

If point b of the notation V_{ab} is specified as ground potential (zero volts), then a single-subscript notation can be employed that provides the voltage at a point with respect to ground. In Fig. 7.26, V_a is the voltage from point a to ground. In this case, it is obviously 10 V since it is right across the source voltage E. Voltage V_b is the voltage from point b to ground. Because it is directly across the 4 Ω resistor, $V_b = +4$ V.

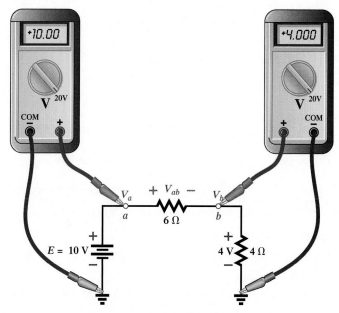

FIG. 7.26
Defining the use of single-subscript notation for voltage levels.

In summary,

the single-subscript notation V_a specifies the voltage at point a with respect to ground (zero volts). If the voltage is less than zero volts, a negative sign must be associated with the magnitude of V_a.

When hooking up a meter to measure the voltage at a point to ground, connect the red or positive lead to the point in question and the black or negative lead to ground (or whatever reference is desired).

For single-subscript voltages, the negative lead of a voltmeter is always connected to the ground or reference potential.

The connections for measuring voltages V_a and V_b of Fig. 7.26 are included in the figure. Since both have a potential higher than the ground or reference level, the reading is positive for both.

General Comments

Since point a of the double-subscript notation V_{ab} is assumed to be at the point of higher potential, the following important relationship can be defined:

$$\boxed{V_{ab} = V_a - V_b} \tag{7.1}$$

In words, Equation (7.1) states that voltage V_{ab} is equal to the voltage at point a less the voltage at point b. In Fig. 7.26, for example,

$$V_{ab} = V_a - V_b = 10\text{ V} - 4\text{ V} = \textbf{6 V}$$

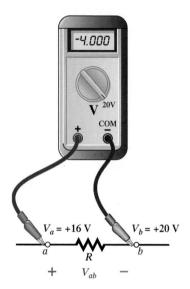

$V_a = +16\text{ V}$ $V_b = +20\text{ V}$

a   *R*   *b*

$+$ V_{ab} $-$

FIG. 7.27
Example 7.7.

EXAMPLE 7.7 Find voltage V_{ab} for the situation of Fig. 7.27.

Solution: Since point a is at a lower potential, we know that V_{ab} must have a negative sign. Using Eq. (7.1):

$$V_{ab} = V_a - V_b = 16\text{ V} - 20\text{ V} = \textbf{-4 V}$$

as confirmed by the meter reading.

EXAMPLE 7.8 Find voltage V_a for the configuration of Fig. 7.28.

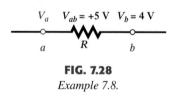

V_a $V_{ab} = +5\text{ V}$ $V_b = 4\text{ V}$

a   *R*   *b*

FIG. 7.28
Example 7.8.

Solution: Applying Eq. (7.1):

$$V_{ab} = V_a - V_b$$

so that

$$V_a = V_{ab} + V_b = 5\text{ V} + 4\text{ V} = \textbf{9 V}$$

EXAMPLE 7.9 Find voltage V_{ab} for the configuration of Fig. 7.29.

Solution: Applying Eq. (7.1):

$$V_{ab} = V_a - V_b = 20\text{ V} - (-15\text{ V}) = 20\text{ V} + 15\text{ V} = \textbf{35 V}$$

as verified by the meter reading.

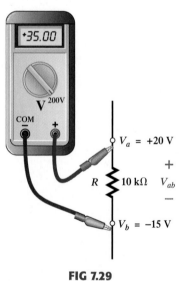

$V_a = +20\text{ V}$

$+$

R $10\text{ k}\Omega$ V_{ab}

$-$

$V_b = -15\text{ V}$

FIG 7.29
Example 7.9.

Note that in Example 7.9, the voltage at point b is below ground level, requiring that the net difference in voltage be the sum of that above ground and that below ground level. It also points out the need to be careful with signs when applying the equation. As shown in Fig. 7.30, this represents a drop in voltage of 35 V. In some ways, it's like going from a positive checking account of $20 to owing $15; the total expenditure is $35.

7.6 APPLYING THE SPECIAL NOTATION

We will now apply the notation introduced in the previous section to a few networks. In each example the network is redrawn to ensure that the special notation is correctly understood.

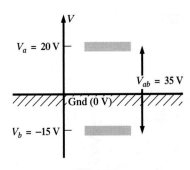

FIG. 7.30
The impact of positive and negative voltages on the total voltage drop.

EXAMPLE 7.10 Determine voltage V_x for the circuit of Fig. 7.31.

Solution: Inserting the standard symbol for the voltage source will result in the circuit of Fig. 7.32. In this case, the desired voltage V_x (the voltage from that point to ground) is the same as voltage V_2 across resistor R_2.

Applying the voltage divider rule:

$$V_x = V_2 = \frac{R_2 E}{R_1 + R_2} = \frac{(2\ \Omega)(24\ \text{V})}{2\ \Omega + 4\ \Omega} = \textbf{8 V}$$

EXAMPLE 7.11 Determine voltage V_a for the circuit of Fig. 7.33.

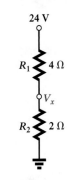

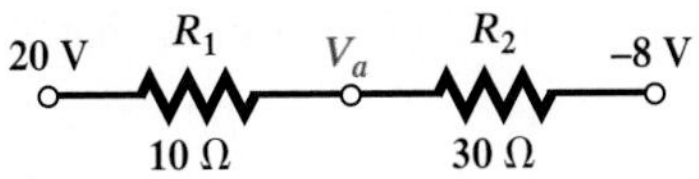

FIG. 7.31
Circuit for Example 7.10.

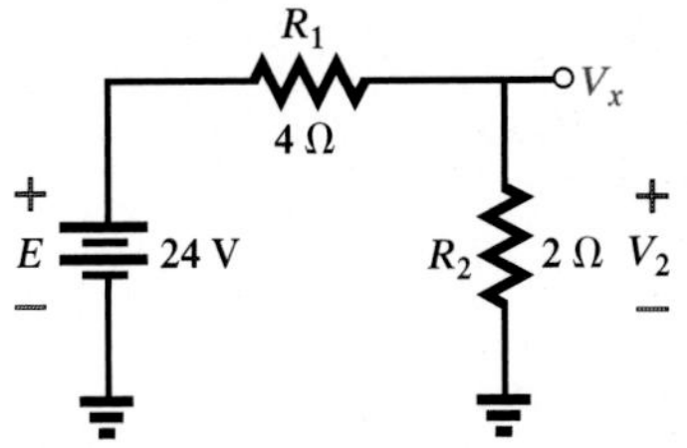

R_1 V_a R_2
20 V ——WW——•——WW—— −8 V
        10 Ω   30 Ω

FIG. 7.33
Circuit for Example 7.11.

Solution: Redrawing the network will result in the configuration of Fig. 7.34. Note that two voltage sources are actually supporting each other. In order to determine V_a, the voltage across R_1 or R_2 must first be determined.

Using Ohm's law:

$$I_1 = \frac{E_1 + E_2}{R_1 + R_2} = \frac{20\ \text{V} + 8\ \text{V}}{10\ \Omega + 30\ \Omega} = \frac{28\ \text{V}}{40\ \Omega} = 0.7\ \text{A}$$

and

$$V_1 = I_1 R_1 = (0.7\ \text{A})(10\ \Omega) = 7\ \text{V}$$

Applying Kirchhoff's voltage law around the indicated loop will result in

$$+E_1 - V_1 - V_a = 0$$

and

$$V_a = E_1 - V_1 = 20\ \text{V} - 7\ \text{V} = \textbf{13 V}$$

FIG. 7.32
Circuit of Fig. 7.31 redrawn with the standard voltage source symbol.

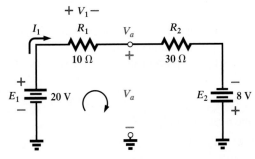

FIG. 7.34
Circuit of Fig. 7.33 redrawn with standard voltage source symbols.

229

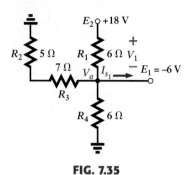

FIG. 7.35

Network for Example 7.12.

EXAMPLE 7.12 For the network of Fig.7.35:

a. Determine voltages V_a and V_1.

b. Find the source current for voltage source E_1.

Solutions:

a. It would indeed be difficult to analyze the network in the form of Fig. 7.35 with the symbolic notation for the sources and the reference or ground connection in the upper left-hand corner of the diagram. However, when the network is redrawn as shown in Fig. 7.36, the unknowns and the relationship between branches become significantly clearer. Note the common connection of the grounds and the replacing of the terminal notation by the actual supplies.

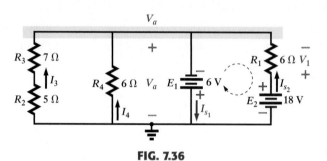

FIG. 7.36

Network of Fig. 7.35 redrawn to better define unknown quantities.

Since the voltages are in parallel, it is now clear that

$$V_a = -E_1 = -6\,\text{V}$$

The minus sign simply reflects the fact that for single-subscript notation, point a is defined as positive with respect to ground and that is not the case here.

Voltage V_1 is surrounded by two voltage sources, so it can be determined by an application of Kirchhoff's voltage law as follows:

$$-E_1 + V_1 - E_2 = 0$$

and
$$V_1 = E_1 + E_2 = 18\,\text{V} + 6\,\text{V} = \textbf{24 V}$$

b. The source current for source E_1 can be determined by an application of Kirchhoff's current law to node a as follows:

$$\Sigma I_i = \Sigma I_o$$
$$I_3 + I_4 + I_{s_2} = I_{s_1}$$

and
$$I_{s_1} = I_3 + I_4 + I_{s_2} = \frac{E_1}{R_2 + R_3} + \frac{E_1}{R_4} + \frac{V_1}{R_1}$$

$$= \frac{6\,\text{V}}{5\,\Omega + 7\,\Omega} + \frac{6\,\text{V}}{6\,\Omega} + \frac{24\,\text{V}}{6\,\Omega} = 0.5\,\text{A} + 1\,\text{A} + 4\,\text{A} = \textbf{5.5 A}$$

The network of the next example is actually an example of a complex network—one in which there are no series or parallel elements. However, we will be able to find the desired unknowns because of the strategic placement of voltage sources.

EXAMPLE 7.13 For the network of Fig. 7.37:

a. Determine voltages V_a, V_b, and V_c.

b. Find voltages V_{ac} and V_{bc}.

c. Find current I_2.

d. Find the source current I_{s_3}.

e. Insert voltmeters to measure voltages V_a and V_{bc}.

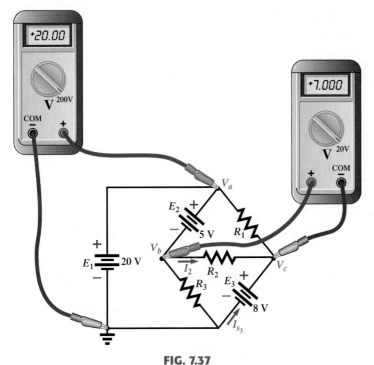

FIG. 7.37

Complex network for Example 7.13.

Solutions:

a. The network is redrawn in Fig. 7.38 to clearly indicate the arrangement between elements.

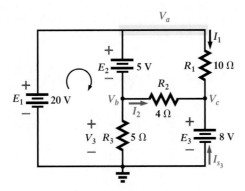

FIG. 7.38

*Network of Fig. 7.37 redrawn to better define
a path toward the desired unknowns.*

First, it is clear that voltage V_a is directly across voltage source E_1. Therefore,

$$V_a = E_1 = \textbf{20 V}$$

The same is true for voltage V_c, which is directly across the voltage source E_3. Therefore,

$$V_c = E_3 = \textbf{8 V}$$

In order to find voltage V_b, which is actually the voltage across R_3, we must apply Kirchhoff's voltage law around loop 1 as follows:

$$+E_1 - E_2 - V_3 = 0$$

and $V_3 = E_1 - E_2 = 20\text{ V} - 5\text{ V} = 15\text{ V}$

and $V_b = V_3 = \textbf{15 V}$

b. Voltage V_{ac}, which is actually the voltage across resistor R_1, can then be determined as follows:

$$V_{ac} = V_a - V_c = 20\text{ V} - 8\text{ V} = \textbf{12 V}$$

Similarly, voltage V_{bc}, which is actually the voltage across resistor R_2, can then be determined as follows:

$$V_{bc} = V_b - V_c = 15 \text{ V} - 8 \text{ V} = \mathbf{7 \text{ V}}$$

c. Current I_2 can be determined using Ohm's law:

$$I_2 = \frac{V_2}{R_2} = \frac{V_{bc}}{R_2} = \frac{7 \text{ V}}{4 \text{ }\Omega} = \mathbf{1.75 \text{ A}}$$

d. The source current I_{s_3} can be determined using Kirchhoff's current law at node c:

$$\Sigma I_i = \Sigma I_o$$

$$I_1 + I_2 + I_{s_3} = 0$$

and

$$I_{s_3} = -I_1 - I_2 = -\frac{V_1}{R_1} - I_2$$

with

$$V_1 = V_{ac} = V_a - V_c = 20 \text{ V} - 8 \text{ V} = 12 \text{ V}$$

so that

$$I_{s_3} = -\frac{12 \text{ V}}{10 \text{ }\Omega} - 1.75 \text{ A} = -1.2 \text{ A} - 1.75 \text{ A} = \mathbf{-2.95 \text{ A}}$$

revealing that current is actually being forced through source E_3 in a direction opposite to that shown in Fig. 7.37.

e. Both meters have a positive reading as shown in Fig. 7.37.

7.7 LADDER NETWORKS

A three-section **ladder network** appears in Fig. 7.39. The reason for the terminology should be quite obvious from its similarities with a ladder laid to rest on its side. The network can be analyzed in one of two ways. In the first method, the reduce and return approach is employed. In the second method, a letter symbol is assigned to the desired unknown, which is then carried back to establish a direct relationship with the applied voltage.

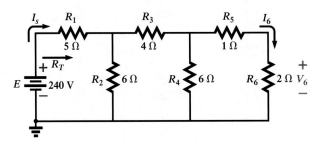

FIG. 7.39
Ladder network.

Method 1

Using the reduce and return approach, we first find that resistors R_5 and R_6 are in series, with a total resistance of 3 Ω, as shown in Fig. 7.40. Resistor R_4 is then in parallel with this series combination, resulting in a parallel combination of 2 Ω. The resulting equivalent of 2 Ω is then in

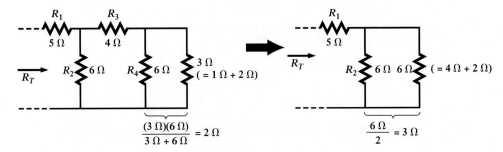

FIG. 7.40
Working back to the source to determine R_T for the network of Fig. 7.39.

series with resistor R_3 which, when combined, results in a total of 6 Ω. The resulting parallel combination of 6 Ω resistors results in a 3 Ω equivalent in series with R_1, as shown in Fig. 7.41.

The total resistance is then

$$R_T = 5\,\Omega + 3\,\Omega = 8\,\Omega$$

and the source current is

$$I_s = \frac{E}{R_T} = \frac{240\ \text{V}}{8\ \Omega} = 30\ \text{A}$$

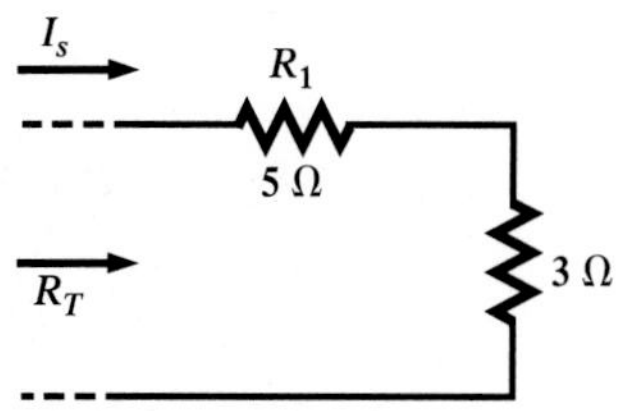

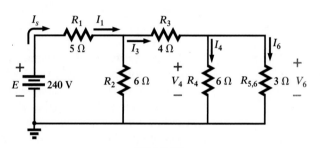

FIG. 7.41

Calculating R_T and I_s.

Working our way back to V_6 using Fig. 7.42, we find

$$I_1 = I_s$$

and

$$I_3 = \frac{I_1}{2} = \frac{30\ \text{A}}{2} = 15\ \text{A}$$

Finally, using Fig. 7.43,

$$I_6 = \frac{(6\ \Omega)I_3}{6\ \Omega + 3\ \Omega} = \frac{6}{9}(15\ \text{A}) = 10\ \text{A}$$

and

$$V_6 = I_6 R_6 = (10\ \text{A})(2\ \Omega) = \textbf{20 V}$$

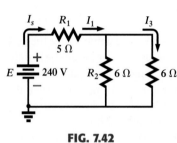

FIG. 7.42

Working back toward V_6.

FIG. 7.43

Calculating V_6.

Method 2

As noted above, this method involves choosing a variable name for the desired unknown and then carrying that unknown back to the voltage source using the basic laws of electric circuits. Resistor values are used in the process to provide the simplest format for the relationships between and among various variables of the network. The following steps should clarify the procedure.

Starting with the desired voltage V_6 of Fig. 7.44, we know that

$$V_6 = \frac{R_6 V_4}{R_6 + R_5} = \frac{(2\ \Omega)V_4}{2\ \Omega + 1\ \Omega} = \frac{2}{3}V_4$$

However,

$$V_4 = I_4 R_4$$

so

$$V_6 = \frac{2}{3}V_4 = \frac{2}{3}(I_4 R_4) = \frac{2}{3}I_4(6\ \Omega) = (4\ \Omega)I_4$$

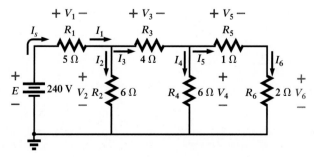

FIG. 7.44

Applying Method 2 to the ladder network of Fig. 7.39.

or

$$I_4 = \frac{V_6}{4\ \Omega}$$

with

$$I_3 = I_4 + I_6 = \frac{V_6}{4\ \Omega} + \frac{V_6}{2\ \Omega} = (0.75\ \text{S})V_6$$

and

$$V_3 = I_3 R_3 = [0.75\ \text{S}\ (V_6)](4\ \Omega) = 3V_6$$

with

$$V_2 = V_3 + V_4 = 3V_6 + \frac{3}{2}V_6 = 4.5V_6$$

and

$$I_2 = \frac{V_2}{R_2} = \frac{4.5V_6}{6\ \Omega} = (0.75\ \text{S})V_6$$

Therefore,

$$I_s = I_2 + I_3 = (0.75\ \text{S})V_6 + (0.75\ \text{S})V_6 = (1.5\ \text{S})V_6$$

and

$$V_1 = I_1 R_1 = I_s R_1 = [(1.5\ \text{S})V_6](5\ \Omega) = 7.5V_6$$

with

$$E = V_1 + V_2 = 7.5V_6 + 4.5V_6 = 12V_6$$

so that

$$V_6 = \frac{E}{12} = \frac{240\ \text{V}}{12} = \mathbf{20\ V}$$

as obtained using Method 1.

Mathcad

The ladder network provides an excellent opportunity to demonstrate the power of a mathematical software package such as Mathcad. The entries for the analysis follow exactly the same procedure as performing the analysis in a longhand manner. First, the values of the parameters must be defined as shown on the top row of Fig. 7.45. Then the resistors are combined in the same manner as described for Method 1. The process of working back to the desired unknown is also the same as described for Method 1.

The wonderful thing about Mathcad is that this sequence can be put in memory and called for as the need arises for a network with different values. Simply redefine the parameters on line 1, and all the new values for the important parameters of the network will be displayed almost immediately.

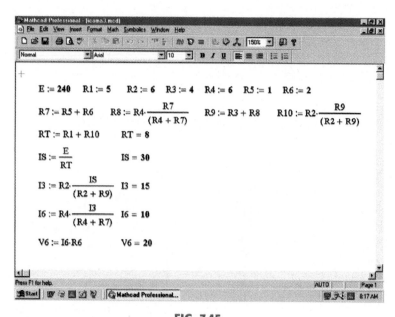

FIG. 7.45

Using Mathcad to analyze the ladder network of Fig. 7.39.

The term *loaded* appearing in the heading of this section refers to the application of an element, network, or system to a supply that will draw current from the supply. In other words,

the loading down of a system is the process of introducing elements that will draw current from the system. The heavier the current, the greater the loading effect.

Recall from Section 5.8 that the application of a load can affect the terminal voltage of a supply due to the internal resistance.

No-Load Conditions

Through a voltage divider network such as appearing in Fig. 7.46, a number of different terminal voltages can be made available from a single supply. Instead of having a single supply of 120 V, we now have terminal voltages of 100 V and 60 V available—a wonderful result for such a simple network. However, as with all things, there can be a down side. The disadvantage here is that the applied resistive loads can have values too close to those making up the voltage divider network.

In general,

for a voltage divider supply to be effective, the applied resistive loads should be significantly larger than the resistors appearing in the voltage divider network.

To demonstrate the validity of the above statement, let us now examine the effect of applying resistors with values very close to those of the voltage divider network.

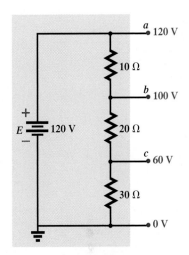

FIG. 7.46
Unloaded voltage divider supply.

Loaded Conditions

In Fig. 7.47, resistors of 20 Ω have been connected to each of the terminal voltages. Note that this value is equal to one of the resistors in the voltage divider network and very close to the other two.

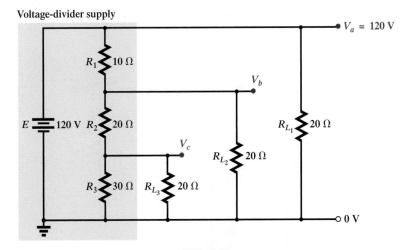

FIG. 7.47
Voltage divider supply with loads equal to the average value of the resistive elements that make up the supply.

Voltage V_a is unaffected by the load R_{L_1} since the load is in parallel with the supply voltage E. The result is $V_a = 120$ V, which is the same as the no-load level. To determine V_b, we must first note that R_3 and R_{L_3} are in parallel and $R'_3 = R_3 \| R_{L_3} = 30\ \Omega \| 20\ \Omega = 12\ \Omega$. The parallel combination

$$R'_2 = (R_2 + R'_3) \| R_{L_2} = (20\ \Omega + 12\ \Omega) \| 20\ \Omega = 32\ \Omega \| 20\ \Omega = 12.31\ \Omega$$

Applying the voltage divider rule gives

$$V_b = \frac{(12.31\ \Omega)(120\ \text{V})}{12.31\ \Omega + 10\ \Omega} = \mathbf{66.21\ V}$$

versus 100 V under no-load conditions.

Voltage V_c is

$$V_c = \frac{(12 \ \Omega)(66.21 \ V)}{12 \ \Omega + 20 \ \Omega} = \textbf{24.83 V}$$

versus 60 V under no-load conditions.

The effect of load resistors close in value to the resistor employed in the voltage divider network is, therefore, to decrease significantly some of the terminal voltages.

If the load resistors are changed to the 1 kΩ level, the terminal voltages will all be relatively close to the no-load values. The analysis is similar to the above, with the following results:

$$V_a = \textbf{120 V} \qquad V_b = \textbf{98.88 V} \qquad V_c = \textbf{58.63 V}$$

If we compare current drains established by the applied loads, we find for the network of Fig. 7.47 that

$$I_{L_2} = \frac{V_{L_2}}{R_{L_2}} = \frac{66.21 \ V}{20 \ \Omega} = 3.31 \ A$$

and for the 1 kΩ level,

$$I_{L_2} = \frac{98.88 \ V}{1 \ k\Omega} = 98.88 \ mA < 0.1 \ A$$

As noted above in the highlighted statement, the greater the current drain, the greater the change in terminal voltage with the application of the load. This is certainly verified by the fact that I_{L_2} is about 33.5 times larger with the 20 Ω loads.

The next example is a design exercise. The voltage and current ratings of each load are provided, along with the terminal ratings of the supply. The required voltage divider resistors must be found.

EXAMPLE 7.14 Determine R_1, R_2, and R_3 for the voltage divider supply of Fig. 7.48. Can 2 W resistors be used in the design?

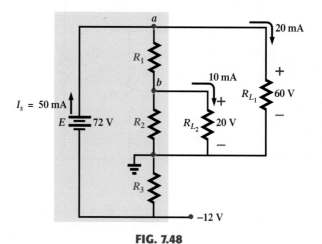

FIG. 7.48
Voltage divider supply for Example 7.14.

Solution: R_3:

$$R_3 = \frac{V_{R_3}}{I_{R_3}} = \frac{V_{R_3}}{I_s} = \frac{12 \ V}{50 \ mA} = \textbf{240 } \boldsymbol{\Omega}$$

$$P_{R_3} = (I_{R_3})^2 R_3 = (50 \ mA)^2 \ 240 \ \Omega = 0.6 \ W < 2 \ W$$

R_1: Applying Kirchhoff's current law to node a:

$$I_s - I_{R_1} - I_{L_1} = 0$$

and
$$I_{R_1} = I_s - I_{L_1} = 50\text{ mA} - 20\text{ mA} = 30\text{ mA}$$

$$R_1 = \frac{V_{R_1}}{I_{R_1}} = \frac{V_{L_1} - V_{L_2}}{I_{R_1}} = \frac{60\text{ V} - 20\text{ V}}{30\text{ mA}} = \frac{40\text{ V}}{30\text{ mA}} = \mathbf{1.33\text{ k}\Omega}$$

$$P_{R_1} = (I_{R_1})^2 R_1 = (30\text{ mA})^2\,1.33\text{ k}\Omega = 1.197\text{ W} < 2\text{ W}$$

R_2: Applying Kirchhoff's current law at node b:

$$I_{R_1} - I_{R_2} - I_{L_2} = 0$$

and
$$I_{R_2} = I_{R_1} - I_{L_2} = 30\text{ mA} - 10\text{ mA} = 20\text{ mA}$$

$$R_2 = \frac{V_{R_2}}{I_{R_2}} = \frac{20\text{ V}}{20\text{ mA}} = \mathbf{1\text{ k}\Omega}$$

$$P_{R_2} = (I_{R_2})^2 R_2 = (20\text{ mA})^2\,1\text{ k}\Omega = 0.4\text{ W} < 2\text{ W}$$

Since P_{R_1}, P_{R_2}, and P_{R_3} are less than 2 W, 2 W resistors can be used for the design.

7.9 POTENTIOMETER LOADING

For the unloaded potentiometer of Fig. 7.49, the output voltage is determined by the voltage divider rule, with R_T in the figure representing the total resistance of the potentiometer. Too often it is assumed that the voltage across a load connected to the wiper arm is determined solely by the potentiometer and the effect of the load can be ignored. This is definitely not the case, as is demonstrated in the next few paragraphs.

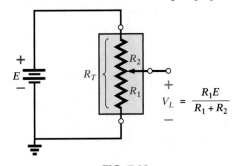

FIG. 7.49
Unloaded potentiometer.

When a load is applied as shown in Fig. 7.50, the output voltage V_L is now a function of the magnitude of the load applied since R_1 is not as shown in Fig. 7.49 but is instead the parallel combination of R_1 and R_L.

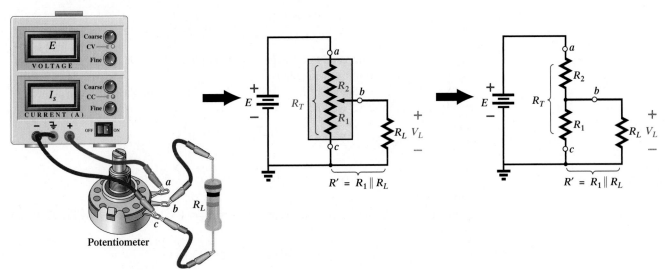

FIG. 7.50
Loaded potentiometer.

The output voltage is now

$$V_L = \frac{R'E}{R' + R_2} \quad \text{with } R' = R_1 \| R_L \qquad \textbf{(7.2)}$$

If it is desired to have good control of the output voltage V_L through the controlling dial, knob, screw, or whatever, it is advisable to choose a load or potentiometer that satisfies the following relationship:

$$R_L \gg R_T \qquad \textbf{(7.3)}$$

In general,

when hooking up a load to a potentiometer, be sure that the load resistance far exceeds the maximum terminal resistance of the potentiometer if good control of the output voltage is desired.

For example, let's disregard Eq. (7.3) and choose a 1 MΩ potentiometer with a 100 Ω load and set the wiper arm to 1/10 the total resistance, as shown in Fig. 7.51. Then

$$R' = 100 \text{ k}\Omega \| 100 \text{ }\Omega = 99.9 \text{ }\Omega$$

and

$$V_L = \frac{99.9 \text{ }\Omega(10 \text{ V})}{99.9 \text{ }\Omega + 900 \text{ k}\Omega} \cong 0.001 \text{ V} = 1 \text{ mV}$$

which is extremely small compared to the expected level of 1 V.

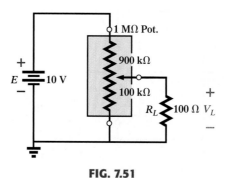

FIG. 7.51

Loaded potentiometer with $R_L \ll R_T$.

In fact, if we move the wiper arm to the midpoint,

$$R' = 500 \text{ k}\Omega \| 100 \text{ }\Omega = 99.98 \text{ }\Omega$$

and

$$V_L = \frac{(99.98 \text{ }\Omega)(10 \text{ V})}{99.98 \text{ }\Omega + 500 \text{ k}\Omega} \cong 0.002 \text{ V} = 2 \text{ mV}$$

which is negligible compared to the expected level of 5 V. Even at $R_1 = 900 \text{ k}\Omega$, V_L is simply 0.01 V, or 1/1000 of the available voltage.

Using the reverse situation of $R_T = 100 \text{ }\Omega$ and $R_L = 1 \text{ M}\Omega$ and the wiper arm at the 1/10 position, as in Fig. 7.52, we find

$$R' = 10 \text{ }\Omega \| 1 \text{ M}\Omega \cong 10 \text{ }\Omega$$

and

$$V_L = \frac{10 \text{ }\Omega(10 \text{ V})}{10 \text{ }\Omega + 90 \text{ }\Omega} = 1 \text{ V}$$

as desired.

For the lower limit (worst-case design) of $R_L = R_T = 100 \text{ }\Omega$, as defined by Eq. (7.3) and the halfway position of Fig. 7.50,

$$R' = 50 \text{ }\Omega \| 100 \text{ }\Omega = 33.33 \text{ }\Omega$$

and

$$V_L = \frac{33.33 \text{ }\Omega(10 \text{ V})}{33.33 \text{ }\Omega + 50 \text{ }\Omega} \cong 4 \text{ V}$$

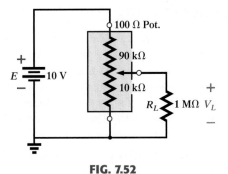

FIG. 7.52
Loaded potentiometer with $R_L \gg R_T$.

It may not be the ideal level of 5 V, but at least 40% of the voltage E has been achieved at the halfway position rather than the 0.02% obtained with $R_L = 100\ \Omega$ and $R_T = 1\ \text{M}\Omega$.

In general, therefore, try to establish a situation for potentiometer control in which Equation (7.3) is satisfied to the highest degree possible.

Someone might suggest that we make R_T as small as possible to bring the percent result as close to the ideal as possible. Keep in mind, however, that the potentiometer has a power rating, and for networks such as Fig. 7.52, $P_{max} \cong E^2/R_T = (10\ \text{V})^2/100\ \Omega = 1\ \text{W}$. If R_T is reduced to $10\ \Omega$, $P_{max} = (10\ \text{V})^2/10\ \Omega = 10\ \text{W}$, which would require a *much larger* unit.

EXAMPLE 7.15 Find voltages V_1 and V_2 for the loaded potentiometer of Fig. 7.53.

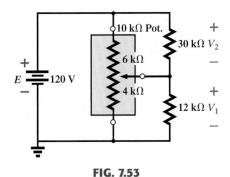

FIG. 7.53
Loaded potentiometer for Example 7.15.

Solution: Ideal (no load):

$$V_1 = \frac{4\ \text{k}\Omega(120\ \text{V})}{10\ \text{k}\Omega} = \textbf{48 V}$$

$$V_2 = \frac{6\ \text{k}\Omega(120\ \text{V})}{10\ \text{k}\Omega} = \textbf{72 V}$$

Loaded:

$$R' = 4\ \text{k}\Omega \parallel 12\ \text{k}\Omega = 3\ \text{k}\Omega$$

$$R'' = 6\ \text{k}\Omega \parallel 30\ \text{k}\Omega = 5\ \text{k}\Omega$$

$$V_1 = \frac{3\ \text{k}\Omega(120\ \text{V})}{8\ \text{k}\Omega} = \textbf{45 V}$$

$$V_2 = \frac{5\ \text{k}\Omega(120\ \text{V})}{8\ \text{k}\Omega} = \textbf{75 V}$$

The ideal and loaded voltage levels are so close that the design can be considered a good one for the applied loads. A slight variation in the position of the wiper arm will establish the ideal voltage levels across the two loads.

7.10 OPEN AND SHORT CIRCUITS

Open circuits and short circuits can often cause more confusion and difficulty in the analysis of a system than standard series or parallel configurations. This will become more obvious in the chapters to follow when we apply some of the methods and theorems.

An **open circuit** is simply two isolated terminals not connected by an element of any kind, as shown in Fig. 7.54. Since a path for conduction does not exist, the current associated with an open circuit must always be zero. The voltage across the open circuit, however, can be any value, as determined by the system it is connected to. In summary, therefore,

an open circuit can have a potential difference (voltage) across its terminals, but the current is always zero amperes.

In Fig. 7.55, an open circuit exists between terminals *a* and *b*. As shown in the figure, the voltage across the open-circuit terminals is the supply voltage, but the current is zero due to the absence of a complete circuit.

Some practical examples of open circuits and their impact are provided in Fig. 7.56. In Fig. 7.56(a), the excessive current demanded by the circuit caused a fuse to fail, creating an open circuit that reduced the current to zero amperes. However, it is important to note that **the full applied voltage is now across the open circuit,** so you must be careful when changing the fuse. If there is a main breaker ahead of the fuse, it should be thrown first to remove the possibility of getting a shock. This situation clearly reveals the benefit of circuit breakers: You can simply reset the breaker without having to get near the hot wires.

In Fig. 7.56(b), the pressure plate at the bottom of the bulb cavity in a flashlight was bent when the flashlight was dropped. An open circuit now exists between the contact point of

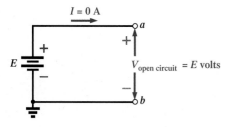

FIG. 7.54

Defining an open circuit.

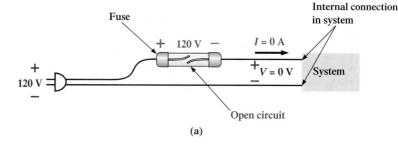

FIG. 7.55

Demonstrating the characteristics of an open circuit.

(a)

(b)

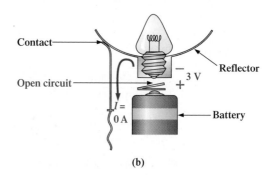

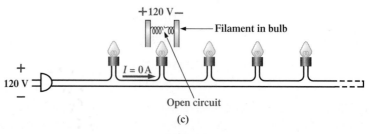

(c)

FIG. 7.56

Examples of open circuits.

the bulb and the plate connected to the batteries. The current has dropped to zero amperes, but the 3 V provided by the series batteries appears across the open circuit. The situation can be corrected by simply placing a flat-edge screwdriver under the plate and bending it toward the bulb.

Finally, in Fig. 7.56(c), the filament in a bulb in a series connection has opened due to excessive current or old age, creating an open circuit that knocks out all the bulbs in the series configuration. Again, the current has dropped to zero amperes, but the full 120 V will appear across the contact points of the bad bulb. For situations such as this, **you should remove the plug from the wall before changing the bulb.**

A **short circuit** is a very low resistance, direct connection between two terminals of a network, as shown in Fig. 7.57. The current through the short circuit can be any value, as determined by the system it is connected to, but the voltage across the short circuit will always be zero volts because the resistance of the short circuit is assumed to be essentially zero ohms and $V = IR = I(0\ \Omega) = 0\ V$.

In summary, therefore,

a short circuit can carry a current of a level determined by the external circuit, but the potential difference (voltage) across its terminals is always zero volts.

In Fig. 7.58(a), the current through the 2 Ω resistor is 5 A. If a short circuit should develop across the 2 Ω resistor, the total resistance of the parallel combination of the 2 Ω resistor and

the short (of essentially zero ohms) will be $2\ \Omega \parallel 0\ \Omega = \dfrac{(2\ \Omega)(0\ \Omega)}{2\ \Omega + 0\ \Omega} = 0\ \Omega$, as indicated in

Fig. 7.58(b), and the current will rise to very high levels, as determined by Ohm's law:

$$I = \frac{E}{R} = \frac{10\ V}{0\ \Omega} \rightarrow \infty\ A$$

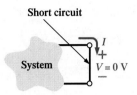

FIG. 7.57
Defining a short circuit.

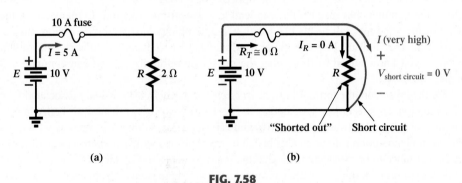

(a)   (b)

FIG. 7.58
Demonstrating the effect of a short circuit on current levels.

The effect of the 2 Ω resistor has effectively been "shorted out" by the low-resistance connection. The maximum current is now limited only by the circuit breaker or fuse in series with the source.

Some practical examples of short circuits and their impact are provided in Fig. 7.59. In Fig. 7.59(a), a hot (the feed) wire wrapped around a screw became loose for a variety of reasons, and it is touching the return connection. A short-circuit connection between the two terminals has been established that could result in a very heavy current and a possible fire hazard. Hopefully, the breaker will "pop" and the circuit will be deactivated. Problems such as this is one of the reasons aluminum wires (cheaper and lighter than copper) are not permitted in residential or industrial wiring. Aluminum is more sensitive to temperature than copper and would expand and contract due to the heat developed by the current passing through the wire. Eventually, this expansion and contraction could loosen the screw, and a wire under some torsional stress from the installation could move and make contact as shown in Fig. 7.59(a). Aluminum is still used in large panels as a bus-bar connection, but in this case it is bolted down.

In Fig. 7.59(b), the wires of an iron have started to twist and crack due to excessive currents or long-term use of the iron. Once the insulation breaks down, the twisting can cause the two wires to touch and establish a short circuit and hopefully a circuit breaker or fuse will quickly disconnect the circuit. Often, it is not the wire of the iron that causes the problem but a cheap extension cord with the wrong gage wire. Be aware that you cannot tell the capacity of an extension cord by its outside jacket. It may have a thick orange covering but have a very thin wire inside. Check the gage on the wire the next time you buy an extension cord, and be sure that it is at least #14 gage, with #12 being the better choice for high-current appliances.

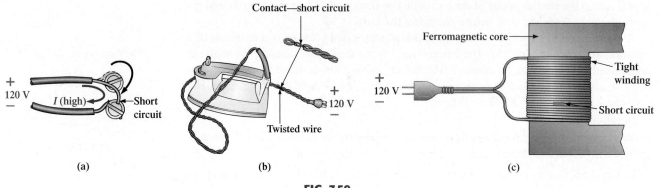

FIG. 7.59

Examples of short circuits.

Finally, in Fig. 7.59(c), we have the windings in a transformer or motor for residential or industrial use. The windings are wound so tightly together with such a thin coating of insulation that it is possible with age and use for the insulation to break down and short out the windings. In many cases, shorts can develop, but a short will simply reduce the number of effective windings in the unit. The tool or appliance may still work, but with less strength or rotational speed. If you notice such a change in the response, you should check the windings because a short can lead to a dangerous situation. In many cases, the state of the windings can be checked with a simple ohmmeter reading. If a short has occurred, the length of usable wire in the winding has been reduced, and the resistance will drop. If you know ahead of time what the resistance normally is, you can compare and make a judgment.

For the layperson, the terminology, *short circuit* or *open circuit* is usually associated with dire situations such as power loss, smoke, or fire. However, in network analysis both can play an integral role in determining specific parameters about a system. Most often, however, if a short-circuit condition is to be established, it is accomplished with a *jumper*—a lead of negligible resistance to be connected between the points of interest. Establishing an open circuit simply requires making sure that the terminals of interest are isolated from each other.

EXAMPLE 7.16   Determine voltage V_{ab} for the network of Fig. 7.60.

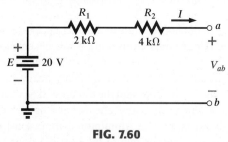

FIG. 7.60

Network for Example 7.16.

Solution: The open circuit requires that I be zero amperes. The voltage drop across both resistors is therefore zero volts since $V = IR = (0)R = 0$ V. Applying Kirchhoff's voltage law around the closed loop,

$$V_{ab} = E = \mathbf{20\ V}$$

EXAMPLE 7.17 Determine voltages V_{ab} and V_{cd} for the network of Fig. 7.61.

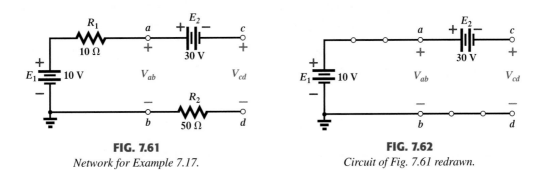

FIG. 7.61

Network for Example 7.17.

FIG. 7.62

Circuit of Fig. 7.61 redrawn.

Solution: The current through the system is zero amperes due to the open circuit, resulting in a 0 V drop across each resistor. Both resistors can therefore be replaced by short circuits, as shown in Fig. 7.62. Voltage V_{ab} is then directly across the 10 V battery, and

$$V_{ab} = E_1 = \textbf{10 V}$$

Voltage V_{cd} requires an application of Kirchhoff's voltage law:

$$+E_1 - E_2 - V_{cd} = 0$$

or

$$V_{cd} = E_1 - E_2 = 10 \text{ V} - 30 \text{ V} = \textbf{−20 V}$$

The negative sign in the solution simply indicates that the actual voltage V_{cd} has the opposite polarity of that appearing in Fig. 7.61.

EXAMPLE 7.18 Determine the unknown voltage and current for each network of Fig. 7.63.

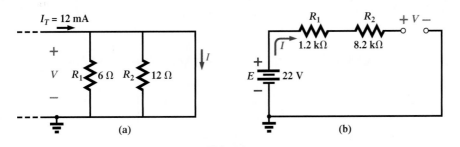

FIG. 7.63

Networks for Example 7.18.

Solution: For the network of Fig. 7.63(a), the current I_T will take the path of least resistance, and since the short-circuit condition at the end of the network is the least-resistance path, all the current will pass through the short circuit. This conclusion can be verified using the current divider rule. The voltage across the network is the same as that across the short circuit and will be zero volts, as shown in Fig. 7.64(a).

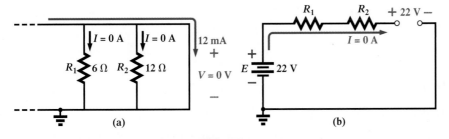

FIG. 7.64

Solutions to Example 7.18.

For the network of Fig. 7.63(b), the open-circuit condition requires that the current be zero amperes. The voltage drops across the resistors must therefore be zero volts, as determined by Ohm's law [$V_R = IR = (0)R = 0$ V], with the resistors simply acting as a connection from the supply to the open circuit. The result is that the open-circuit voltage will be $E = 22$ V, as shown in Fig. 7.64(a).

EXAMPLE 7.19 Calculate the current I and the voltage V for the network of Fig. 7.65.

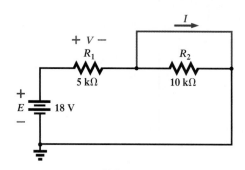

FIG. 7.65
Network for Example 7.19.

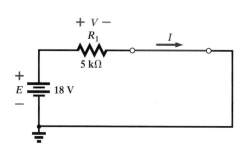

FIG. 7.66
Network of Fig. 7.65 redrawn.

Solution: The 10 kΩ resistor has been effectively shorted out by the jumper, resulting in the equivalent network of Fig. 7.66. Using Ohm's law,

$$I = \frac{E}{R_1} = \frac{18 \text{ V}}{5 \text{ k}\Omega} = \textbf{3.6 mA}$$

and

$$V = E = \textbf{18 V}$$

EXAMPLE 7.20 Determine V and I for the network of Fig. 7.67 if resistor R_2 is shorted out.

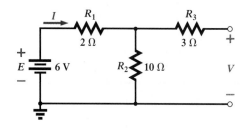

FIG. 7.67
Network for Example 7.20.

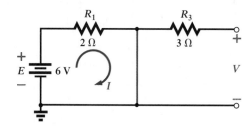

FIG. 7.68
Network of Fig. 7.67 with R_2 replaced by a jumper.

Solution: The redrawn network appears in Fig. 7.68. The current through the 3 Ω resistor is zero due to the open circuit, causing all the current I to pass through the jumper. Since $V_{3\Omega} = IR = (0)R = 0$ V, the voltage V is directly across the short, and

$$V = \textbf{0 V}$$

with

$$I = \frac{E}{R_1} = \frac{6 \text{ V}}{2 \text{ }\Omega} = \textbf{3 A}$$

7.11 APPLICATION

Boosting a Car Battery

Although boosting a car battery may initially appear to be a simple application of parallel networks, it is really a series-parallel operation that is worthy of some investigation. As indicated in Chapter 2, every dc supply has some internal resistance. For the typical 12 V lead-acid car battery, the resistance is quite small—in the milliohm range. In most cases the low internal resistance will ensure that most of the voltage (or power) is delivered to the load and not lost on the internal resistance. In Fig. 7.69, battery #2 has discharged because the lights were left on for three hours during a movie. Fortunately, a friend who made sure his own lights were out has a fully charged battery #1 and a good set of 16 ft cables with #6 gage stranded wire and well-designed clips. The investment in a good set of cables with sufficient length and heavy wire is a wise one, particularly if you live in a cold climate. Flexibility, as provided by stranded wire, is also a very desirable characteristic under some conditions. Be sure to check the gage of the wire and not just the thickness of the insulating jacket. You get what you pay for, and the copper is the most expensive part of the cables. Too often the label says "heavy-duty," but the wire is too high a gage number.

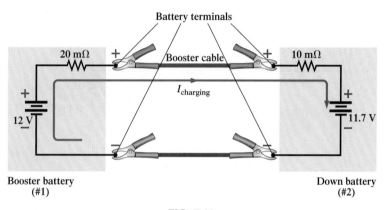

FIG. 7.69

Boosting a car battery.

The proper sequence of events in boosting a car is often a function of whom you speak to or what information you read. For safety sake some people recommend that the car with the good battery be turned off when making the connections. This, however, can create an immediate problem if the "dead" battery is in such a bad state that when it is hooked up to the good battery, it will immediately drain it to the point that neither car will start. With this in mind, it does make some sense to leave the car running to ensure that the charging process continues until the starting of the disabled car is initiated. **Because accidents do happen, it is strongly recommended that the person making the connections wear the proper type of protective eye equipment. Take sufficient time to be sure that you know which are the positive and negative terminals for both cars.** If it's not immediately obvious, keep in mind that the negative or ground side is usually connected to the chassis of the car with a relatively short, heavy wire. When you are sure of which are the positive and negative terminals, first connect one of the red wire clamps of the booster cables to the positive terminal of the discharged battery—all the while being sure that the other red clamp is not touching the battery or car. Then connect the other end of the red wire to the positive terminal of the fully charged battery. Next, connect one end of the black cable of the booster cables to the negative terminal of the booster battery, and finally connect the other end of the black cable to the engine block of the stalled vehicle (not the negative post of the dead battery) away from the carburetor, fuel lines, or moving parts of the car. Lastly, have someone maintain a constant idle speed in the car with the good battery as you start the car with the bad battery. After the vehicle starts, remove the cables in the reverse order starting with the cable connected to the engine block. Always be careful to ensure that clamps don't touch the battery or chassis of the car or get near any moving parts. Some people feel that the car with the good battery should charge the bad battery for 5 to 10 minutes before starting the disabled car so the disabled car will be essentially using its own battery in the starting process. Keep in mind that the instant the booster cables are connected, the booster car will be making a concerted effort to charge both its own battery and the drained battery. At starting it will then be asked to supply a heavy current to start the other car. It's a pretty heavy load to put on a single battery. For the situation of Fig. 7.69, the voltage of battery #2 is less than that of battery #1, and the charging current will flow as shown. The resistance in series with the boosting battery is more because of the long length of the booster cable to the other car. The current is limited only by the series milliohm resistors of the batteries, but the voltage difference is so small that the starting current will be in safe range for the cables involved. The initial charging current will be $I = (12\text{ V} - 11.7\text{ V})/(20\text{ m}\Omega + 10\text{ m}\Omega) = 0.3\text{ V}/30\text{ m}\Omega = 10$ A. At starting, the current levels will be as shown in Fig. 7.70 for the resistance levels and battery voltages assumed. At starting, an internal resistance for the starting circuit of $0.1\ \Omega = 100\text{ m}\Omega$ is assumed. Note that the battery of the disabled car has now charged up to 11.8 V with an associated increase in its power level. The presence of two batteries requires that the analysis wait for the methods to be introduced in the next chapter.

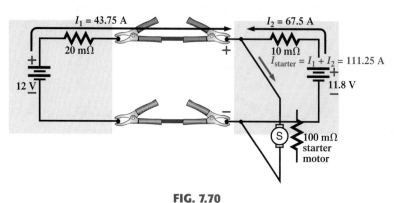

FIG. 7.70
Current levels at starting.

Note also that the current drawn from the starting circuit for the disabled car is over 100 A and that the majority of the starting current is provided by the battery being charged. In essence, therefore, the majority of the starting current is coming from the disabled car. The good battery has provided an initial charge to the bad battery and has provided the additional current necessary to start the car. But, in total, it is the battery of the disabled car that is the primary source of the starting current. For this very reason, it is advised to let the charging action continue for 5 or 10 minutes before starting the car. If the disabled car is in really bad shape with a voltage level of only 11 V, the resulting levels of current will reverse, with the good battery providing 68.75 A and the bad battery only 37.5 A. Quite obviously, therefore, the worse the condition of the dead battery, the heavier the drain on the good battery. A point can also be reached where the bad battery is in such bad shape that it cannot accept a good charge or provide its share of the starting current. The result can be continuous cranking of the down car without starting, and thus possible damage to the battery of the running car due to the enormous current drain. Once the car is started and the booster cables are removed, the car with the discharged battery will continue to run because the alternator will carry the load (charging the battery and providing the necessary dc voltage) after ignition.

The above discussion was all rather straightforward, but let's investigate what might happen if it is a dark and rainy night, you get rushed, and you hook up the cables incorrectly as shown in Fig. 7.71. The result is two series-aiding batteries and a very low resistance path. The

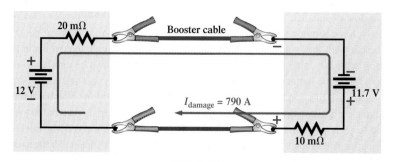

FIG. 7.71
Current levels if the booster battery is improperly connected.

resulting current can then theoretically be extremely high [$I = (12 \text{ V} + 11.7 \text{ V})/30 \text{ m}\Omega = 23.7 \text{ V}/30 \text{ m}\Omega = 790 \text{ A}$], perhaps permanently damaging the electrical system of both cars and, worst of all, causing an explosion that might seriously injure someone. It is therefore very important that you treat the process of boosting a car with great care. Find that flashlight, double-check the connections, and be sure that everyone is clear when you start that car.

Before leaving the subject, we should point out that getting a boost from a tow truck results in a somewhat different situation: The connections to the battery in the truck are very secure; the cable from the truck is a heavy wire with thick insulation; the clamps are also quite large and make an excellent connection with your battery; and the battery is heavy-duty for this type of expected load. The result is less internal resistance on the supply side and a heavier current from the truck battery. In this case, the truck is really starting the disabled car, which simply reacts to the provided surge of power.

7.12   COMPUTER ANALYSIS

PSpice

Voltage Divider Supply   PSpice will now be used to verify the results of Example 7.14. The calculated resistor values will be substituted and the voltage and current levels checked to see if they match the handwritten solution.

As shown in Fig. 7.72, the network is drawn as in earlier chapters using only the tools described thus far—in one way, a practice exercise for everything learned about the **Capture Lite Edition.** Note in this case that rotating the first resistor sets everything up for the remaining resistors. Further, it is a nice advantage that you can place one resistor after another without going to the **End Mode** option. Be especially careful with the placement of the ground, and be sure that **0/SOURCE** is used. Note also that resistor R_1 of Fig. 7.72 was entered as 1.333 kΩ rather than 1.33 kΩ as in Example 7.14. When running the program, we found that the computer solutions were not a perfect match to the longhand solution to the level of accuracy desired unless this change was made.

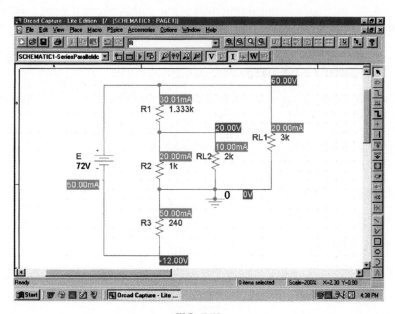

FIG. 7.72
Using PSpice to verify the results of Example 7.14.

Since all the voltages are to ground, the voltage across R_{L_1} is 60 V; across R_{L_2}, 20 V; and across R_3, -12 V. The currents are also an excellent match with the handwritten solution, with $I_E = 50$ mA, $I_{R_1} = 30$ mA, $I_{R_2} = 20$ mA, $I_{R_3} = 50$ mA, $I_{R_{L2}} = 10$ mA, and $I_{R_{L1}} = 20$ mA. For the display of Fig. 7.72, the **W** option was disabled to permit concentrating on the voltage and current levels. There is again an exact match with the longhand solution.

CHAPTER SUMMARY

SERIES-PARALLEL NETWORKS

- A series-parallel configuration is one that is formed by a combination of series and parallel elements.
- The first steps in the analysis of series-parallel networks should include a careful determination of which path to follow and which elements are in series and which are in parallel.
- Redrawing the network a number of times can be very helpful in the analysis of series-parallel networks.

NOTATION

- The potential at every ground symbol of a network, no matter how many grounds there are, is 0 V.
- All the ground symbols of a network can be connected together or left separate without affecting the behavior of the network.
- In double-subscript notation, the first subscript is assumed to be at the point of higher potential. If this is not the case, a negative sign must be associated with the magnitude of the voltage.
- Single-subscript notation specifies the voltage at a point with respect to ground.

VOLTAGE DIVIDER SUPPLY

- The loading of a system is the introduction of components that will draw current from the supply.
- For a voltage divider supply to be effective, the applied resistive loads should be significantly larger than the resistors appearing in the voltage divider network.
- When hooking up a load to a potentiometer, be sure that the load resistance far exceeds the maximum terminal resistance of the potentiometer if good control of the output voltage is desired.

OPEN AND SHORT CIRCUITS

- An open circuit can have a potential difference across its terminals, but the current is always zero amperes.
- A short circuit can carry a current of a level determined by the external circuit, but the voltage across its terminals is always zero volts.

Important Equations

$$V_{ab} = V_a - V_b$$

Potentiometer loading:

$$R_L \gg R_T$$

PROBLEMS

SECTION 7.2 Series-Parallel Networks

1. Which elements (individual elements, not combinations of elements) of the networks of Fig. 7.73 are in series? Which are in parallel? As a check on your assumptions, be sure that the elements in series have the same current and that the elements in parallel have the same voltage. Restrict your decisions to single elements, not combinations of elements.

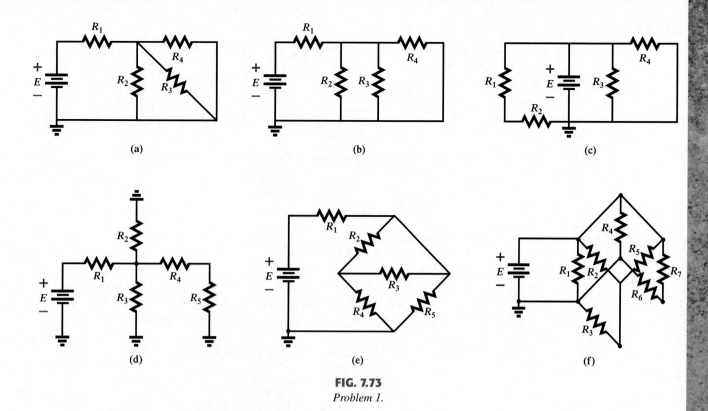

FIG. 7.73

Problem 1.

2. Determine R_T for the networks of Fig. 7.74.

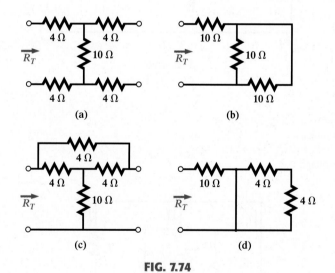

FIG. 7.74

Problem 2.

SECTION 7.4 Descriptive Examples

3. For the network of Fig. 7.75:
   a. Does $I_s = I_5 = I_6$? Explain.
   b. If $I_s = 10$ A and $I_1 = 4$ A, find I_2.
   c. Does $I_1 + I_2 = I_3 + I_4$? Explain.
   d. If $V_2 = 8$ V and $E = 14$ V, find V_3.
   e. If $R_1 = 4 \ \Omega$, $R_2 = 2 \ \Omega$, $R_3 = 4 \ \Omega$, and $R_4 = 6 \ \Omega$, what is R_T?
   f. If all the resistors of the configuration are 20 Ω, what is the source current if the applied voltage is 20 V?
   g. Using the values of part (f), find the power delivered by the battery and the power absorbed by the total resistance R_T.

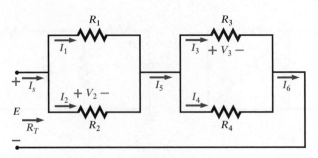

FIG. 7.75

Problem 3.

4. For the network of Fig. 7.76:
   a. Find the total resistance R_T.
   b. Find the source current I_s and currents I_2 and I_3.
   c. Find current I_5.
   d. Find voltages V_2 and V_4.

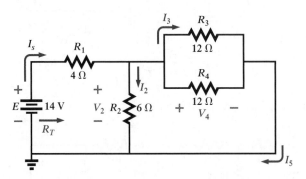

FIG. 7.76

Problem 4.

5. For the network of Fig. 7.77:
   a. Determine R_T.
   b. Find I_s, I_1, and I_2.
   c. Find voltage V_4.

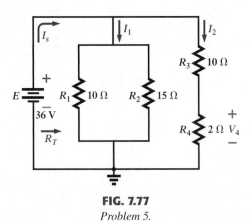

FIG. 7.77

Problem 5.

6. For the circuit board of Fig. 7.78:
   a. Find the total resistance R_T of the configuration.
   b. Find the current drawn from the supply if the applied voltage is 48 V.
   c. Find the reading of the applied voltmeter.

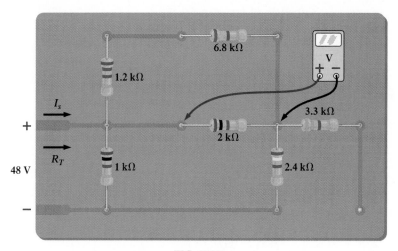

FIG. 7.78
Problem 6.

*7. For the network of Fig. 7.79:
  a. Find currents I_s, I_2, and I_6.
  b. Find voltages V_1 and V_5.
  c. Find the power delivered to the 3 kΩ resistor.

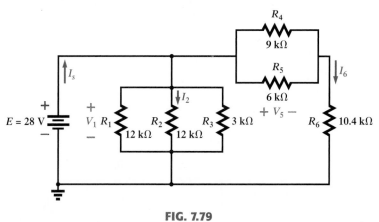

FIG. 7.79
Problem 7.

*8. For the series-parallel configuration of Fig. 7.80:
  a. Find the source current I_s.
  b. Find currents I_3 and I_9.
  c. Find current I_8.
  d. Find voltage V_x.

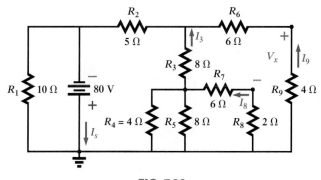

FIG. 7.80
Problem 8.

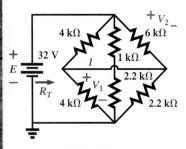

FIG. 7.81
Problem 9.

*9. For the network of Fig. 7.81:
    a. Determine R_T.
    b. Find voltage V_1.
    c. Find voltage V_2.
    d. Find the current I with direction.

*10. For the network of Fig. 7.82:
    a. Find the total resistance seen by the source.
    b. Find voltages V_1 and V_4.
    c. Find current I_3 with direction.
    d. Determine current I_6 with direction.
    e. Determine current I_7 with direction.

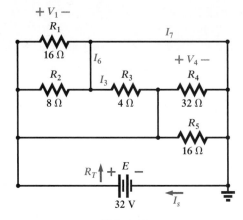

FIG. 7.82
Problem 10.

*11. For the configuration of Fig. 7.83:
    a. Find currents I_2, I_6, and I_8.
    b. Find voltages V_4 and V_8.

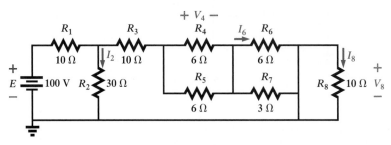

FIG. 7.83
Problem 11.

*12. For the network of Fig. 7.84, find resistance R_3 if the current through it is 2 A.

*13. If all the resistances of the cube of Fig. 7.85 are 20 Ω, what is the total resistance? (*Hint:* Make some basic assumptions about current division through the cube.)

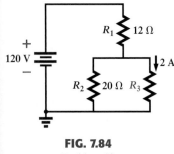

FIG. 7.84
Problem 12.

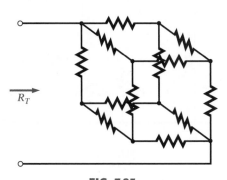

FIG. 7.85
Problem 13.

14. Given the voltmeter reading of 27 V in Fig. 7.86:
   a. Is the network operating correctly?
   b. If not, what could be the cause of the unexpected reading?

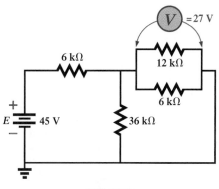

FIG. 7.86

Problem 14.

15. Given the network of Fig. 7.87, composed of standard values:
   a. Determine currents I_1 and I_2.
   b. Find voltages V_a, V_b, and V_{ab}.

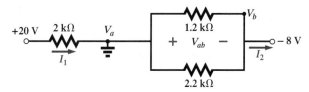

FIG. 7.87

Problem 15.

*16. For the network of Fig. 7.88:
   a. Determine currents I_s, I_1, I_3, and I_4.
   b. Calculate V_a and V_{bc}.

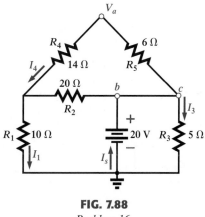

FIG. 7.88

Problem 16.

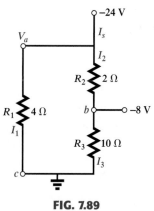

FIG. 7.89

Problem 17.

17. For the network of Fig. 7.89:
   a. Find the magnitude and direction of currents I_s, I_1, I_2, and I_3.
   b. Find voltages V_a and V_{bc}.

18. For the network of Fig. 7.90:
 a. Determine current I_1.
 b. Calculate currents I_2 and I_3.
 c. Determine voltage levels V_a and V_b.

19. For the network of Fig. 7.91, composed of standard values:
 a. Determine the current I.
 b. Find the voltage V.

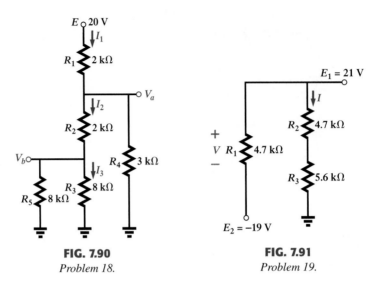

FIG. 7.90
Problem 18.

FIG. 7.91
Problem 19.

20. Determine the current I and the voltages V_a, V_b, and V_{ab} for the network of Fig. 7.92.

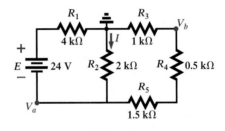

FIG. 7.92
Problem 20.

21. For the ladder network of Fig. 7.93:
 a. Find the current I.
 b. Find current I_7.
 c. Determine voltages V_3, V_5, and V_7.
 d. Calculate the power delivered to R_7, and compare it to the power delivered by the 240 V supply.

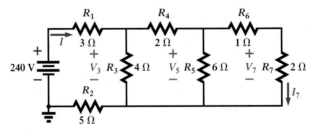

FIG. 7.93
Problem 21.

*22. For the ladder network of Fig. 7.94:
 a. Determine R_T.
 b. Calculate I_s.
 c. Find I_8.

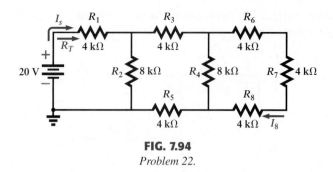

FIG. 7.94
Problem 22.

*23. Determine the power delivered to the 10 Ω load of Fig. 7.95.

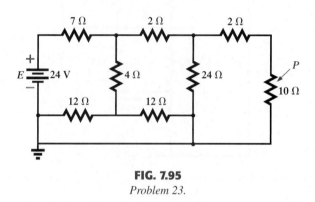

FIG. 7.95
Problem 23.

SECTION 7.8 Voltage Divider Supply (Unloaded and Loaded)

24. For the configuration of Fig. 7.96, composed of standard values:
   a. Find voltages V_a, V_b, and V_c.
   b. Apply a 10 kΩ load in parallel with the 4.7 kΩ resistor, and find the new value of V_c. How does it compare to the value calculated in part (a)? Why?
   c. Change the 4.7 kΩ resistor to a 100 kΩ resistor, and determine voltage V_c. How does it compare to the value of part (a)? Why?

25. Given the voltage divider supply of Fig. 7.97:
   a. Determine the supply voltage E.
   b. Find the load resistors R_{L_2} and R_{L_3}.
   c. Determine the voltage divider resistors R_1, R_2, and R_3.

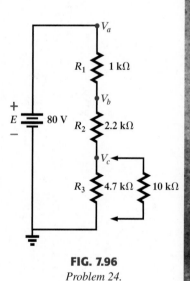

FIG. 7.96
Problem 24.

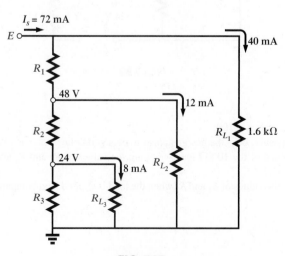

FIG. 7.97
Problem 25.

*26. Determine the voltage divider supply resistors for the configuration of Fig. 7.98. Also determine the required wattage rating for each resistor, and compare their levels.

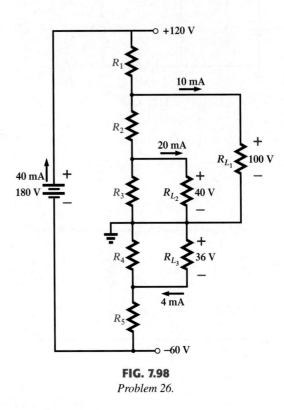

FIG. 7.98
Problem 26.

SECTION 7.9   Potentiometer Loading

27. For the system of Fig. 7.99:
  a. Find the value of R_1 (a portion of the total potentiometer resistance) to establish 24 V at the output as shown.
  b. What will the voltage be across a 1 kΩ load if applied as shown?
  c. Repeat part (b) for an applied 10 kΩ load.
  d. Comment on the results of parts (a) and (b).

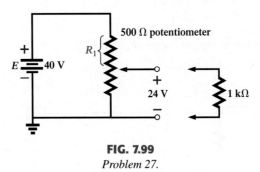

FIG. 7.99
Problem 27.

28. For the system of Fig. 7.100:
  a. At first exposure, does the design appear to be a good one?
  b. In the absence of the 10 kΩ load, what are the values of R_1 and R_2 necessary to establish 3 V across R_2?
  c. Determine the values of R_1 and R_2 when the load is applied, and compare them to the results of part (b).

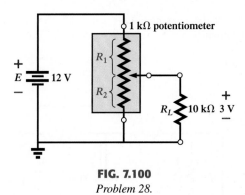

FIG. 7.100
Problem 28.

29. For the potentiometer of Fig. 7.101:
    a. What are voltages V_{ab} and V_{bc} with no load applied?
    b. What are voltages V_{ab} and V_{bc} with the indicated loads applied?
    c. What is the power dissipated by the potentiometer under the loaded conditions of Fig. 7.101?
    d. What is the power dissipated by the potentiometer with no loads applied? Compare this answer to the results of part (c).

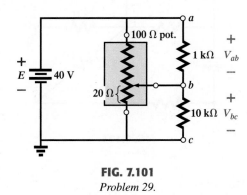

FIG. 7.101
Problem 29.

SECTION 7.10 Open and Short Circuits

30. For the network of Fig. 7.102:
    a. Determine I_s and V_L.
    b. Determine I_s if R_L is shorted out.
    c. Determine V_L if R_L is replaced by an open circuit.

31. For the network of Fig. 7.103:
    a. Determine the open-circuit voltage V_L.
    b. If the 2.2 kΩ resistor is short circuited, what is the new value of V_L?
    c. Determine V_L if the 4.7 kΩ resistor is replaced by an open circuit.

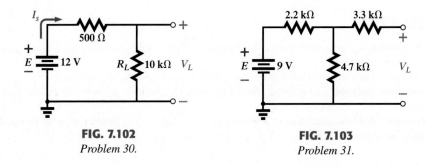

FIG. 7.102
Problem 30.

FIG. 7.103
Problem 31.

*32. For the network of Fig. 7.104, determine:
   a. the short-circuit currents I_1 and I_2.
   b. voltages V_1 and V_2.
   c. the source current I_s.

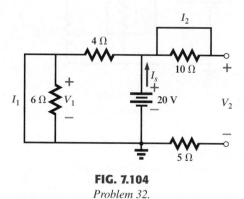

FIG. 7.104
Problem 32.

33. Determine the voltage V and the current I for the network of Fig. 7.105.

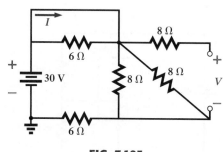

FIG. 7.105
Problem 33.

34. For the network of Fig. 7.106:
   a. Determine voltage V_{ab}. (*Hint:* Just use Kirchhoff's voltage law.)
   b. Calculate the current I.

*35. For the network of Fig. 7.107:
   a. Determine the current I.
   b. Calculate the open-circuit voltage V.

SECTION 7.12   Computer Analysis

36. Using PSpice or Electronics Workbench, verify the results of Example 7.2.

37. Using PSpice or Electronics Workbench, confirm the solutions of Example 7.5.

38. Using PSpice or Electronics Workbench, verify the results of Example 7.13.

39. Using PSpice or Electronics Workbench, find voltage V_6 of Fig. 7.39.

40. Using PSpice or Electronics Workbench, find voltages V_1 and V_2 of Fig. 7.53.

GLOSSARY

Ladder network   A network that consists of a cascaded set of series-parallel combinations and has the appearance of a ladder.

Open circuit   The absence of a direct connection between two points in a network.

Series-parallel network   A network consisting of a combination of both series and parallel branches.

Short circuit   A direct connection of low resistive value that can significantly alter the behavior of an element or system.

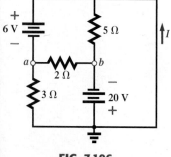

FIG. 7.106
Problem 34.

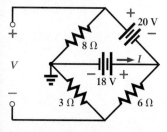

FIG. 7.107
Problem 35.

Methods of Analysis and Selected Topics (dc)

OBJECTIVES

- Become aware of the characteristics of a current source and learn how to respond to it in a network analysis.
- Be able to convert from a voltage source to a current source, and vice versa.
- Understand how to apply branch-current analysis and learn how to properly interpret the results.
- Be able to apply mesh analysis using the general and format approaches, and be able to properly interpret the results.
- Understand how to apply nodal analysis using the general and format approaches, and be able to properly interpret the results.
- Be able to apply mesh and nodal analysis to networks with ideal sources—that is, networks without a specified level of internal resistance.
- Become familiar with the bridge configuration and the conditions that will establish a balanced condition.
- Be able to perform a Δ-Y or a Y-Δ conversion and use the conversion effectively in a network analysis.
- Begin to appreciate how computer software can be an effective aid in solving complex networks and performing the necessary complex mathematical operations.

8.1 INTRODUCTION

The circuits described in previous chapters had only one source or two or more sources in series or parallel. The step-by-step procedures outlined in those chapters cannot be applied if the sources are not in series or parallel. **There will be an interaction of sources that will not permit the reduction techniques used to find quantities such as the total resistance and the source current.**

For such situations, methods of analysis have been developed that allow us to approach, in a systematic manner, networks with any number of sources in any arrangement. To our benefit, **the methods to be introduced can also be applied to networks with only one source or to networks in which sources are in series or parallel.**

The methods to be introduced in this chapter include branch-current analysis, mesh analysis, and nodal analysis. Each can be applied to the same network, although usually one is more appropriate than the other. The "best" method cannot be defined by a strict set of rules but can be determined only after developing an understanding of the relative advantages of each.

Before considering the first of the methods, we will examine current sources in detail because they will appear throughout the analyses to follow. The chapter will conclude with an investigation of a complex network called the *bridge configuration,* followed by the use of Δ-Y and Y-Δ conversions to analyze such configurations.

8.2 CURRENT SOURCES

In previous chapters, the voltage source was the only source appearing in the circuit analysis. This was primarily due to the fact that voltage sources such as the battery and supply are the most common in our daily lives and in the laboratory environment.

We must now turn our attention to a second type of source called the **current source** which will appear throughout the analyses to appear in this chapter. Although current sources are available as laboratory supplies (introduced in Chapter 2), they appear extensively in the modeling of electronic devices such as the transistor. Their characteristics and their impact on the currents and voltages of a network must therefore be clearly understood if electronic systems are to be properly investigated.

The current source is often described as the *dual* of the voltage source. Just as a battery provides a fixed voltage to a network, a current source establishes a fixed current in the branch in which it is located. Further, the current through a battery is a function of the network to which it is applied, just as the voltage across a current source is a function of the connected network. The term *dual* is applied to any two elements in which the traits of one variable can be interchanged with the traits of another. This is certainly true for the current and voltage of the two types of sources.

The symbol for a current source appears in Fig. 8.1(a). The arrow indicates the direction in which it is supplying current to the branch in which it is located. The result is a current equal to the source current through the series resistor. In Fig. 8.1(b), we find that the voltage across a current source is determined by the polarity of the voltage drop caused by the current source. For single-source networks, it will always have the polarity of Fig. 8.1(b), but for multisource networks it can have either polarity.

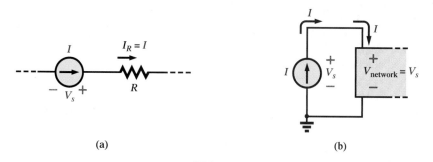

(a)   (b)

FIG. 8.1

Introducing the current source symbol.

In general, therefore,

a current source determines the direction and magnitude of the current in the branch in which it is located.

Furthermore,

the magnitude and the polarity of the voltage across a current source are each a function of the network to which the voltage is applied.

A few examples will demonstrate the similarities between solving for the source current of a voltage source and the terminal voltage of a current source. All the rules and laws developed in the previous chapter still apply, so it is simply a matter of keeping track of what we are looking for and properly understanding the characteristics of each source.

The simplest possible configuration with a current source appears in the first example.

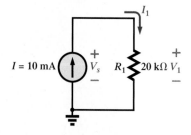

FIG. 8.2

Circuit for Example 8.1.

EXAMPLE 8.1 Find the source voltage, the voltage V_1, and current I_1 for the circuit of Fig. 8.2.

Solution: Since the current source establishes the current in the branch in which it is located, the current I_1 must equal I, and

$$I_1 = I = \textbf{10 mA}$$

The voltage across R_1 is then determined by Ohm's law:

$$V_1 = I_1R_1 = (10 \text{ mA})(20 \text{ }\Omega) = \textbf{200 V}$$

Since resistor R_1 and the current source are in parallel, the voltage across each must be the same, and

$$V_s = V_1 = \textbf{200 V}$$

with the polarity shown.

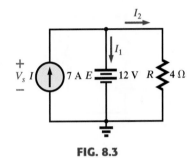

FIG. 8.3

Network for Example 8.2.

EXAMPLE 8.2   Find the voltage V_s and currents I_1 and I_2 for the network of Fig. 8.3.

Solution: This is an interesting problem because it has both a current source and a voltage source. For each source, the dependent (a function of something else) variable will be determined. That is, for the current source, V_s must be determined, and for the voltage source, I_s must be determined.

Since the current source and voltage source are in parallel,

$$V_s = E = \textbf{12 V}$$

Further, since the voltage source and resistor R are in parallel,

$$V_R = E = 12 \text{ V}$$

and

$$I_2 = \frac{V_R}{R} = \frac{12 \text{ V}}{4 \text{ }\Omega} = \textbf{3 A}$$

The current I_1 of the voltage source can then be determined by applying Kirchhoff's current law at the top of the network as follows:

$$\Sigma I_i = \Sigma I_o$$

$$I = I_1 + I_2$$

and

$$I_1 = I - I_2 = 7 \text{ A} - 3 \text{ A} = \textbf{4 A}$$

EXAMPLE 8.3   Determine the current I_1 and the voltage V_s for the network of Fig. 8.4.

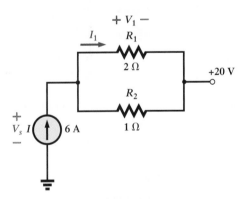

FIG. 8.4

Network for Example 8.3.

Solution: First note that the current in the branch with the current source must be 6 A, no matter what the magnitude of the voltage source to the right. In other words, the currents of the network are defined by I, R_1, and R_2. However, the voltage across the current source will be directly affected by the magnitude and polarity of the applied source.

Using the current divider rule:

$$I_1 = \frac{R_2I}{R_2 + R_1} = \frac{(1 \text{ }\Omega)(6 \text{ A})}{1 \text{ }\Omega + 2 \text{ }\Omega} = \frac{1}{3}(6 \text{ A}) = \textbf{2 A}$$

The voltage V_1:

$$V_1 = I_1 R_1 = (2 \text{ A})(2 \text{ } \Omega) = 4 \text{ V}$$

Applying Kirchhoff's voltage rule to determine V_s:

$$+V_s - V_1 - 20 \text{ V} = 0$$

and

$$V_s = V_1 + 20 \text{ V} = 4 \text{ V} + 20 \text{ V} = \textbf{24 V}$$

In particular, note the polarity of the voltage V_s as determined by the network.

8.3 SOURCE CONVERSIONS

The current source appearing in the previous section is called an *ideal source* due to the absence of any internal resistance. In reality, all sources—whether they are voltage sources or current sources—have some internal resistance in the relative positions shown in Figs. 8.5 and 8.6. For the voltage source, if $R_s = \infty \text{ } \Omega$, or if it is so small compared to any series resistors that it can be ignored, then we have an "ideal" voltage source for all practical purposes. For the current source, since the resistor R_s is in parallel, if $R_s = \infty \text{ } \Omega$, or if it is large enough compared to any parallel resistive elements that it can be ignored, then we have an "ideal" current source.

Unfortunately, however, **ideal sources cannot be converted from one type to another.** That is, a voltage source cannot be converted to a current source, and vice versa—**the internal resistance must be present.** If the voltage source of Fig. 8.5 is to be equivalent to the source of Fig. 8.6, any load connected to the sources such as R_L in the figures should receive the same current, voltage, and power from each configuration. In other words, if the source were enclosed in a container, the load R_L would not know which source it was connected to.

This type of equivalence is established using the equations appearing in Fig. 8.7. First note that the resistance is the same in each configuration—a nice advantage. For the voltage source equivalent, the voltage is determined by a simple application of Ohm's law to the current source: $E = IR_s$. For the current source equivalent, the current is again determined by a simple application of Ohm's law to the voltage source: $I = E/R_s$. At first glance, it all seems too simple, but Example 8.4 will verify the results.

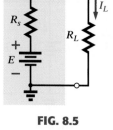

FIG. 8.5

Practical voltage source.

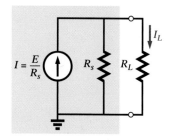

FIG. 8.6

Practical current source.

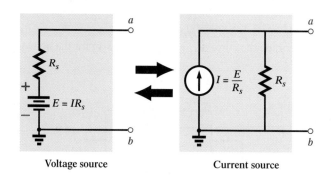

Voltage source   Current source

FIG. 8.7

Source conversion.

It is important to realize, however, that

the equivalence between a current source and a voltage source exists only at their external terminals.

The internal characteristics of each are quite different.

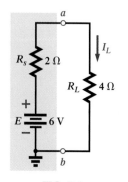

FIG. 8.8

Practical voltage source and load for Example 8.4.

EXAMPLE 8.4 For the circuit of Fig. 8.8:

a. Determine the current I_L.
b. Convert the voltage source to a current source.
c. Using the resulting current source of part (b), calculate the current through the load resistor, and compare your answer to the result of part (a).

a. Applying Ohm's law:

$$I_L = \frac{E}{R_s + R_L} = \frac{6\text{ V}}{2\ \Omega + 4\ \Omega} = \frac{6\text{ V}}{6\ \Omega} = \mathbf{1\text{ A}}$$

b. Using Ohm's law again:

$$I = \frac{E}{R_s} = \frac{6\text{ V}}{2\ \Omega} = \mathbf{3\text{ A}}$$

and the equivalent source appears in Fig. 8.9 with the load reapplied.

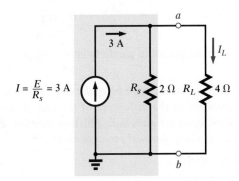

FIG. 8.9

Equivalent current source and load for the voltage source of Fig. 8.8.

c. Using the current divider rule:

$$I_L = \frac{R_s I}{R_s + R_L} = \frac{(2\ \Omega)(3\text{ A})}{2\ \Omega + 4\ \Omega} = \frac{1}{3}(3\text{ A}) = \mathbf{1\text{ A}}$$

We find that the current I_L is the same for the voltage source as it was for the equivalent current source—the sources are therefore equivalent.

As demonstrated in Figs. 8.5 and 8.6 and in Example 8.4, note that

a source and its equivalent will establish current in the same direction through the applied load.

In Example 8.4 note that both sources pressure or establish current up through the circuit to establish the same direction for the load current I_L and the same polarity for the voltage V_L.

EXAMPLE 8.5 Determine current I_2 for the network of Fig. 8.10.

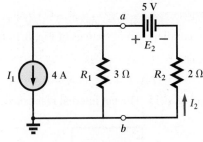

FIG. 8.10

Two-source network for Example 8.5.

Solution: Although it may appear that the network cannot be solved using methods introduced thus far, one source conversion, as shown in Fig. 8.11, will result in a simple series circuit. It does not make sense to convert the voltage source to a current source because you would lose the current I_2 in the redrawn network. Note the polarity for the equivalent voltage source as determined by the current source.

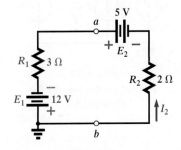

FIG. 8.11

Network of Fig. 8.10 following the conversion of the current source to a voltage source.

For the source conversion:

$$E_1 = I_1 R_1 = (4 \text{ A})(3 \ \Omega) = 12 \text{ V}$$

and

$$I_2 = \frac{E_1 + E_2}{R_1 + R_2} = \frac{12 \text{ V} + 5 \text{ V}}{3 \ \Omega + 2 \ \Omega} = \frac{17 \text{ V}}{5 \ \Omega} = \textbf{3.4 A}$$

8.4 CURRENT SOURCES IN PARALLEL

We found that voltage sources of different terminal voltages could not be placed in parallel because of a violation of Kirchhoff's voltage law. Similarly,

current sources of different values cannot be placed in series due to a violation of Kirchhoff's current law.

However, current sources can be placed in parallel just as voltage sources can be placed in series. In general,

two or more current sources in parallel can be replaced by a single current source having a magnitude determined by the difference of the sum of the currents in one direction and the sum in the opposite direction. The new parallel internal resistance is the total resistance of the resulting parallel resistive elements.

Consider the following examples.

EXAMPLE 8.6 Reduce the parallel current sources of Fig. 8.12 to a single current source.

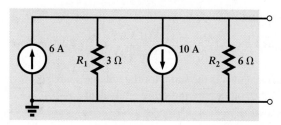

FIG. 8.12
Parallel current sources for Example 8.6.

Solution: The net source current is

$$I = 10 \text{ A} - 6 \text{ A} = \textbf{4 A}$$

with the direction of the larger.

The net internal resistance is the parallel combination of resistors R_1 and R_2:

$$R_s = 3 \ \Omega \ \| \ 6 \ \Omega = \textbf{2} \ \boldsymbol{\Omega}$$

The reduced equivalent appears in Fig. 8.13.

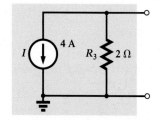

FIG. 8.13
Reduced equivalent for the configuration of Fig. 8.12.

EXAMPLE 8.7 Reduce the parallel current sources of Fig. 8.14 to a single current source.

Solution: The net current is

$$I = 7 \text{ A} + 4 \text{ A} - 3 \text{ A} = \textbf{8 A}$$

with the direction shown in Fig. 8.15. The net internal resistance remains the same.

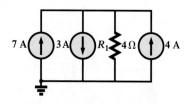

FIG. 8.14
Parallel current sources for Example 8.7.

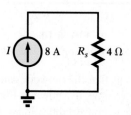

FIG. 8.15
Reduced equivalent for Fig. 8.14.

EXAMPLE 8.8   Reduce the network of Fig. 8.16 to a single current source, and calculate the current through R_L.

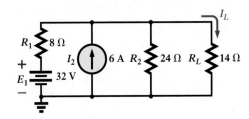

FIG. 8.16
Parallel sources for Example 8.8.

Solution: In this example, the voltage source will first be converted to a current source as shown in Fig. 8.17.

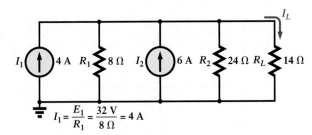

FIG. 8.17
Network of Fig. 8.16 after converting the voltage source to a current source.

Combining current sources:

$$I = I_1 + I_2 = 4\,\text{A} + 6\,\text{A} = \mathbf{10\,A}$$

with

$$R_s = R_1 \| R_2 = 8\,\Omega \| 24\,\Omega = \mathbf{6\,\Omega}$$

as shown in Fig. 8.18.
Applying the current divider rule:

$$I_L = \frac{R_s I}{R_s + R_L} = \frac{(6\,\Omega)(10\,\text{A})}{6\,\Omega + 14\,\Omega} = \frac{60}{20}\,\text{A} = \mathbf{3\,A}$$

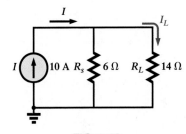

FIG. 8.18
Redrawn network of Fig. 8.17 after combining sources.

8.5   BRANCH-CURRENT ANALYSIS

Before examining the details of the first important method of analysis, let us examine the network of Fig. 8.19 to be sure that you understand the need for these special methods.

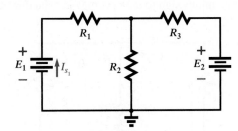

FIG. 8.19
Demonstrating the need for an approach such as branch-current analysis.

Initially, it might appear that we can use the reduce and return approach to work our way back to the source E_1 and calculate the source current I_{s_1}. Unfortunately, however, the series elements R_3 and E_2 cannot be combined because they are different types of elements. A further examination of the network reveals that there are no two like elements that are in series or parallel. No combination of elements can be performed, and it is clear that another approach must be defined.

The first approach to be introduced is called the **branch-current method** because we will define and solve for the currents of each branch of the network. From experience it has been found that the best way to introduce this method and understand its application is to follow a series of steps such as listed below. Each step will be carefully defined in the examples to follow.

Branch-Current Analysis:

1. *Assign a distinct current of arbitrary direction to each branch of the network.*
2. *Add the polarities for each voltage drop as determined by the direction of each branch current.*
3. *Apply Kirchhoff's voltage law around each closed, independent loop of the network.*
4. *Apply Kirchhoff's current law to the minimum number of nodes (junction points) that will include all the branch currents of the network.*
5. *Solve the resulting equations for the assumed branch currents.*

A few descriptive examples follow.

EXAMPLE 8.9   Determine the current in each branch of the network of Fig. 8.20 using branch-current analysis.

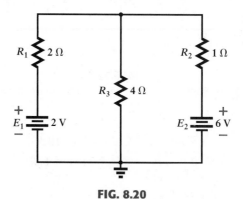

FIG. 8.20
Network to be analyzed in Example 8.9.

Solution:

Step 1:   First, we must determine the number of branches in the network. By definition, a branch is any series combination of elements with the same current. In Fig. 8.21, E_1 and R_1 are in series to form a branch that will have the defined current I_1. The direction was chosen simply because the voltage source E_1 appears to be pressuring current in this direction. The final results may indicate that this branch current actually has the opposite direction, but nonetheless the magnitude of the result will be correct, with a minus sign to indicate the opposite direction.

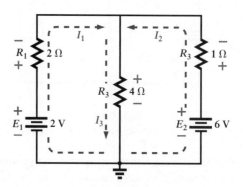

FIG. 8.21
Defining the branch currents and voltage polarities for the network of Fig. 8.20.

For similar reasons, a branch current I_2 was defined, and finally the branch current I_3 was defined for the branch containing resistor R_3. The direction of I_3 was the result of noting that both I_1 and I_2 enter the junction at the top of the network, and it seemed appropriate to define I_3 as leaving. Each branch now has a defined branch current, and we can continue to the next step.

Step 2: For each resistor, simply enter the polarities as defined by the assumed direction of the branch current as shown in Fig. 8.21. Take particular note that the polarities of the voltage sources are unaffected by the branch currents.

Step 3: So far the steps have been pretty mechanical, but now we need to do some thinking. Before applying Kirchhoff's voltage law, we must determine how many times we need to apply the law and which paths would be best.

The best way is to simply **note the number of "windows" in the network.** The network of Fig. 8.20 has a definite similarity to the two-window configuration of Fig. 8.22(a). Since it has two "windows," we only need to apply the law twice. For the configurations of Fig. 8.22(b), the law would have to be applied three times; for Fig. 8.22(c), four times; and so on. If the solution is approached in this way, all the information carried by the network that is necessary for a solution will be available. **Any additional applications of the law will simply carry information already included in the other equations.**

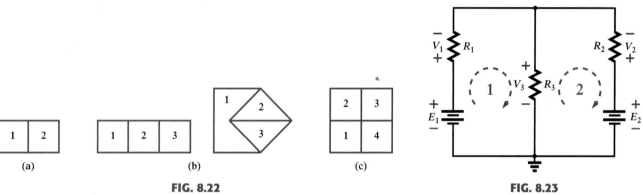

FIG. 8.22

Determining the number of independent closed loops for the application of Kirchhoff's voltage law.

FIG. 8.23

Applying Kirchhoff's voltage law to the two closed loops of Fig. 8.21.

For this example, Kirchhoff's law will be applied to the two windows of Fig. 8.23 in the directions indicated.

Loop 1: Applying Kirchhoff's voltage law in a clockwise direction starting at the negative terminal of E_1 will result in

$$\Sigma_{\circlearrowright} V = 0$$

$$+E_1 - V_1 - V_3 = 0$$

Loop 2: Applying Kirchhoff's voltage law in a clockwise direction starting at the base of resistor R_3 will result in

$$\Sigma_{\circlearrowright} V = 0$$

$$+V_3 + V_2 - E_2 = 0$$

Note for this latter application of Kirchhoff's voltage law that the drop across R_3 results in a positive sign for V_3. Also note the negative sign assigned to the voltage source E_2. One cannot assume that voltage drops across resistors will have negative signs and that the voltage across a supply will always have a positive sign. When applying branch-current analysis, it is important to ignore what the element is and concentrate solely on the polarities. In some respects, maintain a mechanical approach to the writing of the equations.

Substituting

$$V_1 = I_1 R_1 \qquad V_2 = I_2 R_2 \qquad V_3 = I_3 R_3$$

will result in the following for the two equations:

$$E_1 - I_1 R_1 - I_3 R_3 = 0$$

$$I_3 R_3 + I_2 R_2 - E_2 = 0$$

Rewritten:

$$I_1R_1 + I_3R_3 = E_1$$
$$I_2R_2 + I_3R_3 = E_2$$

We now have two equations but three unknowns (I_1, I_2, and I_3), so the currents cannot be determined. We need a third equation linking the branch currents. It will be provided by the next step.

Step 4: The question now is, How many times does Kirchhoff's current law have to be applied? **The minimum number is 1 less than the number of independent nodes (connection points) of the network.** In Fig. 8.24(a), which has a similar appearance to Fig. 8.20, we find that the network has two well-defined nodes. Subtracting 1 as dictated above will result in the need for one application of Kirchhoff's current law. The same would be true for Fig. 8.24(b) because it also has only two defined nodes. The configuration of Fig. 8.24(b) has three defined nodes and would require two applications of the law. In Fig. 8.24(c), the four independent nodes would require three applications of Kirchhoff's current law.

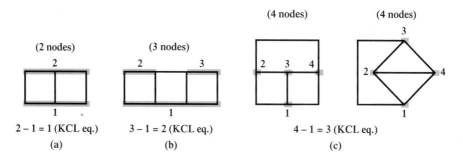

FIG. 8.24

Defining the number of Kirchhoff current law equations required.

As concluded above, the law must be applied once for the network of Fig. 8.20. Using node *a* of Fig. 8.22:

$$\Sigma I_i = \Sigma I_o$$
$$I_1 + I_2 = I_3$$

Step 5: Now we have three equations and three unknowns:

$$I_1R_1 + \quad 0 + I_3R_3 = E_1$$
$$0 + I_2R_2 + I_3R_3 = E_2$$
$$\underline{I_1 + I_2 \quad - I_3 \quad = 0}$$

This step is a purely mathematical one that can be solved in a number of ways, including calculators and computers. A longhand approach would require the use of determinants as discussed in Appendix C.

The number of equations can be reduced to two by simply solving for one variable in one of the equations and substituting into the other two. This would eliminate one variable for the moment and permit the use of second-order determinants. However, let us first pursue working with three simultaneous equations and third-order determinants.

Third-Order Determinants: Substituting the resistor values and the values of the voltage sources will result in the following equations:

$$I_1(2\,\Omega) + 0 + I_3(4\,\Omega) = 2\text{ V}$$
$$0 + I_2 + I_3(4\,\Omega) = 6\text{ V}$$
$$\underline{I_1 \quad + I_2 - I_3 \quad = 0}$$

Ignoring the units of measurement for the moment, we have

$$2I_1 + 0 + 4I_3 = 2$$
$$0 + I_2 + 4I_3 = 6$$
$$\underline{I_1 + I_2 - I_3 = 0}$$

$$I_1 = \frac{\begin{vmatrix} 2 & 0 & 4 \\ 6 & 1 & 4 \\ 0 & 1 & -1 \end{vmatrix}}{\begin{vmatrix} 2 & 0 & 4 \\ 0 & 1 & 4 \\ 1 & 1 & -1 \end{vmatrix} = D} = -1\,A$$

$$I_2 = \frac{\begin{vmatrix} 2 & 2 & 4 \\ 0 & 6 & 4 \\ 1 & 0 & -1 \end{vmatrix}}{D} = 2\,A$$

$$I_3 = \frac{\begin{vmatrix} 2 & 0 & 2 \\ 0 & 1 & 6 \\ 1 & 1 & 0 \end{vmatrix}}{D} = 1\,A$$

The impact of the results will be discussed in a moment. For now, however, let us first investigate the other mathematical approaches.

Second-Order Determinants: We will now proceed with the second option, which was to reduce the set to two simultaneous equations by substituting the information of one into the other two. Since Kirchhoff's current law equation is the "cleanest" of the three, let us solve for current I_3 in the equation and substitute into the other two as follows:

$$I_3 = I_1 + I_2$$
$$2I_1 + 4I_3 = 2$$
$$2I_1 + 4(I_1 + I_2) = 2$$
$$2I_1 + 4I_1 + 4I_2 = 2$$

and
$$6I_1 + 4I_2 = 2$$

Similarly,

$$I_2 + 4I_3 = 6$$
$$I_2 + 4(I_1 + I_2) = 6$$
$$I_2 + 4I_1 + 4I_2 = 6$$

and
$$4I_1 + 5I_2 = 6$$

The result is the following two equations with two unknowns:

$$6I_1 + 4I_2 = 2$$
$$4I_1 + 5I_2 = 6$$

One more substitution of one equation into the other would result in one equation and one unknown, and either I_1 or I_2 could be obtained. However, applying second-order determinants (Appendix C):

$$I_1 = \frac{\begin{vmatrix} 2 & 4 \\ 6 & 5 \end{vmatrix}}{\begin{vmatrix} 6 & 4 \\ 4 & 5 \end{vmatrix} = D} = -1\,A \qquad \text{and} \qquad I_2 = \frac{\begin{vmatrix} 6 & 2 \\ 4 & 6 \end{vmatrix}}{D} = 2\,A$$

and we have the same results for I_1 and I_2.

Current I_3 can be found by simply substituting into Kirchhoff's current law equation as follows:

$$I_3 = I_1 + I_2 = -1\,A + 2\,A = 1\,A$$

Before looking at the use of the calculator or computer software, let us be sure that the results are clearly understood. Currents I_1, I_2, and I_3 just determined are the actual branch currents of the network, as shown in Fig. 8.25. The negative sign in the solution for I_1 simply means that it has the opposite direction as shown in Fig. 8.25. Once the actual current directions are inserted, the various voltages and power levels can be determined. The resulting polarities for the voltages across the resistors of Fig. 8.20 appear in Fig. 8.25. Everything we need to know about the network of Fig. 8.20 is now known—any other properties can be determined quite easily.

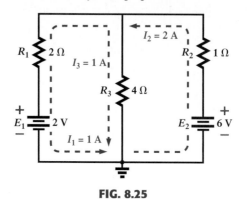

FIG. 8.25

Defining the results of the branch-current analysis.

Mathematical Techniques: Today's scientific calculators are programmed to solve second-order and third-order determinants in a very efficient and powerful way. It takes only a few minutes to learn the particulars for your calculator, leaving you with the ability to solve the most complex network. Using the TI-86 calculator, the format for the solution of the third-order determinant for I_1 would appear as shown in Fig. 8.26.

det[[2,0,4][6,1,4][0,1,–1]]/det[[2,0,4][0,1,4][1,1,–1]] (ENTER)   –1

FIG. 8.26

Using the TI-86 calculator to determine current I_1 from a third-order determinant.

The **det** (determinant) is obtained from a **MATH** listing under the **MATRX** menu. The first set of brackets defines the first row of the determinant, while the second set defines the second row, and so on. A comma separates the entries for each row.

The format for using the calculator to solve the second-order determinant for I_1 would appear as shown in Fig. 8.27. Using Mathcad, the printout would appear as shown in Fig. 8.28.

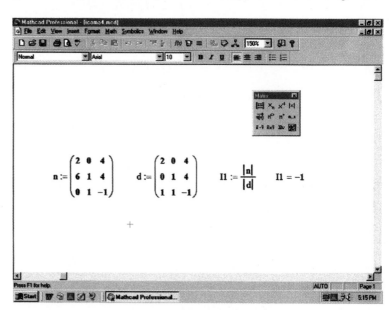

det[[2,4][6,5]]/det[[6,4][4,5]] (ENTER)   –1

FIG. 8.27

Using the TI-86 calculator to determine current I_1 from a second-order determinant.

FIG. 8.28

Using Mathcad to verify the numerical calculations of Example 8.9.

Priorities do not permit a detailed discussion of the application of the Mathcad software package. Let it be said, however, that it would take about five minutes to demonstrate how the entries are made once Mathcad is up and running.

When you consider the time required to perform a longhand calculation, the time involved in learning to use either the calculator or the software program is time well spent, and you will appreciate it more and more as the network complexity increases.

Example 8.9 may have seemed lengthy because of all the discussions about procedure. The next example will demonstrate that the method can be applied quickly once the procedure is understood.

EXAMPLE 8.10

a. Apply branch-current analysis to the network of Fig. 8.29.
b. Determine the voltage across each resistor.

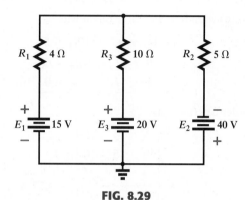

FIG. 8.29
Three-source network for Example 8.10.

Solutions: **Part (a):**

Steps 1–3: This configuration has three sources, but the two-window configuration will still be analyzed in much the same way. For each branch the direction of the branch current was chosen to match the "pressure" of the source as shown in Fig. 8.30. Polarities were then added for the resistive elements in Fig. 8.31, and Kirchhoff's voltage law was applied to each "window" in the clockwise direction.

Loop 1: $\qquad +E_1 - V_1 + V_3 - E_3 = 0$

or $\qquad\qquad -V_1 + V_3 = E_3 - E_1$

Loop 2: $\qquad +E_3 - V_3 - V_2 + E_2 = 0$

or $\qquad\qquad V_2 + V_3 = E_2 + E_3$

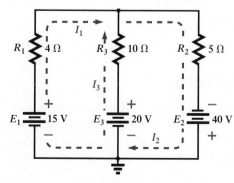

FIG. 8.30
Defining the branch currents for the network of Fig. 8.29.

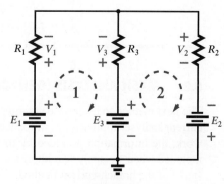

FIG. 8.31
Applying Kirchhoff's voltage law to the "windows" of the network.

Substituting:
$$V_1 = I_1R_1 = I_1(4\,\Omega)$$
$$V_2 = I_2R_2 = I_2(5\,\Omega)$$
$$V_3 = I_3R_3 = I_3(10\,\Omega)$$

which will result in
$$-I_1(4\,\Omega) + I_3(10\,\Omega) = 20\,\text{V} - 15\,\text{V}$$
$$I_2(5\,\Omega) + I_3(10\,\Omega) = 40\,\text{V} + 20\,\text{V}$$

and the following if units are removed:
$$-4I_1 + 10I_3 = 5$$
$$5I_2 + 10I_3 = 60$$

Step 4: Applying Kirchhoff's current law at node *a* of Fig. 8.30 will result in
$$\Sigma I_i = \Sigma I_o$$
$$I_1 + I_3 = I_2$$

and we have three equations and three unknowns:
$$-4I_1 + 0 \quad + 10I_3 = 5$$
$$0 + 5I_2 + 10I_3 = 60$$
$$I_1 - I_2 + \quad I_3 = 0$$

Using third-order determinants:

$$I_1 = \frac{\begin{vmatrix} 5 & 0 & 10 \\ 60 & 5 & 10 \\ 0 & -1 & 1 \end{vmatrix}}{\begin{vmatrix} -4 & 0 & 10 \\ 0 & 5 & 10 \\ 1 & -1 & 1 \end{vmatrix} = D} = \textbf{4.77 A}$$

with

$$I_2 = \frac{\begin{vmatrix} -4 & 5 & 10 \\ 0 & 60 & 10 \\ 1 & 0 & 1 \end{vmatrix}}{D} = \textbf{7.18 A}$$

and $\quad I_3 = I_2 - I_1 = 7.18\,\text{A} - 4.77\,\text{A} = \textbf{2.41 A}$

The positive signs for all the currents indicate that the assumed directions of Fig. 8.30 were correct.

Part (b): The voltage across each resistor is then determined using Ohm's law as follows:

$$V_1 = I_1R_1 = (4.77\,\text{A})(4\,\Omega) = \textbf{19.1 V}$$
$$V_2 = I_2R_2 = (7.18\,\text{A})(5\,\Omega) = \textbf{35.9 V}$$
$$V_3 = I_3R_3 = (2.41\,\text{A})(10\,\Omega) = \textbf{24.1 V}$$

8.6 MESH ANALYSIS (GENERAL APPROACH)

The next method to be described—**mesh analysis**—is actually an extension of the branch-current analysis approach just introduced. By defining a unique array of currents to the network, the information provided by the application of Kirchhoff's current law is already included when we apply Kirchhoff's voltage law. In other words, there is no need to apply step 4 of the branch-current method.

The currents to be defined are called **mesh** or **loop currents.** The two terms will be used interchangeably. In Fig. 8.32(a), a network with two "windows" has had two mesh currents defined. Note that each forms a closed "loop" around the inside of each window; these loops

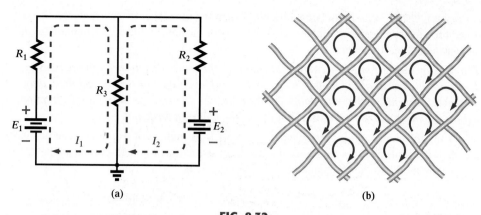

FIG. 8.32

Defining the mesh (loop) current: (a) two-"window" network; (b) mesh fence analogy.

are similar to the loops defined in the wire mesh fence of Fig. 8.25(b)—hence the use of the term *mesh* for the loop currents. We will find that

the number of mesh currents required to analyze a network will equal the number of "windows" of the configuration.

The defined mesh currents can initially be a little confusing because it would appear that two currents have been defined for resistor R_3. There is no problem with E_1 and R_1, which have only current I_1, or with E_2 and R_2, which have only current I_2. However, defining the current through R_3 may seem a little troublesome. Actually, it is quite straightforward. The current through R_3 will simply be the difference between I_1 and I_2, with the direction of the larger. This will be clearly demonstrated in the examples to follow.

Because mesh currents can result in more than one current through an element, branch-current analysis was introduced first. Branch-current analysis is the straightforward application of the basic laws of electric circuits. Mesh analysis employs a maneuver ("trick," if you prefer) that removes the need to apply Kirchhoff's current law.

Before applying mesh analysis to a network, **you must first be sure that only voltage sources are present.** If a current source is present, it must first be converted to a voltage source as described in Section 8.3. If the current source is ideal, that is, if it does not have a parallel internal resistance, the analysis will have to wait for Section 8.10.

The steps required to perform mesh analysis are listed below. The details of applying each will be covered in detail in the examples to follow.

Mesh Analysis:

1. *Assign a current in the clockwise direction to each independent, closed loop of the network.*
2. *Within each loop, insert the polarities for each resistor as determined by the assumed direction of the loop current.*
3. *Apply Kirchhoff's voltage law around each closed loop in the clockwise direction.*
4. *Solve the resulting equations for the assumed loop currents.*

Before we attack the first example, it should be pointed out that the assignment of loop currents is accomplished most effectively by simply placing a loop current in each "window" of the network. This will ensure that the equations are independent and that you have the correct number. A variety of other paths can be chosen, but this approach will ensure that the information carried by any one loop equation is not included in another loop equation or combination of loop equations. This is essentially the crux of the terminology *independent.*

Note in steps 1 and 3 that the clockwise direction is specified for the loop currents and the writing of Kirchhoff's voltage law. This was done for two reasons. First, it eliminates the need to wonder about which direction to take with each problem. Since it doesn't matter which direction is chosen, this approach eliminates the need to think about it at all. Second, we will find in the next section that by choosing the clockwise direction, we can establish a shorthand method for writing the equations that will speed up the process and avoid a lot of errors.

Applying step 3 can initially be a little trying in the sense that two currents may be defined for a resistor, and the voltage across the resistor must be included in Kirchhoff's voltage law equation. In general,

if a resistor has two or more currents through it, the total current through the resistor is the mesh current of the loop in which Kirchhoff's voltage law is being applied, plus the mesh currents of the other loops passing through the resistor in the same direction, minus the mesh currents passing through in the opposite direction.

Again, the above will become clearer as we work through a few examples. The first example is the same network examined during the introduction of branch-current analysis, so the two methods can be compared.

EXAMPLE 8.11 Using mesh analysis, find the current through each branch of Fig. 8.33.

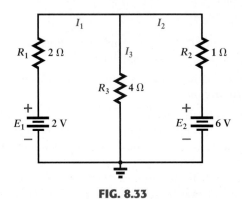

FIG. 8.33
Network to be explored in Example 8.11.

Solution:

Step 1: Since the network has two "windows," two mesh currents must be defined as shown in Fig. 8.34. A third mesh current (I_3) could have been defined as shown in the figure, but the information carried in this path about the network will already be included in the other paths. In other words, by limiting our choice to the "window" approach, we have defined the independent set—there is no need to think further about the issue.

Step 2: This step is quite mechanical. For each resistor, simply insert the polarities of the voltage drop caused by each mesh current. Put the polarities inside the window as shown in Fig. 8.35 because there will be two sets of polarities for resistor R_3: one set defined by each mesh current.

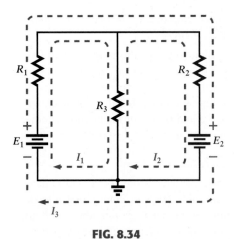

FIG. 8.34
Defining the mesh currents for the network of Fig. 8.33.

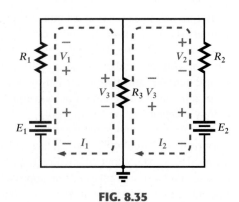

FIG. 8.35
Inserting the polarities for each voltage as defined by the mesh currents.

When applying Kirchhoff's voltage law around each loop, keep in mind that it is
the sequence of polarities that will define whether a plus or a minus sign is applied, not whether it is a supply or a resistor.** In addition, when writing the equations for one loop, **concentrate solely on that loop and forget about the other loops for the moment.**

Loop 1: Applying Kirchhoff's voltage law in the clockwise direction to loop number 1 of Fig. 8.35 will result in the following equation:

$$E_1 - V_1 - V_3 = 0$$

The voltage across resistor R_1 is determined by

$$V_1 = I_1R_1$$

However, for V_3, we must recall the statement provided earlier for the voltage across a resistor with two mesh currents: The net current is determined by taking the loop current of the loop of interest, and then subtracting the other loop current if it has the opposite direction and adding it if it has the same direction. In this example, currents I_1 and I_2 have opposite directions, so the voltage across V_3 is determined by

$$V_3 = (I_1 - I_2)R_3$$

Substituting into Kirchhoff's voltage law equation will result in

$$E_1 - I_1R_1 - (I_1 - I_2)R_3 = 0$$

Loop 2: Applying Kirchhoff's voltage law in the clockwise direction to loop number 2 of Fig. 8.35 will result in the following equation:

$$-V_3 - V_2 - E_2 = 0$$

The voltage across R_2 is simply

$$V_2 = I_2R_2$$

and for R_3 the following:

$$V_3 = (I_2 - I_1)R_3$$

resulting in the following equation:

$$-(I_2 - I_1)R_3 - I_2R_2 - E_2 = 0$$

Substituting values into the two equations will result in

$$2\,\text{V} - I_1(2\,\Omega) - (I_1 - I_2)(4\,\Omega) = 0$$
$$-(I_2 - I_1)(4\,\Omega) - I_2(1\,\Omega) - 6\,\text{V} = 0$$

Dropping units, rearranging, and multiplying through will result in

$$-2I_1 - 4I_1 + 4I_2 = -2$$
$$-4I_2 + 4I_1 - I_2 = 6$$

or

$$6I_1 - 4I_2 = 2$$
$$-4I_1 + 5I_2 = -6$$

The result is two equations and two unknowns rather than the three simultaneous equations obtained using branch-current analysis. Kirchhoff's current law equation has been substituted into the other two equations using the mesh-analysis approach.

Step 4: Applying determinants:

$$I_1 = \frac{\begin{vmatrix} 2 & -4 \\ -6 & 5 \end{vmatrix}}{\begin{vmatrix} 6 & -4 \\ -4 & 5 \end{vmatrix}} = \frac{(2)(5) - (-4)(-6)}{(6)(5) - (-4)(-4)} = \frac{10 - 24}{30 - 16} = \frac{-14}{14} = -1\,\text{A}$$

$$I_2 = \frac{\begin{vmatrix} 6 & 2 \\ -4 & -6 \end{vmatrix}}{14} = \frac{(6)(-6) - (2)(-4)}{14} = \frac{-36 + 8}{14} = \frac{-28}{14} = -2\,\text{A}$$

Now, the question remains as to the meaning of the results. First, the negative signs indicate that currents I_1 and I_2 actually have the directions indicated in Fig. 8.36. The current through E_1 and R_1 is therefore 1 A in the direction shown in the figure. For E_2 and R_2, the current is 2 A in the direction shown. For resistor R_3, the magnitude of the current is simply the difference since they have opposite directions. The direction is that of the larger, which in this case is I_2. The current I_3 is therefore 1 A in the direction shown in Fig. 8.36.

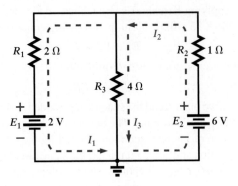

FIG. 8.36

Actual direction for the branch currents of the network of Fig. 8.33.

Again, the solution to Example 8.11 may appear lengthy and time-consuming. Keep in mind, however, that this was the first exposure to the method, and supporting comments had to be made. The flow of the examples to follow will demonstrate the positive features of this approach.

EXAMPLE 8.12 Using mesh analysis, find the current through resistor R_3 of Fig. 8.37.

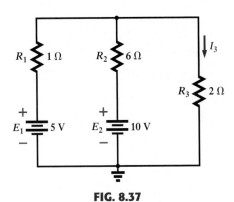

FIG. 8.37

Network for Example 8.12.

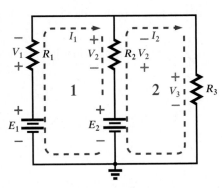

FIG. 8.38

Network of Fig. 8.37 with mesh currents and polarities defined.

Solution:

Steps 1–2: The mesh currents and the resulting polarities are indicated in Fig. 8.38.

Step 3: Applying Kirchhoff's voltage law in the clockwise direction for *loop 1* will result in

$$E_1 - V_1 - V_2 - E_2 = 0$$

Substituting $V_1 = I_1 R_1$ and $V_2 = (I_1 - I_2)R_2$

will result in $E_1 - I_1 R_1 - (I_1 - I_2)R_2 - E_2 = 0$

or $-I_1 R_1 - I_1 R_2 + I_2 R_2 = E_2 - E_1$

and $I_1(R_1 + R_2) - I_2 R_2 = E_1 - E_2$

For *loop 2*, Kirchhoff's voltage law will result in

$$E_2 - V_2 - V_3 = 0$$

Substituting $V_2 = (I_2 - I_1)R_2$ and $V_3 = I_2 R_3$

will result in $E_2 - (I_2 - I_1)R_2 - I_2 R_3 = 0$

or $\qquad E_2 - I_2R_2 + I_1R_2 - I_2R_3 = 0$

and $\qquad -I_1R_2 + I_2(R_2 + R_3) = E_2$

Substituting values into the two resulting equations:

$$I_1(1\,\Omega + 6\,\Omega) - I_2(6\,\Omega) = 5\,\text{V} - 10\,\text{V}$$

$$-I_1(6\,\Omega) + I_2(6\,\Omega + 2\,\Omega) = 10\,\text{V}$$

and removing units and rearranging will yield

$$7I_1 - 6I_2 = -5$$
$$\underline{-6I_1 + 8I_2 = 10}$$

Step 4: The only current asked for is the current through resistor R_3. Therefore, the application of determinants can be limited solely to finding current I_2 since $I_3 = I_2$.

$$I_3 = I_2 = \frac{\begin{vmatrix} 7 & -5 \\ -6 & 10 \end{vmatrix}}{\begin{vmatrix} 7 & -6 \\ -6 & 8 \end{vmatrix}} = \frac{(7)(10) - (-5)(-6)}{(7)(8) - (-6)(-6)} = \frac{70 - 30}{56 - 36} = \frac{40}{20} = \textbf{2 A}$$

We find that the mesh current I_2 has the same direction as the actual current I_3.

EXAMPLE 8.13

a. Using mesh analysis, determine the source current and current I_6 for the ladder network of Fig. 8.39. (This is the same ladder network appearing as Fig. 7.39.)
b. Insert ammeters to measure the mesh currents.

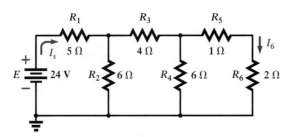

FIG. 8.39
Ladder network for Example 8.13.

Solutions: Recall that in the introduction to this chapter, we noted that all the methods to be introduced in this chapter can be applied to networks with only one source. The ladder network of Fig. 8.39 presents an excellent opportunity to demonstrate the versatility of this approach.

Part (a):

Steps 1–2: Since the configuration has three "windows," three mesh currents were defined in Fig. 8.40. The polarities of the voltage drops across each resistor were then added as shown in the same figure.

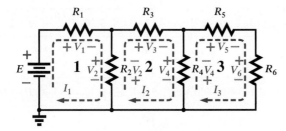

FIG. 8.40
Network of Fig. 8.39 with mesh currents and polarities defined.

Step 3: Three "windows" will require three applications of Kirchhoff's voltage law. For *loop 1,* the result is

$$E - V_1 - V_2 = 0$$

Substituting $\quad V_1 = I_1R_1 \quad$ and $\quad V_2 = (I_1 - I_2)R_2$

will result in $\quad E - I_1R_1 - (I_1 - I_2)R_2 = 0$

which becomes $\quad -I_1R_1 - I_1R_2 + I_2R_2 = E$

and $\quad I_1(R_1 + R_2) - I_2R_2 = E$

Applying Kirchhoff's voltage law to *loop 2:*

$$-V_2 - V_3 - V_4 = 0$$

Substituting $\quad V_2 = (I_2 - I_1)R_2 \qquad V_3 = I_3R_3 \qquad V_4 = (I_2 - I_3)R_4$

will result in $\quad -(I_2 - I_1)R_2 - I_2R_3 - (I_2 - I_3)R_4 = 0$

which becomes $\quad -I_2R_2 + I_1R_2 - I_2R_3 - I_2R_4 + I_3R_4 = 0$

and $\quad I_2(R_2 + R_3 + R_4) - I_1R_2 - I_3R_4 = 0$

Applying Kirchhoff's voltage law to *loop 3:*

$$-V_4 - V_5 - V_6 = 0$$

Substituting $\quad V_4 = (I_3 - I_2)R_4 \qquad V_5 = I_3R_5 \qquad V_6 = I_3R_6$

will result in $\quad -(I_3 - I_2)R_4 - I_3R_5 - I_3R_6 = 0$

which becomes $\quad -I_3R_4 + I_2R_4 - I_3R_5 - I_3R_6 = 0$

and $\quad I_3(R_4 + R_5 + R_6) - I_2R_4 = 0$

The result is three equations and three unknowns. Substituting values yields

$$I_1(5\,\Omega + 6\,\Omega) - I_2(6\,\Omega) = 24\text{ V}$$
$$I_2(6\,\Omega + 4\,\Omega + 6\,\Omega) - I_1(6\,\Omega) - I_3(6\,\Omega) = 0$$
$$I_3(6\,\Omega + 1\,\Omega + 2\,\Omega) - I_2(6\,\Omega) = 0$$

Removing the units and rearranging terms will result in

$$11I_1 - 6I_2 + 0 = 24$$
$$-6I_1 + 16I_2 - 6I_3 = 0$$
$$0 - 6I_2 + 9I_3 = 0$$

Step 4: The source current I_s is equal to I_1, and current I_6 is equal to I_3. Applying third-order determinants to find I_1 will result in

$$I_1 = \frac{\begin{vmatrix} 24 & -6 & 0 \\ 0 & 16 & -6 \\ 0 & -6 & 9 \end{vmatrix}}{\begin{vmatrix} 11 & -6 & 0 \\ -6 & 16 & -6 \\ 0 & -6 & 9 \end{vmatrix}}$$

and the sequence of Fig. 8.41(a) using the TI-86 calculator.

det[[24,–6,0][0,16,–6][0,–6,9]]/det[[11,–6,0][–6,16,–6][0,–6,9]] (ENTER)   3

(a)

det[[11,–6,24][–6,16,0][0,–6,0]]/det[[11,–6,0][–6,16,–6][0,–6,9]] (ENTER)   1

(b)

FIG. 8.41
Determining (a) I_s and (b) I_6 using the TI-86 calculator.

Therefore, $I_s = 3\,\text{A}$

as before.

Since $I_6 = I_3$, we will use the calculator to find I_3 also as shown in Fig. 8.41(b).

$$I_6 = 1\,\text{A}$$

as before.

Part (b): The meters have been inserted in Fig. 8.42.

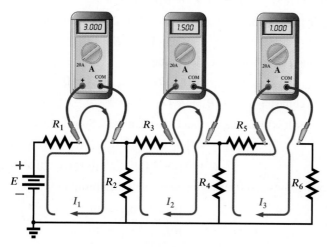

FIG. 8.42

Inserting ammeters in Fig. 8.39 to measure the mesh currents.

EXAMPLE 8.14 Using mesh analysis, find the current through resistor R_2 for the network of Fig. 8.43.

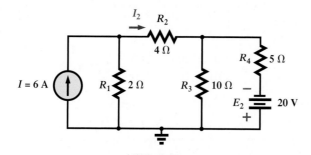

FIG. 8.43

Network for Example 8.14.

Solution: In order to apply mesh analysis using the techniques described thus far, we must first convert the current sources to voltage sources. Performing the conversion will result in the configuration of Fig. 8.44.

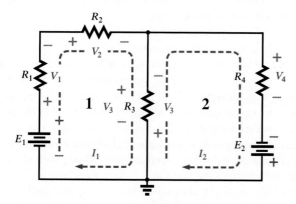

FIG. 8.44

Network of Fig. 8.43 redrawn with a voltage source and mesh currents defined.

Steps 1–3: The result is a two-"window" network in which two mesh currents are defined and the polarities entered.

Applying Kirchhoff's voltage law to *loop 1* will result in

$$E_1 - V_{R_1} - V_{R_2} - V_{R_3} = 0$$

Substituting Ohm's law for each voltage:

$$E_1 - I_1R_1 - I_1R_2 - (I_1 - I_2)R_3 = 0$$

which becomes $\quad I_1(R_1 + R_2 + R_3) - I_2R_3 = E_1$

Applying Kirchhoff's voltage law to *loop 2* will result in

$$-V_{R_3} - V_{R_4} + E_2 = 0$$

Substituting Ohm's law for each voltage:

$$-(I_2 - I_1)R_3 - I_2R_4 + E_2 = 0$$

which becomes $\quad I_2(R_3 + R_4) - I_1R_3 = E_2$

The result is two equations and two unknowns. Substituting values will result in

$$I_1(2\ \Omega + 4\ \Omega + 10\ \Omega) - I_2(10\ \Omega) = 12\ \text{V}$$
$$I_2(10\ \Omega + 5\ \Omega) - I_1(10\ \Omega) = 20\ \text{V}$$

Removing units and rearranging terms will then result in

$$16I_1 - 10I_2 = 12$$
$$-10I_1 + 15I_2 = 20$$

Step 4: Since we are interested only in the current through resistor R_2, we must solve for only the mesh current I_1.

Applying second-order determinants:

$$I_1 = \frac{\begin{vmatrix} 12 & -10 \\ 20 & 15 \end{vmatrix}}{\begin{vmatrix} 16 & -10 \\ -10 & 15 \end{vmatrix}} = \frac{(12)(15) - (-10)(20)}{(16)(15) - (-10)(-10)}$$

$$= \frac{180 + 200}{240 - 100} = \frac{380}{140} = \textbf{2.71 A}$$

and we find

$$I_{R_2} = I_1 = \textbf{2.71 A}$$

8.7 MESH ANALYSIS (FORMAT APPROACH)

In previous sections, branch-current analysis provided a solution for the branch currents of a network using both Kirchhoff's voltage and current law. Mesh analysis saved us some time by including Kirchhoff's current law in the definition of mesh currents. This section will now take us a step further by providing a route to writing the mesh equations that is quick and direct and will usually result in fewer errors.

The question may arise, If this method is that direct, why was so much time spent introducing the methods of the last two sections? The answer is really quite simple: The method to be introduced here is quite mechanical in a number of ways. You proceed in a manner that is dictated by the results of the last section. In other words, this approach is not developed through applications of the basic laws of electric circuits, but simply in recognition of the ways various elements of the network appear in the network equations. In essence, we are working back from the results to find a quick way to write the equations. **In any learning process it is important that you completely understand the source of the equations before using mechanical methods void of any application of the basic laws of electric circuits.** You must be able to defend an approach and be able to respond if a particular network does not fit the criteria for application of mechanical methods.

Let us now investigate how we obtain the mesh equations in a manner that most students find very satisfying. We will use the network of Fig. 8.45, which is the same network appearing in Fig. 8.33 that was used to introduce mesh analysis.

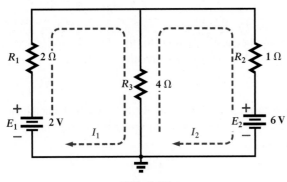

FIG. 8.45
Introducing the format approach.

The resulting equations were

$$6I_1 - 4I_2 = 2$$
$$-4I_1 + 5I_2 = -6$$

which can be rearranged as

$$6I_1 - 4I_2 = 2$$
$$5I_2 - 4I_1 = -6$$

and expanded as

$$\begin{array}{ccc} \textbf{Col. 1} & \textbf{Col. 2} & \textbf{Col. 3} \\ (2+4)I_1 & -4I_2 & = 2 \\ (4+1)I_2 & -4I_1 & = -6 \end{array}$$

If we now carefully examine the resulting equations, a format begins to surface that can be used to write the equations for other networks. First, note that column 1 for both equations is composed of a loop current times the resistors it passes through in the network. The second column includes the effect of the resistors that have additional loop currents passing through them. For each equation, it is the product of the other loop current times the mutual resistor. Note also that they are both subtracted from the first term of the equation. Column 3 is composed of the algebraic sum of the voltage sources touched by the loop current of column 1. They are assigned a positive sign if supporting that loop current, and a negative sign if not.

All of the above statements are applicable only if all of the loop currents are applied in the clockwise direction. Since all of the comments above will apply to any series-parallel or complex network, we have the foundation for the following procedure for the format approach.

Mesh Analysis (Format Approach):

1. *Assign a current in the clockwise direction to each independent, closed loop of the network.*
2. *Column 1 is the product of the loop current of interest and the sum of the resistors through which it passes.*
3. *The mutual terms are always subtracted from the first column. The mutual terms include all the resistors that have more than one loop current passing through them. They are formed by the product of the resistive element and the other loop current.*
4. *The column to the right of the equals sign is the algebraic sum of the voltage sources through which the loop current of interest passes. Positive signs are assigned to those that support the direction of the assigned loop current. Negative signs are assigned to those that oppose it.*
5. *Solve the resulting simultaneous equations for the desired loop currents.*

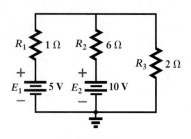

FIG. 8.46

Network for Example 8.15 (the same network as in Fig. 8.37).

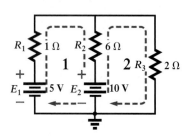

FIG. 8.47

Defining the mesh currents for Fig. 8.46.

Because step 4 refers only to voltage sources, all current sources must first be converted to voltage sources before applying the format approach. A few examples will clarify the procedure defined by the above steps.

EXAMPLE 8.15   Write the mesh equations for the network of Fig. 8.46 (the same network analyzed in Example 8.12). Compare your results.

Solution:

Steps 1–2:   The loop currents are defined in Fig. 8.47. We find that the loop current I_1 passes through resistors R_1 and R_2. The loop current I_2 passes through resistors R_2 and R_3. The result is

Col. 1

$$I_1(R_1 + R_2)$$
$$I_2(R_2 + R_3)$$

Step 3:   For the loop current I_1, the only resistor with a mutual loop current is R_2. For the loop current I_2, the only mutual term is also R_2. Forming the product and subtracting from the first column entries will result in

Col. 1   **Col. 2**

$$I_1(R_1 + R_2) - I_2R_2$$
$$I_2(R_2 + R_3) - I_1R_2$$

Note in each case that the *other* loop current is subtracted from the first term and that voltage sources are not considered mutual terms.

Step 4:   For loop 1, the voltage source E_1 supports the direction of I_1, while the voltage source E_2 opposes it. For loop 2, the voltage source E_2 supports the direction of I_2, resulting in the following equations:

$$I_1(R_1 + R_2) - I_2R_2 = E_1 - E_2$$
$$I_2(R_2 + R_3) - I_1R_2 = E_2$$

Note that the only voltage sources appearing in each equation are those "touched" by the loop current of interest.

Substituting values:

$$I_1(1\,\Omega + 6\,\Omega) - I_2(6\,\Omega) = 5\,V - 10\,V$$
$$I_2(6\,\Omega + 2\,\Omega) - I_1(6\,\Omega) = 10\,V$$

and dropping units:

$$7I_1 - 6I_2 = -5$$
$$8I_2 - 6I_1 = 10$$

or:

$$7I_1 - 6I_2 = -5$$
$$-6I_1 + 8I_2 = 10$$

which matches the results of Example 8.12.

There is no question that the process is short and sweet—very mechanical with very little thinking involved. As mentioned earlier, it is certainly nice to obtain the equations so quickly, but it is even more rewarding, because of our earlier efforts, to know that they are correct and are supported by the basic laws of circuit analysis.

EXAMPLE 8.16 Write the mesh equations for the network of Fig. 8.48 (the same network appearing in Example 8.13). Compare your solutions.

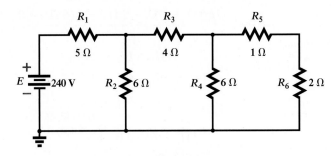

FIG. 8.48
Ladder network for Example 8.16.

Solution:

Steps 1–2: The loop currents are defined in Fig. 8.49. The mesh current I_1 passes through R_1 and R_2, while the mesh current I_2 passes through R_2, R_3, and R_4. Finally, the mesh current I_3 passes through R_4, R_5, and R_6.

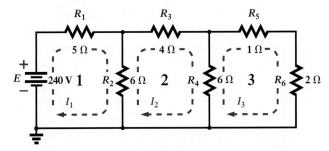

FIG. 8.49
Defining the mesh currents for the network of Fig. 8.48.

The result is

Col. 1

$I_1(R_1 + R_2)$

$I_2(R_2 + R_3 + R_4)$

$\underline{I_3(R_4 + R_5 + R_6)}$

Step 3: For the loop current I_1, the only resistor with a mutual loop current is R_2. The second loop, however, has two mutual terms due to R_2 and R_4. The third loop has only one mutual term established by R_4. Forming the product and subtracting from the first-column entries will result in

Col. 1   **Col. 2**

$I_1(R_1 + R_2)$ $- I_2R_2$

$I_2(R_2 + R_3 + R_4) - I_1R_2 - I_3R_4$

$\underline{I_3(R_4 + R_5 + R_6) - I_2R_4}$

Step 4: For loop 1, the voltage source supports the mesh current I_1, so it is placed to the right of the equals sign with a plus (+) sign. There are no sources in the remaining loops, so a zero is placed to the right of the equals sign for each equation.

The result is three equations and three unknowns:

$$I_1(R_1 + R_2) - I_2R_2 = E$$
$$I_2(R_2 + R_3 + R_4) - I_1R_2 - I_3R_4 = 0$$
$$\underline{I_3(R_4 + R_5 + R_6) - I_2R_4 = 0}$$

Substituting values:

$$I_1(5\,\Omega + 6\,\Omega) - I_2(6\,\Omega) = 24$$
$$I_2(6\,\Omega + 4\,\Omega + 6\,\Omega) - I_1(6\,\Omega) - I_3(6\,\Omega) = 0$$
$$I_3(6\,\Omega + 1\,\Omega + 2\,\Omega) - I_2(6\,\Omega) = 0$$

Dropping units:

$$11I_1 - 6I_2 = 24$$
$$16I_2 - 6I_1 - 6I_3 = 0$$
$$9I_3 - 6I_2 = 0$$

or:

$$11I_1 - 6I_2 + 0 = 24$$
$$-6I_1 + 6I_2 - 6I_3 = 0$$
$$0 - 6I_2 + 9I_3 = 0$$

These results match those of Example 8.13 and were found in less than half the time.

EXAMPLE 8.17   Write the mesh equations for the network of Fig. 8.50.

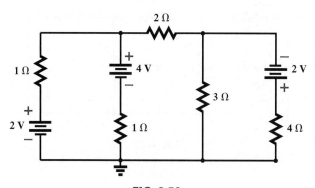

FIG. 8.50
Three-loop network for Example 8.17.

Solution: This time, since we have had some practice, we will go directly to the equations rather break the analysis up into steps. The three clockwise mesh currents are defined in Fig. 8.51, and the resulting equations are

Loop 1: $I_1(1\,\Omega + 1\,\Omega) - I_2(1\,\Omega)$ $\qquad\qquad\quad = 2\,V - 4\,V$

Loop 2: $I_2(1\,\Omega + 2\,\Omega + 3\,\Omega) - I_1(1\,\Omega) - I_3(3\,\Omega) = 4\,V$

Loop 3: $I_3(3\,\Omega + 4\,\Omega) - I_2(3\,\Omega)$ $\qquad\qquad\quad = 2\,V$

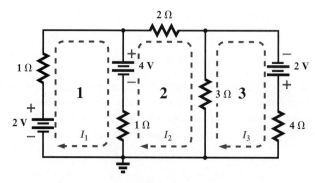

FIG. 8.51
Defining the mesh currents for the network of Fig. 8.50.

Removing units and rearranging terms will result in

$$2I_1 - 1I_2 + 0 = -2$$
$$-1I_1 + 6I_2 - 3I_3 = 4$$
$$0 - 3I_2 + 7I_3 = 2$$

A check of the results can be made if the equations are rewritten as shown below to permit drawing the dashed diagonals:

$$\begin{array}{ccc} c & b & a \\ 2I_1 & - 1I_2 & + 0 & = -2 \\ b & & \\ -1I_1 & + 6I_2 & - 3I_3 = 4 \\ a & & \\ 0 & - 3I_2 & + 7I_3 = 2 \end{array}$$

Note that the coefficients of the *a* diagonal at the top match those of the *a* diagonal on the bottom. The same is true for the two *b* diagonals. This will be true for any number of "windows" for the network if clockwise mesh currents are selected. It provides a quick check of the results before proceeding with the calculations.

EXAMPLE 8.18 Find the current through the 10 Ω resistor of the network of Fig. 8.52.

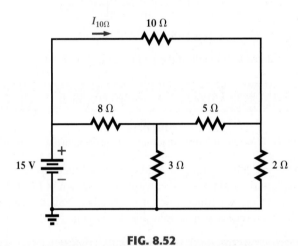

FIG. 8.52
Three-loop network for Example 8.18.

Solution: The three clockwise mesh currents are defined in Fig. 8.53.

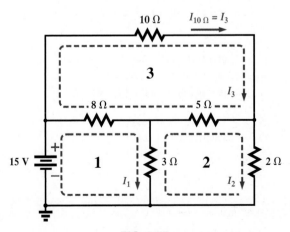

FIG. 8.53
Defining the mesh currents for the network of Fig. 8.52.

The resulting mesh equations are

Loop 1: $I_1(8\ \Omega + 3\ \Omega) - I_2(3\ \Omega) - I_3(8\ \Omega)$ = 15 V

Loop 2: $I_2(3\ \Omega + 5\ \Omega + 2\ \Omega) - I_1(3\ \Omega) - I_3(5\ \Omega)$ = 0

Loop 3: $I_3(8\ \Omega + 10\ \Omega + 5\ \Omega) - I_1(8\ \Omega) - I_2(5\ \Omega) = 0$

Removing units and gathering terms will result in the following standard matrix form:

$$11I_1 - 3I_2 - 8I_3 = 15$$
$$-3I_1 + 10I_2 - 5I_3 = 0$$
$$-8I_1 - 5I_2 + 23I_3 = 0$$

Checking diagonals again, we find that the coefficient of I_3 on the first row is the same as that for I_1 in the third row—a match. Then we find that the coefficient of I_2 in the first row is the same as that for I_1 in the second row. Finally, the coefficient for I_3 in the second row is the same as for I_2 in the last row. Everything seems to check on the diagonals, so it seems wise to proceed with the required mathematics.

Since the current through the 10 Ω resistor is the mesh current I_3, we will solve for this current using determinants as follows:

$$I_3 = I_{10\Omega} = \frac{\begin{vmatrix} 11 & -3 & 15 \\ -3 & 10 & 0 \\ -8 & -5 & 0 \end{vmatrix}}{\begin{vmatrix} 11 & -3 & -8 \\ -3 & 10 & -5 \\ -8 & -5 & 23 \end{vmatrix}} = \textbf{1.220 A}$$

Using the TI-86 calculator will result in the sequence of Fig. 8.54.

det[[11,–3,15][–3,10,0][–8,–5,0]]/det[[11,–3,–8][–3,10,–5][–8,–5,23]] (ENTER) | 1.220

FIG. 8.54

Using the TI-86 calculator to determine I_3 for the network of Fig. 8.53.

Mathcad Solution: For this example of Mathcad, rather than simply solve the resulting determinant, we will go right to the set of equations. As shown in Fig. 8.55, a **Guess** value for each variable must first be defined when using Mathcad. Such a guess helps the computer begin the iteration process as it searches for a solution.

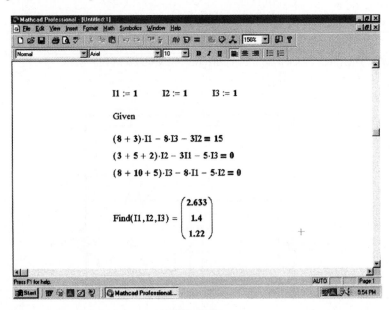

FIG. 8.55

Using Mathcad to verify the numerical calculations of Example 8.18.

Next, as shown, the word **Given** must be entered to tell the computer that the defining equations will follow. Finally, each equation must be carefully entered and set equal to the constant on the right side using the **Ctrl=** operation.

The results are then obtained with the **Find(I1,I2,I3)** expression and an equals sign. As shown, the results will appear immediately with a satisfactory degree of accuracy. When you consider that the equations were left in their roughest form and could be entered in a few minutes, the fact that the results will appear almost immediately is a clear demonstration of the power of the Mathcad software.

8.8 NODAL ANALYSIS (GENERAL APPROACH)

The methods introduced thus far have all been to find the currents of the network. We will now turn our attention to **nodal analysis**—a method that will provide the nodal voltages of a network, that is, the voltage from the various **nodes** (junction points) of the network to ground. The method will be developed through the use of Kirchhoff's current law in much the same manner as Kirchhoff's voltage law was used to develop the mesh analysis approach.

Although it is not a requirement, we will make it a policy to make ground our reference node and assign it a potential level of zero volts. All the other voltage levels will then be found with respect to this reference level. For a network of N nodes, by assigning one as our reference node, we will have $(N - 1)$ nodes for which the voltage must be determined. In other words,

the number of nodes for which the voltage must be determined using nodal analysis is 1 less than the total number of nodes.

The result of the above is $(N - 1)$ nodal voltages that need to be determined, requiring that $(N - 1)$ independent equations be written to find the nodal voltages. In other words,

the number of equations required to solve for all the nodal voltages of a network is 1 less than the total number of independent nodes.

Since each equation will be the result of an application of Kirchhoff's current law, Kirchhoff's current law will have to be applied $(N - 1)$ times for each network.

Recall that the procedure defined for mesh analysis in the previous chapter required that only voltage sources be present. For nodal analysis, the procedure to be described here requires that **only current sources can be present.** For instances where the conversion from one form to the other cannot be made because the internal resistance of the voltage source is not provided, the maneuvers of Section 8.10 must be applied.

Nodal analysis, like mesh analysis, can be applied by a series of carefully defined steps. The examples to follow will explain each step in detail.

Nodal Analysis:

1. *Determine the number of nodes (junctions of two or more branches) for the network.*
2. *Pick a reference node (normally the ground connection), and label each of the remaining nodes with a subscripted label such as V_1, V_2, and so on.*
3. *Apply Kirchhoff's current law at each node except the reference node. For each application of Kirchhoff's current law, assume that each of the unknown currents leaves the node (this removes the concern about direction; a minus sign will appear in the solution if incorrectly chosen).*
4. *Solve the resulting equations for the nodal voltages.*

The four steps seem quite straightforward—certainly no more daunting than those encountered for mesh analysis. A few examples, with each step covered in detail, should remove most questions about application of these steps. Initially, it will take some practice to set up and write Kirchhoff's current law correctly for each node, but the advantage of assuming that all the unknown currents leave a junction will soon become obvious. It is essentially the same type of advantage as occurs when choosing the clockwise direction for all the mesh currents when mesh analysis is applied.

EXAMPLE 8.19 Apply nodal analysis to the network of Fig. 8.56.

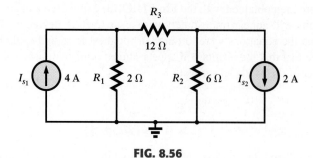

FIG. 8.56
Network for Example 8.19.

Solution:

Steps 1–2: As shown in Fig. 8.57, this network has three nodes, that is, three points in the network where the branches are tied together. The bottom node, connected to ground, is defined as the reference node, and the other two are labeled as shown. Since there are two nodes defined by subscripted voltages, two equations will be required to solve for the nodal voltages.

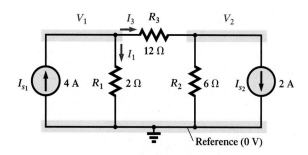

FIG. 8.57
Network of Fig. 8.56 with the nodal voltages and current directions for V_1 defined.

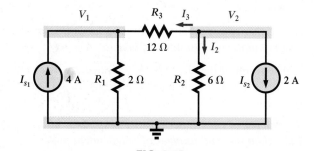

FIG. 8.58
Network of Fig. 8.56 with nodal voltages and current directions for V_2 defined.

Step 3: The current directions have been added to Fig. 8.57 for the application of Kirchhoff's current law to the node defined as V_1. Note, as required by step 3, that currents I_1 and I_3 are defined as leaving the node.

Applying Kirchhoff's current law to node V_1 will result in

$$\Sigma I_i = \Sigma I_o$$
$$I_{s_1} = I_1 + I_3$$

Note that the direction of the current source (like the polarity for voltage sources when mesh analysis was applied) is unaffected by the analysis.

Since the voltage across R_1 is the subscripted voltage V_1, the current I_1 is defined by

$$I_1 = \frac{V_1}{R_1}$$

The voltage across R_3 is equal to $V_1 - V_2$, resulting in I_3 being defined by

$$I_3 = \frac{V_1 - V_2}{R_3}$$

The voltage across R_3 is $V_1 - V_2$ rather than $V_2 - V_1$ because of the chosen direction for I_3. Remember that when current passes through a resistor, it establishes a drop in voltage across the resistor.

Substituting into Kirchhoff's current law equation above will result in

$$I_{s_1} = \frac{V_1}{R_1} + \frac{V_1 - V_2}{R_3}$$

Expanding and rearranging will then result in

$$V_1\left(\frac{1}{R_1} + \frac{1}{R_3}\right) - V_2\left(\frac{1}{R_3}\right) = I_{s_1}$$

For node V_2, the currents are redefined as shown in Fig. 8.58. Note again that the currents are defined as leaving the node.

Applying Kirchhoff's current law to node V_2 will result in

$$\Sigma I_i = \Sigma I_o$$
$$0 = I_3 + I_2 + I_{s_2}$$

The voltage across resistor R_2 is the subscripted voltage V_2, so the current I_2 is defined by

$$I_2 = \frac{V_2}{R_2}$$

Due to the **chosen direction** for I_3, the voltage across R_3 is $V_2 - V_1$, and the current is defined by

$$I_3 = \frac{V_2 - V_1}{R_3}$$

Substituting into Kirchhoff's current law equation will result in

$$0 = \frac{V_2 - V_1}{R_3} + \frac{V_2}{R_2} + I_{s_2}$$

Expanding and rearranging yields

$$V_2\left(\frac{1}{R_2} + \frac{1}{R_3}\right) - V_1\left(\frac{1}{R_3}\right) = -I_{s_2}$$

resulting in two equations and two unknowns:

$$V_1\left(\frac{1}{R_1} + \frac{1}{R_3}\right) - V_2\left(\frac{1}{R_3}\right) = I_{s_1}$$
$$\underline{V_2\left(\frac{1}{R_2} + \frac{1}{R_3}\right) - V_1\left(\frac{1}{R_3}\right) = -I_{s_2}}$$

Substituting values will result in

$$V_1\left(\frac{1}{2\,\Omega} + \frac{1}{12\,\Omega}\right) - V_2\left(\frac{1}{12\,\Omega}\right) = 4\text{ A}$$
$$\underline{V_2\left(\frac{1}{6\,\Omega} + \frac{1}{12\,\Omega}\right) - V_1\left(\frac{1}{12\,\Omega}\right) = -2\text{ A}}$$

which, after we drop units and work through the mathematics, can be written as

$$0.583V_1 - 0.083V_2 = 4$$
$$\underline{-0.083V_1 + 0.25V_2 = -2}$$

Step 4: Using determinants:

$$V_1 = \frac{\begin{vmatrix} 4 & -0.083 \\ -2 & 0.25 \end{vmatrix}}{\begin{vmatrix} 0.583 & -0.083 \\ -0.083 & 0.25 \end{vmatrix}} = \frac{(4)(0.25) - (-0.083)(-2)}{(0.583)(0.25) - (-0.083)(-0.083)}$$

$$= \frac{1 - 0.166}{0.146 - 0.007} = \frac{0.834}{0.139} = \textbf{6 V}$$

$$V_2 = \frac{\begin{vmatrix} 0.583 & 4 \\ -0.083 & -2 \end{vmatrix}}{0.139} = \frac{(0.583)(-2) - (4)(-0.083)}{0.139}$$

$$= \frac{-1.166 + 0.332}{0.139} = \frac{-0.834}{0.139} = \textbf{-6 V}$$

We now have to take a moment to be sure that we understand the significance of the results. Since $V_1 = 6$ V is a positive quantity, the voltage at V_1 is positive with respect to ground, as shown in Fig. 8.59. The current I_1 can then be determined from

$$I_1 = \frac{V_1}{R_1} = \frac{6 \text{ V}}{2 \text{ }\Omega} = \textbf{3 A}$$

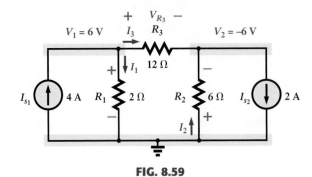

FIG. 8.59
Reviewing the results of Example 8.19.

Since V_2 is a negative quantity, the voltage across R_2 has the polarity shown in Fig. 8.59. Therefore, the current I_2 has the direction shown in Fig. 8.59 and is equal to

$$I_2 = \frac{V_2}{R_2} = \frac{6 \text{ V}}{6 \text{ }\Omega} = \textbf{1 A}$$

The voltage across R_3 is now $V_1 - V_2 = 6$ V $- (-6$ V$) = 12$ V, and the current through R_3 has the direction shown in Fig. 8.59. Its value is

$$I_3 = \frac{V_3}{R_3} = \frac{12 \text{ V}}{12 \text{ }\Omega} = \textbf{1 A}$$

We now know all the currents and voltages of the network, and the analysis is complete.

EXAMPLE 8.20   For the network of Fig. 8.60:

a. Find the nodal voltages.
b. Insert voltmeters to measure the nodal voltages.

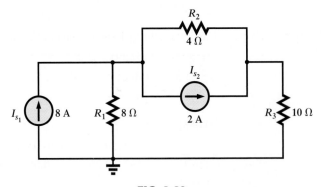

FIG. 8.60
Network for Example 8.20.

Solutions: **Part (a):**

Steps 1–2:   The nodal voltages have been defined in Fig. 8.61 with the current directions for the application of Kirchhoff's current law to node V_1.
Step 3:   Applying Kirchhoff's current law to node V_1 will result in

$$\Sigma I_i = \Sigma I_o$$

$$I_{s_1} = I_1 + I_2 + I_{s_2}$$

Since the voltage across R_1 is V_1, current I_1 can be defined by

$$I_1 = \frac{V_1}{R_1}$$

The direction of I_2 requires that V_1 be at a higher potential than V_2, so current I_2 is defined by

$$I_2 = \frac{V_1 - V_2}{R_2}$$

Substituting into Kirchhoff's current law equation will result in

$$I_{s_1} = \frac{V_1}{R_1} + \frac{V_1 - V_2}{R_2} + I_{s_2}$$

Expanding and rearranging yields

$$V_1\left(\frac{1}{R_1} + \frac{1}{R_2}\right) - V_2\left(\frac{1}{R_2}\right) = I_{s_1} - I_{s_2}$$

and we have one equation for the solution.

For nodal voltage V_2, the currents must be redefined as shown in Fig. 8.62.

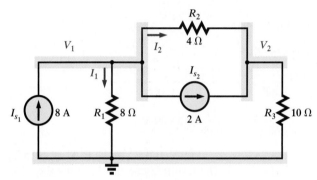

FIG. 8.61

Network of Fig. 8.60 with the nodal voltages and current directions for V_1 defined.

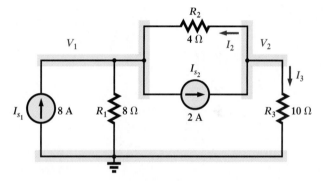

FIG. 8.62

Network of Fig. 8.60 with the nodal voltages and current directions for V_2 defined.

Applying Kirchhoff's current law to node V_2 will result in

$$\Sigma I_i = \Sigma I_o$$

$$I_{s_2} = I_2 + I_3$$

with current I_3 defined by

$$I_3 = \frac{V_2}{R_3}$$

and current I_2 by

$$I_2 = \frac{V_2 - V_1}{R_2}$$

Substituting into Kirchhoff's current law equation will result in

$$I_{s_2} = \frac{V_2 - V_1}{R_2} + \frac{V_2}{R_3}$$

Expanding and rearranging yields

$$V_2\left(\frac{1}{R_2} + \frac{1}{R_3}\right) - V_1\left(\frac{1}{R_2}\right) = I_{s_2}$$

and we have two equations and two unknowns:

$$V_1\left(\frac{1}{R_1} + \frac{1}{R_2}\right) - V_2\left(\frac{1}{R_2}\right) = I_{s_1} - I_{s_2}$$

$$V_2\left(\frac{1}{R_2} + \frac{1}{R_3}\right) - V_1\left(\frac{1}{R_2}\right) = I_{s_2}$$

Substituting values:

$$V_1\left(\frac{1}{8\ \Omega} + \frac{1}{4\ \Omega}\right) - V_2\left(\frac{1}{4\ \Omega}\right) = 8\ \text{A} - 2\ \text{A}$$

$$V_2\left(\frac{1}{4\ \Omega} + \frac{1}{10\ \Omega}\right) - V_1\left(\frac{1}{4\ \Omega}\right) = 2\ \text{A}$$

which, after we drop units and work through the mathematics, can be written as

$$0.375V_1 - 0.25V_2 = 6$$
$$-0.25V_1 + 0.35V_2 = 2$$

Step 4: Applying determinants:

$$V_1 = \frac{\begin{vmatrix} 6 & -0.25 \\ 2 & 0.35 \end{vmatrix}}{\begin{vmatrix} 0.375 & -0.25 \\ -0.25 & 0.35 \end{vmatrix}} = \frac{(6)(0.35) - (-0.25)(2)}{(0.375)(0.35) - (-0.25)(-0.25)}$$

$$= \frac{2.1 + 0.5}{0.131 - 0.063} = \frac{2.6}{0.068} = \textbf{38.24 V}$$

$$V_2 = \frac{\begin{vmatrix} 0.375 & 6 \\ -0.25 & 2 \end{vmatrix}}{0.068} = \frac{(0.375)(2) - (6)(-0.25)}{0.068}$$

$$= \frac{0.75 + 1.5}{0.068} = \frac{2.25}{0.068} = \textbf{33.09 V}$$

We now know that $V_1 = 38.24$ V, $V_2 = 33.09$ V, and $V_{R_2} = V_1 - V_2 = 5.15$ V, with the polarities shown in Fig. 8.63. The resulting currents have the direction shown in the same figure with the following magnitudes:

$$I_1 = \frac{V_1}{R_1} = \frac{38.24\ \text{V}}{8\ \Omega} = \textbf{4.78 A}$$

$$I_2 = \frac{V_1 - V_2}{R_2} = \frac{5.15\ \text{V}}{4\ \Omega} = \textbf{1.29 A}$$

$$I_3 = \frac{V_2}{R_3} = \frac{33.09\ \text{V}}{10\ \Omega} = \textbf{3.31 A}$$

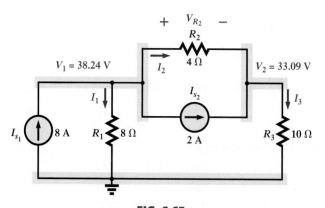

FIG. 8.63
Reviewing the results of Example 8.20.

Part (b): The meters have been inserted in Fig. 8.64 to measure the nodal voltages.

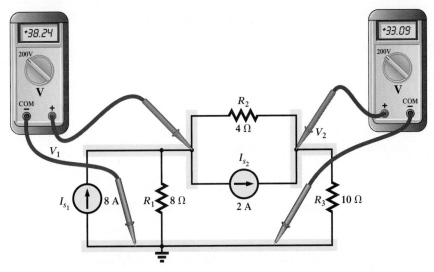

FIG. 8.64

Measuring the nodal voltages for the network of Fig. 8.60.

EXAMPLE 8.21 Using nodal analysis, find voltage V_6 for the ladder network of Fig. 8.65. Note that it is the same ladder network appearing in Examples 8.13 and 8.16 to provide a basis for comparison.

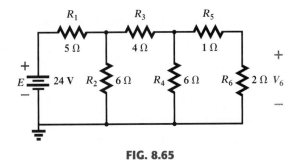

FIG. 8.65

Ladder network for Example 8.21.

Solution:

Steps 1–2: The voltage source is converted to a current source as shown in Fig. 8.66. The result is that resistors R_1 and R_2 are now in parallel and could be combined. There is no need to combine them, however, since the nodal analysis can handle the two parallel elements without a change in procedure.

Since we are interested in voltage V_6, we will define that point as a third node V_3 rather than combine resistors R_5 and R_6 into one series equivalent resistor. The

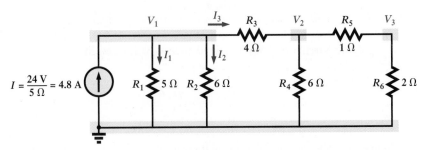

FIG. 8.66

Network of Fig. 8.65 following the source conversion and the definition of the nodal voltages and current directions for V_1.

nodes are defined in Fig. 8.66 along with the current directions for applying Kirchhoff's current law to node V_1.

Step 3: Applying Kirchhoff's current law to node V_1 will result in

$$\Sigma I_i = \Sigma I_o$$

$$I = I_1 + I_2 + I_3$$

or

$$I = \frac{V_1}{R_1} + \frac{V_1}{R_2} + \frac{V_1 - V_2}{R_3}$$

Combining and rearranging terms will result in

$$V_1 \left(\frac{1}{R_1} + \frac{1}{R_2} + \frac{1}{R_3} \right) - \frac{V_2}{R_3} = I$$

For node V_2, the currents must be defined as appearing in Fig. 8.67.

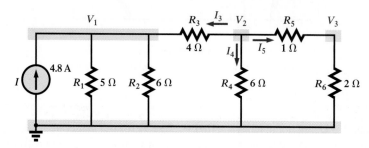

FIG. 8.67

Network of Fig. 8.65 with the nodal voltages and current directions for V_2 defined.

Applying Kirchhoff's current law to node V_2 will result in

$$\Sigma I_i = \Sigma I_o$$

$$0 = I_3 + I_4 + I_5$$

$$0 = \frac{V_2 - V_1}{R_3} + \frac{V_2}{R_4} + \frac{V_2 - V_3}{R_5}$$

Combining and rearranging terms will result in

$$V_2 \left(\frac{1}{R_3} + \frac{1}{R_4} + \frac{1}{R_5} \right) - V_1 \left(\frac{1}{R_3} \right) - V_3 \left(\frac{1}{R_5} \right) = 0$$

For node V_3, the currents are defined as appearing in Fig. 8.68.

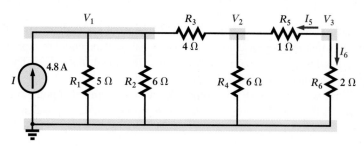

FIG. 8.68

Network of Fig. 8.65 with the nodal voltages and current directions for V_3 defined.

Applying Kirchhoff's current law to node V_3 will result in

$$\Sigma I_i = \Sigma I_o$$

$$0 = I_5 + I_6$$

$$0 = \frac{V_3 - V_2}{R_5} + \frac{V_3}{R_6}$$

Combining and rearranging terms will result in

$$V_3 \left(\frac{1}{R_5} + \frac{1}{R_6} \right) - V_2 \left(\frac{1}{R_5} \right) = 0$$

The result is three equations and three unknowns:

$$V_1\left(\frac{1}{R_1} + \frac{1}{R_2} + \frac{1}{R_3}\right) - \frac{V_2}{R_3} = I$$

$$V_2\left(\frac{1}{R_3} + \frac{1}{R_4} + \frac{1}{R_5}\right) - V_1\left(\frac{1}{R_3}\right) - V_3\left(\frac{1}{R_5}\right) = 0$$

$$V_3\left(\frac{1}{R_5} + \frac{1}{R_6}\right) - V_2\left(\frac{1}{R_5}\right) = 0$$

Substituting values:

$$V_1\left(\frac{1}{5\ \Omega} + \frac{1}{6\ \Omega} + \frac{1}{4\ \Omega}\right) - V_2\left(\frac{1}{4\ \Omega}\right) = 4.8\ \text{A}$$

$$V_2\left(\frac{1}{4\ \Omega} + \frac{1}{6\ \Omega} + \frac{1}{1\ \Omega}\right) - V_1\left(\frac{1}{4\ \Omega}\right) - V_3\left(\frac{1}{1\ \Omega}\right) = 0$$

$$V_3\left(\frac{1}{1\ \Omega} + \frac{1}{2\ \Omega}\right) - V_2\left(\frac{1}{1\ \Omega}\right) = 0$$

which, after we drop units and work through the mathematics, will result in the following matrix form:

$$
\begin{aligned}
0.617V_1 - 0.25V_2 + 0 &= 4.8 \\
-0.25V_1 + 1.417V_2 - 1V_3 &= 0 \\
0 - 1V_2 + 1.5V_3 &= 0
\end{aligned}
$$

Step 4: Since we are interested only in voltage V_6, we will use determinants to find V_3:

$$V_6 = V_3 = \frac{\begin{vmatrix} 0.617 & -0.25 & 4.8 \\ -0.25 & 1.417 & 0 \\ 0 & -1 & 0 \end{vmatrix}}{\begin{vmatrix} 0.617 & -0.25 & 0 \\ -0.25 & 1.417 & -1 \\ 0 & 1 & 1.5 \end{vmatrix}} = \textbf{2 V}$$

Applying the TI-86 calculator will result in the listing of Fig. 8.69, which is essentially equal to the solution of 2.0 V in the earlier examples.

---

det[[0.617,−0.25,4.8][−0.25,1.417,0][0,−1,0]]/det[[0.617,−0.25,0][−0.25,1.417,−1][0,−1,1.5]] (ENTER) 1.998

FIG. 8.69
Using the TI-86 calculator to find voltage V_3 for the network of Fig. 8.66.

8.9 NODAL ANALYSIS (FORMAT APPROACH)

Now that the fundamental steps leading to the nodal equations have been described in detail, we can turn our attention to a shorthand method for writing the equations. The procedure will match that introduced for mesh analysis in a number of ways. Mutual terms will again be present, and all the sources will be relegated to the other side of the equals sign. The general approach was first introduced to provide the equations that reveal how this shorthand method was developed. Further, the format approach is very mechanical in nature, with no verification that the results are correct. We can be sure that the approach is valid only because the results match those obtained in the previous section using the basic laws of electric circuits.

One major requirement of the format approach is that only current sources be present. The first step must therefore be to convert all voltage sources to current sources. If this is impossible because the internal resistance of the source was not provided, or if the source is considered ideal, it will be necessary to turn to the methods of Section 8.10.

To introduce the method, let us investigate the resulting equations for Example 8.19 in the previous section, repeated below for convenience. The network appears in Fig. 8.70.

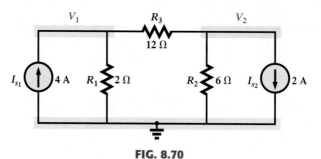

$$V_1\left(\frac{1}{R_1}+\frac{1}{R_3}\right) \qquad -V_2\left(\frac{1}{R_3}\right) \qquad = I_{s_1}$$

$$V_2\left(\frac{1}{R_2}+\frac{1}{R_3}\right) \qquad -V_1\left(\frac{1}{R_3}\right) \qquad = -I_{s_2}$$

FIG. 8.70
Network used to define the format approach to nodal analysis.

First note that column 1 contains a node of interest times the conductances ($1/R$) that are connected to that node. The second column is subtracted from the first and is equal to the product of the conductance connecting the two nodes and the other nodal voltage. The current sources connected to each node appear at the right of the equals sign and are given a plus sign if they supply current to the node and a negative sign if they draw current from the node. The above is essentially the basis for the format approach to be introduced in this section.

The details of all the steps listed below will be covered in detail in the examples to follow.

Nodal Analysis (Format Approach):

1. *Choose a reference node, and assign a subscripted nodal voltage to the remaining nodes of the network.*
2. *The number of required equations is equal to the number of resulting independent subscripted nodal voltages. Column 1 of each equation is formed by multiplying the subscripted nodal voltage of interest by the sum of the conductances connected to that node.*
3. *Column 2 contains the mutual terms which are always subtracted from the entries of the first column. Each mutual term is the product of the mutual conductance (the conductance connected between the two nodes) and the other nodal voltages.*
4. *The column to the right of the equals sign is the algebraic sum of the current sources connected to the node of interest. A current source is assigned a positive sign if it supplies current to the node, and a negative sign if it draws current from the node.*
5. *Solve the resulting simultaneous equations for the nodal voltages.*

Each step will be discussed in detail in the examples to follow.

EXAMPLE 8.22 Write the nodal equations for the network of Fig. 8.71, and compare your results with the solution to Example 8.20.

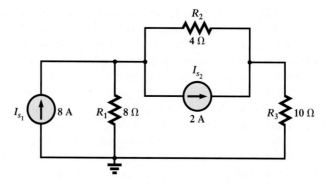

FIG. 8.71
Network to be analyzed in Example 8.22.

Solution:

Step 1: The network is redrawn as shown in Fig. 8.72 with the defined nodes.

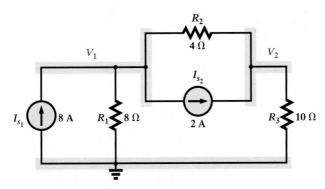

FIG. 8.72
Network of Fig. 8.71 with the defined nodal voltages.

Step 2: The first column is established by first multiplying the nodal voltage V_1 by the sum of the conductances that touch the node. In this case, resistors R_1 and R_2 are connected to the node and therefore are included as conductances $G_1 = 1/R_1$ and $G_2 = 1/R_2$. For nodal voltage V_2, resistors R_2 and R_3 are connected to the node and are included as conductance levels $G_2 = 1/R_2$ and $G_3 = 1/R_3$.

Col. 1

$$V_1\left(\frac{1}{R_1} + \frac{1}{R_2}\right)$$

$$V_2\left(\frac{1}{R_2} + \frac{1}{R_3}\right)$$

Step 3: There is a mutual conductance $1/R_2$ between nodes V_1 and V_2 that will appear in both equations as a mutual term subtracted from first column. The term is simply the product of the mutual conductance and the **other** node for each equation:

Col. 1   **Col. 2**

$$V_1\left(\frac{1}{R_1} + \frac{1}{R_2}\right) - V_2\left(\frac{1}{R_2}\right)$$

$$V_2\left(\frac{1}{R_2} + \frac{1}{R_3}\right) - V_1\left(\frac{1}{R_2}\right)$$

Step 4: For node V_1, the source current I_{s_1} supplies current to the node and is therefore included as a positive quantity to the right of the equals sign. The source I_{s_2} draws current from the same node and is therefore included with a negative sign. For node V_2, the only current source connected to the node is I_{s_2} which is supplying current to the node and thus will appear as a positive quantity:

Col. 1   **Col. 2**   **Col. 3**

$$V_1\left(\frac{1}{R_1} + \frac{1}{R_2}\right) - V_2\left(\frac{1}{R_2}\right) = I_{s_1} - I_{s_2}$$

$$V_2\left(\frac{1}{R_2} + \frac{1}{R_3}\right) - V_1\left(\frac{1}{R_2}\right) = I_{s_2}$$

The resulting equations are an exact match with the equations obtained in Example 8.20.

EXAMPLE 8.23 Write the nodal equations for the network of Fig. 8.73, and compare your results with the solution in Example 8.21.

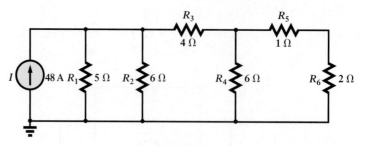

FIG. 8.73

Ladder network for Example 8.23.

Solution:

Step 1: The nodal voltages are defined in Fig. 8.74.

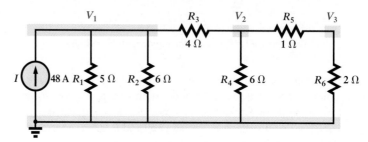

FIG. 8.74

Network of Fig. 8.73 with defined nodal voltages.

Step 2: For nodal voltage V_1, the resistors touching the node are R_1, R_2, and R_3, so the sum of their conductance values is multiplied by the nodal voltage to form the first column of the first equation. For nodal voltage V_2, the resistors connected to the node are R_3, R_4, and R_5, so the sum of their conductance values is multiplied by the nodal voltage to start the second equation. Finally, for nodal voltage V_3, the resistors connected to the node are R_5 and R_6, so the sum of their conductance values is multiplied by the nodal voltage to begin the third equation:

Col. 1

$$V_1\left(\frac{1}{R_1} + \frac{1}{R_2} + \frac{1}{R_3}\right)$$

$$V_2\left(\frac{1}{R_3} + \frac{1}{R_4} + \frac{1}{R_5}\right)$$

$$V_3\left(\frac{1}{R_5} + \frac{1}{R_6}\right)$$

Step 3: The mutual term between nodes V_1 and V_2 is the conductance formed by resistor R_3. The product of the conductance level and the other node will therefore appear with a negative sign in the equation for V_1. Similarly, the product of the same conductance and the other nodal voltage will appear in the second equation with a minus sign. Node V_2, however, has a second mutual term with nodal voltage V_3, so the product formed by the conductance $1/R_5$ and nodal voltage V_3 must also appear with a negative sign in the second equation. For node V_3, the only mutual term is the conductance formed by $1/R_5$, so the product of this conductance and nodal voltage V_2 will appear with a negative sign in the third equation:

Col. 1   **Col. 2**

$$V_1\left(\frac{1}{R_1} + \frac{1}{R_2} + \frac{1}{R_3}\right) - V_2\left(\frac{1}{R_3}\right)$$

$$V_2\left(\frac{1}{R_3} + \frac{1}{R_4} + \frac{1}{R_5}\right) - V_1\left(\frac{1}{R_3}\right) - V_3\left(\frac{1}{R_5}\right)$$

$$V_3\left(\frac{1}{R_5} + \frac{1}{R_6}\right) \qquad - V_2\left(\frac{1}{R_5}\right)$$

Step 4: The current source is connected only to node V_1, so it will appear only in the first equation. It has a positive sign because it supplies current to the node. A zero is entered for the other two equations to define the right side of the equation.

Col. 1	Col. 2	Col. 3

$$V_1\left(\frac{1}{R_1} + \frac{1}{R_2} + \frac{1}{R_3}\right) - V_2\left(\frac{1}{R_3}\right) \qquad\qquad = I$$

$$V_2\left(\frac{1}{R_3} + \frac{1}{R_4} + \frac{1}{R_5}\right) - V_1\left(\frac{1}{R_3}\right) - V_3\left(\frac{1}{R_5}\right) = 0$$

$$V_3\left(\frac{1}{R_5} + \frac{1}{R_6}\right) \qquad\quad - V_2\left(\frac{1}{R_5}\right) \qquad\quad = 0$$

The resulting equations are an exact match with those obtained in Example 8.21.

EXAMPLE 8.24 Using nodal analysis, find the voltage across resistor R_5 of Fig. 8.75.

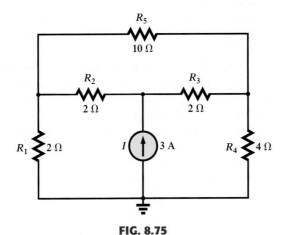

FIG. 8.75
Network to be analyzed in Example 8.24.

Solution: Although an ideal current source appears in the network (no internal resistance is provided), nodal analysis can be applied because it is a current source. The same can be said for ideal voltage sources when mesh analysis is applied. If the source were an ideal voltage source, we would have to turn to the methods of Section 8.10 for a solution. The nodal voltages are defined in Fig. 8.76.

Since we have now had some practice with the approach, let us write the entire equation for V_1. First, we find that resistors R_1, R_2, and R_5 are connected to node V_1. The first

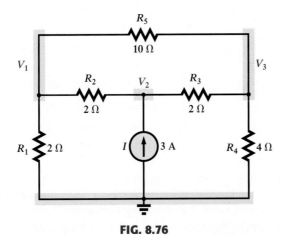

FIG. 8.76
Network of Fig. 8.75 with nodal voltages defined.

component of the equation will therefore be the product of nodal voltage V_1 and the sum of the conductance levels of the three resistors. Next, we find that the conductance $1/R_2$ is acting as a mutual term with nodal voltage V_2. The result is a negative term equal to the product of the conductance and nodal voltage V_2. Now, we must be careful because the conductance $1/R_5$ is creating a mutual term with nodal voltage V_3. Therefore, there will be two mutual terms as shown below. Finally, we find that the current source is not connected to node V_1, so a zero is entered to the right of the equals sign, resulting in the following equation:

$$V_1\left(\frac{1}{R_1} + \frac{1}{R_2} + \frac{1}{R_5}\right) - V_2\left(\frac{1}{R_2}\right) - V_3\left(\frac{1}{R_5}\right) = 0$$

For nodal voltage V_2, resistors R_2 and R_3 are connected to the node, so the sum of their conductance levels is multiplied by the nodal voltage to form the first term of the next equation. Two mutual terms will appear again because the conductance $1/R_2$ forms a bridge with V_1, and the conductance $1/R_3$ forms a bridge with V_3. Both terms are added with a negative sign as shown below. Finally, we must add the effect of the current source connected to the node. Because it is supplying current to the node, the source current will appear with a positive sign to the right of the equals sign:

$$V_2\left(\frac{1}{R_2} + \frac{1}{R_3}\right) - V_1\left(\frac{1}{R_2}\right) - V_3\left(\frac{1}{R_3}\right) = I$$

Finally, for node V_3, we find that resistors R_3, R_4, and R_5 are connected to the node, so the first term is the sum of their conductance levels multiplied by the nodal voltage. Next, we again find two mutual terms due to the conductance link of $1/R_3$ with V_2 and the conductance link of $1/R_5$ with V_1. They both appear with a negative sign as shown below. Since there are no source connections to node V_3, a zero is entered to the right of the equals sign:

$$V_3\left(\frac{1}{R_3} + \frac{1}{R_4} + \frac{1}{R_5}\right) - V_2\left(\frac{1}{R_3}\right) - V_1\left(\frac{1}{R_5}\right) = 0$$

The result is three equations and three unknowns:

$$V_1\left(\frac{1}{R_1} + \frac{1}{R_2} + \frac{1}{R_5}\right) - V_2\left(\frac{1}{R_2}\right) - V_3\left(\frac{1}{R_5}\right) = 0$$

$$V_2\left(\frac{1}{R_2} + \frac{1}{R_3}\right) - V_1\left(\frac{1}{R_2}\right) - V_3\left(\frac{1}{R_3}\right) = I$$

$$V_3\left(\frac{1}{R_3} + \frac{1}{R_4} + \frac{1}{R_5}\right) - V_2\left(\frac{1}{R_3}\right) - V_1\left(\frac{1}{R_5}\right) = 0$$

Substituting values will result in

$$V_1\left(\frac{1}{2\,\Omega} + \frac{1}{2\,\Omega} + \frac{1}{10\,\Omega}\right) - V_2\left(\frac{1}{2\,\Omega}\right) - V_3\left(\frac{1}{10\,\Omega}\right) = 0$$

$$V_2\left(\frac{1}{2\,\Omega} + \frac{1}{2\,\Omega}\right) - V_1\left(\frac{1}{2\,\Omega}\right) - V_3\left(\frac{1}{2\,\Omega}\right) = 3\text{ A}$$

$$V_3\left(\frac{1}{2\,\Omega} + \frac{1}{4\,\Omega} + \frac{1}{10\,\Omega}\right) - V_2\left(\frac{1}{2\,\Omega}\right) - V_1\left(\frac{1}{10\,\Omega}\right) = 0$$

Dropping units and rearranging will result in the following matrix form:

$$1.1V_1 - 0.5V_2 - 0.1V_3 = 0$$
$$-0.5V_1 + 1V_2 - 0.5V_3 = 3$$
$$-0.1V_1 - 0.5V_2 + 0.85V_3 = 0$$

Checking diagonals of the matrix, we find that the multiplier of V_2 in row 1 is the same as that for V_1 in row 2, and that the multiplier for V_3 in row 1 matches that of V_1 in row 3. Further, the multiplier for V_3 in row 2 matches that of V_2 in row 3. Since the diagonals check, there is an excellent chance that the equations were properly generated, and we should proceed with the mathematical analysis.

Defining V_{R_5} as $V_1 - V_3$ will require that we solve for both V_1 and V_3. Using the TI-86 calculator for V_1 will result in the display of Fig. 8.77. Using the calculator for V_3 will result in the display of Fig. 8.78.

det[[0,−0.5,−0.1][3,1,−0.5][0,−0.5,0.85]]/det[[1.1,−0.5,−0.1][−0.5,1,−0.5][−0.1,−0.5,0.85]] (ENTER) 3.677

FIG. 8.77
Using the TI-86 calculator to determine V_1 for the network of Fig. 8.76.

det[[1.1,−0.5,0][−0.5,1,3][−0.1,−0.5,0]]/det[[1.1,−0.5,−0.1][−0.5,1,−0.5][−0.1,−0.5,0.85]] (ENTER) 4.645

FIG. 8.78
Using the TI-86 calculator to determine V_3 for the network of Fig. 8.76.

The voltage across R_5 is therefore

$$V_{R_5} = V_1 - V_3 = 3.677 \text{ V} - 4.645 \text{ V} = \textbf{−0.968 V}$$

with the minus sign simply indicating that the polarity is opposite to that assumed since V_3 is greater than V_1.

Mathcad Solution: As shown in Fig. 8.79, Mathcad can also be used to work directly with the nodal equations just developed. There is no need to apply determinants once the equations are entered. Simply enter **Find(V1,V2,V3)**, and the results for all three voltages will appear immediately—now that is an amazing savings in time. Simply write the equations in a few minutes using the format method, and obtain your solution using Mathcad in a few minutes more—that is fast. Try doing the entire problem longhand and see how long it takes. The values of 1 for each of the nodal voltages at the top of the printout is just a **Guess** value required by Mathcad. The **Given** is required to inform the computer that the simultaneous equations necessary to find the solution are now being entered.

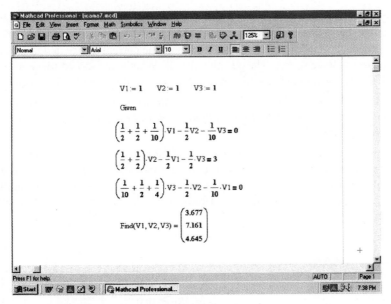

FIG. 8.79
Using Mathcad to verify the mathematical calculations of Example 8.24.

8.10 SPECIAL CASES FOR MESH AND NODAL ANALYSIS

On occasion, you might run into a snag when trying to apply mesh or nodal analysis. The most common cause is the presence of an ideal source that does not include an internal resistance to permit a conversion from one form to the other. Recall that our discussion of mesh analysis required that all the sources be voltage sources, and nodal analysis required that all the sources be current sources. This section will provide two ways to get around this problem. The first, using approximate conversions, is easier to apply, but it requires a certain amount of diligence regarding the level of accuracy in the mathematical equations. The second method, using a supermesh current or a supernode, is "more accurate" but requires a unique set of maneuvers to apply.

Note also that the first method can use the format approach while the second cannot.

Mesh Analysis

Approximate Conversions For discussion purposes, let us assume that mesh analysis is to be applied to the network of Fig. 8.80. The current source cannot be converted to a voltage source because the current source does not have a parallel resistor (which may be the internal resistance of the supply) to apply in the conversion equations.

Leaving the network as is, the mesh currents were defined as shown in Fig. 8.81. If we now tried to apply mesh analysis, how would the voltage across the current source be included in the mesh equations? It could be defined as simply V_s, but this would add another unknown and the need for an additional equation relating the mesh currents to the source voltage.

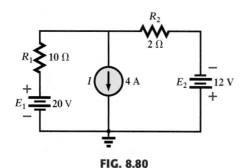

FIG. 8.80

Network with an ideal current source.

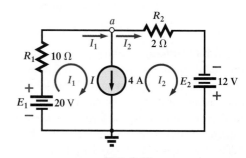

FIG. 8.81

Defining the mesh currents for the network of Fig. 8.80.

The fact that the current source appears without a parallel internal resistance suggests that the internal resistance is so large that it can be ignored. The solution, therefore, might be to simply place a large resistance in parallel with the current source and proceed with a conversion to a voltage source. The question then is, What value should we choose for this resistance so that we don't change the overall behavior of the network? In general,

inserting a resistor in parallel with an ideal current source that is at least 10 times larger than any resistor in the neighboring network will not adversely affect the behavior of the network.

For improved accuracy, you may want to use a factor greater than 10:1, but be aware that this will require holding on to a higher level of accuracy throughout the calculations. For this example, the largest resistor is 10 Ω, so a resistor of 100 Ω was placed in parallel with the current source as shown in Fig. 8.82. Converting to a voltage source will result in the network of Fig. 8.83.

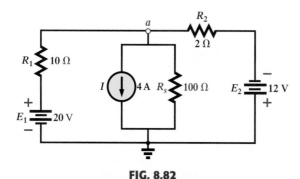

FIG. 8.82

Introducing a source resistor of 100 Ω in parallel with the current source of Fig. 8.81.

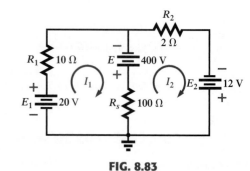

FIG. 8.83

Network of Fig. 8.82 following the conversion of the current source to a voltage source.

Applying the format approach to mesh analysis will result in the following equations:

$$I_1(10\ \Omega + 100\ \Omega) - I_2(100\ \Omega) = 20\ \text{V} + 400\ \text{V}$$

$$I_2(100\ \Omega + 2\ \Omega) - I_1(100\ \Omega) = -400\ \text{V} + 12\ \text{V}$$

or

$$110I_1 - 100I_2 = 420$$

$$-100I_1 + 102I_2 = -388$$

Using determinants, the mesh current through E_1 and R_1 will be

$$I_1 = \frac{\begin{vmatrix} 420 & -100 \\ -388 & 102 \end{vmatrix}}{\begin{vmatrix} 110 & -100 \\ -100 & 102 \end{vmatrix}} = \frac{4.04}{1.22} = \mathbf{3.31\ A}$$

a result that is a very close match with the exact solution of 3.33 A.

The result of the above is that anytime you encounter an ideal current source in a network to which you want to apply mesh analysis, simply introduce a large resistance in parallel with the current source, and continue as before with the format approach.

Let us now examine the second approach and compare solutions.

Supermesh Current   For this method, the mesh currents are first defined as shown in Fig. 8.81. Next the current source is removed (the network redrawn) as shown in Fig. 8.84, and Kirchhoff's voltage law is applied to the resulting network. The single path, which now includes the effects of two mesh currents (for what appears to be a series circuit), is referred to as the path of a **supermesh current.** It is certainly a "super something" in the sense that we have two mesh currents defined for the same series path and we know that the current in a series circuit must be the same everywhere. Remember, however, that this is an approach—a "trick" or "maneuver," if you prefer—to obtain equations that will provide a correct solution.

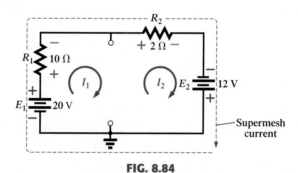

FIG. 8.84
Defining the supermesh current.

Applying Kirchhoff's voltage law around the supermesh loop defined in Fig. 8.84 will result in the following equation using both of the defined mesh currents:

$$20\ \text{V} - I_1(10\ \Omega) - I_2(2\ \Omega) + 12\ \text{V} = 0$$

or

$$10I_1 + 2I_2 = 32$$

The result is one equation and two unknowns. The second equation is obtained through an application of Kirchhoff's current law at node a of Fig. 8.81 (noting the directions of I_1 and I_2 near the node):

$$I_1 = I + I_2 = 4\ \text{A} + I_2$$

The result is two equations and two unknowns:

$$10I_1 + 2I_2 = 32$$

$$I_1 - I_2 = 4$$

Applying determinants will then result in

$$I_1 = \frac{\begin{vmatrix} 32 & 2 \\ 4 & -1 \end{vmatrix}}{\begin{vmatrix} 10 & 2 \\ 1 & -1 \end{vmatrix}} = \frac{40}{12} = \textbf{3.33 A}$$

an exact solution that is a very close match with the approximate solution obtained above.

As a second example using the supermesh current approach, let us determine the mesh currents for the network of Fig. 8.85. At first glance we find that the sources each have a parallel resistance, permitting the conversion to voltage sources. However, once the network is solved, we would have to work backwards to find the current through resistors R_1 and R_2. Using mesh analysis on the network as it is will also give us some practice in using the supermesh current approach.

The mesh currents were defined in Fig. 8.85. If we remove the current sources, the network of Fig. 8.86 will result, and Kirchhoff's current law can be applied as follows to the supermesh path:

$$-V_1 - V_3 - V_2 = 0$$

However,
$$V_1 = (I_2 - I_1)R_1$$

with
$$V_3 = I_2R_3 \quad \text{and} \quad V_2 = (I_2 - I_3)R_2$$

so that
$$-(I_2 - I_1)R_1 - I_2R_3 - (I_2 - I_3)R_2 = 0$$

or
$$I_1R_1 - I_2(R_1 + R_2 + R_3) + I_3R_2 = 0$$

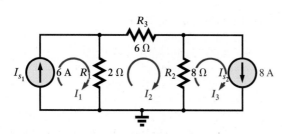

FIG. 8.85

Network to be analyzed using the supermesh approach.

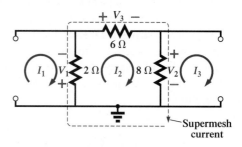

FIG. 8.86

Defining the mesh current for the network of Fig. 8.85.

Substituting values and removing units will result in

$$2I_1 - 16I_2 + 8I_3 = 0$$

From Fig. 8.85, we can see that the mesh currents pass right through the current sources, so

$$I_1 = I_{s_1} = 6 \text{ A} \quad \text{and} \quad I_3 = I_{s_2} = 8 \text{ A}$$

which, when substituted into the above equation, will result in

$$2(6 \text{ A}) - 16I_2 + 8(8 \text{ A}) = 0$$

$$12 \text{ A} - 16I_2 + 64 \text{ A} = 0$$

$$I_2 = \frac{76}{16} \text{ A} = \textbf{4.75 A}$$

Then
$$I_{R_1}(\downarrow) = I_1 - I_2 = 6 \text{ A} - 4.75 \text{ A} = \textbf{1.25 A}$$

and
$$I_{R_2}(\uparrow) = I_3 - I_2 = 8 \text{ A} - 4.75 \text{ A} = \textbf{3.25 A}$$

Now we can take a look at the impact of an ideal voltage source on the application of nodal analysis.

Nodal Analysis

Approximate Conversions For nodal analysis, the problem is usually an ideal voltage source that cannot be converted to a current source. The appearance of an ideal voltage source suggests that the series internal resistance is so small that it can be ignored. The solution, therefore, may be to simply add a small resistance in series with the voltage source so that the conversion can be made. However, the question is, How small do we make the resistance so that it will not affect the overall response of the network? In general,

inserting a resistor in series with an ideal voltage source that is at least 10 times smaller than any resistor in the neighboring network will not adversely affect the behavior of the network.

Take special note of the words *at least* in the above statement. The larger the factor, the closer the solution will be to the exact solution. This will demonstrated in the example to follow.

The network of Fig. 8.87 has an ideal voltage source between the two defined nodes. Since the smallest resistor is 2 Ω, a resistor 10 times smaller would be 0.2 Ω. To improve our accuracy level and make the mathematics somewhat easier, we will use 0.1 Ω, as shown in Fig. 8.88.

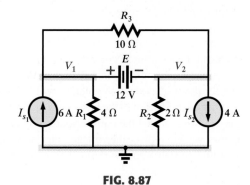

FIG. 8.87
Network with an ideal voltage source.

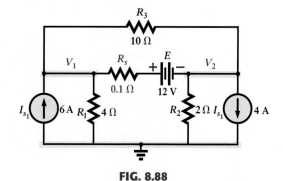

FIG. 8.88
Network of Fig. 8.87 with the addition of a source resistance in series with the ideal voltage source.

Converting to a current source will result in the network of Fig. 8.89. Applying the format approach to nodal analysis will result in the following equations:

$$V_1\left(\frac{1}{R_1} + \frac{1}{R_s} + \frac{1}{R_3}\right) - V_2\left(\frac{1}{R_s}\right) - V_2\left(\frac{1}{R_3}\right) = I_{s_1} + I_s$$

$$V_2\left(\frac{1}{R_2} + \frac{1}{R_s} + \frac{1}{R_3}\right) - V_1\left(\frac{1}{R_s}\right) - V_1\left(\frac{1}{R_3}\right) = -I_{s_2} - I_s$$

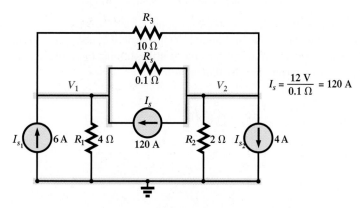

FIG. 8.89
Network of Fig. 8.88 after converting the voltage source to a current source.

Substituting values:

$$V_1\left(\frac{1}{4\,\Omega} + \frac{1}{0.1\,\Omega} + \frac{1}{10\,\Omega}\right) - V_2\left(\frac{1}{0.1\,\Omega}\right) - V_2\left(\frac{1}{10\,\Omega}\right) = 6\,A + 120\,A$$

$$V_2\left(\frac{1}{2\,\Omega} + \frac{1}{0.1\,\Omega} + \frac{1}{10\,\Omega}\right) - V_1\left(\frac{1}{0.1\,\Omega}\right) - V_1\left(\frac{1}{10\,\Omega}\right) = -4\,A - 120\,A$$

Combining terms and rearranging:

$$10.35V_1 - 10.1V_2 = 126$$
$$-10.1V_1 + 10.6V_2 = -124$$

Using determinants:

$$V_1 = \frac{\begin{vmatrix} 126 & -10.1 \\ -124 & +10.6 \end{vmatrix}}{\begin{vmatrix} 10.35 & -10.1 \\ -10.1 & +10.6 \end{vmatrix}} = \mathbf{10.909\ V} \qquad V_2 = \frac{\begin{vmatrix} 10.35 & 126 \\ -10.1 & -124 \end{vmatrix}}{\begin{vmatrix} 10.35 & -10.1 \\ -10.1 & +10.6 \end{vmatrix}} = \mathbf{-1.429\ V}$$

which are reasonably close to the exact solutions of $V_1 = 10.667$ V and $V_2 = -1.333$ V.

For the future, therefore, if a good approximation to the actual voltages is desired, simply add a very small resistor in series with the source, and continue as before with the format approach. Now for the second approach to nodal analysis.

Supernode For mesh analysis we had the supermesh current. For nodal analysis a **supernode** is defined. When we introduced the supermesh current, we found that two mesh currents were defined for the same series configuration—a contradiction to the basic laws of electric circuits. We will now find that for the supernode, two voltages will be defined for a parallel combination of elements—another contradiction to the basic laws of electric circuits. For ideal current sources in mesh analysis, we replaced the current sources by open circuits. For ideal voltage sources in nodal analysis, we will replace the voltage sources by short-circuit equivalents as shown in Fig. 8.90. Note how the supernode is defined in the figure.

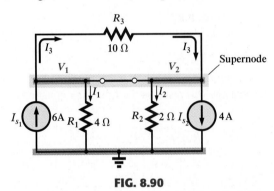

FIG. 8.90

Defining the supernode for the network of Fig. 8.87.

Since Kirchhoff's current law must now be applied to the supernode, the current for each branch must be defined as shown in Figure 8.90. In particular, note that current I_3 will leave the supernode at V_1 and return to the supernode at V_2. It must therefore appear twice when applying Kirchhoff's current law, as shown below:

$$\Sigma I_i = \Sigma I_o$$
$$I_{s_1} + I_3 = I_1 + I_2 + I_{s_2} + I_3$$

or

$$I_1 + I_2 = I_{s_1} - I_{s_2}$$

Since

$$I_1 = \frac{V_1}{R_1} \quad \text{and} \quad I_2 = \frac{V_2}{R_2}$$

we find

$$\frac{V_1}{R_1} + \frac{V_2}{R_2} = I_{s_1} - I_{s_2}$$

Now, returning to Fig. 8.87, we also know that

$$V_1 - V_2 = E$$

The result is two equations and two unknowns:

$$\frac{V_1}{R_1} + \frac{V_2}{R_2} = I_{s_1} - I_{s_2}$$
$$V_1 - V_2 = E$$

Substituting values, rearranging, and dropping units will then result in

$$0.25V_1 + 0.5V_2 = 2$$
$$V_1 - V_2 = 12$$

Solving for V_1 and V_2 using determinants will result in

$$V_1 = \frac{\begin{vmatrix} 2 & 0.5 \\ 12 & -1 \end{vmatrix}}{\begin{vmatrix} 0.25 & 0.5 \\ 1 & -1 \end{vmatrix}} = \textbf{10.667 V} \qquad V_2 = \frac{\begin{vmatrix} 0.25 & 2 \\ 1 & 12 \end{vmatrix}}{\begin{vmatrix} 0.25 & 0.5 \\ 1 & -1 \end{vmatrix}} = \textbf{-1.333 V}$$

which are the exact solutions for the network of Fig. 8.87.

The result is that we now have two approaches for networks with ideal sources that present problems for the method to be applied. One approach is quite straightforward but also provides approximate results that are a function of the chosen series resistor for the source. The other approach is more difficult to apply but provides an exact solution.

8.11 BRIDGE NETWORKS

This section introduces the **bridge network,** a configuration with a multitude of applications in both the dc and the ac environments. You will encounter this configuration early in electronics courses in the discussion of rectifying circuits employed in converting a varying signal to one of a steady nature (such as dc). It belongs in the family of **complex networks** because no two elements of the configuration are in series or parallel.

The bridge network can appear in one of the three forms appearing in Fig. 8.91. The network of Fig. 8.91(c) is also called a *symmetrical lattice* network if $R_2 = R_3$ and $R_1 = R_4$. Figure 8.91(c) is also an excellent example of how a planar (one surface) network can appear to be nonplanar (overlapping surfaces).

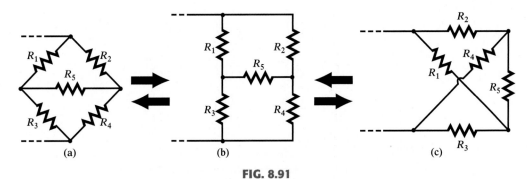

FIG. 8.91

Various formats for a bridge network.

The bridge configuration can be analyzed using either mesh or nodal analysis. For practice, let us apply both methods to the network of Fig. 8.92. Using the format approach for mesh analysis, the following equations can be written directly from the network:

$$I_1(3\,\Omega + 4\,\Omega + 2\,\Omega) - I_2(4\,\Omega) - I_3(2\,\Omega) = 20\ \text{V}$$
$$I_2(4\,\Omega + 5\,\Omega + 2\,\Omega) - I_1(4\,\Omega) - I_3(5\,\Omega) = 0$$
$$I_3(2\,\Omega + 5\,\Omega + 1\,\Omega) - I_1(2\,\Omega) - I_2(5\,\Omega) = 0$$

which will result in the following:

$$9I_1 - 4I_2 - 2I_3 = 20$$
$$-4I_1 + 11I_2 - 5I_3 = 0$$
$$-2I_1 - 5I_2 + 8I_3 = 0$$

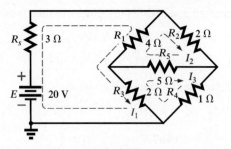

FIG. 8.92

Assigning the mesh currents to the network of Fig. 8.91.

Note the satisfaction of the diagonal criteria introduced for mesh and nodal analysis. The multiplier for I_2 in the first row is the same as that for I_1 in the second row, and the multiplier for I_3 in the first row is the same as that for I_1 in the third row. Finally, the multiplier for I_3 in the second row is the same as that for I_2 in the third row.

Applying determinants, we will find that

$$I_1 = \textbf{4 A}$$

$$I_2 = \textbf{2.667 A}$$

$$I_3 = \textbf{2.667 A}$$

It is particularly interesting that the mesh currents I_2 and I_3 are the same. The effect will be for them to cancel out for the resistor R_5, and

$$I_{R_5} = I_2 - I_3 = 2.667 \text{ A} - 2.667 \text{ A} = \textbf{0 A}$$

If the current through the bridge arm is 0 A, the bridge network is said to be balanced.

We will discuss the balanced condition in detail later in this section.

If the voltage source is converted to a current source as shown in Fig. 8.93, we can use the format approach to nodal analysis and obtain the following equations. Again, they were obtained directly from the network with little difficulty:

$$V_1\left(\frac{1}{3\ \Omega} + \frac{1}{4\ \Omega} + \frac{1}{2\ \Omega}\right) - V_2\left(\frac{1}{4\ \Omega}\right) - V_3\left(\frac{1}{2\ \Omega}\right) = 6.667 \text{ A}$$

$$V_2\left(\frac{1}{4\ \Omega} + \frac{1}{2\ \Omega} + \frac{1}{5\ \Omega}\right) - V_1\left(\frac{1}{4\ \Omega}\right) - V_3\left(\frac{1}{5\ \Omega}\right) = 0$$

$$V_3\left(\frac{1}{5\ \Omega} + \frac{1}{2\ \Omega} + \frac{1}{1\ \Omega}\right) - V_2\left(\frac{1}{5\ \Omega}\right) - V_1\left(\frac{1}{2\ \Omega}\right) = 0$$

which will result in the following:

$$1.083V_1 - 0.25V_2 - 0.5V_3 = 6.667 \text{ A}$$
$$-0.25V_1 + 0.950V_2 - 0.2V_3 = 0$$
$$-0.5V_1 - 0.2V_2 + 1.7V_3 = 0$$

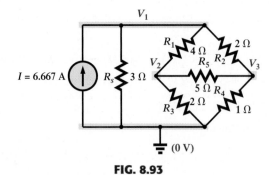

FIG. 8.93

Defining the nodal voltages for the network of Fig. 8.91.

$$V_1 = \mathbf{8\ V}$$
$$V_2 = \mathbf{2.667\ V}$$
$$V_3 = \mathbf{2.667\ V}$$

Since $V_2 = V_3$, the voltage across resistor R_5 will be

$$V_{R_5} = V_2 - V_3 = 2.667\ \text{V} - 2.667\ \text{V} = \mathbf{0\ V}$$

If the voltage across the bridge arm is 0 V, the bridge network is said to be balanced.

Obviously, if the current through R_5 is 0 A, the voltage across the resistor will be 0 V also as determined by Ohm's law. In every sense of the word, therefore, the bridge network of Fig. 8.92 is balanced. All you would have to do is change one of the resistor values, and the bridge would no longer be balanced.

Since the current through the bridge arm is 0 A, an open circuit can be used to replace resistor R_5, and the network can be redrawn as shown in Fig. 8.94(a). The network can then be redrawn as shown in Fig. 8.94(b) and analyzed using series-parallel techniques.

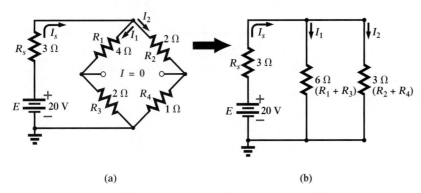

(a)   (b)

FIG. 8.94

Substituting the open-circuit equivalent for the balance arm of a balanced bridge.

However, since the voltage across the bridge arm is 0 V, the bridge arm can also be replaced by a short-circuit equivalent as shown in Fig. 8.95(a). The network can then be redrawn as shown in Fig. 8.95(b) and analyzed using series-parallel techniques.

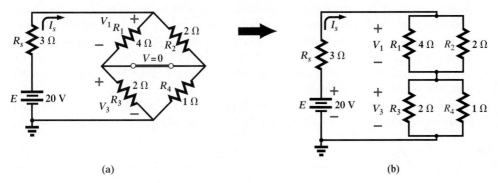

(a)   (b)

FIG. 8.95

Substituting the short-circuit equivalent for the balance arm of a balanced bridge.

When balanced, therefore, the network can be analyzed in the form of Fig. 8.92, Fig. 8.94, or Fig. 8.95, and the same voltage and current will be determined for each element of the network.

The question then remains as to what relationship between resistor values will establish the balanced condition. The answer is quite simple and one that is not hard to remember. If the following ratio of resistors is satisfied, the bridge is balanced:

$$\boxed{\dfrac{R_1}{R_3} = \dfrac{R_2}{R_4}} \qquad\qquad \textbf{(8.1)}$$

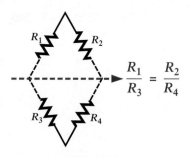

FIG. 8.96

A visual approach to remembering the balance condition.

The relationship can be remembered by simply making a mental image of Fig. 8.96, which shows that the ratios are the same as they appear in the diagram. That is, in the diagram, R_1 is above R_3, and R_2 is above R_4. If this relationship is *not satisfied*, the bridge network is *not balanced*, and a current will exist through R_5, causing a voltage drop across it.

EXAMPLE 8.25 The bridge of Fig. 8.97 is balanced. Determine the unknown resistance R_{unk}.

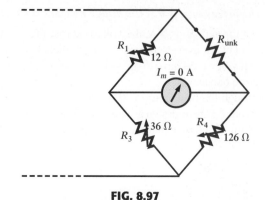

FIG. 8.97

Using a balanced bridge network to determine an unknown resistance (Example 8.25).

Solution: Applying Eq. (8.1):

$$\frac{R_1}{R_3} = \frac{R_2}{R_4} \Rightarrow \frac{12\ \Omega}{36\ \Omega} = \frac{R_{unk}}{126\ \Omega}$$

and

$$R_{unk} = 126\ \Omega \left(\frac{12\ \Omega}{36\ \Omega} \right) = \mathbf{42\ \Omega}$$

8.12 Δ-Y AND Y-Δ CONVERSIONS

Although we now have the ability to analyze almost any network using mesh or nodal analysis, there are times when it helps to change the network from one type to another. In fact, the configurations to be discussed in this section are such that the replacement process can change a network from one that is complex to one that can be solved using series-parallel techniques.

Two circuit configurations that often account for these difficulties are the **wye (Y)** and **delta (Δ)** configurations, depicted in Fig. 8.98(a). They are also referred to as the **tee (T)** and **pi (π)** configurations, respectively, as indicated in Fig. 8.98(b). Note that the pi configuration is actually an inverted delta.

In Fig. 8.99, the Y configuration can be changed to the Δ configuration (actually an inverted delta), and vice versa, without the network connected to points a, b, and c having any

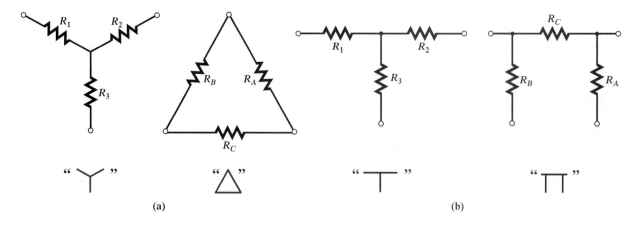

(a)   (b)

FIG. 8.98

The Y (T) and Δ (π) configurations.

idea the change was made. That is, if points *a*, *b*, and *c* are three points of a network, *either* configuration can be inserted between the three points without changing any of the voltages or currents in the surrounding network. The trick is to know the resistor values that will accomplish this type of replacement.

For the Y configuration of Fig. 8.99 to be equivalent to the Δ configuration between points *a*, *b*, and *c*, the resistors of the Y configuration must be determined from the resistors of the Δ using the following equations:

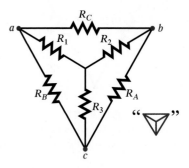

$$R_1 = \frac{R_B R_C}{R_A + R_B + R_C}$$

$$R_2 = \frac{R_A R_C}{R_A + R_B + R_C}$$ **(8.2)**

$$R_3 = \frac{R_A R_B}{R_A + R_B + R_C}$$

FIG. 8.99

Introducing the concept of Δ-Y or Y-Δ conversions.

In particular, note that

each resistor of the Y is equal to the product of the resistors in the two closest branches of the Δ divided by the sum of the resistors in the Δ.

For the Δ configuration of Fig. 8.99 to be equivalent to the Y configuration between points *a*, *b*, and *c*, the resistors of the Δ configuration must be determined from the resistors of the Y using the following equations:

$$R_A = \frac{R_1 R_2 + R_1 R_3 + R_2 R_3}{R_1}$$

$$R_B = \frac{R_1 R_2 + R_1 R_3 + R_2 R_3}{R_2}$$ **(8.3)**

$$R_C = \frac{R_1 R_2 + R_1 R_3 + R_2 R_3}{R_3}$$

In particular, note that

each resistor of the Δ is equal to the sum of the possible product combinations of the Y taken two at a time, divided by the resistance of the Δ farthest from the resistor to be determined.

There are some special situations to be aware of:

If all the resistors of the Δ or Y are the same, then the equivalent Y or Δ, respectively, will have three equal values.

If two of the resistors of the Δ or Y are the same, then two of the equivalent Y or Δ, respectively, will be the same.

If all the resistors of one or the other form are the same, the conversion equations for the resistors are as follows:

$$R_Y = \frac{R_\Delta}{3}$$ **(8.4)**

$$R_\Delta = 3R_Y$$

EXAMPLE 8.26 Convert the Y configuration of Fig. 8.100 to a Δ configuration.

Solution: Using Eq. (8.3):

$$R_A = \frac{R_1 R_2 + R_1 R_3 + R_2 R_3}{R_1} = \frac{(18\ \Omega)(18\ \Omega) + (18\ \Omega)(6\ \Omega) + (18\ \Omega)(6\ \Omega)}{18\ \Omega}$$

$$= \frac{540}{18}\Omega = \mathbf{30\ \Omega}$$

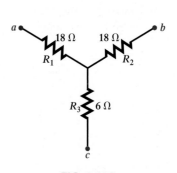

FIG. 8.100

Y configuration to be converted to a Δ configuration in Example 8.26.

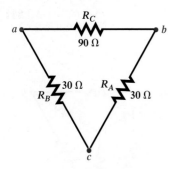

FIG. 8.101
Equivalent Δ configuration for the Y configuration of Fig. 8.100.

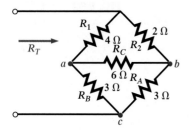

FIG. 8.102
Bridge network for Example 8.27.

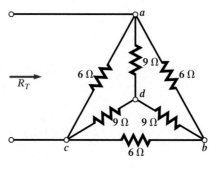

FIG. 8.104
Replacing the Δ of Fig. 8.103 by its equivalent Y configuration.

$$R_B = \frac{R_1R_2 + R_1R_3 + R_2R_3}{R_2} = \frac{540}{18} \ \Omega$$

$$= \mathbf{30 \ \Omega}$$

$$R_C = \frac{R_1R_2 + R_1R_3 + R_2R_3}{R_3} = \frac{540}{6} \ \Omega$$

$$= \mathbf{90 \ \Omega}$$

The equivalent network between points *a*, *b*, and *c* appears in Fig. 8.101. Note that since two of the resistors of the Y configuration are equal, two resistors of the equivalent Δ are also equal—a good check on the calculations.

EXAMPLE 8.27 Given the bridge configuration of Fig. 8.102, find the total resistance R_T.

Solution: Here we have a complex configuration in which no two elements are in series or parallel. The methods of Chapter 7 cannot be applied. However, if we replace the Δ configuration at the bottom of the bridge by a Y configuration, resistors will then be in series and parallel and can be combined.

Since all the resistors of the lower Δ of Fig. 8.103 are the same, the three resistors of the Y will also be the same and determined by Eq. (8.4). That is,

$$R_Y = \frac{R_\Delta}{3} = \frac{9 \ \Omega}{3} = 3 \ \Omega$$

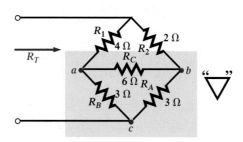

FIG. 8.103
Defining the Δ of Fig. 8.102 to be converted to a Y configuration.

Substituting the equivalent Y configuration in the network will result in the configuration of Fig. 8.104. Resistors R_1 and R'_1 and R_2 and R'_2 are now in series. The result is

$$R_1 + R'_1 = 4 \ \Omega + 3 \ \Omega = 7 \ \Omega$$

$$R_2 + R'_2 = 2 \ \Omega + 3 \ \Omega = 5 \ \Omega$$

and

$$R_T = 7 \ \Omega \| 5 \ \Omega + R'_3 = 2.917 \ \Omega + 3 \ \Omega = \mathbf{5.917 \ \Omega}$$

EXAMPLE 8.28 Find the total resistance R_T of the configuration of Fig. 8.105.

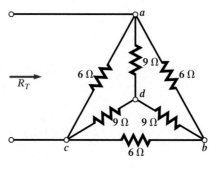

FIG. 8.105
Complex network for Example 8.28.

Solution: Again, although six resistors are present, no two are in series or parallel. The only option is to use a Δ-Y or a Y-Δ conversion; either path will yield the same result.

Using a Y-Δ conversion will result in the configuration of Fig. 8.106(a). Since all the resistors of the Y are the same, we can use Eq. (8.3):

$$R_\Delta = 3R_Y = 3(9\ \Omega) = 27\ \Omega$$

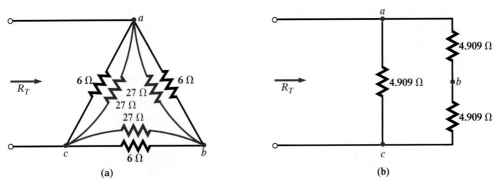

FIG. 8.106

Inserting the Δ equivalent for the Y configuration of Fig. 8.105.

Each branch of one Δ is in parallel with the corresponding branch of the other Δ. The parallel combination for each branch is

$$R'_T = 6\ \Omega\ \|\ 27\ \Omega = \frac{(6\ \Omega)(27\ \Omega)}{6\ \Omega + 27\ \Omega} = 4.909\ \Omega$$

resulting in the configuration of Fig. 8.106(b). The total resistance is then

$$R_T = R'_T\ \|\ (R'_T + R'_T) = R'_T\ \|\ 2R'_T$$

$$= \frac{(R'_T)(2R'_T)}{R'_T + 2R'_T} = \frac{2R'^2_T}{3R'^2_T} = \frac{2}{3}R'_T$$

$$= \frac{2}{3}(4.909\ \Omega) = \mathbf{3.273\ \Omega}$$

If we decided to convert the Δ configuration to a Y, the network of Fig. 8.107(a) would result. Each branch of one Y is now in parallel with the corresponding branch of the other Y with a parallel resistance of

$$R'_T = 2\ \Omega\ \|\ 9\ \Omega = \frac{(2\ \Omega)(9\ \Omega)}{2\ \Omega + 9\ \Omega} = 1.636\ \Omega$$

resulting in the configuration of Fig. 8.107(b).

The total resistance is then

$$R_T = R'_T + R'_T = 2\ R'_T = 2(1.636\ \Omega) = \mathbf{3.273\ \Omega}$$

which is a match with the above solution.

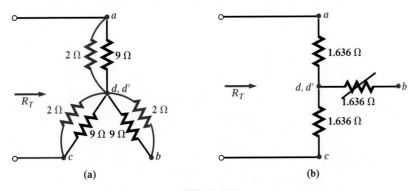

FIG. 8.107

Inserting the Y equivalent for the Δ configuration of Fig. 8.105.

8.13 APPLICATIONS

Constant Current Alarm Systems

The basic components of an alarm system employing a constant current supply are provided in Fig. 8.108. The design is improved over that provided in Chapter 5 in the sense that it is less sensitive to changes in resistance in the circuit due to heating, humidity, changes in the length of the line to the sensors, and so on. The 1.5 kΩ rheostat (total resistance between points *a* and *b*) is adjusted to ensure a current of 5 mA through the single-series security circuit. The adjustable rheostat is necessary to compensate for variations in the total resistance of the circuit introduced by the resistance of the wire, sensors, sensing relay, and milliammeter. The milliammeter is included to set the rheostat and ensure a current of 5 mA.

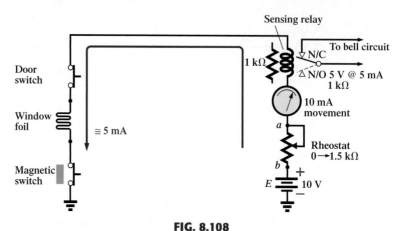

FIG. 8.108
Constant current alarm system.

If any of the sensors should open, the current through the entire circuit will drop to zero, the coil of the relay will release the plunger, and contact will be made with the N/C position of the relay. This action will complete the circuit for the bell circuit, and the alarm will sound. For the future, keep in mind that switch positions for a relay are always shown with no power to the network, resulting in the N/C position of Fig. 8.108. When power is applied, the switch will have the position indicated by the dashed line. That is, various factors, such as a change in resistance of any of the elements due to heating, humidity, and so on, would cause the applied voltage to redistribute itself and create a sensitive situation. With an adjusted 5 mA, the loading can change, but the current will always be 5 mA and the chance of tripping reduced. Take note of the fact that the relay is rated as 5 V at 5 mA, indicating that in the on state the voltage across the relay is 5 V and the current through the relay 5 mA. Its internal resistance is therefore 5 V/5 mA = 1 kΩ in this state.

A more advanced alarm system using a constant current is provided in Fig. 8.109. In this case, an electronic system employing a single transistor, biasing resistors, and a dc battery are establishing a current of 4 mA through the series sensor circuit connected to the positive side of an operational amplifier (op-amp). Although the transistor and op-amp devices may be new to you, they will be discussed in detail in your electronics courses—you do not

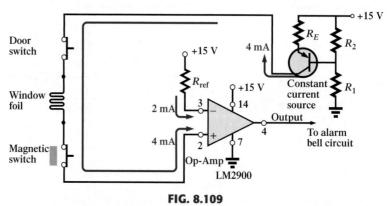

FIG. 8.109
Constant current alarm system with electronic components.

need to be aware of the details of their operation for now. Suffice it to say for the moment that the transistor in this application is being used not as an amplifier but as part of a design to establish a constant current through the circuit. The op-amp is a very useful component of numerous electronic systems, and it has important terminal characteristics established by a variety of components internal to its design.

The LM2900 operational amplifier of Fig. 8.109 is one of four found in the dual-in-line integrated circuit package appearing in Fig. 8.110(a). Pins 2, 3, 4, 7, and 14 were used for the design of Fig. 8.109. Note in Fig. 8.110(b) the number of elements required to establish the desired terminal characteristics—the details of which will be investigated in your electronics courses.

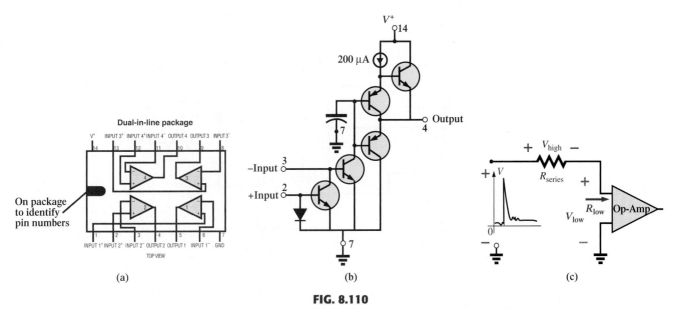

FIG. 8.110

LM2900 operational amplifier: (a) dual-in-line package (DIP); (b) components; (c) impact of low input impedance.

In Fig. 8.109, the designed 15 V dc supply, biasing resistors, and transistor in the upper right-hand corner of the schematic establish a constant 4 mA current through the circuit. It is referred to as a *constant current source* because the current will remain fairly constant at 4 mA even though there may be moderate variations in the total resistance of the series sensor circuit connected to the transistor. Following the 4 mA through the circuit, we find that it enters terminal 2 (the positive side of the input) of the op-amp. A second current of 2 mA, called the *reference current,* is established by the 15 V source and resistance R and enters terminal 3 (the negative side of the input) of the op-amp. The reference current of 2 mA is necessary to establish a current for the 4 mA current of the network to be compared against. As long as the 4 mA current exists, the operational amplifier will provide a "high" output voltage that exceeds 13.5 V, with a typical level of 14.2 V (according to the specification sheet for the op-amp). However, if the sensor current drops from 4 mA to a level below the reference level of 2 mA, the op-amp will respond with a "low" output voltage that is typically about 0.1 V. The output of the operational amplifier will then signal the alarm circuit about the disturbance. Note from the above that it is not necessary for the sensor current to drop to 0 mA to signal the alarm circuit—just a variation around the reference level that appears unusual.

One very important characteristic of this particular op-amp is that the input impedance to the op-amp is relatively low. This feature is important because you don't want alarm circuits reacting to every voltage spike or turbulence that comes down the line because of external switching action or outside forces such as lightning. In Fig. 8.110(c), for instance, if a high voltage should appear at the input to the series configuration, most of the voltage will be absorbed by the series resistance of the sensor circuit rather than traveling across the input terminals of the operational amplifier—thus preventing a false output and an activation of the alarm.

Schematic with Nodal Voltages

When an investigator is presented with a system that is down or not operating properly, one of the first options is to check the system's specified voltages on the schematic. These specified voltage levels are actually the nodal voltages determined in this chapter. *Nodal voltage* is simply a special term for a voltage measured from that point to ground. The technician will attach the negative or lower-potential lead to the ground of the network (often the chassis) and then place the positive or higher-potential lead on the specified points of the network to check the nodal voltages. If they match, it is a good sign that that section of the system is operating properly. If one or more fail to match the given values, the problem area can usually be identified. Be aware that a reading of -15.87 V is significantly different from an expected reading of $+16$ V if the leads have been properly attached. Although the actual numbers seem close, the difference is actually more than 30 V. You must expect some deviation from the given value as shown, but always be very sensitive to the resulting sign of the reading.

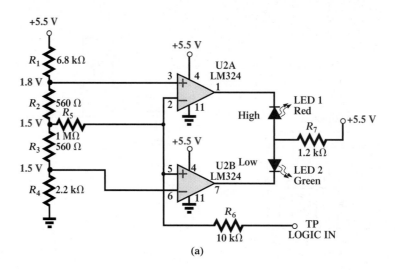

(a)

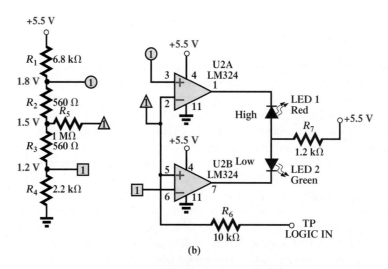

(b)

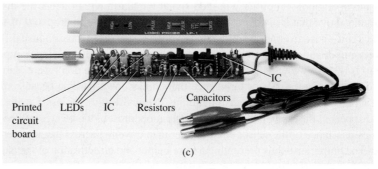

(c)

FIG. 8.111

Logic probe: (a) schematic with nodal voltages; (b) network with global connections;
(c) photograph of commercially available unit.

The schematic of Fig. 8.111(a) includes the nodal voltages for a logic probe used to measure the input and output states of integrated circuit logic chips. In other words, the probe determines whether the measured voltage is one of two states: high or low (often referred to as "on" or "off" or 1 or 0). If the LOGIC IN terminal of the probe is placed on a chip at a location where the voltage is between 0 and 1.2 V, the voltage is considered a low level, and the green LED will light. (LEDs are light-emitting semiconductor diodes that will emit light when current is passed through them.) If the measured voltage is between 1.8 V and 5 V, the reading is considered high, and the red LED will light. Any voltage between 1.2 V and 1.8 V is considered a "floating level" and is an indication that the system being measured is not operating correctly. Note that the reference levels mentioned above are established by the voltage divider network to the right of the schematic. The op-amps employed are of such high input impedance that their loading on the voltage divider network can be ignored and the voltage divider network considered a network unto itself. Even though three 5.5 V dc supply voltages are indicated on the diagram, be aware that all three points are connected to the same supply. The other voltages provided (the nodal voltages) are the voltage levels that should be present from that point to ground if the system is working properly.

The op-amps are used to sense the difference between the reference at points 3 and 6 and the voltage picked up in LOGIC IN. Any difference will result in an output that will light either the green or the red LED. Be aware, because of the direct connection, that the voltage at point 3 is the same as shown by the nodal voltage to the left, or 1.8 V. Likewise, the voltage at point 6 is 1.2 V for comparison with the voltages at points 5 and 2, which reflect the measured voltage. If the input voltage happened to be 1.0 V, the difference between the voltages at points 5 and 6 would be 0.2 V, which ideally would appear at point 7. This low potential at point 7 would result in a current flowing from the much higher 5.5 V dc supply through the green LED, causing it to light and indicating a low condition. By the way, LEDs, like diodes, permit current through them only in the direction of the arrow in the symbol. Also note that the voltage at point 6 must be higher than that at point 5 for the output to turn on the LED. The same is true for point 2 over point 3, which reveals why the red LED does not light when the 1.0 V level is measured.

Oftentimes it is impractical to draw the full network as shown in Fig. 8.111(b) because there are space limitations or because the same voltage divider network is used to supply other parts of the system. In such cases, you must recognize that points having the same shape are connected, and the number in the figure reveals how many connections are made to that point.

A photograph of the outside and inside of a commercially available logic probe is provided in Fig. 8.111(c). Note the increased complexity of the system because of the variety of functions that the probe can perform.

8.14 COMPUTER ANALYSIS

PSpice

The bridge network of Fig. 8.93 will now be analyzed using PSpice to be sure that it is in the balanced state. The only component that has not been introduced in earlier chapters is the dc current source. It can be obtained by first selecting the **Place a part** key and then the **SOURCE** library. Scrolling the **Part List** will result in the option **IDC.** A left click of **IDC** followed by **OK** will result in a dc current source whose direction is toward the bottom of the screen. One left click of the mouse (to make it red—active), followed by a right click of the mouse, will result in a listing having a **Mirror Vertically** option. Selecting that option will flip the source and give it the direction of Fig. 8.93.

The remaining parts of the PSpice analysis are pretty straightforward, with the results of Fig. 8.112 matching those obtained in the analysis of Fig. 8.93. The voltage across the current source is 8 V positive to ground, and the voltage at either end of the bridge arm is at 2.667 V. The voltage across R_5 is obviously 0 V for the level of accuracy displayed, and the current is of such a small magnitude compared to the other current levels of the network that it can essentially be considered 0 A. Note also for the balanced bridge that the current through R_1 equals that through R_3, and the current through R_2 equals that through R_4.

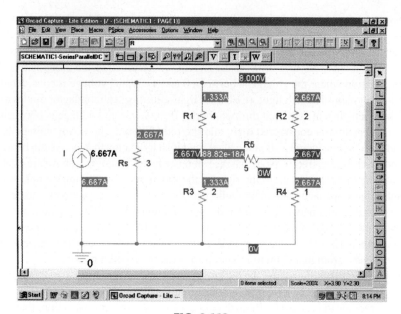

FIG. 8.112

Applying PSpice to the bridge network of Fig. 8.93.

Electronics Workbench

Electronics Workbench will now be used to verify the results of Example 8.18. All the elements of creating the schematic of Fig. 8.113 have been presented in earlier chapters and will be omitted here to demonstrate how little documentation is now necessary to carry you through a fairly complex network.

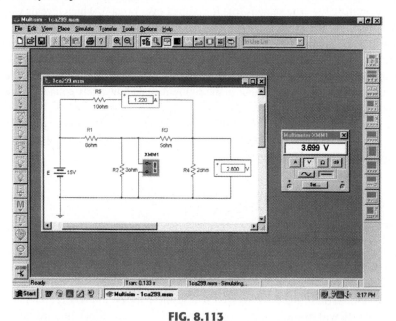

FIG. 8.113

Using Electronics Workbench to verify the results of Example 8.18.

For the analysis, both indicators and a meter will be used to display the desired results. An **A** indicator in the **H** position was used for the current through R_5, and a **V** indicator in the **V** position was used for the voltage across R_2. A multimeter in the voltmeter mode was placed to read the voltage across R_4. The ammeter is reading the mesh or loop current for that branch, and the two voltmeters are displaying the nodal voltages of the network.

After simulation, the results displayed are an exact match with those of Example 8.18.

CHAPTER SUMMARY
CURRENT SOURCES

- A current source determines the direction and magnitude of the current in the branch in which it is located.

- The magnitude and the polarity of the voltage across a current source are each a function of the network to which they are applied.
- The equivalence between a current source and a voltage source exists only at their external terminals.
- Current sources of different values cannot be placed in series due to a violation of Kirchhoff's current law.

BRANCH-CURRENT ANALYSIS

- The number of equations generated using branch-current analysis must equal the number of unknown branch currents. The number of "windows" determines how often Kirchhoff's voltage law must be applied. Kirchhoff's current law must be applied a sufficient number of times to include all the branch currents.

MESH ANALYSIS

- Mesh analysis includes the application of Kirchhoff's current law in the definition of the mesh currents. The number of mesh currents equals the number of "windows" for the network. Kirchhoff's voltage law must be applied around every independent closed loop of the network.
- An approximate solution to a network with an ideal current source can be obtained by placing a relatively large resistor in parallel with the current source and then performing a source conversion.

NODAL ANALYSIS

- Nodal analysis is the application of Kirchhoff's current law to $(N - 1)$ nodes, where the network is defined to have N independent nodes. The nodal voltages are always with respect to a chosen reference level.
- An approximate solution to a network with an ideal voltage source can be obtained by placing a relatively small resistor in series with the voltage source and then performing a source conversion.

BRIDGE NETWORKS

- If a bridge network is balanced, the current through, and the voltage across, the balance arm resistor is zero. The balance equation will have the resistors in the same relative positions as in the bridge network.

Δ-Y OR Y-Δ CONVERSION

- A Δ-Y or Y-Δ conversion is the replacement of one configuration by the other.
- Remember the factor 3 if all the resistors of the Δ or Y configuration are the same, with the Δ resistor being the larger of two.

Important Equations

Source conversion:

$$I = \frac{E}{R_s} \qquad E = IR_p \qquad R_s = R_p$$

Balance condition (bridge configuration):

$$\frac{R_1}{R_3} = \frac{R_2}{R_4}$$

Δ-Y conversion:

$$R_1 = \frac{R_B R_C}{R_A + R_B + R_C} \qquad R_2 = \frac{R_A R_C}{R_A + R_B + R_C} \qquad R_3 = \frac{R_A R_B}{R_A + R_B + R_C}$$

Y-Δ conversion:

$$R_A = \frac{R_1R_2 + R_1R_3 + R_2R_3}{R_1} \qquad R_B = \frac{R_1R_2 + R_1R_3 + R_2R_3}{R_2} \qquad R_C = \frac{R_1R_2 + R_1R_3 + R_2R_3}{R_3}$$

$$R_Y = \frac{R_\Delta}{3} \qquad R_\Delta = 3R_Y$$

PROBLEMS

SECTION 8.2 Current Sources

1. For the network of Fig. 8.114:
   a. Determine currents I_2 and I_3.
   b. Find voltage V_1.
   c. Find the voltage across the source V_s.

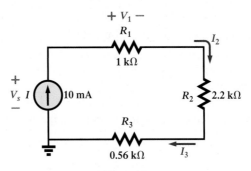

FIG. 8.114
Problem 1.

2. For the network of Fig. 8.115:
   a. Find current I_2.
   b. Calculate voltage V_2.
   c. Find the source voltage V_s.

3. Find voltage V_s (with polarity) across the ideal current source of Fig. 8.116.

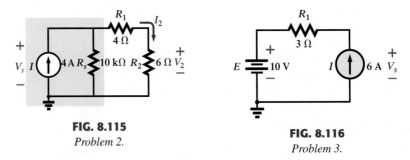

FIG. 8.115
Problem 2.

FIG. 8.116
Problem 3.

4. For the network of Fig. 8.117:
   a. Find voltage V_s.
   b. Calculate current I_2.
   c. Find the source current I_s.

5. Find voltage V_3 and current I_2 for the network of Fig. 8.118.

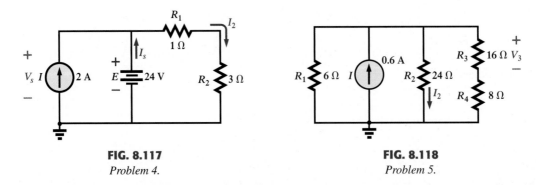

FIG. 8.117
Problem 4.

FIG. 8.118
Problem 5.

6. Convert the voltage sources of Fig. 8.119 to current sources.

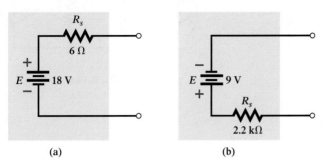

(a)   (b)

FIG. 8.119
Problem 6.

7. Convert the current sources of Fig. 8.120 to voltage sources.

8. For the network of Fig. 8.121:
  a. Find the current through the 2 Ω resistor.
  b. Convert the current source and the 4 Ω resistor to a voltage source, and again solve for the current in the 2 Ω resistor. Compare the results of parts (a) and (b).

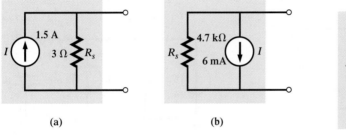

(a)   (b)

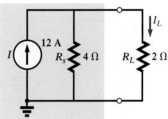

FIG. 8.120   **FIG. 8.121**
Problem 7.   *Problem 8.*

9. For the network of Fig. 8.122:
  a. Convert the current source to a voltage source.
  b. Find current I_2.
  c. Find the source current I_s.
  d. Find the source voltage V_s.

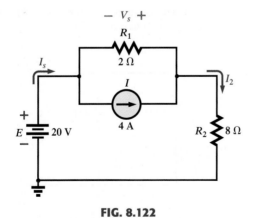

FIG. 8.122
Problem 9.

SECTION 8.4   Current Sources in Parallel

10. For the network of Fig. 8.123:
  a. Replace all the current sources by a single current source.
  b. Find the source voltage V_s.

11. Find voltage V_2 and current I_1 for the network of Fig. 8.124.

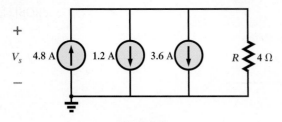

FIG. 8.123

Problem 10.

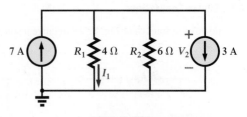

FIG. 8.124

Problem 11.

12. For the network of Fig. 8.125:
   a. Convert the voltage source to a current source.
   b. Combine the two current sources into a single current source.
   c. Find the current through the 2 Ω resistor.

13. For the network of Fig. 8.126:
   a. Convert the voltage source to a current source.
   b. Reduce the network to a single current source, and determine voltage V_1.
   c. Using the results of part (b), determine V_2.
   d. Calculate current I_2.

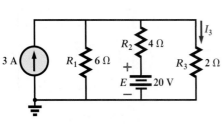

FIG. 8.125

Problem 12.

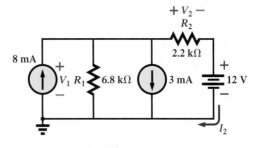

FIG. 8.126

Problem 13.

SECTION 8.5   Branch-Current Analysis

14. Using branch-current analysis, find the magnitude and direction for the current through each resistor for the network of Fig. 8.127.

15. Using branch-current analysis, find the magnitude and direction for the current through each resistor for the network of Fig. 8.128 (all standard values).

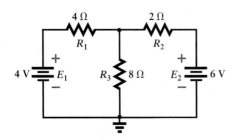

FIG. 8.127

Problems 14, 20, 26, and 65.

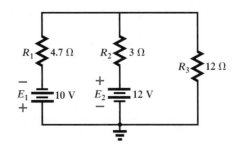

FIG. 8.128

Problems 15, 21, and 27.

16. For the single-source network of Fig. 8.129 (all standard values), use branch-current analysis to find the current through resistor R_2.

*17. Using branch-current analysis, find the magnitude and direction for the current through each resistor for the network of Fig. 8.130 (all standard values).

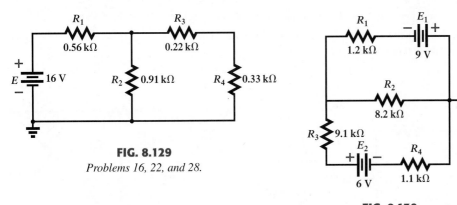

FIG. 8.129
Problems 16, 22, and 28.

FIG. 8.130
Problems 17, 23, and 29.

*18. For the network of Fig. 8.131:
  a. Using branch-current analysis, write the equations necessary to solve for the branch currents.
  b. Using Kirchhoff's current law to relate branch currents, reduce the number of equations to three.
  c. Rewrite the equations in a format that can be solved using third-order determinants.
  d. Solve for the branch current through resistor R_3.

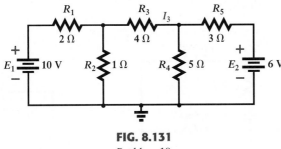

FIG. 8.131
Problem 18.

*19. For the transistor configuration of Fig. 8.132:
  a. Solve for currents I_B, I_C, and I_E using the fact that $V_{BE} = 0.7$ V and $V_{CE} = 8$ V as shown on the figure.
  b. Find voltages V_B, V_C, and V_E.
  c. What is the ratio of output current I_C to input current I_B? Did the transistor increase the current level?

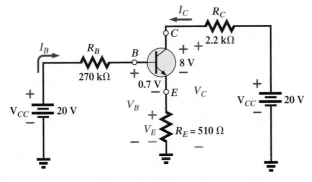

FIG. 8.132
Problem 19.

SECTION 8.6 Mesh Analysis (General Approach)

20. Using mesh analysis, determine the current through each resistor of Fig. 8.127.

21. Using mesh analysis, determine the current through each resistor of Fig. 8.128.

22. Using mesh analysis, determine the current through each resistor of Fig. 8.129.

*23. Using mesh analysis, determine the current through each resistor of Fig. 8.130.

*24. For the networks of Fig. 8.133, write the mesh equations, and, using determinants, solve for the current through resistor R_4 of each network.

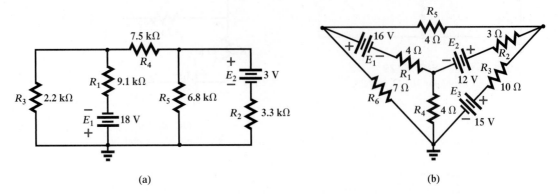

FIG. 8.133
Problems 24 and 30.

*25. For the networks of Fig. 8.134, write the mesh equations, and, using determinants, solve for the loop currents of each network.

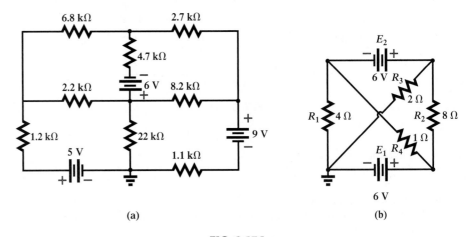

FIG. 8.134
Problems 25 and 31.

SECTION 8.7   Mesh Analysis (Format Approach)

26. Using the format approach to mesh analysis, find the current through each resistor of Fig. 8.129.

27. Using the format approach to mesh analysis, find the current through each resistor of Fig. 8.128.

28. Using the format approach to mesh analysis, find the current through each resistor of Fig. 8.129.

*29. Using the format approach to mesh analysis, find the current through each resistor of Fig. 8.130.

*30. Using the format approach to mesh analysis, find the current through resistor R_4 for each network of Fig. 8.133.

*31. Using the format approach to mesh analysis, find the mesh currents for each network of Fig. 8.134.

SECTION 8.8   Nodal Analysis (General Approach)

32. Find the nodal voltages for the network of Fig. 8.135. Then find the voltage across each resistor.

33. Find the nodal voltages for the network of Fig. 8.136 (all standard values). Then find the voltage across each resistor.

34. Find the nodal voltages for the network of Fig. 8.137. Then find the voltage across each resistor.

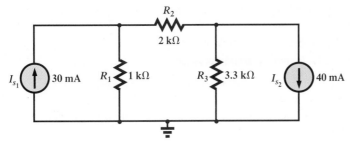

FIG. 8.135
Problems 32, 39, and 67.

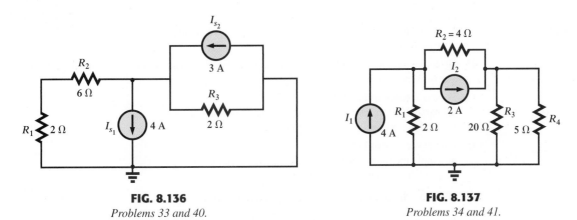

FIG. 8.136
Problems 33 and 40.

FIG. 8.137
Problems 34 and 41.

*35. For the network of Fig. 8.138:
  a. Convert the voltage source to a current source.
  b. Find the nodal voltages.
  c. Find the voltage across each resistor.

*36. Find the nodal voltages for the network of Fig. 8.139. Then find the voltage across each resistor.

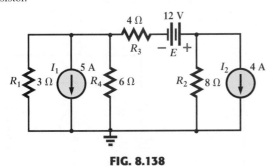

FIG. 8.138
Problems 35 and 42.

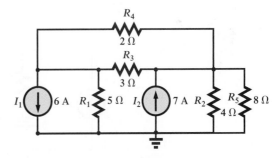

FIG. 8.139
Problems 36 and 43.

*37. For the network of Fig. 8.140:
  a. Find the nodal voltages.
  b. Find the voltage across the current source.

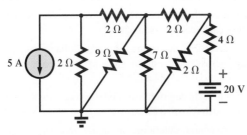

FIG. 8.140
Problems 37, 44, and 68.

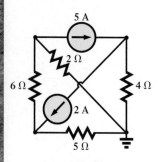

FIG. 8.141
Problems 38 and 45.

38. For the network of Fig. 8.141:
   a. Find the nodal voltages.
   b. Find the voltage across each current source.

SECTION 8.9 Nodal Analysis (Format Approach)

39. Using the format approach, find the nodal voltages for the network of Fig. 8.135. Then find the voltage across each resistor.

40. Using the format approach, find the nodal voltages for the network of Fig. 8.136. Then find the voltage across each resistor.

41. Using the format approach, find the nodal voltages for the network of Fig. 8.137. Then find the voltage across each resistor.

42. For the network of Fig. 8.138:
   a. Convert the voltage source to a current source.
   b. Apply the format approach to determine the nodal voltages.
   c. Find the voltage across each resistor.

43. Using the format approach, find the nodal voltages for the network of Fig. 8.139. Then find the voltage across each resistor.

44. For the network of Fig. 8.140:
   a. Apply the format approach to determine the nodal voltages.
   b. Find the voltage across the current source.

45. For the network of Fig. 8.141:
   a. Apply the format approach to determine the nodal voltages.
   b. Find the voltage across each current source.

SECTION 8.10 Special Cases for Mesh and Nodal Analysis

46. Using the supermesh approach, determine the current through each element of the network of Fig. 8.142 (all standard values).

47. Using the approximate conversion approach, determine the current through each element of Fig. 8.142. If the conversion is performed, compare your results with the solution of Problem 46.

48. Using the supermesh approach, determine the current through each element of the network of Fig. 8.143.

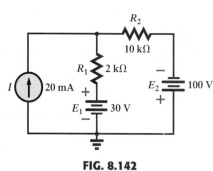

FIG. 8.142
Problems 46 and 47.

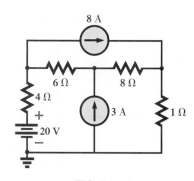

FIG. 8.143
Problems 48 and 49.

49. Using the approximate conversion approach, determine the current through each element of Fig. 8.143. If the conversion is performed, compare your results with the solution of Problem 48.

50. Using the supernode approach, determine the nodal voltages for the network of Fig. 8.144. Then find the voltage across each resistive element.

51. Using approximate conversion, determine the nodal voltages for the network of Fig. 8.144. Then find the voltage across each resistive element. If the conversion is performed, compare your results with the solution of Problem 50.

52. Using the supernode approach, determine the nodal voltages for the network of Fig. 8.145. Then find the voltage across each resistive element.

53. Using approximate conversion, determine the nodal voltages for the network of Fig. 8.145. Then find the voltage across each resistive element. If the conversion is performed, compare your results with the solution of Problem 52.

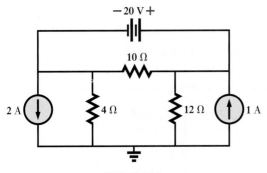

FIG. 8.144
Problems 50 and 51.

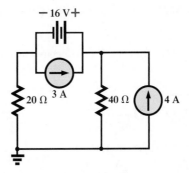

FIG. 8.145
Problems 52 and 53.

SECTION 8.11   Bridge Networks

54. For the bridge network of Fig. 8.146:
   a. Write the mesh equations using the format or general approach.
   b. Determine the current through R_5.
   c. Is the bridge balanced?
   d. Is Equation (8.1) satisfied?

55. For the bridge network of Fig. 8.146:
   a. Write the nodal equations using the format or general approach.
   b. Determine the voltage across R_5.
   c. Is the bridge balanced?
   d. Is Equation (8.1) satisfied?

56. For the bridge network of Fig. 8.147:
   a. Is the bridge balanced?
   b. Determine the current through R_5 using the most direct route.
   c. Is Equation (8.1) satisfied?

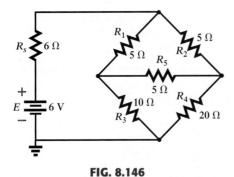

FIG. 8.146
Problems 54 and 55.

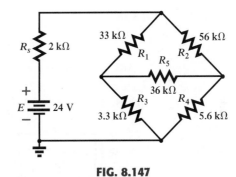

FIG. 8.147
Problem 56.

*57. Using either mesh or nodal analysis, determine the current through the source resistor R_s for the network of Fig. 8.148. Discuss why you chose one method over the other. Is the bridge balanced?

*58. Using either mesh or nodal analysis, determine the voltage across resistor R_s for the network of Fig. 8.149. Discuss why you chose one method over the other. Is the bridge balanced?

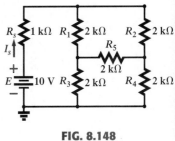

FIG. 8.148
Problem 57.

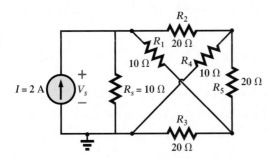

FIG. 8.149
Problems 58 and 66.

SECTION 8.12 Δ-Y and Y-Δ Conversions

59. Using a Δ-Y or Y-Δ conversion, find the current I for the network of Fig. 8.150.

60. Using a Δ-Y or Y-Δ conversion, find the current I for the network of Fig. 8.151.

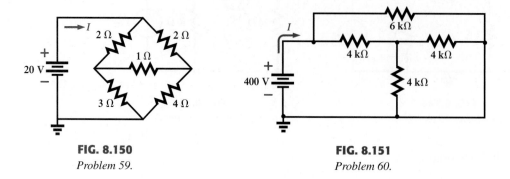

FIG. 8.150

Problem 59.

FIG. 8.151

Problem 60.

*61. Using a Δ-Y or Y-Δ conversion, find the current I for the network of Fig. 8.152.

*62. a. Replace the T configuration of Fig. 8.153 (composed of 6 kΩ resistors) with a π configuration.

    b. Solve for the source current I_s.

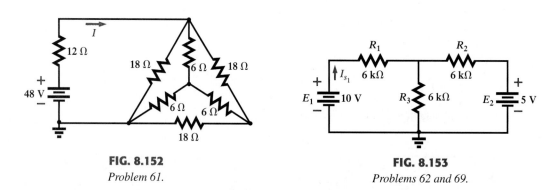

FIG. 8.152

Problem 61.

FIG. 8.153

Problems 62 and 69.

*63. a. Replace the π configuration of Fig. 8.154 (composed of 3 kΩ resistors) with a T configuration.

    b. Solve for the source current I_{s_1}.

*64. Using Δ-Y or Y-Δ conversions, determine the total resistance of the network of Fig. 8.155.

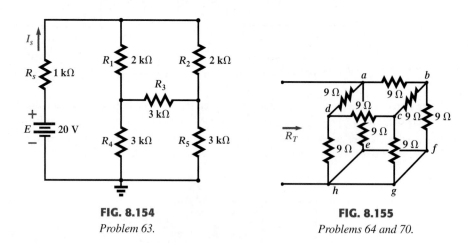

FIG. 8.154

Problem 63.

FIG. 8.155

Problems 64 and 70.

SECTION 8.14 Computer Analysis

65. Using PSpice or EWB, find the branch currents for the network of Fig. 8.127.

66. Using PSpice or EWB, find the current through each element of Fig. 8.149, and determine if the bridge is balanced.

67. Using PSpice or EWB, find the nodal voltages for the network of Fig. 8.135.

68. Using PSpice or EWB, find the nodal voltages for the network of Fig. 8.140.

69. Using PSpice or EWB, find the voltage across each element of Fig. 8.153.

*70. Using PSpice or EWB, find the current through each element of Fig. 8.155.

GLOSSARY

Branch-current method A technique for determining the branch currents of a multiloop network.

Bridge network A network configuration typically having a diamond appearance in which no two elements are in series or parallel.

Current sources Sources that supply a fixed current to a network and have a terminal voltage dependent on the network to which they are applied.

Delta (Δ), pi (π) configuration A network structure that consists of three branches and has the appearance of the Greek letter delta (Δ) or pi (π).

Determinants method A mathematical technique for finding the unknown variables of two or more simultaneous linear equations.

Mesh analysis A technique for determining the mesh (loop) currents of a network that results in a reduced set of equations compared to the branch-current method.

Mesh (loop) current A labeled current assigned to each distinct closed loop of a network that can, individually or in combination with other mesh currents, define all of the branch currents of a network.

Nodal analysis A technique for determining the nodal voltages of a network.

Node A junction of two or more branches in a network.

Supermesh current A current defined in a network with ideal current sources that permits the use of mesh analysis.

Supernode A node defined in a network with ideal voltage sources that permits the use of nodal analysis.

Wye (Y), tee (T) configuration A network structure that consists of three branches and has the appearance of the capital letter Y or T.

Network Theorems

OBJECTIVES

- Become familiar with the superposition theorem and its unique ability to separate the impact of each source on the quantity of interest.
- Be able to apply Thévenin's theorem to reduce any two-terminal, series-parallel network with any number of sources to a single voltage source and series resistor.
- Understand how to apply the maximum power transfer theorem to determine the maximum power to a load.
- Be aware of the differences between the dc operating efficiency of a system and the conditions resulting in maximum power transfer to a load.
- Be able to apply Norton's theorem to reduce any two-terminal, series-parallel network with any number of sources to a single current source and parallel resistor.
- Understand how to use instrumentation to determine the Thévenin and Norton equivalent circuits from an active network.
- Be able to apply computer methods to determine the parameters of a Thévenin or a Norton equivalent circuit.
- Become aware of how to apply the superposition theorem, Thévenin's theorem, and the maximum power theorem using PSpice and EWB.

9.1 INTRODUCTION

This chapter will introduce a number of theorems that have application throughout the field of electricity and electronics. Not only can they be used to solve networks such as encountered in the previous chapter, but they also provide an opportunity to determine the impact of a particular source or element on the response of the entire system. In most cases, the network to be analyzed and the mathematics required to find the solution are simplified. All of the theorems will be appear again in the analysis of ac networks. In fact, the application of each theorem to ac networks will be very similar in content to that found in this chapter.

The first theorem to be introduced is the superposition theorem, followed by Thévenin's theorem, Norton's theorem, and the maximum power transfer theorem. Each theorem will be covered in detail to be sure that you can apply it with confidence in the chapters to follow.

9.2 SUPERPOSITION THEOREM

The **superposition theorem** is unquestionably one of the most powerful in this field. It has such widespread application that people often apply it without recognizing that their maneuvers are valid only because of this theorem.

In general, the theorem can be used to do the following:

1. *Analyze networks such as introduced in the last chapter that have two or more sources that are not in series or parallel.*
2. *Reveal the effect of each source on a particular quantity of interest.*
3. *For sources of different types (such as dc and ac which affect the parameters of the network in a different manner), apply a separate analysis for each type, with the total result simply the algebraic sum of the results.*

The first two areas of application will be described in detail in this section. The last will be left for the discussion of the superposition theorem in the ac portion of the text.

The superposition theorem states the following:

The current through, or voltage across, any element of a network is equal to the algebraic sum of the currents or voltages produced independently by each source.

In other words, this theorem allows us to find a solution for a current or voltage using **only one source at a time.** Once we have the solution for each source, we can combine the results to obtain the total solution. The term *algebraic* appears in the above theorem statement because the currents resulting from the sources of the network can have different directions, just as the resulting voltages can have opposite polarities.

If we are to consider the effects of each source, the other sources obviously must be removed. Setting a voltage source to zero volts is like placing a short circuit across its terminals. Therefore,

when removing a voltage source from a network schematic, replace it with a direct connection (short circuit) of zero ohms. Any internal resistance associated with the source must remain in the network.

Setting a current source to zero amperes is like replacing it with an open circuit. Therefore,

when removing a current source from a network schematic, replace it by an open circuit of infinite ohms. Any internal resistance associated with the souce must remain in the network.

The above statements are illustrated in Fig. 9.1.

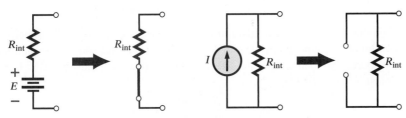

FIG. 9.1
Removing a voltage source and a current source to permit the application of the superposition theorem.

Since the effect of each source will be determined independently, the number of networks to be analyzed will equal the number of sources.

If a particular current of a network is to be determined, the contribution to that current will have to be determined for **each source.** When the effect of each source has been determined, those currents in the same direction will be added, and those having the opposite direction will be subtracted; the algebraic sum is being determined. The total result will be the direction of the larger sum and the magnitude of the difference.

Similarly, if a particular voltage of a network is to be determined, the contribution to that voltage must be determined for each source. When the effect of each source has been determined, those voltages with the same polarity will be added, and those with the opposite polarity will be subtracted; the algebraic sum is being determined. The total result will have the polarity of the larger sum and the magnitude of the difference.

Superposition cannot be applied to power effects because the power is related to the square of the voltage across a resistor or the current through a resistor. The squared term will result in a nonlinear (not a straight line but a curve) relationship between the power and the determining current or voltage. For example, doubling the current through a resistor will not double the power to the resistor (as defined by a linear relationship) but, in fact, will increase it by a factor of 4 (due to the squared term). Tripling the current will increase the power level by a factor of 9. Example 9.3 will demonstrate the differences between a linear and a nonlinear relationship.

A few examples will clarify how sources are removed and total solutions obtained.

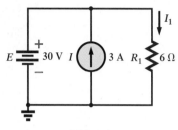

FIG. 9.2

Two-source network to be analyzed using the superposition theorem in Example 9.1.

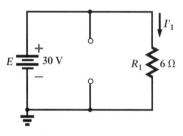

FIG. 9.3

Determining the effect of the 30 V supply on the current I_1 of Fig. 9.2.

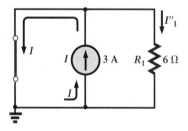

FIG. 9.4

Determining the effect of the 3 A current source on the current I_1 of Fig. 9.2.

EXAMPLE 9.1 Using the superposition theorem, determine current I_1 for the network of Fig. 9.2.

Solution: Since two sources are present, there will be two networks to be analyzed. First let us determine the effects of the voltage source by setting the current source to zero amperes as shown in Fig. 9.3. Note that the resulting current is defined as I'_1 because it is the current through resistor R_1 due to the voltage source only.

Due to the open circuit, resistor R_1 is in series (and, in fact, in parallel) with the voltage source E. The voltage across the resistor is the applied voltage, and current I'_1 is determined by

$$I'_1 = \frac{V_1}{R_1} = \frac{E}{R_1} = \frac{30 \text{ V}}{6 \text{ }\Omega} = 5 \text{ A}$$

Now for the contribution due to the current source. Setting the voltage source to zero volts will result in the network of Fig. 9.4, which presents us with an interesting situation. The current source has been replaced with a short-circuit equivalent that is directly across the current source and resistor R_1. Since the source current will take the path of least resistance, it will choose the zero ohm path of the inserted short-circuit equivalent, and the current through R_1 will be zero amperes. This is clearly demonstrated by an application of the current divider rule as follows:

$$I''_1 = \frac{R_{sc}I}{R_{sc} + R_1} = \frac{(0 \text{ }\Omega)I}{0 \text{ }\Omega + 6 \text{ }\Omega} = 0 \text{ A}$$

Since I'_1 and I''_1 have the same defined direction in Figs. 9.3 and 9.4, the total current is defined by

$$I_1 = I'_1 + I''_1 = 5 \text{ A} + 0 \text{ A} = \mathbf{5 \text{ A}}$$

Although this has been an excellent introduction to the application of the superposition theorem, it should be immediately clear in Fig. 9.2 that the voltage source is in parallel with the current source and load resistor R_1, so the voltage across each must be 30 V. The result is that I_1 must be determined solely by

$$I_1 = \frac{V_1}{R_1} = \frac{E}{R_1} = \frac{30 \text{ V}}{6 \text{ }\Omega} = \mathbf{5 \text{ A}}$$

EXAMPLE 9.2 Using the superposition theorem, determine the current through the 12 Ω resistor of Fig. 9.5. Note that this is a two-source network of the type examined in the previous chapter when we applied branch-current analysis and mesh analysis.

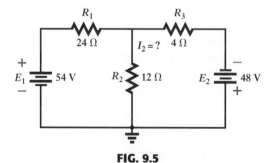

FIG. 9.5

Using the superposition theorem to determine the current through the 12 Ω resistor (Example 9.2).

Solution: Considering the effects of the 54 V source will require replacing the 48 V source by a short-circuit equivalent as shown in Fig. 9.6. As shown in the figure, the result is that the 12 Ω and 4 Ω resistors are in parallel.

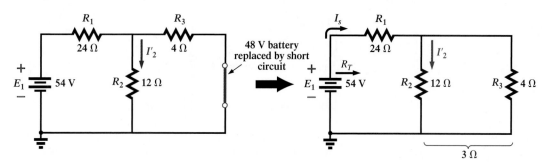

FIG. 9.6

Using the superposition theorem to determine the effect of the 54 V voltage source on current I_2 of Fig. 9.5.

The total resistance seen by the source is therefore

$$R_T = R_1 + R_2 \| R_3 = 24 \, \Omega + 12 \, \Omega \| 4 \, \Omega = 24 \, \Omega + 3 \, \Omega = 27 \, \Omega$$

and the source current is

$$I_s = \frac{E_1}{R_T} = \frac{54 \text{ V}}{27 \, \Omega} = 2 \text{ A}$$

Using the current divider rule will result in the contribution to I_2 due to the 54 V source:

$$I'_2 = \frac{R_3 I_s}{R_3 + R_2} = \frac{(4 \, \Omega)(2 \text{ A})}{4 \, \Omega + 12 \, \Omega} = 0.5 \text{ A}$$

If we now replace the 54 V source by a short-circuit equivalent, the network of Fig. 9.7 will result. The result is a parallel connection for the 12 Ω and 24 Ω resistors as shown in the figure.

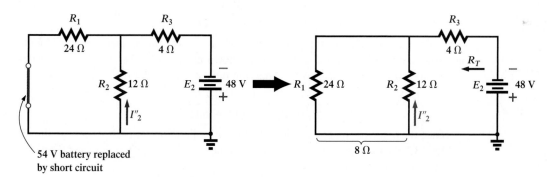

FIG. 9.7

Using the superposition theorem to determine the effect of the 48 V voltage source on current I_2 of Fig. 9.5.

Therefore, the total resistance seen by the 48 V source is

$$R_T = R_3 + R_2 \| R_1 = 4 \, \Omega + 12 \, \Omega \| 24 \, \Omega = 4 \, \Omega + 8 \, \Omega = 12 \, \Omega$$

and the source current is

$$I_s = \frac{E_2}{R_T} = \frac{48 \text{ V}}{12 \, \Omega} = 4 \text{ A}$$

Applying the current divider rule will result in

$$I''_2 = \frac{R_1 (I_s)}{R_1 + R_2} = \frac{(24 \, \Omega)(4 \text{ A})}{24 \, \Omega + 12 \, \Omega} = 2.667 \text{ A}$$

It is now important to realize that current I_2 due to each source has a different direction, as shown in Fig. 9.8. The net current will therefore be the difference of the two and the direction of the larger as follows:

$$I_2 = I''_2 - I'_2 = 2.667 \text{ A} - 0.5 \text{ A} = \textbf{2.167 A}$$

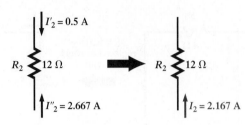

FIG. 9.8

Using the results of Figs. 9.6 and 9.7 to determine current I_2 for the network of Fig. 9.5.

Using Figs. 9.6 and 9.7 in Example 9.2, the other currents of the network could have been determined with little added effort. That is, we could determine all the branch currents of the network, matching an application of the branch-current analysis or mesh analysis approach. In general, therefore, not only can the superposition theorem provide a complete solution for the network, but it will also reveal the effect of each source on the desired quantity.

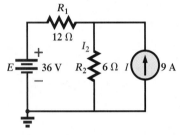

FIG. 9.9

Network to be analyzed in Example 9.3 using the superposition theorem.

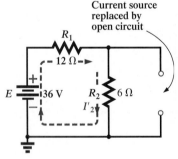

FIG. 9.10

Replacing the 9 A current source of Fig. 9.9 by an open circuit to determine the effect of the 36 V voltage source on current I_2.

EXAMPLE 9.3

a. Using the superposition theorem, determine the current through resistor R_2 for the network of Fig. 9.9.
b. Demonstrate that the superposition theorem is not applicable to power levels.

Solutions:

a. In order to determine the effect of the 36 V voltage source, the current source must be replaced by an open-circuit equivalent as shown in Fig. 9.10. The result is a simple series circuit with a current equal to

$$I'_2 = \frac{E}{R_T} = \frac{E}{R_1 + R_2} = \frac{36 \text{ V}}{12 \text{ }\Omega + 6 \text{ }\Omega} = \frac{36 \text{ V}}{18 \text{ }\Omega} = 2 \text{ A}$$

Examining the effect of the 9 A current source will require replacing the 36 V voltage source by a short-circuit equivalent as shown in Fig. 9.11. The result is a parallel combination of resistors R_1 and R_2. Applying the current divider rule will result in

$$I''_2 = \frac{R_1(I)}{R_1 + R_2} = \frac{(12 \text{ }\Omega)(9 \text{ A})}{12 \text{ }\Omega + 6 \text{ }\Omega} = 6 \text{ A}$$

Since the contribution to current I_2 has the same direction for each source, as shown in Fig. 9.12, the total solution for current I_2 will be the sum of the currents established by the two sources. That is,

$$I_2 = I'_2 + I''_2 = 2 \text{ A} + 6 \text{ A} = \textbf{8 A}$$

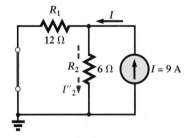

FIG. 9.11

Replacing the 36 V voltage source by a short-circuit equivalent to determine the effect of the 9 A current source on current I_2.

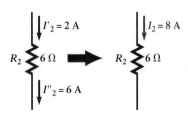

FIG. 9.12

Using the results of Figs. 9.10 and 9.11 to determine current I_2 for the network of Fig. 9.9.

b. Using Fig. 9.10 and the results obtained, the power delivered to the 6 Ω resistor is

$$P_1 = (I'_2)^2(R_2) = (2 \text{ A})^2(6 \text{ }\Omega) = \textbf{24 W}$$

Using Fig. 9.11 and the results obtained, the power delivered to the 6 Ω resistor is

$$P_2 = (I''_2)^2(R_2) = (6 \text{ A})^2(6 \text{ }\Omega) = \textbf{216 W}$$

Using the total results of Fig. 9.12, the power delivered to the 6 Ω resistor is

$$P_T = I_2^2 R_2 = (8 \text{ A})^2(6 \text{ }\Omega) = \textbf{384 W}$$

It is now quite clear that the power delivered to the 6 Ω resistor using the total current of 8 A is not equal to the sum of the power levels due to each source independently. That is,

$$P_1 + P_2 = 24 \text{ W} + 216 \text{ W} = 240 \text{ W} \neq P_T = 384 \text{ W}$$

To expand on the above conclusion and further demonstrate what is meant by a *nonlinear relationship,* the power to the 6 Ω resistor versus current through the 6 Ω resistor was plotted in Fig. 9.13. Note that the curve is not a straight line but one whose rise gets steeper with increase in current level.

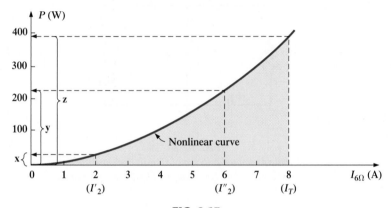

FIG. 9.13

Plotting power versus current for the 6 Ω resistor of Fig. 9.9.

Recall from Fig. 9.11 that the power level was 24 W for a current of 2 A developed by the 36 V voltage source as shown in Fig. 9.13. From Fig. 9.12, we found that the current level was 6 A for a power level of 216 W as shown in Fig. 9.13. Using the total current of 8 A, we find that the power level is 384 W as shown in Fig. 9.13. Quite clearly, the sum of power levels due to the 2 A and 6 A current levels does not equal that due to the 8 A level. That is,

$$x + y \neq z$$

Now, the relationship between the voltage across a resistor and the current through a resistor is a linear (straight line) one as shown in Fig. 9.14, with

$$c = a + b$$

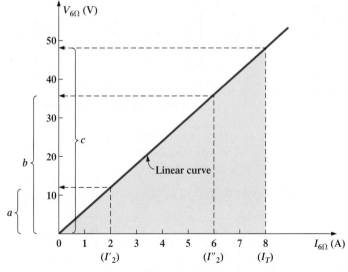

FIG. 9.14

Plotting voltage versus current for the 6 Ω resistor of Fig. 9.9.

Note that the voltage level established at the 2 A level, when added to that established at the 6 A level, will add up to that established by the 8 A level. The superposition theorem, therefore, is applicable to the voltage across an element or the current through the element.

In total, therefore, simply remember that

the superposition theorem is applicable to voltage and current levels, but not power levels.

EXAMPLE 9.4   As a last example of the superposition theorem, we will find the voltage across the 2 Ω resistor of Fig. 9.15.

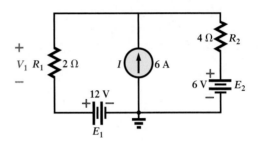

FIG. 9.15
Network to be examined in Example 9.4 to demonstrate that the superposition theorem can also be applied to determine voltage levels.

Solution: Since three sources are present, three distinct networks must be analyzed. Starting with voltage source E_1 will require that voltage source E_2 be replaced by a short-circuit equivalent, and the current source by an open-circuit equivalent, as shown in Fig. 9.16.

A series circuit results with a current of

$$I'_1 = \frac{E_1}{R_T} = \frac{E_1}{R_1 + R_2} = \frac{12 \text{ V}}{2 \Omega + 4 \Omega} = \frac{12 \text{ V}}{6 \Omega} = 2 \text{ A}$$

and $\qquad V'_1 = I'_1 R_1 = (2 \text{ A})(2 \Omega) = 4 \text{ V}$

Considering the effects of the 6 V supply will result in the circuit of Fig. 9.17.

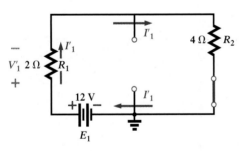

FIG. 9.16
Determining the effect of E_1 on voltage V_1.

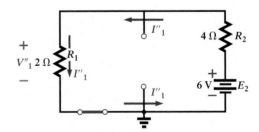

FIG. 9.17
Determining the effect of E_2 on voltage V_1.

Again, due to the open-circuit equivalent for the current source, a series circuit results, and the current is

$$I''_1 = \frac{E_2}{R_T} = \frac{E_2}{R_1 + R_2} = \frac{6 \text{ V}}{2 \Omega + 4 \Omega} = \frac{6 \text{ V}}{6 \Omega} = 1 \text{ A}$$

and $\qquad V''_1 = I''_1 R_1 = (1 \text{ A})(2 \Omega) = 2 \text{ V}$

Next, both voltage sources are replaced by a short-circuit equivalent, and the parallel network of Fig. 9.18 will result.

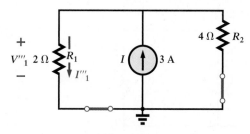

FIG. 9.18

Determining the effect of I on voltage V₁.

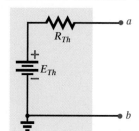

FIG. 9.20

Thévenin equivalent circuit.

Applying the current divider rule will result in

$$I'''_1 = \frac{R_2 I}{R_2 + R_1} = \frac{(4\,\Omega)(6\,\text{A})}{4\,\Omega + 2\,\Omega} = 4\,\text{A}$$

and

$$V'''_1 = I'''_1 R_1 = (4\,\text{A})(2\,\Omega) = 8\,\text{V}$$

Finally, the three contributions to voltage V_1 appear alongside the $2\,\Omega$ resistor in Fig. 9.19. The result is

$$V_1 = -V'_1 + V''_1 + V'''_1 = -4\,\text{V} + 2\,\text{V} + 8\,\text{V} = \textbf{6 V}$$

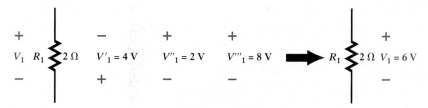

FIG. 9.19

Determining the resultant voltage for V₁ of Fig. 9.15.

That completes the introduction to the superposition theorem for now. The theorem will surface again when we analyze ac networks to show that the theorem is also applicable to sources of a totally different nature—a real blessing when we examine electronic systems.

9.3 THÉVENIN'S THEOREM

The next theorem to be introduced, **Thévenin's theorem,** is probably one of the most interesting in that it permits the reduction of complex networks to a simpler form for analysis and design.

In general, the theorem can be used to do the following:

1. *Analyze networks with sources that are not in series or parallel.*
2. *Reduce the number of components required to establish the same characteristics at the output terminals.*
3. *Investigate the effect of changing a particular component on the behavior of a network without having to analyze the entire network after each change.*

All three areas of application will be demonstrated in the examples to follow.

Thévenin's theorem states the following:

Any two-terminal dc network can be replaced by an equivalent circuit consisting solely of a voltage source and a series resistor as shown in Fig. 9.20.

The theorem was developed by Commandant Leon-Charles Thévenin in 1883 as described in Fig. 9.21.

To demonstrate the power of the theorem, consider the fairly complex network of Fig. 9.22(a) with its two sources and series-parallel connections. The theorem states that the entire network inside the blue shaded area can be replaced by one voltage source and one resistor as shown in Fig. 9.22(b). If the replacement is done properly, the load resistor R_L will not know whether it is connected to the network of Fig. 9.22(a) or Fig. 9.22(b). The value of R_L can be changed to any value, and the voltage, current, or

Courtesy of the Bibliothèque, École Polytechnique, Paris, France.

FIG. 9.21

Leon-Charles Thévenin.

French (Meaux, Paris)
(1857–1927)
Telegraph Engineer,
Commandant and Educator
École Polytechnique and École Supérieure de Télégraphie

Although active in the study and design of telegraphic systems (including underground transmission), cylindrical condensers (capacitors), and electromagnetism, he is best known for a theorem first presented in the French *Journal of Physics—Theory and Applications* in 1883. It appeared under the heading of "Sur un nouveau théorème d'électricité dynamique" ("On a new theorem of dynamic electricity") and was originally referred to as the *equivalent generator theorem.* There is some evidence that a similar theorem was introduced by Hermann von Helmholtz in 1853. However, Professor Helmholtz applied the theorem to animal physiology and not to communication or generator systems, and therefore he has not received the credit in this field that he might deserve. In the early 1920s, AT&T did some pioneering work using the equivalent circuit and may have initiated the reference to the theorem as simply Thévenin's theorem. In fact, Edward L. Norton, an engineer at AT&T at the time, introduced a current source equivalent of the Thévenin equivalent currently referred to as the Norton equivalent circuit. As an aside, Commandant Thévenin was an avid skier and in fact was commissioner of an international ski competition in Chamonix, France, in 1912.

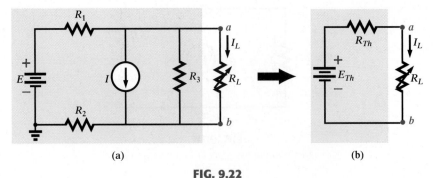

FIG. 9.22

Substituting the Thévenin equivalent circuit for a complex network.

power to the load resistor will be the same for each configuration. Now, that is a pretty powerful statement—one that will be verified in the examples to follow.

The question then remains as to how to determine the proper value of Thévenin voltage and resistance. In general, finding the Thévenin resistance value is quite straightforward. Finding the Thévenin voltage can be more of a challenge and, in fact, may require the use of the superposition theorem or one of the methods described in Chapter 8.

Fortunately, there is a series of steps that will lead to the proper value of each parameter. Although a few of the steps may seem trivial at first, they can become quite important when the network becomes complex. The steps are as follows:

1. *Remove the portion of the network across which the Thévenin equivalent circuit is to be found.*
2. *Mark the resulting two terminals with some special notation.*
3. *Determine R_{Th} by first setting all the sources to zero (voltage sources are replaced by short circuits, and current sources by open circuits) and finding the resulting resistance between the two marked terminals.*
4. *Calculate the Thévenin voltage by first reestablishing each source and finding the open-circuit voltage between the two marked terminals.*
5. *Draw the Thévenin equivalent circuit, and reattach the portion of the network previously removed between the two terminals of the equivalent circuit.*

The examples to follow will provide a detailed explanation of the application of each step.

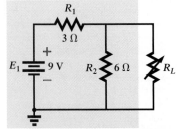

FIG. 9.23

Finding the Thévenin equivalent circuit for the network in the shaded region (Example 9.5).

EXAMPLE 9.5 Find the Thévenin equivalent circuit for the shaded portion of the network of Fig. 9.23. Then find the current through R_L for values of 2 Ω, 10 Ω, and 100 Ω.

Solution:

Steps 1–2: As shown in Fig. 9.24, the variable resistor has been removed from the diagram, and the two resulting terminals have been labeled.

Step 3: Replacing the voltage source by a short-circuit equivalent will result in the configuration of Fig. 9.25(a). The Thévenin resistance is then determined between points *a* and *b*. It may take a little time to adjust to this step, because the resis-

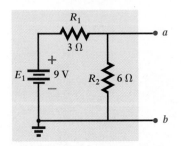

FIG. 9.24

Identifying the terminals of particular importance when applying Thévenin's theorem.

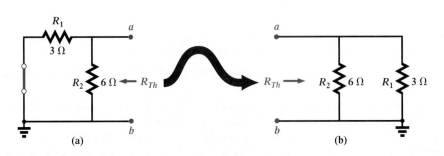

FIG. 9.25

Determining R_{Th} for the network of Fig. 9.23.

tance is not determined looking into the network from the source as occurred in the previous chapters. It is now determined between the very important terminals *a* and *b*. In the early stages of development, it might be easier to redraw the network as shown in Fig. 9.25(b) so that the resistance to be determined is found by looking into the network from the left rather than from the right.

The Thévenin resistance is then

$$R_{Th} = R_2 \| R_1 = 6 \ \Omega \| 3 \ \Omega = 2 \ \Omega$$

Step 4: Replacing the source will result in the configuration of Fig. 9.26. The important Thévenin voltage is then measured between the same two marked terminals *a* and *b*. In this case, the Thévenin voltage is the same as that across resistor R_2 since points *a* and *b* are connected directly to the ends of the 6 Ω resistor.

Since we have a series configuration, the Thévenin voltage can be found by one simple application of the voltage divider rule as follows:

$$E_{Th} = \frac{R_2 E_1}{R_2 + R_1} = \frac{(6 \ \Omega)(9 \ V)}{6 \ \Omega + 3 \ \Omega} = \mathbf{6 \ V}$$

Before continuing, take a moment to appreciate that the Thévenin voltage is the open-circuit voltage between the two marked terminals. Recall that an open circuit can have any voltage across it, but the current must be zero. Therefore, all the source current in Fig. 9.26 must pass through the 6 Ω resistor. The result, as pointed out above, is a series configuration.

Step 5: The last step is to simply draw the resulting Thévenin equivalent circuit as shown in Fig. 9.27 and reattach the variable load resistor.

The general equation for the current through the load resistor of Fig. 9.27 is

$$I_L = \frac{E_{Th}}{R_{Th} + R_L}$$

The only quantity in the equation that will change as R_L changes is R_L. Both E_{Th} and R_{Th} will remain fixed as shown below for the three values of load resistance:

$$I_L = \frac{E_{Th}}{R_{Th} + R_L} = \frac{6 \ V}{2 \ \Omega + 2 \ \Omega} = \frac{6 \ V}{4 \ \Omega} = \mathbf{1.5 \ A}$$

$$I_L = \frac{E_{Th}}{R_{Th} + R_L} = \frac{6 \ V}{2 \ \Omega + 10 \ \Omega} = \frac{6 \ V}{12 \ \Omega} = \mathbf{0.5 \ A}$$

$$I_L = \frac{E_{Th}}{R_{Th} + R_L} = \frac{6 \ V}{2 \ \Omega + 100 \ \Omega} = \frac{6 \ V}{102 \ \Omega} = \mathbf{58.82 \ mA}$$

One true advantage of Thévenin's theorem is now evident: Once the equivalent circuit has been determined, it is relatively easy to find the new load current for each load value. If it were not for the equivalent circuit, **the entire network would have to be analyzed for each load value,** and that would take a great deal more time. The chance of error is also greater when compared to using the one equation derived above.

EXAMPLE 9.6 Find the Thévenin equivalent circuit for the shaded area of the network of Fig. 9.28.

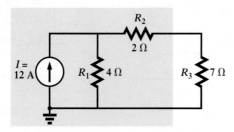

FIG. 9.28
Finding the Thévenin equivalent circuit for the network in the shaded area (Example 9.6).

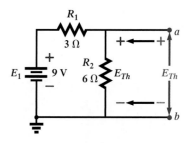

FIG. 9.26
Determining E_{Th} for the network of Fig. 9.23.

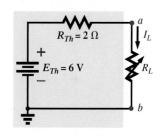

FIG. 9.27
Substituting the Thévenin equivalent circuit for the network external to R_L in Fig. 9.23.

Solution:

Steps 1–2: Resistor R_3 was removed and the resulting terminals were carefully marked as shown in Fig. 9.29.

Step 3: When the current source is replaced by an open-circuit equivalent as shown in Fig. 9.30(a), we find that resistors R_2 and R_1 are in series as shown in Fig. 9.30(b). The Thévenin resistance between the two marked terminals of the redrawn circuit of Fig. 9.30(b) is then

$$R_{Th} = R_2 + R_1 = 2\,\Omega + 4\,\Omega = \mathbf{6\,\Omega}$$

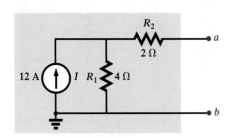

FIG. 9.29

Establishing the terminals of particular interest for the network of Fig. 9.28.

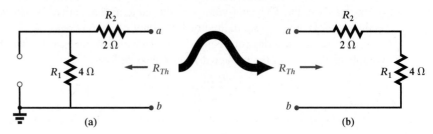

FIG. 9.30

Determining R_{Th} for the network of Fig. 9.28.

Step 4: The source is returned to the configuration, and the Thévenin voltage is defined in Fig. 9.31. For this case it is particularly important that you correctly understand the characteristics of an open circuit. As shown in Fig. 9.31, the current of the open circuit must be zero amperes; in addition, because the 2 Ω resistor is in series with the open circuit, its current must be also zero amperes. The result is that the voltage across the 2 Ω resistor must be zero volts also as determined by Ohm's law. Since the voltage drop across the 2 Ω resistor is zero volts, the voltage at the top of the 4 Ω resistor must be the same as at point *a*. Similarly, the voltage at the bottom of the 4 Ω resistor must be the same as at point *b*. The result is that the Thévenin voltage is the same as across the 4 Ω resistor. Due to the open circuit, the current source and the 4 Ω resistor are in series (and in parallel). The current through the 4 Ω resistor must therefore be the source current, and the Thévenin voltage is determined by

$$E_{Th} = V_1 = I_1 R_1 = IR_1 = (12\text{ A})(4\,\Omega) = \mathbf{48\ V}$$

If we chose to simply apply Kirchhoff's voltage law around the closed loop formed by the resistors and the open circuit, we would obtain

$$+V_1 - V_2 - E_{Th} = 0$$

and

$$E_{Th} = V_1 - V_2 = V_1 - 0\text{ V}$$

so that

$$E_{Th} = V_1$$

Step 5: In Fig. 9.32, the Thévenin circuit was constructed and resistor R_3 was reattached. The voltage, current, or power to the 7 Ω resistor can now be found with little difficulty.

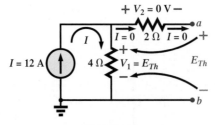

FIG. 9.31

Determining E_{Th} for the network of Fig. 9.28.

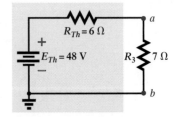

FIG. 9.32

Substituting the Thévenin equivalent circuit in the network external to resistor R_3 of Fig. 9.28.

The next example will demonstrate that Thévenin's theorem can be used to solve networks with sources that are not in series or parallel. In fact, this example will examine the same network as appeared in Example 8.9 of the branch-current analysis.

EXAMPLE 9.7 Using Thévenin's theorem, find the current through resistor R_3 of Fig. 9.33. Check to see if it matches the 1 A solution of Example 8.9.

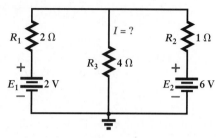

FIG. 9.33

Using Thévenin's theorem to find the current through the 4 Ω resistor (Example 9.7).

Solution:

Steps 1–2: The 4 Ω resistor was removed and the remaining terminals were marked as shown in Fig. 9.34.

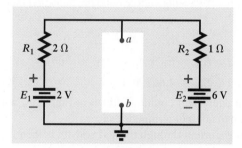

FIG. 9.34

Network of Fig. 9.33 after removing the 4 Ω resistor and marking the terminals.

Step 3: In Fig. 9.35(a), both sources have been replaced by their short-circuit equivalent, resulting in the parallel combination of elements appearing in Fig. 9.35(b). The Thévenin resistance is then

$$R_{Th} = R_2 \parallel R_1 = 1\ \Omega \parallel 2\ \Omega = \frac{(1\ \Omega)(2\ \Omega)}{1\ \Omega + 2\ \Omega} = \frac{2}{3}\ \Omega = \mathbf{0.667\ \Omega}$$

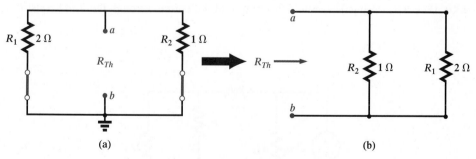

(a)

(b)

FIG. 9.35

Determining R_{Th} for the configuration of Fig. 9.34.

Step 4: The Thévenin voltage is identified in Fig. 9.36 as a voltage that extends across the E_1 and R_1 or the E_2 and R_2 combination. The result is that the voltage across one of the resistors will have to be determined and Kirchhoff's voltage law applied.

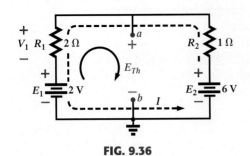

FIG. 9.36

Determining E_{Th} for the configuration of Fig. 9.34.

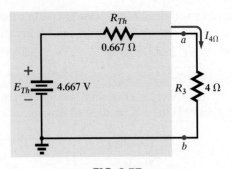

FIG. 9.37

Constructing the Thévenin equivalent circuit and finding current $I_{4\Omega}$.

Due to the open circuit defining the Thévenin voltage, all the elements of the network form a series configuration. Since E_2 is the larger of the two sources, the current will have a counterclockwise direction. Using Ohm's law:

$$I = \frac{E_2 - E_1}{R_2 + R_1} = \frac{6\ \text{V} - 2\ \text{V}}{1\ \Omega - 2\ \Omega} = \frac{4\ \text{V}}{3\ \Omega} = 1\tfrac{1}{3}\ \text{A}$$

The voltage across resistor R_1 is then

$$V_1 = I_1 R_1 = I R_1 = (1\tfrac{1}{3}\ \text{A})(2\ \Omega) = 2.667\ \text{V}$$

Applying Kirchhoff's voltage law around the indicated loop (the left window) will result in

$$+E_1 + V_1 - E_{Th} = 0$$

and

$$E_{Th} = E_1 + V_1 = 2\ \text{V} + 2.667\ \text{V}$$

so that

$$E_{Th} = \mathbf{4.667\ V}$$

Step 5: The Thévenin equivalent circuit is drawn in Fig. 9.37, with the 4 Ω resistor inserted between the output terminals.

The resulting current through the 4 Ω resistor is then

$$I_{4\Omega} = \frac{E_{Th}}{R_{Th} + R_3} = \frac{4.667\ \text{V}}{0.667\ \Omega + 4\ \Omega} = \frac{4.667\ \text{V}}{4.667\ \Omega} = \mathbf{1\ A}$$

which matches the result of Example 8.9.

The next example will include two sources that are not in series or parallel. The complexity of the network is such that the superposition theorem will be used to find the Thévenin voltage.

EXAMPLE 9.8 Using Thévenin's theorem, find the current through the 22 Ω resistor of Fig. 9.38.

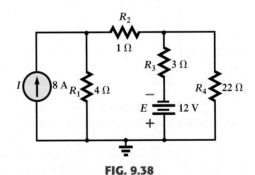

FIG. 9.38

Network to be analyzed in Example 9.8.

Steps 1–2: The 22 Ω resistor was removed and the resulting terminals were marked as shown in Fig. 9.39. The network to be replaced by a single voltage source and a series resistor appears in the shaded region.

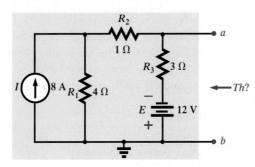

FIG. 9.39

Removing resistor R_4 from Fig. 9.38 and defining the
Thévenin equivalent circuit to be found.

Step 3: The current source was replaced by an open circuit and the voltage source by a short circuit to permit the determination of the Thévenin resistance as shown in Fig. 9.40(a). For clarity, the resulting configuration is redrawn as shown in Fig. 9.40(b). It is certainly interesting that the removal of the current source resulted in a series connection between resistors R_1 and R_2.

The Thévenin resistance is determined by

$$R_{Th} = R_3 \| (R_1 + R_2) = 3\,\Omega \| (4\,\Omega + 1\,\Omega) = 3\,\Omega \| 5\,\Omega = \frac{(3\,\Omega)(5\,\Omega)}{3\,\Omega + 5\,\Omega} = \mathbf{1.875\ \Omega}$$

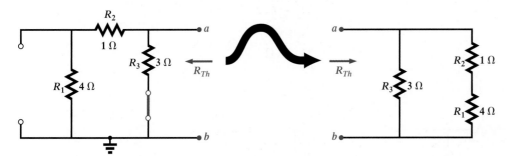

FIG. 9.40

Determining R_{Th} for the network of Fig. 9.39.

Step 4: Due to the presence of two sources, the superposition theorem will be used to find the Thévenin open-circuit voltage. First the voltage source will be removed as shown in Fig. 9.41 to determine the contribution of the current source to the Thévenin voltage. As shown in the figure, the Thévenin voltage is directly across the 3 Ω resistor. The current source supplies a current that will divide between the 4 Ω resistor and the series combination of the 1 Ω and 3 Ω resistors.

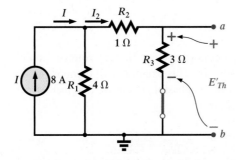

FIG. 9.41

Finding the contribution to E_{Th} by the 8 A current source
using the superposition theorem.

Applying the current divider rule will result in

$$I_2 = \frac{R_1 I}{R_1 + R_2 + R_3} = \frac{(4\ \Omega)(8\ \text{A})}{4\ \Omega + 1\ \Omega + 3\ \Omega} = \frac{32}{8}\ \text{A} = 4\ \text{A}$$

Then, using Ohm's law,

$$E'_{Th} = I_3 R_3 = I_2 R_3 = (4\ \text{A})(3\ \Omega) = 12\ \text{V}$$

In Fig. 9.42, the current source was removed to determine the contribution of the voltage source to the Thévenin voltage. The result is a series circuit with a current I as shown in the figure. The direction is established by the polarity of the source. The Thévenin voltage is across the series combination of the voltage source and the 3 Ω resistor.

The current is determined by

$$I = \frac{E}{R_1 + R_2 + R_3} = \frac{12\ \text{V}}{4\ \Omega + 1\ \Omega + 3\ \Omega} = \frac{12\ \text{V}}{8\ \Omega} = 1.5\ \text{A}$$

and the voltage across R_3 by

$$V_3 = I_3 R_3 = I R_3 = (1.5\ \text{A})(3\ \Omega) = 4.5\ \text{V}$$

Applying Kirchhoff's voltage law around the loop indicated in Fig. 9.42 will result in

$$-E + V_3 - E''_{Th} = 0$$
$$E''_{Th} = -E + V_3 = -12\ \text{V} + 4.5\ \text{V} = -7.5\ \text{V}$$

The total Thévenin voltage is

$$E_{Th} = E'_{Th} + E''_{Th} = 12\ \text{V} - 7.5\ \text{V} = \textbf{4.5 V}$$

Step 5: The Thévenin equivalent circuit is drawn in Fig. 9.43 with the 22 Ω resistor connected between its terminals. The resulting current through the 22 Ω resistor is

$$I_{R_4} = \frac{E_{Th}}{R_{Th} + R_4} = \frac{4.5\ \text{V}}{1.875\ \Omega + 22\ \Omega}$$

$$= \frac{4.5\ \text{V}}{23.875\ \Omega} = \textbf{188.5 mA}$$

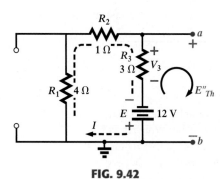

FIG. 9.42

Finding the contribution to E_{Th} by the 12 V voltage source using the superposition theorem.

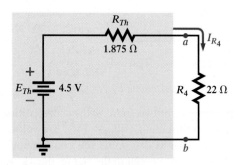

FIG. 9.43

Drawing the Thévenin equivalent circuit and inserting resistor R_4 between its output terminals.

In the next example, the shorthand notation for voltage sources was employed to reduce the complexity of the network. This example will demonstrate how important it can be to redraw the network as often as possible to ensure that the correct Thévenin resistance and voltage source are determined.

EXAMPLE 9.9 Determine the Thévenin equivalent circuit for the network in the shaded area of Fig. 9.44.

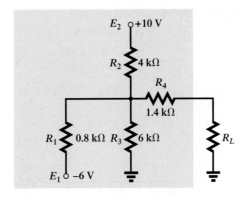

FIG. 9.44

Network to be analyzed in Example 9.9.

Solution:

Steps 1–2: Resistor R_L has been removed and the resulting terminals have been marked as shown in Fig. 9.45. Note the friendlier appearance of the network after it has been redrawn in the familiar series-parallel fashion.

Step 3: Both sources have been removed and the Thévenin resistance has been defined in Fig. 9.46. We now have a parallel combination of resistors R_1, R_2, and R_3, which in turn are in series with resistor R_4.

The Thévenin resistance is

$$R_{Th} = R_4 + R_1 \| R_2 \| R_3 = 1.4 \text{ k}\Omega + 0.8 \text{ k}\Omega \| 4 \text{ k}\Omega \| 6 \text{ k}\Omega$$
$$= 1.4 \text{ k}\Omega + 0.6 \text{ k}\Omega = \mathbf{2 \text{ k}\Omega}$$

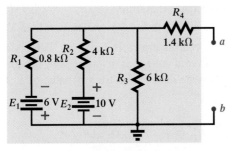

FIG. 9.45

Identifying the terminals of particular interest and redrawing the network of Fig. 9.44.

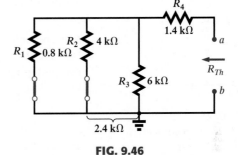

FIG. 9.46

Determining R_{Th} for the network of Fig. 9.44.

Step 4: Since two sources are present, the superposition theorem will again be applied. In most cases there is more than one way to find the Thévenin voltage. For instance, in Fig. 9.45, both of the voltage sources and their series resistances could be converted to current sources. They could then be reduced to one current source and the three parallel resulting resistors combined. The result would be a simpler network to deal with and a quick solution for the Thévenin resistance and voltage. The approach we are using, however, is just as effective and gives us a chance to note how elements form simple series or parallel combinations when sources are removed.

In Fig. 9.47, the voltage source E_2 was removed and the Thévenin voltage defined. Since resistor R_4 is in series with the open circuit defining the Thévenin voltage, the current through resistor R_4 must be zero amperes. The result is that V_4 must also be zero volts. Therefore, the voltage across resistor R_3 is the Thévenin voltage. However, the polarity of the voltage source E_1 will establish a voltage across R_3 that is opposite the defined Thévenin voltage.

Voltage V_3 is across the parallel combination of the 4 kΩ and 6 kΩ resistors. Their parallel resistance is

$$R' = R_2 \| R_3 = 4 \text{ k}\Omega \| 6 \text{ k}\Omega = 2.4 \text{ k}\Omega$$

Applying the voltage divider rule will result in

$$V_3 = \frac{R'E_1}{R' + R_1} = \frac{(2.4 \text{ k}\Omega)(6 \text{ V})}{2.4 \text{ k}\Omega + 0.8 \text{ k}\Omega} = \frac{14.4 \text{ V}}{3.2} = 4.5 \text{ V}$$

with
$$E'_{Th} = -V_3 = -4.5 \text{ V}$$

The impact of removing the 6 V source is shown in Fig. 9.48. The resulting network is very similar to that of Fig. 9.47, with a similar sequence of calculations to determine V_3. The parallel combination of the 0.8 kΩ and 6 kΩ resistors is

$$R'' = R_1 \| R_3 = 0.8 \text{ k}\Omega \| 6 \text{ k}\Omega = 0.706 \text{ k}\Omega$$

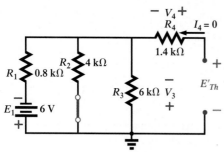

FIG. 9.47

Determining the contribution to E_{Th} from the source E_1 for the network of Fig. 9.44.

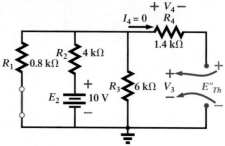

FIG. 9.48

Determining the contribution to E_{Th} from the source E_2 for the network of Fig. 9.44.

Applying the voltage divider rule yields

$$V_3 = \frac{R''E_2}{R'' + R_2} = \frac{(0.706 \text{ k}\Omega)(10 \text{ V})}{0.706 \text{ k}\Omega + 4 \text{ k}\Omega} = \frac{7.06 \text{ V}}{4.706} = 1.5 \text{ V}$$

with
$$E''_{Th} = V_3 = 1.5 \text{ V}$$

The resulting Thévenin voltage is therefore

$$E_{Th} = E'_{Th} + E''_{Th} = -4.5 \text{ V} + 1.5 \text{ V} = \mathbf{-3 \text{ V}}$$

Step 5: The Thévenin equivalent circuit is drawn in Fig. 9.49 with resistor R_L replaced. Note the polarity of the Thévenin voltage as determined by the above solution. The current through R_L can now be determined for any value of R_L by simply using Ohm's law as follows:

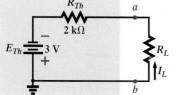

FIG. 9.49

Substituting the Thévenin equivalent circuit for the network external to resistor R_L of Fig. 9.44.

$$I_L = \frac{E_{Th}}{R_{Th} + R_L} = \frac{3 \text{ V}}{2 \text{ k}\Omega + R_L}$$

There is no need to repeat an entire analysis of the network for each value of R_L—a real time saver.

Example 9.10 is one of the more difficult problems to which to apply Thévenin's theorem. It requires a careful redrawing of the network, with particular attention to "holding on to" the marked terminals. Note also that it is a complex network without a single series or parallel combination.

EXAMPLE 9.10 Find the Thévenin equivalent circuit for the network appearing in the shaded area of Fig. 9.50.

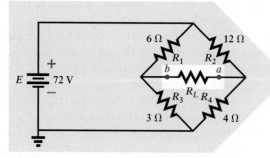

FIG. 9.50

Bridge network to be analyzed in Example 9.10.

Steps 1–2: Resistor R_L was removed and the resulting terminals were marked as shown in Fig. 9.51.

Step 3: In Fig. 9.52(a), the source has been removed and the Thévenin resistance defined. In particular, note that points c and c' of the network are now connected by the short-circuit equivalent of the source. If we hold on to terminals a and b, we can flip the top portion of the network down as shown in Fig. 9.52(b) and connect points c and c'. The result is that R_1 and R_3 are in parallel, as well as R_2 and R_4. The two parallel combinations are then in series, resulting in

$$R_{Th} = R_1 \| R_3 + R_2 \| R_4 = 6\,\Omega \| 3\,\Omega + 12\,\Omega \| 4\,\Omega = 2\,\Omega + 3\,\Omega = \mathbf{5\,\Omega}$$

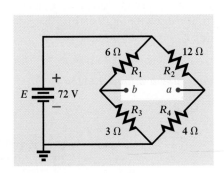

FIG. 9.51

Identifying the terminals of particular interest for the network of Fig. 9.50.

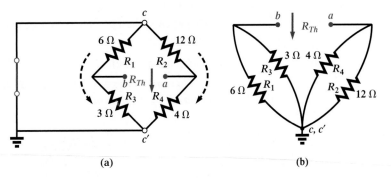

(a)   (b)

FIG. 9.52

Solving for R_{Th} for the network of Fig. 9.50.

Step 4: To determine the Thévenin voltage, the network was drawn in the more familiar rectangular shape shown in Fig. 9.53. As shown in the figure, we now have a parallel network with 72 V across each branch.

Voltage V_1 can be determined using the voltage divider rule as follows:

$$V_1 = \frac{R_1 E}{R_1 + R_3} = \frac{(6\,\Omega)(72\,\text{V})}{6\,\Omega + 3\,\Omega} = \frac{432\,\text{V}}{9} = 48\,\text{V}$$

Voltage V_2 can also be determined using the voltage divider rule as follows:

$$V_2 = \frac{R_2 E}{R_2 + R_4} = \frac{(12\,\Omega)(72\,\text{V})}{12\,\Omega + 4\,\Omega} = \frac{864\,\text{V}}{16} = 54\,\text{V}$$

Applying Kirchhoff's voltage law around the indicated loop will result in

$$+V_1 - V_2 + E_{Th} = 0$$

$$E_{Th} = V_2 - V_1 = 54\,\text{V} - 48\,\text{V} = \mathbf{6\,V}$$

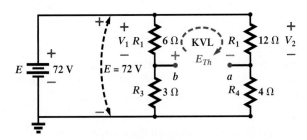

FIG. 9.53

Determining E_{Th} for the network of Fig. 9.50.

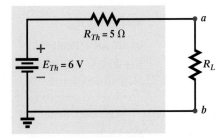

FIG. 9.54

Substituting the Thévenin equivalent circuit for the network external to resistor R_L of Fig. 9.50.

Step 5: The resulting Thévenin equivalent circuit has been drawn in Fig. 9.54, with resistor R_L replaced between its terminals. Resistor R_L will respond in the same way to the reduced circuit of Fig. 9.54 as to the complex network of Fig. 9.50.

Experimental Procedures

Now that the analytical procedure has been described in detail and a sense for the Thévenin impedance and voltage established, it is time to investigate how both quantities can be determined using an experimental procedure.

Even though the Thévenin resistance is usually the easiest to determine analytically, the Thévenin voltage is often the easiest to determine experimentally, and therefore it will be examined first.

Measuring E_{Th}   Since the network of Fig. 9.55(a) (from Example 9.8) has an equivalent Thévenin circuit such as appearing in Fig. 9.55(b), the open-circuit Thévenin voltage can be determined by simply placing a voltmeter on the output terminals of Fig. 9.55(a) as shown. This is due to the fact that the open circuit of Fig. 9.55(b) dictates that the current through and the voltage across the Thévenin resistance must be zero. The result for Fig. 9.55(b) is that

$$V_{oc} = E_{Th} = \textbf{4.5 V}$$

In general, therefore,

the Thévenin voltage is determined by simply connecting a voltmeter to the output terminals of the network.

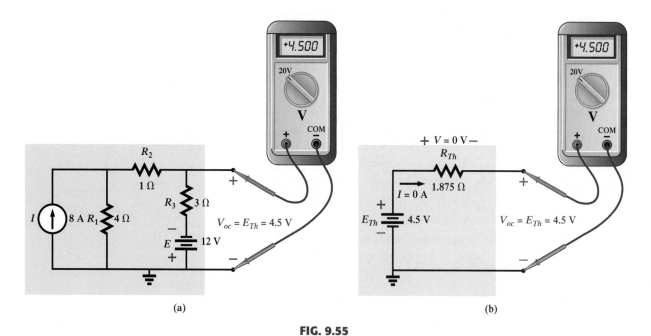

FIG. 9.55

Measuring the Thévenin voltage with a voltmeter: (a) actual network; (b) Thévenin equivalent.

Measuring R_{Th}

USING AN OHMMETER:

In Fig. 9.56, the sources of Fig. 9.55(a) have been set to zero and an ohmmeter has been applied to measure the Thévenin resistance. In Fig. 9.55(b), it is clear that if the Thévenin voltage is set to zero volts, the ohmmeter will read the Thévenin resistance directly.

In general, therefore,

the Thévenin resistance can be measured by simply setting all the sources to zero and measuring the resistance at the output terminals.

It is important to remember, however, that ohmmeters cannot be used on live circuits, and you cannot set a voltage source by putting a short circuit across it—it causes instant damage. The source must be set to zero or removed entirely and replaced by a direct connection. For the current source, the open-circuit condition must be clearly established; otherwise, the measured resistance will be incorrect. For most situations, it is usually best to simply remove the sources and replace them by the appropriate equivalent.

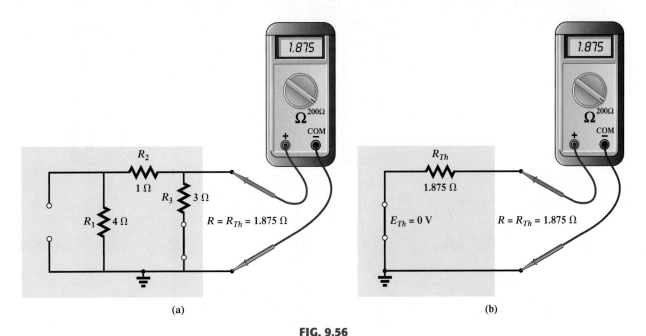

FIG. 9.56

Measuring R_{Th} with an ohmmeter: (a) actual network; (b) Thévenin equivalent.

USING A POTENTIOMETER:

If we use a potentiometer to measure the Thévenin resistance, the sources can be left as is. For this reason alone, this approach is one of the more popular. In Fig. 9.57(a), a potentiometer has been connected across the output terminals of the network to establish the condition appearing in Fig. 9.57(b) for the Thévenin equivalent. If the resistance of the potentiometer is now adjusted so that the voltage across the potentiometer is one-half the measured Thévenin voltage, the Thévenin resistance must match that of the potentiometer. Recall that for a series circuit, the applied voltage will divide equally across two equal series resistors.

If the potentiometer is then disconnected and the resistance measured with an ohmmeter as shown in Fig. 9.57(c), the ohmmeter will display the Thévenin resistance of the network. In general, therefore,

the Thévenin resistance can be measured by applying a potentiometer to the output terminals and varying the resistance until the output voltage is one-half the measured Thévenin voltage. The resistance of the potentiometer is the Thévenin resistance for the network.

USING THE SHORT-CIRCUIT CURRENT:

The Thévenin resistance can also be determined by placing a short circuit across the output terminals and finding the current through the short circuit. Since ammeters ideally have zero internal ohms between their terminals, hooking up an ammeter as shown in Fig. 9.58(a) will have the effect of both hooking up a short circuit across the terminals and measuring the resulting current. The same ammeter was connected across the Thévenin equivalent circuit in Fig. 9.58(b).

On a practical level, it is assumed, of course, that the internal resistance of the ammeter is approximately zero ohms in comparison to the other resistors of the network. It is also important to be sure that the resulting current does not exceed the maximum current for the chosen ammeter scale.

In Fig. 9.58(b), since the short-circuit current is

$$I_{sc} = \frac{E_{Th}}{R_{Th}}$$

the Thévenin resistance can be determined by

$$R_{Th} = \frac{E_{Th}}{I_{sc}}$$

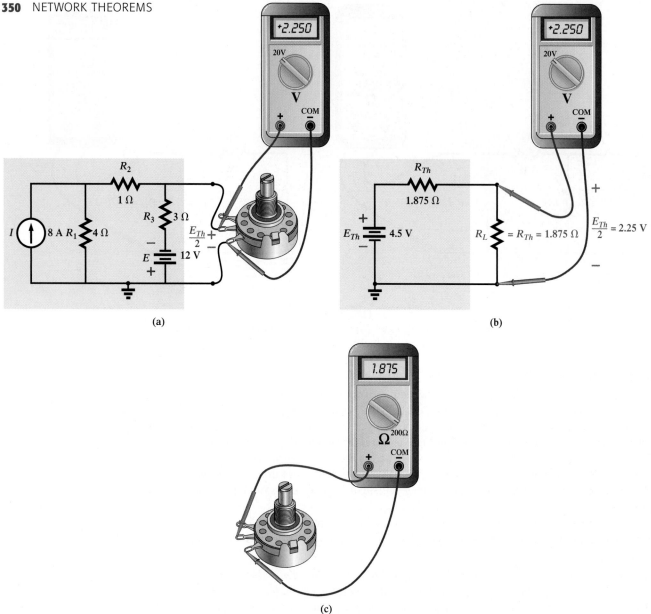

FIG. 9.57

Using a potentiometer to determine R_{Th}: (a) actual network; (b) Thévenin equivalent; (c) measuring R_{Th}.

In general, therefore,

the Thévenin resistance can be determined by hooking up an ammeter across the output terminals to measure the short-circuit current and then using the open-circuit voltage to calculate the Thévenin resistance in the following manner:

$$R_{Th} = \frac{V_{oc}}{I_{sc}} \qquad\qquad (9.1)$$

As a result, we have three ways to measure the Thévenin resistance of a configuration. Because of the concern about setting the sources to zero in the first procedure and the concern about current levels in the last, the second method is often chosen.

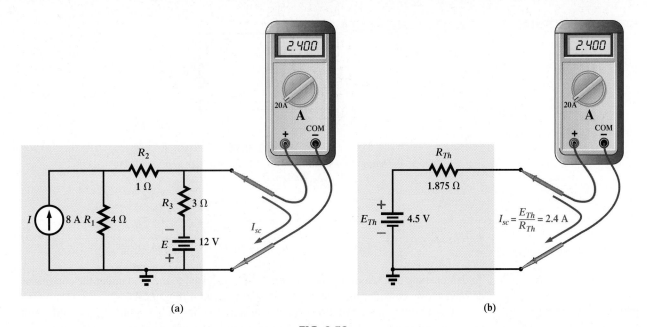

FIG. 9.58

Determing R_{Th} using the short-circuit current: (a) actual network; (b) Thévenin equivalent.

9.4 MAXIMUM POWER TRANSFER THEOREM

It is often important to know the following:

What load should be applied to a system to ensure that the load is receiving maximum power from the system?

and, conversely:

For a particular load, what conditions should be imposed on the source to ensure that it will deliver the maximum power available?

Even if a load cannot be set at the value that would result in maximum power transfer, it is often helpful to have some idea of the value that will draw maximum power so that you can compare it to the load at hand. For instance, if a design calls for a load of 100 Ω, to ensure that the load receives maximum power, using a resistor of 1 Ω or 1 kΩ would result in a power transfer that would be much less than the maximum possible. However, using a load of 82 Ω or 120 Ω would probably result in a fairly good level of power transfer.

Fortunately, the process of finding the load that will receive maximum power from a particular system is quite straightforward due to the **maximum power transfer theorem,** which states the following:

A load will receive maximum power from a network when its resistance is exactly equal to the Thévenin resistance of the network applied to the load. That is,

$$R_L = R_{Th} \qquad\qquad (9.2)$$

In other words, for the Thévenin equivalent circuit of Fig. 9.59, when the load is set equal to the Thévenin resistance, the load will receive maximum power from the network.

Using Fig. 9.59 with $R_L = R_{Th}$, the maximum power delivered to the load can be determined by first finding the current:

$$I_L = \frac{E_{Th}}{R_{Th} + R_L} = \frac{E_{Th}}{R_{Th} + R_{Th}} = \frac{E_{Th}}{2R_{Th}}$$

Then substitute into the power equation:

$$P_L = I_L^2 R_L = \left(\frac{E_{Th}}{2R_{Th}}\right)^2 (R_{Th}) = \frac{E_{Th}^2 R_{Th}}{4R_{Th}^2}$$

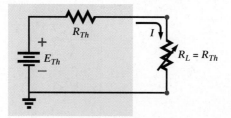

FIG. 9.59

Defining the conditions for maximum power to a load using the Thévenin equivalent circuit.

and

$$P_{L_{max}} = \frac{E_{Th}^2}{4R_{Th}}$$

(9.3)

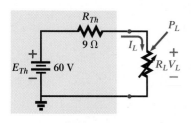

FIG. 9.60

Thévenin equivalent circuit to be used to validate the maximum power transfer theorem.

To demonstrate that maximum power is indeed transferred to the load under the conditions defined above, consider the Thévenin equivalent circuit of Fig. 9.60.

Before getting into detail, however, if you were to guess what value of R_L would result in maximum power transfer to R_L, you might think that the smaller the value of R_L, the better, because the current would reach a maximum when it is squared in the power equation. The problem is, however, that in the equation $P_L = I_L^2 R_L$, the load resistance is a multiplier. As it gets smaller, it will form a smaller product. Then again, you might suggest larger values of R_L, because the output voltage would increase and power is determined by $P_L = V_L^2/R_L$. This time, however, the load resistance is in the denominator of the equation and will cause the resulting power to decrease. A balance must obviously be made between the load resistance and the resulting current or voltage. In the description to follow we will show that

maximum power transfer occurs when the load voltage and current are one-half of their maximum possible values.

For the circuit of Fig. 9.60, the current through the load is determined by

$$I_L = \frac{E_{Th}}{R_{Th} + R_L} = \frac{60\text{ V}}{9\ \Omega + R_L}$$

The voltage is determined by

$$V_L = \frac{R_L E_{Th}}{R_L + R_{Th}} = \frac{R_L(60\text{ V})}{R_L + R_{Th}}$$

and the power by

$$P_L = I_L^2 R_L = \left(\frac{60\text{ V}}{9\ \Omega + R_L}\right)^2 (R_L) = \frac{3600 R_L}{(9\ \Omega + R_L)^2}$$

If we tabulate the three quantities versus a range of values for R_L from 0.1 Ω to 30 Ω, we obtain the results appearing in Table 9.1. Note in particular that when R_L is equal to the Thévenin resistance of 9 Ω, the power has a maximum value of 100 W, the current is 3.33 A or one-half its maximum value of 6.60 A (as would result with a short circuit across the output terminals), and the voltage across the load is 30 V or one-half its maximum value of 60 V (as would result with an open circuit across its output terminals). As you can see, there is no question that maximum power is transferred to the load when the load equals the Thévenin value.

The power to the load versus the range of resistor values is provided in Fig. 9.61. Note in particular that for values of load resistance less than the Thévenin value, the change is dramatic as it approaches the peak value. However, for values greater than the Thévenin value, the drop is a great deal more gradual. This is important because it tells you the following:

If the load applied is less than the Thévenin resistance, the power to the load will drop off rapidly as it gets smaller. However, if the applied load is greater than the Thévenin resistance, the power to the load will not drop off as rapidly as it increases.

TABLE 9.1

R_L (Ω)	P_L (W)		I_L (A)		V_L (V)	
0.1	4.35		6.59		0.66	
0.2	8.51		6.52		1.30	
0.5	19.94		6.32		3.16	
1	36.00		6.00		6.00	
2	59.50		5.46		10.91	
3	75.00		5.00		15.00	
4	85.21		4.62		18.46	
5	91.84		4.29		21.43	
6	96.00		4.00		24.00	
7	98.44	Increase	3.75	Decrease	26.25	Increase
8	99.65	↓	3.53	↓	28.23	↓
9 (R_{Th})	100.00 (Maximum)		3.33 ($I_{max}/2$)		30.00 ($E_{Th}/2$)	
10	99.72		3.16		31.58	
11	99.00		3.00		33.00	
12	97.96		2.86		34.29	
13	96.69		2.73		35.46	
14	95.27		2.61		36.52	
15	93.75		2.50		37.50	
16	92.16		2.40		38.40	
17	90.53		2.31		39.23	
18	88.89		2.22		40.00	
19	87.24		2.14		40.71	
20	85.61		2.07		41.38	
25	77.86		1.77		44.12	
30	71.00		1.54		46.15	
40	59.98		1.22		48.98	
100	30.30		0.55		55.05	
500	6.95	Decrease	0.12	Decrease	58.94	Increase
1000	3.54	↓	0.06		59.47	↓

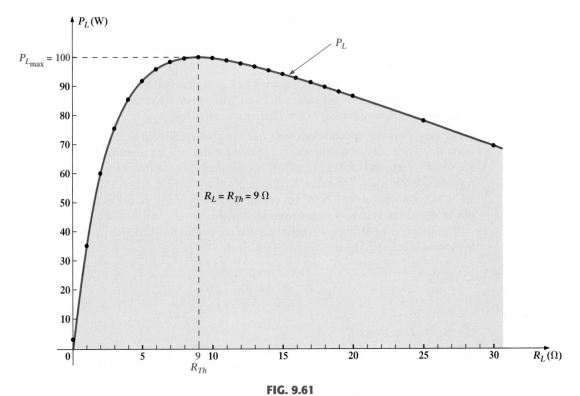

FIG. 9.61
P_L versus R_L for the network of Fig. 9.60.

For instance, the power to the load is at least 90 W for the range of about 4.5 Ω to 9 Ω below the peak value, but it is at least the same level for a range of about 9 Ω to 18 Ω above the peak value. The range below the peak is 4.5 Ω, while the range above the peak is almost twice as much at 9 Ω. As mentioned above, if maximum transfer conditions cannot be established, at least we now know from Fig. 9.61 that any resistance relatively close to the Thévenin value will result in a strong transfer of power. More distant values such as 1 Ω or 100 Ω will result in much lower levels.

It is particularly interesting to plot the power to the load versus load resistance using a log scale, as shown in Fig. 9.62. Logarithms will be discussed in detail in Chapter 17, but for now notice that the spacing between values of R_L is not linear, but the distance between powers of ten (such as 0.1 and 1, 1 and 10, and 10 and 100) are all equal. The advantage of the log scale is that a wide frequency range can be plotted on a relatively small graph.

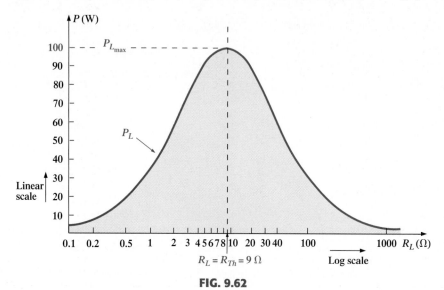

FIG. 9.62

P_L versus R_L for the network of Fig. 9.60.

Note in Fig. 9.62 that a smooth, bell-shaped curve results that is symmetrical about the Thévenin resistance of 9 Ω. At 0.1 Ω, the power has dropped to about the same level as appearing at 1000 Ω, and at 1 Ω and 100 Ω, the power has dropped to the neighborhood of 30 W.

Although all of the above discussion centers on the power to the load, it is important to remember the following:

The total power delivered by a supply such as E_{Th} is absorbed by both the Thévenin equivalent resistance and the load resistance. Any power delivered by the source that does not get to the load is lost to the Thévenin resistance.

Under maximum power conditions, only half the power delivered by the source gets to the load. Now, that sounds disastrous, but remember that we are starting out with a fixed Thévenin voltage and resistance, and the above simply tells us that we must make the two resistance levels equal if we want maximum power to the load. On an efficiency basis, we are working at only a 50% level, but we are content because **we are getting maximum power out of the system we were presented with.**

The dc operating efficiency is defined as the ratio of the power delivered to the load (P_L) to the power delivered by the source (P_s). That is,

$$\eta\% = \frac{P_L}{P_s} \times 100\% \tag{9.4}$$

For the situation where $R_L = R_{Th}$,

$$\eta\% = \frac{I_L^2 R_L}{I_L^2 R_T} \times 100\% = \frac{R_L}{R_T} \times 100\% = \frac{R_{Th}}{R_{Th} + R_{Th}} \times 100\%$$

$$= \frac{R_{Th}}{2R_{Th}} \times 100\% = \frac{1}{2} \times 100\% = \mathbf{50\%}$$

For the circuit of Fig. 9.60, if we plot the efficiency of operation versus load resistance, we obtain the plot of Fig. 9.63, which clearly shows that the efficiency continues to rise to a 100% level as R_L gets larger. Note in particular that the efficiency is 50% when $R_L = R_{Th}$.

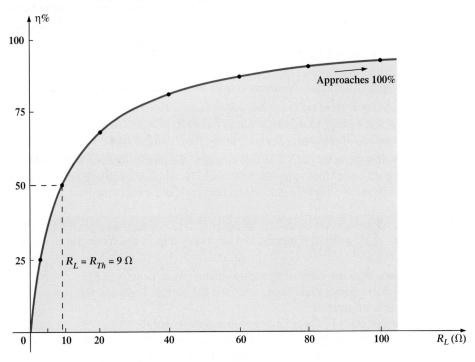

FIG. 9.63

Efficiency of operation versus increasing values of R_L.

To ensure that you completely understand the effect of the maximum power transfer theorem and the efficiency criteria, consider the circuit of Fig. 9.64 where the load resistance is set at 100 Ω and the power to the Thévenin resistance and to the load are calculated as follows:

$$I_L = \frac{E_{Th}}{R_{Th} + R_L} = \frac{60 \text{ V}}{9 \text{ Ω} + 100 \text{ Ω}} = \frac{60 \text{ V}}{109 \text{ Ω}} = 550.5 \text{ mA}$$

with $\qquad P_{R_{Th}} = I_L^2 R_{Th} = (550.5 \text{ mA})^2 (9 \text{ Ω}) \cong \textbf{2.73 W}$

and $\qquad P_L = I_L^2 R_L = (550.5 \text{ mA})^2 (100 \text{ Ω}) \cong \textbf{30.3 W}$

The results clearly show that most of the power supplied by the battery is getting to the load—a desirable attribute on an efficiency basis. However, the power getting to the load is only 30.3 W compared to the 100 W obtained under maximum power conditions. In general, therefore, the following guidelines apply:

If efficiency is the overriding factor, then the load should be much larger than the internal resistance of the supply. If maximum power transfer is desired and efficiency less of a concern, then the conditions dictated by the maximum power transfer theorem should be applied.

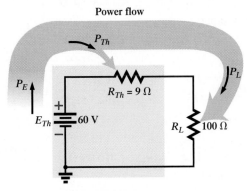

Power flow

FIG. 9.64

Examining a circuit with high efficiency but a relatively low level of power to the load.

A relatively low efficiency of 50% can be tolerated in situations where power levels are relatively low, such as in a wide variety of electronic systems, where maximum power transfer for the given system is usually more important. However, when large power levels are involved, such as at generating plants, efficiencies of 50% could not be tolerated. In fact, a great deal of expense and research is dedicated to raising power generating and transmission efficiencies a few percentage points. Raising an efficiency level of a 10 MkW power plant from 94% to 95% (a 1% increase) can save 0.1 MkW, or 100 million watts, of power—an enormous saving.

In all of the above discussions, the effect of changing the load was discussed for a fixed Thévenin resistance. Looking at the situation from a different viewpoint,

if the load resistance is fixed and does not match the applied Thévenin equivalent resistance, then some effort should be made (if possible) to redesign the system so that the Thévenin equivalent resistance is closer to the fixed applied load.

In other words, if a designer faces a situation where the load resistance is fixed, he/she should investigate whether the supply section should be replaced or redesigned to create a closer match of resistance levels to produce higher levels of power to the load.

EXAMPLE 9.11 A dc generator, battery, and laboratory supply are connected to resistive load R_L in Fig. 9.65.

a. For each, determine the value of R_L for maximum power transfer to R_L.
b. Under maximum power conditions, what are the current level and the power to the load for each configuration?
c. What is the efficiency of operation for each supply in part (b)?
d. If a load of 1 kΩ were applied to the laboratory supply, what would the power delivered to the load be? Compare your answer to the level of part (b). What is the level of efficiency?

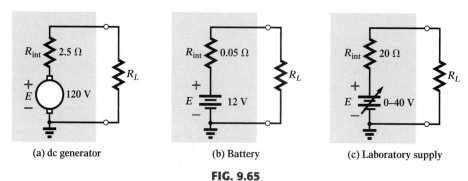

(a) dc generator   (b) Battery   (c) Laboratory supply

FIG. 9.65

Three types of dc supply for Example 9.11: (a) generator; (b) battery; (c) laboratory supply.

Solutions:

a. For the dc generator,

$$R_L = R_{Th} = R_{int} = \textbf{2.5 } \boldsymbol{\Omega}$$

For the 12 V car battery,

$$R_L = R_{Th} = R_{int} = \textbf{0.05 } \boldsymbol{\Omega}$$

For the dc laboratory supply,

$$R_L = R_{Th} = R_{int} = \textbf{20 } \boldsymbol{\Omega}$$

b. For the dc generator,

$$P_{L_{max}} = \frac{E_{Th}^2}{4R_{Th}} = \frac{E^2}{4R_{int}} = \frac{(120 \text{ V})^2}{4(2.5 \text{ }\Omega)} = \textbf{1.44 kW}$$

For the 12 V car battery,

$$P_{L_{max}} = \frac{E_{Th}^2}{4R_{Th}} = \frac{E^2}{4R_{int}} = \frac{(12 \text{ V})^2}{4(0.05 \text{ }\Omega)} = \textbf{720 W}$$

For the dc laboratory supply,

$$P_{L_{max}} = \frac{E_{Th}^2}{4R_{Th}} = \frac{E^2}{4R_{int}} = \frac{(40 \text{ V})^2}{4(20 \text{ }\Omega)} = \textbf{20 W}$$

c. They are all operating under a 50% efficiency level because $R_L = R_{Th}$.

d. The power to the load is determined as follows:

$$I_L = \frac{E}{R_{int} + R_L} = \frac{40 \text{ V}}{20 \text{ }\Omega + 1000 \text{ }\Omega} = \frac{40 \text{ V}}{1020 \text{ }\Omega} = 39.22 \text{ mA}$$

and

$$P_L = I_L^2 R_L = (39.22 \text{ mA})^2(1000 \text{ }\Omega) = \textbf{1.538 W}$$

The power level is significantly less than the 20 W achieved in part (b). The efficiency level is

$$\eta\% = \frac{P_L}{P_s} \times 100\% = \frac{1.538 \text{ W}}{EI_s} \times 100\% = \frac{1.538 \text{ W}}{(40 \text{ V})(39.22 \text{ mA})} \times 100\%$$

$$= \frac{1.538 \text{ W}}{1.569 \text{ W}} \times 100\% = \textbf{98.02\%}$$

which is markedly higher than achieved under maximum power conditions—albeit at the expense of the power level.

EXAMPLE 9.12 The analysis of a transistor network resulted in the reduced equivalent of Fig. 9.66.

a. Find the load resistance that will result in maximum power transfer to the load, and find the maximum power delivered.
b. If the load were changed to 68 kΩ, would you expect a fairly high level of power transfer to the load based on the results of part (a)? What would the new power level be? Is your initial assumption verified?
c. If the load were changed to 8.2 kΩ, would you expect a fairly high level of power transfer to the load based on the results of part (a)? What would the new power level be? Is your initial assumption verified?

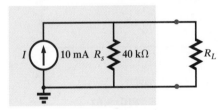

FIG. 9.66
Reduced equivalent network obtained from a
transistor amplifier design (Example 9.12).

Solutions:

a. Replacing the current source by an open-circuit equivalent will result in

$$R_{Th} = R_s = 40 \text{ k}\Omega$$

Restoring the current source and finding the open-circuit voltage at the output terminals will result in

$$E_{Th} = V_{oc} = IR_s = (10 \text{ mA})(40 \text{ k}\Omega) = 400 \text{ V}$$

For maximum power transfer to the load,

$$R_L = R_{Th} = \textbf{40 k}\Omega$$

with a maximum power level of

$$P_{L_{max}} = \frac{E_{Th}^2}{4R_{Th}} = \frac{(400 \text{ V})^2}{4(40 \text{ k}\Omega)} = \textbf{1 W}$$

b. Yes, because the 68 kΩ load is greater (note Fig. 9.61) than the 40 kΩ load, but relatively close in magnitude.

$$I_L = \frac{E_{Th}}{R_{Th} + R_L} = \frac{400 \text{ V}}{40 \text{ k}\Omega + 68 \text{ k}\Omega} = \frac{400 \text{ V}}{108 \text{ k}\Omega} \cong 3.7 \text{ mA}$$

$$P_L = I_L^2 R_L = (3.7 \text{ mA})^2(68 \text{ k}\Omega) \cong \textbf{0.93 W}$$

Yes, the power level of 0.93 W compared to the 1 W level of part (a) verifies the assumption.

c. No, 8.2 kΩ is quite a bit less (note Fig. 9.61) than the 40 kΩ value.

$$I_L = \frac{E_{Th}}{R_{Th} + R_L} = \frac{400 \text{ V}}{40 \text{ k}\Omega + 8.2 \text{ k}\Omega} = \frac{400 \text{ V}}{48.2 \text{ k}\Omega} \cong 8.3 \text{ mA}$$

$$P_L = I_L^2 R_L = (8.3 \text{ mA})^2 (8.2 \text{ k}\Omega) \cong \textbf{0.57 W}$$

Yes, the power level of 0.57 W compared to the 1 W level of part (a) verifies the assumption.

EXAMPLE 9.13   In Fig. 9.67, a fixed load of 16 Ω is applied to a 48 V supply with an internal resistance of 36 Ω.

a. For the conditions of Fig. 9.67, what is the power delivered to the load and lost to the internal resistance of the supply?
b. If the designer has some control over the internal resistance level of the supply, what value should he/she make it for maximum power to the load? What is the maximum power to the load? How does it compare to the level obtained in part (a)?
c. Without making a single calculation, if the designer could change the internal resistance to 22 Ω or 8.2 Ω, which value would result in more power to the load? Verify your conclusion by calculating the power to the load for each value.

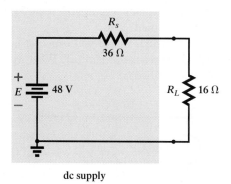

FIG. 9.67

dc supply with a fixed 16 Ω load (Example 9.13).

Solutions:

a. $$I_L = \frac{E}{R_s + R_L} = \frac{48 \text{ V}}{36 \text{ }\Omega + 16 \text{ }\Omega} = \frac{48 \text{ V}}{52 \text{ }\Omega} = 923.1 \text{ mA}$$

$$P_{R_s} = I_L^2 R_s = (923.1 \text{ mA})^2 (36 \text{ }\Omega) = \textbf{30.68 W}$$

$$P_L = I_L^2 R_L = (923.1 \text{ mA})^2 (16 \text{ }\Omega) = \textbf{13.63 W}$$

b. Be careful here. The quick response is to make the source resistance R_s equal to the load resistance to satisfy the criteria of the maximum power transfer theorem. However, this is a totally different type of problem from what was examined earlier in this section. If the load is fixed, the smaller the source resistance R_s, the more of the applied voltage will reach the load and the less will be lost in the internal series resistor. In fact, the source resistance should be made as small as possible. If zero ohms were possible for R_s, the voltage across the load would be the full supply voltage, and the power delivered to the load would equal

$$P_L = \frac{V_L^2}{R_L} = \frac{(48 \text{ V})^2}{16 \text{ }\Omega} = \textbf{144 W}$$

which is more than 10 times the value with a source resistance of 36 Ω.

c. Again, forget the impact of Fig. 9.61: The smaller the source resistance, the greater the power to the fixed 16 Ω load. Therefore, the 8.2 Ω resistance level will result in a higher power transfer to the load than the 22 Ω resistor, as shown on the next page.

For $R_s = 8.2 \, \Omega$:

$$I_L = \frac{E}{R_s + R_L} = \frac{48 \text{ V}}{8.2 \, \Omega + 16 \, \Omega} = \frac{48 \text{ V}}{24.2 \, \Omega} = 1.983 \text{ A}$$

and

$$P_L = I_L^2 R_L = (1.983 \text{ A})^2 (16 \, \Omega) \cong \mathbf{62.92 \text{ W}}$$

For $R_s = 22 \, \Omega$:

$$I_L = \frac{E}{R_s + R_L} = \frac{48 \text{ V}}{22 \, \Omega + 16 \, \Omega} = \frac{48 \text{ V}}{38 \, \Omega} = 1.263 \text{ A}$$

and

$$P_L = I_L^2 R_L = (1.263 \text{ A})^2 (16 \, \Omega) \cong \mathbf{25.52 \text{ W}}$$

EXAMPLE 9.14 Given the network of Fig. 9.68, find the value of R_L for maximum power to the load, and find the maximum power to the load.

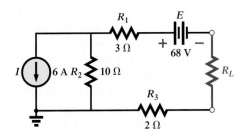

FIG. 9.68

Applying the maximum power transfer theorem to a network with two sources (Example 9.14).

Solution: The Thévenin resistance is determined from Fig. 9.69.

$$R_{Th} = R_1 + R_2 + R_3 = 3 \, \Omega + 10 \, \Omega + 2 \, \Omega = 15 \, \Omega$$

so that

$$R_L = R_{Th} = \mathbf{15 \, \Omega}$$

The Thévenin voltage is determined using Fig. 9.70, where

$$V_1 = V_3 = 0 \text{ V} \quad \text{and} \quad V_2 = I_2 R_2 = I R_2 = (6 \text{ A})(10 \, \Omega) = 60 \text{ V}$$

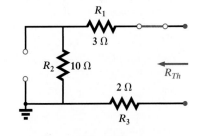

FIG. 9.69

Determining R_{Th} for the network external to resistor R_L of Fig. 9.68.

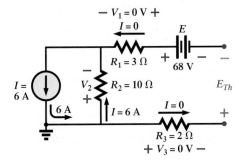

FIG. 9.70

Determining E_{Th} for the network external to resistor R_L of Fig. 9.68.

Applying Kirchhoff's voltage law:

$$-V_2 - E + E_{Th} = 0$$

and

$$E_{Th} = V_2 + E = 60 \text{ V} + 68 \text{ V} = \mathbf{128 \text{ V}}$$

with the maximum power equal to

$$P_{L_{max}} = \frac{E_{Th}^2}{4 R_{Th}} = \frac{(128 \text{ V})^2}{4(15 \, \Omega)} = \mathbf{273.07 \text{ W}}$$

9.5 NORTON'S THEOREM

Thévenin's theorem clearly demonstrates that a two-terminal dc network can be replaced by a single voltage source and resistor in a series combination. Interestingly enough, that same configuration can also be replaced by a single current source and resistor in a parallel combination.

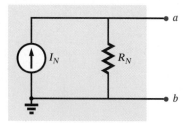

FIG. 9.71

Norton equivalent circuit.

FIG. 9.72

Dr. Edward L. Norton.

American (Rockland, Maine; Summit, New Jersey)
(1898–1983)
Electrical Engineer, Scientist, Inventor
Department Head: Bell Laboratories
Fellow: Acoustical Society and Institute of Radio Engineers

Although interested primarily in communications circuit theory and the transmission of data at high speeds over telephone lines, Edward L. Norton is best remembered for development of the dual of Thévenin's equivalent circuit, currently referred to as *Norton's equivalent circuit.* In fact, Norton and his associates at AT&T in the early 1920s are recognized as some of the first to perform pioneering work applying Thévenin's equivalent circuit and who referred to this concept simply as Thévenin's theorem. In 1926 he proposed the equivalent circuit using a current source and parallel resistor to assist in the design of recording instrumentation that was primarily current driven. He began his telephone career in 1922 with the Western Electric Company's Engineering Department, which later became Bell Laboratories. His areas of active research included network theory, acoustical systems, electromagnetic apparatus, and data transmission. A graduate of MIT and Columbia University, he held nineteen patents on his work.

The current source equivalent is a result of **Norton's theorem,** which states the following:

Any two-terminal dc network can be replaced by an equivalent circuit consisting of a current source and a parallel resistor, as shown in Fig. 9.71.

A brief biography of Dr. Edward L. Norton, the developer of this theorem, appears in Fig. 9.72.

The Norton equivalent circuit, like the Thévenin equivalent, can be found by following a series of steps that are similar in many respects to those applied to determine the Thévenin equivalent circuit. In fact, the first three steps are exactly the same, because the Thévenin and Norton resistances have the same value. The steps are as follows:

1. *Remove the portion of the network across which the Norton equivalent circuit is to be found.*
2. *Mark the terminals of the remaining two-terminal network.*
3. *Find R_N by first setting all sources to zero (voltage sources are replaced with short circuits, and current sources with open circuits) and then finding the resistance between the two marked terminals. (If a source has an internal resistance, it must remain as part of the network when the Norton resistance is determined.)*
4. *Determine I_N by first returning all of the sources to their original positions and finding the short-circuit current between the two marked terminals.*
5. *Draw the Norton equivalent circuit with the portion of the network previously removed replaced between the terminals of the Norton equivalent circuit.*

In this series of steps, only step 4 is different from the steps for Thévenin's theorem. It will take some practice to determine the correct Norton current, in much the same way patience had to be exercised to find the correct open-circuit Thévenin voltage. A few examples, however, will clarify the approach.

EXAMPLE 9.15 Find the Norton equivalent circuit for the network applied to resistor R_L in Fig. 9.73.

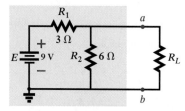

FIG. 9.73

Finding the Norton equivalent circuit for the shaded area of the network (Example 9.15).

Solution:

Steps 1–2: Resistor R_L was removed and the resulting terminals were marked as shown in Fig. 9.74.

Step 3: Replacing the voltage source by a short-circuit equivalent will result in the configuration of Fig. 9.75. The result is a parallel combination of resistors R_2 and R_1 so that the Norton resistance is determined by

$$R_N = R_2 \| R_1 = 6\ \Omega \| 3\ \Omega = \mathbf{2\ \Omega}$$

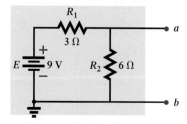

FIG. 9.74

Identifying the terminals of particular interest for the network of Fig. 9.73.

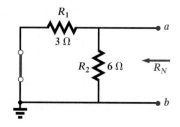

FIG. 9.75

Determining R_N for the configuration of Fig. 9.74.

Step 4: Replacing the voltage source and placing a short circuit between the marked terminals will result in the network of Fig. 9.76. The current through that short circuit must now be determined to find the Norton current. Since the applied short circuit is in parallel with resistor R_2, all the current entering the junction of resistor R_2 and the short circuit will take the path of least resistance, which is that of the short circuit. The current through resistor R_2 will be zero amperes, resulting in

$$V_2 = I_2R_2 = (0 \text{ A})(6 \text{ }\Omega) = 0 \text{ V}$$

Applying Kirchhoff's voltage law around the closed loop will result in

$$E - V_1 - V_2 = 0$$

or

$$V_1 = E - V_2 = 9 \text{ V} - 0 \text{ V} = 9 \text{ V}$$

and

$$I_1 = \frac{V_1}{R_1} = \frac{9 \text{ V}}{3 \text{ }\Omega} = 3 \text{ A}$$

Current I_1 will then continue through the short circuit and define the level of I_N. That is,

$$I_N = I_1 = \textbf{3 A}$$

Step 5: The Norton equivalent circuit is drawn as shown in Fig. 9.77, with the load resistor replaced between its terminals.

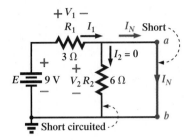

FIG. 9.76

Determining I_N for the network of Fig. 9.74.

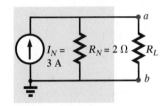

FIG. 9.77

Substituting the Norton equivalent circuit for the network external to resistor R_L of Fig. 9.73.

Before looking at a few more examples, we should point out that the Norton equivalent circuit can be obtained directly from the Thévenin equivalent circuit using a simple source conversion as introduced in Section 8.3. The conversion equations appear in Fig. 9.78.

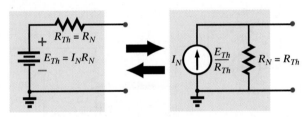

FIG. 9.78

Converting between Thévenin and Norton equivalent circuits.

EXAMPLE 9.16

a. Find the Norton equivalent circuit for the network external to the 9 Ω resistor in Fig. 9.79.
b. Find the Thévenin voltage for the shaded area of the network of Fig. 9.79.

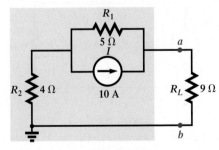

FIG. 9.79

Determining the Norton equivalent circuit for the shaded area of the network (Example 9.16).

c. Convert the Norton equivalent circuit obtained in part (a) to a Thévenin equivalent circuit using the conversion equations of Fig. 9.78. Compare the resulting Thévenin voltage to that obtained in part (b).

Solutions: **Part (a):**

Steps 1–2: The 9 Ω resistor is removed and the terminals are marked in Fig. 9.80.

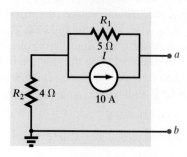

FIG. 9.80

Identifying the terminals of particular interest for the network of Fig. 9.79.

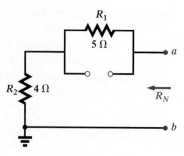

FIG. 9.81

Determining R_N for the network of Fig. 9.80.

Step 3: The current source is removed, resulting in a series connection between resistors R_1 and R_2 as shown in Fig. 9.81. The Norton resistance is

$$R_N = R_1 + R_2 = 5\,\Omega + 4\,\Omega = \mathbf{9\,\Omega}$$

Step 4: The current source is replaced and a short circuit placed between the marked terminals as shown in Fig. 9.82. The network is redrawn in the same figure to show that the Norton current is the same as that through the 4 Ω resistor. Using the current divider rule:

$$I_N = I_2 = \frac{R_1 I}{R_1 + R_2} = \frac{(5\,\Omega)(10\text{ A})}{5\,\Omega + 4\,\Omega} = \frac{50}{9}\text{ A} = \mathbf{5.556\text{ A}}$$

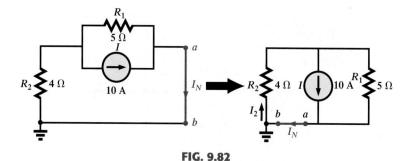

FIG. 9.82

Determining I_N for the network of Fig. 9.80.

Step 5: The Norton equivalent circuit is redrawn in Fig. 9.83, with the 9 Ω resistor replaced between its terminals.

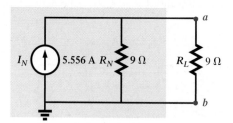

FIG. 9.83

Substituting the Norton equivalent circuit for the network external to resistor R_L of Fig. 9.79.

Part (b): The Thévenin voltage is defined by Fig. 9.84. Due to the open circuit, all the source current passes through resistor R_1. The current through the 4 Ω resistor must be zero amperes since it is in series with the open circuit. The voltage across the 5 Ω resistor is

$$V_1 = IR_1 = (10\text{ A})(5\,\Omega) = 50\text{ V}$$

and

$$E_{Th} = V_1 = \mathbf{50\text{ V}}$$

Note that the polarities of the voltage across the 4 Ω resistor match those of the defined Thévenin voltage, so a plus sign is associated with the result.

Part (c): Using the conversion equations will result in

$$R_{Th} = R_N = \textbf{9 } \boldsymbol{\Omega}$$
$$E_{Th} = I_N R_N = (5.556 \text{ A})(9 \text{ }\Omega) = \textbf{50 V}$$

and the Thévenin equivalent circuit of Fig. 9.85. The Thévenin voltage of Fig. 9.85 is the same as that obtained in part (b).

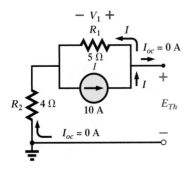

FIG. 9.84
Determining E_{Th} for the network of Fig. 9.80.

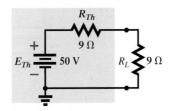

FIG. 9.85
Thévenin equivalent circuit for the network of Fig. 9.79.

Experimental Procedures

Measuring R_N Since $R_N = R_{Th}$, the experimental procedures described for the Thévenin equivalent circuit can be applied here also.

Measuring I_N The Norton current can be determined by simply placing an ammeter across the output terminals of the network as shown in Fig. 9.86(a) for the network of Fig. 9.73. This is due to the fact that, ideally, the ammeter has zero internal resistance and appears as a short circuit to the configuration. The same ammeter was connected to the Norton equivalent circuit in Fig. 9.86(b) to show that the current measured is actually the Norton current. In Fig. 9.86(b), the Norton current will choose the path of zero resistance offered by the ammeter.

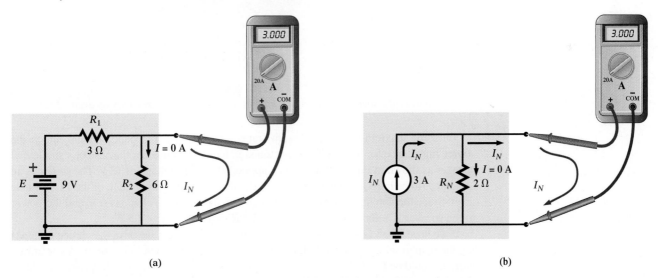

FIG. 9.86
Measuring I_N with an ammeter: (a) actual network; (b) Norton equivalent network.

9.6 APPLICATION

Speaker System

One of the most common applications of the maximum power transfer theorem introduced in this chapter is to speaker systems. An audio amplifier (an amplifier with a frequency range matching the typical range of the human ear) with an output impedance of 8 Ω is shown in Fig. 9.87(a). *Impedance* is a term applied to opposition in ac networks—for the moment think of it as a resistance level. We can also think of impedance as the internal resistance of the source which is normally shown in series with the source voltage as shown in the same figure. Every speaker has an internal resistance that can be represented as shown in Fig. 9.87(b) for a standard 8 Ω speaker. Figure 9.87(c) is a photograph of a commercially available 8 Ω woofer (for very low frequencies). The primary purpose of the following discussion is to shed some light on how the audio power can be distributed and which approach would be the most effective.

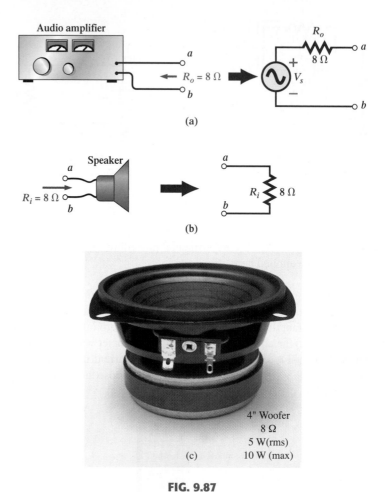

FIG. 9.87

Components of a speaker system: (a) amplifier; (b) speaker; (c) commercially available unit.

Since the maximum power theorem states that the load impedance should match the source impedance for maximum power transfer, let us first consider the case of a single 8 Ω speaker as shown in Fig. 9.88(a) with an applied amplifier voltage of 12 V. The source is an ac source (to be introduced in Chapter 12). For the moment, simply think of the 12 V as a dc level. Since the applied voltage will split equally, the speaker voltage is 6 V, and the power to the speaker is a maximum value of $P = V^2/R = (6 \text{ V})^2/8 \text{ Ω} = 4.5 \text{ W}$.

If we have two 8 Ω speakers that we would like to hook up, we have the choice of hooking them up in series or in parallel. For the series configuration of Fig. 9.88(b), the resulting current would be $I = E/R = 12 \text{ V}/24 \text{ Ω} = 500 \text{ mA}$, and the power to each speaker would be $P = I^2R = (500 \text{ mA})^2(8 \text{ Ω}) = 2 \text{ W}$, which is a drop of over 50% from the maximum output level of 4.5 W. If the speakers are hooked up in parallel as shown in Fig. 9.88(c), the total resistance of the parallel combination is 4 Ω, and the voltage across each speaker as determined by the voltage divider rule will be 4 V. The power to each speaker is $P = V^2/R = (4 \text{ V})^2/8 \text{ Ω} = 2 \text{ W}$ which, interestingly enough, is the same power delivered to each speaker whether in series or in parallel. However, the parallel arrangement is normally chosen for a variety of reasons. First, when the speakers are connected in parallel, if a wire should become disconnected from one of the speakers due simply to the vibration caused by the emitted sound, the other speakers will still be operating—perhaps not at maximum efficiency,

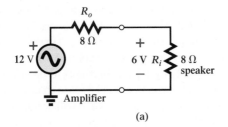

(a)

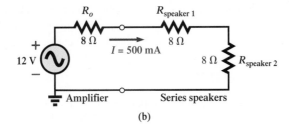

(b)

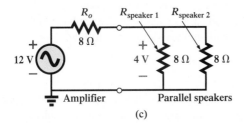

(c)

FIG. 9.88

Speaker connections: (a) single unit; (b) in series; (c) in parallel.

but they will still be operating. If in series they would all fail to operate. A second reason relates to the general wiring procedure. When all of the speakers are in parallel, from various parts of a room all the red wires can be connected together and all the black wires together. If the speakers are in series, and if you are presented with a bundle of red and black wires in the basement, you would first have to determine which wires go with which speakers.

Speakers are also available with input impedances of 4 Ω and 16 Ω. If you know that the output impedance is 8 Ω, purchasing either two 4 Ω speakers or two 16 Ω speakers would result in maximum power to the speakers as shown in Fig. 9.89. The 16 Ω speakers would be connected in parallel and the 4 Ω speakers in series to establish a total load impedance of 8 Ω.

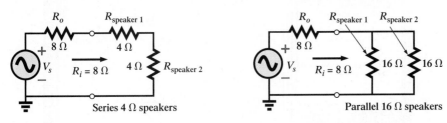

FIG. 9.89

Applying 4 Ω and 16 Ω speakers to an amplifier with an output impedance of 8 Ω.

In any case, always try to match the total resistance of the speaker load to the output resistance of the supply. Yes, a 4 Ω speaker can be placed in series with a parallel combination of 8 Ω speakers for maximum power transfer from the supply since the total resistance will be 8 Ω. However, the power distribution will not be equal, with the 4 Ω speaker receiving 2.25 W and the 8 Ω speakers each 1.125 W for a total of 4.5 W. The 4 Ω speaker is therefore receiving twice the audio power of the 8 Ω speakers, and this difference may cause distortion or imbalance in the listening area.

All speakers have maximum and minimum levels. A 50 W speaker is rated for a maximum output power of 50 W and will provide that level on demand. However, in order to function properly, it will probably need to be operating at least at the 1 W to 5 W level. A 100 W speaker typically needs between 5 W and 10 W of power to operate properly. It is also important to realize that power levels less than the rated value (such as 40 W for the 50 W speaker) will not result in an increase in distortion, but simply in a loss of volume. However, distortion will result if you exceed the rated power level. For example, if we apply 2.5 W to a 2 W speaker, we will definitely have distortion. However, applying 1.5 W will simply result in less volume. A rule of thumb regarding audio levels states that the human ear can sense changes in audio level only if you double the applied power [a 3 dB increase; decibels (dB) will be introduced in Chapter 17]. The doubling effect is always with respect to the initial level. For instance, if the original level were 2 W, you would have to go to 4 W to notice the change. If starting at 10 W, you would have to go to 20 W to appreciate the increase in volume. An exception to the above is at very low power levels or very high power levels. For instance, a change from 1 W to 1.5 W may be discernible, just as a change from 50 W to 80 W may be noticeable.

9.7 COMPUTER ANALYSIS

Once the mechanics of applying a software package or language are understood, the opportunity to be creative and innovative presents itself. Through years of exposure and trial-and-error experiences, professional programmers develop a catalog of innovative techniques that are not only functional but very interesting and truly artistic in nature. Now that some of the basic operations associated with PSpice have been introduced, a few innovative maneuvers will be made in the examples to follow.

PSpice

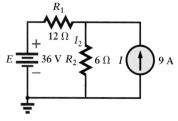

FIG. 9.90

Network from Example 9.3 to be analyzed using PSpice.

Superposition Let us now apply superposition to the network of Fig. 9.90, which appeared earlier as Fig. 9.9 in Example 9.3, to permit a comparison of resulting solutions. The current through R_2 is to be determined. Using methods described in earlier chapters for the application of PSpice, the network of Fig. 9.91 will result to determine the effect of the 36 V voltage source. Note that both **VDC** and **IDC** (flipped vertically) appear in the network. The current source, however, was set to zero simply by selecting the source and changing its value to 0 A in the **Display Properties** dialog box.

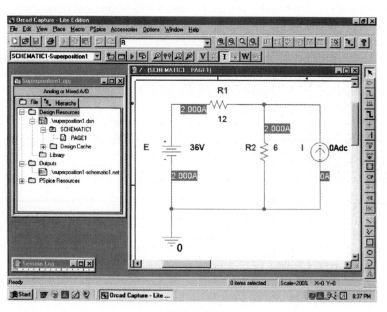

FIG. 9.91

Using PSpice to determine the contribution of the 36 V voltage source to the current through R_2.

Following simulation, the network and values appearing in Fig. 9.91 will result. The current through the 6 Ω resistor is 2 A due solely to the 36 V voltage source. Although direction is not indicated, it is fairly obvious in this case. For those cases where it is not obvious, the voltage levels can be displayed, and the direction would be from the point of high potential to the point of lower potential.

For the effects of the current source, the voltage source is set to 0 V as shown in Fig. 9.92. The resulting current is then 6 A through R_2, with the same direction as the contribution due to the voltage source.

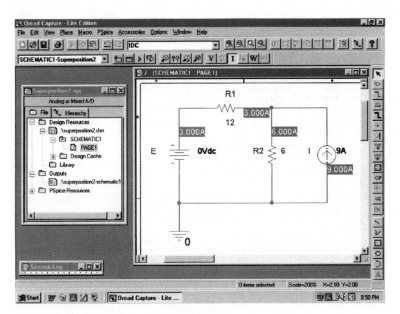

FIG. 9.92

Using PSpice to determine the contribution of the 9 A current source to the current through R_2.

The resulting current for resistor R_2 is the sum of the two currents: $I_T = 2\,\text{A} + 6\,\text{A} = 8\,\text{A}$, as determined in Example 9.3.

Thévenin's Theorem The application of Thévenin's theorem requires an interesting maneuver to determine the Thévenin resistance. It is a maneuver, however, that has application beyond Thévenin's theorem whenever a resistance level is required. The network to be analyzed appears in Fig. 9.93 and is the same one analyzed in Example 9.9 (Fig. 9.44).

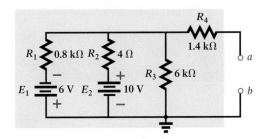

FIG. 9.93

Using PSpice to determine the Thévenin network for the same network appearing in Example 9.9.

Since PSpice is not set up to measure resistance levels directly, a 1 A current source can be applied as shown in Fig. 9.94, and Ohm's law can be used to determine the magnitude of the Thévenin resistance in the following manner:

$$|R_{Th}| = \left|\frac{V_s}{I_s}\right| = \left|\frac{V_s}{1\,\text{A}}\right| = |V_s| \qquad (9.5)$$

In Eq. (9.5), since $I_s = 1\,\text{A}$, the magnitude of R_{Th} in ohms is the same as the magnitude of the voltage V_s (in volts) across the current source. The result is that when the voltage across the current source is displayed, it can be read as ohms rather than volts.

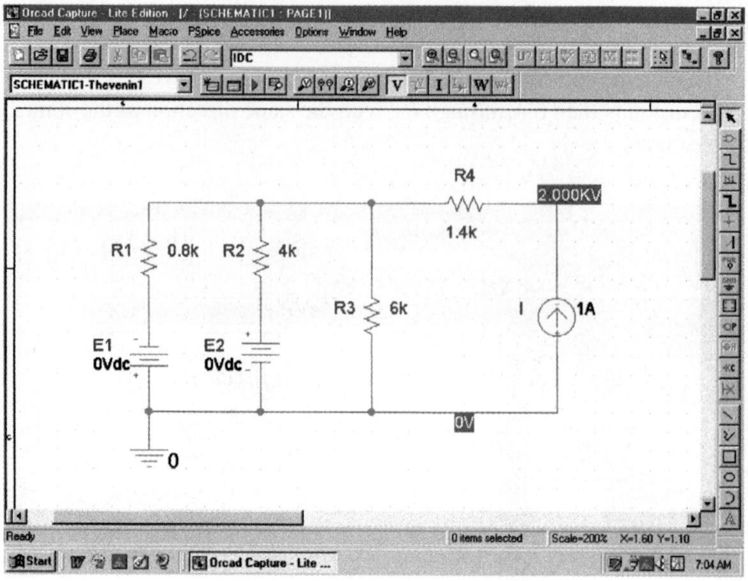

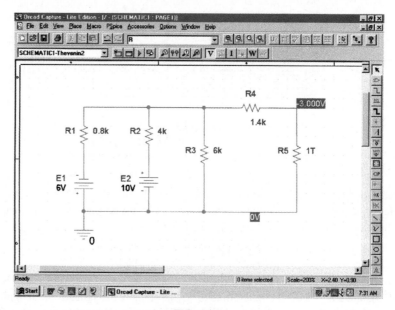

FIG. 9.94

Using PSpice to determine the Thévenin resistance of a network through the application of a 1 A current source.

When PSpice is applied, the network will appear as shown in Fig. 9.94. The voltage source E_1 and the current source are flipped using a right click on the source and using the **Mirror Vertically** option. Both voltage sources are set to zero through the **Display Properties** dialog box obtained by double-clicking on the source symbol. The result of the **Bias Point** simulation is 2 kV across the current source. The Thévenin resistance is therefore 2 kΩ between the two terminals of the network to the left of the current source (to match the results of Example 9.9). In total, by setting the voltage sources to 0 V, we have dictated that the voltage is the same at both ends of the voltage source, replicating the effect of a short-circuit connection between the two points.

For the open-circuit Thévenin voltage between the terminals of interest, the network must be constructed as shown in Fig. 9.95. The resistance of 1 T (= 1 million MΩ) is considered large enough to represent an open circuit, permitting an analysis of the network using PSpice. PSpice does not recognize floating nodes and would generate an error signal if a connection were not made from the top right node to ground. Both voltage sources are now set on their prescribed values, and a simulation will result in 3 V across the 1 T resistor. The open-circuit Thévenin voltage is therefore 3 V which agrees with the solution of Example 9.9.

FIG. 9.95

Using PSpice to determine the Thévenin voltage for a network using a very large resistance value to represent the open-circuit condition between the terminals of interest.

Maximum Power Transfer The procedure for plotting a quantity versus a parameter of the network will now be introduced. In this case it will be the output power versus values of load resistance to verify the fact that maximum power will be delivered to the load when its value equals the series Thévenin resistance. A number of new steps will be introduced, but keep in mind that the method has broad application beyond Thévenin's theorem and is therefore well worth the learning process.

The circuit to be analyzed appears in Fig. 9.96. The circuit is constructed in exactly the same manner as described earlier except for the value of the load resistance. Begin the process by starting a **New Project** called **MaxPower,** and build the circuit of Fig. 9.96. For the moment hold off on setting the value of the load resistance.

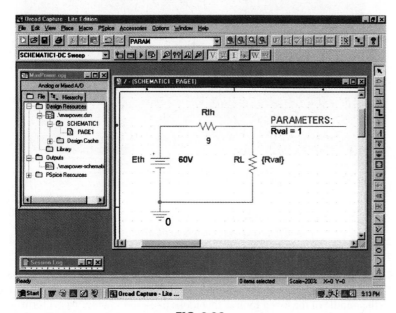

FIG. 9.96

Using PSpice to plot the power to R_L for a range of values for R_L.

The first step will be to establish the value of the load resistance as a variable since it will not be assigned a fixed value. Double-click on the value of **RL** to obtain the **Display Properties** dialog box. For **Value,** type in **{Rval}** and click in place. The brackets (*not* parentheses) are required, but the variable does not have to be called **Rval**—it is the choice of the user. Next select the **Place part** key to obtain the **Place Part** dialog box. If you are not already in the **Libraries** list, choose **Add Library** and add **SPECIAL** to the list. Select the **SPECIAL** library and scroll the **Part List** until **PARAM** appears. Select it; then click **OK** to obtain a rectangular box next to the cursor on the screen. Select a spot near **Rval** and deposit the rectangle. The result is **PARAMETERS:** as shown in Fig. 9.96.

Next double-click on **PARAMETERS:** to obtain a **Property Editor** dialog box which should have **SCHEMATIC1:PAGE1** in the second column from the left. Now select the **New Column** option from the top list of choices to obtain the **Add New Column** dialog box. Enter the **Name:Rval** and **Value:1** followed by an **OK** to leave the dialog box. The result is a return to the **Property Editor** dialog box but with **Rval** and its value (below **Rval**) added to the horizontal list. Now, select **Rval/1** by clicking on the **Rval** to surround **Rval** by a dashed line and add a black background around the **1.** Choose **Display** to produce the **Display Properties** dialog box, and select **Name and Value** followed by **OK.** Then exit the **Property Editor** dialog box (**X**) to obtain the screen of Fig. 9.96. Note that now the first value (1 Ω) of **Rval** is displayed.

We are now ready to set up the simulation process. Select the **New Simulation Profile** key to obtain the **New Simulation** dialog box. Enter **DC Sweep** under **Name** followed by **Create.** The **Simulation Settings-DC Sweep** dialog box will appear. After selecting **Analysis,** select **DC Sweep** under the **Analysis type** heading. Then leave the **Primary Sweep** under the **Options** heading, and select **Global parameter** under the **Sweep variable.** The **Parameter name** should then be entered as **Rval.** For the **Sweep type,** the **Start value** should be 1 Ω; but if we use 1 Ω, the curve to be generated will start at 1 Ω, leaving a blank from 0 to 1 Ω. The curve will look incomplete. To get around this problem, we will select 0.001 Ω as the **Start value** (very close to 0 Ω), and 30.001 Ω as the **End value,** with an **Increment** of 1 Ω. The values of **RL** will therefore be 0.001 Ω, 1.001 Ω, 2.001 Ω, and so on, although the plot will look as if the values were 0 Ω, 1 Ω, 2 Ω, and so on. Click **OK,** and select the **Run PSpice** key to obtain the display of Fig. 9.97.

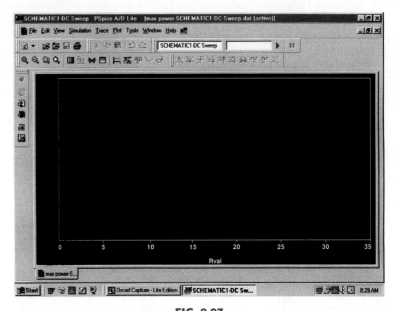

FIG. 9.97

Plot resulting from the dc sweep of R_L for the network of Fig. 9.96 before defining the parameters to be displayed.

First note that there are no plots on the graph and that the graph extends to 35 Ω rather than 30 Ω as desired. It did not respond with a plot of power versus **RL** because we have not defined the plot of interest for the computer. This is done by selecting the **Add Trace** key (the key in the middle of the lower toolbar that looks like a red sawtooth waveform) or **Trace-Add Trace** from the top menu bar. Either choice will result in the **Add Traces** dialog box. The most important region of this dialog box is the **Trace Expression** listing at the bottom. The desired trace can be typed in directly, or the quantities of interest can be chosen from the list of **Simulation Output Variables** and deposited in the **Trace Expression** listing. Since we are interested in the power to **RL** for the chosen range of values for **RL**, W(RL) is selected in the listing; it will then appear as the **Trace Expression.** Click **OK,** and the plot of Fig. 9.98 will appear. Originally, the plot extended from 0 Ω to 35 Ω. We reduced the range to 0 Ω to 30 Ω by selecting **Plot-Axis Settings-X Axis-User Defined 0 to 30-OK.**

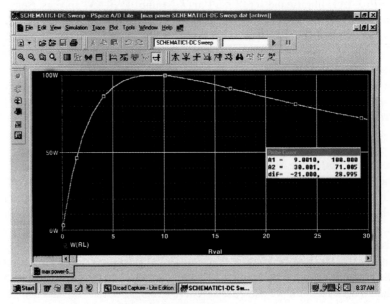

FIG. 9.98

A plot of the power delivered to R_L in Fig. 9.96 for a range of values for R_L extending from 0 Ω to 30 Ω.

Select the **Toggle cursor** key (which looks like a red curve passing through the origin of a graph), and then left-click the mouse. A vertical line and a horizontal line will appear, with the vertical line controlled by the position of the cursor. Moving the cursor to the peak value will result in **A1** = 9.0010 as the *x* value and 100.000 W as the *y* value, as shown in the **Probe Cursor** box at the right of the screen. A second cursor can be generated by a right click of the mouse, which was set at **RL** = 30.001 Ω to result in a power of 71.005 W. Notice also that the plot generated appears as a listing at the bottom left of the screen as **W(RL).**

Before leaving the subject, we should mention that the power to **RL** can be determined in more ways than one from the **Add Traces** dialog box. For example, first enter a minus sign because of the resulting current direction through the resistor, and then select **V2(RL)** followed by the multiplication of **I(RL).** The following expression will appear in the **Trace Expression box: V2(RL)*I(RL),** which is an expression having the basic power format of $P = V * I$. Click **OK,** and the same power curve of Fig. 9.98 will appear. Other quantities, such as the voltage across the load and the current through the load, can be plotted against **RL** by simply following the sequence **Trace-Delete All Traces-Trace-Add Trace-V1(RL)** or **I(RL).**

CHAPTER SUMMARY

SUPERPOSITION THEOREM

- The current through, or voltage across, any element of a network is equal to the algebraic sum of the currents or voltages produced independently by each source.
- To remove the impact of a voltage source, use a short-circuit equivalent. To remove the impact of a current source, use an open-circuit equivalent.
- When the superposition theorem is applied, the number of networks to be analyzed is equal to the number of sources present.
- The superposition theorem cannot be applied to power effects because of the nonlinear relationship between the current or voltage of an element and the power to that element.

THÉVENIN'S THEOREM

- Thévenin's theorem permits the reduction of a two-terminal, series-parallel dc network with any number of sources or components to a single voltage source in series with a resistor.
- The Thévenin voltage is the open-circuit voltage between the two terminals of the network to be reduced to the Thévenin form. The Thévenin resistance is the resistance between the same two terminals with all the sources set to zero.
- The Thévenin voltage can be measured by simply placing a voltmeter across the terminals of interest.
- The Thévenin resistance can be found by applying an ohmmeter across the terminals of interest after the effects of all the sources have been removed.

MAXIMUM POWER TRANSFER THEOREM

- The maximum power transfer theorem states that maximum power will be delivered to a load when it equals the Thévenin resistance of the network.
- For a fixed load, the smaller the Thévenin resistance, the larger the power delivered to the load.
- The dc efficiency of a system is the ratio of the power delivered to a load divided by the power supplied by the source.

NORTON'S THEOREM

- Norton's equivalent circuit can be determined by simply converting the Thévenin equivalent circuit to a current source equivalent.
- The Norton current is the short-circuit current between the terminals of interest.
- The Norton resistance is the same as the Thévenin resistance and is found in exactly the same manner.
- Using instruments, the Norton resistance is found in the same way as the Thévenin resistance. The Norton current is determined by placing an ammeter across the terminals of interest.

Important Equations

$$R_{Th} = \frac{E_{Th}}{I_{sc}} = \frac{V_{oc}}{I_{sc}}$$

$$R_L = R_{Th} \quad \text{(maximum power transfer)}$$

$$P_{L_{max}} = \frac{E_{Th}^2}{4R_{Th}}$$

$$\eta\% = \frac{P_L}{P_s} \times 100\%$$

$$R_N = R_{Th}$$

$$I_N = \frac{E_{Th}}{R_{Th}} = \frac{V_{oc}}{R_N}$$

PROBLEMS

SECTION 9.2   Superposition Theorem

1. Using superposition, determine the reading of the meter in Fig. 9.99.

2. For the network of Fig. 9.100:
   a. Using superposition, find current I_1.
   b. Find the power delivered to R_1 for each source.
   c. Find the power delivered to R_1 using the total current from part (a).
   d. Based on the results of parts (b) and (c), is superposition applicable to power effects? Explain.

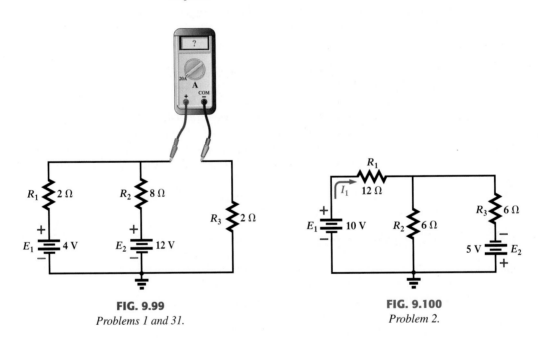

FIG. 9.99
Problems 1 and 31.

FIG. 9.100
Problem 2.

3. Using superposition, find current I_1 for the network of Fig. 9.101.

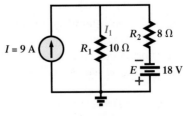

FIG. 9.101
Problem 3.

4. Using superposition, find current I_4 for the network of Fig. 9.102.

5. Using superposition, find current I_1 for the network of Fig. 9.103.

6. Using superposition, find voltage V_2 for the network of Fig. 9.104.

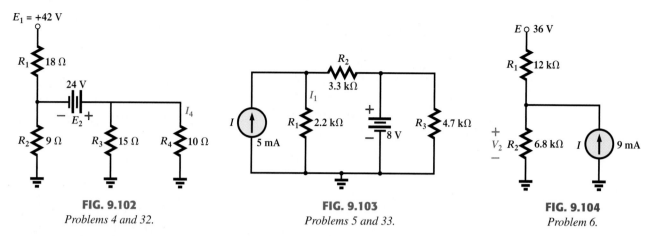

FIG. 9.102
Problems 4 and 32.

FIG. 9.103
Problems 5 and 33.

FIG. 9.104
Problem 6.

SECTION 9.3 Thévenin's Theorem

7. For the network of Fig. 9.105:
   a. Find the Thévenin equivalent circuit external to resistor R_L.
   b. Find the current through R_L when R_L is 2 Ω, 30 Ω, and 100 Ω.

8. For the network (of standard values) of Fig. 9.106:
   a. Find the Thévenin equivalent circuit external to resistor R_L.
   b. Find the power delivered to R_L when R_L is 1.2 kΩ and 6.8 kΩ.

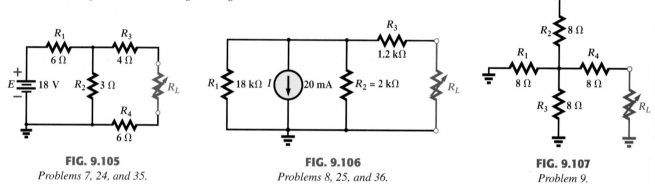

FIG. 9.105
Problems 7, 24, and 35.

FIG. 9.106
Problems 8, 25, and 36.

FIG. 9.107
Problem 9.

9. a. Find the Thévenin equivalent circuit for the network external to resistor R_L in Fig. 9.107.
   b. Find the power delivered to R_L when R_L is 10 Ω and 100 Ω.

10. For each set of measurements appearing in Fig. 9.108, determine the Thévenin equivalent circuit.

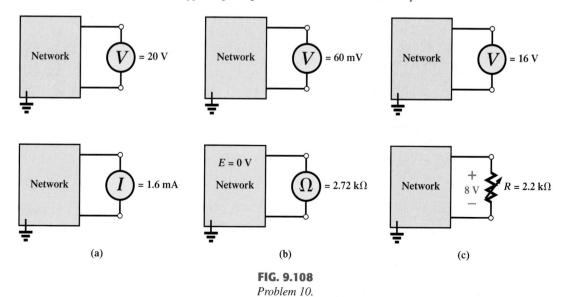

(a)

(b)

(c)

FIG. 9.108
Problem 10.

11. Find the Thévenin equivalent circuit for the network external to resistor R_L in Fig. 9.109.

12. Find the Thévenin equivalent circuit for the network external to resistor R_L in Fig. 9.110.

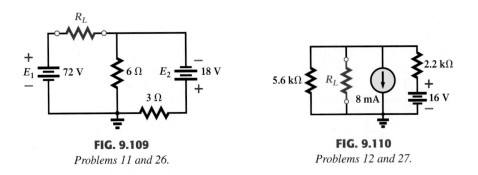

FIG. 9.109
Problems 11 and 26.

FIG. 9.110
Problems 12 and 27.

***13.** Find the Thévenin equivalent circuit for the network external to resistor R_L in Fig. 9.111.

***14.** Find the Thévenin equivalent circuit for the portion of the network of Fig. 9.112 external to points a and b.

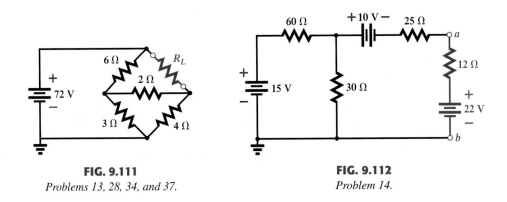

FIG. 9.111
Problems 13, 28, 34, and 37.

FIG. 9.112
Problem 14.

***15.** Determine the Thévenin equivalent circuit for the network external to resistor R_L in Fig. 9.113.

***16.** Determine the Thévenin equivalent circuit for the network external to resistor R_L in Fig. 9.114.

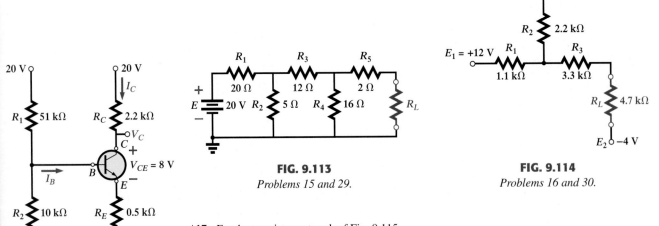

FIG. 9.115
Problem 17.

FIG. 9.113
Problems 15 and 29.

FIG. 9.114
Problems 16 and 30.

***17.** For the transistor network of Fig. 9.115:
    a. Find the Thévenin equivalent circuit for the portion of the network to the left of the base (B) terminal.
    b. Using the fact that $I_C = I_E$ and $V_{CE} = 8$ V, determine the magnitude of I_E.
    c. Using the results of parts (a) and (b), calculate the base current I_B if $V_{BE} = 0.7$ V.
    d. What is the voltage V_C?

18. For the network of Fig. 9.116:
   a. Determine the Thévenin equivalent circuit external to resistor R_L.
   b. Find the value of R_L that will ensure maximum power to R_L.
   c. Find the maximum power that can be delivered to R_L.

*19. For the network of Fig. 9.117:
   a. Determine the Thévenin equivalent circuit external to resistor R_L.
   b. Find the value of R_L that will ensure maximum power to R_L.
   c. Find the maximum power that can be delivered to R_L.
   d. What is the operating efficiency with R_L set as in part (b)?
   e. What is the operating efficiency with R_L set equal to one-half the value of part (b)?

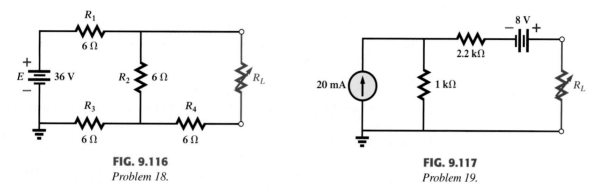

FIG. 9.116
Problem 18.

FIG. 9.117
Problem 19.

20. For the network of Fig. 9.118:
   a. Determine the value of R_L for maximum power to R_L.
   b. Determine the maximum power to R_L.
   c. Plot a curve of power to R_L versus R_L for R_L equal to ¼, ½, 1, 1¼, 1½, 1¾, and 2 times the value obtained in part (a).

*21. Find resistance R_1 of Fig. 9.119 such that resistor R_1 will receive maximum power. Think!

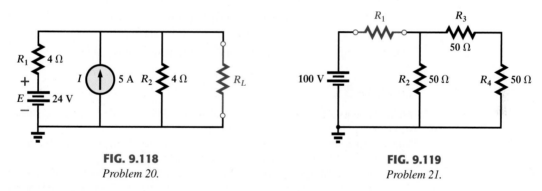

FIG. 9.118
Problem 20.

FIG. 9.119
Problem 21.

*22. For the network of Fig. 9.120:
   a. Determine the value of R_2 for maximum power to R_4.
   b. Is there a general statement that can be made about situations such as those presented here and in Problem 21?

*23. For the network of Fig. 9.121, determine the level of R that will ensure maximum power to the 100 Ω resistor.

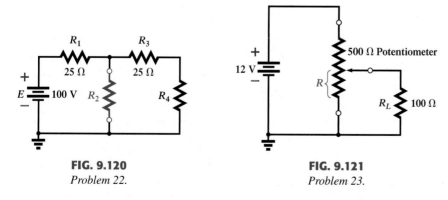

FIG. 9.120
Problem 22.

FIG. 9.121
Problem 23.

SECTION 9.5   Norton's Theorem

24. For the network of Fig. 9.105:
    a. Find the Norton equivalent circuit external to resistor R_L.
    b. Convert the Norton equivalent circuit to a Thévenin equivalent circuit. Compare this solution to that of Problem 7 if the results are available.

25. For the network of Fig. 9.106:
    a. Find the Norton equivalent circuit external to resistor R_L.
    b. Convert the Norton equivalent circuit to a Thévenin equivalent circuit. Compare this solution to that of Problem 8 if the results are available.

26. Find the Norton equivalent circuit for the network of Fig. 9.109.

27. Find the Norton equivalent circuit for the network of Fig. 9.110.

*28. For the network of Fig. 9.111:
    a. Find the Norton equivalent circuit external to resistor R_L.
    b. Convert the Norton equivalent circuit to a Thévenin equivalent circuit. Compare this solution to that of Problem 13 if the results are available.

*29. For the network of Fig. 9.113:
    a. Find the Norton equivalent circuit external to resistor R_L.
    b. Convert the Norton equivalent circuit to a Thévenin equivalent circuit. Compare this solution to that of Problem 15 if the results are available.

*30. For the network of Fig. 9.114:
    a. Find the Norton equivalent circuit external to resistor R_L.
    b. Convert the Norton equivalent circuit to a Thévenin equivalent circuit. Compare this solution to that of Problem 16 if the results are available.

SECTION 9.7   Computer Analysis

31. Using PSpice or EWB, find current I_3 for the network of Fig. 9.99.

32. Using PSpice or EWB, find current I_4 for the network of Fig. 9.102.

33. Using PSpice or EWB, find E_{Th} and R_{Th} for the network of Fig. 9.103.

34. Using PSpice or EWB, find E_{Th} and R_{Th} for the network of Fig. 9.111.

35. For the network of Fig. 9.105, plot the power to R_L for a suitable range of values for R_L.

36. Using PSpice or EWB, find I_N and R_N for the network of Fig. 9.106.

37. Using PSpice or EWB, find I_N and R_N for the network of Fig. 9.111.

GLOSSARY

Maximum power transfer theorem   A theorem used to determine the load resistance necessary to ensure maximum power transfer to the load.

Norton's theorem   A theorem that permits the reduction of any two-terminal linear dc network to one having a single current source and parallel resistor.

Superposition theorem   A network theorem that permits considering the effects of each source independently. The resulting current and/or voltage is the algebraic sum of the currents and/or voltages developed by each source independently.

Thévenin's theorem   A theorem that permits the reduction of any two-terminal, linear dc network to one having a single voltage source and series resistor.

Capacitors

OBJECTIVES

- Become aware of the characteristics of an electric field.
- Be able to calculate the capacitance of a capacitor and understand the impact of each parameter.
- Become aware of the various types of capacitors and learn how to determine their value and rating from the provided labeling.
- Become familiar with the transient response of a capacitor and learn how to plot the response of the voltage and current.
- Learn how to incorporate initial values for the voltage of a capacitor in the transient response.
- Become aware of how Thévenin's theorem can help define the response of a capacitor in a series-parallel network.
- Be able to determine and plot the current of a capacitor given the voltage across the capacitor.
- Understand how to combine capacitors in series or in parallel.
- Develop some familiarity with the use of capacitors in practical applications.
- Be able to set up a transient response using PSpice or EWB.

10.1 INTRODUCTION

Thus far, the resistor has been the only network component appearing in our analyses. In this chapter, we introduce the **capacitor,** which will have a significant impact on the types of networks that you will be able to design and analyze. Like the resistor, it is a two-terminal device, but its characteristics are totally different from those of a resistor. In fact, **the capacitor will display its true characteristics only when a change in the voltage or current is made in the network.** Also, if you recall, all the power delivered to a resistor is dissipated in the form of heat. An ideal capacitor, however, will simply store the energy delivered to it in a form that can be returned to the system.

Although the basic construction of capacitors is actually quite simple, it is a component that opens the door to all types of practical applications, extending from touch pads to sophisticated control systems. A few applications will be introduced and discussed in detail later in this chapter.

10.2 THE ELECTRIC FIELD

Recall from Chapter 2 that a force of attraction exists between two oppositely charged bodies such as shown in Fig. 10.1(a). The strength of this attraction can be determined using **Coulomb's law:** $F = kQ_1Q_2/r^2$ [Eq. (2.1)], where Q_1 and Q_2 are the charges in coulombs

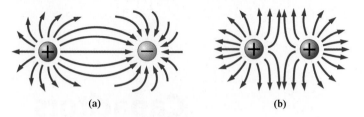

FIG. 10.1

Electric flux distribution: (a) opposite charges; (b) like charges.

(*C*) and *r* is the distance in meters between the charges. Between the two bodies, there exists an **electric field,** the strength and the distribution of which are represented by **electric flux lines** that extend from the positive charge to the negative charge as shown in Fig. 10.1(a). In general,

electric flux lines always extend from a positively charged body to a negatively charged body. They always leave or terminate at an angle perpendicular to the charged surface, and they never intersect.

Note in Fig. 10.1(a) that the electric flux lines form a direct path from the positive to the negative charge. Since electric flux lines always strive to take the shortest path from one charged body to another, there is a pressure on the system to bring the two charges together. If two bodies of the same charge are brought together as shown in Fig. 10.1(b), there is a force of repulsion between the two bodies that can also be determined using Coulomb's law. The resulting pattern for the electric flux lines clearly displays this repulsion by showing no interest in the neighboring charge.

Electric flux lines also reveal the **strength of the electric field** in a particular region through the fact that

the denser the flux lines (the more flux lines per unit area), the stronger the electric field is in that region.

In both parts of Fig. 10.1, the flux lines are tighter together when they are closer to the charge than they are once they leave the vicinity of the charge. The result is a stronger electric field near the charge, just as a magnet has more pulling power the closer you bring an object, such as a nail, to the magnet. The farther away you are from a charge, the wider the distance between flux lines and the weaker the field.

10.3 CAPACITANCE

Thus far, we have examined only isolated positive and negative spherical charges, but the description can be extended to charged surfaces of any shape and size. In Fig. 10.2, for example, two parallel plates of a material such as aluminum (the most commonly used metal

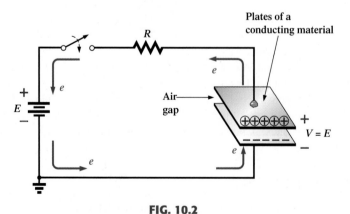

FIG. 10.2

Fundamental charging circuit.

in the construction of capacitors) have been connected through a switch to a resistor and a battery. The instant the switch is closed, electrons on the surface of the top plate and in the wire connected to the plate will be attracted to the positive terminal of the supply. At the same time and rate, electrons will be repelled by the negative terminal of the supply toward the bottom plate. In time, the result will be a layer of positive ions on the top plate and a similar layer of electrons on the lower plate, as shown in the figure. The transfer of electrons will continue until the voltage across the capacitor is the same as the applied voltage. This is due to the fact that the two plates have all the characteristics of an open circuit—**there is no direct connection between the plates** and a current of zero amperes between the plates. The plates will then sit with the charge distribution just established until the switch is opened.

This element, constructed simply of two conducting surfaces separated by the air gap, is called a **capacitor.**

Capacitance is a measure of a capacitor's ability to store charge on its plates—in other words, its storage capacity.

In addition,

the higher the capacitance of a capacitor, the greater the amount of charge stored on the plates for the same applied voltage.

The unit of measure applied to capacitors is the farad (abbreviated F), named after an English scientist, Michael Faraday, who did extensive research in the field (Fig. 10.3). In particular,

a capacitor has a capacitance of 1 F if 1 C of charge (6.242 × 10^18 electrons) is deposited on the plates by a potential difference of 1 V across its plates.

The relationship between the applied voltage, the charge on the plates, and the capacitance level is defined by the following equation:

$$\boxed{C = \frac{Q}{V}} \qquad \begin{aligned} C &= \text{farads (F)} \\ Q &= \text{coulombs (C)} \\ V &= \text{volts (V)} \end{aligned} \qquad \textbf{(10.1)}$$

The equation clearly reveals that for the same voltage (V), the greater the charge (Q) on the plates (in the numerator of the equation), and the higher the capacitance level (C).

If we write the equation in the following form:

$$\boxed{Q = CV} \qquad \text{(coulombs, C)} \qquad \textbf{(10.2)}$$

it becomes obvious through the product relationship that the higher the capacitance (C) or applied voltage (V), the greater the charge on the plates.

EXAMPLE 10.1

a. If 82.4×10^{14} electrons are deposited on the negative plate of a capacitor by an applied voltage of 60 V, find the capacitance of the capacitor.
b. If 40 V are applied across a 470 μF capacitor, find the charge on the plates.

Solutions:

a. First the number of coulombs of charge must be found as follows:

$$82.4 \times 10^{14} \ \text{electrons} \left(\frac{1 \ \text{C}}{6.242 \times 10^{18} \ \text{electrons}} \right) = 1.32 \ \text{mC}$$

and then

$$C = \frac{Q}{V} = \frac{1.32 \ \text{mC}}{60 \ \text{V}} = \textbf{22} \ \boldsymbol{\mu}\textbf{F} \quad \text{(a standard value)}$$

b. Applying Eq. (10.2):

$$Q = CV = (470 \ \mu\text{F})(40 \ \text{V}) = \textbf{18.8 mC}$$

Courtesy of the Smithsonian Institution, Photo No. 51,147.

FIG. 10.3
Michael Faraday.

English (London)
(1791–1867)
Chemist and Electrical Experimenter
Honorary Doctorate,
Oxford University, 1832

An experimenter with no formal education, he began his research career at the Royal Institute in London as a laboratory assistant. Intrigued by the interaction between electrical and magnetic effects, he discovered *electromagnetic induction,* demonstrating that electrical effects can be generated from a magnetic field (the birth of the generator as we know it today). He also discovered *self-induced currents* and introduced the concept of *lines and fields of magnetic force.* Having received over one hundred academic and scientific honors, he became a Fellow of the Royal Society in 1824 at the young age of 32.

In Fig. 10.4(a), the electric field distribution is represented by the electric flux lines from the positive to the negative layers of charge. Note the **fringing** that occurs at the edges as the flux lines originating from the points farthest away from the negative plate strive to complete the connection. This fringing, which has the effect of reducing the net capacitance somewhat, can be ignored for most applications. Ideally, and the way we will assume the distribution to be in this text, the electric flux distribution would appear as shown in Fig. 10.4(b), where all the flux lines are equally distributed and "fringing" does not occur.

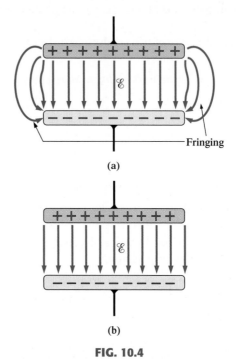

(a)

(b)

FIG. 10.4

Electric flux distribution between the plates of a capacitor:
(a) including fringing; (b) ideal.

The **electric field strength** between the plates is determined by the voltage across the plates and the distance between the plates as follows:

$$\boxed{\mathscr{E} = \frac{V}{d}} \qquad \begin{aligned} \mathscr{E} &= \text{volts/m (V/m)} \\ V &= \text{volts (V)} \\ d &= \text{meters (m)} \end{aligned} \qquad \textbf{(10.3)}$$

Take particular note of the fact that the distance between the plates is measured in meters, not centimeters or inches.

It is interesting to note that the equation for the electric field strength is determined by two factors only: the applied voltage and the distance between the plates. The charge on the plates does not appear in the equation; nor does the size of the capacitor or the plate material. This fact is one that permits us to change the capacitance of a capacitor by placing certain materials between the plates, as shown in Fig. 10.5.

In Fig. 10.5(a), two plates are separated by an air gap and have layers of charge on the plates as established by the applied voltage and the distance between the plates. The electric field strength is $\mathscr{E}_1$ as defined by Eq. (10.3). In Fig. 10.5(b), a slice of mica is introduced which, through an alignment of cells within the dielectric, will establish an electric field $\mathscr{E}_2$ that will oppose electric field $\mathscr{E}_1$. The effect is to try to reduce the electric field strength between the plates. However, Equation (10.3) states that the electric field strength **must be** the value established by the applied voltage and the distance between the plates. This condition is maintained by placing more charge on the plates, thereby increasing the electric field strength between the plates to a level that will cancel out the opposing electric field introduced by the mica sheet. The net result is an increase in charge on the plates and an increase in the capacitance level as established by Eq. (10.1).

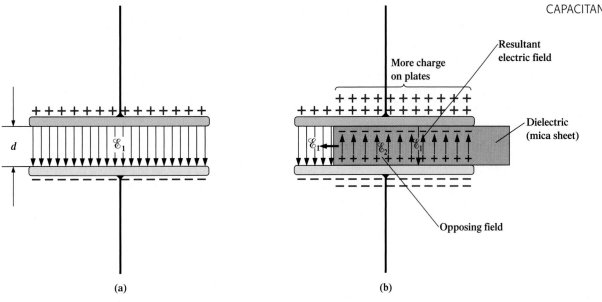

FIG. 10.5

*Demonstrating the effect of inserting a dielectric between the plates of a capacitor:
(a) air capacitor; (b) dielectric being inserted.*

Different materials placed between the plates will establish different amounts of additional charge on the plates. All, however, must be insulators and must have the ability to set up an electric field within the structure. A list of common materials appears in Table 10.1 using air as the reference level of 1.* All of these materials are referred to as **dielectrics,** the "di" for *opposing,* and the "electric" from *electric field.* The symbol ϵ_r in Table 10.1 is called the **relative permittivity** (or **dielectric constant**). The term **permittivity** is applied as a measure of how easily a material will "permit" the establishment of an electric field in the material. The relative permittivity compares the permittivity of a material to that of air. For instance, Table 10.1 reveals that mica, with a relative

TABLE 10.1

Relative permittivity (dielectric constant) ϵ_r of various dielectrics.

Dielectric	ϵ_r (Average Values)
Vacuum	1.0
Air	1.0006
Teflon®	2.0
Paper, paraffined	2.5
Rubber	3.0
Polystyrene	3.0
Oil	4.0
Mica	5.0
Porcelain	6.0
Bakelite®	7.0
Aluminum oxide	7
Glass	7.5
Tantalum oxide	30
Ceramics	20–7500
Barium-strontium titanite (ceramic)	7500.0

*Although there is a difference in dielectric characteristics between air and a vacuum, the difference is so small that air is commonly used as the reference level.

TABLE 10.2

Dielectric strength of some dielectric materials.

Dielectric	Dielectric Strength (Average Value) in Volts/Mil
Air	75
Barium-strontium titanite (ceramic)	75
Ceramics	75–1000
Porcelain	200
Oil	400
Bakelite®	400
Rubber	700
Paper, paraffined	1300
Teflon®	1500
Glass	3000
Mica	5000

permittivity of 5, will "permit" the establishment of an opposing electric field in the material five times better than in air. Note the ceramic material at the bottom of the chart with a relative permittivity of 7500—a relative permittivity that makes it a very special dielectric in the manufacture of capacitors.

Defining ϵ_o as the permittivity of air, the relative permittivity of a material with a permittivity ϵ is defined by Eq. (10.4):

$$\epsilon_r = \frac{\epsilon}{\epsilon_o} \qquad \text{(dimensionless)} \qquad \textbf{(10.4)}$$

Note that ϵ_r, which (as mentioned previously) is often called the **dielectric constant,** is a dimensionless quantity because it is a ratio of similar quantities. However, permittivity does have the units of farads/meter (F/m) and is 8.85×10^{-12} F/m for air. Although the relative permittivity for the air we breathe is listed as 1.006, a value of 1 is normally used for the relative permittivity of air.

For every dielectric there is a potential that, if applied across the dielectric, will break down the bonds within it and cause current to flow through it. The voltage required per unit length is an indication of its **dielectric strength** and is called the **breakdown voltage.** When breakdown occurs, the capacitor has characteristics very similar to those of a conductor. A typical example of dielectric breakdown is lightning, which occurs when the potential between the clouds and the earth is so high that charge can pass from one to the other through the atmosphere (the dielectric). The average dielectric strengths for various dielectrics are tabulated in volts/mil in Table 10.2 (1 mil = 1/1000 inch).

One of the important parameters of a capacitor is the **maximum working voltage.** It defines the maximum voltage that can be placed across the capacitor on a continuous basis without damaging it or changing its characteristics. For most capacitors it is the dielectric strength that defines the maximum working voltage.

10.4 CAPACITORS

Capacitor Construction

We are now aware of the basic components of a capacitor: conductive plates, separation, and dielectric. However, the question remains as to how all these factors interact to determine the capacitance of a capacitor. **Larger plates** permit an increased area for the storage of charge, so the area of the plates should be in the numerator of the defining equation. **The smaller the distance between the plates,** the larger the capacitance so this factor should appear in the numerator of the equation. Finally, since **higher levels of permittivity** result in higher levels of capacitance, the factor ϵ should appear in the numerator of the defining equation. The effect of each of these parameters is defined in Fig. 10.6.

The result is the following general equation for capacitance:

$$C = \epsilon \frac{A}{d} \qquad \begin{array}{l} C = \text{farads (F)} \\ \epsilon = \text{permittivity (F/m)} \\ A = \text{m}^2 \\ d = \text{m} \end{array} \qquad \textbf{(10.5)}$$

If we substitute Eq. (10.4) for the permittivity of the material, we obtain the following equation for the capacitance:

$$C = \epsilon_o \epsilon_r \frac{A}{d} \qquad \text{(farads, F)} \qquad \textbf{(10.6)}$$

or if we substitute the known value for the permittivity of air, we obtain the following useful equation:

$$C = 8.85 \times 10^{-12} \epsilon_r \frac{A}{d} \qquad \text{(farads, F)} \qquad \textbf{(10.7)}$$

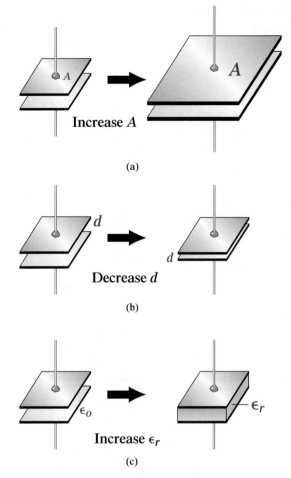

FIG. 10.6

Factors that increase capacitance: (a) increase in area of plates;
(b) decrease in distance between plates; (c) introduction of a dielectric with a high ϵ_r.

It is important to note in Eq. (10.7) that the area of the plates (actually the area of only one plate) is in meters squared (m^2); the distance between the plates is measured in meters; and the numerical value of ϵ_r is simply taken from Table 10.1.

You should also be aware that most capacitors are in the μF, nF, or pF range, not the 1 F or greater range. A 1 F capacitor is typically larger than a 1 gallon container, requiring that the housing for the system be quite large. Most capacitors in electronic systems are the size of a thumbnail or smaller.

The capacitance of a capacitor using a dielectric divided by the capacitance that it would have with a dielectric of air gives the following result:

$$\frac{C = \epsilon \dfrac{A}{d}}{C_o = \epsilon_o \dfrac{A}{d}} \Rightarrow \frac{C}{C_o} = \frac{\epsilon}{\epsilon_o} = \epsilon_r$$

and

$$\boxed{C = \epsilon_r C_o} \qquad\qquad (10.8)$$

The result is that

the capacitance of a capacitor with a dielectric having a relative permittivity of ϵ_r is ϵ_r times the capacitance using air as the dielectric.

The next few examples will review the concepts and equations just presented.

EXAMPLE 10.2   In Fig. 10.7, if each air capacitor in the left-hand column is changed to the type appearing in the right-hand column, find the new capacitance level. For each change, the other factors remain the same.

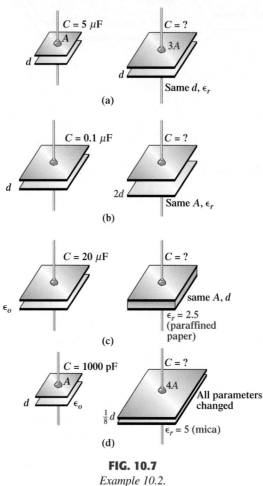

FIG. 10.7
Example 10.2.

Solutions:

a. The area has increased by a factor of three, providing more space for the storage of charge on each plate. Since the area appears in the numerator of the capacitance equation, the capacitance will increase by a factor of three. That is,

$$C = 3(C_i) = 3(5\ \mu F) = \mathbf{15\ \mu F}$$

b. In this case the area stayed the same, but the distance between the plates was increased by a factor of two. Increasing the distance reduces the capacitance level, so the resulting capacitance is one-half of what it was before. That is,

$$C = \frac{1}{2}(0.1\ \mu F) = \mathbf{0.05\ \mu F}$$

c. In this case the area and the distance between the plates were maintained, but a dielectric of paraffined (waxed) paper was added between the plates. Since the permittivity appears in the numerator of the capacitance equation, the capacitance will increase by a factor determined by the relative permittivity. That is,

$$C = \epsilon_r C_o = 2.5(20\ \mu F) = \mathbf{50\ \mu F}$$

d. In this case we have a multitude of changes happening at the same time. However, solving the problem is simply a matter of determining whether the change will increase or decrease the capacitance and then placing the multiplying factor in the numerator or denominator of the equation. The increase in area by a factor of four will produce a multiplier of four in the numerator, as shown in the equation below. Reducing the distance by a factor of 1/8 will increase the capacitance by its inverse, or a factor of eight. Inserting the mica dielectric will increase the capacitance by a factor of five. The result is

$$C = (5)\frac{4}{(1/8)}(C_i) = 160(1000 \text{ pF}) = \textbf{0.16 } \boldsymbol{\mu}\textbf{F}$$

In the next example, the dimensions of an air capacitor will be provided and the capacitance is to be determined. The example clearly reveals the importance of knowing the units of each factor of the equation. Failing to make a conversion to the proper set of units would most likely result in a meaningless result, even if the proper equation were used and the mathematics properly executed.

EXAMPLE 10.3 For the capacitor of Fig. 10.8:

a. Find the capacitance.
b. Find the strength of the electric field between the plates if 48 V are applied across the plates.
c. Find the charge on each plate.

Solutions:

a. First, the area and the distance between the plates must be converted to the MKS system as required by Eq. (10.7):

$$d = \frac{1}{32}\text{ in.}\left(\frac{1 \text{ m}}{39.37 \text{ in.}}\right) = 0.794 \text{ mm}$$

and $$A = (2 \text{ in.})(2 \text{ in.})\left(\frac{1 \text{ m}}{39.37 \text{ in.}}\right)\left(\frac{1 \text{ m}}{39.37 \text{ in.}}\right) = 2.581 \times 10^{-3} \text{ m}^2$$

Eq. (10.7):

$$C = 8.85 \times 10^{-12}\epsilon_r\frac{A}{d} = 8.85 \times 10^{-12}(1)\frac{(2.581 \times 10^{-3} \text{ m}^2)}{0.794 \text{ mm}} = \textbf{28.8 pF}$$

b. The electric field between the plates is determined by Eq. (10.3):

$$\mathscr{E} = \frac{V}{d} = \frac{48 \text{ V}}{0.794 \text{ mm}} = \textbf{60.5 kV/m}$$

c. The charge on the plates is determined by Eq. (10.2):

$$Q = CV = (28.8 \text{ pF})(48 \text{ V}) = \textbf{1.38 nC}$$

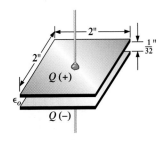

FIG. 10.8
Air capacitor for Example 10.3.

In the next example, we will insert a ceramic dielectric between the plates of the air capacitor of Fig. 10.8 and see how it affects the capacitance level, electric field, and charge on the plates.

EXAMPLE 10.4

a. Insert a ceramic dielectric with an ϵ_r of 250 between the plates of the capacitor of Fig. 10.8. Then determine the new level of capacitance. Compare your results to the solution of Example 10.3.
b. Find the resulting electric field strength between the plates, and compare your answer to the result of Example 10.3.
c. Determine the charge on each of the plates, and compare your answer to the result of Example 10.3.

Solutions:

a. Using Eq. (10.8), the new capacitance level is

$$C = \epsilon_r C_o = (250)(28.8 \text{ pF}) = \textbf{7200 pF} = \textbf{7.2 nF} = \textbf{0.0072 } \boldsymbol{\mu}\textbf{F}$$

which is **significantly higher** than the level of Example 10.3.

b. $\mathscr{E} = \dfrac{V}{d} = \dfrac{48\ V}{0.794\ mm} = \mathbf{60.5\ kV/m}$

Since the applied voltage and the distance between the plates did not change, **the electric field between the plates remains the same.**

c. $Q = CV = (7200\ pF)(48\ V) = \mathbf{345.6\ nC} = \mathbf{0.3456\ \mu C}$

We now know that the insertion of a dielectric between the plates will increase the amount of charge stored on the plates. In Example 10.4, since the relative permittivity increased by a factor of 250, the charge on the plates **increased by the same amount.**

EXAMPLE 10.5   Find the maximum voltage that can be applied across the capacitor of Example 10.4 if the dielectric strength is 80 V/mil.

Solution:

$$d = \frac{1}{32}\ \text{in.}\left(\frac{1000\ \text{mils}}{1\ \text{in.}}\right) = 31.25\ \text{mils}$$

and

$$V_{max} = 31.25\ \text{mils}\left(\frac{80\ V}{\text{mil}}\right) = \mathbf{2.5\ kV}$$

although the provided working voltage may be only 2 kV to provide a margin of safety.

Types of Capacitors

Capacitors, like resistors, can be listed under two general headings: **fixed** and **variable.** The symbol for the fixed capacitor appears in Fig. 10.9(a). Note that the curved side is normally connected to ground or to the point of lower dc potential. The symbol for variable capacitors appears in Fig. 10.9(b).

(a)   (b)

FIG. 10.9

Symbols for the capacitor:
(a) fixed; (b) variable.

Fixed Capacitors   Fixed-type capacitors come in all shapes and sizes. However,

in general, for the same type of construction and dielectric, the larger the capacitor, the greater is the capacitance.

In Fig. 10.10(a), the 10,000 μF electrolytic capacitor is significantly larger than the 1 μF capacitor. However, it is certainly not 10,000 times larger. For the polyester-film type of Fig. 10.10(b), the 2.2 μF capacitor is significantly larger than the 0.01 μF capacitor, but again it is not 220 times larger. The 22 μF tantalum capacitor of Fig. 10.10(c) is about 6 times larger than the 1.5 μF capacitor, even though the capacitance level is about 15 times higher. It is particularly interesting to note that due to the difference in dielectric and construction, the 22 μF tantalum capacitor is significantly smaller than the 2.2 μF polyester-film capacitor, and much smaller than 1/5 the size of the 100 μF electrolytic capacitor. The relatively large 10,000 μF electrolytic capacitor is normally used for high-power applications, such as in power supplies and high-output speaker systems. All the others may appear in any commercial electronic system.

The increase in size is due primarily to the effect of area and thickness of the dielectric on the capacitance level. There are a number of ways to increase the area without making the capacitor too large. One is to lay out the plates and the dielectric in long, narrow strips and then roll them all together, as shown in Fig. 10.11(a). The dielectric (remember that it has the characteristics of an insulator) between the conducting strips ensures the strips will never touch. Of course, the dielectric must be the type that can be rolled without breaking up. Depending on how the materials are wrapped, the capacitor can take on a cylindrical or a rectangular, box-type shape.

A second popular method is to stack the plates and the dielectrics, as shown in Fig. 10.11(b). The area is now a multiple of the number of dielectric layers. This construction is very popular for smaller capacitors. A third method is to use the dielectric to establish the

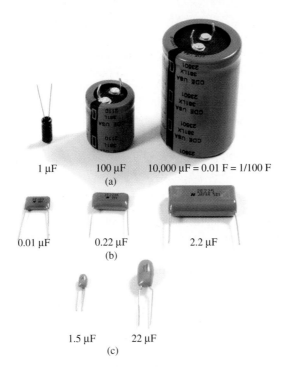

FIG. 10.10

Demonstrating that, in general, for each type of construction, the size of a capacitor will increase with the capacitance value: (a) electrolytic; (b) polyester-film; (c) tantalum.

body shape [a cylinder in Fig. 10.11(c)]. Then simply insert a rod for the positive plate, and coat the surface of the cylinder to form the negative plate, as shown in Fig. 10.11(c). Although the resulting "plates" are not the same in construction or surface area, the effect is to provide a large surface area for storage (the density of electric field lines will be different on the two "plates"), although the resulting distance factor may be larger than desired. Using a dielectric with a high ϵ_r, however, will compensate for the increased distance between the plates.

There are other variations of the above to increase the area factor, but the three depicted in Fig. 10.11 are the most popular.

The next controllable factor is the distance between the plates. This factor, however, is very sensitive to how thin the dielectric can be made, with natural concerns for the fact that the working voltage (the breakdown voltage) will drop as the gap decreases. Some of the thinnest dielectrics are simply oxide coatings on one of the conducting surfaces (plates). A

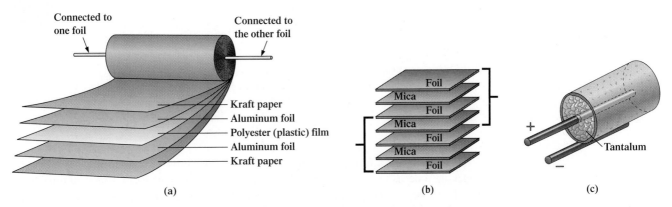

FIG. 10.11

Three ways to increase the area of a capacitor: (a) rolling; (b) stacking; (c) insertion.

very thin polyester material, such as Mylar®, Teflon®, or even paper with a paraffin coating, provides a thin sheet of material that can easily be wrapped for increased areas. Materials such as mica and some ceramic materials can be made only so thin before crumbling or breaking down under stress.

The last factor is the dielectric, for which there is a wide range of possibilities. However, following factors will greatly influence which dielectric is used:

The level of capacitance desired
The resulting size
The possibilities for rolling, stacking, and so on
Temperature sensitivity
Working voltage

The range of relative permittivities is enormous, as shown in Table 10.2, but all the factors listed above must be considered in the construction process.

In general, the most common fixed capacitors are the **electrolytic, film, polyester, foil, Mylar®, ceramic, mica, dipped,** and **oil.**

The **electrolytic capacitors** of Fig. 10.12 are easy to identify by their shape and the fact that they usually have a polarity marking on the body (although special-application electrolytics are available that are not polarized). Few capacitors have a polarity marking, but those that do must be connected with the negative terminal connected to ground or to the point of lower potential. The markings often used to denote the positive terminal or plate include +, □, and Δ. In general, electrolytic capacitors offer some of the highest capacitance values available, although their working voltage levels are limited. Typical values range from **0.1 μF to 15,000 μF,** with working voltages from **5 V to 450 V.** The basic construction employs the rolling process of Fig. 10.11(a) using a roll of aluminum foil coated on one side with aluminum oxide—the aluminum being the positive plate, and the oxide the dielectric. A layer of paper or gauze saturated with an electrolyte (a solution or paste that forms the conducting medium between the electrodes of the capacitor) is placed over the aluminum-oxide coating of the positive plate. Another layer of aluminum without the oxide coating is then placed over this layer to assume the role of the negative plate. In most cases, the negative plate is connected directly to the aluminum container, which then serves as the negative terminal for external connections. Because of the size of the roll of aluminum foil, the overall size of the electrolytic capacitor is greater than most.

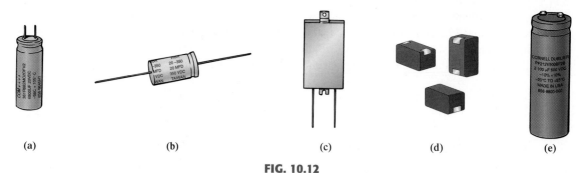

(a)   (b)   (c)   (d)   (e)

FIG. 10.12
Various types of electrolytic capacitors: (a) miniature radial leads; (b) axial leads;
(c) flatpack; (d) surface-mount; (e) screw-in terminals.
(Courtesy of Cornell-Dubilier.)

Film, polyester, foil, polypropylene, Teflon®, or **Mylar® capacitors** use a rolling or stacking process to increase the surface area, as shown in Fig. 10.13. The resulting shape can be either round or rectangular, with radial or axial leads. The typical range for such capacitors is **100 pF to 10 μF,** with units available up to 100 μF. The name of the unit defines the type of dielectric employed. Working voltages can extend from a **few volts to 2000 V,** depending on the type of unit.

(a) (b) (c) (d)

FIG. 10.13

(a) Film/foil polyester radial lead; (b) metalized polyester-film axial lead;
(c) surface-mount polyester-film; (d) polypropylene-film, radial lead.

Ceramic capacitors (often called **disc capacitors**) use a ceramic dielectric, as shown in Fig. 10.14(a), to utilize the excellent ϵ_r values and high working voltages associated with a number of ceramic materials. Stacking can also be applied to increase the surface area. An example of the disc variety appears in Fig. 10.14(b). Ceramic capacitors typically range in value from **10 pF to 0.047 μF,** with high working voltages that can reach **as high as 10 kV.**

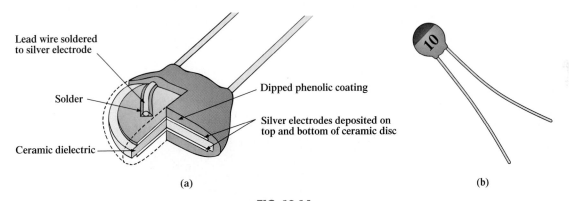

Lead wire soldered
to silver electrode

Solder

Ceramic dielectric

Dipped phenolic coating

Silver electrodes deposited on
top and bottom of ceramic disc

(a) (b)

FIG. 10.14

Ceramic (disc) capacitor: (a) construction; (b) appearance.

Mica capacitors use a mica dielectric that can be monolithic (single chip) or stacked. The relatively small size of monolithic mica chip capacitors is demonstrated in Fig. 10.15(a), with their placement shown in Fig. 10.15(b). A variety of high-voltage mica paper capacitors are displayed in Fig. 10.15(c). Mica capacitors typically range in value from **2 pF to several microfarads,** with working voltages **up to 20 kV.**

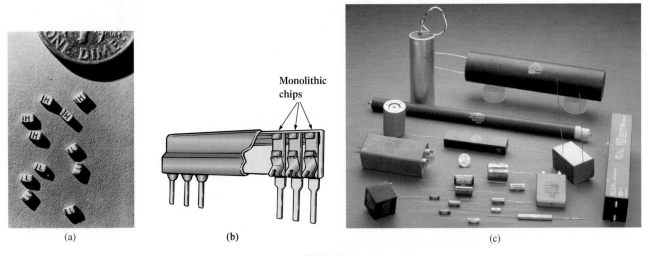

Monolithic
chips

(a) (b) (c)

FIG. 10.15

Mica capacitors: (a) and (b) surface-mount monolithic chips; (c) high-voltage/temperature mica paper capacitors.
[(a) and (b) courtesy of Vishay Intertechnology, Inc.; (c) courtesy of Custom Electronics, Inc.]

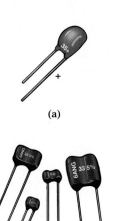

(a)

(b)

FIG. 10.16
Dipped capacitors: (a) polarized tantalum; (b) nonpolarized mica.

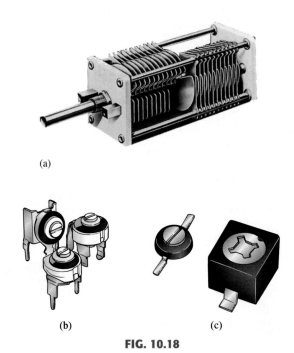

(a)

(b)   **(c)**

FIG. 10.18
Variable capacitors: (a) air; (b) air trimmer; (c) ceramic dielectric compression trimmer.
[(a) courtesy of James Millen Manufacturing Co.; (c) courtesy of Sprague-Goodman, Inc.]

FIG. 10.17
Oil-filled, metallic oval case snubber capacitor (the snubber removes unwanted voltage spikes).
(Courtesy of Cornell-Dubilier.)

Dipped capacitors are made by dipping the dielectric (tantalum or mica) into a conductor in a molten state to form a thin, conductive sheet on the dielectric. Due to the presence of an electrolyte in the manufacturing process, dipped tantalum capacitors require a polarity marking to ensure that the positive plate is always at a higher potential than the negative plate, as shown in Fig. 10.16(a). A series of small positive signs is typically applied to the casing near the positive lead. A group of nonpolarized, mica dipped capacitors are shown in Fig. 10.16(b). They typically range in value from **0.1 μF to 680 μF,** but with lower working voltages ranging from **6 V to 50 V.**

Most **oil capacitors** such as appearing in Fig. 10.17 are used for industrial applications such as welding, high-voltage power supplies, surge protection, and power-factor correction (Chapter 18). They can provide capacitance levels extending from **0.001 μF all the way up to 10,000 μF,** with working voltages **up to 150 kV.** Internally, there are a number of parallel plates sitting in a bath of oil or oil-impregnated material (the dielectric).

Variable Capacitors All the parameters of Eq. (10.7) can be changed to some degree to create a **variable capacitor.** For example, in Fig. 10.18(a), the capacitance of the variable air capacitor is changed by turning the shaft at the end of the unit. By turning the shaft, you control the amount of common area between the plates: The less common area there is, the less capacitance there is. In Fig. 10.18(b), we have a much smaller **air trimmer capacitor.** It works under the same principle, but the rotating blades are totally hidden inside the structure. In Fig. 10.18(c), the capacitance level is changed by changing the distance between the plates. Another type, called a **ceramic trimmer capacitor,** permits varying the capacitance by changing the common area as above or by applying pressure to the ceramic plate to reduce the distance between the plates.

Leakage Current

We would naturally like to think of capacitors as ideal elements. Unfortunately, however, this is not the case. Up to this point, we have assumed that the insulating characteristics of dielectrics would prevent any flow of charge between the plates unless the breakdown voltage was exceeded. In reality, however, dielectrics are not perfect insulators, and they do carry a few free electrons in their atomic structure.

When a voltage is applied across a capacitor, a **leakage current** will be established between the plates. This current is usually so small that it can be ignored for the application under investigation. The availability of free electrons to support current flow is represented by a large parallel resistor in the equivalent circuit for a capacitor as shown in Fig. 10.19(a). If we apply 10 V across a capacitor with an internal resistance of 1000 MΩ, the current will be 0.01 μA—a level that can be ignored for most applications.

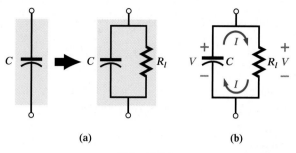

(a)   (b)

FIG. 10.19

Leakage current: (a) including the leakage resistance in the equivalent model for a capacitor; (b) internal discharge of a capacitor due to the leakage current.

The real problem associated with leakage currents is not evident until you ask the capacitors to sit in a charged state for long periods of time. As shown in Fig. 10.19(b), the voltage ($V = Q/C$) across a charged capacitor will also appear across the parallel leakage resistance and will establish a discharge current through the resistor. In time, the capacitor will be totally discharged. Capacitors such as the electrolytic that have high leakage currents (a leakage resistance of 0.5 MΩ is typical) usually have a limited shelf life due to this internal discharge characteristic. Ceramic, tantalum, and mica capacitors typically have unlimited shelf life due to leakage resistances in excess of 1000 MΩ. Thin-film capacitors have lower levels of leakage resistances that result in some concern about shelf life.

Temperature Effects: ppm

Every capacitor is temperature sensitive, with the nameplate capacitance level specified at room temperature. Depending on the type of dielectric, increasing or decreasing temperatures can cause either a drop or a rise in capacitance. If temperature is a concern for a particular application, the manufacturer will provide a temperature plot, such as shown in Fig. 10.20, or a **ppm/°C** (parts per million per degree Celsius) rating for the capacitor. Note in Fig. 10.20 the 0% variation from the nominal (nameplate) value at 25°C (room temperature). At 0°C (freezing), it has dropped 20%, while at 100°C (the boiling point of water), it has dropped 70%—a factor to consider for some applications.

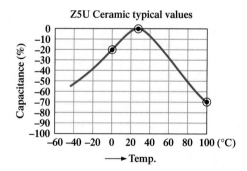

FIG. 10.20

Variation of capacitor value with temperature.

As an example using the ppm level, consider a 100 μF capacitor with a **temperature coefficient** or **ppm** of −150 ppm/°C. It is important to note the negative sign in front of the ppm value, because it reveals that the capacitance will drop with increase in temperature. It takes a moment to fully appreciate a term such as *parts per million*. In equation form, 150 parts per million can be written as

$$\frac{150}{1{,}000{,}000} \times$$

If we then multiply this term by the capacitor value, we can obtain the change in capacitance for each 1°C change in temperature. That is,

$$\frac{150}{1,000,000}(100 \ \mu\text{F})/°\text{C} = \mathbf{0.015 \ \mu\text{F}/°\text{C}} = \mathbf{15,000 \ pF/°\text{C}}$$

If the temperature should rise 25°C, the capacitance will decrease by

$$\frac{15,000 \ \text{pF}}{°\cancel{\text{C}}}(25°\cancel{\text{C}}) = 0.375 \ \mu\text{F}$$

changing the capacitance level to

$$100 \ \mu\text{F} - 0.375 \ \mu\text{F} = \mathbf{99.625 \ \mu\text{F}}$$

Capacitor Labeling

Due to the small size of some capacitors, various marking schemes have been adopted to provide the capacitance level, the tolerance, and, if possible, the working voltage. In general, however, as pointed out above, **the size of the capacitor is the first indicator of its value.** In fact, for most marking schemes there is no indication as to whether it is in μF or pF. It is assumed that you can make that judgment purely from the size. The smaller units are typically in pF, and the larger units in μF. Unless indicated by an **n** or **N,** most units are not provided in nF. On larger μF units, the value can often be printed on the jacket with the tolerance and working voltage. However, smaller units need to use some form of abbreviation as shown in Fig. 10.21. For very small units such as appearing in Fig. 10.21(a) with only two numbers, the value is recognized immediately as in pF with the **K** an indicator of a $\pm 10\%$ tolerance level. Too often the K is read as a multiplier of 10^3, and the capacitance is read as 20,000 pF or 20 nF rather than the actual 20 pF.

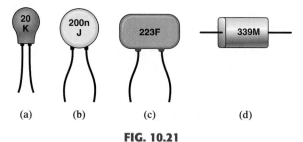

| (a) | (b) | (c) | (d) |

FIG. 10.21

Various marking schemes for small capacitors.

For the unit of Fig. 10.21(b), there was room for a lowercase **n** to represent a multiplier of 10^{-9}, resulting in a value of 200 nF. To avoid unnecessary confusion, the letters used for tolerance do not include **N** or **U** or **P,** so the presence of any of these letters in upper- or lowercase normally refers to the multiplier level. The **J** appearing on the unit of Fig. 10.20(b) represents a $\pm 5\%$ tolerance level. For the capacitor of Fig. 10.21(c), the first two numbers are the numerical value of the capacitor, while the third number is the power of the multiplier (or number of zeros to be added to the first two numbers). The question then remains as to whether the units are μF or pF. With the 223 representing a number of 22,000, the units are certainly not μF, because the unit is too small for such a large capacitance. It is a 22,000 pF = 22 nF capacitor. The **F** represents a $\pm 1\%$ tolerance level. Multipliers of 0.01 use an 8 for the third digit, while multipliers of 0.1 use a 9. The capacitor of Fig. 10.21(d) is a $33 \times 0.1 = 3.3 \ \mu$F capacitor with a tolerance of $\pm 20\%$ as defined by the capital letter **M.** The capacitance is not 3.3 pF because the unit is too large; again, the factor of size is very helpful in making a judgment about the capacitance level.

Measurement and Testing of Capacitors

The capacitance of a capacitor can be read directly using a meter such as the Universal LCR Meter of Fig. 10.22. If you set the meter on **C** for *capacitance,* it will automatically choose the most appropriate unit of measurement for the element, that is, F, μF, nF, or pF. Note the polarity markings on the meter for capacitors that have a specified polarity.

The best check is to use a meter such as the type appearing in Fig. 10.22. However, if it is unavailable, an ohmmeter can be used to determine whether the dielectric is still in good

FIG. 10.22

Digital reading capacitance meter.
(Image provided by B & K Precision Corporation.)

working order or whether it has deteriorated due to age or use (especially for paper and electrolytics). As the dielectric breaks down, the insulating qualities of the material will decrease to the point where the resistance between the plates drops to a relatively low level. To use an ohmmeter, be sure that the capacitor is fully discharged by placing a lead directly across its terminals. Then hook up the meter (paying attention to the polarities if the unit is polarized) as shown in Fig. 10.23, and note whether the resistance has dropped to a relatively low value (0 to a few kilohms). If so, the capacitor should be discarded. You may find that the reading will change when the meter is first connected. This change is due to the charging of the capacitor by the internal supply of the ohmmeter. In time the capacitor will settle down, however, and the correct reading can be observed. Typically, it should pin at the highest level on the megohm scales or indicate OL on a digital meter.

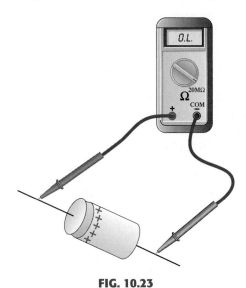

FIG. 10.23
Checking the dielectric of an electrolytic capacitor.

The above ohmmeter test is not all-inclusive, since some capacitors will exhibit the breakdown characteristics only when a large voltage is applied. The test, however, does help isolate capacitors in which the dielectric has deteriorated.

Standard Capacitor Values

The most common capacitors employ the same numerical multipliers encountered for resistors.

The vast majority are available with 5%, 10%, or 20% tolerances. There are capacitors available, however, with tolerances of 1%, 2%, or 3%, if you are willing to pay the price. Typical values include **0.1 μF, 0.15 μF, 0.22 μF, 0.33 μF, 0.47 μF, 0.68 μF; or 1 μF, 1.5 μF, 2.2 μF, 3.3 μF, 4.7 μF, 6.8 μF; and 10 pF, 22 pF, 33 pF, 100 pF;** and so on.

10.5   TRANSIENTS IN CAPACITIVE NETWORKS: THE CHARGING PHASE

The placement of charge on the plates of a capacitor does not occur instantaneously. Instead, it occurs over a period of time determined by the components of the network. The charging phase, the phase during which charge is deposited on the plates, can be described by reviewing the response of the simple series circuit of Fig. 10.2. The circuit has been redrawn in Fig. 10.24 with the symbol for a fixed capacitor.

Recall that the instant the switch is closed, electrons are drawn from the top plate and deposited on the bottom plate by the battery, resulting in a net positive charge on the top plate and a negative charge on the bottom plate. The transfer of electrons is very rapid at first, slowing down as the potential across the plates approaches the applied voltage of the battery. Eventually, when the voltage across the capacitor equals the applied voltage, the transfer of electrons will cease, and the plates will have a net charge determined by $Q = CV_C = CE$. This period of time during which charge is being deposited on the plates is called **a transient period because it is defined for a specific period of time—it does not last forever.**

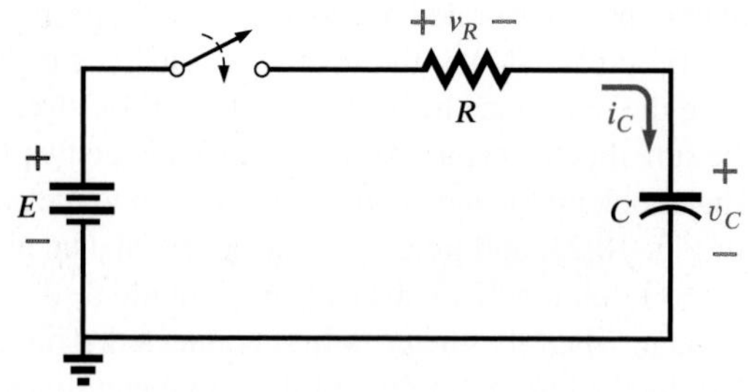

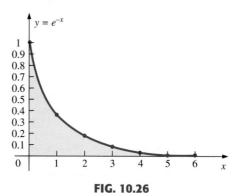

FIG. 10.24

Basic R-C charging network.

Since the voltage across the plates is directly related to the charge on the plates by $V = Q/C$, a plot of the voltage across the capacitor will have the same shape as a plot of the charge on the plates over time. As shown in Fig. 10.25, the voltage across the capacitor is zero volts when the switch is closed ($t = 0$ s). It then builds up very quickly at first since charge is being deposited at a very high rate of speed. As time moves on, the charge is deposited at a slower rate, and the change in voltage drops off with time. The voltage continues to grow, but at a much slower rate. Eventually, as the voltage across the plates approaches the applied voltage, the charging rate is very slow, until finally the voltage across the plates is equal to the applied voltage—the transient phase has passed.

Fortunately, the waveform of Fig. 10.25 from beginning to end can be described using the mathematical function e^{-x}. It is an exponential function that will decrease with time, as shown in Fig. 10.26. If we subsitite 0 for x, we obtain e^{-0} which by definition is 1, as shown in Table 10.3 and on the plot of Fig. 10.26. Table 10.3 clearly reveals that as x increases, the function e^{-x} decreases in magnitude until it is very close to zero after $x = 5$. As noted in Table 10.3, the exponential factor $e^1 = e = 2.71828$.

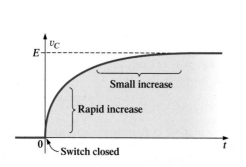

FIG. 10.25

Transient response of v_C for the network of Fig. 10.24.

FIG. 10.26

Plot of e^x versus x.

You may wonder how this function can help us if it decreases with time and the curve for the voltage across the capacitor increases with time. We simply place the exponential in the proper mathematical form as shown in Eq. (10.9):

$$v_C = E(1 - e^{-t/\tau}) \Big|_{\text{charging}} \qquad \text{(volts, V)} \qquad \textbf{(10.9)}$$

First note in Eq. (10.9) that the voltage v_C **is written in lowercase (not capital) italic to point out that it is a function that will change with time**—it is not a constant. The exponent of the exponential function is no longer just x, but now is time (t) divided by a constant τ, the Greek letter *tau*. The quantity τ is defined by

$$\tau = RC \qquad \text{(time, s)} \qquad \textbf{(10.10)}$$

The factor τ, called the **time constant** of the network, has the units of time as shown below using some of the basic equations introduced earlier in this text:

$$\tau = RC = \left(\frac{V}{I}\right)\left(\frac{Q}{V}\right) = \left(\frac{\cancel{V}}{\cancel{Q}/t}\right)\left(\frac{\cancel{Q}}{\cancel{V}}\right) = t \ (\text{seconds})$$

TABLE 10.3

Selected values of e^{-x}

$x = 0$	$e^{-x} = e^{-0} = \dfrac{1}{e^0} = \dfrac{1}{1} = 1$
$x = 1$	$e^{-1} = \dfrac{1}{e} = \dfrac{1}{2.71828\ldots} = 0.3679$
$x = 2$	$e^{-2} = \dfrac{1}{e^2} = 0.1353$
$x = 5$	$e^{-5} = \dfrac{1}{e^5} = 0.00674$
$x = 10$	$e^{-10} = \dfrac{1}{e^{10}} = 0.0000454$
$x = 100$	$e^{-100} = \dfrac{1}{e^{100}} = 3.72 \times 10^{-44}$

A plot of Eq. (10.9) will result in the curve of Fig. 10.27 whose shape is an exact match with that of Fig. 10.25.

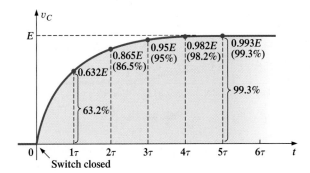

FIG. 10.27

Plotting the equation $v_C = E(1 - e^{-t/\tau})$ versus time (t).

In Eq. (10.9), if we substitute $t = 0$ s, we find that

$$e^{-t/\tau} = e^{-0/\tau} = e^{-0} = \frac{1}{e^0} = \frac{1}{1} = 1$$

and

$$v_C = E(1 - e^{-t/\tau}) = E(1 - 1) = \textbf{0 V}$$

as appearing in the plot of Fig. 10.27.

It is important to realize at this point that the plot of Fig. 10.27 is not against simply time but against τ, the time constant of the network. If we want to know the voltage across the plates after one time constant, we simply plug $t = 1\tau$ into Eq. (10.9). The result is

$$e^{-t/\tau} = e^{-\tau/\tau} = e^{-1} \cong 0.368$$

and

$$v_C = E(1 - e^{-t/\tau}) = E(1 - 0.368) = \textbf{0.632}E$$

as shown in Fig. 10.27.

At $t = 2\tau$:

$$e^{-t/\tau} = e^{-2\tau/\tau} = e^{-2} \cong 0.135$$

and

$$v_C = E(1 - e^{-t/\tau}) = E(1 - 0.135) = \textbf{0.865}E$$

as shown in Fig. 10.27.

As the number of time constants increases, the voltage across the capacitor does indeed approach the applied voltage.

At $t = 5\tau$:

$$e^{-t/\tau} = e^{-5\tau/\tau} = e^{-5} \cong 0.007$$

and

$$v_C = E(1 - e^{-t/\tau}) = E(1 - 0.007) = \textbf{0.993}E \cong E$$

In fact, we can conclude from the results just obtained that

the voltage across a capacitor in a dc network is essentially equal to the applied voltage after five time constants of the charging phase have passed.

or, in more general terms,

the transient or charging phase of a capacitor has essentially ended after five time constants.

It is indeed fortunate that the same exponential function can be used to plot the current of the capacitor versus time. When the switch is first closed, the flow of charge or current jumps very quickly to a value limited by the applied voltage and the circuit resistance, as shown in Fig. 10.28. The rate of deposit, and hence the current, then decreases quite rapidly, until eventually charge is not being deposited on the plates and the current drops to zero amperes.

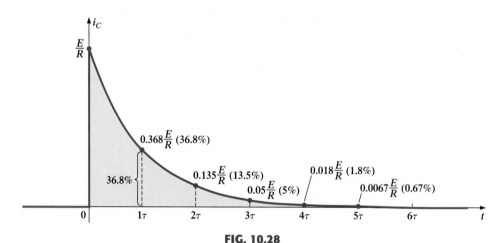

FIG. 10.28

Plotting the equation $i_C = \dfrac{E}{R}e^{-t/\tau}$ versus time (t).

The equation for the current is the following:

$$\boxed{i_C = \frac{E}{R}e^{-t/\tau}}_{\text{charging}} \qquad \text{(amperes, A)} \qquad \textbf{(10.11)}$$

In Fig. 10.24, the current (conventional flow) has the direction shown since electrons flow in the opposite direction.

At $t = 0$ s:

$$e^{-t/\tau} = e^{-0} = 1$$

and

$$i_C = \frac{E}{R}e^{-t/\tau} = \frac{E}{R}(1) = \frac{E}{R}$$

At $t = 1\tau$:

$$e^{-t/\tau} = e^{-\tau/\tau} = e^{-1} \cong 0.368$$

and

$$i_C = \frac{E}{R}e^{-t/\tau} = \frac{E}{R}(0.368) = \textbf{0.368}\frac{E}{R}$$

In general, Figure 10.28 clearly reveals that

the current of a capacitive dc network is essentially zero amperes after five time constants of the charging phase have passed.

It is also important to recognize that

during the charging phase, the major change in voltage and current occurs during the first time constant.

The voltage across the capacitor reaches about 63.2% (about 2/3) of its final value, whereas the current drops to 36.8% (about 1/3) of its peak value. During the next time constant, the voltage increases only about 23.3%, whereas the current drops to 13.5%. The first time constant is therefore a very dramatic time for the changing parameters. Between the fourth and fifth time constants, the voltage will increase only about 1.2%, whereas the current will drop to less than 1% of its peak value.

Returning to Figs. 10.27 and 10.28, it is particularly interesting to note that when the voltage across the capacitor reaches the applied voltage E, the current will drop to zero amperes, as reviewed in Fig. 10.29. These conditions match those of an open circuit, permitting the following conclusion:

A capacitor can be replaced by an open-circuit equivalent once the charging phase in a dc network has passed.

This conclusion will be particularly useful when analyzing dc networks that have been sitting for a long period of time or have passed the transient phase that normally occurs when a system is first turned on.

A similar conclusion can be reached if we consider the instant the switch is closed in the circuit of Fig. 10.24. Referring to Figs. 10.27 and 10.28 again, we find that the current is a peak value at $t = 0$ s, whereas the voltage across the capacitor is 0 V, as shown in the equivalent circuit of Fig. 10.30. The result is that

a capacitor has the characteristics of a short-circuit equivalent at the instant the switch is closed in an uncharged series R-C circuit.

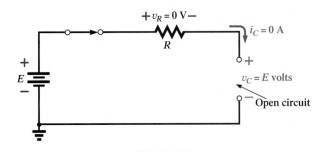

FIG. 10.29

Demonstrating that a capacitor has the characteristics of an open circuit after the charging phase has passed.

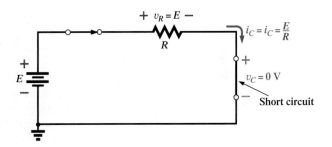

FIG. 10.30

Revealing the short-circuit equivalent for the capacitor that occurs when the switch is first closed.

In Eq. (10.9), the time constant τ will always have some value because some resistance is always present in a capacitive network. In some cases, the value of τ may be very small, but five times that value of τ, no matter how small, must therefore always exist; it cannot be zero. The result is the following very important conclusion:

The voltage across a capacitor cannot change instantaneously.

In fact, we can take this statement a step further by saying that the capacitance of a network is a measure of how much it will oppose a change in voltage in a network. The larger the capacitance, the larger the time constant, and the longer it will take the voltage across the capacitor to reach the applied value. This can prove very helpful when lightning arresters and surge suppressors are designed to protect equipment from unexpected high surges in voltage.

Since the resistor and the capacitor of Fig. 10.24 are in series, the current through the resistor is the same as that associated with the capacitor. The voltage across the resistor can be determined by simply using Ohm's law in the following manner:

$$v_R = i_R R = i_C R$$

so that

$$v_R = \left(\frac{E}{R} e^{-t/\tau}\right) R$$

and

$$\boxed{v_R = E e^{-t/\tau}}_{\text{charging}} \qquad \text{(volts, V)} \qquad \textbf{(10.12)}$$

A plot of the voltage as provided in Fig. 10.31 has the same shape as that for the current because they are related by the constant R. Note, however, that the voltage across the resistor starts at a level of E volts because the voltage across the capacitor is zero volts and Kirchhoff's voltage law must always be satisfied. When the capacitor has reached the applied voltage, the voltage across the resistor must drop to zero volts for the same reason. Always remember that

Kirchhoff's voltage law is applicable at any instant of time for any type of voltage in any type of network.

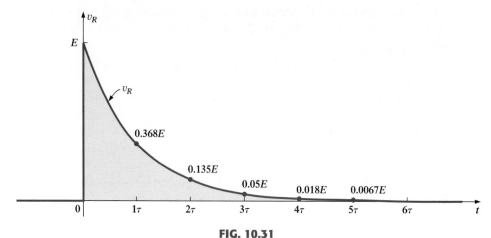

FIG. 10.31

Plotting the equation $v_R = Ee^{-t/\tau}$ versus time (t).

Using the Calculator to Solve Exponential Functions

Before looking at an example, we will first discuss the use of the calculator with exponential functions. The process is actually quite simple for a number such as $e^{-1.2}$. Simply select the 2nd function key, followed by the function e^x. Then insert the (−) sign from the numerical keyboard (not the mathematical functions), and insert the number 1.2 followed by ENTER to obtain the result of 0.301, as shown in Fig. 10.32. The use of the computer software program Mathcad will be demonstrated in a later example.

FIG. 10.32

Calculator key strokes to determine $e^{-1.2}$.

EXAMPLE 10.6 For the circuit of Fig. 10.33:

a. Find the mathematical expression for the transient behavior of v_C, i_C, and v_R if the switch is closed at $t = 0$ s.
b. Plot the waveform of v_C versus the time constant of the network.
c. Plot the waveform of v_C versus time.
d. Plot the waveforms of i_C and v_R versus the time constant of the network.
e. What is the value of v_C at $t = 20$ ms?
f. On a practical basis, how much time must pass before we can assume that the charging phase has passed?
g. When the charging phase has passed, how much charge is sitting on the plates?

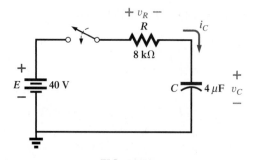

FIG. 10.33

Transient network for Example 10.6.

h. If the capacitor has a leakage current of 10,000 MΩ, what is the initial leakage current? Once the capacitor is separated from the circuit, how long will it take to totally discharge, assuming a linear (unchanging) discharge rate?

Solutions:

a. The time constant of the network is

$$\tau = RC = (8 \text{ k}\Omega)(4 \text{ }\mu\text{F}) = 32 \text{ ms}$$

resulting in the following mathematical equations:

$$v_C = E(1 - e^{-t/\tau}) = \textbf{40 V}(\textbf{1} - e^{-t/32\text{ms}})$$

$$i_C = \frac{E}{R}e^{-t/\tau} = \frac{40 \text{ V}}{8 \text{ k}\Omega}e^{-t/32\text{ms}} = \textbf{5 mA}e^{-t/32\text{ms}}$$

$$v_R = Ee^{-t/\tau} = \textbf{40 V}e^{-t/32\text{ms}}$$

b. The resulting plot appears in Fig. 10.34.

c. The horizontal scale will now be against time rather than time constants, as shown in Fig. 10.35. The plot points in Fig. 10.35 were taken from Fig. 10.34.

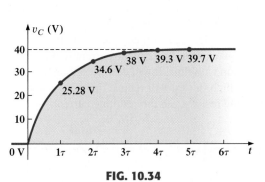

FIG. 10.34

v_C versus time for the charging network of Fig. 10.33.

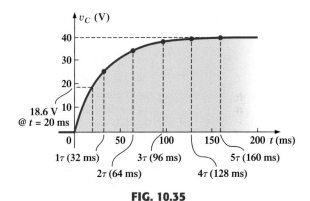

FIG. 10.35

Plotting the waveform of Fig. 10.34 versus time (t).

d. Both plots appear in Fig. 10.36.

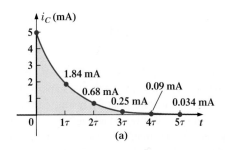

FIG. 10.36

i_C and v_R for the charging network of Fig. 10.33.

e. Substituting the time $t = 20$ ms will result in the following for the exponential part of the equation:

$$e^{-t/\tau} = e^{-20\text{ms}/32\text{ms}} = e^{-0.625} = 0.535 \quad \text{(using a calculator)}$$

so that $v_C = 40 \text{ V}(1 - e^{-t/32\text{ms}}) = 40 \text{ V}(1 - 0.535)$

$$= (40 \text{ V})(0.465) = \textbf{18.6 V} \quad \text{(as verified by Fig. 10.35)}$$

f. Assuming a full charge in five time constants will result in

$$5\tau = 5(32 \text{ ms}) = \textbf{160 ms} = \textbf{0.16 s}$$

g. Using Eq. (10.2):

$$Q = CV = (4 \text{ }\mu\text{F})(40 \text{ V}) = \textbf{160 }\mu\textbf{C}$$

h. Using Ohm's law:

$$I_{\text{leakage}} = \frac{40 \text{ V}}{10,000 \text{ M}\Omega} = 4 \text{ nA}$$

Finally, the basic equation $I = Q/t$ will result in

$$t = \frac{Q}{I} = \frac{160 \,\mu\text{C}}{4 \text{ nA}} = (40,000 \text{ s})\left(\frac{1 \text{ min}}{60 \text{ s}}\right)\left(\frac{1 \text{ h}}{60 \text{ min}}\right) = \textbf{11.11 h}$$

10.6 TRANSIENTS IN CAPACITIVE NETWORKS: THE DISCHARGING PHASE

We will now investigate how to discharge a capacitor while exerting some control on how long the discharge time will be. You can, of course, place a lead directly across a capacitor to discharge it very quickly—and possibly cause a visible spark. For larger capacitors such as found in TV sets, this procedure should not be attempted because of the high voltages involved—unless, of course, you are trained in the maneuver.

In Fig. 10.37(a), a second contact for the switch was added to the circuit of Fig. 10.24 to permit a controlled discharge of the capacitor. With the switch in position 1, we have the charging network described in the last section. Following the full charging phase, if we move the switch to position 1, the capacitor can be discharged through the resulting circuit of Fig. 10.37(b). In Fig. 10.37(b), the voltage across the capacitor will appear directly across the resistor to establish a discharge current. Initially the current will jump to a relatively high value; then it will begin to drop with time. It will drop with time because charge is leaving the plates of the capacitor, which in turn will reduce the voltage across the capacitor and thereby the voltage across the resistor and the resulting current.

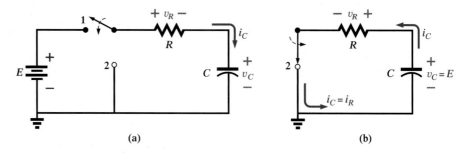

(a)   (b)

FIG. 10.37

(a) Charging network; (b) discharging configuration.

Before looking at the wave shapes for each quantity of interest, note that the current i_C has now reversed direction as shown in Fig. 10.37(b). As shown in parts (a) and (b) of Fig. 10.37, the voltage across the capacitor does not reverse polarity, but the current reverses direction. We will show the reversals on the resulting plots by sketching the waveforms in the negative regions of the graph. In all the waveforms it is interesting and, of course, helpful that all the mathematical expressions use the same e^{-x} factor appearing during the charging phase.

For the voltage across the capacitor that is decreasing with time, the mathematical expression is the following:

$$\boxed{v_C = Ee^{-t/\tau}}_{\text{discharging}} \tag{10.13}$$

For this circuit, the time constant τ is defined by the same equation as employed for the charging phase. That is,

$$\boxed{\tau = RC}_{\text{discharging}} \tag{10.14}$$

Since the current will decrease with time, it will have a similar format:

$$\boxed{i_C = \frac{E}{R}e^{-t/\tau}}_{\text{discharging}} \tag{10.15}$$

For the configuration of Fig. 10.37(b), since $v_R = v_C$ (in parallel), the equation for the voltage v_R has the same format:

$$v_R = Ee^{-t/\tau} \bigg|_{\text{discharging}}$$

(10.16)

The complete discharge will occur, for all practical purposes, in five time constants. If the switch is moved between terminals 1 and 2 every five time constants, the wave shapes of Fig. 10.38 will result for v_C, i_C, and v_R. For each curve the current directions and voltage polarities are as defined by the configurations of Fig. 10.37. Note, as pointed out above, that the current reverses direction during the discharge phase.

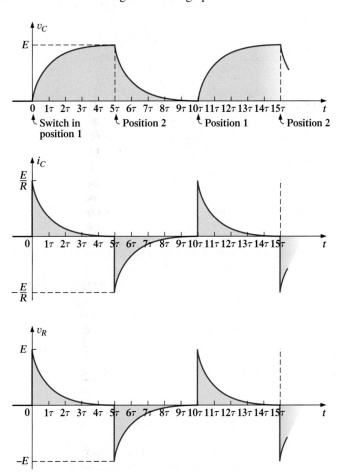

FIG. 10.38

v_C, i_C, and v_R for 5τ switching between contacts of Fig. 10.37(a).

The discharge rate does not have to equal the charging rate if a different switching arrangement is used. In fact, Example 10.8 will demonstrate how to change the discharge rate.

EXAMPLE 10.7 Using the values of Example 10.6, plot the waveforms for v_C and i_C resulting from switching between contacts 1 and 2 of Fig. 10.37 every five time constants.

Solution: The time constant is the same for the charging and discharging phases. That is,

$$\tau = RC = (8 \text{ k}\Omega)(4 \text{ μF}) = 32 \text{ ms}$$

For the discharge phase, the equations are

$$v_C = Ee^{-t/\tau} = \textbf{40 V}e^{-t/\textbf{32ms}}$$

$$i_C = -\frac{E}{R}e^{-t/\tau} = -\frac{40 \text{ V}}{8 \text{ k}\Omega}e^{-t/32\text{ms}} = \textbf{-5 mA}e^{-t/\textbf{32 ms}}$$

$$v_R = v_C = \textbf{40 V}e^{-t/\textbf{32ms}}$$

A continuous plot for the charging and discharging phases appears in Fig. 10.39.

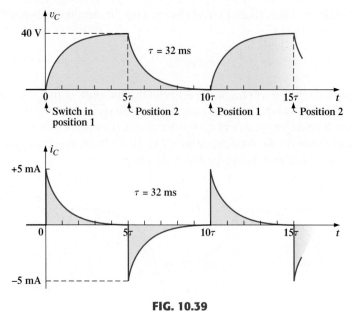

FIG. 10.39

v_C and i_C for the network of Fig. 10.37(a) with the values of Example 10.6.

The Effect of τ on the Response

In Example 10.7, if the value of τ were changed by changing the resistance, or the capacitor, or both, **the resulting waveforms would appear the same because they were plotted against the time constant of the network.** If they were plotted against time, there could be a dramatic change in the appearance of the resulting plots. In fact, on an oscilloscope, an instrument designed to display such waveforms, the plots are against time, and the change will be immediately apparent. In Fig. 10.40(a), the waveforms of Fig. 10.39 for v_C and i_C were plotted against time. In Fig. 10.40(b), the capacitance was decreased to 1 μF which will reduce the time constant to 8 ms. Note the dramatic effect on the appearance of the waveform.

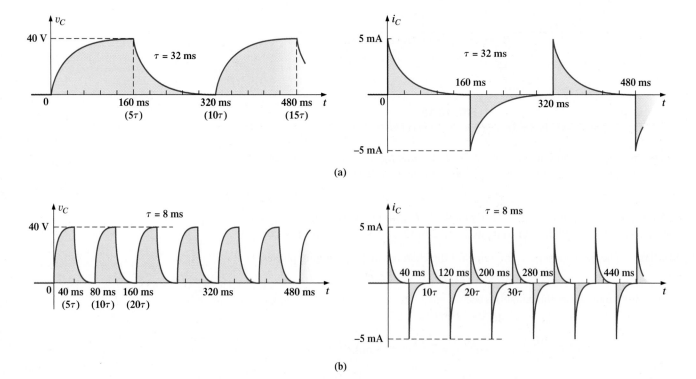

FIG. 10.40

Plotting v_C and i_C versus time in ms: (a) $\tau = 32$ ms; (b) $\tau = 8$ ms.

For a fixed-resistance network, the effect of increasing the capacitance is clearly demonstrated in Fig. 10.41. The larger the capacitance, and hence the time constant, the longer it takes the capacitor to charge up—there is more charge to be stored. The same effect can be created by holding the capacitance constant and increasing the resistance, but now the longer time is due to the lower currents that are a result of the higher resistance.

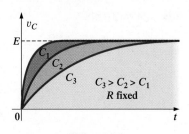

FIG. 10.41

Effect of increasing values of C (with R constant) on the charging curve for v_C.

EXAMPLE 10.8 For the circuit of Fig. 10.42:

a. Find the mathematical expressions for the transient behavior of the voltage v_C and the current i_C if the capacitor was initially uncharged and the switch is thrown into position 1 at $t = 0$ s.
b. Find the mathematical expressions for the voltage v_C and the current i_C if the switch is moved to position 2 at $t = 10$ ms. (Assume that the leakage resistance of the capacitor is infinite ohms; that is, there is no leakage current.)
c. Find the mathematical expressions for the voltage v_C and the current i_C if the switch is thrown into position 3 at $t = 20$ ms.
d. Plot the waveforms obtained in parts (a)–(c) on the same time axis using the defined polarities of Fig. 10.42.

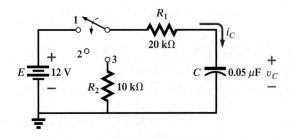

FIG. 10.42

Network to be analyzed in Example 10.8.

Solutions:

a. *Charging phase:*

$$\tau = R_1 C = (20 \text{ k}\Omega)(0.05 \text{ }\mu\text{F}) = 1 \text{ ms}$$

$$v_C = E(1 - e^{-t/\tau}) = \mathbf{12 \text{ V}(1 - e^{-t/1ms})}$$

$$i_C = \frac{E}{R_1}e^{-t/\tau} = \frac{12 \text{ V}}{20 \text{ k}\Omega}e^{-t/1ms} = \mathbf{0.6 \text{ mA}e^{-t/1ms}}$$

b. *Storage phase*: At 10 ms, a period of time equal to 10τ has passed, permitting the assumption that the capacitor is fully charged. Since $R_{leakage} = \infty \text{ }\Omega$, the capacitor will hold its charge indefinitely. The result is that both v_C and i_C will remain at a fixed value of

$$v_C = \mathbf{12 \text{ V}}$$

$$i_C = \mathbf{0 \text{ A}}$$

c. *Discharge phase* (using 20 ms as the new $t = 0$ s for the equations): The new time constant is

$$\tau' = RC = (R_1 + R_2)C = (20 \text{ k}\Omega + 10 \text{ k}\Omega)(0.05 \text{ }\mu\text{F}) = 1.5 \text{ ms}$$

$$v_C = Ee^{-t/\tau'} = \mathbf{12 \text{ V}e^{-t/1.5ms}}$$

$$i_C = -\frac{E}{R}e^{-t/\tau'} = -\frac{E}{R_1 + R_2}e^{-t/\tau'}$$

$$= \frac{12 \text{ V}}{20 \text{ k}\Omega + 10 \text{ k}\Omega}e^{-t/1.5ms} = \mathbf{-0.4 \text{ mA}e^{-t/1.5ms}}$$

d. See Fig. 10.43.

403

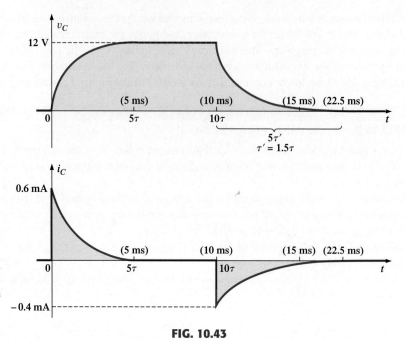

FIG. 10.43

v_C and i_C for the network of Fig. 10.42.

EXAMPLE 10.9 For the network of Fig. 10.44:

a. Find the mathematical expression for the transient behavior of the voltage across the capacitor if the switch is thrown into position 1 at $t = 0$ s.
b. Find the mathematical expression for the transient behavior of the voltage across the capacitor if the switch is moved to position 2 at $t = 1\tau$.
c. Plot the resulting waveform for the voltage v_C as determined by parts (a) and (b).
d. Repeat parts (a)–(c) for the current i_C.

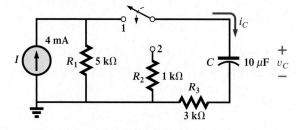

FIG. 10.44

Network to be analyzed in Example 10.9.

Solutions:

a. Converting the current source to a voltage source will result in the configuration of Fig. 10.45 for the charging phase.

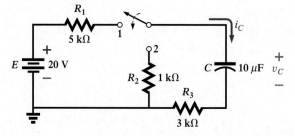

FIG. 10.45

Network of Fig. 10.44 redrawn after making the source conversion.

For the source conversion: $E = IR = (4\text{ mA})(5\text{ k}\Omega) = 20\text{ V}$

and $R_s = R_p = 5\text{ k}\Omega$

$$\tau = RC = (R_1 + R_3)C = (5\text{ k}\Omega + 3\text{ k}\Omega)(10\ \mu\text{F}) = 80\text{ ms}$$

$$v_C = E(1 - e^{-t/\tau}) = \mathbf{20\ V(1 - e^{-t/80ms})}$$

b. With the switch in position 2, the network will appear as shown in Fig. 10.46. The voltage at 1τ can be found by using the fact that the voltage is 63.2% of its final value of 20 V, so that $0.632(20\text{ V}) = 12.64\text{ V}$. Or you can simply substitute into the derived equation as follows:

$$e^{-t/\tau} = e^{-\tau/\tau} = e^{-1} = 0.368$$

and $v_C = 20\text{ V}(1 - e^{-t/80ms}) = 20\text{ V}(1 - 0.368)$

$$= (20\text{ V})(0.632) = 12.64\text{ V}$$

Using this voltage as the starting point and simply substituting into the discharge equation will result in

$$\tau' = RC = (R_2 + R_3)C = (1\text{ k}\Omega + 3\text{ k}\Omega)(10\ \mu\text{F}) = 40\text{ ms}$$

$$v_C = Ee^{-t/\tau'} = \mathbf{12.64\ Ve^{-t/40ms}}$$

c. See Fig. 10.47.

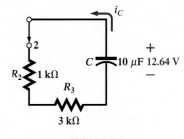

FIG. 10.46
Network of Fig. 10.45 when the switch is thrown into position 2 at t = 1τ.

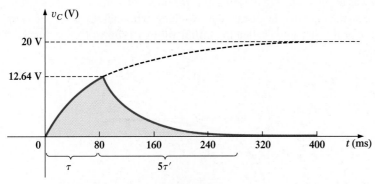

FIG. 10.47
v_C *for the network of Fig. 10.45.*

d. The charging equation for the current is

$$i_C = \frac{E}{R}e^{-t/\tau} = \frac{E}{R_1 + R_3}e^{-t/\tau} = \frac{20\text{ V}}{8\text{ k}\Omega}e^{-t/80ms} = \mathbf{2.5\ mAe^{-t/80ms}}$$

which, at $t = 80\text{ ms}$, will result in

$$i_C = 2.5\text{ mA}e^{-80ms/80ms} = 2.5\text{ mA}e^{-1} = (2.5\text{ mA})(0.368) = 0.92\text{ mA}$$

When the switch is moved to position 2, the 12.64 V across the capacitor will appear across the resistor to establish a current of $12.64\text{ V}/4\text{ k}\Omega = 3.16\text{ mA}$. Substituting into the discharge equation with $V_i = 12.64\text{ V}$ and $\tau' = 40\text{ ms}$ yields

$$i_C = -\frac{V_i}{R_2 + R_3}e^{-t/\tau'} = -\frac{12.64\text{ V}}{1\text{ k}\Omega + 3\text{ k}\Omega}e^{-t/40ms}$$

$$= -\frac{12.64\text{ V}}{4\text{ k}\Omega}e^{-t/40ms} = \mathbf{-3.16\ mAe^{-t/40ms}}$$

The equation has a minus sign because the direction of the discharge current is opposite to that defined for the current in Fig. 10.45. The resulting plot appears in Fig. 10.48.

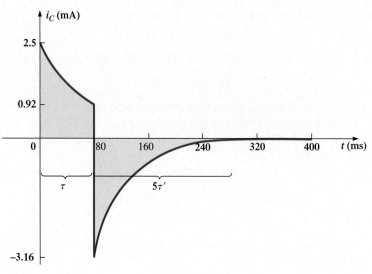

FIG. 10.48

i_C for the network of Fig. 10.45.

10.7 INITIAL CONDITIONS

In all the examples examined in the previous sections, the capacitor was uncharged before the switch was thrown. We will now examine the effect of a charge, and therefore a voltage ($V = Q/C$), on the plates at the instant the switching action takes place. The voltage across the capacitor at this instant is called the **initial value,** as shown for the general waveform of Fig. 10.49.

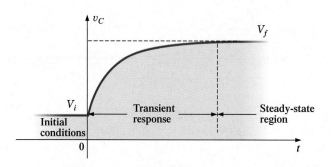

FIG. 10.49

Defining the regions associated with a transient response.

Once the switch is thrown, the transient phase will commence until a leveling off occurs after five time constants. This region of relatively fixed value that follows the transient response is called the **steady-state region,** and the resulting value is called the **steady-state** or **final value.** The steady-state value is found by simply substituting the open-circuit equivalent for the capacitor and finding the voltage across the plates. Using the transient equation developed in the previous section, an equation for the voltage v_C can be written for the entire time interval of Fig. 10.49. That is, for the transient period, the voltage will rise from V_i (previously 0 V) to a final value of V_f. Therefore,

$$v_C = E(1 - e^{-t/\tau}) = (V_f - V_i)(1 - e^{-t/\tau})$$

Adding the starting value of V_i to the equation will result in

$$v_C = V_i + (V_f - V_i)(1 - e^{-t/\tau})$$

However, by multiplying through and rearranging terms:

$$v_C = V_i + V_f - V_f e^{-t/\tau} - V_i + V_i e^{-t/\tau}$$
$$= V_f - V_f e^{-t/\tau} + V_i e^{-t/\tau}$$

we find

$$v_C = V_f + (V_i - V_f)e^{-t/\tau}$$ **(10.17)**

Now that the equation has been developed, it is important to recognize that

Equation (10.17) is a universal equation for the transient response of a capacitor.

That is, it can be used whether or not the capacitor has an initial value. If the initial value is 0 V as it was in all the previous examples, simply set V_i equal to zero in the equation, and the desired equation will result. The final value is simply the voltage across the capacitor when the open-circuit equivalent is substituted.

EXAMPLE 10.10   The capacitor of Fig. 10.50 has an initial voltage of 4 V.

a. Find the mathematical expression for the voltage across the capacitor once the switch is closed.
b. Find the mathematical expression for the current during the transient period.
c. Sketch the waveform for each from initial value to final value.

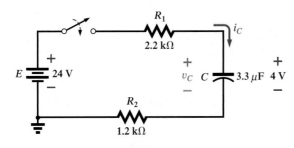

FIG. 10.50
Charging network for Example 10.10.

Solutions:

a. Substituting the open-circuit equivalent for the capacitor will result in a final or steady-state voltage v_C of 24 V.
   The time constant is determined by

$$\tau = (R_1 + R_2)C$$
$$= (2.2\ k\Omega + 1.2\ k\Omega)(3.3\ \mu F) = 11.22\ ms$$

with $5\tau = 56.1\ ms$

   Applying Eq. (10.17):

$$v_C = V_f + (V_i - V_f)e^{-t/\tau} = 24\ V + (4\ V - 24\ V)e^{-t/11.22ms}$$

and $v_C = \textbf{24 V} - \textbf{20 V}e^{-t/\textbf{11.22ms}}$

b. Since the voltage across the capacitor is constant at 4 V prior to the closing of the switch, the current (whose level is sensitive only to changes in voltage across the capacitor) must have an initial value of 0 mA. At the instant the switch is closed, the voltage across the capacitor cannot change instantaneously, so the voltage across the resistive elements at this instant is the applied voltage less the initial voltage across the capacitor. The resulting peak current is

$$I_m = \frac{E - V_C}{R_1 + R_2} = \frac{24\ V - 4\ V}{2.2\ k\Omega + 1.2\ k\Omega} = \frac{20\ V}{3.4\ k\Omega} = 5.88\ mA$$

   The current will then decay (with the same time constant as the voltage v_C) to zero because the capacitor is approaching its open-circuit equivalence.
   The equation for i_C is therefore:

$$i_C = \textbf{5.88 mA}e^{-t/\textbf{11.22 ms}}$$

c. See Fig. 10.51. The initial and final values of the voltage were drawn first, and then the transient response was included between these levels. For the current, the waveform begins and ends at zero, with the peak value having a sign sensitive to the defined direction of i_C in Fig. 10.50.

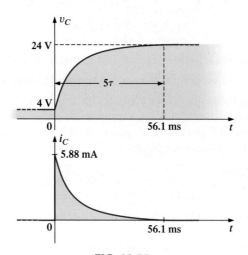

FIG. 10.51

v_C and i_C for the network of Fig. 10.50.

Let us now test the validity of the equation for v_C by substituting $t = 0$ s to reflect the instant the switch is closed.

$$e^{-t/\tau} = e^{-0} = 1$$

and $\qquad v_C = 24\,\text{V} - 20\,\text{V}e^{-t/\tau} = 24\,\text{V} - 20\,\text{V} = 4\,\text{V}$

When $t > 5\tau$,

$$e^{-t/\tau} \cong 0$$

and $\qquad v_C = 24\,\text{V} - 20\,\text{V}e^{-t/\tau} = 24\,\text{V} - 0\,\text{V} = 24\,\text{V}$

Equation (10.17) can also be applied to the discharge phase by simply applying the correct levels of V_i and V_f.

For the discharge pattern of Fig. 10.52, $V_f = 0$ V, and Equation (10.17) becomes

$$v_C = V_f + (V_i - V_f)e^{-t/\tau} = 0\,\text{V} + (V_i - 0\,\text{V})e^{-t/\tau}$$

and $\qquad \boxed{v_C = V_i e^{-t/\tau}}_{\text{discharging}}$ **(10.18)**

Substituting $V_i = E$ volts will result in Eq. (10.13).

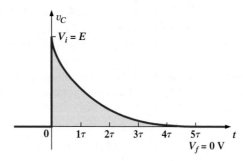

FIG. 10.52

Defining the parameters of Eq. (10.17) for the discharge phase.

10.8 INSTANTANEOUS VALUES

On occasion it will be necessary to determine the voltage or current at a particular instant of time that is not an integral multiple of τ, as in the previous sections. For example, if

$$v_C = 20\,\text{V}(1 - e^{-t/(2\times10^{-3})})$$

the voltage v_C may be required at $t = 5$ ms, which does not correspond to a particular value of τ. Figure 10.27 reveals that $(1 - e^{-t/\tau})$ is approximately 0.93 at $t = 5$ ms $= 2.5\tau$, resulting in $v_C = 20(0.93) = 18.6$ V. Additional accuracy can be obtained simply by substituting $v = 5$ ms into the equation and solving for v_C using a calculator or table to determine $e^{-2.5}$. Thus,

$$v_C = 20 \text{ V}(1 - e^{-5ms/2ms}) = (20 \text{ V})(1 - e^{-2.5}) = (20 \text{ V})(1 - 0.082)$$
$$= (20 \text{ V})(0.918) = \textbf{18.36 V}$$

The results are close, but accuracy beyond the tenths place is suspect using Fig. 10.27. The above procedure can also be applied to any other equation introduced in this chapter for currents or other voltages.

There are also occasions when the time to reach a particular voltage or current is required. The procedure is complicated somewhat by the use of natural logs ($\log_e$, or ln), but today's calculators are equipped to handle the operation with ease.

For example, solving for t in the equation

$$v_C = V_f + (V_i - V_f)e^{-t/\tau}$$

will result in

$$t = \tau(\log_e)\frac{(V_i - V_f)}{(v_C - V_f)} \tag{10.19}$$

For example, suppose that

$$v_C = 20 \text{ V}(1 - e^{-t/2ms})$$

and the time t to reach 10 V is desired. Since $V_i = 0$ V, and $V_f = 20$ V, we have

$$t = \tau(\log_e)\frac{(V_i - V_f)}{(v_C - V_f)} = (2 \text{ ms})(\log_e)\frac{(0 \text{ V} - 20 \text{ V})}{(10 \text{ V} - 20 \text{ V})}$$

$$= (2 \text{ ms})\left[\log_e\left(\frac{-20 \text{ V}}{-10 \text{ V}}\right)\right] = (2 \text{ ms})(\log_e 2) = (2 \text{ ms})(0.693)$$

$$= \textbf{1.386 ms}$$

The calculator key strokes appear in Fig. 10.53.

FIG. 10.53
Key strokes in a TI-86 calculator to determine $(2 \text{ ms})(\log_e 2)$.

For the discharge equation,

$$v_C = Ee^{-t/\tau} = V_i(e^{-t/\tau}) \qquad \text{with } V_f = 0 \text{ V}$$

Using Eq. (10.19):

$$t = \tau(\log_e)\frac{(V_i - V_f)}{(v_C - V_f)} = \tau(\log_e)\frac{(V_i - 0 \text{ V})}{(v_C - 0 \text{ V})}$$

and

$$t = \tau \log_e \frac{V_i}{v_C} \tag{10.20}$$

For the current equation,

$$i_C = \frac{E}{R}e^{-t/\tau} \qquad I_i = \frac{E}{R} \qquad I_f = 0 \text{ A}$$

and

$$t = \log_e \frac{I_i}{i_C} \tag{10.21}$$

Using Mathcad to Perform Transient Analysis

It is time to see how Mathcad can be applied to the transient analysis described in this chapter. For the first equation described in Section 10.5, the value of t must be defined before the expression is written, or the value can simply be inserted in the equation. The former approach is often better, because changing the defined value of t will result in an immediate change in the result. In other words, the value can be used for further calculations. In Fig. 10.54, the value of t was defined as 5 ms. The equation was then entered using the e function from the **Calculator** palette obtained from **View-Toolbars-Calculator.** Be sure to insert a multiplication operator between the initial 20 and the main left bracket. Also, be careful that the control bracket is in the correct place before placing the right bracket to enclose the equation. It will take some practice to ensure that the insertion bracket is in the correct place before entering a parameter, but in time you will find that it is a fairly direct procedure. The -3 is placed using the shift operator over the number **6** on the standard keyboard. The result is displayed by simply entering v again, followed by an equals sign. The result for $t = 1$ ms can now be obtained by simply changing the defined value for t. The result of 7.869 V will appear immediately.

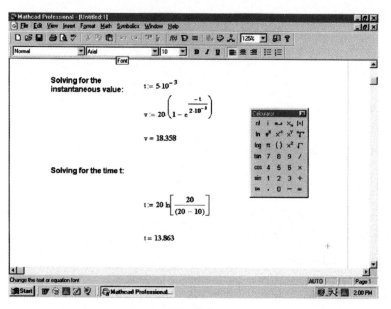

FIG. 10.54

Applying Mathcad to the transient R-C equations.

For the example above, the equation for t can be entered directly as shown in the bottom of Fig. 10.54. The **ln** from the **Calculator** is for a base e calculation, whereas **log** is for a base 10 calculation. The result will appear the instant the equals sign is placed after the t on the bottom line.

The text you see on the screen to define each operation is obtained from **Insert-Text Region**. Then simply type in the text material. The boldface was obtained by clicking on the text material and swiping the text to establish a black background. Then select **B** from the toolbar, and the boldface will appear.

10.9 THÉVENIN EQUIVALENT: $\tau = R_{Th}C$

Occasions will arise in which the network does not have the simple series form of Fig. 10.24. It will then be necessary first to find the Thévenin equivalent circuit for the network external to the capacitive element. E_{Th} will then be the source voltage E of Eqs. (10.9) through (10.21), and R_{Th} will be the resistance R. The time constant is then $\tau = R_{Th}C$.

EXAMPLE 10.11   For the network of Fig. 10.55:

a. Find the mathematical expression for the transient behavior of the voltage v_C and the current i_C following the closing of the switch (position 1 at $t = 0$ s).

b. Find the mathematical expression for the voltage v_C and the current i_C as a function of time if the switch is thrown into position 2 at $t = 9$ ms.

c. Draw the resultant waveforms of parts (a) and (b) on the same time axis.

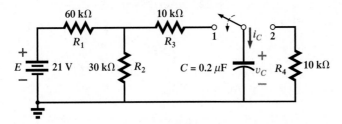

FIG. 10.55

Network for Example 10.11.

Solutions:

a. Applying Thévenin's theorem to the 0.2 μF capacitor, we obtain Fig. 10.56.

$$R_{Th} = R_1\|R_2 + R_3 = \frac{(60\text{ k}\Omega)(30\text{ k}\Omega)}{90\text{ k}\Omega} + 10\text{ k}\Omega$$

$$= 20\text{ k}\Omega + 10\text{ k}\Omega = 30\text{ k}\Omega$$

$$E_{Th} = \frac{R_2 E}{R_2 + R_1} = \frac{(30\text{ k}\Omega)(21\text{ V})}{30\text{ k}\Omega + 60\text{ k}\Omega} = \frac{1}{3}(21\text{ V}) = 7\text{ V}$$

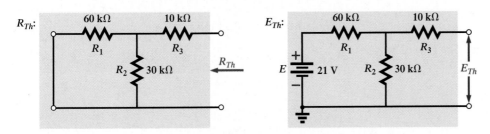

FIG. 10.56

Applying Thévenin's theorem to the network of Fig. 10.55.

The resultant Thévenin equivalent circuit with the capacitor replaced is shown in Fig. 10.57.

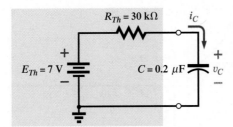

FIG. 10.57

Substituting the Thévenin equivalent for the network of Fig. 10.55.

Using Eq. (10.17) with $V_f = E_{Th}$ and $V_i = 0$ V, we find that

$$v_C = V_f + (V_i - V_f)e^{-t/\tau}$$

becomes $\qquad v_C = E_{Th} + (0\text{ V} - E_{Th})e^{-t/\tau}$

or $\qquad v_C = E_{Th}(1 - e^{-t/\tau})$

with $\qquad \tau = RC = (30\text{ k}\Omega)(0.2\ \mu\text{F}) = 6$ ms

Therefore, $\qquad v_C = \mathbf{7\ V(1 - e^{-t/6\ ms})}$

For the current i_C:

$$i_C = \frac{E_{Th}}{R}e^{-t/RC} = \frac{7 \text{ V}}{30 \text{ k}\Omega}e^{-t/6\text{ms}}$$

$$= \mathbf{0.223 \text{ mA}e^{-t/6\text{ms}}}$$

b. At $t = 9$ ms,

$$v_C = E_{Th}(1 - e^{-t/\tau}) = 7 \text{ V}(1 - e^{-(9 \times 10^{-3})/(6 \times 10^{-3})})$$
$$= (7 \text{ V})(1 - e^{-1.5}) = (7 \text{ V})(1 - 0.223)$$
$$= (7 \text{ V})(0.777) = 5.44 \text{ V}$$

and

$$i_C = \frac{E_{Th}}{R}e^{-t/\tau} = 0.233 \text{ mA}e^{-1.5}$$

$$= (0.233 \times 10^{-3})(0.233) = 0.052 \times 10^{-3} = 0.052 \text{ mA}$$

Using Eq. (10.17) with $V_f = 0$ V and $V_i = 5.44$ V, we find that

$$v_c = V_f + (V_i - V_f)e^{-t/\tau'}$$

becomes

$$v_C = 0 \text{ V} + (5.44 \text{ V} - 0 \text{ V})e^{-t/\tau'}$$
$$= (5.44 \text{ V})e^{-t/\tau'}$$

with

$$\tau' = R_4C = (10 \text{ k}\Omega)(0.2 \text{ }\mu\text{F}) = 2 \text{ ms}$$

and

$$v_C = \mathbf{5.4 \text{ V}e^{-t/2\text{ms}}}$$

By Eq. (10.15):

$$I_i = \frac{5.44 \text{ V}}{10 \text{ k}\Omega} = 0.054 \text{ mA}$$

and

$$i_C = I_i e^{-t/\tau} = \mathbf{-0.54 \text{ mA}e^{-t/2\text{ms}}}$$

c. See Fig. 10.58.

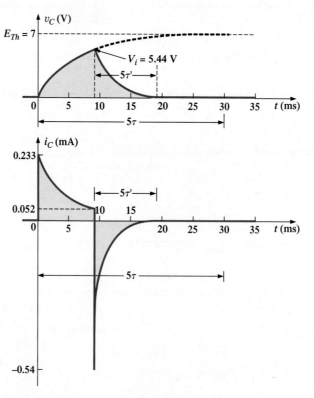

FIG. 10.58

The resulting waveforms for the network of Fig. 10.55.

EXAMPLE 10.12 The capacitor of Fig. 10.59 is initially charged to 40 V. Find the mathematical expression for v_C after the closing of the switch. Plot the waveform for v_C.

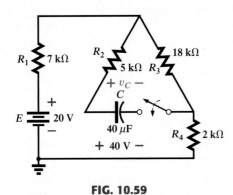

FIG. 10.59

Network to be analyzed in Example 10.12.

Solution: The network is redrawn in Fig. 10.60.

E_{Th}:

$$E_{Th} = \frac{R_3 E}{R_3 + R_1 + R_4} = \frac{(18 \text{ k}\Omega)(120 \text{ V})}{18 \text{ k}\Omega + 7 \text{ k}\Omega + 2 \text{ k}\Omega} = 80 \text{ V}$$

R_{Th}:

$$R_{Th} = (5 \text{ k}\Omega + 18 \text{ k}\Omega) \,\|\, (7 \text{ k}\Omega + 2 \text{ k}\Omega)$$
$$= 5 \text{ k}\Omega + 6 \text{ k}\Omega = 11 \text{ k}\Omega$$

Therefore, $\qquad\qquad V_i = 40 \text{ V} \quad \text{and} \quad V_f = 80 \text{ V}$

and $\qquad\qquad \tau = R_{Th}C = (11 \text{ k}\Omega)(40 \text{ }\mu\text{F}) = 0.44 \text{ s}$

Eq. (10.17): $\qquad v_C = V_f + (V_i - V_f)e^{-t/\tau}$
$$= 80 \text{ V} + (40 \text{ V} - 80 \text{ V})e^{-t/0.44s}$$

and $\qquad\qquad \boldsymbol{v_C = 80 \text{ V} - 40 \text{ V}e^{-t/0.44s}}$

The waveform appears as Fig. 10.61.

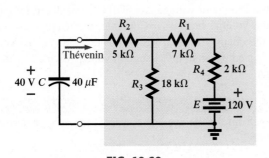

FIG. 10.60

Network of Fig. 10.59 redrawn.

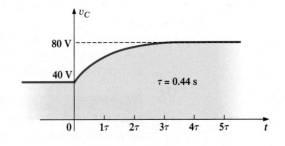

FIG. 10.61

v_C *for the network of Fig. 10.59.*

10.10 THE CURRENT i_C

There is a very special relationship between the current of a capacitor and the voltage across it. For the resistor, it is defined by Ohm's law: $i_R = v_R/R$. The current through and the voltage across the resistor are related by a constant R—a very simple direct linear relationship. For the capacitor, it is the more complex relationship defined by

$$i_C = C\frac{dv_C}{dt} \qquad\qquad \textbf{(10.22)}$$

The factor C reveals that the higher the capacitance, the greater the resulting current. Intuitively, this relationship makes sense, because higher capacitance levels result in increased levels of stored charge, providing a source for increased current levels. The second term, dv_C/dt, is sensitive to the **rate of change** of v_C with time. The function dv_C/dt is called the **derivative** (calculus) of the voltage v_C with respect to time t. The faster the voltage v_C changes with time, the larger will be the factor dv_C/dt, and the larger will be the resulting current i_C. That is why the current jumps to its maximum of E/R in a charging circuit the instant the switch is closed. At that instant, if you look at the charging curve for v_C, the voltage is **changing** at its greatest rate. As it approaches its final value, the rate of change decreases, and, as confirmed by Eq. (10.22), the level of current decreases.

Take special note of the following:

The capacitive current is directly related to the rate of change of the voltage across the capacitor, not the levels of voltage involved.

For example, the current of a capacitor **will be greater** when the voltage changes from 1 V to 10 V in 1 ms than when it changes from 10 V to 100 V in 1 s; in fact, it will be 100 times more.

If the voltage fails to change over time, then

$$\frac{dv_C}{dt} = 0$$

and

$$i_C = C\frac{dv_C}{dt} = C(0) = 0 \text{ A}$$

In an effort to develop a clearer understanding of Eq. (10.22), let us calculate the **average current** associated with a capacitor for various voltages impressed across the capacitor. The average current is defined by the equation

$$\boxed{i_{C_{\text{av}}} = C\frac{\Delta v_C}{\Delta t}} \tag{10.23}$$

where Δ indicates a finite (measurable) change in voltage or time. The instantaneous current can be derived from Eq. (10.23) by letting Δt become vanishingly small; that is,

$$i_{C_{\text{inst}}} = \lim_{\Delta t \to 0} C\frac{\Delta v_C}{\Delta t} = C\frac{dv_C}{dt}$$

In the following example, the change in voltage Δv_C will be considered for each slope of the voltage waveform. If the voltage increases with time, the average current is the change in voltage divided by the change in time, with a positive sign. If the voltage decreases with time, the average current is again the change in voltage divided by the change in time, but with a negative sign.

EXAMPLE 10.13 Find the waveform for the average current if the voltage across a 2 μF capacitor is as shown in Fig. 10.62.

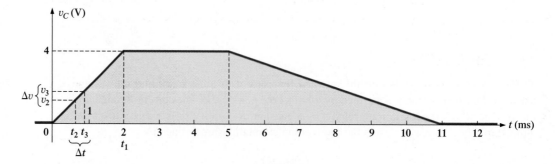

FIG. 10.62
v_C for Example 10.13.

a. From 0 ms to 2 ms, the voltage increases linearly from 0 V to 4 V; the change in voltage $\Delta v = 4\,V - 0 = 4\,V$ (with a positive sign since the voltage increases with time). The change in time $\Delta t = 2\,ms - 0 = 2\,ms$, and

$$i_{C_{av}} = C\frac{\Delta v_C}{\Delta t} = (2 \times 10^{-6}\,F)\left(\frac{4\,V}{2 \times 10^{-3}\,s}\right)$$

$$= 4 \times 10^{-3}\,A = \textbf{4 mA}$$

b. From 2 ms to 5 ms, the voltage remains constant at 4 V; the change in voltage $\Delta v = 0$. The change in time $\Delta t = 3\,ms$, and

$$i_{C_{av}} = C\frac{\Delta v_C}{\Delta t} = C\frac{0}{\Delta t} = \textbf{0 mA}$$

c. From 5 ms to 11 ms, the voltage decreases from 4 V to 0 V. The change in voltage Δv is, therefore, $4\,V - 0 = 4\,V$ (with a negative sign since the voltage is decreasing with time). The change in time $\Delta t = 11\,ms - 5\,ms = 6\,ms$, and

$$i_{C_{av}} = C\frac{\Delta v_C}{\Delta t} = -(2 \times 10^{-6}\,F)\left(\frac{4\,V}{6 \times 10^{-3}\,s}\right)$$

$$= -1.33 \times 10^{-3}\,A = \textbf{-1.33 mA}$$

d. From 11 ms on, the voltage remains constant at 0 and $\Delta v = 0$, so $i_{C_{av}} = 0\,mA$. The waveform for the average current for the impressed voltage is as shown in Fig. 10.63.

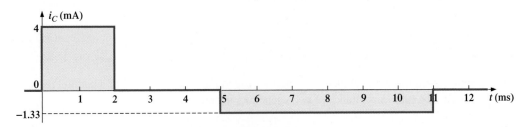

FIG. 10.63
The resulting current i_C for the applied voltage of Fig. 10.62.

Note in Example 10.13 that, in general, the steeper the slope, the greater the current, and when the voltage fails to change, the current is zero. In addition, the average value is the same as the instantaneous value at any point along the slope over which the average value was found. For example, if the interval Δt is reduced from $0 \rightarrow t_1$ to $t_2 - t_3$, as noted in Fig. 10.62, $\Delta v / \Delta t$ is still the same. In fact, no matter how small the interval Δt, the slope will be the same, and therefore the current $i_{C_{av}}$ will be the same. If we consider the limit as $\Delta t \rightarrow 0$, the slope will still remain the same, and therefore $i_{C_{av}} = i_{C_{inst}}$ at any instant of time between 0 and t_1. The same can be said about any portion of the voltage waveform that has a constant slope.

An important point to be gained from this discussion is that it is not the magnitude of the voltage across a capacitor that determines the current but rather how quickly the voltage *changes* across the capacitor. An applied steady dc voltage of 10,000 V would (ideally) not create any flow of charge (current), but a change in voltage of 1 V in a very brief period of time could create a significant current.

The method described above is only for waveforms with straight-line (linear) segments. For nonlinear (curved) waveforms, a method of calculus (differentiation) must be employed.

10.11 CAPACITORS IN SERIES AND IN PARALLEL

Capacitors, like resistors, can be placed in series and in parallel. Increasing levels of capacitance can be obtained by placing capacitors in parallel, while decreasing levels can be obtained by placing capacitors in series.

For capacitors in series, the charge is the same on each capacitor (Fig. 10.64):

$$Q_T = Q_1 = Q_2 = Q_3 \qquad \textbf{(10.24)}$$

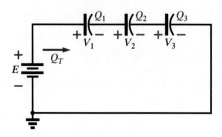

FIG. 10.64
Series capacitors.

Applying Kirchhoff's voltage law around the closed loop gives

$$E = V_1 + V_2 + V_3$$

However,

$$V = \frac{Q}{C}$$

so that

$$\frac{Q_T}{C_T} = \frac{Q_1}{C_1} + \frac{Q_2}{C_2} + \frac{Q_3}{C_3}$$

Using Eq. (10.24) and dividing both sides by Q yields

$$\frac{1}{C_T} = \frac{1}{C_1} + \frac{1}{C_2} + \frac{1}{C_3} \qquad \textbf{(10.25)}$$

which is similar to the manner in which we found the total resistance of a parallel resistive circuit. The total capacitance of two capacitors in series is

$$C_T = \frac{C_1 C_2}{C_1 + C_2} \qquad \textbf{(10.26)}$$

The voltage across each capacitor of Fig. 10.63 can be found by first recognizing that

$$Q_T = Q_1$$

or

$$C_T E = C_1 V_1$$

Solving for V_1:

$$V_1 = \frac{C_T E}{C_1}$$

and substituting for C_T:

$$V_1 = \left(\frac{1/C_1}{1/C_1 + 1/C_2 + 1/C_3} \right) E \qquad \textbf{(10.27)}$$

A similar equation will result for each capacitor of the network.

For capacitors in parallel, as shown in Fig. 10.65, the voltage is the same across each capacitor, and the total charge is the sum of that on each capacitor:

$$Q_T = Q_1 + Q_2 + Q_3 \qquad \textbf{(10.28)}$$

However,

$$Q = CV$$

Therefore,

$$C_T E = C_1 V_1 = C_2 V_2 = C_3 V_3$$

but

$$E = V_1 = V_2 = V_3$$

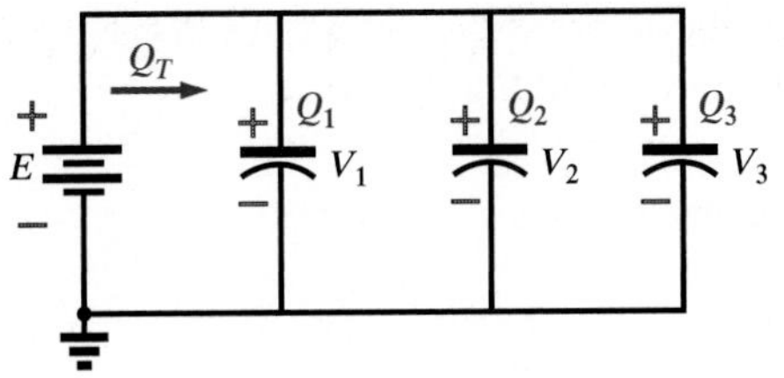

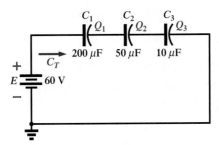

FIG. 10.65
Parallel capacitors.

Thus,

$$\boxed{C_T = C_1 + C_2 + C_3}$$ **(10.29)**

which is similar to the manner in which the total resistance of a series circuit is found.

EXAMPLE 10.14 For the circuit of Fig. 10.66:

a. Find the total capacitance.
b. Determine the charge on each plate.
c. Find the voltage across each capacitor.

FIG. 10.66
Series capacitors for Example 10.14.

Solutions:

a. $\dfrac{1}{C_T} = \dfrac{1}{C_1} + \dfrac{1}{C_2} + \dfrac{1}{C_3}$

$\qquad = \dfrac{1}{200 \times 10^{-6}\,\text{F}} + \dfrac{1}{50 \times 10^{-6}\,\text{F}} + \dfrac{1}{10 \times 10^{-6}\,\text{F}}$

$\qquad = 0.005 \times 10^{6} + 0.02 \times 10^{6} + 0.1 \times 10^{6}$

$\qquad = 0.125 \times 10^{6}$

and $C_T = \dfrac{1}{0.125 \times 10^{6}} = \textbf{8 } \boldsymbol{\mu}\textbf{F}$

b. $Q_T = Q_1 = Q_2 = Q_3$

$\qquad = C_T E = (8 \times 10^{-6}\,\text{F})(60\,\text{V}) = \textbf{480 } \boldsymbol{\mu}\textbf{C}$

c. $V_1 = \dfrac{Q_1}{C_1} = \dfrac{480 \times 10^{-6}\,\text{C}}{200 \times 10^{-6}\,\text{F}} = \textbf{2.4 V}$

$\quad V_2 = \dfrac{Q_2}{C_2} = \dfrac{480 \times 10^{-6}\,\text{C}}{50 \times 10^{-6}\,\text{F}} = \textbf{9.6 V}$

$\quad V_3 = \dfrac{Q_3}{C_3} = \dfrac{480 \times 10^{-6}\,\text{C}}{10 \times 10^{-6}\,\text{F}} = \textbf{48.0 V}$

and $\quad E = V_1 + V_2 + V_3 = 2.4\,\text{V} + 9.6\,\text{V} + 48\,\text{V} = \textbf{60 V}$ (checks)

EXAMPLE 10.15 For the network of Fig. 10.67:

a. Find the total capacitance.
b. Determine the charge on each plate.
c. Find the total charge.

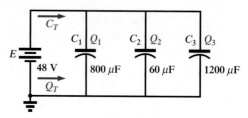

FIG. 10.67
Parallel network for Example 10.15.

Solutions:

a. $C_T = C_1 + C_2 + C_3 = 800 \ \mu\text{F} + 60 \ \mu\text{F} + 1200 \ \mu\text{F} = \textbf{2060} \ \boldsymbol{\mu}\textbf{F}$

b. $Q_1 = C_1 E = (800 \times 10^{-6} \ \text{F})(48 \ \text{V}) = \textbf{38.4 mC}$

$Q_2 = C_2 E = (60 \times 10^{-6} \ \text{F})(48 \ \text{V}) = \textbf{2.88 mC}$

$Q_3 = C_3 E = (1200 \times 10^{-6} \ \text{F})(48 \ \text{V}) = \textbf{57.6 mC}$

c. $Q_T = Q_1 + Q_2 + Q_3 = 38.4 \ \text{mC} + 2.88 \ \text{mC} + 57.6 \ \text{mC} = \textbf{98.88 mC}$

EXAMPLE 10.16 Find the voltage across and the charge on each capacitor for the network of Fig. 10.68.

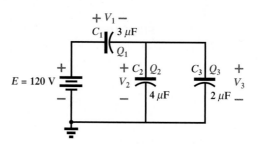

FIG. 10.68
Series-parallel network for Example 10.16.

Solution:

$$C'_T = C_2 + C_3 = 4 \ \mu\text{F} + 2 \ \mu\text{F} = 6 \ \mu\text{F}$$

$$C_T = \frac{C_1 C'_T}{C_1 + C'_T} = \frac{(3 \ \mu\text{F})(6 \ \mu\text{F})}{3 \ \mu\text{F} + 6 \ \mu\text{F}} = 2 \ \mu\text{F}$$

$$Q_T = C_T E = (2 \times 10^{-6} \ \text{F})(120 \ \text{V}) = \textbf{240} \ \boldsymbol{\mu}\textbf{C}$$

An equivalent circuit (Fig. 10.69) has

$$Q_T = Q_1 = Q'_T$$

and, therefore,

$$Q_1 = \textbf{240} \ \boldsymbol{\mu}\textbf{C}$$

and

$$V_1 = \frac{Q_1}{C_1} = \frac{240 \times 10^{-6} \ \text{C}}{3 \times 10^{-6} \ \text{F}} = \textbf{80 V}$$

$$Q'_T = 240 \ \mu\text{C}$$

Therefore,

$$V'_T = \frac{Q'_T}{C'_T} = \frac{240 \times 10^{-6} \ \text{C}}{6 \times 10^{-6} \ \text{F}} = \textbf{40 V}$$

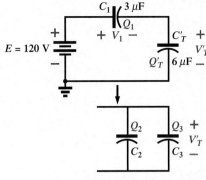

FIG. 10.69
Reduced equivalent for the network of Fig. 10.68.

and
$$Q_2 = C_2 V'_T = (4 \times 10^{-6}\,\text{F})(40\,\text{V}) = \textbf{160}\ \boldsymbol{\mu}\textbf{C}$$
$$Q_3 = C_3 V'_T = (2 \times 10^{-6}\,\text{F})(40\,\text{V}) = \textbf{80}\ \boldsymbol{\mu}\textbf{C}$$

EXAMPLE 10.17 Find the voltage across and the charge on capacitor C_1 of Fig. 10.70 after it has charged up to its final value.

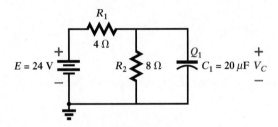

FIG. 10.70
Network for Example 10.17.

Solution: As previously discussed, the capacitor is effectively an open circuit for dc after charging up to its final value (Fig. 10.71).
  Therefore,

$$V_C = \frac{(8\,\Omega)(24\,\text{V})}{4\,\Omega + 8\,\Omega} = \textbf{16 V}$$

$$Q_1 = C_1 V_C = (20 \times 10^{-6}\,\text{F})(16\,\text{V}) = \textbf{320}\ \boldsymbol{\mu}\textbf{C}$$

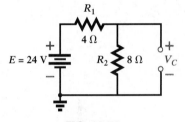

FIG. 10.71
*Determining the final (steady-state)
value for v_C.*

EXAMPLE 10.18 Find the voltage across and the charge on each capacitor of the network of Fig. 10.72(a) after each has charged up to its final value.

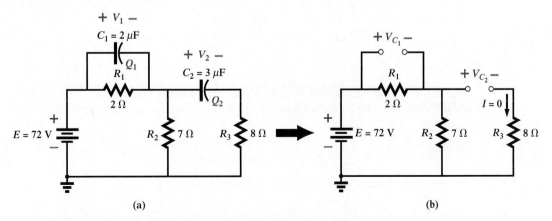

(a)   (b)

FIG. 10.72
Networks for Example 10.18.

Solution: See Fig. 10.72(b):

$$V_{C_2} = \frac{(7\ \Omega)(72\ \text{V})}{7\ \Omega + 2\ \Omega} = \textbf{56 V}$$

$$V_{C_1} = \frac{(2\ \Omega)(72\ \text{V})}{2\ \Omega + 7\ \Omega} = \textbf{16 V}$$

$$Q_1 = C_1 V_{C_1} = (2 \times 10^{-6}\ \text{F})(16\ \text{V}) = \textbf{32}\ \boldsymbol{\mu}\textbf{C}$$

$$Q_2 = C_2 V_{C_2} = (3 \times 10^{-6}\ \text{F})(56\ \text{V}) = \textbf{168}\ \boldsymbol{\mu}\textbf{C}$$

10.12 ENERGY STORED BY A CAPACITOR

An ideal capacitor does not dissipate any of the energy supplied to it. It stores the energy in the form of an electric field between the conducting surfaces. A plot of the voltage, current, and power to a capacitor during the charging phase is shown in Fig. 10.73. The power curve can be obtained by finding the product of the voltage and current at selected instants of time and connecting the points obtained. **The energy stored is represented by the shaded area under the power curve.** Using calculus, we can determine the area under the curve:

$$W_C = \frac{1}{2} CE^2$$

In general,

$$\boxed{W_C = \frac{1}{2} CV^2} \quad \text{(J)} \tag{10.30}$$

where V is the steady-state voltage across the capacitor. In terms of Q and C,

$$W_C = \frac{1}{2} C \left(\frac{Q}{C} \right)^2$$

or

$$\boxed{W_C = \frac{Q^2}{2C}} \quad \text{(J)} \tag{10.31}$$

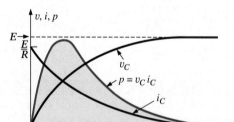

FIG. 10.73
Plotting the power to a capacitive element during the transient phase.

EXAMPLE 10.19 For the network of Fig. 10.72(a), determine the energy stored by each capacitor.

Solution: For C_1:

$$W_C = \frac{1}{2} CV^2$$

$$= \frac{1}{2} (2 \times 10^{-6}\ \text{F})(16\ \text{V})^2 = (1 \times 10^{-6})(256) = \textbf{256}\ \boldsymbol{\mu}\textbf{J}$$

For C_2:

$$W_C = \frac{1}{2}CV^2$$

$$= \frac{1}{2}(3 \times 10^{-6}\,\text{F})(56\,\text{V})^2 = (1.5 \times 10^{-6})(3136) = \mathbf{4704\ \mu J}$$

Due to the squared term, the energy stored increases rapidly with increasing voltages.

10.13 STRAY CAPACITANCES

In addition to the capacitors discussed so far in this chapter, there are **stray capacitances** that exist not through design but simply because two conducting surfaces are relatively close to each other. Two conducting wires in the same network will have a capacitive effect between them, as shown in Fig. 10.74(a). In electronic circuits, capacitance levels exist between conducting surfaces of the transistor, as shown in Fig. 10.74(b). As mentioned earlier, in Chapter 11 we will discuss another element called the *inductor,* which has capacitive effects between the windings [Fig. 10.74(c)]. Stray capacitances can often lead to serious errors in system design if they are not considered carefully.

10.14 APPLICATION

This Application section for capacitors includes a description of the operation of one of the less expensive, throwaway cameras that have become so popular today. Additional examples of the use of capacitors will appear in the chapters to follow.

Flash Lamp

The basic circuitry for the flash lamp of the popular, inexpensive, throwaway camera of Fig. 10.75 is provided in Fig. 10.76, with the physical circuitry appearing in Fig. 10.77. The labels added to Fig. 10.77 identify broad areas of the design and some individual components. The major components of the electronic circuitry include a large 160 μF, 330 V, polarized electrolytic capacitor to store the necessary charge for the flash lamp, a flash lamp to generate the required light, a dc battery of 1.5 V, a chopper network to generate a dc voltage in excess of 300 V, and a trigger network to establish a few thousand volts for a very short period of time to fire the flash lamp. As shown in Fig. 10.76, there are both a 22 nF capacitor in the trigger network and a third capacitor of 470 pF in the high-frequency oscillator of the chopper network. In particular, note that the size of each capacitor is directly related to its capacitance level. It should certainly be of some interest that a single source of energy of only 1.5 V dc can be converted to one of a few thousand volts (albeit for a very short period of time) to fire the flash lamp. In fact, that single, small battery has sufficient power for the entire run of film through the camera. Always keep in mind that energy is related to power and time by $W = Pt = (VI)t$. That is, a high level of voltage can be generated for a defined energy level as long as the factors I and t are sufficiently small.

When you first use the camera, you are directed to press the flash button on the face of the camera and wait for the flash-ready light to come on. As soon as the flash button is depressed, the full 1.5 V of the dc battery are applied to an electronic network (a variety of networks can perform the same function) that will generate an oscillating waveform of very high frequency (with a high repetitive rate) as shown in Fig. 10.76. The high-frequency transformer will then significantly increase the magnitude of the generated voltage and will pass it on to a half-wave rectification system (introduced in earlier chapters), resulting in a dc voltage of about 300 V across the 160 μF capacitor to charge the capacitor (as determined by $Q = CV$). Once the 300 V level is reached, the lead marked "sense" in Fig. 10.76 will feed the information back to the oscillator and will turn it off until the output dc voltage drops to a low threshold level. When the capacitor is fully charged, the neon light in

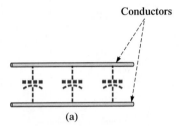

Conductors

(a)

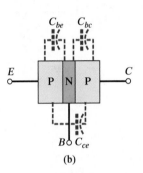

(b)

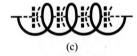

(c)

FIG. 10.74

Examples of stray capacitance.

FIG. 10.75

Flash camera.

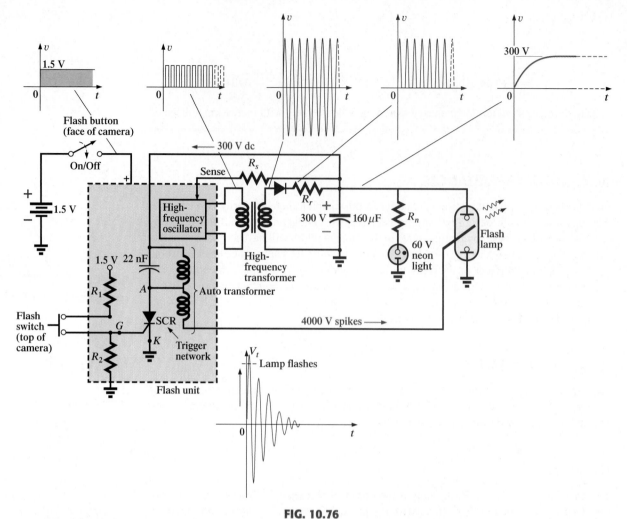

FIG. 10.76

Basic circuitry of flash camera of Fig. 10.75.

parallel with the capacitor will turn on (labeled "flash-ready lamp" on the camera) to let you know that the camera is ready to use. The entire network, from the 1.5 V dc level to the final 300 V level, is called a *dc-dc converter*. The terminology *chopper network* comes from the fact that the applied dc voltage of 1.5 V was chopped up into one that changes level at a very high frequency so that the transformer can perform its function.

Even though the camera may use a 60 V neon light, the neon light and series resistor R_n must have a full 300 V across the branch before the neon light will turn on. Neon lights are simply bulbs with a neon gas that will support conduction when the voltage across the terminals reaches a sufficiently high level. There is no filament, or hot wire as in a light bulb, but simply conduction through the gaseous medium. For new cameras, the first charging sequence may take 12 s to 15 s. Succeeding charging cycles may only take some 7 s or 8 s because the capacitor will still have some residual charge on its plates. If the flash unit is not used, the neon light will begin to drain the 300 V dc supply with a drain current in microamperes. As the terminal voltage drops, there will come a point where the neon light will turn off. For the unit of Fig. 10.75, it takes about 15 min before the light turns off. Once off, the neon light will no longer drain the capacitor, and the terminal voltage of the capacitor will remain fairly constant. Eventually, however, the capacitor will discharge due to its own leakage current, and the terminal voltage will drop to very low levels. The discharge process is very rapid when the flash unit is used, causing the terminal voltage to drop very quickly ($V = Q/C$) and, through the feedback-sense connection signal, causing the oscillator to start up again and recharge the capacitor. You may have noticed when using a camera of this type that once the camera has its initial charge, there is no need to press the charge button between pictures—it is done automatically. However, if the camera sits for a long period of time, the charge button will have to be depressed again; but you will find that the charge time is only 3 s or 4 s due to the residual charge on the plates of the capacitor.

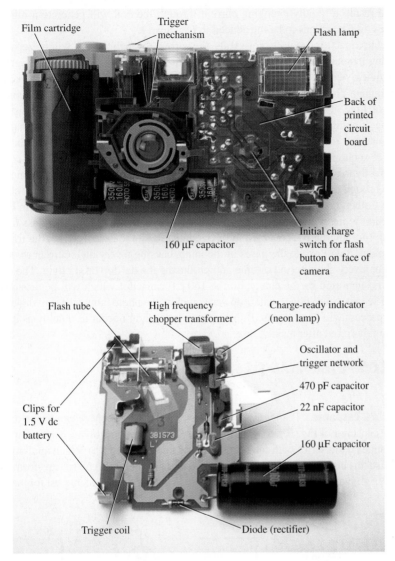

Film cartridge

Trigger
mechanism

Flash lamp

Back of
printed
circuit
board

160 µF capacitor

Initial charge
switch for flash
button on face of
camera

Flash tube

High frequency
chopper transformer

Charge-ready indicator
(neon lamp)

Oscillator and
trigger network

470 pF capacitor

22 nF capacitor

Clips for
1.5 V dc
battery

160 µF capacitor

Trigger coil

Diode (rectifier)

FIG. 10.77
Internal construction of flash camera of Fig. 10.75.

The 300 V across the capacitor are insufficient to fire the flash lamp. Additional circuitry, called the *trigger network,* must be incorporated to generate the few thousand volts necessary to fire the flash lamp. The resulting high voltage is one reason that there is a CAUTION note on each camera regarding the high internal voltages generated and the possibility of electrical shock if the camera is opened.

The thousands of volts required to fire the flash lamp require a discussion that introduces elements and concepts beyond the current level of the text. However, this description is sensitive to this fact and should be looked upon as simply a first exposure to some of the interesting possibilities available from the right mix of elements. When the flash switch at the bottom left of Fig. 10.75 is closed, it will establish a connection between resistors R_1 and R_2. Through a voltage divider action, a dc voltage will appear at the gate (G) terminal of the SCR (silicon-controlled rectifier—a device whose state is controlled by the voltage at the gate terminal). This dc voltage will turn the SCR "on" and will establish a very low resistance path (like a short circuit) between its anode (A) and cathode (K) terminals. At this point the trigger capacitor, which is connected directly to the 300 V sitting across the capacitor, will rapidly charge to 300 V because it now has a direct, low-resistance path to ground through the SCR. Once it reaches 300 V, the charging current in this part of the network will drop to 0 A, and the SCR will open up again since it is a device that needs a steady current in the anode circuit to stay on. The capacitor then sits across the parallel coil (with no connection to ground through the SCR) with its full 300 V and begins to quickly discharge through the coil because the only resistance in the circuit affecting the time constant is the resistance of the parallel coil. As a result, a rapidly changing current through the coil will generate a high voltage across the coil for reasons to be introduced in Chapter 11.

When the capacitor decays to zero volts, the current through the coil will be zero amperes, but a strong magnetic field has been established around the coil. This strong magnetic field will then quickly collapse, establishing a current in the parallel network that will recharge the capacitor again. This continual exchange between the two storage elements will continue for a period of time, depending on the resistance in the circuit. The more the resistance, the shorter the "ringing" of the voltage at the output. This action of the energy "flying back" to the other element is the basis for the "flyback" effect that is frequently used to generate high dc voltages such as needed in TVs. In Fig. 10.76, you will find that the trigger coil is connected directly to a second coil to form an autotransformer (a transformer with one end connected). Through transformer action, the high voltage generated across the trigger coil will be increased further, resulting in the 4000 V necessary to fire the flash lamp. Note in Fig. 10.77 that the 4000 V are applied to a grid that actually lies on the surface of the glass tube of the flash lamp (not internally connected to or in contact with the gases). When the trigger voltage is applied, it will excite the gases in the lamp, causing a very high current to develop in the bulb for a very short period of time and producing the desired bright light. The current in the lamp is supported by the charge on the 160 μF capacitor which will be dissipated very quickly. The capacitor voltage will drop very quickly, the photo lamp will shut down, and the charging process will begin again. If the entire process didn't occur as quickly as it does, the lamp would burn out after a single use.

10.15 COMPUTER ANALYSIS

PSpice

Transient *RC* Response   PSpice will now investigate the transient response for the voltage across the capacitor of Fig. 10.78. In all the examples in the text involving a transient response, a switch appeared in series with the source as shown in Fig. 10.79(a). When applying PSpice, we establish this instantaneous change in voltage level by applying a pulse waveform as shown in Fig. 10.79(b) with a pulse width (*PW*) longer than the period (5τ) of interest for the network.

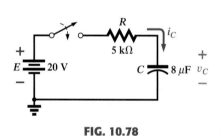

FIG. 10.78

Circuit to be analyzed using PSpice.

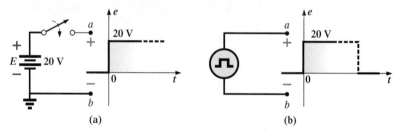

FIG. 10.79

Establishing a switching dc voltage level: (a) series dc voltage-switch combination;
(b) PSpice pulse option.

A pulse source is obtained through the sequence **Place part** key-**Libraries-SOURCE-VPULSE-OK.** Once in place, the label and all the parameters can be set by simply double-clicking on each to obtain the **Display Properties** dialog box. As you scroll down the list of attributes, you will see the following parameters defined by Fig. 10.80:

> **V1** is the initial value.
> **V2** is the pulse level.
> **TD** is the delay time.
> **TR** is the rise time.
> **TF** is the fall time.
> **PW** is the pulse width at the V_2 level.
> **PER** is the period of the waveform.

All the parameters have been set as shown on the schematic of Fig. 10.81 for the network of Fig. 10.78. Since rise and fall times of 0 s are unrealistic from a practical standpoint, 0.1 ms was chosen for each in this example. Further, since $\tau = RC = (5 \text{ k}\Omega)(8 \mu\text{F}) = 20$ ms and $5\tau = 200$ ms, a pulse width of 500 ms was selected. The period was simply chosen as twice the pulse width.

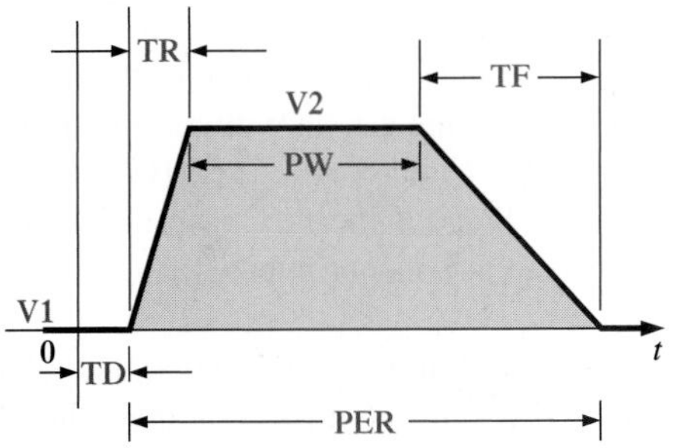

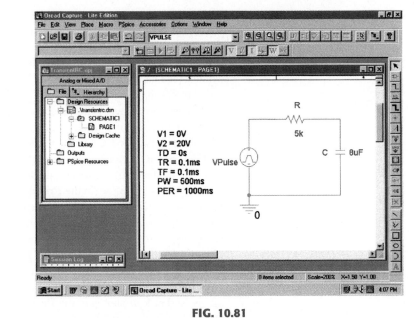

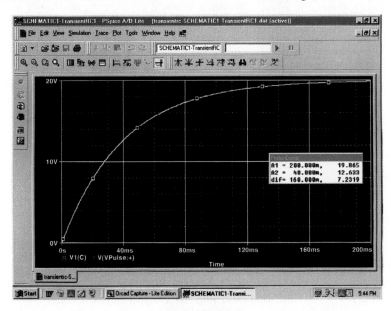

FIG. 10.80

*The defining parameters of PSpice **VPULSE**.*

FIG. 10.81

Using PSpice to investigate the transient response of the series R-C circuit of Fig. 10.78.

Now for the simulation process. First the **New Simulation Profile** key is selected to obtain the **New Simulation** dialog box in which **TransientRC** is inserted for the **Name** and **Create** is chosen to leave the dialog box. The **Simulation Settings-Transient RC** dialog box will result, and under **Analysis,** the **Time Domain (Transient)** option is chosen under **Analysis type.** The **Run to time** is set at 200 ms so that only the first five time constants will be plotted. The **Start saving data after** option will be 0 s to ensure that the data are collected immediately. The **Maximum step size** is 1 ms to provide sufficient data points for a good plot. Click **OK,** and we are ready to select the **Run PSpice** key. The result will be a graph without a plot (since it has not been defined yet) and an *x*-axis that extends from 0 s to 200 ms as defined above.

To obtain a plot of the voltage across the capacitor versus time, the following sequence is applied: **Add Trace** key-**Add Traces** dialog box-**V1(C)-OK**, and the plot of Fig. 10.82 will result. The color and thickness of the plot and the axis can be changed by placing the cursor on the plot line and performing a right click. A list will appear in which **Properties** should be selected; then a **Trace Properties** dialog box will appear in which the color and thickness of the line can be changed. Since the plot is against a black background, a better printout occurred when yellow was selected and the line made thicker as shown in Fig. 10.82.

FIG. 10.82

*Transient response for the voltage across the capacitor of Fig. 10.78 when **VPulse** is applied.*

Next, the cursor can be put on the axis, and another right click will allow you to make the axis yellow and thicker for a better printout. For comparison, it seemed appropriate to plot the applied pulse signal also. This is accomplished by going back to **Trace** and selecting **Add Trace** followed by **V(Vpulse:+)** and **OK.** Now both waveforms appear on the same screen, as shown in Fig. 10.82. In this case, the plot was left a greenish tint so that it could be distinguished from the axis and the other plot. Note that it follows the left axis to the top and travels across the screen at 20 V.

If you want the magnitude of either plot at any instant, simply select the **Toggle cursor** key. Then click on **V1(C)** at the bottom left of the screen. A box will appear around **V1(C)** that will reveal the spacing between the dots of the cursor on the screen. This is important when more than one cursor is used. By moving the cursor to 200 ms, we find that the magnitude **(A1)** is 19.865 V (in the **Probe Cursor** dialog box), clearly showing how close it is to the final value of 20 V. A second cursor can be placed on the screen with a right click; then click on the same **V1(C)** on the bottom of the screen. The box around **V1(C)** cannot show two boxes, but the spacing and the width of the lines of the box have definitely changed. There is no box around the **Pulse** symbol since this was not selected, although it could have been selected by either cursor.

If we now move the second cursor to one time constant of 40 ms, we find that the voltage is 12.633 V as shown in the **Probe Cursor** dialog box. This confirms that the voltage should be 63.2% of its final value of 20 V in one time constant (0.632×20 V = 12.4 V). Two separate plots could have been obtained by going to **Plot-Add Plot to Window** and then using the trace sequence again.

Average Capacitive Current   As an exercise in using the pulse source, and to verify our analysis of the average current for a purely capacitive network, the description to follow will verify the results of Example 10.13. For the pulse waveform of Fig. 10.62, the parameters of the pulse supply appear in Fig. 10.83. Note that the rise time is now 2 ms, starting at 0 s, and the fall time is 6 ms. The period was set at 15 ms to permit monitoring the current after the pulse had passed.

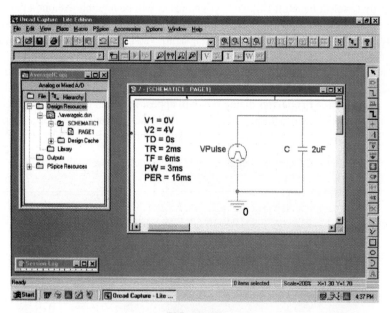

FIG. 10.83
Using PSpice to verify the results of Example 10.13.

Simulation is initiated by first selecting the **New Simulation Profile** key to obtain the **New Simulation** dialog box in which **AverageIC** is entered as the **Name. Create** is then chosen to obtain the **Simulation Settings-AverageIC** dialog box. **Analysis** is selected, and **Time Domain(Transient)** is chosen under the **Analysis type** options. The **Run to time** is set at 15 ms to encompass the period of interest, and the **Start saving data after** is set at 0 s to ensure data points starting at $t = 0$ s. The **Maximum step size** is selected from 15 ms/1000 = 15 μs to ensure 1000 data points for the plot. Click **OK,** and select the **Run PSpice** key. A window will appear with a horizontal scale that extends from 0 to 15 ms as defined above. Then the

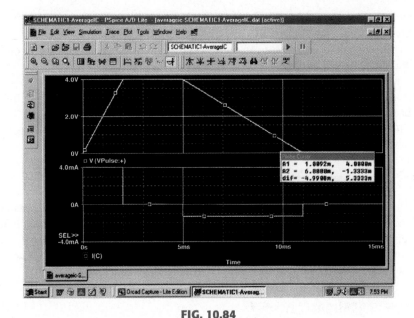

FIG. 10.84

The applied pulse and resulting current for the 2 μF capacitor of Fig. 10.83.

Add Trace key is selected and **I(C)** chosen to appear in the **Trace Expression** below. Click **OK,** and the plot of **I(C)** appears in the bottom of Fig. 10.84.

This time, it would be nice to see the pulse waveform in the same window but as a separate plot. Therefore, continue with **Plot-Add Plot to Window-Trace-Add Trace-V(Vpulse:+)-OK,** and both plots will appear as shown in Fig. 10.84. The cursors can now be used to measure the resulting average current levels. First, select the **I(C)** plot to move the **SEL**≫ notation to the lower plot. The **SEL**≫ defines which plot for multiplot screens is active. Then select the **Toggle cursor** key and left-click on the **I(C)** plot to establish the crosshairs of the cursor. Set the interval at 1 ms, and the magnitude **A1** is displayed as 4 mA. Right-click on the same plot, and a second cursor will result that can be placed at 6 ms to get a response of −1.33 mA **(A2)** as expected from Example 10.13. Both plots were again made yellow with a wider line by right-clicking on the curve and choosing **Properties.**

CHAPTER SUMMARY

ELECTRIC FIELD

- Electric field lines always extend from a positively charged body to a negatively charged body; they always leave or terminate at an angle perpendicular to the charged surface; and they never intersect.
- The denser the flux lines, the stronger the electric field is in that region.
- The energy delivered to a capacitor is stored in the form of an electric field between the plates.

CAPACITANCE

- Capacitance is a measure of a capacitor's ability to store charge on its plates.
- Dielectrics are insulators designed to increase the capacitance of a capacitor by setting up an opposing electric field between the plates.
- The larger the area or permittivity of the dielectric, and the smaller the distance between the plates, the larger the capacitance of the capacitor.
- In general, the larger the capacitor, the larger the capacitance. The smallest capacitors are typically in pF, and the largest in μF.

TRANSIENTS

- In a dc network, the voltage across a capacitor is essentially equal to the applied voltage after five time constants of the charging phase have passed.

- In a dc network, the current of a capacitor is essentially zero amperes after five time constants of the charging phase have passed.
- In a dc network, a capacitor can be replaced by an open-circuit equivalent once the charging phase has passed.
- The voltage across a capacitor cannot change instantaneously.
- Thévenin's theorem can be very helpful in defining the transient response for a capacitor in a series-parallel dc network.
- The current of a capacitor is directly related to the rate of change of voltage across the capacitor.

SERIES AND PARALLEL CAPACITORS

- The total capacitance of capacitors in series is found in the same manner as the resistance of resistors in parallel, whereas the total capacitance of capacitors in parallel is found in the same manner as the resistance of resistors in series.
- For capacitors in series, the charge is the same on each capacitor. For capacitors in parallel, the total charge is the sum of the charges on the capacitors.

Important Equations

$$C = \frac{Q}{V}$$

$$\mathscr{E} = \frac{V}{d}$$

$$C = \epsilon\frac{A}{d} = \epsilon_o\epsilon_r\frac{A}{d} = 8.85 \times 10^{-12}\epsilon_r\frac{A}{d}$$

$$C = \epsilon_r C_o$$

Charging:

$$\tau = RC$$

$$v_C = E(1 - e^{-t/\tau}) \qquad i_C = \frac{E}{R}e^{-t/\tau} \qquad v_R = Ee^{-t/\tau}$$

Discharging:

$$v_C = Ee^{-t/\tau} \qquad i_C = \frac{E}{R}e^{-t/\tau} \qquad v_R = Ee^{-t/\tau}$$

Initial conditions:

$$v_C = V_f + (V_i - V_f)e^{-t/\tau}$$

$$i_C = C\frac{dv_C}{dt} \qquad i_{C_{av}} = C\frac{\Delta v_C}{\Delta t}$$

Series capacitors:

$$Q_T = Q_1 = Q_2 = Q_3$$

$$\frac{1}{C_T} = \frac{1}{C_1} + \frac{1}{C_2} + \frac{1}{C_3}$$

$$C_T = \frac{C_1 C_2}{C_1 + C_2}$$

Parallel capacitors:

$$Q_T = Q_1 + Q_2 + Q_3$$

$$C_T = C_1 + C_2 + C_3$$

Energy:

$$W_C = \frac{1}{2}CV^2$$

PROBLEMS

SECTION 10.3 Capacitance

1. Find the capacitance of a parallel plate capacitor if 1200 μC of charge are deposited on its plates when 10 V are applied across the plates.

2. How much charge is deposited on the plates of a 2 μF capacitor if 60 V are applied across the capacitor?

3. Find the electric field strength between the plates of a parallel plate capacitor if 100 mV are applied across the plates and the plates are 2 mm apart.

*4. Repeat Problem 3 if the plates are separated by 4 mils.

5. A 4 μF parallel plate capacitor has 160 μC of charge on its plates. If the plates are 5 mm apart, find the electric field strength between the plates.

6. A parallel plate air capacitor has a capacitance of 5 μF. Find the new capacitance if:
  a. The distance between the plates is doubled (everything else remains the same).
  b. The area of the plates is doubled (everything else remains the same as for the 5 μF level).
  c. A dielectric with a relative permittivity of 20 is inserted between the plates (everything else remains the same as for the 5 μF level).
  d. A dielectric is inserted with a relative permittivity of 4, and the area is reduced to 1/3 and the distance to 1/4 of their original dimensions.

7. Find the capacitance of a parallel plate capacitor if the area of each plate is 0.075 m^2 and the distance between the plates is 1.77 mm. The dielectric is air.

8. Repeat Problem 7 if the dielectric is paraffin-coated paper.

9. Find the distance in mils between the plates of a 2 μF capacitor if the area of the plates is 0.2 m^2 and the dielectric is transformer oil.

10. The capacitance of a capacitor with a dielectric of air is 1200 pF. When a dielectric is inserted between the plates, the capacitance increases to 0.006 μF. Of what material is the dielectric made?

11. The plates of a parallel plate air capacitor are 0.2 mm apart and have an area of 0.08 m^2, with 200 V applied across the plates.
  a. Determine the capacitance.
  b. Find the electric field intensity between the plates.
  c. Find the charge on each plate if the dielectric is air.

12. A sheet of Bakelite® 0.2 mm thick and having an area of 0.08 m^2 is inserted between the plates of Problem 11.
  a. Find the electric field strength between the plates.
  b. Determine the charge on each plate.
  c. Determine the capacitance.

13. The capacitance of capacitor with a dielectric of air is 20 nF. What is the capacitance if a dielectric of mica is inserted between the plates? All the other dimensions remain the same.

*14. Find the maximum voltage that can be applied across a parallel plate capacitor of 6 nF if the area of one plate is 0.02 m^2 and the dielectric is mica. Assume a linear relationship between the dielectric strength and the thickness of the dielectric.

*15. Find the distance in micrometers between the plates of a parallel plate mica capacitor if the maximum voltage that can be applied across the capacitor is 1200 V. Assume a linear relationship between the breakdown strength and the thickness of the dielectric.

16. A 22 μF capacitor has -200 ppm/°C at room temperature of 20°C. What is the capacitance if the temperature increases to 100°C, the boiling point of water?

17. What is the capacitance of a small teardrop capacitor labeled 40 J? What is the range of expected values as established by the tolerance?

18. A large, flat, mica capacitor is labeled 220M. What are the capacitance and the expected range of values guaranteed by the manufacturer?

19. A small flat disc ceramic capacitor is labeled 333K. What are the capacitance level and the expected range of values?

SECTION 10.5 **Transients in Capacitive Networks: The Charging Phase**

20. For the R-C circuit of Fig. 10.85, composed of standard values:
  a. Determine the time constant of the circuit.
  b. Write the mathematical expression for the voltage v_C following the closing of the switch at $t = 0$ s.
  c. Determine the voltage v_C after one, three, and five time constants.
  d. Write the mathematical equations for the current i_C and the voltage v_R.
  e. Sketch the waveforms for v_C and i_C.

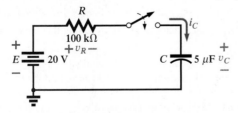

FIG. 10.85
Problems 20 and 21.

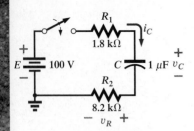

FIG. 10.86
Problem 24.

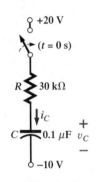

FIG. 10.87
Problem 25.

21. Repeat Problem 20 for $R = 1$ MΩ, and compare the results.

22. Given the voltage $v_C = 60$ mV$(1 - e^{-t/5ms})$:
  a. What is the time constant?
  b. What is the voltage at $t = 2$ ms?
  c. What is the voltage at $t = 100$ ms?

23. The voltage across a 10 μF capacitor in a series R-C circuit is $v_C = 12$ V$(1 - e^{-t/40ms})$.
  a. On a practical basis, how much time must pass before the charging phase has passed?
  b. What is the resistance of the circuit?
  c. What is the voltage at $t = 20$ ms?
  d. What is the voltage at 10 time constants?
  e. Under steady-state conditions, how much charge is on the plates?
  f. If the leakage resistance is 1000 MΩ, how long will it take (in hours) the capacitor to discharge if we assume that the discharge rate is constant throughout the discharge period?

24. For the R-C circuit of Fig. 10.86, composed of standard values:
  a. Determine the time constant of the circuit.
  b. Write the mathematical expressions for the voltage v_C following the closing of the switch.
  c. Determine v_C after one, three, and five time constants.
  d. Write the mathematical expressions for the current i_C and the voltage v_{R_2}.
  e. Sketch the waveforms for v_C and i_C.

*25. For the R-C circuit of Fig. 10.87, composed of standard values:
  a. Determine the time constant of the circuit.
  b. Write the mathematical expressions for the voltages v_C and v_R and the current i_C.
  c. Sketch the waveforms for v_C, v_R, and i_C.

SECTION 10.6 Transients in Capacitive Networks: The Discharging Phase

26. For the R-C circuit of Fig. 10.88, composed of standard values:
  a. Determine the time constant of the circuit when the switch is thrown into position 1.
  b. Find the mathematical expression for the voltage across the capacitor and the current after the switch is thrown into position 1.
  c. Determine the magnitude of the voltage v_C and the current i_C the instant the switch is thrown into position 2 at $t = 1$ s.
  d. Determine the mathematical expression for the voltage v_C and the current i_C for the discharge phase.
  e. Plot the waveforms of v_C and i_C for a period of time extending from 0 to 2 s from when the switch was thrown into position 1.

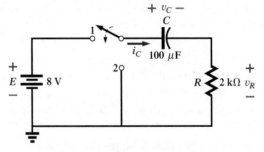

FIG. 10.88
Problem 26.

27. For the network of Fig. 10.89, composed of standard values:
  a. Write the mathematical expressions for the voltages v_C and v_{R_1} and the current i_C after the switch is thrown into position 1.
  b. Find the values of v_C, v_{R_1}, and i_C when the switch is moved to position 2 at $t = 100$ ms.
  c. Write the mathematical expressions for the voltages v_C and v_R and the current i_C if the switch is moved to position 3 at $t = 200$ ms.
  d. Plot the waveforms of v_C, v_{R_2}, and i_C for the time period extending from 0 to 300 ms.

28. Repeat Problem 27 for a capacitance of 20 μF assuming that the leakage resistance of the capacitor is ∞ Ω.

29. For the network of Fig. 10.90, composed of standard values:
  a. Find the mathematical expressions for the voltage v_C and the current i_C when the switch is thrown into position 1.
  b. Find the mathematical expressions for the voltage v_C and the current i_C if the switch is thrown into position 2 at a time equal to five time constants of the charging circuit.
  c. Plot the waveforms of v_C and i_C for a period of time extending from 0 to 30 μs.

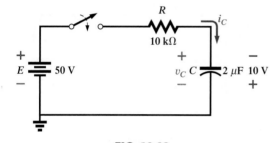

FIG. 10.89
Problems 27 and 28.

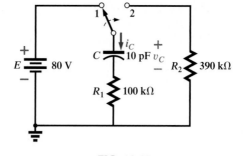

FIG. 10.90
Problem 29.

30. The 1000 μF capacitor of Fig. 10.91 is charged to 6 V. To discharge the capacitor before further use, a wire with a resistance of 2 mΩ is placed across the capacitor.
   a. How long will it take to discharge the capacitor?
   b. What is the peak value of the current?
   c. Based on the answer to part (b), is a spark expected when contact is made with both ends of the capacitor?

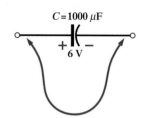

FIG. 10.91
Problem 30.

SECTION 10.7 Initial Conditions

31. The capacitor of Fig. 10.92 is initially charged to 10 V with the polarity shown.
   a. Write the expression for the voltage v_C after the switch is closed.
   b. Write the expression for the current i_C after the switch is closed.
   c. Plot the results of parts (a) and (b).

32. The capacitor of Fig. 10.93 is initially charged to 40 V before the switch is closed. Write the expressions for the voltages v_C and v_R and the current i_C following the closing of the switch. Plot the resulting waveforms.

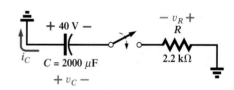

FIG. 10.92
Problem 31.

FIG. 10.93
Problem 32.

*33. The capacitor of Fig. 10.94 is initially charged to 20 V with the polarity shown. Write the expressions for the voltage v_C and the current i_C following the closing of the switch. Plot the resulting waveforms.

*34. The capacitor of Fig. 10.95 is initially charged to 12 V with the polarity shown.
   a. Find the mathematical expressions for the voltage v_C and the current i_C when the switch is closed.
   b. Sketch the waveforms of v_C and i_C.

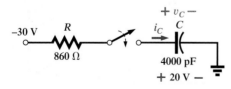

FIG. 10.94
Problem 33.

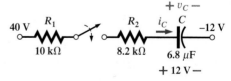

FIG. 10.95
Problem 34.

SECTION 10.8 Instantaneous Values

35. Given the expression $v_C = 12\ \text{V}(1 - e^{-t/20\mu s})$:
   a. Determine v_C at $t = 10\ \mu$s.
   b. Determine v_C at $t = 10\tau$.
   c. Find the time t for v_C to reach 6 V.
   d. Find the time t for v_C to reach 11.98 V.

36. For the network of Fig. 10.96, V_L must be 8 V before the system is activated. If the switch is closed at $t = 0$ s, how long will it take for the system to be activated?

*37. Design the network of Fig. 10.97 such that the system will turn on 10 s after the switch is closed.

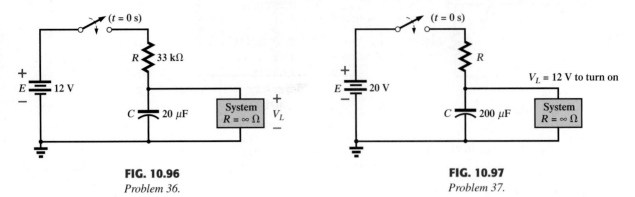

FIG. 10.96

Problem 36.

FIG. 10.97

Problem 37.

38. For the circuit of Fig. 10.98:
  a. Find the time required for v_C to reach 60 V following the closing of the switch.
  b. Calculate the current i_C at the instant $v_C = 60$ V.
  c. Determine the power delivered by the source at the instant $t = 2\tau$.

*39. For the network of Fig. 10.99:
  a. Calculate v_C, i_C, and v_{R_1} at 0.5 s and 1 s after the switch makes contact with position 1.
  b. The network sits in position 1 10 min before the switch is moved to position 2. How long after making contact with position 2 will it take for the current i_C to drop to 8 μA? How much *longer* will it take for v_C to drop to 10 V?

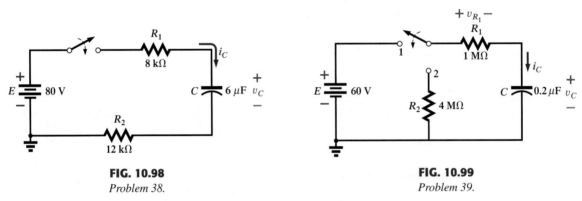

FIG. 10.98

Problem 38.

FIG. 10.99

Problem 39.

40. For the system of Fig. 10.100, using a DMM with a 10 MΩ internal resistance in the voltmeter mode:
  a. Determine the voltmeter reading one time constant after the switch is closed.
  b. Find the current i_C two time constants after the switch is closed.
  c. Calculate the time that must pass after the closing of the switch for the voltage v_C to be 50 V.

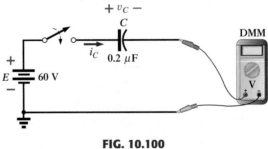

FIG. 10.100

Problem 40.

SECTION 10.9 Thévenin Equivalent: $\tau = R_{Th}C$

41. For the circuit of Fig. 10.101:
  a. Find the mathematical expressions for the transient behavior of the voltage v_C and the current i_C following the closing of the switch.
  b. Sketch the waveforms of v_C and i_C.

42. The capacitor of Fig. 10.102 is initially charged to 2 V with the polarity shown.
   a. Write the mathematical expressions for the voltage v_C and the current i_C when the switch is closed.
   b. Sketch the waveforms of v_C and i_C.

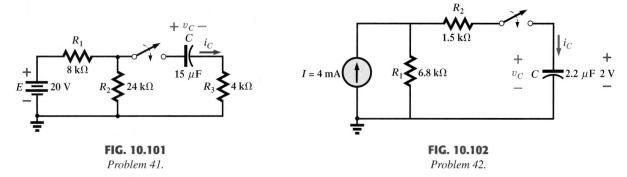

FIG. 10.101
Problem 41.

FIG. 10.102
Problem 42.

43. The capacitor of Fig. 10.103 is initially charged to 4 V with the polarity shown.
   a. Write the mathematical expressions for the voltage v_C and the current i_C when the switch is closed.
   b. Sketch the waveforms of v_C and i_C.

44. For the circuit of Fig. 10.104:
   a. Find the mathematical expressions for the transient behavior of the voltage v_C and the current i_C following the closing of the switch.
   b. Sketch the waveforms of v_C and i_C.

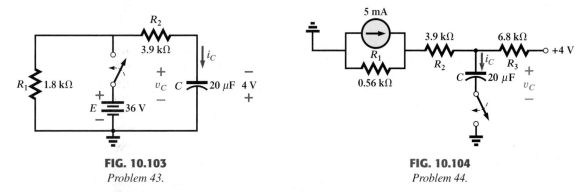

FIG. 10.103
Problem 43.

FIG. 10.104
Problem 44.

*45. The capacitor of Fig. 10.105 is initially charged to 3 V with the polarity shown.
   a. Write the mathematical expressions for the voltage v_C and the current i_C when the switch is closed.
   b. Sketch the waveforms of v_C and i_C.

46. For the system of Fig. 10.106, using a DMM with a 10 MΩ internal resistance in the voltmeter mode:
   a. Determine the voltmeter reading four time constants after the switch is closed.
   b. Find the time that must pass before i_C drops to 3 μA.
   c. Find the time that must pass after the closing of the switch for the voltage across the meter to reach 10 V.

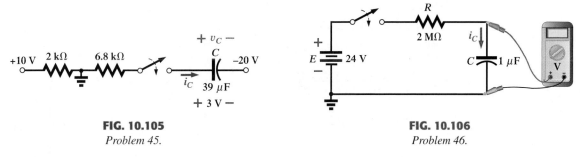

FIG. 10.105
Problem 45.

FIG. 10.106
Problem 46.

SECTION 10.10 The Current i_C

47. Find the waveform for the average current if the voltage across the 2 μF capacitor is as shown in Fig. 10.107.

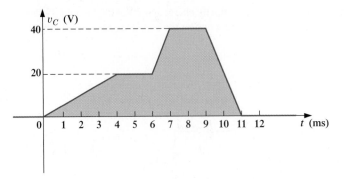

FIG. 10.107
Problem 47.

48. Find the waveform for the average current if the voltage across the 4.7 μF capacitor is as shown in Fig. 10.108.

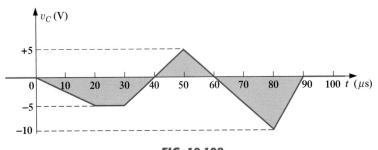

FIG. 10.108
Problem 48.

49. Given the waveform of Fig. 10.109 for the current of a 20 μF capacitor, sketch the waveform of the voltage v_C across the capacitor if $v_C = 0$ V at $t = 0$ s.

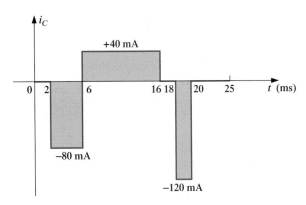

FIG. 10.109
Problem 49.

SECTION 10.11   Capacitors in Series and in Parallel

50. Find the total capacitance C_T for the circuit of Fig. 10.110.
51. Find the total capacitance C_T for the circuit of Fig. 10.111.

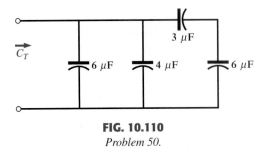

FIG. 10.110
Problem 50.

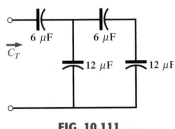

FIG. 10.111
Problem 51.

52. Find the voltage across and the charge on each capacitor for the circuit of Fig. 10.112.

53. Find the voltage across and the charge on each capacitor for the circuit of Fig. 10.113.

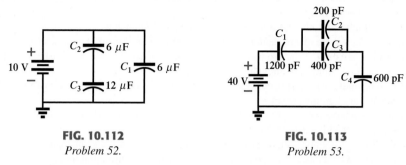

FIG. 10.112
Problem 52.

FIG. 10.113
Problem 53.

54. For the configuration of Fig. 10.114, determine the voltage across each capacitor and the charge on each capacitor.

55. For the configuration of Fig. 10.115, determine the voltage across each capacitor and the charge on each capacitor.

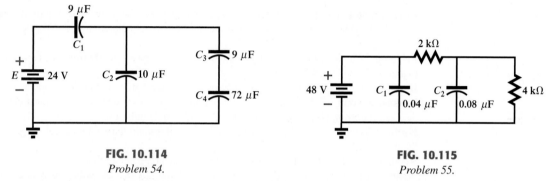

FIG. 10.114
Problem 54.

FIG. 10.115
Problem 55.

SECTION 10.12 Energy Stored by a Capacitor

56. Find the energy stored by a 120 pF capacitor with 12 V across its plates.

57. If the energy stored by a 6 μF capacitor is 1200 J, find the charge Q on each plate of the capacitor.

58. For the network of Fig. 10.116:
  a. Determine the energy stored by each capacitor under steady-state conditions.
  b. Repeat part (a) if the capacitors are in series.

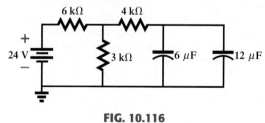

FIG. 10.116
Problem 58.

*59. An electronic flashgun has a 1000 μF capacitor that is charged to 100 V.
  a. How much energy is stored by the capacitor?
  b. What is the charge on the capacitor?
  c. When the photographer takes a picture, the flash fires for 1/2000 s. What is the average current through the flashtube?
  d. Find the power delivered to the flashtube.
  e. After a picture is taken, the capacitor has to be recharged by a power supply that delivers a maximum current of 10 mA. How long will it take to charge the capacitor?

SECTION 10.15 Computer Analysis

60. Using PSpice or EWB, verify the results of Example 10.6.

61. Using the initial condition operator, verify the results of Example 10.8 for the charging phase using PSpice or EWB.

62. Using PSpice or EWB, verify the results for v_C during the charging phase of Example 10.11.

63. Using PSpice or EWB, verify the results of Problem 47.

GLOSSARY

Average current   The current defined by a linear (straight line) change in voltage across a capacitor for a specific period of time.

Breakdown voltage   Another term for *dielectric strength,* listed below.

Capacitance   A measure of a capacitor's ability to store charge; measured in farads (F).

Capacitive time constant   The product of resistance and capacitance that establishes the required time for the charging and discharging phases of a capacitive transient.

Capacitive transient   The waveforms for the voltage and current of a capacitor that result during the charging and discharging phases.

Capacitor   A fundamental electrical element having two conducting surfaces separated by an insulating material and having the capacity to store charge on its plates.

Coulomb's law   An equation relating the force between two like or unlike charges.

Derivative   The instantaneous change in a quantity at a particular instant in time.

Dielectric   The insulating material between the plates of a capacitor that can have a pronounced effect on the charge stored on the plates of a capacitor.

Dielectric constant   Another term for *relative permittivity,* listed below.

Dielectric strength   An indication of the voltage required for unit length to establish conduction in a dielectric.

Electric field strength   The force acting on a unit positive charge in the region of interest.

Electric flux lines   Lines drawn to indicate the strength and direction of an electric field in a particular region.

Fringing   An effect established by flux lines that do not pass directly from one conducting surface to another.

Initial value   The steady-state voltage across a capacitor before a transient period begins.

Leakage current   The current that will result in the total discharge of a capacitor if the capacitor is disconnected from the charging network for a sufficient length of time.

Permittivity   A measure of how well a dielectric will *permit* the establishment of flux lines within the dielectric.

Relative permittivity   The permittivity of a material compared to that of air.

Steady-state region   A period of time defined by the fact that the voltage across a capacitor has reached a level that, for all practical purposes, remains constant.

Stray capacitance   Capacitances that exist not through design but simply because two conducting surfaces are relatively close to each other.

Surge voltage   That voltage level that a capacitor can withstand for a very short period of time.

Temperature coefficient   An indication of how much the capacitance value of a capacitor will change with change in temperature.

Time constant   A period of time defined by the parameters of the network that will define how long the transient behavior of the voltage or current of a capacitor will last.

Transient period   That period of time where the voltage across a capacitor or the current of a capacitor will change in value at a rate determined by the time constant of the network.

Working voltage   That voltage level at which a capacitor can perform its function without concern about breakdown or change in characteristics.

Inductors

- Become aware of the characteristics of a magnetic field.
- Be able to calculate the inductance of an inductor and understand the impact of each parameter.
- Become aware of the various types of inductors and learn how to determine their value and rating from the provided labeling.
- Understand the relationship between the current through and the voltage across a coil.
- Become familiar with the transient response of an inductor and learn how to plot the response of the voltage and current.
- Learn how to incorporate initial values for the current of an inductor in the transient response.
- Become aware of how Thévenin's theorem can help define the response of an inductor in a series-parallel network.
- Understand how to combine inductors in series and in parallel.
- Develop some familiarity with the use of inductors in practical applications.
- Be able to set up the transient response using PSpice or EWB.

11.1 INTRODUCTION

Three basic components make up the majority of electrical/electronic systems in use today. They include the resistor and the capacitor, which have already been introduced, and the **inductor,** to be examined in detail in this chapter. In many ways, the inductor is the dual of the capacitor in the sense that what is true for the voltage of one is applicable to the current of the other, and vice versa. In fact, a number of the sections in this chapter will parallel those of the previous chapter on the capacitor. Like the capacitor, **the inductor will exhibit its true characteristics only when a change in voltage or current is made in the network.**

Recall in the last chapter that a capacitor can be replaced by an open-circuit equivalent under steady-state conditions. It will be demonstrated in this chapter that an inductor can be replaced by a short-circuit equivalent under steady-state conditions. Finally, we recognize that resistors dissipate the power delivered to them in the form of heat, while capacitors store the energy delivered to them in the form of an electric field. Inductors, in the ideal sense, are like capacitors in that they also store the energy delivered to them—but in the form of a magnetic field.

11.2 THE MAGNETIC FIELD

Magnetism plays an integral part in almost every electrical device used today in industry, research, or the home. Generators, motors, transformers, circuit breakers, televisions, computers, tape recorders, and telephones all employ magnetic effects to perform a variety of important tasks.

The compass, used by Chinese sailors as early as the second century A.D., relies on a **permanent magnet** for indicating direction. A permanent magnet is made of a material, such as steel or iron, that will remain magnetized for long periods of time without the need for an external source of energy.

In 1820, the Danish physicist Hans Christian Oersted discovered that the needle of a compass would deflect if brought near a current-carrying conductor. For the first time it was demonstrated that electricity and magnetism were related, and in the same year the French physicist André-Marie Ampère performed experiments in this area and developed what is presently known as **Ampère's circuital law.** In subsequent years, men such as Michael Faraday, Karl Friedrich Gauss, and James Clerk Maxwell continued to experiment in this area and developed many of the basic concepts of **electromagnetism**—magnetic effects induced by the flow of charge, or current.

In the region surrounding a permanent magnet, there exists a magnetic field, which can be represented by **magnetic flux lines** similar to electric flux lines. Magnetic flux lines, however, do not have origins or terminating points as do electric flux lines **but exist in continuous loops,** as shown in Fig. 11.1. The symbol for magnetic flux is the Greek letter Φ (phi).

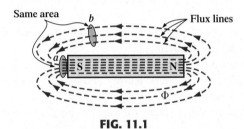

FIG. 11.1

Flux distribution for a permanent magnet.

The magnetic flux lines radiate from the north pole to the south pole, returning to the north pole through the metallic bar. Note the equal spacing between the flux lines within the core and the symmetric distribution outside the magnetic material. These are additional properties of magnetic flux lines in homogeneous materials (that is, materials having uniform structure or composition throughout). It is also important to realize that the continuous magnetic flux line will strive to occupy as small an area as possible. This will result in magnetic flux lines of minimum length between the unlike poles, as shown in Fig. 11.2. The strength of a magnetic field in a particular region is directly related to the density of flux lines in that region. In Fig. 11.1, for example, the magnetic field strength at a is twice that at b since twice as many magnetic flux lines are associated with the perpendicular plane at a than at b. Recall from childhood experiments that the strength of permanent magnets is always stronger near the poles.

If unlike poles of two permanent magnets are brought together, the magnets will attract, and the flux distribution will be as shown in Fig. 11.2. If like poles are brought together, the magnets will repel, and the flux distribution will be as shown in Fig. 11.3.

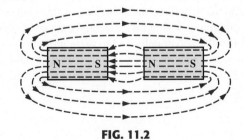

FIG. 11.2

Flux distribution for two adjacent, opposite poles.

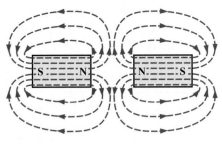

FIG. 11.3

Flux distribution for two adjacent, like poles.

If a nonmagnetic material, such as glass or copper, is placed in the flux paths surrounding a permanent magnet, there will be an almost unnoticeable change in the flux distribution (Fig. 11.4). However, if a magnetic material, such as soft iron, is placed in the flux path, the flux lines will pass through the soft iron rather than the surrounding air because flux lines pass with greater ease through magnetic materials than through air. This principle is put to use in the shielding of sensitive electrical elements and instruments that can be affected by stray magnetic fields (Fig. 11.5).

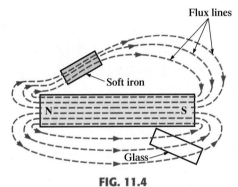

FIG. 11.4

Effect of a ferromagnetic sample on the flux distribution of a permanent magnet.

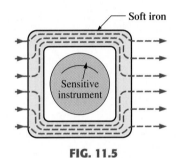

FIG. 11.5

Effect of a magnetic shield on the flux distribution.

As indicated previously, a magnetic field (represented by concentric magnetic flux lines, as in Fig. 11.6) is present around every wire that carries an electric current. The direction of the magnetic flux lines can be found simply by placing the thumb of the *right* hand in the direction of *conventional* current flow and noting the direction of the fingers. (This method is commonly called the *right-hand rule*.) If the conductor is wound in a single-turn coil (Fig. 11.7), the resulting flux will flow in a common direction through the center of the coil. A coil of more than one turn would produce a magnetic field that would exist in a continuous path through and around the coil (Fig. 11.8).

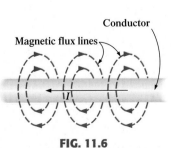

FIG. 11.6

Magnetic flux lines around a current-carrying conductor.

FIG. 11.7

Flux distribution of a single-turn coil.

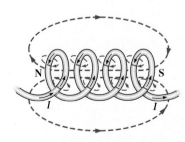

FIG. 11.8

Flux distribution of a current-carrying coil.

The flux distribution of the coil is quite similar to that of the permanent magnet. The flux lines leaving the coil from the left and entering to the right simulate a north and a south pole, respectively. The principal difference between the two flux distributions is that the flux lines are more concentrated for the permanent magnet than for the coil. Also, since **the strength of a magnetic field is determined by the density of the flux lines,** the coil has a weaker field strength. The field strength of the coil can be effectively increased by placing certain materials, such as iron, steel, or cobalt, within the coil to increase the flux density within the coil. By increasing the field strength with the addition of the core, we have devised an *electromagnet* (Fig. 11.9) that, in addition to having all the properties of a permanent magnet, also has a field strength that can be varied by changing one of the component values (current, turns, and so on). Of course, current must pass through the coil of the electromagnet in order for magnetic flux to be developed, whereas there is no need for the coil or current in the permanent magnet. The direction of flux lines can be determined for the electromagnet (or in any core with a wrapping of turns) by placing the fingers of the right hand in the direction of current flow around the core. The thumb will then point in the direction of the north pole of the induced magnetic flux, as demonstrated in Fig. 11.10(a). A cross section of the same electromagnet is included as Fig. 11.10(b) to introduce the convention for directions perpendicular to the page. The cross and the dot refer to the tail and the head of the arrow, respectively.

FIG. 11.9
Electromagnet.

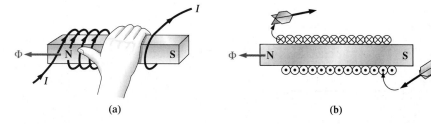

(a)　　　　　　　　　　　　　(b)

FIG. 11.10
Determining the direction of flux for an electromagnet: (a) method; (b) notation.

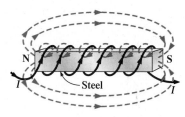

FIG. 11.11
Wilhelm Eduard Weber.

German (Wittenberg, Göttingen)
(1804–91)
Physicist
Professor of Physics,
University of Göttingen

An important contributor to the establishment of a system of *absolute units* for the electrical sciences, which was beginnning to become a very active area of research and development. Established a definition of electric current in an electromagnetic system based on the magnetic field produced by the current. He was politically active and, in fact, was dismissed from the faculty of the University of Göttingen for protesting the suppression of the constitution by the King of Hanover in 1837. However, he found other faculty positions and eventually returned to Göttingen as director of the astronomical observatory. He received honors from England, France, and Germany, including the Copley Medal of the Royal Society.

In the MKS system of units, magnetic flux is measured in **webers (Wb)** as derived from the surname of Wilhelm Eduard Weber (Fig. 11.11). The applied symbol is the capital Greek letter phi, Φ. The number of flux lines per unit area, called the **flux density,** is denoted by the capital letter B and is measured in **teslas (T)** to honor the efforts of Nikola Tesla, a scientist of the late 1800s (Fig. 11.12).

In equation form:

$$B = \frac{\Phi}{A} \qquad \begin{aligned} B &= \text{Wb/m}^2 = \text{teslas (T)} \\ \Phi &= \text{webers (Wb)} \\ A &= \text{m}^2 \end{aligned} \qquad \textbf{(11.1)}$$

where Φ is the number of flux lines passing through the area A of Fig. 11.13. The flux density at point a in Fig. 11.1 is twice that at b because twice as many flux lines pass through the same area.

In Eq. (11.1), the equivalence is given by

$$\boxed{1 \text{ tesla} = 1 \text{ T} = 1 \text{ Wb/m}^2} \qquad \textbf{(11.2)}$$

which states in words that if 1 weber of magnetic flux passes through an area of 1 square meter, the flux density is 1 tesla.

The flux density of an electromagnet is directly related to the number of turns of and the current through the coil. The product of the two, called the **magnetomotive force,** is measured in **ampere-turns (At)** as defined by

$$\boxed{\mathscr{F} = NI} \qquad \text{(ampere-turns, At)} \qquad \textbf{(11.3)}$$

In other words, if you increase the number of turns around a core and/or increase the current through the coil, the magnetic field strength will also increase. In many ways, the magnetomotive force for magnetic circuits is similar to the applied voltage in an electric circuit. Increasing either one will result in an increase in the desired effect: magnetic flux for magnetic circuits and current for electric circuits.

Another factor that will affect the magnetic field strength is the type of core employed. Materials in which magnetic flux lines can readily be set up are said to be **magnetic** and to have a high **permeability.** Again, note the similarity with the word "permit" used to describe permittivity for the dielectrics of capacitors. Similarly, the permeability (represented by the Greek letter mu, μ) of a material is a measure of the ease with which magnetic flux lines can be established in the material.

Just as there is a specific value for the permittivity of air, there is a specific number associated with the permeability of air:

$$\mu_o = 4\pi \times 10^{-7} \text{ Wb/A·m} \tag{11.4}$$

Practically speaking, the permeability of all nonmagnetic materials, such as copper, aluminum, wood, glass, and air, is the same as that for free space. Materials that have permeabilities slightly less than that of free space are said to be **diamagnetic,** and those with permeabilities slightly greater than that of free space are said to be **paramagnetic.** Magnetic materials, such as iron, nickel, steel, cobalt, and alloys of these metals, have permeabilities hundreds and even thousands of times that of free space. Materials with these very high permeabilities are referred to as **ferromagnetic.**

The ratio of the permeability of a material to that of free space is called its **relative permeability;** that is,

$$\mu_r = \frac{\mu}{\mu_o} \tag{11.5}$$

In general, for ferromagnetic materials, $\mu_r \geq 100$, and for nonmagnetic materials, $\mu_r = 1$.

You might now expect to see a table of values for μ to match the provided table for permittivity levels of specific dielectrics. Unfortunately, however, such a table cannot be provided because **relative permeability is a function of the operating conditions.** If you change the magnetomotive force applied, the level of μ can vary between extreme limits. At one level of magnetomotive force, the permeability of a material can be 10 times that at another level.

An instrument designed to measure flux density in gauss (CGS system) appears in Fig. 11.14. Appendix F reveals that $1 \text{ T} = 10^4$ gauss. The magnitude of the reading appearing on the face of the meter in Fig. 11.14 is therefore

$$28.2159 \text{ gauss} \left(\frac{1 \text{ T}}{10^4 \text{ gauss}} \right) = 2.822 \times 10^{-3} \text{ T}$$

FIG. 11.14

Digital display gaussmeter.
(Courtesy of Walker LDJ Scientific Inc. www.walkerldjscientific.com)

Although our emphasis in this chapter is to introduce the parameters that will affect the nameplate data of an inductor, the use of magnetics has widespread application in the electrical/electronics industry, as shown by a few areas of application in Fig. 11.15.

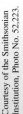
Courtesy of the Smithsonian Institution, Photo No. 52.223.

FIG. 11.12

Nikola Tesla.

Croatian-American (Smiljan, Paris, Colorado Springs, New York City) **(1856–1943)**
Electrical Engineer and Inventor
Recipient of the Edison Medal in 1917

Often regarded as one of the most innovative and inventive individuals in the history of the sciences. He was the first to introduce the *alternating-current machine,* removing the need for commutator bars of dc machines. After emigrating to the United States in 1884, he sold a number of his patents on *ac machines, transformers,* and *induction coils* (including the *Tesla coil* as we know it today) to the Westinghouse Electric Company. Some say that his most important discovery was made at his laboratory in Colorado Springs, where in 1900 he discovered *terrestrial stationary waves.* The range of his discoveries and inventions is too extensive to list here but extends from lighting systems to *polyphase power systems* to a *wireless world broadcasting system.*

FIG. 11.13

Defining the flux density B.

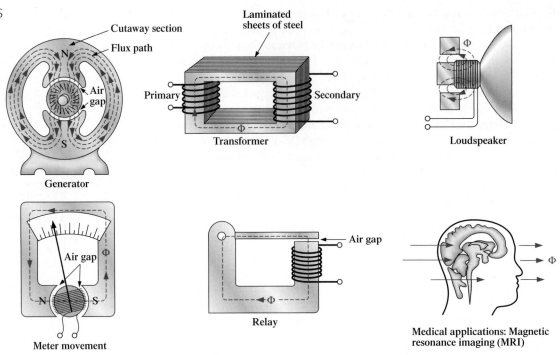

FIG. 11.15

Some areas of application of magnetic effects.

11.3 INDUCTANCE

In the previous section, it was clearly demonstrated that sending a current through a coil of wire, with or without a core, will establish a magnetic field through and surrounding the unit. This component, of rather simple construction as shown in Fig. 11.16, is called an **inductor** (often referred to as a **coil**). Its **inductance** level will determine the strength of the magnetic field around the coil due to an applied current. The higher the inductance level, the greater the strength of the magnetic field. In total, therefore,

inductors are designed to set up a strong magnetic field linking the unit, whereas capacitors are designed to set up a strong electric field between the plates.

Inductance is measured in **henries (H),** after the American physicist Joseph Henry (Fig. 11.17). However, just as the farad is too large a unit for most applications, most inductors are of the millihenry (mH) or microhenry (μH) range.

In Chapter 10, 1 farad was defined as a capacitance level that would result in 1 coulomb of charge on the plates due to the application of 1 volt across the plates. For inductors, 1 henry is the inductance level that will result in a voltage of 1 volt across the coil due to a change in current of 1 A/s through the coil.

Inductor Construction

In the previous chapter we found that capacitance is sensitive to the area of the plates, the distance between the plates, and the dielectric employed. The level of inductance has similar construction sensitivities in that it is dependent on the area within the coil, the length of the unit, and the permeability of the core material, but it is also sensitive to the number of turns of wire in the coil as dictated by Eq. (11.6) and defined in Fig. 11.18 for two of the most popular shapes:

N turns

FIG. 11.16

Basic construction of an inductor.

$$L = \frac{\mu N^2 A}{l}$$

μ = permeability (Wb/A·m)
N = number of turns (t)
A = m²
l = m
L = henries (H)

(11.6)

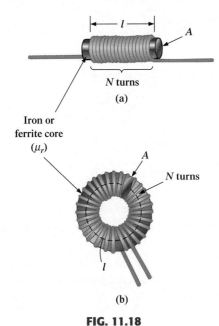

(a)

Iron or
ferrite core
(μ_r)

N turns

A

(b)

FIG. 11.18
Defining the parameters for Eq. (11.6).

FIG. 11.17
Joseph Henry.

American (Albany, NY; Princeton,
NJ)
(1797–1878)
Physicist and Mathematician
Professor of Natural Philosophy,
Princeton University

In the early 1800s the title Professor
of Natural Philosophy was applied
to educators in the sciences. As a
student and teacher at the Albany
Academy, Henry performed extensive research in the area of electromagnetism. He improved the design
of *electromagnets* by insulating the
coil wire to permit a tighter wrap on
the core. One of his earlier designs
was capable of lifting 3600 pounds.
In 1832 he discovered and delivered
a paper on *self-induction.* This was
followed by the construction of an
effective *electric telegraph transmitter and receiver* and extensive
research on the oscillatory nature of
lightning and discharges from a
Leyden jar. In 1845 he was ap-
pointed the first Secretary of the
Smithsonian.

First note that since the turns are squared in the equation, the number of turns is a big
factor. However, also keep in mind that the more turns, the bigger the unit. If the wire is
made too thin in order to get more windings on the core, the rated current of the inductor is
limited. Since higher levels of permeability will result in higher levels of magnetic flux,
permeability should, and does, appear in the numerator of the equation. Increasing the area
of the core or decreasing the length will also increase the inductance level. The effect of
each parameter is defined in Fig. 11.19.

Substituting $\mu = \mu_r \mu_o$ for the permeability will result in Eq. (11.7), which is very similar to the equation for the capacitance of a capacitor:

$$L = \frac{\mu_r \mu_o N^2 A}{l}$$

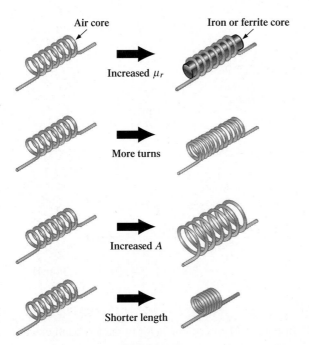

Air core

Iron or ferrite core

Increased μ_r

More turns

Increased A

Shorter length

FIG. 11.19
Factors that will increase the inductance level of an inductor (coil).

or
$$L = 4\pi \times 10^{-7} \frac{\mu_r N^2 A}{l} \qquad \text{(henries, H)}$$
(11.7)

If we break out the relative permeability as follows:

$$L = \mu_r \left(\frac{\mu_o N^2 A}{l} \right)$$

we obtain the following useful equation:

$$\boxed{L = \mu_r L_o}$$
(11.8)

which is very similar to the equation $C = \epsilon_r C_o$. Equation (11.8) clearly states the following:

The inductance of an inductor with a ferromagnetic core is μ_r times the inductance obtained with an air core.

Although Equation (11.6) is approximate at best, the equations for the inductance of a wide variety of coils can be found in reference handbooks. Most of the equations are mathematically more complex than Eq. (11.6), but the impact of each factor is the same in each equation.

EXAMPLE 11.1   For the air-core coil of Fig. 11.20:

a. Find the inductance.
b. Find the inductance if a metallic core with $\mu_r = 2000$ is inserted in the coil.

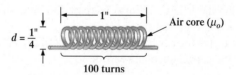

$d = \frac{1}{4}''$ 1''   Air core (μ_o)

100 turns

FIG. 11.20
Air-core coil for Example 11.1.

Solutions:

a. $d = \dfrac{1}{4} \text{in.} \left(\dfrac{1 \text{ m}}{39.37 \text{ in.}} \right) = 6.35 \text{ mm}$

$A = \dfrac{\pi d^2}{4} = \dfrac{\pi (6.35 \text{ mm})^2}{4} = 31.7 = \mu\text{m}^2$

$l = 1 \text{ in.} \left(\dfrac{1 \text{ m}}{39.37 \text{ in.}} \right) = 25.4 \text{ mm}$

$L = 4\pi \times 10^{-7} \dfrac{\mu_r N^2 A}{l}$

$\quad = 4\pi \times 10^{-7} \dfrac{(1)(100 \text{ t})^2 (31.7 \ \mu\text{m}^2)}{25.4 \text{ mm}} = \mathbf{15.68 \ \mu H}$

b. Eq. (11.8): $L = \mu_r L_o = (2000)(15.68 \ \mu\text{H}) = \mathbf{31.36 \ mH}$

EXAMPLE 11.2   In Fig. 11.21, if each inductor in the left-hand column is changed to the type appearing in the right-hand column, find the new inductance level. For each change, assume that the other factors remain the same.

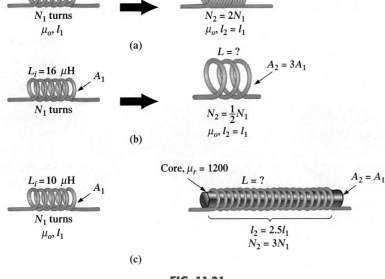

FIG. 11.21

Inductors for Example 11.2.

Solutions:

a. The only change was the number of turns, but it is a squared factor, resulting in

$$L = (2)^2 L_i = (4)(20 \ \mu\text{H}) = \textbf{80} \ \boldsymbol{\mu}\textbf{H}$$

b. In this case, the area is three times the original size, and the number of turns is 1/2. Since the area is in the numerator, it increases the inductance by a factor of three. The drop in the number of turns will reduce the inductance by a factor of $(1/2)^2 = 1/4$. Therefore,

$$L = (3)\left(\frac{1}{4}\right) L_i = \frac{3}{4}(16 \ \mu\text{H}) = \textbf{12} \ \boldsymbol{\mu}\textbf{H}$$

c. Both μ and the number of turns have increased, although the increase in the number of turns is squared. The increased length will reduce the inductance. Therefore,

$$L = \frac{(3)^2 (1200)}{2.5} L_i = (4.32 \times 10^3)(10 \ \mu\text{H}) = \textbf{43.2 mH}$$

Types of Inductors

Inductors, like capacitors and resistors, can be categorized under the general headings **fixed** or **variable.** The symbol for a fixed air-core inductor is provided in Fig. 11.22(a), for an inductor with a ferromagnetic core in Fig. 11.22(b), for a tapped coil in Fig. 11.22(c), and for a variable inductor in Fig. 11.22(d).

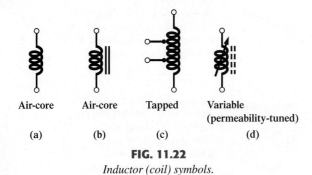

Air-core   Air-core   Tapped   Variable
(permeability-tuned)

(a)   (b)   (c)   (d)

FIG. 11.22

Inductor (coil) symbols.

Fixed   Fixed-type inductors come in all shapes in sizes. However,

in general, the size of an inductor is determined primarily by the type of construction, the core employed, or the current rating.

In Fig. 11.23(a), the 10 μH and 1 mH coils are about the same size because a thinner wire was used for the 1 mH coil to permit more turns in the same space. The result, however, is a drop in rated current from 10 A to only 1.3 A. If the wire of the 10 μH coil had been used to

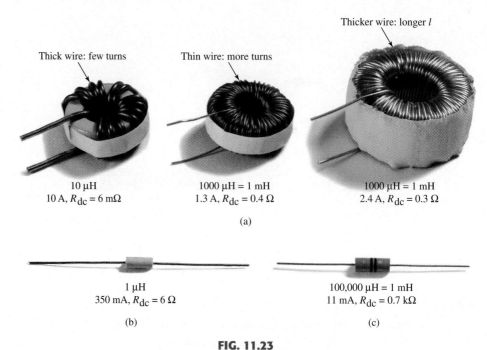

Thicker wire: longer *l*

Thick wire: few turns

Thin wire: more turns

10 μH
10 A, $R_{dc} = 6$ mΩ

1000 μH = 1 mH
1.3 A, $R_{dc} = 0.4$ Ω

1000 μH = 1 mH
2.4 A, $R_{dc} = 0.3$ Ω

(a)

1 μH
350 mA, $R_{dc} = 6$ Ω

100,000 μH = 1 mH
11 mA, $R_{dc} = 0.7$ kΩ

(b)

(c)

FIG. 11.23
Relative sizes of different types of inductors: (a) toroid, high-current;
(b) phenolic (resin or plastic core); (c) ferrite core.

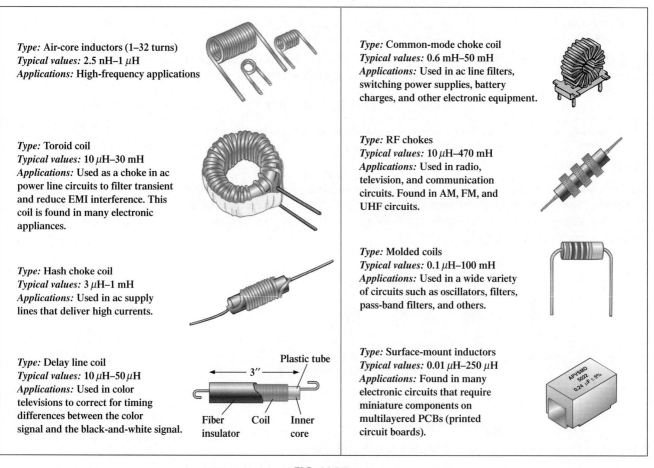

Type: Air-core inductors (1–32 turns)
Typical values: 2.5 nH–1 μH
Applications: High-frequency applications

Type: Toroid coil
Typical values: 10 μH–30 mH
Applications: Used as a choke in ac power line circuits to filter transient and reduce EMI interference. This coil is found in many electronic appliances.

Type: Hash choke coil
Typical values: 3 μH–1 mH
Applications: Used in ac supply lines that deliver high currents.

Type: Delay line coil
Typical values: 10 μH–50 μH
Applications: Used in color televisions to correct for timing differences between the color signal and the black-and-white signal.

Plastic tube

← 3″ →

Fiber
insulator

Coil

Inner
core

Type: Common-mode choke coil
Typical values: 0.6 mH–50 mH
Applications: Used in ac line filters, switching power supplies, battery charges, and other electronic equipment.

Type: RF chokes
Typical values: 10 μH–470 mH
Applications: Used in radio, television, and communication circuits. Found in AM, FM, and UHF circuits.

Type: Molded coils
Typical values: 0.1 μH–100 mH
Applications: Used in a wide variety of circuits such as oscillators, filters, pass-band filters, and others.

Type: Surface-mount inductors
Typical values: 0.01 μH–250 μH
Applications: Found in many electronic circuits that require miniature components on multilayered PCBs (printed circuit boards).

FIG. 11.24
Typical areas of application for inductive elements.

make the 1 mH coil, the resulting coil would have been many times the size of the 10 μH coil. The impact of the wire thickness is clearly revealed by the 1 mH coil at the far right of Fig. 11.23(a), where a thicker wire was used to raise the rated current level from 1.3 A to 2.4 A. Even though the inductance level is the same, the size of the toroid is 4 or 5 times greater.

The phenolic inductor (using a nonferromagnetic core of resin or plastic) of Fig. 11.23(b) is quite small for its level of inductance. We must assume that it has a high number of turns of very thin wire. Note, however, that the use of a very thin wire has resulted in a relatively low current rating of only 350 mA (0.35 A). The use of a ferrite (ferromagnetic) core in the inductor of Fig. 11.23(c) has resulted in an amazingly high level of inductance for its size. However, the wire is so thin that the current rating is only 11 mA = 0.011 A. Note that for all the inductors, the dc resistance of the inductor increases with decrease in the thickness of the wire. The 10 μH toroid has a dc resistance of only 6 mΩ, whereas the dc resistance of the 100 mH ferrite inductor is 700 Ω—a price to be paid for the smaller size and high inductance level.

A number of different types of fixed inductive elements are displayed in Fig. 11.24, including their typical range of values and common areas of application. Recognize that, in general, based on the discussion of construction in recent paragraphs, it is fairly easy to identify an inductive element. There certainly is a similarity between the shape of a molded film resistor and the shape of an inductor. However, careful examination of the typical shapes of each will reveal some differences, such as the ridges at each end of a resistor which do not appear on most inductors.

Variable   A number of variable inductors are depicted in Fig. 11.25. In each case the inductance is changed by turning the slot at the end of the core to move it into and out of the unit. The further in the core is, the more the ferromagnetic material is part of the magnetic circuit, and the higher are the magnetic field strength and the inductance level.

Practical Equivalent Inductors

Inductors, like capacitors, are not ideal. Associated with every inductor is a resistance determined by the resistance of the turns of wire (the thinner the wire, the greater the resistance for the same material) and by the core losses (radiation and skin effect, eddy current and hysteresis losses—all discussed in more advanced texts). There is also some stray capacitance due to the capacitance between the current-carrying turns of wire of the coil. Recall that capacitance will appear whenever there are two conducting surfaces separated by an insulator, such as air, and when those wrappings are fairly tight and are parallel. Both elements are included in the equivalent circuit of Fig. 11.26. For most applications in this text, the capacitance can be ignored, resulting in the equivalent model of Fig. 11.27. The resistance R_l will play an important part in some areas (such as resonance, discussed in Chapter 17), because the resistance can extend from a few ohms to a few hundred ohms depending on the construction. For this chapter, the inductor will be considered an ideal element and the series resistance will be dropped from Fig. 11.27.

FIG. 11.25

Variable inductors with a typical range of values from 1 μH to 100 μH; commonly used in oscillators and various R-F circuits such as CB transceivers, televisions, and radios.

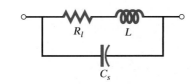

FIG. 11.26

Complete equivalent model for an inductor.

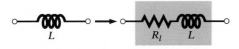

FIG. 11.27

Practical equivalent model for an inductor.

Inductor Labeling

Because some inductors are larger in size, their nameplate value can often be printed on the body of the element. However, on smaller units, there may not be enough room to print the actual value, so an abbreviation is used that is fairly straightforward to follow. First, realize that **the microhenry (μH) is the fundamental unit of measurement for this marking.** Most manuals will list the inductance value in μH even if the value must be reported as 470,000 μH rather than as 470 mH. If the label reads 223K, the third number (3) is the power to be applied to the first two. **The K is not from kilo,** representing a power of three, but is used to denote a tolerance of $\pm$10% as described for capacitors. The resulting number of 22,000 is in μH for reasons described above, so the 223K unit is a 22,000 μH or 22 mH inductor. The letters J and M indicate a tolerance of $\pm$5% and $\pm$20%, respectively.

For molded inductors, a color-coding system very similar to that used for resistors is used. The major difference is that **the resulting value is always in μH,** and a wide band at the beginning of the labeling is an MIL ("meets military standards") indicator. Always read the colors in sequence, starting with the band closest to one end as shown in Fig. 11.28.

The standard values for inductors employ the same numerical values and multipliers used with resistors and capacitors. In general, therefore, expect to find inductors with the following multipliers: **1 μH, 1.5 μH, 2.2 μH, 3.3 μH, 4.7 μH, 6.8 μH, 10 μH,** and so on.

Color Code Table			
Color[1]	**Significant Figure**	**Multiplier**[2]	**Inductance Tolerance (%)**
Black	0	1	
Brown	1	10	
Red	2	100	
Orange	3	1000	
Yellow	4		
Green	5		
Blue	6		
Violet	7		
Gray	8		
White	9		
None[2]			$\pm$20
Silver			$\pm$10
Gold	Decimal point		$\pm$5

[1] Indicates body color.
[2] The multiplier is the factor by which the two significant figures are multiplied to yield the nominal inductance value.

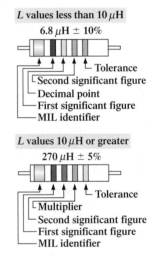

Cylindrical molded choke coils are marked with five colored bands. A wide silver band, located at one end of the coil, identifies military radio-frequency coils. The next three bands indicate the inductance in microhenries, and the fourth band is the tolerance.

Color coding is in accordance with the color code table, shown on the left. If the first or second band is gold, it represents the decimal point for inductance values less than 10. Then the following two bands are significant figures. For inductance values of 10 or more, the first two bands represent significant figures, and the third is the multiplier.

FIG. 11.28
Molded inductor color coding.

Measurement and Testing of Inductors

The inductance of an inductor can be read directly using a meter such as the Universal LCR Meter of Fig. 11.29 which also appeared in Chapter 10 on capacitors. Set the meter to L for inductance, and the meter will automatically choose the most appropriate unit of measurement for the element, that is, H, mH, μH, or pH.

The best check is to use a meter such as appearing in Fig. 11.29. However, if such a meter is unavailable, an ohmmeter can be used to check whether a short has developed between the windings or whether an open circuit has developed. The open-circuit possibility is easy to check because a reading of infinite ohms or very high resistance will result. The short-circuit condition is harder to check because the resistance of many good inductors is relatively small, and the shorting of a few windings may not adversely affect the total resistance. Of course, if

FIG. 11.29
Digital reading inductance meter.
(Image provided by B & K Precision Corporation.)

you are aware of the typical resistance of the coil, you can compare it to the measured value. A short between the windings and the core can be checked by simply placing one lead of the ohmmeter on one wire (perhaps a terminal) and the other on the core itself. A reading of zero ohms reveals a short between the two that may be due to a breakdown in the insulation jacket around the wire resulting from excessive currents, environmental conditions, or simply old age and cracking.

11.4   THE VOLTAGE v_L

Before analyzing the response of inductive elements to an applied dc voltage, we must introduce a number of laws and equations that will affect the transient response.

The first, referred to as **Faraday's law of electromagnetic induction,** is one of the most important in this field because it enables us to establish ac and dc voltages with a generator. If we move a conductor (any material with conductor characteristics as defined in Chapter 2) through a magnetic field so that it cuts magnetic lines of flux as shown in Fig. 11.30, a voltage will be induced across the conductor that can be measured with a sensitive voltmeter. That's all it takes, and, in fact, the faster you move the conductor through the magnetic flux, the greater will be the induced voltage. The same effect can be produced if you hold the conductor still and move the magnetic field across the conductor. It is interesting to note that the direction in which you move the conductor through the field will determine the polarity of the induced voltage. Also, if you move the conductor through the field at right angles to the magnetic flux, you will generate the maximum induced voltage. Moving the conductor parallel with the magnetic flux lines will result in an induced voltage of zero volts since magnetic lines of flux will not be crossed.

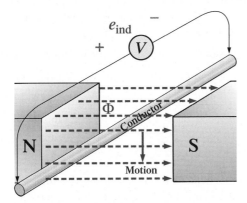

FIG. 11.30

Generating an induced voltage by moving a conductor through a magnetic field.

If we now go a step further and move a coil of N turns through the magnetic field as shown in Fig. 11.31, a voltage will be induced across the coil as determined by **Faraday's law:**

$$e = N\frac{d\phi}{dt} \qquad \text{(volts, V)} \tag{11.9}$$

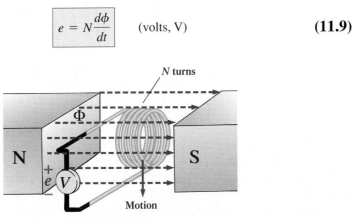

FIG. 11.31

Demonstrating Faraday's law.

The greater the number of turns or the faster the coil is moved through the magnetic flux pattern, the greater will be the induced voltage. The term $d\phi/dt$ is the differential change in magnetic flux through the coil at a particular instant in time. If the magnetic flux passing through a coil remains constant—no matter how strong the magnetic field—the term will be zero, and the induced voltage zero volts. It doesn't matter whether the changing flux is due to moving the magnetic field or moving the coil in the vicinity of a magnetic field: The only requirement is that the flux linking (passing through) the coil changes with time. Before the coil passes through the magnetic poles, the induced voltage is zero because there are no magnetic flux lines passing through the coil. As the coil enters the flux pattern, the number of flux lines cut per instant of time increases until it peaks at the center of the poles. The induced voltage will then decrease with time as it leaves the magnetic field.

This important phenomenon can now be applied to the inductor of Fig. 11.32, which is simply an extended version of the coil of Fig. 11.31. In Section 11.2, we found that the magnetic flux linking the coil of N turns with a current I has the distribution shown in Fig. 11.32. If the current through the coil increases in magnitude, the flux linking the coil will also increase. We just learned through Faraday's law, however, that a coil in the vicinity of a changing magnetic flux will have a voltage induced across it. The result is that a voltage is induced across the coil of Fig. 11.32 due to the **change in current through the coil.**

It is very important to note in Fig. 11.32 that the polarity of the induced voltage across the coil is such that it opposes the increasing level of current in the coil. In other words, the changing current through the coil induces a voltage across the coil that is opposing the applied voltage that establishes the increase in current in the first place. The quicker the change in current through the coil, the greater will be the opposing induced voltage to squelch the attempt of the current to increase in magnitude. The "choking" action of the coil is the reason inductors or coils are often referred to as **chokes.** This effect is a result of an important law referred to as **Lenz's law,** which states that

an induced effect is always such as to oppose the cause that produced it.

The inductance of a coil is also a measure of the change in flux linking the coil due to a change in current through the coil. That is,

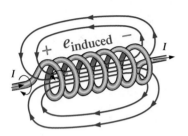

FIG. 11.32

Demonstrating the effect of Lenz's law.

$$\boxed{L = N\frac{d\phi}{di_L}} \qquad \text{(henries, H)} \qquad \textbf{(11.10)}$$

The equation reveals that the greater the number of turns or the greater the change in flux linking the coil due to a particular change in current, the greater will be the level of inductance. In other words, coils with smaller levels of inductance will generate smaller changes in flux linking the coil for the same change in current through the coil. If the inductance level is very small, there will be almost no change in flux linking the coil, and the induced voltage across the coil will be very small. In fact, if we now write Eq. (11.9) in the following form:

$$e = N\frac{d\phi}{dt} = \left(N\frac{d\phi}{di_L}\right)\left(\frac{di_L}{dt}\right)$$

and substitute Eq. (11.10), we obtain

$$\boxed{e_L = L\frac{di_L}{dt}} \qquad \text{(volts, V)} \qquad \textbf{(11.11)}$$

which relates the voltage across a coil to the number of turns of the coil and the change in current through the coil.

When induced effects are employed in the generation of voltages such as those from dc or ac generators, the symbol e is applied to the induced voltage. However, in network analysis, the voltage induced across an inductor will always have a polarity that opposes the applied voltage (like the voltage across a resistor). Therefore, the following notation will be used for the induced voltage across an inductor:

$$\boxed{v_L = L\frac{di_L}{dt}} \qquad \text{(volts, V)} \qquad \textbf{(11.12)}$$

The equation clearly states that

the larger the inductance and/or the faster the current through a coil changes, the larger will be the induced voltage across the coil.

If the current through the coil fails to change with time, the induced voltage across the coil will be zero. We will find in the next section that for dc applications, when the transient phase has passed, $di_L/dt = 0$, and the induced voltage across the coil will be

$$v_L = L\frac{di_L}{dt} = L(0) = 0 \text{ V}$$

The duality that exists between inductive and capacitive elements is now abundantly clear. Simply interchange the voltages and currents of Eq. (11.12), and interchange the inductance and capacitance. The following equation for the current of a capacitor will result:

$$v_L = L\frac{di_L}{dt}$$
$$i_C = C\frac{dv_C}{dt}$$

We are now at a point where we have all the background relationships necessary to investigate the transient behavior of inductive elements.

11.5 *R-L* TRANSIENTS: THE STORAGE PHASE

It is indeed helpful and fortunate that there are a great number of similarities between the analyses of inductive and capacitive networks. That is, what was true for the voltage of a capacitor will also be true for the current of an inductor, and what was true for the current of a capacitor can be matched in many ways by the voltage of an inductor. The storage waveforms will have the same shape, and time constants will be defined for each configuration. In many ways, therefore, this section will be very close in content to Section 10.5 for the charging of a capacitor—providing an opportunity to reinforce concepts introduced earlier, yet still enabling you to learn more about the behavior of inductive elements.

The circuit of Fig. 11.33 will be used to describe the storage phase. Note that it is the same circuit used to describe the charging phase of capacitors, with a simple replacement of the capacitor by an ideal inductor. Throughout the analysis, it is important to keep in mind that energy is stored in the form of an electric field between the plates of a capacitor. For inductors, on the other hand, energy is stored in the form of a magnetic field linking the coil.

At the instant the switch is closed, the choking action of the coil will prevent an instantaneous change in current through the coil, resulting in $i_L = 0$ A as shown in Fig. 11.34(a). The absence of a current through the coil and circuit at the instant the switch is closed will result in zero volts across the resistor as determined by $v_R = i_R R = i_L R = (0 \text{ A})R = 0$ V, as shown in Fig. 11.34(c). Applying Kirchhoff's voltage law around the closed loop will result in E volts across the coil at the instant the switch is closed, as shown in Fig. 11.34(b).

Initially, the current will increase very rapidly as shown in Fig. 11.34(a) and then at a much slower rate as it approaches its steady-state value determined by the parameters of the network (E/R). The voltage across the resistor will rise at the same rate because $v_R = i_R R = i_L R$. Since the voltage across the coil is sensitive to the rate of change of current through the coil, the voltage will be at or near its maximum value early in the storage phase. Finally, when the current reaches its steady-state value of E/R amperes, the current through the coil will cease to change, and the voltage across the coil will drop to zero volts. At any instant of time, the voltage across the coil can be determined using Kirchhoff's voltage law in the following manner: $v_L = E - v_R$.

As mentioned earlier, it is indeed fortunate that the waveforms for the inductor have the same shape as obtained for capacitive networks. We are now familiar with the mathematical

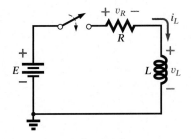

FIG. 11.33
Basic R-L transient network.

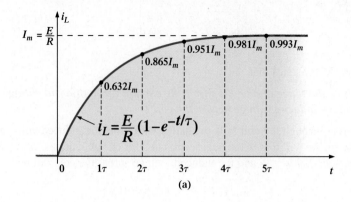

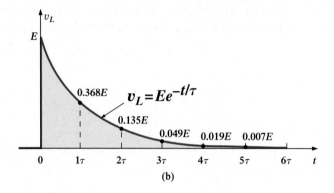

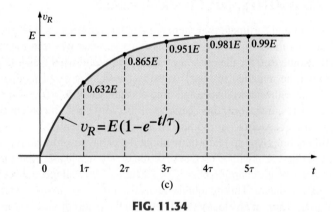

FIG. 11.34

i_L, v_L, and v_R for the circuit of Fig. 11.33 following the closing of the switch.

format and feel comfortable calculating the quantities of interest using a calculator or computer.

The equation for the transient response of the current through an inductor is the following:

$$i_L = \frac{E}{R}(1 - e^{-t/\tau}) \qquad \text{(amperes, A)} \tag{11.13}$$

with the time constant now defined by

$$\tau = \frac{L}{R} \qquad \text{(seconds, s)} \tag{11.14}$$

It is important to note that Equation (11.14) is a ratio of parameters rather than a product as used for capacitive networks, yet the units used are still seconds (for time).

Our experience with the factor $(1 - e^{-t/\tau})$ verifies the level of 63.2% for the inductor current after one time constant, 86.5% after two time constants, and so on. If we keep R constant and increase L, the ratio L/R will increase, and the rise time of 5τ will increase as shown in Fig. 11.35 for increasing levels of L. The change in transient response is expected because the higher the inductance level, the greater the choking action on the changing current level, and the longer it will take to reach steady-state conditions.

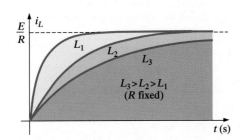

FIG. 11.35

Effect of L on the shape of the i_L storage waveform.

The equation for the voltage across the coil is the following:

$$\boxed{v_L = Ee^{-t/\tau}} \quad \text{(volts, V)} \qquad \textbf{(11.15)}$$

and for the voltage across the resistor:

$$\boxed{v_R = E(1 - e^{-t/\tau})} \quad \text{(volts, V)} \qquad \textbf{(11.16)}$$

As mentioned earlier, the shape of the response curve for the voltage across the resistor must match that of the current i_L since $v_R = i_R R = i_L R$.

Since the waveforms are similar to those obtained for capacitive networks, we will assume that

the storage phase has passed and steady-state conditions have been established once a period of time equal to five time constants has occurred.

In addition, since $\tau = L/R$ will always have some numerical value, even though it may be very small at times, the transient period of 5τ will always have some numerical value. Therefore,

the current cannot change instantaneously in an inductive network.

If we examine the conditions that exist at the instant the switch is closed, we find that the voltage across the coil is E volts, although the current is zero amperes as shown in Fig. 11.36. In essence, therefore,

the inductor takes on the characteristics of an open circuit at the instant the switch is closed.

However, if we consider the conditions that exist when steady-state conditions have been established, we find that the voltage across the coil is zero volts and the current is a maximum value of E/R amperes as shown in Fig. 11.37. In essence, therefore,

the inductor takes on the characteristics of a short circuit when steady-state conditions have been established.

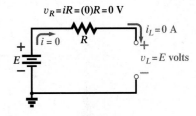

FIG. 11.36

Circuit of Fig. 11.33 the instant the switch is closed.

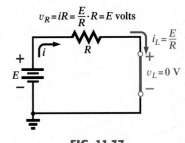

FIG. 11.37

Circuit of Fig. 11.33 under steady-state conditions.

EXAMPLE 11.3 Find the mathematical expressions for the transient behavior of i_L and v_L for the circuit of Fig. 11.38 if the switch is closed at $t = 0$ s. Sketch the resulting curves.

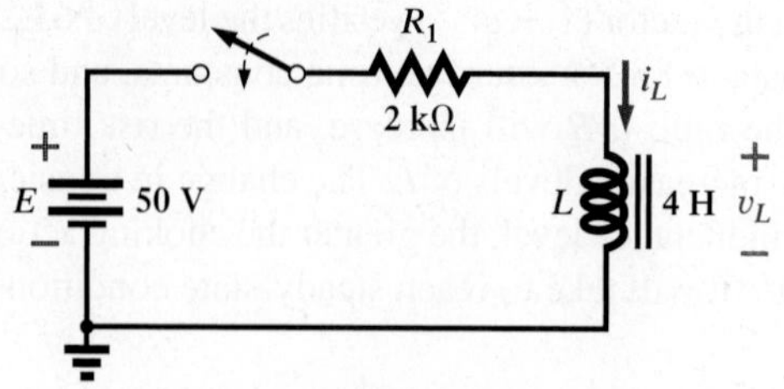

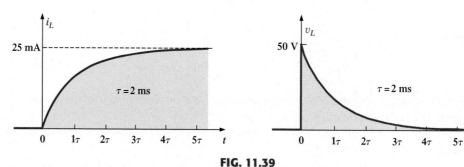

FIG. 11.38

Series R-L circuit for Example 11.3.

Solution: First the time constant is determined:

$$\tau = \frac{L}{R_1} = \frac{4 \text{ H}}{2 \text{ k}\Omega} = 2 \text{ ms}$$

Then the maximum or steady-state current is

$$I_m = \frac{E}{R_1} = \frac{50 \text{ V}}{2 \text{ k}\Omega} = 25 \times 10^{-3} \text{ A} = 25 \text{ mA}$$

Substituting into Eq. (11.13):

$$i_L = \mathbf{25 \text{ mA}(1 - e^{-t/2ms})}$$

Using Eq. (11.15):

$$v_L = \mathbf{50 \text{ V}}e^{-t/2ms}$$

The resulting waveforms appear in Fig. 11.39.

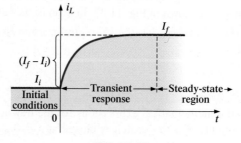

FIG. 11.39

i_L and v_L for the network of Fig. 11.38.

11.6 INITIAL CONDITIONS

This section will parallel Section 10.7 ("Initial Values: Capacitors") on the effect of initial values on the transient phase. Since the current through a coil cannot change instantaneously, the current through a coil will begin the transient phase at the initial value established by the network (note Fig. 11.40) before the switch was closed. It will then pass through the transient phase until it reaches the steady-state (or final) level after about five time constants. The steady-state level of the inductor current can be found by simply substituting its short-circuit equivalent (or R_l for the practical equivalent) and finding the resulting current through the element.

FIG. 11.40

Defining the three phases of a transient waveform.

Using the transient equation developed in the previous section, an equation for the current i_L can be written for the entire time interval of Fig. 11.40; that is,

$$i_L = I_i + (I_f - I_i)(1 - e^{-t/\tau})$$

with $(I_f - I_i)$ representing the total change during the transient phase. However, by multiplying through and rearranging terms:

$$i_L = I_i + I_f - I_f e^{-t/\tau} - I_i + I_i e^{-t/\tau}$$
$$= I_f - I_f e^{-t/\tau} + I_i e^{-t/\tau}$$

we find

$$\boxed{i_L = I_f + (I_i - I_f)e^{-t/\tau}} \qquad \textbf{(11.17)}$$

If you are required to draw the waveform for the current i_L from initial value to final value, start by drawing a line at the initial value and steady-state levels, and then add the transient response (sensitive to the time constant) between the two levels. The following example will clarify the procedure.

EXAMPLE 11.4   The inductor of Fig. 11.41 has an initial current level of 4 mA in the direction shown. (Specific methods to establish the initial current will be presented in the sections and problems to follow.)

a. Find the mathematical expression for the current through the coil once the switch is closed.
b. Find the mathematical expression for the voltage across the coil during the same transient period.
c. Sketch the waveform for each from initial value to final value.

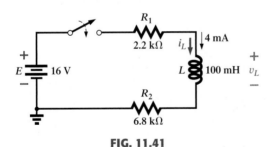

FIG. 11.41
Series R-L circuit for Example 11.4.

Solutions:

a. Substituting the short-circuit equivalent for the inductor will result in a final or steady-state current determined by Ohm's law:

$$I_f = \frac{E}{R_1 + R_2} = \frac{16\ V}{2.2\ k\Omega + 6.8\ k\Omega} = \frac{16\ V}{9\ k\Omega} = 1.78\ mA$$

The time constant is determined by

$$\tau = \frac{L}{R_T} = \frac{100\ mH}{2.2\ k\Omega + 6.8\ k\Omega} = \frac{100\ mH}{9\ k\Omega} = 11.11\ \mu s$$

Applying Eq. (11.17):

$$i_L = I_f + (I_i - I_f)e^{-t/\tau} = 1.78\ mA + (4\ mA - 1.78\ mA)e^{-t/11.11\mu s}$$
$$= \textbf{1.78 mA} + \textbf{2.22 mA}e^{-t/11.11\mu s}$$

b. Since the current through the inductor is constant at 4 mA prior to the closing of the switch, the voltage (whose level is sensitive only to changes in current through the coil) must have an initial value of 0 V. At the instant the switch is closed, the current through the coil cannot change instantaneously, so the current through the resistive elements will be 4 mA. The resulting peak voltage at $t = 0$ s can then be found using Kirchhoff's voltage law as follows:

$$V_m = E - V_{R_1} - V_{R_2} = 16\,V - (4\,mA)(2.2\,k\Omega) - (4\,mA)(6.8\,k\Omega)$$
$$= 16\,V - 8.8\,V - 27.2\,V = 16\,V - 36\,V = -20\,V$$

Note the minus sign to indicate that the polarity of the voltage v_L is opposite to the defined polarity of Fig. 11.41.

The voltage will then decay (with the same time constant as the current i_L) to zero because the inductor is approaching its short-circuit equivalence.

The equation for v_L is therefore

$$v_L = -20\,Ve^{-t/11.11\mu s}$$

c. See Fig. 11.42. The initial and final values of the current were drawn first, and then the transient response was included between these levels. For the voltage, the waveform begins and ends at zero, with the peak value having a sign sensitive to the defined polarity of v_L in Fig. 11.41.

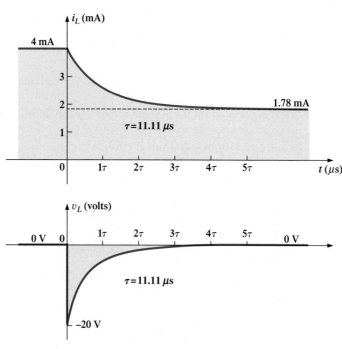

FIG. 11.42
i_L and v_L for the network of Fig. 11.41.

Let us now test the validity of the equation for i_L by substituting $t = 0$ s to reflect the instant the switch is closed.

$$e^{-t/\tau} = e^{-0} = 1$$

and $i_L = 1.78\,mA + 2.22\,mAe^{-t/\tau} = 1.78\,mA + 2.22\,mA = 4\,mA$

When $t > 5\tau$, $e^{-t/\tau} \cong 0$

and $i_L = 1.78\,mA + 2.22\,mAe^{-t/\tau} = 1.78\,mA$

11.7 R-L TRANSIENTS: THE RELEASE PHASE

In the analysis of R-C circuits, we found that the capacitor could hold its charge and store energy in the form of an electric field for a period of time determined by the leakage factors. In R-L circuits, the energy is stored in the form of a magnetic field established by the current through the coil. Unlike the capacitor, however, an isolated inductor cannot continue to store energy, because the absence of a closed path would cause the current to drop to zero, releasing the energy stored in the form of a magnetic field. If the series R-L circuit of Fig. 11.43 had reached steady-state conditions and the switch were quickly opened, a spark would probably occur across the contacts due to the rapid change in current from a maximum of E/R to zero amperes. The change in current di/dt of the equation $v_L = L(di/dt)$ would establish a high voltage v_L across the coil that, in conjunction with the applied voltage E, would appear across the points of the switch. This is the same mechanism as applied in the ignition system of a car to ignite the fuel in the cylinder. Some 25,000 V are generated by the rapid decrease in ignition coil current that occurs when the switch in the system is opened. (In older systems, the "points" in the distributor served as the switch.) This inductive reaction is significant when you consider that the only independent source in a car is a 12 V battery.

If opening the switch to move it to another position will cause such a rapid discharge in stored energy, how can the decay phase of an R-L circuit be analyzed in much the same manner as for the R-C circuit? The solution is to use a network such as that appearing in Fig. 11.44(a). When the switch is closed, the voltage across resistor R_2 is E volts, and the R-L branch will respond in the same manner as described above, with the same waveforms and levels. A Thévenin network of E in parallel with R_2 would simply result in the source as shown in Fig. 11.44(b), since R_2 would be shorted out by the short-circuit replacement of the voltage source E when the Thévenin resistance was determined.

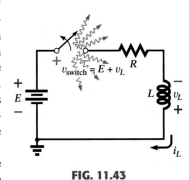

FIG. 11.43

Demonstrating why a high voltage will develop across the points of a switch in series with a coil in the steady-state mode.

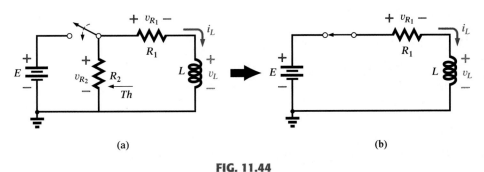

(a)　　　　　　　　　　　(b)

FIG. 11.44

(a) Initiating the storage phase for an inductor by closing the switch; (b) equivalent network for the storage phase.

After the storage phase has passed and steady-state conditions are established, the switch can be opened without the sparking effect or rapid discharge due to resistor R_2, which provides a complete path for the current i_L. In fact, for clarity the discharge path is isolated in Fig. 11.45. The voltage v_L across the inductor will reverse polarity and have a magnitude determined by

$$v_L = -(v_{R_1} + v_{R_2})$$

(11.18)

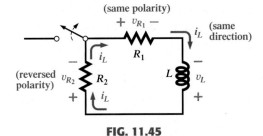

FIG. 11.45

Network of Fig. 11.44 the instant the switch is opened.

Recall that the voltage across an inductor can change instantaneously but the current cannot. The result is that the current i_L must maintain the same direction and magnitude as shown in Fig. 11.45. Therefore, the instant after the switch is opened, i_L is still $I_m = E/R_1$, and

$$v_L = -(v_{R_1} + v_{R_2}) = -(i_1 R_1 + i_2 R_2)$$

$$= -i_L(R_1 + R_2) = -\frac{E}{R_1}(R_1 + R_2) = -\left(\frac{R_1}{R_1} + \frac{R_2}{R_1}\right)E$$

and

$$\boxed{v_L = -\left(1 + \frac{R_2}{R_1}\right)E} \quad \text{switch opened}$$ (11.19)

which is bigger than E volts by the ratio R_2/R_1. In other words, when the switch is opened, the voltage across the inductor will reverse polarity and drop instantaneously from E to $-[1 + (R_2/R_1)]E$ volts.

As an inductor releases its stored energy, the voltage across the coil will decay to zero in the following manner:

$$\boxed{v_L = -V_i e^{-t/\tau'}}$$ (11.20)

with

$$V_i = \left(1 + \frac{R_2}{R_1}\right)E$$

and

$$\tau' = \frac{L}{R_T} = \frac{L}{R_1 + R_2}$$

The current will decay from a maximum of $I_m = E/R_1$ to zero.

Using Eq. (11.17):

$$I_i = \frac{E}{R_1} \quad \text{and} \quad I_f = 0 \text{ A}$$

so that

$$i_L = I_f + (I_i - I_f)e^{-t/\tau'} = 0 \text{ A} + \left(\frac{E}{R_1} - 0 \text{ A}\right)e^{-t/\tau'}$$

and

$$\boxed{i_L = \frac{E}{R_1}e^{-t/\tau'}}$$ (11.21)

with

$$\tau' = \frac{L}{R_1 + R_2}$$

The mathematical expression for the voltage across either resistor can then be determined using Ohm's law:

$$v_{R_1} = i_{R_1} R_1 = i_L R_1 = \frac{E}{R_1} R_1 e^{-t/\tau'}$$

and

$$\boxed{v_{R_1} = E e^{-t/\tau'}}$$ (11.22)

The voltage v_{R_1} has the same polarity as during the storage phase since the current i_L has the same direction. The voltage v_{R_2} is expressed as follows using the defined polarity of Fig. 11.44:

$$v_{R_2} = -i_{R_2} R_2 = -i_L R_2 = -\frac{E}{R_1} R_2 e^{-t/\tau'}$$

and

$$\boxed{v_{R_2} = -\frac{R_2}{R_1} E e^{-t/\tau'}}$$ (11.23)

EXAMPLE 11.5 Resistor R_2 was added to the network of Fig. 11.38 as shown in Fig. 11.46.

a. Find the mathematical expressions for i_L, v_L, v_{R_1}, and v_{R_2} for five time constants of the storage phase.
b. Find the mathematical expressions for i_L, v_L, v_{R_1}, and v_{R_2} if the switch is opened after five time constants of the storage phase.
c. Sketch the waveforms for each voltage and current for both phases covered by this example. Use the defined polarities of Fig. 11.45.

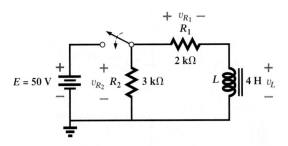

FIG. 11.46
Network to be examined in Example 11.5.

Solutions:

a. From Example 11.3:

$$i_L = \mathbf{25\ mA}(1 - e^{-t/2ms})$$

$$v_L = \mathbf{50\ V}e^{-t/2ms}$$

$$v_{R_1} = i_{R_1}R_1 = i_L R_1$$

$$= \left[\frac{E}{R_1}(1 - e^{-t/\tau})\right]R_1$$

$$= E(1 - e^{-t/\tau})$$

and
$$v_{R_1} = \mathbf{50\ V}(1 - e^{-t/2ms})$$

$$v_{R_2} = E = \mathbf{50\ V}$$

b. $\tau' = \dfrac{L}{R_1 + R_2} = \dfrac{4\ H}{2\ k\Omega + 3\ k\Omega} = \dfrac{4\ H}{5 \times 10^3\ \Omega} = 0.8 \times 10^{-3}\ s = 0.8\ ms$

By Eqs. (11.19) and (11.20):

$$V_i = \left(1 + \frac{R_2}{R_1}\right)E = \left(1 + \frac{3\ k\Omega}{2\ k\Omega}\right)(50\ V) = 125\ V$$

and
$$v_L = -V_i e^{-t/\tau'} = \mathbf{-125\ V}e^{-t/-0.8ms}$$

By Eq. (11.21):

$$I_m = \frac{E}{R_1} = \frac{50\ V}{2\ k\Omega} = 25\ mA$$

and
$$i_L = I_m e^{-t/\tau'} = \mathbf{25\ mA}e^{-t/0.8ms}$$

By Eq. (11.22):

$$v_{R_1} = E e^{-t/\tau'} = \mathbf{50\ V}e^{-t/0.8ms}$$

By Eq. (11.23):

$$v_{R_2} = -\frac{R_2}{R_1}E e^{-t/\tau'} = -\frac{3\ k\Omega}{2\ k\Omega}(50\ V)e^{-t/\tau'} = \mathbf{-75\ V}e^{-t/0.8ms}$$

c. See Fig. 11.47.

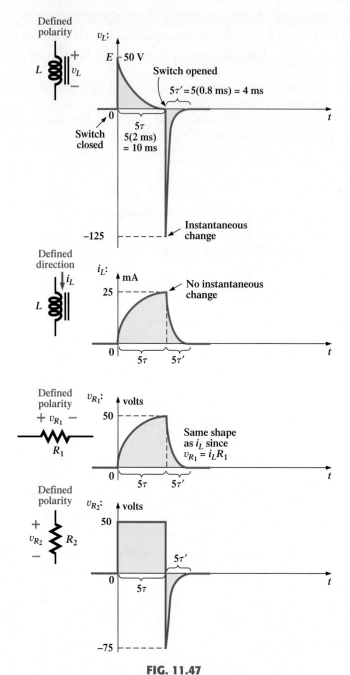

FIG. 11.47
The various voltages and the current for the network of Fig. 11.46.

In the preceding analysis, it was assumed that steady-state conditions were established during the charging phase and $I_m = E/R_1$, with $v_L = 0$ V. However, if the switch of Fig. 11.44 is opened before i_L reaches its maximum value, the equation for the decaying current of Fig. 11.44 must change to

$$i_L = I_i e^{-t/\tau'} \tag{11.24}$$

where I_i is the starting or initial current. The voltage across the coil would be defined by the following:

$$v_L = -V_i e^{-t/\tau'} \tag{11.25}$$

with $\qquad\qquad\qquad V_i = I_i(R_1 + R_2)$

11.8   THÉVENIN EQUIVALENT: $\tau = L/R_{Th}$

In Chapter 10 ("Capacitors"), we found that there are occasions when the circuit does not have the basic form of Fig. 11.33. The solution is to find the Thévenin equivalent circuit before proceeding in the manner described in this chapter. Consider the following example.

EXAMPLE 11.6   For the network of Fig. 11.48:

a. Find the mathematical expression for the transient behavior of the current i_L and the voltage v_L after the closing of the switch ($I_i = 0$ mA).
b. Draw the resultant waveform for each.

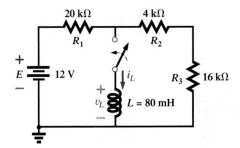

FIG. 11.48
R-L network for Example 11.6.

Solutions:

a. Applying Thévenin's theorem to the 80 mH inductor (Fig. 11.49) yields

$$R_{Th} = \frac{R}{N} = \frac{20 \text{ k}\Omega}{2} = 10 \text{ k}\Omega$$

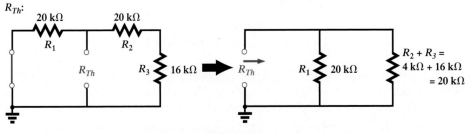

FIG. 11.49
Determining R_{Th} for the network of Fig. 11.48.

Applying the voltage divider rule (Fig. 11.50), we obtain

$$E_{Th} = \frac{(R_2 + R_3)E}{R_1 + R_2 + R_3}$$

$$= \frac{(4 \text{ k}\Omega + 16 \text{ k}\Omega)(12 \text{ V})}{20 \text{ k}\Omega + 4 \text{ k}\Omega + 16 \text{ k}\Omega} = \frac{(20 \text{ k}\Omega)(12 \text{ V})}{40 \text{ k}\Omega} = 6 \text{ V}$$

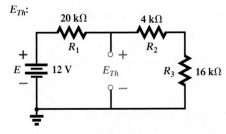

FIG. 11.50
Determining E_{Th} for the network of Fig. 11.48.

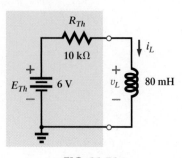

FIG. 11.51

Substituting the Thévenin equivalent circuit for the network of Fig. 11.48.

The Thévenin equivalent circuit is shown in Fig. 11.51. Using Eq. (11.13):

$$i_L = \frac{E_{Th}}{R}(1 - e^{-t/\tau})$$

$$\tau = \frac{L}{R_{Th}} = \frac{80 \times 10^{-3}\ \text{H}}{10 \times 10^3\ \Omega} = 8 \times 10^{-6}\ \text{s} = 8\ \mu\text{s}$$

$$I_m = \frac{E_{Th}}{R_{Th}} = \frac{6\ \text{V}}{10 \times 10^3\ \Omega} = 0.6 \times 10^{-3}\ \text{A} = 0.6\ \text{mA}$$

and $i_L = \mathbf{0.6\ mA(1 - e^{-t/8\mu s})}$

Using Eq. (11.15):

$$v_L = E_{Th}e^{-t/\tau}$$

so that $v_L = \mathbf{6\ Ve^{-t/8\mu s}}$

b. See Fig. 11.52.

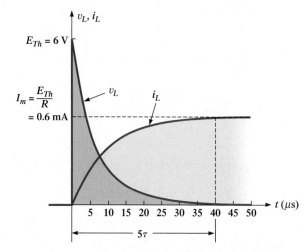

FIG. 11.52

The resulting waveforms for i_L and v_L for the network of Fig. 11.48.

EXAMPLE 11.7   Switch S_1 of Fig. 11.53 has been closed for a long time. At $t = 0$ s, S_1 is opened at the same instant that S_2 is closed to avoid an interruption in current through the coil.

a. Find the initial current through the coil. Pay particular attention to its direction.
b. Find the mathematical expression for the current i_L following the closing of switch S_2.
c. Sketch the waveform for i_L.

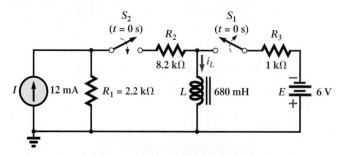

FIG. 11.53

R-L network for Example 11.7.

Solutions:

a. Using Ohm's law, the initial current through the coil is determined by

$$I_i = -\frac{E}{R_3} = -\frac{6\ \text{V}}{1\ \text{k}\Omega} = -6\ \text{mA}$$

b. Applying Thévenin's theorem:

$$R_{Th} = R_1 + R_2 = 2.2 \text{ k}\Omega + 8.2 \text{ k}\Omega = 10.4 \text{ k}\Omega$$

$$E_{Th} = IR_1 = (12 \text{ mA})(2.2 \text{ k}\Omega) = 26.4 \text{ V}$$

The Thévenin equivalent network appears in Fig. 11.54.

The steady-state current can then be determined by substituting the short-circuit equivalent for the inductor:

$$I_f = \frac{E}{R_{Th}} = \frac{26.4 \text{ V}}{10.4 \text{ k}\Omega} = 2.54 \text{ mA}$$

The time constant is

$$\tau = \frac{L}{R_{Th}} = \frac{680 \text{ mH}}{10.4 \text{ k}\Omega} = 65.39 \text{ } \mu s$$

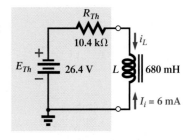

FIG. 11.54
Thévenin equivalent circuit for the network of Fig. 11.53 for $t \geq 0$ s.

Applying Eq. (11.17):

$$i_L = I_f + (I_i - I_f)e^{-t/\tau}$$

$$= 2.54 \text{ mA} + (-6 \text{ mA} - 2.54 \text{ mA})e^{-t/65.39\mu s}$$

$$= \mathbf{2.54 \text{ mA} - 8.54 \text{ mA}}e^{-t/65.39\mu s}$$

c. Note Fig. 11.55.

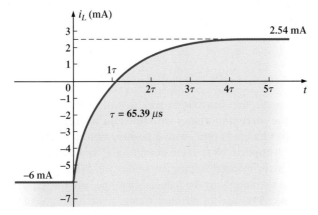

FIG. 11.55
The current i_L for the network of Fig. 11.53.

11.9 INSTANTANEOUS VALUES

The development presented in Section 10.8 for capacitive networks can also be applied to *R-L* networks to determine instantaneous voltages, currents, and time. The instantaneous values of any voltage or current can be determined by simply inserting *t* into the equation and using a calculator or table to determine the magnitude of the exponential term.

The similarity between the equations

$$v_C = V_f + (V_i - V_f)e^{-t/\tau}$$

and

$$i_L = I_f + (I_i - I_f)e^{-t/\tau}$$

results in a derivation of the following for *t* that is identical to that used to obtain Eq. (10.18):

$$t = \tau \log_e \frac{(I_i - I_f)}{(i_L - I_f)} \qquad \text{(seconds, s)} \qquad \textbf{(11.26)}$$

For the other form, the equation $v_C = Ee^{-t/\tau}$ is a close match with $v_L = Ee^{-t/\tau} = V_i e^{-t/\tau}$, permitting a derivation similar to that employed for Eq. (10.19):

$$t = \tau \log_e \frac{V_i}{v_L} \qquad \text{(seconds, s)} \qquad \textbf{(11.27)}$$

For the voltage v_R, $V_i = 0$ V and $V_f = EV$ since $v_R = E(1 - e^{-t/\tau})$. Solving for t yields

$$t = \tau \log_e\left(\frac{E}{E - v_R}\right)$$

or

$$t = \tau \log_e\left(\frac{V_f}{V_f - v_R}\right) \qquad \text{(seconds, s)} \qquad \textbf{(11.28)}$$

11.10 AVERAGE INDUCED VOLTAGE: $v_{L_{av}}$

In an effort to develop some feeling for the impact of the derivative in an equation, the average value was defined for capacitors in Section 10.10, and a number of plots for the current were developed for an applied voltage. For inductors, a similar relationship exists between the induced voltage across a coil and the current through the coil. For inductors, the average induced voltage is defined by

$$v_{L_{av}} = L\frac{\Delta i_L}{\Delta t} \qquad \text{(volts, V)} \qquad \textbf{(11.29)}$$

where Δ indicates a finite (measurable) change in current or time. Equation (11.12) for the instantaneous voltage across a coil can be derived from Eq. (11.29) by letting V_L become vanishingly small. That is,

$$v_{L_{inst}} = \lim_{\Delta t \to 0} L\frac{\Delta i_L}{\Delta t} = L\frac{di_L}{dt}$$

In the following example, the change in current Δi_L will be considered for each slope of the current waveform. **If the current increases with time, the average current is the change in current divided by the change in time, with a positive sign. If the current decreases with time, a negative sign is applied**. Note in the example that the faster the current changes with time, the greater will be the induced voltage across the coil. When making the necessary calculations, do not forget to multiply by the inductance of the coil. Larger inductances result in increased levels of induced voltage for the same change in current through the coil.

EXAMPLE 11.8 Find the waveform for the average voltage across the coil if the current through a 4 mH coil is as shown in Fig. 11.56.

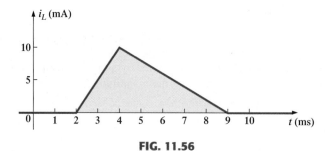

FIG. 11.56
Current i_L to be applied to a 4 mH coil in Example 11.8.

Solutions:

a. *0 to 2 ms:* Since there is no change in current through the coil, there is no voltage induced across the coil. That is,

$$v_L = L\frac{\Delta i}{\Delta t} = L\frac{0}{\Delta t} = \textbf{0 V}$$

b. *2 ms to 4 ms:*

$$v_L = L\frac{\Delta i}{\Delta t} = (4 \times 10^{-3}\,\text{H})\left(\frac{10 \times 10^{-3}\,\text{A}}{2 \times 10^{-3}\,\text{s}}\right) = 20 \times 10^{-3}\,\text{V} = \textbf{20 mV}$$

c. *4 ms to 9 ms:*

$$v_L = L\frac{\Delta i}{\Delta t} = (-4 \times 10^{-3}\,\text{H})\left(\frac{10 \times 10^{-3}\,\text{A}}{5 \times 10^{-3}\,\text{s}}\right) = -8 \times 10^{-3}\,\text{V} = \mathbf{-8\,mV}$$

d. *9 ms to ∞:*

$$v_L = L\frac{\Delta i}{\Delta t} = L\frac{0}{\Delta t} = \mathbf{0\,V}$$

The waveform for the average voltage across the coil is shown in Fig. 11.57. Note from the curve that

the voltage across the coil is not determined solely by the magnitude of the change in current through the coil (Δi), but by the rate of change of current through the coil ($\Delta i/\Delta t$).

A similar statement was made for the current of a capacitor due to a change in voltage across the capacitor.

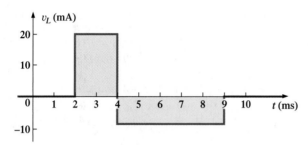

FIG. 11.57
Voltage across a 4 mH coil due to the current of Fig. 11.56.

A careful examination of Fig. 11.57 will also reveal that the area under the positive pulse from 2 ms to 4 ms equals the area under the negative pulse from 4 ms to 9 ms. In Section 11.13, we will find that the area under the curves represents the energy stored or released by the inductor. From 2 ms to 4 ms, the inductor is storing energy, whereas from 4 ms to 9 ms, the inductor is releasing the energy stored. For the full period zero to 10 ms, energy has simply been stored and released; there has been no dissipation as experienced for the resistive elements. Over a full cycle, both the ideal capacitor and inductor do not consume energy but simply store and release it in their respective forms.

11.11 INDUCTORS IN SERIES AND IN PARALLEL

Inductors, like resistors and capacitors, can be placed in series or in parallel. Increasing levels of inductance can be obtained by placing inductors in series, while decreasing levels can be obtained by placing inductors in parallel.

For inductors in series, the total inductance is found in the same manner as the total resistance of resistors in series (Fig. 11.58):

$$\boxed{L_T = L_1 + L_2 + L_3 + \cdots + L_N} \qquad \textbf{(11.30)}$$

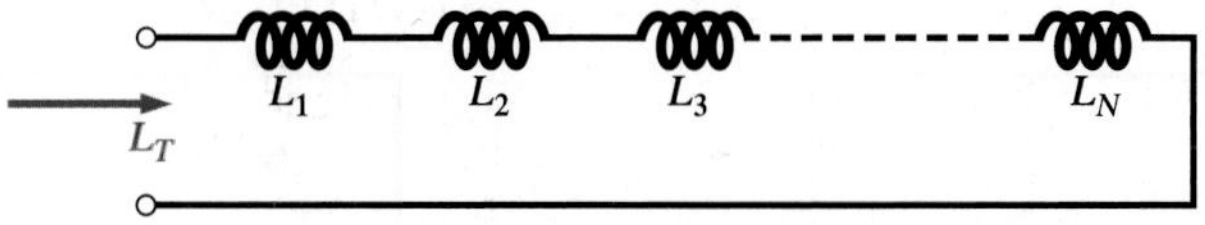

FIG. 11.58
Inductors in series.

For inductors in parallel, the total inductance is found in the same manner as the total resistance of resistors in parallel (Fig. 11.59):

$$\boxed{\frac{1}{L_T} = \frac{1}{L_1} + \frac{1}{L_2} + \frac{1}{L_3} + \cdots + \frac{1}{L_N}} \qquad \textbf{(11.31)}$$

For two inductors in parallel,

$$L_T = \frac{L_1 L_2}{L_1 + L_2}$$ **(11.32)**

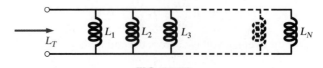

FIG. 11.59
Inductors in parallel.

EXAMPLE 11.9 Reduce the network of Fig. 11.60 to its simplest form.

Solution: Inductors L_2 and L_3 are equal in value and they are in parallel, resulting in an equivalent parallel value of

$$L'_T = \frac{L}{N} = \frac{1.2\text{ H}}{2} = 0.6\text{ H}$$

The resulting 0.6 H is then in parallel with the 1.8 H inductor, and

$$L''_T = \frac{(L'_T)(L_4)}{L'_T + L_4} = \frac{(0.6\text{ H})(1.8\text{ H})}{0.6\text{ H} + 1.8\text{ H}} = 0.45\text{ H}$$

Inductor L_1 is then in series with the equivalent parallel value, and

$$L_T = L_1 + L''_T = 0.56\text{ H} + 0.45\text{ H} = \mathbf{1.01\text{ H}}$$

The reduced equivalent network appears in Fig. 11.61.

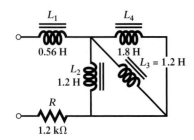

FIG. 11.60
R-L network for Example 11.9.

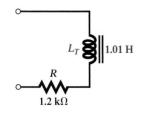

FIG. 11.61
Terminal equivalent of the network of Fig. 11.60.

11.12 STEADY-STATE CONDITIONS

We found in Section 11.5 that, for all practical purposes, an ideal (ignoring internal resistance and stray capacitances) inductor can be replaced by a short-circuit equivalent once steady-state conditions have been established. Recall that the term *steady state* implies that the voltage and current levels have reached their final resting value and will no longer change unless a change is made in the applied voltage or circuit configuration. For all practical purposes, we have assumed, and will continue to assume, that steady-state conditions have been established after five time constants of the storage or release phase have passed.

For the circuit of Fig. 11.62(a), for example, if we assume that steady-state conditions have been established, the inductor can be removed and replaced by a short-circuit equivalent as shown in Fig. 11.62(b). The short-circuit equivalent will short-out the 3 Ω resistor, and current I_1 will determined by

$$I_1 = \frac{E}{R_1} = \frac{10\text{ V}}{2\text{ Ω}} = \mathbf{5\text{ A}}$$

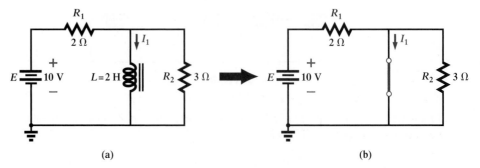

(a)   **(b)**

FIG. 11.62
Substituting the short-circuit equivalent for the inductor for t > 5τ.

For the circuit of Fig. 11.63(a), the steady-state equivalent will appear as shown in Fig. 11.63(b). This time, resistor R_1 is shorted out, and resistors R_2 and R_3 now appear in parallel. The result is

$$I = \frac{E}{R_2 \| R_3} = \frac{21 \text{ V}}{2 \text{ } \Omega} = \mathbf{10.5 \text{ A}}$$

Applying the current divider rule yields

$$I_1 = \frac{R_3 I}{R_3 + R_2} = \frac{(6 \text{ } \Omega)(10.5 \text{ A})}{6 \text{ } \Omega + 3 \text{ } \Omega} = \frac{63}{9} \text{ A} = \mathbf{7 \text{ A}}$$

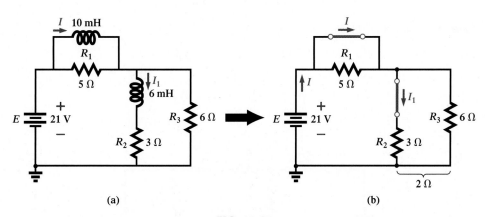

(a)   (b)

FIG. 11.63

Establishing the equivalent network for t > 5τ.

In the examples to follow, it is assumed that steady-state conditions have been established.

EXAMPLE 11.10   Find the current I_L and the voltage V_C for the network of Fig. 11.64.

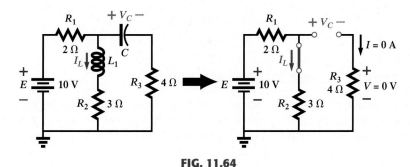

FIG. 11.64

Network for Example 11.10.

Solution:

$$I_L = \frac{E}{R_1 + R_2} = \frac{10 \text{ V}}{5 \text{ } \Omega} = \mathbf{2 \text{ A}}$$

$$V_C = \frac{R_2 E}{R_2 + R_1} = \frac{(3 \text{ } \Omega)(10 \text{ V})}{3 \text{ } \Omega + 2 \text{ } \Omega} = \mathbf{6 \text{ V}}$$

EXAMPLE 11.11   Find currents I_1 and I_2 and voltages V_1 and V_2 for the network of Fig. 11.65.

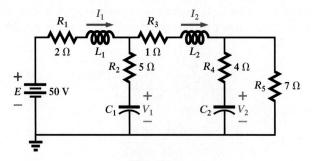

FIG. 11.65

Network for Example 11.11.

Solution: Note Fig. 11.66:

$$I_1 = I_2$$

$$= \frac{E}{R_1 + R_3 + R_5} = \frac{50 \text{ V}}{2 \text{ }\Omega + 1 \text{ }\Omega + 7 \text{ }\Omega} = \frac{50 \text{ V}}{10 \text{ }\Omega} = \mathbf{5 \text{ A}}$$

$$V_2 = I_2 R_5 = (5 \text{ A})(7 \text{ }\Omega) = \mathbf{35 \text{ V}}$$

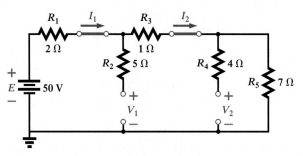

FIG. 11.66

Substituting the short-circuit equivalents for the inductors and the open-circuit equivalents for the capacitor for t > 5τ for the network of Fig. 11.65.

Applying the voltage divider rule yields

$$V_1 = \frac{(R_3 + R_5)E}{R_1 + R_3 + R_5} = \frac{(1 \text{ }\Omega + 7 \text{ }\Omega)(50 \text{ V})}{2 \text{ }\Omega + 1 \text{ }\Omega + 7 \text{ }\Omega} = \frac{(8 \text{ }\Omega)(50 \text{ V})}{10 \text{ }\Omega} = \mathbf{40 \text{ V}}$$

11.13 ENERGY STORED BY AN INDUCTOR

The ideal inductor, like the ideal capacitor, does not dissipate the electrical energy supplied to it. It stores the energy in the form of a magnetic field. A plot of the voltage, current, and power to an inductor is shown in Fig. 11.67 during the buildup of the magnetic field sur-

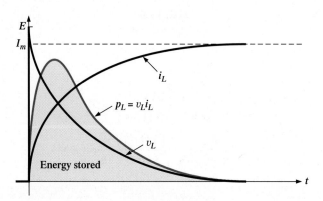

FIG. 11.67

The power curve for an inductive element under transient conditions.

rounding the inductor. The energy stored is represented by the shaded area under the power curve. Using calculus, we can show that the evaluation of the area under the curve yields

$$W_{\text{stored}} = \frac{1}{2}LI_m^2 \qquad \text{(joules, J)} \qquad \qquad \textbf{(11.33)}$$

EXAMPLE 11.12 Find the energy stored by the inductor in the circuit of Fig. 11.68 when the current through it has reached its final value.

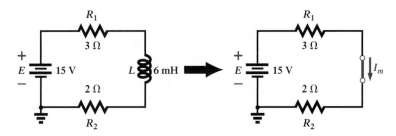

FIG. 11.68

Network for Example 11.12.

Solution:

$$I_m = \frac{E}{R_1 + R_2} = \frac{15\ \text{V}}{3\ \Omega + 2\ \Omega} = \frac{15\ \text{V}}{5\ \Omega} = 3\ \text{A}$$

$$W_{\text{stored}} = \frac{1}{2}LI_m^2 = \frac{1}{2}(6 \times 10^{-3}\ \text{H})(3\ \text{A})^2 = \frac{54}{2} \times 10^{-3}\ \text{J} = \textbf{27 mJ}$$

11.14 APPLICATIONS

Camera Flash Lamp

The inductor played an important role in the camera flash lamp circuitry described in the Application section of Chapter 10 on capacitors. For the camera it was the important component that resulted in the high spike voltage across the trigger coil, which was then magnified by the autotransformer action of the secondary to generate the 4000 V necessary to ignite the flash lamp. Recall that the capacitor in parallel with the trigger coil charged up to 300 V using the low-resistance path provided by the SCR. However, once the capacitor was fully charged, the short-circuit path to ground provided by the SCR was removed, and the capacitor immediately started to discharge through the trigger coil. Since the only resistance in the time constant for the inductive network is the relatively low resistance of the coil itself, the current through the coil grew at a very rapid rate. A significant voltage was then developed across the coil as defined by Eq. (11.12): $v_L = L(di/dt)$. This voltage was in turn increased by transformer action to the secondary coil of the autotransformer, and the flash lamp was ignited. That high voltage generated across the trigger coil will also appear directly across the capacitor of the trigger network. The result is that it will begin to charge up again until the generated voltage across the coil drops to zero volts. However, when it does drop, the capacitor will again discharge through the coil, establish another charging current through the coil, and again develop a voltage across the coil. The high-frequency exchange of energy between the coil and capacitor is called *flyback* because of the "flying back" of energy from one storage element to the other. It will begin to decay with time because of the resistive elements in the loop. The more resistance, the more quickly it will die out. If the capacitor-inductor pairing were isolated and "tickled" along the way with the application of a dc voltage, the high frequency-generated voltage across the coil could be maintained and put to good use. In fact, it is this flyback effect that is used to generate a steady dc voltage (using rectification to convert the oscillating waveform to one of a steady dc nature) that is commonly used in TVs.

Household Dimmer Switch

Inductors can be found in a wide variety of common electronic circuits in the home. The typical household dimmer uses an inductor to protect the other components and the applied load from "rush" currents—currents that increase at very high rates and often to excessively high levels. This feature is particularly important for dimmers since they are most commonly used to control the light intensity of an incandescent lamp. At "turn on," the resistance of incandescent lamps is typically very low, and relatively high currents may flow for short periods of time until the filament of the bulb heats up. The inductor is also effective in blocking high-frequency noise (RFI) generated by the switching action of the triac in the dimmer. A capacitor is also normally included from line to neutral to prevent any voltage spikes from affecting the operation of the dimmer and the applied load (lamp, etc.) and to assist with the suppression of RFI disturbances.

A photograph of one of the most common dimmers is provided in Fig. 11.69(a), with an internal view shown in 11.69(b). The basic components of most commercially available dimmers appear in the schematic of Fig. 11.69(c). In this design, a 14.5 μH inductor is used in the choking capacity described above, with a 0.068 μF capacitor for the "bypass" operation. Note the size of the inductor with its heavy wire and large ferromagnetic core and the relatively large size of the two 0.068 μF capacitors. Both suggest that they are designed to absorb high-energy disturbances.

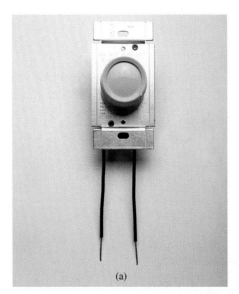

(a)

(b)

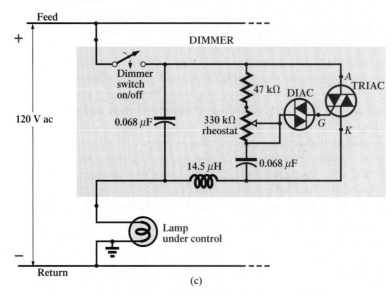

(c)

FIG. 11.69

Dimmer control: (a) external appearance; (b) internal construction; (c) schematic.

The general operation of a dimmer is shown in Fig. 11.70. The controlling network is in series with the lamp and will essentially act as an impedance (like resistance—to be introduced in Chapter 15) that can vary between very low and very high levels: very low impedance levels resembling a short circuit so that the majority of the applied voltage appears across the lamp [Fig. 11.70(a)] and very high impedances approaching open circuit where very little voltage appears across the lamp [Fig. 11.70(b)]. Intermediate levels of impedance will control the terminal voltage of the bulb accordingly. For instance, if the controlling network has a very high impedance (open-circuit equivalent) through half the cycle, as shown in Fig. 11.70(c), the brightness of the bulb will be less than full voltage but not 50% due to the non-linear relationship between the brightness of a bulb and the applied voltage. There is also a lagging effect present in the actual operation of the dimmer, but this subject will have to wait until leading and lagging networks are examined in the ac chapters.

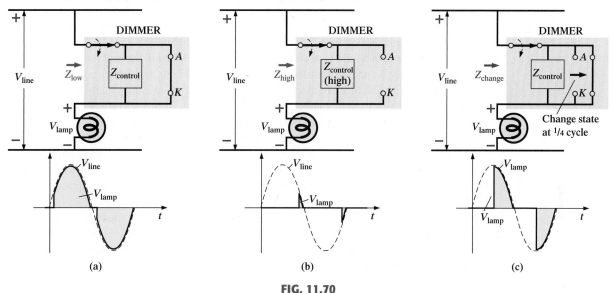

FIG. 11.70

Basic operation of the dimmer of Fig. 11.69: (a) full voltage to the lamp;
(b) approaching the cutoff point for the bulb; (c) reduced illumination of the lamp.

The controlling knob, slide, or whatever other method is used on the face of the switch to control the light intensity is connected directly to the rheostat in the branch parallel to the triac. Its setting will determine when the voltage across the capacitor reaches a sufficiently high level to turn on the diac (a bidirectional diode) and establish a voltage at the gate (G) of the triac to turn it on. When it does, it will establish a very low resistance path from the anode (A) to the cathode (K), and the applied voltage will appear directly across the lamp. During the period the SCR is off, its terminal resistance between anode and cathode will be very high and can be approximated by an open circuit. During this period the applied voltage will not reach the load (lamp). During such intervals the impedance of the parallel branch containing the rheostat, fixed resistor, and capacitor is sufficiently high compared to the load that it can also be ignored, completing the open-circuit equivalent in series with the load. Note the placement of the elements in the photograph of Fig. 11.69(b) and the fact that the metal plate to which the triac is connected is actually a heat sink for the device. The on/off switch is in the same housing as the rheostat. The total design is certainly well planned to maintain a relatively small size for the dimmer.

Since the effort here is simply to control the amount of power getting to the load, the question is often asked, Why don't we simply use a rheostat in series with the lamp? The question is best answered by examining Fig. 11.71, which shows a rather simple network with a rheostat in series with the lamp. At full wattage, a 60 W bulb on a 120 V line theoretically has an internal resistance of $R = V^2/P$ (from the equation $P = V^2/R$) = $(120 \text{ V})^2/60 \text{ W} = 240 \text{ }\Omega$. Although the resistance is sensitive to the applied voltage, we will assume this level for the following calculations.

If we consider the case where the rheostat is set for the same level as the bulb, as shown in Fig. 11.71, there will be 60 V across the rheostat and the bulb. The power to each element will then be $P = V^2/R = (60 \text{ V})^2/240 \text{ }\Omega = 15 \text{ W}$. The bulb is certainly quite dim, but the rheostat inside the dimmer switch would be dissipating 15 W of power on a continuous

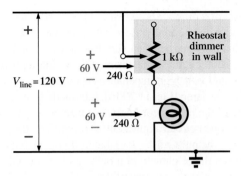

FIG. 11.71
Direct rheostat control of the brightness of a 60 W bulb.

basis. When you consider the size of a 2 W potentiometer in your laboratory, you can imagine the size rheostat you would need for 15 W, not to mention the purchase cost, although the biggest concern would probably be all the heat developed in the walls of the house. You would certainly be paying for electric power that would not be performing a useful function. Also, if you had four dimmers set at the same level, you would actually be wasting sufficient power to fully light another 60 W bulb.

On occasion, especially when the lights are set very low by the dimmer, a faint "singing" can sometimes be heard from the light bulb. This effect will sometimes occur when the conduction period of the dimmer is very small. The short, repetitive voltage pulse applied to the bulb will set the bulb into a condition that could be likened to a resonance (Chapter 17) state. The short pulses are just enough to heat up the filament and its supporting structures, and then the pulses are removed to allow the filament to cool down again for a longer period of time. This repetitive heating and cooling cycle can set the filament in motion, and the "singing" can be heard in a quiet environment. Incidentally, the longer the filament, the louder the "singing." A further condition for this effect is that the filament be in the shape of a coil and not a straight wire so that the "slinky" effect can develop.

11.15 COMPUTER ANALYSIS

PSpice

Transient *R-L* Response The computer analysis will begin with a transient analysis of the network of parallel inductive elements in Fig. 11.72. The inductors are taken from the **ANALOG** library in the **Place Part** dialog box. As noted on Fig. 11.72, the inductor is dis-

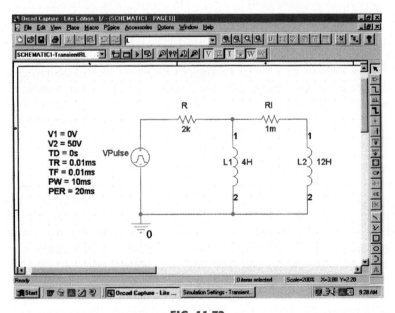

FIG. 11.72
Using PSpice to obtain the transient response of a parallel inductive network due to an applied pulse of 50 V.

played with its terminal identification which is helpful for identifying nodes when calling for specific output plots and values. In general, when an element is first placed on a schematic, the number **1** is assigned to the left end on a horizontal display and to the top on a vertical display. Similarly, the number **2** is assigned to the right end of an element in the horizontal position and to the bottom in the vertical position. Be aware, however, that the option **Rotate** rotates the element in the CCW direction, so taking a horizontal resistor to the vertical position requires three rotations to get the number **1** to the top again.

In previous chapters you may have noted that a number of the outputs were taken from terminal 2 because a single rotation placed this terminal at the top of the vertical display. Also note in Fig. 11.72 the need for a series resistor R_l within the parallel loop of inductors. In PSpice, inductors must have a series resistor to reflect real-world conditions. The chosen value of 1 mΩ is so small, however, that it will not affect the response of the system. For **VPulse,** the rise time was selected as 0.01 ms and the pulse width was chosen as 10 ms because the time constant of the network is $\tau = L_T/R = (4 \text{ H} \parallel 12 \text{ H})/2 \text{ k}\Omega = 1.5 \text{ ms}$ and $5\tau = 7.5 \text{ ms}$.

The simulation is the same as applied when obtaining the transient response of capacitive networks. In condensed form, the sequence to obtain a plot of the voltage across the coils versus time is as follows: **New SimulationProfile** key-**TransientRL-Create-TimeDomain(Transient)-Run to time:**10ms-**Start saving data after:**0s and **Maximum step size:**5 μs-**OK-Run PSpice** key-**Add Trace** key-**V1(L2)-OK.** The result is in the trace appearing in the bottom of Fig. 11.73. A maximum step size of 5 μs was chosen to ensure that it was less than the rise or fall times of 10 μs. Note that the voltage across the coil jumps to the 50 V level almost immediately; then it decays to 0 V in about 8 ms. A plot of the total current through the parallel coils can be obtained through **Plot-Plot to Window-Add Trace** key-**I(R)-OK,** resulting in the trace appearing at the top of Fig. 11.73. When the trace first appeared, the vertical scale extended from 0 A to 40 mA even though the maximum value of i_R was 25 mA. To bring the maximum value to the top of the graph, **Plot** was selected followed by **Axis Settings-Y Axis-User Defined-0A to 25mA-OK.**

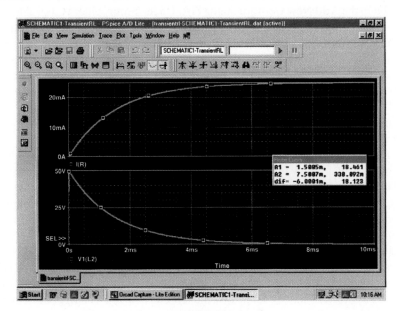

FIG. 11.73

The transient response of v_L and i_R for the network of Fig. 11.72.

For values, the voltage plot was selected, **SEL >>,** followed by the **Toggle cursor** key and a click on the screen to establish the crosshairs. The left-click cursor was set on one time constant to reveal a value of 18.461 V for **A1** (about 36.8% of the maximum as defined by the exponential waveform). The right-click cursor was set at 7.5 ms or five time constants, resulting in a relatively low value of 0.338 V for **A2**.

Transient Response with Initial Conditions The next application will verify the results of Example 11.4 which has an initial condition associated with the inductive element. **VPULSE** is again employed with the parameters appearing in Fig. 11.74. Since $\tau = L/R =$

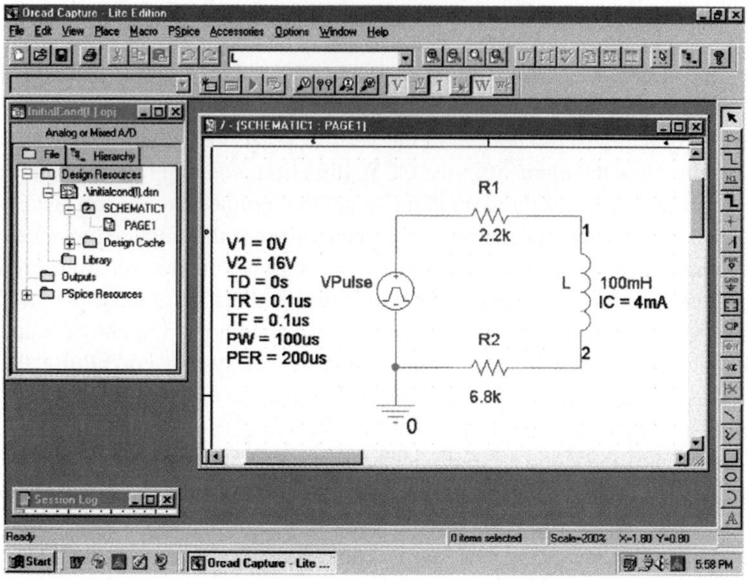

FIG. 11.74

Using PSpice to determine the transient response for a circuit in which the inductive element has an initial condition.

$100 \text{ mH}/(2.2 \text{ k}\Omega + 6.8 \text{ k}\Omega) = 100 \text{ mH}/9 \text{ k}\Omega = 11.11 \text{ } \mu\text{s}$ and $5\tau = 55.55 \text{ } \mu\text{s}$, the pulse width (**PW**) was set to 100 μs. The rise and fall times were set at $100 \text{ } \mu\text{s}/1000 = 0.1 \text{ } \mu\text{s}$. Note again that the labels **1** and **2** appear with the inductive element.

Setting the initial conditions for the inductor requires a procedure that has not been described as yet. First double-click on the inductor symbol to obtain the **Property Editor** dialog box. Then select **Parts** at the bottom of the dialog box, and select **New Column** to obtain the **Add New Column** dialog box. Under **Name,** enter **IC** (an abbreviation for "initial condition"—not "capacitive current") followed by the initial condition of 4 mA under **Value;** then click **OK.** The **Property Editor** dialog box will appear again, but now the initial condition appears as a **New Column** in the horizontal listing dedicated to the inductive element. Now select **Display** to obtain the **Display Properties** dialog box, and under **Display Format** choose **Name and Value** so that both **IC** and **4mA** will appear. Click **OK,** and we return to the **Property Editor** dialog box. Finally, click on **Apply,** and exit the dialog box (**X**). The result is the display of Fig. 11.74 for the inductive element.

Now for the simulation. First select the **New Simulation Profile** key, insert the name **InitialCond(L),** and follow up with **Create.** Then in the **Simulation Settings** dialog box, select **Time Domain(Transient)** for the **Analysis type** and **General Settings** for the **Options.** The **Run to time** should be 200 μs so that we can see the full effect of the pulse source on the transient response. The **Start saving data after** should remain at 0 s, and the **Maximum step size** should be $200 \text{ } \mu\text{s}/1000 = 200 \text{ ns}$. Click **OK** and then select the **Run PSpice** key. The result will be a screen with an *x*-axis extending from 0 to 200 μs. Selecting **Trace** to get to the **Add Traces** dialog box and then selecting **I(L)** followed by **OK** will result in the display of Fig. 11.75. The plot for **I(L)** clearly starts at the initial value of 4 mA and then decays to 1.78 mA as defined by the left-click cursor. The right-click cursor reveals that the current has dropped to 0.222 μA (essentially 0 A) after the pulse source has dropped to 0 V for 100 μs. The **VPulse** source was placed in the same figure through **Plot-Add Plot to Window-Trace-Add Trace-V(VPulse:+)-OK** to permit a comparison between the applied voltage and the resulting inductor current.

Electronics Workbench (EWB)

The transient response of an *R-L* network can also be obtained using EWB. The circuit to be examined appears in Fig. 11.76 with a pulse voltage source to simulate the closing of a switch at $t = 0$ s. The source, referred to as **PULSE_VOLTAGE_SOURCE** in the **Source** listing, is the near the bottom left of the **Sources** parts bin. When selected, it will appear with a label, an initial voltage, a step voltage, and a frequency. All can be changed

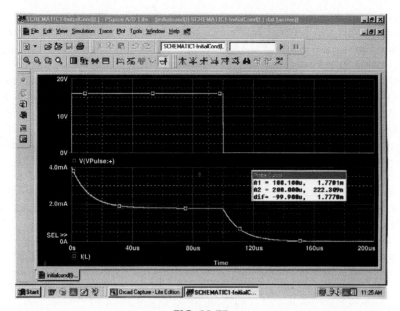

FIG. 11.75

A plot of the applied pulse and resulting current for the circuit of Fig. 11.73.

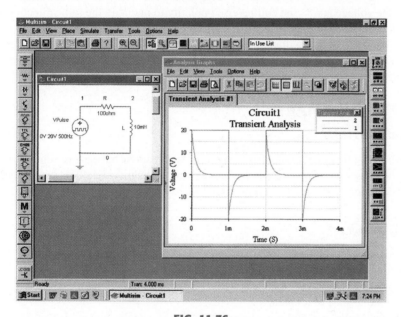

FIG. 11.76

Using EWB to obtain the transient response for an inductive circuit.

by simply double-clicking on the source symbol to obtain the **Pulse Voltage** dialog box. The **Pulse Value** will be set at 20 V as shown in Fig. 11.76, and the **Delay Time** to 0 s. The **Rise Time** and **Fall Time** will both remain at the default levels of 1 ns. For our analysis we want a **Pulse Width** that is at least twice the 5τ transient period of the circuit. For the chosen values of R and L, $\tau = L/R = 10$ mH/100 Ω = 0.1 ms = 100 μs. The transient period of 5τ is therefore 500 μs or 0.5 ms. Thus, a **Pulse Width** of 1 ms would seem appropriate with a **Period** of 2 ms. The result is a frequency of $f = 1/T = 1/2$ ms = 500 Hz. When all have been set and selected, the parameters of the pulse source will appear as shown in Fig. 11.76. Next the resistor, inductor, and ground are placed on the screen to complete the circuit.

This time we will want to see the node names so that we can call for them when we set up the simulation process. This is accomplished through **Options-Preferences-Show node names.** In this case we have two—one at the positive terminal of the supply (**1**) and the other at the top end of the inductor (**2**) representing the voltage across the inductor.

The simulation process is initiated by the following sequence: **Simulate-Analyses-Transient Analysis.** The result is the **Transient Analysis** dialog box in which **Analysis**

Parameters is chosen first. Under **Parameters,** use 0 s as the **Start time** and 4 ms as the **End time** so that we get two full cycles of the applied voltage. After enabling the **Maximum time step settings (TMAX),** we set the **Minimum number of time points** at 1000 to obtain a reasonably good plot during the rapidly changing transient period.

Next, the **Output variables** section must be selected and the program told which voltage and current levels we are interested in. On the left side of the dialog box is a list of **Variables** that have been defined for the circuit. On the right is a list of **Selected variables for analysis.** In between you see a **Plot during simulation** or **Remove.** To move a variable from the left to the right column, simply select it in the left column and choose **Plot during simulation.** It will then appear in the right column. For our purposes it seems appropriate that we plot both the applied voltage and the voltage across the coil, so **1** and **2** were moved to the right column. Then **Simulate** is selected, and a window titled **Analysis Graphs** will appear with the selected plots as shown in Fig. 11.76. Click on the **Show/Hide Grid** key (a red grid on a black axis), and the grid lines will appear. Then selecting the **Show/Hide Legend** key on the immediate right will result in the small **Transient Anal** dialog box that will identify the color that goes with each nodal voltage. In our case, blue is the color of the applied voltage, and red is the color of the voltage across the coil.

The source voltage appears as expected with its transition to 20 V, 50% duty cycle, and the period of 2 ms. The voltage across the coil jumped immediately to the 20 V level and then began its decay to 0 V in about 0.5 ms as predicted. When the source voltage dropped to zero, the voltage across the coil reversed polarity to maintain the same direction of current in the inductive circuit. Remember that for a coil, the voltage can change instantaneously, but the inductor will choke any instantaneous change in current. By reversing its polarity, the voltage across the coil ensures the same polarity of voltage across the resistor and therefore the same direction of current through the coil and circuit.

CHAPTER SUMMARY
MAGNETIC FIELD

- Magnetic flux lines do not have origins or terminations but exist in continuous loops.
- The denser the magnetic flux lines, the stronger the magnetic field in that region.
- The energy delivered to an inductor is stored in the form of a magnetic field.

INDUCTANCE

- Inductance is a measure of an inductor's ability to store energy in the form of a magnetic field.
- Using cores of ferromagnetic materials with high levels of permeability will increase the inductance of an inductor.
- The larger the permeability, the number of turns, and the core area of an inductor, and the shorter its length, the greater the inductance.
- In general, the larger the inductor, the greater the inductance level. The smallest inductors are typically in picohenries (pH), and the largest in millihenries (mH).

TRANSIENTS

- In a dc network, the current of an inductor is essentially equal to the maximum steady-state value after five time constants of the storage phase have passed.
- The voltage across an inductor is essentially zero volts after five time constants of the storage phase have passed.
- An ideal inductor can be replaced by a short-circuit equivalent once the storage phase has passed.
- The current through a coil cannot change instantaneously.
- Thévenin's theorem can be very helpful in defining the transient response for an inductor in a series-parallel dc network.
- The voltage across a coil is directly related to the rate of change of current through the coil.

- The total inductance of inductors in series is found in the same manner as the total resistance of resistors in series, whereas the total inductance of inductors in parallel is found in the same manner as the total resistance of resistors in parallel.

Important Equations

$$B = \frac{\Phi}{A}$$

$$\mathscr{F} = NI$$

$$L = \frac{\mu N^2 A}{l} = \frac{\mu_o \mu_r N^2 A}{l} = \frac{4\pi \times 10^{-7} \mu_r N^2 A}{l}$$

$$L = \mu_r L_o$$

Storage phase:

$$\tau = \frac{L}{R}$$

$$i_L = \frac{E}{R}(1 - e^{-t/\tau}) \qquad v_L = Ee^{-t/\tau} \qquad v_R = E(1 - e^{-t/\tau})$$

Release phase:

$$v_L = -V_i e^{-t/\tau'} \qquad V_i = \left(1 + \frac{R_2}{R_1}\right)E \qquad \tau' = \frac{L}{R_1 + R_2}$$

$$i_L = \frac{E}{R_1}e^{-t/\tau'}$$

$$v_{R_1} = Ee^{-t/\tau'} \qquad v_{R_2} = -\frac{R_2}{R_1}Ee^{-t/\tau'}$$

$$v_L = L\frac{di_L}{dt} \qquad v_{L_{av}} = L\frac{\Delta i_L}{\Delta t}$$

Series inductors:

$$L_T = L_1 + L_2 + L_3 + \cdots + L_N$$

Parallel inductors:

$$\frac{1}{L_T} = \frac{1}{L_1} + \frac{1}{L_2} + \frac{1}{L_3} + \cdots + \frac{1}{L_N}$$

$$L_T = \frac{L_1 L_2}{L_1 + L_2}$$

Energy:

$$W_{stored} = \frac{1}{2}LI_m^2$$

PROBLEMS

SECTION 11.2 The Magnetic Field

1. For the electromagnet of Fig. 11.77:
   a. Find the flux density in Wb/m^2.
   b. What is the flux density in teslas?
   c. What is the applied magnetomotive force?
   d. What would the reading of the meter of Fig. 11.14 read in gauss?

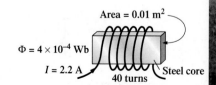

Area = 0.01 m²
$\Phi = 4 \times 10^{-4}$ Wb
$I = 2.2$ A
40 turns
Steel core

FIG. 11.77
Problem 1.

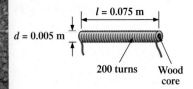

FIG. 11.78
Problems 2 and 3.

300 turns $A = 1.5 \times 10^{-4} \text{ m}^2$

Air core

$l = 0.1$ m

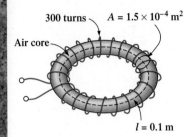

FIG. 11.79
Problem 4.

2. For the inductor of Fig. 11.78, find the inductance L in henries.

3. Repeat Problem 2 with $l = 4$ in. and $d = \frac{1}{4}$ in.

4. For the inductor of Fig. 11.79, find the inductance L in henries.

5. An air-core inductor has a total inductance of 5 mH.
   a. What is the inductance if the only change is to increase the number of turns by a factor of three?
   b. What is the inductance if the only change is to increase the length by a factor of three?
   c. What is the inductance if the area is doubled, the length cut in half, and the number of turns doubled?
   d. What is the inductance if the area, length, and number of turns are cut in half and a ferro-magnetic core with a μ_r of 1500 is inserted?

6. What are the inductance and the range of expected values for an inductor with the following label?
   a. 123J
   b. 47K

SECTION 11.4   The Voltage v_L

7. If the flux linking a coil of 50 turns changes at a rate of 0.085 Wb/s, what is the induced voltage across the coil?

8. Determine the rate of change of flux linking a coil if 20 V are induced across a coil of 40 turns.

9. How many turns does a coil have if 42 mV are induced across the coil by a change in flux of 0.003 Wb/s?

10. Find the voltage induced across a coil of 5 H if the rate of change of current through the coil is:
    a. 0.5 A/s.
    b. 60 mA/s.
    c. 0.04 A/ms.

11. Find the induced voltage across a 50 mH inductor if the current through the coil changes at a rate of 0.1 mA/μs.

SECTION 11.5   R-L Transients: The Storage Phase

12. For the circuit of Fig. 11.80:
    a. Determine the time constant.
    b. Write the mathematical expression for the current i_L after the switch is closed.
    c. Repeat part (b) for v_L and v_R.
    d. Determine i_L and v_L at one, three, and five time constants.
    e. Sketch the waveforms of i_L, v_L, and v_R.

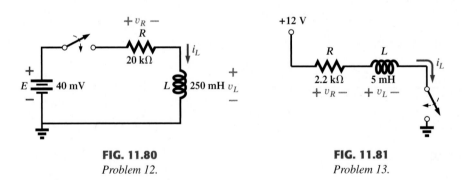

FIG. 11.80
Problem 12.

FIG. 11.81
Problem 13.

13. For the circuit of Fig. 11.81:
    a. Determine τ.
    b. Write the mathematical expression for the current i_L after the switch is closed at $t = 0$ s.
    c. Write the mathematical expression for v_L and v_R after the switch is closed at $t = 0$ s.
    d. Determine i_L and v_L at $t = 1\tau$, 3τ, and 5τ.
    e. Sketch the waveforms of i_L, v_L, and v_R for the storage phase.

*14. For the circuit of Fig. 11.82:
    a. Determine the time constant.
    b. Write the mathematical expression for the voltage v_L and the current i_L using the defined polarities and direction.
    c. Sketch the waveforms of v_L and i_L.

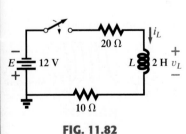

FIG. 11.82
Problem 14.

15. For the circuit of Fig. 11.83:
   a. Write the mathematical expressions for the current i_L and the voltage v_L following the closing of the switch. Note the magnitude and the direction of the initial current.
   b. Sketch the waveform of i_L and v_L for the entire period from initial value to steady-state level.

16. In this problem the effect of reversing the initial current will be investigated. The circuit of Fig. 11.84 is the same as that appearing in Fig. 11.83, with the only change being the direction of the initial current.
   a. Write the mathematical expressions for the current i_L and the voltage v_L following the closing of the switch. Take careful note of the defined polarity for v_L and the direction for i_L.
   b. Sketch the waveform of i_L and v_L for the entire period from initial value to steady-state level.
   c. Compare the results with those of Problem 15.

17. For the network of Fig. 11.85:
   a. Write the mathematical expressions for the current i_L and the voltage v_L following the closing of the switch. Note the magnitude and the direction of the initial current.
   b. Sketch the waveform of i_L and v_L for the entire period from initial value to steady-state level.

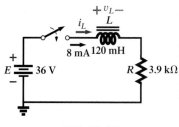

FIG. 11.83
Problems 15 and 42.

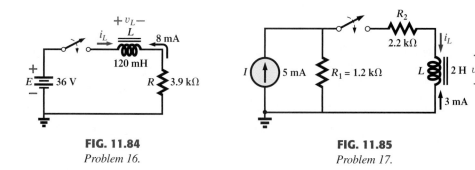

FIG. 11.84
Problem 16.

FIG. 11.85
Problem 17.

SECTION 11.7   *R-L* Transients: The Release Phase

18. For the network of Fig. 11.86:
   a. Determine the mathematical expressions for the current i_L and the voltage v_L when the switch is closed.
   b. Repeat part (a) if the switch is opened after a period of five time constants has passed.
   c. Sketch the waveforms of parts (a) and (b) on the same set of axes.

*19. For the network of Fig. 11.87:
   a. Determine the mathematical expressions for the current i_L and the voltage v_L following the closing of the switch.
   b. Repeat part (a) if the switch is opened at $t = 1\ \mu s$.
   c. Sketch the waveforms of parts (a) and (b) on the same set of axes.

*20. For the network of Fig. 11.88:
   a. Write the mathematical expression for the current i_L and the voltage v_L following the closing of the switch.
   b. Determine the mathematical expressions for i_L and v_L if the switch is opened after a period of five time constants has passed.
   c. Sketch the waveforms of i_L and v_L for the time periods defined by parts (a) and (b).
   d. Sketch the waveform for the voltage across R_2 for the same period of time encompassed by i_L and v_L. Take careful note of the defined polarities and directions of Fig. 11.88.

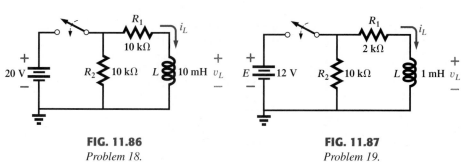

FIG. 11.86
Problem 18.

FIG. 11.87
Problem 19.

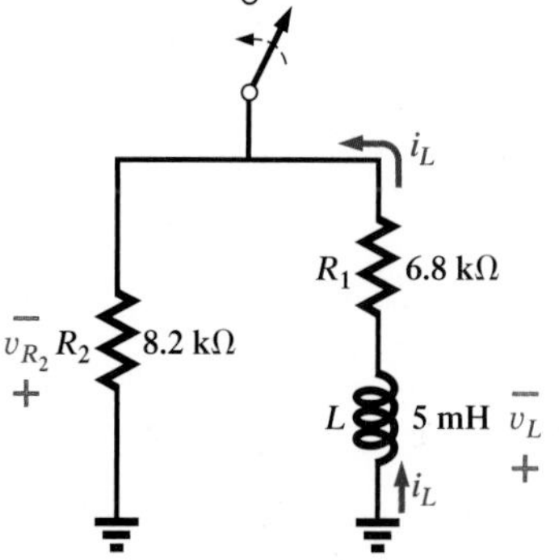

FIG. 11.88
Problem 20.

SECTION 11.8 Thévenin Equivalent: $\tau = L/R_{Th}$

21. For Fig. 11.89:
  a. Determine the mathematical expressions for i_L and v_L following the closing of the switch.
  b. Determine i_L and v_L after one time constant.

22. For Fig. 11.90:
  a. Determine the mathematical expressions for i_L and v_L following the closing of the switch.
  b. Determine i_L and v_L at $t = 100$ ns.

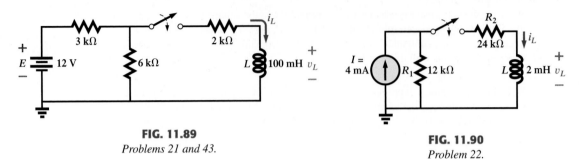

FIG. 11.89
Problems 21 and 43.

FIG. 11.90
Problem 22.

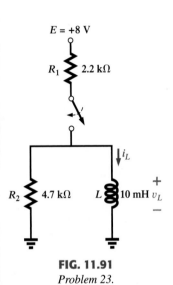

FIG. 11.91
Problem 23.

***23.** For Fig. 11.91:
  a. Determine the mathematical expressions for i_L and v_L following the closing of the switch.
  b. Calculate i_L and v_L at $t = 10\ \mu s$.
  c. Write the mathematical expressions for the current i_L and the voltage v_L if the switch is opened at $t = 10\ \mu s$.
  d. Sketch the waveforms of i_L and v_L for parts (a) and (c).

***24.** For the network of Fig. 11.92, the switch is closed at $t = 0$ s.
  a. Determine v_L at $t = 25$ ms.
  b. Find v_L at $t = 1$ ms.
  c. Calculate v_{R_1} at $t = 1\tau$.
  d. Find the time required for the current i_L to reach 100 mA.

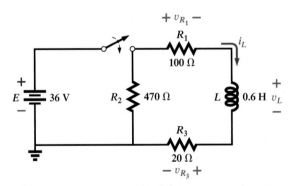

FIG. 11.92
Problem 24.

***25.** The switch in Fig. 11.93 has been open for a long time. It is then closed at $t = 0$ s.
  a. Write the mathematical expression for the current i_L and the voltage v_L after the switch is closed.
  b. Sketch the waveform of i_L and v_L from the initial value to the steady-state level.

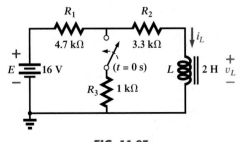

FIG. 11.93
Problem 25.

*26. The switch of Fig. 11.94 has been closed for a long time. It is then opened at $t = 0$ s.
   a. Write the mathematical expression for the current i_L and the voltage v_L after the switch is opened.
   b. Sketch the waveform of i_L and v_L from initial value to the steady-state level.

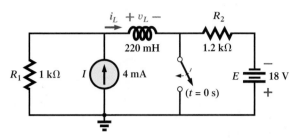

FIG. 11.94
Problem 26.

SECTION 11.9   Instantaneous Values

27. Given $i_L = 100$ mA$(1 - e^{-t/20 \text{ ms}})$:
   a. Determine i_L at $t = 1$ ms.
   b. Determine i_L at $t = 100$ ms.
   c. Find the time t when i_L will equal 50 mA.
   d. Find the time t when i_L will equal 99 mA.

28. The network of Fig. 11.95 employs a DMM with an internal resistance of 10 MΩ in the voltmeter mode. The switch is closed at $t = 0$ s.
   a. Find the voltage across the coil the instant after the switch is closed.
   b. What is the final value of the current i_L?
   c. How much time must pass before i_L reaches 10 μA?
   d. What is the voltmeter reading at $t = 12$ μs?

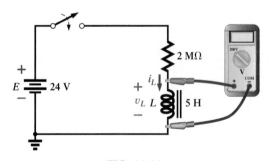

FIG. 11.95
Problem 28.

SECTION 11.10   Average Induced Voltage: $v_{L_{\text{av}}}$

29. Find the waveform for the voltage induced across a 200 mH coil if the current through the coil is as shown in Fig. 11.96.

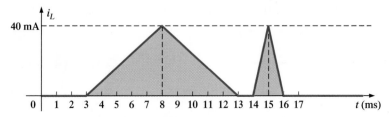

FIG. 11.96
Problem 29.

30. Find the waveform for the voltage induced across a 5 mH coil if the current through the coil is as shown in Fig. 11.97.

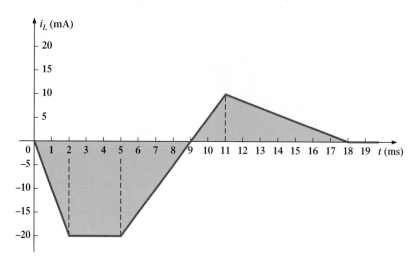

FIG. 11.97
Problem 30.

*31. Find the waveform for the current of a 10 mH coil if the voltage across the coil follows the pattern of Fig. 11.98. The current i_L is 4 mA at $t = 0$ s.

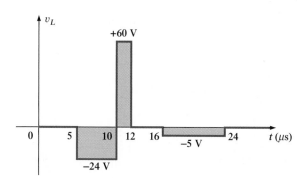

FIG. 11.98
Problem 31.

SECTION 11.11   Inductors in Series and in Parallel

32. Find the total inductance of the circuits of Fig. 11.99.

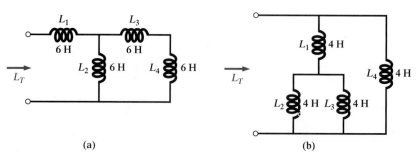

(a)   (b)

FIG. 11.99
Problem 32.

33. Reduce the networks of Fig. 11.100 to the fewest elements.

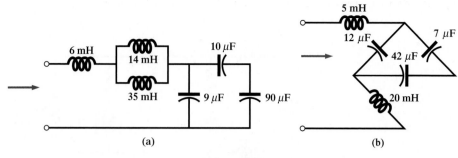

FIG. 11.100
Problem 33.

*34. For the network of Fig. 11.101:
  a. Write the mathematical expressions for the voltages v_L and v_R and the current i_L if the switch is closed at $t = 0$ s.
  b. Sketch the waveforms of v_L, v_R, and i_L.

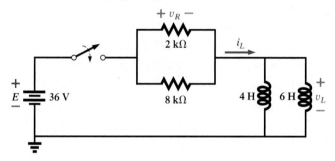

FIG. 11.101
Problem 34.

*35. For the network of Fig. 11.102:
  a. Write the mathematical expressions for the voltage v_L and the current i_L if the switch is closed at $t = 0$ s.
  b. Sketch the waveforms of v_L and i_L.

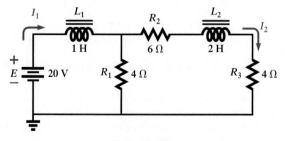

FIG. 11.102
Problem 35.

SECTION 11.12 Steady-State Conditions

36. Find the steady-state currents I_1 and I_2 for the network of Fig. 11.103.

37. Find the steady-state currents and voltages for the network of Fig. 11.104.

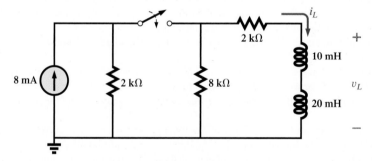

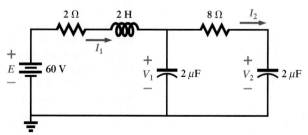

FIG. 11.103
Problem 36.

FIG. 11.104
Problem 37.

38. Find the steady-state currents and voltages for the network of Fig. 11.105.

39. Find the steady-state currents and voltages for the network of Fig. 11.106.

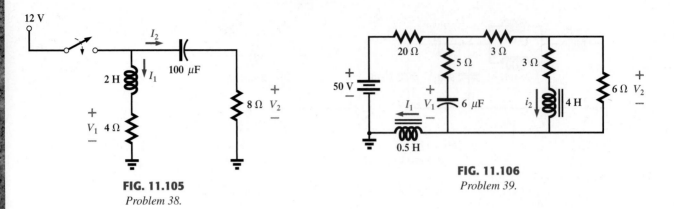

FIG. 11.105
Problem 38.

FIG. 11.106
Problem 39.

SECTION 11.15   Computer Analysis

40. Using PSpice or Electronics Workbench, verify the results of Example 11.3.

41. Using PSpice or Electronics Workbench, verify the results of Example 11.4.

42. Using PSpice or Electronics Workbench, find the solution to Problem 15.

43. Using PSpice or Electronics Workbench, find the solution to Problem 21.

44. Using PSpice or Electronics Workbench, verify the results of Example 11.8.

GLOSSARY

Ampère's circuital law   A law establishing the fact that the algebraic sum of the rises and drops of the mmf around a closed loop of a magnetic circuit is equal to zero.

Choke   A term often applied to an inductor, due to the ability of an inductor to resist a change in current through it.

Diamagnetic materials   Materials that have permeabilities slightly less than that of free space.

Domain   A group of magnetically aligned atoms.

Electromagnetism   Magnetic effects introduced by the flow of charge or current.

Faraday's law   A law relating the voltage induced across a coil to the number of turns in the coil and the rate at which the flux linking the coil is changing.

Ferromagnetic materials   Materials having permeabilities hundreds and thousands of times greater than that of free space.

Flux density (B)   A measure of the flux per unit area perpendicular to a magnetic flux path. It is measured in teslas (T) or webers per square meter (Wb/m^2).

Hysteresis   The lagging effect between the flux density of a material and the magnetizing force applied.

Inductance (L)   A measure of the ability of a coil to oppose any change in current through the coil and to store energy in the form of a magnetic field in the region surrounding the coil.

Inductor (coil)   A fundamental element of electrical systems constructed of numerous turns of wire around a ferromagnetic core or an air core.

Lenz's law   A law stating that an induced effect is always such as to oppose the cause that produced it.

Magnetic flux lines   Lines of a continuous nature that reveal the strength and direction of a magnetic field.

Magnetizing force (H)   A measure of the magnetomotive force per unit length of a magnetic circuit.

Magnetomotive force (mmf) ($\mathscr{F}$)   The "pressure" required to establish magnetic flux in a ferromagnetic material. It is measured in ampere-turns (At).

Paramagnetic materials   Materials that have permeabilities slightly greater than that of free space.

Permanent magnet   A material such as steel or iron that will remain magnetized for long periods of time without the aid of external means.

Permeability (μ)   A measure of the ease with which magnetic flux can be established in a material. It is measured in Wb/Am.

Relative permeability (μ_r)   The ratio of the permeability of a material to that of free space.

Reluctance ($\mathscr{R}$)   A quantity determined by the physical characteristics of a material that will provide an indication of the "reluctance" of that material to the setting up of magnetic flux lines in the material. It is measured in rels or At/Wb.

Sinusoidal Alternating Waveforms

OBJECTIVES

- Become familiar with the characteristics of a sinusoidal waveform.
- Be able to determine the phase relationship between two sinusoidal waveforms.
- Understand how to use the oscilloscope to make phase measurements.
- Be able to find the average value of a waveform.
- Become familiar with the effective (rms) value of a sinusoidal voltage or current.
- Understand the use of instruments to measure ac quantities.
- Become familiar with the use of the PSpice and EWB software packages for the analysis of ac networks.

12.1 INTRODUCTION

The analysis thus far has been limited to dc networks, networks in which the currents or voltages are fixed in magnitude except for transient effects. We will now turn our attention to the analysis of networks in which the magnitude of the source varies in a set manner. Of particular interest is the time-varying voltage that is commercially available in large quantities and is commonly called the *ac voltage.* (The letters *ac* are an abbreviation for *alternating current.*) To be absolutely rigorous, the terminology *ac voltage* or *ac current* is not sufficient to describe the type of signal we will be analyzing. Each waveform of Fig. 12.1 is an **alternating waveform** available from commercial supplies. The term *alternating* indicates only that the waveform alternates between two prescribed levels in a set time sequence (Fig. 12.1). To be absolutely correct, the term *sinusoidal, square-wave,* or *triangular* must also be applied.

The pattern of particular interest is the **sinusoidal ac waveform** for the voltage of Fig. 12.1. Since this type of signal is encountered in the vast majority of instances, the abbreviated phrases *ac voltage* and *ac current* are commonly applied without confusion. For the other patterns of Fig. 12.1, the descriptive term is always present, but frequently the *ac* abbreviation is dropped, resulting in the designation *square-wave* or *triangular* waveforms.

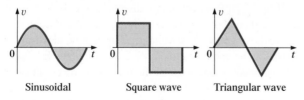

Sinusoidal   Square wave   Triangular wave

FIG. 12.1

Alternating waveforms.

One of the important reasons for concentrating on the sinusoidal ac voltage is that it is the voltage generated by utilities throughout the world. Other reasons include its application throughout electrical, electronic, communication, and industrial systems. In addition, the chapters to follow will reveal that the waveform itself has a number of characteristics that will result in a unique response when it is applied to basic electrical elements. The wide range of theorems and methods introduced for dc networks will also be applied to sinusoidal ac systems. Although the application of sinusoidal signals will raise the required math level, once the notation given in Chapter 13 is understood, **most of the concepts introduced in the dc chapters can be applied to ac networks with a minimum of added difficulty.**

12.2 SINUSOIDAL ac VOLTAGE CHARACTERISTICS AND DEFINITIONS

Generation

Sinusoidal ac voltages are available from a variety of sources. The most common source is the typical home outlet, which provides an ac voltage that originates at a power plant; such a power plant is most commonly fueled by water power, oil, gas, or nuclear fusion. In each case, an **ac generator** (also called an *alternator*), as shown in Fig. 12.2(a), is the primary component in the energy-conversion process. The power to the shaft developed by one of the energy sources listed will turn a *rotor* (constructed of alternating magnetic poles) inside a set of windings housed in the *stator* (the stationary part of the dynamo) and will induce a voltage across the windings of the stator, as defined by Faraday's law:

$$e = N\frac{d\phi}{dt}$$

Through proper design of the generator, a sinusoidal ac voltage is developed that can be transformed to higher levels for distribution through the power lines to the consumer. For isolated locations where power lines have not been installed, portable ac generators [Fig. 12.2(b)] are available that run on gasoline. As in the larger power plants, however, an ac generator is an integral part of the design.

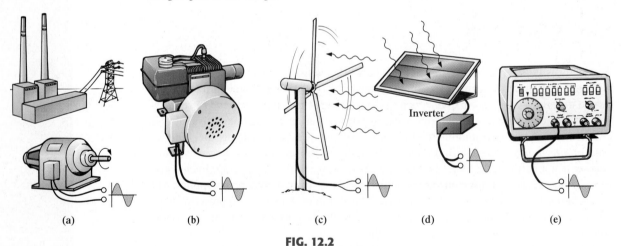

(a)	(b)	(c)	(d)	(e)

FIG. 12.2

Various sources of ac power: (a) generating plant; (b) portable ac generator; (c) wind-power station; (d) solar panel; (e) function generator.

In an effort to conserve our natural resources, wind power and solar energy are receiving increasing interest from various districts of the world that have such energy sources available in level and duration that make the conversion process viable. The turning propellers of the wind-power station [Fig. 12.2(c)] are connected directly to the shaft of an ac generator to provide the ac voltage described above. Through light energy absorbed in the form of *photons,* solar cells [Fig. 12.2(d)] can generate dc voltages. Through an electronic package called an *inverter,* the dc voltage can be converted to one of a sinusoidal nature. Boats, recreational vehicles (RVs), and so on, make frequent use of the inversion process in isolated areas.

Sinusoidal ac voltages with characteristics that can be controlled by the user are available from **function generators,** such as the one in Fig. 12.2(e). By setting the various switches and controlling the position of the knobs on the face of the instrument, you can make available sinusoidal voltages of different peak values and different repetition rates.

The function generator plays an integral role in the investigation of the variety of theorems, methods of analysis, and topics to be introduced in the chapters that follow.

Definitions

The sinusoidal waveform of Fig. 12.3 with its additional notation will now be used as a model in defining a few basic terms. These terms, however, can be applied to any alternating waveform. It is important to remember, as you proceed through the various definitions, that the vertical scaling is in volts or amperes and the horizontal scaling is in units of time.

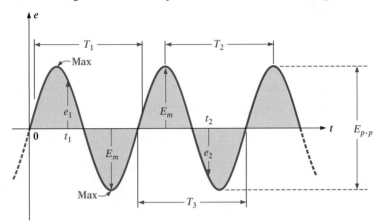

FIG. 12.3
Important parameters for a sinusoidal voltage.

Waveform: **The path traced by a quantity,** such as the voltage in Fig. 12.3, plotted as a function of some variable such as time (as above), position, degrees, radians, temperature, and so on.

Instantaneous value: The magnitude of a waveform at **any instant of time;** denoted by lowercase letters (e_1, e_2 in Fig. 12.3).

Peak amplitude: **The maximum value** of a waveform as measured from its *average,* or *mean,* value, denoted by uppercase letters [such as E_m (Fig. 12.3) for sources of voltage and V_m for the voltage drop across a load]. For the waveform of Fig. 12.3, the average value is zero volts, and E_m is as defined by the figure.

Peak value: **The maximum instantaneous value** of a function as measured from the zero volt level. For the waveform of Fig. 12.3, the peak amplitude and peak value are the same, since the average value of the function is zero volts.

Peak-to-peak value: Denoted by $E_{p\text{-}p}$ or $V_{p\text{-}p}$ (as shown in Fig. 12.3), **the full voltage between positive and negative peaks** of the waveform, that is, the sum of the magnitude of the positive and negative peaks.

Periodic waveform: A waveform that continually **repeats itself** after the same time interval. The waveform of Fig. 12.3 is a periodic waveform.

Period (T): **The time interval between successive repetitions** of a periodic waveform (the period $T_1 = T_2 = T_3$ in Fig. 12.3), as long as successive *similar points* of the periodic waveform are used in determining T.

Cycle: The portion of a waveform contained in **one period of time.** The cycles within T_1, T_2, and T_3 of Fig. 12.3 may appear different in Fig. 12.4, but they are all bounded by one period of time and therefore satisfy the definition of a cycle.

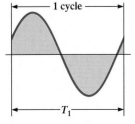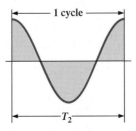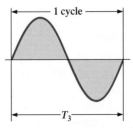

FIG. 12.4
Defining the cycle and period of a sinusoidal waveform.

Frequency (f): The number of cycles that occur in 1 s. The frequency of the waveform of Fig. 12.5(a) is 1 cycle per second, and for Fig. 12.5(b), 2½ cycles per second. If a waveform of similar shape had a period of 0.5 s [Fig. 12.5(c)], the frequency would be 2 cycles per second.

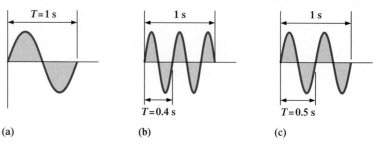

(a)   (b)   (c)

FIG. 12.5

Demonstrating the effect of a changing frequency on the period of a sinusoidal waveform.

The unit of measure for frequency is the *hertz* (Hz), where

$$1 \text{ hertz (Hz)} = 1 \text{ cycle per second (c/s)} \qquad \textbf{(12.1)}$$

The unit hertz is derived from the surname of Heinrich Rudolph Hertz (Fig. 12.6), who did original research in the area of alternating currents and voltages and their effect on the basic *R*, *L*, and *C* elements. The frequency standard for North America is 60 Hz, whereas for Europe it is predominantly 50 Hz.

As with all standards, any variation from the norm will cause difficulties. In 1993, Berlin, Germany, received all its power from eastern plants, whose output frequency was varying between 50.03 Hz and 51 Hz. The result was that clocks were gaining as much as 4 minutes a day. Alarms went off too soon, VCRs clicked off before the end of the program, and so on, requiring that clocks be continually reset. In 1994, however, when power was linked with the rest of Europe, the precise standard of 50 Hz was reestablished and everyone was on time again.

Courtesy of the Smithsonian Institution, Photo No. 66,606.

FIG. 12.6
Heinrich Rudolph Hertz.

German (Hamburg, Berlin, Karlsruhe)
(1857–94)
Physicist
Professor of Physics, Karlsruhe Polytechnic and University of Bonn

Spurred on by the earlier predictions of the English physicist James Clerk Maxwell, Heinrich Hertz produced *electromagnetic waves* in his laboratory at the Karlsruhe Polytechnic while in his early 30s. The rudimentary *transmitter* and *receiver* were in essence the first to broadcast and receive radio waves. He was able to measure the *wavelength* of the electromagnetic waves and confirmed that the *velocity of propagation* is in the same order of magnitude as light. In addition, he demonstrated that the *reflective* and *refractive* properties of electromagnetic waves are the same as those for heat and light waves. It was indeed unfortunate that such an ingenious, industrious individual should pass away at the very early age of 37 due to a bone disease.

EXAMPLE 12.1 For the sinusoidal waveform of Fig. 12.7:

a. What is the peak value?
b. What is the instantaneous value at 0.3 s and 0.6 s?
c. What is the peak-to-peak value of the waveform?
d. What is the period of the waveform?
e. How many cycles are shown?
f. What is the frequency of the waveform?

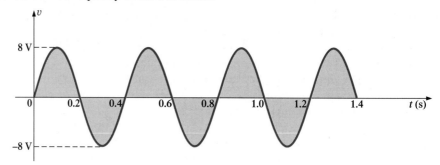

FIG. 12.7
Example 12.1.

Solutions:

a. **8 V.**
b. At 0.3 s, **−8 V;** at 0.6 s, **0 V.**
c. **16 V.**
d. **0.4 s.**
e. **3.5 cycles.**
f. **2.5 cycles/second,** or **2.5 Hz.**

12.3 FREQUENCY SPECTRUM

Using a log scale (to be described in detail in Chapter 17), a frequency spectrum from 1 Hz to 1000 GHz can be scaled off on the same axis, as shown in Fig. 12.8. A number of terms in the various spectrums are probably familiar to you from everyday experiences. Note that the audio range (human ear) extends from only 15 Hz to 20 kHz, but the transmission of radio signals can occur between 3 kHz and 300 GHz. The uniform process of defining the intervals of the radio-frequency spectrum from VLF to EHF is quite evident from the length

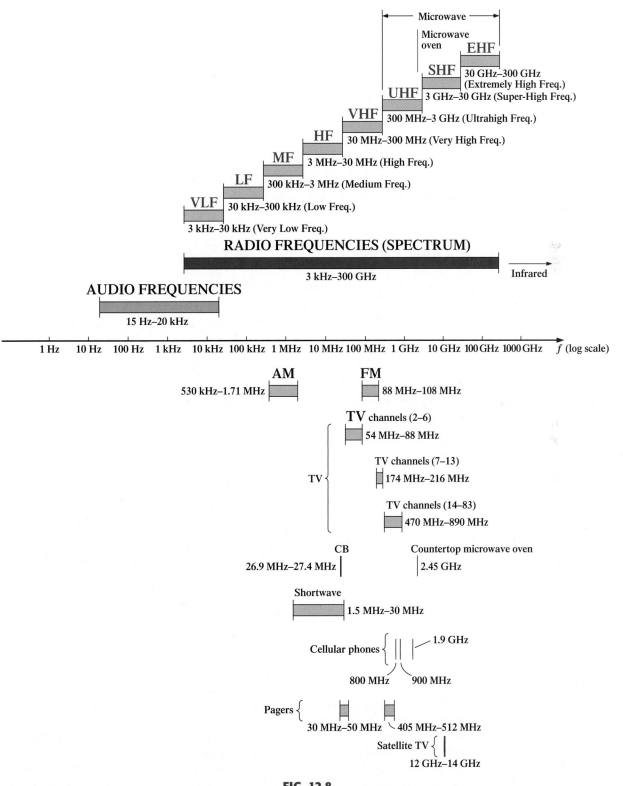

FIG. 12.8

Areas of application for specific frequency bands.

of the bars in the figure (although keep in mind that it is a log scale, so the frequencies encompassed within each segment are quite different). Other frequencies of particular interest (TV, CB, microwave, and so on), are also included for reference purposes. Although it is numerically easy to talk about frequencies in the megahertz and gigahertz range, keep in mind that a frequency of 100 MHz, for instance, represents a sinusoidal waveform that passes through 100,000,000 cycles in only 1 s—an incredible number when we compare it to the 60 Hz of our conventional power sources. The Intel® Pentium® 4 chip manufactured by Intel can run at speeds of 2 GHz. Imagine a product able to handle 2 billion instructions per second—an incredible achievement.

Since the frequency is inversely related to the period—that is, as one increases, the other decreases by an equal amount—the two can be related by the following equation:

$$f = \frac{1}{T} \qquad \begin{array}{l} f = \text{Hz} \\ T = \text{seconds (s)} \end{array}$$

(12.2)

or

$$T = \frac{1}{f}$$

(12.3)

EXAMPLE 12.2 Find the period of a periodic waveform with a frequency of

a. 60 Hz.
b. 1000 Hz.

Solutions:

a. $T = \dfrac{1}{f} = \dfrac{1}{60 \text{ Hz}} \cong 0.01667 \text{ s or } \textbf{16.67 ms}$

(a recurring value since 60 Hz is so prevalent)

b. $T = \dfrac{1}{f} = \dfrac{1}{1000 \text{ Hz}} = 10^{-3} \text{ s} = \textbf{1 ms}$

EXAMPLE 12.3 Determine the frequency of the waveform of Fig. 12.9.

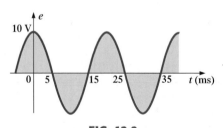

FIG. 12.9
Example 12.3.

Solution: From the figure, $T = (25 \text{ ms} - 5 \text{ ms}) = 20 \text{ ms}$, and

$$f = \frac{1}{T} = \frac{1}{20 \times 10^{-3} \text{ s}} = \textbf{50 Hz}$$

In Fig. 12.10, the seismogram resulting from a seismometer near an earthquake is displayed. Prior to the disturbance, the waveform has a relatively steady level, but as the event is about to occur, the frequency begins to increase along with the amplitude. Finally, the earthquake occurs, and **the frequency and the amplitude increase dramatically.** In other words, the relative frequencies can be determined simply by looking at the tightness of the waveform and the associated period. The change in amplitude is immediately obvious from the resulting waveform. The fact that the earthquake lasts for only a few minutes is clear from the horizontal scale.

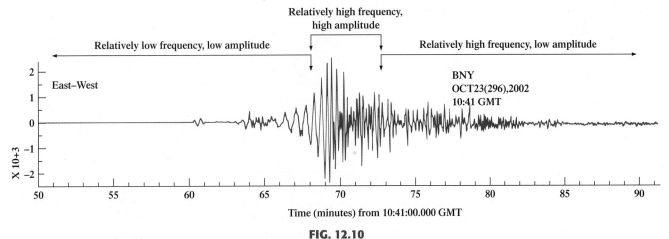

Relatively high frequency, high amplitude

Relatively low frequency, low amplitude

Relatively high frequency, low amplitude

East–West

BNY
OCT23(296),2002
10:41 GMT

X 10+3

Time (minutes) from 10:41:00.000 GMT

FIG. 12.10

*Seismogram from station BNY (Binghamton University) in New York due to magnitude 6.7 earthquake
in Central Alaska that occurred at 63.62°N, 148.04°W, with a depth of 10 km, on Wednesday, October 23, 2002.*

Defined Polarities and Direction

You may be wondering how a polarity for a voltage or a direction for a current can be established if the waveform moves back and forth from the positive to the negative region. For a period of time, a voltage will have one polarity, while for the next equal period it will reverse. To take care of this problem, a positive sign is applied if the voltage is above the axis, as shown in Fig. 12.11(a). For a current source the direction in the symbol corresponds with the positive region of the waveform, as shown in Fig. 12.11(b).

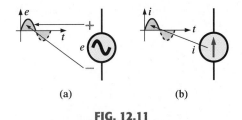

(a)   (b)

FIG. 12.11

(a) Sinusoidal ac voltage sources; (b) sinusoidal current sources.

For any quantity that will not change with time, an uppercase letter such as V or I will be employed. For expressions that are time dependent or that represent a particular instant of time, a lowercase letter such as e or i will be used.

The need for defining polarities and current direction will become quite obvious when we consider multisource ac networks. Note in the last sentence the absence of the term *sinusoidal* before the phrase *ac networks*. This phrase will be used to an increasing degree as we progress; *sinusoidal* is to be understood unless otherwise indicated.

12.4 THE SINUSOIDAL WAVEFORM

The terms defined in the previous section can be applied to any type of periodic waveform, whether smooth or discontinuous. The sinusoidal waveform is of particular importance, however, since it lends itself readily to the mathematics and the physical phenomena associated with electric circuits. Consider the power of the following statement:

***The sinusoidal waveform is the only alternating waveform whose shape is unaffected
by the response characteristics of R, L, and C elements.***

In other words, if the voltage across (or current through) a resistor, coil, or capacitor is sinusoidal in nature, the resulting current (or voltage, respectively) for each will also have sinusoidal characteristics, as shown in Fig. 12.12. If any other alternating waveform such as a square wave or a triangular wave were applied, such would not be the case.

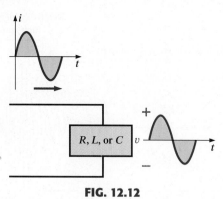

FIG. 12.12

The sine wave is the only alternating waveform whose shape is not altered by the response characteristics of a pure resistor, inductor, or capacitor.

The unit of measurement for the horizontal axis can be **time** (as appearing in the figures thus far), **degrees,** or **radians.** *Degrees* is a familiar term, but the measure *radian* needs to be defined. If we mark off a portion of the circumference of a circle by a length equal to the radius of the circle, as shown in Fig. 12.13, the angle resulting is called **1 radian.** The result is

$$1 \text{ rad} = 57.296° \cong 57.3° \qquad (12.4)$$

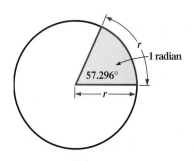

FIG. 12.13

Defining the radian.

where 57.3° is the usual approximation applied.

In one full circle there are 2π radians, as shown in Fig. 12.14. That is,

$$2\pi \text{ rad} = 360° \qquad (12.5)$$

so that $2\pi = 2(3.142) = 6.28$

and $2\pi(57.3°) = 6.28(57.3°) = 359.84° \cong 360°$

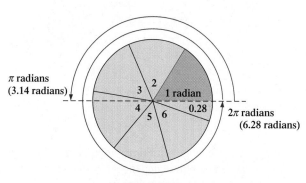

FIG. 12.14

There are 2π radians in one full circle of 360°.

A number of electrical formulas contain a multiplier of π. For this reason, it is sometimes preferable to measure angles in radians rather than in degrees.

The quantity π is the ratio of the circumference of a circle to its diameter.

π has been determined to an extended number of places, primarily in an attempt to see if a repetitive sequence of numbers appears. It does not. A sampling of the effort appears below:

$$\pi = 3.14159 \ 26535 \ 89793 \ 23846 \ 26433 \ \dots$$

Although the approximation $\pi \cong 3.14$ is often applied, all the calculations in this text will use the π function as provided on all scientific calculators.

For 180° and 360°, the two units of measurement are related as shown in Fig. 12.14. The conversion equations between the two are the following:

$$\text{Radians} = \left(\frac{\pi}{180°}\right) \times (\text{degrees}) \qquad (12.6)$$

$$\text{Degrees} = \left(\frac{180°}{\pi}\right) \times (\text{radians}) \qquad (12.7)$$

Applying these equations, we find

$$\textbf{90°:} \quad \text{Radians} = \frac{\pi}{180°}(90°) = \frac{\pi}{2} \textbf{ rad}$$

$$\textbf{30°:} \quad \text{Radians} = \frac{\pi}{180°}(30°) = \frac{\pi}{6} \textbf{ rad}$$

$$\frac{\pi}{3} \textbf{ rad:} \quad \text{Degrees} = \frac{180°}{\pi}\left(\frac{\pi}{3}\right) = \textbf{60°}$$

$$\frac{3\pi}{2} \textbf{ rad:} \quad \text{Degrees} = \frac{180°}{\pi}\left(\frac{3\pi}{2}\right) = \textbf{270°}$$

For comparison purposes, two sinusoidal voltages were plotted in Fig. 12.15 using degrees and radians as the units of measurement for the horizontal axis.

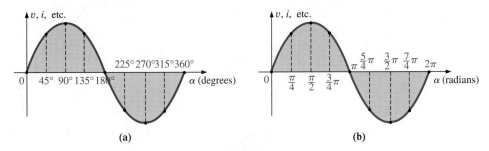

FIG. 12.15

Plotting a sine wave versus (a) degrees and (b) radians.

It is of particular interest that the sinusoidal waveform can be derived from the length of the *vertical projection* of a radius vector rotating in a uniform circular motion about a fixed point. Starting as shown in Fig. 12.16(a) and plotting the amplitude (above and below zero) on the coordinates drawn to the right [Figs. 12.16(b) through (i)], we will trace a complete sinusoidal waveform after the radius vector has completed a 360° rotation about the center.

The velocity with which the radius vector rotates about the center, called the **angular velocity,** can be determined from the following equation:

$$\text{Angular velocity} = \frac{\text{distance (degrees or radians)}}{\text{time (seconds)}} \qquad (12.8)$$

Substituting into Eq. (12.8) and assigning the Greek letter omega (ω) to the angular velocity, we have

$$\omega = \frac{\alpha}{t} \qquad (12.9)$$

and

$$\alpha = \omega t \qquad (12.10)$$

Since ω is typically provided in radians per second, the angle α obtained using Eq. (12.10) is usually in radians. If α is required in degrees, Equation (12.7) must be applied. The importance of remembering the above will become obvious in the examples to follow.

In Fig. 12.16, the time required to complete one revolution is equal to the period (T) of the sinusoidal waveform of Fig. 12.16(i). The radians subtended in this time interval are 2π. Substituting, we have

$$\omega = \frac{2\pi}{T} \qquad \text{(rad/s)} \qquad (12.11)$$

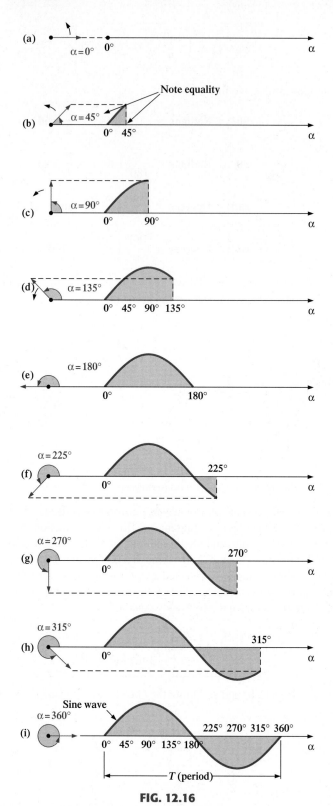

FIG. 12.16

Generating a sinusoidal waveform through the vertical projection of a rotating vector.

In words, this equation states that the smaller the period of the sinusoidal waveform of Fig. 12.16(i), or the smaller the time interval before one complete cycle is generated, the greater must be the angular velocity of the rotating radius vector. Certainly this statement agrees with what we have learned thus far. We can now go one step further and apply the fact that the frequency of the generated waveform is inversely related to the period of the waveform; that is, $f = 1/T$. Thus,

$$\omega = 2\pi f \quad \text{(rad/s)} \tag{12.12}$$

This equation states that the higher the frequency of the generated sinusoidal waveform, the higher must be the angular velocity. Equations (12.11) and (12.12) are verified somewhat by Fig. 12.17, where for the same radius vector, $\omega = 100$ rad/s and 500 rad/s.

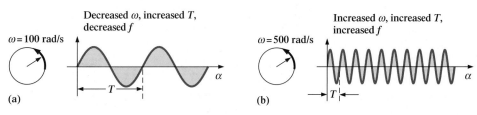

FIG. 12.17

Demonstrating the effect of ω on the frequency and period.

EXAMPLE 12.4 Determine the angular velocity of a sine wave having a frequency of 60 Hz.

Solution:

$$\omega = 2\pi f = (2\pi)(60\ \text{Hz}) \cong \mathbf{377\ rad/s}$$

(a recurring value due to 60 Hz predominance)

EXAMPLE 12.5 Determine the frequency and period of the sine wave of Fig. 12.17(b).

Solution: Since $\omega = 2\pi/T$,

$$T = \frac{2\pi}{\omega} = \frac{2\pi\ \text{rad}}{500\ \text{rad/s}} = \frac{2\pi\ \text{rad}}{500\ \text{rad/s}} = \mathbf{12.57\ ms}$$

and

$$f = \frac{1}{T} = \frac{1}{12.57 \times 10^{-3}\ \text{s}} = \mathbf{79.58\ Hz}$$

EXAMPLE 12.6 Given $\omega = 200$ rad/s, determine how long it will take the sinusoidal waveform to pass through an angle of 90°.

Solution: Eq. (12.10): $\alpha = \omega t$, and

$$t = \frac{\alpha}{\omega}$$

However, α must be substituted as $\pi/2$ (=90°) since ω is in radians per second:

$$t = \frac{\alpha}{\omega} = \frac{\pi/2\ \text{rad}}{200\ \text{rad/s}} = \frac{\pi}{400}\ \text{s} = \mathbf{7.85\ ms}$$

EXAMPLE 12.7 Find the angle through which a sinusoidal waveform of 60 Hz will pass in a period of 5 ms.

Solution: Eq. (12.11): $\alpha = \omega t$, or

$$\alpha = 2\pi f t = (2\pi)(60\ \text{Hz})(5 \times 10^{-3}\ \text{s}) = \mathbf{1.885\ rad}$$

If not careful, you might be tempted to interpret the answer as 1.885°. However,

$$\alpha\ (°) = \frac{180°}{\pi\ \text{rad}}(1.885\ \text{rad}) = \mathbf{108°}$$

12.5 GENERAL FORMAT FOR THE SINUSOIDAL VOLTAGE OR CURRENT

The basic mathematical format for the sinusoidal waveform is

$$\boxed{A_m \sin \alpha} \tag{12.13}$$

where A_m is the peak value of the waveform and α is the unit of measure for the horizontal axis, as shown in Fig. 12.18.

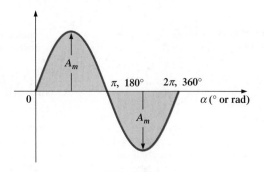

FIG. 12.18

Basic sinusoidal function.

The equation $\alpha = \omega t$ states that the angle α through which the rotating vector of Fig. 12.16 will pass is determined by the angular velocity of the rotating vector and the length of time the vector rotates. For example, for a particular angular velocity (fixed ω), the longer the radius vector is permitted to rotate (that is, the greater the value of t), the greater will be the number of degrees or radians through which the vector will pass. Relating this statement to the sinusoidal waveform, for a particular angular velocity, the longer the time, the greater the number of cycles shown. For a fixed time interval, the greater the angular velocity, the greater the number of cycles generated.

Due to Eq. (12.10), the general format of a sine wave can also be written

$$\boxed{A_m \sin \omega t} \tag{12.14}$$

with ωt as the horizontal unit of measure.

For electrical quantities such as current and voltage, the general format is

$$i = I_m \sin \omega t = I_m \sin \alpha$$

$$e = E_m \sin \omega t = E_m \sin \alpha$$

where the capital letters with the subscript m represent the amplitude, and the lowercase letters i and e represent the instantaneous value of current or voltage, respectively, at any time t. This format is particularly important because it presents the sinusoidal voltage or current as a function of time, which is the horizontal scale for the oscilloscope. Recall that the horizontal sensitivity of a scope is in time per division, not degrees per centimeter.

EXAMPLE 12.8   Given $e = 5 \sin \alpha$, determine e at $\alpha = 40°$ and $\alpha = 0.8\pi$.

Solution: For $\alpha = 40°$,

$$e = 5 \sin 40° = 5(0.6428) = \textbf{3.214 V}$$

For $\alpha = 0.8\pi$,

$$\alpha\,(°) = \frac{180°}{\pi}(0.8\pi) = 144°$$

and

$$e = 5 \sin 144° = 5(0.5878) = \textbf{2.939 V}$$

The angle at which a particular voltage level is attained can be determined by rearranging the equation

$$e = E_m \sin \alpha$$

in the following manner:

$$\sin \alpha = \frac{e}{E_m}$$

which can be written

$$\alpha = \sin^{-1} \frac{e}{E_m} \qquad\qquad \textbf{(12.15)}$$

Similarly, for a particular current level,

$$\alpha = \sin^{-1} \frac{i}{I_m} \qquad\qquad \textbf{(12.16)}$$

EXAMPLE 12.9

a. Determine the angle at which the magnitude of the sinusoidal function $v = 10 \sin 377t$ is 4 V.
b. Determine the time at which the magnitude is attained.

Solutions:

a. Eq. (12.15):

$$\alpha_1 = \sin^{-1} \frac{v}{E_m} = \sin^{-1} \frac{4\ \text{V}}{10\ \text{V}} = \sin^{-1} 0.4 = \textbf{23.578°}$$

However, Figure 12.19 reveals that the magnitude of 4 V (positive) will be attained at two points between 0° and 180°. The second intersection is determined by

$$\alpha_2 = 180° - 23.578° = \textbf{156.422°}$$

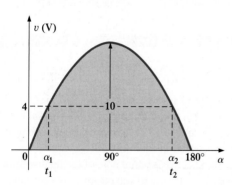

FIG. 12.19
Example 12.9.

In general, therefore, keep in mind that Equations (12.15) and (12.16) will provide an angle with a magnitude between 0° and 90°.

b. Eq. (12.10): $\alpha = \omega t$, and so $t = \alpha/\omega$. However, α must be in radians. Thus,

$$\alpha\ (\text{rad}) = \frac{\pi}{180°}(23.578°) = 0.412\ \text{rad}$$

and

$$t_1 = \frac{\alpha}{\omega} = \frac{0.412\ \text{rad}}{377\ \text{rad/s}} = \textbf{1.09 ms}$$

For the second intersection,

$$\alpha\ (\text{rad}) = \frac{\pi}{180°}(156.422°) = 2.73\ \text{rad}$$

$$t_2 = \frac{\alpha}{\omega} = \frac{2.73\ \text{rad}}{377\ \text{rad/s}} = \textbf{7.24 ms}$$

Calculator Operations

Both sin and $\sin^{-1}$ are available on all scientific calculators. You can also use them to work with the angle in degrees or radians without having to make a conversion from one form to the other. That is, if the angle is in radians and the mode setting is for radians, you can enter the radian measure directly.

In Example 12.8, the voltage at an angle of 40° is determined by the calculator sequence of Fig. 12.20 **(with the mode set for angles in degrees).**

FIG. 12.20

Finding e = 5 sin 40° using the calculator in the degree mode.

Setting **the mode to radians** will permit the sequence of Fig. 12.21 to determine the voltage at 0.8π.

FIG. 12.21

Finding e = 5 sin 0.8π using the calculator in the radian mode.

Finally, the angle in degrees for α_1 in part (a) of Example 12.9 can be determined by the sequence of Fig. 12.22 with the mode set in degrees, whereas the angle in radians for part (a) of Example 12.9 can be determined by the sequence of Fig. 12.23 with the mode set in radians.

FIG. 12.22

Finding $\alpha_1 = sin^{-1}(4/10)$ using the calculator in the degree mode.

FIG. 12.23

Finding $\alpha_1 = sin^{-1}(4/10)$ using the calculator in the radian mode.

The sinusoidal waveform can also be plotted against *time* on the horizontal axis. The time period for each interval can be determined from $t = \alpha/\omega$, but the most direct route is simply to find the period T from $T = 1/f$ and break it up into the required intervals. This latter technique will be demonstrated in Example 12.10.

Before reviewing the example, take special note of the relative simplicity of the mathematical equation that can represent a sinusoidal waveform. Any alternating waveform whose characteristics differ from those of the sine wave cannot be represented by a single term, but may require two, four, six, or perhaps an infinite number of terms to be represented accurately.

EXAMPLE 12.10   Sketch $e = 10 \sin 314t$ with the abscissa

a. angle (α) in degrees.
b. angle (α) in radians.
c. time (t) in seconds.

Solutions:

a. See Fig. 12.24. (Note that no calculations are required.)

b. See Fig. 12.25. (Once the relationship between degrees and radians is understood, there is again no need for calculations.)

c. See Fig. 12.26.

$$360°: \quad T = \frac{2\pi}{\omega} = \frac{2\pi}{314} = 20 \text{ ms}$$

$$180°: \quad \frac{T}{2} = \frac{20 \text{ ms}}{2} = 10 \text{ ms}$$

$$90°: \quad \frac{T}{4} = \frac{20 \text{ ms}}{4} = 5 \text{ ms}$$

$$30°: \quad \frac{T}{12} = \frac{20 \text{ ms}}{12} = 1.67 \text{ ms}$$

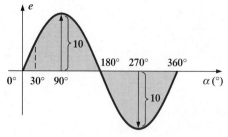

FIG. 12.24

Example 12.10, horizontal axis in degrees.

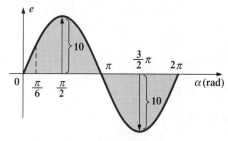

FIG. 12.25

Example 12.10, horizontal axis in radians.

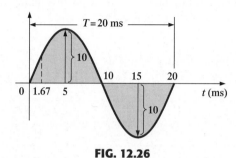

FIG. 12.26

Example 12.10, horizontal axis in milliseconds.

EXAMPLE 12.11   Given $i = 6 \times 10^{-3} \sin 1000t$, determine i at $t = 2$ ms.

Solution:

$$\alpha = \omega t = 1000t = (1000 \text{ rad/s})(2 \times 10^{-3} \text{ s}) = 2 \text{ rad}$$

$$\alpha \, (°) = \frac{180°}{\pi \text{ rad}}(2 \text{ rad}) = 114.59°$$

$$i = (6 \times 10^{-3})(\sin 114.59°) = (6 \text{ mA})(0.9093) = \mathbf{5.46 \text{ mA}}$$

12.6   PHASE RELATIONS

Thus far, we have considered only sine waves that have maxima at $\pi/2$ and $3\pi/2$, with a zero value at 0, π, and 2π, as shown in Fig. 12.25. If the waveform is shifted to the right or left of $0°$, the expression becomes

$$A_m \sin(\omega t \pm \theta) \tag{12.17}$$

where θ is the angle in degrees or radians that the waveform has been shifted.

If the waveform passes through the horizontal axis with a *positive-going* (increasing with time) slope *before* $0°$, as shown in Fig. 12.27, the expression is

$$A_m \sin(\omega t + \theta) \tag{12.18}$$

At $\omega t = \alpha = 0°$, the magnitude is determined by $A_m \sin \theta$. If the waveform passes through the horizontal axis with a positive-going slope *after* $0°$, as shown in Fig. 12.28, the expression is

$$A_m \sin(\omega t - \theta) \tag{12.19}$$

Finally, at $\omega t = \alpha = 0°$, the magnitude is $A_m \sin(-\theta)$, which, by a trigonometric identity, is $-A_m \sin \theta$.

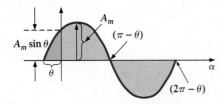

FIG. 12.27

Defining the phase shift for a sinusoidal function that crosses the horizontal axis with a positive slope before 0°.

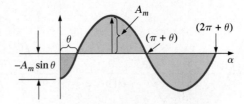

FIG. 12.28

Defining the phase shift for a sinusoidal function that crosses the horizontal axis with a positive slope after 0°.

If the waveform crosses the horizontal axis with a positive-going slope 90° ($\pi/2$) sooner, as shown in Fig. 12.29, it is called a *cosine wave;* that is,

$$\sin(\omega t + 90°) = \sin\left(\omega t + \frac{\pi}{2}\right) = \cos \omega t \qquad \textbf{(12.20)}$$

or

$$\sin \omega t = \cos(\omega t - 90°) = \cos\left(\omega t - \frac{\pi}{2}\right) \qquad \textbf{(12.21)}$$

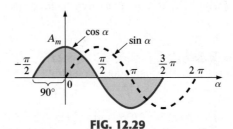

FIG. 12.29

Phase relationship between a sine wave and a cosine wave.

The terms **leading** and **lagging** are used to indicate the relationship between two sinusoidal waveforms of the **same frequency** plotted on the same set of axes. In Fig. 12.29, the cosine curve is said to *lead* the sine curve by 90°, and the sine curve is said to *lag* the cosine curve by 90°. The 90° is referred to as the phase angle between the two waveforms. In language commonly applied, the waveforms are *out of phase* by 90°. Note that the phase angle between the two waveforms is measured between those two points on the horizontal axis through which each passes with the **same slope.** If both waveforms cross the axis at the same point with the same slope, they are *in phase.*

The geometric relationship between various forms of the sine and cosine functions can be derived from Fig. 12.30. For instance, starting at the $+\sin \alpha$ position, we find that $+\cos \alpha$ is an additional 90° in the counterclockwise direction. Therefore, $\cos \alpha = \sin(\alpha + 90°)$. For $-\sin \alpha$ we must travel 180° in the counterclockwise (or clockwise) direction so that $-\sin \alpha = \sin(\alpha \pm 180°)$, and so on, as listed below:

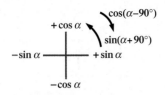

FIG. 12.30

Graphic tool for finding the relationship between specific sine and cosine functions.

$$\begin{aligned} \cos \alpha &= \sin(\alpha + 90°) \\ \sin \alpha &= \cos(\alpha - 90°) \\ -\sin \alpha &= \sin(\alpha \pm 180°) \\ -\cos \alpha &= \sin(\alpha + 270°) = \sin(\alpha - 90°) \\ &\text{etc.} \end{aligned} \qquad \textbf{(12.22)}$$

In addition, note that

$$\begin{aligned} \sin(-\alpha) &= -\sin \alpha \\ \cos(-\alpha) &= \cos \alpha \end{aligned} \qquad \textbf{(12.23)}$$

500

If a sinusoidal expression appears as

$$e = -E_m \sin \omega t$$

the negative sign is associated with the sine portion of the expression, not the peak value E_m. In other words, the expression, if not for convenience, would be written

$$e = E_m(-\sin \omega t)$$

Since

$$-\sin \omega t = \sin(\omega t \pm 180°)$$

the expression can also be written

$$e = E_m \sin(\omega t \pm 180°)$$

revealing that a negative sign can be replaced by a 180° change in phase angle (+ or −); that is,

$$e = -E_m \sin \omega t = E_m \sin(\omega t + 180°) = E_m \sin(\omega t - 180°)$$

A plot of each will clearly show their equivalence. There are, therefore, two correct mathematical representations for the functions.

The **phase relationship** between two waveforms indicates which one leads or lags the other, and by how many degrees or radians.

EXAMPLE 12.12 What is the phase relationship between the sinusoidal waveforms of each of the following sets?

a. $v = 10 \sin(\omega t + 30°)$
    $i = 5 \sin(\omega t + 70°)$

b. $i = 15 \sin(\omega t + 60°)$
    $v = 10 \sin(\omega t - 20°)$

c. $i = 2 \cos(\omega t + 10°)$
    $v = 3 \sin(\omega t - 10°)$

d. $i = -\sin(\omega t + 30°)$
    $v = 2 \sin(\omega t + 10°)$

e. $i = -2 \cos(\omega t - 60)$
    $v = 3 \sin(\omega t - 150°)$

Solutions:

a. See Fig. 12.31.
    i leads v by 40°, or v lags i by 40°.

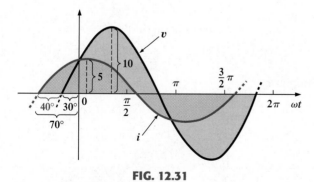

FIG. 12.31
Example 12.12(a): i leads v by 40°.

b. See Fig. 12.32.
    i leads v by 80°, or v lags i by 80°.

c. See Fig. 12.33.

$$i = 2 \cos(\omega t + 10°) = 2 \sin(\omega t + 10° + 90°)$$
$$= 2 \sin(\omega t + 100°)$$

i leads v by 110°, or v lags i by 110°.

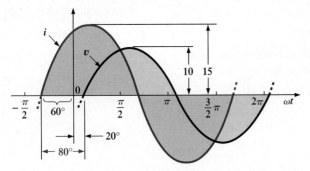

FIG. 12.32

Example 12.12(b): i leads v by 80°.

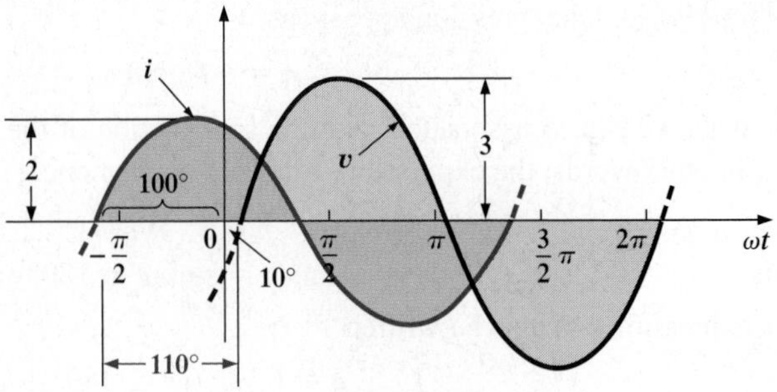

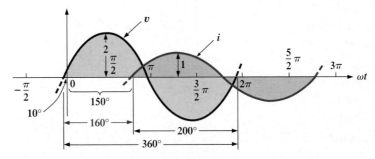

FIG. 12.33

Example 12.12(c): i leads v by 110°.

d. See Fig. 12.34.

$$-\sin(\omega t + 30°) = \sin(\omega t + 30° - 180°) \quad \overset{\text{Note}}{\swarrow}$$
$$= \sin(\omega t - 150°)$$

v leads i by 160°, or i lags v by 160°.
Or using

$$-\sin(\omega t + 30°) = \sin(\omega t + 30° + 180°) \quad \overset{\text{Note}}{\swarrow}$$
$$= \sin(\omega t + 210°)$$

i leads v by 200°, or v lags i by 200°.

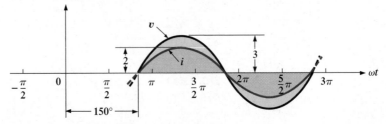

FIG. 12.34

Example 12.12(d): v leads i by 160°.

e. See Fig. 12.35.

$$i = -2\cos(\omega t - 60°) = 2\cos(\omega t - 60° - 180°) \quad \overset{\text{By choice}}{\swarrow}$$
$$= 2\cos(\omega t - 240°)$$

However, $\qquad\qquad \cos\alpha = \sin(\alpha + 90°)$

so that $\qquad 2\cos(\omega t - 240°) = 2\sin(\omega t - 240° + 90°)$
$$= 2\sin(\omega t - 150°)$$

v and i are in phase.

FIG. 12.35

Example 12.12(e): v and i are in phase.

The Oscilloscope

The **oscilloscope** is an instrument that will display the sinusoidal alternating waveform in a way that will permit the reviewing of all of the waveform's characteristics. In some ways, the screen and the dials give an oscilloscope the appearance of a small TV, but remember that **it can display only what you feed into it.** You can't turn it on and ask for a sine wave, a square wave, and so on; it must be connected to a source or an active circuit to pick up the desired waveform.

The screen has a standard appearance, with 10 horizontal divisions and 8 vertical divisions. **The distance between divisions is 1 cm** on the vertical and horizontal scales, providing you with an excellent opportunity to become aware of the length of 1 cm. **The vertical scale is set to display voltage levels, whereas the horizontal scale is always in units of time.** The **vertical sensitivity control** will set the voltage level for each division, whereas the **horizontal sensitivity control** will set the time associated with each division. In other words, if the vertical sensitivity is set at 1 V/div., each division will display a 1 V swing, so that a total vertical swing of 8 divisions will represent 8 V peak-to-peak. If the horizontal control is set on 10 μs/div., 4 divisions will equal a time period of 40 μs. In general, therefore, do not expect the oscilloscope display to present a sinusoidal voltage versus degrees or radians; it will always be against time. Further, the vertical scale can never be in units of amperes; it is always a voltage sensitivity.

EXAMPLE 12.13 Find the period, frequency, and peak value of the sinusoidal waveform appearing on the screen of the oscilloscope in Fig. 12.36. Note the sensitivities provided on the figure.

Solution: One cycle spans 4 divisions. Therefore, the period is

$$T = 4 \text{ div.} \left(\frac{50 \ \mu s}{\text{div.}} \right) = \textbf{200} \ \boldsymbol{\mu} \textbf{s}$$

and the frequency is

$$f = \frac{1}{T} = \frac{1}{200 \times 10^{-6} \text{ s}} = \textbf{5 kHz}$$

The vertical height above the horizontal axis encompasses 2 divisions. Therefore,

$$V_m = 2 \text{ div.} \left(\frac{0.1 \text{ V}}{\text{div.}} \right) = \textbf{0.2 V}$$

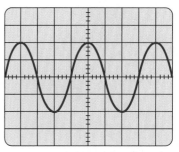

Vertical sensitivity = 0.1 V/div.
Horizontal sensitivity = 50 μs/div.

FIG. 12.36
Example 12.13.

An oscilloscope can also be used to make phase measurements between two sinusoidal waveforms. Virtually all modern day laboratory oscilloscopes have the dual-trace option, that is, the ability to show two waveforms at the same time. It is important to remember, however, that both waveforms will and must have the same frequency. The hookup procedure for using an oscilloscope to measure phase angles is covered in detail in Section 14.11. However, the equation for determining the phase angle can be introduced using Fig. 12.37.

First, note that each sinusoidal function **has the same frequency,** permitting the use of either waveform to determine the period. For the waveform chosen in Fig. 12.37, the period encompasses 5 divisions at 0.2 ms/div. The phase shift between the waveforms (irrespective of which is leading or lagging) is 2 divisions. Since the full period represents a cycle of 360°, the following ratio [from which Equation (12.24) can be derived] can be formed:

$$\frac{360°}{T \text{ (no. of div.)}} = \frac{\theta}{\text{phase shift (no. of div.)}}$$

and

$$\theta = \frac{\text{phase shift (no. of div.)}}{T \text{ (no. of div.)}} \times 360° \qquad \textbf{(12.24)}$$

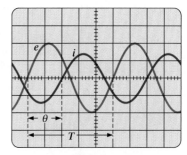

Vertical sensitivity = 2 V/div.
Horizontal sensitivity = 0.2 ms/div.

FIG. 12.37
Finding the phase angle between waveforms using a dual-trace oscilloscope.

Substituting into Eq. (12.24) will result in

$$\theta = \frac{(2 \text{ div.})}{(5 \text{ div.})} \times 360° = \textbf{144°}$$

and *e* leads *i* by 144°.

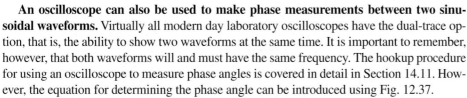

12.7 AVERAGE VALUE

Even though the concept of the **average value** is an important one in most technical fields, its true meaning is often misunderstood. In Fig. 12.38(a), for example, the average height of the sand may be required to determine the volume of sand available. The average height of the sand is that height obtained if the distance from one end to the other is maintained while the sand is leveled off, as shown in Fig. 12.38(b). The area under the mound of Fig. 12.38(a) will then equal the area under the rectangular shape of Fig. 12.38(b) as determined by $A = b \times h$. Of course, the depth (into the page) of the sand must be the same for Fig. 12.38(a) and (b) for the preceding conclusions to have any meaning.

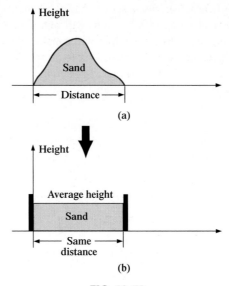

FIG. 12.38

Defining average value.

In Fig. 12.38, the distance was measured from one end to the other. In Fig. 12.39(a), the distance extends beyond the end of the original pile of Fig. 12.38. The situation could be one where a landscaper would like to know the average height of the sand if spread out over a distance such as defined in Fig. 12.39(a). The result of an increased distance is as shown in Fig. 12.39(b). The average height has decreased compared to Fig. 12.38. Quite obviously, therefore, the longer the distance, the lower is the average value.

If the distance parameter includes a depression, as shown in Fig. 12.40(a), some of the sand will be used to fill the depression, resulting in an even lower average value for the land-

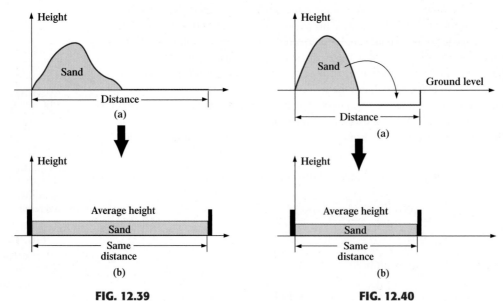

FIG. 12.39

Effect of distance (length) on average value.

FIG. 12.40

Effect of depressions (negative excursions) on average value.

scaper, as shown in Fig. 12.40(b). For a sinusoidal waveform, the depression would have the same shape as the mound of sand (over one full cycle), resulting in an average value at ground level (or zero volts for a sinusoidal voltage over one full period).

After traveling a considerable distance by car, some drivers like to calculate their average speed for the entire trip. This is usually done by dividing the miles traveled by the hours required to drive that distance. For example, if a person traveled 225 mi in 5 h, the average speed was 225 mi/5 h, or 45 mi/h. This same distance may have been traveled at various speeds for various intervals of time, as shown in Fig. 12.41.

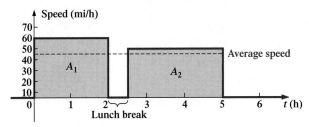

FIG. 12.41

Plotting speed versus time for an automobile excursion.

By finding the total area under the curve for the 5 h and then dividing the area by 5 h (the total time for the trip), we obtain the same result of 45 mi/h; that is,

$$\text{Average speed} = \frac{\text{area under curve}}{\text{length of curve}} \qquad \textbf{(12.25)}$$

$$\text{Average speed} = \frac{A_1 + A_2}{5 \text{ h}} = \frac{(60 \text{ mi/h})(2 \text{ h}) + (50 \text{ mi/h})(2.5 \text{ h})}{5 \text{ h}}$$

$$= \frac{225}{5} \text{ mi/h} = \textbf{45 mi/h}$$

Equation (12.25) can be extended to include any variable quantity, such as current or voltage, if we let G denote the average value, as follows:

$$G \text{ (average value)} = \frac{\text{algebraic sum of areas}}{\text{length of curve}} \qquad \textbf{(12.26)}$$

The **algebraic** sum of the areas must be determined, since some area contributions will be from below the horizontal axis. Areas above the axis will be assigned a positive sign, and those below, a negative sign. A positive average value will then be above the axis, and a negative value, below.

The average value of *any* current or voltage is the value indicated on a dc meter. In other words, over a complete cycle, the average value is the equivalent dc value. In the analysis of electronic circuits to be considered in a later course, both dc and ac sources of voltage will be applied to the same network. It will then be necessary to know or determine the dc (or average value) and ac components of the voltage or current in various parts of the system.

EXAMPLE 12.14 Determine the average value of the waveforms of Fig. 12.42.

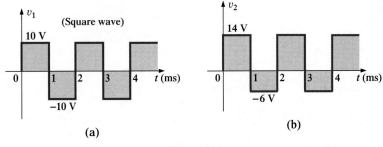

FIG. 12.42

Example 12.14.

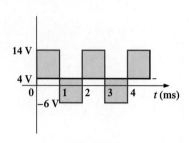

FIG. 12.43

*Defining the average value for the
waveform of Fig. 12.42(b).*

Solutions:

a. By inspection, the area above the axis equals the area below over one cycle, resulting in an average value of zero volts. Using Eq. (12.26):

$$G = \frac{(10 \text{ V})(1 \text{ ms}) - (10 \text{ V})(1 \text{ ms})}{2 \text{ ms}} = \frac{0}{2 \text{ ms}} = \textbf{0 V}$$

b. Using Eq. (12.26):

$$G = \frac{(14 \text{ V})(1 \text{ ms}) - (6 \text{ V})(1 \text{ ms})}{2 \text{ ms}} = \frac{14 \text{ V} - 6 \text{ V}}{2} = \frac{8 \text{ V}}{2} = \textbf{4 V}$$

as shown in Fig. 12.43.

In reality, the waveform of Fig. 12.42(b) is simply the square wave of Fig. 12.42(a) with a dc shift of 4 V; that is,

$$v_2 = v_1 + 4 \text{ V}$$

EXAMPLE 12.15 Find the average values of the following waveforms over one full cycle:

a. Fig. 12.44.
b. Fig. 12.45.

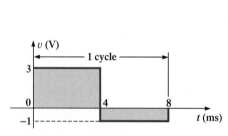

FIG. 12.44

Example 12.15(a).

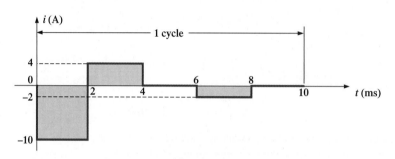

FIG. 12.45

Example 12.15(b).

Solutions:

a. $$G = \frac{+(3 \text{ V})(4 \text{ ms}) - (1 \text{ V})(4 \text{ ms})}{8 \text{ ms}} = \frac{12 \text{ V} - 4 \text{ V}}{8} = \textbf{1 V}$$

Note Fig. 12.46.

b. $$G = \frac{-(10 \text{ V})(2 \text{ ms}) + (4 \text{ V})(2 \text{ ms}) - (2 \text{ V})(2 \text{ ms})}{10 \text{ ms}}$$

$$= \frac{-20 \text{ V} + 8 \text{ V} - 4 \text{ V}}{10} = -\frac{16 \text{ V}}{10} = \textbf{-1.6 V}$$

Note Fig. 12.47.

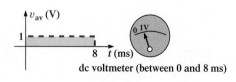

dc voltmeter (between 0 and 8 ms)

FIG. 12.46

*The response of a dc meter to the waveform
of Fig. 12.44.*

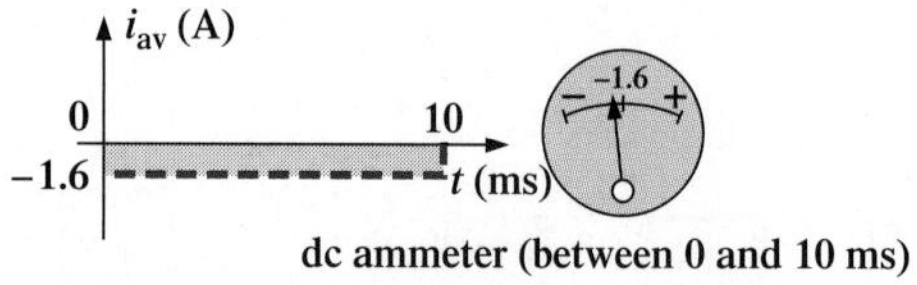

dc ammeter (between 0 and 10 ms)

FIG. 12.47

*The response of a dc meter to the waveform
of Fig. 12.45.*

We found the areas under the curves in Example 12.15 by using a simple geometric formula. If we should encounter a sine wave or any other unusual shape, however, we must find the area by some other means. We can obtain a good approximation of the area by attempting to reproduce the original wave shape using a number of small rectangles or other familiar shapes, the area of which we already know through simple geometric formulas. For example,

the area of the positive (or negative) pulse of a sine wave is $2A_m$.

Approximating this waveform by two triangles (Fig. 12.48), we obtain (using *area = 1/2 base × height* for the area of a triangle) a rough idea of the actual area:

$$\text{Area shaded} = 2\left(\frac{1}{2}bh\right) = 2\left[\left(\frac{1}{2}\right)\overbrace{\left(\frac{\pi}{2}\right)}^{b}\overbrace{(A_m)}^{h}\right] = \frac{\pi}{2}A_m \cong 1.58A_m$$

A closer approximation might be a rectangle with two similar triangles (Fig. 12.49):

$$\text{Area} = A_m\frac{\pi}{3} + 2\left(\frac{1}{2}bh\right) = A_m\frac{\pi}{3} + \frac{\pi}{3}A_m = \frac{2}{3}\pi A_m = 2.094A_m$$

which is certainly close to the actual area. If an infinite number of forms were used, an exact answer of $2A_m$ could be obtained. For irregular waveforms, this method can be especially useful if data such as the average value are desired.

The procedure of calculus that gives the exact solution $2A_m$ is known as *integration*. Integration is presented here only to make the method recognizable to you; it is not necessary to be proficient in its use to continue with this text. It is a useful mathematical tool, however, and should be learned. Finding the area under the positive pulse of a sine wave using integration, we have

$$\text{Area} = \int_0^\pi A_m \sin \alpha \, d\alpha$$

where $\int$ is the sign of integration, 0 and π are the limits of integration, $A_m \sin \alpha$ is the function to be integrated, and $d\alpha$ indicates that we are integrating with respect to α.

Integrating, we obtain

$$\text{Area} = A_m[-\cos \alpha]_0^\pi$$
$$= -A_m(\cos \pi - \cos 0°)$$
$$= -A_m[-1 - (+1)] = -A_m(-2)$$

$$\boxed{\text{Area} = 2A_m} \qquad (12.27)$$

Since we know the area under the positive (or negative) pulse, we can easily determine the average value of the positive (or negative) region of a sine wave pulse by applying Eq. (12.26):

$$G = \frac{2A_m}{\pi}$$

and

$$\boxed{G = 0.637A_m} \qquad (12.28)$$

For the waveform of Fig. 12.50,

$$G = \frac{(2A_m/2)}{\pi/2} = \frac{2A_m}{\pi} \qquad \text{(The average is the same as for a full pulse.)}$$

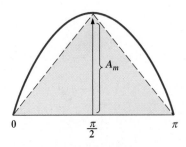

FIG. 12.48

Approximating the shape of the positive pulse of a sinusoidal waveform with two right triangles.

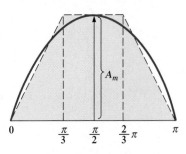

FIG. 12.49

A better approximation for the shape of the positive pulse of a sinusoidal waveform.

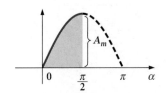

FIG. 12.50

Finding the average value of one-half the positive pulse of a sinusoidal waveform.

EXAMPLE 12.16 Determine the average value of the sinusoidal waveform of Fig. 12.51.

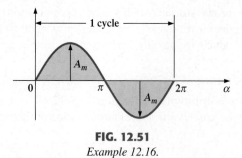

FIG. 12.51
Example 12.16.

Solution: By inspection it is fairly obvious that
the average value of a pure sinusoidal waveform over one full cycle is zero.
Eq. (12.26):

$$G = \frac{+2A_m - 2A_m}{2\pi} = \textbf{0 V}$$

EXAMPLE 12.17 Determine the average value of the waveform of Fig. 12.52.

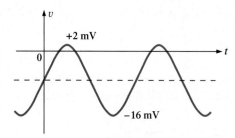

FIG. 12.52
Example 12.17.

Solution: The peak-to-peak value of the sinusoidal function is 16 mV + 2 mV = 18 mV. The peak amplitude of the sinusoidal waveform is, therefore, 18 mV/2 = 9 mV. Counting down 9 mV from 2 mV (or 9 mV up from −16 mV) results in an average or dc level of −7 mV, as noted by the dashed line of Fig. 12.52.

EXAMPLE 12.18 Determine the average value of the waveform of Fig. 12.53.

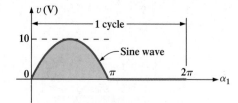

FIG. 12.53
Example 12.18.

Solution:

$$G = \frac{2A_m + 0}{2\pi} = \frac{2(10 \text{ V})}{2\pi} \cong \textbf{3.18 V}$$

EXAMPLE 12.19 For the waveform of Fig. 12.54, determine whether the average value is positive or negative, and determine its approximate value.

v (mV)

10 mV

0

t

FIG. 12.54

Example 12.19.

Solution: From the appearance of the waveform, the average value is positive and in the vicinity of 2 mV. Occasionally, judgments of this type will have to be made.

Instrumentation

The dc level or average value of any waveform can be found using a digital multimeter (DMM) or an **oscilloscope.** For purely dc circuits, simply set the DMM on dc, and read the voltage or current levels. Oscilloscopes are limited to voltage levels using the sequence of steps listed below:

1. First choose GND from the DC-GND-AC option list associated with each vertical channel. The GND option blocks any signal to which the oscilloscope probe may be connected from entering the oscilloscope and responds with just a horizontal line. Set the resulting line in the middle of the vertical axis on the horizontal axis, as shown in Fig. 12.55(a).

2. Apply the oscilloscope probe to the voltage to be measured (if not already connected), and switch to the DC option. If a dc voltage is present, the horizontal line will shift up or down, as demonstrated in Fig. 12.55(b). Multiplying the shift by the vertical sensitivity will result in the dc voltage. An upward shift is a positive voltage (higher potential at the red or positive lead of the oscilloscope), while a downward shift is a negative voltage (lower potential at the red or positive lead of the oscilloscope).

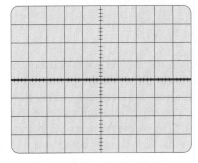

(a)

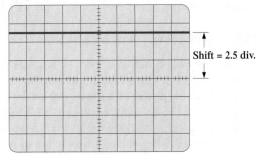

Shift = 2.5 div.

Vertical sensitivity = 50 mV/div.

(b)

FIG. 12.55

Using the oscilloscope to measure dc voltages: (a) setting the GND condition; (b) the vertical shift resulting from a dc voltage when shifted to the DC option.

In general,

$$\boxed{V_{dc} = (\text{vertical shift in div.}) \times (\text{vertical sensitivity in V/div.})} \quad \textbf{(12.29)}$$

For the waveform of Fig. 12.55(b),

$$V_{dc} = (2.5 \text{ div.})(50 \text{ mV/div.}) = \textbf{125 mV}$$

The oscilloscope can also be used to measure the dc or average level of any waveform using the following sequence:

1. Using the GND option, reset the horizontal line to the middle of the screen.

2. Switch to AC (all dc components of the signal to which the probe is connected will be blocked from entering the oscilloscope—only the alternating, or chang-

ing, components will be displayed). Note the location of some definitive point on the waveform, such as the bottom of the half-wave rectified waveform of Fig. 12.56(a); that is, note its position on the vertical scale. For the future, **whenever you use the AC option, keep in mind that the computer will distribute the waveform above and below the horizontal axis such that the average value is zero;** that is, the area above the axis will equal the area below.

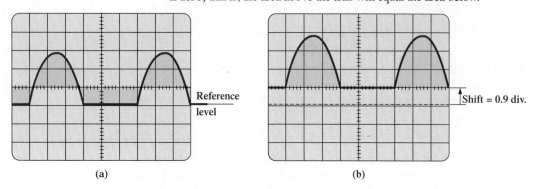

(a)
(b)

FIG. 12.56

Determining the average value of a nonsinusoidal waveform using the oscilloscope:
(a) vertical channel on the ac mode; (b) vertical channel on the dc mode.

3. Then switch to DC (to permit both the dc and the ac components of the waveform to enter the oscilloscope), and note the shift in the chosen level of part 2, as shown in Fig. 12.56(b). Equation (12.29) can then be used to determine the dc or average value of the waveform. For the waveform of Fig. 12.56(b), the average value is about

$$V_{av} = V_{dc} = (0.9 \text{ div.})(5 \text{ V/div.}) = \mathbf{4.5 \text{ V}}$$

The procedure outlined above can be applied to any alternating waveform such as the one in Fig. 12.54. In some cases the average value may require moving the starting position of the waveform under the AC option to a different region of the screen or choosing a higher voltage scale. By simply choosing the appropriate scale, you can enable DMMs to read the average or dc level of any waveform.

12.8 EFFECTIVE (rms) VALUES

This section will begin to relate dc and ac quantities with respect to the power delivered to a load. It will help us **determine the amplitude of a sinusoidal ac current required to deliver the same power as a particular dc current.** The question frequently arises, How is it possible for a sinusoidal ac quantity to deliver a net power if, over a full cycle, the net current in any one direction is zero (average value = 0)? It would almost appear that the power delivered during the positive portion of the sinusoidal waveform is withdrawn during the negative portion, and since the two are equal in magnitude, the net power delivered is zero. However, understand that **irrespective of direction,** current of any magnitude through a resistor will deliver power *to that resistor.* In other words, during the positive or negative portions of a sinusoidal ac current, power is being delivered at **each instant of time** to the resistor. The power delivered at each instant will, of course, vary with the magnitude of the sinusoidal ac current, but there will be a net flow during either the positive or the negative pulses with a net flow over the full cycle. The net power flow will equal twice that delivered by either the positive or the negative regions of sinusoidal quantity.

A fixed relationship between ac and dc voltages and currents can be derived from the experimental setup shown in Fig. 12.57. A resistor in a water bath is connected by switches to a dc and an ac supply. If switch 1 is closed, a dc current I, determined by the resistance R and battery voltage E, will be established through the resistor R. The temperature reached by the water is determined by the dc power dissipated in the form of heat by the resistor.

If switch 2 is closed and switch 1 left open, the ac current through the resistor will have a peak value of I_m. The temperature reached by the water is now determined by the ac power dissipated in the form of heat by the resistor. The ac input is varied until the temperature is the same as that reached with the dc input. When this is accomplished, the average electrical power delivered to the resistor R by the ac source is the same as that delivered by the dc source.

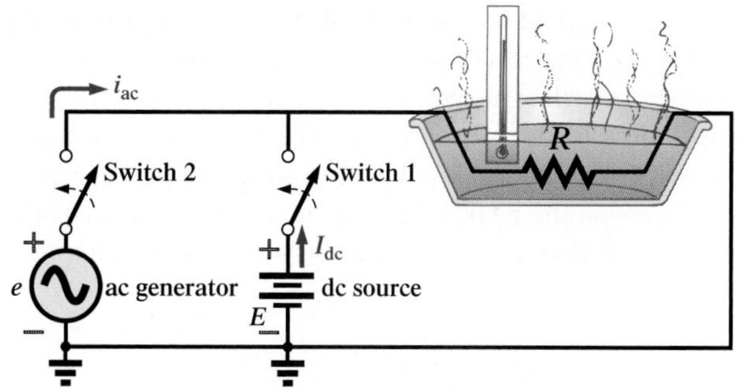

FIG. 12.57

An experimental setup to establish a relationship between dc and ac quantities.

The power delivered by the ac supply at any instant of time is

$$P_{ac} = (i_{ac})^2 R = (I_m \sin \omega t)^2 R = (I_m^2 \sin^2 \omega t)R$$

However,

$$\sin^2 \omega t = \frac{1}{2}(1 - \cos 2\omega t) \qquad \text{(trigonometric identity)}$$

Therefore,

$$P_{ac} = I_m^2 \left[\frac{1}{2}(1 - \cos 2\omega t) \right] R$$

and

$$P_{ac} = \frac{I_m^2 R}{2} - \frac{I_m^2 R}{2} \cos 2\omega t \qquad \text{(12.30)}$$

The *average power* delivered by the ac source is just the first term, since the average value of a cosine wave is zero even though the wave may have twice the frequency of the original input current waveform. Equating the average power delivered by the ac generator to that delivered by the dc source,

$$P_{av(ac)} = P_{dc}$$

$$\frac{I_m^2 R}{2} = I_{dc}^2 R$$

and

$$I_{dc} = \frac{I_m}{\sqrt{2}} = 0.707 I_m$$

which, in words, states that

the equivalent dc value of a sinusoidal current or voltage is $1/\sqrt{2}$ or 0.707 of its peak value.

The equivalent dc value is called the **effective value** of the sinusoidal quantity.

As a simple numerical example, it would require an ac current with a peak value of $\sqrt{2}(10) = 14.14$ A to deliver the same power to the resistor in Fig. 12.57 as a dc current of 10 A. The effective value of any quantity plotted as a function of time can be found by using the following equation derived from the experiment just described.

Calculus format:

$$I_{eff} = \sqrt{\frac{\int_0^T i^2(t)\, dt}{T}} \qquad \text{(12.31)}$$

which means:

$$I_{eff} = \sqrt{\frac{\text{area } (i^2(t))}{T}} \qquad \text{(12.32)}$$

In words, Equations (12.31) and (12.32) state that to find the effective value, the function $i(t)$ must first be squared. After $i(t)$ is squared, the area under the curve is found by integration. It is then divided by T, the length of the cycle or the period of the waveform, to obtain the average or *mean* value of the squared waveform. The final step is to take the *square root* of the mean value. This procedure gives us another designation for the effective value, the **root-mean-square (rms) value.** In fact, since the rms term is the most commonly used in the educational and industrial communities, it will be used throughout this text.

The relationship between the peak value and the effective or rms value is the same for voltages, resulting in the following set of relationships for the examples and text material to follow:

$$I_{rms} = \frac{1}{\sqrt{2}} I_m = 0.707 I_m$$

$$E_{rms} = \frac{1}{\sqrt{2}} E_m = 0.707 E_m$$

(12.33)

Similarly,

$$I_m = \sqrt{2} I_{rms} = 1.414 I_{rms}$$

$$E_m = \sqrt{2} E_{rms} = 1.414 E_{rms}$$

(12.34)

EXAMPLE 12.20 Find the rms values of the sinusoidal waveform in each part of Fig. 12.58.

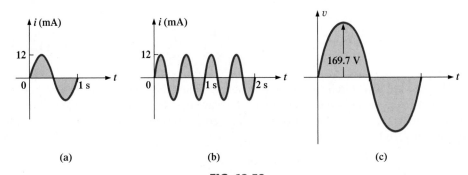

FIG. 12.58
Example 12.20.

Solution: For part (a), $I_{rms} = 0.707(12 \times 10^{-3} \text{ A}) =$ **8.484 mA.** For part (b), again $I_{rms} =$ **8.484 mA.** Note that frequency did not change the effective value in (b) compared to (a). For part (c), $V_{rms} = 0.707(169.73 \text{ V}) \cong$ **120 V,** the same as available from a home outlet.

EXAMPLE 12.21 The 120 V dc source of Fig. 12.59(a) delivers 3.6 W to the load. Determine the peak value of the applied voltage (E_m) and the current (I_m) if the ac source [Fig. 12.59(b)] is to deliver the same power to the load.

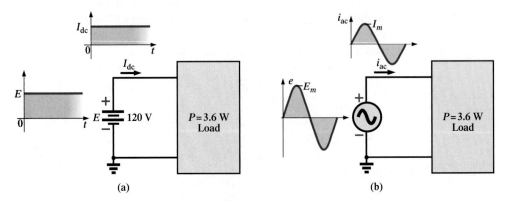

FIG. 12.59
Example 12.21.

$$P_{dc} = V_{dc}I_{dc}$$

and
$$I_{dc} = \frac{P_{dc}}{V_{dc}} = \frac{3.6 \text{ W}}{120 \text{ V}} = 30 \text{ mA}$$

$$I_m = \sqrt{2}I_{dc} = (1.414)(30 \text{ mA}) = \textbf{42.42 mA}$$

$$E_m = \sqrt{2}E_{dc} = (1.414)(120 \text{ V}) = \textbf{169.68 V}$$

EXAMPLE 12.22 Find the effective or rms value of the waveform of Fig. 12.60.

Solution: v^2 (Fig. 12.61):

$$V_{rms} = \sqrt{\frac{(9)(4) + (1)(4)}{8}} = \sqrt{\frac{40}{8}} = \textbf{2.236 V}$$

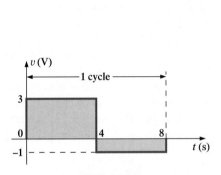

FIG. 12.60
Example 12.22.

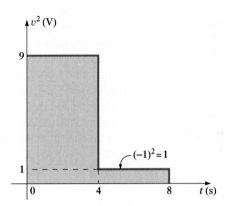

FIG. 12.61
The squared waveform of Fig. 12.60.

EXAMPLE 12.23 Calculate the rms value of the voltage of Fig. 12.62.

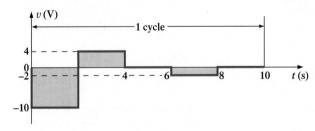

FIG. 12.62
Example 12.23.

Solution: v^2 (Fig. 12.63):

$$V_{rms} = \sqrt{\frac{(100)(2) + (16)(2) + (4)(2)}{10}} = \sqrt{\frac{240}{10}} = \textbf{4.899 V}$$

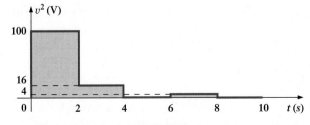

FIG. 12.63
The squared waveform of Fig. 12.62.

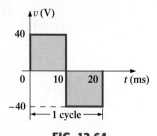

FIG. 12.64

Example 12.24.

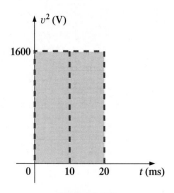

FIG. 12.65

The squared waveform of
Fig. 12.64.

EXAMPLE 12.24 Determine the average and rms values of the square wave of Fig. 12.64.

Solution: By inspection, the average value is zero.

v^2 (Fig. 12.65):

$$V_{rms} = \sqrt{\frac{(1600)(10 \times 10^{-3}) + (1600)(10 \times 10^{-3})}{20 \times 10^{-3}}}$$

$$= \sqrt{\frac{32,000 \times 10^{-3}}{20 \times 10^{-3}}} = \sqrt{1600} = \textbf{40 V}$$

(the maximum value of the waveform of Fig. 12.65).

The waveforms appearing in these examples are the same as those used in the examples on the average value. It might prove interesting to compare the rms and average values of these waveforms.

The rms values of sinusoidal quantities such as voltage or current will be represented by E and I. These symbols are the same as those used for dc voltages and currents. **To avoid confusion, the peak value of a waveform will always have a subscript m associated with it: $I_m \sin \omega t$.** *Caution:* When finding the rms value of the positive pulse of a sine wave, note that the squared area is *not* simply $(2A_m)^2 = 4A_m^2$; it must be found by a completely new integration. This will always be the case for any waveform that is not rectangular.

A unique situation arises if a waveform has both a dc and an ac component that may be due to a source such as the one in Fig. 12.66. The combination appears frequently in the analysis of electronic networks where both dc and ac levels are present in the same system.

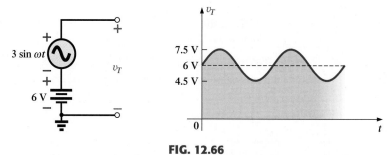

FIG. 12.66

Generation and display of a waveform having a dc and an ac component.

The question arises, What is the rms value of the voltage v_T? One might be tempted to simply assume that it is the sum of the rms values of each component of the waveform; that is, $V_{T_{rms}} = 0.7071(1.5 \text{ V}) + 6 \text{ V} = 1.06 \text{ V} + 6 \text{ V} = 7.06 \text{ V}$. However, the rms value is actually determined by

$$\boxed{V_{rms} = \sqrt{V_{dc}^2 + V_{ac(rms)}^2}} \qquad \textbf{(12.35)}$$

which for the waveform of Fig. 12.66 is

$$V_{rms} = \sqrt{(6 \text{ V})^2 + (1.06 \text{ V})^2} = \sqrt{37.124} \text{ V} \cong \textbf{6.1 V}$$

This result is noticeably less than the solution of 7.06 V.

Instrumentation

It is important to note whether the DMM in use is a true rms meter or simply a meter where the average value is calibrated (as described in the next section) to indicate the rms level. **A true rms meter will read the effective value of any waveform (such as Figs. 12.54 and 12.66) and is not limited to only sinusoidal waveforms.** Since the label *true rms* is normally not placed on the face of the meter, it is prudent to check the manual if waveforms other than purely sinusoidal are to be encountered. For any type of rms meter, be sure to check the manual for its frequency range of application. For most it is less than 1 kHz.

12.9 ac METERS AND INSTRUMENTS

The d'Arsonval movement employed in dc meters can also be used to measure sinusoidal voltages and currents if the *bridge rectifier* of Fig. 12.67 is placed between the signal to be measured and the average reading movement.

The bridge rectifier, composed of four diodes (electronic switches), will convert the input signal of zero average value to one having an average value sensitive to the peak value of the input signal. The conversion process is well described in most basic electronics texts. Fundamentally, conduction is permitted through the diodes in such a manner as to convert the sinusoidal input of Fig. 12.68(a) to one having the appearance of Fig. 12.68(b). The negative portion of the input has been effectively "flipped over" by the bridge configuration. The resulting waveform of Fig. 12.68(b) is called a *full-wave rectified waveform.*

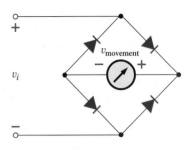

FIG. 12.67
Full-wave bridge rectifier.

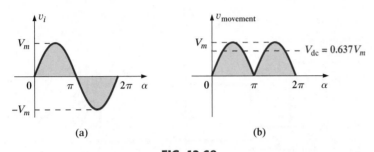

(a)   (b)

FIG. 12.68
(a) Sinusoidal input; (b) full-wave rectified signal.

The zero average value of Fig. 12.68(a) has been replaced by a pattern having an average value determined by

$$G = \frac{2V_m + 2V_m}{2\pi} = \frac{4V_m}{2\pi} = \frac{2V_m}{\pi} = 0.637V_m$$

The movement of the pointer will therefore be directly related to the peak value of the signal by the factor 0.637.

Forming the ratio between the rms and dc levels will result in

$$\frac{V_{rms}}{V_{dc}} = \frac{0.707V_m}{0.637V_m} \cong 1.11$$

revealing that the scale indication is 1.11 times the dc level measured by the movement; that is,

$$\boxed{\text{Meter indication} = 1.11 \text{ (dc or average value)}}_{\text{full-wave}} \qquad \textbf{(12.36)}$$

Some ac meters use a half-wave rectifier arrangement that results in the waveform of Fig. 12.69, which has half the average value of Fig. 12.68(b) over one full cycle. The result is

$$\boxed{\text{Meter indication} = 2.22 \text{ (dc or average value)}}_{\text{half-wave}} \qquad \textbf{(12.37)}$$

A second movement, called the **electrodynamometer movement** (Fig. 12.70), can measure both ac and dc quantities without a change in internal circuitry. The movement can, in fact, read the effective value of any periodic or nonperiodic waveform because a reversal in current direction reverses the fields of both the stationary and the movable coils, so the deflection of the pointer is always up-scale.

The **VOM**, introduced in Chapter 2, can be used to measure both dc and ac voltages using a d'Arsonval movement and the proper switching networks. That is, when the meter is used for dc measurements, the dial setting will establish the proper series resistance for the

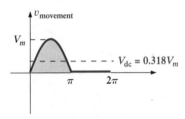

FIG. 12.69
Half-wave rectified signal.

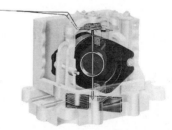

FIG. 12.70
Electrodynamometer movement.
(Courtesy of Schlumberger Technology Corp.)

chosen scale and will permit the appropriate dc level to pass directly to the movement. For ac measurements, the dial setting will introduce a network that employs a full- or half-wave rectifier to establish a dc level. As discussed above, each setting is properly calibrated to indicate the desired quantity on the face of the instrument.

EXAMPLE 12.25 Determine the reading of each meter for each situation of Fig. 12.71(a) and (b).

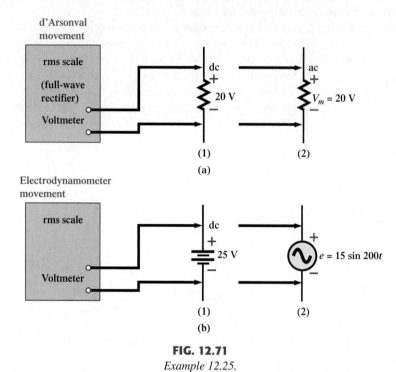

FIG. 12.71
Example 12.25.

Solution: For Fig. 12.71(a), situation (1): By Eq. (12.36),

$$\text{Meter indication} = 1.11(20 \text{ V}) = \textbf{22.2 V}$$

For Fig. 12.71(a), situation (2):

$$V_{\text{rms}} = 0.707 V_m = 0.707(20 \text{ V}) = \textbf{14.14 V}$$

For Fig. 12.71(b), situation (1):

$$V_{\text{rms}} = V_{\text{dc}} = \textbf{25 V}$$

For Fig. 12.71(b), situation (2):

$$V_{\text{rms}} = 0.707 V_m = 0.707(15 \text{ V}) \cong \textbf{10.6 V}$$

Most DMMs employ a full-wave rectification system to convert the input ac signal to one with an average value. In fact, for the VOM of Fig. 2.27, the same scale factor of Eq. (12.36) is employed; that is, the average value is scaled up by a factor of 1.11 to obtain the rms value. In digital meters, however, there are no moving parts such as in the d'Arsonval or electrodynamometer movements to display the signal level. Rather, the average value is sensed by a multiprocessor integrated circuit (IC), which in turn determines which digits should appear on the digital display.

Digital meters can also be used to measure nonsinusoidal signals, but the scale factor of each input waveform must first be known (normally provided by the manufacturer in the operator's manual). For instance, the scale factor for an average responding DMM on the ac rms scale will produce an indication for a square-wave input that is 1.11 times the peak value. For a triangular input, the response is 0.555 times the peak value. Obviously, for a sine wave input, the response is 0.707 times the peak value.

For any instrument, it is always good practice to read (if only briefly) the operator's manual if it appears that you will use the instrument on a regular basis.

For frequency measurements, the **frequency counter** of Fig. 12.72 provides a digital readout of sine, square, and triangular waves from 1 Hz to 1.3 GHz. Note the relative simplicity of the panel and the high degree of accuracy available. The temperature-compensated, crystal-controlled time base is stable to ±1 part per million per year.

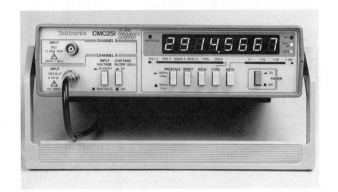

FIG. 12.72
Frequency counter. Tektronix CMC251 1.3 GHz multifunction counter.
(Photo courtesy of Tektronix, Inc.)

The AEMC® Clamp Meter of Fig. 12.73 is an instrument that can measure alternating current in the ampere range without having to open the circuit. The loop is opened by squeezing the "trigger"; then it is placed around the current-carrying conductor. Through transformer action, the level of current in rms units will appear on the appropriate scale. The Model 501 is auto-ranging (that is, each scale changes automatically) and can measure dc or ac currents up to 400 mA. Through the use of additional leads, it can also be used as a voltmeter (up to 400 V, dc or ac) and an ohmmeter (from zero to 400 Ω).

One of the most versatile and important instruments in the electronics industry is the **oscilloscope,** which has already been introduced in this chapter. It provides a display of the waveform on a cathode-ray tube to permit the detection of irregularities and the determination of quantities such as magnitude, frequency, period, dc component, and so on. The digital oscilloscope of Fig. 12.74 can display four waveforms at the same time. You use menu buttons to set the vertical and horizontal scales by choosing from selections appearing on the screen. The TDS model of Fig. 12.74 can display, store, and analyze the amplitude, time, and distribution of amplitude over time. It is also completely portable due to its battery-capable design.

A student accustomed to watching TV might be confused when first introduced to an oscilloscope. There is, at least initially, an assumption that the oscilloscope is generating the

FIG. 12.73
Clamp-on ammeter and voltmeter.
(Courtesy of AEMC® Instruments. Used with permission.)

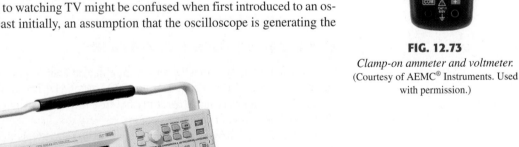

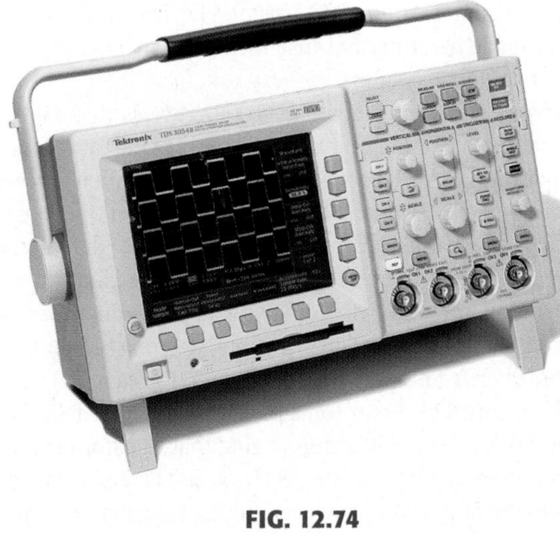

FIG. 12.74
Four-channel digital phosphor oscilloscope. Tektronix TDS3000B series oscilloscope.
(Photo courtesy of Tektronix, Inc.)

waveform on the screen—much like a TV broadcast. However, it is important to clearly understand that

an oscilloscope displays only those signals generated elsewhere and connected to the input terminals of the oscilloscope. The absence of an external signal will simply result in a horizontal line on the screen of the scope.

On most modern-day oscilloscopes, there is a switch or knob with the choice DC/GND/AC, as shown in Fig. 12.75(a), that is often ignored or treated too lightly in the early stages of scope utilization. The effect of each position is fundamentally as shown in Fig. 12.75(b). **In the DC mode the dc and ac components of the input signal can pass directly to the display. In the AC mode the dc input is blocked by the capacitor, but the ac portion of the signal can pass through to the screen. In the GND position the input signal is prevented from reaching the scope display by a direct ground connection,** which reduces the scope display to a single horizontal line.

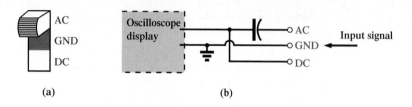

(a)   (b)

FIG. 12.75

AC-GND-DC switch for the vertical channel of an oscilloscope.

Before we leave the subject of ac meters and instrumentation, it is important to understand that

an ohmmeter cannot be used to measure the ac reactance or impedance of an element or system even though reactance and impedance are measured in ohms.

Recall that ohmmeters cannot be used on energized networks—the power must be shut off or disconnected. For an inductor, if the ac power is removed, the reactance of the coil will simply be the dc resistance of the windings because the applicable frequency would be 0 Hz. For a capacitor, if the ac power is removed, the reactance of the capacitor will simply be the leakage resistance of the capacitor. In general, therefore, always keep in mind that **ohmmeters can read only the dc resistance of an element or network, and only after the applied power has been removed.**

12.10 APPLICATIONS

(120 V at 60 Hz) versus (220 V at 50 Hz)

In North and South America the most common available ac supply is 120 V at 60 Hz, while in Europe and the Eastern countries it is 220 V at 50 Hz. The choices of rms value and frequency were obviously made carefully because they have such an important impact on the design and operation of so many systems.

The fact that the frequency difference is only 10 Hz reveals that there was agreement on the general frequency range that should be used for power generation and distribution. History suggests that the question of frequency selection originally focused on the frequency that would not exhibit *flicker in the incandescent lamps* available in those days. Technically, however, there really wouldn't be a noticeable difference between 50 and 60 cycles per second based on this criterion. Another important factor in the early design stages was the effect of frequency on the size of transformers, which play a major role in power generation and distribution. Working through the fundamental equations for transformer design, you will find that *the size of a transformer is inversely proportional to frequency.* The result is that transformers operating at 50 Hz must be larger (on a purely mathematical basis about 17% larger) than those operating at 60 Hz. You will therefore find that transformers designed for the international market where they can operate on 50 Hz or 60 Hz are designed around the 50 Hz frequency. On the other side of the coin, however, higher frequencies result in increased concerns about arcing, increased losses in the transformer core due to eddy current and hysteresis losses, and skin effect phenomena. Somewhere in the discussion we must consider the fact that 60 Hz is an exact multiple of 60 seconds in a minute and 60 minutes in an hour. On the other side of the coin, however, a 60 Hz signal has a period of 16.67 ms (an awkward number), but the period of a 50 Hz signal is exactly 20 ms. Since accurate timing is such a critical part of our technological design, was this a significant motive in the final choice? There is also the question about whether the 50 Hz is a result of the close affinity of this value to the metric system. Keep in mind that powers of ten are all powerful in the metric system, with 100 cm in a meter, 100°C the boiling point of water, and so on. Note that 50 Hz is exactly half of this special number. All in all, it would seem that both sides have an argument that would be worth defending. However, in the final analysis, we must also wonder whether the difference is simply political in nature.

The difference in voltage between North America and Europe is a different matter entirely in the sense that the difference is close to 100%. Again, however, there are valid arguments for both sides. There is no question that larger voltages such as 220 V *raise safety issues* beyond those raised by voltages of 120 V. However, when higher voltages are supplied, there is less current in the wire for the same power demand, permitting the use of smaller conductors—a real money saver. In addition, motors, compressors, and so on, found in common home appliances and throughout the industrial community *can be smaller in size.* Higher voltages, however, also bring back the concern about arcing effects, insulation requirements, and, due to real safety concerns, higher installation costs. In general, however, international travelers are prepared for most situations if they have a transformer that can convert from their home level to that of the country they plan to visit. Most equipment (not clocks, of course) can run quite well on 50 Hz or 60 Hz for most travel periods. For any unit not operating at its design frequency, it will simply have to "work a little harder" to perform the given task. The major problem for the traveler is not the transformer itself but the wide variety of plugs used from one country to another. Each country has its own design for the "female" plug in the wall. For a three-week tour, this could mean as many as 6 to 10 different plugs of the type shown in Fig. 12.76. For a 120 V, 60 Hz supply, the plug is quite standard in appearance with its two spade leads (and possible ground connection).

In any event, both the 120 V at 60 Hz and the 220 V at 50 Hz are obviously meeting the needs of the consumer. It is a debate that could go on at length without an ultimate victor.

FIG. 12.76
Variety of plugs for a 220 V, 50 Hz connection.

Safety Concerns (High Voltages and dc versus ac)

Be aware that any "live" network should be treated with a calculated level of respect. Electricity in its various forms is not to be feared but should be employed with some awareness of its potentially dangerous side effects. It is common knowledge that electricity and water do not mix (never use extension cords or plug in TVs or radios in the bathroom) because a full 120 V in a layer of water of any height (from a shallow puddle to a full bath) can be *lethal.* However, other effects of dc and ac voltages are less known. In general, as the voltage and current increase, your concern about safety should increase exponentially. For instance, under dry conditions, most human beings can survive a 120 V ac shock such as obtained when changing a light bulb, turning on a switch, and so on. Most electricians have experienced such a jolt many times in their careers. However, ask an electrician to relate how it feels to hit 220 V, and the response (if he or she has been unfortunate to have had such an experience) will be totally different. How often have you heard of a back-hoe operator hitting a 220 V line and having a fatal heart attack? Remember, the operator is sitting in a metal container on a damp ground which provides an excellent path for the resulting current to flow from the line to ground. If only for a short period of time, with the best environment (rubber-sole shoes, and so on), in a situation where you can quickly escape the situation, most human beings can also survive a 220 V shock. However, as mentioned above, it is one you will not quickly forget. For voltages beyond 220 V rms, the chances of survival go down exponentially with increase in voltage. It takes only about 10 mA of steady current through the heart to put it in defibrillation. In general, therefore, always be sure that the power is disconnected when working on the repair of electrical equipment. Don't assume that throwing a wall switch will disconnect the power. Throw the main circuit breaker and test the lines with a voltmeter before working on the system. Since voltage is a two-point phenomenon, be sure to work with only one line at at time—accidents happen!

You should also be aware that the reaction to dc voltages is quite different from that to ac voltages. You have probably seen in movies or comic strips that people are often unable to let go of a *hot* wire. This is evidence of the most important difference between the two types of voltages. As mentioned above, if you happen to touch a "hot" 120 V ac line, you will probably get a good sting, but *you can let go.* If it happens to be a "hot" 120 V dc line, you will probably not be able to let go, and a fatality could occur. Time plays an important role when this happens, because the longer you are subjected to the dc voltage, the more the resistance in the body decreases until a fatal current can be established. The reason that we can let go of an ac line is best demonstrated by carefully examining the 120 V rms, 60 Hz voltage in Fig. 12.77. Since the voltage is oscillating, there is a period of time when the voltage is near zero or less than, say, 20 V, and is reversing in direction. Although this time interval is very short, it appears every 8.3 ms and provides a window to *let go.*

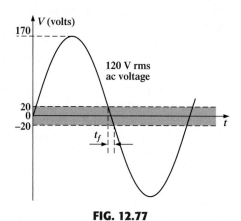

FIG. 12.77
Interval of time when sinusoidal voltage is near zero volts.

Now that we are aware of the additional dangers of dc voltages, it is important to mention that under the wrong conditions, dc voltages as low as 12 V such as from a car battery can be quite dangerous. If you happen to be working on a car under wet conditions, or if you are sweating badly for some reason or, worse yet, wearing a wedding ring that may have moisture and body salt underneath, touching the positive terminal may initiate the process whereby the body resistance begins to drop and serious injury could take place. It is one of the reasons you seldom see a professional electrician wearing any rings or jewelry—it is just not worth the risk.

Before leaving this topic of safety concerns, you should also be aware of the dangers of high-frequency supplies. We are all aware of what 2.45 GHz at 120 V can do to a meat product in a microwave oven, and it is therefore very important that the seal around the oven be as tight as possible. However, don't ever assume that anything is absolutely perfect in design—so don't make it a habit to view the cooking process in the microwave 6 in. from the door on a continuing basis. Find something else to do, and check the food only when the cooking process is complete. If you ever visit the Empire State Building, you will notice that you are unable to get close to the antenna on the dome due to the high-frequency signals being emitted with a great deal of power. Also note the large KEEP OUT signs near radio transmission towers for local radio stations. Standing within 10 ft of an AM transmitter working at 540 kHz would bring on disaster. Simply holding (not to be tried!) a fluorescent bulb near the tower could make it light up due to the excitation of the molecules inside the bulb.

In total, therefore, treat any situation with high ac voltages or currents, high-energy dc levels, and high frequencies with added care.

12.11 COMPUTER ANALYSIS

PSpice

OrCAD Capture offers a variety of ac voltage and current sources. However, for the purposes of this text, the voltage source **VSIN** and the current source **ISIN** are the most appropriate because they have a list of attributes that will cover current areas of interest.

Under the library **SOURCE,** a number of others are listed, but they don't have the full range of the above, or they are dedicated to only one type of analysis. On occasion, **ISRC** will be used because it has an arrow symbol like that appearing in the text, and it can be used for dc, ac, and some transient analyses. The symbol for **ISIN** is simply a sine wave which utilizes the plus-and-minus sign ($\pm$) to indicate direction. The sources **VAC, IAC, VSRC,** and **ISRC** are fine if the magnitude and the phase of a specific quantity are desired or if a transient plot against frequency is desired. However, they will not provide a transient response against time even if the frequency and the transient information are provided for the simulation.

For all of the sinusoidal sources, the magnitude (**VAMPL**) is the peak value of the waveform, not the rms value. This will become clear when a plot of a quantity is desired and the magnitude calculated by PSpice is the peak value of the transient response. However, for a purely steady-state ac response, the magnitude provided can be the rms value, and the output read as the rms value. Only when a plot is desired will it be clear that PSpice is accepting every ac magnitude as the peak value of the waveform. Of course, the phase angle is the same whether the magnitude is the peak or the rms value.

Before examining the mechanics of getting the various sources, remember that

Transient Analysis provides an ac or a dc output versus time, while AC Sweep is used to obtain a plot versus frequency.

To obtain any of the sources listed above, apply the following sequence: **Place part** key-**Place Part** dialog box-**Source**-(enter type of source). Once the source is selected the ac source **VSIN** will appear on the schematic with **VOFF, VAMPL,** and **FREQ.** Always specify **VOFF** as 0 V (unless a specific value is part of the analysis), and provide a value for the amplitude and frequency. The remaining quantities of **PHASE, AC, DC, DF,** and **TD** can be entered by double-clicking on the source symbol to obtain the **Property Editor,** although **PHASE, DF** (damping factor), and **TD** (time delay) do have a default of 0 s. To add a phase angle, simply click on **PHASE,** enter the phase angle in the box below, and then select **Apply.** If you want to display a factor such as a phase angle of 60°, simply click on **PHASE** followed by **Display** to obtain the **Display Properties** dialog box. Then choose **Name and Value** followed by **OK** and **Apply,** and leave the **Properties Editor** dialog box (×) to see **PHASE=60** next to the **VSIN** source. The next chapter will include the use of the ac source in a simple circuit.

Electronics Workbench

For EWB, the ac voltage source is available from two sources—the **Sources** parts bin and the **Function Generator.** The major difference between the two is that the phase angle can be set when using the **Sources** parts bin, whereas it cannot be set using the **Function Generator.**

Under **Sources,** the ac voltage source is the fourth option down on the left column of the toolbar. When selected and placed, it will display the default values for the amplitude, frequency, and phase. All the parameters of the source can be changed by double-clicking on the source symbol to obtain the **AC Voltage** dialog box. The **Voltage Amplitude** and **Voltage RMS** are interlinked so that when you change one, the other will change accordingly. For the **1V** default value, the rms value is automatically listed as **0.71** (not 0.7071 because of the hundredths-place accuracy). Note that the unit of measurement is controlled by the scrolls to the right of the default label and cannot be set by typing in the desired unit of measurement. The label can be changed by simply switching the **Label** heading and inserting the desired label. After all the changes have been made in the **AC Voltage** dialog box, click **OK,** and all the changes will appear next to the ac voltage source symbol. In Fig. 12.78, the label was changed to **Vs** and the amplitude to 10 V while the frequency and phase angle were left with their default values. It is particularly important to realize that

for any frequency analysis (that is, where the frequency will change), the AC Magnitude of the ac source must be set under Analysis Setup in the AC Voltage dialog box. Failure to do so will create results linked to the default values rather than the value set under the Value heading.

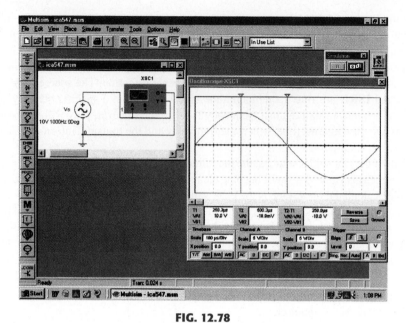

FIG. 12.78

Using the oscilloscope to display the sinusoidal ac voltage source available in the
*Electronics Workbench **Sources** tool bin.*

To view the sinusoidal voltage set in Fig. 12.78, an oscilloscope can be selected from the **Instrument** toolbar at the right of the screen. It is the fourth option down and has the appearance shown in Fig. 12.78 when selected. Note that it is a dual-channel oscilloscope with an **A** channel and a **B** channel. It has a ground (**G**) connection and a trigger (**T**) connection. The connections for viewing the ac voltage source on the **A** channel are provided in Fig. 12.78. Note that the trigger control is also connected to the **A** channel for sync control. The screen appearing in Fig. 12.78 can be displayed by double-clicking on the oscilloscope symbol on the screen. It has all the major controls of a typical laboratory oscilloscope. When you select **Simulate-Run** or select **1** on the **Simulate Switch,** the ac voltage will appear on the screen. Changing the **Time base** to 100 μs/div. will result in the display of Fig. 12.78 since there are 10 divisions across the screen and 10(100 μs) = 1 ms (the period of the applied signal). Changes in the **Time base** are made by simply clicking on the default value to obtain the scrolls in the same box. It is important to remember, however, that

changes in the oscilloscope setting or any network should not be made until the simulation is ended by disabling the Simulate-Run option or placing the Simulate switch in the 0 mode.

The options within the time base are set by the scroll bars and cannot be changed—again they match those typically available on a laboratory oscilloscope. The vertical sensitivity of the **A** channel was automatically set by the program at 5 V/div. to result in two vertical boxes for the peak value as shown in Fig. 12.78. Note the **AC** and **DC** key pads below Channel **A.** Since there is no dc component in the applied signal, either one will result in the same display. The **Trigger** control is set on the positive transition at a level of 0 V. The **T1** and **T2** refer to the cursor positions on the horizontal time axis. By simply clicking on the small red triangle at the top of the red line at the far left edge of the screen and dragging the triangle, you can move the vertical red line to any position along the axis. In Fig. 12.78, it was moved to the peak value of the waveform at ¼ of the total period or 0.25 ms = 250 μs. Note the value of **T1** (250.3 μs) and the corresponding value of **VA1** (10.0 V). By moving the other cursor with a blue triangle at the top to ½ the total period or 0.5 ms = 500 μs, we find that the value at **T2** (500.3 μs) is −18.9 mV (**VA2**), which is approximately 0 V for a waveform with a peak value of 10 V. The accuracy is controlled by the number of data points called for in the simulation setup. The more data points, the higher the likelihood of a higher degree of accuracy for the desired quantity. However, an increased number of data points will also extend the running time of the simulation. The third display box to the right gives the difference between **T2** and **T1** as 250 μs and difference between their magnitudes (**VA2–VA1**) as −10 V, with the negative sign appearing because **VA1** is greater than **VA2.**

As mentioned above, an ac voltage can also be obtained from the **Function Generator** appearing as the second option down on the **Instrument** toolbar. Its symbol appears in Fig. 12.79 with positive, negative, and ground connections. Double-click on the generator graphic symbol, and the **Function Generator-XFG1** dialog box will appear in which selections can be made. For this example, the sinusoidal waveform is chosen. The **Frequency** is set at 1 kHz, the **Amplitude** is set at 10 V, and the **Offset** is left at 0 V. Note that there is no option to set the phase angle as was possible for the source above. Double-clicking on the oscilloscope will generate the oscilloscope on which a **Timebase** of 100 μs/div. can be set again with a vertical sensitivity of 5 V/div. Select **1** on the **Simulate** switch, and the waveform of Fig. 12.79 will appear. Choosing **Singular** under **Trigger** will result in a fixed display; then set the **Simulate** switch on **0** to end the simulation. Placing the cursors in the same position shows that the waveforms for Figs. 12.78 and 12.79 are the same.

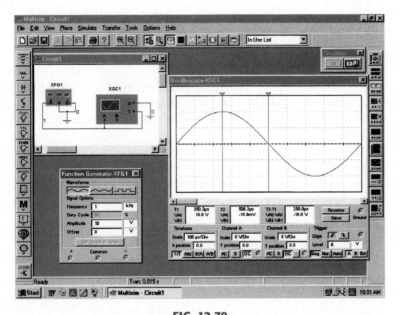

FIG. 12.79

Using the function generator to place a sinusoidal ac voltage waveform on the screen of the oscilloscope.

For most of the EWB analyses to appear in this text, the **AC_VOLTAGE_SOURCE** under **Sources** will be employed. However, with such a limited introduction to EWB, it seemed appropriate to introduce the use of the **Function Generator** because of its close linkage to the laboratory experience.

CHAPTER SUMMARY

SINUSOIDAL WAVEFORM

- The period of a waveform is defined as the time between similar points of the waveform.
- One cycle of a waveform will appear within one period of the waveform.
- The higher the frequency, the smaller the period of the waveform, and the faster the waveform will oscillate between the positive and negative regions.
- There are 2π radians in 360°.
- Angular velocity is the velocity with which a vector will rotate about its fixed end.

PHASE RELATIONS

- If a sinusoidal waveform crosses the axis before 0°, the phase angle is entered as a positive value in the sinusoidal expression. If it crosses after 0°, it is entered as a negative value.

- A negative sign in front of a sinusoidal expression can be absorbed in the sinusoidal expression by simply adding 180° to or subtracting 180° from the initial phase angle.

AVERAGE VALUE

- The average value of a waveform is the algebraic sum of the areas above and below the axis in a defined period of time.
- The average value of a sinusoidal waveform is zero over one full period.
- The area under one pulse of a sinusoidal waveform is twice the peak value.

EFFECTIVE (rms) VALUE

- The effective or rms value of a voltage or current waveform is the equivalent dc value.
- For a sinusoidal waveform, the effective or rms value will always be 70.71% of the peak value.
- The terminology *rms* ("root-mean-square") defines the mathematical procedure for finding the effective value of any waveform.

ac METERS AND INSTRUMENTS

- If a d'Arsonval average reading movement is used in an ac instrument, first the incoming signal must be changed to one that has an average value; then a calibration factor must be incorporated to ensure that the pointer is indicating the correct level for the applied signal.
- VOMs typically use d'Arsonval movements to measure ac quantities, whereas digital meters typically use an integrated circuit (IC).
- Dual-trace oscillocopes are the most effective way to measure the phase angle between two sinusoidal waveforms.

Important Equations

$$f = \frac{1}{T}$$

$$1 \text{ rad} \cong 57.3°$$

$$2\pi \text{ rad} = 360°$$

$$\text{Radians} = \left(\frac{\pi}{180°}\right) \times (\text{degrees})$$

$$\text{Degrees} = \left(\frac{180°}{\pi}\right) \times (\text{radians})$$

$$\omega = 2\pi f = \frac{2\pi}{T}$$

$$\cos \alpha = \sin(\alpha + 90°)$$

$$-\sin \alpha = \sin(\alpha \pm 180°)$$

$$\sin(-\alpha) = -\sin \alpha \qquad \cos(-\alpha) = \cos \alpha$$

$$G \text{ (average value)} = \frac{\text{algebraic sum of areas}}{\text{length of curve}}$$

$$I_{\text{eff}} = \sqrt{\frac{\text{area } (i^2(t))}{T}}$$

$$I_{\text{rms}} = \frac{1}{\sqrt{2}} I_m = 0.707 I_m$$

$$E_m = \sqrt{2} E_{\text{rms}} = 1.414 E_{\text{rms}}$$

$$V_{\text{rms}} = \sqrt{V_{\text{dc}}^2 + V_{\text{ac(rms)}}^2}$$

PROBLEMS

SECTION 12.2   Sinusoidal ac Voltage Characteristics and Definitions

1. For the sinusoidal waveform of Fig. 12.80:
   a. What is the peak value?
   b. What is the instantaneous value at 15 ms and at 20 ms?
   c. What is the peak-to-peak value of the waveform?
   d. What is the period of the waveform?
   e. How many cycles are shown?

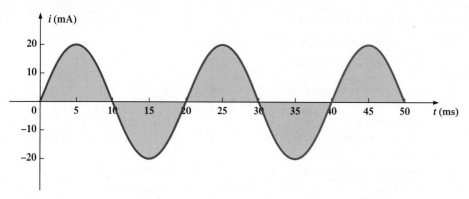

FIG. 12.80
Problem 1.

2. For the square-wave signal of Fig. 12.81:
   a. What is the peak value?
   b. What is the instantaneous value at 5 μs and at 11 μs?
   c. What is the peak-to-peak value of the waveform?
   d. What is the period of the waveform?
   e. How many cycles are shown?

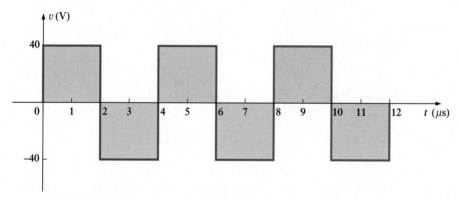

FIG. 12.81
Problem 2.

3. For the periodic waveform of Fig. 12.82:
   a. What is the peak value?
   b. What is the instantaneous value at 3 μs and at 9 μs?
   c. What is the peak-to-peak value of the waveform?
   d. What is the period of the waveform?
   e. How many cycles are shown?

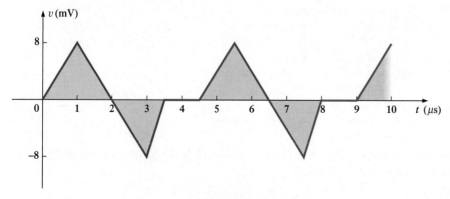

FIG. 12.82
Problem 3.

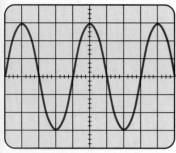

Vertical sensitivity = 50 mV/div.
Horizontal sensitivity = 10 μs/div.

FIG. 12.83

Problem 9.

SECTION 12.3   Frequency Spectrum

4. Find the period of a periodic waveform whose frequency is
   a. 25 Hz.   **b.** 40 MHz.
   c. 25 kHz.   **d.** 1 Hz.

5. Find the frequency of a repeating waveform whose period is
   a. 1/60 s.   **b.** 0.01 s.
   c. 40 ms.   **d.** 25 μs.

6. If a periodic waveform has a frequency of 20 Hz, how long (in seconds) will it take to complete five cycles?

7. Find the period of a sinusoidal waveform that completes 80 cycles in 24 ms.

8. What is the frequency of a periodic waveform that completes 42 cycles in 6 s?

9. For the oscilloscope pattern of Fig. 12.83:
   a. Determine the peak amplitude.
   b. Find the period.
   c. Calculate the frequency.
   Redraw the oscilloscope pattern if a +25 mV dc level were added to the input waveform.

SECTION 12.4   The Sinusoidal Waveform

10. Convert the following degrees to radians:
   a. 45°   **b.** 60°
   c. 270°   **d.** 170°

11. Convert the following radians to degrees:
   a. $\pi/4$ **b.** $\pi/6$
   c. $\frac{1}{10}\pi$ **d.** 0.6π

12. Find the angular velocity of a waveform with a period of
   a. 2 s.   **b.** 0.3 ms.
   c. 4 μs.   **d.** $\frac{1}{26}$ s.

13. Find the angular velocity of a waveform with a frequency of
   a. 50 Hz.   **b.** 600 Hz.
   c. 2 kHz.   **d.** 0.004 MHz.

14. Find the frequency and period of sine waves having an angular velocity of
   a. 754 rad/s.   **b.** 8.4 rad/s.
   c. 6000 rad/s.   **d.** $\frac{1}{16}$ rad/s.

*15. Given $f = 60$ Hz, determine how long it will take the sinusoidal waveform to pass through an angle of 45°.

*16. If a sinusoidal waveform passes through an angle of 30° in 5 ms, determine the angular velocity of the waveform.

SECTION 12.5   General Format for the Sinusoidal Voltage or Current

17. Find the amplitude and frequency of the following waves:
   a. 20 sin 377t **b.** 5 sin 754t
   c. 10^6 sin 10,000t **d.** −6.4 sin 942t

18. Sketch 5 sin 754t with the abscissa
   a. angle in degrees.   **b.** angle in radians.
   c. time in seconds.

*19. Sketch −7.6 sin 43.6t with the abscissa
   a. angle in degrees.   **b.** angle in radians.
   c. time in seconds.

20. If $e = 300$ sin 157t, how long (in seconds) does it take this waveform to complete 1/2 cycle?

21. Given $i = 0.5$ sin α, determine i at $\alpha = 72°$.

22. Given $v = 20$ sin α, determine v at $\alpha = 1.2\pi$.

*23. Given $v = 30 \times 10^{-3}$ sin α, determine the angles at which v will be 6 mV.

*24. If $v = 40$ V at $\alpha = 30$ and $t = 1$ ms, determine the mathematical expression for the sinusoidal voltage.

25. Sketch sin(377t + 60°) with the abscissa
   a. angle in degrees.   **b.** angle in radians.
   c. time in seconds.

26. Sketch the following waveforms:
   a. 50 sin(ωt + 0°)   **b.** 5 sin(ωt + 60°)
   c. 2 cos(ωt + 10°)   **d.** −20 sin(ωt + 2°)

27. Write the analytical expression for the waveforms of Fig. 12.84 with the phase angle in degrees.

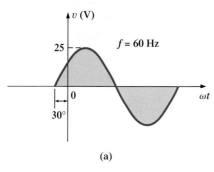

(a)

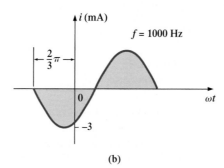

(b)

FIG. 12.84
Problem 27.

28. Write the analytical expression for the waveforms of Fig. 12.85 with the phase angle in degrees.

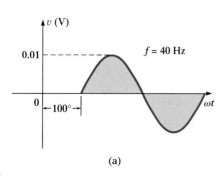

(a)

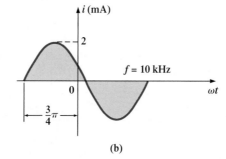

(b)

FIG. 12.85
Problem 28.

29. Find the phase relationship between the following waveforms:
$$v = 4 \sin(\omega t + 50°)$$
$$i = 6 \sin(\omega t + 40°)$$

30. Find the phase relationship between the following waveforms:
$$v = 25 \sin(\omega t - 80°)$$
$$i = 5 \times 10^{-3} \sin(\omega t - 10°)$$

31. Find the phase relationship between the following waveforms:
$$v = 0.2 \sin(\omega t - 60°)$$
$$i = 0.1 \sin(\omega t + 20°)$$

*32. Find the phase relationship between the following waveforms:
$$v = 2 \cos(\omega t - 30°)$$
$$i = 5 \sin(\omega t + 60°)$$

*33. Find the phase relationship between the following waveforms:
$$v = -4 \cos(\omega t + 90°)$$
$$i = -2 \sin(\omega t + 10°)$$

*34. The sinusoidal voltage $v = 200 \sin(2\pi\ 1000t + 60°)$ is plotted in Fig. 12.86. Determine the time t_1 when the waveform crosses the axis.

*35. The sinusoidal current $i = 4 \sin(50,000t - 40°)$ is plotted in Fig. 12.87. Determine the time t_1 when the waveform crosses the axis.

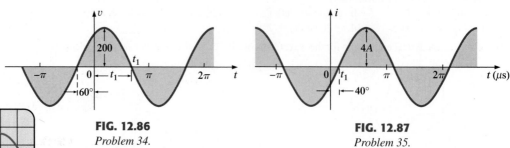

FIG. 12.86
Problem 34.

FIG. 12.87
Problem 35.

36. For the oscilloscope display of Fig. 12.88:
   a. Determine the period of the waveform.
   b. Determine the frequency of each waveform.
   c. Find the rms value of each waveform.
   d. Determine the phase shift between the two waveforms and determine which leads and which lags.

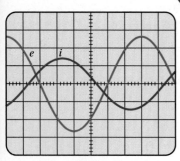

Vertical sensitivity = 0.5 V/div.
Horizontal sensitivity = 1 ms/div.

FIG. 12.88
Problem 36.

SECTION 12.7 Average Value

37. Find the average value of the periodic waveform of Fig. 12.89.

38. Find the average value of the periodic waveform of Fig. 12.90.

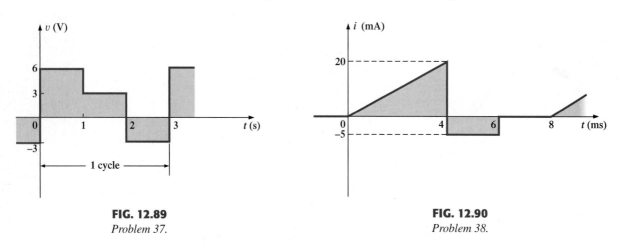

FIG. 12.89
Problem 37.

FIG. 12.90
Problem 38.

39. Find the average value of the periodic waveform of Fig. 12.91.

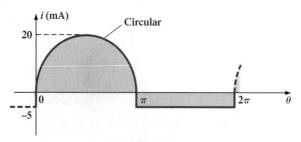

FIG. 12.91
Problem 39.

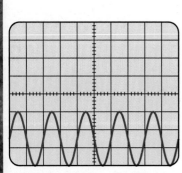

Vertical sensitivity = 10 mV/div.
Horizontal sensitivity = 0.2 ms/div.

FIG. 12.92
Problem 40.

40. For the waveform of Fig. 12.92:
   a. Determine the period.
   b. Find the frequency.
   c. Determine the average value.
   d. Sketch the resulting oscilloscope display if the vertical channel is switched from DC to AC.

*41. For the waveform of Fig. 12.93:
  a. Determine the period.
  b. Find the frequency.
  c. Determine the average value.
  d. Sketch the resulting oscilloscope display if the vertical channel is switched from DC to AC.

SECTION 12.8   Effective (rms) Values

42. Find the rms values of the following sinusoidal waveforms:
  a. $v = 140 \sin(377t + 60°)$
  b. $i = 6 \times 10^{-3} \sin(2\pi\ 1000t)$
  c. $v = 40 \times 10^{-6} \sin(2\pi\ 5000t + 30°)$

43. Write the sinusoidal expressions for voltages and currents having the following rms values at a frequency of 60 Hz with zero phase shift:
  a. 10 V   **b.**   50 mA
  c. 2 kV

44. Find the rms value of the periodic waveform of Fig. 12.94 over one full cycle.

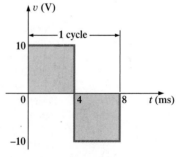

Vertical sensitivity = 10 mV/div.
Horizontal sensitivity = 10 μs/div.

FIG. 12.93
Problem 41.

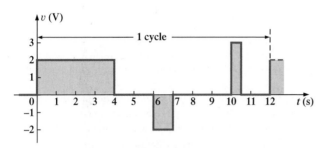

FIG. 12.94
Problem 44.

45. Find the rms value of the periodic waveform of Fig. 12.95 over one full cycle.

46. What are the average and rms values of the square wave of Fig. 12.96?

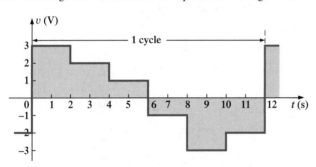

FIG. 12.95
Problem 45.

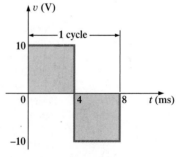

FIG. 12.96
Problem 46.

*47. For each waveform of Fig. 12.97, determine the period, frequency, average value, and rms value.

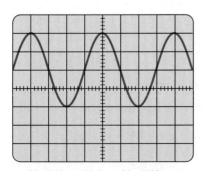

Vertical sensitivity = 20 mV/div.
Horizontal sensitivity = 10 μs/div.
(a)

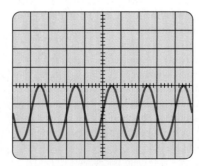

Vertical sensitivity = 0.2 V/div.
Horizontal sensitivity = 50 μs/div.
(b)

FIG. 12.97
Problem 47.

SECTION 12.9 ac Meters and Instruments

48. Determine the reading of the meter for each situation of Fig. 12.98.

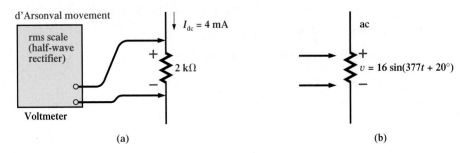

FIG. 12.98
Problem 48.

GLOSSARY

Alternating waveform   A waveform that oscillates above and below a defined reference level.

Angular velocity   The velocity with which a radius vector projecting a sinusoidal function rotates about its center.

Average value   The level of a waveform defined by the condition that the area enclosed by the curve above this level is exactly equal to the area enclosed by the curve below this level.

Clamp Meter®   A clamp-type instrument that will permit noninvasive current measurements and that can be used as a conventional voltmeter or ohmmeter.

Cycle   A portion of a waveform contained in one period of time.

Effective value   The equivalent dc value of any alternating voltage or current.

Electrodynamometer meters   Instruments that can measure both ac and dc quantities without a change in internal circuitry.

Frequency (*f*)   The number of cycles of a periodic waveform that occur in 1 second.

Frequency counter   An instrument that will provide a digital display of the frequency or period of a periodic time-varying signal.

Instantaneous value   The magnitude of a waveform at any instant of time, denoted by lowercase letters.

Lagging waveform   A waveform that crosses the time axis at a point in time later than another waveform of the same frequency.

Leading waveform   A waveform that crosses the time axis at a point in time ahead of another waveform of the same frequency.

Oscilloscope   An instrument that will display, through the use of a cathode-ray tube, the characteristics of a time-varying signal.

Peak amplitude   The maximum value of a waveform as measured from its average, or mean, value, denoted by uppercase letters.

Peak-to-peak value   The magnitude of the total swing of a signal from positive to negative peaks. The sum of the absolute values of the positive and negative peak values.

Peak value   The maximum value of a waveform, denoted by uppercase letters.

Period (*T*)   The time interval between successive repetitions of a periodic waveform.

Periodic waveform   A waveform that continually repeats itself after a defined time interval.

Phase relationship   An indication of which of two waveforms leads or lags the other, and by how many degrees or radians.

Radian (rad)   A unit of measure used to define a particular segment of a circle. One radian is approximately equal to $57.3°$; 2π rad are equal to $360°$.

Root-mean-square (rms) value   The root-mean-square or effective value of a waveform.

Sinusoidal ac waveform   An alternating waveform of unique characteristics that oscillates with equal amplitude above and below a given axis.

VOM   A multimeter with the capability to measure resistance and both ac and dc levels of current and voltage.

Waveform   The path traced by a quantity, plotted as a function of some variable such as position, time, degrees, temperature, and so on.

The Basic Elements
and Phasors

OBJECTIVES

- Be able to add and subtract sinusoidal voltages or currents.
- Understand how to use the phasor format to add and subtract sinusoidal waveforms.
- Become familiar with complex numbers and the rectangular and polar formats.
- Learn how to perform all the basic mathematical operations using complex numbers.
- Be able to use phasor notation to apply Kirchhoff's voltage and current laws to ac networks.
- Become proficient in the use of the calculator or Mathcad to work with complex numbers.
- Understand the response of resistors, inductors, and capacitors to an ac signal.
- Be able to apply phasor notation to the analysis of the basic elements.
- Understand how to calculate the real power to a resistive element and the reactive power to an inductive or a capacitive element.
- Become aware of how the resistance of a resistor or the reactance of a capacitor or an inductor will change with frequency.
- Be able to use PSpice or EWB to analyze the response of the basic elements to an ac signal.

13.1 INTRODUCTION

In this chapter we will tackle the problem of adding and subtracting sinusoidal waveforms, and we will investigate the response of the basic elements to an ac signal. The early sections will have a mathematical orientation that may seem tedious and overdone, but these sections are necessary in order for us to continue with the material to appear in later chapters. In time, when you can obtain results with a minimum of effort and time, you will appreciate the benefits associated with learning the process.

The notation to be introduced will allow us to use all the laws, theorems, and methods of analysis introduced for dc networks. The result is that the coverage of ac networks will parallel the material covered for dc circuits. That is, the sequence of material covered will be the same, with the only major difference being the mathematical operations.

13.2 ADDING AND SUBTRACTING SINUSOIDAL WAVEFORMS

In the analysis of dc circuits, you learned that currents and voltages must be added and subtracted to apply laws such as Kirchhoff's current and voltage laws. Such operations didn't present any particular problem because we were dealing with fixed numbers that may or may not have had a sign to worry about. For ac circuits, however, the currents and voltages will be **sinusoidal,** will **vary with time,** and may be **out of phase.** If the currents and voltages were all in phase, the approach would be quite straightforward: Just add or subtract the peak values. However, we will find in the chapters to follow that the sinusoidal voltages will often be out of phase, so some other approach must be developed.

In Fig. 13.1, there are two out-of-phase sinusoidal voltages of the same frequency that have to be added. At this stage it is important to realize the following:

The methods to be described in this chapter are only for sinusoidal waveforms of the same frequency.

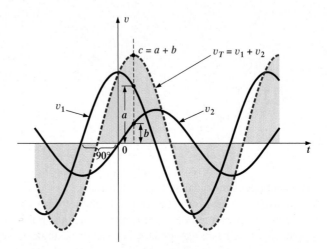

FIG. 13.1

Adding two sinusoidal waveforms on a point-by-point basis.

Voltage v_1 in Fig. 13.1 has a peak value at $t = 0$ s while the other voltage is passing through the origin. Using the terminology of the previous chapter, voltage v_1 is **leading** voltage v_2 by 90°. If both voltages were carefully plotted on graph paper using the same vertical and horizontal scales, the two waveforms could be added at a number of points along the horizontal axis, as demonstrated by $c = a + b$ at one point in the figure. Connecting the resulting data points would result in the waveform v_T. Note that in some regions the sum would be positive while in others it would be negative. However,

when you are adding or subtracting sinusoidal waveforms of the same frequency, the resulting waveform will also be sinusoidal with the same frequency.

This summing of points along the horizontal axis would be a long and tedious process, with results that would be a direct function of the accuracy with which the plots were drawn and added. Accuracy levels beyond tenths place would be difficult to obtain.

An approach that takes considerably less time and effort uses a rotating vector (appearing in Fig. 12.16 of Section 12.4) to generate the sinusoidal waveform. In Fig. 13.2, the waveforms of Fig. 13.1 were given peak values and were carefully plotted. Voltage $v_1 = 2 \sin(\omega t + 90°)$, and voltage $v_2 = 1 \sin \omega t$. Looking back at Fig. 12.16, we recognize that the vector representing voltage v_2 would be at an angle of 0° as shown in Fig. 13.2(a) the instant the waveform is at $t = 0$ s ($\alpha = 0°$) for the waveform of Fig. 13.2(b). The vector representing waveform v_1 would be at 90° as shown in Fig. 13.2(a) at the **exact same instant.** The result is two vectors in Fig. 13.2(a) that are 90° apart and that represent the two sinusoidal waveforms at the instant $t = 0$ s or $\alpha = 0°$. In other words, **the vectors of Fig. 13.2(a) are a "snapshot" of the rotating vectors of Fig. 12.16 when $t = 0$ s or $\alpha = 0°$.** The radius vectors appearing in Fig. 13.2(a) are called **phasors** when applied to electric circuits.

If we now add the two vectors of Fig. 13.2(a) to obtain the vector at 2.24 V, we will find that it is the peak value of the sinusoidal waveform representing the sum $v_T = v_1 + v_2$. The angle of 63.43° associated with the resultant vector is the phase angle of the sum. We can then return to Fig. 13.2(b), scale off the peak value, insert the phase shift, and quickly plot the resultant voltage as appearing in Fig. 13.2(b). This procedure is certainly much faster than adding the two waveforms point by point, and it can provide accuracy levels to the thousandths place.

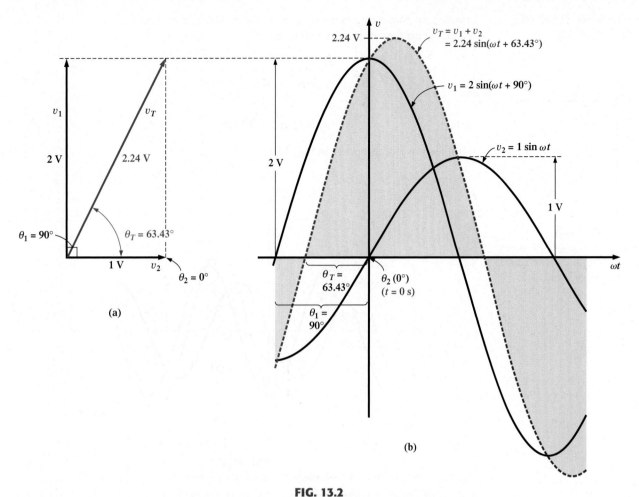

FIG. 13.2

(a) The phasor representation of the sinusoidal waveforms of Fig. 13.2(b); (b) finding the sum of two sinusoidal waveforms of v_1 and v_2.

It is now clear from the above example that if we represent sinusoidal waveforms by their peak value and phase angle, the sum or difference of the two waveforms can be determined using vector algebra. Note in Fig. 13.2(a) that the angle associated with each vector is actually the phase angle associated with the standard format for a sinusoidal waveform. The phasor notation for the sinusoidal function is defined by the following:

$$v_1 = V_m \sin(\omega t + \theta_1) \Rightarrow \mathbf{V}_1 = V_m \angle \theta_1$$
$$i_1 = I_m \sin(\omega t + \theta_2) \Rightarrow \mathbf{I}_1 = I_m \angle \theta_2$$

$$(13.1)$$

Any pattern that includes one or more phasors such as appearing in Fig. 13.2(a) is called a **phasor diagram.**

In general, therefore,

finding the sum or the difference of two sinusoidal waveforms simply requires putting each in the phasor format and performing the required vector algebra.

EXAMPLE 13.1 Find the sum of the two sinusoidal currents i_1 and i_2 if

$$i_1 = 3 \sin \omega t$$

and

$$i_2 = 4 \sin(\omega t - 90°)$$

Solution: Placing both in phasor form:

$$\mathbf{I}_1 = 3 \text{ A } \angle 0°$$

$$\mathbf{I}_2 = 4 \text{ A } \angle -90°$$

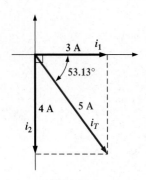

FIG. 13.3

*Determining the magnitude and the
phase angle for the current equal to
the sum of i_1 and i_2.*

Drawing the phasor diagram will result in the display of Fig. 13.3.

Adding the two vectorially will result in

$$|I_T| = \sqrt{(3\text{ A})^2 + (4\text{ A})^2} = 5\text{ A}$$

and

$$\theta = \tan^{-1}\frac{4}{3} = 53.13°$$

so that

$$\mathbf{I}_T = \mathbf{5\text{ A}} \angle \mathbf{-53.13°}$$

The three waveforms appear in Fig. 13.4.

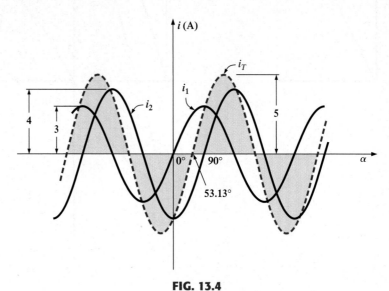

FIG. 13.4

Plotting i_T, i_1, and i_2 for Example 13.1.

13.3 COMPLEX NUMBERS

The procedure described above for finding the sum or the difference of two sinusoidal waveforms is fine if the waveforms happen to have a phase shift of 0°, 90°, or 180°. However, if the phase angle between the waveforms is some number between those just listed, you must turn to a mathematical approach called **complex algebra;** otherwise, the mathematics can get relatively confusing and difficult.

A complex number defines a point in a two-dimensional plane established by two axes at 90° to one another (in quadrature).

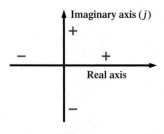

FIG. 13.5

*Defining the real and imaginary
axes of a complex plane.*

That point can also define a vector drawn from the origin to the point. The horizontal axis is called the **real axis,** and the vertical axis is called the **imaginary axis.** Both are labeled in Fig. 13.5. Every number from 0 to $\pm\infty$ can be represented by some point on the real axis. Before this system of complex numbers was developed, mathematicians believed that any number not on the axis could not exist—hence the term *imaginary.*

In the complex plane, the horizontal or real axis represents all positive numbers to the right of the imaginary axis, and all negative numbers to the left of the imaginary axis. All positive imaginary numbers are represented above the real axis, and all negative imaginary numbers, below the real axis. The symbol j (sometimes i) is used to denote the imaginary component.

Two forms are used to represent a complex number: **rectangular** and **polar.** Each can define the vector drawn from the origin to a point in the two-dimensional plane. Once we become familiar with the two forms and the mathematics associated with each, we will be able to perform algebraic operations such as addition and subtraction of any two or more sinusoidal voltages or currents with ease and with confidence in our results. It is a session in mathematics that must be examined in depth because the operations to be described will appear throughout the remaining chapters of the text.

Rectangular Form

The format for the **rectangular form** is

$$\mathbf{C} = X + jY \qquad (13.2)$$

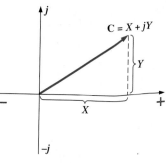

FIG. 13.6
Defining the rectangular form.

as shown in Fig. 13.6. The letter **C** was chosen from the word "complex." The **boldface** notation is for **any number with magnitude and direction.** The *italic* is for magnitude only.

EXAMPLE 13.2   Sketch the following complex numbers in the complex plane:

a. $\mathbf{C} = 3 + j4$
b. $\mathbf{C} = 0 - j6$
c. $\mathbf{C} = -10 - j20$

Solutions:

a. See Fig. 13.7.
b. See Fig. 13.8.
c. See Fig. 13.9.

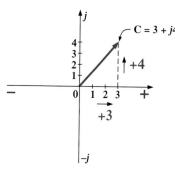

FIG. 13.7
Example 13.2(a).

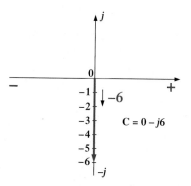

FIG. 13.8
Example 13.2(b).

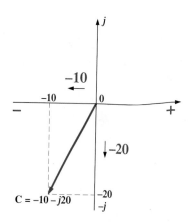

FIG. 13.9
Example 13.2(c).

Polar Form

The format for the **polar form** is

$$\mathbf{C} = Z \angle \theta \qquad (13.3)$$

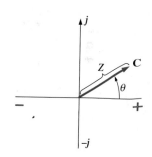

FIG. 13.10
Defining the polar form.

where Z indicates magnitude only and θ is **always measured counterclockwise (CCW)** from the *positive real axis*, as shown in Fig. 13.10. Angles measured in the clockwise direction from the positive real axis must have a negative sign associated with them.

A negative sign in front of the polar form has the effect shown in Fig. 13.11. Note that it results in a complex number directly opposite the complex number with a positive sign.

$$-\mathbf{C} = -Z \angle \theta = Z \angle \theta \pm 180° \qquad (13.4)$$

EXAMPLE 13.3   Sketch the following complex numbers in the complex plane:

a. $\mathbf{C} = 5 \angle 30°$
b. $\mathbf{C} = 7 \angle -120°$
c. $\mathbf{C} = -4.2 \angle 60°$

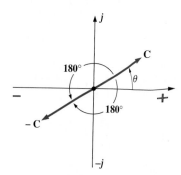

FIG. 13.11
Demonstrating the effect of a negative sign on the polar form.

Solutions:

a. See Fig. 13.12.
b. See Fig. 13.13.
c. See Fig. 13.14.

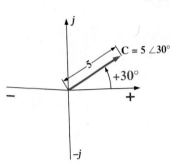

FIG. 13.12
Example 13.3(a).

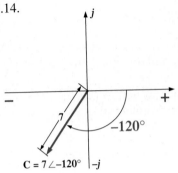

FIG. 13.13
Example 13.3(b).

$$C = -4.2 \angle 60° = 4.2 \angle 60° + 180°$$
$$= 4.2 \angle +240°$$

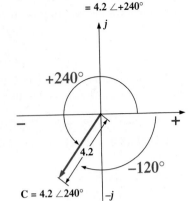

FIG. 13.14
Example 13.3(c).

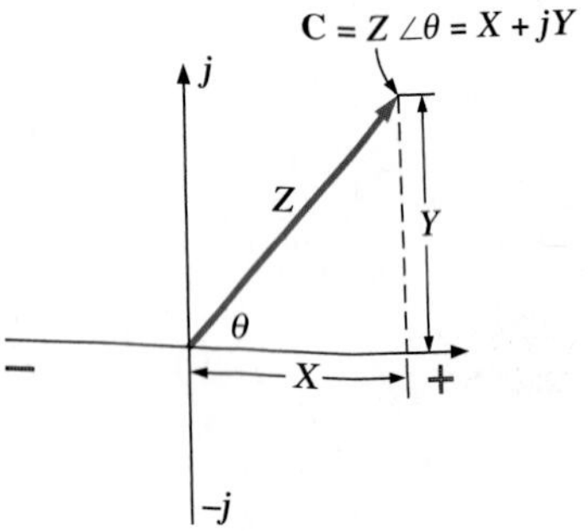

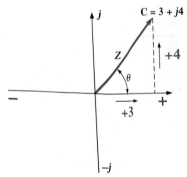

FIG. 13.15
Conversion between forms.

Conversion between Forms

The two forms are related by the following equations, as illustrated in Fig. 13.15.

Rectangular to Polar

$$Z = \sqrt{X^2 + Y^2} \qquad \textbf{(13.5)}$$

$$\theta = \tan^{-1} \frac{Y}{X} \qquad \textbf{(13.6)}$$

Polar to Rectangular

$$X = Z \cos \theta \qquad \textbf{(13.7)}$$

$$Y = Z \sin \theta \qquad \textbf{(13.8)}$$

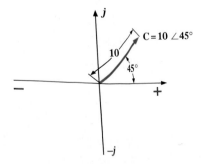

FIG. 13.16
Example 13.4.

EXAMPLE 13.4 Convert the following from rectangular to polar form:

$$\mathbf{C} = 3 + j4 \qquad \text{(Fig. 13.16)}$$

Solution:

$$Z = \sqrt{(3)^2 + (4)^2} = \sqrt{25} = 5$$

$$\theta = \tan^{-1}\left(\frac{4}{3}\right) = 53.13°$$

and $$\mathbf{C} = 5 \angle \mathbf{53.13°}$$

EXAMPLE 13.5 Convert the following from polar to rectangular form:

$$\mathbf{C} = 10 \angle 45° \qquad \text{(Fig. 13.17)}$$

Solution:

$$X = 10 \cos 45° = (10)(0.707) = 7.07$$

$$Y = 10 \sin 45° = (10)(0.707) = 7.07$$

and $$\mathbf{C} = \mathbf{7.07} + j\,\mathbf{7.07}$$

FIG. 13.17
Example 13.5.

If the complex number should appear in the second, third, or fourth quadrant, simply convert it in that quadrant, and carefully determine the proper angle to be associated with the magnitude of the vector.

EXAMPLE 13.6 Convert the following from rectangular to polar form:

$$C = -6 + j\,3 \qquad \text{(Fig. 13.18)}$$

Solution:

$$Z = \sqrt{(6)^2 + (3)^2} = \sqrt{45} = 6.71$$

$$\beta = \tan^{-1}\left(\frac{3}{6}\right) = 26.57°$$

$$\theta = 180° - 26.57° = 153.43°$$

and

$$C = 6.71 \angle 153.43°$$

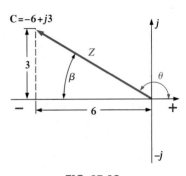

FIG. 13.18
Example 13.6.

EXAMPLE 13.7 Convert the following from polar to rectangular form:

$$C = 10 \angle 230° \qquad \text{(Fig. 13.19)}$$

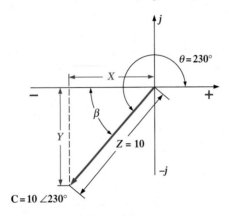

FIG. 13.19
Example 13.7.

Solution:

$$X = Z \cos \beta = 10 \cos(230° - 180°) = 10 \cos 50° = (10)(0.6428) = 6.428$$
$$Y = Z \sin \beta = 10 \sin 50° = (10)(0.7660) = 7.660$$

and

$$C = -6.428 - j\,7.660$$

13.4 MATHEMATICAL OPERATIONS WITH COMPLEX NUMBERS

Complex numbers lend themselves readily to the basic mathematical operations of addition, subtraction, multiplication, and division. A few basic rules and definitions must be understood before we consider these operations.

Definitions

The Symbol *j* Let us first examine the symbol *j* associated with imaginary numbers. By definition,

$$\boxed{j = \sqrt{-1}} \qquad \qquad \textbf{(13.9)}$$

Thus,

$$\boxed{j^2 = -1} \qquad \qquad \textbf{(13.10)}$$

and
$$j^3 = j^2 j = -1j = -j$$

with
$$j^4 = j^2 j^2 = (-1)(-1) = +1$$
$$j^5 = j$$

and so on. Further,

$$\frac{1}{j} = (1)\left(\frac{1}{j}\right) = \left(\frac{j}{j}\right)\left(\frac{1}{j}\right) = \frac{j}{j^2} = \frac{j}{-1}$$

and

$$\boxed{\frac{1}{j} = -j} \tag{13.11}$$

The Complex Conjugate The **conjugate** or **complex conjugate** of a complex number can be found by simply changing the sign of the imaginary part in the rectangular form or by using the negative of the angle of the polar form. For example, the conjugate of

$$\mathbf{C} = 2 + j3$$

is
$$2 - j3$$

as shown in Fig. 13.20. The conjugate of

$$\mathbf{C} = 2 \angle 30°$$

is
$$2 \angle -30°$$

as shown in Fig. 13.21.

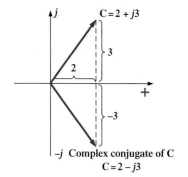

$C = 2 + j3$

FIG. 13.20

Defining the complex conjugate of a complex number in rectangular form.

The Reciprocal The **reciprocal** of a complex number is 1 divided by the complex number. For example, the reciprocal of

$$\mathbf{C} = X + jY$$

is
$$\frac{1}{X + jY}$$

and of $Z \angle \theta$,

$$\frac{1}{Z \angle \theta}$$

We are now prepared to consider the four basic operations of *addition, subtraction, multiplication,* and *division* with complex numbers.

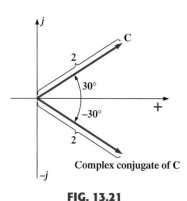

FIG. 13.21

Defining the complex conjugate of a complex number in polar form.

Addition and Subtraction

The addition or subtraction of complex numbers will normally be performed in the rectangular form because adding or subtracting complex numbers in the polar form requires that they have the same angle or that they be 180° out of phase. If the numbers are presented in polar form, it is simply a matter of converting to the rectangular form before performing the required operation.

For all the mathematical operations, the resulting equations will be boxed off and numbered. However, in general, **it is best to simply remember the required operations in the written form** rather than trying to memorize the equation. For instance, in words,

the addition or subtraction of two complex numbers requires that the real and imaginary parts be worked on independently.

For example, if

$$\mathbf{C_1} = \pm X_1 \pm jY_1 \quad \text{and} \quad \mathbf{C_2} = \pm X_2 \pm jY_2$$

then

$$\boxed{\mathbf{C_1} + \mathbf{C_2} = (\pm X_1 \pm X_2) + j(\pm Y_1 \pm Y_2)} \tag{13.12}$$

and

$$\boxed{\mathbf{C_1} - \mathbf{C_2} = [\pm X_1 - (\pm X_2)] + j[\pm Y_1 - (\pm Y_2)]} \tag{13.13}$$

As indicated above, it is usually best to simply remember the boldface statement rather than memorizing Eqs. (13.12) and (13.13). A few examples will demonstrate the process.

EXAMPLE 13.8

a. Add $\mathbf{C}_1 = 2 + j4$ and $\mathbf{C}_2 = 3 + j1$.
b. Add $\mathbf{C}_1 = 3 + j6$ and $\mathbf{C}_2 = -6 + j3$.

Solutions:

a. By Eq. (13.12):

$$\mathbf{C}_1 + \mathbf{C}_2 = (2 + 3) + j(4 + 1) = \mathbf{5 + j5}$$

Note Fig. 13.22. The alternative method is

$$
\begin{array}{c}
2 + j\,4 \\
3 + j\,1 \\
\hline
\downarrow \qquad \downarrow \\
\mathbf{5 + j\,5}
\end{array}
$$

b. By Eq. (13.12):

$$\mathbf{C}_1 + \mathbf{C}_2 = (3 - 6) + j(6 + 3) = \mathbf{-3 + j9}$$

Note Fig. 13.23. The alternative method is

$$
\begin{array}{c}
3 + j\,6 \\
-6 + j\,3 \\
\hline
\downarrow \qquad \downarrow \\
\mathbf{-3 + j\,9}
\end{array}
$$

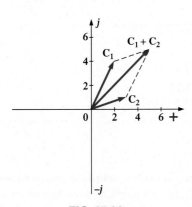

FIG. 13.22
Example 13.8(a).

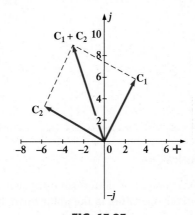

FIG. 13.23
Example 13.8(b).

EXAMPLE 13.9

a. Subtract $\mathbf{C}_2 = 1 + j4$ from $\mathbf{C}_1 = 4 + j6$.
b. Subtract $\mathbf{C}_2 = -2 + j5$ from $\mathbf{C}_1 = +3 + j3$.

Solutions:

a. By Eq. (13.13):

$$\mathbf{C}_1 - \mathbf{C}_2 = (4 - 1) + j(6 - 4) = \mathbf{3 + j2}$$

Note Fig. 13.24. The alternative method is

$$
\begin{array}{c}
4 + j\,6 \\
-(1 + j\,4) \\
\hline
\downarrow \qquad \downarrow \\
\mathbf{3 + j\,2}
\end{array}
$$

b. By Eq. (13.13):

$$\mathbf{C}_1 - \mathbf{C}_2 = [3 - (-2)] + j(3 - 5) = \mathbf{5 - j2}$$

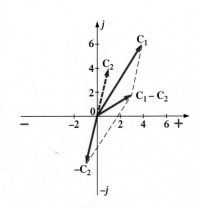

FIG. 13.24
Example 13.9(a).

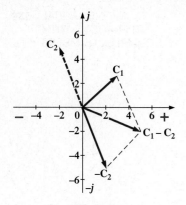

FIG. 13.25

Example 13.9(b).

Note Fig. 13.25. The alternative method is

$$
\begin{array}{r}
3 + j\,3 \\
-(-2 + j\,5) \\
\hline
\downarrow \qquad \downarrow \\
\mathbf{5 - j\,2}
\end{array}
$$

Addition or subtraction cannot be performed in polar form unless the complex numbers have the same angle θ or unless they differ only by multiples of 180°.

EXAMPLE 13.10

a. $2 \angle 45° + 3 \angle 45° = \mathbf{5 \angle 45°}$
Note Fig. 13.26.

b. $2 \angle 0° - 4 \angle 180° = \mathbf{6 \angle 0°}$
Note Fig. 13.27.

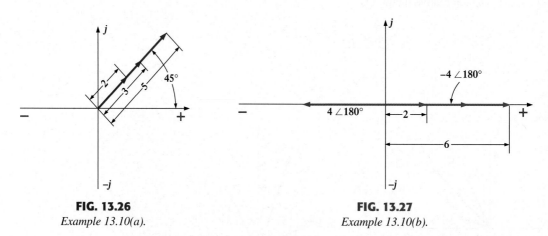

FIG. 13.26

Example 13.10(a).

FIG. 13.27

Example 13.10(b).

Multiplication

Multiplication can be performed using either form, although **it is usually much easier to perform the operation in the polar form.** In some cases it might be advantageous to convert to the polar form before performing the operation, but this choice is usually dependent on the complex numbers. For the rectangular form, the resulting equation will again be generated, but in most cases, it is best to simply work with the components of the complex numbers rather than memorize the equation.

Given

$$\mathbf{C}_1 = X_1 + j\,Y_1 \quad \text{and} \quad \mathbf{C}_2 = X_2 + j\,Y_2$$

then $\mathbf{C}_1 \cdot \mathbf{C}_2$:

$$
\begin{array}{r}
X_1 + j\,Y_1 \\
X_2 + j\,Y_2 \\
\hline
X_1 X_2 + j\,Y_1 X_2 \\
+ j\,X_1 Y_2 + j^2\,Y_1 Y_2 \\
\hline
X_1 X_2 + j\,(X_1 Y_1 X_2 + X_1 Y_2) + Y_1 Y_2(-1)
\end{array}
$$

and

$$\boxed{\mathbf{C}_1 \cdot \mathbf{C}_2 = (X_1 X_2 - Y_1 Y_2) + j\,(Y_1 X_2 + X_1 Y_2)} \tag{13.14}$$

In *polar* form, the magnitudes are multiplied and the angles added algebraically. For example, for

$$\mathbf{C}_1 = Z_1 \angle \theta_1 \quad \text{and} \quad \mathbf{C}_2 = Z_2 \angle \theta_2$$

we write

$$\boxed{\mathbf{C}_1 \cdot \mathbf{C}_2 = Z_1 Z_2 \ \underline{/\theta_1 + \theta_2}} \tag{13.15}$$

EXAMPLE 13.11

a. Find $\mathbf{C}_1 \cdot \mathbf{C}_2$ if

$$\mathbf{C}_1 = 2 + j3 \quad \text{and} \quad \mathbf{C}_2 = 5 + j10$$

b. Find $\mathbf{C}_1 \cdot \mathbf{C}_2$ if

$$\mathbf{C}_1 = -2 - j3 \quad \text{and} \quad \mathbf{C}_2 = +4 - j6$$

Solutions:

a. Using the format above, we have

$$\mathbf{C}_1 \cdot \mathbf{C}_2 = [(2)(5) - (3)(10)] + j[(3)(5) + (2)(10)] = \mathbf{-20 + j\,35}$$

b. Without using the format, we obtain

$$
\begin{array}{r}
-2 - j\,3 \\
+4 - j\,6 \\
\hline
-8 - j\,12 \\
+ j\,12 + j^2 18 \\
\hline
-8 + j(-12 + 12) - 18
\end{array}
$$

and

$$\mathbf{C}_1 \cdot \mathbf{C}_2 = \mathbf{-26} = \mathbf{26 \angle 180°}$$

EXAMPLE 13.12

a. Find $\mathbf{C}_1 \cdot \mathbf{C}_2$ if

$$\mathbf{C}_1 = 5 \angle 20° \quad \text{and} \quad \mathbf{C}_2 = 10 \angle 30°$$

b. Find $\mathbf{C}_1 \cdot \mathbf{C}_2$ if

$$\mathbf{C}_1 = 2 \angle -40° \quad \text{and} \quad \mathbf{C}_2 = 7 \angle +120°$$

Solutions:

a. $\mathbf{C}_1 \cdot \mathbf{C}_2 = (5 \angle 20°)(10 \angle 30°) = (5)(10) \underline{/20° + 30°} = \mathbf{50 \angle 50°}$

b. $\mathbf{C}_1 \cdot \mathbf{C}_2 = (2 \angle -40°)(7 \angle +120°) = (2)(7) \underline{/-40° + 120°} = \mathbf{14 \angle +80°}$

To multiply a complex number in rectangular form by a real number requires that both the real part and the imaginary part be multiplied by the real number. For example,

$$(10)(2 + j3) = 20 + j30$$

and

$$50 \angle 0°(0 + j6) = j300 = 300 \angle 90°$$

Division

Division can also be performed using either form, although **it is usually much easier to perform the operation in the polar form.** In some cases it might be advantageous to convert to the polar form before performing the operation, but this choice is usually dependent on the complex numbers. For the rectangular form the resulting equation will again be generated, but in most cases, it is best to simply work with the components of the complex numbers rather than memorize the equation.

To divide two complex numbers in *rectangular* form, multiply the numerator and the denominator by the conjugate of the denominator and the resulting real and imaginary parts collected. That is, if

$$\mathbf{C}_1 = X_1 + jY_1 \quad \text{and} \quad \mathbf{C}_2 = X_2 + jY_2$$

then

$$\frac{\mathbf{C}_1}{\mathbf{C}_2} = \frac{(X_1 + jY_1)(X_2 - jY_2)}{(X_2 + jY_2)(X_2 - jY_2)}$$

$$= \frac{(X_1 X_2 + Y_1 Y_2) + j(X_2 Y_1 - X_1 Y_2)}{X_2^2 + Y_2^2}$$

and
$$\frac{\mathbf{C}_1}{\mathbf{C}_2} = \frac{X_1X_2 + Y_1Y_2}{X_2^2 + Y_2^2} + j\frac{X_2Y_1 - X_1Y_2}{X_2^2 + Y_2^2}$$
(13.16)

In *polar* form, division is accomplished by simply dividing the magnitude of the numerator by the magnitude of the denominator and subtracting the angle of the denominator from that of the numerator. That is, for

$$\mathbf{C}_1 = Z_1 \angle\theta_1 \quad \text{and} \quad \mathbf{C}_2 = Z_2 \angle\theta_2$$

we write

$$\frac{\mathbf{C}_1}{\mathbf{C}_2} = \frac{Z_1}{Z_2}\,\underline{/\theta_1 - \theta_2}$$
(13.17)

EXAMPLE 13.13

a. Find $\mathbf{C}_1/\mathbf{C}_2$ if $\mathbf{C}_1 = 1 + j4$ and $\mathbf{C}_2 = 4 + j5$.
b. Find $\mathbf{C}_1/\mathbf{C}_2$ if $\mathbf{C}_1 = -4 - j8$ and $\mathbf{C}_2 = +6 - j1$.

Solutions:

a. By Eq. (13.16):

$$\frac{\mathbf{C}_1}{\mathbf{C}_2} = \frac{(1)(4) + (4)(5)}{4^2 + 5^2} + j\frac{(4)(4) - (1)(5)}{4^2 + 5^2}$$

$$= \frac{24}{41} + \frac{j11}{41} \cong \mathbf{0.585 + j\,0.268}$$

b. Using an alternative method, we obtain

$$
\begin{array}{r}
-4 - j8 \\
+6 + j1 \\
\hline
-24 - j48 \\
-j4 - j^2 8 \\
\hline
-24 - j52 + 8 = -16 - j52
\end{array}
$$

$$
\begin{array}{r}
+6 - j1 \\
+6 + j1 \\
\hline
36 + j6 \\
-j6 - j^2 1 \\
\hline
36 + 0 + 1 \quad = 37
\end{array}
$$

and
$$\frac{\mathbf{C}_1}{\mathbf{C}_2} = \frac{-16 - j52}{37} = \mathbf{-0.432 - j1.405}$$

To divide a complex number in rectangular form by a real number, you must divide both the real part and the imaginary part by the real number. For example,

$$\frac{8 + j10}{2} = 4 + j5$$

and
$$\frac{6.8 - j0}{2} = 3.4 - j0 = 3.4 \angle 0°$$

EXAMPLE 13.14

a. Find $\mathbf{C}_1/\mathbf{C}_2$ if $\mathbf{C}_1 = 15 \angle 10°$ and $\mathbf{C}_2 = 2 \angle 7°$.
b. Find $\mathbf{C}_1/\mathbf{C}_2$ if $\mathbf{C}_1 = 8 \angle 120°$ and $\mathbf{C}_2 = 16 \angle -50°$.

a. $\dfrac{\mathbf{C}_1}{\mathbf{C}_2} = \dfrac{15\angle 10°}{2\angle 7°} = \dfrac{15}{2}\,\underline{/10° - 7°} = \mathbf{7.5\angle 3°}$

b. $\dfrac{\mathbf{C}_1}{\mathbf{C}_2} = \dfrac{8\angle 120°}{16\angle -50°} = \dfrac{8}{16}\,\underline{/120° - (-50°)} = \mathbf{0.5\angle 170°}$

We obtain the **reciprocal** in the rectangular form by multiplying the numerator and the denominator by the complex conjugate of the denominator:

$$\frac{1}{X + jY} = \left(\frac{1}{X + jY}\right)\left(\frac{X - jY}{X - jY}\right) = \frac{X - jY}{X^2 + Y^2}$$

and

$$\boxed{\frac{1}{X + jY} = \frac{X}{X^2 + Y^2} - j\frac{Y}{X^2 + Y^2}} \qquad (13.18)$$

In polar form, the reciprocal is

$$\boxed{\frac{1}{Z\angle\theta} = \frac{1}{Z}\angle -\theta} \qquad (13.19)$$

A concluding example using the four basic operations follows.

EXAMPLE 13.15 Perform the following operations, leaving the answer in polar or rectangular form:

a. $\dfrac{(2 + j3) + (4 + j6)}{(7 + j7) - (3 - j3)} = \dfrac{(2 + 4) + j(3 + 6)}{(7 - 3) + j(7 + 3)}$

$= \dfrac{(6 + j9)(4 - j10)}{(4 + j10)(4 - j10)}$

$= \dfrac{[(6)(4) + (9)(10)] + j[(4)(9) - (6)(10)]}{4^2 + 10^2}$

$= \dfrac{114 - j24}{116} = \mathbf{0.983 - j0.207}$

b. $\dfrac{(50\angle 30°)(5 + j5)}{10\angle -20°} = \dfrac{(50\angle 30°)(7.07\angle 45°)}{10\angle -20°} = \dfrac{353.5\angle 75°}{10\angle -20°}$

$= 35.35\,\underline{/75° - (-20°)} = \mathbf{35.35\angle 95°}$

c. $\dfrac{(2\angle 20°)^2(3 + j4)}{8 - j6} = \dfrac{(2\angle 20°)(2\angle 20°)(5\angle 53.13°)}{10\angle -36.87°}$

$= \dfrac{(4\angle 40°)(5\angle 53.13°)}{10\angle -36.87°} = \dfrac{20\angle 93.13°}{10\angle -36.87°}$

$= 2\,\underline{/93.13° - (-36.87°)} = \mathbf{2.0\angle 130°}$

d. $3\angle 27° - 6\angle -40° = (2.673 + j1.362) - (4.596 - j3.857)$

$= (2.673 = 4.596) + j(1.362 + 3.857)$

$= \mathbf{-1.923 + j5.219}$

FIG. 13.28
TI-86 scientific calculator.
(Courtesy of Texas
Instruments, Inc.)

FIG. 13.29
*Converting 3 − j4 to polar form
using the TI-86 calculator
[Example 13.16(a)].*

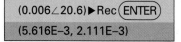

FIG. 13.30
*Converting 0.006 ∠20.6° to
rectangular form using the TI-86
calculator [Example 13.16(b)].*

Using a Calculator with Complex Numbers

The process of converting from one form to another or working through a lengthy series of calculations with complex numbers can be very time-consuming and often frustrating if you lose a minus sign or a decimal point, thereby making the solution meaningless. Fortunately, technologists of today have calculators and computer software to make the process measurably easier with higher degrees of reliability and accuracy. In this section, we will look at the use of the TI-86 scientific calculator and leave the computer software Mathcad for the computer section (Section 13.11) of the chapter.

The TI-86 calculator of Fig. 13.28 is only one of numerous calculators that can convert from one form to another and perform lengthy calculations with complex numbers in a concise, neat form. Not all of the details of using a specific calculator will be included here because each has its own format and sequence of steps. However, the basic operations with the TI-86 will be included primarily to demonstrate the ease with which the conversions can be made and the format for more complex operations.

For the TI-86 calculator, you must first call up the 2nd function CPLX from the keyboard, which results in a menu at the bottom of the display including conj, real, imag, abs, and angle. If we choose the key MORE , ▶ Rec and ▶ Pol will appear as options (for the conversion process). To convert from one form to another, simply enter the current form in brackets, with a comma between components for the rectangular form and an angle symbol for the polar form. Follow this form with the operation to be performed, and press the ENTER key— the result will appear on the screen in the desired format.

EXAMPLE 13.16 This example is for demonstration purposes only. It is not expected that all readers will have a TI-86 calculator. The sole purpose of the example is to demonstrate the power of today's calculators.

Using the TI-86 calculator, perform the following conversions:

a. $3 - j4$ to polar form.
b. $0.006 \angle 20.6°$ to rectangular form.

Solutions:

a. The TI-86 display for part (a) appears as Fig. 13.29.
b. The TI-86 display for part (b) appears as Fig. 13.30.

EXAMPLE 13.17 Using the TI-86 calculator, perform the desired operations required in part (c) of Example 13.15, and compare solutions.

Solution: You must now be aware of the hierarchy of mathematical operations. In other words, in which sequence will the calculator perform the desired operations? In most cases, the sequence is the same as that used in longhand calculations, although you must become adept at setting up the parentheses to ensure the correct order of operations. For this example, the TI-86 display appears as Fig. 13.31, which is a perfect match with the earlier solution.

FIG. 13.31
*Performing the operations to determine the solution
to Example 13.15(c) using the TI-86 calculator.*

13.5 APPLYING KIRCHHOFF'S LAWS USING PHASOR NOTATION

We are now prepared to apply Kirchhoff's laws to any sinusoidal waveform with any phase angle. There is no longer any need to be concerned that the waveforms have a quadrature relationship. Apply each law in the same way that it would be applied to dc circuits. Then simply work with the phasor notation for each to determine the desired solution. The use of phasor notation to analyze ac networks was first introduced by Dr. Charles Steinmetz in 1897 (Fig. 13.32).

EXAMPLE 13.18 Given the series configuration of Fig. 13.33, find the applied voltage if

$$v_1 = 50 \sin(\omega t + 30°) \quad \text{and} \quad v_2 = 30 \sin(\omega t + 60°)$$

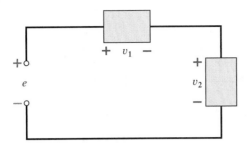

FIG. 13.33
Example 13.18.

Solution: Applying Kirchhoff's voltage law will result in the following:

$$e = v_1 + v_2$$

Converting each from **the time to the phasor domain** will result in

$$v_1 = 50 \sin(\omega t + 30°) \Rightarrow \mathbf{V}_1 = 50 \text{ V } \angle 30°$$
$$v_2 = 30 \sin(\omega t + 60°) \Rightarrow \mathbf{V}_2 = 30 \text{ V } \angle 60°$$

Converting from the polar to the rectangular form will result in

$$\mathbf{V}_1 = 50 \text{ V } \angle 30° = 43.30 \text{ V} + j\, 25.0 \text{ V}$$
$$\mathbf{V}_2 = 30 \text{ V } \angle 60° = 15.0 \text{ V} + j\, 25.98 \text{ V}$$

Then
$$\mathbf{E} = \mathbf{V}_1 + \mathbf{V}_2 = (43.30 \text{ V} + j\, 25.0 \text{ V}) + (15.0 \text{ V} + j\, 25.98 \text{ V})$$
$$= 43.30 \text{ V} + 15.0 \text{ V} + j(25.0 \text{ V} + 25.98 \text{ V})$$
$$= 58.3 \text{ V} + j\, 50.98 \text{ V}$$

Converting back to the polar form will result in

$$\mathbf{E} = 58.3 \text{ V} + j\, 50.98 \text{ V} = 77.45 \text{ V } \angle 41.17°$$

Then converting back to the time domain yields

$$e = \mathbf{77.45 \sin(\omega t + 41.17°)}$$

The resulting waveforms all appear in Fig. 13.34.

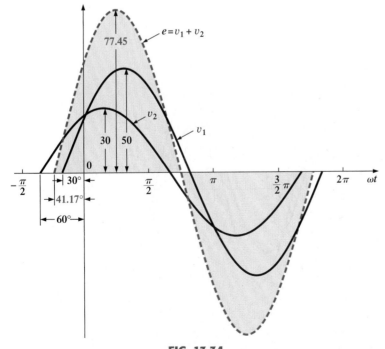

FIG. 13.34
Solution to Example 13.18.

EXAMPLE 13.19 Determine current i_2 for the network of Fig. 13.35 if

$$i_T = 120 \times 10^{-3} \sin(\omega t + 60°) \quad \text{and} \quad i_1 = 80 \times 10^{-3} \sin(\omega t + 80°)$$

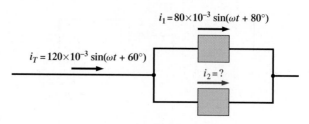

FIG. 13.35
Example 13.19.

Solution: Applying Kirchhoff's current law will result in the following:

$$i_T = i_1 + i_2 \quad \text{or} \quad i_2 = i_T - i_1$$

Converting each from the time to the phasor domain will result in

$$i_T = 120 \times 10^{-3} \sin(\omega t + 60°) \Rightarrow \mathbf{I}_T = 120 \text{ mA} \angle 60°$$

$$i_1 = 80 \times 10^{-3} \sin(\omega t + 80°) \Rightarrow \mathbf{I}_1 = 80 \text{ mA} \angle 80°$$

Converting from the polar to the rectangular form will result in

$$\mathbf{I}_T = 120 \text{ mA} \angle 60° = 60 \text{ mA} + j\,103.92 \text{ mA}$$

$$\mathbf{I}_1 = 80 \text{ mA} \angle 80° = 13.89 \text{ mA} + j\,78.79 \text{ mA}$$

Then

$$\mathbf{I}_2 = \mathbf{I}_T - \mathbf{I}_1$$

$$= (60 \text{ mA} + j\,103.92 \text{ mA}) - (13.89 \text{ mA} + j\,78.79 \text{ mA})$$

$$= 46.11 \text{ mA} + j\,25.13 \text{ mA}$$

Converting back to the polar form will result in

$$\mathbf{I}_2 = 46.11 \text{ mA} + j\,25.13 \text{ mA} = 52.51 \text{ mA} \angle 28.59°$$

Then converting back to the time domain yields

$$i_2 = \mathbf{52.51 \times 10^{-3} \sin(\omega t + 28.59°)}$$

The resulting waveforms all appear in Fig. 13.36.

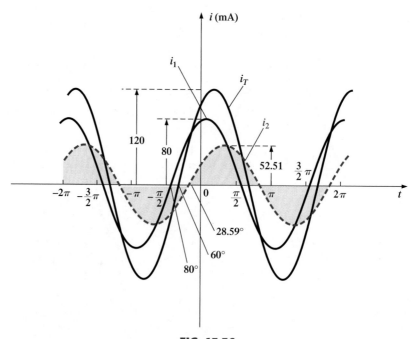

FIG. 13.36
Sketching i_T, i_1, and i_2 for Example 13.19.

Now that we are comfortable with performing some basic calculations with sinusoidal waveforms, it is time to determine the response of the basic elements to such a signal. Initially, there is often some concern about the fact that the sinusoidal voltage has exactly the same pattern above the axis as it has below. It would suggest that over one full period, the power delivered during the positive cycle is returned during the negative cycle. In other words, how can a waveform that seems to reverse polarity over one full cycle, with equal areas above and below the axis, deliver anything?

For resistors, the answer lies in the fact that

at every instant of time other than when $i_R = 0\,A$ (or $v_R = 0\,V$), power is being delivered to the resistor irrespective of the direction of the current through (or the polarity of the voltage across) the resistor.

To demonstrate the above, let us examine in some detail the application of an ac voltage to the resistor of Fig. 13.37. At the instant the applied voltage has a peak value of 8 V, the current through the resistor is 4 A, and the power delivered to the resistor is 32 W as shown in the figure. When the applied voltage drops to half its peak value, the current is 2 A, and the power is 8 W as also shown in the figure. At the instant the applied voltage is 0 V, the current is 0 A, and the power delivered is 0 W. In total, therefore, at every instant of the positive portion of the applied voltage, power is being delivered to the resistor.

For a purely resistive element, the power delivered is dissipated in the form of heat to the surrounding medium.

Electric heaters use a high-resistance wire that will glow red-hot due to relatively high currents to generate the desired heat.

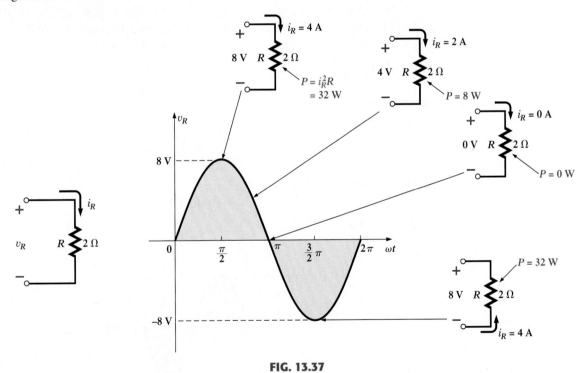

FIG. 13.37

Demonstrating that power is delivered at every instant of a sinusoidal voltage waveform (except $v_R = 0\,V$).

At the negative peak of the applied voltage, the polarity across the resistor has reversed as shown in Fig. 13.37, but the current through the resistor is still 4 A, and the power to the resistor is again 32 W. The change in direction of current did not affect the fact that power can still be delivered to that resistor at that instant in time. In fact, during the negative cycle, the power delivered to the resistor will be the same as that delivered during the positive cycle.

If we now plot the resulting current and applied voltage on the same graph, we obtain the display of Fig. 13.38. First note that both peak at the same time and pass through zero at the same instant. Using the terminology of Chapter 12, we can conclude that

for resistive elements, the applied voltage and the resulting current are in phase.

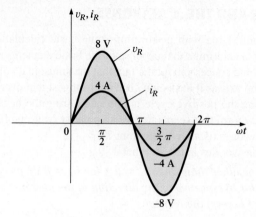

FIG. 13.38

The voltage across a resistor is in phase with the current through the resistor.

The peak values of the current and voltage are related by the following:

$$I_m = \frac{V_m}{R}$$

(13.20)

or

$$V_m = I_m R$$

(13.21)

EXAMPLE 13.20 The voltage across a resistor is indicated. Find the sinusoidal expression for the current if the resistor is 10 Ω. Sketch the curves for v and i.

a. $v = 100 \sin 377t$
b. $v = 25 \sin(377t + 60°)$

Solutions:

a. By Eq. (13.20):

$$I_m = \frac{V_m}{R} = \frac{100 \text{ V}}{10 \text{ Ω}} = 10 \text{ A}$$

(v and i are in phase), resulting in

$$i = \mathbf{10 \sin 377}t$$

The curves are sketched in Fig. 13.39.

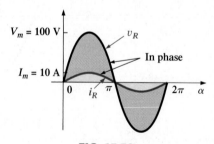

FIG. 13.39

Example 13.20(a).

b. By Eq. (13.20):

$$I_m = \frac{V_m}{R} = \frac{25 \text{ V}}{10 \text{ Ω}} = 2.5 \text{ A}$$

(v and i are in phase), resulting in

$$i = \mathbf{2.5 \sin(377}t + \mathbf{60°)}$$

The curves are sketched in Fig. 13.40.

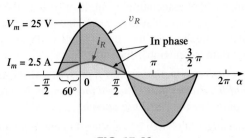

FIG. 13.40
Example 13.20(b).

EXAMPLE 13.21 The current through a 5 Ω resistor is given. Find the sinusoidal expression for the voltage across the resistor for $i = 40 \sin(377t + 30°)$.

Solution: By Eq. (13.21):

$$V_m = I_m R = (40 \text{ A})(5 \text{ Ω}) = 200 \text{ V}$$

(v and i are in phase), resulting in

$$v = \mathbf{200 \sin(377t + 30°)}$$

Using Phasor Notation

For a single resistive element, the analysis just concluded was quite straightforward once the basic equations and the phase relationship were established. **For more complex networks, however, an approach must be developed that carries the information about the phase relationship in the mechanics of the mathematical operations.** In other words, the approach will remove the need to remember the phase relationship. For a single resistive element, it is not difficult to remember that the voltage and the current are in phase, but we will find that this is not the case for inductive and capacitive elements. Further, any network with a combination of resistive and reactive elements will result in phase relationships that can be any value from 0° to 360°.

The approach will be developed using the phasor notation introduced in the previous sections. It will take a moment to develop the required equations, but be assured that the approach is quite straightforward and the results will prove invaluable in the analyses to follow.

Using the phasor notation, the voltage across a resistor can be written as

$$v_R = V_m \sin \omega t \Rightarrow \mathbf{V}_R = V_R \angle 0°$$

with

$$V_R = V_{R(\text{rms})} = 0.707V_m$$

It is important to note in the above representation that

the rms (effective) value of a voltage or current is used in the phasor notation when applied to the analysis of electric circuits.

This is due primarily to the fact that ac meters measure the rms value, and using this value permits a direct correlation with the laws and theorems of dc networks.

Applying Ohm's law:

$$\mathbf{I}_R = \frac{V_R \angle 0°}{R \angle \theta_R} = \frac{V_R}{R} \underline{/0° - \theta_R} = \frac{V}{R} \angle -\theta_R$$

For uniformity in format, an angle θ_R has been associated with the resistive element.

Since we now know that v_R and i_R are in phase and the angle associated with i_R must equal 0°, it follows that θ_R must also be 0°. Substituting $\theta_R = 0°$, we find, using phasor algebra, that

$$\mathbf{I}_R = \frac{V_R \angle 0°}{R \angle 0°} = \frac{V_R}{R} \underline{/0° - 0°} = \frac{V}{R} \angle 0°$$

Converting the result for i_R back to the time domain will result in

$$i_R = \sqrt{2}\left(\frac{V_R}{R}\right) \sin \omega t$$

The result of the above is that associating an angle of 0° with the magnitude of a resistive element will take care of the fact that the voltage and the current are in phase. Simply perform the required phasor algebra, and the correct angle will result for the voltage or current. However, it is important to remember the following:

The vector form for a resistive element does not represent a sinusoidal waveform with an initial phase angle of 0°. Rather, it is simply a mechanism to ensure the correct angle for the applied voltage or the resulting current.

From this point on, an angle of 0° will be employed in the following polar format to ensure the proper phase relationship between the voltage and the current of a resistor:

$$\boxed{\mathbf{Z}_R = R \angle 0°} \qquad \text{(ohms, } \Omega) \tag{13.22}$$

The boldface roman quantity $\mathbf{Z}_R$, having both magnitude and an associated angle, is referred to as the **impedance** of a resistive element. It is measured in ohms and is a measure of how much the resistive element will **impede** the flow of charge through the resistor due to an applied ac voltage. The above format will prove a very useful tool when the networks become more complex and phase relationships become less obvious. However, at the risk of being repetitious, remember that the format of Eq. (13.22) does not define a phasor; it is simply a vector form for resistive elements in an ac network. The term *phasor* is reserved for quantities such as voltage or current that vary with time.

EXAMPLE 13.22 Repeat Example 13.20 using the complex number format for the resistive element. Compare your results.

Solutions:

a. Writing the applied voltage in phasor form:

$$v_R = 100 \sin 377t \Rightarrow \mathbf{V}_R = V_R \angle 0°$$
$$= (0.707)(100 \text{ V}) \angle 0° = 70.7 \text{ V} \angle 0°$$

Applying Ohm's law:

$$\mathbf{I}_R = \frac{\mathbf{V}_R}{\mathbf{Z}_R} = \frac{70.7 \text{ V} \angle 0°}{10 \ \Omega \ \angle 0°} = 7.07 \text{ A} \angle 0°$$

Returning to the time domain:

$$i_R = \sqrt{2}(7.07) \sin 377t = \mathbf{10 \sin 377} t$$

which is the same result as in Example 13.20(a).

b. Writing the applied voltage in phasor form:

$$v_R = 25 \sin(377t + 60°) \Rightarrow \mathbf{V}_R = (0.707)(25 \text{ V}) \angle 60° = 17.68 \text{ V} \angle 60°$$

Applying Ohm's law:

$$\mathbf{I}_R = \frac{\mathbf{V}_R}{\mathbf{Z}_R} = \frac{17.68 \text{ V} \angle 60°}{10 \ \Omega \ \angle 0°} = 1.768 \text{ A} \angle 60°$$

Returning to the time domain:

$$i_R = \sqrt{2}(1.768) \sin(377t + 60°) = \mathbf{2.5 \sin(377} t + \mathbf{60°)}$$

matching the result of Example 13.20(b).

EXAMPLE 13.23 The current through a 2 kΩ resistor is $i_R = 40 \times 10^{-3} \sin(1000t + 42°)$. Find the mathematical expression for the voltage across the resistor.

Solution: The phasor form for the current is

$$\mathbf{I}_R = I_R \angle \theta = (0.707)(40 \text{ mA}) \angle 42° = 28.28 \text{ mA} \angle 42°$$

Applying Ohm's law:

$$\mathbf{V}_R = \mathbf{I}_R \mathbf{Z}_R = (28.28 \text{ mA} \angle 42°)(2 \text{ k}\Omega \angle 0°) = 56.56 \text{ V} \angle 42°$$

Returning to the time domain:

$$v_R = \sqrt{2}(56.56)\sin(1000t + 42°) = \mathbf{80\sin(1000t + 42°)}$$

Note in both Examples 13.22 and 13.23 that it is completely unnecessary to remember that the voltage and current are in phase; the information was carried along by the notation. The application of Ohm's law remains the same as it appeared in the dc chapters of this text. The only difference is the necessity to use the special notation to include the in-phase relationship between the voltage and the current.

Before continuing in a similar fashion with the inductive element, note that

the frequency or angular velocity is not part of the phasor notation.

It is always assumed that the frequency is the same for all the voltages and currents of a network and can simply be included wherever necessary.

In the analysis of networks, it is often helpful to have a **phasor diagram,** which shows at a glance the *magnitudes* and *phase relations* among the various quantities within the network. For example, the phasor diagram of the circuit considered in Example 13.22 is shown in Fig. 13.41. In both cases, it is immediately obvious that v_R and i_R are in phase since they both have the same phase angle.

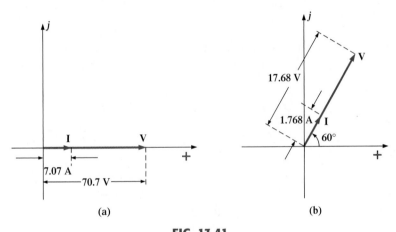

(a) (b)

FIG. 13.41

Phasor diagrams for Example 13.22.

13.7 INDUCTORS AND THE ac RESPONSE

The discussion to follow will be for pure inductors only, with no associated internal resistance due to the turns of wire nor stray capacitances between the conducting leads. The effect of these parameters will be discussed as the need arises later in the text. For the pure inductor, the voltage across the element and the current through the element are determined by Eq. (11.12), repeated here for convenience:

$$v_L = L\frac{di_L}{dt}$$

Let's write the equation for the voltage across a resistor as $v_R = R \cdot i_R$ rather than as $v_R = i_R \cdot R$ and insert the equation for the voltage across a coil directly below:

$$v_R = R(i_R)$$

$$v_L = L\underbrace{\frac{d}{dt}}_{\text{opposition}}(i_L)$$

Since the current appears in the numerator of each equation, the multiplying factor is the opposition level for each element. For the resistor it is simply the resistance R, but for the inductor, it is the product of the inductance of the coil and the instantaneous rate of change in current through the coil.

The **larger the inductance,** therefore, the greater the opposition factor for an inductor, and the larger the voltage across a coil for the same change in current through the coil. This statement supports a conclusion of Chapter 11 stating that the larger the inductance, the greater the induced voltage across a coil due to a change in current through the coil.

The factor d/dt is the derivative (calculus) of the function at a particular instant of time. It provides a measure of how rapidly the function is changing at that instant. As the frequency of the current through the coil increases, the **rate of change** of the current at any point on the curve will increase as shown in Fig. 13.42 at $t = 0$ s and $i_L = I_m/2$ for two currents of different frequencies. This higher transition rate will continue throughout the waveform, leaving us with the fact that the current changes at a much quicker rate for sinusoidal waveforms with higher frequencies. The opposition presented by the coil will therefore be **significantly more for high-frequency currents.** This statement also agrees with a conclusion of Chapter 11 that inductors act like "chokes" in that they oppose rapid changes in current through the coil. The quicker the rate of change, the greater the opposition.

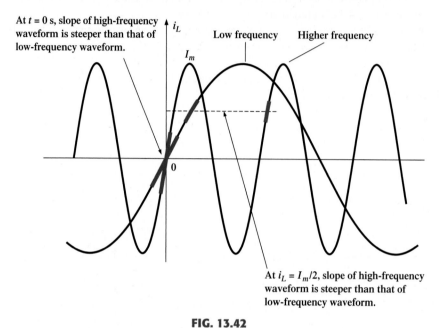

FIG. 13.42

Demonstrating that the rate of change of a sinusoidal waveform with a high frequency will always be more than one with a lower frequency.

In total, therefore, the opposition of an inductor to the flow of charge, or current, through the inductor is directly related to the inductance of the inductor and the frequency of the current. The opposition of an inductor is called its **reactance,** from the phase "react against," and is determined by the following:

$$X_L = 2\pi f L = \omega L \qquad \text{(ohms, } \Omega\text{)} \qquad \textbf{(13.23)}$$

It is important to understand that

reactance is quite different from resistance in that all the electrical energy transferred to an ideal inductor is not dissipated but simply stored in the form of a magnetic field.

Note that Equation (13.23) is directly related to the angular velocity rather than simply the applied frequency. Both quantities define the oscillation rate of the sinusoidal waveform, however, and are related by a constant 2π that would result if Equation (13.23) were determined using a purely mathematical derivation.

For a pure inductor, the peak values of the current and the voltage are related by an Ohm's law relationship:

$$V_m = I_m X_L \qquad \textbf{(13.24)}$$

and

$$I_m = \frac{V_m}{X_L} \qquad \textbf{(13.25)}$$

The phase relationship between the voltage across a coil and the current through the coil can be determined by examining Fig. 13.43. It defines the regions where the rate of change of the current is a maximum and a minimum. If we plot the instantaneous change in current at various points along the curve, the lower plot will result, clearly indicating that when one curve is a maximum, the other is a minimum, and vice versa. The resulting waveform is a sinusoidal function because the rate of change $\Delta v/\Delta t$ will change in a sinusoidal manner. Of major importance is the fact that the resulting waveform leads the defining current waveform by 90°. Since the lower waveform defines the shape of the voltage across an inductor, we can conclude that

the voltage across an inductor leads the current through the inductor by 90°.

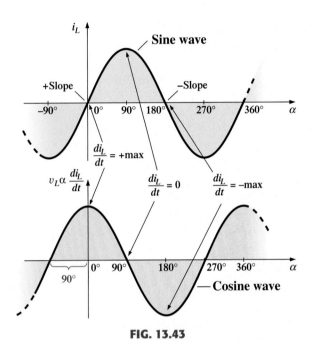

FIG. 13.43

*Demonstrating how the derivative in the equation for the voltage across an inductor
results in a 90° phase shift between v_L and i_L.*

EXAMPLE 13.24 The current through a 0.1 H coil is provided. Find the sinusoidal expression for the voltage across the coil. Sketch the v and i curves.

a. $i = 10 \sin 377t$
b. $i = 7 \sin(377t - 70°)$

Solutions:

a. Eq. (13.23): $X_L = \omega L = (377 \text{ rad/s})(0.1 \text{ H}) = 37.7 \ \Omega$
   Eq. (13.24): $V_m = I_m X_L = (10 \text{ A})(37.7 \ \Omega) = 377 \text{ V}$

   and we know that for a coil, v leads i by 90°. Therefore,

$$v = \mathbf{377 \sin(377t + 90°)}$$

   The curves are sketched in Fig. 13.44.

b. X_L remains at 37.7 Ω.

$$V_m = I_m X_L = (7 \text{ A})(37.7 \ \Omega) = 263.9 \text{ V}$$

   and we know that for a coil, v leads i by 90°. Therefore,

$$v = 263.9 \sin(377t - 70° + 90°)$$

   and $\qquad\qquad v = \mathbf{263.9 \sin(377t + 20°)}$

   The curves are sketched in Fig. 13.45.

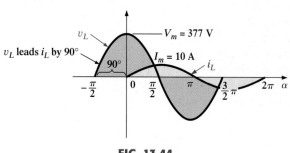

FIG. 13.44
Example 13.24(a).

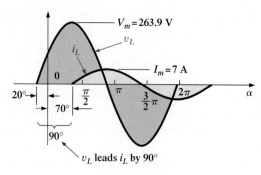

FIG. 13.45
Example 13.24(b).

EXAMPLE 13.25 The voltage across a 0.5 H coil is provided below. What is the sinusoidal expression for the current?

$$v = 100 \sin 20t$$

Solution:

$$X_L = \omega L = (20 \text{ rad/s})(0.5 \text{ H}) = 10 \, \Omega$$

Eq. (13.25):
$$I_m = \frac{V_m}{X_L} = \frac{100 \text{ V}}{10 \, \Omega} = 10 \text{ A}$$

and we know that i lags v by 90°. Therefore,

$$i = \mathbf{10 \sin(20t - 90°)}$$

Using Phasor Notation

The analysis will now proceed in the same manner that was applied to the resistor for vector representation of the impedance of an inductor. The method will incorporate the phase shift between the voltage across a coil and the current through the coil.

The phasor notation for the current is defined by

$$i_L = I_m \sin \omega t \Rightarrow \mathbf{I}_L = I_L \angle 0°$$

where
$$I_L = I_{L(\text{rms})} = 0.707 I_m$$

Applying Ohm's law to an inductive element will result in

$$\mathbf{V}_L = I_L \angle 0° \cdot X_L \angle \theta_L = IX_L \underline{/0° + \theta_L} = IX_L \angle \theta_L$$

using an angle θ_L for the inductive component.

Since v_L must lead i_L by 90°, θ_L must equal 90°. Substituting $\theta_L = 90°$, we will obtain

$$\mathbf{V}_L = I_L \angle 0° \cdot X_L \angle 90° = IX_L \angle 90°$$

which in the time domain is

$$v_L = \sqrt{2}(IX_L) \sin(\omega t + 90°)$$

From this point on, if we simply associate an angle of 90° with the reactance of the coil, the phase angle between the applied voltage or current and between the resulting current or voltage, respectively, will be taken care of.

For the future, the impedance of a coil will be defined by

$$\boxed{\mathbf{Z}_L = X_L \angle 90°} \qquad (\text{ohms}, \, \Omega) \qquad \mathbf{(13.26)}$$

As with $\mathbf{Z}_R$, remember that the format of Eq. (13.26) does not define a sinusoidal function. It is simply a vector representation of the impedance of an inductor that ensures that the phase angle associated with the voltage or current is correct.

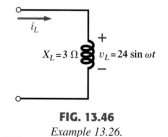

FIG. 13.46
Example 13.26.

EXAMPLE 13.26 Using complex algebra, find the current i_L for the circuit of Fig. 13.46. Sketch the v_L and i_L curves.

Solution: First, the reactance of the coil must be determined:

$$X_L = \omega L = (1000 \text{ rad/s})(3 \text{ mH}) = 3 \text{ } \Omega$$

Placing the applied voltage in phasor form:

$$v_L = 24 \sin \omega t \Rightarrow \mathbf{V}_L = (0.707)(24 \text{ V}) \angle 0° = 16.968 \text{ V} \angle 0°$$

Applying Ohm's law:

$$\mathbf{I}_L = \frac{\mathbf{V}_L}{\mathbf{Z}_L} = \frac{V_L \angle \theta}{X_L \angle 90°} = \frac{16.968 \text{ V} \angle 0°}{3 \text{ } \Omega \angle 90°} = 5.656 \text{ A} \angle -90°$$

Converting back to the time domain:

$$i_L = \sqrt{2}(5.656) \sin(\omega t - 90°) = \mathbf{8.0 \sin(\omega t - 90°)}$$

The waveforms appear in Fig. 13.47.

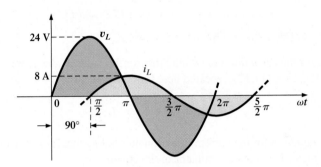

FIG. 13.47
Waveforms for Example 13.26.

EXAMPLE 13.27 Using complex algebra, find the voltage v_L for the circuit of Fig. 13.48. Sketch the curves for v_L and i_L.

Solution: In this example, the impedance is provided as

$$\mathbf{Z}_L = X_L \angle 90° = 4 \text{ } \Omega \angle 90°$$

Following through, we obtain

$$i_L = 5 \sin(\omega t + 30°) \Rightarrow \mathbf{I}_L = (0.707)(5 \text{ A}) \angle 30° = 3.535 \text{ A} \angle 30°$$

and

$$\mathbf{V}_L = \mathbf{I}_L \cdot \mathbf{Z}_L = (I \angle \theta)(X_L \angle 90°) = (3.535 \text{ A} \angle 30°)(4 \text{ } \Omega \angle 90°)$$
$$= 14.140 \text{ V} \angle 120°$$

so that

$$v_L = \sqrt{2}(14.140) \sin(\omega t + 120°) = \mathbf{20 \sin(\omega t + 120°)}$$

The resulting waveforms appear in Fig. 13.49.

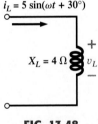

FIG. 13.48
Example 13.27.

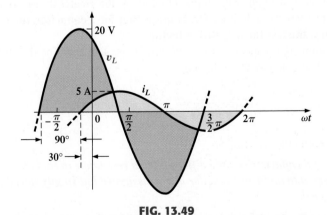

FIG. 13.49
Waveforms for Example 13.27.

The phasor diagrams for Examples 13.26 and 13.27 appear in Fig. 13.50. Both clearly reveal that the voltage leads the current by 90°.

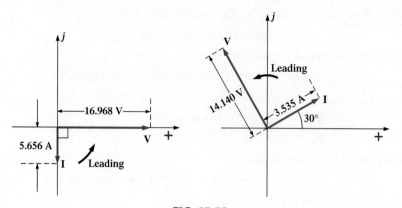

FIG. 13.50

Phasor diagrams for Examples 13.26 and 13.27.

13.8 CAPACITORS AND THE ac RESPONSE

The discussion for the capacitor will parallel that for the inductor due to the similarities between Eqs. (10.21) and (11.12), repeated here for convenience:

$$i_C = C\frac{dv_C}{dt} \qquad v_L = L\frac{di_L}{dt}$$

If we write the equation for the current of a capacitor below that for a resistor, we find that the opposition factor is in the denominator of the equations:

$$i_R = \frac{(v_R)}{R}$$

$$i_C = C\underbrace{\frac{d}{dt}}_{\substack{1 \\ \text{opposition}}}(v_C)$$

The result is that the opposition of a capacitor is proportional to 1 over the product of the capacitance value and the rate of change of the voltage across the capacitor. That is,

$$\text{Opposition} \propto \frac{1}{C\dfrac{d}{dt}}$$

The larger the capacitance of a capacitor, therefore, the smaller the opposition to the flow of charge or current. This conclusion appears valid because the larger the capacitance, the more the available charge for conduction (for the same voltage across the plates). The previous section gave us some insight into the impact of the derivative term in the equation. That is, the higher the applied frequency, the greater the rate of change of the waveform as it passes through time. **The result is that for a capacitor, the higher the applied frequency, the less the opposition factor.**

The opposition of a capacitor in an ac circuit is, like that for the inductor, called **reactance** and is defined by the following:

$$\boxed{X_C = \frac{1}{2\pi fC} = \frac{1}{\omega C}} \qquad \text{(ohms, } \Omega\text{)} \tag{13.27}$$

Again, it is important to realize that

the reactance of a capacitor is quite different from resistance in that all the electrical energy transferred to an ideal capacitor is not dissipated but simply stored in the form of an electric field.

For the capacitor, the peak values of the current and voltage are related by an Ohm's law relationship:

$$\boxed{I_m = \frac{V_m}{X_C}} \tag{13.28}$$

and
$$V_m = I_m X_C \qquad \text{(13.29)}$$

The phase relationship between the voltage across a capacitor and the current can be determined by examining Fig. 13.51. The regions of maximum and minimum instantaneous change have been identified on the waveform for the applied voltage. Since the current of a capacitor is directly proportional to the derivative of the applied voltage, the current will have the peaks and values appearing in the lower figure. The result, as shown, is that

the current of a capacitor leads the voltage across the capacitor by 90°.

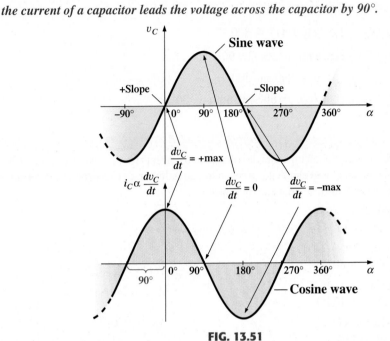

FIG. 13.51

*Demonstrating how the derivative in the equation for the current of a capacitor results
in a 90° phase shift between v_C and i_C.*

EXAMPLE 13.28 The voltage across a 1 μF capacitor is provided below. What is the sinusoidal expression for the current? Sketch the v and i curves.

$$v = 30 \sin 400t$$

Solution: Eq. (13.27):

$$X_C = \frac{1}{\omega C} = \frac{1}{(400 \text{ rad/s})(1 \times 10^{-6} \text{ F})} = \frac{10^6 \, \Omega}{400} = 2500 \, \Omega$$

Eq. (13.28): $\quad I_m = \dfrac{V_m}{X_C} = \dfrac{30 \text{ V}}{2500 \, \Omega} = 0.0120 \text{ A} = 12 \text{ mA}$

and we know that for a capacitor, i leads v by 90°. Therefore,

$$i = 12 \times 10^{-3} \sin(400t + 90°)$$

The curves are sketched in Fig. 13.52.

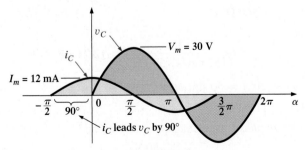

FIG. 13.52

Waveforms for Example 13.28.

EXAMPLE 13.29 The current through a 100 μF capacitor is given. Find the sinusoidal expression for the voltage across the capacitor.

$$i = 40 \sin(500t + 60°)$$

Solution:

$$X_C = \frac{1}{\omega C} = \frac{1}{(500 \text{ rad/s})(100 \times 10^{-6} \text{ F})} = \frac{10^6 \text{ } \Omega}{5 \times 10^4} = \frac{10^2 \text{ } \Omega}{5} = 20 \text{ } \Omega$$

$$V_m = I_m X_C = (40 \text{ A})(20 \text{ } \Omega) = 800 \text{ V}$$

and we know that for a capacitor, v lags i by 90°. Therefore,

$$v = 800 \sin(500t + 60° - 90°)$$

and

$$v = \textbf{800 sin}(\textbf{500}t - \textbf{30°})$$

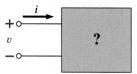

FIG. 13.53
Example 13.30.

EXAMPLE 13.30 For the following pairs of voltages and currents for the system of Fig. 13.53, determine whether the enclosed element is a capacitor, an inductor, or a resistor, and determine the value of C, L, or R if sufficient data are provided:

a. $v = 100 \sin(\omega t + 40°)$
$i = 20 \sin(\omega t + 40°)$

b. $v = 1000 \sin(377t + 10°)$
$i = 5 \sin(377t - 80°)$

c. $v = 500 \sin(157t + 30°)$
$i = 1 \sin(157t + 120°)$

d. $v = 50 \cos(\omega t + 20°)$
$i = 5 \sin(\omega t + 110°)$

Solutions:

a. Since v and i are *in phase*, the element is a *resistor*, and

$$R = \frac{V_m}{I_m} = \frac{100 \text{ V}}{20 \text{ A}} = \textbf{5 } \boldsymbol{\Omega}$$

b. Since v *leads* i by 90°, the element is an *inductor*, and

$$X_L = \frac{V_m}{I_m} = \frac{1000 \text{ V}}{5 \text{ A}} = 200 \text{ } \Omega$$

so that $X_L = \omega L = 200 \text{ } \Omega$ or

$$L = \frac{200 \text{ } \Omega}{\omega} = \frac{200 \text{ } \Omega}{377 \text{ rad/s}} = \textbf{0.531 H}$$

c. Since i *leads* v by 90°, the element is a *capacitor*, and

$$X_C = \frac{V_m}{I_m} = \frac{500 \text{ V}}{1 \text{ A}} = 500 \text{ } \Omega$$

so that $X_C = \dfrac{1}{\omega C} = 500 \text{ } \Omega$ or

$$C = \frac{1}{\omega 500 \text{ } \Omega} = \frac{1}{(157 \text{ rad/s})(500 \text{ } \Omega)} = \textbf{12.74 } \boldsymbol{\mu}\textbf{F}$$

d. $v = 50 \cos(\omega t + 20°) = 50 \sin(\omega t + 20° + 90°)$
$= 50 \sin(\omega t + 110°)$

Since v and i are *in phase*, the element is a *resistor*, and

$$R = \frac{V_m}{I_m} = \frac{50 \text{ V}}{5 \text{ A}} = \textbf{10 } \boldsymbol{\Omega}$$

Using Phasor Notation

Proceeding as described for a resistive and an inductive element, we define the phasor notation for the voltage across a capacitor by

$$v_C = V_m \sin \omega t \Rightarrow \mathbf{V}_C = V_C \angle 0°$$

where

$$V_C = V_{C(\text{rms})} = 0.707V_m$$

Applying Ohm's law will result in

$$\mathbf{I}_C = \frac{V_C \angle 0°}{X_C \angle \theta_C} = \frac{V_C}{X_C} \underline{/0° - \theta_C} = \frac{V_C}{X_C} \angle -\theta_C$$

using the angle θ_C for the capacitive element.

Since i_C leads v_C by 90°, θ_C must have an angle of +90° associated with it. To satisfy this condition, θ_C must equal $-90°$. Substituting $\theta_C = -90°$ yields

$$\mathbf{I}_C = \frac{V_C \angle 0°}{X_C \angle -90°} = \frac{V_C}{X_C} \underline{/0° - (-90°)} = \frac{V_C}{X_C} \angle 90°$$

Thus, in the time domain,

$$i_C = \sqrt{2}\left(\frac{V_C}{X_C}\right) \sin(\omega t + 90°)$$

The fact that $\theta_C = -90°$ will now be used in the following polar format for capacitive reactance to ensure the proper phase relationship between the voltage and the current of a capacitor:

$$\boxed{\mathbf{Z}_C = X_C \angle -90°} \qquad (\text{ohms, } \Omega) \qquad \textbf{(13.30)}$$

The boldface roman quantity $\mathbf{Z}_C$ has both magnitude and an associated angle. Referred to as the **impedance** of a capacitive element, it is measured in ohms and is a measure of how much the capacitive element will control or "impede" the level of current for a capacitor. The above format, like that defined for the resistor and the inductor, will prove a very useful tool in the analysis of ac networks. Again, be aware that $\mathbf{Z}_C$ is not a phasor quantity representing a sinusoidal function. It is simply a representation for the impedance of a capacitor that simplifies the analysis.

EXAMPLE 13.31   Using phasor notation, find the current i_C for the circuit of Fig. 13.54. Sketch the curves for v_C and i_C.

Solution:

$$v_C = 15 \sin \omega t \Rightarrow \mathbf{V}_C = (0.707)(15 \text{ V}) \angle 0° = 10.605 \text{ V} \angle 0°$$

$$\mathbf{I}_C = \frac{\mathbf{V}_C}{\mathbf{Z}_C} = \frac{10.605 \text{ V} \angle 0°}{2 \Omega \angle -90°} = 5.303 \text{ A} \angle 90°$$

and $i_C = \sqrt{2}(5.303) \sin(\omega t + 90°) = \textbf{7.5 } \sin(\boldsymbol{\omega t + 90°})$

The curves appear in Fig. 13.55.

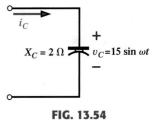

FIG. 13.54
Example 13.31.

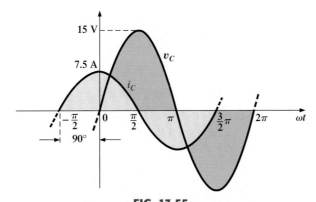

FIG. 13.55
Waveforms for Example 13.31.

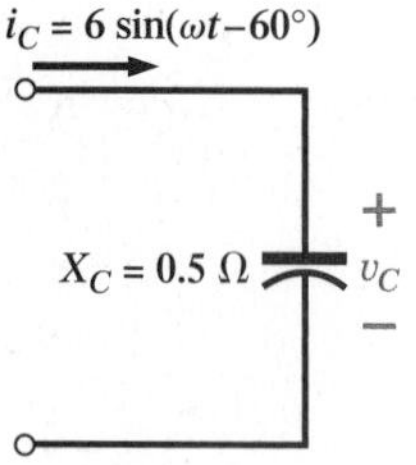

$i_C = 6 \sin(\omega t - 60°)$

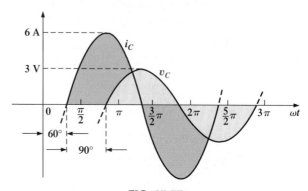

$X_C = 0.5 \, \Omega$ + v_C −

FIG. 13.56
Example 13.32.

EXAMPLE 13.32 Using phasor notation, find the voltage v_C for the circuit of Fig. 13.56. Sketch the curves for v_C and i_C.

Solution:

$$i_C = 6 \sin(\omega t - 60°) \Rightarrow \mathbf{I}_C = (0.707)(6 \text{ A}) \angle{-60°} = 4.242 \text{ A} \angle{-60°}$$

$$\mathbf{V}_C = \mathbf{I}_C \cdot \mathbf{Z}_C = (4.242 \text{ A} \angle{-60°})(0.5 \, \Omega \angle{-90°}) = 2.121 \text{ V} \angle{-150°}$$

and $\quad v_C = \sqrt{2}\,(2.121) \sin(\omega t - 150°) = \mathbf{3.0 \sin(\omega t - 150°)}$

The curves appear in Fig. 13.57.

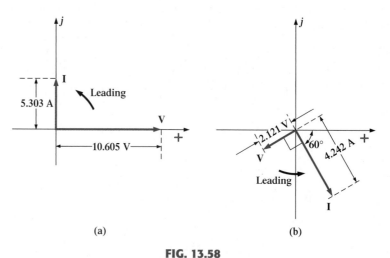

FIG. 13.57
Waveforms for Example 13.32.

The phasor diagrams for the circuits of Examples 13.31 and 13.32 are shown in Fig. 13.58. Both clearly indicate that the current i_C leads the voltage v_C by 90°.

(a)   (b)

FIG. 13.58
Phasor diagrams for Examples 13.31 and 13.32.

13.9 POWER AND THE BASIC ELEMENTS

It has already been pointed out that resistors dissipate all the energy delivered to them, whether the energy is dc, ac, or any type of signal. We are now aware that for **ideal** inductors or capacitors, the energy delivered is not dissipated but simply stored in the form of a magnetic field or an electric field, respectively, ready to be returned to the electrical system when called for by the design. This section will develop an equation for the ac power delivered to resistive elements. It will also illustrate that even though reactive elements do not dissipate power, they do draw power from the source at particular instants of time.

The Resistor *R*

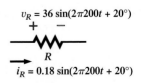

The equation for the power to the resistor of Fig. 13.59 has the same format as that used for dc circuits. That is,

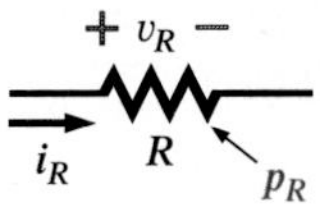

$$p_R = v_R \cdot i_R \qquad (13.31)$$

FIG. 13.59
Determining the ac power delivered to a purely resistive load.

If we plot the product of the two quantities at each instant of time, we will obtain the power waveform of Fig. 13.60, which has twice the frequency of the applied voltage or current. First, note that v_R and i_R are in phase because the element is resistive and the resulting power curve is sinusoidal in nature. The first positive peak of the power curve results when both v_R and i_R are at their positive peak values. The second peak of the power curve occurs when v_R and i_R are at their negative peaks. It is positive because the product of two negative quantities is positive. If we form the product of v_R and i_R at their peak values, we find the peak value of the power curve:

$$V_m \cdot I_m = (\sqrt{2}V_{\text{rms}})(\sqrt{2}I_{\text{rms}}) = 2V_{\text{rms}}I_{\text{rms}} = 2VI$$

so that the peak value of the sinusoidal function defining the power curve is simply the product of the **effective** or **rms values** of the voltage and current (*VI*).

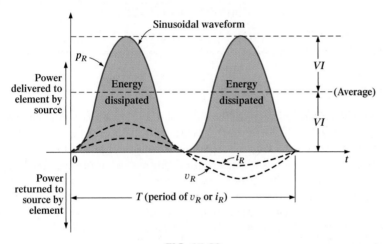

FIG. 13.60
Power versus time for a purely resistive load.

The most important fact to note in Fig. 13.60 is that the entire power curve is above the horizontal axis, indicating that all the power delivered is dissipated by the resistive element. Anything below the axis would have been returned to the source. As shown in the figure, the average value of the power curve is a product of the effective values of the voltage and the current. This product, called the **average** or **real power** delivered to a resistor, can be determined by any of the following equations (using rms values):

$$P_R = V_R I_R = I_R^2 R = \frac{V_R^2}{R} \qquad \text{(watts, W)} \qquad (13.32)$$

In particular, note the similarity of these equations to the equations used for dc circuits. The only difference is that the voltages and currents are rms values rather than dc values.

The area under the curves is the energy dissipated by the resistive element. It can be determined by using the equation $W = Pt$ and substituting the time for one full period of the applied voltage or current. That is,

$$W_R = P_R \cdot T \qquad \text{(joules, J)} \qquad (13.33)$$

EXAMPLE 13.33 For the resistor of Fig. 13.61:

a. Determine the power delivered to the resistor.
b. Find the energy dissipated over one full cycle of the applied voltage.

$v_R = 36 \sin(2\pi 200 t + 20°)$

$i_R = 0.18 \sin(2\pi 200 t + 20°)$

FIG. 13.61
Example 13.33.

Solutions:

a. The rms values:

$$V_R = (0.707)(36 \text{ V}) = 25.45 \text{ V}$$
$$I_R = (0.707)(0.18 \text{ A}) = 0.127 \text{ A}$$
$$P_R = V_R I_R = (25.45 \text{ V})(0.127 \text{ A}) = \textbf{3.23 W}$$

b. Using Eq. (13.33):

$$W_R = P_R \cdot T$$

with

$$T = \frac{1}{f} = \frac{1}{200 \text{ Hz}} = 5 \text{ ms}$$

and

$$W_R = (3.23 \text{ W})(5 \times 10^{-3} \text{ s}) = \textbf{16 mJ}$$

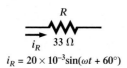

$$i_R = 20 \times 10^{-3}\sin(\omega t + 60°)$$

FIG. 13.62

Example 13.34.

EXAMPLE 13.34 Determine the power delivered to the resistor of Fig. 13.62.

Solution: The rms value:

$$I_R = (0.707)(20 \text{ mA}) = 14.14 \text{ mA}$$

and

$$P_R = I_R^2 R = (14.14 \text{ mA})^2 \, 33 \, \Omega = \textbf{6.598 mW}$$

The Inductor L

If we proceed in the same manner for the inductor of Fig. 13.63, we will obtain the curves of Fig. 13.64. In this case, due to the 90° phase shift, there are regions where either the voltage or the current will be negative, resulting in a negative product for the power level. The resulting power curve has a sinusoidal pattern but **twice the frequency** of the applied voltage or current. That is, for every cycle of the voltage or current, there are two cycles of the power curve. Most important, however, is the fact that **the power curve has equal areas above and below the axis for one full period of the applied signal.** In other words, over one full cycle, the energy absorbed is equal to that returned; there is **no net dissipation.**

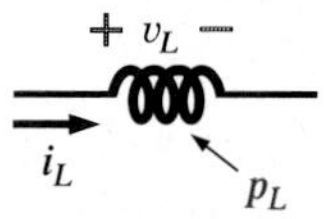

FIG. 13.63

Finding the power delivered to an inductive element.

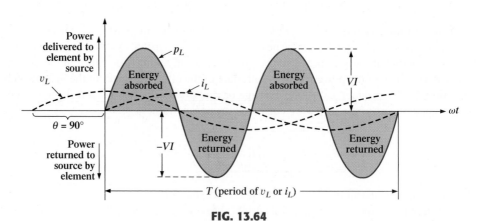

FIG. 13.64

The power curve for a purely inductive load.

Even though the average value of the waveform is zero watts and there is no net dissipation, it is important to realize that at instants such as t_1, power is being delivered to the inductor. However, as pointed out above, the energy transmitted is simply being converted to another form and is not being dissipated as heat (as is the case for resistive elements). Thus, power companies must transmit power at that instant even though the power won't be dissipated, and they must supply the needed current for the provided voltage. The additional demand must be met by a power plant, requiring that the plant provide more power than will be dissipated. This requirement taxes the system, so that an additional cost must be passed on to the consumer. The net result is that a consumer's bill will be affected by the reactive elements in the load. In Chapter 15, a topic called *power-factor correction* will be introduced; this factor brings a consumer's bill more in line with the total power dissipated.

For inductors, a quantity called **reactive power** has been defined. It has the symbol Q_L and is equal to the peak value of the power curve of Fig. 13.64. That is,

$$Q_L = V_L I_L = I_L^2 X_L = \frac{V_L^2}{X_L} \quad \text{(VAR)} \qquad \textbf{(13.34)}$$

It is calculated using the rms values of the voltage and current and is measured in units of **volt-ampere-reactive,** or **VAR.** Simply remember for future applications that it represents the peak value of the power curve for a purely inductive element.

The energy stored or released during one pulse of the power curve can be determined using the following equation, which is very similar to that employed for dc circuits:

$$W_L = L I_L^2 \quad \text{(joules, J)} \qquad \textbf{(13.35)}$$

EXAMPLE 13.35   For the inductor of Fig. 13.65:

a. Find the reactive power level.
b. Calculate the energy stored or released by one pulse of the power curve.

Solutions:

a. $I_{L(\text{rms})} = (0.707)(16 \text{ A}) = 11.31 \text{ A}$

$X_L = \omega L = (377 \text{ rad/s})(100 \text{ mH}) = 37.7 \ \Omega$

$Q_L = I_L^2 X_L = (11.31 \text{ A})^2 \cdot 37.7 \ \Omega = \textbf{4.82 kVAR}$

b. $W_L = L I_L^2 = (100 \text{ mH})(11.31 \text{ A})^2 = \textbf{12.79 J}$

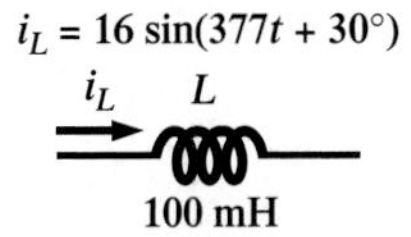

$i_L = 16 \sin(377t + 30°)$

$\overset{i_L}{\longrightarrow} \quad L$
000
100 mH

FIG. 13.65
Example 13.35.

EXAMPLE 13.36   Find the reactive power level for the inductor of Fig. 13.66.

Solution: The effective value is

$$V_L = (0.707)(32 \text{ V}) = 22.62 \text{ V}$$

Therefore,

$$Q_L = \frac{V_L^2}{X_L} = \frac{(22.62 \text{ V})^2}{4 \ \Omega} = \textbf{127.92 VAR}$$

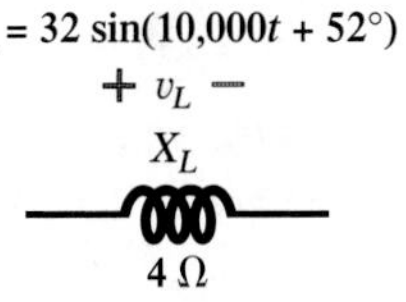

$v_L = 32 \sin(10,000t + 52°)$
$+ \ v_L \ -$
X_L
000
4 Ω

FIG. 13.66
Example 13.36.

The Capacitor C

Since the ideal capacitor and inductor are purely reactive elements, the results obtained for the capacitor here will be a close match to the resulting curves and equations obtained for the inductor. Finding the power curve for the capacitor of Fig. 13.67 will result in the sinusoidal waveform of Fig. 13.68. Again, the curve has twice the frequency of the applied voltage or current and a peak value (like the inductor) equal to the product of the rms values of the voltage and the current. The primary difference between this power curve and that of the inductor is that there is a 180° phase shift between the two. Whereas the power curve for the inductor is positive for the first quarter-cycle, the power curve for the inductor is negative (but the same shape). This difference will surface as an important characteristic when we discuss power-level corrections in Chapter 15. Note again that the power curve has the same area under the curve above the horizontal axis as it has below, revealing that the energy transmitted is returned within the same cycle and **the net dissipation is zero watts.**

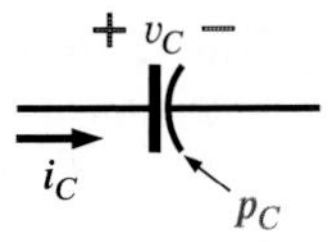

FIG. 13.67
Finding the power delivered to a capacitive element.

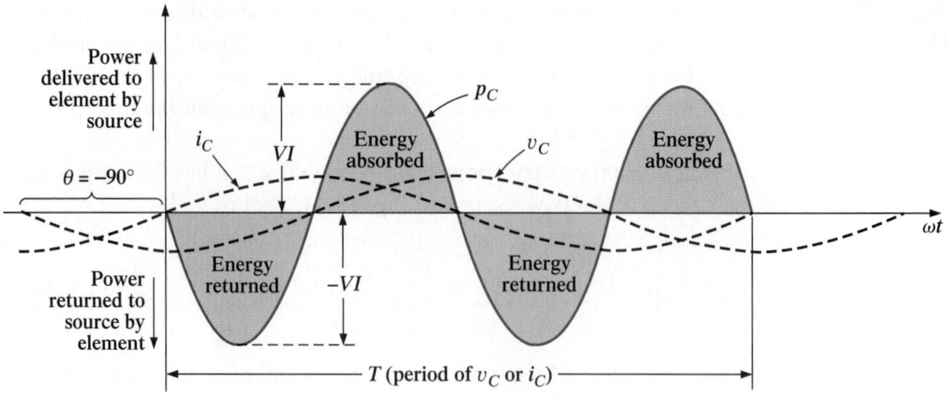

FIG. 13.68

The power curve for a purely capacitive load.

The average value of the power curve is again zero watts, but again there are periods of time when power is being delivered to the capacitor. Thus, we need to define a reactive power level using the symbol Q_C that is equal to the peak value of the curve. That is,

$$Q_C = V_C I_C = I_C^2 X_C = \frac{V_C^2}{X_C} \quad \text{(VAR)} \qquad \textbf{(13.36)}$$

The reactive power to a capacitor is also measured in **volt-ampere-reactive (VAR)** units which match those applied to the inductive element.

The energy stored or released during one pulse of the power curve can be determined using the following equation which is very similar to that used for dc circuits:

$$W_C = CV_C^2 \quad \text{(joules, J)} \qquad \textbf{(13.37)}$$

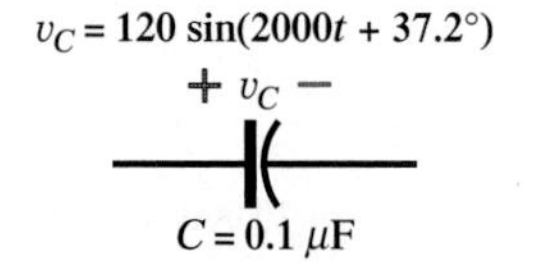

$v_C = 120 \sin(2000t + 37.2°)$

$+ \; v_C \; -$

$C = 0.1 \; \mu F$

FIG. 13.69

Example 13.37.

EXAMPLE 13.37 For the capacitor of Fig. 13.69:

a. Find the reactive power level.
b. Calculate the energy stored by one pulse of the power curve.

Solutions:

a. Effective value: $V_C = (0.707)(120 \text{ V}) = 84.84 \text{ V}$

$$\text{Reactance level:} \quad X_C = \frac{1}{\omega C} = \frac{1}{(2000 \text{ rad/s})(0.1 \; \mu F)} = 5 \text{ k}\Omega$$

$$Q_C = \frac{V_C^2}{X_C} = \frac{(84.84 \text{ V})^2}{5 \text{ k}\Omega} = \textbf{1.44 VAR}$$

b. $W_C = CV_C^2 = (0.1 \; \mu F)(84.84 \text{ V})^2 = \textbf{719.78} \; \boldsymbol{\mu}\textbf{J}$

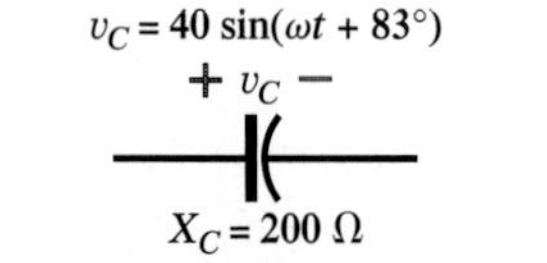

$v_C = 40 \sin(\omega t + 83°)$

$+ \; v_C \; -$

$X_C = 200 \; \Omega$

FIG. 13.70

Example 13.38.

EXAMPLE 13.38 Find the reactive power level for the capacitor of Fig. 13.70.

Solution:

$$V_{C(rms)} = 0.707(40 \text{ V}) = 28.28 \text{ V}$$

$$Q_C = \frac{V_C^2}{X_C} = \frac{(28.28 \text{ V})^2}{200 \; \Omega} = \textbf{4 VAR}$$

13.10 FREQUENCY RESPONSE OF THE BASIC ELEMENTS

Thus far, each description has been for a set frequency, resulting in a fixed level of impedance for each of the basic elements. We must now investigate how a change in frequency will affect the impedance level of the basic elements. It is an important consideration because most signals other than those provided by a power plant contain a variety of frequency levels. The last section made it quite clear that the reactance of an inductor or a capacitor is sensitive to the applied frequency. However, the question is, How will these reactance levels change if we steadily increase the frequency from a very low level to a much higher level?

Although we would like to think of every element as ideal, it is important to realize that **every commercial element available today will not respond in an ideal fashion for the full range of possible frequencies.** That is, each element is such that for a particular range of frequencies, it will perform in an essentially ideal manner. However, there will always be a range of frequencies in which the performance will vary from the ideal. **Fortunately, the designer is aware of these limitations and will take them into account in the design.**

The discussion will begin with a look at the response of the **ideal elements—a response that will be assumed for the remaining chapters of this text and one that can be assumed for any initial investigation of a network.** This discussion will be followed by a look at the factors that cause an element to deviate from an ideal response as frequency levels become too low or high.

Ideal Response

The Resistor R For an ideal resistor, you can assume that **frequency will have absolutely no effect on the impedance level,** as shown by the response of Fig. 13.71. Note that at 5 kHz or 20 kHz, the resistance of the resistor remains at 22 Ω; there is no change whatsoever. **For the analyses to follow in this text,** the resistance level will remain as the nameplate value, no matter what frequency is applied.

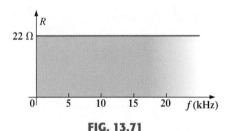

FIG. 13.71
R versus f for the range of interest.

The Inductor L For the ideal inductor, the equation for the reactance can be written as follows to isolate the frequency term in the equation. The result is a constant times the frequency variable that will change as we move down the horizontal axis of a plot:

$$X_L = \omega L = 2\pi f L = (2\pi L)f = kf \quad \text{with } k = 2\pi L$$

The resulting equation can be compared directly with the equation for a straight line:

$$y = mx + b = kf + 0 = kf$$

where $b = 0$ and the slope is k or $2\pi L$. X_L is the y variable, and f is the x variable, as shown in Fig. 13.72. Since the inductance determines the slope of the curve, the higher the inductance, the steeper the straight-line plot as shown in Fig. 13.72 for two levels of inductance.

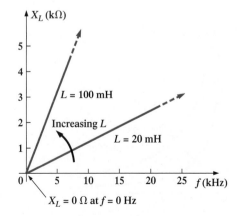

FIG. 13.72
X_L versus frequency.

In particular, note that at $f = 0$ Hz, the reactance of each plot is zero ohms as determined by substituting $f = 0$ Hz into the basic equation for the reactance of an inductor:

$$X_L = 2\pi f L = 2\pi(0 \text{ Hz})L = 0 \text{ }\Omega$$

Since a reactance of zero ohms corresponds with the characteristics of a short circuit, we can conclude that

at a frequency of 0 Hz, an inductor takes on the characteristics of a short circuit, as shown in Fig. 13.73.

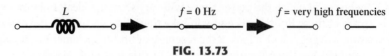

FIG. 13.73
Effect of low and high frequencies on the circuit model of an inductor.

As shown in Fig. 13.73, as the frequency increases, the reactance increases, until it reaches an extremely high level at very high frequencies. The result is that

at very high frequencies, the characteristics of an inductor approach those of an open circuit, as shown in Fig. 13.73.

The inductor, therefore, is capable of handling impedance levels that cover the entire range, from zero ohms to infinite ohms, changing at a **steady rate** determined by the inductance level. The higher the inductance, the faster it will approach the open-circuit equivalent.

The Capacitor C   For the capacitor, the equation for the reactance

$$X_C = \frac{1}{2\pi f C}$$

can be written as

$$X_C f = \frac{1}{2\pi C} = k \quad \text{(a constant)}$$

which matches the basic format for a hyberbola:

$$yx = k$$

where X_C is the y variable, f the x variable, and k a constant equal to $1/(2\pi C)$.

Hyberbolas have the shape appearing in Fig. 13.74 for two levels of capacitance. Note that the higher the capacitance, the closer the curve approaches the vertical and horizontal axes at low and high frequencies.

At or near 0 Hz, the reactance of any capacitor is extremely high, as determined by the basic equation for capacitance:

$$X_C = \frac{1}{2\pi f c} = \frac{1}{2\pi(0 \text{ Hz})C} \Rightarrow \infty \text{ }\Omega$$

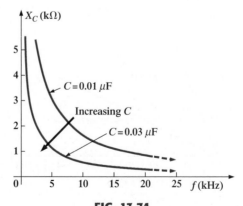

FIG. 13.74
X_C versus frequency.

The result is that

at or near 0 Hz, the characteristics of a capacitor approach those of an open circuit, as shown in Fig. 13.75.

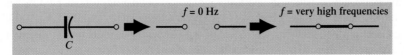

FIG. 13.75
Effect of low and high frequencies on the circuit model of a capacitor.

As the frequency increases, the reactance approaches a value of zero ohms. The result is that

at very high frequencies, a capacitor takes on the characteristics of a short circuit, as shown in Fig. 13.75.

It is important to note in Fig. 13.74 that the reactance drops very rapidly as the frequency increases. It is not a gradual drop as encountered for the rise in inductive reactance. In addition, the reactance sits at a fairly low level for a broad range of frequencies. In general, therefore, recognize that for capacitive elements, the change in reactance level can be dramatic with a relatively small change in frequency level.

Finally, recognize the following:

As frequency increases, the reactance of an inductive element increases while that of a capacitor decreases, with one approaching an open-circuit equivalent as the other approaches a short-circuit equivalent.

Practical Response

The Resistor R   In the manufacturing process, every resistive element inherits some stray capacitance levels and lead inductances. For most applications the levels are so low that their effects can be ignored. However, as the frequency extends beyond a few megahertz, it may be necessary to be aware of their effects. For instance, a number of carbon composition resistors have the frequency response appearing in Fig. 13.76. The 100 Ω resistor is essentially stable up to about 300 MHz, whereas the 100 kΩ resistor starts to drop off at about 15 MHz. In general, therefore, this type of carbon composition resistor has the ideal characteristics of Fig. 13.71 for frequencies up to about 15 MHz. For frequencies of 100 Hz, 1 kHz, 150 kHz, and so on, the resistor can be considered ideal.

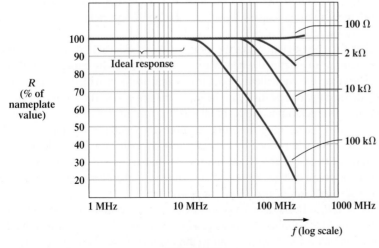

FIG. 13.76
Typical resistance-versus-frequency curves for carbon composition resistors.

The Inductor L   In reality, inductance can be affected by frequency, temperature, and current. A true equivalent for an inductor appears in Fig. 13.77. The series resistance R_s represents the copper losses (resistance of the many turns of thin copper wire); the eddy current

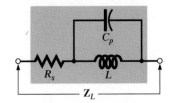

FIG. 13.77
Practical equivalent for an inductor.

losses (losses due to small circular currents in the core when an ac voltage is applied); and the hysteresis losses (losses due to core losses created by the rapidly reversing field in the core). The capacitance C_p is the stray capacitance that exists between the windings of the inductor.

For most inductors, the construction is usually such that the larger the inductance, the lower the frequency at which the parasitic elements become important. That is, for inductors in the millihenry range (which is very typical), frequencies approaching 100 kHz can have an effect on the ideal characteristics of the element. For inductors in the microhenry range, a frequency of 1 MHz may introduce negative effects. This is not to suggest that the inductors lose their effect at these frequencies but more that they can no longer be considered ideal (purely inductive elements).

Figure 13.78 is a plot of the magnitude of the impedance Z_L of Fig. 13.77 versus frequency. Note that up to about 2 MHz, the impedance increases almost linearly with frequency, clearly suggesting that the 100 μH inductor is essentially ideal. However, above 2 MHz all the factors contributing to R_s will start to increase, while the reactance due to the capacitive element C_p will be more pronounced. The dropping level of capacitive reactance will begin to have a shorting effect across the windings of the inductor and will reduce the overall inductive effect. Eventually, if the frequency continues to increase, the capacitive effects will overcome the inductive effects, and the element will actually begin to behave in a capacitive fashion. Note the similarities of this region with the curves of Fig. 13.75. Also, note that decreasing levels of inductance (available with fewer turns and therefore lower levels of C_p) will not demonstrate the degrading effect until higher frequencies are applied.

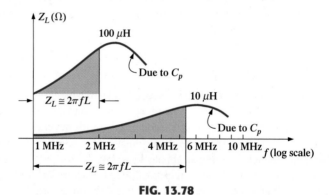

FIG. 13.78

Z_L versus frequency for the practical inductor equivalent of Fig. 13.77.

In general, therefore, the frequency of application for a coil becomes important at increasing frequencies. Inductors lose their ideal characteristics and, in fact, begin to act as capacitive elements with increasing losses at very high frequencies.

The Capacitor C   The capacitor, like the inductor, is not ideal at higher frequencies. In fact, a transition point can be defined where the characteristics of the capacitor will actually be inductive. The complete equivalent model for a capacitor is provided in Fig. 13.79. The resistance R_s, defined by the resistivity of the dielectric (typically 10^{12} $\Omega \cdot$ m or better) and the case resistance, will determine the level of leakage current to expect during the discharge cycle. In other words, a charged capacitor can discharge both through the case and through the dielectric at a rate determined by the resistance of each path. Depending on the capacitor, the discharge time can extend from a few seconds for some electrolytic capacitors to hours (paper) or perhaps days (polystyrene). Inversely, therefore, electrolytics obviously have much lower levels of R_s than paper or polystyrene.

The resistance R_p reflects the energy lost as the atoms continually realign themselves in the dielectric due to the applied alternating ac voltage. Molecular friction is present due to the motion of the atoms as they respond to the alternating applied electric field. Interestingly enough, however, the relative permittivity will decrease with increasing frequencies but will eventually take a complete turnaround and begin to increase at very high frequencies. The inductance L_s includes the inductance of the capacitor leads and any inductive effects introduced by the design of the capacitor. Be aware that the inductance of the leads is about 0.05 μH per centimeter or 0.2 μH for a capacitor with two 2 cm leads—a level that can be important at high frequencies. As for the inductor, the capacitor will behave quite

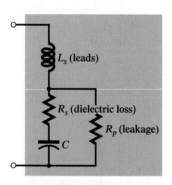

FIG. 13.79

Practical equivalent for a capacitor.

ideally for the low- and mid-frequency range, as shown by the plot of Fig. 13.80 for a 0.01 μF metalized film capacitor with 2 cm leads. As the frequency increases, however, and the reactance X_s becomes larger, a frequency will eventually be reached where the reactance of the coil equals that of the capacitor (a resonant condition to be described in Chapter 17). Any additional increase in frequency will simply result in X_s being greater than X_C, and the element will behave like an inductor.

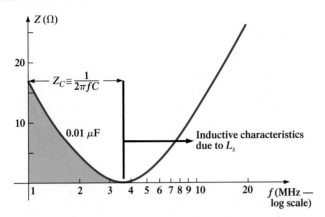

FIG. 13.80

Impedance characteristics of a 0.01 μF metalized film capacitor versus frequency.

In general, therefore, the frequency of application is important for capacitive elements because there comes a point with increasing frequency when the element will take on inductive characteristics. It also points out that the frequency of application defines the type of capacitor (or inductor) that would be applied: Electrolytics are limited to frequencies up to perhaps 10 kHz, while ceramic or mica can handle frequencies beyond 10 MHz.

The expected temperature range of operation can have an important impact on the type of capacitor chosen for a particular application. Electrolytics, tantalum, and some high-k ceramic capacitors are very sensitive to colder temperatures. In fact, most electrolytics lose 20% of their room-temperature capacitance at 0°C (freezing). Higher temperatures (up to 100°C or 212°F) seem to have less impact in general than colder temperatures, but high-k ceramics can lose up to 30% of their capacitance level at 100°C compared to room temperature. With exposure and experience, you will learn the type of capacitor employed for each application, and concern will arise only when very high frequencies, extreme temperatures, or very high currents or voltages are encountered.

EXAMPLE 13.39 At what frequency will the reactance of a 200 mH inductor match the resistance level of a 5 kΩ resistor?

Solution: The resistance remains constant at 5 kΩ for the frequency range of the inductor. Therefore,

$$R = 5000\ \Omega = X_L = 2\pi fL = 2\pi Lf$$
$$= 2\pi(200 \times 10^{-3}\text{ H})f = 1.257f$$

and
$$f = \frac{5000\text{ Hz}}{1.257} \cong \textbf{3.98 kHz}$$

EXAMPLE 13.40 At what frequency will an inductor of 5 mH have the same reactance as a capacitor of 0.1 μF?

Solution:

$$X_L = X_C$$
$$2\pi fL = \frac{1}{2\pi fC}$$
$$f^2 = \frac{1}{4\pi^2 LC}$$

and
$$f = \frac{1}{2\pi\sqrt{LC}} = \frac{1}{2\pi\sqrt{(5 \times 10^{-3}\text{ H})(0.1 \times 10^{-6}\text{ F})}}$$

$$= \frac{1}{2\pi\sqrt{5 \times 10^{-10}}} = \frac{1}{(2\pi)(2.236 \times 10^{-5})} = \frac{10^5\text{ Hz}}{14.05} \cong \mathbf{7.12\text{ kHz}}$$

13.11 COMPUTER ANALYSIS

The computer analysis section of this chapter will include a description of using Mathcad to work with complex numbers and using PSpice and EWB to set up the response of the basic elements. In some respects the coverage may seem more extensive than necessary, but this is the first application of computer methods to ac circuits, and the use of each software package should receive some attention.

Mathcad

You may be somewhat reluctant to investigate the use of Mathcad with complex numbers due to the availability of the versatile TI-86 calculator. However, Mathcad has advantages when you are working with complex networks. The procedure is quite straightforward and can be learned in a few minutes. We will continue to use j when defining a number in rectangular form even though the Mathcad solution will always appear with the letter i. You can change the solution to the j format by going to the **Format** menu, but for this presentation it was decided to use the default operators as much as possible.

When entering j to define the imaginary component of a complex number, be sure to enter it as $1j$; but do not put a multiplication operator between the 1 and the j. Just type 1 and then j. In addition, place the j after the constant rather than before as in the text material.

When Mathcad operates on an angle, it will assume that the angle is in radians and not degrees. Further, all results will appear in radians rather than degrees.

The first operation to be developed is the conversion from rectangular to polar form. In Fig. 13.81, the rectangular number $4 + j\,3$ is being converted to polar form using Mathcad. First X and Y are defined using the colon operator. Next the equation for the magnitude of the polar form is written in terms of the two variables just defined. The magnitude of the polar form is then revealed by writing the variable again and using the equals sign. It will take some practice, but be careful when writing the equation for Z in the sense that you should pay particular attention to the location of the bracket before performing the next operation. The resulting magnitude of 5 is as expected.

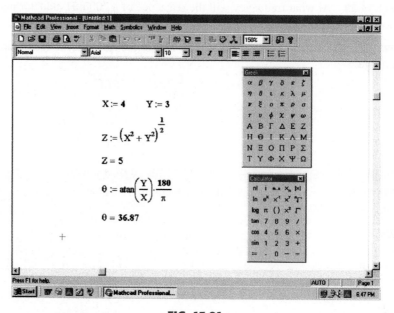

FIG. 13.81

Using Mathcad to convert from rectangular to polar form.

For the angle, the sequence **View-Toolbars-Greek** is first applied to obtain the **Greek** toolbar appearing in Fig. 13.81. It can be moved to any location by simply clicking on the blue at the top of the toolbar and dragging it to the preferred location. Then 0 is selected from the toolbar as the variable to be defined. The $\tan^{-1}\theta$ is obtained through the sequence **Insert-*f(x)*-Insert Function** dialog box-**trigonometric-atan-OK** in which *Y/X* is inserted. Then bring the controlling bracket to the outside of the entire expression, and multiply by the ratio of $180/\pi$ with π selected from the **Calculator** toolbar (available from the same sequence used to obtain the **Greek** toolbar). The multiplication by the last factor of the equation will ensure that the angle is in degrees. Selecting θ again followed by an equals sign will result in the correct angle of 36.87° as shown in Fig. 13.81.

We will now look at two forms for the polar form of a complex number. The first is defined by the basic equations introduced in this chapter, while the second uses a special format. For all the Mathcad analyses in this text, the latter format will be employed. First the magnitude of the polar form is defined, followed by the conversion of the angle of 60° to radians by multiplying by the factor $\pi/180$ as shown in Fig. 13.82. In this example, the resulting angular measure is $\pi/3$ radians. Next the rectangular format is defined by a real part $X = Z \cos\theta$ and by an imaginary part $Y = Z \sin\theta$. Both the cos and the sin are obtained by the sequence **Insert-*f(x)*-trigonometric-cos**(or **sin**)**-OK**. Note the multiplication by *j* which was actually entered as 1*j*. Entering *C* again followed by an equals sign will result in the correct conversion shown in Fig. 13.82.

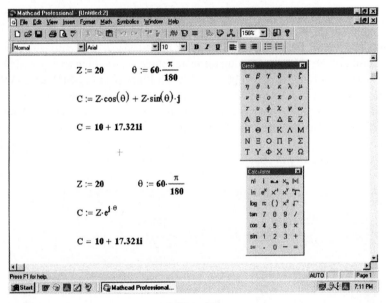

FIG. 13.82

Using Mathcad to convert from polar to rectangular form.

The next format is based on the mathematical relationship $e^{j\theta} = \cos\theta + j\sin\theta$. Both Z and θ are as defined above, but now the complex number is written as shown in Fig. 13.82 using the notation just introduced. Note that both Z and θ are part of this defining form. The e^x is obtained directly from the **Calculator** toolbar. Remember to enter the *j* as 1*j* without a multiplication sign between the 1 and the *j*. However, there is a multiplication operator placed between the *j* and θ. When entered again followed by an equals sign, the rectangular form appears to match the above results. As mentioned above, it is this latter format that will be used throughout the text due to its cleaner form and more direct entering path.

The last example using Mathcad will be a confirmation of the results of Example 13.15(b) as shown in Fig. 13.83. The three complex numbers are first defined as shown. Then the equation for the desired result is entered using C_4, and finally the results are called for.

Note the relative simplicity of the equation for C_4 now that all the other variables have been defined. As shown, however, the immediate result is in the rectangular form using the magnitude feature from the calculator and the **arg** function from **Insert-*f(x)*-Complex Numbers-arg.** In the chapters to follow, there will be a number of other examples on the use of Mathcad with complex numbers.

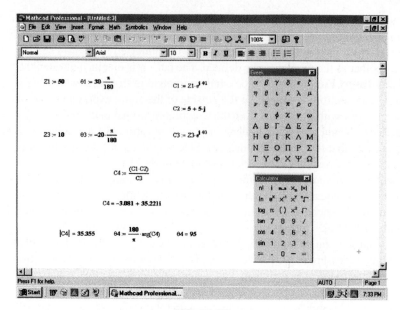

FIG. 13.83

Using Mathcad to confirm the results of Example 13.15(b).

PSpice

Capacitors and the ac Response   The simplest of ac capacitive circuits will now be analyzed to introduce the process of setting up an ac source and running an ac transient simulation. The ac source of Fig. 13.84 is obtained through **Place part** key-**SOURCE-VSIN-OK.** The name or value of any parameter can be changed by simply double-clicking on the parameter on the display or by double-clicking on the source symbol to get the **Property Editor** dialog box. Within the dialog box the values appearing in Fig. 13.84 were set, and under **Display, Name and Value** were selected. After you have selected **Apply** and exited the dialog box, the parameters will appear as shown in the figure.

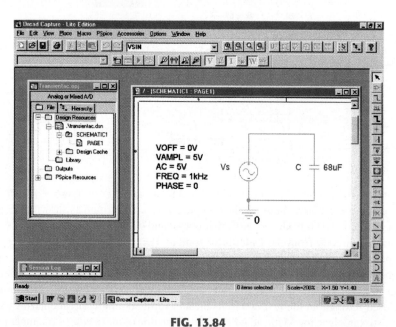

FIG. 13.84

Using PSpice to analyze the response of a capacitor to a sinusoidal ac signal.

The simulation process is initiated by selecting the **New Simulation Profile** and under **New Simulation** entering **Transientac** for the **Name** followed by **Create.** In the **Simulation Settings** dialog box, **Analysis** is selected and **Time Domain(Transient)** is chosen under **Analysis type.** The **Run to time** will be set at 3 ms to permit a display of three cycles of the sinusoidal waveforms ($T = 1/f = 1/1000$ Hz $= 1$ ms). The **Start saving data**

after will be left at 0 s, and the **Maximum step size** will be 3 ms/1000 = 3 μs. Clicking **OK** and then selecting the **Run PSpice** icon will result in a plot having a horizontal axis that extends from 0 to 3 ms.

Now we have to tell the computer which waveforms we are interested in. First, we should take a look at the applied ac source by selecting **Trace-Add Trace-V(Vs:+)** followed by **OK.** The result is the sweeping ac voltage in the bottom region of the screen of Fig. 13.85. Note that it has a peak value of 5 V, and three cycles appear in the 3 ms time frame.

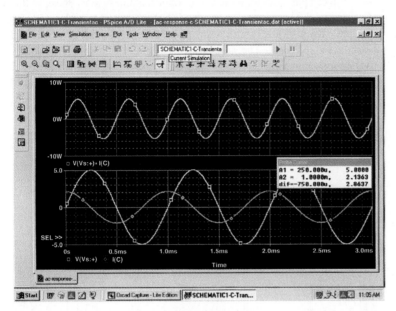

FIG. 13.85

A plot of the voltage, current, and power for the capacitor of Fig. 13.84.

The current for the capacitor can be added by selecting **Trace-Add Trace** and choosing **I(C)** followed by **OK.** The resulting waveform for **I(C)** appears at a 90° phase shift from the applied voltage, with the current leading the voltage (the current has already peaked as the voltage crosses the 0 V axis). Since the peak value of each plot is in the same magnitude range, the 5 appearing on the vertical scale can be used for both. A theoretical analysis would result in $X_C = 2.34 \ \Omega$, and the peak value of $I_C = E/X_C = 5 \ V/2.34 \ \Omega = 2.136$ A, as shown in Fig. 13.85.

For interest sake, and a little bit of practice, let us obtain the curve for the power delivered to the capacitor over the same time period. First select **Plot-Add Plot to Window-Trace-Add Trace** to obtain the **Add Traces** dialog box. Then choose **V(Vs:+),** follow it with a * for multiplication, and finish by selecting **I(C).** The result is the expression **V(Vs:+)*I(C)** of the power format: $p = vi$. Click **OK,** and the power plot at the top of Fig. 13.85 will appear. Note that over the full three cycles, the area above the axis equals the area below; there is no net transfer of power over the 3 ms period. Note also that the power curve is sinusoidal (which is quite interesting) with a frequency twice that of the applied signal. Using the cursor control, we can determine that the maximum power (peak value of the sinusoidal waveform) is 5.34 W. The cursors, in fact, have been added to the lower curves to show the peak value of the applied sinusoid and the resulting current.

After selecting the **Toggle cursor** icon, left-click the mouse to surround the **V(Vs:+)** at the bottom of the plot with a dashed line to show that the cursor is providing the levels of that quantity. When placed at 1/4 of the total period of 250 μs **(A1),** the peak value is exactly 5 V as shown in the **Probe Cursor** dialog box. Placing the cursor over the symbol next to **I(C)** at the bottom of the plot and right-clicking the mouse will assign the right cursor to the current. Placing it at exactly 1 ms **(A2)** will result in a peak value of 2.136 A to match the solution above. To further distinguish between the voltage and current waveforms, the color and the width of the lines of the traces were changed. Place the cursor right on the plot line and perform a right click. Then the **Properties** option appears. When **Properties** is selected, a **Trace Properties** dialog box will appear in which the yellow color can be selected and the width widened to improve the visibility on the black background. Note that yellow was chosen for **Vs** and green for **I(C).** Note also that the axis and the grid have been changed to a more visible color using the same procedure.

Electronics Workbench (EWB)

Since PSpice reviewed the response of a capacitive element to an ac voltage, Electronics Workbench will repeat the analysis for an inductive element. The ac voltage source was derived from the **Sources** parts bin with the values appearing in Fig. 13.86 set in the **AC Voltage** dialog box. Since the transient response of EWB is limited to a plot of voltage versus time, a plot of the current of the circuit will require the addition of a resistor of 1 Ω in series with the inductive element. The magnitude of the current through the resistor and, of course, the series inductor will then be determined by

$$|i_R| = \left|\frac{v_R}{R}\right| = \left|\frac{v_R}{1\,\Omega}\right| = |v_R| = |i_L|$$

revealing that the current will have the same peak value as the voltage across the resistor due to the division by 1.

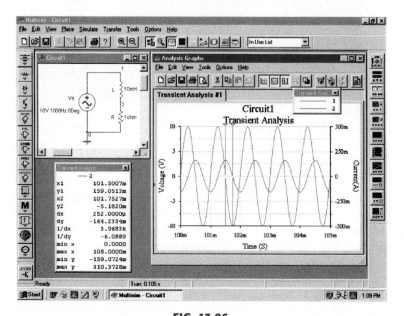

FIG. 13.86

Using EWB to review the response of an inductive element to a sinusoidal ac signal.

When viewed on the graph, it can simply be considered a plot of the current. In actuality, all inductors require a series resistance, so the 1 Ω resistor serves an important dual purpose. The 1 Ω resistance is also so small compared to the reactance of the coil at the 1 kHz frequency that its effect on the total impedance or voltage across the coil can be ignored.

Once the circuit has been constructed, the sequence **Simulate-Analyses-Transient Analysis** will result in a **Transient Analysis** dialog box in which the **Start time** is set at 0 s and the **End time** at 105 ms. The 105 ms was set as the **End time** to give the network 100 ms to settle down in its steady-state mode and 5 ms for five cycles in the output display. The **Minimum number of time points** was set at 10,000 to ensure a good display for the rapidly changing waveforms.

Next the **Output variables** heading was chosen within the dialog box, and nodes **1** and **2** were moved from the **Variables in Circuit** to **Selected variables** for analysis using the **Plot during simulation** key pad. Choosing **Simulate** will then result in a waveform that extends from 0 s to 105 ms. Even though we plan to save only the response that occurs after 100 ms, the computer is unaware of our interest, and it plots the response for the entire period. This is corrected by selecting the **Properties** key pad in the toolbar at the top of the graph (it looks like a tag and pencil) to obtain the **Graph Properties** dialog box. Selecting **Bottom Axis** will permit setting the **Range** from a **Minimum of 0.100s=100ms** to a **Maximum of**

0.105s=105ms. Click **OK,** and the time period of Fig. 13.86 will be displayed. The grid structure is added by selecting the **Show/Hide Grid** key pad, and the color associated with each nodal voltage will be displayed if we choose the **Show/Hide Legend** key next to it.

The scale for the plot of i_L can be improved by first going to **Traces** and setting the **Trace** to the number **2** representing the voltage across the 1 Ω resistor. When **2** is selected, the **Color** displayed will automaticaly change to blue. In the **Y Range,** select **Right Axis** followed by **OK.** Then select the **Right Axis** heading, and enter **Current(A)** for the **Label,** enable **Axis,** change the **Pen Size** to 1, and change the **Range** from −500 mA to +500 mA. Finally, set the **Total Ticks** at 8 with **Minor Ticks** at 2 to match the **Left Axis,** and leave the box with an **OK.** The plot of Fig. 13.86 will result.

Take immediate note of the new axis on the right and the **Current(A)** label. We can now see that the current has a peak of about 160 mA. For more detail on the peak values, simply click on the **Show/Hide Cursors** key pad on the top toolbar. A **Transient Analysis** dialog box will appear with a **1** and a red line to indicate that it is working on the full source voltage at node **1.** To switch to the current curve (the blue curve), simply bring the cursor to any point on the blue curve and perform a left click. A blue line and the number **2** will appear at the heading of the **Transient Analysis** dialog box. Clicking on the **1** in the small inverted arrow at the top will allow you to drag the vertical red line to any horizontal point on the graph. As shown in Fig. 13.86, when the cursor is set on 101.5 ms **(x1),** the peak value of the current curve is 159.05 mA **(y1).** A second cursor appears in blue with a number **2** in the inverted arrowhead that can also be moved with a left click on the number **2** at the top of the line. If set at 101.75 ms **(x2),** it has a minimum value of −5.18 mA **(y2),** the smallest value available for the calculated data points. Note that the difference between horizontal time values **dx** = 252 μs = 0.25 ms which is 1/4 of the period of the wave (at 1 ms).

CHAPTER SUMMARY

SINUSOIDAL WAVEFORMS

- Sinusoidal waveforms can be added graphically or by representing each by a radius vector and adding the vectors using vector algebra.
- The addition or subtraction of two sinusoidal waveforms of the same frequency will result in a sinusoidal waveform of the same frequency.

COMPLEX NUMBERS

- Addition and subtraction are performed in the rectangular form unless the phase angles are the same or 180° out of phase in the polar form.
- Multiplication and division can be performed in both the rectangular and the polar forms.
- The symbol j is equal to the square root of −1, with j^2 equal to −1.

PHASOR NOTATION

- Using the phasor notation, two sinusoidal waveforms can be added or their difference found using phasor (complex) algebra.
- When applied to electric circuits, the magnitude of a phasor quantity is the effective or rms value of the voltage or current. The angle is the phase angle of the voltage or current.
- The vector form for the resistance of a resistor, or for the impedance of an inductor or a capacitor, does not represent a sinusoidal function.

POWER

- The power delivered to a resistive element is dissipated in the form of heat.
- The average or real power delivered to a purely inductive or capacitive element is zero. However, a reactive power level is associated with each which is found using equations having the same format as those used for the resistive element.

FREQUENCY RESPONSE

- Ideally, the resistance of a resistor is independent of the frequency applied. In reality, however, the resistance level will change with the application of very high frequencies.
- The impedance of an ideal inductive element will increase in a straight-line fashion as the frequency increases. On a practical basis, however, the reactance will drop off at high frequencies due to the stray capacitance between the turns of wire.
- The impedance of any ideal capacitor will be very high at low frequencies and will drop off dramatically as the frequency increases. Practically speaking, however, the impedance will eventually increase due to the inherent inductance in the equivalent circuit.
- At very low frequencies, an inductor can be replaced by a short-circuit equivalent, and the capacitor by an open-circuit equivalent. At very high frequencies, the reverse is true.

Important Equations

$$\mathbf{C} = X + jY = Z \angle \theta$$

$$Z = \sqrt{X^2 + Y^2} \qquad \theta = \tan^{-1}\frac{Y}{X}$$

$$X = Z\cos\theta \qquad Y = Z\sin\theta$$

$$j = \sqrt{-1} \qquad j^2 = -1 \qquad \frac{1}{j} = -j$$

$$\mathbf{C_1} \cdot \mathbf{C_2} = (X_1X_2 - Y_1Y_2) + j(Y_1X_2 + X_1Y_2) = Z_1Z_2 \underline{/\theta_1 + \theta_2}$$

$$\frac{\mathbf{C_1}}{\mathbf{C_2}} = \frac{X_1X_2 + Y_1Y_2}{X_2^2 + Y_2^2} + j\frac{X_2Y_1 - X_1Y_2}{X_2^2 + Y_2^2} = \frac{Z_1}{Z_2}\underline{/\theta_1 - \theta_2}$$

$$\frac{1}{X + jY} = \frac{X}{X^2 + Y^2} - j\frac{Y}{X^2 + Y^2}$$

$$v_1 = V_m\sin(\omega t + \theta_1) \Rightarrow \mathbf{V_1} = 0.707V_m \angle\theta$$

$$\mathbf{Z}_R = R \angle 0° \qquad I_m = \frac{V_m}{R}$$

$$X_L = \omega L \qquad \mathbf{Z}_L = X_L \angle 90° \qquad I_m = \frac{V_m}{X_L}$$

$$X_C = \frac{1}{\omega C} \qquad \mathbf{Z}_C = X_C \angle -90° \qquad I_m = \frac{V_m}{X_C}$$

$$P_R = V_R I_R = I_R^2 R = \frac{V_R^2}{R} \qquad W_R = P \cdot T$$

$$Q_L = V_L I_L = I_L^2 X_L = \frac{V_L^2}{X_L} \qquad W_L = LI_L^2$$

$$Q_C = V_C I_C = I_C^2 X_C = \frac{V_C^2}{X_C} \qquad W_C = CV_C^2$$

PROBLEMS

SECTION 13.2 Adding and Subtracting Sinusoidal Waveforms

1. **a.** Using graph paper, add the two sinusoidal waveforms $v_1 = 10\sin\omega t$ and $v_2 = 20\sin\omega t$.
   b. Using the phasor form, add the two waveforms of part (a), and compare your answer with the results of part (a).

*2. **a.** Using graph paper, add the two sinusoidal waveforms $i_1 = 5\sin\omega t$ and $i_2 = 10\sin(\omega t + 45°)$.
   b. Using the phasor form, add the two waveforms of part (a), and compare your answer with the results of part (a).

*3. a. Using graph paper, find the difference $v_T = v_1 - v_2$ if $v_1 = 12 \sin \omega t$ and $v_2 = 4 \sin(\omega t - 90°)$.
  b. Using the phasor form, find the difference of the two waveforms of part (a), and compare your answer with the results of part (a).

SECTION 13.3   Complex Numbers

4. Sketch the following in the complex plane:
  a. $\mathbf{C} = 8 + j12$ **b.** $\mathbf{C} = -8 + j8$
  c. $\mathbf{C} = 4 - j10$

5. Sketch the following in the complex plane:
  a. $\mathbf{C} = 14 \angle 20°$ **b.** $\mathbf{C} = 7 \angle 120°$
  c. $\mathbf{C} = 0.5 \angle -80°$

6. Convert the following from rectangular to polar form:
  a. $4 + j3$ **b.** $3.5 + j16$
  c. $100 + j800$ **d.** $7.6 - j9$
  e. $-8 + j4$ **f.** $-15 - j60$
  g. $-2400 + j3600$ **h.** $5 \times 10^{-3} - j25 \times 10^{-3}$

7. Convert the following from polar to rectangular form:
  a. $6 \angle 30°$ **b.** $7400 \angle 70°$
  c. $4 \times 10^{-4} \angle 8°$ **d.** $0.0093 \angle 23°$
  e. $65 \angle 150°$ **f.** $6320 \angle -35°$
  g. $7.52 \angle -125°$ **h.** $0.008 \angle 310°$

SECTION 13.4   Mathematical Operations with Complex Numbers

8. Perform the following operations (addition and subtraction):
  a. $(4.2 + j6.8) + (7.6 + j0.2)$
  b. $(142 + j7) + (9.8 + j42) + (0.1 + j0.9)$
  c. $(4 \times 10^{-6} + j76) + (7.2 \times 10^{-7} - j5)$
  d. $(9.8 + j6.2) - (4.6 + j4.6)$
  e. $(167 + j243) - (-42.3 - j68)$
  f. $(-36.0 + j78) - (-4 - j6) + (10.8 - j72)$
  g. $6 \angle 20° + 8 \angle 80°$
  h. $42 \angle 45° + 62 \angle 60° - 70 \angle 120°$

9. Perform the following operations (multiplication):
  a. $(2 + j3)(6 + j8)$
  b. $(7.8 + j1)(4 + j2)(7 + j6)$
  c. $(2 \angle 60°)(4 \angle 22°)$
  d. $(6.9 \angle 8°)(7.2 \angle -72°)$
  e. $(540 \angle -20°)(-5 \angle 180°)(6.2 \angle 0°)$

10. Perform the following operations (division):
  a. $(42 \angle 10°)/(7 \angle 60°)$
  b. $(4360 \angle -20°)/(40 \angle 210°)$
  c. $(8 + j8)/(2 + j2)$
  d. $(8 + j42)/(-6 + j60)$
  e. $(-4.5 - j6)/(0.1 - j0.4)$

*11. Perform the following operations (express your answers in rectangular form):

  a. $\dfrac{(4 + j3) + (6 - j8)}{(3 + j3) - (2 + j3)}$

  b. $\dfrac{8 \angle 60°}{(2 \angle 0°) + (100 + j100)}$

  c. $\dfrac{(6 \angle 20°)(120 \angle -40°)(3 + j4)}{2 \angle -30°}$

  d. $\dfrac{(0.4 \angle 60°)^2(300 \angle 40°)}{3 + j9}$

  e. $\left(\dfrac{1}{(0.02 \angle 10°)^2}\right)\left(\dfrac{2}{j}\right)^3\left(\dfrac{1}{6^2 - j\sqrt{900}}\right)$

SECTION 13.5 Applying Kirchhoff's Laws Using Phasor Notation

12. Express the following in phasor form:

    a. $\sqrt{2}(100) \sin(\omega t + 30°)$
    b. $100 \sin(\omega t - 90°)$
    c. $42 \sin(377t + 0°)$
    d. $3.6 \times 10^{-6} \cos(754t - 20)$

13. Express the following phasor currents and voltages as sine waves if the frequency is 60 Hz:

    a. $\mathbf{I} = 40 \text{ A } \angle 20°$
    b. $\mathbf{V} = 120 \text{ V } \angle 0°$
    c. $\mathbf{I} = 8 \times 10^{-3} \text{ A } \angle 120°$
    d. $\mathbf{V} = \dfrac{6000}{\sqrt{2}} \text{ V } \angle -180°$

14. For the system of Fig. 13.87, find the sinusoidal expression for the unknown voltage v_a if

$$e_{in} = 60 \sin(377t + 20°)$$

    and $\qquad\qquad v_b = 20 \sin 377t$

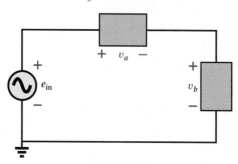

FIG. 13.87
Problem 14.

15. For the system of Fig. 13.88, find the sinusoidal expression for the unknown current i_1 if

$$i_s = 20 \times 10^{-6} \sin(\omega t + 90°)$$

    and $\qquad\qquad i_2 = 6 \times 10^{-6} \sin(\omega t - 60°)$

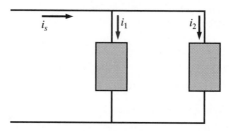

FIG. 13.88
Problem 15.

16. Find the sinusoidal expression for the applied voltage e for the system of Fig. 13.89 if

$$v_a = 60 \sin(\omega t + 30°)$$
$$v_b = 30 \sin(\omega t - 30°)$$

    and $\qquad\qquad v_c = 40 \sin(\omega t + 120°)$

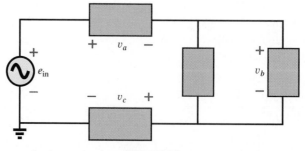

FIG. 13.89
Problem 16.

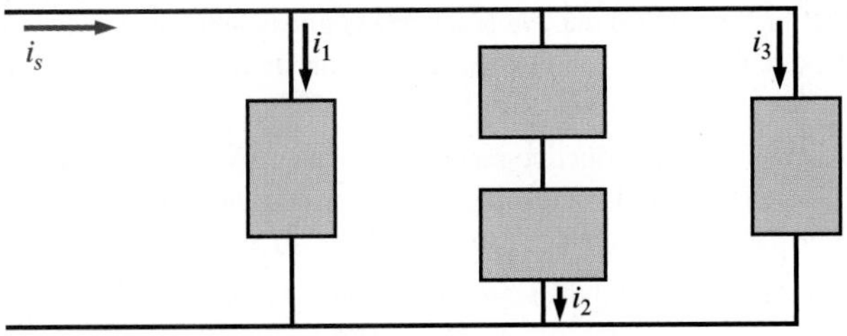

FIG. 13.90
Problem 17.

17. Find the sinusoidal expression for the current i_s for the system of Fig. 13.90 if

$$i_1 = 6 \times 10^{-3} \sin(377t + 180°)$$
$$i_2 = 8 \times 10^{-3} \sin 377t$$

and $\quad i_3 = 2i_2$

SECTION 13.6 Resistors and the ac Response

18. The voltage across a 4.7 Ω resistor is indicated. Find the sinusoidal expression for the current. In addition, sketch the v and i sinusoidal waveforms on the same set of axes.
    a. 150 sin 377t   **b.** 30 sin(377t + 20°)

19. The current through a 2.2 kΩ resistor is indicated. Find the sinusoidal expression for the voltage. In addition, sketch the i and v sinusoidal waveforms on the same set of axes.
    a. 0.03 sin 754t   **b.** $2 \times 10^{-3} \sin(400t - 120°)$

*20. The voltage across a 9.1 Ω resistor is indicated. Find the sinusoidal expression for the current. In addition, sketch the v and i sinusoidal waveforms on the same set of axes.
    a. 40 cos(ωt + 10°)   **b.** −80 sin(ωt + 40°)

*21. The current through a 3.3 kΩ resistor is indicated. Find the sinusoidal expression for the voltage. In addition, sketch the i and v sinusoidal waveforms on the same set of axes.
    a. $6 \times 10^{-6} \sin(ωt - 90°)$ **b.** −0.004 cos(ωt + 60°)

22. **a.** Using phasor notation, find the voltage across a 5.6 Ω resistor if the current is
    $$i_R = 2 \sin(ωt + 30°)$$
    b. Sketch the phasor diagram of the voltage and current.

23. **a.** Using phasor notation, find the current through a 1.2 kΩ resistor if the voltage across the resistor is $v_R = 64 \sin(ωt + 80°)$.
    b. Sketch the phasor diagram of the voltage and current.

SECTION 13.7 Inductors and the ac Response

24. Determine the inductive reactance (in ohms) of a 2 H coil for
    a. dc
    and for the following frequencies:
    b. 25 Hz   **c.** 2 kHz

25. Determine the inductance of a coil that has a reactance of
    a. 20 Ω at $f = 2$ Hz.   **b.** 1000 Ω at $f = 60$ Hz.
    c. 5.28 kΩ at $f = 1000$ Hz.

26. Determine the frequency at which a 10 H inductance has the following inductive reactances:
    a. 50 Ω   **b.** 15.7 kΩ

27. The current through a 20 Ω inductive reactance is given. What is the sinusoidal expression for the voltage? Sketch the i and v sinusoidal waveforms on the same set of axes.
    a. $i = 5 \sin ωt$ **b.** $i = 0.4 \sin(ωt + 60°)$

28. The current through a 0.1 H coil is given. What is the sinusoidal expression for the voltage?
    a. 0.3 sin 30t   **b.** $6 \times 10^{-3} \sin 377t$

29. The voltage across a 50 Ω inductive reactance is given. What is the sinusoidal expression for the current? Sketch the v and i sinusoidal waveforms on the same set of axes.
    a. 50 sin ωt   **b.** 30 sin(ωt + 20°)

30. The voltage across a 0.2 H coil is provided. What is the sinusoidal expression for the current?
    a. 1.5 sin 60t   **b.** 0.016 sin(t + 4°)

*31. The current through a 16 Ω inductive reactance is given. What is the sinusoidal expression for the voltage? Sketch the i and v sinusoidal waveforms on the same set of axes.
   a. $i = -6 \sin(\omega t - 30°)$ **b.** $i = 3 \cos(\omega t + 10°)$

*32. The voltage across a 32 Ω inductive reactance is given. What is the sinusoidal expression for the current? Sketch the v and i sinusoidal waveforms on the same set of axes.
   a. $40 \cos(\omega t + 70°)$ **b.** $-80 \cos(377t - 40°)$

33. **a.** Using phasor notation, find the sinusoidal expression for the voltage across a 10 mH coil if the current is $i_L = 50 \times 10^{-3} \sin(1000t + 40°)$.
   b. Sketch the phasor diagram of the voltage and current.

34. **a.** Using phasor notation, find the sinusoidal expression for the current through a 0.2 H coil if the voltage across the coil is $v_L = 16 \sin(2\pi\, 500t + 60°)$.
   b. Sketch the phasor diagram of the voltage and current.

SECTION 13.8   Capacitors and the ac Response

35. Determine the capacitive reactance (in ohms) of a 5 μF capacitor for
   a. dc
   and for the following frequencies:
   b. 60 Hz   **c.** 1.8 kHz

36. Determine the capacitance (in microfarads) if a capacitor has a reactance of
   a. 250 Ω at $f = 60$ Hz.   **b.** 55 Ω at $f = 312$ Hz.
   c. 10 Ω at $f = 25$ Hz.

37. Determine the frequency at which a 50 μF capacitor has the following capacitive reactances:
   a. 684 Ω   **b.** 2 kΩ

38. The voltage across a 2.5 Ω capacitive reactance is given. What is the sinusoidal expression for the current? Sketch the v and i sinusoidal waveforms on the same set of axes.
   a. $100 \sin \omega t$ **b.** $0.4 \sin(\omega t + 20°)$

39. The voltage across a 1 μF capacitor is given. What is the sinusoidal expression for the current?
   a. $30 \sin 200t$ **b.** $90 \sin 377t$

40. The current through a 10 Ω capacitive reactance is given. Write the sinusoidal expression for the voltage. Sketch the i and v sinusoidal waveforms on the same set of axes.
   a. $i = 50 \sin \omega t$ **b.** $i = 40 \sin(\omega t + 60°)$

41. The current through a 0.5 μF capacitor is given. What is the sinusoidal expression for the voltage?
   a. $i = 0.20 \sin 300t$ **b.** $i = 7 \times 10^{-3} \sin 377t$

*42. The voltage across a 1.8 kΩ capacitive reactance is given. What is the sinusoidal expression for the current? Sketch the v and i sinusoidal waveforms on the same set of axes.
   a. $8 \cos(\omega t + 10°)$ **b.** $-70 \sin(\omega t + 40°)$

*43. For the following pairs of voltages and currents, indicate whether the element involved is a capacitor, an inductor, or a resistor, and find the value of R, L, or C if sufficient data are provided:

   a. $v = 550 \sin(377t + 40°)$
   $i = 11 \sin(377t - 50°)$

   b. $v = 36 \sin(754t + 80°)$
   $i = 4 \sin(754t + 170°)$

   c. $v = 10.5 \sin(\omega t + 13°)$
   $i = 1.5 \sin(\omega t + 13°)$

*44. For the following pairs of voltages and currents, indicate whether the element involved is a resistor, an inductor, or a capacitor, and find the value of the R, L, or C if sufficient data are provided:

   a. $v = 2000 \sin \omega t$ **b.** $v = 80 \sin(157t + 150°)$
   $i = 5 \cos \omega t$ $i = 2 \sin(157t + 60°)$

   c. $v = 35 \sin(\omega t - 20°)$
   $i = 7 \cos(\omega t - 110°)$

SECTION 13.9   Power and the Basic Elements

45. Find the power delivered to a 2.2 Ω resistor by a current
$$i_R = 4 \sin(\omega t + 60°)$$

46. Find the power delivered to a 1.2 kΩ resistor if the voltage across the resistor is
$$v_R = 22 \sin(377t + 10°)$$

47. a. Find the power delivered to a resistor of unknown value if the voltage across the resistor is

$$v_R = 68 \sin(2\pi\ 1000t + 80°)$$

and the current through the resistor is

$$i_R = 200 \times 10^{-3} \sin(2\pi\ 1000t + 80°)$$

b. Find the energy dissipated during one full cycle of the applied voltage.

48. For a 200 mH inductor with a voltage $v_L = 20 \sin(4000t + 60°)$ across it:
   a. Find the real power delivered to the inductor over one full cycle of the power curve.
   b. Find the reactive power to the inductor.
   c. Find the energy stored during one positive pulse of the power curve.

49. The current through a 5 mH coil is

$$i_L = 20 \times 10^{-3} \sin(600t - 60°)$$

and the voltage across the coil is

$$v_L = 60 \times 10^{-3}(600t + 30°)$$

   a. Find the real power delivered to the inductor over one full cycle of the power curve.
   b. Find the reactive power to the inductor.
   c. Find the energy stored during one positive pulse of the power curve.

50. The current of a 3.3 μF capacitor is

$$i_C = 6 \times 10^{-3} \sin(1000t - 30°)$$

   a. Find the real power delivered to the capacitor over one full cycle of the power curve.
   b. Find the reactive power to the capacitor.
   c. Find the energy stored during one positive pulse of the power curve.

51. The current of a 68 μF capacitor is

$$i_C = 20 \times 10^{-3} \sin(800t + 40°)$$

and the voltage across it is

$$v_C = 48 \times 10^{-3} \sin(800t - 50°)$$

   a. Find the real power delivered to the capacitor over one full cycle of the power curve.
   b. Find the reactive power to the capacitor.
   c. Find the energy stored during one positive pulse of the power curve.

SECTION 13.10   Frequency Response of the Basic Elements

52. Plot X_L versus frequency for a 5 mH coil using a frequency range of zero to 100 kHz on a linear scale.

53. Plot X_C versus frequency for a 1 μF capacitor using a frequency range of zero to 10 kHz on a linear scale.

54. At what frequency will the reactance of a 1 μF capacitor equal the resistance of a 2 kΩ resistor?

55. The reactance of a coil equals the resistance of a 10 kΩ resistor at a frequency of 5 kHz. Determine the inductance of the coil.

56. Determine the frequency at which a 1 μF capacitor and a 10 mH inductor will have the same reactance.

57. Determine the capacitance required to establish a capacitive reactance that will match that of a 2 mH coil at a frequency of 50 kHz.

SECTION 13.11   Computer Analysis

58. Using PSpice, repeat the analysis provided for the capacitor of Fig. 13.84 for an inductive element of 5 mH. Use the same applied voltage.

59. Using PSpice, repeat the frequency analysis provided for the capacitor of Fig. 13.84 for a resistance of 1.2 kΩ. Use the same applied voltage.

60. Using PSpice, plot the product of the voltage across the capacitor and the current of the capacitor for the capacitor of Fig. 13.84 using the same applied voltage. What does the resulting curve demonstrate about the power delivered to a capacitor over one full cycle of the applied voltage?

61. Using EWB, repeat the frequency analysis provided for the inductor of Fig. 13.86 for a 100 μF capacitor. Use the same applied voltage.

62. Using EWB, repeat the frequency analysis provided for the inductor of Fig. 13.86 for a 2.2 kΩ resistor. Use the same applied voltage.

GLOSSARY

Apparent power   The power delivered to a load without consideration of the effects of a power-factor angle of the load. It is determined solely by the product of the terminal voltage and current of the load.

Average or **real power**   The power delivered to and dissipated by the load over a full cycle.

Complex conjugate   A complex number defined by simply changing the sign of an imaginary component of a complex number in the rectangular form.

Complex number   A number that represents a point in a two-dimensional plane located with reference to two distinct axes. It defines a vector drawn from the origin to that point.

Impedance   The opposition of an ac network to the flow of charge.

Phasor   A radius vector that has a constant magnitude at a fixed angle from the positive real axis and that represents a sinusoidal voltage or current in the vector domain.

Phasor diagram   A "snapshot" of the phasors that represent a number of sinusoidal waveforms at $t = 0$.

Polar form   A method of defining a point in a complex plane that includes a single magnitude to represent the distance from the origin, and an angle to reflect the counterclockwise distance from the positive real axis.

Reactance   The opposition of an inductor or a capacitor to the flow of charge that results in the continual exchange of energy between the circuit and magnetic field of an inductor or the electric field of a capacitor.

Reactive power   The power associated with reactive elements that provides a measure of the energy associated with setting up the magnetic and electric fields of inductive and capacitive elements, respectively.

Reciprocal   A format defined by 1 divided by the complex number.

Rectangular form   A method of defining a point in a complex plane that includes the magnitude of the real component and the magnitude of the imaginary component, the latter component being defined by an associated letter j.

Series and Parallel
ac Circuits

OBJECTIVES

- Become familiar with the characteristics of a series and a parallel ac circuit.
- Be able to find the total impedance of a series and a parallel ac circuit, and sketch the impedance and admittance diagrams of each.
- Learn how to find all the currents and voltages of a series and a parallel ac circuit.
- Be able to sketch a phasor diagram for all the voltages and currents of a series or parallel ac network.
- Be able to apply Kirchhoff's current law and voltage law to an ac network.
- Learn how to apply the voltage divider rule and current divider rule to ac networks.
- Be able to find the power delivered to any series or parallel ac network and become familiar with the power factor of a network.
- Understand how to determine the frequency response of a series or parallel ac network.
- Learn how to use an oscilloscope to determine the phase angle between any two voltages or currents in a network.
- Become familiar with the use of PSpice and EWB to analyze the response of an ac network to a sinusoidal ac voltage and current.

14.1 INTRODUCTION

The content of this chapter will be a pleasant surprise in that we will be able to apply all the laws and rules learned for dc circuits to ac circuits by simply using the phasor notation introduced in the previous chapter. There is no need to learn an entirely different set of laws for ac networks. Simply use the phasor notation for all the voltages and currents, and use the vector notation for the impedances. Then proceed as before. The mathematics will be a bit more complex due to the need to worry about angles, but, with the calculator or computer to perform the mathematical calculations, this additional requirement will not be much of a stumbling block.

14.2 SERIES CONFIGURATION

In Fig. 14.1, a number of elements have been placed in series. Each impedance can be a resistor, an inductor, or a capacitor. As for dc circuits,

the total impedance of series ac elements is simply the sum of the individual impedances.

That is,

$$\mathbf{Z}_T = \mathbf{Z}_1 + \mathbf{Z}_2 + \mathbf{Z}_3 + \cdots + \mathbf{Z}_N \qquad \textbf{(14.1)}$$

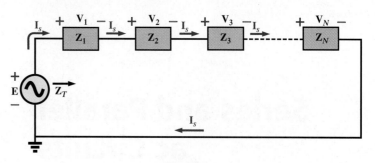

FIG. 14.1

Series impedances.

As shown by the current I_s throughout the circuit of Fig. 14.1,

the current is the same at every point in a series ac circuit.

The current is determined by Ohm's law:

$$I_s = \frac{E}{Z_T} \qquad (14.2)$$

and the voltage across each element by

$$V_1 = I_s Z_1 \qquad V_2 = I_s Z_2 \qquad V_3 = I_s Z_3 \qquad \cdots \qquad V_N = I_s Z_N \qquad (14.3)$$

Kirchhoff's voltage law can then be applied in the same manner as for dc circuits. However, keep in mind that we are now dealing with the algebraic manipulation of quantities that have both magnitude and direction.

In the clockwise direction for Fig. 14.1,

$$+E - V_1 - V_2 - V_3 - \cdots - V_N = 0$$

or

$$E = V_1 + V_2 + V_3 + \cdots + V_N \qquad (14.4)$$

As an example in the use of the above equations, let us take a look at the series *R-L-C* circuit of Fig. 14.2. If we plot the impedance of each quantity on the same set of axes, we obtain the diagram of Fig. 14.3. Note that the angle associated with the resistor is zero degrees, with an angle of 90° for the inductive element and −90° for the capacitive element as derived from

$$Z_R = R \angle 0° \qquad Z_L = X_L \angle 90° \qquad Z_C = X_C \angle -90°$$

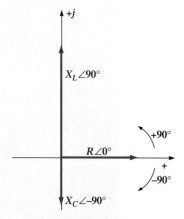

FIG. 14.3

Impedance diagram for a series R-L-C circuit.

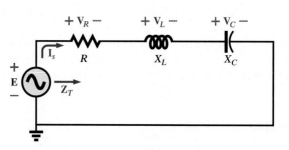

FIG. 14.2

Series R-L-C ac circuit.

The result is an **impedance diagram** for the circuit that can be used to find the total impedance using vector algebra. In Fig. 14.4, since X_L and X_C are 180° apart or in direct opposition, the net reactance is their difference. This difference is then combined with the resistive element to determine the total impedance Z_T.

The current is then determined by

$$\mathbf{I}_s = \frac{\mathbf{E}}{\mathbf{Z}_T}$$

and the voltage across each element by

$$\mathbf{V}_R = \mathbf{I}_s\mathbf{Z}_R \qquad \mathbf{V}_L = \mathbf{I}_s\mathbf{Z}_L \qquad \mathbf{V}_C = \mathbf{I}_s\mathbf{Z}_C$$

Since power is dissipated only by the resistive element, the total power delivered by the source (P_s) or absorbed by the circuit is simply

$$P_s = P_R = V_R I_R = I_R^2 R = \frac{V_R^2}{R} \qquad\qquad (14.5)$$

A few examples will clarify the use of the above equations.

EXAMPLE 14.1 Draw the impedance diagram for the circuit of Fig. 14.5, and find the total impedance.

Solution: As indicated by Fig. 14.6, the input impedance can be found graphically from the impedance diagram by properly scaling the real and imaginary axes and finding the length of the resultant vector Z_T and angle θ_T. Or, by using vector algebra, we obtain

$$\mathbf{Z}_T = \mathbf{Z}_1 + \mathbf{Z}_2 = R \angle 0° + X_L \angle 90°$$
$$= R + jX_L = 4\ \Omega + j8\ \Omega = \mathbf{8.944\ \Omega\ \angle 63.43°}$$

EXAMPLE 14.2 Determine the input impedance to the series network of Fig. 14.7. Draw the impedance diagram.

Solution:

$$\mathbf{Z}_T = \mathbf{Z}_1 + \mathbf{Z}_2 + \mathbf{Z}_3 = R \angle 0° + X_L \angle 90° + X_C \angle -90°$$
$$= R + jX_L - jX_C = R + j(X_L - X_C) = 6\ \Omega + j(10\ \Omega - 12\ \Omega)$$
$$= 6\ \Omega - j2\ \Omega = \mathbf{6.325\ \Omega\ \angle -18.43°}$$

The impedance diagram appears in Fig. 14.8.

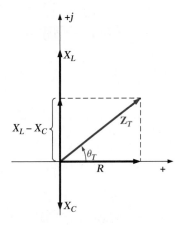

FIG. 14.4
Finding the total impedance of a series R-L-C circuit in which $X_L > X_C$.

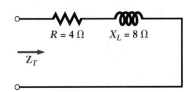

FIG. 14.5
Example 14.1.

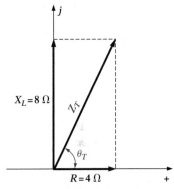

FIG. 14.6
Impedance diagram for Example 14.1.

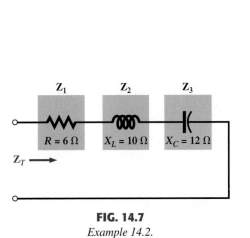

FIG. 14.7
Example 14.2.

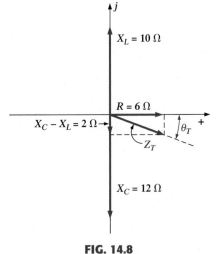

FIG. 14.8
Impedance diagram for Example 14.2.

Note that in Example 14.2, series inductive and capacitive reactances are in direct opposition. For the circuit of Fig. 14.7, if the inductive reactance were equal to the capacitive reactance, the input impedance would be purely resistive. We will have more to say about this particular condition in a later chapter.

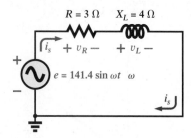

FIG. 14.9

Series R-L circuit for Example 14.3.

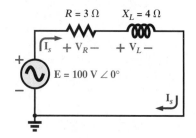

FIG. 14.10

Applying phasor notation to the network of Fig. 14.9.

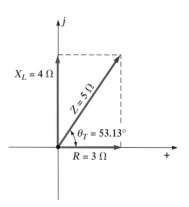

FIG. 14.11

Impedance diagram for the series R-L circuit of Fig. 14.9.

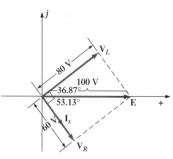

FIG. 14.12

Phasor diagram for the series R-L circuit of Fig. 14.9.

EXAMPLE 14.3   For the series *R-L* circuit of Fig. 14.9:

a. Redraw the circuit using phasor notation.
b. Draw the impedance diagram and find the total impedance.
c. Calculate the current $\mathbf{I}_s$.
d. Find the voltages $\mathbf{V}_R$ and $\mathbf{V}_L$.
e. Sketch the phasor diagram for the circuit.
f. Verify Kirchhoff's voltage law around the closed path.
g. Calculate the power dissipated by the circuit.
h. Write the current and voltages in the sinusoidal or time domain, and sketch the waveforms.

Solutions:

a. $e = 141.4 \sin \omega t \Rightarrow \mathbf{E} = (0.707)(141.4 \text{ V}) \angle 0° = 100 \text{ V} \angle 0°$
Note Fig. 14.10.

b. The impedance diagram appears in Fig. 14.11. Using complex algebra:

$$\mathbf{Z}_T = \mathbf{Z}_1 + \mathbf{Z}_2 = 3 \text{ } \Omega \angle 0° + 4 \text{ } \Omega \angle 90° = 3 \text{ } \Omega + j\,4 \text{ } \Omega$$
$$= \mathbf{5 \text{ } \Omega \angle 53.13°}$$

c. Applying Ohm's law:

$$\mathbf{I}_s = \frac{\mathbf{E}}{\mathbf{Z}_T} = \frac{100 \text{ V} \angle 0°}{5 \text{ } \Omega \angle 53.13°} = 20 \text{ A} \angle -53.13°$$

d. $\mathbf{V}_R = \mathbf{I}_s \mathbf{Z}_R = (20 \text{ A} \angle -53.13°)(3 \text{ } \Omega \angle 0°) = \mathbf{60 \text{ V} \angle -53.13°}$

$\mathbf{V}_L = \mathbf{I}_s \mathbf{Z}_L = (20 \text{ A} \angle -53.13°)(4 \text{ } \Omega \angle 90°) = \mathbf{80 \text{ V} \angle 36.87°}$

e. Note Fig. 14.12. In particular, note in Fig. 14.12 that the voltage across the resistor is in phase with the current through the resistor (the same direction for the phasors). Also note that the voltage across the coil leads the current through the coil by 90°.

f. Applying Kirchhoff's voltage law around the closed path in the clockwise direction will result in

$$+\mathbf{E} - \mathbf{V}_R - \mathbf{V}_L = 0 \quad \text{or} \quad \mathbf{E} = \mathbf{V}_R + \mathbf{V}_L$$

Using phasor algebra:

$$\mathbf{V}_R = 60 \text{ V} \angle -53.13° = 36 \text{ V} - j\,48 \text{ V}$$
$$\mathbf{V}_L = 80 \text{ V} \angle +36.87° = 64 \text{ V} + j\,48 \text{ V}$$

and $\mathbf{E} = \mathbf{V}_R + \mathbf{V}_L = (36 \text{ V} - j\,48 \text{ V}) + (64 \text{ V} + j\,48 \text{ V}) = 100 \text{ V} + j\,0$

$$\mathbf{E} = \mathbf{100 \text{ V} \angle 0°}$$

which is verified graphically by Fig. 14.12.

g. Power will be dissipated only by the resistive element, resulting in

$$P_s = P_R = V_R I_R = (60 \text{ V})(20 \text{ A}) = \mathbf{1200 \text{ W}}$$

or $$P_R = I_R^2 R = (20 \text{ A})^2 (3 \text{ } \Omega) = \mathbf{1200 \text{ W}}$$

or $$P_R = \frac{V_R^2}{R} = \frac{(60 \text{ V})^2}{3 \text{ } \Omega} = \mathbf{1200 \text{ W}}$$

h. In the time domain:

$$i = \sqrt{2}(20) \sin(\omega t - 53.13°) = \mathbf{28.28 \sin(\omega t - 53.13°)}$$
$$v_R = \sqrt{2}(60) \sin(\omega t - 53.13°) = \mathbf{84.84 \sin(\omega t - 53.13°)}$$
$$v_L = \sqrt{2}(80) \sin(\omega t + 36.87°) = \mathbf{113.12 \sin(\omega t + 36.87°)}$$

Note Fig. 14.13.

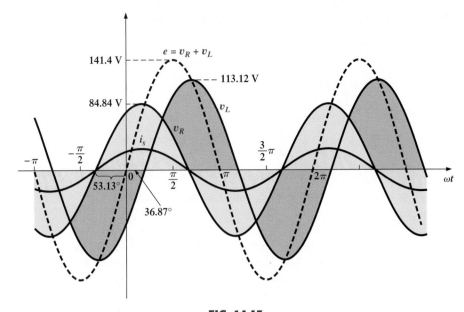

FIG. 14.13

Plot of e, v_R, v_L, and i for the circuit of Fig. 14.9.

EXAMPLE 14.4 For the series *R-L-C* circuit of Fig. 14.14:

a. Redraw the circuit using phasor notation.
b. Draw the impedance diagram and find the total impedance.
c. Calculate the current $\mathbf{I}_s$.
d. Find the voltages $\mathbf{V}_R$, $\mathbf{V}_L$, and $\mathbf{V}_C$.
e. Sketch the phasor diagram for the circuit.
f. Verify Kirchhoff's voltage law around the closed path.
g. Calculate the average power supplied by the source.
h. Write the current and voltages in the sinusoidal or time domain, and sketch the wave-forms of i_s, v_R, v_L, and v_C.

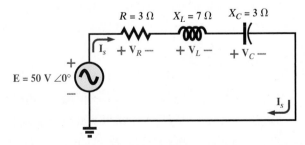

FIG. 14.14

Series R-L-C ac circuit for Example 14.4.

Solutions:

a. $e = 70.7 \sin \omega t \Rightarrow \mathbf{E} = (0.707)(70.7 \text{ V}) \angle 0° = 50 \text{ V} \angle 0°$
   Note Fig. 14.15.

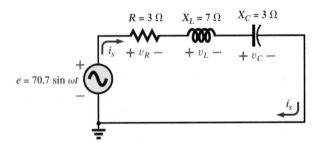

FIG. 14.15

Applying phasor notation to the circuit of Fig. 14.14.

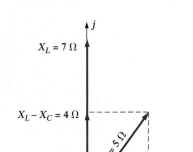

$X_L = 7\,\Omega$

$X_L - X_C = 4\,\Omega$

$Z_T = 5\,\Omega$

$\theta_T = 53.13°$

$R = 3\,\Omega$

$X_C = 3\,\Omega$

FIG. 14.16
*Impedance diagram for the series
R-L-C circuit of Fig. 14.14.*

b. The impedance diagram appears in Fig. 14.16. Using complex algebra:

$$\mathbf{Z}_T = \mathbf{Z}_R + \mathbf{Z}_L + \mathbf{Z}_C$$
$$= 3\,\Omega\,\angle 0° + 7\,\Omega\,\angle 90° + 3\,\Omega\,\angle{-90°} = 3\,\Omega + j\,7\,\Omega - j\,3\,\Omega$$
$$= 3\,\Omega + j\,4\,\Omega = \mathbf{5\,\Omega\,\angle 53.13°}$$

c. Applying Ohm's law:

$$\mathbf{I}_s = \frac{\mathbf{E}}{\mathbf{Z}_T} = \frac{50\text{ V}\,\angle 0°}{5\,\Omega\,\angle 53.13°} = \mathbf{10\text{ A}\,\angle{-53.13°}}$$

d. $\mathbf{V}_R = \mathbf{I}_s\mathbf{Z}_R = (10\text{ A}\,\angle{-53.13°})(3\,\Omega\,\angle 0°) = \mathbf{30\text{ V}\,\angle{-53.13°}}$

$\mathbf{V}_L = \mathbf{I}_s\mathbf{Z}_L = (10\text{ A}\,\angle{-53.13°})(7\,\Omega\,\angle 90°) = \mathbf{70\text{ V}\,\angle 36.87°}$

$\mathbf{V}_C = \mathbf{I}_s\mathbf{Z}_C = (10\text{ A}\,\angle{-53.13°})(3\,\Omega\,\angle{-90°}) = \mathbf{30\text{ V}\,\angle{-143.13°}}$

e. Note Fig. 14.17. In particular, note in Fig. 14.17 that the voltage across the resistor is in phase with the current, the voltage across the coil leads the current by 90°, and the current for the capacitor leads the voltage across the capacitor by 90°.

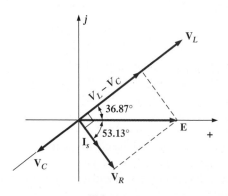

FIG. 14.17
Phasor diagram for the series R-L-C circuit of Fig. 14.14.

f. Applying Kirchhoff's voltage law around the closed path in the clockwise direction will result in

$$+\mathbf{E} - \mathbf{V}_R - \mathbf{V}_L - \mathbf{V}_C = 0 \quad \text{or} \quad \mathbf{E} = \mathbf{V}_R + \mathbf{V}_L + \mathbf{V}_C$$

Using phasor algebra:

$$50\text{ V}\,\angle 0° = 30\text{ V}\,\angle{-53.13°} + 70\text{ V}\,\angle 36.87° + 30\text{ V}\,\angle{-143.13°}$$
$$= (18\text{ V} - j\,24\text{ V}) + (56\text{ V} + j\,42\text{ V}) + (-24\text{ V} - j\,18\text{ V})$$
$$= (18\text{ V} + 56\text{ V} - 24\text{ V}) + j\,(-24\text{ V} + 42\text{ V} - 18\text{ V})$$
$$= 50\text{ V} + j\,0 = \mathbf{50\text{ V}\,\angle 0°} \quad \text{(checks)}$$

g. The power supplied by the source equals that dissipated by the circuit. The result is

$$P_s = P_R = I_R^2\,R = (10\text{ A})^2\,(3\,\Omega) = \mathbf{300\text{ W}}$$

h. In the time domain:

$$i_s = \sqrt{2}(10)\sin(\omega t - 53.13°) = \mathbf{14.14\sin(\omega t - 53.13°)}$$
$$v_R = \sqrt{2}(30)\sin(\omega t - 53.13°) = \mathbf{42.42\sin(\omega t - 53.13°)}$$
$$v_L = \sqrt{2}(70)\sin(\omega t + 36.87°) = \mathbf{98.98\sin(\omega t + 36.87°)}$$
$$v_C = \sqrt{2}(30)\sin(\omega t - 143.13°) = \mathbf{42.42\sin(\omega t - 143.13°)}$$

Note Fig. 14.18.

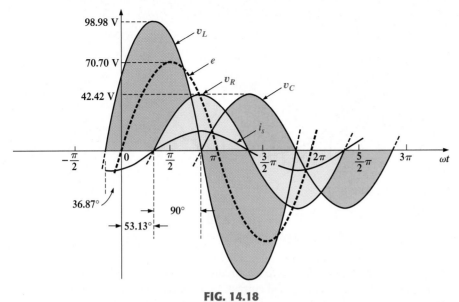

FIG. 14.18

Waveforms for the series R-L-C circuit of Fig. 14.14.

14.3 AVERAGE POWER AND THE POWER FACTOR

In the preceding discussion and examples, the power delivered to a network was determined by finding the power delivered to the resistive elements. In this section we will find that the power to any network can be found if we know simply the applied voltage, the resulting current, and the phase angle between the two. There is no need to be aware of the internal structure of the network and the current through each resistive element—we need to use only a terminal calculation. Rather than merely state the equation, which can be derived using calculus, the approach here will be to first examine a series of loading situations so that when the equation is revealed, it will have some theoretical support.

In Fig. 14.19, the current resulting from the applied voltage is provided. Since the two are in phase, the terminal characteristics of the network are resistive. A plot of the power delivered using the equation $p = vi$ will result in the pattern of Fig. 14.20, which is similar to the plot appearing in Chapter 13.

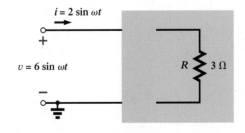

FIG. 14.19

The voltage and current for a purely resistive load.

Note in Fig. 14.20 that the current and voltage are in phase, and the power curve has twice the frequency of the voltage or current. The average value of the power curve is simply the product of the rms values of the voltage and current as derived in Chapter 13. The peak value of the power curve and the average power level are indicated on the figure for the applied voltage. In general, therefore,

the average or real power to a network composed solely of resistive elements can be found by simply finding the product of the rms values of the applied voltage and the resulting current.

That is, there is no need to find the power delivered to each individual resistor and add them up—just use the terminal voltage and current.

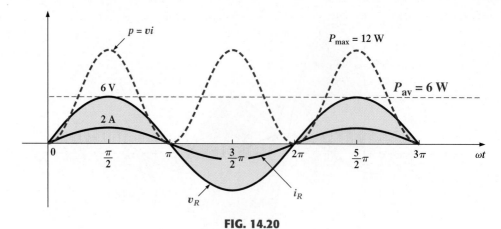

FIG. 14.20
Power curve for the resistive load of Fig. 14.19.

If we now add inductive elements to the circuit as shown in Fig. 14.21, a phase shift will result between the applied voltage and current. In this case, the *R-L* combination was chosen to maintain the same level of current and to introduce a phase shift of 30°. The resulting patterns for the voltage, current, and power are provided in Fig. 14.22. First note the phase shift and the fact that the power curve will not peak at the peak value of the voltage or current. The power curve is sensitive to the magnitude of the product of the voltage and current; it is not sensitive to simply the value of one or the other. Also note that even though the peak value of the voltage and current are the same, the peak value of the power curve and the average or real power delivered has dropped. The introduction of a phase shift be-

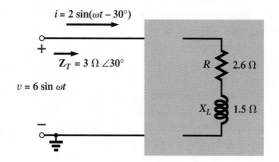

FIG. 14.21
*A circuit with the same magnitude for the total impedance as
in Fig. 14.19 but with a phase shift of 30°.*

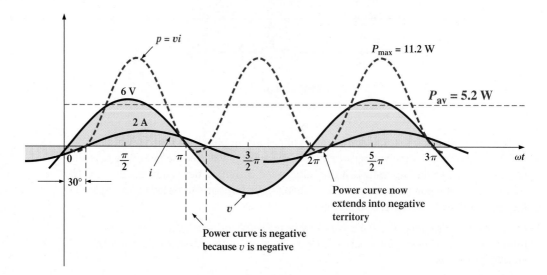

FIG. 14.22
Power curve for an R-L load that causes a phase shift of 30° between v and i.

tween the applied voltage and the resulting current has caused a decrease in the power de-livered—quite obviously the phase angle between the voltage and current will affect the power level and must appear in the equation for the power delivered to any ac system.

If we now introduce additional inductance into the network, the phase angle by which the voltage will lead the current will increase. The *R-L* combination chosen is such as to maintain the same level of current but also to introduce a phase shift of 60°. In other words, throughout this analysis we are trying to maintain the same level of applied voltage and re-sulting current, and concentrate solely on the effect of the phase angle between the two. The resulting curves for the voltage, current, and power in Fig. 14.23 reveal that the peak value and average value of the power curve continue to drop. We can therefore conclude that the power level will continue to decrease as the phase angle increases.

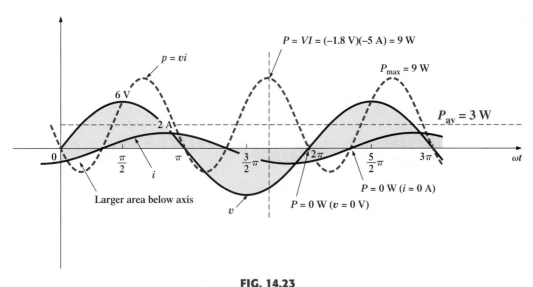

FIG. 14.23

Power curve for an R-L load that causes a phase shift of 60° between v and i.

Let us now go to the extreme of making the network look essentially inductive, as shown in Fig. 14.24, but maintain the same level of voltage and current. The resulting power curve will appear as shown in Fig. 14.25. Quite clearly, when the phase angle is increased to its maximum value of 90°, the power curve will continue to have a peak value, but now the av-erage value is zero. There is no net power transfer to the network.

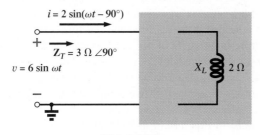

FIG. 14.24

A purely inductive load establishing the same current as in Figs. 14.19 and 14.21.

In summary, therefore, as the phase angle went from 0° to 90°, the power level contin-ued to drop—a characteristic similar to the cosine of the same angle:

At 0°, $\cos 0° = 1$, the maximum value.
At 30°, $\cos 30° = 0.866$.
At 60°, $\cos 60° = 0.500$.
At 90°, $\cos 90° = 0$.

Note in Fig. 14.22 that the average power level is 0.866 times the maximum level of 6 W, whereas in Fig. 14.23 the average power level is 0.500 times the maximum value.

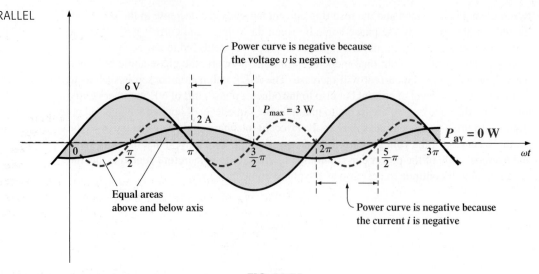

FIG. 14.25

Power curve to a purely inductive load.

A direct correspondence with the cosine function results in the following format for the power equation:

$$P = VI \cos \theta_T \qquad \text{(watts, W)} \qquad \textbf{(14.6)}$$

where V and I are the rms values at the terminals of interest and θ_T is the phase angle between them. There is no need to worry about whether the phase angle is leading or lagging. **The equation simply calls for the magnitude of the phase angle.**

The factor $\cos \theta_T$, called the **power factor of the network,** has the symbol F_p. That is,

$$F_p = \cos \theta_T \qquad \text{(constant, unitless)} \qquad \textbf{(14.7)}$$

Since the angle θ_T is actually the angle associated with the total impedance of a network, **the angle and the resulting power factor give us an immediate indication of whether the network is primarily resistive or reactive in nature.** If the angle is close to $0°$, or the power factor close to 1, the network is essentially resistive in nature. Most of the power delivered to the network will be dissipated. If the angle is close to $90°$, or the power factor close to 0, the network is primarily reactive. A very small part of the power delivered to the network will be dissipated, and most will be stored in the form of a magnetic or an electric field to be returned to the network when called for by the design.

It is interesting to note that if the resistive and reactive components of the total impedance are equal, resulting in an angle of $45°$ and a power factor of 0.7071, 70.71% of the power will still be absorbed by the network rather than the expected level of 50%. The reason is that $\cos \theta$ at $45°$ is not half its value at $90°$. In fact, the reactive component would have to be greater than 1.7 times the resistive component to bring the power level down to 50% of the maximum value.

Although the angle θ_T in Eqs. (14.6) and (14.7) is not affected by whether the phase angle is leading or lagging, the terms **leading** and **lagging** are applied to the power factor. They are defined by the current through the load. If the current leads the voltage across the load, the load has a **leading power factor.** If the current lags the voltage across the load, the load has a **lagging power factor.**

In other words,

inductive networks have lagging power factors, and capacitive networks have leading power factors.

Let us write the basic power equation $P = VI_s \cos \theta_T$ as follows:

$$\cos \theta_T = \frac{P}{VI_s}$$

and then perform the following substitutions from the basic series ac circuit:

$$\cos\theta = \frac{I_s^2 R}{VI} = \frac{I_s R}{V} = \frac{R}{V/I_s} = \frac{R}{Z_T}$$

We find that

$$\boxed{F_p = \cos\theta_T = \frac{R}{Z_T}} \qquad (14.8)$$

confirming that the angle θ_T is in fact the angle associated with the total impedance of a network.

EXAMPLE 14.5   Given the voltage and current for the system of Fig. 14.26:

a. Determine the power factor of the system and the power delivered.
b. Is the system more resistive than reactive, or vice versa?

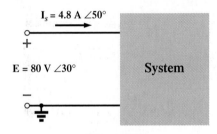

FIG. 14.26
Example 14.5.

Solutions:

a. $F_p = \cos\theta_T = \cos 20° = \mathbf{0.940}$

   $P = VI_s \cos\theta_T = (80\text{ V})(4.8\text{ A})\cos 20°$

   $\quad = (384)(0.940) = \mathbf{360.96\ W}$

b. Since the current leads the voltage by 20°, the system is capacitive. A power factor of 0.940 clearly indicates that the network is primarily resistive in nature.

EXAMPLE 14.6   Given the R-C circuit of Fig. 14.27:

a. Find the reactance of the capacitor.
b. Calculate the total impedance.
c. Find the current in phasor form.
d. What is the power factor of the circuit?
e. Is the circuit primarily resistive or reactive?
f. Find the power delivered to the circuit.
g. Write the sinusoidal expression for the current.

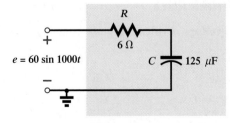

FIG. 14.27
Example 14.6.

Solutions:

a. $X_C = \dfrac{1}{\omega C} = \dfrac{1}{(1000 \text{ rad/s})(125 \ \mu\text{F})} = \mathbf{8 \ \Omega}$

b. $\mathbf{Z}_T = \mathbf{Z}_R + \mathbf{Z}_C$

$\qquad = 6 \ \Omega \ \angle 0° + 8 \ \Omega \ \angle -90° = 6 \ \Omega - j8 \ \Omega = \mathbf{10 \ \Omega \ \angle -53.13°}$

c. $\mathbf{E} = (0.707)(60 \text{ V}) \ \angle 0° = 42.42 \text{ V} \ \angle 0°$

and $\qquad\qquad \mathbf{I}_s = \dfrac{\mathbf{E}}{\mathbf{Z}_T} = \dfrac{42.42 \text{ V} \ \angle 0°}{10 \ \Omega \ \angle -53.13°} = \mathbf{4.242 \text{ A} \ \angle 53.13°}$

d. $F_p = \cos \theta_T = \cos 53.13° = \mathbf{0.6} \quad$ or $\quad F_p = \dfrac{R}{Z_T} = \dfrac{6 \ \Omega}{10 \ \Omega} = \mathbf{0.6}$

e. Since F_p is relatively low at 0.6, the circuit is primarily capacitive—a result verified by the fact that $X_C > R$.

f. $P = EI_s \cos \theta_T = (42.42 \text{ V})(4.242 \text{ A}) \cos 53.13°$

$\qquad = (179.95)(0.6) = \mathbf{107.97 \text{ W}}$

g. $i_s = \sqrt{2}(4.242) \sin(1000t + 53.13°) = \mathbf{6 \sin(1000t + 53.13°)}$

EXAMPLE 14.7 For the following applied voltage and resulting current:

$$v = 100 \sin(\omega t + 40°)$$
$$i = 20 \sin(\omega t - 40°)$$

a. Find the power factor of the network and indicate whether it is leading or lagging.
b. Is the system primarily resistive or reactive?
c. Find the average power delivered to the system.
d. Find the average power dissipated by the system.

Solutions:

a. The voltage leads the current by a total of 80°. The power factor is

$$F_p = \cos \theta_T = \cos 80° = \mathbf{0.174}$$

The voltage leads the current in an inductive network, resulting in a **lagging power factor.**

b. Since the phase angle is 80° and the power factor is 0.174, the network is **highly reactive.**

c. The average power delivered is found as follows:

$\qquad V_{\text{rms}} = 0.707(100 \text{ V}) = 70.70 \text{ V}$

$\qquad I_{\text{rms}} = 0.707(20 \text{ A}) = 14.14 \text{ A}$

$\qquad\qquad P = VI_s \cos \theta_T = (70.70 \text{ V})(14.14 \text{ A}) \cos 80° = \mathbf{173.6 \text{ W}}$

d. The purpose of this question was to emphasize that the total average power delivered will always equal the total average power dissipated. Therefore, the average power dissipated is also **173.6W.**

14.4 VOLTAGE DIVIDER RULE

The basic format for the **voltage divider rule** in ac circuits is exactly the same as that for dc circuits:

$$\boxed{\mathbf{V}_x = \dfrac{\mathbf{Z}_x \mathbf{E}}{\mathbf{Z}_T}} \qquad\qquad (14.9)$$

where $\mathbf{V}_x$ is the voltage across one or more elements in series that have total impedance $\mathbf{Z}_x$, $\mathbf{E}$ is the total voltage appearing across the series circuit, and $\mathbf{Z}_T$ is the total impedance of the series circuit.

EXAMPLE 14.8   Using the voltage divider rule, find the voltage across each element of the circuit of Fig. 14.28.

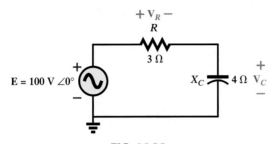

FIG. 14.28
Circuit for Example 14.8.

Solution:

$$\mathbf{V}_C = \frac{\mathbf{Z}_C\mathbf{E}}{\mathbf{Z}_C + \mathbf{Z}_R} = \frac{(4\ \Omega\ \angle-90°)(100\ V\ \angle0°)}{4\ \Omega\ \angle-90° + 3\ \Omega\ \angle0°} = \frac{400\ \angle-90°}{3 - j\,4}$$

$$= \frac{400\ \angle-90°}{5\ \angle-53.13°} = \mathbf{80\ V\ \angle-36.87°}$$

$$\mathbf{V}_R = \frac{\mathbf{Z}_R\mathbf{E}}{\mathbf{Z}_C + \mathbf{Z}_R} = \frac{(3\ \Omega\ \angle0°)(100\ V\ \angle0°)}{5\ \Omega\ \angle-53.13°} = \frac{300\ \angle0°}{5\ \angle-53.13°}$$

$$= \mathbf{60\ V\ \angle+53.13°}$$

EXAMPLE 14.9   Using the voltage divider rule, find the unknown voltages $\mathbf{V}_R$, $\mathbf{V}_L$, $\mathbf{V}_C$, and $\mathbf{V}_1$ for the circuit of Fig. 14.29.

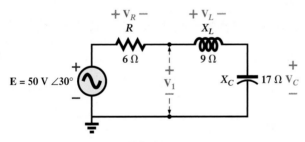

FIG. 14.29
Circuit for Example 14.9.

Solution:

$$\mathbf{V}_R = \frac{\mathbf{Z}_R\mathbf{E}}{\mathbf{Z}_R + \mathbf{Z}_L + \mathbf{Z}_C} = \frac{(6\ \Omega\ \angle0°)(50\ V\ \angle30°)}{6\ \Omega\ \angle0° + 9\ \Omega\ \angle90° + 17\ \Omega\ \angle-90°}$$

$$= \frac{300\ \angle30°}{6 + j\,9 - j\,17} = \frac{300\ \angle30°}{6 - j\,8}$$

$$= \frac{300\ \angle30°}{10\ \angle-53.13°} = \mathbf{30\ V\ \angle83.13°}$$

Calculator Solution: The calculation for $\mathbf{V}_R$ provides an excellent opportunity to demonstrate the power of today's calculators. Using the notation of the TI-86 calculator, the above calculation and the result appear in Fig. 14.30.

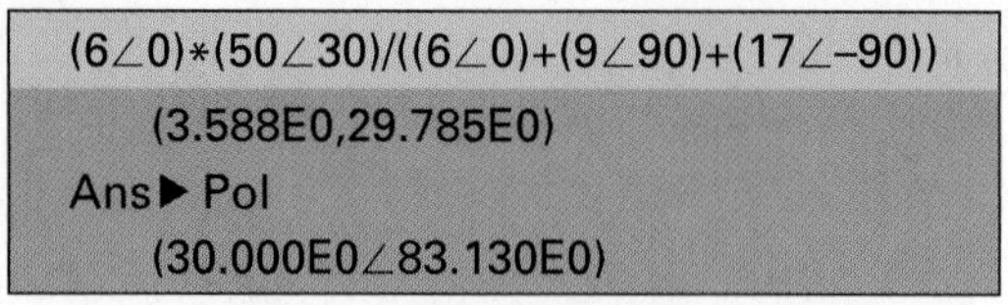

$$(6\angle0)*(50\angle30)/((6\angle0)+(9\angle90)+(17\angle-90))$$

$$(3.588E0, 29.785E0)$$

Ans▶ Pol

$$(30.000E0\angle83.130E0)$$

FIG. 14.30

Calculator solution for $\mathbf{V}_R$ of Example 14.9.

$$\mathbf{V}_L = \frac{\mathbf{Z}_L\mathbf{E}}{\mathbf{Z}_T} = \frac{(9\ \Omega\ \angle90°)(50\ \text{V}\ \angle30°)}{10\ \Omega\ \angle-53.13°} = \frac{450\ \text{V}\ \angle120°}{10\ \angle-53.13°} = \mathbf{45\ V}\ \angle\mathbf{173.13°}$$

$$\mathbf{V}_C = \frac{\mathbf{Z}_C\mathbf{E}}{\mathbf{Z}_T} = \frac{(17\ \Omega\ \angle-90°)(50\ \text{V}\ \angle30°)}{10\ \Omega\ \angle-53.13°} = \frac{850\ \text{V}\ \angle-60°}{10\ \angle-53°}$$

$$= \mathbf{85\ V}\ \angle\mathbf{-6.87°}$$

$$\mathbf{V}_1 = \frac{(\mathbf{Z}_L + \mathbf{Z}_C)\mathbf{E}}{\mathbf{Z}_T} = \frac{(9\ \Omega\ \angle90°\ +\ 17\ \Omega\ \angle-90°)(50\ \text{V}\ \angle30°)}{10\ \Omega\ \angle-53.13°}$$

$$= \frac{(8\ \angle-90°)(50\ \angle30°)}{10\ \angle-53.13°}$$

$$= \frac{400\ \angle-60°}{10\ \angle-53.13°} = \mathbf{40\ V}\ \angle\mathbf{-6.87°}$$

EXAMPLE 14.10   For the circuit of Fig. 14.31:

a. Calculate $\mathbf{I}_s$, $\mathbf{V}_R$, $\mathbf{V}_L$, and $\mathbf{V}_C$ in phasor form.
b. Calculate the total power factor.
c. Calculate the average power delivered to the circuit.
d. Draw the phasor diagram.
e. Obtain the phasor sum of $\mathbf{V}_R$, $\mathbf{V}_L$, and $\mathbf{V}_C$, and show that it equals the input voltage $\mathbf{E}$.
f. Find $\mathbf{V}_R$ and $\mathbf{V}_C$ using the voltage divider rule.

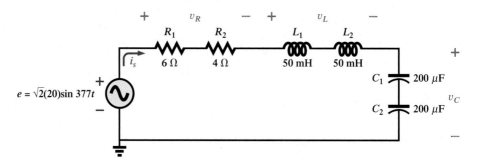

FIG. 14.31

Circuit for Example 14.10.

Solutions:

a. Combining common elements and finding the reactance of the inductor and capacitor, we obtain

$$R_T = 6\ \Omega\ +\ 4\ \Omega\ =\ 10\ \Omega$$

$$L_T = 50\ \text{mH}\ +\ 50\ \text{mH}\ =\ 100\ \text{mH}\ =\ 0.1\ \text{H}$$

$$C_T = \frac{200\ \mu\text{F}}{2} = 100\ \mu\text{F}$$

$$X_L = \omega L = (377\ \text{rad/s})(0.1\ \text{H}) = 37.70\ \Omega$$

$$X_C = \frac{1}{\omega C} = \frac{1}{(377\ \text{rad/s})(100\ \times\ 10^{-6}\ \text{F})} = \frac{10^6\ \Omega}{37,700} = 26.53\ \Omega$$

Redrawing the circuit using phasor notation results in Fig. 14.32.

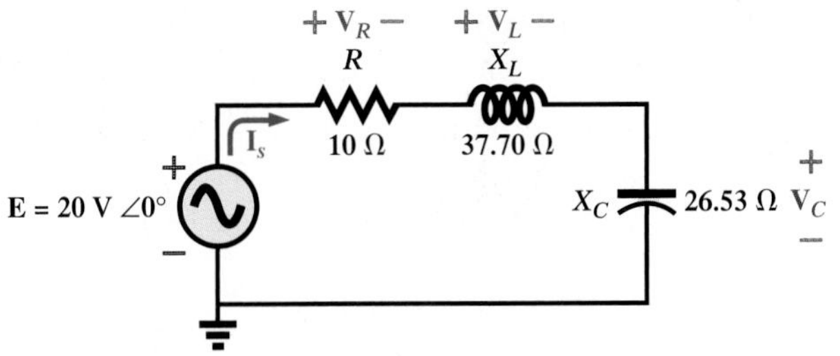

FIG. 14.32

Circuit of Fig. 14.31 using phasor notation.

For the circuit of Fig. 14.32,

$$\mathbf{Z}_T = R \angle 0° + X_L \angle 90° + X_C \angle -90°$$
$$= 10\ \Omega + j\,37.70\ \Omega - j\,26.53\ \Omega$$
$$= 10\ \Omega + j\,11.17\ \Omega = \mathbf{15\ \Omega\ \angle 48.16°}$$

The current $\mathbf{I}_s$ is

$$\mathbf{I}_s = \frac{\mathbf{E}}{\mathbf{Z}_T} = \frac{20\ \text{V}\ \angle 0°}{15\ \Omega\ \angle 48.16°} = \mathbf{1.33\ A\ \angle -48.16°}$$

The voltage across the resistor, inductor, and capacitor can be found using Ohm's law:

$$\mathbf{V}_R = \mathbf{I}_s\mathbf{Z}_R = (I\angle\theta)(R\angle 0°) = (1.33\ \text{A}\ \angle -48.16°)(10\ \Omega\ \angle 0°)$$
$$= \mathbf{13.30\ V\ \angle -48.16°}$$

$$\mathbf{V}_L = \mathbf{I}_s\mathbf{Z}_L = (I\angle\theta)(X_L\angle 90°) = (1.33\ \text{A}\ \angle -48.16°)(37.70\ \Omega\ \angle 90°)$$
$$= \mathbf{50.14\ V\ \angle 41.84°}$$

$$\mathbf{V}_C = \mathbf{I}_s\mathbf{Z}_C = (I\angle\theta)(X_C\angle -90°) = (1.33\ \text{A}\ \angle -48.16°)(26.53\ \Omega\ \angle -90°)$$
$$= \mathbf{35.28\ V\ \angle -138.16°}$$

b. The total power factor, determined by the angle between the applied voltage $\mathbf{E}$ and the resulting current $\mathbf{I}_s$, is 48.16°:

$$F_p = \cos\theta_T = \cos 48.16° = \mathbf{0.667\ lagging}$$

or

$$F_p = \cos\theta_T = \frac{R}{Z_T} = \frac{10\ \Omega}{15\ \Omega} = \mathbf{0.667\ lagging}$$

c. The total power in watts delivered to the circuit is

$$P_T = EI\cos\theta = (20\ \text{V})(1.33\ \text{A})(0.667) = \mathbf{17.74\ W}$$

d. The phasor diagram appears in Fig. 14.33.

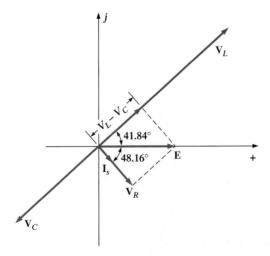

FIG. 14.33

Phasor diagram for the circuit of Fig. 14.31.

e. The phasor sum of $\mathbf{V}_R$, $\mathbf{V}_L$, and $\mathbf{V}_C$ is

$$\mathbf{E} = \mathbf{V}_R + \mathbf{V}_L + \mathbf{V}_C$$

$$= 13.30 \text{ V} \angle -48.16° + 50.14 \text{ V} \angle 41.84° + 35.28 \text{ V} \angle -138.16°$$

$$= 13.30 \text{ V} \angle -48.16° + 14.86 \text{ V} \angle 41.84°$$

Therefore,

$$E = \sqrt{(13.30 \text{ V})^2 + (14.86 \text{ V})^2} = \mathbf{20 \text{ V}}$$

and

$$\theta_E = \mathbf{0°} \qquad \text{(from the phasor diagram)}$$

and

$$\mathbf{E} = 20 \angle 0°$$

f.

$$\mathbf{V}_R = \frac{\mathbf{Z}_R \mathbf{E}}{\mathbf{Z}_T} = \frac{(10 \text{ } \Omega \angle 0°)(20 \text{ V} \angle 0°)}{15 \text{ } \Omega \angle 48.16°} = \frac{200 \text{ V} \angle 0°}{15 \angle 48.16°}$$

$$= \mathbf{13.3 \text{ V}} \angle \mathbf{-48.16°}$$

$$\mathbf{V}_C = \frac{\mathbf{Z}_C \mathbf{E}}{\mathbf{Z}_T} = \frac{(26.5 \text{ } \Omega \angle -90°)(20 \text{ V} \angle 0°)}{15 \text{ } \Omega \angle 48.16°} = \frac{530.6 \text{ V} \angle -90°}{15 \angle 48.16°}$$

$$= \mathbf{35.37 \text{ V}} \angle \mathbf{-138.16°}$$

14.5   FREQUENCY RESPONSE FOR SERIES ac CIRCUITS

Thus far, the analysis has been for a fixed frequency, resulting in a fixed value for the reactance of an inductor or a capacitor. We will now examine how the response of a series circuit will change as the frequency changes. We will assume ideal elements throughout the discussion so that the response of each element will be as shown in Fig. 14.34. Each response of Fig. 14.34 was discussed in detail in Chapter 13.

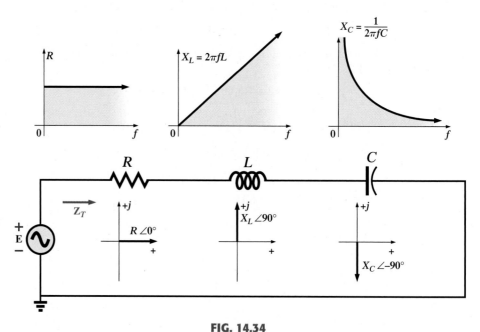

FIG. 14.34

Reviewing the frequency response of the basic elements.

When considering elements in series, remember that the total impedance is the sum of the individual elements and that the reactance of an inductor is in direct opposition to that of a capacitor. For Fig. 14.34, we are first aware that the resistance will remain fixed for the full range of frequencies: It will always be there, but, more importantly, its magnitude will not change. The inductor, however, will provide increasing levels of impedance as the frequency increases, while the capacitor will provide lower levels of impedance.

We are also aware from Chapter 13 that the inductor has a short-circuit equivalence at $f = 0$ Hz or very low frequencies, while the capacitor is nearly an open circuit for the same frequency range. For very high frequencies, the capacitor approaches the short-circuit equivalence, and the inductor approaches the open-circuit equivalence.

In general, therefore, if we encounter a series *R-L-C* circuit at very low frequencies, we can assume that the capacitor, with its very large impedance, will be the predominant factor. If the circuit is just an *R-L* series circuit, the impedance may be determined primarily by the resistive element since the reactance of the inductor is so small. As the frequency increases, the reactance of the coil will increase to the point where it will totally outshadow the impedance of the resistor. For an *R-L-C* combination, as the frequency increases, the reactance of the capacitor will begin to approach a short-circuit equivalence, and the total impedance will be determined primarily by the inductive element. At very high frequencies, for an *R-C* series circuit, the total impedance will eventually approach that of the resistor since the impedance of the capacitor is dropping off so quickly.

In total, therefore,

when encountering a series ac circuit of any combination of elements, always use the idealized response of each element to establish some feeling for how the circuit will respond as the frequency changes.

Once you have a logical, overall sense for what the response will be, you can spend the time working out the details.

Series *R-C* ac Circuit

As an example of establishing the frequency response of a circuit, consider the series *R-C* circuit of Fig. 14.35. As noted next to the source, the frequency range of interest is from 0 to 20 kHz. A great deal of detail will be provided for this particular combination so that obtaining the response of a series *R-L* or *R-L-C* combination will be quite straightforward.

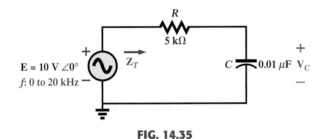

FIG. 14.35
Determining the frequency response of a series R-C circuit.

Since the resistance will remain fixed at 5 kΩ for the full frequency range, and the total impedance is the sum of the impedances, it is immediately obvious that the lowest possible impedance is 5 kΩ. The highest impedance, however, is dependent on the capacitive element since its impedance at very low frequencies is extremely high. At very low frequencies we can conclude, without a single calculation, that the impedance is determined primarily by the impedance of the capacitor. At the highest frequencies we can assume that the reactance of the capacitor has dropped to such low levels that the impedance of the combination will approach that of the resistance.

The frequency at which the reactance of the capacitor will drop to that of the resistor can be determined by simply setting the reactance of the capacitor equal to that of the resistor as follows:

$$X_C = \frac{1}{2\pi f_1 C} = R$$

Solving for the frequency yields

$$f_1 = \frac{1}{2\pi RC}$$

This significant point appears in the frequency plots of Fig. 14.36. Substituting values, we find that it occurs at

$$f_1 = \frac{1}{2\pi RC} = \frac{1}{2\pi(5 \text{ k}\Omega)(0.01 \text{ }\mu\text{F})} \cong 3.18 \text{ kHz}$$

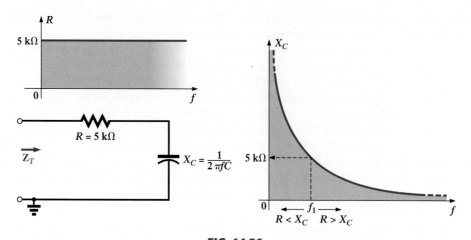

FIG. 14.36
The frequency response for the individual elements of Fig. 14.35.

We now know that for frequencies greater than f_1, $R > X_C$ and that for frequencies less than f_1, $X_C > R$, as shown in Fig. 14.36.

A plot of the total impedance of the series combination can be determined by finding the total impedance and writing it in the following form:

$$\mathbf{Z}_T = R - jX_C$$

and
$$\mathbf{Z}_T = Z_T \angle \theta_T = \sqrt{R^2 + X_C^2} \angle -\tan^{-1}\frac{X_C}{R}$$ **(14.10)**

Therefore,
$$Z_T = \sqrt{R^2 + X_C^2}$$

and
$$\theta_T = -\tan^{-1}\frac{X_C}{R}$$

For a plot of the magnitude of the impedance, all that you have to do is calculate Z_T at each frequency for the frequency range of interest. At each frequency the resistance will remain fixed at 5 kΩ while the reactance will be determined by $1/(2\pi f C)$. For example, at $f = 100$ Hz,

$$X_C = \frac{1}{2\pi f C} = \frac{1}{2\pi(100 \text{ Hz})(0.01 \ \mu\text{F})} = 159.16 \text{ k}\Omega$$

and
$$Z_T = \sqrt{R^2 + X_C^2} = \sqrt{(5 \text{ k}\Omega)^2 + (159.16 \text{ k}\Omega)^2}$$
$$= 159.24 \text{ k}\Omega \cong X_C \quad (\gg R = 5 \text{ k}\Omega)$$

confirming the comment above that the capacitor will be the predominant factor at low frequencies.

If a series of frequencies were substituted into the impedance equation, the plot of Fig. 14.37 would result. Note how the circuit switches from one of a capacitive nature to one with resistive characteristics as the frequency increases. Also note the small change in impedance level beyond 10 kHz as the curve flattens out. The effect of the capacitor is almost nonexistent after 10 kHz. For the frequency range of 10 kHz to 20 kHz, it would not be a bad approximation to replace the series R-L combination of Fig. 14.35 by simply a 5 kΩ resistor.

At $f = 100$ Hz, the phase angle is determined by

$$\theta_T = -\tan^{-1}\frac{X_C}{R} = -\tan^{-1}\frac{159.16 \text{ k}\Omega}{5 \text{ k}\Omega} = -\tan^{-1} 31.83 = -88.2°$$

A plot of θ_T would result in the pattern of Fig. 14.38. Note that the phase angle is close to $-90°$ at low frequencies because the circuit is primarily capacitive. However, at frequencies above 10 kHz, the phase is close to 0°, revealing that the total impedance is essentially resistive in nature.

If a plot of the voltage across the capacitor versus frequency were desired, the voltage divider rule could be applied as follows:

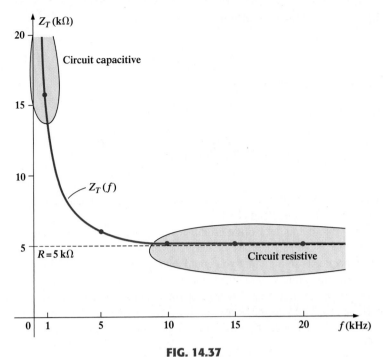

FIG. 14.37

The magnitude of the input impedance versus frequency for the circuit of Fig. 14.35.

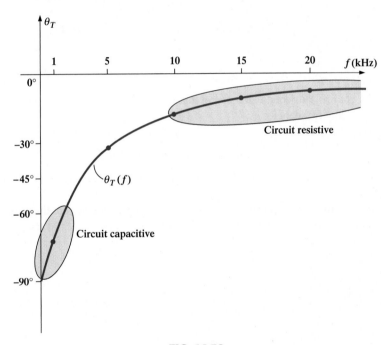

FIG. 14.38

The phase angle of the input impedance (θ_T) versus frequency for the circuit of Fig. 14.35.

$$\mathbf{V}_C = \frac{\mathbf{Z}_C \mathbf{E}}{\mathbf{Z}_R + \mathbf{Z}_C}$$

$$= \frac{(X_C \angle -90°)(E \angle 0°)}{R - jX_C} = \frac{X_C E \angle -90°}{R - jX_C}$$

$$= \frac{X_C E \angle -90°}{\sqrt{R^2 + X_C^2} \; \angle -\tan^{-1} X_C/R}$$

or $\qquad \mathbf{V}_C = V_C \angle \theta_C = \dfrac{X_C E}{\sqrt{R^2 + X_C^2}} \; \underline{/-90° + \tan^{-1}(X_C/R)}$

The magnitude of $\mathbf{V}_C$ is therefore determined by

$$V_C = \frac{X_C E}{\sqrt{R^2 + X_C^2}} \qquad (14.11)$$

and the phase angle θ_C by which $\mathbf{V}_C$ leads $\mathbf{E}$ is given by

$$\theta_C = -90° + \tan^{-1}\frac{X_C}{R} = -\tan^{-1}\frac{R}{X_C} \qquad (14.12)$$

To determine the frequency response, X_C must be calculated for each frequency of interest and inserted into Eqs. (14.11) and (14.12).

Before going into detail, let us first establish some idea of what to expect by reviewing the basic response of each element as the frequency increases. At low frequencies the reactance of the capacitor is so high that you could expect most or all of the applied voltage to appear across the capacitor. At very high frequencies the reactance of a capacitor is so small (it approaches the short-circuit equivalent) that you could assume that the voltage across the capacitor will approach 0 V as the series resistor picks up most of the applied voltage. Now we will verify our conclusions by considering frequencies of 1 kHz and 20 kHz.

At 1 kHz:

$$X_C = \frac{1}{2\pi f C} = \frac{1}{(2\pi)(1 \times 10^3 \text{ Hz})(0.01 \times 10^{-6} \text{ F})} \cong \mathbf{15.92 \text{ k}\Omega}$$

$$\sqrt{R^2 + X_C^2} = \sqrt{(5 \text{ k}\Omega)^2 + (15.92 \text{ k}\Omega)^2} \cong 16.69 \text{ k}\Omega$$

and

$$V_C = \frac{X_C E}{\sqrt{R^2 + X_C^2}} = \frac{(15.92 \text{ k}\Omega)(10)}{16.69 \text{ k}\Omega} = \mathbf{9.54 \text{ V}}$$

These results confirm our conclusion above that the capacitor will pick up most of the applied voltage at low frequencies.

At 20 kHz:

$$X_C \cong 796 \text{ } \Omega \quad \text{with } V_C = 1.57 \text{ V}$$

again confirming that X_C will be much less than R and that the voltage V_C will begin to approach 0 V.

The resulting plot for V_C appears in Fig. 14.39. It is interesting to note that it has almost the same shape as a curve for X_C against frequency. The curve for V_R versus frequency

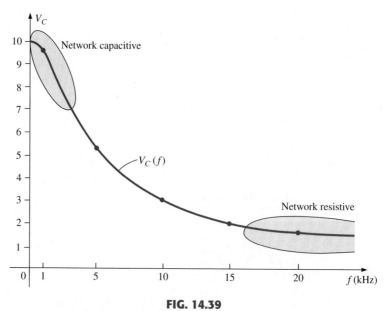

FIG. 14.39
The magnitude of the voltage V_C versus frequency for the circuit of Fig. 14.35.

would start at 0 V and rise to 10 V, with a shape similar to the transient rises of Chapters 10 and 11. In any event, at every frequency, the vector sum of $\mathbf{V}_C$ and $\mathbf{V}_R$ must equal the applied voltage of 10 V to satisfy Kirchhoff's voltage law.

EXAMPLE 14.11   For the series R-L circuit of Fig. 14.40:

a. Determine the frequency at which $X_L = R$.
b. Develop a mental image of the change in total impedance with frequency without resorting to a single calculation.
c. Find the total impedance at $f = 100$ Hz and 40 kHz, and compare your answer with the assumptions of part (b).
d. Plot the curve of V_L versus frequency.
e. Find the phase angle of the total impedance at $f = 40$ kHz. Can the circuit be considered inductive at this frequency? Why?

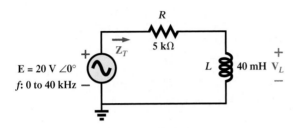

FIG. 14.40
Circuit for Example 14.11.

Solutions:

a. $X_L = 2\pi f_1 L = R$

and $\qquad\qquad f_1 = \dfrac{R}{2\pi L} = \dfrac{2\text{ k}\Omega}{2\pi(40\text{ mH})} = \textbf{7957.7 Hz}$

b. At low frequencies, $R > X_L$ and the impedance will be very close to that of the resistor, or 2 kΩ. As the frequency increases, X_L will increase to a point where it will be the predominant factor. The result is that the curve will start almost horizontal at 2 kΩ and then increase linearly to very high levels.

c. $Z_T = R + j X_L = Z_T \angle\theta_T = \sqrt{R^2 + X_L^2}\;\angle\tan^{-1}\dfrac{X_L}{R}$

At $f = 100$ Hz:

$$X_L = 2\pi fL = 2\pi(100\text{ Hz})(40\text{ mH}) = 25.13\ \Omega$$

and $\qquad Z_T = \sqrt{R^2 + X_L^2} = \sqrt{(2\text{ k}\Omega)^2 + (25.13\ \Omega)^2}$

$$= 2000.16\ \Omega \cong R$$

At $f = 40$ kHz:

$$X_L = 2\pi fL = 2\pi(40\text{ kHz})(40\text{ mH}) \cong 10.05\text{ k}\Omega$$

and $\quad Z_T = \sqrt{R^2 + X_L^2} = \sqrt{(2\text{ k}\Omega)^2 + (10.05\text{ k}\Omega)^2} = 10.25\text{ k}\Omega \cong X_L$

Both calculations support the conclusions of part (b).

d. Applying the voltage divider rule:

$$\mathbf{V}_L = \dfrac{\mathbf{Z}_L\mathbf{E}}{\mathbf{Z}_T}$$

From part (c), we know that at 100 Hz, $Z_T \cong R$ so that $V_R \cong 20$ V and $V_L \cong 0$ V. Part (c) revealed that at 40 kHz, $Z_T \cong X_L$ so that $V_L \cong 20$ V and $V_R \cong 0$ V. The result is two plot points for the curve of Fig. 14.41.

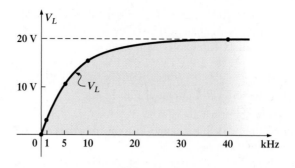

FIG. 14.41

Plotting V_L versus f for the series R-L circuit of Fig. 14.40.

At 1 kHz: $$X_L = 2\pi fL \cong 0.25 \text{ k}\Omega$$

and $$\mathbf{V}_L = \frac{(0.25 \text{ k}\Omega \angle 90°)(20 \text{ V} \angle 0°)}{2 \text{ k}\Omega + j\,0.25 \text{ k}\Omega} = \mathbf{2.48 \text{ V} \angle 82.87°}$$

At 5 kHz: $$X_L = 2\pi fL \cong 1.26 \text{ k}\Omega$$

and $$\mathbf{V}_L = \frac{(1.26 \text{ k}\Omega \angle 90°)(20 \text{ V} \angle 0°)}{2 \text{ k}\Omega + j\,1.26 \text{ k}\Omega} = \mathbf{10.68 \text{ V} \angle 57.79°}$$

At 10 kHz: $$X_L = 2\pi fL \cong 2.5 \text{ k}\Omega$$

and $$\mathbf{V}_L = \frac{(2.5 \text{ k}\Omega \angle 90°)(20 \text{ V} \angle 0°)}{2.5 \text{ k}\Omega + j\,2.5 \text{ k}\Omega} = \mathbf{15.63 \text{ V} \angle 38.66°}$$

The complete plot appears in Fig. 14.41.

e. $\theta_T = \tan^{-1} \dfrac{X_L}{R} = \tan^{-1} \dfrac{10.05 \text{ k}\Omega}{2 \text{ k}\Omega} = \mathbf{78.75°}$

The angle θ_T is closing in on the 90° of a purely inductive network. Therefore, yes, the network can be considered quite inductive at a frequency of 40 kHz.

14.6 SUMMARY: SERIES ac CIRCUITS

The following is a review of important conclusions that can be derived from the discussion and examples of the previous sections. The list is not all-inclusive, but it does emphasize some of the conclusions that should be carried forward in future analyses of ac systems.

For series ac circuits with reactive elements:

1. *The total impedance will be frequency dependent.*
2. *The impedance of any one element can be greater than the total impedance of the network.*
3. *The inductive and capacitive reactances are always in direct opposition on an impedance diagram.*
4. *Depending on the frequency applied, the same circuit can be either predominantly inductive or predominantly capacitive.*
5. *At lower frequencies the capacitive elements will usually have the most impact on the total impedance, while at high frequencies the inductive elements will usually have the most impact.*
6. *The magnitude of the voltage across any one element can be greater than the applied voltage.*
7. *The magnitude of the voltage across an element compared to the other elements of the circuit is directly related to the magnitude of its impedance; that is, the larger the impedance of an element, the larger the magnitude of the voltage across the element.*
8. *The voltages across a coil or capacitor are always in direct opposition on a phasor diagram.*
9. *The current is always in phase with the voltage across the resistive elements, lags the voltage across all the inductive elements by 90°, and leads the voltage across all the capacitive elements by 90°.*

10. *The larger the resistive element of a circuit compared to the net reactive impedance, the closer the power factor is to unity.*

14.7 PARALLEL CONFIGURATION

The analysis of parallel ac networks can take one of two directions. One approach is to continue to treat parallel networks in the same manner as described for dc circuits, whereas the other requires the introduction of terms defined in a way similar to that of conductance for dc circuits. Since the terminology most frequently applied to the input characteristics of a system, device, and so on, is in terms of impedance elements, our approach will be to continue to use impedance terms throughout the analysis. Before we leave the subject, however, the terms employed in the other approach will be introduced and their use demonstrated so that if you encounter them in the future, you will recognize them.

For the parallel combination of elements in Fig. 14.42, the total impedance is determined using Eq. (14.13) which has the same format as Eq. (6.1). The only difference is that the equation is now in terms of impedances rather than just resistive elements.

$$\frac{1}{\mathbf{Z}_T} = \frac{1}{\mathbf{Z}_1} + \frac{1}{\mathbf{Z}_2} + \frac{1}{\mathbf{Z}_3} + \cdots + \frac{1}{\mathbf{Z}_N} \qquad \textbf{(14.13)}$$

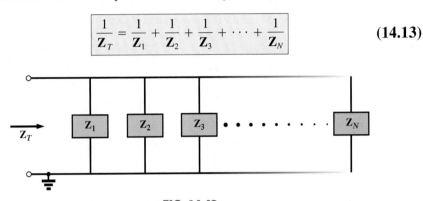

FIG. 14.42
Parallel impedances.

Equation (6.2) will then appear as

$$\mathbf{Z}_T = \cfrac{1}{\cfrac{1}{\mathbf{Z}_1} + \cfrac{1}{\mathbf{Z}_2} + \cfrac{1}{\mathbf{Z}_3} + \cdots + \cfrac{1}{\mathbf{Z}_N}} \qquad \textbf{(14.14)}$$

For N equal impedances, with $\mathbf{Z}_1$ in parallel, the equation becomes

$$\mathbf{Z}_T = \frac{\mathbf{Z}_1}{N} \qquad \textbf{(14.15)}$$

When using Eqs. (14.13) through (14.15), always be aware that each impedance element in the equation can be a resistor, an inductor, or a capacitor, requiring that the angle associated with each be included in the calculations.

For two impedances in parallel, the equation becomes

$$\mathbf{Z}_T = \frac{\mathbf{Z}_1 \mathbf{Z}_2}{\mathbf{Z}_1 + \mathbf{Z}_2} \qquad \textbf{(14.16)}$$

matching the format of Eq. (6.5).

For parallel dc circuits, the smallest resistor has the most impact on the resulting total resistance, and the total resistance is always less than the smallest parallel resistor. For ac networks, however, since there is a mix of impedances with different associated angles, such simplistic general conclusions cannot be made. However, it can be stated that

the smallest parallel resistor will still have the most impact on determining the resistive component of the total impedance, whereas the smallest reactive element will have the most impact on determining the reactive component of the total impedance.

In addition,

the applied frequency will affect the terminal characteristics of an ac network with reactive elements. In fact, a change in frequency may change the terminal characteristics from inductive to capacitive, or vice versa.

Since we are interested in the terminal impedance of a network, the impedance diagram can be used to define the resistive and reactive components. The impedance diagram will tell us at a glance whether the parallel ac network has a general resistive or reactive appearance and whether it is inductive or capacitive.

In fact,

the total impedance of a parallel ac network will define an equivalent series combination of elements with the same total impedance.

In addition,

if the parallel network is resistive and inductive, the resulting impedance and series combination of elements will also be resistive and inductive. The same applies to parallel resistive and capacitive networks.

A series of examples should clarify the above and validate some of the conclusions.

EXAMPLE 14.12 Find the total impedance of the parallel network of Fig. 14.43.

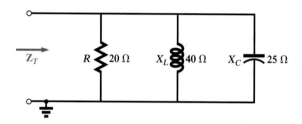

FIG. 14.43
Network for Examples 14.12 and 14.13.

Solution: Applying Eq. (14.14):

$$\mathbf{Z}_T = \cfrac{1}{\cfrac{1}{\mathbf{Z}_1} + \cfrac{1}{\mathbf{Z}_2} + \cfrac{1}{\mathbf{Z}_3}} = \cfrac{1}{\cfrac{1}{20\ \Omega\ \angle 0°} + \cfrac{1}{40\ \Omega\ \angle 90°} + \cfrac{1}{25\ \Omega\ \angle -90°}}$$

$$= \cfrac{1}{0.05\ \text{S}\ \angle 0° + 0.025\ \text{S}\ \angle -90° + 0.04\ \text{S}\ \angle 90°} = \cfrac{1}{0.05\ \text{S} + j\,0.015\ \text{S}}$$

$$= \cfrac{1}{0.158\ \text{S}\ \angle 71.57°} = \mathbf{6.33\ \Omega\ \angle -71.57°}$$

EXAMPLE 14.13 Find the total impedance of the parallel network of Fig. 14.43 if the resistor is changed to 1 Ω.

Solution: Applying Eq. (14.14):

$$\mathbf{Z}_T = \cfrac{1}{\cfrac{1}{\mathbf{Z}_1} + \cfrac{1}{\mathbf{Z}_2} + \cfrac{1}{\mathbf{Z}_3}} = \cfrac{1}{\cfrac{1}{1\ \Omega\ \angle 0°} + \cfrac{1}{40\ \Omega\ \angle 90°} + \cfrac{1}{25\ \Omega\ \angle -90°}}$$

$$= \cfrac{1}{1\ \text{S}\ \angle 0° + 0.025\ \text{S}\ \angle -90° + 0.04\ \text{S}\ \angle 90°} = \cfrac{1}{1\ \text{S} + j\,0.015\ \text{S}}$$

$$= \cfrac{1}{1.000\ \text{S}\ \angle 0.86°} = \mathbf{1\ \Omega\ \angle -0.86°}$$

Note that in Example 14.13, the resistive component of the total impedance is considerably less than the value for the solution of Example 14.12, and, in fact, the total impedance is essentially equal to that of the resistor—the smallest parallel impedance.

EXAMPLE 14.14 Find the total impedance of the parallel reactive elements of Fig. 14.44.

Solution: Applying Eq. (14.16):

$$\mathbf{Z}_T = \frac{\mathbf{Z}_1\mathbf{Z}_2}{\mathbf{Z}_1 + \mathbf{Z}_2} = \frac{(20\ \Omega\ \angle 90°)(12\ \Omega\ \angle -90°)}{+j\ 20\ \Omega\ -\ j\ 12\ \Omega}$$

$$= \frac{240\ \Omega\ \angle 0°}{+j\ 8} = \frac{240\ \Omega\ \angle 0°}{8\ \angle 90°} = \mathbf{30\ \Omega\ \angle -90°}$$

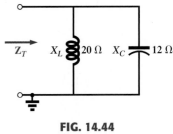

FIG. 14.44
Example 14.14.

In Example 14.14, since the capacitive element has the least impedance, the total impedance will be capacitive. For parallel reactive elements, the reactive element with the least reactance will always be the predominant factor.

EXAMPLE 14.15 For the network of Fig. 14.45:

a. Find the total impedance.
b. Draw the impedance diagram.
c. Draw the equivalent series combination of elements.

Solutions:

a. $\mathbf{Z}_T = \dfrac{\mathbf{Z}_1\mathbf{Z}_2}{\mathbf{Z}_1 + \mathbf{Z}_2} = \dfrac{(2\ \text{k}\Omega\ \angle 0°)(4\ \text{k}\Omega\ \angle 90°)}{2\ \text{k}\Omega\ +\ j\ 4\ \text{k}\Omega} = \dfrac{8\ \text{k}\Omega\ \angle 90°}{4.47\ \angle 63.4°}$

$= \mathbf{1.79\ k\Omega\ \angle 26.6°} = \mathbf{1.6\ k\Omega\ +\ j\ 0.8\ k\Omega}$

b. See Fig. 14.46.

c. See Fig. 14.47. This verifies that a parallel *R-L* network will have a series *R-L* equivalent.

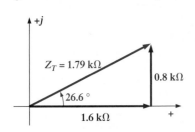

FIG. 14.46
*Impedance diagram for the parallel
network of Fig. 14.45.*

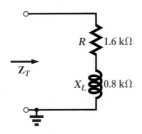

FIG. 14.47
*Series equivalent circuit for the
parallel network of Fig. 14.45.*

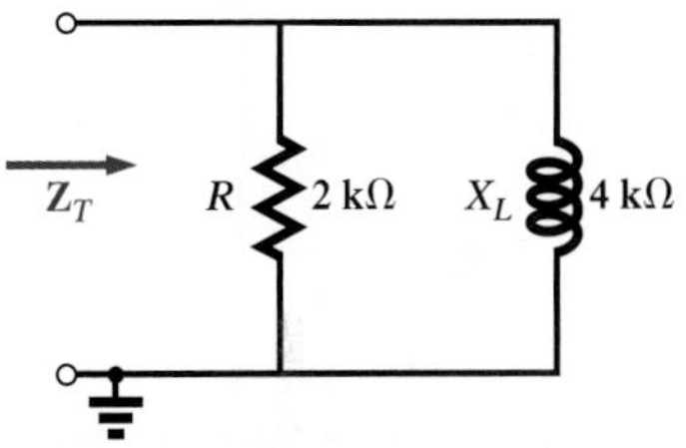

FIG. 14.45
Network for Example 14.15.

EXAMPLE 14.16 For the network of Fig. 14.48:

a. Find the total impedance.
b. Draw the impedance diagram.
c. Draw the equivalent series combination of elements.

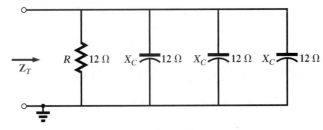

FIG. 14.48
Network for Example 14.16.

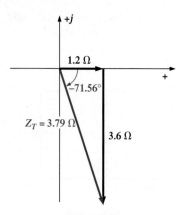

FIG. 14.49

Impedance diagram for the network of Fig. 14.48.

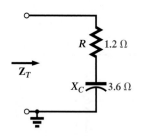

FIG. 14.50

Series equivalent circuit for the parallel network of Fig. 14.48.

Solutions:

a. $\mathbf{Z'}_T = \dfrac{12\ \Omega\ \angle{-90°}}{3} = 4\ \Omega\ \angle{-90°}$

and $\mathbf{Z}_T = \dfrac{\mathbf{Z}_R\mathbf{Z'}_T}{\mathbf{Z}_R + \mathbf{Z'}_T} = \dfrac{(12\ \Omega\ \angle 0°)(4\ \Omega\ \angle{-90°})}{12\ \Omega - j\,4\ \Omega} = \dfrac{48\ \Omega\ \angle{-90°}}{12.65\ \angle{-18.44°}}$

$= \mathbf{3.79\ \Omega\ \angle{-71.56°}} = \mathbf{1.2\ \Omega - j\,3.6\ \Omega}$

b. See Fig. 14.49.

c. See Fig. 14.50. In this case, the parallel *R-C* network has a series *R-C* equivalent network.

The analysis of parallel ac circuits will proceed in exactly the same manner as described for dc circuits. The voltage is the same across the parallel elements, and the current through each branch is determined using Ohm's law. The source current can be determined through an application of Ohm's law or Kirchhoff's current law.

For the parallel ac network of Fig. 14.51, the source current can be determined using the total impedance and Ohm's law as follows:

$$\mathbf{I}_s = \frac{\mathbf{E}}{\mathbf{Z}_T}$$

(14.17)

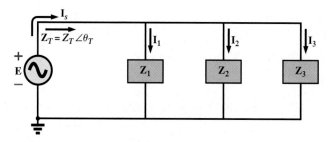

FIG. 14.51

Parallel ac network.

The current through each branch is determined by

$$\mathbf{I}_1 = \frac{\mathbf{E}}{\mathbf{Z}_1} \qquad \mathbf{I}_2 = \frac{\mathbf{E}}{\mathbf{Z}_2} \qquad \mathbf{I}_3 = \frac{\mathbf{E}}{\mathbf{Z}_3}$$

(14.18)

with the source current related to the branch currents through an application of Kirchhoff's current law:

$$\mathbf{I}_s = \mathbf{I}_1 + \mathbf{I}_2 + \mathbf{I}_3 + \cdots + \mathbf{I}_N$$

(14.19)

The power delivered to the network can be determined by summing the powers delivered to all the resistive elements using any one of the following equations:

$$P_R = V_R I_R = I_R^2 R = \frac{V_R^2}{R}$$

(14.20)

or using Eq. (14.6) and the parameters defined in Fig. 14.51:

$$P_T = E I_s \cos \theta_T$$

(14.21)

A few examples will clarify the use of the above equations.

EXAMPLE 14.17 For the parallel *R-L* network of Fig. 14.52:

a. Find the total impedance.
b. Draw the impedance diagram.
c. Find the source current.
d. Find the current through each branch.
e. Draw the phasor diagram.
f. Calculate the power dissipated by the resistor.
g. Find the power delivered by the source.
h. Find the sinusoidal expressions for the voltage and the currents, and plot the values on a time scale.

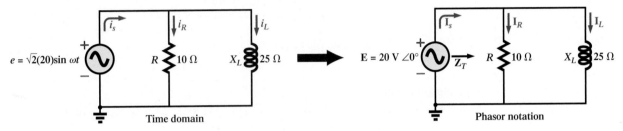

FIG. 14.52
Network for Example 14.17.

Solutions:

a. $\mathbf{Z}_T = \dfrac{\mathbf{Z}_R \mathbf{Z}_L}{\mathbf{Z}_R + \mathbf{Z}_L} = \dfrac{(10\ \Omega\ \angle 0°)(25\ \Omega\ \angle 90°)}{10\ \Omega\ +\ j\ 25\ \Omega} = \dfrac{250\ \Omega\ \angle 90°}{26.93\ \angle 68.2°}$

$= \mathbf{9.28\ \Omega\ \angle 21.8°} = \mathbf{8.62\ \Omega\ +\ j\ 3.45\ \Omega}$

b. See Fig. 14.53.

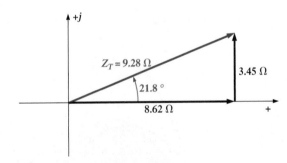

FIG. 14.53
Impedance diagram for the parallel network of Fig. 14.52.

c. Applying Ohm's law:

$$\mathbf{I}_s = \frac{\mathbf{E}}{\mathbf{Z}_T} = \frac{20\ V\ \angle 0°}{9.28\ \Omega\ \angle 21.8°} = \mathbf{2.16\ A\ \angle -21.8°}$$

d. Applying Ohm's law:

$$\mathbf{I}_R = \frac{\mathbf{E}}{\mathbf{Z}_R} = \frac{20\ V\ \angle 0°}{10\ \Omega\ \angle 0°} = \mathbf{2\ A\ \angle 0°}$$

$$\mathbf{I}_L = \frac{\mathbf{E}}{\mathbf{Z}_L} = \frac{20\ V\ \angle 0°}{25\ \Omega\ \angle 90°} = \mathbf{0.8\ A\ \angle -90°}$$

e. See Fig. 14.54. Note that the source current lags the applied voltage by 53.13° due to the inductive network. Note also that the current through the resistor is in phase with the voltage across it, whereas the current through the coil lags the applied voltage by 90°.

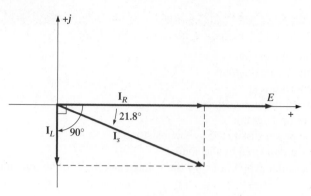

FIG. 14.54

Phasor diagram for the network of Fig. 14.52.

    f. The power dissipated by the resistor is

$$P_T = I_R^2 R = (2 \text{ A})^2 (10 \text{ }\Omega) = \textbf{40 W}$$

    g. Applying Eq. (14.6):

$$P_T = EI_s \cos \theta_T = (20 \text{ V})(2.16 \text{ A}) \cos 21.8° = 43.2(0.928) = \textbf{40 W}$$

As expected, the result of part (f) matches that of part (g).

    h. Converting to the time domain:

$$e = \sqrt{2}(20) \sin \omega t = \textbf{28.28 sin } \boldsymbol{\omega t}$$
$$i_s = \sqrt{2}(2.16) \sin(\omega t - 21.8°) = \textbf{3.05 sin}(\boldsymbol{\omega t} - \textbf{21.8°})$$
$$i_R = \sqrt{2}(2) \sin \omega t = \textbf{2.828 sin } \boldsymbol{\omega t}$$
$$i_L = \sqrt{2}(0.8) \sin(\omega t - 90°) = \textbf{1.13 sin}(\boldsymbol{\omega t} - \textbf{90°})$$

See Fig. 14.55. Note again that the applied voltage is in phase with the current i_R and leads i_L by 90°. Note also that at any instant of time on the horizontal axis, the sum of the currents i_R and i_L equals the source current i_s.

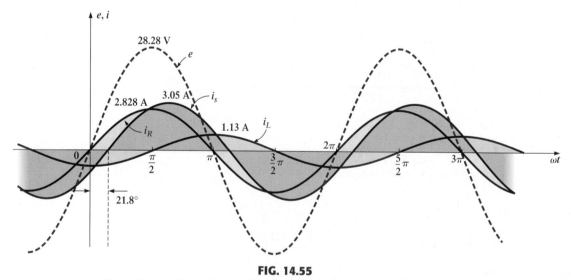

FIG. 14.55

Sinusoidal waveforms for the voltage and currents of the network of Fig. 14.52.

EXAMPLE 14.18 For the parallel *R-L-C* network of Fig. 14.56:

    a. By inspection, is the network inductive or capacitive? Why?
    b. Find the total impedance and draw the impedance diagram.
    c. What is the power factor of the network, and is it leading or lagging?
    d. Find the current through each branch of the network.
    e. Using Kirchhoff's current law, find the source current.

f. Using Ohm's law and the total impedance, calculate the source current, and compare your answer to the result of part (e).
g. Draw the phasor diagram.
h. Find the power delivered to the network.
i. Write the voltage and currents in the time domain, and plot them all on the same time axis.

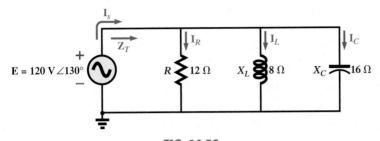

FIG. 14.56
Network for Example 14.18.

Solutions:

a. The network is inductive because $X_L < X_C$.

b. $\mathbf{Z}_T = \dfrac{1}{\dfrac{1}{\mathbf{Z}_R} + \dfrac{1}{\mathbf{Z}_L} + \dfrac{1}{\mathbf{Z}_C}} = \dfrac{1}{\dfrac{1}{12\,\Omega\,\angle 0°} + \dfrac{1}{8\,\Omega\,\angle 90°} + \dfrac{1}{16\,\Omega\,\angle -90°}}$

$= \dfrac{1}{0.083\text{ S }\angle 0° + 0.125\text{ S }\angle -90° + 0.063\text{ S }\angle 90°}$

$= \dfrac{1}{0.083\text{ S} - j\,0.062\text{ S}} = \dfrac{1}{0.104\text{ S }\angle -36.76°}$

$= \mathbf{9.615\,\Omega\,\angle 36.76°} = \mathbf{7.7\,\Omega + j\,5.75\,\Omega}$

See Fig. 14.57.

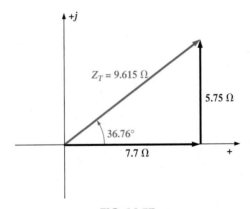

FIG. 14.57
Impedance diagram for the network of Fig. 14.56.

c. $F_p = \cos\theta_T = \cos 36.76° = \mathbf{0.8}$ which is a **lagging power factor** because the network is inductive.

d. Applying Ohm's law:

$$\mathbf{I}_R = \frac{\mathbf{E}}{\mathbf{Z}_R} = \frac{120\text{ V }\angle 30°}{12\,\Omega\,\angle 0°} = \mathbf{10\text{ A }\angle 30°}$$

$$\mathbf{I}_L = \frac{\mathbf{E}}{\mathbf{Z}_L} = \frac{120\text{ V }\angle 30°}{8\,\Omega\,\angle 90°} = \mathbf{15\text{ A }\angle -60°}$$

$$\mathbf{I}_C = \frac{\mathbf{E}}{\mathbf{Z}_C} = \frac{120\text{ V }\angle 30°}{16\,\Omega\,\angle -90°} = \mathbf{7.5\text{ A }\angle 120°}$$

e. Applying Kirchhoff's current law:

$$\mathbf{I}_s = \mathbf{I}_R + \mathbf{I}_L + \mathbf{I}_C$$

$$= 10\text{ A } \angle 30° + 15\text{ A } \angle -60° + 7.5\text{ A } \angle 120°$$

$$= (8.66\text{ A} + j\,5\text{ A}) + (7.5\text{ A} - j\,12.99\text{ A}) + (-3.75\text{ A} + j\,6.495\text{ A})$$

$$= 12.\,41\text{ A} - j\,1.495\text{ A} = \mathbf{12.5\text{ A } \angle -6.87°}$$

f. Applying Ohm's law:

$$\mathbf{I}_s = \frac{\mathbf{E}}{\mathbf{Z}_T} = \frac{120\text{ V } \angle 30°}{9.615\ \Omega\ \angle 36.76°} = \mathbf{12.48\text{ A } \angle -6.76°}$$

The difference between this result and that of part (e) is due solely to the accuracy carried through the calculations.

g. See Fig. 14.58. Note that the current through the resistor continues to be in phase with the applied voltage, whereas the applied voltage leads the current through the coil by 90°. The current i_C leads the voltage by 90° but is 180° out of phase with the current through the coil (as depicted on the phasor diagram).

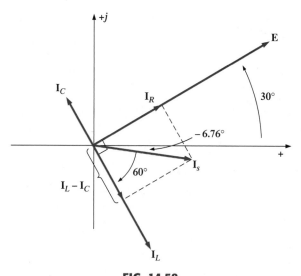

FIG. 14.58
Phasor diagram for the network of Fig. 14.56.

h. The power delivered to the network is

$$P_T = I_R^2 R = (10\text{ A})^2(12\ \Omega) = \mathbf{1200\text{ W}}$$

or

$$P_T = EI_s \cos \theta_T = (120\text{ V})(12.5\text{ A}) \cos 36.76° = \mathbf{1200\text{ W}}$$

i. In the time domain:

$$e = \sqrt{2}(120) \sin(\omega t + 30°) = \mathbf{169.7 \sin(\omega t + 30°)}$$
$$i_s = \sqrt{2}(12.5) \sin(\omega t - 6.87°) = \mathbf{17.7 \sin(\omega t - 6.87°)}$$
$$i_R = \sqrt{2}(10) \sin(\omega t + 30°) = \mathbf{14.1 \sin(\omega t + 30°)}$$
$$i_L = \sqrt{2}(15) \sin(\omega t - 60°) = \mathbf{21.2 \sin(\omega t - 60°)}$$
$$i_C = \sqrt{2}(7.5) \sin(\omega t + 120°) = \mathbf{10.6 \sin(\omega t + 120°)}$$

See Fig. 14.59. The applied voltage is clearly in phase with the current through the resistor, but it lags the current i_C by 90° and leads the current i_L by 90°. A careful examination of the plot at $t = 0$ s will reveal that the source current does, in fact, equal the sum of the other currents at every instant of time. The inductor current has the highest peak value since the inductor has the smallest impedance level, whereas the current i_C has the lowest peak value because it is the highest impedance level.

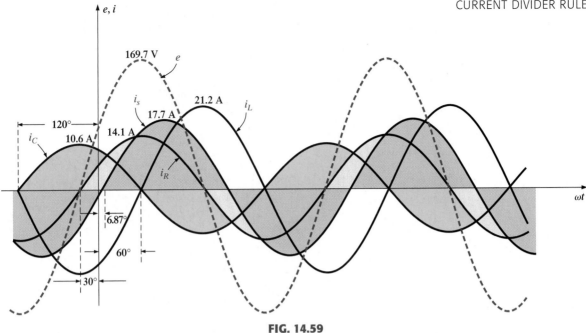

FIG. 14.59

Sinusoidal waveforms for the voltage and currents of Fig. 14.56.

The above examples should make it quite clear that the analysis of ac networks is the same as for dc networks, with the only major difference being the use of impedance parameters rather than just resistance levels. The format of every equation and the application of every important rule or law are exactly the same. This similarity will continue to be demonstrated in the chapters to follow.

14.8   CURRENT DIVIDER RULE

The basic format for the **current divider rule** in ac circuits is exactly the same as that for dc circuits; that is, for two parallel branches with impedances $\mathbf{Z}_1$ and $\mathbf{Z}_2$ as shown in Fig. 14.60,

$$\mathbf{I}_1 = \frac{\mathbf{Z}_2\mathbf{I}_T}{\mathbf{Z}_1 + \mathbf{Z}_2} \quad \text{or} \quad \mathbf{I}_2 = \frac{\mathbf{Z}_1\mathbf{I}_T}{\mathbf{Z}_1 + \mathbf{Z}_2} \qquad (14.22)$$

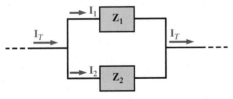

FIG. 14.60

Applying the current divider rule.

EXAMPLE 14.19   Using the current divider rule, find the current through each impedance of Fig. 14.61.

Solution:

$$\mathbf{I}_R = \frac{\mathbf{Z}_L\mathbf{I}_T}{\mathbf{Z}_R + \mathbf{Z}_L} = \frac{(4\ \Omega\ \angle 90°)(20\ \text{A}\ \angle 0°)}{3\ \Omega\ \angle 0° + 4\ \Omega\ \angle 90°} = \frac{80\ \text{A}\ \angle 90°}{5\ \angle 53.13°}$$

$$= \mathbf{16\ A\ \angle 36.87°}$$

$$\mathbf{I}_L = \frac{\mathbf{Z}_R\mathbf{I}_T}{\mathbf{Z}_R + \mathbf{Z}_L} = \frac{(3\ \Omega\ \angle 0°)(20\ \text{A}\ \angle 0°)}{5\ \Omega\ \angle 53.13°} = \frac{60\ \text{A}\ \angle 0°}{5\ \angle 53.13°}$$

$$= \mathbf{12\ A\ \angle -53.13°}$$

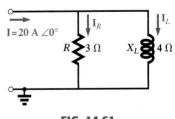

FIG. 14.61

Example 14.19.

EXAMPLE 14.20 Using the current divider rule, find the current through each parallel branch of Fig. 14.62.

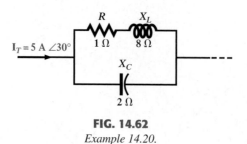

FIG. 14.62
Example 14.20.

Solution:

$$\mathbf{I}_{R\text{-}L} = \frac{\mathbf{Z}_C\mathbf{I}_T}{\mathbf{Z}_C + \mathbf{Z}_{R\text{-}L}} = \frac{(2\ \Omega\ \angle{-}90°)(5\ \text{A}\ \angle 30°)}{-j\,2\ \Omega + 1\ \Omega + j\,8\ \Omega} = \frac{10\ \text{A}\ \angle{-}60°}{1 + j\,6}$$

$$= \frac{10\ \text{A}\ \angle{-}60°}{6.083\ \angle 80.54°} \cong \mathbf{1.644\ A\ \angle{-}140.54°}$$

$$\mathbf{I}_C = \frac{\mathbf{Z}_{R\text{-}L}\mathbf{I}_T}{\mathbf{Z}_{R\text{-}L} + \mathbf{Z}_C} = \frac{(1\ \Omega + j\,8\ \Omega)(5\ \text{A}\ \angle 30°)}{6.08\ \Omega\ \angle 80.54°}$$

$$= \frac{(8.06\ \angle 82.87°)(5\ \text{A}\ \angle 30°)}{6.08\ \angle 80.54°} = \frac{40.30\ \text{A}\ \angle 112.87°}{6.083\ \angle 80.54°}$$

$$= \mathbf{6.625\ A\ \angle 32.33°}$$

14.9 FREQUENCY RESPONSE OF PARALLEL ELEMENTS

Recall that for elements in series, the total impedance is the direct sum of the impedances of each element, and the largest real or imaginary component has the most impact on the total impedance. For parallel elements, it is important to remember that **the smallest parallel resistor or the smallest parallel reactance will have the most impact on the real or imaginary component of the total impedance.**

In Fig. 14.63, the frequency response has been included for each element of a parallel *R-L-C* combination. At very low frequencies, the impedance of the coil will be less than that of the resistor or capacitor, resulting in an inductive network in which the reactance of the inductor will have the most impact on the total impedance. As the frequency in-

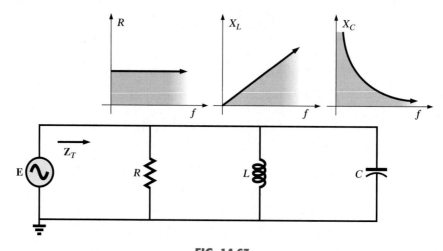

FIG. 14.63
Frequency response for parallel R-L-C elements.

creases, the impedance of the inductor will increase while the impedance of the capacitor will decrease. Depending on the components chosen, it is possible that the reactance of the capacitor will drop to a point where it will equal the impedance of the coil before either one reaches the resistance level.

Therefore, it is impossible to make too many broad statements about the effect of each element as the frequency increases. In general, however, for very low frequencies, we can assume that a parallel R-L-C network will be inductive as described above, and at very high frequencies it will be capacitive since X_C will drop to very low levels. In-between, a point will result at which X_L will equal X_C and where X_L or X_C will equal R. The frequencies at which these events occur, however, will depend on the elements chosen and the frequency range of interest. In general, however, keep in mind that the smaller the resistance or reactance, the greater its impact on the total impedance of a parallel system.

R-L Parallel ac Network

Let us now investigate the frequency response of the parallel ac network of Fig. 14.64 using specific values for the elements. As shown in the figure, the reactance of the inductor will be less than that of the resistor for all frequencies less than f_2. The result is that the network is more inductive than resistive for frequencies less than f_2. For frequencies greater than f_2, the inductor has the highest impedance, and the network will begin to appear more and more resistive.

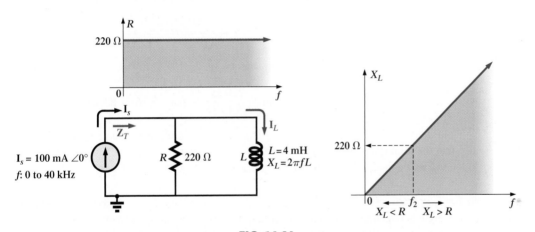

FIG. 14.64

The frequency response of the individual elements of a parallel R-L network.

Frequency f_2 can be found by simply setting the two impedances equal to each other as follows and solving for the frequency:

$$X_L = 2\pi f_2 L = R$$

and

$$f_2 = \frac{R}{2\pi L} \qquad (14.23)$$

For the network of Fig. 14.64,

$$f_2 = \frac{R}{2\pi L} = \frac{220\ \Omega}{2\pi(4\ \text{mH})} = \textbf{8.754 kHz}$$

An equation for the input impedance of the combination for any frequency can be found as follows:

$$\mathbf{Z}_T = \frac{\mathbf{Z}_R \mathbf{Z}_L}{\mathbf{Z}_R + \mathbf{Z}_L} = \frac{(R\angle 0°)(X_L \angle 90°)}{R + jX_L} = \frac{RX_L \angle 90°}{\sqrt{R^2 + X_L^2}\ \underline{/\tan^{-1} X_L/R}}$$

$$= \frac{RX_L}{\sqrt{R^2 + X_L^2}}\ \underline{/90° - \tan^{-1} X_L/R}$$

However,
$$\theta_T = 90° - \tan^{-1}\frac{X_L}{R} = \tan^{-1}\frac{R}{X_L}$$

so that
$$\mathbf{Z}_T = \frac{RX_L}{\sqrt{R^2 + X_L^2}}\ \underline{/\tan^{-1} R/X_L}$$

with the magnitude of $\mathbf{Z}_T$ given by

$$Z_T = \frac{RX_L}{\sqrt{R^2 + X_L^2}} \qquad\qquad (14.24)$$

and the angle of θ_T given by

$$\theta_T = \tan^{-1}\frac{R}{X_L} \qquad\qquad (14.25)$$

At f = 1 kHz   At a frequency of 1 kHz, which is much less than frequency f_2 calculated above, we would expect that the reactance of the inductor would be much less than that of the resistor so that the network will take on inductive characteristics.
Substituting:

$$X_L = 2\pi fL = 2\pi(1\text{ kHz})(4\text{ mH}) = 25.12\ \Omega$$

and
$$Z_T = \frac{RX_L}{\sqrt{R^2 + X_L^2}} = \frac{(220\ \Omega)(25.12\ \Omega)}{\sqrt{(220\ \Omega)^2 + (25.12\ \Omega)^2}}$$

$$= \mathbf{24.96\ \Omega} \cong X_L \ll R$$

with
$$\theta_T = \tan^{-1}\frac{R}{X_L} = \tan^{-1}\frac{220\ \Omega}{25.12\ \Omega} = \tan^{-1} 8.76 = \mathbf{83.49°}$$

Note in the above calculations that X_L is indeed much less than the resistance R. Note also that the total impedance is very close to the value of X_L and that the angle is very close to the 90° associated with a purely inductive load.

At f = f_2 = 8.754 kHz   At a frequency equal to f_2, we would expect X_L to equal R, and the phase angle between the two to be 45°. Since we expect the inductor to eventually look like an open circuit due to its very high impedance, we would expect the total impedance to approach that of the resistor since its impedance is frequency independent.
Substituting:

$$X_L = 2\pi fL = 2\pi(8.754\text{ kHz})(4\text{ mH}) = 220\ \Omega$$

and
$$Z_T = \frac{RX_L}{\sqrt{R^2 + X_L^2}} = \frac{(220\ \Omega)(220\ \Omega)}{\sqrt{(220\ \Omega)^2 + (220\ \Omega)^2}} = \mathbf{155.56\ \Omega}$$

with
$$\theta_T = \tan^{-1}\frac{R}{X_L} = \tan^{-1}\frac{220\ \Omega}{220\ \Omega} = \tan^{-1} 1 = \mathbf{45°}$$

At f = 20 kHz   At a frequency more than twice f_2, we find that the total impedance is beginning to approach that of the resistor R as the coil approaches its open-circuit equivalent. The total impedance is now

$$\mathbf{Z}_T = \mathbf{201.53\ \Omega\ \angle23.65°}$$

At f = 40 kHz   The chosen frequency is now almost five times f_2, and the total impedance is very close to that of the resistor with

$$\mathbf{Z}_T = \mathbf{214.91\ \Omega\ \angle12.35°}$$

A plot of Z_T versus frequency is provided in Fig. 14.65. Note the regions where the network has inductive and resistive characteristics. In the mid-region, one element is predominant but not with the strength of the defined regions.

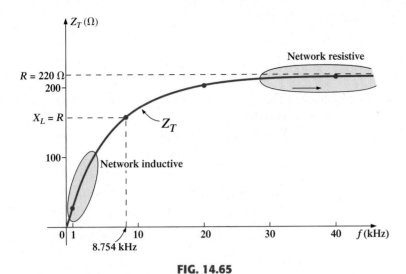

FIG. 14.65

The magnitude of the input impedance versus frequency for the network of Fig. 14.64.

A plot of the phase angle of the total impedance versus frequency will result in the plot of Fig. 14.66, where it is again clear that with a phase angle near 90°, the network is predominantly inductive at low frequencies. At high frequencies, it approaches 0° and has a predominantly resistive total impedance.

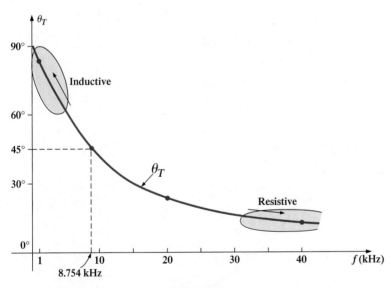

FIG. 14.66

The phase angle of the input impedance versus frequency for the network of Fig. 14.64.

We know from prior experience that current always seeks the path of least resistance. Since we know that the inductor has the lower level of impedance at low frequencies, we would expect the current through the coil to be higher at low frequencies and to drop as the frequency increases.

Applying the current divider rule:

$$\mathbf{I}_L = \frac{\mathbf{Z}_R \mathbf{I}_s}{\mathbf{Z}_R + \mathbf{Z}_L} = \frac{(R \angle 0°)(I_s \angle 0°)}{R + jX_L} = \frac{RI_s \angle 0°}{\sqrt{R^2 + X_L^2} \;\underline{/\tan^{-1} X_L/R}}$$

so that

$$\mathbf{I}_L = \frac{RI_s}{\sqrt{R^2 + X_L^2}} \;\underline{/-\tan^{-1} X_L/R}$$

The result is that the magnitude of the current through the coil is determined by

$$I_L = \frac{RI_s}{\sqrt{R^2 + X_L^2}} \qquad\qquad \textbf{(14.26)}$$

and the phase angle by

$$\boxed{\theta_L = -\tan^{-1}\frac{X_L}{R}} \tag{14.27}$$

At $f = 0$ Hz, the short-circuit equivalent is appropriate for the inductor, and the source current will equal that of the inductor. That is,

$$X_L = 2\pi f L = 2\pi(0 \text{ Hz})L = 0 \ \Omega$$

and

$$\mathbf{I}_L = \mathbf{I}_s = \textbf{100 mA} \angle \textbf{0°}$$

At $f = 1$ kHz,

$$X_L = 2\pi f L = 2\pi(1 \text{ kHz})(4 \text{ mH}) = 25.12 \ \Omega$$

and

$$I_L = \frac{RI_s}{\sqrt{R^2 + X_L^2}} = \frac{(220 \ \Omega)(100 \text{ mA})}{\sqrt{(220 \ \Omega)^2 + (25.12 \ \Omega)^2}}$$

$$= \textbf{99.35 mA} \cong I_s$$

with

$$\theta_L = -\tan^{-1}\frac{X_L}{R} = -\tan^{-1}\frac{25.12 \ \Omega}{220 \ \Omega} = -\tan^{-1} 0.114 = \textbf{−6.51°}$$

and

$$\mathbf{I}_L = \textbf{99.35 mA} \angle \textbf{−6.51°} \cong \mathbf{I}_s$$

resulting in a current very close to the source current.

The magnitude of the current and phase angle for various other frequencies is the following:

At $f = 8.754$ kHz: $I_L = \textbf{70.71 mA} \angle \textbf{−45°}$

At $f = 20$ kHz: $I_L = \textbf{40.11 mA} \angle \textbf{−66.35°}$

At $f = 40$ kHz: $I_L = \textbf{21.38 mA} \angle \textbf{−77.65°}$

resulting in the plot of Fig. 14.67 for the magnitude of the current I_L.

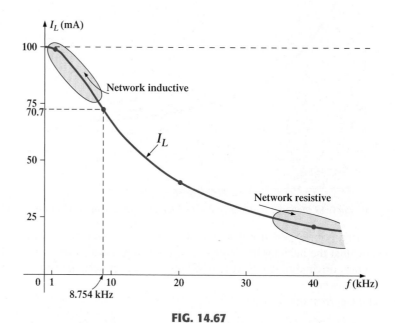

FIG. 14.67

The magnitude of the current $\mathbf{I}_L$ versus frequency for the parallel R-L network of Fig. 14.64.

Clearly, the current I_L is a maximum at low frequencies and drops as the reactance of the coil increases. When $X_L = R$, the current I_L will equal $(0.7071)(100 \text{ mA}) = 70.71$ mA as defined by Eq. (14.26).

The analysis of a parallel R-C circuit will proceed in a very similar manner, except for a reversal of the impact of the low and high frequencies. That is, the capacitor will be the predominant factor at high frequencies due its very low impedance, with the magnitude of the total impedance approaching that of the capacitor rather than that of the resistor. The current I_C will be essentially zero amperes at very low frequencies, increasing exponentially to the source current as the frequency increases.

For a parallel *R-L-C* network, the total impedance will start at very low levels due to the inductive element and then peak at some value, depending on the values of the inductor, capacitor, and resistor. The total impedance will then start dropping off toward zero ohms again as the frequency increases due to the impact of the parallel capacitive element. The current I_L will start off at high levels and then drop off to very low levels as the reactance of the coil increases with frequency. For the capacitor, the current will start off at very low levels and then increase rapidly as the reactance of the capacitor drops exponentially.

Plotting the response of *R-L* and *R-L-C* networks will appear as problems at the end of the chapter.

14.10 SUMMARY: PARALLEL ac NETWORKS

The following is a review of important conclusions that can be derived from the discussion and examples of the previous sections. The list is not all-inclusive, but it does emphasize some of the conclusions that should be carried forward in the future analyses of ac systems.

For parallel ac networks with reactive elements:

 1. *The total impedance is frequency dependent.*
 2. *The impedance of any one element can be less than the total impedance. (Recall that for dc circuits the total resistance must always be less than the smallest parallel resistor.)*
 3. *The smallest parallel resistor or smallest parallel reactance will have the most impact on the real or imaginary component of the total impedance.*
 4. *Depending on the frequency applied, the same network can be either predominantly inductive or predominantly capacitive.*
 5. *At lower frequencies the inductive elements will usually have the most impact on the total impedance, while at high frequencies the capacitive elements will usually have the most impact.*
 6. *The magnitude of the current through any one branch can be greater than the source current.*
 7. *The magnitude of the current through an element, compared to the other elements of the network, is directly related to the magnitude of its impedance; that is, the smaller the impedance of an element, the larger the magnitude of the current through the element.*
 8. *The current through a coil is always in direct opposition with the current through a capacitor on a phasor diagram.*
 9. *The applied voltage is always in phase with the current through the resistive elements, leads the voltage across all the inductive elements by 90°, and lags the current through all capacitive elements by 90°.*
10. *The smaller the resistive element of a network compared to the net reactive component, the closer the power factor is to unity.*

14.11 PHASE MEASUREMENTS

Measuring the phase angle between quantities is one of the most important functions that an oscilloscope can perform. It is an operation that must be performed carefully, however, or the incorrect result will be obtained or the equipment damaged. Whenever you are using the dual-trace capability of an oscilloscope, the most important thing to remember is that

both channels of a dual-trace oscilloscope must be connected to the same ground.

Measuring Z_T and θ_T

For ac parallel networks restricted to **resistive loads,** the total impedance can be found in the same manner as described for dc circuits: Simply remove the source and place an ohmmeter across the network terminals. However,

for parallel ac networks with reactive elements, the total impedance cannot be measured with an ohmmeter.

An experimental procedure must be defined that will permit determining the magnitude and the angle of the terminal impedance.

The phase angle between the applied voltage and the resulting source current is one of the most important because (a) it is also the phase angle associated with the total impedance; (b) it provides an instant indication of whether a network is resistive or reactive; (c) it reveals whether a network is inductive or capacitive; and (d) it can be used to find the power delivered to the network.

In Fig. 14.68, a resistor has been added to the configuration between the source and the network to permit measuring the current and finding the phase angle between the applied voltage and the source current.

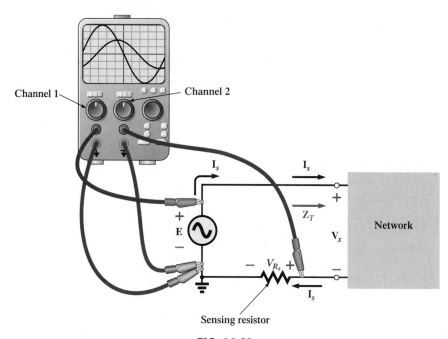

FIG. 14.68

Using an oscilloscope to measure Z_T and θ_T.

At the frequency of interest, the applied voltage will establish a voltage across the sensing resistor which can be displayed by one channel of the dual-trace oscilloscope. In Fig. 14.68, channel 1 is displaying the applied voltage, and channel 2 the voltage across the sensing resistor. Sensitivities for each channel are chosen to establish the waveforms appearing on the screen in Fig. 14.69. As emphasized above, note that both channels have the same ground connection. In fact, the need for a common ground connection is the only reason that the sensing resistor was not connected to the positive side of the supply. Since oscilloscopes will display only voltages versus time, the peak value of the source current must be found using Ohm's law. **Since the voltage across a resistor and the current through the resistor are in phase, the phase angle between the two voltages will be the same as that between the applied voltage and the resulting source current.**

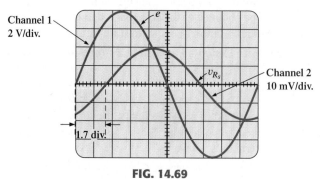

FIG. 14.69

e and v_{R_s} for the configuration of Fig. 14.68.

Using the sensitivities, the peak value of the applied voltage is

$$E_m = (4 \text{ div.})\,(2 \text{ V/div.}) = 8 \text{ V}$$

while the peak value of the voltage across the sensing resistor is

$$V_{R_s(\text{peak})} = (2 \text{ div.})(10 \text{ mV/div.}) = 20 \text{ mV}$$

Using Ohm's law, the peak value of the current is

$$I_{s(\text{peak})} = \frac{V_{R_s(\text{ peak})}}{R_s} = \frac{20 \text{ mV}}{10 \text{ }\Omega} = 2 \text{ mA}$$

The sensing resistor is chosen small enough so that the voltage across the sensing resistor is small enough to permit the approximation $\mathbf{V}_x = \mathbf{E} - \mathbf{V}_{R_s} \cong \mathbf{E}$. The magnitude of the input impedance is then

$$Z_T = \frac{V_x}{I_s} \cong \frac{E}{I_s} = \frac{8 \text{ V}}{2 \text{ mA}} = \mathbf{4 \text{ k}\Omega}$$

with the phase angle determined using Eq. (12.24):

$$\frac{\theta}{1.7 \text{ div.}} = \frac{180°}{5 \text{ div.}}$$

and

$$\theta = \frac{1.7}{5}(180°) = \mathbf{61.2°}$$

Therefore, the total impedance is

$$\mathbf{Z}_T = \mathbf{4 \text{ k}\Omega \ \angle 61.2° = 1.93 \text{ k}\Omega + j\, 3.51 \text{ k}\Omega = R + j\, X_L}$$

which is equivalent to the series combination of a 1.93 kΩ resistor and an inductor with a reactance of 3.51 kΩ (at the frequency of interest).

Measuring the Phase Angle between Various Voltages

In Fig. 14.70, an oscilloscope is being used to find the phase relationship between the applied voltage and the voltage across the inductor. Note again that each channel shares the same ground connection. The resulting pattern appears in Fig. 14.71 with the chosen sensitivities. This time both channels have the same sensitivity, resulting in the following peak values for the voltages:

$$E_m = (3 \text{ div.})\,(2 \text{ V/div.}) = \mathbf{6 \text{ V}}$$

$$V_{L(\text{peak})} = (1.6 \text{ div.})\,(2 \text{ V/div.}) = \mathbf{3.2 \text{ V}}$$

The phase angle is determined as follows:

$$\frac{\theta}{1 \text{ div.}} = \frac{360°}{8 \text{ div.}}$$

$$\theta = \frac{1}{8}(360°) = \mathbf{45°}$$

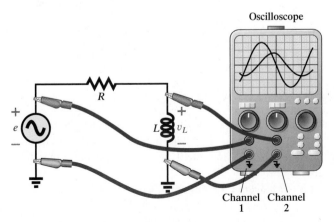

FIG. 14.70

Determining the phase relationship between e and v_L.

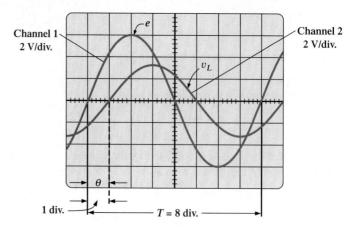

FIG. 14.71

Determining the phase angle between e and v_L for the configuration of Fig. 14.70.

If the phase relationship between e and v_R is desired, the oscilloscope **cannot be connected** as shown in Fig. 14.72. **The grounds of each channel are internally connected in the oscilloscope,** forcing point b to have the same potential as point a. The result would be a direct connection between points a and b that would short-out the inductive element. If the inductive element is the predominant factor in controlling the level of the current, the current in the circuit could rise to dangerous levels and damage the oscilloscope or supply. The easiest way to find the phase relationship between e and v_R would be to simply interchange the positions of the resistor and the inductor and proceed as before.

For the parallel network of Fig. 14.73, the phase relationship between two of the branch currents, i_R and i_L, can be determined using a sensing resistor, as shown in the figure. The value of the sensing resistor is chosen small enough in comparison to the value of the series inductive reactance to ensure that it will not affect the general response of the network. Channel 1 will display the voltage v_R, and channel 2 the voltage v_{R_s}. Since v_R is in phase with i_R, and v_{R_s} is in phase with i_L, the phase relationship between v_R and v_{R_s} will be the same as between i_R and i_L. The peak value of each current can be found through a simple application of Ohm's law.

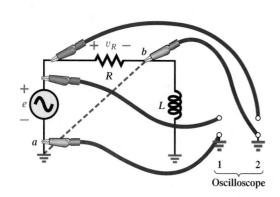

FIG. 14.72

An improper phase-measurement connection.

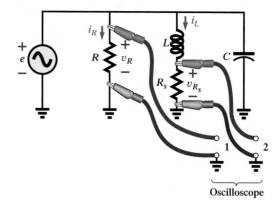

FIG. 14.73

Determining the phase relationship between i_R and i_L.

14.12 ADMITTANCE AND SUSCEPTANCE

The analysis of parallel ac networks can also take place using parameters defined in a similar manner to conductance; that is, $G = 1/R$. In the basic equation for the total impedance of parallel ac elements, the division into 1 occurs so frequently that special names were given to the ratios.

For instance, for the inductor, the ratio of $1/X_L$ is defined as

$$\boxed{B_L = \frac{1}{X_L}} \quad \text{(siemens, S)} \qquad \textbf{(14.28)}$$

where the term B_L is called the **susceptance** of the inductor and is measured in **siemens (S).** The higher the susceptance, the greater the **susceptiblility** to the flow of charge, or current, through the element.

For the capacitor, the ratio $1/X_C$ is defined as

$$B_C = \frac{1}{X_C} \qquad \text{(siemens, S)} \qquad \textbf{(14.29)}$$

with B_C called the **susceptance** of the capacitor which is also measured in **siemens.**

For the parallel ac network of Fig. 14.74, the ratio $1/\mathbf{Z}$ is called **admittance** because it provides an indication of how easily a parallel branch will **admit** the flow of charge or current. It is assigned the symbol **Y** and is also measured in siemens.

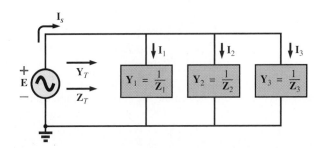

FIG. 14.74

Parallel ac network with admittance parameters defined.

If we take the equation for the total impedance:

$$\frac{1}{\mathbf{Z}_T} = \frac{1}{\mathbf{Z}_1} + \frac{1}{\mathbf{Z}_2} + \frac{1}{\mathbf{Z}_3}$$

and substitute the admittance identification, we obtain the following:

$$\mathbf{Y}_T = \mathbf{Y}_1 + \mathbf{Y}_2 + \mathbf{Y}_3 \qquad \text{(siemens, S)} \qquad \textbf{(14.30)}$$

which takes on a simpler form than that for the impedance equation. However, remember that each term is 1 over the impedance of that element.

For each element, the admittance parameter can now be defined by

$$\mathbf{Y}_L = \frac{1}{\mathbf{Z}_L} = \frac{1}{X_L \angle 90°} = \frac{1}{X_L} \angle -90°$$

and

$$\mathbf{Y}_L = B_L \angle -90° \qquad \textbf{(14.31)}$$

$$\mathbf{Y}_C = \frac{1}{\mathbf{Z}_C} = \frac{1}{X_C \angle -90°} = \frac{1}{X_C} \angle 90°$$

and

$$\mathbf{Y}_C = B_C \angle 90° \qquad \textbf{(14.32)}$$

$$\mathbf{Y}_R = \frac{1}{\mathbf{Z}_R} = \frac{1}{R \angle 0°} = \frac{1}{R} \angle 0°$$

and

$$\mathbf{Y}_R = G \angle 0° \qquad \textbf{(14.33)}$$

Since each admittance parameter has a magnitude and an angle, the parameters can be plotted on a real–imaginary plane as shown in Fig. 14.75. The result is an **admittance diagram** that is very similar to the impedance diagram introduced earlier.

Using the above defined parameters, a parallel ac network can now be analyzed by first finding the susceptance of each parallel reactive element and the conductance of each parallel resistive element. Then sum up the admittance associated with each using Eq. (14.30).

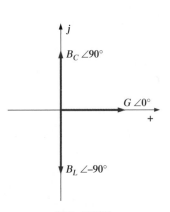

FIG. 14.75

Admittance diagram.

The result will provide the input admittance of the network which will provide an indication of how easily the network will "admit" the flow of charge or current.

The larger the admittance, the greater the resulting source current for the same applied voltage.

The total admittance can be found in the same manner as described for the impedance diagram. In this case, however, **the angle associated with the total admittance is the angle by which the source current leads the applied source voltage.** Recall that the angle associated with the total impedance of a network is the angle by which the applied voltage leads the resulting source current. Be aware from the admittance diagram that the susceptance of an inductor is in direct opposition to the susceptance of a capacitor, and their difference is used in the determination of the total admittance.

Some of the basic equations applied to parallel ac networks can then be modified as follows:

$$I_s = \frac{E}{Z_T} = EY_T \qquad \textbf{(14.34)}$$

The following equations are used to determine the current through each branch:

$$I_1 = \frac{E}{Z_1} = EY_1 \qquad I_2 = \frac{E}{Z_2} = EY_2 \qquad I_3 = \frac{E}{Z_3} = EY_3 \qquad \textbf{(14.35)}$$

However, the application of Kirchhoff's current law results in the same equation:

$$I_s = I_1 + I_2 + I_3 \qquad \textbf{(14.36)}$$

The power equation remains

$$P_T = EI_s \cos \theta_T \qquad \textbf{(14.37)}$$

since it is not sensitive to whether θ_T is a leading or a lagging phase angle, and the magnitude of the angle associated with Z_T is the same as that associated with Y_T.

The power factor is determined by

$$F_p = \cos \theta_T = \frac{G}{Y_T} \qquad \textbf{(14.38)}$$

The approach that you use to analyze parallel ac networks is simply a matter of choice—use whichever approach you are more comfortable with. The results for the currents, voltages, and power distribution will be the same. In one case, we use the impedance parameters for the series and the parallel configurations. In the other, we use the admittance parameters because the equations initially have a cleaner appearance. In any case, you should be familiar with each and the terms introduced in this section.

The next example will clarify the use of susceptance levels and admittance parameters to solve for the unknowns of a parallel ac network.

EXAMPLE 14.21   For the parallel *R-C* network of Fig. 14.76:

a. Find the total admittance and impedance.
b. Sketch the admittance and impedance diagrams.
c. Find the source current and the current through each branch.
d. Draw the phasor diagram.
e. Write the sinusoidal expression for the applied voltage and resulting currents.
f. Compare this approach to that of Section 14.7.

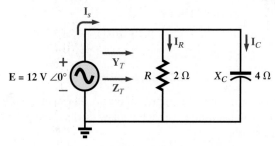

FIG. 14.76

Network for Example 14.21.

Solutions:

a. The total admittance is

$$\mathbf{Y}_R = G \angle 0° = \frac{1}{R} \angle 0° = \frac{1}{2\ \Omega} \angle 0° = 0.5\ \text{S} \angle 0°$$

$$\mathbf{Y}_C = B_C \angle 90° = \frac{1}{X_C} \angle 90° = \frac{1}{4\ \Omega} \angle 90° = 0.25\ \text{S} \angle 90°$$

and $\qquad \mathbf{Y}_T = \mathbf{Y}_R + \mathbf{Y}_C = \mathbf{0.5\ S} + j\mathbf{0.25\ S} = \mathbf{0.559\ S} \angle\mathbf{26.57°}$

The total impedance is

$$\mathbf{Z}_T = \frac{1}{\mathbf{Y}_T} = \frac{1}{0.559\ \text{S} \angle 26.57°} = \mathbf{1.789\ \Omega} \angle\mathbf{-26.57°} = \mathbf{1.6\ \Omega} - j\mathbf{0.8\ \Omega}$$

b. See Fig. 14.77.

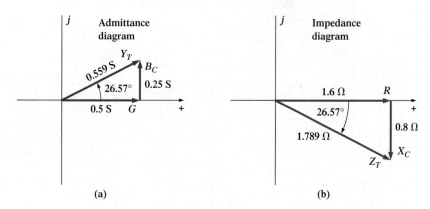

(a) (b)

FIG. 14.77

(a) Admittance diagram and (b) impedance diagram for Fig. 14.76.

c. Applying Ohm's law:

$$\mathbf{I}_s = \mathbf{EY}_T = (12\ \text{V} \angle 0°)(0.559\ \text{S} \angle 26.57°) = \mathbf{6.71\ A} \angle\mathbf{26.57°}$$
$$\mathbf{I}_R = \mathbf{EY}_R = (12\ \text{V} \angle 0°)(0.5\ \text{S} \angle 0°) = \mathbf{6\ A} \angle\mathbf{0°}$$
$$\mathbf{I}_C = \mathbf{EY}_C = (12\ \text{V} \angle 0°)(0.25\ \text{S} \angle 90°) = \mathbf{3\ A} \angle\mathbf{90°}$$

d. See Fig. 14.78.

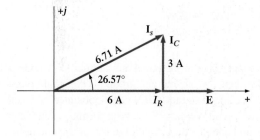

FIG. 14.78

Phasor diagram for Fig. 14.76.

e. $e = \sqrt{2}(12) \sin \omega t = \mathbf{16.97 \sin \omega t}$

   $i_s = \sqrt{2}(6.71) \sin(\omega t + 26.57°) = \mathbf{9.49 \sin(\omega t + 26.57°)}$

   $i_R = \sqrt{2}(6) \sin \omega t = \mathbf{8.48 \sin \omega t}$

   $i_C = \sqrt{2}(3) \sin(\omega t + 90°) = \mathbf{4.24 \sin(\omega t + 90°)}$

f. The only significant difference in approach occurs at the beginning, when $\mathbf{Y}_T$ or $\mathbf{Z}_T$ is determined. Once you begin to apply Ohm's law or draw the phasor diagram, the two methods are very similar.

14.13 APPLICATION

Home Wiring

An expanded view of house wiring is provided in Fig. 14.79 to permit a discussion of the entire system. The house panel has been included with the "feed" and the important grounding mechanism. In addition, a number of typical circuits found in the home have been included to provide a sense for the manner in which the total power is distributed.

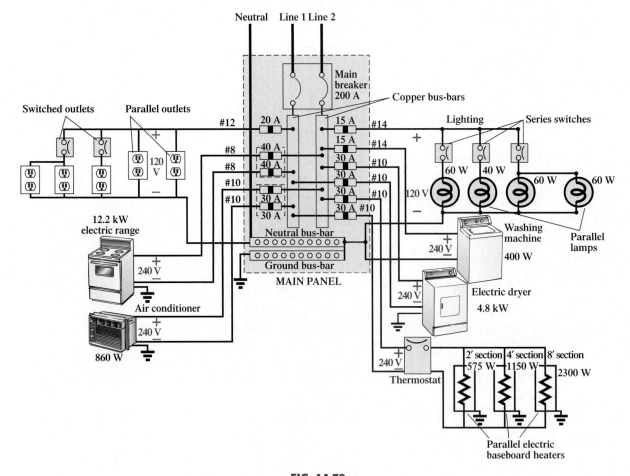

FIG. 14.79

Home wiring diagram.

First note how the copper bars in the panel are laid out to provide both 120 V and 240 V (sometimes 208 V). Between any one bar and ground is the single-phase 120 V supply. However, the bars have been arranged so that 240 V can be obtained between two vertical adjacent bars using a double-gang circuit breaker. When time permits, examine your own panel (but do not remove the cover), and note the dual circuit breaker arrangement for the 240 V supply.

For appliances such as fixtures and heaters that have a metal casing, the ground wire is connected to the metal casing to provide a direct path to ground path for a "shorting" or errant current. For outlets and such that do not have a conductive casing, the ground lead is connected to a point on the outlet that distributes to all important points of the outlet.

Note the series arrangement between the thermostat and the heater, but the parallel arrangement between heaters on the same circuit. In addition, note the series connection of switches to lights in the upper-right corner, but the parallel connection of lights and outlets. Due to high current demand, the air conditioner, heaters, and electric stove have 30 A breakers. Keep in mind that the total current does not equal the product of the two (or 60 A) since each breaker is in a line and the same current will flow through each breaker.

In general, you now have a surface understanding of the general wiring in your home. You may not be a qualified, licensed electrician, but at least you should now be able to converse with some intelligence about the system.

14.14 COMPUTER ANALYSIS

PSpice

Series *R-L-C* Circuit   The *R-L-C* network of Fig. 14.15 will now be analyzed using OrCAD Capture. Since the inductive and capacitive reactances cannot be entered onto the screen, the associated inductive and capacitive levels were first determined as follows:

$$X_L = 2\pi f L \Rightarrow L = \frac{X_L}{2\pi f} = \frac{7\ \Omega}{2\pi(1\ \text{kHz})} = 1.114\ \text{mH}$$

$$X_C = \frac{1}{2\pi f C} \Rightarrow C = \frac{1}{2\pi f X_C} = \frac{1}{2\pi(1\ \text{kHz})3\ \Omega} = 53.05\ \mu\text{F}$$

The values were then entered into the schematic as shown in Fig. 14.80. For the ac source, the sequence is **Place part** icon-**SOURCE-VSIN-OK** with **VOFF** set at 0 V, **VAMPL** set at 70.7 V (the peak value of the applied sinusoidal source in Fig. 14.15), and **FREQ** = 1 kHz. If we double-click on the source symbol, the **Property Editor** will appear, confirming the above choices and showing that **DF** = 0 s, **PHASE** = 0°, and **TD** = 0 s as set by the default levels. We are now ready to do an analysis of the circuit for the fixed frequency of 1 kHz.

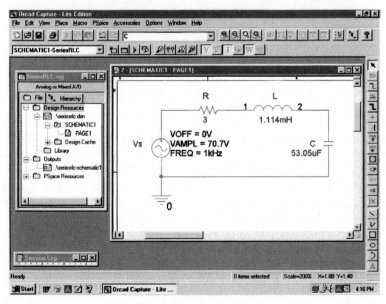

FIG. 14.80
Using PSpice to analyze a series R-L-C ac circuit.

The simulation process is initiated by first selecting the **New Simulation Profile** icon and inserting **SeriesRLC** as the **Name** followed by **Create**. The **Simulation Settings** dialog box will now appear, and since we are continuing to plot the results against time, the **Time Domain(Transient)** option is selected under **Analysis type.** Since the period of each **cycle** of the applied source is 1 ms, the **Run to time** will be set at 5 ms so that five cycles will appear. The **Start saving data after** will be left at 0 s even though there will be an oscillatory period for the reactive elements before the circuit settles down. The **Maximum step size** will be set at 5 ms/1000 = 5 μs. Finally **OK** is selected followed by the **Run PSpice** key. The result will be a blank screen with an *x*-axis extending from 0 s to 5 ms.

The first quantity of interest is the current through the circuit, so **Trace-Add-Trace** is selected followed by **I(R)** and **OK.** The resulting plot of Fig. 14.81 clearly shows that there is a period of storing and discharging of the reactive elements before a steady-state level is established. It would appear that after 3 ms, steady-state conditions have been essentially established. Select the **Toggle cursor** key, and left-click the mouse; a cursor will result that can be moved along the axis near the maximum value around 1.4 ms. In fact, the cursor reveals a maximum value of 16.4 A which exceeds the steady-state solution by over 2 A. A right click of the mouse will establish a second cursor on the screen that can be placed near the steady-state peak around 4.4 ms. The resulting peak value is about 14.15 A which is a match with the longhand solution for Fig. 14.15. We will therefore assume that steady-state conditions have been established for the circuit after 4 ms.

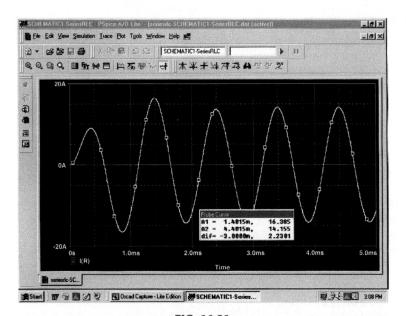

FIG. 14.81

A plot of the current for the circuit of Fig. 14.80 showing the transition from the transient state to the steady-state response.

Let us now add the source voltage through **Trace-Add Trace-V(Vs:+)-OK** to obtain the multiple plot at the bottom of Fig. 14.82. For the voltage across the coil, the sequence **Plot-Add Plot to Window-Trace-Add Trace-V(L:1)-V(L:2)** will result in the plot appearing at the top of Fig. 14.82. Take special note of the fact that the **Trace Expression** is **V(L:1)−V(L:2)** rather than just **V(L:1)** because **V(L:1)** would be the voltage from that point to ground which would include the voltage across the capacitor. In addition, the − sign between the two comes from the **Functions or Macros** list at right of the **Add Traces** dialog box.

Finally, since we know that the waveforms are fairly steady after 3 ms, let us cut away the waveforms before 3 ms with **Plot-Axis Settings-X axis-User Defined-3ms to 5ms-OK** to obtain the two cycles of Fig. 14.82. Now you can clearly see that the peak value of the voltage across the coil is 100 V to match the analysis of Fig. 14.15. It is also clear that the applied voltage leads the input current by an angle that can be determined using the cursors. First activate the cursor option by selecting the cursor key (a red plot through the origin) in the second toolbar down from the menu bar. Then select **V(Vs:+)** at the bottom left of the screen with a left click of the mouse, and set it at the point where the applied voltage passes through the horizontal axis with a positive slope. The result is **A1** = 4 ms at −4.243 μV $\cong$ 0 V. Then select **I(R)** at the bottom left of the screen with a right click of the mouse, and place it at the point where the current waveform passes through the horizontal axis with a positive slope. The result is **A2** = 4.15 ms at −55.15 mA = 0.55 A $\cong$ 0 A (compared to a peak value of 14.14 A). At the bottom of the **Probe Cursor** dialog box, the time difference is 147.24 μs.

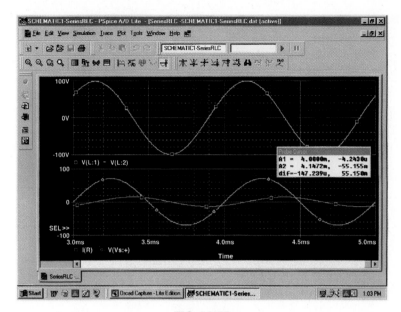

FIG. 14.82

A plot of the steady-state response (t > 3 ms) for v_L, v_s, and i for the circuit of Fig. 14.80.

Now set up the ratio

$$\frac{147.24 \; \mu s}{1000 \; \mu s} = \frac{\theta}{360°}$$

$$\theta = 52.99°$$

The phase angle by which the applied voltage leads the source is 52.99° which is very close to the theoretical solution of 53.13°. Increasing the number of data points for the plot would have increased the accuracy level and brought the results closer to 53.13°.

Electronics Workbench (EWB)

We will now examine the response of a network versus frequency rather than time using the network of Fig. 14.64 which now appears on the schematic of Fig. 14.83. The ac current source appears as **AC_CURRENT_SOURCE** in the **Sources** tool bin next to the ac voltage source. Note that the current source was given an amplitude of 1 A to establish a magnitude match between the response of the voltage across the network and the impedance of the network. That is,

$$|Z_T| = \left|\frac{V_s}{I_s}\right| = \left|\frac{V_s}{1 \; A}\right| = |V_s|$$

Before applying computer methods, we should develop a rough idea of what to expect so that we have something to which to compare the computer solution. At very high frequencies such as 1 MHz, the impedance of the inductive element will be about 25 kΩ which when placed in parallel with the 220 Ω will look like an open circuit. The result is that as the frequency gets very high, we should expect the impedance of the network to approach the 220 Ω level of the resistor. In addition, since the network will take on resistive characteristics at very high frequencies, the angle associated with the input impedance should also approach 0 Ω. At very low frequencies the reactance of the inductive element will be much less than the 220 Ω of the resistor, and the network will take on inductive characteristics. In fact, at, say, 10 Hz, the reactance of the inductor is only about 0.25 Ω which is very close to a short-circuit equivalent compared to the parallel 220 Ω resistor. The result is that the impedance of the network is very close to 0 Ω at very low frequencies. Again, since the inductive effects are so strong at low frequencies, the phase angle associated with the input impedance should be very close to 90°.

Now for the computer analysis. The current source, the resistor element, and the inductor are all placed and connected using procedures described in detail in earlier chapters. However, there is one big difference this time that the user must be aware of: Since the out-

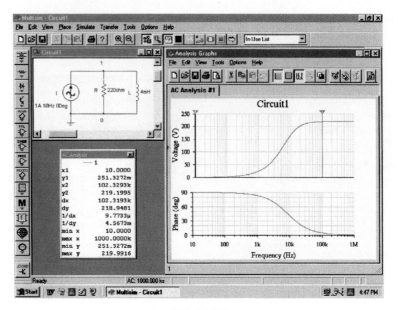

FIG. 14.83

Obtaining an impedance plot for a parallel R-L network using EWB.

put will be plotted versus frequency, the **Analysis Setup** heading must be selected in the **AC Current** dialog box for the current source. When selected, **the AC Magnitude** must be set to the value of the ac source. In this case, the default level of **1A** matches that of the applied source, so we were set even if we failed to check the setting. In the future, however, a voltage or current source may be used that does not have an amplitude of 1, and proper entries must be made to this listing.

For the simulation the sequence **Simulate-Analyses-AC Analysis** is first applied to obtain the **AC Analysis** dialog box. The **Start frequency** will be set at **10 Hz** so that we have entries at very low frequencies, and the **Stop frequency** will be set at **1MHz** so that we have data points at the other end of the spectrum. The **Sweep type** can remain **Decade,** but the number of points per decade will be 1000 so that we obtain a detailed plot. The **Vertical scale** will be set on **Linear.** Within **Output variables** we find that only one node, **1,** is defined. Shifting it over to the **Selected variables for analysis** column using the **Plot during simulation** key pad and then hitting the **Simulate** key will result in the two plots of Fig. 14.83. The **Show/Hide Grid** key was selected to place the grid on the graph, and the **Show/Hide Cursors** key was selected to place the **AC Analysis** dialog box appearing in Fig. 14.83.

Since two graphs are present, we must define the one we are working on by clicking on the **Voltage** or **Phase** heading on the left side of each plot. A small red arrow will appear when selected to keep us aware of the active plot. When setting up the cursors, be sure that you have activated the correct plot. When the red cursor is moved to 10 Hz **(x1),** we find that the voltage across the network is only 0.251 V **(y1),** resulting in an input impedance of only 0.25 Ω—quite small and matching our theoretical prediction. In addition, note that the phase angle is essentially at 90° in the other plot, confirming our other assumption above—a totally inductive network. If we set the blue cursor near 100 kHz **(x2** = 102.3 kHz), we find that the impedance at 219.2 Ω **(y2)** is closing in on the resistance of the parallel resistor of 220 Ω, again confirming the preliminary analysis above. As noted in the bottom of the **AC Analysis** box, the maximum value of the voltage is 219.99 Ω or essentially 220 Ω at 1 MHz. Before leaving the plot, note the advantages of using a log axis when you want a response over a wide frequency range.

CHAPTER SUMMARY
SERIES CONFIGURATION

- The total impedance of series ac elements is the sum of the individual impedances.
- The current is the same at every point in a series ac circuit.
- The applied voltage equals the sum of the voltage drops around an ac closed path.
- The angle associated with the total impedance is the same as that between the applied voltage and the resulting source current.

- The smaller the angle associated with the total impedance, the more resistive the circuit.
- In an impedance diagram, the inductive and capacitive reactances are always opposing elements.
- The phasor diagram shows at a glance the relative magnitude of each quantity and the phase relationship between the voltages and current.
- The power to a series circuit can be found by simply summing the power to all the resistive elements or using the general equation for power that includes the phase angle between the applied voltage and the resulting current.
- For any series ac circuit, the largest resistor or largest reactance will have the most impact on the total resistance or reactance, respectively.
- When examining the frequency response, keep in mind that, ideally, the resistance does not change with frequency. On the other hand, the reactance of the inductor increases linearly with frequency, whereas the reactance of the capacitor decreases exponentially with frequency.

PARALLEL CONFIGURATION

- The total impedance of a parallel ac network is found in the same way as the total resistance for a dc network, except that you are now working with magnitudes and angles, not just magnitudes.
- The total impedance of two parallel impedances is the product divided by their sum.
- The angle associated with the total impedance is the same as that between the applied voltage and the resulting current.
- The smaller the angle associated with the total impedance, the more resistive the network.
- The voltage is the same across parallel ac elements.
- Ohm's law can be used to find all the currents of a parallel ac network.
- Kirchhoff's current law can be applied in the same manner as applied to dc circuits.
- The total power to a parallel ac network can be found in the same manner as applied to series ac circuits. The power-factor angle for parallel networks will reveal at a glance whether the network is resistive or reactive. The larger the power factor, the more resistive the total impedance.
- For parallel ac networks, the smallest parallel resistive or reactive element will have the most impact on the total resistance or reactance, respectively.
- For the frequency response of parallel elements, the response of each element is the same as that for series ac elements. The only difference is that the element with the smallest impedance will be the predominant factor.

PHASE MEASUREMENTS

- When using a dual-trace scope, always be sure that there is a common ground for the two channels.
- A sensing resistor can be placed in series with any inductor or capacitor to determine the phase relationship between the current through each and a voltage or current at some other point in the network. This approach works because of the in-phase relationship between the voltage across a resistor and the current through the resistor.
- Always choose a sensing resistor that will have an impedance significantly less than that of any element in series with the sensing resistor.

Important Equations

Series Configuration:

$$\mathbf{Z}_T = \mathbf{Z}_1 + \mathbf{Z}_2 + \mathbf{Z}_3 + \cdots + \mathbf{Z}_N$$

$$\mathbf{I}_s = \frac{\mathbf{E}}{\mathbf{Z}_T}$$

$$\mathbf{V}_1 = \mathbf{I}_s \mathbf{Z}_1 \qquad \mathbf{V}_N = \mathbf{I}_s \mathbf{Z}_N$$

$$\mathbf{E} = \mathbf{V}_1 + \mathbf{V}_2 + \mathbf{V}_3 + \cdots + \mathbf{V}_N$$

$$P_T = \Sigma P_R = EI_s \cos \theta_T$$

$$F_p = \cos \theta_T = \frac{R}{Z_T}$$

$$\mathbf{V}_x = \frac{\mathbf{Z}_x \mathbf{E}}{\mathbf{Z}_T}$$

Parallel Configuration:

$$Z_T = \cfrac{1}{\cfrac{1}{Z_1} + \cfrac{1}{Z_2} + \cfrac{1}{Z_3} + \cdots + \cfrac{1}{Z_N}}$$

$$I_1 = \frac{E}{Z_1} = EY_1 \qquad I_N = \frac{E}{Z_N} = EY_N$$

$$I_s = I_1 + I_2 + I_3 + \cdots + I_N$$

$$Z_T = \frac{Z_1}{N}$$

$$P_T = \Sigma P_R = EI_s \cos \theta_T$$

$$Z_T = \frac{Z_1 Z_2}{Z_1 + Z_2}$$

$$F_p = \cos \theta_T = \frac{G}{Y_T}$$

$$I_s = \frac{E}{Z_T} = EY_T$$

$$I_1 = \frac{Z_2 I_T}{Z_1 + Z_2} \qquad I_2 = \frac{Z_1 I_T}{Z_1 + Z_2}$$

PROBLEMS

SECTION 14.2   Series Configuration

1. Determine the total impedance of the circuits of Fig. 14.84. Express your answer in rectangular and polar form, and draw the impedance diagram.

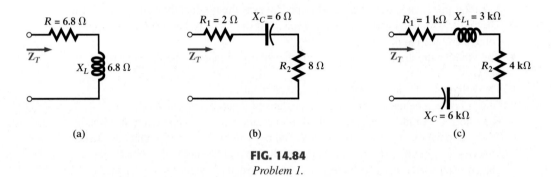

FIG. 14.84
Problem 1.

2. Find the type of the series circuit elements, and their impedance in ohms, that must be in the closed container of Fig. 14.85 for the indicated voltages and currents to exist at the input terminals. That is, find the simplest series circuit that will satisfy the indicated conditions.

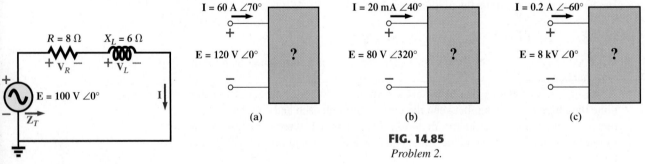

FIG. 14.85
Problem 2.

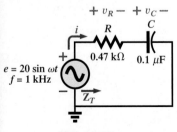

FIG. 14.86
Problems 3 and 34.

3. For the circuit of Fig. 14.86:
   a. Find the total impedance Z_T in polar form.
   b. Draw the impedance diagram.
   c. Find the current I and the voltages V_R and V_L in phasor form.
   d. Draw the phasor diagram of the voltages E, V_R, and V_L, and the current I.
   e. Verify Kirchhoff's voltage law around the closed loop.
   f. Find the average power delivered to the circuit.
   g. Find the sinusoidal expressions for the voltages and current if the frequency is 60 Hz.
   h. Plot the waveforms for the voltages and current on the same set of axes.

4. Given the network of Fig. 14.87:
   a. Determine Z_T.
   b. Find I.
   c. Calculate V_R and V_L.

FIG. 14.87
Problems 4 and 36.

5. For the circuit of Fig. 14.88:
   a. Find the total impedance $\mathbf{Z}_T$ in polar form.
   b. Draw the impedance diagram.
   c. Find the value of C in microfarads and L in henries.
   d. Find the current $\mathbf{I}$ and the voltages $\mathbf{V}_R$, $\mathbf{V}_L$, and $\mathbf{V}_C$ in phasor form.
   e. Draw the phasor diagram of the voltages $\mathbf{E}$, $\mathbf{V}_R$, $\mathbf{V}_L$, and $\mathbf{V}_C$, and the current $\mathbf{I}$.
   f. Verify Kirchhoff's voltage law around the closed loop.
   g. Find the average power delivered to the circuit.
   h. Find the sinusoidal expressions for the voltages and current.
   i. Plot the waveforms for the voltages and current on the same set of axes.

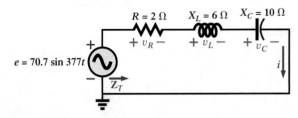

FIG. 14.88
Problem 5.

6. Using the oscilloscope reading of Fig. 14.89, determine the resistance R.

*7. Using the DMM current reading and the oscilloscope measurement of Fig. 14.90:
   a. Determine the inductance L.
   b. Find the resistance R.

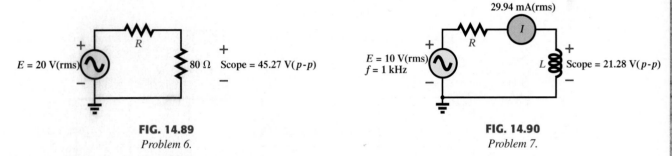

FIG. 14.89
Problem 6.

FIG. 14.90
Problem 7.

*8. Using the oscilloscope reading of Fig. 14.91, determine the capacitance C.

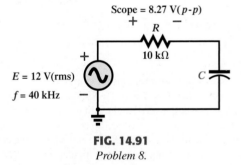

FIG. 14.91
Problem 8.

SECTION 14.3   Average Power and the Power Factor

9. Given the R-L circuit of Fig. 14.92:
   a. Find the reactance of the inductor.
   b. Calculate the total impedance of the network and draw the impedance diagram.
   c. Find the current in phasor form.
   d. What is the power factor of the circuit?
   e. Find the power delivered to the circuit.
   f. Find the reactive power to the inductive element.

10. For the R-L-C circuit of Fig. 14.93:
   a. Find the total impedance of the circuit.
   b. Find the current in phasor form.
   c. Find the power factor of the circuit. Is it leading or lagging?
   d. Find the power delivered to the circuit.

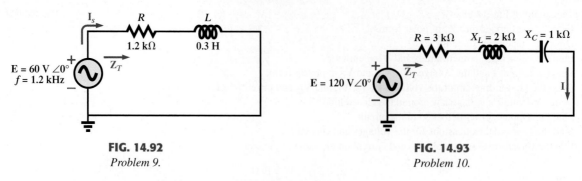

FIG. 14.92
Problem 9.

FIG. 14.93
Problem 10.

SECTION 14.4   Voltage Divider Rule

11. Calculate the voltages $\mathbf{V}_1$ and $\mathbf{V}_2$ for the circuit of Fig. 14.94 in phasor form using the voltage divider rule.

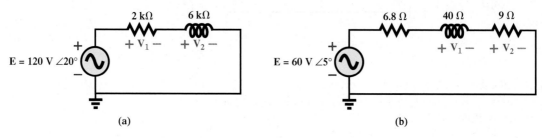

(a)   **(b)**

FIG. 14.94
Problem 11.

12. Calculate the voltages $\mathbf{V}_1$ and $\mathbf{V}_2$ for the circuit of Fig. 14.95 in phasor form using the voltage divider rule.

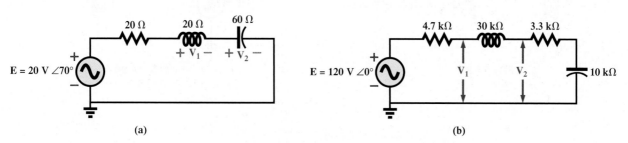

(a)   **(b)**

FIG. 14.95
Problem 12.

*13. For the circuit of Fig. 14.96:
  a. Determine $\mathbf{I}$, $\mathbf{V}_R$, and $\mathbf{V}_C$ in phasor form using Ohm's law.
  b. Calculate the total power factor, and indicate whether it is leading or lagging.
  c. Calculate the average power delivered to the circuit. Is the power factor leading or lagging?
  d. Draw the impedance diagram.
  e. Draw the phasor diagram of the voltages $\mathbf{E}$, $\mathbf{V}_R$, and $\mathbf{V}_C$, and the current $\mathbf{I}$.
  f. Find the voltages $\mathbf{V}_R$ and $\mathbf{V}_C$ using the voltage divider rule, and compare them with the results of part (a) above.
  g. Draw the equivalent series circuit of the above as far as the total impedance and the current i are concerned.

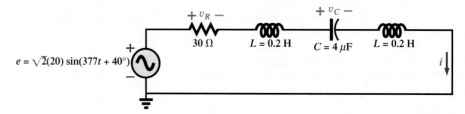

FIG. 14.96
Problems 13 and 37.

14. An electrical load has a power factor of 0.8 lagging. It dissipates 8 kW at a voltage of 200 V. Calculate the impedance of this load in rectangular coordinates.

*15. Find the series element or elements that must be in the enclosed container of Fig. 14.97 to satisfy the following conditions:
   a. Average power to circuit = 300 W.
   b. Circuit has a lagging power factor.

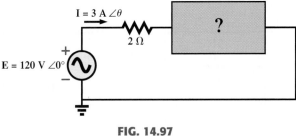

FIG. 14.97
Problem 15.

SECTION 14.5   Frequency Response for Series ac Circuits

*16. For the circuit of Fig. 14.98:
   a. Plot Z_T and θ_T versus frequency for a frequency range of zero to 20 kHz.
   b. Plot V_L versus frequency for the frequency range of part (a).
   c. Plot θ_L versus frequency for the frequency range of part (a).
   d. Plot V_R versus frequency for the frequency range of part (a).

*17. For the circuit of Fig. 14.99:
   a. Plot Z_T and θ_T versus frequency for a frequency range of zero to 10 kHz.
   b. Plot V_C versus frequency for the frequency range of part (a).
   c. Plot θ_C versus frequency for the frequency range of part (a).
   d. Plot V_R versus frequency for the frequency range of part (a).

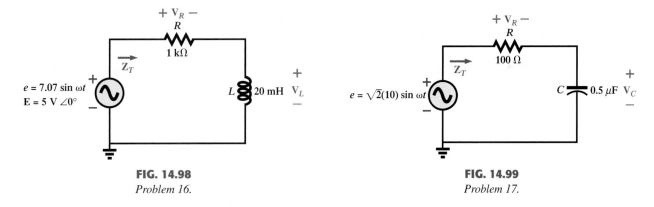

FIG. 14.98	**FIG. 14.99**
Problem 16.	*Problem 17.*

*18. For the series *R-L-C* circuit of Fig. 14.100:
   a. Plot Z_T and θ_T versus frequency for a frequency range of zero to 20 kHz in increments of 1 kHz.
   b. Plot V_C (magnitude only) versus frequency for the same frequency range of part (a).
   c. Plot I (magnitude only) versus frequency for the same frequency range of part (a).

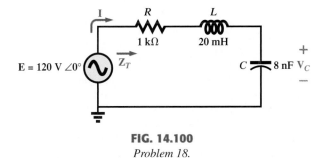

FIG. 14.100
Problem 18.

SECTION 14.7 Parallel Configuration

19. Find the total impedance of the circuits of Fig. 14.101 in rectangular form.

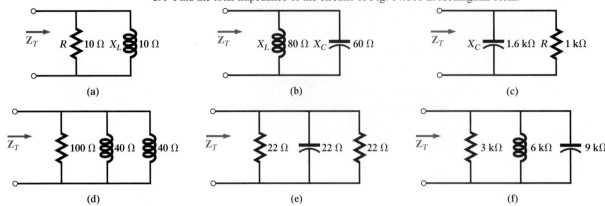

FIG. 14.101
Problem 19.

*20. Repeat Problem 2 for the parallel circuit elements that must be in the closed container for the same voltage and current to exist at the input terminals. (Find the simplest parallel circuit that will satisfy the conditions indicated.)

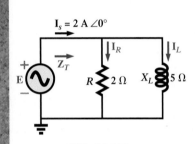

FIG. 14.102
Problem 21.

21. For the circuit of Fig. 14.102:
  a. Find the total impedance Z_T in polar form.
  b. Draw the impedance diagram.
  c. Find the voltage **E** and the currents I_R and I_L in phasor form.
  d. Draw the phasor diagram of the currents I_s, I_R, and I_L, and the voltage **E.**
  e. Verify Kirchhoff's current law at one node.
  f. Find the average power delivered to the circuit.
  g. Find the power factor of the circuit, and indicate whether it is leading or lagging.
  h. Find the sinusoidal expressions for the currents and voltage if the frequency is 60 Hz.
  i. Plot the waveforms for the currents and voltage on the same set of axes.

22. For the circuit of Fig. 14.103:
  a. Find the total impedance Z_T in polar form.
  b. Draw the impedance diagram.
  c. Find the currents I_s, I_R, and I_C in phasor form.
  d. Draw the phasor diagram of the currents I_s, I_R, and I_C, and the voltage **E.**
  e. Verify Kirchhoff's current law at one node.
  f. Find the average power delivered to the circuit.
  g. Find the power factor of the circuit, and indicate whether it is leading or lagging.
  h. Find the sinusoidal expressions for the currents and voltage if the frequency is 60 Hz.
  i. Plot the waveforms for the currents and voltage on the same set of axes.

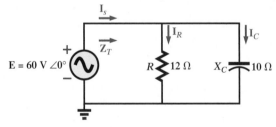

FIG. 14.103
Problems 22 and 35.

23. For the circuit of Fig. 14.104:
  a. Find the total impedance Z_T in polar form.
  b. Draw the impedance diagram.
  c. Find the value of C in microfarads and L in henries.
  d. Find the voltage **E** and currents I_R, I_L, and I_C in phasor form.
  e. Draw the phasor diagram of the currents I_s, I_R, I_L, and I_C, and the voltage **E.**
  f. Verify Kirchhoff's current law at one node.
  g. Find the average power delivered to the circuit.
  h. Find the power factor of the circuit, and indicate whether it is leading or lagging.
  i. Find the sinusoidal expressions for the currents and voltage.
  j. Plot the waveforms for the currents and voltage on the same set of axes.

24. Repeat Problem 23 for the circuit of Fig. 14.105, replacing **E** with **I**$_s$ in part (d).

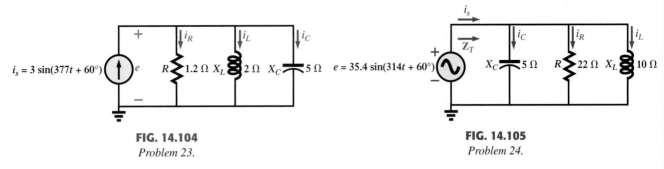

FIG. 14.104

Problem 23.

FIG. 14.105

Problem 24.

SECTION 14.8   Current Divider Rule

25. Calculate the currents **I**$_1$ and **I**$_2$ of Fig. 14.106 in phasor form using the current divider rule.

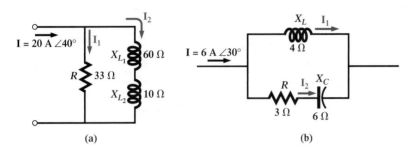

(a)

(b)

FIG. 14.106

Problem 25.

SECTION 14.9   Frequency Response of Parallel Elements

*26. For the parallel *R-C* network of Fig. 14.107:
  a. Plot Z_T and θ_T versus frequency for a frequency range of zero to 20 kHz.
  b. Plot V_C versus frequency for the frequency range of part (a).
  c. Plot I_R versus frequency for the frequency range of part (a).

*27. For the parallel *R-L* network of Fig. 14.108:
  a. Plot Z_T and θ_T versus frequency for a frequency range of zero to 10 kHz.
  b. Plot I_L versus frequency for the frequency range of part (a).
  c. Plot I_R versus frequency for the frequency range of part (a).

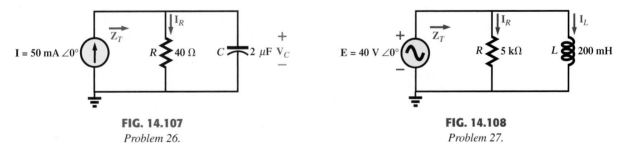

FIG. 14.107

Problem 26.

FIG. 14.108

Problem 27.

*28. For the parallel *R-L-C* network of Fig. 14.109:
  a. Plot Z_T and θ_T for a frequency range of zero to 20 kHz.
  b. Plot V_C versus frequency for the frequency range of part (a).
  c. Plot I_L versus frequency for the frequency range of part (a).

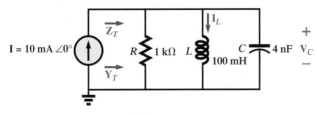

FIG. 14.109

Problem 28.

SECTION 14.11   Phase Measurements

29. For the circuit of Fig. 14.110, show the connections and determine the phase relationship between the following using a dual-trace oscilloscope. The circuit can be reconstructed differently for each part, but do not use sensing resistors. Show all connections on a redrawn diagram.
    a. e and v_C
    b. e and i_s
    c. e and v_L

30. For the network of Fig. 14.111, determine the phase relationship between the following using a dual-trace oscilloscope. The network must remain as constructed in Fig. 14.111, but sensing resistors can be introduced. Show all connections on a redrawn diagram.
    a. e and v_{R_2}
    b. e and i_s
    c. i_L and i_C

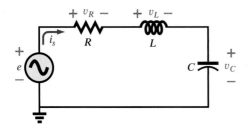

FIG. 14.110
Problem 29.

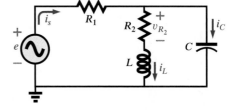

FIG. 14.111
Problem 30.

31. For the oscilloscope traces of Fig. 14.112:
    a. Determine the phase relationship between the waveforms, and indicate which one leads or lags.
    b. Determine the peak-to-peak and rms values of each waveform.
    c. Find the frequency of each waveform.

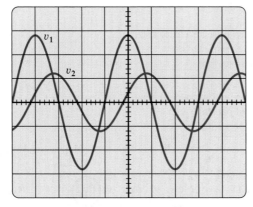

Vertical sensitivity = 0.5 V/div.
Horizontal sensitivity = 0.2 ms/div.
(I)

Vertical sensitivity = 2 V/div.
Horizontal sensitivity = 10 μs/div.
(II)

FIG. 14.112
Problem 31.

SECTION 14.12   Admittance and Susceptance

32. For the network of Fig. 14.113:
    a. Find the total admittance $\mathbf{Y}_T$ in polar form.
    b. Draw the admittance diagram.
    c. Find the total impedance $\mathbf{Z}_T$ in polar form.
    d. Draw the impedance diagram.
    e. Find the currents $\mathbf{I}_R$ and $\mathbf{I}_L$ using Ohm's law.
    f. Calculate the source current using the total admittance.
    g. Find the average power delivered to and the power factor of the network.

33. For the network of Fig. 14.114:
    a. Find the total admittance $\mathbf{Y}_T$ in polar form.
    b. Draw the admittance diagram.
    c. Find the currents $\mathbf{I}_R$, $\mathbf{I}_L$, and $\mathbf{I}_C$ using Ohm's law.

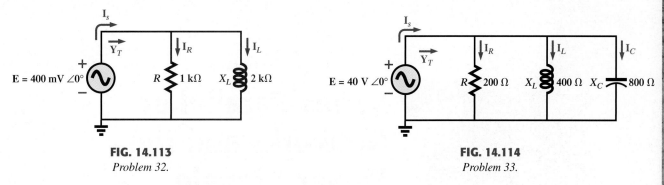

FIG. 14.113
Problem 32.

FIG. 14.114
Problem 33.

   d. Calculate I_s using the total admittance.

   e. Find I_s using Kirchhoff's current law [and the results of part (c)], and compare your answer with the solution for part (d).

   f. Calculate the power factor of the network, and determine whether it is leading or lagging.

   g. Find the power dissipated by the network.

SECTION 14.14 Computer Analysis

34. Using PSpice or EWB, for the network of Fig. 14.86 (use $f = 1$ kHz):

   a. Determine the rms values of the voltages V_R and V_L and the current **I**.

   b. Plot v_R, v_L, and i versus time on separate plots.

   c. Place e, v_R, v_L, and i on the same plot, and label them accordingly.

35. Using PSpice or EWB, for the network of Fig. 14.103:

   a. Determine the rms values of the currents I_s, I_R, and I_L.

   b. Plot i_s, i_R, and i_L versus time on separate plots.

   c. Place e, i_s, i_R, and i_L on the same plot, and label them accordingly.

36. Using PSpice or EWB, for the network of Fig. 14.87:

   a. Plot the impedance of the network versus frequency from zero to 10 kHz.

   b. Plot the current i versus frequency for the frequency range zero to 10 kHz.

*37. Using PSpice or EWB, for the network of Fig. 14.96:

   a. Find the rms values of the voltages v_R and v_C at a frequency of 1 kHz.

   b. Plot v_C versus frequency for the frequency range zero to 10 kHz.

   c. Plot the phase angle between e and i for the frequency range zero to 10 kHz.

GLOSSARY

Admittance   A measure of how easily a network will "admit" the passage of current through that system. It is measured in siemens, abbreviated S, and is represented by the capital letter Y.

Admittance diagram   A vector display that clearly depicts the magnitude of the admittance of the conductance, capacitive susceptance, and inductive susceptance, and the magnitude and angle of the total admittance of the system.

Current divider rule   A method by which the current through either of two parallel branches can be determined in an ac network without first finding the voltage across the parallel branches.

Equivalent circuits   For every series ac network there is a parallel ac network (and vice versa) that will be "equivalent" in the sense that the input current and impedance are the same.

Impedance diagram   A vector display that clearly depicts the magnitude of the impedance of the resistive, reactive, and capacitive components of a network, and the magnitude and angle of the total impedance of the system.

Parallel ac circuits   A connection of elements in an ac network in which all the elements have two points in common. The voltage is the same across each element.

Phasor diagram   A vector display that provides at a glance the magnitude and phase relationships among the various voltages and currents of a network.

Series ac configuration   A connection of elements in an ac network in which no two impedances have more than one terminal in common and the current is the same through each element.

Susceptance   A measure of how "susceptible" an element is to the passage of current through it. It is measured in siemens. abbreviated S, and is represented by the capital letter B.

Voltage divider rule   A method through which the voltage across one element of a series of elements in an ac network can be determined without first having to find the current through the elements.

Series-Parallel ac Networks and the Power Triangle

OBJECTIVES

- Be able to solve complex series-parallel ac networks.
- Understand how to use the calculator or Mathcad to perform the complex mathematical equations that might result in the analysis of series-parallel ac networks.
- Be comfortable with the analysis of ac ladder networks.
- Understand how the real, reactive, and apparent power levels are related on a vector diagram.
- Develop confidence in finding the total average, reactive, and apparent power levels of a network.
- Be able to find the total power factor of any network and be able to determine whether the power factor is leading or lagging.
- Clearly understand the concept of power-factor correction and the impact it has on the supply current and the utility bill of a large manufacturing plant.
- Appreciate the importance of the concept of grounding, including why it is such an important part of any system design.

15.1 INTRODUCTION

This chapter will clearly demonstrate the similarities between analyzing ac circuits (at a fixed frequency) and dc circuits if the elements are replaced by impedance notation. All the steps leading to the analysis of series-parallel dc circuits will be exactly the same for ac circuits. The similarities are so strong that the content of the next section is simply a series of examples to demonstrate the analysis process.

Although the primary emphasis has been on the average power delivered to a network, it was pointed out in Chapter 13 that at specific instances the power drain from the source will be affected by the energy being delivered to the reactive elements. That additional drain is a real concern for the supplier and big users of electricity, often resulting in a need to improve the power factor of the load. Since most industrial loads are inductive, the use of capacitors to improve the power factor will be discussed in some detail.

Finally, the need to be fully aware of the impact of proper grounding techniques in the laboratory and in residential and industrial appliances will be considered. It is a topic often taken too likely and one that can cause some real problems if ignored.

15.2 SERIES-PARALLEL ac NETWORKS

Before examining the details of this section, it might be wise to review Chapter 7 on series-parallel dc circuits, since the examples and problems are already familiar and the approach to be applied in this section will be quite similar. Throughout the analyses, note the use of **block impedances** to reduce the complexity of the network and the common practice of redrawing the network as often as possible. Both steps help reveal the similarities between the analysis of dc and ac networks.

Recall that in the dc chapter, a series of steps were provided to guide you through this type of analysis. For ac networks the steps take the following form:

1. *Redraw the network, employing block impedances to combine obvious series and parallel elements, which will reduce the network to one that clearly reveals the fundamental structure of the system.*
2. *Study the problem and make a brief mental sketch of the overall approach you plan to use. Doing this may result in time- and energy-saving shortcuts. In some cases, a lengthy, drawn-out analysis may not be necessary. A single application of a fundamental law of circuit analysis may result in the desired solution.*
3. *After the overall approach has been determined, it is usually best to consider each branch involved in your method independently before tying them together in series-parallel combinations. In most cases, work back from the obvious series and parallel combinations to the source to determine the total impedance of the network. The source current can then be determined, and the path back to specific unknowns can be defined. As you progress back to the source, continually define those unknowns that have not been lost in the reduction process. It will save time when you have to work back through the network to find specific quantities.*
4. *When you have arrived at a solution, check to see that it is reasonable by considering the magnitudes of the energy source and the elements in the circuit. If it is not reasonable, either solve the network using another approach, or check over your work very carefully. At this point a computer solution can be an invaluable asset in the validation process.*

EXAMPLE 15.1 For the network of Fig. 15.1:

a. Calculate $\mathbf{Z}_T$.
b. Determine $\mathbf{I}_s$.
c. Calculate $\mathbf{V}_R$ and $\mathbf{V}_C$.
d. Find $\mathbf{I}_C$.
e. Compute the power delivered.
f. Find F_p of the network.

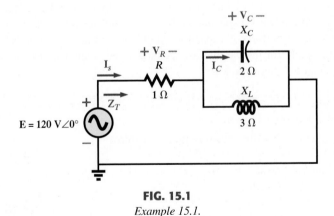

FIG. 15.1

Example 15.1.

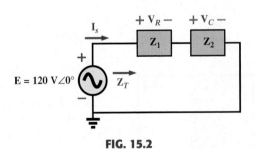

FIG. 15.2

Network of Fig. 15.1 after assigning the block impedances.

Solutions:

a. As suggested in the introduction, the network has been redrawn with block impedances, as shown in Fig. 15.2. The impedance $\mathbf{Z}_1$ is simply the resistor R of 1 Ω, and $\mathbf{Z}_2$ is the parallel combination of X_C and X_L. The network now clearly reveals that it is fundamentally a series circuit, suggesting a direct path toward the total impedance and the source current. As also noted in the introduction, for many such problems you must work back to the source to find first the total impedance and then the source current. When the unknown quantities are found in terms of these subscripted impedances, the numerical values can then be substituted to find the magnitude and phase angle of the unknown. In other words, try to find the desired solution solely in terms of the subscripted impedances before substituting numbers. This approach will usually enhance the clarity of the chosen path toward a solution while saving time and preventing careless calculation errors. Note also in Fig. 15.2 that all the unknown quantities except $\mathbf{I}_C$ have been preserved, meaning that we can use Fig. 15.2 to determine these quantities rather than having to return to the more complex network of Fig. 15.1.

The total impedance is defined by

$$\mathbf{Z}_T = \mathbf{Z}_1 + \mathbf{Z}_2$$

with

$$\mathbf{Z}_1 = R \angle 0° = 1 \; \Omega \; \angle 0°$$

$$\mathbf{Z}_2 = \mathbf{Z}_C \parallel \mathbf{Z}_L = \frac{(X_C \angle -90°)(X_L \angle 90°)}{-j \, X_C + j \, X_L} = \frac{(2 \; \Omega \; \angle -90°)(3 \; \Omega \; \angle 90°)}{-j \, 2 \; \Omega + j \, 3 \; \Omega}$$

$$= \frac{6 \; \Omega \; \angle 0°}{j \, 1} = \frac{6 \; \Omega \; \angle 0°}{1 \; \angle 90°} = 6 \; \Omega \; \angle -90°$$

and

$$\mathbf{Z}_T = \mathbf{Z}_1 + \mathbf{Z}_2 = 1 \; \Omega - j \, 6 \; \Omega = \textbf{6.08} \; \boldsymbol{\Omega} \; \angle \textbf{-80.54°}$$

b. $$\mathbf{I}_s = \frac{\mathbf{E}}{\mathbf{Z}_T} = \frac{120 \; \text{V} \; \angle 0°}{6.08 \; \Omega \; \angle -80.54°} = \textbf{19.74 A} \; \angle \textbf{80.54°}$$

c. Referring to Fig. 15.2, we find that $\mathbf{V}_R$ and $\mathbf{V}_C$ can be found by a direct application of Ohm's law:

$$\mathbf{V}_R = \mathbf{I}_s \mathbf{Z}_1 = (19.74 \; \text{A} \; \angle 80.54°)(1 \; \Omega \; \angle 0°) = \textbf{19.74 V} \; \angle \textbf{80.54°}$$

$$\mathbf{V}_C = \mathbf{I}_s \mathbf{Z}_2 = (19.74 \; \text{A} \; \angle 80.54°)(6 \; \Omega \; \angle -90°)$$

$$= \textbf{118.44 V} \; \angle \textbf{-9.46°}$$

d. Now that $\mathbf{V}_C$ is known, the current $\mathbf{I}_C$ can also be found using Ohm's law.

$$\mathbf{I}_C = \frac{\mathbf{V}_C}{\mathbf{Z}_C} = \frac{118.44 \; \text{V} \; \angle -9.46°}{2 \; \Omega \; \angle -90°} = \textbf{59.22 A} \; \angle \textbf{80.54°}$$

e. $P_{\text{del}} = I_s^2 R = (19.74 \; \text{A})^2 (1 \; \Omega) = \textbf{389.67 W}$

f. $F_p = \cos \theta = \cos 80.54° = \textbf{0.164 leading}$

The fact that the total impedance has a negative phase angle (revealing that $\mathbf{I}_s$ leads $\mathbf{E}$) is a clear indication that the network is capacitive in nature and therefore has a leading power factor. The fact that the network is capacitive can be determined from the original network by first realizing that, for the parallel L-C elements, the smaller impedance predominates and results in an R-C network.

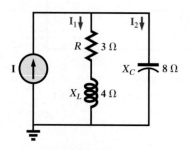

FIG. 15.3

Network for Example 15.2.

EXAMPLE 15.2 For the network of Fig. 15.3:

a. If $\mathbf{I}$ is 50 A $\angle 30°$, calculate $\mathbf{I}_1$ using the current divider rule.
b. Repeat part (a) for $\mathbf{I}_2$.
c. Verify Kirchhoff's current law at one node.

Solutions:

a. Redrawing the circuit as in Fig. 15.4, we have

$$\mathbf{Z}_1 = R + jX_L = 3 \; \Omega + j \, 4 \; \Omega = 5 \; \Omega \; \angle 53.13°$$

$$\mathbf{Z}_2 = -j X_C = -j \, 8 \; \Omega = 8 \; \Omega \; \angle -90°$$

Using the current divider rule yields

$$\mathbf{I}_1 = \frac{\mathbf{Z}_2 \mathbf{I}}{\mathbf{Z}_2 + \mathbf{Z}_1} = \frac{(8 \; \Omega \; \angle -90°)(50 \; \text{A} \; \angle 30°)}{(-j \, 8 \; \Omega) + (3 \; \Omega + j \, 4 \; \Omega)} = \frac{400 \; \angle -60°}{3 - j \, 4}$$

$$= \frac{400 \; \angle -60°}{5 \; \angle -53.13°} = \textbf{80 A} \; \angle \textbf{-6.87°}$$

b. $\mathbf{I}_2 = \dfrac{\mathbf{Z}_1\mathbf{I}}{\mathbf{Z}_2 + \mathbf{Z}_1} = \dfrac{(5\ \Omega\ \angle 53.13°)(50\ \text{A}\ \angle 30°)}{5\ \Omega\ \angle -53.13°} = \dfrac{250\ \angle 83.13°}{5\ \angle -53.13°}$

$\qquad\qquad = \mathbf{50\ A\ \angle 136.26°}$

c. $\qquad\qquad \mathbf{I} = \mathbf{I}_1 + \mathbf{I}_2$

$\quad 50\ \text{A}\ \angle 30° = 80\ \text{A}\ \angle -6.87° + 50\ \text{A}\ \angle 136.26°$

$\qquad\qquad = (79.43 - j\,9.57) + (-36.12 + j\,34.57)$

$\qquad\qquad = 43.31 + j\,25.0$

$\quad \mathbf{50\ A\ \angle 30°} = \mathbf{50\ A\ \angle 30°}\quad \text{(checks)}$

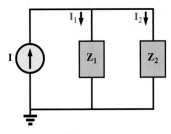

FIG. 15.4
*Network of Fig. 15.3 after assigning
the block impedances.*

EXAMPLE 15.3 For the network of Fig. 15.5:

a. Calculate the voltage $\mathbf{V}_C$ using the voltage divider rule.
b. Calculate the current $\mathbf{I}_s$.

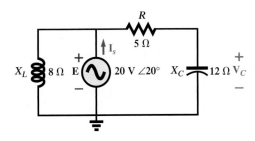

FIG. 15.5
Network for Example 15.3.

Solutions:

a. The network is redrawn as shown in Fig. 15.6, with

$$\mathbf{Z}_1 = 5\ \Omega = 5\ \Omega\ \angle 0°$$

$$\mathbf{Z}_2 = -j\,12\ \Omega = 12\ \Omega\ \angle -90°$$

$$\mathbf{Z}_3 = +j\,8\ \Omega = 8\ \Omega\ \angle 90°$$

Since $\mathbf{V}_C$ is desired, we will not combine R and X_C into a single block impedance. Note also that Fig. 15.6 clearly reveals that $\mathbf{E}$ is the total voltage across the series combination of $\mathbf{Z}_1$ and $\mathbf{Z}_2$, permitting the use of the voltage divider rule to calculate $\mathbf{V}_C$. In addition, note that all the currents necessary to determine $\mathbf{I}_s$ have been preserved in Fig. 15.6, revealing that there is no need to ever return to the network of Fig. 15.5—everything is defined by Fig. 15.6.

$$\mathbf{V}_C = \dfrac{\mathbf{Z}_2\mathbf{E}}{\mathbf{Z}_1 + \mathbf{Z}_2} = \dfrac{(12\ \Omega\ \angle -90°)(20\ \text{V}\ \angle 20°)}{5\ \Omega - j\,12\ \Omega} = \dfrac{240\ \text{V}\ \angle -70°}{13\ \angle -67.38°}$$

$$= \mathbf{18.46\ V}\ \angle\mathbf{-2.62°}$$

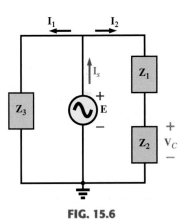

FIG. 15.6
*Network of Fig. 15.5 after assigning
the block impedances.*

b. $\mathbf{I}_1 = \dfrac{\mathbf{E}}{\mathbf{Z}_3} = \dfrac{20\ \text{V}\ \angle 20°}{8\ \Omega\ \angle 90°} = 2.5\ \text{A}\ \angle -70°$

$\mathbf{I}_2 = \dfrac{\mathbf{E}}{\mathbf{Z}_1 + \mathbf{Z}_2} = \dfrac{20\ \text{V}\ \angle 20°}{13\ \Omega\ \angle -67.38°} = 1.54\ \text{A}\ \angle 87.38°$

and

$$\mathbf{I}_s = \mathbf{I}_1 + \mathbf{I}_2$$

$$= 2.5\ \text{A}\ \angle -70° + 1.54\ \text{A}\ \angle 87.38°$$

$$= (0.86 - j\,2.35) + (0.07 + j\,1.54)$$

$$= 0.93 - j\,0.81 = \mathbf{1.23\ A}\ \angle\mathbf{-41.05°}$$

EXAMPLE 15.4 For Fig. 15.7:

a. Calculate the current $\mathbf{I}_s$.
b. Find the voltage $\mathbf{V}_{ab}$.

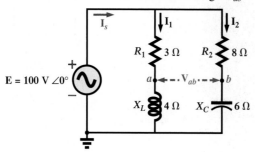

FIG. 15.7

Circuit for Example 15.4.

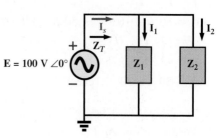

FIG. 15.8

Network of Fig. 15.7 after assigning the block impedances.

Solutions:

a. Redrawing the circuit as in Fig. 15.8, we obtain

$$\mathbf{Z}_1 = R_1 + jX_L = 3\,\Omega + j\,4\,\Omega = 5\,\Omega\,\angle53.13°$$
$$\mathbf{Z}_2 = R_2 - jX_C = 8\,\Omega - j\,6\,\Omega = 10\,\Omega\,\angle-36.87°$$

In this case, voltage $\mathbf{V}_{ab}$ is lost in the redrawn network, but currents $\mathbf{I}_1$ and $\mathbf{I}_2$ remain defined for future calculations necessary to determine $\mathbf{V}_{ab}$. Figure 15.8 clearly reveals that the total impedance can be found using the equation for two parallel impedances:

$$\mathbf{Z}_T = \frac{\mathbf{Z}_1\mathbf{Z}_2}{\mathbf{Z}_1 + \mathbf{Z}_2} = \frac{(5\,\Omega\,\angle53.13°)(10\,\Omega\,\angle-36.87°)}{(3\,\Omega + j\,4\,\Omega) + (8\,\Omega - j\,6\,\Omega)}$$

$$= \frac{50\,\Omega\,\angle16.26°}{11 - j\,2} = \frac{50\,\Omega\,\angle16.26°}{11.18\,\angle-10.30°}$$

$$= \mathbf{4.472\,\Omega\,\angle26.56°}$$

and

$$\mathbf{I}_s = \frac{\mathbf{E}}{\mathbf{Z}_T} = \frac{100\,\text{V}\,\angle0°}{4.472\,\Omega\,\angle26.56°} = \mathbf{22.36\,A\,\angle-26.56°}$$

b. By Ohm's law,

$$\mathbf{I}_1 = \frac{\mathbf{E}}{\mathbf{Z}_1} = \frac{100\,\text{V}\,\angle0°}{5\,\Omega\,\angle53.13°} = \mathbf{20\,A\,\angle-53.13°}$$

$$\mathbf{I}_2 = \frac{\mathbf{E}}{\mathbf{Z}_2} = \frac{100\,\text{V}\,\angle0°}{10\,\Omega\,\angle-36.87°} = \mathbf{10\,A\,\angle36.87°}$$

Returning to Fig. 15.7, we have

$$\mathbf{V}_{R_1} = \mathbf{I}_1\mathbf{Z}_{R_1} = (20\,\text{A}\,\angle-53.13°)(3\,\Omega\,\angle0°) = \mathbf{60\,V\,\angle-53.13°}$$
$$\mathbf{V}_{R_2} = \mathbf{I}_2\mathbf{Z}_{R_2} = (10\,\text{A}\,\angle+36.87°)(8\,\Omega\,\angle0°) = \mathbf{80\,V\,\angle+36.87°}$$

Instead of using the two steps just shown, we could have determined $\mathbf{V}_{R_1}$ or $\mathbf{V}_{R_2}$ in one step using the voltage divider rule:

$$\mathbf{V}_{R_1} = \frac{(3\,\Omega\,\angle0°)(100\,\text{V}\,\angle0°)}{3\,\Omega\,\angle0° + 4\,\Omega\,\angle90°} = \frac{300\,\text{V}\,\angle0°}{5\,\angle53.13°} = \mathbf{60\,V\,\angle-53.13°}$$

To find $\mathbf{V}_{ab}$, Kirchhoff's voltage law must be applied around the loop (Fig. 15.9) consisting of the 3 Ω and 8 Ω resistors. By Kirchhoff's voltage law,

$$\mathbf{V}_{ab} + \mathbf{V}_{R_1} - \mathbf{V}_{R_2} = 0$$

or

$$\mathbf{V}_{ab} = \mathbf{V}_{R_2} - \mathbf{V}_{R_1}$$

$$= 80\,\text{V}\,\angle36.87° - 60\,\text{V}\,\angle-53.13°$$

$$= (64 + j\,48) - (36 - j\,48)$$

$$= 28 + j\,96 = \mathbf{100\,V\,\angle73.74°}$$

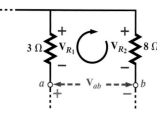

FIG. 15.9

Determining the voltage $\mathbf{V}_{ab}$ for the network of Fig. 15.7.

EXAMPLE 15.5   The network of Fig. 15.10 is frequently encountered in the analysis of transistor networks. The transistor equivalent circuit includes a current source $\mathbf{I}$ and an output impedance R_o. The resistor R_C is a biasing resistor to establish specific dc conditions, and the resistor R_i represents the loading of the next stage. The coupling capacitor is designed to be an open circuit for dc and to have as low an impedance as possible for the frequencies of interest to ensure that $\mathbf{V}_L$ is a maximum value. The frequency range of the example includes the entire audio (hearing) spectrum from 100 Hz to 20 kHz.

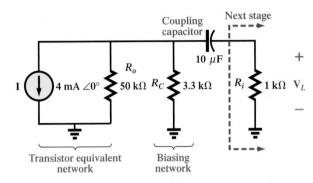

FIG. 15.10
Basic transistor amplifier.

The purpose of this example is to demonstrate that, for the full audio range, the effect of the capacitor can be ignored. It performs its function as a dc blocking agent but permits the ac to pass through with little disturbance.

a. Determine $\mathbf{V}_L$ for the network of Fig. 15.10 at a frequency of 100 Hz.
b. Repeat part (a) at a frequency of 20 kHz.
c. Compare the results of parts (a) and (b).

Solutions:

a. The network is redrawn with subscripted impedances in Fig. 15.11.

$$\mathbf{Z}_1 = 50 \text{ k}\Omega \ \angle 0° \ \| \ 3.3 \text{ k}\Omega \ \angle 0° = 3.096 \text{ k}\Omega \ \angle 0°$$

$$\mathbf{Z}_2 = R_i - jX_C$$

At $f = 100$ Hz: $X_C = \dfrac{1}{2\pi fC} = \dfrac{1}{2\pi(100 \text{ Hz})(10 \ \mu\text{F})} = 159.16 \ \Omega$

and
$$\mathbf{Z}_2 = 1 \text{ k}\Omega - j \ 159.16$$

Current divider rule:

$$\mathbf{I}_L = \frac{\mathbf{Z}_1 \mathbf{I}}{\mathbf{Z}_1 + \mathbf{Z}_2} = \frac{(3.096 \text{ k}\Omega \ \angle 0°)(4 \text{ mA} \ \angle 0°)}{3.096 \text{ k}\Omega + 1 \text{ k}\Omega - j \ 159.16 \ \Omega}$$

$$= \frac{12.384 \text{ A} \ \angle 0°}{4096 - j \ 159.16} = \frac{12.384 \text{ A} \ \angle 0°}{4099 \ \angle -2.225°} = 3.021 \text{ mA} \ \angle 2.225°$$

and
$$\mathbf{V}_L = \mathbf{I}_L \mathbf{Z}_R$$

$$= (3.021 \text{ mA} \ \angle 2.225°)(1 \text{ k}\Omega \ \angle 0°)$$

$$= \mathbf{3.021 \text{ V} \ \angle 2.225°}$$

b. At $f = 20$ kHz: $X_C = \dfrac{1}{2\pi fC} = \dfrac{1}{2\pi(20 \text{ kHz})(10 \ \mu\text{F})} = 0.796 \ \Omega$

Note the dramatic change in X_C with frequency. Obviously, the higher the frequency, the better the short-circuit approximation for X_C for ac conditions.

$$\mathbf{Z}_2 = 1 \text{ k}\Omega - j \ 0.796 \ \Omega$$

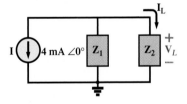

FIG. 15.11
Network of Fig. 15.10 following the assignment of the block impedances.

Current divider rule:

$$I_L = \frac{\mathbf{Z}_1\mathbf{I}}{\mathbf{Z}_1 + \mathbf{Z}_2} = \frac{(3.096 \text{ k}\Omega \angle 0°)(4 \text{ mA} \angle 0°)}{3.096 \text{ k}\Omega + 1 \text{ k}\Omega - j\,0.796 \,\Omega}$$

$$= \frac{12.384 \text{ A} \angle 0°}{4096 - j\,0.796 \,\Omega} = \frac{12.384 \text{ A} \angle 0°}{4096 \angle -0.011°}$$

$$= 3.023 \text{ mA} \angle 0.011°$$

and

$$\mathbf{V}_L = \mathbf{I}_L\mathbf{Z}_R$$

$$= (3.023 \text{ mA} \angle 0.011°)(1 \text{ k}\Omega \angle 0°)$$

$$= \mathbf{3.023 \text{ V} \angle 0.011°}$$

c. The results clearly indicate that the capacitor had little effect on the frequencies of interest. In addition, note that most of the supply current reached the load for the typical parameters employed.

EXAMPLE 15.6 For the network of Fig. 15.12:

a. Determine the current **I**.
b. Find the voltage **V**.

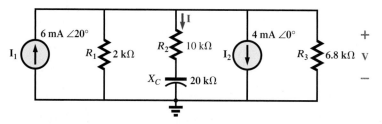

FIG. 15.12
Network for Example 15.6.

Solutions:

a. The rules for parallel current sources are the same for dc and ac networks. That is, the equivalent current source is their sum or difference (as phasors). Therefore,

$$\mathbf{I}_T = 6 \text{ mA} \angle 20° - 4 \text{ mA} \angle 0° = 5.638 \text{ mA} + j\,2.052 \text{ mA} - 4 \text{ mA}$$

$$= 1.638 \text{ mA} + j\,2.052 \text{ mA} = 2.626 \text{ mA} \angle 51.402°$$

Redrawing the network using block impedances will result in the network of Fig. 15.13, where

$$\mathbf{Z}_1 = 2 \text{ k}\Omega \angle 0° \,\|\, 6.8 \text{ k}\Omega \angle 0° = 1.545 \text{ k}\Omega \angle 0°$$

and

$$\mathbf{Z}_2 = 10 \text{ k}\Omega - j\,20 \text{ k}\Omega = 22.361 \text{ k}\Omega \angle -63.435°$$

Note that **I** and **V** are still defined in Fig. 15.13.

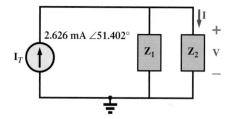

FIG. 15.13
Network of Fig. 15.12 following the assignment of the subscripted impedances.

Current divider rule:

$$\mathbf{I} = \frac{\mathbf{Z}_1\mathbf{I}_T}{\mathbf{Z}_1 + \mathbf{Z}_2} = \frac{(1.545 \text{ k}\Omega \angle 0°)(2.626 \text{ mA} \angle 51.402°)}{1.545 \text{ k}\Omega - 10 \text{ k}\Omega - j\,20 \text{ k}\Omega}$$

$$= \frac{4.057 \text{ A} \angle 51.402°}{11.545 \times 10^3 - j\,20 \times 10^3} = \frac{4.057 \text{ A} \angle 51.402°}{23.093 \times 10^3 \angle -60.004°}$$

$$= \mathbf{0.176 \text{ mA} \angle 111.406°}$$

b. $\mathbf{V} = \mathbf{I}\mathbf{Z}_2 = (0.176 \text{ mA } \angle 111.406°)(22.36 \text{ k}\Omega \angle -63.435°) = \mathbf{3.936 \text{ V }} \angle 47.971°$

EXAMPLE 15.7   For the network of Fig. 15.14:

a. Compute $\mathbf{I}$.
b. Find $\mathbf{I}_1$, $\mathbf{I}_2$, and $\mathbf{I}_3$.
c. Verify Kirchhoff's current law by showing that
$$\mathbf{I} = \mathbf{I}_1 + \mathbf{I}_2 + \mathbf{I}_3$$
d. Find the total impedance of the circuit.

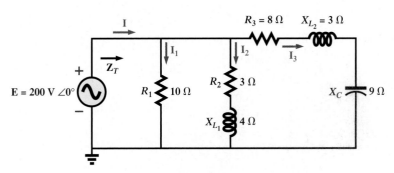

FIG. 15.14
Network for Example 15.7.

Solutions:

a. Redrawing the circuit as in Fig. 15.15 reveals a strictly parallel network where

$$\mathbf{Z}_1 = R_1 = 10 \ \Omega \ \angle 0°$$
$$\mathbf{Z}_2 = R_2 + j X_{L_1} = 3 \ \Omega + j \ 4 \ \Omega$$
$$\mathbf{Z}_3 = R_3 + j X_{L_2} - j X_C = 8 \ \Omega + j \ 3 \ \Omega - j \ 9 \ \Omega = 8 \ \Omega - j \ 6 \ \Omega$$

The total admittance is

$$\mathbf{Y}_T = \mathbf{Y}_1 + \mathbf{Y}_2 + \mathbf{Y}_3$$

$$= \frac{1}{\mathbf{Z}_1} + \frac{1}{\mathbf{Z}_2} + \frac{1}{\mathbf{Z}_3} = \frac{1}{10 \ \Omega} + \frac{1}{3 \ \Omega + j \ 4 \ \Omega} + \frac{1}{8 \ \Omega - j \ 6 \ \Omega}$$

$$= 0.1 \text{ S} + \frac{1}{5 \ \Omega \ \angle 53.13°} + \frac{1}{10 \ \Omega \ \angle -36.87°}$$

$$= 0.1 \text{ S} + 0.2 \text{ S} \ \angle -53.13° + 0.1 \text{ S} \ \angle 36.87°$$

$$= 0.1 \text{ S} + 0.12 \text{ S} - j \ 0.16 \text{ S} + 0.08 \text{ S} + j \ 0.06 \text{ S}$$

$$= 0.3 \text{ S} - j \ 0.1 \text{ S} = 0.316 \text{ S} \ \angle -18.435°$$

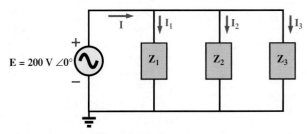

FIG. 15.15
Network of Fig. 15.14 following the assignment of the subscripted impedances.

Calculator Solution: The above mathematical exercise presents an excellent opportunity to demonstrate the power of some of today's calculators. Using the TI-86, the above operation would appear as follows on the display:

$$1/(10,0) + 1/(3,4) + 1/(8, -6)$$

with the result:

$$(300.000E{-}3, -100.000E{-}3)$$

Converting to polar form:

Ans ▶ Pol

$$(316.228E{-}3 \angle{-}18.435E0)$$

The current **I**:

$$\mathbf{I} = \mathbf{E}\mathbf{Y}_T = (200 \text{ V} \angle 0°)(0.316 \text{ S} \angle{-}18.435°)$$
$$= \mathbf{63.2 \text{ A}} \angle{-}\mathbf{18.435°}$$

b. Since the voltage is the same across parallel branches,

$$\mathbf{I}_1 = \frac{\mathbf{E}}{\mathbf{Z}_1} = \frac{200 \text{ V} \angle 0°}{10 \text{ Ω} \angle 0°} = \mathbf{20 \text{ A}} \angle \mathbf{0°}$$

$$\mathbf{I}_2 = \frac{\mathbf{E}}{\mathbf{Z}_2} = \frac{200 \text{ V} \angle 0°}{5 \text{ Ω} \angle 53.13°} = \mathbf{40 \text{ A}} \angle{-}\mathbf{53.13°}$$

$$\mathbf{I}_3 = \frac{\mathbf{E}}{\mathbf{Z}_3} = \frac{200 \text{ V} \angle 0°}{10 \text{ Ω} \angle{-}36.87°} = \mathbf{20 \text{ A}} \angle{+}\mathbf{36.87°}$$

c. $\mathbf{I} = \mathbf{I}_1 + \mathbf{I}_2 + \mathbf{I}_3$

$$60 - j\,20 = 20 \angle 0° + 40 \angle{-}53.13° + 20 \angle{+}36.87°$$
$$= (20 + j\,0) + (24 - j\,32) + (16 + j\,12)$$
$$\mathbf{60 - j\,20} = \mathbf{60 - j\,20} \quad \text{(checks)}$$

d. $\mathbf{Z}_T = \dfrac{1}{\mathbf{Y}_T} = \dfrac{1}{0.316 \text{ S} \angle{-}18.435°} = \mathbf{3.165 \text{ Ω}} \angle\mathbf{18.435°}$

EXAMPLE 15.8   For the network of Fig. 15.16:

a. Calculate the total impedance $\mathbf{Z}_T$.
b. Compute **I**.
c. Find the total power factor.
d. Calculate $\mathbf{I}_1$ and $\mathbf{I}_2$.
e. Find the average power delivered to the circuit.

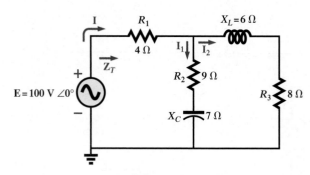

FIG. 15.16

Network for Example 15.8.

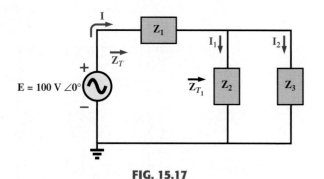

FIG. 15.17

*Network of Fig. 15.16 following the assignment of the
subscripted impedances.*

Solutions:

a. Redrawing the circuit as in Fig. 15.17, we have

$$\mathbf{Z}_1 = R_1 = 4 \text{ Ω} \angle 0°$$

$$\mathbf{Z}_2 = R_2 - j\,X_C = 9 \text{ Ω} - j\,7 \text{ Ω} = 11.40 \text{ Ω} \angle{-}37.87°$$

$$\mathbf{Z}_3 = R_3 + j\,X_L = 8 \text{ Ω} + j\,6 \text{ Ω} = 10 \text{ Ω} \angle{+}36.87°$$

Notice that all the desired quantities were conserved in the redrawn network. The total impedance:

$$\mathbf{Z}_T = \mathbf{Z}_1 + \mathbf{Z}_{T_1} = \mathbf{Z}_1 + \frac{\mathbf{Z}_2\mathbf{Z}_3}{\mathbf{Z}_2 + \mathbf{Z}_3}$$

$$= 4\,\Omega + \frac{(11.4\,\Omega\,\angle-37.87°)(10\,\Omega\,\angle36.87°)}{(9\,\Omega - j\,7\,\Omega) + (8\,\Omega + j\,6\,\Omega)}$$

$$= 4\,\Omega + \frac{114\,\Omega\,\angle-1.00°}{17.03\,\angle-3.37°} = 4\,\Omega + 6.69\,\Omega\,\angle2.37°$$

$$= 4\,\Omega + 6.68\,\Omega + j\,0.28\,\Omega = 10.68\,\Omega + j\,0.28\,\Omega$$

$$= \mathbf{10.684\,\Omega\,\angle1.5°}$$

Mathcad Solution: The complex algebra just presented in detail provides an excellent opportunity to practice our Mathcad skills with complex numbers. Remember that the j must follow the numerical value of the imaginary part and **is not** multiplied by the numerical value. Simply type in the numerical value and then j. Also recall that unless you make a global change in the format, an i will appear with the imaginary part of the solution. As shown in Fig. 15.18, each impedance is first defined with **Shift:**. Then each impedance is entered in sequence on the same line or succeeding lines.

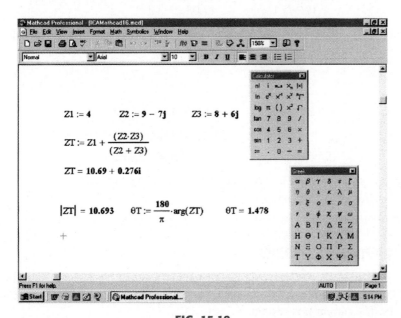

FIG. 15.18
Using Mathcad to determine the total impedance for the network of Fig. 15.16.

Next, the equation for the total impedance is defined using the brackets to ensure that the bottom summation is carried out before the division and also to provide the same format to the equation as appearing above. Then enter **ZT**, select the equals sign key, and the rectangular form for the total impedance will appear as shown.

The polar form can be obtained by first going to the **Calculator** toolbar to obtain the magnitude operation and inserting **ZT** as shown in Fig. 15.18. Then selecting the equals sign will result in the magnitude of 10.693 Ω. The angle is obtained by first going to the **Greek** toolbar and picking up theta, entering **T**, and defining the variable. The π comes from the **Calculator** toolbar, and the **arg( )** from **Insert-*f(x)*- Function Name-arg.** Finally the variable is written again and the equals sign selected to obtain an angle of 1.478°. The computer solution of 10.693 Ω ∠1.478° is an excellent verification of the theoretical solution of 10.684 Ω ∠1.5°.

Calculator Solution: Another opportunity to demonstrate the versatility of the calculator! For the above operation, however, you must be aware of the priority of the mathematical operations, as demonstrated in the calculator display below. In most cases, the operations are performed in the same order they would be performed longhand.

$$(4,0)+((9,-7)+(8,6))^{-1}*(11.4 \angle -37.87)(10 \angle 36.87) \boxed{\text{ENTER}}$$
$$(10.689\text{E}0,276.413\text{E}-3)$$
$$\text{Ans} \blacktriangleright \text{Pol} \boxed{\text{ENTER}}$$
$$(10.692\text{E}0 \angle 1.481\text{E}0)$$

b. $\mathbf{I} = \dfrac{\mathbf{E}}{\mathbf{Z}_T} = \dfrac{100 \text{ V} \angle 0°}{10.684 \text{ }\Omega \angle 1.5°} = \mathbf{9.36 \text{ A} \angle -1.5°}$

c. $F_p = \cos \theta_T = \dfrac{R}{Z_T} = \dfrac{10.68 \text{ }\Omega}{10.684 \text{ }\Omega} \cong \mathbf{1}$

(essentially resistive, which is interesting, considering the complexity of the network)

d. Current divider rule:

$$\mathbf{I}_2 = \frac{\mathbf{Z}_2\mathbf{I}}{\mathbf{Z}_2 + \mathbf{Z}_3} = \frac{(11.40 \text{ }\Omega \angle -37.87°)(9.36 \text{ A} \angle -1.5°)}{(9 \text{ }\Omega - j\,7 \text{ }\Omega) + (8 \text{ }\Omega + j\,6 \text{ }\Omega)}$$

$$= \frac{106.7 \text{ A} \angle -39.37°}{17 - j\,1} = \frac{106.7 \text{ A} \angle -39.37°}{17.03 \angle -3.37°} = \mathbf{6.27 \text{ A} \angle -36°}$$

Applying Kirchhoff's current law (rather than another application of the current divider rule) yields

$$\mathbf{I}_1 = \mathbf{I} - \mathbf{I}_2$$

or $\qquad \mathbf{I} = \mathbf{I}_1 - \mathbf{I}_2$

$$= (9.36 \text{ A} \angle -1.5°) - (6.27 \text{ A} \angle -36°)$$
$$= (9.36 \text{ A} - j\,0.25 \text{ A}) - (5.07 \text{ A} - j\,3.69 \text{ A})$$
$$\mathbf{I}_1 = 4.29 \text{ A} + j\,3.44 \text{ A} = \mathbf{5.5 \text{ A} \angle 38.72°}$$

e. $P_T = EI \cos \theta_T = (100 \text{ V})(9.36 \text{ A}) \cos 1.5°$
$\qquad = (936)(0.99966) = \mathbf{935.68 \text{ W}}$

15.3 LADDER NETWORKS

Ladder networks were discussed in some detail in Chapter 7. This section will simply apply the first method described in Section 7.7 to the general sinusoidal ac ladder network of Fig. 15.19. The current $\mathbf{I}_6$ is desired.

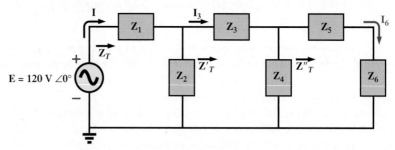

FIG. 15.19

Defining an approach to the analysis of ladder networks.

Impedances $\mathbf{Z}_T$, $\mathbf{Z}'_T$, and $\mathbf{Z}''_T$ and currents $\mathbf{I}_1$ and $\mathbf{I}_3$ are defined in Fig. 15.19:

$$\mathbf{Z}''_T = \mathbf{Z}_5 + \mathbf{Z}_6$$

and $\qquad \mathbf{Z}'_T = \mathbf{Z}_3 + \mathbf{Z}_4 \| \mathbf{Z}''_T$

with $\qquad \mathbf{Z}_T = \mathbf{Z}_1 + \mathbf{Z}_2 \| \mathbf{Z}'_T$

Then
$$\mathbf{I} = \frac{\mathbf{E}}{\mathbf{Z}_T}$$

and
$$\mathbf{I}_3 = \frac{\mathbf{Z}_2\mathbf{I}}{\mathbf{Z}_2 + \mathbf{Z}'_T}$$

with
$$\mathbf{I}_6 = \frac{\mathbf{Z}_4\mathbf{I}_3}{\mathbf{Z}_4 + \mathbf{Z}''_T}$$

15.4 THE POWER TRIANGLE

It was pointed out in Chapter 13 that the power delivered to a resistive element is dissipated and power delivered to inductive or capacitive elements is simply stored and returned to the network when called for by the network design. This storage of energy by the reactive components will require an increase in current demand from the supply that will cause increased expense for a power plant. The result is that the electric bill for an industrial complex is sensitive not only to the power dissipated. It also is sensitive to the effect of the reactive elements.

For the series R-L circuit of Fig. 15.20(a), the impedance diagram appears in Fig. 15.20(b). The total impedance is $\mathbf{Z}_T = 50\ \Omega\ \angle 53.13°$, with a resulting source current of

$$\mathbf{I}_s = \frac{\mathbf{E}}{\mathbf{Z}_T} = 2.4\ A\ \angle -53.13°$$

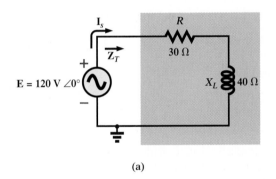
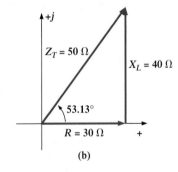

(a)   (b)

FIG. 15.20

Series R-L circuit: (a) circuit diagram; (b) impedance diagram.

The power to the resistive element is

$$P_R = I_R^2 R = (2.4\ A)^2(30\ \Omega) = 172.8\ W$$

while the reactive power of the inductive element is

$$Q_L = I_L^2 X_L = (2.4\ A)^2(40\ \Omega) = 230.4\ VAR$$

If we multiply each component of the impedance diagram by the current squared, as shown in Fig. 15.21, the equation for the power associated with each element will result with the values shown. In particular, note that the hypotenuse of the vector diagram is determined

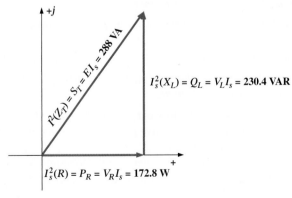

FIG. 15.21

Making the transition from an impedance diagram to a power diagram.

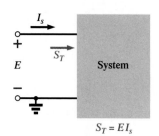

FIG. 15.22

*Defining the apparent power level
delivered to a system.*

solely by the product of the applied voltage and the resulting current. This product, as defined by the terminal values of Fig. 15.22, is given the symbol S and the units volt-ampere (VA) and is called the **apparent power** because initially it would seem "apparent" that the power delivered to the system is determined by the product of the applied voltage and the resulting source current.

In summary:

$$S = EI_s \quad \text{(volt-ampere, VA)} \tag{15.1}$$

However, since

$$I_s = \frac{E}{Z_T} \quad \text{and} \quad E = I_s Z_T$$

$$S = I_s^2 Z_T \quad \text{(VA)} \tag{15.2}$$

and

$$S = \frac{E^2}{Z_T} \quad \text{(VA)} \tag{15.3}$$

Further, since the average or real power to a load is defined by

$$P = EI_s \cos \theta_T$$

and

$$S = EI_s$$

we have the following helpful relationship between the real and apparent power of a system:

$$P = S \cos \theta_T \tag{15.4}$$

The power factor can then be determined by

$$F_p = \cos \theta_T = \frac{P}{S} \tag{15.5}$$

In total, therefore, if we associate an angle of zero degrees with the real power dissipated by a network, and an angle of 90° with the inductive reactive power, the apparent power can be found using the Pythagorean theorem. For a capacitive load, an angle of −90° would be employed as shown in Fig. 15.23 for all three possible components.

The smaller the reactive component in Fig. 15.21, the closer the apparent power will be to the real power component, or the closer the apparent power will be to the power dissipated. For large industrial plants that pay for the apparent power delivered, the smaller the reactive component, the closer the apparent power will be to the real power, and the smaller the electric bill.

The smaller the angle between the real component and the apparent power, the closer the power factor $F_p = \cos \theta_T$ will be to unity and a purely resistive load. In the next section, a subject called **power-factor correction** will be discussed as a means to reducing the utility bills to large consumers.

One of the most interesting things about electric circuits is the fact that

no matter how complex a network may be, the total dissipated power is simply the sum of the powers delivered to each resistive element.

In other words, **there is no need to worry about how the resistors are connected in the network.** Simply find the power to each and add them up.

Similarly, for the total reactive power,

the total reactive power of any complex configuration is simply the algebraic sum of the reactive powers to all the reactive elements.

In other words, **there is again no need to worry about how the elements are connected in the network.** Simply add up all the reactive power levels to the inductive elements and subtract those from the capacitive elements.

The total apparent power will always be related to the total resistive and reactive power levels through the right triangle of Fig. 15.24 and the Pythagorean theorem:

$$S_T^2 = P_T^2 + Q_T^2 \tag{15.6}$$

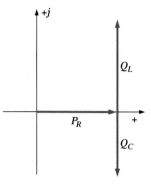

FIG. 15.23

*Displaying the real and reactive
power of a network.*

or
$$S_T = \sqrt{P_T^2 + Q_T^2} \qquad \textbf{(15.7)}$$

The total number of watts, volt-ampere reactive, and volt-ampere, and the power factor of any system, can be found using the following procedure:

1. *Find the real power and reactive power for each branch of the circuit.*
2. *The total real power of the system (P_T) is then the sum of the average power delivered to each branch.*
3. *The total reactive power (Q_T) is the difference between the reactive power of the inductive loads and that of the capacitive loads.*
4. *The total apparent power is $S_T = \sqrt{P_T^2 + Q_T^2}$.*
5. *The total power factor is P_T/S_T.*

There are two important points in the above tabulation. First, the total apparent power must be determined from the total average and reactive powers and *cannot* be determined from the apparent powers of each branch. Second, and more important, it is *not necessary* to consider the series-parallel arrangement of branches. In other words, the total real, reactive, or apparent power is independent of whether the loads are in series, parallel, or series-parallel. The following examples will demonstrate the relative ease with which all of the quantities of interest can be found.

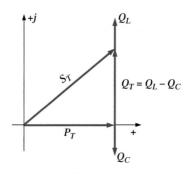

FIG. 15.24
Displaying the components of a power triangle.

EXAMPLE 15.9 Find the total number of watts, volt-ampere reactive, and volt-ampere, and the power factor F_p, of the network in Fig. 15.25. Draw the power triangle and find the current in phasor form.

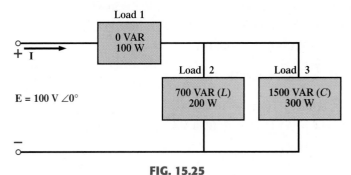

FIG. 15.25
Network for Example 15.9.

Solution: Construct a table such as shown in Table 15.1.

TABLE 15.1

Load	W	VAR	VA
1	100	0	100
2	200	700 (L)	$\sqrt{(200)^2 + (700)^2}$ = 728.0
3	300	1500 (C)	$\sqrt{(300)^2 + (1500)^2}$ = 1529.71
	$P_T = \textbf{600}$ Total power dissipated	$Q_T = \textbf{800}$ (C) Resultant reactive power of network	$S_T = \sqrt{(600)^2 + (800)^2} = \textbf{1000}$ (Note that $S_T \neq$ sum of each branch: $1000 \neq 100 + 728 + 1529.71$)

Thus,
$$F_p = \frac{P_T}{S_T} = \frac{600 \text{ W}}{1000 \text{ VA}} = \textbf{0.6 leading } (\textbf{C})$$

The power triangle is shown in Fig. 15.26.

Since $S_T = VI = 1000$ VA, $I = 1000$ VA/100 V = 10 A; and since θ of $\cos \theta = F_p$ is the angle between the input voltage and current:

$$\textbf{I} = \textbf{10 A} \angle \textbf{+53.13°}$$

The plus sign is associated with the phase angle since the circuit is predominantly capacitive.

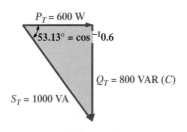

FIG. 15.26
Power triangle for Example 15.9.

EXAMPLE 15.10

a. Find the total number of watts, volt-ampere reactive, and volt-ampere, and the power factor F_p, for the network of Fig. 15.27.
b. Sketch the power triangle.

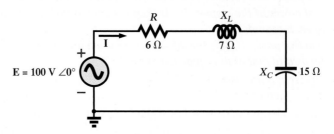

FIG. 15.27

Network for Example 15.10.

Solutions:

a. $\quad \mathbf{I} = \dfrac{\mathbf{E}}{\mathbf{Z}_T} = \dfrac{100 \text{ V } \angle 0°}{6 \text{ }\Omega + j \, 7 \text{ }\Omega - j \, 15 \text{ }\Omega} = \dfrac{100 \text{ V } \angle 0°}{10 \text{ }\Omega \angle -53.13°}$

$\quad = 10 \text{ A } \angle 53.13°$

$\mathbf{V}_R = (10 \text{ A } \angle 53.13°)(6 \text{ }\Omega \angle 0°) = 60 \text{ V } \angle 53.13°$

$\mathbf{V}_L = (10 \text{ A } \angle 53.13°)(7 \text{ }\Omega \angle 90°) = 70 \text{ V } \angle 143.13°$

$\mathbf{V}_C = (10 \text{ A } \angle 53.13°)(15 \text{ }\Omega \angle -90°) = 150 \text{ V } \angle -36.87°$

$P_T = EI \cos \theta = (100 \text{ V})(10 \text{ A}) \cos 53.13° = \mathbf{600 \ W}$

$\quad = I^2 R = (10 \text{ A})^2 (6 \text{ }\Omega) = \mathbf{600 \ W}$

$\quad = \dfrac{V_R^2}{R} = \dfrac{(60 \text{ V})^2}{6} = \mathbf{600 \ W}$

$S_T = EI = (100 \text{ V})(10 \text{ A}) = \mathbf{1000 \ VA}$

$\quad = I^2 Z_T = (10 \text{ A})^2 (10 \text{ }\Omega) = \mathbf{1000 \ VA}$

$\quad = \dfrac{E^2}{Z_T} = \dfrac{(100 \text{ V})^2}{10 \text{ }\Omega} = \mathbf{1000 \ VA}$

$Q_T = EI \sin \theta = (100 \text{ V})(10 \text{ A}) \sin 53.13° = \mathbf{800 \ VAR}$

$\quad = Q_C - Q_L$

$\quad = I^2 (X_C - X_L) = (10 \text{ A})^2 (15 \text{ }\Omega - 7 \text{ }\Omega) = \mathbf{800 \ VAR}$

$\quad = \dfrac{V_C^2}{X_C} - \dfrac{V_L^2}{X_L} = \dfrac{(150 \text{ V})^2}{15 \text{ }\Omega} - \dfrac{(70 \text{ V})^2}{7 \text{ }\Omega}$

$\quad = 1500 \text{ VAR} - 700 \text{ VAR} = \mathbf{800 \ VAR}$

$E_p = \dfrac{P_T}{S_T} = \dfrac{600 \text{ W}}{1000 \text{ VA}} = \mathbf{0.6 \ leading} \ (C)$

b. The power triangle is as shown in Fig. 15.28.

FIG. 15.28

Power triangle for Example 15.10.

EXAMPLE 15.11 For the system of Fig. 15.29:

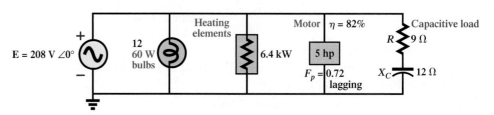

FIG. 15.29

Example 15.11.

a. Find the average power, apparent power, reactive power, and F_p for each branch.

b. Find the total number of watts, volt-ampere reactive, and volt-ampere, and the power factor F_p, of the system. Sketch the power triangle.
c. Find the source current I.

Solutions:

a. *Bulbs:*

Total dissipation of applied power

$$P_1 = 12(60 \text{ W}) = \textbf{720 W}$$
$$Q_1 = \textbf{0 VAR}$$
$$S_1 = P_1 = \textbf{720 VA}$$
$$F_{p1} = \textbf{1}$$

Heating elements:

Total dissipation of applied power

$$P_2 = \textbf{6.4 kW}$$
$$Q_2 = \textbf{0 VAR}$$
$$S_2 = P_2 = \textbf{6.4 kVA}$$
$$F_{p2} = \textbf{1}$$

Motor:

$$\eta = \frac{P_o}{P_i} \rightarrow P_i = \frac{P_o}{\eta} = \frac{5(746 \text{ W})}{0.82} = \textbf{4548.78 W} = P_3$$

$$F_p = \textbf{0.72 lagging}$$

$$P_3 = S_3 \cos \theta \rightarrow S_3 = \frac{P_3}{\cos \theta} = \frac{4548.78 \text{ W}}{0.72} = \textbf{6317.75 VA}$$

Also, $\theta = \cos^{-1} 0.72 = 43.95°$, so that

$$Q_3 = S_3 \sin \theta = (6317.75 \text{ VA})(\sin 43.95°)$$
$$= (6317.75 \text{ VA})(0.694) = \textbf{4384.71 VAR } (\textbf{\textit{L}})$$

Capacitive load:

$$\mathbf{I} = \frac{\mathbf{E}}{\mathbf{Z}} = \frac{208 \text{ V } \angle 0°}{9 \ \Omega - j \ 12 \ \Omega} = \frac{208 \text{ V } \angle 0°}{15 \ \Omega \ \angle -53.13°} = 13.87 \text{ A } \angle 53.13°$$

$$P_4 = I^2 R = (13.87 \text{ A})^2 \cdot 9 \ \Omega = \textbf{1731.39 W}$$

$$Q_4 = I^2 X_C = (13.87 \text{ A})^2 \cdot 12 \ \Omega = \textbf{2308.52 VAR } (\textbf{\textit{C}})$$

$$S_4 = \sqrt{P_4^2 + Q_4^2} = \sqrt{(1731.39 \text{ W})^2 + (2308.52 \text{ VAR})^2}$$
$$= \textbf{2885.65 VA}$$

$$F_p = \frac{P_4}{S_4} = \frac{1731.39 \text{ W}}{2885.65 \text{ VA}} = \textbf{0.6 leading}$$

b. $P_T = P_1 + P_2 + P_3 + P_4$

$$= 720 \text{ W} + 6400 \text{ W} + 4548.78 \text{ W} + 1731.39 \text{ W}$$
$$= \textbf{13,400.17 W}$$

$Q_T = \pm Q_1 \pm Q_2 \pm Q_3 \pm Q_4$

$$= 0 + 0 + 4384.71 \text{ VAR } (L) - 2308.52 \text{ VAR } (C)$$
$$= \textbf{2076.19 VAR } (\textbf{\textit{L}})$$

$S_T = \sqrt{P_T^2 + Q_T^2} = \sqrt{(13{,}400.17 \text{ W})^2 + (2076.19 \text{ VAR})^2}$

$$= \textbf{13,560.06 VA}$$

$$F_p = \frac{P_T}{S_T} = \frac{13.4 \text{ kW}}{13{,}560.06 \text{ VA}} = \textbf{0.988 lagging}$$

$$\theta = \cos^{-1} 0.988 = 8.89°$$

Note Fig. 15.30.

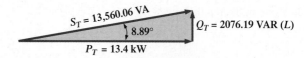

FIG. 15.30
Power triangle for Example 15.11.

c. $S_T = EI \rightarrow I = \dfrac{S_T}{E} = \dfrac{13{,}559.89 \text{ VA}}{208 \text{ V}} = 65.19 \text{ A}$

Lagging power factor: **E** leads **I** by 8.89°, and

$$\mathbf{I} = \mathbf{65.19 \text{ A}} \angle -\mathbf{8.89°}$$

EXAMPLE 15.12 An electrical device is rated 5 kVA, 100 V at a 0.6 power-factor lag. What is the impedance of the device in rectangular coordinates?

Solution:

$$S = EI = 5000 \text{ VA}$$

Therefore,

$$I = \frac{5000 \text{ VA}}{100 \text{ V}} = 50 \text{ A}$$

For $F_p = 0.6$, we have

$$\theta = \cos^{-1} 0.6 = 53.13°$$

Since the power factor is lagging, the circuit is predominantly inductive, and **I** lags **E**. Or, for $\mathbf{E} = 100$ V $\angle 0°$,

$$\mathbf{I} = 50 \text{ A} \angle -53.13°$$

However,

$$\mathbf{Z}_T = \frac{\mathbf{E}}{\mathbf{I}} = \frac{100 \text{ V} \angle 0°}{50 \text{ A} \angle -53.13°} = 2 \text{ } \Omega \angle 53.13° = \mathbf{1.2 \text{ } \Omega + j \text{ } 1.6 \text{ } \Omega}$$

which is the impedance of the circuit of Fig. 15.31.

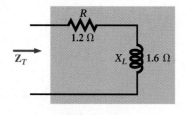

FIG. 15.31
Example 15.12.

15.5 POWER-FACTOR CORRECTION

The design of any power transmission system is very sensitive to the magnitude of the current in the lines as determined by the applied loads. Increased currents result in increased power losses (by a squared factor since $P = I^2 R$) in the transmission lines due to the resistance of the lines. Heavier currents also require larger conductors, increasing the amount of copper needed for the system, and, quite obviously, they require increased generating capacities by the utility company.

Every effort must therefore be made to keep current levels at a minimum. Since the line voltage of a transmission system is fixed, the apparent power is directly related to the current level. In turn, the smaller the net apparent power, the smaller the current drawn from the supply. Minimum current is therefore drawn from a supply when $S = P$ and $Q_T = 0$. Note the effect of decreasing levels of Q_T on the length (and magnitude) of S in Fig. 15.32 for the same real power. Note also that the power-factor angle approaches zero degrees and F_p approaches 1, revealing that the network is appearing more and more resistive at the input terminals.

The process of introducing reactive elements to bring the power factor closer to unity is called **power-factor correction.** Since most loads are inductive, the process normally involves introducing elements with capacitive terminal characteristics having the sole purpose of improving the power factor.

In Fig. 15.33(a), for instance, an inductive load is drawing a current I_L that has a real and an imaginary component. In Fig. 15.33(b), a capacitive load was added in parallel with the original load to raise the power factor of the total system to the unity power-factor level.

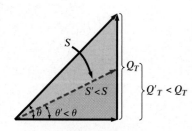

FIG. 15.32
Demonstrating the impact of power-factor correction on the power triangle of a network.

Note that by placing all the elements in parallel, the load still receives the same terminal voltage and draws the same current I_L. In other words, the load is unaware of and unconcerned about whether it is hooked up as shown in Fig. 15.33(a) or Fig. 15.33(b).

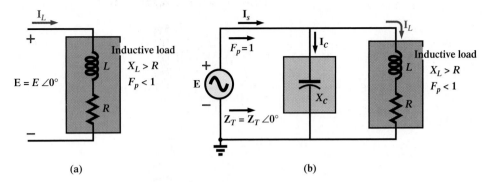

FIG. 15.33

Demonstrating the impact of a capacitive element on the power factor of a network.

Solving for the source current in Fig. 15.33(b):

$$\mathbf{I}_s = \mathbf{I}_C + \mathbf{I}_L$$
$$= j\,I_C + I_L(\text{real component}) + j\,I_L(\text{imaginary component})$$
$$= I_L(\text{real component}) + j[I_C + I_L\,(\text{imaginary component})]$$

If X_C is chosen such that $I_C = I_L(\text{imaginary component})$, then

$$\mathbf{I}_s = I_L(\text{real component}) + j\,(0) = I_L(\text{real component})\,\angle 0°$$

The result is a source current whose magnitude is simply equal to the real part of the load current, which can be considerably less than the magnitude of the load current of Fig. 15.33(a). In addition, since the phase angle associated with both the applied voltage and the source current is the same, the system appears "resistive" at the input terminals, and all of the power supplied is absorbed, creating maximum efficiency for a generating utility.

EXAMPLE 15.13   A 5 hp motor with a 0.6 lagging power factor and an efficiency of 92% is connected to a 208 V, 60 Hz supply.

a. Establish the power triangle for the load.
b. Determine the power-factor capacitor that must be placed in parallel with the load to raise the power factor to unity.
c. Determine the change in supply current from the uncompensated to the compensated system.
d. Find the network equivalent of the above, and verify the conclusions.

Solutions:

a. Since 1 hp = 746 W,

$$P_o = 5 \text{ hp} = 5(746 \text{ W}) = 3730 \text{ W}$$

and $\qquad P_i\,(\text{drawn from the line}) = \dfrac{P_o}{\eta} = \dfrac{3730 \text{ W}}{0.92} = \mathbf{4054.35 \text{ W}}$

Also, $\qquad\qquad\qquad\qquad F_P = \cos\theta = 0.6$

and $\qquad\qquad\qquad\qquad \theta = \cos^{-1} 0.6 = 53.13°$

Applying $\qquad\qquad\qquad\qquad \tan\theta = \dfrac{Q_L}{P_i}$

we obtain $\qquad Q_L = P_i \tan\theta = (4054.35 \text{ W}) \tan 53.13°$

$$= \mathbf{5405.8 \text{ VAR }}(L)$$

and $\qquad S = \sqrt{P_i^2 + Q_L^2} = \sqrt{(4054.35 \text{ W})^2 + (5405.8 \text{ VAR})^2}$

$$= \mathbf{6757.25 \text{ VA}}$$

The power triangle appears in Fig. 15.34.

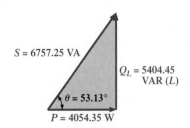

FIG. 15.34

Initial power triangle for the load of Example 15.13.

b. A net unity power-factor level is established by introducing a capacitive reactive power level of 5405.8 VAR to balance Q_L. Since

$$Q_C = \frac{V^2}{X_C}$$

then

$$X_C = \frac{V^2}{Q_C} = \frac{(208 \text{ V})^2}{5405.8 \text{ VAR }(C)} = 8 \text{ }\Omega$$

and

$$C = \frac{1}{2\pi f X_C} = \frac{1}{(2\pi)(60 \text{ Hz})(8 \text{ }\Omega)} = \textbf{331.6 }\mu\textbf{F}$$

c. At $0.6F_p$,

$$S = VI = 6757.25 \text{ VA}$$

and

$$I = \frac{S}{V} = \frac{6757.25 \text{ VA}}{208 \text{ V}} = \textbf{32.49 A}$$

At unity F_p,

$$S = VI = 4054.35 \text{ VA}$$

and

$$I = \frac{S}{V} = \frac{4054.35 \text{ VA}}{208 \text{ V}} = \textbf{19.49 A}$$

producing a 40% reduction in supply current.

d. For the motor, the angle by which the applied voltage leads the current is

$$\theta = \cos^{-1} 0.6 = 53.13°$$

and $P = EI_m \cos \theta = 4054.35$ W, from above, so that

$$I_m = \frac{P}{E \cos \theta} = \frac{4054.35 \text{ W}}{(208 \text{ V})(0.6)} = \textbf{32.49 A} \quad \text{(as above)}$$

resulting in $\quad \textbf{I}_m = 32.49 \text{ A } \angle{-53.13°}$

Therefore,

$$\textbf{Z}_m = \frac{\textbf{E}}{\textbf{I}_m} = \frac{208 \text{ V } \angle 0°}{32.49 \text{ A } \angle{-53.13°}} = 6.4 \text{ }\Omega \angle 53.13° = 3.84 \text{ }\Omega + j \, 5.12 \text{ }\Omega$$

as shown in Fig. 15.35(a).

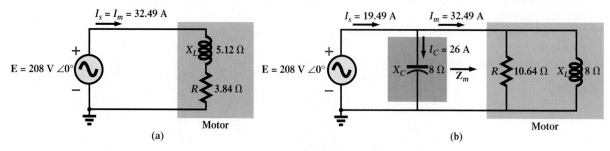

FIG. 15.35

Demonstrating the impact of power-factor corrections on the source current.

The equivalent parallel load is determined from

$$\textbf{Y} = \frac{1}{\textbf{Z}} = \frac{1}{6.4 \text{ }\Omega \angle 53.13°}$$

$$= 0.156 \text{ S } \angle{-53.13°} = 0.094 \text{ S} - j \, 0.125 \text{ S}$$

$$= \frac{1}{10.64 \text{ }\Omega} + \frac{1}{j \, 8 \text{ }\Omega}$$

as shown in Fig. 15.35(b).

It is now clear that the effect of the 8 Ω inductive reactance can be compensated for by a parallel capacitive reactance of 8 Ω using a power-factor correction capacitor of 332 μF.

Since

$$\mathbf{Y}_T = \frac{1}{-j\,X_C} + \frac{1}{R} + \frac{1}{+j\,X_L} = \frac{1}{R}$$

$$I_s = EY_T = E\left(\frac{1}{R}\right) = (208\text{ V})\left(\frac{1}{10.64\ \Omega}\right) = \mathbf{19.54\ A} \qquad \text{(as above)}$$

In addition, the magnitude of the capacitive current can be determined as follows:

$$I_C = \frac{E}{X_C} = \frac{208\text{ V}}{8\ \Omega} = \mathbf{26\ A}$$

15.6 POWER METERS

The digital-display power meter of Fig. 15.36 employs a sophisticated electronic package to sense the voltage and current levels and, through the use of an analog-to-digital conversion unit, display the proper digits on the display. It is capable of providing a digital readout for distorted nonsinusoidal waveforms, and it can provide the phase power, total power, apparent power, reactive power, and power factor. It can also measure currents up to 500 A, voltages up to 600 V, and frequencies from 30 Hz to 1000 Hz.

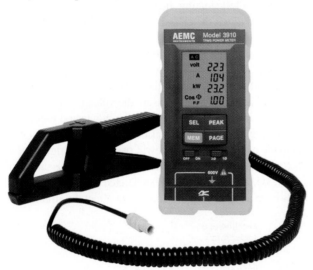

FIG. 15.36
Digital single-phase and three-phase power meter.
(Courtesy of AEMC® Instruments, Foxborough, MA.)

The power quality analyzer of Fig. 15.37 can also display the real, reactive, and apparent power levels along with the power factor. However, it has a broad range of other options, including providing the harmonic content of up to 51 terms for the voltage, current, and power. The power range extends from 250 W to 2.5 MW, and the current can be read up to 1000 A. The meter can also be used to measure resistance levels from 500 Ω to 30 MΩ, capacitance levels from 50 nF to 500 μF, and temperature in both °C and °F.

15.7 GROUNDING

Although usually treated too lightly in most introductory electrical or electronics texts, the impact of the ground connection and how it can provide a measure of safety to a design are very important topics. **Ground potential is 0 V at every point in a network that has a ground symbol.** Since all points are at the same potential, they can all be connected together, but for purposes of clarity most are left isolated on a large schematic. On a schematic the voltage levels provided are always with respect to ground. A system can therefore be checked quite rapidly by simply connecting the black lead of the voltmeter to the ground connection and placing the red lead at the various points where the typical operating voltage is provided. A close match normally implies that that portion of the system is operating properly.

There are various types of grounds whose use depends on the application. An *earth ground* is one that is connected directly to the earth by a low-impedance connection. The entire surface

FIG. 15.37
Power quality analyzer capable of displaying the power in watts, the current in amperes, and the voltage in volts.
(Courtesy of Fluke Corporation. Reproduced with Permission.)

of the earth is defined to have a potential of 0 V. It is the same at every point because there are sufficient conductive agents in the soil such as water and electrolytes to ensure that any difference in voltage on the surface is equalized by a flow of charge between the two points. Every home has an earth ground, usually established by a long conductive rod driven into the ground and connected to the power panel. The electrical code requires a direct connection from earth ground to the cold-water pipes of a home for safety reasons. A "hot" wire touching a cold-water pipe draws sufficient current because of the low-impedance ground connection to throw the breaker. Otherwise, people in the bathroom could pick up the voltage when they touch the cold-water faucet, thereby risking bodily harm. Because water is a conductive agent, any area of the home with water, such as a bathroom or the kitchen, is of particular concern. Most electrical systems are connected to earth ground primarily for safety reasons. All the power lines in a laboratory, at industrial locations, or in the home are connected to earth ground.

A second type is referred to as a *chassis ground*, which may be *floating* or connected directly to an earth ground. A chassis ground simply stipulates that the chassis has a reference potential for all points of the network. If the chassis is not connected to earth potential (0 V), it is said to be *floating* and can have any other reference voltage for the other voltages to be compared to. For instance, if the chassis is sitting at 120 V, all measured voltages of the network will be referenced to this level. A reading of 32 V between a point in the network and the chassis ground will therefore actually be at 152 V with respect to earth potential. Most high-voltage systems are not left floating, however, because of loss of the safety factor. For instance, if someone should touch the chassis and be standing on a suitable ground, the full 120 V would fall across that individual.

Grounding can be particularly important when working with numerous pieces of measuring equipment in the laboratory. For instance, the supply and oscilloscope in Fig. 15.38(a) are each connected directly to an earth ground through the negative terminal of each. If the oscilloscope is connected as shown in Fig. 15.38(a) to measure the voltage V_{R_1}, a dangerous situation will develop. The grounds of each piece of equipment are connected together through the earth ground, and they effectively short out the resistor. Since the resistor is the primary current-controlling element in the network, the current will rise to a very high level and possibly damage the instruments or cause dangerous side effects. In this case, the supply or scope should be used in the floating mode, or the resistors interchanged as shown in Fig. 15.38(b). In Fig. 15.38(b), the grounds have a common point and do not affect the structure of the network.

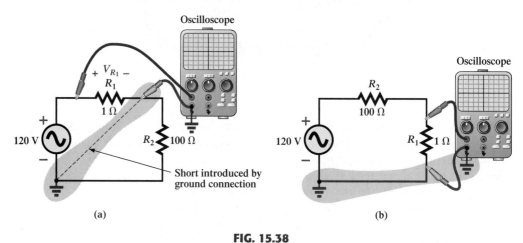

FIG. 15.38

Demonstrating the effect of the oscilloscope ground on the measurement of the voltage across resistor R_1.

The National Electrical Code requires that the "hot" (or *feeder*) line that carries current to a load be *black*, and the line (called the *neutral*) that carries the current back to the supply be *white*. Three-wire conductors have a ground wire that must be *green* or *bare,* which will ensure a common ground but which is not designed to carry current. The components of a three-prong extension cord and wall outlet are shown in Fig. 15.39. Note that on both fixtures, the connection to the hot lead is smaller than the return leg and that the ground connection is partially circular.

The complete wiring diagram for a household outlet is shown in Fig. 15.40. Note that the current through the ground wire is zero and that both the return wire and the ground wire are connected to an earth ground. The full current to the loads flows through the feeder and return lines.

The importance of the ground wire in a three-wire system can be demonstrated by the toaster in Fig. 15.41, rated 1200 W at 120 V. From the power equation $P = EI$, the current

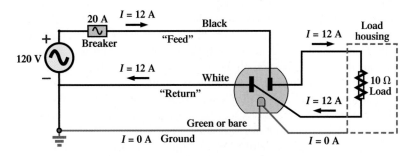

FIG. 15.39

Three-wire conductors: (a) extension cord; (b) home outlet.

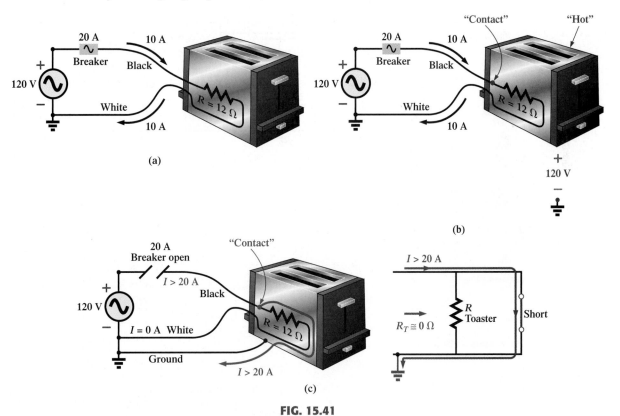

FIG. 15.40

Complete wiring diagram for a household outlet with a 10 Ω load.

FIG. 15.41

Demonstrating the importance of a properly grounded appliance: (a) ungrounded; (b) ungrounded and undesirable contact; (c) grounded appliance with undesirable contact.

drawn under normal operating conditions is $I = P/E = 1200$ W/120 V = 10 A. If a two-wire line were employed as shown in Fig. 15.41(a), the 20 A breaker would be quite comfortable with the 10 A current, and the system would perform normally. However, if abuse to the feeder (or return line) caused it to become frayed and to touch the metal housing of the toaster, the situation depicted in Fig. 15.41(b) would result. The housing would become "hot," yet the breaker would not "pop" because the current would still be the rated 10 A. A dangerous condition would exist because anyone touching the toaster would feel the full 120 V to ground. If the ground wire were attached to the chassis as shown in Fig. 15.41(c), a low-resistance path would be created between the short-circuit point and ground, and the current would jump to very high levels. The breaker would "pop," and the user would be warned that a problem exists.

Although the above discussion does not cover all possible areas of concern with proper grounding or introduce all the nuances associated with the effect of grounds on a system's performance, it should provide an awareness of the importance of understanding its impact.

15.8 APPLICATION

GFI (Ground Fault Interrupter)

The National Electric Code, the "bible" for all electrical contractors, now requires that ground fault interrupter (GFI) outlets be used in any area where water and dampness could result in serious injury, such as in bathrooms, pools, marinas, and so on. The outlet looks like any other except that it has a reset button and a test button in the center of the unit as shown in Fig. 15.42(a). Its primary difference between an ordinary outlet is that it will shut the power off much more quickly than the breaker all the way down in the basement could. You may still feel a shock with a GFI outlet, but the current will cut off so quickly (in a few milliseconds) that a person in normal health should not receive a serious electrical injury. Whenever in doubt about its use, remember that the cost is such that it should be installed. It works just as a regular outlet does, but it provides an increased measure of safety.

The basic operation is best described by the simple network of Fig. 15.42(b). The protection circuit separates the power source from the outlet itself. Note in Fig. 15.42(b) the

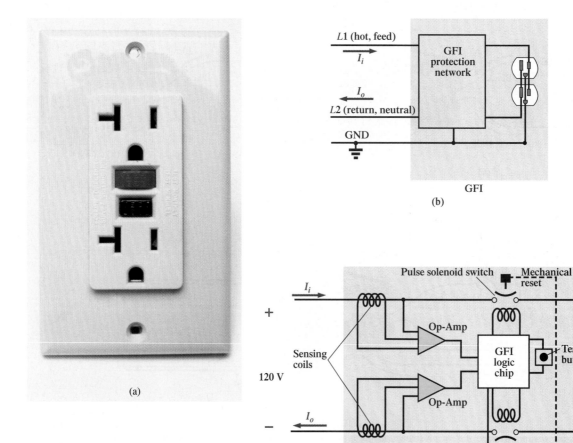

FIG. 15.42
GFI outlet: (a) wall-mounted appearance; (b) basic operation; (c) schematic.

importance of grounding the protection circuit to the central ground of the establishment (a water pipe, ground bar, and so on, connected to the main panel). In general, the outlet will be grounded to the same connection. Basically, the network shown in Fig. 15.42(b) senses both the current entering (I_i) and the current leaving (I_o) and provides a direct connection to the outlet when they are equal. If a fault should develop, such as caused by someone touching the hot leg while standing on a wet floor, the return current will be less than the feed current (just a few milliamperes is enough). The protection circuitry will sense this difference, establish an open circuit in the line, and cut off the power to the outlet.

In Fig. 15.43(a), you can see the feed and return lines passing through the sensing coils. The two sensing coils are separately connected to the printed circuit board. There are two

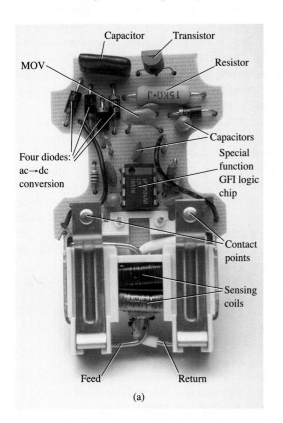

(a)

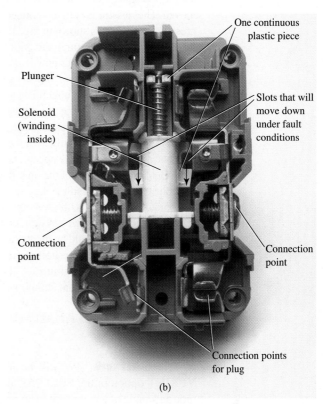

(b)

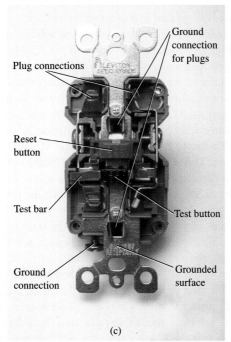

(c)

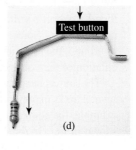

(d)

FIG. 15.43

GFI construction; (a) sensing coils; (b) solenoid control (bottom view); (c) grounding (top view); (d) test bar.

pulse control switches in the line and a return to establish an open circuit under errant conditions. The two contacts in Fig. 15.43(a) are the contacts that provide conduction to the outlet. When a fault develops, another set of similar contacts in the housing will slide away, providing the desired open-circuit condition. The separation is created by the solenoid appearing in Fig. 15.43(b). When the solenoid is energized due to a fault condition, it will pull the plunger toward the solenoid, compressing the spring. At the same time, the slots in the lower plastic piece (connected directly to the plunger) will shift down, causing a disconnect by moving the structure inserted in the slots. The test button is connected to the brass bar across the unit in Fig. 15.43(c) below the reset button. When pressed, it will place a large resistor between the line and ground to "unbalance" the line and cause a fault condition. When the button is released, the resistor will be separated from the line, and the unbalanced condition will be removed. The resistor is actually connected directly to one end of the bar and moves down with pressure on the bar as shown in Fig. 15.43(d). Note in Fig. 15.43(c) that the metal ground connection passes right through the entire unit and that it is connected to the ground terminal of an applied plug. Also note how it is separated from the rest of the network with the plastic housing. Although this unit appears simple on the outside and is relatively small in size, it is beautifully designed and contains a great deal of technology and innovation.

Before leaving the subject, note the logic chip in the center of Fig. 15.43(a) and the various sizes of capacitors and resistors. Note also the four diodes in the upper-left region of the circuit board used as a bridge rectifier for the ac-to-dc conversion process. The transistor is the black element with the half-circle appearance. It is part of the driver circuit for the controlling solenoid. Because of the size of the unit, there wasn't a lot of room to provide the power to quickly open the circuit. The result is the use of a pulse circuit to control the motion of the controlling solenoid. In other words, the solenoid is pulsed for a short period of time to cause the required release. If the design used a system that would hold the circuit open on a continuing basis, the power requirement would be greater and the size of the coil larger. A small coil can handle the required power pulse for a short period of time without any long-term damage.

As mentioned earlier, if unsure, then install a GFI. It provides a measure of safety—at a very reasonable cost—that should not be ignored.

15.9 COMPUTER ANALYSIS

PSpice

ac Bridge Network We will be using Example 15.4 to demonstrate the power of the **VPRINT** option in the **SPECIAL** library. It permits a direct determination of the magnitude and angle of any voltage in an ac network. Similarly, the **IPRINT** option does the same for ac currents. In Example 15.4, the ac voltages across R_1 and R_2 were first determined, and then Kirchhoff's voltage law was applied to determine the voltage between the two known points. Since PSpice is designed primarily to determine the voltage at a point with respect to ground, the network of Fig. 15.7 is entered as shown in Fig. 15.44 to permit a direct calculation of the voltages across R_1 and R_2.

The source and network elements are entered using a procedure that has been demonstrated several times in previous chapters, although for the **AC Sweep** analysis to be performed in this example, the source must carry an **AC** level also. Fortunately, it is the same as **VAMPL** as shown in Fig. 15.44. It is introduced into the source description by double-clicking on the source symbol to obtain the **Property Editor** dialog box. The AC column is selected and the 100 V entered in the box below. Then **Display** is selected and **Name and Value** chosen. Click **OK** followed by **Apply,** and you can exit the dialog box. The result is AC = 100 V added to the source description on the diagram and in the system. Using the reactance values of Fig. 15.7, the values for L and C were determined using a frequency of 1 kHz. The voltage across R_1 and R_2 can be determined using the **Trace** command in the same manner as described in the previous chapter or by using the **VPRINT** option. Both approaches will be discussed in this section because they have application to any ac network.

The **VPRINT** option is under the **SPECIAL** library in the **Place Part** dialog box. Once selected, the printer symbol will appear on the screen next to the cursor, and it can be placed near the point of interest. Once the printer symbol is in place, a double-click on it will result in the **Property Editor** dialog box. Scrolling from left to right, type the word **ok** under **AC, MAG**, and **PHASE**. When each is active, the **Display** key should be selected and the option **Name and Value** chosen followed by **OK.** When all the entries have been made, choose **Apply** and

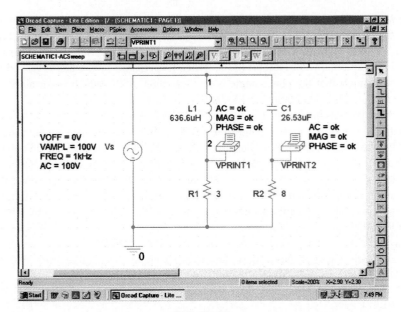

FIG. 15.44

*Determining the voltage across R_1 and R_2 using the **VPRINT** option of a PSpice analysis.*

exit the dialog box. The result appears in Fig. 15.44 for the two applications of the **VPRINT** option. If you prefer, **VPRINT1** and **VPRINT2** can be added to distinguish between the two when you review the output data. This is accomplished by returning to the **Property Editor** dialog box for each by double-clicking on the printer symbol of each and selecting **Value** and then **Display** followed by **Value Only.** We are now ready for the simulation.

The simulation is initiated by selecting the **New Simulation Profile** icon and entering **ACSweep** as the **Name.** Then select **Create** to bring up the **Simulation Settings** dialog box. This time, we want to analyze the network at 1 kHz but are not interested in plots against time. Thus, the **AC Sweep/Noise** option will be selected under **Analysis type** in the **Analysis** section. An **AC Sweep Type** region will then appear in the dialog asking for the **Start Frequency.** Since we are interested in the response at only one frequency, the **Start** and **End Frequency** will be the same: 1 kHz. Since we need only one point of analysis, the **Points/Decade** will be 1. Click **OK,** and the **Run PSpice** icon can be selected. The **SCHEMATIC1** screen will appear, and the voltage across R_1 can be determined by selecting **Trace** followed by **Add Trace** and then **V(R1:1).** The result is the bottom display of Fig. 15.45, with only one plot point at 1 kHz. Since we fixed the frequency of interest at 1 kHz, this is the only frequency with a response.

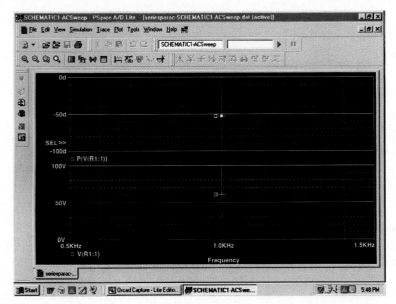

FIG. 15.45

The resulting magnitude and phase angle for the voltage V_{R_1} of Fig. 15.44.

The magnitude of the voltage across R_1 is 60 V to match the longhand solution of Example 15.4. The phase angle associated with the voltage can be determined by the sequence **Plot-Add Plot to Window-Trace-Add Trace-P( )** from the **Functions or Macros** list and then **V(R1:1)** to obtain **P(V(R1:1))** in the **Trace Expression** box. Click **OK,** and the resulting plot shows that the phase angle is just less than $-50°$, which is certainly a close match with the $-53.13°$ obtained in Example 15.4.

The above process made no use of the new **VPRINT** option just introduced. We will now see what this option provides. When the **SCHEMATIC1** window appears after the simulation, the window should be exited using the **X,** and **PSpice** should be selected on the top menu bar of the resulting screen. A list will appear of which **View Output File** is an option. Selecting this option will result in a long list of data about the construction of the network and the results obtained from the simulation. In Fig. 15.46, the portion of the output file listing the resulting magnitude and phase angle for the voltages defined by **VPRINT1** and **VPRINT2** is provided. Note that the voltage across R_1 defined by **VPRINT1** is 60 V at an angle of 53.13°. The voltage across R_2 as defined by **VPRINT2** is 80 V at an angle of 36.87°. Both are exact matches of the solutions of Example 15.4. In the future, therefore, if the **VPRINT** option is used, the results will appear in the output file.

```
82:
83:   ****       AC ANALYSIS                      TEMPERATURE =    27.000 DEG C
84:
85:
86:   *******************************************************************************
87:
88:
89:
90:    FREQ        VM(N00809)   VP(N00809)
91:
92:
93:     1.000E+03   6.000E+01   -5.313E+01
94: □
95: **** 06/30/01 17:45:20 ************** PSpice Lite (Mar 2000) ****************
96:
97:   ** Profile: "SCHEMATIC1-ACSweep"   [ C:\PSpice\seriesparac-SCHEMATIC1-ACSweep.s
     im ]
98:
99:
100:   ****       AC ANALYSIS                      TEMPERATURE =    27.000 DEG C
101:
102:
103:   *******************************************************************************
104:
105:
106:
107:    FREQ        VM(N00717)   VP(N00717)
108:
109:
110:     1.000E+03   8.000E+01   3.687E+01
111:
```

FIG. 15.46

The VPRINT1 (V_{R_1}) and VPRINT2 (V_{R_2}) response for the network of Fig. 15.44.

Now we will determine the voltage across the two branches from point a to point b. Return to **SCHEMATIC1,** and select **Trace** followed by **Add Trace** to obtain the list of **Simulation Output Variables.** Then, by applying Kirchhoff's voltage law around the closed loop, we find that the desired voltage is **V(R1:1)−V(R2:1)** which when followed by **OK** will result in the plot point in the screen in the bottom of Fig. 15.47. Note that it is exactly 100 V as obtained in the longhand solution. The phase angle can then be determined through **Plot-Add Plot to Window-Trace-Add Trace** and creating the expression **P(V(R1:1)−V(R2:1)).**

Remember that the expression can be generated using the lists of **Output variables** and **Functions,** but it can also be simply typed in from the keyboard. However, always be sure that there are as many left parentheses as there are right. Click **OK,** and a solution near $-105°$ appears. A better reading can be obtained by using **Plot-Axis Settings-Y Axis-User Defined** and changing the scale to $-100°$ to $-110°$. The result is the top screen of Fig. 15.47, with an angle closer to $-106.5°$ or 73.5° which is very close to the theoretical solution of 73.74°.

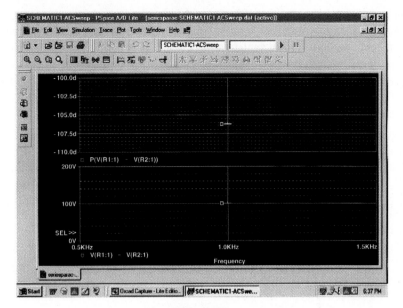

FIG. 15.47

The PSpice reponse for the voltage between the two points above resistors R_1 and R_2.

Electronics Workbench (EWB)

Electronics Workbench will now be used to determine the voltage across the last element of the ladder network of Fig. 15.48. The mathematical content of this chapter would certainly suggest that this analysis would be a lengthy exercise in complex algebra, with one mistake (a single sign or an incorrect angle) enough to invalidate the results. However, it will take only a few minutes to "draw" the network on the screen and only a few seconds to generate the results—results you can usually assume are correct if all the parameters were entered correctly. The results are certainly an excellent check against a longhand solution.

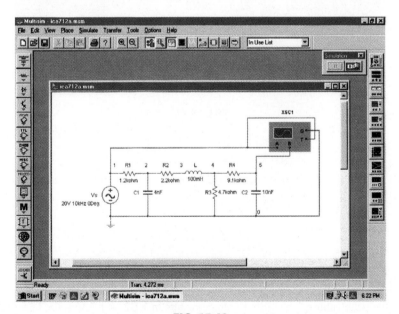

FIG. 15.48

Using the oscilloscope of EWB to determine the voltage across the capacitor C_2.

Our first approach will be to use an oscilloscope to measure the amplitude and phase angle of the output voltage as shown in Fig. 15.48. Note that five nodes are defined, with node 5 the desired voltage. The oscilloscope settings include a **Time base** of 20 μs/div. since the period of the 10 kHz signal is 100 μs. Channel **A** was set on 10 V/div, so that the full 20 V of the applied signal will have a peak value encompassing two divisions. Note that **Channel A** in Fig. 15.48 is connected directly to the source **Vs** and to the

Trigger input for synchronization. Expecting the output voltage to have a smaller amplitude resulted in a vertical sensitivity of 1 V/div. for **Channel B.** The analysis was initiated by placing the **Simulation** switch in the **1** position. It is important to realize that *when simulation is initiated, it will take time for networks with reactive elements to settle down and for the response to reach its steady-state condition. Therefore, it is wise to let a system run for a while after simulation before selecting Single on the oscilloscope to obtain a steady waveform for analysis.*

The resulting plots of Fig. 15.49 clearly show that the applied voltage has an amplitude of 20 V and a period of 100 μs (5 div. at 20 μs/div). The cursors sit ready for use at the left and right edges of the screen. Clicking on the small red arrow (with number 1) at the top of the oscilloscope screen will permit you to drag it to any location on the horizontal axis. As you move the cursor, the magnitude of each waveform will appear in the **T1** box below. By comparing positive slopes through the origin, you should see that the applied voltage is leading the output voltage by an angle that is more than 90°. Setting the cursor at the point where the output voltage on channel B passes through the origin with a positive slope, we find that we cannot achieve exactly 0 V; but $-313.4\ \mu$V $= -0.313$ mV **(VB1)** is certainly very close at 39.7 μs **(T1).**

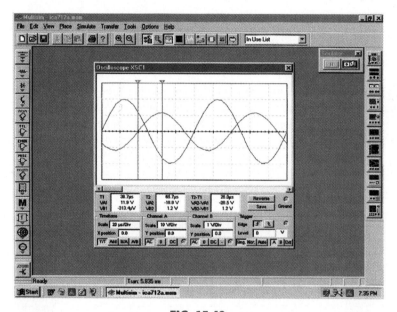

FIG. 15.49

Using EWB to display the applied voltage and the voltage across capacitor C_2 for the network of Fig. 15.48.

Knowing that the applied voltage passed through the origin at 0 μs permits the following calculation for the phase angle:

$$\frac{39.7\ \mu s}{100\ \mu s} = \frac{\theta}{360°}$$

$$\theta = 142.92°$$

with the result that the output voltage has an angle of $-142.92°$ associated with it. The second cursor is found at the right edge of the screen and has a blue color. Selecting it and moving it to the peak value of the output voltage results in **VB2** $= 1.2$ V at 65.7 μs **(T2).** The result of all the above is

$$\mathbf{V}_{C_2} = 1.2\ \text{V}\ \angle -142.92°$$

Our second approach will be to use the **AC Analysis** option under the **Simulate** heading. First, realize that when we were using the oscilloscope as we did above, there was no need to pass through the sequence of dialog boxes to choose the desired analysis. All that was necessary was to simulate using either the switch or the **PSpice-Run** sequence—the oscilloscope was there to measure the output voltage. Remember that the source defined the magnitude of the applied voltage, the frequency, and the phase shift.

This time we will use the sequence **Simulate-Analyses-AC Analysis** to obtain the **AC Analysis** dialog box in which the **Start** and **Stop frequencies** will be 10 kHz and the **Selected variable for analysis** will be node 5 only. Selecting **Simulate** will then result in a magnitude-phase plot with no apparent indicators at 10 kHz. However, this is easily corrected by first selecting one of the plots by clicking on the **Voltage** label at the left of the plot. Then select the **Show/Hide Grid, Show/Hide Legend,** and **Show/Hide Cursors** keys to obtain the cursors, legend, and **AC Analysis** dialog box. Hook on the red cursor and move it to 10 kHz. At that location, and that location only, **x1** will appear as 10 kHz in the dialog box, and **y1** will be 1.1946 as shown in Fig. 15.50. In other words, the cursor has defined the magnitude of the voltage across the output capacitor as 1.1946 V or approximately 1.2 V as obtained above. If you then select the **Phase** curve and repeat the procedure, you will find that at 10 kHz (**x1**) the angle is $-142.15°$ (**y1**) which is very close to the $-142.92°$ obtained above.

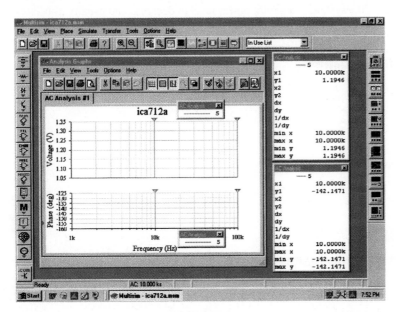

FIG. 15.50

Using the AC Analysis option under EWB to determine the magnitude and phase angle for the voltage V_{C_2} for the network of Fig. 15.48.

In total, therefore, we have two methods to obtain an ac voltage in a network—one by instrumentation and the other through computer methods. Both are valid, although, as expected, the computer approach has a higher level of accuracy.

CHAPTER SUMMARY

SERIES-PARALLEL ac NETWORKS

- Using block impedances to represent individual elements or combinations of elements will often help define the approach that should be used to analyze a particular series-parallel ac network.
- Redrawing the network is often helpful in simplifying the analysis of a complex series-parallel ac network.
- Always check the magnitude and phase angle of your solution to determine whether the solution makes any sense at all based on the components of the network.

THE POWER TRIANGLE

- The apparent power to a system is measured in volt-ampere and is determined solely by the product of the rms values of the applied voltage and the resulting source current.
- The power-factor angle is the same angle as appearing in the total impedance equation.
- The total real, reactive, and apparent power of any system are related by the Pythagorean theorem.

- The power factor of a network is defined by the ratio of the real power over the apparent power.
- No matter how complex a network, the total real power is simply the sum of the real power levels of the network, whereas the total reactive power is the difference between the total inductive and capacitive power levels.

POWER-FACTOR CORRECTION

- The process of introducing reactive elements to bring the network power factor closer to unity is called **power-factor correction.**
- Since most industrial loads are inductive, capacitive elements are normally introduced in parallel with the load to improve the power factor.
- The direct result of power-factor correction is a reduction in the source current, which causes the system's apparent power level to decrease and usually translates into a reduced power bill for the industrial plant.

Important Equations

$$S = EI_s = I_s^2 Z_T = \frac{E^2}{Z_T} \quad \text{(volt-ampere, VA)}$$

$$P = S \cos \theta_T$$

$$F_P = \cos \theta_T = \frac{P}{S}$$

$$S_T = \sqrt{P_T^2 + Q_T^2}$$

PROBLEMS

SECTION 15.2   Series-Parallel ac Networks

1. For the series-parallel network of Fig. 15.51:
   - **a.** Calculate Z_T.
   - **b.** Determine I.
   - **c.** Determine I_1.
   - **d.** Find I_2 and I_3.
   - **e.** Find V_L.

2. For the network of Fig. 15.52:
   - **a.** Find the total impedance Z_T.
   - **b.** Determine the current I_s.
   - **c.** Calculate I_C using the current divider rule.
   - **d.** Calculate V_L using the voltage divider rule.

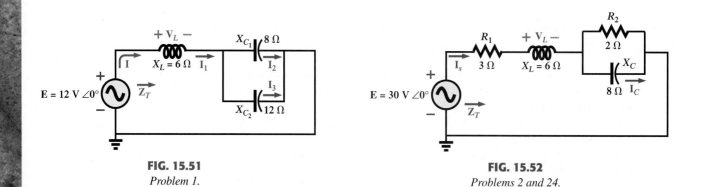

FIG. 15.51
Problem 1.

FIG. 15.52
Problems 2 and 24.

3. For the network of Fig. 15.53:
   - **a.** Find the total impedance Z_T and the total admittance Y_T.
   - **b.** Find the current I_s.
   - **c.** Calculate I_2 using the current divider rule.
   - **d.** Calculate V_C.
   - **e.** Calculate the average power delivered to the network.

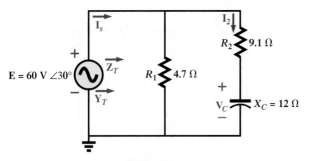

FIG. 15.53
Problem 3.

4. For the network of Fig. 15.54:
  a. Find the current **I**.
  b. Find the voltage $\mathbf{V}_C$.
  c. Find the average power delivered to the network.

*__5.__ For the network of Fig. 15.55:
  a. Find the current $\mathbf{I}_1$.
  b. Calculate the voltage $\mathbf{V}_C$ using the voltage divider rule.
  c. Find the voltage $\mathbf{V}_{ab}$.

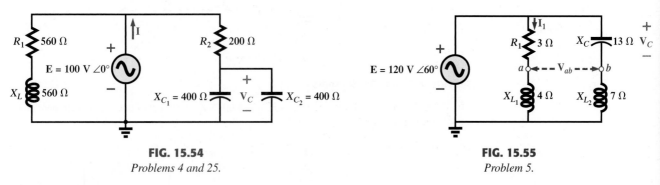

FIG. 15.54
Problems 4 and 25.

FIG. 15.55
Problem 5.

*__6.__ For the network of Fig. 15.56:
  a. Find the current $\mathbf{I}_s$.
  b. Find the voltage $\mathbf{V}_L$.
  c. Calculate the average power delivered to the network.

7. For the network of Fig. 15.57:
  a. Find the total impedance $\mathbf{Z}_T$ and the admittance $\mathbf{Y}_T$.
  b. Find the currents $\mathbf{I}_1$, $\mathbf{I}_2$, and $\mathbf{I}_3$.
  c. Verify Kirchhoff's current law by showing that $\mathbf{I}_s = \mathbf{I}_1 + \mathbf{I}_2 + \mathbf{I}_3$.
  d. Find the power factor of the network, and indicate whether it is leading or lagging.

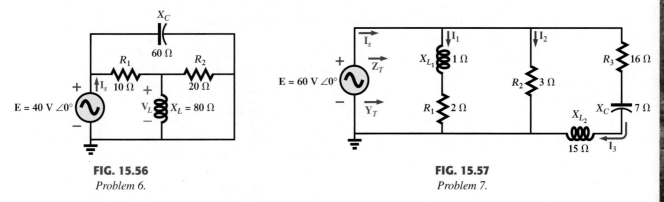

FIG. 15.56
Problem 6.

FIG. 15.57
Problem 7.

*__8.__ For the network of Fig. 15.58:
  a. Find the total impedance $\mathbf{Z}_T$ and the admittance $\mathbf{Y}_T$.
  b. Find the source current $\mathbf{I}_s$ in phasor form.
  c. Find the currents $\mathbf{I}_1$ and $\mathbf{I}_2$ in phasor form.
  d. Find the voltages $\mathbf{V}_1$ and $\mathbf{V}_{ab}$ in phasor form.
  e. Find the average power delivered to the network.
  f. Find the power factor of the network, and indicate whether it is leading or lagging.

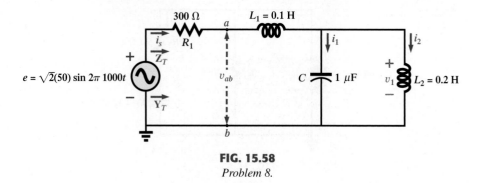

FIG. 15.58
Problem 8.

*9. Find the current $\mathbf{I}$ for the network of Fig. 15.59.

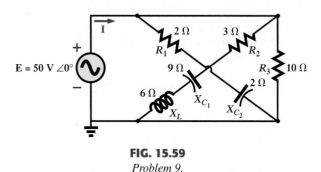

FIG. 15.59
Problem 9.

SECTION 15.3 Ladder Networks

10. Find the current $\mathbf{I}_5$ for the network of Fig. 15.60. Note the effect of one reactive element on the resulting calculations.

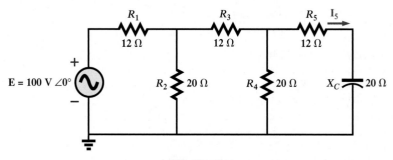

FIG. 15.60
Problems 10 and 26.

11. Find the average power delivered to R_4 in Fig. 15.61.

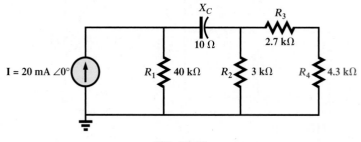

FIG. 15.61
Problem 11.

SECTION 15.4 The Power Triangle

12. For the battery of bulbs (purely resistive) appearing in Fig. 15.62:
   a. Determine the total power dissipation.
   b. Calculate the total reactive and apparent power.
   c. Find the source current I_s.

d. Calculate the resistance of each bulb for the specified operating conditions.

e. Determine the currents I_1 and I_2.

13. For the network of Fig. 15.63:

a. Find the average power delivered to each element.

b. Find the reactive power for each element.

c. Find the apparent power for each element.

d. Find the total number of watts, volt-ampere reactive, and volt-ampere, and the power factor F_p.

e. Sketch the power triangle.

f. Find the energy dissipated by the resistor over one full cycle of the input voltage.

g. Find the energy stored or returned by the capacitor and the inductor over one half-cycle of the power curve for each.

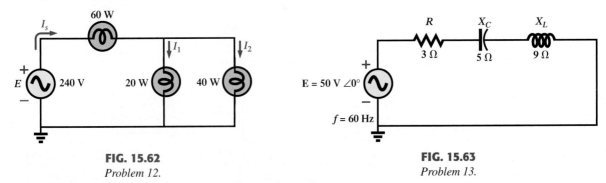

FIG. 15.62
Problem 12.

FIG. 15.63
Problem 13.

14. For the circuit of Fig. 15.64:

a. Find the average, reactive, and apparent power for the 20 Ω resistor.

b. Repeat part (a) for the 10 Ω inductive reactance.

c. Find the total number of watts, volt-ampere reactive, and volt-ampere, and the power factor F_p.

d. Find the current $\mathbf{I}_s$.

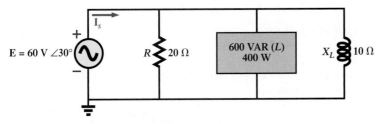

FIG. 15.64
Problem 14.

15. For the network of Fig. 15.65:

a. Find the average power delivered to each element.

b. Find the reactive power for each element.

c. Find the apparent power for each element.

d. Find P_T, Q_T, S_T, and F_p for the system.

e. Sketch the power triangle.

f. Find $\mathbf{I}_s$.

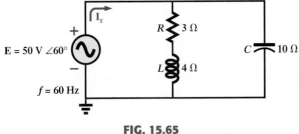

FIG. 15.65
Problem 15.

***16.** For the system of Fig. 15.66:

a. Find the total number of watts, volt-ampere reactive, and volt-ampere, and F_p.

b. Find the current $\mathbf{I}_s$.

c. Draw the power triangle.

d. Find the type of elements and their impedance in ohms within each electrical box. (Assume that all elements of a load are in series.)

    e. Verify that the result of part (b) is correct by finding the current $\mathbf{I}_s$ using only the input voltage $\mathbf{E}$ and the results of part (d). Compare the value of $\mathbf{I}_s$ with that obtained for part (b).

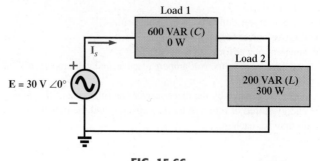

FIG. 15.66
Problem 16.

*17. Repeat Problem 16 for the system of Fig. 15.67.

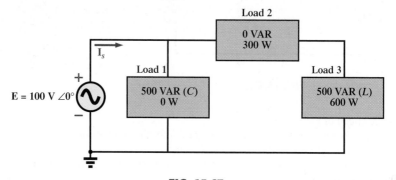

FIG. 15.67
Problem 17.

*18. For the circuit of Fig. 15.68:
    a. Find the total number of watts, volt-ampere reactive, and volt-ampere, and F_p.
    b. Find the current $\mathbf{I}_s$.
    c. Find the types of elements and their impedance in each box. (Assume that the elements within each box are in series.)

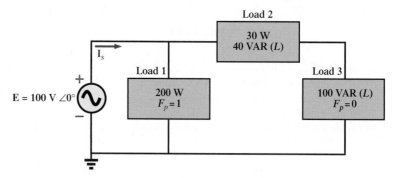

FIG. 15.68
Problem 18.

SECTION 15.5   Power-Factor Correction

*19. The lighting and motor loads of a small factory establish a 10 kVA power demand at a 0.7 lagging power factor on a 208 V, 60 Hz supply.
    a. Establish the power triangle for the load.
    b. Determine the power-factor capacitor that must be placed in parallel with the load to raise the power factor to unity.
    c. Determine the change in supply current from the uncompensated to the compensated system.
    d. Repeat parts (b) and (c) if the power factor is increased to 0.9.

20. The load on a 120 V, 60 Hz supply is 5 kW (resistive), 8 kVAR (inductive), and 2 kVAR (capacitive).
    a. Find the total kilovolt-amperes.
    b. Determine the F_p of the combined loads.
    c. Find the current drawn from the supply.
    d. Calculate the capacitance necessary to establish a unity power factor.

e. Find the current drawn from the supply at unity power factor, and compare it to the uncompensated level.

21. The loading of a factory on a 1000 V, 60 Hz system includes:

20 kW heating (unity power factor)
10 kW (P_i) induction motors (0.7 lagging power factor)
5 kW lighting (0.85 lagging power factor)

a. Establish the power triangle for the total loading on the supply.
b. Determine the power-factor capacitor required to raise the power factor to unity.
c. Determine the change in supply current from the uncompensated to the compensated system.

SECTION 15.6   Wattmeters and Power-Factor Meters

22. a. A wattmeter is connected with its current coil as shown in Fig. 15.69 and with the potential coil across points *f-g*. What does the wattmeter read?
b. Repeat part (a) with the potential coil (*PC*) across *a-b, b-c, a-c, a-d, c-d, d-e,* and *f-e*.

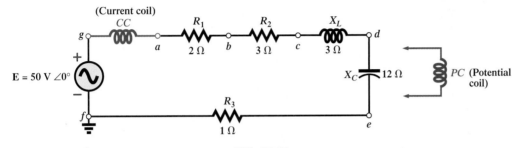

FIG. 15.69
Problem 22.

23. The voltage source of Fig. 15.70 delivers 660 VA at 120 V, with a supply current that lags the voltage by a power factor of 0.6.
a. Determine the voltmeter, ammeter, and wattmeter readings.
b. Find the load impedance in rectangular form.

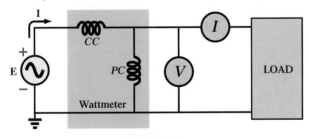

FIG. 15.70
Problem 23.

SECTION 15.9   Computer Analysis

24. Using PSpice or EWB, perform a complete analysis of the network of Fig. 15.52.

25. Using PSpice or EWB, perform a complete analysis of the network of Fig. 15.54.

26. Using PSpice or EWB, perform a complete analysis of the ladder network of Fig. 15.60.

27. Using PSpice or EWB, obtain a plot of the reactive power for a pure capacitor of 636.62 μF at a frequency of 1 kHz for one cycle of the input voltage. Use an applied voltage of $\mathbf{E} = 10$ V $\angle 0°$. On the same graph, plot both the applied voltage and the resulting current.

28. Using PSpice or EWB, plot a curve of the power to a series R-L combination where the resistance and inductive reactance are both 1 kΩ. Apply a voltage of $\mathbf{E} = 2$ V $\angle 0°$. Include curves of the applied voltage and current on the same plot.

GLOSSARY

Ladder network   A repetitive combination of series and parallel branches that has the appearance of a ladder.

Power-factor correction   The addition of reactive components (typically capacitive) to establish a system power factor closer to unity.

Series-parallel ac network   A combination of series and parallel branches in the same network configuration. Each branch may contain any number of elements whose impedance is dependent on the applied frequency.

ac Methods of Analysis and Theorems

OBJECTIVES

- Be able to make a source conversion in the ac domain.
- Understand how to apply mesh analysis to ac networks.
- Be able to apply nodal analysis to an ac network.
- Understand the criteria for balance conditions in an ac bridge network.
- Be able to perform a Δ-Y a or Y-Δ conversion in an ac network.
- Understand how to apply the superposition theorem to an ac network.
- Be able to apply Thévenin's theorem to an ac network.
- Understand how to apply Norton's theorem to an ac network.
- Understand the conditions for maximum power transfer in an ac network.
- Be able to use the calculator and computer software to perform some of the complex mathematical procedures called for when using the above methods of analysis and theorems.

16.1 INTRODUCTION

Through the use of block impedances, the methods of analysis and the theorems introduced for dc networks can be applied to ac networks using the same approach and the same series of steps. In fact, the similarities are so strong that the two chapters (Chapters 8 and 9) devoted to dc circuits will be covered in this one chapter for the ac circuits.

The branch-current method will not be discussed again because it falls within the framework of mesh analysis. In addition to the methods and theorems included in the above list of objectives, the bridge network and Δ-Y and Y-Δ conversions will be discussed in some detail.

16.2 SOURCE CONVERSIONS

When applying the methods and theorems to be discussed, you may have to convert a current source to a voltage source, or vice versa. The conversion process is the same as discussed for dc circuits, except now we will be dealing with phasors and impedances instead of just real numbers and resistors.

The general format and the equations for converting from one form to the other are provided in Fig. 16.1. Note that the series impedance for the voltage source becomes the parallel impedance for the current source with no change in magnitude or angle—we simply move from one position to the other. The source voltages and currents are determined using Ohm's law.

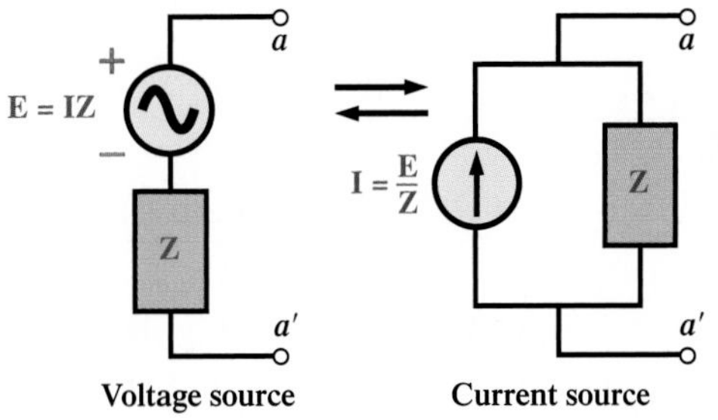

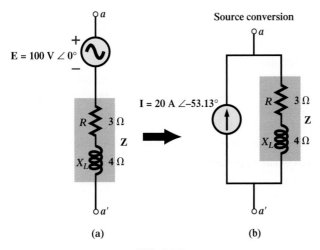

FIG. 16.1

Source conversion.

EXAMPLE 16.1 Convert the voltage source of Fig. 16.2(a) to a current source.

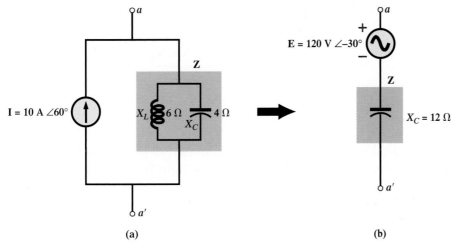

FIG. 16.2

Example 16.1.

Solution:

$$\mathbf{I} = \frac{\mathbf{E}}{\mathbf{Z}} = \frac{100 \text{ V}\angle 0°}{5 \text{ }\Omega \angle 53.13°} = \mathbf{20 \text{ A}} \angle -\mathbf{53.13°} \qquad [\text{Fig. 16.2(b)}]$$

EXAMPLE 16.2 Convert the current source of Fig. 16.3(a) to a voltage source.

FIG. 16.3

Example 16.2.

Solution:

$$Z = \frac{Z_C Z_L}{Z_C + Z_L} = \frac{(X_C \angle -90°)(X_L \angle 90°)}{-j X_C + j X_L}$$

$$= \frac{(4\ \Omega \angle -90°)(6\ \Omega \angle 90°)}{-j\,4\ \Omega + j\,6\ \Omega} = \frac{24\ \Omega \angle 0°}{2 \angle 90°}$$

$$= 12\ \Omega \angle -90° \quad \text{[Fig. 16.3(b)]}$$

$$E = IZ = (10\ A \angle 60°)(12\ \Omega \angle -90°)$$

$$= 120\ V \angle -30° \quad \text{[Fig. 16.3(b)]}$$

16.3 MESH ANALYSIS

Before examining the application of **mesh analysis** to ac networks, you should consider first reviewing the sections on the application of mesh analysis to dc networks, because the content of this section is a very close match with the earlier coverage. The general approach will use the same series of steps listed for dc networks, with the only change in the general procedure being the substitution of impedance for resistance, and admittance for conductance.

As with dc networks, first be sure that only voltage sources are present. If not, a conversion must be performed. If the current source does not have a parallel impedance, then the procedures outlined in Section 8.10 should be applied.

For ac networks, mesh analysis is applied using the following sequence of steps:

Mesh Analysis (General Approach):

1. *Assign a distinct current in the clockwise direction to each independent, closed loop of the network.*
2. *Within each loop insert the polarities for each impedance as determined by the direction of the loop current.*
3. *Apply Kirchhoff's voltage law around each closed loop in the clockwise direction.*
4. *Solve the resulting equations for the assumed loop currents.*

A few examples will review the procedure described earlier in detail for dc networks.

EXAMPLE 16.3 Using the general approach to mesh analysis, find current I_1 in Fig. 16.4.

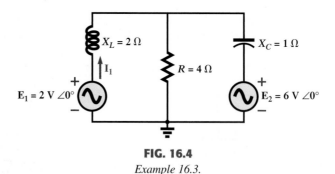

FIG. 16.4
Example 16.3.

Solution: When you are applying these methods to ac circuits, it is good practice to represent the resistors and reactances (or combinations thereof) by subscripted impedances. When the total solution is found in terms of these subscripted impedances, the numerical values can be substituted to find the unknown quantities.

The network is redrawn in Fig. 16.5 with subscripted impedances:

$$\begin{aligned} Z_1 &= +j X_L = +j\,2\ \Omega & E_1 &= 2\ V \angle 0° \\ Z_2 &= R = 4\ \Omega & E_2 &= 6\ V \angle 0° \\ Z_3 &= -j X_C = -j\,1\ \Omega \end{aligned}$$

Steps 1 and 2 are as indicated in Fig. 16.5.

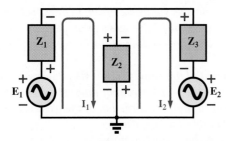

FIG. 16.5

Assigning the mesh currents and subscripted impedances for the network of Fig. 16.4.

Step 3:

$$+\mathbf{E}_1 - \mathbf{I}_1\mathbf{Z}_1 - \mathbf{Z}_2(\mathbf{I}_1 - \mathbf{I}_2) = 0$$

$$-\mathbf{Z}_2(\mathbf{I}_2 - \mathbf{I}_1) - \mathbf{I}_2\mathbf{Z}_3 - \mathbf{E}_2 = 0$$

or

$$\mathbf{E}_1 - \mathbf{I}_1\mathbf{Z}_1 - \mathbf{I}_1\mathbf{Z}_2 + \mathbf{I}_2\mathbf{Z}_2 = 0$$

$$-\mathbf{I}_2\mathbf{Z}_2 + \mathbf{I}_1\mathbf{Z}_2 - \mathbf{I}_2\mathbf{Z}_3 - \mathbf{E}_2 = 0$$

so that

$$\mathbf{I}_1(\mathbf{Z}_1 + \mathbf{Z}_2) - \mathbf{I}_2\mathbf{Z}_2 = \mathbf{E}_1$$

$$\mathbf{I}_2(\mathbf{Z}_2 + \mathbf{Z}_3) - \mathbf{I}_1\mathbf{Z}_2 = -\mathbf{E}_2$$

which are rewritten as

$$\mathbf{I}_1(\mathbf{Z}_1 + \mathbf{Z}_2) - \mathbf{I}_2\mathbf{Z}_2 \qquad = \mathbf{E}_1$$

$$-\mathbf{I}_1\mathbf{Z}_2 + \mathbf{I}_2(\mathbf{Z}_2 + \mathbf{Z}_3) = -\mathbf{E}_2$$

Step 4: Using determinants, we obtain

$$\mathbf{I}_1 = \frac{\begin{vmatrix} \mathbf{E}_1 & -\mathbf{Z}_2 \\ -\mathbf{E}_2 & \mathbf{Z}_2 + \mathbf{Z}_3 \end{vmatrix}}{\begin{vmatrix} \mathbf{Z}_1 + \mathbf{Z}_2 & -\mathbf{Z}_2 \\ -\mathbf{Z}_2 & \mathbf{Z}_2 + \mathbf{Z}_3 \end{vmatrix}}$$

$$= \frac{\mathbf{E}_1(\mathbf{Z}_2 + \mathbf{Z}_3) - \mathbf{E}_2(\mathbf{Z}_2)}{(\mathbf{Z}_1 + \mathbf{Z}_2)(\mathbf{Z}_2 + \mathbf{Z}_3) - (\mathbf{Z}_2)^2}$$

$$= \frac{(\mathbf{E}_1 - \mathbf{E}_2)\mathbf{Z}_2 + \mathbf{E}_1\mathbf{Z}_3}{\mathbf{Z}_1\mathbf{Z}_2 + \mathbf{Z}_1\mathbf{Z}_3 + \mathbf{Z}_2\mathbf{Z}_3}$$

Substituting numerical values yields

$$\mathbf{I}_1 = \frac{(2\ \text{V} - 6\ \text{V})(4\ \Omega) + (2\ \text{V})(-j\,1\ \Omega)}{(+j\,2\ \Omega)(4\ \Omega) + (+j\,2\ \Omega)(-j\,1\ \Omega) + (4\ \Omega)(-j\,1\ \Omega)}$$

$$= \frac{-16 - j\,2}{j\,8 - j^2 2 - j\,4} = \frac{-16 - j\,2}{2 + j\,4} = \frac{16.12\ \text{A}\ \angle -172.87°}{4.47\ \angle 63.43°}$$

$$= \mathbf{3.61\ A\ \angle -236.30°} \quad \text{or} \quad \mathbf{3.61\ A\ \angle 123.70°}$$

Recall from the dc coverage that there is a general approach and a format approach. The format approach permits the writing of the network equations in a very direct fashion that saves both time and energy. The steps applied for the format approach for ac networks are listed below. The only change will be to substitute the term *impedance* wherever the term *resistance* might appear.

Mesh Analysis (Format Approach):

1. *Assign a current in the clockwise direction to each independent, closed loop of the network.*
2. *Column 1 is the product of the loop current of interest and the sum of the impedances through which it passes.*

3. *The mutual terms are always subtracted from the first column. The mutual terms include all the impedances that have more than one loop current passing through them. They are formed by the product of the impedance element and the other loop current.*

4. *The column to the right of the equals sign is the algebraic sum of the voltage sources through which the loop current of interest passes. Positive signs are assigned to those that support the direction of the assigned loop current. Negative signs are assigned to those that oppose it.*

5. *Solve the resulting simultaneous equations for the desired loop currents.*

EXAMPLE 16.4 Using the format approach to mesh analysis, find current I_2 for the network of Fig. 16.6.

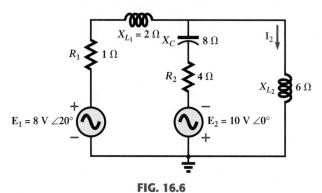

FIG. 16.6
Example 16.4.

Solution: The network is redrawn in Fig. 16.7:

$$\mathbf{Z}_1 = R_1 + j X_{L_1} = 1\,\Omega + j\,2\,\Omega \qquad \mathbf{E}_1 = 8\text{ V }\angle 20°$$
$$\mathbf{Z}_2 = R_2 + j X_C = 4\,\Omega - j\,8\,\Omega \qquad \mathbf{E}_2 = 10\text{ V }\angle 0°$$
$$\mathbf{Z}_3 = +j X_{L_2} = +j\,6\,\Omega$$

Note the reduction in complexity of the problem with the substitution of the subscripted impedances.

Step 1 is as indicated in Fig. 16.7.
Steps 2–4:

$$\mathbf{I}_1(\mathbf{Z}_1 + \mathbf{Z}_2) - \mathbf{I}_2\mathbf{Z}_2 = \mathbf{E}_1 + \mathbf{E}_2$$
$$\mathbf{I}_2(\mathbf{Z}_2 + \mathbf{Z}_3) - \mathbf{I}_1\mathbf{Z}_2 = -\mathbf{E}_2$$

which are rewritten as

$$\mathbf{I}_1(\mathbf{Z}_1 + \mathbf{Z}_2) - \mathbf{I}_2\mathbf{Z}_2 \qquad\qquad = \mathbf{E}_1 + \mathbf{E}_2$$
$$-\,\mathbf{I}_2\,\mathbf{Z}_2 \qquad + \mathbf{I}_2(\mathbf{Z}_2 + \mathbf{Z}_3) = -\mathbf{E}_2$$

Step 5: Using determinants, we have

$$\mathbf{I}_2 = \frac{\begin{vmatrix} \mathbf{Z}_1 + \mathbf{Z}_2 & \mathbf{E}_1 + \mathbf{E}_2 \\ -\mathbf{Z}_2 & -\mathbf{E}_2 \end{vmatrix}}{\begin{vmatrix} \mathbf{Z}_1 + \mathbf{Z}_2 & -\mathbf{Z}_2 \\ -\mathbf{Z}_2 & \mathbf{Z}_2 + \mathbf{Z}_3 \end{vmatrix}}$$

$$= \frac{-(\mathbf{Z}_1 + \mathbf{Z}_2)\mathbf{E}_2 + \mathbf{Z}_2(\mathbf{E}_1 + \mathbf{E}_2)}{(\mathbf{Z}_1 + \mathbf{Z}_2)(\mathbf{Z}_2 + \mathbf{Z}_3) - \mathbf{Z}_2^2}$$

$$= \frac{\mathbf{Z}_2\mathbf{E}_1 - \mathbf{Z}_1\mathbf{E}_2}{\mathbf{Z}_1\mathbf{Z}_2 + \mathbf{Z}_1\mathbf{Z}_3 + \mathbf{Z}_2\mathbf{Z}_3}$$

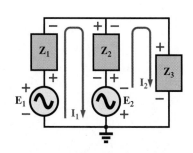

FIG. 16.7
Assigning the mesh currents and subscripted impedances for the network of Fig. 16.6.

$$I_2 = \frac{(4\,\Omega - j\,8\,\Omega)(8\,\text{V}\,\angle 20°) - (1\,\Omega + j\,2\,\Omega)(10\,\text{V}\,\angle 0°)}{(1\,\Omega + j\,2\,\Omega)(4\,\Omega - j\,8\,\Omega) + (1\,\Omega + j\,2\,\Omega)(+j\,6\,\Omega) + (4\,\Omega - j\,8\,\Omega)(+j\,6\,\Omega)}$$

$$= \frac{(4 - j\,8)(7.52 + j\,2.74) - (10 + j\,20)}{20 + (j\,6 - 12) + (j\,24 + 48)}$$

$$= \frac{(52.0 - j\,49.20) - (10 + j\,20)}{56 + j\,30} = \frac{42.0 - j\,69.20}{56 + j\,30} = \frac{80.95\,\text{A}\,\angle -58.74°}{63.53\,\angle 28.18°}$$

$$= \mathbf{1.27\,A\,\angle -86.92°}$$

Calculator Solution: The calculator (TI-86 or equivalent) can be an effective tool in performing the long, laborious calculations involved with the final equation appearing above. However, you must be very careful to use the correct number of brackets and to define by brackets the order of the arithmetic operations as shown in Fig. 16.8.

```
((4,–8)*8(∠20)–(1,2)*(10∠0))/((1,2)*(4,–8)+(1,2)*(0,6)+(4,–8)*(0,6)) (ENTER)
(67.854E–3,–1.272E0)
Ans▶Pol
(1.274E0∠–86.956E0)
```

FIG. 16.8

Determining I_2 for the network of Fig. 16.6 using the TI-86 calculator.

Mathcad Solution: This example provides an excellent opportunity to demonstrate the power of Mathcad. First the impedances and parameters are defined for the equations to follow as shown in Fig. 16.9. Then the **Guess** values of mesh currents I_1 and I_2 are entered. The label **Given** must then be entered followed by the equations for the network. Note that in this example, we are not continuing with the analysis until the matrix is defined—we are working directly from the network equations. Once the equations have been properly entered, **Find(I1,I2)** is entered. Then selecting the equals sign will result in the single-column matrix with the results in rectangular form. Conversion to polar form requires defining a variable **A** and then calling for the magnitude and the angle using the definitions entered earlier in the listing and both the **Calculator** and **Greek** toolbars. The result for I_2 is **1.274 A∠−86.94°** which is an excellent match with the theoretical solution.

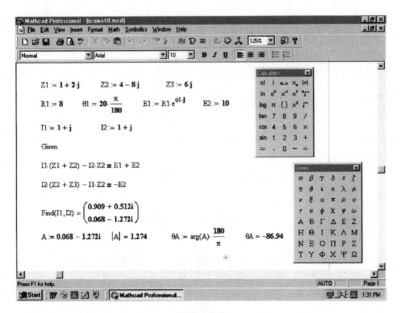

FIG. 16.9

Using Mathcad to verify the results of Example 16.4.

EXAMPLE 16.5 Write the mesh equations for the network of Fig. 16.10. Do not solve them.

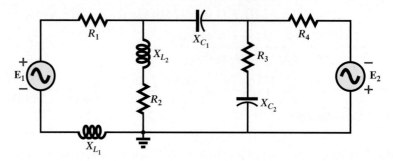

FIG. 16.10
Example 16.5.

Solution: The network is redrawn in Fig. 16.11. Again note the reduced complexity and increased clarity provided by the use of subscripted impedances:

$$\mathbf{Z}_1 = R_1 + j X_{L_1} \qquad \mathbf{Z}_4 = R_3 - j X_{C_2}$$
$$\mathbf{Z}_2 = R_2 + j X_{L_2} \qquad \mathbf{Z}_5 = R_4$$
$$\mathbf{Z}_3 = j X_{C_1}$$

and

$$\mathbf{I}_1(\mathbf{Z}_1 + \mathbf{Z}_2) - \mathbf{I}_2\mathbf{Z}_2 = \mathbf{E}_1$$
$$\mathbf{I}_2(\mathbf{Z}_2 + \mathbf{Z}_3 + \mathbf{Z}_4) - \mathbf{I}_1\mathbf{Z}_2 - \mathbf{I}_3\mathbf{Z}_4 = 0$$
$$\mathbf{I}_3(\mathbf{Z}_4 + \mathbf{Z}_5) - \mathbf{I}_2\mathbf{Z}_4 = \mathbf{E}_2$$

or

$$\mathbf{I}_1(\mathbf{Z}_1 + \mathbf{Z}_2) - \mathbf{I}_2(\mathbf{Z}_2) \qquad\qquad + 0 \qquad\qquad = \mathbf{E}_1$$
$$\mathbf{I}_1\mathbf{Z}_2 \qquad - \mathbf{I}_2(\mathbf{Z}_2 + \mathbf{Z}_3 + \mathbf{Z}_4) + \mathbf{I}_3(\mathbf{Z}_4) \qquad = 0$$
$$0 \qquad - \mathbf{I}_2(\mathbf{Z}_4) \qquad + \mathbf{I}_3(\mathbf{Z}_4 + \mathbf{Z}_5) = \mathbf{E}_2$$

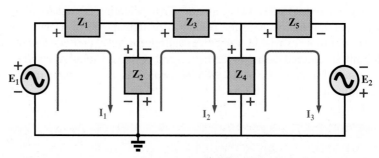

FIG. 16.11
Assigning the mesh currents and subscripted impedances for the network of Fig. 16.10.

EXAMPLE 16.6 Using the format approach, write the mesh equations for the network of Fig. 16.12.

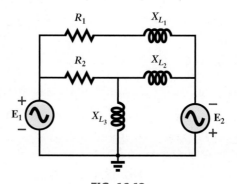

FIG. 16.12
Network for Example 16.6.

Solution: The network is redrawn as shown in Fig. 16.13, where

$$\mathbf{Z}_1 = R_1 + j X_{L_1} \qquad \mathbf{Z}_3 = j X_{L_2}$$
$$\mathbf{Z}_2 = R_2 \qquad\qquad \mathbf{Z}_4 = j X_{L_3}$$

and

$$\mathbf{I}_1(\mathbf{Z}_2 + \mathbf{Z}_4) - \mathbf{I}_2\mathbf{Z}_2 - \mathbf{I}_3\mathbf{Z}_4 = \mathbf{E}_1$$

$$\mathbf{I}_2(\mathbf{Z}_1 + \mathbf{Z}_2 + \mathbf{Z}_3) - \mathbf{I}_1\mathbf{Z}_2 - \mathbf{I}_3\mathbf{Z}_3 = 0$$

$$\mathbf{I}_3(\mathbf{Z}_3 + \mathbf{Z}_4) - \mathbf{I}_2\mathbf{Z}_3 - \mathbf{I}_1\mathbf{Z}_4 = \mathbf{E}_2$$

or

$$\mathbf{I}_1(\mathbf{Z}_2 + \mathbf{Z}_4) - \mathbf{I}_2\mathbf{Z}_2 \qquad\qquad - \mathbf{I}_3\mathbf{Z}_4 \qquad = \mathbf{E}_1$$

$$-\mathbf{I}_1\mathbf{Z}_2 \qquad + \mathbf{I}_2(\mathbf{Z}_1 + \mathbf{Z}_2 + \mathbf{Z}_3) - \mathbf{I}_3\mathbf{Z}_3 \qquad = 0$$

$$-\mathbf{I}_1\mathbf{Z}_4 \qquad\qquad - \mathbf{I}_2\mathbf{Z}_3 \qquad + \mathbf{I}_3(\mathbf{Z}_3 + \mathbf{Z}_4) = \mathbf{E}_2$$

Note the symmetry *about* the diagonal axis; that is, note the location of $-\mathbf{Z}_2$, $-\mathbf{Z}_4$, and $-\mathbf{Z}_3$ off the diagonal.

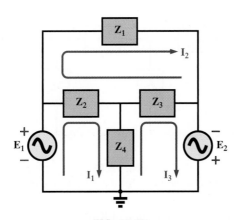

FIG. 16.13
Assigning the mesh currents and subscripted impedances
for the network of Fig. 16.12.

16.4 NODAL ANALYSIS

Before we examine the application of **nodal analysis** to ac networks, a review of the corresponding sections in Chapter 8 is suggested. The only major change here will be the need to work with phasors and impedances rather than dc voltage sources and resistors.

It is important to remember that the network must consist solely of current sources for the steps listed below to be applicable. Otherwise, a source conversion must first be performed or the approach introduced in Section 8.10 applied.

The steps leading to the application of nodal analysis to an ac network are the following:

Nodal Analysis (General Approach):

1. *First determine the number of nodes for the network.*
2. *Pick a reference node (normally the ground connection), and label each of the remaining nodes with a subscripted label such as V_1, V_2, and so on.*
3. *Apply Kirchhoff's current law at each node except the reference node. For each application of Kirchhoff's current law, assume that each of the unknown currents leaves the node (this removes the concern about direction; a minus sign will appear in the solution if the direction is incorrectly chosen).*
4. *Solve the resulting equations for the nodal voltages.*

EXAMPLE 16.7 Determine the voltage across the inductor for the network of Fig. 16.14.

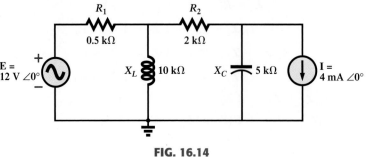

FIG. 16.14
Example 16.7.

Solution:

Steps 1 and 2 are as indicated in Fig. 16.15.

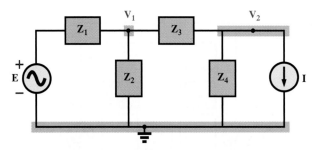

FIG. 16.15
Assigning the nodal voltages and subscripted impedances to the network of Fig. 16.14.

Step 3: Note Fig. 16.16 for the application of Kirchhoff's current law to node $\mathbf{V}_1$:

$$\Sigma\, \mathbf{I}_i = \Sigma\, \mathbf{I}_o \qquad 0 = \mathbf{I}_1 + \mathbf{I}_2 + \mathbf{I}_3$$

$$\frac{\mathbf{V}_1 - \mathbf{E}}{\mathbf{Z}_1} + \frac{\mathbf{V}_1}{\mathbf{Z}_2} + \frac{\mathbf{V}_1 - \mathbf{V}_2}{\mathbf{Z}_3} = 0$$

Rearranging terms:

$$\mathbf{V}_1\left[\frac{1}{\mathbf{Z}_1} + \frac{1}{\mathbf{Z}_2} + \frac{1}{\mathbf{Z}_3}\right] - \mathbf{V}_2\left[\frac{1}{\mathbf{Z}_3}\right] = \frac{\mathbf{E}_1}{\mathbf{Z}_1}$$

Note Fig. 16.17 for the application of Kirchhoff's current law to node $\mathbf{V}_2$.

$$0 = \mathbf{I}_3 + \mathbf{I}_4 + \mathbf{I}$$

$$\frac{\mathbf{V}_2 - \mathbf{V}_1}{\mathbf{Z}_3} + \frac{\mathbf{V}_2}{\mathbf{Z}_4} + \mathbf{I} = 0$$

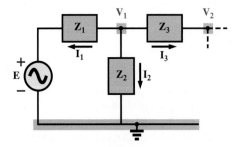

FIG. 16.16
Applying Kirchhoff's current law to node V_1 of Fig. 16.15.

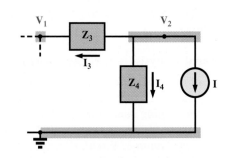

FIG. 16.17
Applying Kirchhoff's current law to node V_2 of Fig. 16.15.

Rearranging terms:

$$\mathbf{V}_2\left[\frac{1}{\mathbf{Z}_3}+\frac{1}{\mathbf{Z}_4}\right]-\mathbf{V}_1\left[\frac{1}{\mathbf{Z}_3}\right]=-\mathbf{I}$$

Grouping equations:

$$\mathbf{V}_1\left[\frac{1}{\mathbf{Z}_1}+\frac{1}{\mathbf{Z}_2}+\frac{1}{\mathbf{Z}_3}\right]-\mathbf{V}_2\left[\frac{1}{\mathbf{Z}_3}\right]=\frac{\mathbf{E}}{\mathbf{Z}_1}$$

$$\mathbf{V}_1\left[\frac{1}{\mathbf{Z}_3}\right]-\mathbf{V}_2\left[\frac{1}{\mathbf{Z}_3}+\frac{1}{\mathbf{Z}_4}\right]=\mathbf{I}$$

$$\frac{1}{\mathbf{Z}_1}+\frac{1}{\mathbf{Z}_2}+\frac{1}{\mathbf{Z}_3}=\frac{1}{0.5\text{ k}\Omega}+\frac{1}{j\,10\text{ k}\Omega}+\frac{1}{2\text{ k}\Omega}=2.5\text{ mS }\angle-2.29°$$

$$\frac{1}{\mathbf{Z}_3}+\frac{1}{\mathbf{Z}_4}=\frac{1}{2\text{ k}\Omega}+\frac{1}{-j\,5\text{ k}\Omega}=0.539\text{ mS }\angle21.80°$$

and

$$\mathbf{V}_1[2.5\text{ mS }\angle-2.29°]-\mathbf{V}_2[0.5\text{ mS }\angle0°]=24\text{ mA }\angle0°$$
$$\mathbf{V}_1[0.5\text{ mS }\angle0°]-\mathbf{V}_2[0.539\text{ mS }\angle21.80°]=4\text{ mA }\angle0°$$

with

$$\mathbf{V}_1=\frac{\begin{vmatrix}24\text{ mA }\angle0° & -0.5\text{ mS }\angle0° \\ 4\text{ mA }\angle0° & -0.539\text{ mS }\angle21.80°\end{vmatrix}}{\begin{vmatrix}2.5\text{ mS }\angle-2.29° & -0.5\text{ mS }\angle0° \\ 0.5\text{ mS }\angle0° & -0.539\text{ mS }\angle21.80°\end{vmatrix}}$$

$$=\frac{(24\text{ mA }\angle0°)(-0.539\text{ mS }\angle21.80°)+(0.5\text{ mS }\angle0°)(4\text{ mA }\angle0°)}{(2.5\text{ mS }\angle-2.29°)(-0.539\text{ mS }\angle21.80°)+(0.5\text{ mS }\angle0°)(0.5\text{ mS }\angle0°)}$$

$$=\frac{-12.94\times10^{-6}\text{ V }\angle21.80°+2\times10^{-6}\text{ V }\angle0°}{-1.348\times10^{-6}\angle19.51°+0.25\times10^{-6}\angle0°}$$

$$=\frac{-(12.01+j\,4.81)\times10^{-6}\text{ V}+2\times10^{-6}\text{ V}}{-(1.271+j\,0.45)\times10^{-6}+0.25\times10^{-6}}$$

$$=\frac{-10.01\text{ V}-j\,4.81\text{ V}}{-1.021-j\,0.45}=\frac{11.106\text{ V }\angle-154.33°}{1.116\angle-156.21°}$$

$$\mathbf{V}_1=\mathbf{9.95\text{ V }\angle1.88°}$$

Mathcad Solution: The length and the complexity of the above mathematical development strongly suggest the use of an alternative approach such as Mathcad. The printout of Fig. 16.18 first defines the letters **k** and **m** to specific numerical values so that the power-of-ten format did not have to be included in the equations. Thus, the results are cleaner and easier to review.

When entering the equations, remember that the *j* is entered as 1*j* **without** the multiplication sign between the 1 and the *j*. A multiplication sign between the two will define the *j* as another variable. Also be sure that the multiplication process is inserted between the nodal variables and the brackets. If an error signal continues to surface, it is often best to simply reenter the entire listing—errors are often not easy to spot simply by looking at the resulting equations.

Finally the results are obtained and converted to polar form for comparison with the theoretical solution. The solution of **9.949 A∠ 1.837°** is a very close confirmation of the long-hand solution.

Before leaving this example, let's look at another method for obtaining the polar form of the solution. The method appears in the bottom of Fig. 16.18. First **deg** is defined as shown, and then **arg** is taken from the **Insert-*f(x)*-Insert Function-arg** sequence. Next **V1** is entered; the result will be in radian form but with a small black rectangle in the place

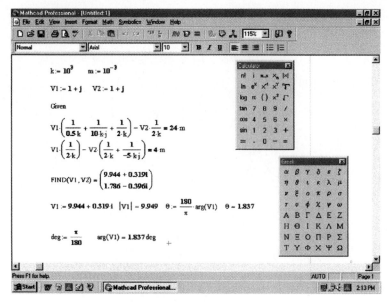

FIG. 16.18

Using Mathcad to verify the results of Example 16.7.

where the units normally appear. Click on that black rectangle, and the bracket will appear and **deg** can be typed. When the equals sign is selected, the angle in degrees will appear.

The format approach to nodal analysis is, as you might expect, the same as introduced for dc networks. The only change is the use of the term *admittance* rather than *conductance*.

Nodal Analysis (Format Approach):

1. *Choose a reference node and assign a subscripted nodal voltage to the remaining nodes of the network.*
2. *The number of required equations is equal to the number of resulting independent subscripted nodal voltages. Column 1 of each equation is formed by multiplying the subscripted nodal voltage of interest by the sum of the admittances connected to that node.*
3. *Column 2 contains the mutual terms which are always subtracted from the entries of the first column. Each mutual term is the product of the mutual admittance (the admittance connected between the two nodes) and the other nodal voltages.*
4. *The column to the right of the equals sign is the algebraic sum of the current sources connected to the node of interest. A current source is assigned a positive sign if it supplies current to the node, and a negative sign if it draws current from the node.*
5. *Solve the resulting simultaneous equations for the nodal voltages.*

EXAMPLE 16.8 Using the format approach to nodal analysis, find the voltage across the 4 Ω resistor in Fig. 16.19.

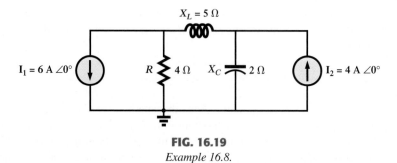

FIG. 16.19

Example 16.8.

Solution: Choosing nodes (Fig. 16.20) and writing the nodal equations, we have

$$\mathbf{Z}_1 = R = 4 \ \Omega \qquad \mathbf{Z}_2 = j \, X_L = j \, 5 \ \Omega \qquad \mathbf{Z}_3 = -j \, X_C = -j \, 2 \ \Omega$$

$$\mathbf{V}_1(\mathbf{Y}_1 + \mathbf{Y}_2) - \mathbf{V}_2(\mathbf{Y}_2) = -\mathbf{I}_1$$
$$\underline{\mathbf{V}_2(\mathbf{Y}_3 + \mathbf{Y}_2) - \mathbf{V}_1(\mathbf{Y}_2) = +\mathbf{I}_2}$$

or

$$\mathbf{V}_1(\mathbf{Y}_1 + \mathbf{Y}_2) - \mathbf{V}_2(\mathbf{Y}_2) \qquad = -\mathbf{I}_1$$
$$\underline{-\mathbf{V}_1(\mathbf{Y}_2) + \mathbf{V}_2(\mathbf{Y}_3 + \mathbf{Y}_2) = +\mathbf{I}_2}$$

$$\mathbf{Y}_1 = \frac{1}{\mathbf{Z}_1} \qquad \mathbf{Y}_2 = \frac{1}{\mathbf{Z}_2} \qquad \mathbf{Y}_3 = \frac{1}{\mathbf{Z}_3}$$

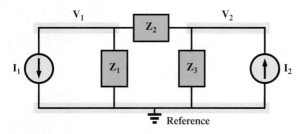

FIG. 16.20
Assigning the nodal voltages and subscripted impedances for the network of Fig. 16.19.

Using determinants yields

$$\mathbf{V}_1 = \frac{\begin{vmatrix} -\mathbf{I}_1 & -\mathbf{Y}_2 \\ +\mathbf{I}_2 & \mathbf{Y}_3 + \mathbf{Y}_2 \end{vmatrix}}{\begin{vmatrix} \mathbf{Y}_1 + \mathbf{Y}_2 & -\mathbf{Y}_2 \\ -\mathbf{Y}_2 & \mathbf{Y}_3 + \mathbf{Y}_2 \end{vmatrix}}$$

$$= \frac{-(\mathbf{Y}_3 + \mathbf{Y}_2)\mathbf{I}_1 + \mathbf{I}_2\mathbf{Y}_2}{(\mathbf{Y}_1 + \mathbf{Y}_2)(\mathbf{Y}_3 + \mathbf{Y}_2) - \mathbf{Y}_2^2}$$

$$= \frac{-(\mathbf{Y}_3 + \mathbf{Y}_2)\mathbf{I}_1 + \mathbf{I}_2\mathbf{Y}_2}{\mathbf{Y}_1\mathbf{Y}_3 + \mathbf{Y}_2\mathbf{Y}_3 + \mathbf{Y}_1\mathbf{Y}_2}$$

Substituting numerical values, we have

$$\mathbf{V}_1 = \frac{-[(1/-j\,2\ \Omega) + (1/j\,5\ \Omega)]6\ \text{A}\ \angle 0° + 4\ \text{A}\ \angle 0°(1/j\,5\ \Omega)}{(1/4\ \Omega)(1/-j\,2\ \Omega) + (1/j\,5\ \Omega)(1/-j\,2\ \Omega) + (1/4\ \Omega)(1/j\,5\ \Omega)}$$

$$= \frac{-(+j\,0.5 - j\,0.2)6\ \angle 0° + 4\ \angle 0°(-j\,0.2)}{(1/-j\,8) + (1/10) + (1/j\,20)}$$

$$= \frac{(-0.3\ \angle 90°)(6\ \angle 0°) + (4\ \angle 0°)(0.2\ \angle -90°)}{j\,0.125 + 0.1 - j\,0.05}$$

$$= \frac{-1.8\ \angle 90° + 0.8\ \angle -90°}{0.1 + j\,0.075}$$

$$= \frac{2.6\ \text{V}\ \angle -90°}{0.125\ \angle 36.87°}$$

$$\mathbf{V}_1 = \mathbf{20.80\ V} \angle \mathbf{-126.87°}$$

Mathcad Solution: For this example we will use the matrix format to find the nodal voltage $\mathbf{V}_1$. First the various parameters of the network are defined including the factor **deg** so that the phase angle will be displayed in degrees. Next the numerator is defined by **n,** and the **Matrix** icon is selected from the **Matrix** toolbar. Within the **Insert Matrix** dialog box, the **Rows** and **Columns** are set as 2 followed by an **OK** to place the 2×2 matrix on the screen. The parameters are then entered as shown in Fig. 16.21 using a left click of the mouse to select the parameter to be entered. Once the numerator is set, the process is repeated to define the numerator. Finally the equation for **V1** is defined, and the result in rectangular form will appear when the equals sign is selected. The magnitude and the angle are then found in polar form as described in earlier sections of this chapter. The results are again a clear confirmation of the theoretical result.

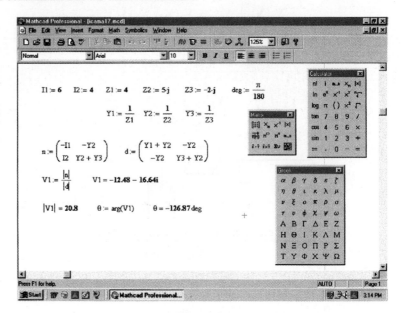

FIG. 16.21
Using Mathcad to verify the results of Example 16.8.

EXAMPLE 16.9   Using the format approach, write the nodal equations for the network of Fig. 16.22.

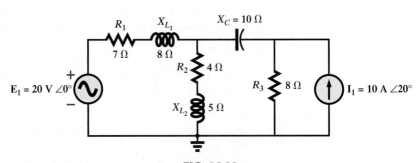

FIG. 16.22
Example 16.9.

Solution: The circuit is redrawn in Fig. 16.23, where

$$\mathbf{Z}_1 = R_1 + jX_{L_1} = 7\ \Omega + j\,8\ \Omega \qquad \mathbf{E}_1 = 20\ \text{V}\ \angle 0°$$

$$\mathbf{Z}_2 = R_2 + jX_{L_2} = 4\ \Omega + j\,5\ \Omega \qquad \mathbf{I}_1 = 10\ \text{A}\ \angle 20°$$

$$\mathbf{Z}_3 = -jX_C = -j\,10\ \Omega$$

$$\mathbf{Z}_4 = R_3 = 8\ \Omega$$

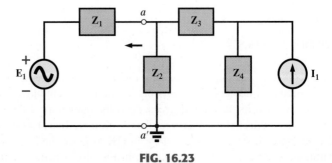

FIG. 16.23
Assigning the subscripted impedances for the network of Fig. 16.22.

Converting the voltage source to a current source and choosing nodes, we obtain Fig. 16.24. Note the neat appearance of the network using the subscripted impedances. Working directly with Fig. 16.22 would be more difficult and could produce errors.

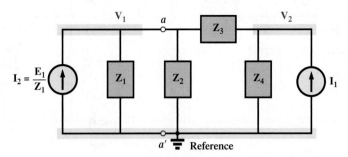

FIG. 16.24

Converting the voltage source of Fig. 16.23 to a current source and defining the nodal voltages.

Write the nodal equations:

$$\mathbf{V}_1(\mathbf{Y}_1 + \mathbf{Y}_2 + \mathbf{Y}_3) - \mathbf{V}_2(\mathbf{Y}_3) = +\mathbf{I}_2$$
$$\mathbf{V}_2(\mathbf{Y}_3 + \mathbf{Y}_4) - \mathbf{V}_1(\mathbf{Y}_3) = +\mathbf{I}_1$$

$$\mathbf{Y}_1 = \frac{1}{\mathbf{Z}_1} \qquad \mathbf{Y}_2 = \frac{1}{\mathbf{Z}_2} \qquad \mathbf{Y}_3 = \frac{1}{\mathbf{Z}_3} \qquad \mathbf{Y}_4 = \frac{1}{\mathbf{Z}_4}$$

which are rewritten as

$$\mathbf{V}_1(\mathbf{Y}_1 + \mathbf{Y}_2 + \mathbf{Y}_3) - \mathbf{V}_2(\mathbf{Y}_3) \qquad = +\mathbf{I}_2$$
$$-\mathbf{V}_1(\mathbf{Y}_3) \qquad \qquad + \mathbf{V}_2(\mathbf{Y}_3 + \mathbf{Y}_4) = +\mathbf{I}_1$$

EXAMPLE 16.10 Write the nodal equations for the network of Fig. 16.25. Do not solve them.

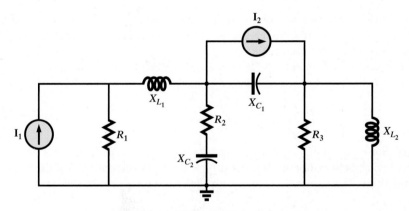

FIG. 16.25

Network for Example 16.10.

Solution: Choose nodes (Fig. 16.26):

$$\mathbf{Z}_1 = R_1 \qquad \mathbf{Z}_2 = j\,X_{L_1} \qquad \mathbf{Z}_3 = R_2 - j\,X_{C_2}$$
$$\mathbf{Z}_4 = -j\,X_{C_1} \qquad \mathbf{Z}_5 = R_3 \qquad \mathbf{Z}_6 = j\,X_{L_2}$$

and write the nodal equations:

$$\mathbf{V}_1(\mathbf{Y}_1 + \mathbf{Y}_2) - \mathbf{V}_2(\mathbf{Y}_2) = +\mathbf{I}_1$$
$$\mathbf{V}_2(\mathbf{Y}_2 + \mathbf{Y}_3 + \mathbf{Y}_4) - \mathbf{V}_1(\mathbf{Y}_2) - \mathbf{V}_3(\mathbf{Y}_4) = -\mathbf{I}_2$$
$$\mathbf{V}_3(\mathbf{Y}_4 + \mathbf{Y}_5 + \mathbf{Y}_6) - \mathbf{V}_2(\mathbf{Y}_4) = +\mathbf{I}_2$$

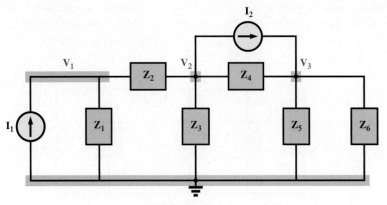

FIG. 16.26

Assigning the nodal voltages and subscripted impedances for the network of Fig. 16.25.

which are rewritten as

$$\mathbf{V}_1(\mathbf{Y}_1 + \mathbf{Y}_2) \quad - \mathbf{V}_2(\mathbf{Y}_2) \qquad\qquad\qquad + 0 \qquad\qquad\qquad = +\mathbf{I}_1$$
$$-\mathbf{V}_1(\mathbf{Y}_2) \qquad + \mathbf{V}_2(\mathbf{Y}_2 + \mathbf{Y}_3 + \mathbf{Y}_4) \quad - \mathbf{V}_3(\mathbf{Y}_4) \qquad\qquad = -\mathbf{I}_2$$
$$0 \qquad\qquad\qquad - \mathbf{V}_2(\mathbf{Y}_4) \qquad\qquad + \mathbf{V}_3(\mathbf{Y}_4 + \mathbf{Y}_5 + \mathbf{Y}_6) = +\mathbf{I}_2$$

$$\mathbf{Y}_1 = \frac{1}{R_1} \qquad\qquad \mathbf{Y}_2 = \frac{1}{j\,X_{L_1}} \qquad \mathbf{Y}_3 = \frac{1}{R_2 - j\,X_{C_2}}$$

$$\mathbf{Y}_4 = \frac{1}{-j\,X_{C_1}} \qquad \mathbf{Y}_5 = \frac{1}{R_3} \qquad \mathbf{Y}_6 = \frac{1}{j\,X_{L_2}}$$

Note the symmetry about the diagonal for Example 16.10 and the examples preceding it in this section.

16.5 BRIDGE NETWORKS

The basic **bridge network** was discussed in some detail in Section 8.11 for dc networks. The discussion in this section will review the changes that occur due to reactive components and sinusoidal voltages and currents.

For the bridge configuration of Fig. 16.27, balance conditions are defined by

$$\boxed{\frac{\mathbf{Z}_1}{\mathbf{Z}_3} = \frac{\mathbf{Z}_2}{\mathbf{Z}_4}} \qquad\qquad \textbf{(16.1)}$$

Although this equation may appear to be exactly the same as that obtained for dc networks, it is very important to realize that **it is not sufficient that the magnitudes of the ratios have the same value. The angles associated with each ratio must also be the same.**

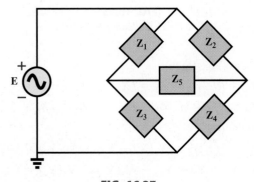

FIG. 16.27

Bridge configuration.

For instance, for the network of Fig. 16.28, it may appear that a balance condition has been established, but this will not be the case because the resulting phase angles are not the same. That is,

$$\frac{\mathbf{Z}_1}{\mathbf{Z}_3} = \frac{\mathbf{Z}_2}{\mathbf{Z}_4}$$

$$\frac{16\ \Omega\ \angle 90°}{4\ \Omega\ \angle 90°} = \frac{8\ \Omega\ \angle 0°}{2\ \Omega\ \angle -90°}$$

$$4\ \angle 0° \neq 4\ \angle 90° \qquad (\text{not balanced})$$

For the **Hay bridge** of Fig. 16.29,

$$\mathbf{Z}_1 = R_1 - j\,X_C \qquad \mathbf{Z}_2 = R_2$$

$$\mathbf{Z}_3 = R_3 \qquad \mathbf{Z}_4 = R_4 + j\,X_L$$

Resistance R_4 and inductance L are to be determined.

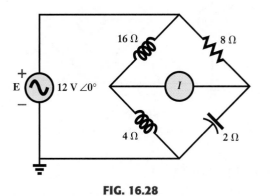

FIG. 16.28
Testing the balance condition.

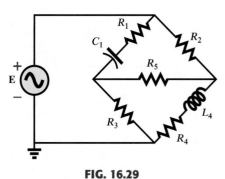

FIG. 16.29
Hay bridge.

Substituting into the modified form Eq. (16.1) will result in

$$\mathbf{Z}_2\mathbf{Z}_3 = \mathbf{Z}_4\mathbf{Z}_1$$

$$R_2R_3 = (R_4 + j\,X_L)(R_1 - j\,X_C)$$

$$R_2R_3 = R_1R_4 + j\,(R_1X_L - R_4X_C) + X_CX_L$$

so that $\qquad R_2R_3 + j\,0 = (R_1\,R_4 + X_CX_L) + j\,(R_1X_L - R_4X_C)$

For the equations to be equal, and to ensure that the current through the galvanometer is zero amperes, the real and imaginary parts on each side of the equals sign must be equal. That is,

$$R_2R_3 = R_1R_4 + X_CX_L$$

and $\qquad\qquad 0 = R_1X_L - R_4X_C$

Substituting $\qquad\qquad X_L = \omega L \quad \text{and} \quad X_C = \dfrac{1}{\omega C}$

will then result in the following balance equations:

$$\boxed{L = \frac{CR_2R_3}{1 + \omega^2 C^2 R_1^2}} \qquad\qquad \textbf{(16.2)}$$

and

$$\boxed{R_4 = \frac{\omega^2 C^2 R_1 R_2 R_3}{1 + \omega^2 C^2 R_1}} \qquad\qquad \textbf{(16.3)}$$

Note that the results are frequency dependent and thus the inductance and resistance levels of the unknown quantities are sensitive to the applied frequency.

For the **Maxwell bridge** of Fig. 16.30, the resulting equations for the unknowns are

$$\boxed{L_4 = C_1R_2R_3} \qquad\qquad \textbf{(16.4)}$$

and
$$R_4 = \frac{R_2 R_3}{R_1}$$ **(16.5)**

For the **capacitance comparison bridge** of Fig. 16.31, the balance equations are

$$C_4 = C_3 \frac{R_1}{R_2}$$ **(16.6)**

and
$$R_4 = R_2 \frac{R_3}{R_1}$$ **(16.7)**

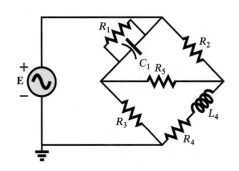

FIG. 16.30
Maxwell bridge.

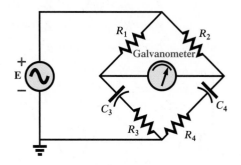

FIG. 16.31
Capacitance comparison bridge.

16.6 Δ-Y AND Y-Δ CONVERSIONS

The Δ-Y, Y-Δ (or π-T, T-π as defined in Section 8.12) conversions for ac circuits will not be derived here since the development corresponds exactly with that for dc circuits. Taking the **Δ-Y configuration** shown in Fig. 16.32, we find the general equations for the impedances of the Y in terms of those for the Δ:

$$\mathbf{Z}_1 = \frac{\mathbf{Z}_B \mathbf{Z}_C}{\mathbf{Z}_A + \mathbf{Z}_B + \mathbf{Z}_C}$$ **(16.8)**

$$\mathbf{Z}_2 = \frac{\mathbf{Z}_A \mathbf{Z}_C}{\mathbf{Z}_A + \mathbf{Z}_B + \mathbf{Z}_C}$$ **(16.9)**

$$\mathbf{Z}_3 = \frac{\mathbf{Z}_A \mathbf{Z}_B}{\mathbf{Z}_A + \mathbf{Z}_B + \mathbf{Z}_C}$$ **(16.10)**

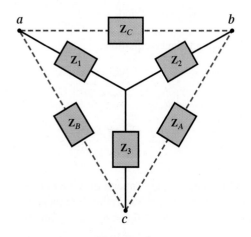

FIG. 16.32
Δ-Y configuration.

For the impedances of the Δ in terms of those for the Y, the equations are

$$\mathbf{Z}_B = \frac{\mathbf{Z}_1\mathbf{Z}_2 + \mathbf{Z}_1\mathbf{Z}_3 + \mathbf{Z}_2\mathbf{Z}_3}{\mathbf{Z}_2}$$ **(16.11)**

$$\mathbf{Z}_A = \frac{\mathbf{Z}_1\mathbf{Z}_2 + \mathbf{Z}_1\mathbf{Z}_3 + \mathbf{Z}_2\mathbf{Z}_3}{\mathbf{Z}_1}$$ **(16.12)**

$$\mathbf{Z}_C = \frac{\mathbf{Z}_1\mathbf{Z}_2 + \mathbf{Z}_1\mathbf{Z}_3 + \mathbf{Z}_2\mathbf{Z}_3}{\mathbf{Z}_3}$$ **(16.13)**

Note that each impedance of the Y is equal to the product of the impedances in the two closest branches of the Δ, divided by the sum of the impedances in the Δ.

Further, the value of each impedance of the Δ is equal to the sum of the possible product combinations of the impedances of the Y, divided by the impedances of the Y farthest from the impedance to be determined.

Drawn in different forms (Fig. 16.33), they are also referred to as the T and π configurations.

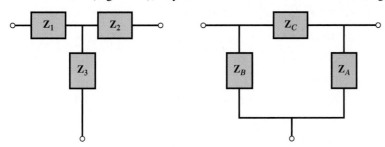

FIG. 16.33
The T and π configurations.

In the study of dc networks, we found that if all of the resistors of the Δ or Y were the same, the conversion from one to the other could be accomplished using the equation

$$R_\Delta = 3R_Y \quad \text{or} \quad R_Y = \frac{R_\Delta}{3}$$

For ac networks,

$$\mathbf{Z}_\Delta = 3\mathbf{Z}_Y \quad \text{or} \quad \mathbf{Z}_Y = \frac{\mathbf{Z}_\Delta}{3}$$ **(16.14)**

Be careful when using this simplified form. It is not sufficient for all the impedances of the Δ or Y to be of the same magnitude: **The angle associated with each must also be the same.**

EXAMPLE 16.11 Find the total impedance $\mathbf{Z}_T$ of the network of Fig. 16.34.

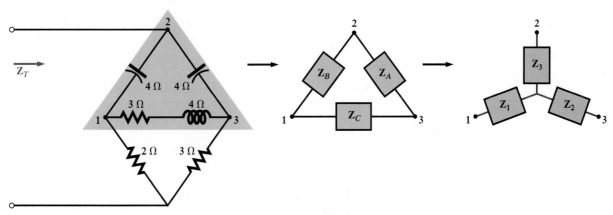

FIG. 16.34
Converting the upper Δ of a bridge configuration to a Y (Example 16.11).

Solution:

$$\mathbf{Z}_B = -j\,4 \qquad \mathbf{Z}_A = -j\,4 \qquad \mathbf{Z}_C = 3 + j\,4$$

$$\mathbf{Z}_1 = \frac{\mathbf{Z}_B\mathbf{Z}_C}{\mathbf{Z}_A + \mathbf{Z}_B + \mathbf{Z}_C} = \frac{(-j\,4\,\Omega)(3\,\Omega + j\,4\,\Omega)}{(-j\,4\,\Omega) + (-j\,4\,\Omega) + (3\,\Omega + j\,4\,\Omega)}$$

$$= \frac{(4\angle{-90°})(5\angle 53.13°)}{3 - j\,4} = \frac{20\angle{-36.87°}}{5\angle{-53.13°}}$$

$$= 4\,\Omega\,\angle 16.13° = 3.84\,\Omega + j\,1.11\,\Omega$$

$$\mathbf{Z}_2 = \frac{\mathbf{Z}_A\mathbf{Z}_C}{\mathbf{Z}_A + \mathbf{Z}_B + \mathbf{Z}_C} = \frac{(-j\,4\,\Omega)(3\,\Omega + j\,4\,\Omega)}{5\,\Omega\,\angle{-53.13°}}$$

$$= 4\,\Omega\,\angle 16.13° = 3.84\,\Omega + j\,1.11\,\Omega$$

Recall from the study of dc circuits that if two branches of the Y or Δ were the same, the corresponding Δ or Y, respectively, would also have two similar branches. In this example, $\mathbf{Z}_A = \mathbf{Z}_B$. Therefore, $\mathbf{Z}_1 = \mathbf{Z}_2$, and

$$\mathbf{Z}_3 = \frac{\mathbf{Z}_A\mathbf{Z}_B}{\mathbf{Z}_A + \mathbf{Z}_B + \mathbf{Z}_C} = \frac{(-j\,4\,\Omega)(-j\,4\,\Omega)}{5\,\Omega\,\angle{-53.13°}}$$

$$= \frac{16\,\Omega\,\angle{-180°}}{5\angle{-53.13°}} = 3.2\,\Omega\,\angle{-126.87°} = -1.92\,\Omega - j\,2.56\,\Omega$$

Replace the Δ by the Y (Fig. 16.35):

$$\mathbf{Z}_1 = 3.84\,\Omega + j\,1.11\,\Omega \qquad \mathbf{Z}_2 = 3.84\,\Omega + j\,1.11\,\Omega$$

$$\mathbf{Z}_3 = -1.92\,\Omega - j\,2.56\,\Omega \qquad \mathbf{Z}_4 = 2\,\Omega$$

$$\mathbf{Z}_5 = 3\,\Omega$$

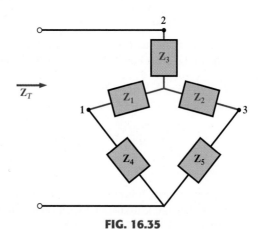

FIG. 16.35

The network of Fig. 16.34 following the substitution of the Y configuration.

Impedances $\mathbf{Z}_1$ and $\mathbf{Z}_4$ are in series:

$$\mathbf{Z}_{T_1} = \mathbf{Z}_1 + \mathbf{Z}_4 = 3.84\,\Omega + j\,1.11\,\Omega + 2\,\Omega = 5.84\,\Omega + j\,1.11\,\Omega$$

$$= 5.94\,\Omega\,\angle 10.76°$$

Impedances $\mathbf{Z}_2$ and $\mathbf{Z}_5$ are in series:

$$\mathbf{Z}_{T_2} = \mathbf{Z}_2 + \mathbf{Z}_5 = 3.84\,\Omega + j\,1.11\,\Omega + 3\,\Omega = 6.84\,\Omega + j\,1.11\,\Omega$$

$$= 6.93\,\Omega\,\angle 9.22°$$

Impedances $\mathbf{Z}_{T_1}$ and $\mathbf{Z}_{T_2}$ are in parallel:

$$\mathbf{Z}_{T_3} = \frac{\mathbf{Z}_{T_1}\mathbf{Z}_{T_2}}{\mathbf{Z}_{T_1} + \mathbf{Z}_{T_2}} = \frac{(5.94\,\Omega\,\angle 10.76°)(6.93\,\Omega\,\angle 9.22°)}{5.84\,\Omega + j\,1.11\,\Omega + 6.84\,\Omega + j\,1.11\,\Omega}$$

$$= \frac{41.16\,\Omega\,\angle 19.98°}{12.68 + j\,2.22} = \frac{41.16\,\Omega\,\angle 19.98°}{12.87\,\angle 9.93°} = 3.198\,\Omega\,\angle 10.05°$$

$$= 3.15\,\Omega + j\,0.56\,\Omega$$

Impedances $\mathbf{Z}_3$ and $\mathbf{Z}_{T_3}$ are in series. Therefore,

$$\mathbf{Z}_T = \mathbf{Z}_3 + \mathbf{Z}_{T_3} = -1.92\ \Omega - j\,2.56\ \Omega + 3.15\ \Omega + j\,0.56\ \Omega$$
$$= 1.23\ \Omega - j\,2.0\ \Omega = \mathbf{2.35\ \Omega\ \angle-58.41°}$$

16.7 SUPERPOSITION THEOREM

The content of this section will parallel that of the superposition theorem in Chapter 9. The only major differences will be the need to work with impedances and ac sources rather than just resistors and fixed dc supplies.

Recall from Chapter 9 that the **superposition theorem** eliminates the need for solving simultaneous linear equations by considering the effects of each source independently. To consider the effects of each source, we must remove the remaining sources. We accomplish this by setting voltage sources to zero (short-circuit representation) and current sources to zero (open-circuit representation). The current through, or voltage across, a portion of the network produced by each source is then added algebraically to find the total solution for the current or voltage.

The superposition theorem is not applicable to power effects in ac networks because we are still dealing with a nonlinear relationship. It can be applied to networks with sources of different frequencies only if the total response for *each* frequency is found independently and the results are expanded in a nonsinusoidal expression.

One of the most frequent applications of the superposition theorem is to electronic systems in which the dc and ac analyses are treated separately and the total solution is the sum of the two. It is an important application of the theorem because the impact of the reactive elements changes dramatically in response to the two types of independent sources. In addition, the dc analysis of an electronic system can often define important parameters for the ac analysis. Example 16.15 will demonstrate the impact of the applied source on the general configuration of the network.

EXAMPLE 16.12 Using the superposition theorem, find the current $\mathbf{I}$ through the 4 Ω reactance (X_{L_2}) of Fig. 16.36.

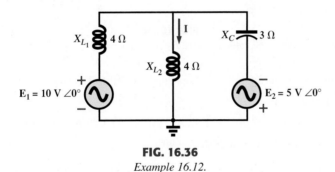

FIG. 16.36
Example 16.12.

Solution: For the redrawn circuit (Fig. 16.37),

$$\mathbf{Z}_1 = +j\,X_{L_1} = j\,4\ \Omega$$
$$\mathbf{Z}_2 = +j\,X_{L_2} = j\,4\ \Omega$$
$$\mathbf{Z}_3 = -j\,X_C = -j\,3\ \Omega$$

Considering the effects of the voltage source $\mathbf{E}_1$ (Fig. 16.38), we have

$$\mathbf{Z}_{2\|3} = \frac{\mathbf{Z}_2\mathbf{Z}_3}{\mathbf{Z}_2 + \mathbf{Z}_3} = \frac{(j\,4\ \Omega)(-j\,3\ \Omega)}{j\,4\ \Omega - j\,3\ \Omega} = \frac{12\ \Omega}{j} = -j\,12\ \Omega = 12\ \Omega\ \angle-90°$$

$$I_{s_1} = \frac{\mathbf{E}_1}{\mathbf{Z}_{2\|3} + \mathbf{Z}_1} = \frac{10\ \text{V}\ \angle0°}{-j\,12\ \Omega + j\,4\ \Omega} = \frac{10\ \text{V}\ \angle0°}{8\ \Omega\ \angle-90°} = 1.25\ \text{A}\ \angle90°$$

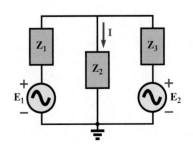

FIG. 16.37
*Assigning the subscripted
impedances to the network of
Fig. 16.36.*

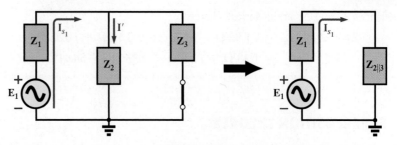

FIG. 16.38
Determining the effect of the voltage source E_1 on the current I of the network of Fig. 16.36.

and
$$\mathbf{I'} = \frac{\mathbf{Z}_3\mathbf{I}_{s_1}}{\mathbf{Z}_2 + \mathbf{Z}_3} \quad \text{(current divider rule)}$$

$$= \frac{(-j\,3\,\Omega)(j\,1.25\,\text{A})}{j\,4\,\Omega - j\,3\,\Omega} = \frac{3.75\,\text{A}}{j\,1} = 3.75\,\text{A}\,\angle-90°$$

Considering the effects of the voltage source $\mathbf{E}_2$ (Fig. 16.39), we have

$$\mathbf{Z}_{1\|2} = \frac{\mathbf{Z}_1}{N} = \frac{j\,4\,\Omega}{2} = j\,2\,\Omega$$

$$\mathbf{I}_{s_2} = \frac{\mathbf{E}_2}{\mathbf{Z}_{1\|2} + \mathbf{Z}_3} = \frac{5\,\text{V}\,\angle0°}{j\,2\,\Omega - j\,3\,\Omega} = \frac{5\,\text{V}\,\angle0°}{1\,\Omega\,\angle-90°} = 5\,\text{A}\,\angle90°$$

and
$$\mathbf{I''} = \frac{\mathbf{I}_{s_2}}{2} = 2.5\,\text{A}\,\angle90°$$

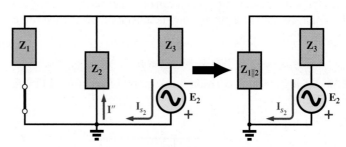

FIG. 16.39
Determining the effect of the voltage source E_2 on the current I of the network of Fig. 16.36.

FIG. 16.40
*Determining the resultant current
for the network of Fig. 16.36.*

The resultant current through the 4 Ω reactance X_{L_2} (Fig. 16.40) is

$$\mathbf{I} = \mathbf{I'} - \mathbf{I''}$$

$$= 3.75\,\text{A}\,\angle-90° - 2.50\,\text{A}\,\angle90° = -j\,3.75\,\text{A} - j\,2.50\,\text{A}$$

$$= -j\,6.25\,\text{A} = \mathbf{6.25\,\text{A}\,\angle-90°}$$

EXAMPLE 16.13 Using superposition, find the current **I** through the 6 Ω resistor of Fig. 16.41.

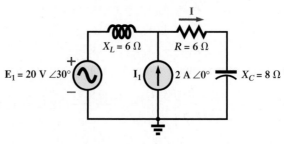

FIG. 16.41
Example 16.13.

Solution: For the redrawn circuit (Fig. 16.42),

$$\mathbf{Z}_1 = j\,6\,\Omega \qquad \mathbf{Z}_2 = 6 - j\,8\,\Omega$$

Consider the effects of the current source (Fig. 16.43). Applying the current divider rule, we have

$$\mathbf{I}' = \frac{\mathbf{Z}_1\mathbf{I}_1}{\mathbf{Z}_1 + \mathbf{Z}_2} = \frac{(j\,6\,\Omega)(2\text{ A})}{j\,6\,\Omega + 6\,\Omega - j\,8\,\Omega} = \frac{j\,12\text{ A}}{6 - j\,2}$$

$$= \frac{12\text{ A }\angle 90°}{6.32\ \angle{-18.43°}} = 1.9\text{ A }\angle 108.43°$$

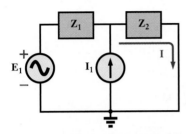

FIG. 16.42

Assigning the subscripted impedances to the network of Fig. 16.41.

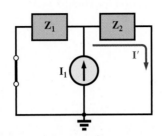

FIG. 16.43

Determining the effect of the current source $\mathbf{I}_1$ on the current $\mathbf{I}$ of the network of Fig. 16.41.

Consider the effects of the voltage source (Fig. 16.44). Applying Ohm's law gives us

$$\mathbf{I}'' = \frac{\mathbf{E}_1}{\mathbf{Z}_T} = \frac{\mathbf{E}_1}{\mathbf{Z}_1 + \mathbf{Z}_2} = \frac{20\text{ V }\angle 30°}{6.32\ \Omega\ \angle{-18.43°}} = 3.16\text{ A }\angle 48.43°$$

The total current through the 6 Ω resistor (Fig. 16.45) is

$$\mathbf{I} = \mathbf{I}' + \mathbf{I}'' = 1.9\text{ A }\angle 108.43° + 3.16\text{ A }\angle 48.43°$$

$$= (-0.60\text{ A} + j\,1.80\text{ A}) + (2.10\text{ A} + j\,2.36\text{ A})$$

$$= 1.50\text{ A} + j\,4.16\text{ A} = \mathbf{4.42\text{ A }\angle 70.2°}$$

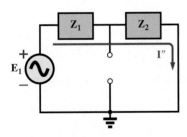

FIG. 16.44

Determining the effect of the voltage source $\mathbf{E}_1$ on the current $\mathbf{I}$ of the network of Fig. 16.41.

FIG. 16.45

Determining the resultant current $\mathbf{I}$ for the network of Fig. 16.41.

EXAMPLE 16.14 Using superposition, find the voltage across the 6 Ω resistor in Fig. 16.41. Check the results against $\mathbf{V}_{6\Omega} = \mathbf{I}(6\,\Omega)$, where $\mathbf{I}$ is the current found through the 6 Ω resistor in Example 16.13.

Solution: For the current source,

$$\mathbf{V}'_{6\Omega} = \mathbf{I}'(6\,\Omega) = (1.9\text{ A }\angle 108.43°)(6\,\Omega) = 11.4\text{ V }\angle 108.43°$$

For the voltage source,

$$\mathbf{V}''_{6\Omega} = \mathbf{I}''(6\,\Omega) = (3.16\text{ A }\angle 48.43°)(6\,\Omega) = 18.96\text{ V }\angle 48.43°$$

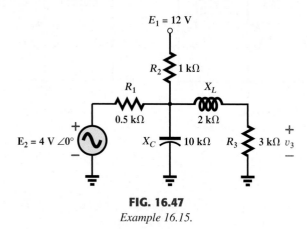

+ V'₆Ω −

+ V''₆Ω −

R
6 Ω

+ V₆Ω −

FIG. 16.46

Determining the resultant voltage V₆Ω for the network of Fig. 16.41 (Example 16.14).

The total voltage across the 6 Ω resistor (Fig. 16.46) is

$$\mathbf{V}_{6\Omega} = \mathbf{V}'_{6\Omega} + \mathbf{V}''_{6\Omega} = 11.4 \text{ V} \angle 108.43° + 18.96 \text{ V} \angle 48.43°$$

$$= (-3.60 \text{ V} + j\,10.82 \text{ V}) + (12.58 \text{ V} + j\,14.18 \text{ V})$$

$$= 8.98 \text{ V} + j\,25.0 \text{ V} = \mathbf{26.5 \text{ V}} \angle \mathbf{70.2°}$$

Checking the result, we have

$$\mathbf{V}_{6\Omega} = \mathbf{I}(6\,\Omega) = (4.42 \text{ A} \angle 70.2°)(6\,\Omega) = \mathbf{26.5 \text{ V}} \angle \mathbf{70.2°} \quad \text{(checks)}$$

EXAMPLE 16.15 For the network of Fig. 16.47, use superposition to determine the sinusoidal expression for voltage v_3.

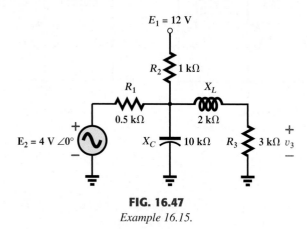

FIG. 16.47

Example 16.15.

Solution: For the dc source, recall that for dc analysis, in the steady state the capacitor can be replaced by an open-circuit equivalent, and the inductor by a short-circuit equivalent. The result is the network of Fig. 16.48.

Resistors R_1 and R_3 are then in parallel, and voltage V_3 can be determined using the voltage divider rule:

$$R' = R_1 \| R_3 = 0.5 \text{ k}\Omega \| 3 \text{ k}\Omega = 0.429 \text{ k}\Omega$$

and

$$V_3 = \frac{R'E_1}{R' + R_2}$$

$$= \frac{(0.429 \text{ k}\Omega)(12 \text{ V})}{0.429 \text{ k}\Omega + 1 \text{ k}\Omega} = \frac{5.148 \text{ V}}{1.429} \cong \mathbf{3.6 \text{ V}}$$

For ac analysis, the dc source is set to zero and the network is redrawn as shown in Fig. 16.49.

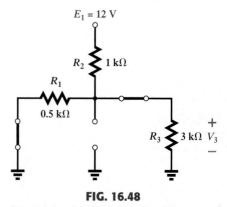

FIG. 16.48

Determining the effect of the dc voltage source E_1 on voltage v_3 of the network of Fig. 16.47.

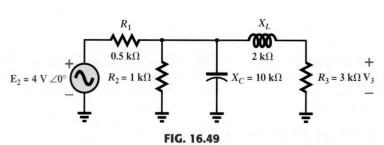

FIG. 16.49

Redrawing the network of Fig. 16.47 to determine the effect of the ac voltage source E_2.

The block impedances are then defined as in Fig. 16.50, and series-parallel techniques are applied as follows:

$$\mathbf{Z}_1 = 0.5 \text{ k}\Omega \angle 0°$$

$$\mathbf{Z}_2 = (R_2 \angle 0° \parallel (X_C \angle -90°)$$

$$= \frac{(1 \text{ k}\Omega \angle 0°)(10 \text{ k}\Omega \angle -90°)}{1 \text{ k}\Omega - j\, 10 \text{ k}\Omega} = \frac{10 \text{ k}\Omega \angle -90°}{10.05 \angle -84.29°}$$

$$= 0.995 \text{ k}\Omega \angle -5.71°$$

$$\mathbf{Z}_3 = R_3 + j\, X_L = 3 \text{ k}\Omega + j\, 2 \text{ k}\Omega = 3.61 \text{ k}\Omega \angle 33.69°$$

Therefore, $\mathbf{Z}_T = \mathbf{Z}_1 + \mathbf{Z}_2 \parallel \mathbf{Z}_3$

$$= 0.5 \text{ k}\Omega + (0.995 \text{ k}\Omega \angle -5.71°) \parallel (3.61 \text{ k}\Omega \angle 33.69°)$$

$$= 1.312 \text{ k}\Omega \angle 1.57°$$

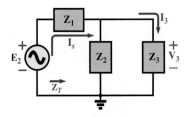

FIG. 16.50
Assigning the subscripted impedances to the network of Fig. 16.49.

Calculator Solution: Performing the above on the TI-86 calculator: will result in the sequence of Fig. 16.51.

$$\mathbf{I}_s = \frac{\mathbf{E}_2}{\mathbf{Z}_T} = \frac{4 \text{ V} \angle 0°}{1.312 \text{ k}\Omega \angle 1.57°} = 3.05 \text{ mA} \angle -1.57°$$

(0.5,0)+((0.995∠−5.71)*(3.61∠33.69))/((0.995∠−5.71)+(3.61∠33.69)) (ENTER)

((1.311E0,35.373E−3)
Ans▶Pol
(1.312E0∠1.545E0)

FIG. 16.51
Calculator sequence for finding the total impedance for the network of Fig. 16.50.

Current divider rule:

$$\mathbf{I}_3 = \frac{\mathbf{Z}_2 \mathbf{I}_s}{\mathbf{Z}_2 + \mathbf{Z}_3} = \frac{(0.995 \text{ k}\Omega \angle -5.71°)(3.05 \text{ mA} \angle -1.57°)}{0.995 \text{ k}\Omega \angle -5.71° + 3.61 \text{ k}\Omega \angle 33.69°}$$

$$= 0.686 \text{ mA} \angle -32.74°$$

with $\mathbf{V}_3 = (I_3 \angle \theta)(R_3 \angle 0°) = (0.686 \text{ mA} \angle -32.74°)(3 \text{ k}\Omega \angle 0°)$

$$= \mathbf{2.06 \text{ V}} \angle \mathbf{-32.74°}$$

The total solution is

$$v_3 = v_3 \text{ (dc)} + v_3 \text{ (ac)} = 3.6 \text{ V} + 2.06 \text{ V} \angle -32.74°$$

$$= \mathbf{3.6 + 2.91 \sin(\omega t - 32.74°)}$$

The result is a sinusoidal voltage having a peak value of 2.91 V riding on an average value of 3.6 V, as shown in Fig. 16.52.

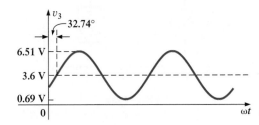

FIG. 16.52
The resultant voltage v_3 for the network of Fig. 16.47.

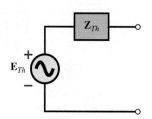

FIG. 16.53

*Thévenin equivalent circuit for
ac networks.*

16.8 THÉVENIN'S THEOREM

Thévenin's theorem, as stated for sinusoidal ac circuits, is changed only to include the term *impedance* instead of *resistance*. That is,

any two-terminal linear ac network can be replaced with an equivalent circuit consisting of a voltage source and an impedance in series, as shown in Fig. 16.53.

Since the reactances of a circuit are frequency dependent, the Thévenin circuit found for a particular network **is applicable only at one frequency.**

The steps required to apply this method to dc circuits are repeated here with changes for sinusoidal ac circuits. As before, the only change is the replacement of the term *resistance* with *impedance*. Again, dependent and independent sources will be treated separately.

1. *Remove the portion of the network across which the Thévenin equivalent circuit is to be found.*
2. *Mark (○, •, and so on) the terminals of the remaining two-terminal network.*
3. *Calculate Z_{Th} by first setting all voltage and current sources to zero (short circuit and open circuit, respectively) and then finding the resulting impedance between the two marked terminals.*
4. *Calculate E_{Th} by first replacing the voltage and current sources and then finding the open-circuit voltage between the marked terminals.*
5. *Draw the Thévenin equivalent circuit with the portion of the circuit previously removed replaced between the terminals of the Thévenin equivalent circuit.*

EXAMPLE 16.16 Find the Thévenin equivalent circuit for the network external to resistor *R* in Fig. 16.54.

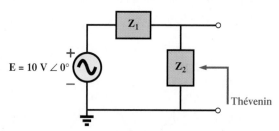

FIG. 16.54

Example 16.16.

Solution:

Steps 1–2 (Fig. 16.55):

$$\mathbf{Z}_1 = jX_L = j\,8\ \Omega \qquad \mathbf{Z}_2 = -jX_C = -j\,2\ \Omega$$

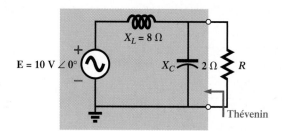

FIG. 16.55

Assigning the subscripted impedances to the network of Fig. 16.54.

Step 3 (Fig. 16.56):

$$\mathbf{Z}_{Th} = \frac{\mathbf{Z}_1\mathbf{Z}_2}{\mathbf{Z}_1 + \mathbf{Z}_2} = \frac{(j\,8\ \Omega)(-j\,2\ \Omega)}{j\,8\ \Omega - j\,2\ \Omega} = \frac{-j^2\,16\ \Omega}{j\,6} = \frac{16\ \Omega}{6\,\angle 90°} = \mathbf{2.67\ \Omega\ \angle{-90°}}$$

Step 4 (Fig. 16.57):

$$\mathbf{E}_{Th} = \frac{\mathbf{Z}_2 \mathbf{E}}{\mathbf{Z}_1 + \mathbf{Z}_2} \qquad \text{(voltage divider rule)}$$

$$= \frac{(-j\,2\,\Omega)(10\text{ V})}{j\,8\,\Omega - j\,2\,\Omega} = \frac{-j\,20\text{ V}}{j\,6} = \mathbf{3.33\text{ V} \angle -180°}$$

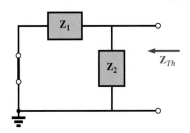

FIG. 16.56
Determining the Thévenin impedance for the network of Fig. 16.54.

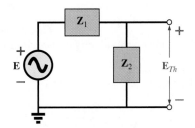

FIG. 16.57
Determining the open-circuit Thévenin voltage for the network of Fig. 16.54.

Step 5: The Thévenin equivalent circuit is shown in Fig. 16.58.

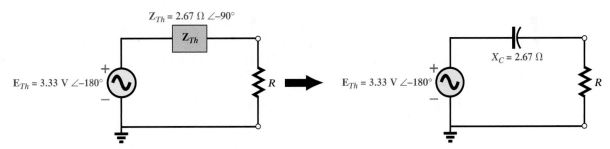

FIG. 16.58
The Thévenin equivalent circuit for the network of Fig. 16.54.

EXAMPLE 16.17 Find the Thévenin equivalent circuit for the network external to branch *a-a'* in Fig. 16.59.

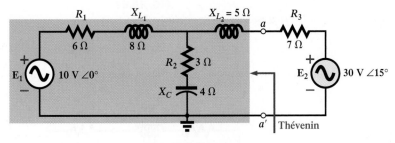

FIG. 16.59
Example 16.17.

Solution:

Steps 1–2 (Fig. 16.60): Note the reduced complexity with subscripted impedances:

$$\mathbf{Z}_1 = R_1 + j\,X_{L_1} = 6\,\Omega + j\,8\,\Omega$$

$$\mathbf{Z}_2 = R_2 - j\,X_C = 3\,\Omega - j\,4\,\Omega$$

$$\mathbf{Z}_3 = +j\,X_{L_2} = j\,5\,\Omega$$

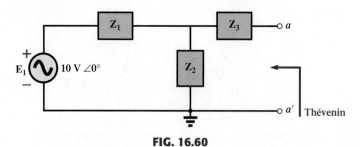

FIG. 16.60

Assigning the subscripted impedances to the network of Fig. 16.59.

Step 3 (Fig. 16.61):

$$\mathbf{Z}_{Th} = \mathbf{Z}_3 + \frac{\mathbf{Z}_1\mathbf{Z}_2}{\mathbf{Z}_1 + \mathbf{Z}_2} = j\,5\,\Omega + \frac{(10\,\Omega\,\angle 53.13°)(5\,\Omega\,\angle -53.13°)}{(6\,\Omega + j\,8\,\Omega) + (3\,\Omega - j\,4\,\Omega)}$$

$$= j\,5 + \frac{50\,\angle 0°}{9 + j\,4} = j\,5 + \frac{50\,\angle 0°}{9.85\,\angle 23.96°}$$

$$= j\,5 + 5.08\,\angle -23.96° = j\,5 + 4.64 - j\,2.06$$

$$= \mathbf{4.64\,\Omega + j\,2.94\,\Omega = 5.49\,\Omega\,\angle 32.36°}$$

Step 4 (Fig. 16.62): Since $a\text{-}a'$ is an open circuit, $\mathbf{I}_{Z_3} = 0$. Then $\mathbf{E}_{Th}$ is the voltage drop across $\mathbf{Z}_2$:

$$\mathbf{E}_{Th} = \frac{\mathbf{Z}_2\mathbf{E}}{\mathbf{Z}_2 + \mathbf{Z}_1} \qquad (\text{voltage divider rule})$$

$$= \frac{(5\,\Omega\,\angle -53.13°)(10\,\text{V}\,\angle 0°)}{9.85\,\Omega\,\angle 23.96°}$$

$$= \frac{50\,\text{V}\,\angle -53.13°}{9.85\,\angle 23.96°} = \mathbf{5.08\,\text{V}\,\angle -77.09°}$$

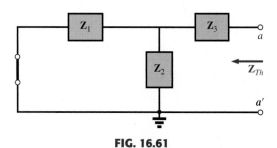

FIG. 16.61

Determining the Thévenin impedance for the network of Fig. 16.59.

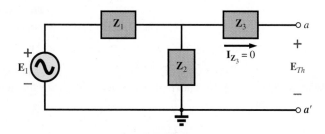

FIG. 16.62

Determining the open-circuit Thévenin voltage for the network of Fig. 16.59.

Step 5: The Thévenin equivalent circuit is shown in Fig. 16.63.

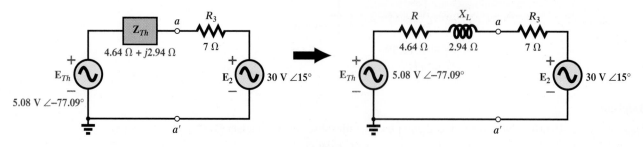

FIG. 16.63

The Thévenin equivalent circuit for the network of Fig. 16.59.

Recall from Chapter 9 that Norton's theorem allows us to replace any two-terminal linear bilateral ac network with an equivalent circuit consisting of a current source and an impedance, as shown in Fig. 16.64. The Norton equivalent circuit, like the Thévenin equivalent circuit, is applicable at only one frequency since the reactances are frequency dependent.

The procedure outlined below to find the Norton equivalent of a sinusoidal ac network is changed (from that in Chapter 9) in only one respect: the replacement of the term *resistance* with the term *impedance*.

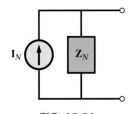

FIG. 16.64

The Norton equivalent circuit for ac networks.

1. *Remove the portion of the network across which the Norton equivalent circuit is to be found.*
2. *Mark (○, •, and so on) the terminals of the remaining two-terminal network.*
3. *Calculate $\mathbf{Z}_N$ by first setting all voltage and current sources to zero (short circuit and open circuit, respectively) and then finding the resulting impedance between the two marked terminals.*
4. *Calculate $\mathbf{I}_N$ by first replacing the voltage and current sources and then finding the short-circuit current between the marked terminals.*
5. *Draw the Norton equivalent circuit with the portion of the circuit previously removed replaced between the terminals of the Norton equivalent circuit.*

The Norton and Thévenin equivalent circuits can be found from each other by using the source transformation shown in Fig. 16.65. The source transformation is applicable for any Thévenin or Norton equivalent circuit determined from a network with any combination of independent or dependent sources.

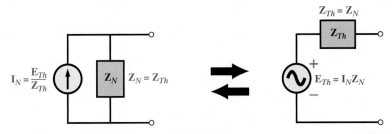

FIG. 16.65

Conversion between the Thévenin and Norton equivalent circuits.

EXAMPLE 16.18 Determine the Norton equivalent circuit for the network external to the 6 Ω resistor of Fig. 16.66.

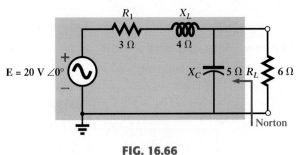

FIG. 16.66

Example 16.18.

Solution:

Steps 1–2 (Fig. 16.67):

$$\mathbf{Z}_1 = R_1 + jX_L = 3\ \Omega + j\,4\ \Omega = 5\ \Omega\ \angle 53.13°$$
$$\mathbf{Z}_2 = -jX_C = -j\,5\ \Omega$$

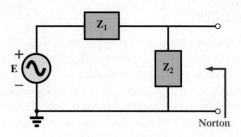

FIG. 16.67

*Assigning the subscripted impedances to the
network of Fig. 16.66.*

Step 3 (Fig. 16.68):

$$\mathbf{Z}_N = \frac{\mathbf{Z}_1\mathbf{Z}_2}{\mathbf{Z}_1 + \mathbf{Z}_2} = \frac{(5\ \Omega\ \angle 53.13°)(5\ \Omega\ \angle -90°)}{3\ \Omega + j\,4\ \Omega - j\,5\ \Omega} = \frac{25\ \Omega\ \angle -36.87°}{3 - j\,1}$$

$$= \frac{25\ \Omega\ \angle -36.87°}{3.16\ \angle -18.43°} = 7.91\ \Omega\ \angle -18.44° = \mathbf{7.50\ \Omega - j\,2.50\ \Omega}$$

Step 4 (Fig. 16.69):

$$\mathbf{I}_N = \mathbf{I}_1 = \frac{\mathbf{E}}{\mathbf{Z}_1} = \frac{20\ \text{V}\ \angle 0°}{5\ \Omega\ \angle 53.13°} = \mathbf{4\ A\ \angle -53.13°}$$

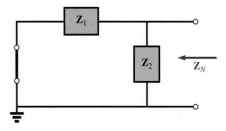

FIG. 16.68

*Determining the Norton impedance for the
network of Fig. 16.66.*

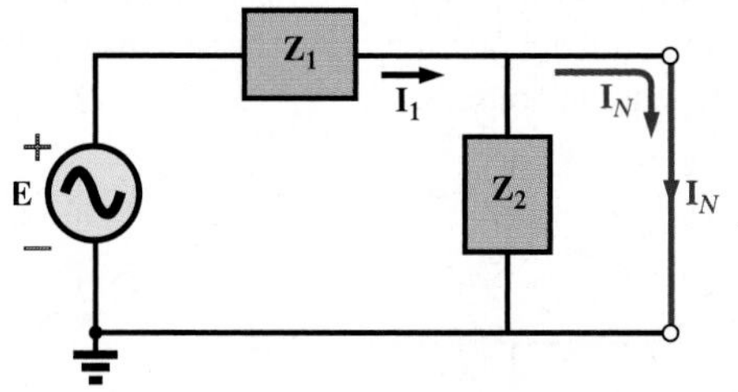

FIG. 16.69

Determining $\mathbf{I}_N$ for the network of Fig. 16.66.

Step 5: The Norton equivalent circuit is shown in Fig. 16.70.

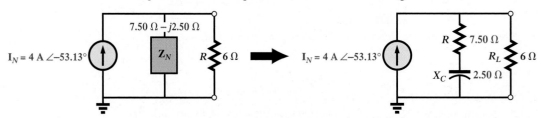

FIG. 16.70

The Norton equivalent circuit for the network of Fig. 16.66.

EXAMPLE 16.19 Find the Norton equivalent circuit for the network external to the 7 Ω capacitive reactance in Fig. 16.71.

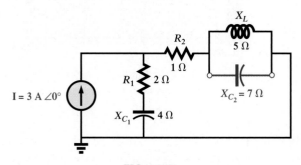

FIG. 16.71

Example 16.19.

Steps 1–2 (Fig. 16.72):

$$\mathbf{Z}_1 = R_1 - j X_{C_1} = 2\,\Omega - j\,4\,\Omega$$

$$\mathbf{Z}_2 = R_2 = 1\,\Omega$$

$$\mathbf{Z}_3 = +j\,X_L = j\,5\,\Omega$$

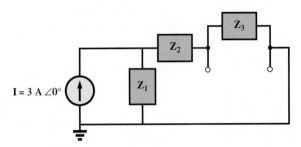

FIG. 16.72

Assigning the subscripted impedances to the network of Fig. 16.71.

Step 3 (Fig. 16.73):

$$\mathbf{Z}_N = \frac{\mathbf{Z}_3(\mathbf{Z}_1 + \mathbf{Z}_2)}{\mathbf{Z}_3 + (\mathbf{Z}_1 + \mathbf{Z}_2)}$$

$$\mathbf{Z}_1 + \mathbf{Z}_2 = 2\,\Omega - j\,4\,\Omega + 1\,\Omega = 3\,\Omega - j\,4\,\Omega = 5\,\Omega\,\angle-53.13°$$

$$\mathbf{Z}_N = \frac{(5\,\Omega\,\angle 90°)(5\,\Omega\,\angle-53.13°)}{j\,5\,\Omega + 3\,\Omega - j\,4\,\Omega} = \frac{25\,\Omega\,\angle 36.87°}{3 + j\,1}$$

$$= \frac{25\,\Omega\,\angle 36.87°}{3.16\,\angle +18.43°}$$

$$= 7.91\,\Omega\,\angle 18.44° = \mathbf{7.50\,\Omega + j\,2.50\,\Omega}$$

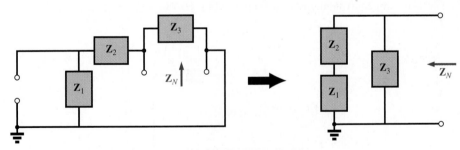

FIG. 16.73

Finding the Norton impedance for the network of Fig. 16.71.

Calculator Solution: Performing the above on the TI-86 calculator, we obtain the sequence of Fig. 16.74.

Step 4 (Fig. 16.75):

$$\mathbf{I}_N = \mathbf{I}_2 = \frac{\mathbf{Z}_1\mathbf{I}}{\mathbf{Z}_1 + \mathbf{Z}_2} \qquad \text{(current divider rule)}$$

$$= \frac{(2\,\Omega - j\,4\,\Omega)(3\,\text{A})}{3\,\Omega - j\,4\,\Omega} = \frac{6\,\text{A} - j\,12\,\text{A}}{5\,\angle-53.13°} = \frac{13.4\,\text{A}\,\angle-63.43°}{5\,\angle-53.13°}$$

$$= \mathbf{2.68\,\text{A}\,\angle -10.3°}$$

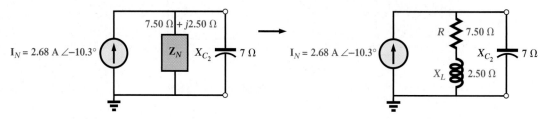

```
((0,5)*((2,–4)+(1,0)))/((0,5)+((2,–4)+(1,0)))
(7.500E0,2.500E0)
Ans▶Pol
(7.906E0∠18.435E0)
```

FIG. 16.74

Determining $\mathbf{Z}_N$ for the network of Fig. 16.71 using the TI-86 calculator.

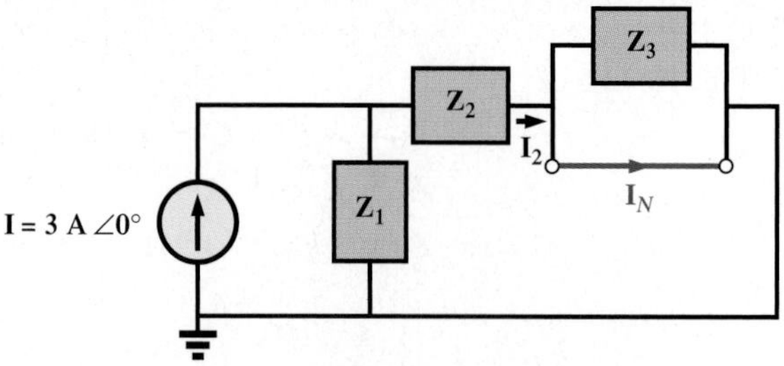

FIG. 16.75

Determining $\mathbf{I}_N$ for the network of Fig. 16.71.

Step 5: The Norton equivalent circuit is shown in Fig. 16.76.

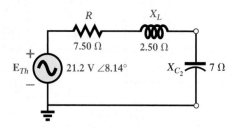

FIG. 16.76

The Norton equivalent circuit for the network of Fig. 16.71.

EXAMPLE 16.20 Find the Thévenin equivalent circuit for the network external to the 7 Ω capacitive reactance in Fig. 16.71.

Solution: Using the conversion between sources (Fig. 16.77), we obtain

$$\mathbf{Z}_{Th} = \mathbf{Z}_N = \mathbf{7.50 \; \Omega \; + \; j \; 2.50 \; \Omega}$$

$$\mathbf{E}_{Th} = \mathbf{I}_N \mathbf{Z}_N = (2.68 \text{ A } \angle -10.3°)(7.91 \; \Omega \; \angle 18.44°)$$

$$= \mathbf{21.2 \; V \; \angle 8.14°}$$

The Thévenin equivalent circuit is shown in Fig. 16.78.

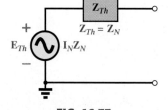

FIG. 16.77

Determining the Thévenin equivalent circuit for the Norton equivalent of Fig. 16.76.

FIG. 16.78

The Thévenin equivalent circuit for the network of Fig. 16.71.

16.10 MAXIMUM POWER TRANSFER THEOREM

When applied to ac circuits, the **maximum power transfer theorem** states that

maximum power will be delivered to a load when the load impedance is the conjugate of the Thévenin impedance across its terminals.

That is, for Fig. 16.79, for maximum power transfer to the load,

$$Z_L = Z_{Th} \quad \text{and} \quad \theta_L = -\theta_{Th_z}$$ **(16.15)**

or, in rectangular form,

$$R_L = R_{Th} \quad \text{and} \quad \pm j X_{\text{load}} = \mp j X_{Th}$$ **(16.16)**

The conditions just mentioned will make the total impedance of the circuit appear purely resistive, as indicated in Fig. 16.80.

$$\mathbf{Z}_T = (R \pm j X) + (R \mp j X)$$

and $$\mathbf{Z}_T = 2R$$ **(16.17)**

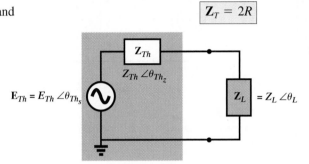

FIG. 16.79
Defining the conditions for maximum power transfer to a load.

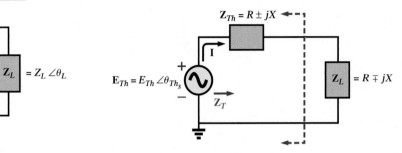

FIG. 16.80
Conditions for maximum power transfer to $\mathbf{Z}_L$.

Since the circuit is purely resistive, the power factor of the circuit under maximum power conditions is 1; that is,

$$F_p = 1 \qquad \text{(maximum power transfer)}$$ **(16.18)**

The magnitude of the current **I** of Fig. 16.80 is

$$I = \frac{E_{Th}}{Z_T} = \frac{E_{Th}}{2R}$$

The maximum power to the load is

$$P_{\text{max}} = I^2 R = \left(\frac{E_{Th}}{2R}\right)^2 R$$

and $$P_{\text{max}} = \frac{E_{Th}^2}{4R}$$ **(16.19)**

EXAMPLE 16.21 Find the load impedance in Fig. 16.81 for maximum power to the load, and find the maximum power.

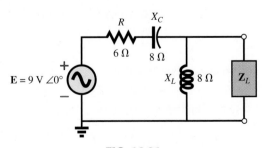

FIG. 16.81
Example 16.21.

Solution: Determine $\mathbf{Z}_{Th}$ [Fig. 16.82(a)]:

$$\mathbf{Z}_1 = R - jX_C = 6 \ \Omega - j8 \ \Omega = 10 \ \Omega \ \angle{-53.13°}$$

$$\mathbf{Z}_2 = +jX_L = j8 \ \Omega$$

$$\mathbf{Z}_{Th} = \frac{\mathbf{Z}_1\mathbf{Z}_2}{\mathbf{Z}_1 + \mathbf{Z}_2} = \frac{(10 \ \Omega \ \angle{-53.13°})(8 \ \Omega \ \angle{90°})}{6 \ \Omega - j8 \ \Omega + j8 \ \Omega} = \frac{80 \ \Omega \ \angle{36.87°}}{6 \ \angle{0°}}$$

$$= 13.33 \ \Omega \ \angle{36.87°} = 10.66 \ \Omega + j8 \ \Omega$$

and

$$\mathbf{Z}_L = 13.3 \ \Omega \ \angle{-36.87°} = \mathbf{10.66 \ \Omega - j8 \ \Omega}$$

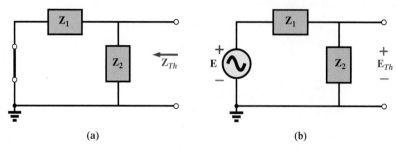

(a)　　　　　　　　　　**(b)**

FIG. 16.82

Determining (a) $\mathbf{Z}_{Th}$ and (b) $\mathbf{E}_{Th}$ for the network external to the load in Fig. 16.81.

To find the maximum power, we must first find $\mathbf{E}_{Th}$ [Fig. 16.82(b)] as follows:

$$\mathbf{E}_{Th} = \frac{\mathbf{Z}_2\mathbf{E}}{\mathbf{Z}_2 + \mathbf{Z}_1} \qquad (\text{voltage divider rule})$$

$$= \frac{(8 \ \Omega \ \angle{90°})(9 \ \text{V}\angle{0°})}{j8 \ \Omega + 6 \ \Omega - j8 \ \Omega} = \frac{72 \ \text{V}\angle{90°}}{6 \ \angle{0°}} = 12 \ \text{V}\angle{90°}$$

Then

$$P_{max} = \frac{E_{Th}^2}{4R} = \frac{(12 \ \text{V})^2}{4(10.66 \ \Omega)} = \frac{144}{42.64} = \mathbf{3.38 \ W}$$

EXAMPLE 16.22 Find the load impedance in Fig. 16.83 for maximum power to the load, and find the maximum power.

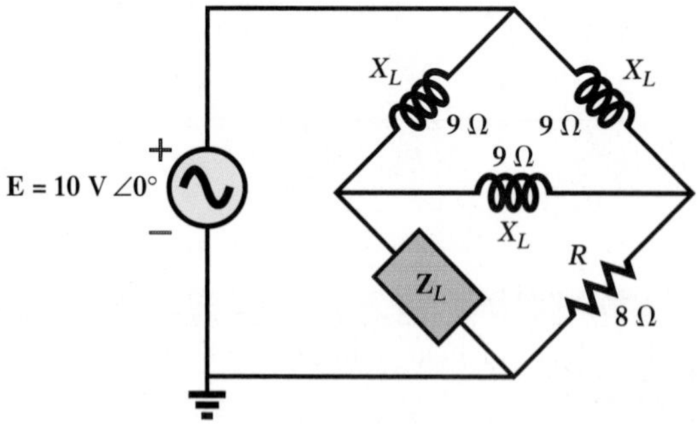

FIG. 16.83

Example 16.22.

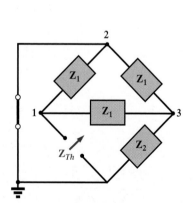

FIG. 16.84

Defining the subscripted impedances for the network of Fig. 16.83.

Solution: First we must find $\mathbf{Z}_{Th}$ (Fig. 16.84).

$$\mathbf{Z}_1 = +jX_L = j9 \ \Omega \qquad \mathbf{Z}_2 = R = 8 \ \Omega$$

Converting from a Δ to a Y (Fig. 16.85), we have

$$\mathbf{Z}'_1 = \frac{\mathbf{Z}_1}{3} = j3 \ \Omega \qquad \mathbf{Z}_2 = 8 \ \Omega$$

The redrawn circuit (Fig. 16.86) shows

$$\mathbf{Z}_{Th} = \mathbf{Z}'_1 + \frac{\mathbf{Z}'_1(\mathbf{Z}'_1 + \mathbf{Z}_2)}{\mathbf{Z}'_1 + (\mathbf{Z}'_1 + \mathbf{Z}_2)}$$

$$= j\,3\,\Omega + \frac{3\,\Omega\,\angle 90°(j\,3\,\Omega + 8\,\Omega)}{j\,6\,\Omega + 8\,\Omega}$$

$$= j\,3 + \frac{(3\,\angle 90°)(8.54\,\angle 20.56°)}{10\,\angle 36.87°}$$

$$= j\,3 + \frac{25.62\,\angle 110.56°}{10\,\angle 36.87°} = j\,3 + 2.56\,\angle 73.69°$$

$$= j\,3 + 0.72 + j\,2.46 = 0.72\,\Omega + j\,5.46\,\Omega$$

and

$$\mathbf{Z}_L = \mathbf{0.72}\,\Omega - j\,\mathbf{5.46}\,\Omega$$

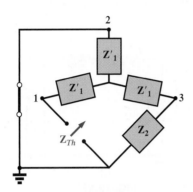

FIG. 16.85

*Substituting the Y equivalent for the
upper Δ configuration of Fig. 16.84.*

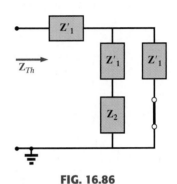

FIG. 16.86

*Determining $\mathbf{Z}_{Th}$ for the network of
Fig. 16.83.*

For $\mathbf{E}_{Th}$, use the modified circuit of Fig. 16.87 with the voltage source replaced in its original position. Since $I_1 = 0$, $\mathbf{E}_{Th}$ is the voltage across the series impedance of $\mathbf{Z}'_1$ and $\mathbf{Z}_2$. Using the voltage divider rule gives us

$$\mathbf{E}_{Th} = \frac{(\mathbf{Z}'_1 + \mathbf{Z}_2)\mathbf{E}}{\mathbf{Z}'_1 + \mathbf{Z}_2 + \mathbf{Z}'_1} = \frac{(j\,3\,\Omega + 8\,\Omega)(10\,\text{V}\angle 0°)}{8\,\Omega + j\,6\,\Omega}$$

$$= \frac{(8.54\,\angle 20.56°)(10\,\text{V}\angle 0°)}{10\,\angle 36.87°} = 8.54\,\text{V}\,\angle -16.31°$$

and

$$P_{\text{max}} = \frac{E_{Th}^2}{4R} = \frac{(8.54\,\text{V})^2}{4(0.72\,\Omega)} = \frac{72.93}{2.88}\,\text{W} = \mathbf{25.32\ W}$$

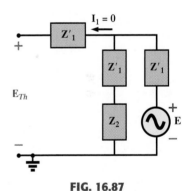

FIG. 16.87

*Finding the Thévenin voltage for the
network of Fig. 16.81.*

If the load resistance is adjustable but the magnitude of the load reactance cannot be set equal to the magnitude of the Thévenin reactance, then the maximum power *that can be delivered* to the load will occur when the load reactance is made as close to the Thévenin reactance as possible and the load resistance is set to the following value:

$$R_L = \sqrt{R_{Th}^2 + (X_{Th} + X_{\text{load}})^2} \qquad \textbf{(16.20)}$$

where each reactance carries a positive sign if inductive and a negative sign if capacitive.

The power delivered will be determined by

$$P = \frac{E_{Th}^2}{4R_{\text{av}}} \qquad \textbf{(16.21)}$$

where
$$R_{av} = \frac{R_{Th} + R_L}{2}$$
(16.22)

The derivation of the above equations is given in Appendix F of the text, and the following example demonstrates their use.

EXAMPLE 16.23 For the network of Fig. 16.88:

a. Determine the value of R_L for maximum power to the load if the load reactance is fixed at 4 Ω.
b. Find the power delivered to the load under the conditions of part (a).
c. Find the maximum power to the load if the load reactance is made adjustable to any value, and compare the result to part (b) above.

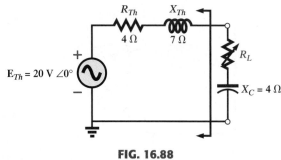

FIG. 16.88
Example 16.23.

Solutions:

a. Eq. (16.20):
$$R_L = \sqrt{R_{Th}^2 + (X_{Th} + X_{load})^2}$$
$$= \sqrt{(4\,\Omega)^2 + (7\,\Omega - 4\,\Omega)^2}$$
$$= \sqrt{16 + 9} = \sqrt{25} = \mathbf{5\,\Omega}$$

b. Eq. (16.22):
$$R_{av} = \frac{R_{Th} + R_L}{2} = \frac{4\,\Omega + 5\,\Omega}{2} = \mathbf{4.5\,\Omega}$$

Eq. (16.21):
$$P = \frac{E_{Th}^2}{4R_{av}} = \frac{(20\text{ V})^2}{4(4.5\,\Omega)} = \frac{400}{18}\text{ W} \cong \mathbf{22.22\ W}$$

c. For $\mathbf{Z}_L = 4\,\Omega - j\,7\,\Omega$,

$$P_{max} = \frac{E_{Th}^2}{4R_{Th}} = \frac{(20\text{ V})^2}{4(4\,\Omega)} = \mathbf{25\ W}$$

exceeding the result of part (b) by 2.78 W.

16.11 COMPUTER ANALYSIS

PSpice

Nodal Analysis The first application of PSpice will be to determine the nodal voltages for the network of Example 16.4 and compare solutions. The network will appear as shown in Fig. 16.89 using elements that were determined from the reactance level at a frequency of 1 kHz. There is no need to continually use 1 kHz. Any frequency will do, but remember to use the chosen frequency to find the network components and when setting up the simulation.

For the current sources, **ISIN** was chosen so that the phase angle could be specified (even though it is 0°), although the symbol does not have the arrow used in the text material. The direction must be recognized as pointing from the + to − sign of the source. That requires that the sources I_1 and I_2 be set as shown in Fig. 16.89. The source I_2 is reversed by using the **Mirror Vertically** option obtained by right-clicking the source symbol on the screen. Setting up the **ISIN** source is the same as that employed with the **VSIN** source. It can be found under the **SOURCE** library, and its attributes are the same as for the **VSIN** source. For each source, **IOFF** is set to 0 A, and the amplitude is the peak value of the source current. The frequency will be the same for each source. Then **VPRINT1** is selected from the **SPECIAL** library and placed to generate the desired nodal voltages. Finally the remaining elements are added to the network as shown in Fig. 16.89. For each source the symbol is double-clicked to generate the **Property Editor** dialog box. **AC** is set at the 6 A level for the I_1 source and at 4 A for the I_2 source, followed by **Display** and **Name and Value** for each. It will appear as shown in Fig. 16.89. A double-click on each **VPRINT1** option will also provide the **Property Editor,** so **OK** can be added under **AC, MAG,** and **PHASE.** For each quantity, **Display** is selected followed by **Name and Value** and **OK.** Then **Value** is selected and **VPRINT1** is displayed as **Value** only. Selecting **Apply** and leaving the dialog box will result in the listing next to each source in Fig. 16.89. For **VPRINT2,** the listing on **Value** must first be changed from **VPRINT1** to **VPRINT2** before selecting **Display** and **Apply.**

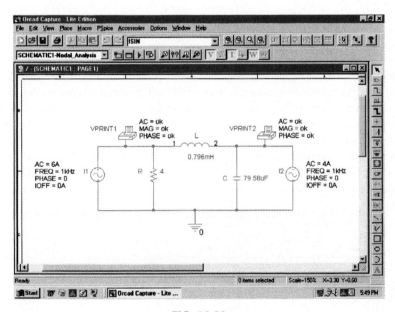

FIG. 16.89

Using PSpice to verify the results of Example 16.4.

Now the **New Simulation Profile** icon is selected and **ACNodal** entered as the **Name** followed by **Create.** In the **Simulation Settings** dialog box, **AC Sweep** is selected, and the **Start Frequency** and **End Frequency** are set at 1 kHz with 1 for the **Points/Decade.** Click **OK,** and select the **Run PSpice** icon; a **SCHEMATIC1** screen will result. Exiting (**X**) will bring us back to the **Orcad Capture** window. Selecting **PSpice** followed by **View Output File** will result in the display of Fig. 16.90, providing exactly the same results as obtained in Example 16.4 with $V_1 = 20.8$ V $\angle-126.9°$. The other nodal voltage is 8.617 V $\angle-15.09°$.

```
 82:
 83:    ****        AC ANALYSIS                        TEMPERATURE =    27.000 DEG C
 84:
 85:
 86:    ***********************************************************************
 87:
 88:
 89:
 90:    FREQ          VM(N01310)  VP(N01310)
 91:
 92:
 93:     1.000E+03    2.080E+01   -1.269E+02
 94: □
 95: **** 07/16/01 17:40:22 ************** PSpice Lite (Mar 2000) ****************
 96:
 97:    ** Profile: "SCHEMATIC1-Nodal_Analysis"  [ C:\PSpice\nodal_analysis-SCHEMATIC1-N
        odal_Analysis.sim ]
 98:
 99:
100:    ****        AC ANALYSIS                        TEMPERATURE =    27.000 DEG C
101:
102:
103:    ***********************************************************************
104:
105:
106:
107:    FREQ          VM(N01383)  VP(N01383)
108:
109:
110:     1.000E+03    8.617E+00   -1.509E+01
111:
```

FIG. 16.90

Output file for the nodal voltages for the network of Fig. 16.89.

Superposition The analysis will begin with the network of Fig. 16.47 from Example 16.15 because it has both an ac and a dc source. You will find that it is not necessary to specify an analysis for each, even though one is essentially an ac sweep and the other is a bias point calculation. When **AC Sweep** is selected, the program will automatically perform the bias calculations and display the results in the output file.

The resulting schematic appears in Fig. 16.91 with **VSIN** and **VDC** as the **SOURCE** voltages. The placement of all the *R-L-C* elements and the dc source should be quite straightforward at this point. For the ac source, be sure to double-click on the source symbol to obtain

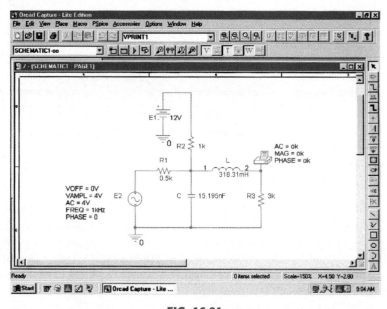

FIG. 16.91

Using PSpice to apply superposition to the network of Fig. 16.47.

the **Property Editor** dialog box. Then set **AC** to 4 V, **FREQ** to 1 kHz, **PHASE** to 0°, **VAMPL** to 4 V, and **VOFF** to 0 V. In each case, choose **Name and Value** under the **Display** heading so that we have a review of the parameters on the screen. Also, be sure to **Apply** before exiting the dialog box. Obtain the **VPRINT1** option from the **SPECIAL** library, place it as shown, and then double-click to obtain its **Property Editor.** The parameters **AC, MAG,** and **PHASE** must then receive the **OK** listing, and **Name and Value** must be applied to each under **Display** before you choose **Apply** and **OK.** The network is then ready for simulation.

After you have selected the **New Simulation Profile** icon, the **New Simulation** dialog box will appear in which **SuperpositionAC** is entered as the **Name.** Following the selection of **Create,** the **Simulation Settings** dialog box will appear in which **AC Sweep/Noise** is selected. The **Start** and **End Frequencies** are both set at 1 kHz, and 1 is entered for the **Points/Decade** request. Click **OK,** and then select the **Run PSpice** key; the **SCHEMATIC1** screen will result with an axis extending from 0.5 kHz to 1.5 kHz. Through the sequence **Trace-Add Trace-V(R3:1)-OK,** the plot point appearing in the bottom of Fig. 16.92 will result. Its value is slightly above the 2 V level and could be read as 2.05 V, which compares very nicely with the hand-calculated solution of 2.06 V. The phase angle can be obtained from **Plot-Add Plot to Window-Trace-Add Trace-P(V(R3:1))** to obtain a phase angle close to −33°. Additional accuracy can be added to the phase plot through the sequence **Plot-Axis Settings-Y Axis-User Defined −40d to −30d-OK**, resulting in the −32.5° reading of Fig. 16.92—again, very close to the hand calculation of −32.74° of Example 16.15. Now, this solution is fine for the ac signal, but it tells us nothing about the dc component.

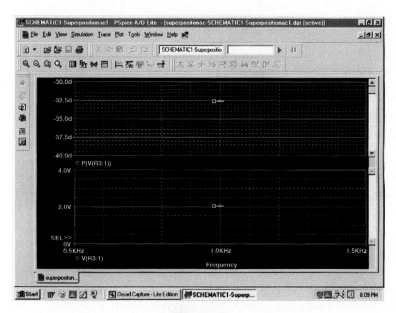

FIG. 16.92
The output results from the simulation of the network of Fig. 16.91.

By exiting the **SCHEMATIC1** screen, we obtain the **Orcad Capture** window on which **PSpice** can be selected followed by **View Output File.** The result is the printout of Fig. 16.93 which has both the dc and the ac solutions. The **SMALL SIGNAL BIAS SOLUTION** includes the nodes of the network and their dc levels. The node numbers are defined under the netlist starting on line 30. In particular, note the dc level of 3.6 V at node **N00676** which is at the top of resistor R_3 in Fig. 16.91. Also note that the dc level of both ends of the inductor is the same value because of the substitution of a short-circuit equivalent for the inductor for dc analysis. The ac solution appears under the **AC ANALYSIS** heading as 2.06 V at −32.66°, which again is a great verification of the results of Example 16.15.

```
52:
53:   ****        SMALL SIGNAL BIAS SOLUTION        TEMPERATURE =    27.000 DEG C
54:
55:
56:   *****************************************************************************
57:
58:
59:
60:   NODE    VOLTAGE        NODE    VOLTAGE        NODE    VOLTAGE        NODE    VOLTAGE
61:
62:
63:  (N00530)    0.0000  (N00596)     3.6000  (N00623)    12.0000  (N00676)     3.6000
64:
65:
66:
67:
68:      VOLTAGE SOURCE CURRENTS
69:      NAME            CURRENT
70:
71:      V_E1          -8.400E-03
72:      V_E2           7.200E-03
73:
74:      TOTAL POWER DISSIPATION    1.01E-01   WATTS
75:
76:  ⊔
77:  **** 07/01/01 20:07:27 ************** PSpice Lite (Mar 2000) ****************
78:
79:   ** Profile: "SCHEMATIC1-Superpositionac1"   [ C:\PSpice\superpositonac-SCHEMATIC
     1-Superpositionac1.sim ]
80:
81:
82:   ****       AC ANALYSIS                       TEMPERATURE =    27.000 DEG C
83:
84:
85:   *****************************************************************************
86:
87:
88:
89:   FREQ          VM(N00676)  VP(N00676)
90:
91:
92:    1.000E+03   2.060E+00  -3.266E+01
93:
```

FIG. 16.93

The output file for the dc (SMALL SIGNAL BIAS SOLUTION) and AC ANALYSIS for the network of Fig. 16.91.

Finally, if a plot of the voltage across resistor R_3 is desired, we must return to the **New Simulation Profile** and enter a new **Name** such as **SuperpositionAC1** followed by **Create fill** in the **Simulation Profile** dialog box. This time, however, we will choose the **Time Domain(Transient)** option so that we can obtain a plot against time. The fact that the source has a defined frequency of 1 kHz will tell the program which frequency to apply. The **Run to time** will be 5 ms, resulting in a five-cycle display of the 1 kHz signal. The **Start saving data after** will remain at 0 s, and the **Maximum step size** will be 5 ms/1000 = 5 μs. Click **OK,** and select the **Run PSpice** icon; the **SCHEMATIC1** screen will result again. This time **Trace-Add Trace-V(R3:1)-OK** will result in the plot of Fig. 16.94 which clearly shows a dc level of 3.6 V. Setting a cursor at $t = 0$ s (**A1**) will result in 3.6 V in the **Probe Cursor** display box. Placing the other cursor at the peak value at 2.34 ms (**A2**) will result in a peak value of about 5.66 V. The difference between the peak and the dc level provides the peak value of the ac signal and is listed as 2.06 V in the same **Probe Cursor** display box. A variety of options have now been introduced to find a particular voltage or current in a network with both dc and ac sources. It is certainly satisfying that they all verify our theoretical solution.

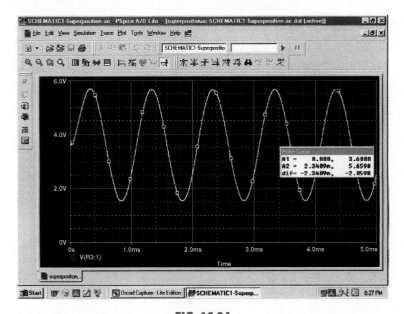

FIG. 16.94

Using PSpice to display the voltage across R_3 for the network of Fig. 16.91.

CHAPTER SUMMARY

SOURCE CONVERSIONS

- The direction of a current source (or the polarity of a voltage source) is determined by the polarity of the voltage source (or the current source, respectively) in its equivalent form.
- The equivalence between a current source and a voltage source exists only at their external terminals.

MESH ANALYSIS

- The number of mesh currents equals the number of windows for the network.
- Kirchhoff's voltage law must be applied around every independent, closed loop of the network.
- For branches with more than one mesh current, a direction for the solution must be chosen and the algebraic sum of the currents determined.
- When applying the format approach to mesh analysis, be sure to subtract all the mutual terms from the first column. Also be sure that the voltage sources in a loop are given a positive sign if supporting the assumed direction of the mesh current, and a negative sign if opposing the direction.

NODAL ANALYSIS

- The number of nodal equations is 1 less than the total number of nodes for the network.
- Kirchhoff's current law is applied to each independent node of the network, assuming that each unknown current leaves the node.
- When applying the format approach to nodal analysis, subtract all mutual terms from the first column. Also be sure that all current sources connected to a node are given a positive sign if supplying current to a node, and a negative sign if drawing current from the node.

BRIDGE NETWORKS

- When applying the balance condition to bridge networks, be sure that the magnitude and the angle of each side of the balance equation are equal.
- If balance conditions exist, any element in the bridge arm can be replaced by a short-circuit or an open-circuit equivalent.

Δ-Y AND Y-Δ CONVERSIONS

- Each impedance of a Y equivalent is equal to the product of the impedances in the two closest branches of the Δ, divided by the sum of the impedances in the Δ.
- Each impedance of the Δ equivalent is equal to the sum of the possible product combinations of the impedances of the Y, divided by the impedance of the Y farthest from the impedance to be determined.
- If all the impedances of the Y or Δ are the same, the impedance of the equivalent is related by a factor of 3, with the Δ values equal to 3 times the Y values. The angle associated with each of the impedances must also be the same for such a conversion.

SUPERPOSITION THEOREM

- When applying the superposition theorem, replace all voltage sources by short-circuit equivalents, and currents sources by open-circuit equivalents.
- Any internal impedance associated with a source must remain when the effects of a source are removed.
- The net current for an element is found by first assuming a direction for the solution and then assigning a positive sign for each current with the same direction, and a negative sign for each current with the opposite direction.
- The net voltage for an element is found by first selecting a polarity for the result and then assigning a positive sign for each voltage with the same polarity, and a negative sign for each voltage with the opposite polarity.

THÉVENIN'S THEOREM

- The Thévenin equivalent for a network is applicable at only one frequency.
- The Thévenin voltage and impedance are always determined between the two terminals of interest. Any internal resistance associated with a source must remain when the Thévenin impedance is determined.
- The Thévenin voltage is an open-circuit voltage, so that the current is zero amperes through any elements in series with the defined Thévenin voltage.
- The application of Thévenin's theorem to a network with dc and ac sources can be separated due to the superposition theorem.

NORTON'S THEOREM

- The Norton equivalent network can be determined from the Thévenin equivalent network using a source conversion.
- The Norton current is the current through a short circuit placed between the terminals of interest. The voltage across any impedance in parallel with the short-circuit branch must be zero volts, and the current zero amperes.
- The Norton impedance is found in the same manner as the Thévenin impedance.

MAXIMUM POWER TRANSFER THEOREM

- Maximum power transfer to a load will occur only if both the magnitude and the angle of the Thévenin impedance and those of the load are equal.
- The maximum power to a load can be determined directly from the Thévenin voltage and the real part of the Thévenin impedance.
- If the load impedance is fixed, the Thévenin impedance or the internal impedance of the source should be the minimum possible for maximum power to the load.

Important Equations

Source conversions:

$$\mathbf{I} = \frac{\mathbf{E}}{\mathbf{Z}} \qquad \mathbf{E} = \mathbf{IZ}$$

Bridge configuration:

$$\frac{\mathbf{Z}_1}{\mathbf{Z}_3} = \frac{\mathbf{Z}_2}{\mathbf{Z}_4}$$

Δ-Y and Y-Δ conversions:

$$\mathbf{Z}_{1(Y)} = \frac{\mathbf{Z}_B \mathbf{Z}_C}{\mathbf{Z}_A + \mathbf{Z}_B + \mathbf{Z}_C}$$

$$\mathbf{Z}_{2(Y)} = \frac{\mathbf{Z}_A \mathbf{Z}_C}{\mathbf{Z}_A + \mathbf{Z}_B + \mathbf{Z}_C}$$

$$\mathbf{Z}_{3(Y)} = \frac{\mathbf{Z}_A \mathbf{Z}_B}{\mathbf{Z}_A + \mathbf{Z}_B + \mathbf{Z}_C}$$

$$\mathbf{Z}_{B(\Delta)} = \frac{\mathbf{Z}_1 \mathbf{Z}_2 + \mathbf{Z}_1 \mathbf{Z}_3 + \mathbf{Z}_2 \mathbf{Z}_3}{\mathbf{Z}_2}$$

$$\mathbf{Z}_{A(\Delta)} = \frac{\mathbf{Z}_1 \mathbf{Z}_2 + \mathbf{Z}_1 \mathbf{Z}_3 + \mathbf{Z}_2 \mathbf{Z}_3}{\mathbf{Z}_1}$$

$$\mathbf{Z}_{C\Delta} = \frac{\mathbf{Z}_1 \mathbf{Z}_2 + \mathbf{Z}_1 \mathbf{Z}_3 + \mathbf{Z}_2 \mathbf{Z}_3}{\mathbf{Z}_3}$$

Equal impedances:

$$\mathbf{Z}_1 = \mathbf{Z}_2 = \mathbf{Z}_3 \quad \text{or} \quad \mathbf{Z}_A = \mathbf{Z}_B = \mathbf{Z}_C$$

$$\mathbf{Z}_\Delta = 3\mathbf{Z}_Y \quad \text{or} \quad \mathbf{Z}_Y = \frac{\mathbf{Z}_\Delta}{3}$$

Maximum power transfer:

$$\mathbf{Z}_L = \mathbf{Z}_{Th} \qquad \theta_L = -\theta_{Th_z}$$

$$\text{or} \quad R_L = R_{Th} \qquad \pm jX_{\text{load}} = \mp jX_{Th}$$

$$P_{\text{max}} = \frac{E_{Th}^2}{4R}$$

PROBLEMS

SECTION 16.2 Source Conversions

1. Convert the voltage sources of Fig. 16.95 to current sources.

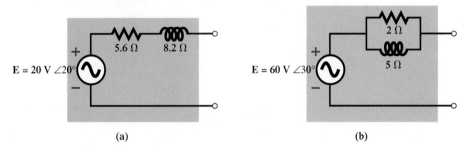

(a)   (b)

FIG. 16.95
Problem 1.

2. Convert the current sources of Fig. 16.96 to voltage sources.

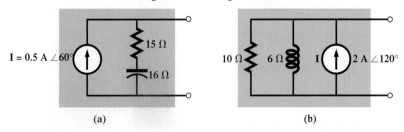

(a)   (b)

FIG. 16.96
Problem 2.

SECTION 16.3 Mesh Analysis

3. Write the mesh equations for the networks of Fig. 16.97. Determine the current through the resistor R.

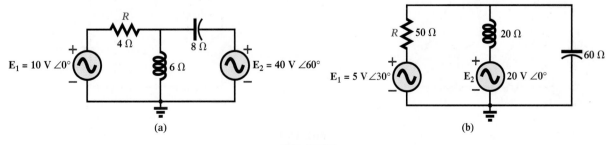

(a)   (b)

FIG. 16.97
Problems 3 and 43.

4. Write the mesh equations for the networks of Fig. 16.98. Determine the current through resistor R_1.

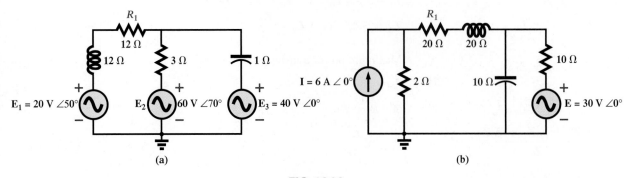

FIG. 16.98
Problems 4 and 9.

*5. Write the mesh equations for the networks of Fig. 16.99. Determine the current through resistor R_1.

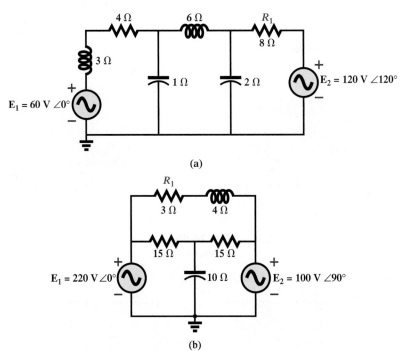

FIG. 16.99
Problems 5, 10, and 44.

*6. Write the mesh equations for the networks of Fig. 16.100. Determine the current through resistor R_1.

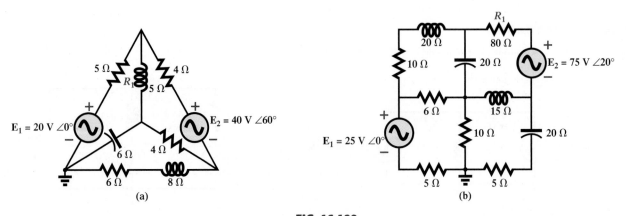

FIG. 16.100
Problems 6, 11, 12, 45, and 47.

7. Determine the nodal voltages for the networks of Fig. 16.101.

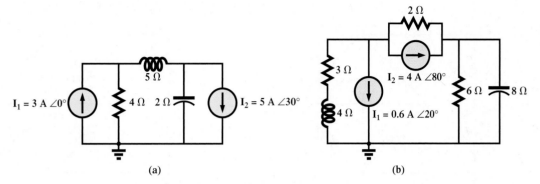

FIG. 16.101
Problems 7 and 46.

8. Determine the nodal voltages for the networks of Fig. 16.102.

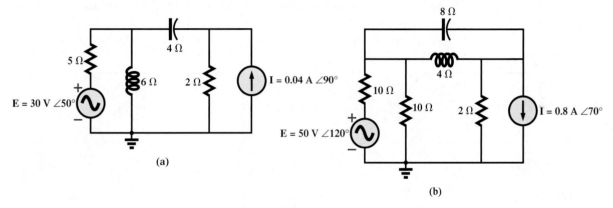

FIG. 16.102
Problem 8.

9. Determine the nodal voltages for the network of Fig. 16.98(b).
10. Determine the nodal voltages for the network of Fig. 16.99(b).
*11. Determine the nodal voltages for the network of Fig. 16.100(a).
*12. Determine the nodal voltages for the network of Fig. 16.100(b).
*13. Determine the nodal voltages for the networks of Fig. 16.103.

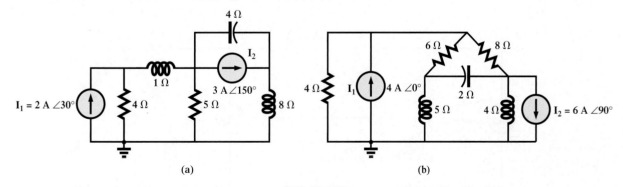

FIG. 16.103
Problems 13 and 48.

SECTION 16.5 Bridge Networks

14. For the bridge network of Fig. 16.104:
  a. Is the bridge balanced?
  b. Using mesh analysis, determine the current through the capacitive reactance.
  c. Using nodal analysis, determine the voltage across the capacitive reactance.

15. For the bridge network of Fig. 16.105:
    a. Is the bridge balanced?
    b. Using mesh analysis, determine the current through the capacitive reactance.
    c. Using nodal analysis, determine the voltage across the capacitive reactance.

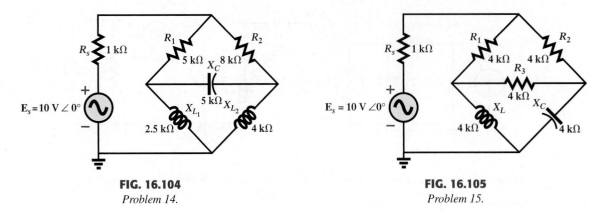

FIG. 16.104
Problem 14.

FIG. 16.105
Problem 15.

16. The Hay bridge of Fig. 16.106 is balanced. Using Eqs. (16.2) and (16.3), determine the unknown inductance L_x and resistance R_x.

17. Determine whether the Maxwell bridge of Fig. 16.107 is balanced ($\omega = 1000$ rad/s).

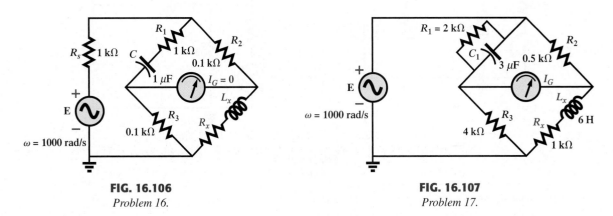

FIG. 16.106
Problem 16.

FIG. 16.107
Problem 17.

*18. Derive the balance equations (16.6) and (16.7) for the capacitance comparison bridge.

SECTION 16.6 Δ-Y and Y-Δ Conversions

19. Using the Δ-Y or Y-Δ conversion, determine the current **I** for the networks of Fig. 16.108.

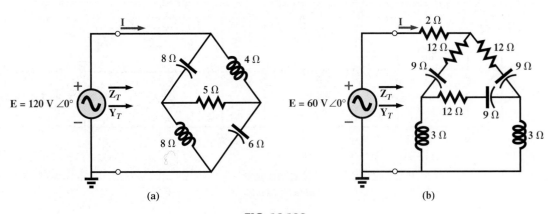

(a)

(b)

FIG. 16.108
Problem 19.

*20. Using the Δ-Y or Y-Δ conversion, determine the current **I** for the networks of Fig. 16.109. (**E** = 100 V ∠0° in each case.)

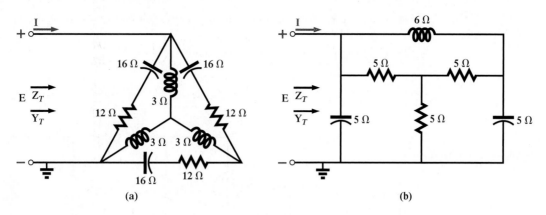

FIG. 16.109
Problem 20.

SECTION 16.7 Superposition Theorem

21. Using superposition, determine the current through the inductance X_L for each network of Fig. 16.110.

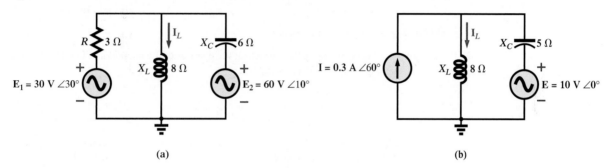

FIG. 16.110
Problem 21.

*22. Using superposition, determine the current $\mathbf{I}_L$ for each network of Fig. 16.111.

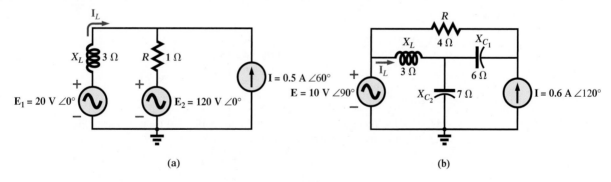

FIG. 16.111
Problem 22.

*23. Using superposition, find the sinusoidal expression for the current i for the network of Fig. 16.112.

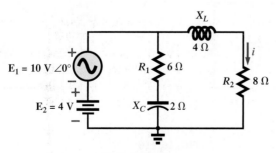

FIG. 16.112
Problems 23, 29, and 36.

*24. Using superposition, find the sinusoidal expression for the voltage v_C for the network of Fig. 16.113.

*25. Using superposition, find the current **I** for the network of Fig. 16.114.

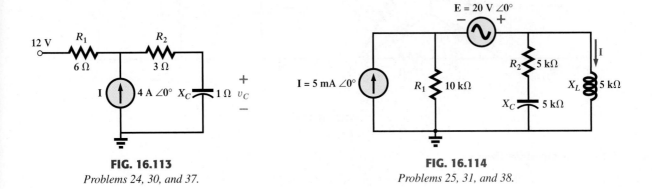

FIG. 16.113

Problems 24, 30, and 37.

FIG. 16.114

Problems 25, 31, and 38.

SECTION 16.8 Thévenin's Theorem

26. Find the Thévenin equivalent circuit for the portions of the networks of Fig. 16.115 external to the elements between points a and b.

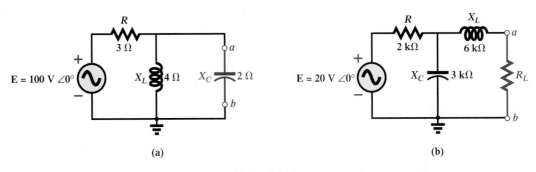

(a)

(b)

FIG. 16.115

Problems 26 and 32.

*27. Find the Thévenin equivalent circuit for the portions of the networks of Fig. 16.116 external to the elements between points a and b.

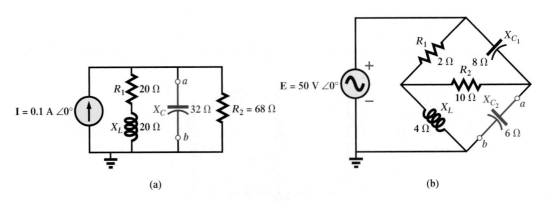

(a)

(b)

FIG. 16.116

Problems 27 and 33.

*28. Find the Thévenin equivalent circuit for the portions of the networks of Fig. 16.117 external to the elements between points a and b.

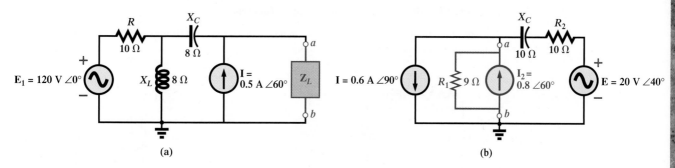

FIG. 16.117
Problems 28 and 34.

*29. For Fig. 16.112:
   a. Find the Thévenin equivalent circuit for the network external to the resistor R_2.
   b. Using the results of part (a), determine the current i.

30. For Fig. 16.113:
   a. Find the Thévenin equivalent circuit for the network external to the capacitor.
   b. Using the results of part (a), determine voltage V_C.

*31. For Fig. 16.114:
   a. Find the Thévenin equivalent circuit for the network external to the inductor.
   b. Using the results of part (a), determine the current I.

SECTION 16.9 Norton's Theorem

32. Find the Norton equivalent circuit for the network external to the elements between a and b for the networks of Fig. 16.115.

33. Find the Norton equivalent circuit for the network external to the elements between a and b for the networks of Fig. 16.116.

34. Find the Norton equivalent circuit for the network external to the elements between a and b for the networks of Fig. 16.117.

*35. Find the Norton equivalent circuit for the portions of the networks of Fig. 16.118 external to the elements between points a and b.

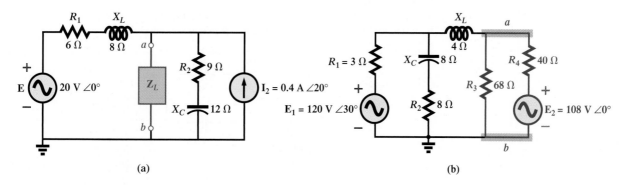

FIG. 16.118
Problem 35.

*36. For Fig. 16.112:
   a. Find the Norton equivalent circuit for the network external to resistor R_2.
   b. Using the results of part (a), determine the current I.

*37. For Fig. 16.113:
   a. Find the Norton equivalent circuit for the network external to the capacitor.
   b. Using the results of part (a), determine the voltage V_C.

*38. For Fig. 16.114:
   a. Find the Norton equivalent circuit for the network external to the inductor.
   b. Using the results of part (a), determine the current **I**.

SECTION 16.10   Maximum Power Transfer Theorem

39. Find the load impedance Z_L for the networks of Fig. 16.119 for maximum power to the load, and find the maximum power to the load.

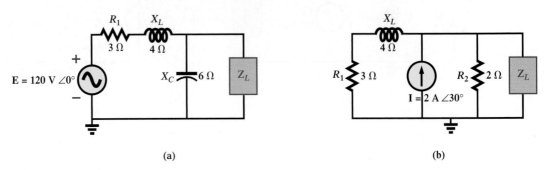

(a)   (b)

FIG. 16.119
Problem 39.

*40. Find the load impedance Z_L for the networks of Fig. 16.120 for maximum power to the load, and find the maximum power to the load.

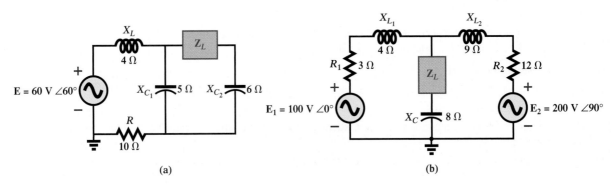

(a)   (b)

FIG. 16.120
Problem 40.

41. For the network of Fig. 16.121.
   a. Determine the value of R_L that will result in maximum power to the load.
   b. Using the results of part (a), determine the maximum power delivered.

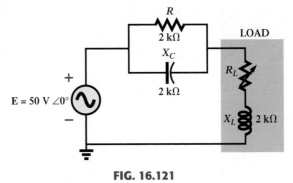

FIG. 16.121
Problem 41.

*42. For the network of Fig. 16.122:
   a. Determine the level of capacitance that will ensure maximum power to the load if the range of capacitance is limited to 1 nF to 5 nF.
   b. Using the results of part (a), determine the value of R_L that will ensure maximum power to the load.
   c. Using the results of parts (a) and (b), determine the maximum power to the load.

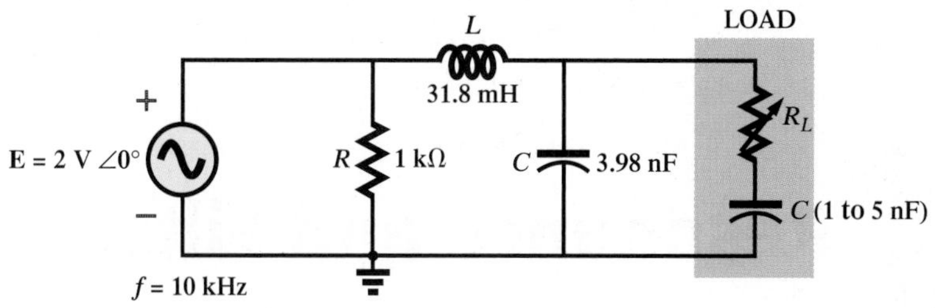

FIG. 16.122
Problem 42.

SECTION 16.12 Computer Analysis

43. Using PSpice or EWB, determine the mesh currents for the network of Fig. 16.97(a).

44. Using PSpice or EWB, determine the mesh currents for the network of Fig. 16.99(a).

__45.__ Using PSpice or EWB, determine the mesh currents for the network of Fig. 16.100.

46. Using PSpice or EWB, determine the nodal voltages for the network of Fig. 16.101(b).

__47.__ Using PSpice or EWB, determine the nodal voltages for the network of Fig. 16.100(a).

__48.__ Using PSpice or EWB, determine the nodal voltages for the network of Fig. 16.103.

49. Using PSpice or EWB, apply superposition to the network of Fig. 16.41. That is, determine the current $\mathbf{I}$ due to each source, and then find the resultant current.

__50.__ Using PSpice or EWB, determine the current $\mathbf{I}_C$ for the network of Fig. 16.47 using schematics and the superposition theorem.

__51.__ Using PSpice or EWB, and using schematics, plot the power to the *R-C* load of Fig. 16.88 for values of R_L from 1 Ω to 10 Ω.

GLOSSARY

Bridge network   A network configuration having the appearance of a diamond in which no two branches are in series or in parallel.

Capacitance comparison bridge   A bridge configuration having a galvanometer in the bridge arm that is used to determine an unknown capacitance and an associated resistance.

Delta (Δ) configuration   A network configuration having the appearance of the capital Greek letter delta.

Dependent (controlled) source   A source whose magnitude and/or phase angle is determined (controlled) by a current or voltage of the system in which it appears.

Hay bridge   A bridge configuration used for measuring the resistance and inductance of coils in cases where the resistance is a small fraction of the reactance of the coil.

Maximum power transfer theorem   A theorem used to determine the load impedance necessary to ensure maximum power to the load.

Maxwell bridge   A bridge configuration used for inductance measurements when the resistance of the coil is large enough not to require a Hay bridge.

Mesh analysis   A method through which the loop (or mesh) currents of a network can be determined. The branch currents of the network can then be determined directly from the loop currents.

Nodal analysis   A method through which the nodal voltages of a network can be determined. The voltage across each element can then be determined through application of Kirchhoff's voltage law.

Norton's theorem   A theorem that permits the reduction of any two-terminal linear ac network to one having a single current source and parallel impedance. The resulting configuration can then be employed to determine a particular current or voltage in the original network or to examine the effects of a specific portion of the network on a particular variable.

Source conversion   The changing of a voltage source to a current source, or vice versa, which will result in the same terminal behavior of the source. In other words, the external network is unaware of the change in sources.

Superposition theorem   A method of network analysis that permits considering the effects of each source independently. The resulting current and/or voltage is the phasor sum of the currents and/or voltages developed by each source independently.

Thévenin's theorem   A theorem that permits the reduction of any two-terminal linear ac network to one having a single voltage source and series impedance. The resulting configuration can then be employed to determine a particular current or voltage in the original network or to examine the effects of a specific portion of the network on a particular variable.

Wye (Y) configuration   A network configuration having the appearance of the capital letter Y.

Resonance and Filters

OBJECTIVES

- Become familiar with the characteristics of a resonance curve.
- Understand the conditions required to establish resonance in a series or a parallel resonant circuit.
- Be able to calculate the resonant frequency, bandwidth, quality factor, and cutoff frequencies for a resonant circuit.
- Understand how the quality factor and the selectivity of a parallel resonant circuit are related.
- Be able to plot the current versus frequency for a series resonant circuit, and the voltage versus frequency for a parallel resonant circuit.
- Become familiar with logarithms and learn to use them effectively when plotting data and comparing voltage and power levels.
- Understand how to interpret logarithmic plots and use the calculator to determine the logarithm and antilogarithm of a number.
- Become familiar with various decibel (dB) levels and learn how to interpret the results for different applications.
- Become familiar with the various types of filters and learn how they control the response of succeeding stages.
- Understand the meaning of *normalized plot* and learn how to establish the dB response of a filter.

17.1 INTRODUCTION

In the field of electronics, there are a number of applications in which the response of the system is sensitive to the applied frequency. In such applications, the output voltage, current, or power is a function of the frequency applied; that is, they are higher for some frequencies and lower for others. In this chapter we will investigate the **tuned** or **resonant circuit,** in which the output is a maximum for a limited range of frequencies as shown in Fig. 17.1.

Note in Fig. 17.1 that the response is a maximum at the frequency f_r, called the **resonant frequency.** The frequencies to the far right or left have very low current and voltage levels and, for all practical purposes, have little effect on the stage to follow. Radio and television receivers have a response curve for each broadcast station of the type indicated in Fig. 17.1. When the receiver is set (or tuned) to a particular station, it is set on or near the frequency f_r of Fig. 17.1. Stations transmitting at frequencies to the far right or left of this resonant frequency are not carried through with significant power to affect the program of interest. The tuning process (setting the dial to f_r) as described above is the reason for the terminology *tuned circuit*. When the response is at or near the maximum, the circuit is said to be in a state of **resonance.**

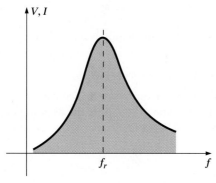

FIG. 17.1
Resonance curve.

The concept of resonance is not limited to electrical or electronic systems. If mechanical impulses are applied to a mechanical system at the proper frequency, the system will enter a state of resonance in which sustained vibrations of very large amplitude will develop. The frequency at which this occurs is called the *natural frequency* of the system. The classic example of this effect was the Tacoma Narrows Bridge built in 1940 over Puget Sound in Washington State. Four months after the bridge, with its suspended span of 2800 ft, was completed, a 42 mi/h pulsating gale set the bridge into oscillations at its natural frequency. The amplitude of the oscillations increased to the point where the main span broke up and fell into the water below. It has since been replaced by the new Tacoma Narrows Bridge, completed in 1950.

The resonant electrical circuit must have both inductance and capacitance. In addition, resistance will always be present due either to the lack of ideal elements or to the control offered on the shape of the resonance curve. When resonance occurs due to the application of the proper frequency (f_r), the energy absorbed by one reactive element is the same as that released by another reactive element within the system. In other words, energy pulsates from one reactive element to the other. Therefore, once an ideal (pure C, L) system has reached a state of resonance, it requires no further reactive power since it is self-sustaining. In a practical circuit, there is some resistance associated with the reactive elements that will result in the eventual "damping" of the oscillations between reactive elements.

The resonant or tuned circuit is the heart of any **filter** design. Filters are *R-L-C* networks designed to permit the passage of particular frequencies to the next stage of a system with minimum loss in voltage, current, or power. They can also be used to block the passage of a particular frequency range to the succeeding stage. Both types, called **pass-band** and **stop-band,** respectively, will be examined in detail in this chapter.

There are two types of resonant circuits: **series** and **parallel.** Each will be considered in some detail in this chapter.

17.2   SERIES RESONANT CIRCUIT

Basic Configuration

A resonant circuit (series or parallel) must have an inductive and a capacitive element. A resistive element will always be present due to the internal resistance of the source (R_s), the internal resistance of the inductor (R_l), and any added resistance to control the shape of the response curve (R_{design}). The basic configuration for the series resonant circuit appears in Fig. 17.2 with the resistive elements listed above all combined into one resistive element R.

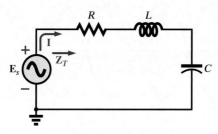

FIG. 17.2
Basic components of a series resonant circuit.

Resonant Frequency

The total impedance of the series combination is the following:

$$Z_T = R + jX_L - jX_C$$

or

$$Z_T = R + j(X_L - X_C)$$

It should be clear from the final expression that a special condition will result when $X_L = X_C$. If substituted into the above equation, the total impedance will reduce to simply that of the resistive element. That is,

$$\boxed{Z_{T_s} = R}\bigg|_{f = f_s} \tag{17.1}$$

The subscript s will be used to define resonance conditions for the series resonant circuit.

Since the impedance will be a minimum value, the current will be a maximum as determined by Ohm's law. That is,

$$\boxed{I_{\max} = \frac{E}{R}}\bigg|_{f = f_s} \tag{17.2}$$

The condition that defined this special result can be used to find the resonant frequency using the following short derivation:

$$X_L = X_C$$

$$2\pi f_s L = \frac{1}{2\pi f_s C}$$

and

$$f_s^2 = \frac{1}{4\pi^2 LC}$$

or

$$f_s = \frac{1}{\sqrt{4\pi^2 LC}}$$

so that

$$\boxed{f_s = \frac{1}{2\pi \sqrt{LC}}} \tag{17.3}$$

In particular, note that **the resistance does not appear in the equation for the resonant frequency.** The resistance, however, is important because it will determine the maximum current at resonance and will affect the shape of the resonance curve, as we shall see later.

If we plot the power curves at resonance for each of the elements, the result will be the curves shown in Fig. 17.3. Note that **at any instant of time** such as t_1, the energy being absorbed by the inductive element is exactly the same as that being released by the capacitor. At every instant of time in the plot, the resistor is dissipating energy, with the peak value occurring when the power curves of the inductor and capacitor intersect on the time axis.

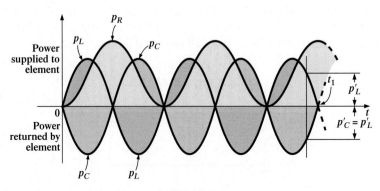

FIG. 17.3

Power curves at resonance for the series resonant circuit.

Total Impedance versus Frequency

If we plot the reactance of the inductor and capacitor on the same set of axes as shown in Fig. 17.4, we find that the reactance of the capacitor is greater than that of the coil until resonance occurs. The network then shifts to one that is primarily inductive in nature. Since the inductive reactance starts out at zero ohms, and the capacitive reactance ends up at zero ohms, the shape of the total impedance curve appearing in Fig. 17.5 will be very much like the capacitive reactance curve at low frequencies and similar to the inductive reactance curve at high frequencies. As shown in Fig. 17.5, the total impedance is its minimum value of $R \ \Omega$ at resonance. Also note that the curve is not completely symmetrical about the resonant frequency. It drops off very quickly at low frequencies and rises more gradually for very high frequencies.

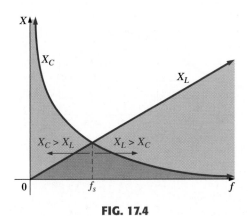

FIG. 17.4

Placing the frequency response of the inductive and capacitive reactance of a series R-L-C circuit on the same set of axes.

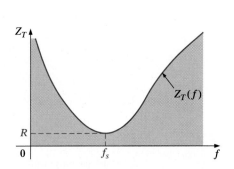

FIG. 17.5

Z_T versus frequency for the series resonant circuit.

Quality Factor (Q)

The **quality factor** Q of a series resonant circuit is defined as the ratio of the reactive power of either the inductor or the capacitor to the average power of the resistor at resonance. That is,

$$Q_s = \frac{\text{reactive power}}{\text{average power}} \qquad \textbf{(17.4)}$$

The quality factor is also an indication of how much energy is placed in storage (continual transfer from one reactive element to the other) compared to that dissipated. The lower the level of dissipation for the same reactive power, the larger the Q_s factor and the more concentrated and intense the region of resonance.

Substituting for an inductive reactance in Eq. (17.4) at resonance gives us

$$Q_s = \frac{I^2 X_L}{I^2 R}$$

and
$$Q_s = \frac{X_L}{R} = \frac{\omega_s L}{R} \qquad \textbf{(17.5)}$$

If the resistance R is just the resistance of the coil (R_l), we can speak of the Q of the coil (Q_l), and

$$Q_s = Q_l = \frac{X_L}{R_l} \qquad \text{(where } R = R_l) \qquad \textbf{(17.6)}$$

Since the quality factor of a coil is typically the information provided by manufacturers of inductors, it is often given the symbol Q without an associated subscript. It would appear from Eq. (17.6) that Q_l will increase linearly with frequency since $X_L = 2\pi fL$. That is, if the frequency doubles, then Q_l will also increase by a factor of 2. This is approximately true for the low range to the mid-range of frequencies such as shown for the coils of Fig. 17.6. Unfortunately, however, as the frequency increases, the effective resistance of the coil will also increase, due primarily to skin effect phenomena, and the resulting Q_l will decrease. In addition, the capacitive effects between the windings will increase, further reducing the Q_l of the coil. For this reason, Q_l must be specified for a particular frequency or frequency range. For wide frequency applications, a plot of Q_l versus frequency is often provided. The maximum Q_l for most commercially available coils is less than 200, with most having a maximum near 100. Note in Fig. 17.6 that for coils of the same type, Q_l drops off more quickly for higher levels of inductance.

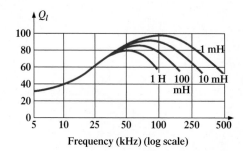

FIG. 17.6

Q_l versus frequency for a series of inductors of similar construction.

If we substitute

$$\omega_s = 2\pi f_s = 2\pi\left(\frac{1}{2\pi\sqrt{LC}}\right) = \frac{1}{\sqrt{LC}}$$

into Eq. (17.5), we obtain

$$Q_s = \frac{1}{R}\sqrt{\frac{L}{C}} \qquad\qquad (17.7)$$

providing Q_s in terms of the circuit parameters.

For series resonant circuits used in communication systems, Q_s is usually greater than 1. By applying the voltage divider rule to the circuit of Fig. 17.2, we obtain

$$V_L = \frac{X_L E}{Z_T} = \frac{X_L E}{R} \qquad (\text{at resonance})$$

and

$$V_{L_s} = Q_s E \qquad\qquad (17.8)$$

or

$$V_C = \frac{X_C E}{Z_T} = \frac{X_C E}{R}$$

and

$$V_{C_s} = Q_s E \qquad\qquad (17.9)$$

Since Q_s is usually greater than 1, the voltage across the capacitor or the inductor of a series resonant circuit can be significantly greater than the input voltage. In fact, in many cases the Q_s is so high that careful design and handling (including adequate insulation) are mandatory with respect to the voltage across the capacitor and inductor.

In the circuit of Fig. 17.7, for example, which is in the state of resonance,

$$Q_s = \frac{X_L}{R} = \frac{480\ \Omega}{6\ \Omega} = 80$$

and

$$V_L = V_C = Q_s E = (80)(10\text{ V}) = \mathbf{800\ V}$$

which is certainly a potential of significant magnitude.

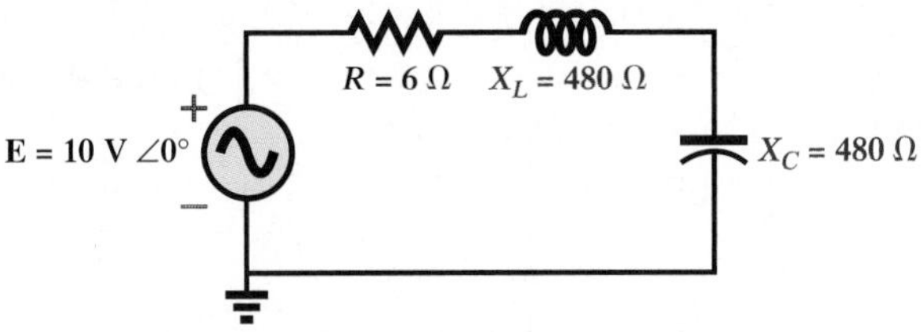

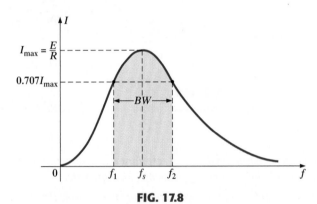

FIG. 17.7

High-Q series resonant circuit.

Selectivity

If we now plot the magnitude of the current $I = E/Z_T$ versus frequency for a *fixed* applied voltage E, we obtain the curve shown in Fig. 17.8, which rises from zero to a maximum value of E/R (where Z_T is a minimum) and then drops toward zero (as Z_T increases) at a slower rate than it rose to its peak value. The curve is actually the inverse of the impedance-versus-frequency curve. Since the Z_T curve is not absolutely symmetrical about the resonant frequency, the curve of the current versus frequency has the same property.

FIG. 17.8

I versus frequency for the series resonant circuit.

There is a definite range of frequencies at which the current is near its maximum value and the impedance is at a minimum. Those frequencies corresponding to 0.707 of the maximum current are called the **band frequencies, cutoff frequencies,** or **half-power frequencies.** They are indicated by f_1 and f_2 in Fig. 17.8. The range of frequencies between the two is referred to as the **bandwidth** (abbreviated *BW*) of the resonant circuit.

Half-power frequencies are the frequencies at which the power delivered is one-half that delivered at the resonant frequency. That is,

$$P_{\text{HPF}} = \frac{1}{2}P_{\text{max}}$$ **(17.10)**

The above condition is derived using the fact that

$$P_{\text{max}} = I^2_{\text{max}}R$$

and $$P_{\text{HPF}} = I^2R = (0.707I_{\text{max}})^2R = (0.5)(I^2_{\text{max}}R) = \frac{1}{2}P_{\text{max}}$$

Since the resonant circuit is adjusted to select a band of frequencies, the curve of Fig. 17.8 is called the **selectivity curve.** The term is derived from the fact that you must be *selective* in choosing the frequency to ensure that it is in the bandwidth. The smaller the bandwidth, the higher the selectivity. The shape of the curve, as shown in Fig. 17.9, depends on each element of the series *R-L-C* circuit. If the resistance is made smaller with a fixed inductance and capacitance, the bandwidth decreases and the selectivity increases. Similarly, if the ratio *L/C* increases with fixed resistance, the bandwidth again decreases with an increase in selectivity.

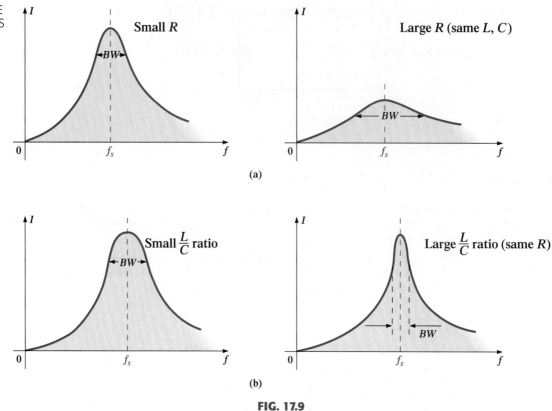

(a)

(b)

FIG. 17.9

Effect of R, L, and C on the selectivity curve for a series resonant circuit: (a) increasing R; (b) increasing L/C ratio.

A small Q_s, therefore, is associated with a resonant curve having a large bandwidth and a small selectivity, while a large Q_s indicates the opposite.

For circuits where $Q_s \geq 10$, a widely accepted approximation is that the resonant frequency bisects the bandwidth and that the resonant curve is symmetrical about the resonant frequency.

These conditions are shown in Fig. 17.10, indicating that the cutoff frequencies are then equidistant from the resonant frequency.

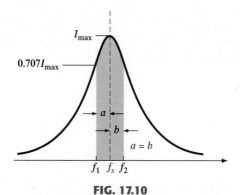

FIG. 17.10

Approximate series resonance curve for $Q_s \geq 10$.

As the quality factor drops below 10, there will be a growing loss in symmetry about the resonant frequency, and the frequency interval between f_2 and f_s will increase over that from f_1 to f_s. However,

for any quality factor greater than 1, a frequently applied approximation is to assume that the symmetry exists in order to obtain some idea of the circuit's response.

For any quality factor, the cutoff frequencies can be determined from the following equations:

$$f_1 = \frac{1}{2\pi}\left[-\frac{R}{2L} + \frac{1}{2}\sqrt{\left(\frac{R}{L}\right)^2 + \frac{4}{LC}}\,\right] \quad (\text{Hz}) \qquad \textbf{(17.11)}$$

$$f_2 = \frac{1}{2\pi}\left[+\frac{R}{2L} + \frac{1}{2}\sqrt{\left(\frac{R}{L}\right)^2 + \frac{4}{LC}}\,\right] \quad (\text{Hz}) \qquad \textbf{(17.12)}$$

The bandwidth is

$$BW = f_2 - f_1 = \text{Eq. (17.11)} - \text{Eq. (17.12)}$$

and

$$BW = f_2 - f_1 = \frac{R}{2\pi L} \quad (\text{Hz}) \qquad \textbf{(17.13)}$$

In terms of the resonant frequency and quality factor, the bandwidth is determined by

$$BW = \frac{f_s}{Q_s} \quad (\text{Hz}) \qquad \textbf{(17.14)}$$

which is a very convenient form because it relates the bandwidth to the Q_s of the circuit. As mentioned earlier, Equation (17.14) verifies that the larger the Q_s, the smaller the bandwidth, and vice versa.

Written in a slightly different form, Equation (17.14) becomes

$$\frac{f_2 - f_1}{f_s} = \frac{1}{Q_s} \qquad \textbf{(17.15)}$$

The ratio $(f_2 - f_1)/f_s$ is sometimes called the *fractional bandwidth,* providing an indication of the width of the bandwidth compared to the resonant frequency.

It can also be shown through mathematical manipulations of the pertinent equations that the resonant frequency is related to the geometric mean of the band frequencies. That is,

$$f_s = \sqrt{f_1 f_2} \qquad \textbf{(17.16)}$$

V_R, V_L, and V_C

Plotting the magnitude (effective value) of the voltages V_R, V_L, and V_C and the current I versus frequency for the series resonant circuit on the same set of axes, we obtain the curves shown in Fig. 17.11. Note that the V_R curve has the same shape as the I curve and a peak

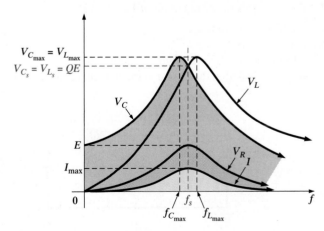

FIG. 17.11

V_R, V_L, V_C, and I versus frequency for a series resonant circuit.

value equal to the magnitude of the input voltage E. The V_C curve builds up slowly at first from a value equal to the input voltage since the reactance of the capacitor is infinite (open circuit) at zero frequency and the reactance of the inductor is zero (short circuit) at this frequency. As the frequency increases, $1/\omega C$ of the equation

$$V_C = IX_C = (I)\left(\frac{1}{\omega C}\right)$$

becomes smaller, but I increases at a rate faster than that at which $1/\omega C$ drops. Therefore, V_C rises and will continue to rise due to the quickly rising current, until the frequency nears resonance. As it approaches the resonant condition, the rate of change of I decreases. When this occurs, the factor $1/\omega C$, which decreased as the frequency rose, will overcome the rate of change of I, and V_C will start to drop. The peak value will occur at a frequency just before resonance. After resonance, both V_C and I drop in magnitude, and V_C approaches zero.

The higher the Q_s of the circuit, the closer $f_{C_{max}}$ will be to f_s, and the closer $V_{C_{max}}$ will be to $Q_s E$. For circuits with $Q_s \geq 10$, $f_{C_{max}} \cong f_s$, and $V_{C_{max}} \cong Q_s E$.

The curve for V_L increases steadily from zero to the resonant frequency since both quantities ωL and I of the equation $V_L = IX_L = (I)(\omega L)$ increase over this frequency range. At resonance, I has reached its maximum value, but ωL is still rising. Therefore, V_L will reach its maximum value after resonance. After reaching its peak value, the voltage V_L will drop toward E since the drop in I will overcome the rise in ωL. It approaches E because X_L will eventually be infinite, and X_C will be zero.

As Q_s of the circuit increases, the frequency $f_{L_{max}}$ drops toward f_s, and $V_{L_{max}}$ approaches $Q_s E$. For circuits with $Q_s \geq 10$, $f_{L_{max}} \cong f_s$, and $V_{L_{max}} \cong Q_s E$.

The V_L curve has a greater magnitude than the V_C curve for any frequency above resonance, and the V_C curve has a greater magnitude than the V_L curve for any frequency below resonance. This again verifies that the series R-L-C circuit is predominantly capacitive from zero to the resonant frequency, and predominantly inductive for any frequency above resonance.

In review:

1. V_C and V_L are at their maximum values at or near resonance (depending on Q_s).
2. At very low frequencies, V_C is very close to the source voltage and V_L is very close to zero volts, whereas at very high frequencies, V_L approaches the source voltage and V_C approaches zero volts.
3. Both V_R and I peak at the resonant frequency and have the same shape.

17.3   EXAMPLES (SERIES RESONANCE)

EXAMPLE 17.1   For the series resonant circuit of Fig. 17.12:

a. Find $\mathbf{I}$, $\mathbf{V}_R$, $\mathbf{V}_L$, and $\mathbf{V}_C$ at resonance.
b. What is the Q_s of the circuit?
c. If the resonant frequency is 5000 Hz, find the bandwidth.
d. What is the power dissipated in the circuit at the half-power frequencies?

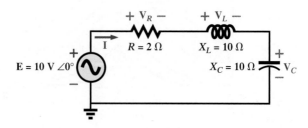

FIG. 17.12
Circuit for Example 17.1.

Solutions:

a. $Z_{T_s} = R = 2 \, \Omega$

$$I = \frac{E}{Z_{T_s}} = \frac{10 \text{ V} \angle 0°}{2 \, \Omega \angle 0°} = \mathbf{5 \text{ A} \angle 0°}$$

$V_R = E = 10 \text{ V} \angle 0°$

$V_L = (I \angle 0°)(X_L \angle 90°) = (5 \text{ A} \angle 0°)(10 \, \Omega \angle 90°) = \mathbf{50 \text{ V} \angle 90°}$

$V_C = (I \angle 0°)(X_C \angle -90°) = (5 \text{ A} \angle 0°)(10 \, \Omega \angle -90°) = \mathbf{50 \text{ V} \angle -90°}$

b. $Q_s = \dfrac{X_L}{R} = \dfrac{10 \, \Omega}{2 \, \Omega} = \mathbf{5}$

c. $BW = f_2 - f_1 = \dfrac{f_s}{Q_s} = \dfrac{5000 \text{ Hz}}{5} = \mathbf{1000 \text{ Hz}}$

d. $P_{\text{HPF}} = \dfrac{1}{2} P_{\text{max}} = \dfrac{1}{2} I_{\text{max}}^2 R = \left(\dfrac{1}{2}\right)(5 \text{ A})^2(2 \, \Omega) = \mathbf{25 \text{ W}}$

EXAMPLE 17.2 The bandwidth of a series resonant circuit is 400 Hz.

a. If the resonant frequency is 4000 Hz, what is the value of Q_s?
b. If $R = 10 \, \Omega$, what is the value of X_L at resonance?
c. Find the inductance L and capacitance C of the circuit.

Solutions:

a. $BW = \dfrac{f_s}{Q_s}$ or $Q_s = \dfrac{f_s}{BW} = \dfrac{4000 \text{ Hz}}{400 \text{ Hz}} = \mathbf{10}$

b. $Q_s = \dfrac{X_L}{R}$ or $X_L = Q_s R = (10)(10 \, \Omega) = \mathbf{100 \, \Omega}$

c. $X_L = 2\pi f_s L$ or $L = \dfrac{X_L}{2\pi f_s} = \dfrac{100 \, \Omega}{2\pi(4000 \text{ Hz})} = \mathbf{3.98 \text{ mH}}$

$X_C = \dfrac{1}{2\pi f_s C}$ or $C = \dfrac{1}{2\pi f_s X_C} = \dfrac{1}{2\pi(4000 \text{ Hz})(100 \, \Omega)} = \mathbf{0.398 \, \mu F}$

EXAMPLE 17.3 A series R-L-C circuit has a series resonant frequency of 12,000 Hz.

a. If $R = 5 \, \Omega$, and if X_L at resonance is 300 Ω, find the bandwidth.
b. Find the cutoff frequencies.

Solutions:

a. $Q_s = \dfrac{X_L}{R} = \dfrac{300 \, \Omega}{5 \, \Omega} = 60$

$BW = \dfrac{f_s}{Q_s} = \dfrac{12,000 \text{ Hz}}{60} = \mathbf{200 \text{ Hz}}$

b. Since $Q_s \geq 10$, the bandwidth is bisected by f_s. Therefore,

$$f_2 = f_s + \dfrac{BW}{2} = 12,000 \text{ Hz} + 100 \text{ Hz} = \mathbf{12,100 \text{ Hz}}$$

and $f_1 = 12,000 \text{ Hz} - 100 \text{ Hz} = \mathbf{11,900 \text{ Hz}}$

EXAMPLE 17.4 For the response curve of Fig. 17.13:

a. Determine the Q_s and the bandwidth.
b. For $C = 101.5$ nF, determine L and R for the series resonant circuit.
c. Determine the applied voltage.

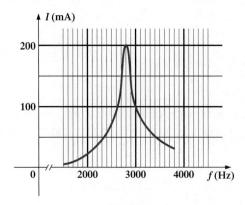

FIG. 17.13
Example 17.4.

Solutions:

a. The resonant frequency is 2800 Hz. At 0.707 times the peak value, or 141.4 mA,

$$BW \cong \textbf{200 Hz}$$

and
$$Q_s = \frac{f_s}{BW} = \frac{2800 \text{ Hz}}{200 \text{ Hz}} = \textbf{14}$$

b. $f_s = \dfrac{1}{2\pi\sqrt{LC}}$ or $L = \dfrac{1}{4\pi^2 f_s^2 C}$

$$= \frac{1}{4\pi^2 (2.8 \times 10^3 \text{ Hz})^2 (101.5 \times 10^{-9} \text{ F})} = \textbf{31.832 mH}$$

$Q_s = \dfrac{X_L}{R}$ or $R = \dfrac{X_L}{Q_s} = \dfrac{2\pi(2800 \text{ Hz})(31.832 \times 10^{-3} \text{ H})}{14} = \textbf{40 } \mathbf{\Omega}$

c. $I_{max} = \dfrac{E}{R}$ or $E = I_{max}R = 200 \text{ mA})(40 \text{ } \Omega) = \textbf{8 V}$

EXAMPLE 17.5 A series *R-L-C* circuit is designed to resonant at $\omega_s = 10^5$ rad/s, have a bandwidth of $0.15\omega_s$, and draw 16 W from a 120 V source at resonance.

a. Determine the value of *R*.
b. Find the bandwidth in hertz.
c. Find the nameplate values of *L* and *C*.
d. Determine the Q_s of the circuit.
e. Determine the fractional bandwidth.

Solutions:

a. At resonance, $V_R = E$ and $P = \dfrac{E^2}{R}$ with $R = \dfrac{E^2}{P} = \dfrac{(120 \text{ V})^2}{16 \text{ W}} = \textbf{900 } \mathbf{\Omega}$

b. $f_s = \dfrac{\omega_s}{2\pi} = \dfrac{10^5 \text{ rad/s}}{2\pi} = 15{,}915.49 \text{ Hz}$

$BW = 0.15f_s = 0.15(15{,}915.49 \text{ Hz}) = \textbf{2387.32 Hz}$

c. Eq. (17.13):

$$BW = \frac{R}{2\pi L} \quad \text{and} \quad L = \frac{R}{2\pi BW} = \frac{900 \text{ } \Omega}{2\pi(2387.32 \text{ Hz})} = \textbf{60 mH}$$

$$f_s = \frac{1}{2\pi\sqrt{LC}} \quad \text{and} \quad C = \frac{1}{4\pi^2 f_s^2 L} = \frac{1}{4\pi^2 (15{,}915.49 \text{ Hz})^2 (60 \times 10^{-3} \text{ H})}$$

$$= \textbf{1.67 nF}$$

d. $Q_s = \dfrac{X_L}{R} = \dfrac{2\pi f_s L}{R} = \dfrac{2\pi(15{,}915.49\ \text{Hz})(60\ \text{mH})}{900\ \Omega} = \textbf{6.67}$

e. $\dfrac{f_2 - f_1}{f_s} = \dfrac{BW}{f_s} = \dfrac{1}{Q_s} = \dfrac{1}{6.67} = \textbf{0.15}$

17.4 PARALLEL RESONANT CIRCUIT

Basic Configuration

The basic format of the series resonant circuit is a series $R\text{-}L\text{-}C$ combination in series with an applied voltage source. The parallel resonant circuit has the basic configuration of Fig. 17.14, which is a parallel $R\text{-}L\text{-}C$ combination in parallel with an applied current source.

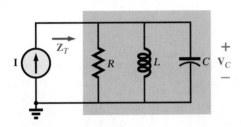

FIG. 17.14

Ideal parallel resonant network.

For the series circuit, the impedance was a minimum at resonance, producing a significant current that resulted in a high output voltage for $\mathbf{V}_C$ and $\mathbf{V}_L$. For the parallel resonant circuit, the impedance is relatively high at resonance, producing a significant voltage for $\mathbf{V}_C$ and $\mathbf{V}_L$ through the Ohm's law relationship $\mathbf{V}_C = \mathbf{I}\mathbf{Z}_T$. For the network of Fig. 17.14, resonance will occur when $X_L = X_C$, and the resonant frequency will have the same format obtained for series resonance.

If the practical equivalent of Fig. 17.14 had the format of Fig. 17.14, the analysis would be as direct and lucid as that experienced for series resonance. However, in the practical world, the internal resistance of the coil must be placed in series with the inductor, as shown in Fig. 17.15. The resistance R_l can no longer be included in a simple series or parallel combination with the source resistance and any other resistance added for design purposes. Even though R_l is usually relatively small in magnitude compared with other resistance and reactance levels of the network, it does have an important impact on the parallel resonant condition, as will be demonstrated in the sections to follow. In other words, the network of Fig. 17.14 is an ideal situation that can be assumed only for specific network conditions.

Our first effort will be to find a parallel network equivalent (at the terminals) for the series $R\text{-}L$ branch of Fig. 17.15 using the following technique:

$$\mathbf{Z}_{R\text{-}L} = R_l + j\,X_L$$

and

$$\mathbf{Y}_{R\text{-}L} = \frac{1}{\mathbf{Z}_{R\text{-}L}} = \frac{1}{R_l + j\,X_L} = \frac{R_l}{R_l^2 + X_L^2} - j\frac{X_L}{R_l^2 + X_L^2}$$

$$= \frac{1}{\dfrac{R_l^2 + X_L^2}{R_l}} + \frac{1}{j\left(\dfrac{R_l^2 + X_L^2}{X_L}\right)} = \frac{1}{R_p} + \frac{1}{j\,X_{L_p}}$$

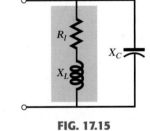

FIG. 17.15

Practical parallel L-C network.

with

$$\boxed{R_p = \frac{R_l^2 + X_L^2}{R_l}} \qquad\qquad \textbf{(17.17)}$$

and

$$\boxed{X_{L_p} = \frac{R_l^2 + X_L^2}{X_L}} \qquad\qquad \textbf{(17.18)}$$

as shown in Fig. 17.16.

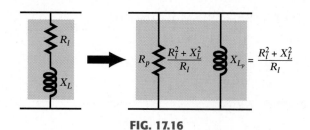

FIG. 17.16

Equivalent parallel network for a series R-L combination.

Redrawing the network of Fig. 17.15 with the equivalent of Fig. 17.16 and a practical current source having an internal resistance R_s will result in the network of Fig. 17.17.

If we define the parallel combination of R_s and R_p by the notation

$$R = R_s \| R_p \tag{17.19}$$

the network of Fig. 17.18 will result. It has the same format as the ideal configuration of Fig. 17.14.

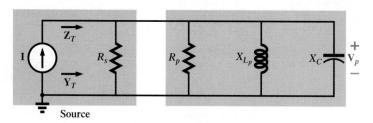

FIG. 17.17

Substituting the equivalent parallel network for the series R-L combination of Fig. 17.15.

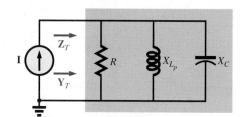

FIG. 17.18

Substituting $R = R_s \| R_p$ for the network of Fig. 17.17.

Resonant Frequency

We are now at a point where the conditions for resonance can be established. For the series configuration, it was simply a matter of satisfying the condition $X_L = X_C$. For the equivalent parallel configuration of Fig. 17.18, it would appear that it was simply a matter of setting $X_{L_p} = X_C$. This will result in the removal of the imaginary component from the total impedance or admittance of the network, and it will also result in an input impedance equal to just R. However, we must remember that **R is not a purely resistive value and is frequency dependent.** The result is that the maximum value of Z_T recognized as the resonant point for parallel resonance will occur at a frequency slightly higher than defined simply by $X_{L_p} = X_C$. Fortunately, however, the two frequencies are so close together that we will make the following general assumption:

The frequency at which resonance occurs is so close to that defined by $X_{L_p} = X_C$ that for most applications the two can be considered equal.

The most direct path toward finding the frequency at which the imaginary component is zero for the network of Fig. 17.18 is to first find the total admittance as follows:

$$\mathbf{Y}_T = \frac{1}{\mathbf{Z}_1} + \frac{1}{\mathbf{Z}_2} + \frac{1}{\mathbf{Z}_3} = \frac{1}{R} + \frac{1}{j X_{L_p}} + \frac{1}{-j X_C}$$

$$= \frac{1}{R} - j\left(\frac{1}{X_{L_p}}\right) + j\left(\frac{1}{X_C}\right)$$

and

$$\mathbf{Y}_T = \frac{1}{R} + j\left(\frac{1}{X_C} - \frac{1}{X_{L_p}}\right) \tag{17.20}$$

For unity power factor, the reactive component must be zero as defined by

$$\frac{1}{X_C} - \frac{1}{X_{L_p}} = 0$$

Therefore,

$$\frac{1}{X_C} = \frac{1}{X_{L_p}}$$

and

$$\boxed{X_{L_p} = X_C}$$ **(17.21)**

Substituting for X_{L_p} yields

$$\boxed{\frac{R_l^2 + X_L^2}{X_L} = X_C}$$ **(17.22)**

The resonant frequency, f_p, can now be determined from Eq. (17.22):

$$\boxed{f_p = \frac{1}{2\pi\sqrt{LC}}\sqrt{1 - \frac{R_l^2 C}{L}}}$$ **(17.23)**

Note that the first part of Eq. (17.23) is exactly the same as the equation for the resonant frequency of a series resonant circuit [Eq. (17.3)]. Although the square-root multiplier will always be less than 1 and will reduce the magnitude of f_p, its effect can often be ignored on an approximate basis. The result is that

on an approximate basis, the resonant frequency of a parallel resonant circuit can be determined using the same equation employed for series resonance. That is,

$$\boxed{f_p \cong \frac{1}{2\pi\sqrt{LC}}}$$ **(17.24)**

Quality Factor (Q)

The quality factor of the parallel resonant circuit continues to be determined by the ratio of the reactive power to the real power. That is,

$$Q_p = \frac{V_p^2/X_{L_p}}{V_p^2/R}$$

where $R = R_s \| R_p$, and V_p is the voltage across the parallel branches. The result is that

$$\boxed{Q_p = \frac{R}{X_{L_p}} = \frac{R_s \| R_p}{X_{L_p}}}$$ **(17.25)**

or, since $X_{L_p} = X_C$ at resonance,

$$\boxed{Q_p = \frac{R_s \| R_p}{X_C}}$$ **(17.26)**

For the ideal current source ($R_s = \infty\ \Omega$), or when R_s is sufficiently large compared to R_p, the equation for Q_p reduces to

$$\boxed{Q_p = \frac{X_L}{R_l} = Q_l}\Big|_{R_s \gg R_p}$$ **(17.27)**

which is simply the quality factor Q_l of the coil.

Selectivity

A plot of the input impedance versus frequency will result in the curve of Fig. 17.19, revealing that the impedance is maximum at resonance and only R_l at $f = 0$ Hz. This result is the total opposite of that for series resonance where the total impedance is a minimum at

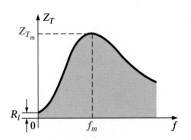

FIG. 17.19

Z_T versus frequency for the parallel resonant circuit.

resonance. At low frequencies, the reactance of the inductor is so small that it predominates in the parallel configuration. At high frequencies, the reactance of the capacitor becomes small enough to predominate.

For the parallel configuration, the parallel voltage across the network is the resonant curve of interest. If the current is a constant as set by the current source, and if the impedance has the shape just described, the voltage across the capacitor will have the shape appearing in Fig. 17.20. At each frequency the voltage across the capacitor is simply the product of the current and the value of the total impedance. The result is a curve with exactly the same shape as that of the impedance curve. Therefore, the peak value of the voltage will occur at the same frequency as the maximum impedance and is determined by Ohm's law: $V_p = IZ_{T_p}$.

All the parameters defined for the series resonant curve can now be defined for the plot of Fig. 17.21. The bandwidth is still defined at 0.707 times the peak value and is related to the resonant frequency and the quality factor by

$$BW = f_2 - f_1 = \frac{f_p}{Q_p} \qquad (17.28)$$

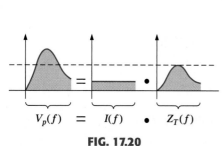

$$V_p(f) = I(f) \bullet Z_T(f)$$

FIG. 17.20

Defining the shape of the $V_p(f)$ curve.

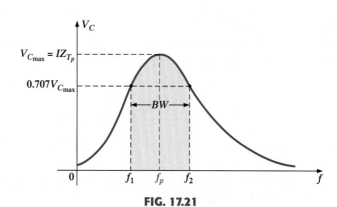

FIG. 17.21

Selectivity curve for V_C for a parallel resonant network.

The cutoff frequencies can be determined from the following expressions:

$$f_1 = \frac{1}{4\pi C}\left[\frac{1}{R} - \sqrt{\frac{1}{R^2} + \frac{4C}{L}}\right] \qquad (17.29)$$

$$f_2 = \frac{1}{4\pi C}\left[\frac{1}{R} + \sqrt{\frac{1}{R^2} + \frac{4C}{L}}\right] \qquad (17.30)$$

Since the term in the brackets of Eq. (17.29) will always be negative, simply associate f_1 with the magnitude of the result.

For most applications,

it is a reasonable approximation to assume that the resonant frequency bisects the bandwidth so that f_2 and f_1 are equidistant from the resonant frequency.

The result is that

$$f_1 \cong f_p - \frac{BW}{2} \qquad (17.31)$$

and

$$f_2 \cong f_p + \frac{BW}{2} \qquad (17.32)$$

In terms of the network parameters, an excellent approximation for the bandwidth can be determined from

$$BW \cong \frac{1}{2\pi}\left[\frac{R_l}{L} + \frac{1}{R_sC}\right] \qquad \textbf{(17.33)}$$

clearly revealing the impact of R_s on the bandwidth. Of course, if $R_s = \infty\,\Omega$ (ideal current source), then the equation reduces to the following simpler format:

$$BW \cong \frac{R_l}{2\pi L}\Bigg|_{R_s = \infty\,\Omega} \qquad \textbf{(17.34)}$$

The effect of the parameters of the network on the shape of the resonant curves are provided in Fig. 17.22. For fixed values of L and C, increasing values of R_l will reduce the quality factor and expand the bandwidth. On the other hand, for fixed values of R_l, increasing levels of the ratio L/C will sharpen the peak and decrease the bandwidth.

The resistance R_p can take on a convenient approximate form if the quality factor of the network is sufficiently high. Given

$$R_p = \frac{R_l^2 + X_L^2}{R_l} = R_l + \frac{X_L^2}{R_l}\left(\frac{R_l}{R_l}\right) = R_l + \frac{X_L^2}{R_l}R_l = R_l + Q_l^2 R_l$$

we find that

$$R_p = (1 + Q_l^2)R_l$$

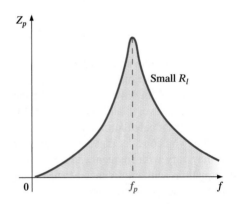
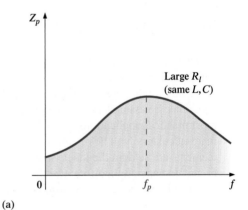

(a)

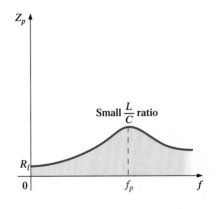
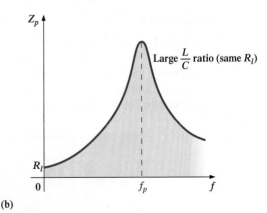

(b)

FIG. 17.22

Effect of R, L, and C on the selectivity curve for a parallel resonant circuit: (a) increasing R_l;
(b) increasing L/C ratio.

Even if Q_l has a relatively low value of 2, the term Q_l^2 is four times larger than the 1 appearing in the brackets. For $Q_l = 10$, it is 100 times larger. The result is that the 1 can be ignored for most applications, and we can conclude the following:

For most applications, it is an excellent approximation to assume that

$$R_p \cong Q_l^2 R_l \tag{17.35}$$

Substituting $Q_l = X_L/R_l$ into Eq. (17.35) will result in the following useful equation:

$$R_p \cong \frac{L}{R_l C} \tag{17.36}$$

The result is that the total impedance at resonance for most applications can be found from the rather simple equation of

$$Z_{T_p} \cong R_s \| R_p \cong R_s \| Q_l^2 R_l \tag{17.37}$$

For an ideal source where $R_s = \infty \, \Omega$, the equation reduces to

$$Z_{T_p} \cong Q_l^2 R_l \tag{17.38}$$

Recall that for series resonant circuits, the peak value of the voltage across the inductor is simply $V_L = Q_s E$ and for the capacitor $V_C = Q_s E$. For the parallel resonant circuit with a reasonably high Q, the current of the capacitor is approximately

$$I_{C_p} \cong Q_l I_{\text{source}} \tag{17.39}$$

and for the inductor

$$I_{L_p} \cong Q_l I_{\text{source}} \tag{17.40}$$

There is no question that the discussion surrounding the above equations is a bit more complex than that offered for the series resonant circuit. A number of approximations were made that might lead to some concern about the accuracy of any investigation using the equations. Be assured, however, that all the approximations are very valid and will work for most practical applications. The differences between an exact calculation and an approximate calculation will be of almost no consequence when you consider that actual values of elements are seldom the label value, and the value may vary with frequency, temperature, or age. In general, therefore, use the approximate equations with confidence because your results will be very close to those you might obtain from the actual network.

17.5 EXAMPLES (PARALLEL RESONANCE)

EXAMPLE 17.6 Given the parallel network of Fig. 17.23, composed of "ideal" elements:

a. Determine the resonant frequency f_p.
b. Find the total impedance at resonance.
c. Calculate the quality factor, bandwidth, and cutoff frequencies f_1 and f_2 of the system.
d. Find the voltage V_C at resonance.
e. Determine the currents I_L and I_C at resonance.

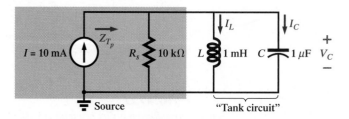

FIG. 17.23
Network for Example 17.6.

a. The fact that R_l is zero ohms results in a very high Q_l $(= X_L/R_l)$.

$$f_p = \frac{1}{2\pi\sqrt{LC}} = \frac{1}{2\pi\sqrt{(1\text{ mH})(1\text{ }\mu F)}} = \textbf{5.03 kHz}$$

b. For the parallel reactive elements:

$$\mathbf{Z}_L\|\mathbf{Z}_C = \frac{(X_L\angle 90°)(X_C\angle -90°)}{+j(X_L - X_C)}$$

However, $X_L = X_C$ at resonance, resulting in a zero in the denominator of the equation and a very high impedance that can be approximated by an open circuit. Therefore,

$$Z_{T_p} = R_s\|\mathbf{Z}_L\|\mathbf{Z}_C = R_s = \textbf{10 k}\Omega$$

c. $Q_p = \dfrac{R_s\|R_p}{X_C}$

$$X_C = \frac{1}{2\pi fC} = \frac{1}{2\pi(5.03\text{ kHz})(1\text{ }\mu F)} = 31.64\text{ }\Omega$$

$$Q_p = \frac{10\text{ k}\Omega\|\infty\text{ }\Omega}{31.64\text{ }\Omega} = \frac{10\text{ k}\Omega}{31.64\text{ }\Omega} = \textbf{316.1}$$

$$BW = \frac{f_p}{Q_p} = \frac{5.03\text{ kHz}}{316.1} = \textbf{15.91 Hz}$$

$$f_1 = f_p - \frac{BW}{2} - 5.03\text{ kHz} - \frac{15.91\text{ Hz}}{2} = \textbf{5022.05 Hz}$$

$$f_2 = f_p + \frac{BW}{2} = 5.03\text{ kHz} + \frac{15.91\text{ Hz}}{2} = \textbf{5037.96 Hz}$$

d. $V_C = IZ_{T_p} = (10\text{ mA})(10\text{ k}\Omega) = \textbf{100 V}$

e. $I_L = Q_lI = (316.1)(10\text{ mA}) = \textbf{3.16 A}$
   $I_C = Q_lI = I_L = \textbf{3.16 A}$

Example 17.6 demonstrated the impact of R_s on the calculations associated with parallel resonance. The source impedance will define the level of input impedance and V_C.

EXAMPLE 17.7 For the parallel resonant circuit of Fig. 17.24 with $R_s = \infty\text{ }\Omega$:

a. Determine f_p.
b. Calculate the maximum impedance and the magnitude of the voltage V_C at f_p.
c. Calculate the bandwidth.

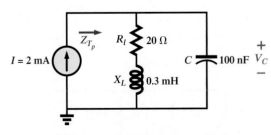

FIG. 17.24
Circuit for Example 17.7.

Solutions:

a. $f_p \cong \dfrac{1}{2\pi\sqrt{LC}} = \dfrac{1}{2\pi\sqrt{(0.3\text{ mH})(100\text{ nF})}} = \textbf{29.06 kHz}$

b. $Q_l = \dfrac{R_s \parallel R_p}{X_C}$

$$R_p \cong \frac{L}{R_l C} = \frac{0.3 \text{ mH}}{(20 \text{ }\Omega)(100 \text{ nF})} = 150 \text{ }\Omega$$

$$X_C = \frac{1}{2\pi f C} = \frac{1}{2\pi (29.06 \text{ kHz})(100 \text{ nF})} = 54.77 \text{ }\Omega$$

$$Q_l = \frac{\infty \text{ }\Omega \parallel 150 \text{ }\Omega}{54.77 \text{ }\Omega} = \frac{150 \text{ }\Omega}{54.77 \text{ }\Omega} = 2.74$$

$$R_p = Z_{T_p} = Q_l^2 R_l = (2.74)^2(20 \text{ }\Omega) = \mathbf{150.15 \text{ }\Omega}$$

$$V_{C_{max}} = I Z_{T_p} = (2 \text{ mA})(150.15 \text{ }\Omega) \cong \mathbf{300 \text{ mV}}$$

c. $BW = \dfrac{f_p}{Q_p} = \dfrac{29.06 \text{ kHz}}{2.74} = \mathbf{10.61 \text{ kHz}}$

Comment: The results obtained in this example represent approximate solutions. If all conditions were pure and ideal, the results would be slightly different if the lengthy, exact methods were employed. To demonstrate that the approximate solutions are an excellent first approximation to the exact solution, the following comparisons were made.

The approximate solution for the resonant frequency was **29.06 kHz.** The frequency at which the impedance is a maximum is **28.58 kHz,** and the frequency determined by ensuring that the imaginary component of the total impedance is zero is **27.06 kHz.** All three results are certainly very close when we consider the range of possibilities. The approximate solution is very close to the frequency of maximum impedance which defines the resonant condition.

The approximate solution for the quality factor is **2.74;** a lengthy analysis would have resulted in **2.55.** Again, the results are very close and essentially still define what to expect from the selectivity curve.

The approximate solution for the total impedance is **150.15 Ω,** very close to a more exact solution of **159.34,** which is a difference of negligible consequence.

The approximate solution for the bandwidth is **10.61 kHz,** while the detailed solution provided exactly the same solution of **10.61 kHz.** Now that is an interesting outcome.

Finally, the approximate solution for the peak voltage is **300 mV,** while the exact solution provided **300.3 mV,** which is another difference of negligible consequence.

Even though the quality factor was quite small at 2.74, the exact and approximate solutions are very close. The fact that higher-Q networks will result in a closer correspondence between the solutions should reinforce the fact that the approximate solutions outlined in this chapter will provide an excellent idea of the response to be expected from the actual network.

EXAMPLE 17.8   The equivalent network for the transistor configuration of Fig. 17.25 is provided in Fig. 17.26.

a.   Find f_p.
b.   Determine Q_p.
c.   Calculate the BW.
d.   Determine V_p at resonance.
e.   Sketch the curve of V_C versus frequency.

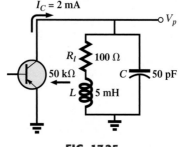

FIG. 17.25
Example 17.8.

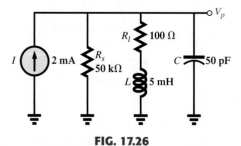

FIG. 17.26
Equivalent network for the transistor configuration of Fig. 17.25.

a. $f_p = \dfrac{1}{2\pi \sqrt{LC}} = \dfrac{1}{2\pi \sqrt{(5 \text{ mH})(50 \text{ pF})}} = \textbf{318.31 kHz}$

$X_L = 2\pi f_s L = 2\pi(318.31 \text{ kHz})(5 \text{ mH}) = 10 \text{ k}\Omega$

$Q_l = \dfrac{X_L}{R_l} = \dfrac{10 \text{ k}\Omega}{100 \text{ }\Omega} = \textbf{100}$

b. $Q_p = \dfrac{R_s \| R_p}{X_L}$

$R_p = Q_l^2 R_l = (100)^2 100 \text{ }\Omega = 1 \text{ M}\Omega$

$Q_p = \dfrac{50 \text{ k}\Omega \| 1 \text{ M}\Omega}{10 \text{ k}\Omega} = \dfrac{47.62 \text{ k}\Omega}{10 \text{ k}\Omega} = \textbf{4.76}$

Note the drop in Q from $Q_l = 100$ to $Q_p = 4.76$ due to R_s.

c. $BW = \dfrac{f_p}{Q_p} = \dfrac{318.31 \text{ kHz}}{4.76} = \textbf{66.87 kHz}$

On the other hand,

$BW = \dfrac{1}{2\pi}\left(\dfrac{R_l}{L} + \dfrac{1}{R_s C}\right) = \dfrac{1}{2\pi}\left[\dfrac{100 \text{ }\Omega}{5 \text{ mH}} + \dfrac{1}{(50 \text{ k}\Omega)(50 \text{ pF})}\right] = \textbf{66.85 kHz}$

which compares very favorably with the above solution.

d. $V_p = IZ_{T_p} = (2 \text{ mA})(R_s \| R_p) = (2 \text{ mA})(47.62 \text{ k}\Omega) = \textbf{95.24 V}$

e. See Fig. 17.27.

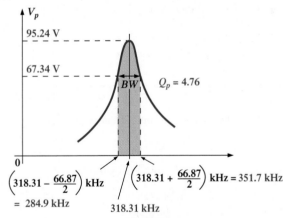

FIG. 17.27

Resonance curve for the network of Fig. 17.26.

EXAMPLE 17.9   Design a parallel resonant circuit to have the response curve of Fig. 17.28 using a 1 mH, 10 Ω inductor and a current source with an internal resistance of 40 kΩ.

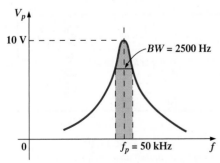

FIG. 17.28

Example 17.9.

Solution:

$$BW = \frac{f_p}{Q_p}$$

Therefore,

$$Q_p = \frac{f_p}{BW} = \frac{50,000 \text{ Hz}}{2500 \text{ Hz}} = \textbf{20}$$

$$X_L = 2\pi f_p L = 2\pi(50 \text{ kHz})(1 \text{ mH}) = 314 \; \Omega$$

and

$$Q_l = \frac{X_L}{R_l} = \frac{314 \; \Omega}{10 \; \Omega} = \textbf{31.4}$$

$$R_p = Q_l^2 R = (31.4)^2(10 \; \Omega) = \textbf{9859.6} \; \boldsymbol{\Omega}$$

$$Q_p = \frac{R}{X_L} = \frac{R_s \| 9859.6 \; \Omega}{314 \; \Omega} = 20 \quad \text{(from above)}$$

so that

$$\frac{(R_s)(9859.6)}{R_s + 9859.6} = 6280$$

resulting in

$$R_s = 17.298 \text{ k}\Omega$$

However, the source resistance was given as 40 kΩ. We must therefore add a parallel resistor (R') that will reduce the 40 kΩ to approximately 17.298 kΩ. That is,

$$\frac{(40 \text{ k}\Omega)(R')}{40 \text{ k}\Omega + R'} = 17.298 \text{ k}\Omega$$

Solving for R':

$$R' = \textbf{30.481 k}\boldsymbol{\Omega}$$

The closest commercial value is **30 kΩ.** At resonance, $X_L = X_C$, and

$$X_C = \frac{1}{2\pi f_p C}$$

$$C = \frac{1}{2\pi f_p X_C} = \frac{1}{2\pi(50 \text{ kHz})(314 \; \Omega)}$$

and

$$C \cong \textbf{0.01} \; \boldsymbol{\mu}\textbf{F} \quad \text{(commercially available)}$$

$$Z_{T_p} = R_s \| Q_l^2 R_l$$

$$= 17.298 \text{ k}\Omega \| 9859.6 \; \Omega$$

$$= 6.28 \text{ k}\Omega$$

with

$$V_p = I Z_{T_p}$$

and

$$I = \frac{V_p}{Z_{T_p}} = \frac{10 \text{ V}}{6.28 \text{ k}\Omega} \cong \textbf{1.6 mA}$$

The network appears in Fig. 17.29.

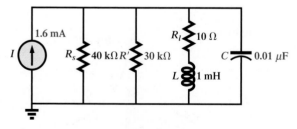

FIG. 17.29
Network designed to meet the criteria of Fig. 17.28.

The use of logarithms in industry is so extensive that a clear understanding of their purpose and use is an absolute necessity. At first exposure, logarithms often appear vague and mysterious due to the mathematical operations required to find the logarithm and antilogarithm using the longhand table approach that is typically taught in mathematics courses. However, almost all of today's scientific calculators have the common and natural log functions, eliminating the complexity of applying logarithms and allowing us to concentrate on the positive characteristics of the function.

Basic Relationships

Let us first examine the relationship between the variables of the logarithmic function. The mathematical expression

$$N = (b)^x$$

states that the number N is equal to the base b taken to the power x. A few examples:

$$100 = (10)^2$$
$$27 = (3)^3$$
$$54.6 = (e)^4 \quad \text{where } e = 2.7183$$

If the question were to find the power x to satisfy the equation

$$1200 = (10)^x$$

the value of x could be determined using logarithms in the following manner:

$$x = \log_{10} 1200 = \mathbf{3.079}$$

revealing that

$$10^{3.079} = 1200$$

Note that the logarithm was taken to the base 10—the number to be taken to the power of x. There is no limitation on the numerical value of the base except that tables and calculators are designed to handle either a base of 10 (common logarithm, $\boxed{\text{LOG}}$) or base $e = 2.7183$ (natural logarithm, $\boxed{\text{LN}}$). In review, therefore,

$$\boxed{\text{If } N = (b)^x, \text{ then } x = \log_b N.} \qquad \textbf{(17.41)}$$

The base to be employed is a function of the area of application. If a conversion from one base to the other is required, the following equation can be applied:

$$\boxed{\log_e x = 2.3 \log_{10} x} \qquad \textbf{(17.42)}$$

The content of this chapter is such that we will concentrate solely on the common logarithm. However, a number of the conclusions are also applicable to natural logarithms.

Some Areas of Application

The following is a short list of the most common applications of the logarithmic function:

1. This chapter will demonstrate that the use of logarithms permits plotting the response of a system for a range of values that may otherwise be impossible or unwieldy with a linear scale.
2. Levels of power, voltage, and so on, can be compared without dealing with very large or very small numbers that often cloud the true impact of the difference in magnitudes.
3. A number of systems respond to outside stimuli in a nonlinear logarithmic manner. The result is a mathematical model that permits a direct calculation of the response of the system to a particular input signal.
4. The response of a cascaded or compound system can be rapidly determined using logarithms if the gain of each stage is known on a logarithmic basis. This characteristic will be demonstrated in an example to follow.

Graphs

Graph paper is available in the **semilog** and **log-log** varieties. Semilog paper has only one log scale, with the other a linear scale. Both scales of log-log paper are log scales. A section of semilog paper appears in Fig. 17.30. Note the linear (evenly spaced-interval) vertical scaling and the repeating intervals of the log scale at multiples of 10.

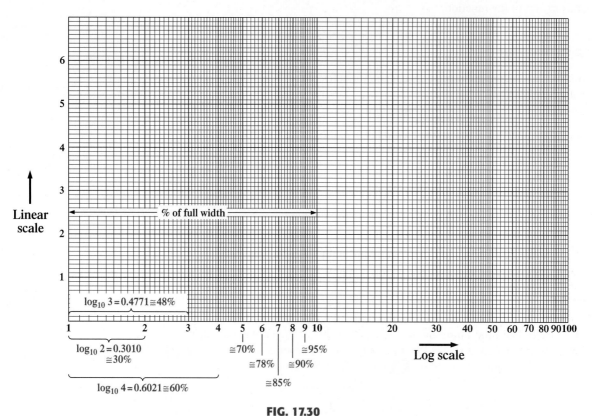

FIG. 17.30

Semilog graph paper.

The spacing of the log scale is determined by taking the common log (base 10) of the number. The scaling starts with 1, since $\log_{10} 1 = 0$. The distance between 1 and 2 is determined by $\log_{10} 2 = 0.3010$, or approximately 30% of the full distance of a log interval, as shown on the graph. The distance between 1 and 3 is determined by $\log_{10} 3 = 0.4771$, or about 48% of the full width. For future reference, keep in mind that almost 50% of the width of one log interval is represented by a 3 rather than by the 5 of a linear scale. In addition, note that the number 5 is about 70% of the full width, and 8 is about 90%. Remembering the percentage of full width of the lines 2, 3, 5, and 8 will be particularly useful when the various lines of a log plot are left unnumbered.

Since

$$\log_{10} 1 = 0$$
$$\log_{10} 10 = 1$$
$$\log_{10} 100 = 2$$
$$\log_{10} 1000 = 3$$
$$\vdots$$

the spacing between 1 and 10, 10 and 100, 100 and 1000, and so on, will be the same on a log scale because the difference between 0 and 1, 1 and 2, or 2 and 3 is fixed. Note in Figs. 17.30 and 17.31 that the horizontal difference between 1 and 10 and between 10 and 100 is the same. In Fig. 17.31 each increase in frequency by a factor 10 results in the same increase in horizontal spacing.

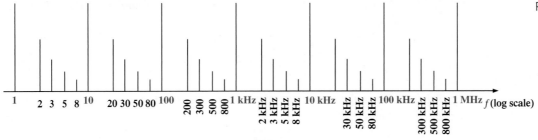

FIG. 17.31
Frequency log scale.

Note also in Figs. 17.30 and 17.31 that the log scale becomes compressed at the high end of each interval. With increasing frequency levels assigned to each interval, a single graph can provide a frequency plot extending from 1 Hz to 1 MHz, as shown in Fig. 17.31, with particular reference to the 30%, 50%, 70%, and 90% levels of each interval.

On many log plots, the tick marks for most of the intermediate levels are omitted because of space constraints. The following equation can be used to determine the logarithmic level at a particular point between known levels using a ruler or simply estimating the distances. The parameters are defined by Fig. 17.32.

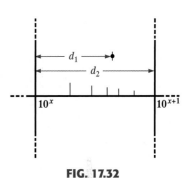

FIG. 17.32
Finding a value on a log plot.

$$\text{Value} = 10^x \times 10^{d_1/d_2} \qquad \textbf{(17.43)}$$

The derivation of Eq. (17.43) is simply an extension of the details regarding distance appearing on Fig. 17.30.

EXAMPLE 17.10 Determine the value of the point appearing on the logarithmic plot of Fig. 17.33 using the measurements made by a ruler (linear).

Solution:

$$\frac{d_1}{d_2} = \frac{7/16''}{3/4''} = \frac{0.438''}{0.750''} = 0.584$$

Using a calculator:

$$10^{d_1/d_2} = 10^{0.584} = 3.837$$

Applying Eq. (17.43):

$$\text{Value} = 10^x \times 10^{d_1/d_2} = 10^2 \times 3.837 = \textbf{383.7}$$

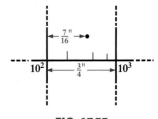

FIG. 17.33
Example 17.10.

17.7 PROPERTIES OF LOGARITHMS

There are a few characteristics of logarithms that should be emphasized:

1. The common or natural logarithm of the number 1 is 0.

$$\log_{10} 1 = 0 \qquad \textbf{(17.44)}$$

just as $10^x = 1$ requires that $x = 0$.

2. The log of any number less than 1 is a negative number.

$$\log_{10} \frac{1}{2} = \log_{10} 0.5 = -0.3$$

$$\log_{10} \frac{1}{10} = \log_{10} 0.1 = -1$$

3. The log of the product of two numbers is the sum of the logs of the numbers.

$$\log_{10} ab = \log_{10} a + \log_{10} b \qquad \textbf{(17.45)}$$

4. *The log of the quotient of two numbers is the log of the numerator minus the log of the denominator.*

$$\log_{10} \frac{a}{b} = \log_{10} a - \log_{10} b \qquad (17.46)$$

5. *The log of a number taken to a power is equal to the product of the power and the log of the number.*

$$\log_{10} a^n = n \log_{10} a \qquad (17.47)$$

Calculator Functions

On most calculators the log of a number is found by simply entering the number and pressing the LOG key. For example,

$$\log_{10} 80 = \boxed{\text{LOG}} \ \boxed{8} \ \boxed{0} \ \boxed{\text{ENTER}}$$

with a display of **1.903.**

For the reverse process, where N, or the antilogarithm, is desired, the function 10^x is employed. On most calculators 10^x appears as a second function above the LOG key. For the case of

$$0.6 = \log_{10} N$$

the following keys are employed:

$$\boxed{\text{2nd}} \ \boxed{10^x} \ \boxed{0} \ \boxed{\cdot} \ \boxed{6} \ \boxed{\text{ENTER}}$$

with a display of **3.981.** Checking, we have $\log_{10} 3.981 = 0.6$.

EXAMPLE 17.11 Evaluate each of the following logarithmic expressions:

a. $\log_{10} 0.004$

b. $\log_{10} 250{,}000$

c. $\log_{10}(0.08)(240)$

d. $\log_{10} \dfrac{1 \times 10^4}{1 \times 10^{-4}}$

e. $\log_{10}(10)^4$

Solutions:

a. **−2.398**

b. **+5.398**

c. $\log_{10}(0.08)(240) = \log_{10} 0.08 + \log_{10} 240 = -1.097 + 2.380 = \mathbf{1.283}$

d. $\log_{10} \dfrac{1 \times 10^4}{1 \times 10^{-4}} = \log_{10} 1 \times 10^4 - \log_{10} 1 \times 10^{-4} = 4 - (-4) = \mathbf{8}$

e. $\log_{10} 10^4 = 4 \log_{10} 10 = 4(1) = \mathbf{4}$

17.8 DECIBELS

Power Gain

Two levels of power can be compared using a unit of measure called the *bel,* which is defined by the following equation:

$$\text{B} = \log_{10} \frac{P_2}{P_1} \qquad \text{(bels)} \qquad (17.48)$$

However, to provide a unit of measure of *lesser* magnitude, a **decibel** is defined, where

$$\boxed{1 \text{ bel} = 10 \text{ decibels (dB)}} \qquad \textbf{(17.49)}$$

The result is the following important equation, which compares power levels P_2 and P_1 in decibels:

$$\boxed{\text{dB} = 10 \log_{10} \frac{P_2}{P_1}} \qquad \text{(decibels, dB)} \qquad \textbf{(17.50)}$$

If the power levels are equal ($P_2 = P_1$), there is no change in power level, and dB = 0. If there is an increase in power level ($P_2 > P_1$), the resulting decibel level is positive. If there is a decrease in power level ($P_2 < P_1$), the resulting decibel level will be negative.

For the special case of $P_2 = 2P_1$, the gain in decibels is

$$\text{dB} = 10 \log_{10} \frac{P_2}{P_1} = 10 \log_{10} 2 = \textbf{3 dB}$$

Therefore, for a speaker system, a 3 dB increase in output would require that the power level be doubled. In the audio industry, it is a generally accepted rule that an increase in sound level is accomplished with 3 dB increments in the output level. In other words, a 1 dB increase is barely detectable, and a 2 dB increase just discernible. A 3 dB increase normally results in a readily detectable increase in sound level. An additional increase in the sound level is normally accomplished by simply increasing the output level another 3 dB If an 8 W system were in use, a 3 dB increase would require a 16 W output, whereas an additional increase of 3 dB (a total of 6 dB) would require a 32 W system as demonstrated by the calculations below:

$$\text{dB} = 10 \log_{10} \frac{P_2}{P_1} = 10 \log_{10} \frac{16}{8} = 10 \log_{10} 2 = \textbf{3 dB}$$

$$\text{dB} = 10 \log_{10} \frac{P_2}{P_1} = 10 \log_{10} \frac{32}{8} = 10 \log_{10} 4 = \textbf{6 dB}$$

For $P_2 = 10P_1$,

$$\text{dB} = 10 \log_{10} \frac{P_2}{P_1} = 10 \log_{10} 10 = 10(1) = \textbf{10 dB}$$

resulting in the unique situation where the power gain has the same magnitude as the decibel level.

For some applications, a reference level is established to permit a comparison of decibel levels from one situation to another. For communication systems a commonly applied reference level is

$$P_{\text{ref}} = 1 \text{ mW} \quad \text{(across a 600 } \Omega \text{ load)}$$

Equation (17.50) is then typically written as

$$\boxed{\text{dB}_m = 10 \log_{10} \frac{P}{1 \text{ mW}} \bigg|_{600 \, \Omega}} \qquad \textbf{(17.51)}$$

Note the subscript m to denote that the decibel level is determined with a reference level of 1 mW.

In particular, for $P = 40$ mW,

$$\text{dB}_m = 10 \log_{10} \frac{40 \text{ mW}}{1 \text{ mW}} = 10 \log_{10} 40 = 10(1.6) = \textbf{16 dB}_m$$

whereas for $P = 4$ W,

$$\text{dB}_m = 10 \log_{10} \frac{4000 \text{ mW}}{1 \text{ mW}} = 10 \log_{10} 4000 = 10(3.6) = \textbf{36 dB}_m$$

Even though the power level has increased by a factor of 4000 mW/40 mW = 100, the dB_m increase is limited to 20 dB_m. In time, the significance of dB_m levels of 16 dB_m and 36 dB_m will generate an immediate appreciation regarding the power levels involved. An increase of 20 dB_m will also be associated with a significant gain in power levels.

Voltage Gain

Decibels are also used to provide a comparison between voltage levels. Substituting the basic power equations $P_2 = V_2^2/R_2$ and $P_1 = V_1^2/R_1$ into Eq. (17.50) will result in

$$dB = 10 \log_{10} \frac{P_2}{P_1} = 10 \log_{10} \frac{V_2^2/R_2}{V_1^2/R_1}$$

$$= 10 \log_{10} \frac{V_2^2/V_1^2}{R_2/R_1} = 10 \log_{10} \left(\frac{V_2}{V_1}\right)^2 - 10 \log_{10} \left(\frac{R_2}{R_1}\right)$$

and

$$dB = 20 \log_{10} \frac{V_2}{V_1} - 10 \log_{10} \frac{R_2}{R_1}$$

For the situation where $R_2 = R_1$, a condition normally assumed when comparing voltage levels on a decibel basis, the second term of the preceding equation will drop out $(\log_{10} 1 = 0)$, and

$$\boxed{dB_v = 20 \log_{10} \frac{V_2}{V_1}} \qquad \text{(dB)} \qquad \textbf{(17.52)}$$

Note the subscript v to define the decibel level obtained.

EXAMPLE 17.12 Find the voltage gain in dB of a system where the applied signal is 2 mV and the output voltage is 1.2 V.

Solution: $dB_v = 20 \log_{10} \frac{V_o}{V_i} = 20 \log_{10} \frac{1.2 \text{ V}}{2 \text{ mV}} = 20 \log_{10} 600 = \textbf{55.56 dB}$

for a voltage gain $A_v = V_o/V_i$ of 600.

EXAMPLE 17.13 If a system has a voltage gain of 36 dB, find the applied voltage if the output voltage is 6.8 V.

Solution:

$$dB_v = 20 \log_{10} \frac{V_o}{V_i}$$

$$36 = 20 \log_{10} \frac{V_o}{V_i}$$

$$1.8 = \log_{10} \frac{V_o}{V_i}$$

From the antilogarithm:

$$\frac{V_o}{V_i} = 63.096$$

and

$$V_i = \frac{V_o}{63.096} = \frac{6.8 \text{ V}}{63.096} = \textbf{107.77 mV}$$

TABLE 17.1

V_o/V_i	$dB = 20 \log_{10}(V_o/V_i)$
1	0 dB
2	6 dB
10	20 dB
20	26 dB
100	40 dB
1,000	60 dB
100,000	100 dB

Table 17.1 compares the magnitude of specific gains to the resulting decibel level. In particular, note that when voltage levels are compared, a doubling of the level results in a change of 6 dB rather than 3 dB as obtained for power levels. In addition, note that an increase in gain from 1 to 100,000 results in a change in decibels that can easily be plotted on a single graph. Also note that doubling the gain (from 1 to 2 and 10 to 20) results in a 6 dB increase in the decibel level, while a change of 10 to 1 (from 1 to 10, 10 to 100, and so on) always results in a 20 dB decrease in the decibel level.

The Human Auditory Response

One of the most frequent applications of the decibel scale is in the communication and entertainment industries. The human ear does not respond in a linear fashion to changes in source power level; that is, a doubling of the audio power level from 1/2 W to 1 W does not result in a doubling of the loudness level for the human ear. In addition, a change from 5 W to 10 W will be received by the ear as the same change in sound intensity as experienced from 1/2 W to 1 W. In other words, the ratio between levels is the same in each case (1 W/0.5 W = 10 W/5 W = 2), resulting in the same decibel or logarithmic change defined by Eq. (17.47). The ear, therefore, responds in a logarithmic fashion to changes in audio power levels.

To establish a basis for comparison between audio levels, a reference level of 0.0002 **microbar** (μbar) was chosen, where 1 μbar is equal to the sound pressure of 1 dyne per square centimeter, or about 1 millionth of the normal atmospheric pressure at sea level. The 0.0002 μbar level is the threshold level of hearing. Using this reference level, the sound pressure level in decibels is defined by the following equation:

$$dB_s = 20 \log_{10} \frac{P}{0.0002 \ \mu\text{bar}} \qquad\qquad (17.53)$$

where P is the sound pressure in microbars.

The decibel levels of Fig. 17.34 are defined by Eq. (17.53). Meters designed to measure audio levels are calibrated to the levels defined by Eq. (17.53) and shown in Fig. 17.34.

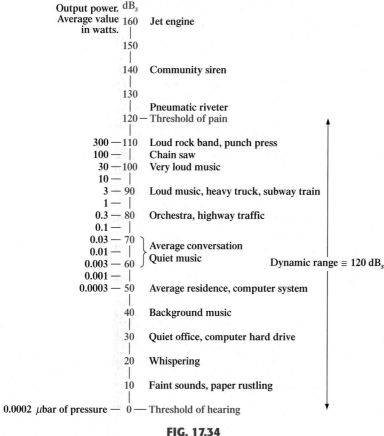

FIG. 17.34

Typical sound levels and their decibel levels.

A common question regarding audio levels is how much the power level of an acoustical source must be increased to double the sound level received by the human ear. The question is not as simple as it first seems due to considerations such as the frequency content of the sound, the acoustical conditions of the surrounding area, the physical characteristics of the surrounding medium, and, of course, the unique characteristics of the human ear. However, a general conclusion can be formulated that has practical value if we note the power levels of an acoustical source as shown in Fig. 17.34. Each power level is associated with a particular decibel level, and a change of 10 dB in the scale corresponds with an increase or a decrease in power by a factor of 10. For instance, a change from 90 dB to 100 dB is associated with a change in wattage from 3 W to 30 W. Through experimentation it has been found that on an average basis the loudness level will double for every 10 dB change in audio level—a conclusion somewhat verified by the examples to the right in Fig. 17.34. Using the fact that a 10 dB change corresponds with a tenfold increase in power level supports the following conclusion (on an approximate basis): Through experimentation it has been found that on an average basis, the loudness level will double for every 10 dB change in audio level.

To double the sound level received by the human ear, the power rating of the acoustical source (in watts) must be increased by a factor of 10.

In other words, doubling the sound level available from a 1 W acoustical source would require moving up to a 10 W source.

Instrumentation

A number of modern VOMs and DMMs have a dB scale designed to provide an indication of power ratios referenced to a standard level of 1 mW at 600 Ω. Since the reading is accurate only if the load has a characteristic impedance of 600 Ω, the 1 mW, 600 reference level is normally printed somewhere on the face of the meter, as shown in Fig. 17.35. The dB scale is usually calibrated to the lowest ac scale of the meter. In other words, when making the dB measurement, choose the lowest ac voltage scale, but read the dB scale. If a higher voltage scale is chosen, a correction factor must be employed that is sometimes printed on the face of the meter but always available in the meter manual. If the impedance is other than 600 Ω or not purely resistive, other correction factors must be used that are normally included in the meter manual.

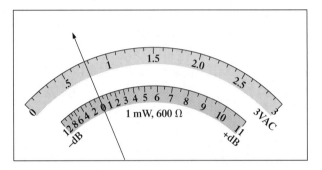

FIG. 17.35

Defining the relationship between a dB scale referenced to 1 mW, 600 Ω and a 3 V rms voltage scale.

Using the basic power equation $P = V^2/R$ will reveal that 1 mW across a 600 Ω load is the same as applying 0.775 V rms across a 600 Ω load; that is,

$$V = \sqrt{PR} = \sqrt{(1 \text{ mW})(600 \text{ }\Omega)} = 0.775 \text{ V}$$

The result is that an analog display will have 0 dB [defining the reference point of 1 mW, dB $= 10 \log_{10} P_2/P_1 = 10 \log_{10} (1 \text{ mW}/1 \text{ mW(ref)}) = 0 \text{ dB}$] and 0.775 V rms on the same pointer projection, as shown in Fig. 17.35. A voltage of 2.5 V across a 600 Ω load would result in a dB level of dB $= 20 \log_{10} V_2/V_1 = 20 \log_{10} 2.5 \text{ V}/0.775 = 10.17 \text{ dB}$, resulting in 2.5 V and 10.17 dB appearing along the same pointer projection. A voltage of less than 0.775 V, such as 0.5 V, will result in a dB level of dB $= 20 \log_{10} V_2/V_1 = 20 \log_{10} 0.5 \text{ V}/0.775 \text{ V}$ $= -3.8 \text{ dB}$, as is also shown on the scale of Fig. 17.35. Although a reading of 10 dB will reveal that the power level is 10 times the reference, don't assume that a reading of 5 dB means that the output level is 5 mW. The 10 : 1 ratio is a special one in logarithmic circles. For the 5 dB level, the power level must be found using the antilogarithm (3.126), which reveals that the power level associated with 5 dB is about 3.1 times the reference, or 3.1 mW. A conversion table is usually provided in the manual for such conversions.

17.9 FILTERS

Any combination of passive (R, L, and C) and/or active (transistors or operational amplifiers) elements designed to select or reject a band of frequencies is called a **filter.** In communication systems, filters are employed to pass those frequencies containing the desired information and to reject the remaining frequencies. In stereo systems, filters can be used to isolate particular bands of frequencies for increased or decreased emphasis by the output acoustical system (amplifier, speaker, and so on). Filters are employed to filter out any unwanted frequencies, commonly called *noise*, due to the nonlinear characteristics of some electronic devices or signals picked up from the surrounding medium. In general, there are two classifications of filters:

1. **Passive filters** are those filters composed of series or parallel combinations of R, L, and C elements such as described for series and parallel resonance.
2. **Active filters** are filters that employ active devices such as transistors and operational amplifiers.

Since this text is limited to passive devices, the analysis of this chapter will be limited to passive filters. In addition, only the most fundamental forms will be examined in the next few sections. The subject of filters is a very broad one that continues to receive extensive research support from industry and the government as new communication systems are developed to meet the demands of increased volume and speed. There are courses and texts devoted solely to the analysis and design of filter systems that can become quite complex and sophisticated. In general, however, all filters belong to the four broad categories of **low-pass, high-pass, pass-band,** and **stop-band,** as depicted in Fig. 17.36. For each form there are critical frequencies that define the regions of pass-bands and stop-bands (often called *reject* bands). Any frequency in the pass-band will pass through to the next stage with at least 70.7% of the maximum output voltage. Recall the use of the 0.707 level to define the bandwidth of a series or parallel resonant circuit (both with the general shape of the pass-band filter).

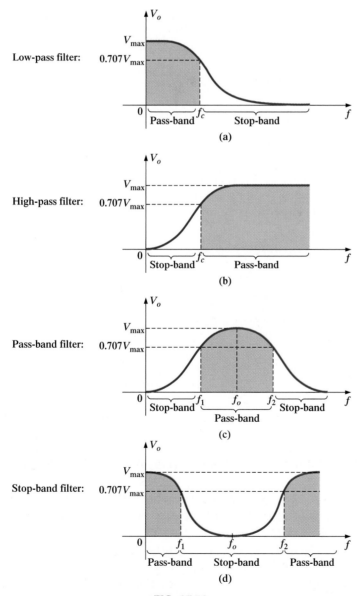

FIG. 17.36
Defining the four broad categories of filters.

For some stop-band filters, the stop-band is defined by conditions other than the 0.707 level. In fact, for many stop-band filters, the condition that $V_o = 1/1000V_{max}$ (corresponding with -60 dB in the discussion to follow) is used to define the stop-band region, with the pass-band continuing to be defined by the 0.707 V level. The resulting frequencies between the two regions are then called the *transition frequencies* and establish the *transition region*.

At least one example of each filter of Fig. 17.36 will be discussed in some detail in the sections to follow. Take particular note of the relative simplicity of some of the designs.

17.10 *R-C* LOW-PASS FILTER

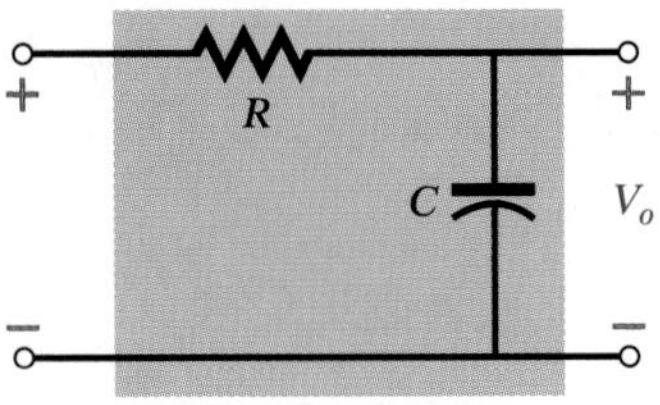

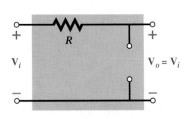

FIG. 17.37
Low-pass filter.

The *R-C* filter, incredibly simple in design, can be used as a low-pass or a high-pass filter. If the output is taken from the capacitor, as shown in Fig. 17.37, it will respond as a low-pass filter. If the positions of the resistor and capacitor are interchanged and the output is taken from the resistor, the response will be that of a high-pass filter.

A glance at Fig. 17.36(a) reveals that the circuit should behave in a manner that will result in a high-level output for low frequencies and a declining level for frequencies above the critical value. Let us first examine the network at the frequency extremes of $f = 0$ Hz and very high frequencies to test the response of the circuit.

At $f = 0$ Hz,

$$X_C = \frac{1}{2\pi fC} = \infty \; \Omega$$

and the open-circuit equivalent can be substituted for the capacitor, as shown in Fig. 17.38, resulting in $\mathbf{V}_o = \mathbf{V}_i$.

At very high frequencies, the reactance is

$$X_C = \frac{1}{2\pi fC} \cong 0 \; \Omega$$

and the short-circuit equivalent can be substituted for the capacitor, as shown in Fig. 17.39, resulting in $\mathbf{V}_o = 0$ V.

A plot of the magnitude of V_o versus frequency will result in the curve of Fig. 17.40. Our next goal is now clearly defined: Find the frequency at which the transition takes place from a pass-band to a stop-band.

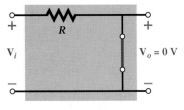

FIG. 17.38
R-C low-pass filter at low frequencies.

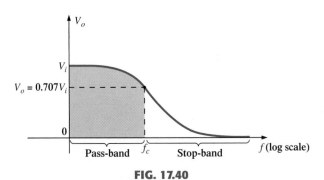

FIG. 17.40
V_o versus frequency for a low-pass R-C filter.

FIG. 17.39
R-C low-pass filter at high frequencies.

For filters, a **normalized plot** is employed more often than the plot of V_o versus frequency of Fig. 17.40.

Normalization is a process whereby a quantity such as voltage, current, or impedance is divided by a quantity of the same unit of measure to establish a dimensionless level of a specific value or range.

A normalized plot in the filter domain can be obtained by dividing the plotted quantity such as V_o of Fig. 17.40 with the applied voltage V_i for the frequency range of interest. Since the maximum value of V_o for the low-pass filter of Fig. 17.37 is V_i, each level of V_o in Fig. 17.40 is divided by the level of V_i. The result is the plot of $A_v = V_o/V_i$ of Fig. 17.41. Note that the maximum value is 1 and the cutoff frequency is defined at the 0.707 level.

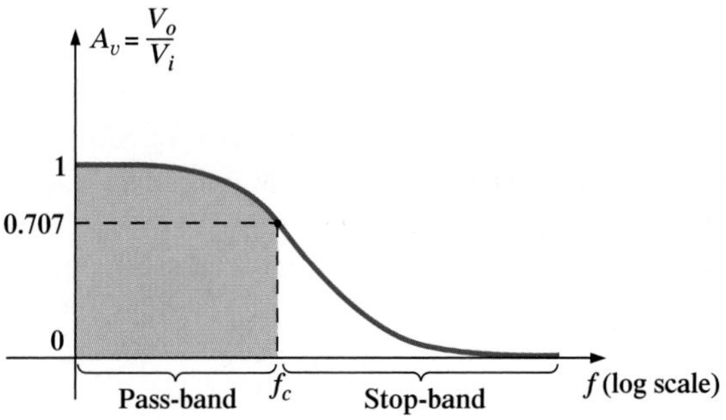

FIG. 17.41
Normalized plot of Fig. 17.40.

At any intermediate frequency, the output voltage $\mathbf{V}_o$ of Fig. 17.37 can be determined using the voltage divider rule:

$$\mathbf{V}_o = \frac{X_C \angle -90° \mathbf{V}_i}{R - jX_C}$$

or

$$\mathbf{A}_v = \frac{\mathbf{V}_o}{\mathbf{V}_i} = \frac{X_C \angle -90°}{R - jX_C} = \frac{X_C \angle -90°}{\sqrt{R^2 + X_C^2}\ \underline{/-\tan^{-1}(X_C/R)}}$$

and

$$\mathbf{A}_v = \frac{\mathbf{V}_o}{\mathbf{V}_i} = \frac{X_C}{\sqrt{R^2 + X_C^2}} \angle -90° + \tan^{-1}\left(\frac{X_C}{R}\right)$$

The magnitude of the ratio V_o/V_i is therefore determined by

$$A_v = \frac{V_o}{V_i} = \frac{X_C}{\sqrt{R^2 + X_C^2}} = \frac{1}{\sqrt{\left(\dfrac{R}{X_C}\right)^2 + 1}} \qquad \textbf{(17.54)}$$

and the phase angle is determined by

$$\theta = -90° + \tan^{-1}\frac{X_C}{R} = -\tan^{-1}\frac{R}{X_C} \qquad \textbf{(17.55)}$$

For the special frequency at which $X_C = R$, the magnitude becomes

$$A_v = \frac{V_o}{V_i} = \frac{1}{\sqrt{\left(\dfrac{R}{X_C}\right)^2 + 1}} = \frac{1}{\sqrt{1 + 1}} = \frac{1}{\sqrt{2}} = 0.707$$

which defines the critical or cutoff frequency of Fig. 17.41.

The frequency at which $X_C = R$ is determined by

$$\frac{1}{2\pi f_c C} = R$$

and

$$f_c = \frac{1}{2\pi RC} \qquad \textbf{(17.56)}$$

The impact of Eq. (17.56) extends beyond its relative simplicity. For any low-pass filter, the application of any frequency less than f_c will result in an output voltage V_o that is at least 70.7% of the maximum. For any frequency above f_c, the output is less than 70.7% of the applied signal.

Solving for $\mathbf{V}_o$ and substituting $\mathbf{V}_i = V_i \angle 0°$ gives

$$\mathbf{V}_o = \left[\frac{X_C}{\sqrt{R^2 + X_C^2}} \angle \theta\right]\mathbf{V}_i = \left[\frac{X_C}{\sqrt{R^2 + X_C^2}} \angle \theta\right]V_i \angle 0°$$

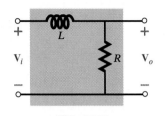

FIG. 17.42
Low-pass R-L filter.

and

$$\mathbf{V}_o = \frac{X_C V_i}{\sqrt{R^2 + X_C^2}} \angle \theta$$

Therefore, the angle θ is the angle by which $\mathbf{V}_o$ leads $\mathbf{V}_i$. Since $\theta = -\tan^{-1} R/X_C$ is always negative (except at $f = 0$ Hz), it is clear that $\mathbf{V}_o$ will always lag $\mathbf{V}_i$, leading to the label **lagging network** for the network of Fig. 17.37.

The low-pass filter response of Fig. 17.36(a) can also be obtained using the R-L combination of Fig. 17.42, with

$$f_c = \frac{R}{2\pi L} \qquad (17.57)$$

In general, however, the R-C combination is more popular due to the smaller size of capacitive elements and the nonlinearities associated with inductive elements. The details of the analysis of the low-pass R-L will be left as an exercise for the reader.

EXAMPLE 17.14 For the low-pass R-C filter of Fig. 17.43:

a. Sketch the output voltage V_o versus frequency using a semilog plot.
b. Determine the voltage V_o at $f = 100$ kHz and 1 MHz, and compare the results to the results obtained from the curve of part (a).
c. Sketch the normalized gain $A_v = V_o/V_i$.

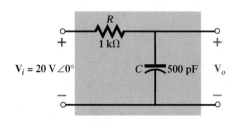

FIG. 17.43
Example 17.14.

Solutions:

a. Eq. (17.56):

$$f_c = \frac{1}{2\pi RC} = \frac{1}{2\pi(1\text{ k}\Omega)(500\text{ pF})} = \mathbf{318.31\text{ kHz}}$$

At f_c, $V_o = 0.707(20\text{ V}) = 14.14$ V. See Fig. 17.44.

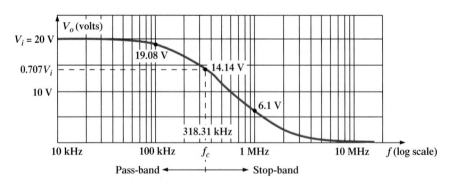

FIG. 17.44
Frequency response for the low-pass R-C network of Fig. 17.43.

b. Eq. (17.54):

$$V_o = \frac{V_i}{\sqrt{\left(\dfrac{R}{X_C}\right)^2 + 1}}$$

At f = 100 kHz:

$$X_C = \frac{1}{2\pi f C} = \frac{1}{2\pi(100 \text{ kHz})(500 \text{ pF})} = 3.18 \text{ k}\Omega$$

and

$$V_o = \frac{20 \text{ V}}{\sqrt{\left(\frac{1 \text{ k}\Omega}{3.18 \text{ k}\Omega}\right)^2 + 1}} = \mathbf{19.08 \text{ V}}$$

At f = 1 MHz:

$$X_C = \frac{1}{2\pi f C} = \frac{1}{2\pi(1 \text{ MHz})(500 \text{ pF})} = 0.32 \text{ k}\Omega$$

and

$$V_o = \frac{20 \text{ V}}{\sqrt{\left(\frac{1 \text{ k}\Omega}{0.32 \text{ k}\Omega}\right)^2 + 1}} = \mathbf{6.1 \text{ V}}$$

Both levels are verified by Fig. 17.44.

c. Dividing every level of Fig. 17.44 by V_i = 20 V will result in the normalized plot of Fig. 17.45.

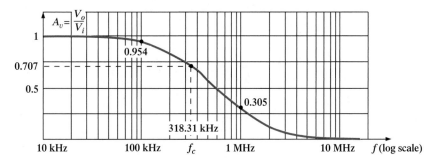

FIG. 17.45

Normalized plot of Fig. 17.44.

17.11 *R-C* HIGH-PASS FILTER

As noted early in Section 17.10, a high-pass *R-C* filter can be constructed by simply reversing the positions of the capacitor and resistor, as shown in Fig. 17.46.

At very high frequencies the reactance of the capacitor is very small, and the short-circuit equivalent can be substituted as shown in Fig. 17.47. The result is that $\mathbf{V}_o = \mathbf{V}_i$.

At f = 0 Hz, the reactance of the capacitor is quite high, and the open-circuit equivalent can be substituted as shown in Fig. 17.48. In this case, $\mathbf{V}_o$ = 0 V.

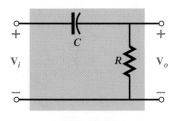

FIG. 17.46

High-pass filter.

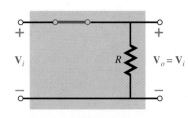

FIG. 17.47

R-C high-pass filter at very high frequencies.

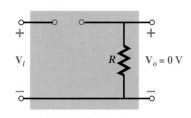

FIG. 17.48

R-C high-pass filter at f = 0 Hz.

A plot of the magnitude versus frequency is provided in Fig. 17.49, with the normalized plot in Fig. 17.50.

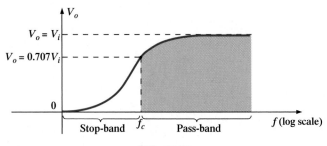

FIG. 17.49

V_o versus frequency for a high-pass R-C filter.

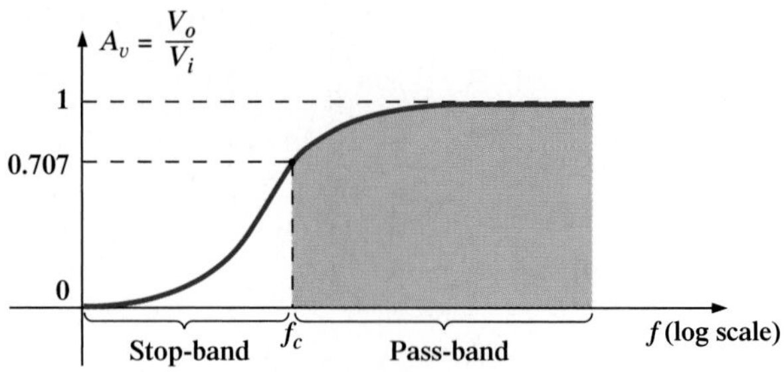

FIG. 17.50

Normalized plot of Fig. 17.49.

At any intermediate frequency, the output voltage can be determined using the voltage divider rule:

$$\mathbf{V}_o = \frac{R \angle 0° \ \mathbf{V}_i}{R - j X_C}$$

or

$$\frac{\mathbf{V}_o}{\mathbf{V}_i} = \frac{R \angle 0°}{R - j X_C} = \frac{R \angle 0°}{\sqrt{R^2 + X_C^2} \angle -\tan^{-1}(X_C/R)}$$

and

$$\frac{\mathbf{V}_o}{\mathbf{V}_i} = \frac{R}{\sqrt{R^2 + X_C^2}} \angle \tan^{-1}(X_C/R)$$

The magnitude of the ratio $\mathbf{V}_o/\mathbf{V}_i$ is therefore determined by

$$A_v = \frac{V_o}{V_i} = \frac{R}{\sqrt{R^2 + X_C^2}} = \frac{1}{\sqrt{1 + \left(\dfrac{X_C}{R}\right)^2}} \qquad \textbf{(17.58)}$$

and the phase angle θ by

$$\theta = \tan^{-1}\frac{X_C}{R} \qquad \textbf{(17.59)}$$

For the frequency at which $X_C = R$, the magnitude becomes

$$\frac{V_o}{V_i} = \frac{1}{\sqrt{1 + \left(\dfrac{X_C}{R}\right)^2}} = \frac{1}{\sqrt{1 + 1}} = \frac{1}{\sqrt{2}} = 0.707$$

as shown in Fig. 17.50.

The frequency at which $X_C = R$ is determined by

$$X_C = \frac{1}{2\pi f_c C} = R$$

and

$$f_c = \frac{1}{2\pi RC} \qquad \textbf{(17.60)}$$

For the high-pass R-C filter, the application of any frequency greater than f_c will result in an output voltage V_o that is at least 70.7% of the magnitude of the input signal. For any frequency below f_c, the output is less than 70.7% of the applied signal.

For the phase angle, high frequencies result in small values of X_C, and the ratio X_C/R will approach zero with $\tan^{-1}(X_C/R)$ approaching 0°, as shown in Fig. 17.51. At low frequencies, the ratio X_C/R becomes quite large, and $\tan^{-1}(X_C/R)$ approaches 90°. For the

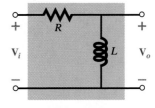

FIG. 17.51

Phase-angle response for the high-pass R-C filter.

case $X_C = R$, $\tan^{-1}(X_C/R) = \tan^{-1} 1 = 45°$. Assigning a phase angle of $0°$ to $\mathbf{V}_i$ such that $\mathbf{V}_i = V_i \angle 0°$, the phase angle associated with $\mathbf{V}_o$ is θ, resulting in $\mathbf{V}_o = V_o \angle \theta$ and revealing that θ is the angle by which $\mathbf{V}_o$ leads $\mathbf{V}_i$. Since the angle θ is the angle by which $\mathbf{V}_o$ leads $\mathbf{V}_i$ throughout the frequency range of Fig. 17.51, the high-pass R-C filter is referred to as a **leading network.**

The high-pass filter response of Fig. 17.50 can also be obtained using the same elements of Fig. 17.42 but interchanging their positions, as shown in Fig. 17.52.

FIG. 17.52

High-pass R-L filter.

EXAMPLE 17.15 Given $R = 20$ kΩ and $C = 1200$ pF:

a. Sketch the normalized plot if the filter is used as both a high-pass and a low-pass filter.
b. Sketch the phase plot for both filters of part (a).
c. Determine the magnitude and phase of $\mathbf{A}_v = \mathbf{V}_o/\mathbf{V}_i$ at $f = \frac{1}{2}f_c$ for the high-pass filter.

Solutions:

a. $f_c = \dfrac{1}{2\pi RC} = \dfrac{1}{(2\pi)(20 \text{ k}\Omega)(1200 \text{ pF})} = \textbf{6631.46 Hz}$

The normalized plots appear in Fig. 17.53.

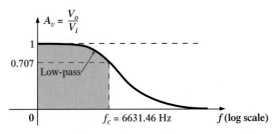

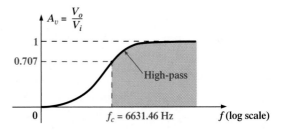

FIG. 17.53

Normalized plots for a low-pass and a high-pass filter using the same elements.

b. The phase plots appear in Fig. 17.54.

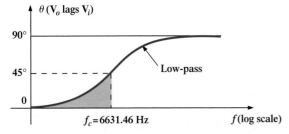

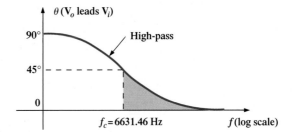

FIG. 17.54

Phase plots for a low-pass and a high-pass filter using the same elements.

c. $f = \dfrac{1}{2}f_c = \dfrac{1}{2}(6631.46 \text{ Hz}) = 3315.73 \text{ Hz}$

$$X_C = \frac{1}{2\pi fC} = \frac{1}{(2\pi)(3315.73 \text{ Hz})(1200 \text{ pF})} \cong 40 \text{ k}\Omega$$

$$A_v = \frac{V_o}{V_i} = \frac{1}{\sqrt{1 + \left(\dfrac{X_C}{R}\right)^2}} = \frac{1}{\sqrt{1 + \left(\dfrac{40 \text{ k}\Omega}{20 \text{ k}\Omega}\right)^2}}$$

$$= \frac{1}{\sqrt{1 + (2)^2}} = \frac{1}{\sqrt{5}} = 0.4472$$

$$\theta = \tan^{-1}\frac{X_C}{R} = \tan^{-1}\frac{40 \text{ k}\Omega}{20 \text{ k}\Omega} = \tan^{-1} 2 = 63.43°$$

and $\qquad\qquad \mathbf{A}_v = \dfrac{\mathbf{V}_o}{\mathbf{V}_i} = \mathbf{0.4472} \angle \mathbf{63.43°}$

17.12 PASS-BAND FILTERS

A number of methods are used to establish the pass-band characteristic of Fig. 17.36(c). One method employs both a low-pass and a high-pass filter in cascade, as shown in Fig. 17.55.

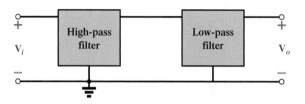

FIG. 17.55

Pass-band filter.

The components are chosen to establish a cutoff frequency for the high-pass filter that is lower than the critical frequency of the low-pass filter, as shown in Fig. 17.56. A frequency f_1 may pass through the low-pass filter but have little effect on V_o due to the reject characteristics of the high-pass filter. A frequency f_2 may pass through the high-pass filter undisturbed but be prohibited from reaching the high-pass filter by the low-pass characteristics. A frequency f_o near the center of the pass-band will pass through both filters with very little degeneration.

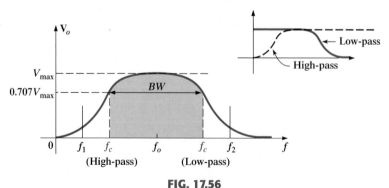

FIG. 17.56

Pass-band characteristics.

The network of Fig. 17.57 in Example 17.16 will generate the characteristics of Fig. 17.56. However, for a circuit such as the one shown in Fig. 17.57, there is a loading between stages at each frequency that will affect the level of $\mathbf{V}_o$. Through proper design, the level of $\mathbf{V}_o$ may be very near the level of $\mathbf{V}_i$ in the pass-band, but it will never equal it exactly. In addition, as the critical frequencies of each filter get closer and closer together to increase the quality factor of the response curve, the peak values within the pass-band will continue to drop. For cases where $V_{o_{max}} \neq V_{i_{max}}$, the bandwidth is defined at 0.707 of the resulting $V_{o_{max}}$.

EXAMPLE 17.16   For the pass-band filter of Fig. 17.57:

a. Determine the critical frequencies for the low- and high-pass filters.
b. Using only the critical frequencies, sketch the response characteristics.
c. Determine the actual value of V_o at the high-pass critical frequency calculated in part (a), and compare it to the level that will define the upper frequency for the pass-band.

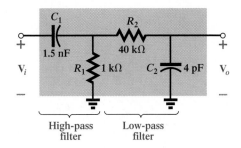

FIG. 17.57
Pass-band filter for Example 17.16.

Solutions:

a. High-pass filter:

$$f_c = \frac{1}{2\pi R_1 C_1} = \frac{1}{2\pi(1 \text{ k}\Omega)(1.5 \text{ nF})} = \mathbf{106.1 \text{ kHz}}$$

Low-pass filter:

$$f_c = \frac{1}{2\pi R_2 C_2} = \frac{1}{2\pi(40 \text{ k}\Omega)(4 \text{ pF})} = \mathbf{994.72 \text{ kHz}}$$

b. In the mid-region of the pass-band at about 500 kHz, an analysis of the network will reveal that $V_o \cong 0.9V_i$ as shown in Fig. 17.58. The bandwidth is therefore defined at a level of $0.707(0.9V_i) = 0.636V_i$ as also shown in Fig. 17.58.

c. At $f = 994.72$ kHz,

$$X_{C_1} = \frac{1}{2\pi f C_1} \cong 107 \text{ } \Omega \quad \text{and} \quad X_{C_2} = \frac{1}{2\pi f C_2} = R_2 = 40 \text{ k}\Omega$$

resulting in the network of Fig. 17.59.

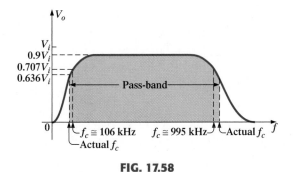

FIG. 17.58
Pass-band characteristics for the filter of Fig. 17.57.

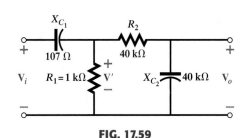

FIG. 17.59
Network of Fig. 17.57 at $f = 994.72$ kHz.

The parallel combination $R_1 \| (R_2 - j X_{C_2})$ is essentially 0.976 kΩ ∠0° because the $R_2 - X_{C_2}$ combination is so large compared to the parallel resistor R_1. Then

$$\mathbf{V'} = \frac{0.976 \text{ k}\Omega \angle 0°(\mathbf{V}_i)}{0.976 \text{ k}\Omega - j\, 0.107 \text{ k}\Omega} \cong 0.994 \mathbf{V}_i \angle 6.26°$$

with
$$\mathbf{V}_o = \frac{(40 \text{ k}\Omega \angle -90°)(0.994\mathbf{V}_i \angle 6.26°)}{40 \text{ k}\Omega - j40 \text{ k}\Omega}$$

$$\mathbf{V}_o \cong 0.703\mathbf{V}_i \angle -39°$$

so that
$$V_o \cong \mathbf{0.703V_i} \qquad \text{at } f = 994.72 \text{ kHz}$$

Since the bandwidth is defined at $0.636V_i$, the upper cutoff frequency will be higher than 994.72 kHz as shown in Fig. 17.58.

The pass-band response can also be obtained using the series and parallel resonant circuits discussed earlier in this chapter. In each case, however, V_o will not be equal to V_i in the pass-band, but a frequency range in which V_o will be equal to or greater than $0.707V_{max}$ can be defined.

For the series resonant circuit of Fig. 17.60, $X_L = X_C$ at resonance, and

$$\boxed{V_{o_{max}} = \frac{R}{R + R_l}V_i \Big|_{f = f_s}} \qquad (17.61)$$

and

$$\boxed{f_s = \frac{1}{2\pi\sqrt{LC}}} \qquad (17.62)$$

with

$$\boxed{Q_s = \frac{X_L}{R + R_l}} \qquad (17.63)$$

and

$$\boxed{BW = \frac{f_s}{Q_s}} \qquad (17.64)$$

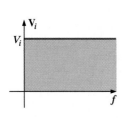
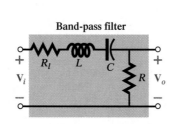
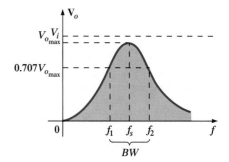

FIG. 17.60

Series resonant pass-band filter.

For the parallel resonant circuit of Fig. 17.61, Z_{T_p} is a maximum value at resonance, and

$$\boxed{V_{o_{max}} = \frac{Z_{T_p}V_i}{Z_{T_p} + R} \Big|_{f = f_p}} \qquad (17.65)$$

with

$$\boxed{Z_{T_p} \cong Q_l^2 R_l} \qquad (17.66)$$

Also,

$$\boxed{f_p \cong \frac{1}{2\pi\sqrt{LC}}} \qquad (17.67)$$

with

$$\boxed{Q_p \cong \frac{X_L}{R_l}} \qquad (17.68)$$

and

$$\boxed{BW = \frac{f_p}{Q_p}} \qquad (17.69)$$

As a first approximation that is acceptable for most practical applications, it can be assumed that the resonant frequency bisects the bandwidth.

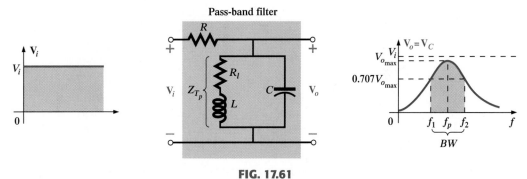

FIG. 17.61

Parallel resonant pass-band filter.

EXAMPLE 17.17 For the series circuit of Fig. 17.62:

a. Determine the frequency response for the voltage V_o.
b. Plot the normalized response $A_v = V_o/V_i$.
c. Plot a normalized response defined by $A'_v = A_v/A_{v_{max}}$.

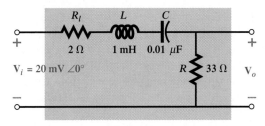

FIG. 17.62

Series resonant pass-band filter for Example 17.17.

Solutions:

a. $f_s = \dfrac{1}{2\pi\sqrt{LC}} = \dfrac{1}{2\pi\sqrt{(1\text{ mH})(0.01\ \mu\text{F})}} = \textbf{50,329.21 Hz}$

$Q_s = \dfrac{X_L}{R + R_l} = \dfrac{2\pi(50{,}329.21\text{ Hz})(1\text{ mH})}{33\ \Omega + 2\ \Omega} = \textbf{9.04}$

$BW = \dfrac{f_s}{Q_s} = \dfrac{50{,}329.21\text{ Hz}}{9.04} = \textbf{5.57 kHz}$

At resonance:

$V_{o_{max}} = \dfrac{RV_i}{R + R_l} = \dfrac{33\ \Omega(V_i)}{33\ \Omega + 2\ \Omega} = 0.943V_i = 0.943(20\text{ mV}) = \textbf{18.86 mV}$

At the cutoff frequencies:

$V_o = (0.707)(0.943V_i) = 0.667V_i = 0.667(20\text{ mV}) = \textbf{13.34 mV}$

Note Fig. 17.63.

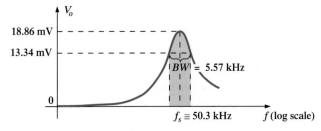

FIG. 17.63

Pass-band response for the network.

b. Dividing all levels of Fig. 17.63 by $V_i = 20$ mV will result in the normalized plot of Fig. 17.64(a).

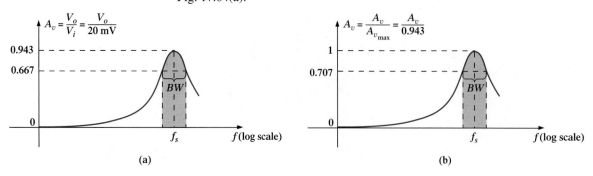

(a)   (b)

FIG. 17.64

Normalized plots for the pass-band filter of Fig. 17.62.

c. Dividing all levels of Fig. 17.64(a) by $A_{v_{max}} = 0.943$ will result in the normalized plot of Fig. 17.64(b).

17.13 STOP-BAND FILTERS

Stop-band filters can also be constructed using a low-pass and a high-pass filter. However, rather than the cascaded configuration used for the pass-band filter, a parallel arrangement is required, as shown in Fig. 17.65. A low-frequency f_1 can pass through the low-pass filter, and a higher-frequency f_2 can use the parallel path, as shown in Figs. 17.65 and 17.66. However, a frequency such as f_o in the reject-band is higher than the low-pass critical frequency and lower than the high-pass critical frequency, and is therefore prevented from contributing to the levels of V_o above $0.707V_{max}$.

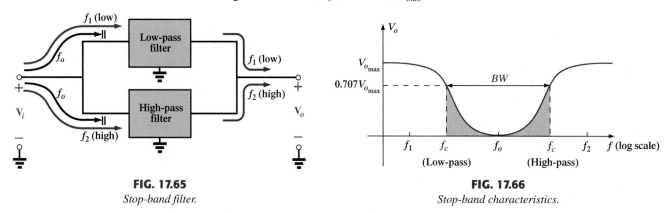

FIG. 17.65   **FIG. 17.66**

Stop-band filter.   *Stop-band characteristics.*

Since the characteristics of a stop-band filter are the inverse of the pattern obtained for the pass-band filters, we can employ the fact that at any frequency, the sum of the magnitudes of the two waveforms to the right of the equals sign in Fig. 17.67 will equal the applied voltage V_i.

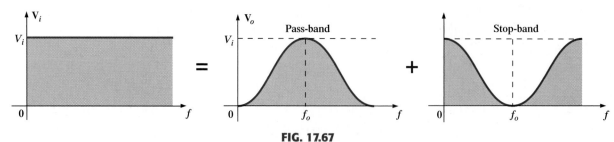

FIG. 17.67

Demonstrating how an applied signal of fixed magnitude can be broken down into a pass-band and stop-band response curve.

For the pass-band filters of Figs. 17.60 and 17.61, therefore, if we take the output from the other series elements as shown in Figs. 17.68 and 17.69, a stop-band characteristic will be obtained, as required by Kirchhoff's voltage law.

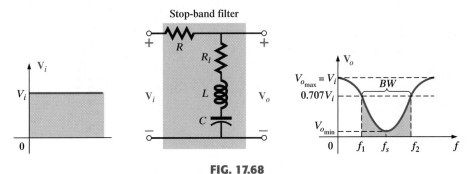

FIG. 17.68

Stop-band filter using a series resonant circuit.

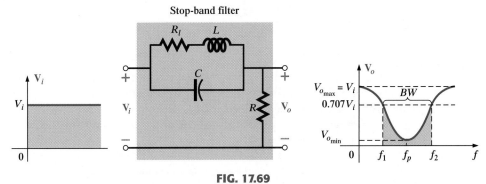

FIG. 17.69

Stop-band filter using a parallel resonant network.

For the series resonant circuit of Fig. 17.68, Equations (17.62) through (17.64) still apply, but now, at resonance,

$$V_{o_{min}} = \frac{R_l V_i}{R_l + R} \qquad \textbf{(17.70)}$$

For the parallel resonant circuit of Fig. 17.69, Equations (17.65) through (17.69) are still applicable, but now, at resonance,

$$V_{o_{min}} = \frac{R V_i}{R + Z_{T_p}} \qquad \textbf{(17.71)}$$

The maximum value of V_o for the series resonant circuit is V_i at the low end due to the open-circuit equivalent for the capacitor, and V_i at the high end due to the high impedance of the inductive element.

For the parallel resonant circuit, at $f = 0$ Hz, the coil can be replaced by a short-circuit equivalent, and the capacitor can be replaced by its open circuit and $V_o = RV_i/(R + R_l)$. At the high-frequency end, the capacitor approaches a short-circuit equivalent, and V_o increases toward V_i.

17.14 DOUBLE-TUNED FILTERS

Some network configurations display both a pass-band and a stop-band characteristic, such as shown in Fig. 17.70. Such networks are called **double-tuned filters.** For the network of Fig. 17.70(a), the parallel resonant circuit will establish a stop-band for the range of frequencies not permitted to establish a significant V_L. The greater part of the applied voltage will appear across the parallel resonant circuit for this frequency range due to its very high impedance compared with R_L. For the pass-band, the parallel resonant circuit is designed to be capacitive (inductive if L_s is replaced by C_s). The inductance L_s is chosen to cancel the effects of the resulting net capacitive reactance at the resonant pass-band frequency of the tank circuit, thereby acting as a series resonant circuit. The applied voltage will then appear across R_L at this frequency.

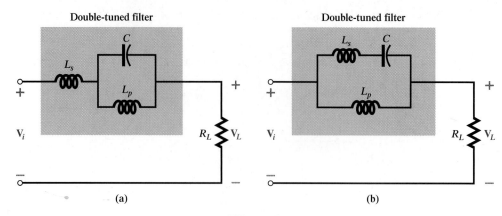

FIG. 17.70

Double-tuned networks.

For the network of Fig. 17.70(b), the series resonant circuit will still determine the pass-band, acting as a very low impedance across the parallel inductor at resonance. At the desired stop-band resonant frequency, the series resonant circuit is capacitive. The inductance L_p is chosen to establish parallel resonance at the resonant stop-band frequency. The high impedance of the parallel resonant circuit will result in a very low load voltage V_L.

For rejected frequencies below the pass-band, the networks should appear as shown in Fig. 17.70. For the reverse situation, L_s in Fig. 17.70(a) and L_p in Fig. 17.70(b) are replaced by capacitors.

EXAMPLE 17.18 For the network of Fig. 17.70(b), determine L_s and L_p for a capacitance C of 500 pF if a frequency of 200 kHz is to be rejected and a frequency of 600 kHz accepted.

Solution: For series resonance, we have

$$f_s = \frac{1}{2\pi\sqrt{LC}}$$

and

$$L_s = \frac{1}{4\pi^2 f_s^2 C} = \frac{1}{4\pi^2(600 \text{ kHz})^2(500 \text{ pF})} = \textbf{140.7 } \boldsymbol{\mu}\textbf{H}$$

At 200 kHz,

$$X_{L_s} = \omega L = 2\pi f_s L_s = (2\pi)(200 \text{ kHz})(140.7 \text{ }\mu\text{H}) = 176.8 \text{ }\Omega$$

and

$$X_C = \frac{1}{\omega C} = \frac{1}{(2\pi)(200 \text{ kHz})(500 \text{ pF})} = 1591.5 \text{ }\Omega$$

For the series elements,

$$j(X_{L_s} - X_C) = j(176.8 \text{ }\Omega - 1591.5 \text{ }\Omega) = -j\,1414.7 \text{ }\Omega = -j\,X'_C$$

At parallel resonance ($Q_l \geq 10$ assumed),

$$X_{L_p} = X'_C$$

and

$$L_p = \frac{X_{L_p}}{\omega} = \frac{1414.7 \text{ }\Omega}{(2\pi)(200 \text{ kHz})} = \textbf{1.13 mH}$$

17.15 dB PLOTS

When commercial literature describes a filter, it usually uses a dB plot of the response versus frequency rather than the normalized plots just described. It is important to understand, however, that the dB plot comes directly from the normalized plot just described.

A dB plot is derived using the following equation:

$$\boxed{A_{v_{dB}} = 20 \log_{10} A_v} \tag{17.72}$$

Note that the equation is taking the logarithm of the voltage gain, not a particular value of gain, and the logarithm is taken to the base 10 (common log). In Fig. 17.71(a), the gain plot of Fig. 17.41 is repeated for the low-pass filter. When the normalized gain is 1, if we apply Eq. (17.72), we find that the result is 0 dB as shown below and in Fig. 17.71(b):

$$A_{v_{dB}} = 20 \log_{10} 1 = 20(0) = \textbf{0 dB}$$

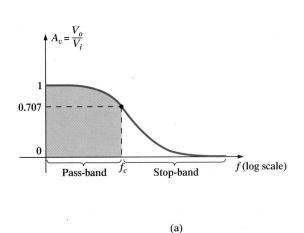

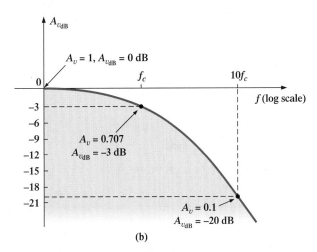

FIG. 17.71

Low-pass filter: (a) A_v versus frequency; (b) $A_{v_{dB}}$ versus frequency.

If we apply Eq. (17.72) to the 0.707 level, we obtain the following as shown in Fig. 17.71(b):

$$A_{v_{dB}} = 20 \log_{10} 0.707 = 20(-0.150) = \textbf{-3 dB}$$

and we can conclude that

at the critical or corner frequency, there will be a 3 dB drop from the maximum value.

As the frequency increases above the critical value, the dB gain will continue to drop as shown in Fig. 17.71(b). When the gain drops to a level of 0.1, the gain in dB is -20 dB. Interestingly, it occurs at a frequency of $10f_c$. The dB plot then approaches a straight line as it moves toward higher frequencies. At the other end of the spectrum, the curve also becomes tangent with the horizontal axis as it moves from the critical frequency to 0 Hz. This tendency toward straight-line segments is the basis for a shorthand method for drawing the dB response.

Through the use of straight-line segments called **Bode plots,** developed by Professor Hendrik Wade Bode (Fig. 17.72), the response of filters of all kinds can be quickly drawn. Priorities do not permit a derivation of the approach, but some of the basics of how to apply the approach will be described. The material will also provide a path toward understanding the commercial literature available on filters and amplifiers.

For a simple unit such as the low-pass filter described above, a Bode plot has two distinct straight-line segments radiating from the critical frequency, as shown in Fig. 17.73. For the region where the gain is 1, a horizontal line is drawn from the critical frequency to 0 Hz as shown. Another straight-line segment would be drawn for frequencies greater than the critical frequency at a negative slope equal to -6 dB/octave where

two frequencies separated by a 2 : 1 ratio are said to be one octave apart.

In other words, the slope of the same line drops off at -20 dB/decade where

two frequencies separated by a 10 : 1 ratio are said to be one decade apart.

Once the two radius vectors are drawn from the critical frequency, the actual curve can be sketched, passing through the -3 dB point below the critical frequency and approaching the two straight-line segments at very low and very high frequencies.

In total, therefore, the actual response curve can be drawn very quickly using these straight-line segments and the critical frequency. Also note that some designs are such that the drop-off for the low-pass filter can be -12 dB/octave or even -18 dB/octave. The greater the drop-off, the faster the filter, ensuring that those frequencies have little impact on the stage to follow.

FIG. 17.72

Hendrik-Wade Bode.

American (Madison, WI; Summit, NJ; Cambridge, MA)
(1905–81)
V.P. at Bell Laboratories
Professor of Systems Engineering, Harvard University

In his early years at Bell Laboratories, Hendrik Bode was involved with *electric filter* and *equalizer design.* He then transferred to the Mathematics Research Group, where he specialized in research pertaining to electrical networks theory and its application to long-distance communication facilities. In 1946 he was awarded the Presidential Certificate of Merit for his work in electronic fire control devices. In addition to the publication of the book *Network Analysis and Feedback Amplifier Design* in 1945, which is considered a classic in its field, he was granted 25 patents in electrical engineering and systems design. Upon retirement, Bode was elected Gordon McKay Professor of Systems Engineering at Harvard University. He was a fellow of the IEEE and American Academy of Arts and Sciences.

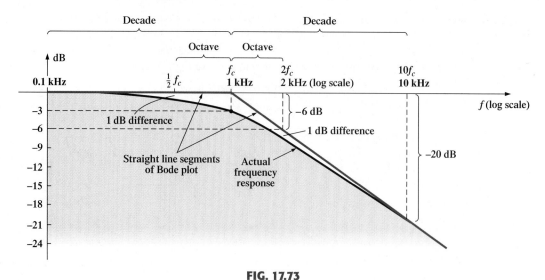

FIG. 17.73

Bode plot for the high-frequency region of a low-pass R-C filter.

For the high-pass filter, the response may appear as shown in Fig. 17.74 using Bode plots to draw the actual response.

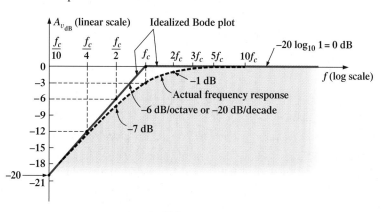

FIG. 17.74

Idealized Bode plot for the low-frequency region of a high-pass filter.

For the band-pass filter, the Bode plots may appear as shown in Fig. 17.75. In this case there are two critical frequencies defined by the $0.707A_v$ level. Each straight-line segment continues to drop off at −6 dB/octave, even though either end of the plot could drop off at a different rate.

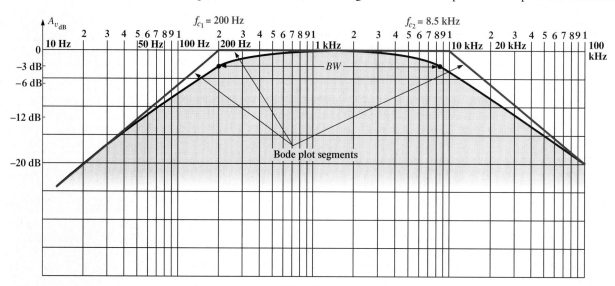

FIG. 17.75

$A_{v_{dB}}$ *versus frequency for a band-pass filter.*

EXAMPLE 17.19   Using Bode plots, draw the dB response for the high-pass filter of Fig. 17.76.

Solution: The critical frequency is

$$f_c = \frac{1}{2\pi RC} = \frac{1}{2\pi(2\text{ k}\Omega)(0.01\ \mu\text{F})} = \textbf{7.96 kHz}$$

On the graph of Fig. 17.77, the critical frequency is first defined and the horizontal line drawn from the critical frequency to very high frequencies. Next a line is drawn with a negative slope from the critical frequency to 0 Hz at a −6 dB/octave rate. Finally, a point 3 dB down from the critical frequency is defined and a curve drawn to meet the Bode plots at 0 Hz and very high frequencies.

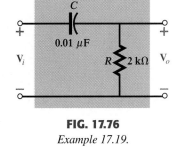

FIG. 17.76
Example 17.19.

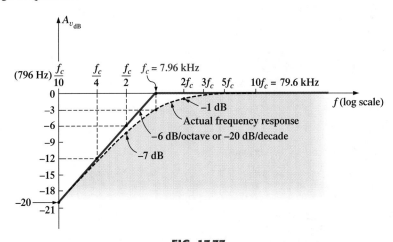

FIG. 17.77
dB gain versus frequency for the high-pass filter of Fig. 17.76.

17.16   APPLICATIONS

Stray Resonance

Stray resonance, like stray capacitance and inductance and unexpected resistance levels, can occur in totally unexpected situations and can severely affect the operation of a system. All that is required to produce stray resonance would be, for example, a level of capacitance introduced by parallel wires or copper leads on a printed circuit board, or simply two parallel conductive surfaces with residual charge and inductance levels associated with any conductor or components such as tape recorder heads, transformers, and so on, that provide the elements necessary for a resonance effect. In fact, this resonance effect is a very common effect in the everyday cassette tape recorder. The play/record head is a coil that can act like an inductor and an antenna. Combine this factor with the stray capacitance and real capacitance in the network to form the tuning network, and the tape recorder with the addition of a semiconductor diode can respond like an AM radio. As you plot the frequency response of any transformer, you will normally find a region where the response has a peaking effect. This peaking is due solely to the inductance of the coils of the transformer and the stray capacitance between the wires.

In general, any time you see an unexpected peaking in the frequency response of an element or a system, it is normally caused by a resonance condition. If the response has a detrimental effect on the overall operation of the system, a redesign may be in order, or a filter can be added that will block the frequencies that result in the resonance condition. Of course, when you add a filter composed of inductors and/or capacitors, you must be careful that you don't add another unexpected resonance condition. It is a problem that can be properly weighed only by constructing the system and exposing it to the full range of tests.

Graphic and Parametric Equalizers

We have all noticed at one time or another that the music we hear in a concert hall doesn't quite sound the same when we play it at home on our entertainment center. Even after we check the

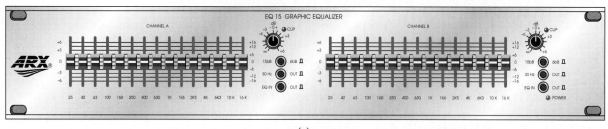

(a)

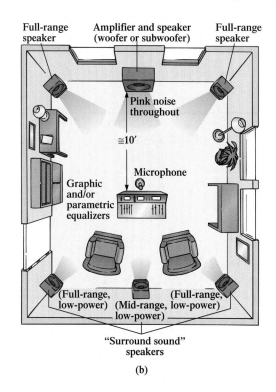

(b)

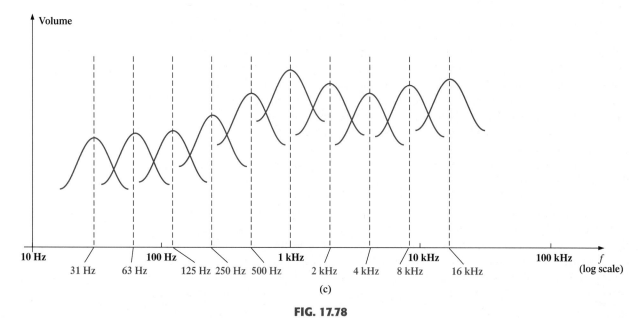

(c)

FIG. 17.78

(a) Dual-channel 15-band "Constant Q" graphic equalizer; (b) setup; (c) frequency response.

[(a) Courtesy of ARX Systems®.]

specifications of the speakers and amplifiers and find that both are nearly perfect (and the most expensive we can afford), the sound is still not what it should be. In general, we are experiencing the effects of the local environmental characteristics on the sound waves. Some typical problems are hard walls or floors (stone, cement) that will make high frequencies sound louder. Curtains and rugs, on the other hand, will absorb high frequencies. The shape of the room and the placement of the speakers and furniture will also affect the sound that reaches our ears. Another criterion is the echo or reflection of sound that will occur in the room. Concert halls are designed very carefully with their vaulted ceilings and curved walls to allow a certain amount of echo. Even the temperature and humidity characteristics of the surrounding air will affect the quality of the sound. It is certainly impossible, in most cases, to redesign your listening area to match a concert hall, but with the proper use of electronic systems you can develop a response that will have all the qualities that you can expect from a home entertainment center.

For a quality system, a number of steps can be taken: *characterization and digital delay (surround sound)* and *proper speaker and amplifier selection and placement.* Characterization is a process whereby a thorough sound absorption check of the room is performed and the frequency response determined. A *graphic equalizer* such as appearing in Fig. 17.78(a) is then used to make the response "flat" for the full range of frequencies. In other words, the room is made to appear as though all the frequencies receive equal amplification in the listening area. For instance, if the room is fully carpeted with full draping curtains, there will be a lot of high-frequency absorption, requiring that the high frequencies have additional amplification to match the sound levels of the mid- and low frequencies. To *characterize* the typical rectangular-shaped room, a setup such as shown in Fig. 17.78(b) may be used. The amplifier and speakers are placed in the center of one wall, with additional speakers in the corners of the room facing the reception area. A mike is then placed in the reception area about 10 ft from the amplifier and centered between the two other speakers. A *pink noise* will then be sent out from a spectrum analyzer (often an integral part of the graphic equalizer) to the amplifier and speakers. Pink noise is actually a square-wave signal whose amplitude and frequency can be controlled. A square-wave signal was chosen because a Fourier breakdown of a square-wave signal will result in a broad range of frequencies for the system to check.

Once the proper volume of pink noise is established, the spectrum analyzer can be used to set the response of each slide band to establish the desired flat response. The center frequencies for the slides of the graphic equalizer of Fig. 17.78(a) are provided in Fig. 17.78(c), along with the frequency response for a number of adjoining frequencies evenly spaced on a logarithmic scale. Note that each center frequency is actually the resonant frequency for that slide. The design is such that each slide can control the volume associated with that frequency, but the bandwidth and frequency response stay fairly constant. A good spectrum analyzer will have each slide set against a decibel (dB) scale. The decibel scale simply establishes a scale for the comparison of audio levels. At a normal listening level, usually a change of about 3 dB is necessary for the audio change to be detectable by the human ear. At low levels of sound, a 2 dB change may be detectable, but at loud sounds probably a 4 dB change would be necessary for the change to be noticed. These are not strict laws but simply rules of thumb commonly used by audio technicians. For the room in question, the mix of settings may be as shown in Fig. 17.78(c). Once set, the slides are not touched again. A flat response has been established for the room for the full audio range so that every sound or type of music is covered.

A *parametric equalizer* such as appearing in Fig. 17.79 is similar to a *graphic equalizer,* but instead of separate controls for the individual frequency ranges, it uses three basic controls over three or four broader frequency ranges. The typical controls—the *gain, center frequency,* and

FIG. 17.79
Six-channel parametric equalizer.
(Courtesy of ARX Systems®.)

bandwidth—are typically available for the *low-*, *mid-*, and *high-frequency* ranges. Each is fundamentally an independent control; that is, a change in one can be made without affecting the other two. For the parametric equalizer of Fig. 17.79, each of the six channels has a frequency control switch which, in conjunction with the $f \times 10$ switch, will give a range of center frequencies from 40 Hz through 16 kHz. It has controls for *BW* ("*Q*") from 3 octaves to 1/20 octave, and ± 18 dB cut and boost.

Some like to refer to the parametric equalizer as a *sophisticated tone control* and will actually use them to enrich the sound after the flat response has been established by the graphic equalizer. The effect achieved with a standard tone control knob is sometimes referred to as "boring" compared to the effect established by a good parametric equalizer, primarily because the former can control only the volume and not the bandwidth or center frequency. In general, graphic equalizers establish the important *flat response* while parametric equalizers are adjusted to provide the *type* and *quality* of sound you like to hear. You can "notch out" the frequencies that bother you and remove tape "hiss" and the "sharpness" often associated with CDs.

One characteristic of concert halls that is more difficult to fake is the fullness of sound that concert halls are able to provide. In the concert hall you have the direct sound from the instruments and the reflection of sound off the walls and the vaulted ceilings, which were all carefully designed expressly for this purpose. Any reflection results in a delay in the sound waves reaching the ear, creating the fullness effect. Through digital delay, speakers can be placed to the back and side of a listener to establish the surround sound effect. In general, the delay speakers are much lower in wattage, with 20 W speakers typically used with a 100 W system. The echo response is one reason that people often like to play their stereos louder than they should for normal hearing. By playing the stereo louder, they create more echo and reflection off the walls, bringing into play some of the fullness heard at concert halls.

It is probably safe to say that any system composed of quality components, a graphic and parametric equalizer, and surround sound will have all the components necessary to have a quality reproduction of the concert hall effect.

17.17 COMPUTER ANALYSIS

PSpice

Series Resonance This chapter provides an excellent opportunity to demonstrate what computer software programs can do for us. Imagine having to plot a detailed resonance curve with all the calculations required for each frequency. At every frequency, the reactance of the inductive and capacitive elements changes, and the phasor operations would have to be repeated—a long and arduous task. However, with PSpice, taking a few moments to enter the circuit and establish the desired simulation will result in a detailed plot in a few seconds that can have plot points every microsecond!

For the first time, the horizontal axis will be in the frequency domain rather than in the time domain as in all the previous plots. For the series resonant circuit of Fig. 17.80, the magnitude of the source was chosen to produce a maximum current of $I = 400$ mV/40 Ω = 10 mA at resonance, and the reactive elements will establish a resonant frequency of

$$f_s = \frac{1}{2\pi\sqrt{LC}} = \frac{1}{2\pi\sqrt{(30\ \text{mH})(0.1\ \mu\text{F})}} \cong \mathbf{2.91\ kHz}$$

The quality factor is

$$Q_l = \frac{X_L}{R_l} = \frac{546.64\ \Omega}{40\ \Omega} \cong \mathbf{13.7}$$

which is relatively high and should give us a nice sharp response.

The bandwidth is

$$BW = \frac{f_s}{Q_l} = \frac{2.91\ \text{kHz}}{13.7} \cong \mathbf{212\ Hz}$$

which will be verified using our cursor options.

For the ac source, **VSIN** was chosen again. All the parameters were set by double-clicking on the source symbol and entering the values in the **Property Editor** dialog box. For each, **Name and Value** was selected under **Display** followed by **Apply** before leaving the dialog box.

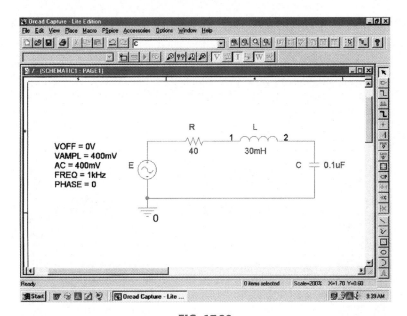

FIG. 17.80

Series resonant circuit to be analyzed using PSpice.

In the **Simulation Settings** dialog box, **AC Sweep/Noise** was selected, and the **Start Frequency** was set at 1 kHz, the **End Frequency** at 10 kHz, and the **Points/Decade** at 10,000. The **Logarithmic scale** and **Decade** settings remain at their default values. The 10,000 for **Points/Decade** was chosen to ensure a number of data points near the peak value. When the **SCHEMATIC1** screen of Fig. 17.81 appears, **Trace-Add Trace-I(R)-OK** will result in a logarithmic plot that peaks just to the left of 3 kHz. The spacing between grid lines on the **X-axis** should be increased, so **Plot-Axis Settings-X Grid-unable Automatic-Spacing-Log-0.1-OK** is implemented.

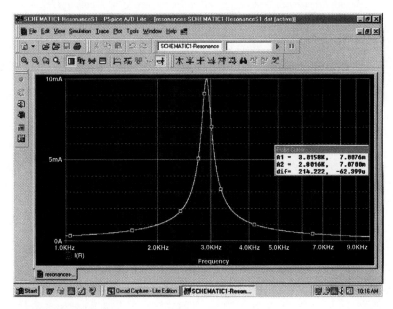

FIG. 17.81

Resonance curve for the current of the circuit of Fig. 17.80.

Next, select the **Toggle cursor** icon, and with a right click of the mouse move the right cursor as close to 7.07 mA as possible (0.707 of the peak value to define the bandwidth) to obtain **A1** with a frequency of 3.02 kHz at a level of 7.01 mA—the best we can do with the 10,000 data points per decade. Now do a left click, and place the left cursor as close to the same level as possible. The result is 2.8 kHz at a level of 7.07 mA for **A2.** The cursors were set in the order above to obtain a positive answer for the difference of the two as appearing in the third line of the **Probe Cursor** box. The resulting 214.22 Hz is an excellent match with the calculated value of 212 Hz.

Electronics Workbench (EWB)

The results of Example 17.8 will now be confirmed using EWB. The network of Fig. 17.26 will appear as shown in Fig. 17.82 after all the elements have been placed as described in earlier chapters. In particular, note that the frequency assigned to the 2 mA ac current source is 100 kHz. Since we have some idea that the resonant frequency is a few hundred kilohertz, it seemed appropriate that the starting frequency for the plot begin at 100 kHz and extend to 1 MHz. Also, be sure that the **AC Magnitude** is set to 2 mA in the **Analysis Setup** within the **AC Current** dialog box.

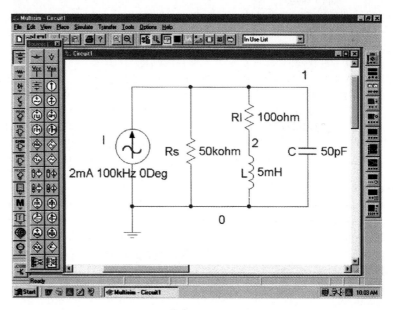

FIG. 17.82

Using Electronics Workbench to confirm the results of Example 17.8.

For simulation, the sequence **Simulate-Analyses-AC Analysis** is first selected to obtain the **AC Analysis** dialog box. The **Start frequency** is set at 100 kHz, and the **Stop frequency** at 1 MHz; **Sweep type** is **Decade; Number of points per decade** is 1000; and the **Vertical scale** is **Linear.** Under **Output variables,** node number 1 is selected as a **Variable for analysis** followed by **Simulate** to run the program. The results are the magnitude and phase plots of Fig. 17.83. Starting with the **Voltage** plot, the **Show/Hide Grid** key, **Show/Hide Legend** key, and **Show/Hide Cursors** key are selected. You will immediately note under the **AC**

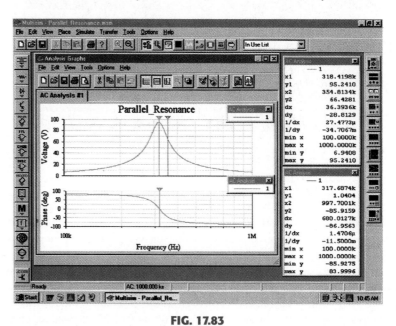

FIG. 17.83

Magnitude and phase plots for the voltage v_c of the network of Fig. 17.82.

Analysis cursor box that the maximum value is 95.24 V and the minimum value is 6.94 V. By moving the cursor until we reach 95.24 V (**y1**), we can find the resonant frequency. As shown in the top cursor dialog box of Fig. 17.83, this is achieved at 318.42 kHz (**x1**). The other (blue) cursor can be used to define the high cutoff frequency for the bandwidth by first calculating the 0.707 level of the output voltage. The result is 0.707(95.24 V) = 67.33 V. The closest we can come to this level with the cursor is 66.43 V (**y2**) which defines a frequency of 354.81 kHz (**x2**). If we now use the red cursor to find the corresponding level below the resonant frequency, we will find a level of 67.49 V (**y1**) at 287.08 kHz (**x1**). The resulting bandwidth is therefore 354.81 kHz − 287.08 kHz = 67.73 kHz.

It would now be interesting to determine the resonant frequency if we define resonance as the frequency that results in a phase angle of 0° for the output voltage. By repeating the process described above for the phase plot, we can set the red cursor as close to 0° as possible. The result is 1.04° (**y1**) at 317.69 kHz (**x1**), clearly revealing that the resonant frequency defined by the phase angle is less than that defined by the peak voltage. However, with a Q_l of about 100, the difference of 0.73 kHz (less than 1 kHz) is not significant. Also note that when the second cursor was set on approximately 1 MHz (997.7 kHz), the phase angle of −85.91° is very close to that of a pure capacitor. The shorting effect of a capacitor at high frequencies has taken over the characteristics of the parallel resonant circuit.

Again, the computer solution was a very close match with the longhand solution of Example 17.8 with a perfect match of 95.24 V for the peak value and only a small difference in bandwidth with 66.87 kHz in Example 17.8 and 67.73 kHz here. For the high cutoff frequency, the computer generated a result of 354.8 kHz, while the theoretical solution was 351.7 kHz. For the low cutoff frequency, the computer responded with 287.08 kHz compared to a theoretical solution of 284 kHz.

CHAPTER SUMMARY

SERIES RESONANCE

- Series resonance occurs when the total impedance is simply the resistance of the circuit and the resulting current is a maximum.
- For series resonance, the resonant frequency is determined solely by the reactive components of the circuit.
- The quality factor (Q) provides an immediate indication of the sharpness of the resonant curve. The higher the Q factor, the smaller the bandwidth and the higher the voltages V_L and V_C at resonance.
- For most applications it is a valid approximation to assume that the resonant frequency bisects the bandwidth.
- The cutoff or half-power frequencies define the bandwidth and the range of frequencies that will ensure that the power delivered to the load is at least half the maximum.

PARALLEL RESONANCE

- For parallel resonance, the input impedance is a maximum at resonance, resulting in a peak value for the voltage across the parallel network.
- The peak value of a resonance curve does not occur at exactly the frequency at which the total network appears resistive with a power factor of 1 because the equivalent resistance is frequency dependent. However, the actual peak is so close to the value defined by $f_p = 1$ that for most applications they are considered equivalent.
- For most applications, as an excellent first approximation, the equation for parallel resonance will be the same as that used for series resonance.
- The quality factor (Q) for a parallel network is sensitive to the source resistance of the current source. For cases where the effect of the source resistance can be ignored, the quality factor is simply that of the coil.
- For most applications, you can assume that the resonant frequency bisects the bandwidth.
- The bandwidth is related to the resonant frequency and the quality factor in the same manner as for series resonant circuits.
- At resonance, the magnitude of the current through the coil or through the capacitor is directly related to the quality factor of the network. The result is usually a current through the parallel branches that is many times greater than the source current.

LOGARITHMS

- The logarithm of a number is a number that, if applied as the power of the base, will result in the number for which the logarithm is being determined.
- The logarithm of 1 to any base is always 0.
- The logarithm of a product of two numbers equals the sum of the logarithm of each number, whereas the logarithm of the division of two numbers is always the logarithm of the numerator minus the logarithm of the denominator.
- The logarithm of a number taken to a power is the power times the logarithm of the number.
- The dB gain between two voltage levels is 20 times the log of the ratio of voltage levels.

FILTERS

- Filters are designed primarily to control the range of frequencies that will be applied to the next stage of a system. Ideally, the gain is 1 for the frequencies to be passed on, and 0 for those to be rejected.
- A low-pass filter will permit the passage of a range of low frequencies and will squelch frequencies above the critical frequency. The critical frequency is the frequency at which the gain drops to 70.7% of the maximum value.
- A high-pass filter is one that will permit the passage of frequencies above the critical frequency and reject those below the critical value.
- A pass-band filter is one that permits the passage of a specific band of frequencies. A stop-band filter is one that rejects a specific range of frequencies but lets all other frequencies pass through.
- When applied to filters, normalization is a process by which the output at every frequency for the range of interest is divided by the peak value.
- A low-pass filter is considered a lagging network because the output voltage lags the applied voltage. A high-pass filter is considered a leading network because the output voltage leads the applied voltage.
- A dB plot for a filter is obtained by applying the basic dB equation to the normalized plot of the filter's response. The dB plot is 0 at the peak value of the filter and -3 dB at the cutoff frequency. A Bode plot at a critical frequency drops off at -6 dB/octave, or -20 dB/decade.

Important Equations

Series resonance:

$$\mathbf{Z}_{T_s} = R$$

$$f_s = \frac{1}{2\pi\sqrt{LC}}$$

$$Q_l = \frac{X_L}{R_l}$$

$$V_{L_s} = Q_s E \qquad V_{C_s} = Q_s E$$

$$f_1 = \frac{1}{2\pi}\left[-\frac{R}{2L} + \frac{1}{2}\sqrt{\left(\frac{R}{L}\right)^2 + \frac{4}{LC}} \right]$$

$$f_2 = \frac{1}{2\pi}\left[+\frac{R}{2L} + \frac{1}{2}\sqrt{\left(\frac{R}{L}\right)^2 + \frac{4}{LC}} \right]$$

$$BW = f_2 - f_1 = \frac{f_s}{Q_s} = \frac{R}{2\pi L}$$

Parallel resonance:

$$R_p = \frac{R_l^2 + X_L^2}{R_l} \qquad X_{L_p} = \frac{R_l^2 + X_L^2}{X_L}$$

$$f_p = \frac{1}{2\pi\sqrt{LC}}\sqrt{1 - \frac{R_l^2 C}{L}} \cong \frac{1}{2\pi\sqrt{LC}}$$

$$Q_p = \frac{R_s \| R_p}{X_{L_p}} \cong \frac{R_s \| R_p}{X_C} = \frac{X_L}{R_l} \qquad (R_s \gg R_p)$$

$$f_1 = \frac{1}{4\pi C}\left[\frac{1}{R} - \sqrt{\frac{1}{R^2} + \frac{4C}{L}} \right]$$

$$f_2 = \frac{1}{4\pi C}\left[\frac{1}{R} + \sqrt{\frac{1}{R^2} + \frac{4C}{L}} \right]$$

$$BW = f_2 - f_1 = \frac{f_p}{Q_p} \cong \frac{R}{2\pi L} \qquad (R_s = \infty\,\Omega)$$

$$Z_{T_p} = R_s \| R_p \cong R_s \| Q_l^2 R_l$$

$$I_{C_p} \cong Q_l I_{\text{source}} \qquad I_{L_p} \cong Q_l I_{\text{source}}$$

Logarithms:

$$x = \log_{10} N \Rightarrow N = b^x$$

Value $= 10^x \times 10^{d_1/d_2}$

$$\log_{10} 1 = 0$$

$$\log_{10} ab = \log_{10} a + \log_{10} b$$

$$\log_{10} \frac{a}{b} = \log_{10} a - \log_{10} b$$

$$\log_{10} a^n = n \log_{10} a$$

Logarithms (*continued*):

$$dB = 10 \log_{10} \frac{P_2}{P_1}$$

$$dB_v = 20 \log_{10} \frac{V_2}{V_1}$$

Filters (Low-Pass/High-Pass):

$$f_c = \frac{1}{2\pi RC}$$

$$A_{v_{dB}} = 20 \log_{10} A_v$$

PROBLEMS

SECTIONS 17.2 AND 17.3   Series Resonant Circuit

1. Find the resonant ω_s and f_s for the series circuit with the following parameters:
   a. $R = 10\ \Omega$, $L = 1$ H, $C = 16\ \mu$F
   b. $R = 300\ \Omega$, $L = 0.5$ H, $C = 0.16\ \mu$F
   c. $R = 20\ \Omega$, $L = 0.28$ mH, $C = 7.46\ \mu$F

2. For the series circuit of Fig. 17.84:
   a. Find the value of X_C for resonance.
   b. Determine the total impedance of the circuit at resonance.
   c. Find the magnitude of the current I.
   d. Calculate the voltages V_R, V_L, and V_C at resonance. How are V_L and V_C related? How does V_R compare to the applied voltage E?
   e. What is the quality factor of the circuit? Is it a high- or low-Q circuit?
   f. What is the power dissipated by the circuit at resonance?

3. For the series circuit of Fig. 17.85:
   a. Find the value of X_L for resonance.
   b. Determine the magnitude of the current I at resonance.
   c. Find the voltages V_R, V_L, and V_C at resonance, and compare their magnitudes.
   d. Determine the quality factor of the circuit. Is it a high- or low-Q circuit?
   e. If the resonant frequency is 5 kHz, determine the value of L and C.
   f. Find the bandwidth of the response if the resonant frequency is 5 kHz.
   g. What are the low and high cutoff frequencies?

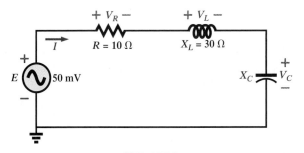

FIG. 17.84

Problem 2.

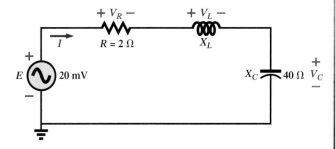

FIG. 17.85

Problem 3.

4. For the circuit of Fig. 17.86:
   a. Find the value of L in millihenries if the resonant frequency is 1800 Hz.
   b. Calculate X_L and X_C. How do they compare?
   c. Find the magnitude of the current I_{rms} at resonance.
   d. Find the power dissipated by the circuit at resonance.
   e. What is the apparent power delivered to the system at resonance?
   f. What is the power factor of the circuit at resonance?
   g. Calculate the Q of the circuit and the resulting bandwidth.
   h. Find the cutoff frequencies, and calculate the power dissipated by the circuit at these frequencies.

5. a. Find the bandwidth of a series resonant circuit having a resonant frequency of 6000 Hz and a Q_s of 15.
   b. Find the cutoff frequencies.

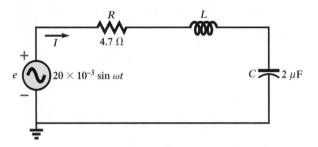

FIG. 17.86
Problem 4.

   c. If the resistance of the circuit at resonance is 3 Ω, what are the values of X_L and X_C in ohms?
   d. What is the power dissipated at the half-power frequencies if the maximum current flowing through the circuit is 0.5 A?

6. A series circuit has a resonant frequency of 10 kHz. The resistance of the circuit is 5 Ω, and X_C at resonance is 200 Ω.
   a. Find the bandwidth.
   b. Find the cutoff frequencies.
   c. Find Q_s.
   d. If the input voltage is 30 V $\angle 0°$, find the voltage across the coil and capacitor in phasor form.
   e. Find the power dissipated at resonance.

7. a. The bandwidth of a series resonant circuit is 200 Hz. If the resonant frequency is 2000 Hz, what is the value of Q_s for the circuit?
   b. If $R = 2$ Ω, what is the value of X_L at resonance?
   c. Find the value of L and C at resonance.
   d. Find the cutoff frequencies.

8. The cutoff frequencies of a series resonant circuit are 5400 Hz and 6000 Hz.
   a. Find the bandwidth of the circuit.
   b. If Q_s is 9.5, find the resonant frequency of the circuit.
   c. If the resistance of the circuit is 2 Ω, find the value of X_L and X_C at resonance.
   d. Find the value of L and C at resonance.

*9. Design a series resonant circuit with an input voltage of 5 V $\angle 0°$ to have the following specifications:

     A peak current of 500 mA at resonance
     A bandwidth of 120 Hz
     A resonant frequency of 8400 Hz

   Find the value of R, L, and C and the cutoff frequencies.

*10. Design a series resonant circuit to have a bandwidth of 400 Hz using a coil with a Q_l of 20 and a resistance of 2 Ω. Find the values of L and C and the cutoff frequencies.

*11. A series resonant circuit is to resonate at $\omega_s = 2\pi \times 10^6$ rad/s and draw 20 W from a 120 V source at resonance. If the fractional bandwidth is 0.16:
   a. Determine the resonant frequency in hertz.
   b. Calculate the bandwidth in hertz.
   c. Determine the values of R, L, and C.
   d. Find the resistance of the coil if $Q_l = 80$.

*12. A series resonant circuit will resonate at a frequency of 1 MHz with a fractional bandwidth of 0.2. If the quality factor of the coil at resonance is 12.5 and its inductance is 100 μH, determine the following:
   a. The resistance of the coil.
   b. The additional resistance required to establish the indicated fractional bandwidth.
   c. The required value of capacitance.

SECTIONS 17.4 AND 17.5   Parallel Resonance

13. For the "ideal" parallel resonant circuit of Fig. 17.87:
   a. Determine the resonant frequency (f_p).
   b. Find the voltage V_C at resonance.
   c. Determine the currents I_L and I_C at resonance.
   d. Find Q_p.

14. For the parallel resonant network of Fig. 17.88:
   a. Calculate f_p.
   b. Determine Q_l.
   c. Calculate X_L and X_C using f_p. How do they compare?
   d. Find the total impedance at resonance.
   e. Calculate V_C at resonance.
   f. Determine the BW.
   g. Calculate I_L and I_C at resonance.

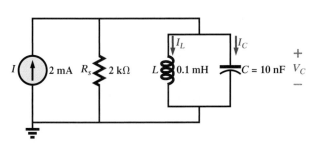

FIG. 17.87
Problem 13.

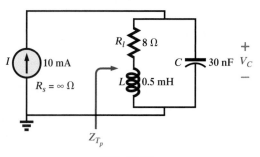

FIG. 17.88
Problem 14.

15. For the network of Fig. 17.89:
   a. Find the value of X_C at resonance.
   b. Find the total impedance Z_{T_p} at resonance.
   c. Find the currents I_L and I_C at resonance.
   d. If the resonant frequency is 20,000 Hz, find the value of L and C at resonance.
   e. Find Q_p and the BW.

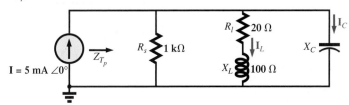

FIG. 17.89
Problem 15.

16. Repeat Problem 15 for the network of Fig. 17.90.

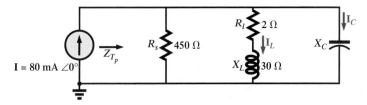

FIG. 17.90
Problem 16.

17. For the network of Fig. 17.91:
   a. Find the resonant frequency.
   b. Find the values of X_L and X_C at resonance (f_p). How do they compare?
   c. Find the impedance Z_{T_p} at resonance (f_p).
   d. Calculate Q_p and the BW.
   e. Find the magnitude of currents I_L and I_C at resonance (f_p).
   f. Calculate the voltage V_C at resonance (f_p).

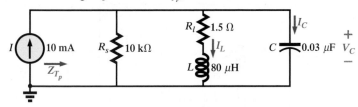

FIG. 17.91
Problems 17 and 63.

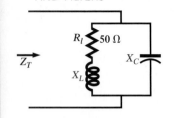

FIG. 17.92

Problem 18.

18. It is desired that the impedance $\mathbf{Z}_T$ of the high-Q circuit of Fig. 17.92 be 50 kΩ $\angle 0°$ at resonance (f_p).
  a. Find the value of X_L.
  b. Compute X_C.
  c. Find the resonant frequency (f_p) if $L = 16$ mH.
  d. Find the value of C.

19. For the network of Fig. 17.93:
  a. Find f_p.
  b. Calculate the magnitude of V_C at resonance (f_p).
  c. Determine the power absorbed at resonance.
  d. Find the BW.

*20. For the network of Fig. 17.94:
  a. Find the value of X_L at resonance.
  b. Find Q_l.
  c. Find the resonant frequency (f_p) if the bandwidth is 1 kHz.
  d. Find the maximum value of the voltage V_C.
  e. Sketch the curve of V_C versus frequency. Indicate its peak value, resonant frequency, and band frequencies.

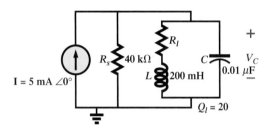

FIG. 17.93

Problem 19.

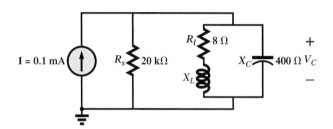

FIG. 17.94

Problem 20.

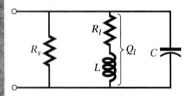

FIG. 17.95

Problem 21.

*21. For the network of Fig. 17.95, the following are specified:

$$f_p = 20 \text{ kHz}$$
$$BW = 1.8 \text{ kHz}$$
$$L = 2 \text{ mH}$$
$$Q_l = 80$$

Find R_s and C.

*22. Design the network of Fig. 17.96 to have the following characteristics:
  a. $BW = 500$ Hz
  b. $Q_p = 30$
  c. $V_{C_{max}} = 1.8$ V

*23. For the parallel resonant circuit of Fig. 17.97:
  a. Determine the resonant frequency.
  b. Find the total impedance at resonance.
  c. Find Q_p.
  d. Calculate the BW.
  e. Repeat parts (a) through (d) for $L = 20$ μH and $C = 20$ nF.
  f. Repeat parts (a) through (d) for $L = 0.4$ mH and $C = 1$ nF.
  g. For the network of Fig. 17.97 and the parameters of parts (e) and (f), determine the ratio L/C.
  h. Do your results confirm the conclusions of Fig. 17.22 for changes in the L/C ratio?

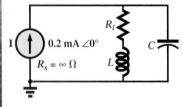

FIG. 17.96

Problem 22.

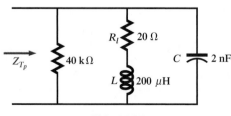

FIG. 17.97

Problem 23.

24. a. Determine the frequencies (in kHz) at the points indicated on the plot of Fig. 17.98(a).
  b. Determine the voltages (in mV) at the points indicated on the plot of Fig. 17.98(b).

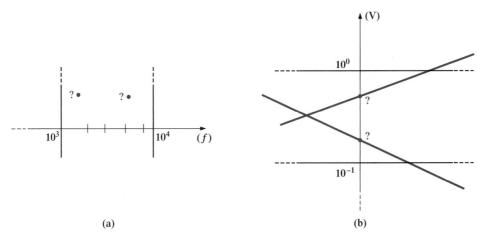

(a)   (b)

FIG. 17.98
Problem 24.

SECTION 17.7 Properties of Logarithms

25. Determine $\log_{10} x$ for each value of x.
  a. 100,000   **b.** 0.0001
  c. 10^8 **d.** 10^{-6}
  e. 20   **f.** 8643.4
  g. 56,000   **h.** 0.318

26. Given $N = \log_{10} x$, determine x for each value of N.
  a. 3   **b.** 12
  c. 0.2   **d.** 0.04
  e. 10   **f.** 3.18
  g. 1.001   **h.** 6.1

27. Determine $\log_e x$ for each value of x.
  a. 100,000   **b.** 0.0001
  c. 20   **d.** 8643.4

  Compare your answers with the solutions to Problem 25.

28. Determine $\log_{10} 48 = \log_{10}(8)(6)$, and compare to $\log_{10} 8 + \log_{10} 6$.

29. Determine $\log_{10} 0.2 = \log_{10} 18/90$, and compare to $\log_{10} 18 - \log_{10} 90$.

30. Verify that $\log_{10} 0.5$ is equal to $-\log_{10} 1/0.5 = -\log_{10} 2$.

31. Find $\log_{10}(3)^3$, and compare with $3 \log_{10} 3$.

SECTION 17.8 Decibels

32. a. Determine the number of bels that relate power levels of $P_2 = 280$ mW and $P_1 = 4$ mW.
  b. Determine the number of decibels for the power levels of part (a), and compare results.

33. A power level of 100 W is 6 dB above what power level?

34. If a 2 W speaker is replaced by one with a 40 W output, what is the increase in decibel level?

35. Determine the dB_m level for an output power of 120 mW.

36. Find the dB_v gain of an amplifier that raises the voltage level from 0.1 mV to 8.4 mV.

37. Find the output voltage of an amplifier if the applied voltage is 20 mV and a dB_v gain of 22 dB is attained.

38. If the sound pressure level is increased from 0.001 μbar to 0.016 μbar, what is the increase in dB_s level?

39. What is the required increase in acoustical power to raise a sound level from that of quiet music to very loud music? Use Fig. 17.34.

40. a. Using semilog paper, plot X_L versus frequency for a 10 mH coil and a frequency range of 100 Hz to 1 MHz. Choose the best vertical scaling for the range of X_L.
  b. Repeat part (a) using log-log graph paper. Compare your results to those of part (a). Which plot is more informative?

c. Using semilog paper, plot X_C versus frequency for a 1 μF capacitor and a frequency range of 10 Hz to 100 kHz. Again choose the best vertical scaling for the range of X_C.
d. Repeat part (a) using log-log graph paper. Compare your results to those of part (c). Which plot is more informative?

41. a. For the meter of Fig. 17.35, find the power delivered to a load for an 8 dB reading.
    b. Repeat part (a) for a −5 dB reading.

SECTION 17.10 *R-C* Low-Pass Filter

42. For the *R-C* low-pass filter of Fig. 17.99:
    a. Sketch $A_v = V_o/V_i$ versus frequency using a log scale for the frequency axis. Determine $A_v = V_o/V_i$ at $0.1f_c$, $0.5f_c$, f_c, $2f_c$, and $10f_c$.
    b. Sketch the phase plot of θ versus frequency, where θ is the angle by which $\mathbf{V}_o$ leads $\mathbf{V}_i$. Determine θ at $f = 0.1f_c$, $0.5f_c$, f_c, $2f_c$, and $10f_c$.

*43. For the network of Fig. 17.100:
    a. Determine V_o at a frequency one octave above the critical frequency.
    b. Determine V_o at a frequency one decade below the critical frequency.
    c. Do the levels of parts (a) and (b) verify the expected frequency plot of V_o versus frequency for the filter?

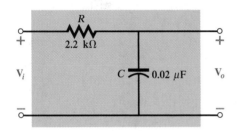

FIG. 17.99
Problem 42.

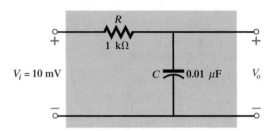

FIG. 17.100
Problem 43.

44. Design an *R-C* low-pass filter to have a cutoff frequency of 500 Hz using a resistor of 1.2 kΩ. Then sketch the resulting magnitude and phase plot for a frequency range of $0.1f_c$ to $10f_c$.

45. For the low-pass filter of Fig. 17.101:
    a. Determine f_c.
    b. Find $A_v = V_o/V_i$ at $f = 0.1f_c$, and compare your answer to the maximum value of 1 for the low-frequency range.
    c. Find $A_v = V_o/V_i$ at $f = 10f_c$, and compare your answer to the minimum value of 0 for the high-frequency range.
    d. Determine the frequency at which $A_v = 0.01$ or $V_o = \frac{1}{100} V_i$.

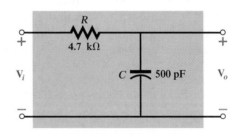

FIG. 17.101
Problem 45.

SECTION 17.11 *R-C* High-Pass Filter

46. For the *R-C* high-pass filter of Fig. 17.102:
    a. Sketch $A_v = V_o/V_i$ versus frequency using a log scale for the frequency axis. Determine $A_v = V_o/V_i$ at f_c, one octave above and below f_c, and one decade above and below f_c.
    b. Sketch the phase plot of θ versus frequency, where θ is the angle by which $\mathbf{V}_o$ leads $\mathbf{V}_i$. Determine θ at the same frequencies noted in part (a).

47. For the network of Fig. 17.103:
    a. Determine $A_v = V_o/V_i$ at $f = f_c$ for the high-pass filter.
    b. Determine $A_v = V_o/V_i$ at two octaves above f_c. Is the rise in V_o significant from the $f = f_c$ level?

c. Determine $A_v = V_o/V_i$ at two decades above f_c. Is the rise in V_o significant from the $f = f_c$ level?

d. If $V_i = 10$ mV, what is the power delivered to R at the critical frequency?

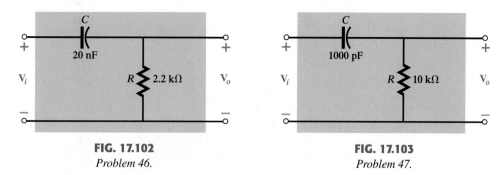

FIG. 17.102
Problem 46.

FIG. 17.103
Problem 47.

48. Design a high-pass R-C filter to have a cutoff or corner frequency of 2 kHz, given a capacitor of 0.1 μF. Choose the closest commercial value for R, and then recalculate the resulting corner frequency. Sketch the normalized gain $A_v = V_o/V_i$ for a frequency range of $0.1f_c$ to $10f_c$.

49. For the high-pass filter of Fig. 17.104:
a. Determine f_c.
b. Find $A_v = V_o/V_i$ at $f = 0.01f_c$, and compare your answer to the minimum level of 0 for the low-frequency region.
c. Find $A_v = V_o/V_i$ at $f = 100f_c$, and compare your answer to the maximum level of 1 for the high-frequency region.
d. Determine the frequency at which $V_o = \frac{1}{2}V_i$.

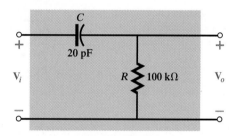

FIG. 17.104
Problems 49 and 65.

SECTION 17.12 Pass-Band Filters

50. For the pass-band filter of Fig. 17.105:
a. Sketch the frequency response of $A_v = V_o/V_i$ against a log scale extending from 10 Hz to 10 kHz.
b. What are the bandwidth and the center frequency?

*51. Design a pass-band filter such as the one appearing in Fig. 17.105 to have a low cutoff frequency of 4 kHz and a high cutoff frequency of 80 kHz.

52. For the pass-band filter of Fig. 17.106:
a. Determine f_s.
b. Calculate Q_s and the BW for V_o.
c. Sketch $A_v = V_o/V_i$ for a frequency range of 1 kHz to 1 MHz.
d. Find the magnitude of V_o at $f = f_s$ and the cutoff frequencies.

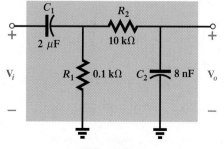

FIG. 17.105
Problems 50 and 51.

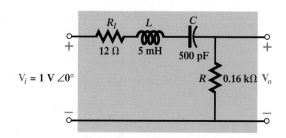

FIG. 17.106
Problem 52.

53. For the pass-band filter of Fig. 17.107:
  a. Determine the frequency response of $A_v = V_o/V_i$ for a frequency range of 100 Hz to 1 MHz.
  b. Find the quality factor Q_p and the BW of the response.

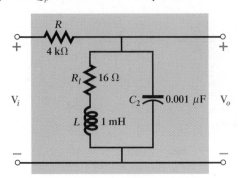

FIG. 17.107
Problems 53 and 66.

SECTION 17.13 Stop-Band Filters

54. For the stop-band filter of Fig. 17.108:
  a. Determine Q_s.
  b. Find the bandwidth and the half-power frequencies.
  c. Sketch the frequency characteristics of $A_v = V_o/V_i$.
  d. What is the effect on the curve of part (c) if a load of 2 kΩ is applied?

55. For the pass-band filter of Fig. 17.109:
  a. Determine Q_p ($R_L = \infty$ Ω, an open circuit).
  b. Sketch the frequency characteristics of $A_v = V_o/V_i$.
  c. Find Q_p (loaded) for $R_L = 100$ kΩ, and indicate the effect of R_L on the characteristics of part (b).
  d. Repeat part (c) for $R_L = 20$ kΩ.

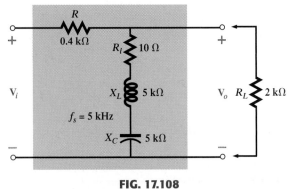

FIG. 17.108
Problem 54.

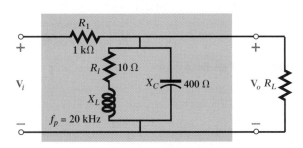

FIG. 17.109
Problem 55.

SECTION 17.14 Double-Tuned Filters

56. For the network of Fig. 17.70(a):
  a. If $L_p = 400$ μH ($Q > 10$), $L_s = 60$ μH, and $C = 120$ pF, determine the rejected and accepted frequencies.
  b. Sketch the response curve for part (a).

57. For the network of Fig. 17.70(b):
  a. If the rejected frequency is 30 kHz and the accepted is 100 kHz, determine the values of L_s and L_p ($Q > 10$) for a capacitance of 200 pF.
  b. Sketch the response curve for part (a).

SECTION 17.15 dB Plots

58. For the high-pass filter of Fig. 17.110:
  a. Sketch the idealized Bode plot for $A_v = V_o/V_i$.
  b. Using the results of part (a), sketch the actual frequency response for the same frequency range.
  c. Determine the decibel level at f_c, $\frac{1}{2}f_c$, $2f_c$, $\frac{1}{10}f_c$, and $10f_c$.

d. Determine the gain $A_v = V_o/V_i$ as $f = f_c, \frac{1}{2}f_c,$ and $2f_c$.

e. Sketch the phase response for the same frequency range.

*59. For the high-pass filter of Fig. 17.111:

a. Sketch the response of the magnitude of V_o (in terms of V_i) versus frequency.

b. Using the results of part (a), sketch the response $A_v = V_o/V_i$ for the same frequency range.

c. Sketch the idealized Bode plot.

d. Sketch the actual response, indicating the dB difference between the idealized and the actual response at $f = f_c, 0.5f_c,$ and $2f_c$.

e. Determine $A_{v_{dB}}$ at $f = 1.5f_c$ from the plot of part (d), and then determine the corresponding magnitude of $A_v = V_o/V_i$.

f. Sketch the phase response for the same frequency range (the angle by which $\mathbf{V}_o$ leads $\mathbf{V}_i$).

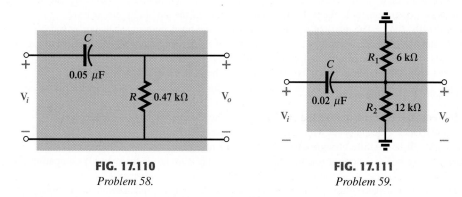

FIG. 17.110
Problem 58.

FIG. 17.111
Problem 59.

60. For the low-pass filter of Fig. 17.112:

a. Sketch the idealized Bode plot for $A_v = V_o/V_i$.

b. Using the results of part (a), sketch the actual frequency response for the same frequency range.

c. Determine the decibel level at $f_c, \frac{1}{2}f_c, 2f_c, \frac{1}{10}f_c,$ and $10f_c$.

d. Determine the gain $A_v = V_o/V_i$ at $f = f_c, \frac{1}{2}f_c,$ and $2f_c$.

e. Sketch the phase response for the same frequency range.

*61. For the low-pass filter of Fig. 17.113:

a. Sketch the response of the magnitude of V_o (in terms of V_i) versus frequency.

b. Using the results of part (a), sketch the response $A_v = V_o/V_i$ for the same frequency range.

c. Sketch the idealized Bode plot.

d. Sketch the actual response indicating the dB difference between the idealized and the actual response at $f = f_c, 0.5f_c,$ and $2f_c$.

e. Determine $A_{v_{dB}}$ at $f = 0.25f_c$ from the plot of part (d), and then determine the corresponding magnitude of $A_v = V_o/V_i$.

f. Sketch the phase response for the same frequency range (the angle by which $\mathbf{V}_o$ leads $\mathbf{V}_i$).

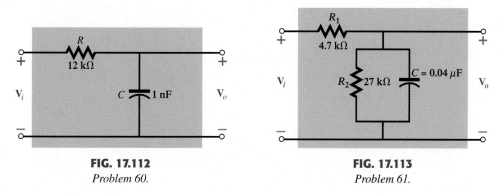

FIG. 17.112
Problem 60.

FIG. 17.113
Problem 61.

SECTION 17.17   Computer Analysis

62. Using PSpice or EWB, verify the results of Example 17.1. That is, show that the resonant frequency is, in fact, 40 kHz, the cutoff frequencies are as calculated, and the bandwidth is 1.85 kHz.

63. Using PSpice or EWB, find f_p for the parallel resonant network of Fig. 17.91, and comment on the resulting bandwidth as it relates to the quality factor of the network.

64. Using PSpice or EWB, verify the results of Example 17.7.

65. Using PSpice or EWB, and using schematics, obtain the magnitude and phase response versus frequency for the network of Fig. 17.104.

66. Using PSpice or EWB, and using schematics, obtain the magnitude and phase response versus frequency for the network of Fig. 17.107.

GLOSSARY

Active filter   A filter that employs active devices such as transistors or operational amplifiers in combination with R, L, and C elements.

Band (cutoff, half-power, corner) frequencies   Frequencies that define the points on the resonance curve that are 0.707 of the peak current or voltage value. In addition, they define the frequencies at which the power transfer to the resonant circuit will be half the maximum power level.

Bandwidth (BW)   The range of frequencies between the band, cutoff, or half-power frequencies.

Bode plot   A plot of the frequency response of a system using straight-line segments called *asymptotes*.

Decibel   A unit of measurement used to compare power levels.

Double-tuned filter   A network having both a pass-band and a stop-band region.

Filter   Networks designed to either pass or reject the transfer of signals at certain frequencies to a load.

High-pass filter   A filter designed to pass high frequencies and reject low frequencies.

Log-log paper   Graph paper with vertical and horizontal log scales.

Low-pass filter   A filter designed to pass low frequencies and reject high frequencies.

Microbar (μbar)   A unit of measurement for sound pressure levels that permits comparing audio levels on a dB scale.

Pass-band (band-pass) filter   A network designed to pass signals within a particular frequency range.

Passive filter   A filter constructed of series, parallel, or series-parallel R, L, and C elements.

Quality factor (Q)   A ratio that provides an immediate indication of the sharpness of the peak of a resonance curve. The higher the Q, the sharper the peak and the more quickly it drops off to the right and left of the resonant frequency.

Resonance   A condition established by the application of a particular frequency (the resonant frequency) to a series or parallel R-L-C network. The transfer of power to the system is a maximum, and, for frequencies above and below, the power transfer drops off to significantly lower levels.

Selectivity   A characteristic of resonant networks directly related to the bandwidth of the resonant system. High selectivity is associated with small bandwidth (high Q's), and low selectivity with larger bandwidths (low Q's).

Semilog paper   Graph paper with one log scale and one linear scale.

Stop-band filter   A network designed to reject (block) signals within a particular frequency range.

Transformers and Three-Phase Systems

OBJECTIVES

- Become aware of the basic construction of an iron-core transformer.
- Be able to calculate the various voltage and current levels of a transformer.
- Understand how to calculate the reflected impedance of a transformer.
- Recognize that because of the high-efficiency characteristics of a transformer, it is assumed that the applied power equals the power to the load.
- Understand the impact of impedance matching on the power to the load.
- Be able to interpret the nameplate data of a transformer.
- Understand why three-phase systems are used for the transfer of power from the plant to the user.
- Be able to calculate all the currents, voltages, and power levels for a Y-Y, Y-Δ, Δ-Δ, or Δ-Y system.
- Become aware of autotransformers and learn why they are used for so many practical applications.
- Be able to use the two-wattmeter or three-wattmeter method to measure the total power to a three-phase load.

18.1 INTRODUCTION

The **transformer** is one of the most commonly used and most useful electrical components in the industry. It is highly efficient and can be used for a variety of applications, including the following:

1. *Raise or lower the voltage or current level.*
2. *Act as an impedance matching device for maximum power transfer.*
3. *Provide isolation between systems for safety or operational reasons.*

Each of the above applications will be examined in this chapter.

The transformer is an integral part of the **three-phase (3-Φ) distribution system** from the power plant to the industrial or residential user. The typical ac generator is designed to develop a single sinusoidal ac voltage for each rotation of the shaft (rotor). Three-phase generators have additional coils in the rotor **to generate three voltages with each rotation of the rotor.** The peak value of each of the three voltages is the same, but they are each out of phase by 120°. Three-phase systems are connected in either a Y or a Δ configuration. Both will be examined in detail in this chapter.

18.2 IRON-CORE TRANSFORMER

Construction

The most common of the transformers available commercially today is the iron-core variety of Fig. 18.1. The core is a ferromagnetic material such as iron or steel to permit the establishment of a high level of magnetic flux in the core. The core is usually laminated sheets of the metal to cut down on the power losses due to eddy currents that can develop in a solid core. Eddy currents are small, circular loop currents that will develop in the core due to the application of an ac voltage. Their negative effect can be reduced considerably by simply putting an insulator between the sheets of ferromagnetic material. The high-resistance path created by the layers of insulation will reduce the power losses to insignificant levels.

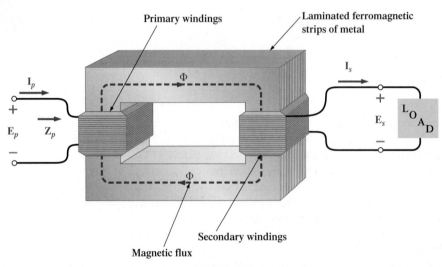

FIG. 18.1

Iron-core transformer.

Voltage Levels

The flux in the core is established by the application of an ac voltage to the primary windings, as shown in Fig. 18.1. The term **primary** is normally applied to the winding to which the source is applied. The current in the primary will establish a flux in the core that will have a strength sensitive to the level of current, the number of turns in the primary winding, and the reluctance level of the ferromagnetic material.

You can find the direction of the resulting flux by simply placing the fingers of your right hand in the direction of the current through the windings and then noting the direction of the thumb.

That direction, of course, will reverse with a change in polarity of the applied voltage and a change in direction of the resulting current in the primary.

Since the strength of the resulting flux is sensitive to the strength of the applied current, the flux in the core will vary in a sinusoidal manner also. That alternating flux will then pass through the core as shown in Fig. 18.1, linking the secondary winding. The term **secondary** is normally applied to the winding to which the load is applied. The changing flux, linking the secondary windings, will develop a voltage across the secondary winding as determined by Faraday's law:

$$e = N\frac{d\phi}{dt}$$

Since the flux varies in a sinusoidal manner, the voltage induced across the secondary will vary in the same manner.

If we assume the ideal situation, in which all the flux developed at the primary will link up with the secondary, the magnitude of the voltages is determined solely by the turns in the windings. That is,

$$\boxed{\frac{\mathbf{E}_p}{\mathbf{E}_s} = \frac{N_p}{N_s}} \qquad \textbf{(18.1)}$$

so that

the voltages of a transformer have the same ratio as the turns of the respective windings.

As we now know, nothing in the real world is ideal. Therefore, some of the flux developed at the primary does not link up with the windings of the secondary, due to an effect called *fringing,* as shown in Fig. 18.2. However, the leakage flux for most ferromagnetic core transformers is so small that it can be ignored. In fact, in general,

the efficiency level of transformers is so high that transformers, for most applications, can be considered to be ideal.

It is an assumption we will make in the discussion to follow.

The turns ratio N_p/N_s is called the **transformation ratio** and is defined by

$$\boxed{a = \frac{N_p}{N_s}} \qquad \textbf{(18.2)}$$

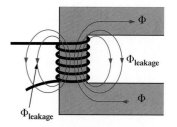

FIG 18.2
Identifying the leakage flux of the primary.

If $a < 1$, the transformer is called a **step-up transformer** because the secondary voltage will be greater than the primary voltage. That is,

$$\frac{E_p}{E_s} = \frac{N_p}{N_s} = a$$

so that

$$E_s = \frac{E_p}{a}$$

and, if $a < 1$, then $E_s > E_p$.

If $a > 1$, the transformer is called a **step-down transformer** because the secondary voltage will be less than the primary voltage. That is,

$$E_p = aE_s$$

and, if $a > 1$, then $\qquad E_p > E_s$

Current Levels

Continuing with the assumption that the transformer is an ideal device, we will find that the current levels are related by

$$\boxed{\frac{\mathbf{I}_p}{\mathbf{I}_s} = \frac{N_s}{N_p}} \qquad \textbf{(18.3)}$$

so that

the currents of a transformer are related by the inverse ratio of the respective turns.

In general, therefore,

for a transformer, we can expect that if the voltage level increases, the current level decreases, and vice versa.

Reflected Impedance

Since

$$\frac{\mathbf{E}_p}{\mathbf{E}_s} = \frac{N_p}{N_s} = a$$

and since

$$\frac{\mathbf{I}_p}{\mathbf{I}_s} = \frac{N_s}{N_p} = \frac{1}{a}$$

if we divide the first equation by the second, we have

$$\frac{\dfrac{E_p}{E_s}}{\dfrac{I_p}{I_s}} = \frac{a}{\dfrac{1}{a}}$$

or

$$\frac{\dfrac{E_p}{I_p}}{\dfrac{E_s}{I_s}} = a^2$$

However, with
$$\mathbf{Z}_p = \frac{E_p}{I_p} \quad \text{and} \quad \mathbf{Z}_s = \frac{E_s}{I_s}$$

we have
$$\frac{\mathbf{Z}_p}{\mathbf{Z}_s} = a^2$$

Finally,
$$\boxed{\mathbf{Z}_p = a^2 \mathbf{Z}_s} \qquad (18.4)$$

revealing that

the impedance appearing at the primary of a transformer is the turns ratio squared times the impedance at the secondary.

Further, since a^2 is a constant, the impedance at the primary will have the same characteristics as the impedance connected to the secondary. For example, if the load is capacitive, the impedance at the primary will also be capacitive, and so on.

Power

For the ideal iron-core transformer,

$$\frac{E_p}{E_s} = a = \frac{I_s}{I_p}$$

or
$$\boxed{E_p I_p = E_s I_s} \qquad (18.5)$$

and
$$\boxed{P_{\text{in}} = P_{\text{out}}} \qquad (18.6)$$

revealing that

for the ideal transformer, the power at the primary is equal to the power at the secondary.

In other words, it is assumed that no power is lost in the core or in the transmission of power from one set of turns to the other. It turns out, however, that the typical efficiency level of a transformer is so high that assuming ideal characteristics is an excellent and appropriate approximation.

The symbol for an iron-core transformer appears in Fig. 18.3 with all the parameters of importance for a transformer.

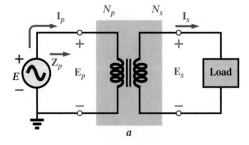

FIG. 18.3
Symbol for an iron-core transformer and the parameters of importance.

EXAMPLE 18.1   For the iron-core transformer of Fig. 18.4:

a. Find the magnitude of the secondary voltage E_s.
b. Find the magnitude of the secondary current I_s.
c. Determine the primary current I_p.
d. What is the turns ratio?
e. Is it a step-up or a step-down transformer?
f. Find the power at the primary and secondary terminals, and compare their levels.
g. Find the input impedance at the primary.

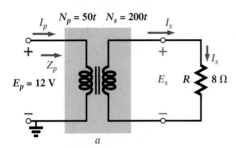

FIG. 18.4
Example 18.1.

Solutions:

a. $\dfrac{E_p}{E_s} = \dfrac{N_p}{N_s}$ and $E_s = \dfrac{N_s}{N_p} E_p$

so that $E_s = \dfrac{200\,\text{t}}{50\,\text{t}}(12\text{ V}) = 4(12\text{ V}) = \textbf{48 V}$

b. $I_s = \dfrac{E_s}{R_L} = \dfrac{48\text{ V}}{8\ \Omega} = \textbf{6 A}$

c. $\dfrac{I_p}{I_s} = \dfrac{N_s}{N_p}$ and $I_p = \dfrac{N_s}{N_p} I_s$

so that $I_p = \dfrac{200\,\text{t}}{50\,\text{t}}(6\text{ A}) = 4(6\text{ A}) = \textbf{24 A}$

d. $a = \dfrac{N_p}{N_s} = \dfrac{50\,\text{t}}{200\,\text{t}} = \dfrac{\textbf{1}}{\textbf{4}}$

e. Since $a < 1$, and $E_s > E_p$, it is a **step-up transformer.**

f. $P_p = E_p I_p = (12\text{ V})(24\text{ A}) = \textbf{288 W}$
   $P_s = E_s I_s = (48\text{ V})(6\text{ A}) = \textbf{288 W}$
   As required by Eq. (18.6):
   $P_p = P_s = \textbf{288 W}$

g. $Z_p = a^2 Z_L = \left(\dfrac{1}{4}\right)^2 (8\ \Omega) = \left(\dfrac{1}{16}\right)(8\ \Omega) = \dfrac{\textbf{1}}{\textbf{2}}\ \Omega$

   or $Z_p = \dfrac{E_p}{I_p} = \dfrac{12\text{ V}}{24\text{ A}} = \dfrac{1}{2}\ \Omega$

EXAMPLE 18.2

a. The source impedance for the supply of Fig. 18.5(a) is 512 Ω, which is a poor match with the 8 Ω input impedance of the speaker. You can expect only that the power delivered to the speaker will be significantly less than the maximum possible level. Determine the power to the speaker under the conditions of Fig. 18.5(a).

b. In Fig. 18.5(b), an audio impedance matching transformer was introduced between the speaker and the source, and it was designed to ensure maximum power to the 8 Ω

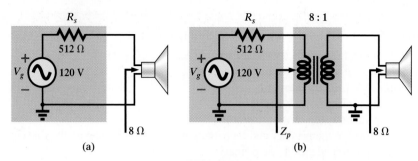

FIG. 18.5
Example 18.2.

speaker. Determine the input impedance of the transformer and the power delivered to
the speaker.

c. Compare the power delivered to the speaker under the conditions of parts (a) and (b).

Solutions:

a. The source current:

$$I_s = \frac{E}{R_T} = \frac{120 \text{ V}}{512\ \Omega + 8\ \Omega} = \frac{120 \text{ V}}{520\ \Omega} = 230.8 \text{ mA}$$

The power to the speaker:

$$P = I^2 R = (230.8 \text{ mA})^2 8\ \Omega = \textbf{426.15 mW} \cong \textbf{0.43 W}$$

or less than one-half watt.

b. $Z_p = a^2 Z_L$

$$a = \frac{N_p}{N_s} = \frac{8}{1} = 8$$

and $$Z_p = (8)^2 8\ \Omega = \textbf{512}\ \boldsymbol{\Omega}$$

which matches that of the source. Maximum power transfer conditions have been established, and the source current is now determined by

$$I_s = \frac{E}{R_T} = \frac{120 \text{ V}}{512\ \Omega + 512\ \Omega} = \frac{120 \text{ V}}{1024\ \Omega} = 117.19 \text{ mA}$$

The power to the primary (which equals that to the secondary for the ideal transformer) is

$$P = I^2 R = (117.19 \text{ mA})^2\ 512\ \Omega = \textbf{7.032 W}$$

The result is not in milliwatts, as obtained above, and exceeds 7 W, which is a significant improvement.

c. Comparing levels, 7.032 W/426.15 mW = 16.5, or more than 16 times the power delivered to the speaker using the impedance matching transformer.

EXAMPLE 18.3 Another important application of the impedance matching capabilities of a
transformer is the matching of the 300 Ω twin line transmission line from a television antenna to
the 75 Ω input impedance of today's televisions (ready-made for the 75 Ω coaxial cable), as
shown in Fig. 18.6. Find the match needed to ensure the strongest signal to the television receiver.

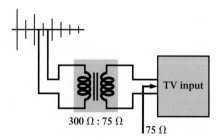

FIG. 18.6
Television impedance matching transformer (Example 18.3).

Solution: Using the equation $Z_p = a^2 Z_L$, we find

$$300 \ \Omega = a^2 (75 \ \Omega)$$

and

$$a = \sqrt{\frac{300 \ \Omega}{75 \ \Omega}} = \sqrt{4} = 2$$

with

$$N_p : N_s = 2 : 1 \qquad \text{(a step-down transformer)}$$

EXAMPLE 18.4   Impedance matching transformers are also quite evident in public address systems, such as the one appearing in the 70.7 V system of Fig. 18.7. Although the system has only one set of output terminals, up to four speakers can be connected to this system (the number is a function of the chosen system). Each 8 Ω speaker is connected to the 70.7 V line through a 10 W audio-matching transformer (defining the frequency range of linear operation).

a. If each speaker of Fig. 18.7 can receive 10 W of power, what is the maximum power drain on the source?
b. For each speaker, determine the impedance seen at the input side of the transformer if each is operating under its full 10 W of power.
c. Determine the turns ratio of the transformers.
d. At 10 W, what are the speaker voltage and current?
e. What is the load seen by the source with one, two, three, or four speakers connected?

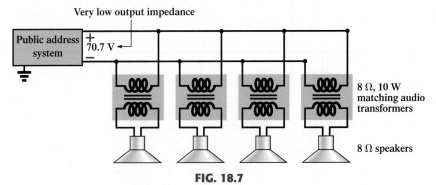

FIG. 18.7
Public address system (Example 18.4).

Solutions:

a. Ideally, the primary power equals the power delivered to the load, resulting in a maximum of **40 W** from the supply.

b. The power at the primary:

$$P_p = V_p I_p = (70.7 \ \text{V}) \, I_p = 10 \ \text{W}$$

and

$$I_p = \frac{10 \ \text{W}}{70.7 \ \text{V}} = 141.4 \ \text{mA}$$

so that

$$Z_p = \frac{V_p}{I_p} = \frac{70.7 \ \text{V}}{141.4 \ \text{mA}} = \mathbf{500 \ \Omega}$$

c. $Z_p = a^2 Z_L \Rightarrow a = \sqrt{\dfrac{Z_p}{Z_L}} = \sqrt{\dfrac{500 \ \Omega}{8 \ \Omega}} = \sqrt{62.5} = \mathbf{7.91} \cong \mathbf{8 : 1}$

d. $V_s = V_L = \dfrac{V_p}{a} = \dfrac{70.7 \ \text{V}}{7.91} = \mathbf{8.94 \ V} \cong \mathbf{9 \ V}$

e. All the speakers are in parallel. Therefore,

One speaker: $R_T = \mathbf{500 \ \Omega}$

Two speakers: $R_T = \dfrac{500 \ \Omega}{2} = \mathbf{250 \ \Omega}$

Three speakers: $\quad R_T = \dfrac{500\ \Omega}{3} = \mathbf{167\ \Omega}$

Four speakers: $\quad R_T = \dfrac{500\ \Omega}{4} = \mathbf{125\ \Omega}$

Even though the load seen by the source will vary with the number of speakers connected, the source impedance is so low (compared to the lowest load of 125 Ω) that the terminal voltage of 70.7 V is essentially constant. This is not the case where the desired result is to match the load to the input impedance; rather, it was to ensure 70.7 V at each primary, no matter how many speakers were connected, and to limit the current drawn from the supply.

Isolation

The transformer is frequently used to isolate one portion of an electrical system from another. *Isolation* implies the absence of any direct physical connection. As a first example of its use as an isolation device, consider the measurement of line voltages on the order of 40,000 V (Fig. 18.8).

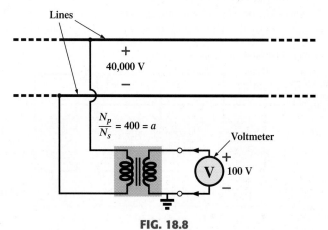

FIG. 18.8

Isolating a high-voltage line from the point of measurement.

To apply a voltmeter across 40,000 V would obviously be a dangerous task due to the possibility of physical contact with the lines when making the necessary connections. By including a transformer in the transmission system as original equipment, one can bring the potential down to a safe level for measurement purposes and can determine the line voltage using the turns ratio. Therefore, the transformer will serve both to isolate and to step down the voltage.

As a second example, consider the application of the voltage v_x to the vertical input of the oscilloscope (a measuring instrument) in Fig. 18.9(a). If the connections are made as shown, and if the generator and oscilloscope have a common ground, the impedance $\mathbf{Z}_2$ has been

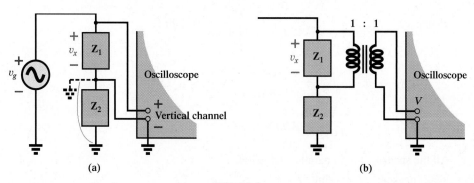

FIG. 18.9

(a) Demonstrating the shorting effect introduced by the grounded side of the vertical channel of an oscilloscope; (b) correcting the situation using an isolation transformer.

effectively shorted out of the circuit by the ground connection of the oscilloscope. The input voltage to the oscilloscope will therefore be meaningless as far as the voltage v_x is concerned. In addition, if $\mathbf{Z}_2$ is the current-limiting impedance in the circuit, the current in the circuit may rise to a level that will cause severe damage to the circuit. If a transformer is used as shown in Fig. 18.9(b), this problem will be eliminated, and the input voltage to the oscilloscope will be v_x.

Nameplate Data

A typical iron-core power transformer rating, included in the **nameplate data** for the transformer, might be the following:

$$5 \text{ kVA} \qquad 2000/100 \text{ V} \qquad 60 \text{ Hz}$$

The 2000 V or the 100 V can be either the primary or the secondary voltage; that is, if 2000 V is the primary voltage, then 100 V is the secondary voltage, and vice versa. The 5 kVA is the apparent power $(S = VI)$ rating of the transformer. If the secondary voltage is 100 V, then the maximum load current is

$$I_L = \frac{S}{V_L} = \frac{5000 \text{ VA}}{100 \text{ V}} = 50 \text{ A}$$

and if the secondary voltage is 2000 V, then the maximum load current is

$$I_L = \frac{S}{V_L} = \frac{5000 \text{ VA}}{2000 \text{ V}} = 2.5 \text{ A}$$

The transformer is rated in terms of the apparent power rather than the average, or real, power for the reason demonstrated by the circuit of Fig. 18.10. Since the current through the load is greater than that determined by the apparent power rating, the transformer may be permanently damaged. Note, however, that since the load is purely capacitive, the average power to the load is zero. Therefore, the wattage rating would be meaningless regarding the ability of this load to damage the transformer.

The transformation ratio of the transformer under discussion can be either of two values. If the secondary voltage is 2000 V, the transformation ratio is $a = N_p/N_s = V_g/V_L = 100 \text{ V}/2000 \text{ V} = 1/20$, and the transformer is a step-up transformer. If the secondary voltage is 100 V, the transformation ratio is $a = N_p/N_s = V_g/V_L = 2000 \text{ V}/100 \text{ V} = 20$, and the transformer is a step-down transformer.

The rated primary current can be determined simply by applying Eq. (18.3):

$$I_p = \frac{I_s}{a}$$

which is equal to $[2.5 \text{ A}/(1/20)] = 50 \text{ A}$ if the secondary voltage is 2000 V, and $(50 \text{ A}/20) = 2.5 \text{ A}$ if the secondary voltage is 100 V.

To explain the necessity for including the frequency in the nameplate data, it can be shown that the primary voltage is related to the applied frequency and the flux in the core by

$$E_p = 4.44 \, f N_p \Phi$$

The B-H curve for the iron core of the transformer appears in Fig. 18.11.

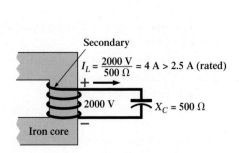

FIG. 18.10

Demonstrating why transformers are rated in kVA rather than kW.

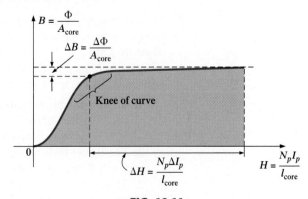

FIG. 18.11

Demonstrating why the frequency of application is important for transformers.

The point of operation on the *B-H* curve for most transformers is at the knee of the curve. If the frequency of the applied signal should drop, and N_p and E_p remain the same, then Φ must increase in magnitude, as determined by the above equation.

$$\Phi\uparrow = \frac{E_p}{4.44f\downarrow N_p}$$

The result is that the flux density *B* will increase, as shown in Fig. 18.11, causing *H* to increase also. The resulting ΔI could cause a very high current in the primary, resulting in possible damage to the transformer.

Autotransformer

The **autotransformer** is a power transformer that, instead of employing the two-circuit principle (complete isolation between coils), has one winding common to both the input and output circuits as shown in Fig. 18.12(b). The induced voltages are related to the turns ratio in the same manner as that described for the two-circuit transformer. If the proper connection is used, a two-circuit power transformer can be employed as an autotransformer. The advantage of using it as an autotransformer is that a larger apparent power can be transformed. This can be demonstrated by the two-circuit transformer of Fig. 18.12(a), shown in Fig. 18.12(b) as an autotransformer.

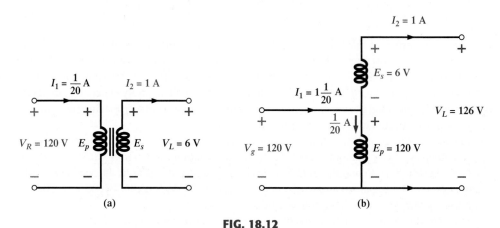

FIG. 18.12

(a) Two-circuit transformer; (b) autotransformer.

For the two-circuit transformer, note that $S = (\frac{1}{20}\,\text{A})(120\,\text{V}) = 6\,\text{VA}$, whereas for the autotransformer, $S = (1\frac{1}{20}\,\text{A})(120\,\text{V}) = 126\,\text{VA}$, which is many times that of the two-circuit transformer. Note also that the current and voltage of each coil are the same as those for the two-circuit configuration. The disadvantage of the autotransformer is obvious: loss of the isolation between the primary and secondary circuits.

18.3 THREE-PHASE SYSTEMS

As mentioned in the introduction, a three-phase generator is one that develops three distinct voltages at the same time—each with the same peak value and each out of phase with the others by 120°.

In general, three-phase systems are preferred over single-phase systems for the transmission of power for many reasons, including the following:

1. Thinner conductors can be used to transmit the same kVA at the same voltage, which reduces the amount of copper required (typically about 25% less) and in turn reduces construction and maintenance costs.
2. The lighter lines are easier to install, and the supporting structures can be less massive and farther apart.
3. Three-phase equipment and motors have preferred running and starting characteristics compared to single-phase systems because of a more even flow of power to the transducer than can be delivered with a single-phase supply.

4. In general, most larger motors are three phase because they are essentially self-starting and do not require a special design or additional starting circuitry.

The frequency generated is determined by the number of poles on the *rotor* (the rotating part of the generator) and the speed with which the shaft is turned. Throughout the United States the line frequency is 60 Hz, whereas in Europe the chosen standard is 50 Hz. Both frequencies were chosen primarily because they can be generated by a relatively efficient and stable mechanical design that is sensitive to the size of the generating systems and the demand that must be met during peak periods. On aircraft and ships the demand levels permit the use of a 400 Hz line frequency.

The three-phase system is used by almost all commercial electric generators. This does not mean that single-phase and two-phase generating systems are obsolete. Most small emergency generators, such as the gasoline type, are one-phase generating systems. The two-phase system is commonly used in servomechanisms, which are self-correcting control systems capable of detecting and adjusting their own operation. Servomechanisms are used in ships and aircraft to keep them on course automatically, or, in simpler devices such as a thermostatic circuit, to regulate heat output. In many cases, however, where single-phase and two-phase inputs are required, they are supplied by one and two phases of a three-phase generating system rather than generated independently.

The number of **phase voltages** that can be produced by a polyphase generator is not limited to three. Any number of phases can be obtained by spacing the windings for each phase at the proper angular position around the stator. Some electrical systems operate more efficiently if more than three phases are used. One such system involves the process of rectification, which is used to convert an alternating output to one having an average, or dc, value. The greater the number of phases, the smoother the dc output of the system.

18.4 THE THREE-PHASE GENERATOR

The three-phase generator of Fig. 18.13(a) has three induction coils placed 120° apart on the stator, as shown symbolically by Fig. 18.13(b). Since the three coils have an equal number of turns, and each coil rotates with the same angular velocity, the voltage induced across each coil will have the same peak value, shape, and frequency. As the shaft of the generator is turned by some external means, the induced voltages e_{AN}, e_{BN}, and e_{CN} will be generated simultaneously, as shown in Fig. 18.14. Note the 120° phase shift between waveforms and the similarities in appearance of the three sinusoidal functions.

In particular, note that

at any instant of time, the algebraic sum of the three phase voltages of a three-phase generator is zero.

This is shown at $\omega t = 0$ in Fig. 18.14, where it is also evident that **when one induced voltage is zero, the other two are 86.6% of their positive or negative maximums. In addition, when any two are equal in magnitude and sign (at $0.5E_m$), the remaining induced voltage has the opposite polarity and a peak value.**

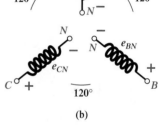

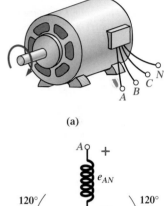

(a)

(b)

FIG. 18.13
(a) Three-phase generator;
(b) induced voltages of a three-phase generator.

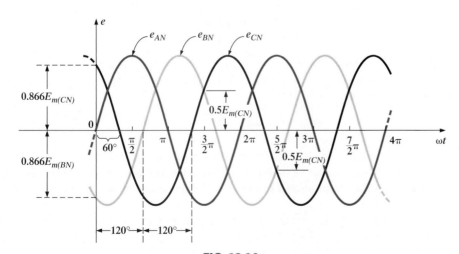

FIG. 18.14
Phase voltages of a three-phase generator.

The sinusoidal expression for each of the induced voltages of Fig. 18.14 is

$$
\begin{aligned}
e_{AN} &= E_{m(AN)} \sin \omega t \\
e_{BN} &= E_{m(BN)} \sin(\omega t - 120°) \\
e_{CN} &= E_{m(CN)} \sin(\omega t - 240°) = E_{m(CN)} \sin(\omega t + 120°)
\end{aligned}
\tag{18.7}
$$

The phasor diagram of the induced voltages is shown in Fig. 18.15, where the effective value of each is determined by

$$
\begin{aligned}
E_{AN} &= 0.707 E_{m(AN)} \\
E_{BN} &= 0.707 E_{m(BN)} \\
E_{CN} &= 0.707 E_{m(CN)}
\end{aligned}
$$

and

$$
\begin{aligned}
\mathbf{E}_{AN} &= E_{AN} \angle 0° \\
\mathbf{E}_{BN} &= E_{BN} \angle -120° \\
\mathbf{E}_{CN} &= E_{CN} \angle +120°
\end{aligned}
$$

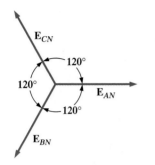

FIG. 18.15

*Phasor diagram for the phase
voltages of a three-phase generator.*

18.5 THE Y-CONNECTED GENERATOR

If the three terminals denoted *N* of Fig. 18.13(b) are connected together, the generator is referred to as a **Y-connected three-phase generator** (Fig. 18.16). As indicated in Fig. 18.16, the Y is inverted for ease of notation and for clarity. The point at which all the terminals are connected is called the **neutral point.** If a conductor is not attached from this point to the load, the system is called a **Y-connected, three-phase, three-wire generator.** If the neutral is connected, the system is a **Y-connected, three-phase, four-wire generator.** The function of the neutral will be discussed in detail when we consider the load circuit.

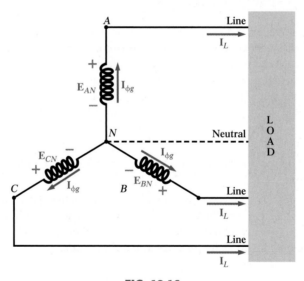

FIG. 18.16

Y-connected generator.

The three conductors connected from *A*, *B*, and *C* to the load are called *lines*. For the Y-connected system, it should be obvious from Fig. 18.16 that the **line current** equals the **phase current** for each phase. That is,

$$
\mathbf{I}_L = \mathbf{I}_{\phi g}
\tag{18.8}
$$

where ϕ is used to denote a phase quantity and *g* is a generator parameter.

The voltage from one line to another is called a **line voltage.** On the phasor diagram (Fig. 18.17), it is the phasor drawn from the end of one phase to another in the counterclockwise direction.

The magnitude of the line voltage of a Y-connected generator is $\sqrt{3}$ times the phase voltage:

$$
E_L = \sqrt{3} E_\phi
\tag{18.9}
$$

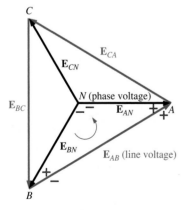

FIG. 18.17

*Line and phase voltages of the
Y-connected three-phase generator.*

with the phase angle between any line voltage and the nearest phase voltage at 30°.

In sinusoidal notation,

$$e_{AB} = \sqrt{2}E_{AB} \sin(\omega t + 30°)$$
$$e_{CA} = \sqrt{2}E_{CA} \sin(\omega t + 150°)$$

and
$$e_{BC} = \sqrt{2}E_{BC} \sin(\omega t + 270°)$$

18.6 PHASE SEQUENCE (Y-CONNECTED GENERATOR)

The **phase sequence** can be determined by the order in which the phasors representing the phase voltages pass through a fixed point on the phasor diagram if the phasors are rotated in a counterclockwise direction. For example, in Fig. 18.18, the phase sequence is *ABC*. However, since the fixed point can be chosen anywhere on the phasor diagram, the sequence can also be written as *BCA* or *CAB*. The phase sequence is quite important in the three-phase distribution of power. In a three-phase motor, for example, if two phase voltages are interchanged, the sequence will change, and the direction of rotation of the motor will be reversed. Other effects will be described when we consider the loaded three-phase system.

The phase sequence can also be described in terms of the line voltages. Drawing the line voltages on a phasor diagram in Fig. 18.19, we are able to determine the phase sequence by again rotating the phasors in the counterclockwise direction. In this case, however, the sequence can be determined by noting the order of the passing first or second subscripts. In the system of Fig. 18.19, for example, the phase sequence of the first subscripts passing point *P* is *ABC*, and the phase sequence of the second subscripts is *BCA*. But we know that *BCA* is equivalent to *ABC*, so the sequence is the same for each. Note that the phase sequence is the same as that of the phase voltages described in Fig. 18.18.

If the sequence is given, the phasor diagram can be drawn by simply picking a reference voltage, placing it on the reference axis, and then drawing the other voltages at the proper angular position. For a sequence of *ACB*, for example, we might choose E_{AB} to be the reference [Fig. 18.20(a)] if we wanted the phasor diagram of the line voltages, or E_{NA} for the phase voltages [Fig. 18.20(b)]. For the sequence indicated, the phasor diagrams would be as shown in Fig. 18.20. In phasor notation,

$$\text{Line voltages} \begin{cases} \mathbf{E}_{AB} = E_{AB} \angle 0° & \text{(reference)} \\ \mathbf{E}_{CA} = E_{CA} \angle -120° \\ \mathbf{E}_{BC} = E_{BC} \angle +120° \end{cases}$$

$$\text{Phase voltages} \begin{cases} \mathbf{E}_{AN} = E_{AN} \angle 0° & \text{(reference)} \\ \mathbf{E}_{CN} = E_{CN} \angle -120° \\ \mathbf{E}_{BN} = E_{BN} \angle +120° \end{cases}$$

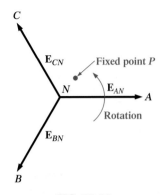

FIG. 18.18
Determining the phase sequence from the phase voltages of a three-phase generator.

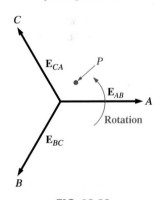

FIG. 18.19
Determining the phase sequence from the line voltages of a three-phase generator.

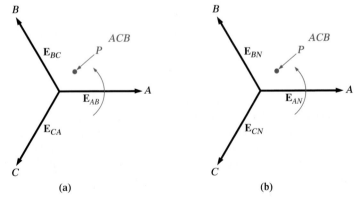

(a)   (b)

FIG. 18.20
Drawing the phasor diagram from the phase sequence.

18.7 THE Y-CONNECTED GENERATOR WITH A Y-CONNECTED LOAD

Loads connected to three-phase supplies are of two types: the Y and the Δ. If a Y-connected load is connected to a Y-connected generator, the system is symbolically represented by Y-Y. The physical setup of such a system is shown in Fig. 18.21.

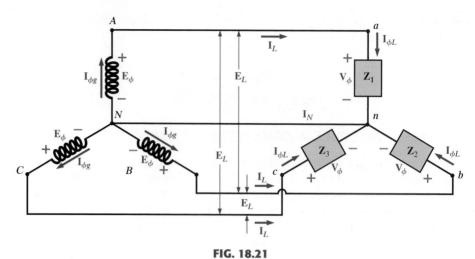

FIG. 18.21

Y-connected generator with a Y-connected load.

If the load is balanced, the **neutral connection** can be removed without affecting the circuit in any manner. That is, if

$$\mathbf{Z}_1 = \mathbf{Z}_2 = \mathbf{Z}_3$$

then I_N will be zero. (This will be demonstrated in Example 18.5.) Note that in order to have a balanced load, the phase angle must also be the same for each impedance—a condition that was unnecessary in dc circuits when we considered balanced systems.

In practice, if a factory, for example, had only balanced, three-phase loads, the absence of the neutral would have no effect since, ideally, the system would always be balanced. The cost would therefore be less since the number of required conductors would be reduced. However, lighting and most other electrical equipment will use only one of the phase voltages, and even if the loading is designed to be balanced (as it should be), there will never be perfect, continuous balancing because lights and other electrical equipment will be turned on and off, upsetting the balanced condition. Therefore, the neutral is necessary to carry the resulting current away from the load and back to the Y-connected generator. This will be demonstrated when we consider unbalanced Y-connected systems.

We shall now examine the *four-wire Y-Y-connected system*. The current passing through each phase of the generator is the same as its corresponding line current, which in turn for a Y-connected load is equal to the current in the phase of the load to which it is attached:

$$\boxed{\mathbf{I}_{\phi g} = \mathbf{I}_L = \mathbf{I}_{\phi L}} \qquad \textbf{(18.10)}$$

For a balanced or an unbalanced load, since the generator and the load have a common neutral point, then

$$\boxed{\mathbf{V}_{\phi} = \mathbf{E}_{\phi}} \qquad \textbf{(18.11)}$$

In addition, since $\mathbf{I}_{\phi L} = \mathbf{V}_{\phi}/\mathbf{Z}_{\phi}$, the magnitude of the current in each phase will be equal for a balanced load and unequal for an unbalanced load. Recall that for the Y-connected generator, the magnitude of the line voltage is equal to $\sqrt{3}$ times the phase voltage. This same relationship can be applied to a balanced or an unbalanced four-wire Y-connected load:

$$\boxed{E_L = \sqrt{3}V_{\phi}} \qquad \textbf{(18.12)}$$

For a voltage drop across a load element, the first subscript refers to the terminal through which the current enters the load element, and the second subscript refers to the terminal from which the current leaves. In other words, the first subscript is, by definition, positive with respect to the second for a voltage drop. Note Fig. 18.22 (Example 18.5), in which the standard double subscripts for a source of voltage and a voltage drop are indicated.

EXAMPLE 18.5 The phase sequence of the Y-connected generator in Fig. 18.22 is *ABC*.

a. Find the phase angles θ_2 and θ_3.
b. Find the magnitude of the line voltages.
c. Find the line currents.
d. Verify that since the load is balanced, $\mathbf{I}_N = 0$.

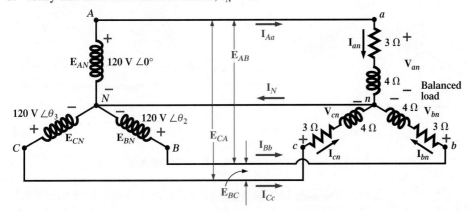

FIG. 18.22
Example 18.5.

Solutions:

a. For an *ABC* phase sequence,

$$\theta_2 = -\mathbf{120°} \quad \text{and} \quad \theta_3 = +\mathbf{120°}$$

b. $E_L = \sqrt{3}E_\phi = (1.73)(120 \text{ V}) = 208 \text{ V}$. Therefore,

$$E_{AB} = E_{BC} = E_{CA} = \mathbf{208 \text{ V}}$$

c. $\mathbf{V}_\phi = \mathbf{E}_\phi$. Therefore,

$$\mathbf{V}_{an} = \mathbf{E}_{AN} \qquad \mathbf{V}_{bn} = \mathbf{E}_{BN} \qquad \mathbf{V}_{cn} = \mathbf{E}_{CN}$$

$$\mathbf{I}_{\phi L} = \mathbf{I}_{an} = \frac{\mathbf{V}_{an}}{\mathbf{Z}_{an}} = \frac{120 \text{ V} \angle 0°}{3 \text{ } \Omega + j \text{ } 4 \text{ } \Omega} = \frac{120 \text{ V} \angle 0°}{5 \text{ } \Omega \angle 53.13°} = 24 \text{ A} \angle -53.13°$$

$$\mathbf{I}_{bn} = \frac{\mathbf{V}_{bn}}{\mathbf{Z}_{bn}} = \frac{120 \text{ V} \angle -120°}{5 \text{ } \Omega \angle 53.13°} = 24 \text{ A} \angle -173.13°$$

$$\mathbf{I}_{cn} = \frac{\mathbf{V}_{cn}}{\mathbf{Z}_{cn}} = \frac{120 \text{ V} \angle +120°}{5 \text{ } \Omega \angle 53.13°} = 24 \text{ A} \angle 66.87°$$

and, since $\mathbf{I}_L = \mathbf{I}_{\phi L}$, $\qquad \mathbf{I}_{Aa} = \mathbf{I}_{an} = \mathbf{24 \text{ A} \angle -53.13°}$

$$\mathbf{I}_{Bb} = \mathbf{I}_{bn} = \mathbf{24 \text{ A} \angle -173.13°}$$

$$\mathbf{I}_{Cc} = \mathbf{I}_{cn} = \mathbf{24 \text{ A} \angle 66.87°}$$

d. Applying Kirchhoff's current law, we have $\mathbf{I}_N = \mathbf{I}_{Aa} + \mathbf{I}_{Bb} + \mathbf{I}_{Cc}$. In rectangular form,

$$\mathbf{I}_{Aa} = 24 \text{ A} \angle -53.13° = 14.40 \text{ A} - j \text{ } 19.20 \text{ A}$$

$$\mathbf{I}_{Bb} = 24 \text{ A} \angle -173.13° = -22.83 \text{ A} - j \text{ } 2.87 \text{ A}$$

$$\mathbf{I}_{Cc} = 24 \text{ A} \angle 66.87° = \underline{9.43 \text{ A} + j \text{ } 22.07 \text{ A}}$$

$$\Sigma(\mathbf{I}_{Aa} + \mathbf{I}_{Bb} + \mathbf{I}_{Cc}) = 0 + j \text{ } 0$$

and $\mathbf{I}_N$ is in fact equal to **zero,** as required for a balanced load.

18.8 THE Y-Δ SYSTEM

There is no neutral connection for the Y-Δ system of Fig. 18.23. Any variation in the impedance of a phase that produces an unbalanced system will simply vary the line and phase currents of the system.

For a balanced load,

$$\boxed{\mathbf{Z}_1 = \mathbf{Z}_2 = \mathbf{Z}_3} \qquad\qquad \textbf{(18.13)}$$

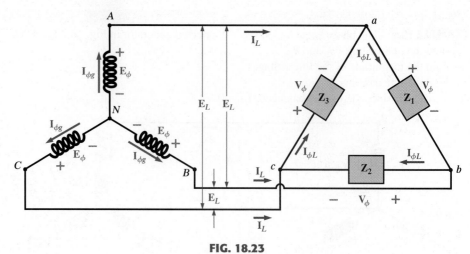

FIG. 18.23

Y-connected generator with a Δ-connected load.

The voltage across each phase of the load is equal to the line voltage of the generator for a balanced or an unbalanced load:

$$\boxed{\mathbf{V}_\phi = \mathbf{E}_L} \qquad (18.14)$$

The relationship between the line currents and phase currents of a balanced Δ load can be found using an approach very similar to that used in Section 18.5 to find the relationship between the line voltages and phase voltages of a Y-connected generator. For this case, however, Kirchhoff's current law is employed instead of Kirchhoff's voltage law.

The results obtained are

$$\boxed{I_L = \sqrt{3}I_\phi} \qquad (18.15)$$

and the phase angle between a line current and the nearest phase current is 30°. A more detailed discussion of this relationship between the line and phase currents of a Δ-connected system can be found in Section 18.9.

For a balanced load, the line currents will be equal in magnitude, as will the phase currents.

EXAMPLE 18.6 For the three-phase system of Fig. 18.24:

a. Find the phase angles θ_2 and θ_3.
b. Find the current in each phase of the load.
c. Find the magnitude of the line currents.

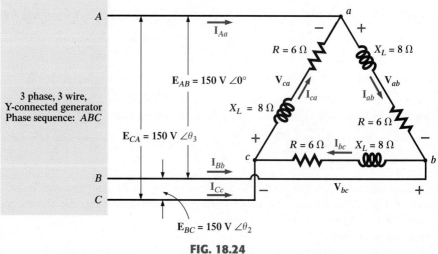

FIG. 18.24

Example 18.6.

Solutions:

a. For an *ABC* sequence,

$$\theta_2 = -120° \quad \text{and} \quad \theta_3 = +120°$$

b. $\mathbf{V}_\phi = \mathbf{E}_L$. Therefore,

$$\mathbf{V}_{ab} = \mathbf{E}_{AB} \qquad \mathbf{V}_{ca} = \mathbf{E}_{CA} \qquad \mathbf{V}_{bc} = \mathbf{E}_{BC}$$

The phase currents are

$$\mathbf{I}_{ab} = \frac{\mathbf{V}_{ab}}{\mathbf{Z}_{ab}} = \frac{150 \text{ V} \angle 0°}{6\,\Omega + j\,8\,\Omega} = \frac{150 \text{ V} \angle 0°}{10\,\Omega \angle 53.13°} = \mathbf{15 \text{ A} \angle -53.13°}$$

$$\mathbf{I}_{bc} = \frac{\mathbf{V}_{bc}}{\mathbf{Z}_{bc}} = \frac{150 \text{ V} \angle -120°}{10\,\Omega \angle 53.13°} = \mathbf{15 \text{ A} \angle -173.13°}$$

$$\mathbf{I}_{ca} = \frac{\mathbf{V}_{ca}}{\mathbf{Z}_{ca}} = \frac{150 \text{ V} \angle +120°}{10\,\Omega \angle 53.13°} = \mathbf{15 \text{ A} \angle 66.87°}$$

c. $I_L = \sqrt{3}I_\phi = (1.73)(15 \text{ A}) = 25.95 \text{ A}$. Therefore,

$$I_{Aa} = I_{Bb} = I_{Cc} = \mathbf{25.95 \text{ A}}$$

18.9 THE Δ-CONNECTED GENERATOR

If we rearrange the coils of the generator in Fig. 18.25(a) as shown in Fig. 18.25(b), the system is referred to as a **three-phase, three-wire, Δ-connected ac generator.** In this system, the phase and line voltages are equivalent and equal to the voltage induced across each coil of the generator. That is,

$$\left.\begin{array}{lll}\mathbf{E}_{AB} = \mathbf{E}_{AN} & \text{and} & e_{AN} = \sqrt{2}E_{AN} \sin \omega t \\ \mathbf{E}_{BC} = \mathbf{E}_{BN} & \text{and} & e_{BN} = \sqrt{2}E_{BN} \sin(\omega t - 120°) \\ \mathbf{E}_{CA} = \mathbf{E}_{CN} & \text{and} & e_{CN} = \sqrt{2}E_{CN} \sin(\omega t + 120°)\end{array}\right\} \begin{array}{l}\text{Phase} \\ \text{sequence} \\ ABC\end{array}$$

or

$$\boxed{\mathbf{E}_L = \mathbf{E}_{\phi g}} \tag{18.16}$$

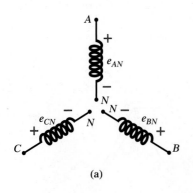

(a)

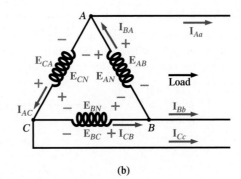

(b)

FIG. 18.25

Δ-connected generator.

Note that only one voltage (magnitude) is available instead of the two available in the Y-connected system.

Unlike the line current for the Y-connected generator, the line current for the Δ-connected system is not equal to the phase current. Through an application of Kirchhoff's current law at a node of Fig. 18.25, the following important relationship can be derived:

$$\boxed{I_L = \sqrt{3}I_{\phi g}} \tag{18.17}$$

with the phase angle between a line current and the nearest phase current at 30°. The phasor diagram of the currents is shown in Fig. 18.26.

Even though the line and phase voltages of a Δ-connected system are the same, it is standard practice to describe the phase sequence in terms of the line voltages. The method used is the same as that described for the line voltages of the Y-connected generator. For example, the phasor diagram of the line voltages for a phase sequence *ABC* is shown in Fig. 18.27. In drawing such a diagram, you must take care to have the sequence of the first and second subscripts the same. In phasor notation,

$$\mathbf{E}_{AB} = E_{AB} \angle 0°$$
$$\mathbf{E}_{BC} = E_{BC} \angle -120°$$
$$\mathbf{E}_{CA} = E_{CA} \angle 120°$$

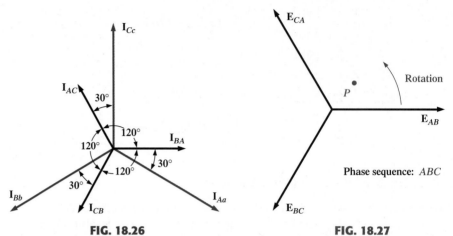

FIG. 18.26

The phasor diagram of the currents of a three-phase, Δ-connected generator.

FIG. 18.27

Determining the phase sequence for a Δ-connected, three-phase generator.

18.10 THE Δ-Δ, Δ-Y THREE-PHASE SYSTEMS

The basic equations necessary to analyze either of the two systems (Δ-Δ, Δ-Y) have been presented at least once in this chapter. Therefore, we will proceed directly to two descriptive examples, one with a Δ-connected load and one with a Y-connected load.

EXAMPLE 18.7 For the Δ-Δ system shown in Fig. 18.28:

a. Find the phase angles θ_2 and θ_3 for the specified phase sequence.
b. Find the current in each phase of the load.
c. Find the magnitude of the line currents.

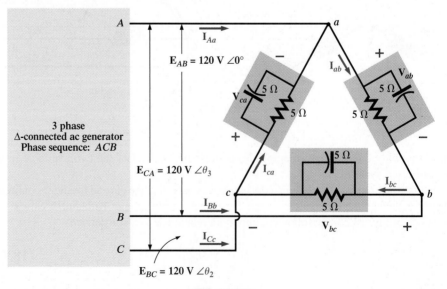

FIG. 18.28

Example 18.7: Δ-Δ system.

Solutions:

a. For an *ACB* phase sequence,

$$\theta_2 = +120° \quad \text{and} \quad \theta_3 = -120°$$

b. $\mathbf{V}_\phi = \mathbf{E}_L$. Therefore,

$$\mathbf{V}_{ab} = \mathbf{E}_{AB} \qquad \mathbf{V}_{ca} = \mathbf{E}_{CA} \qquad \mathbf{V}_{bc} = \mathbf{E}_{BC}$$

The phase currents are

$$\mathbf{I}_{ab} = \frac{\mathbf{V}_{ab}}{\mathbf{Z}_{ab}} = \frac{120 \text{ V } \angle 0°}{\dfrac{(5 \ \Omega \ \angle 0°)(5 \ \Omega \ \angle -90°)}{5 \ \Omega - j \ 5 \ \Omega}} = \frac{120 \text{ V } \angle 0°}{\dfrac{25 \ \Omega \ \angle -90°}{7.071 \ \angle -45°}}$$

$$= \frac{120 \text{ V } \angle 0°}{3.54 \ \Omega \ \angle -45°} = \textbf{33.9 A } \angle \textbf{45°}$$

$$\mathbf{I}_{bc} = \frac{\mathbf{V}_{bc}}{\mathbf{Z}_{bc}} = \frac{120 \text{ V } \angle 120°}{3.54 \ \Omega \ \angle -45°} = \textbf{33.9 A } \angle \textbf{165°}$$

$$\mathbf{I}_{ca} = \frac{\mathbf{V}_{ca}}{\mathbf{Z}_{ca}} = \frac{120 \text{ V } \angle -120°}{3.54 \ \Omega \ \angle -45°} = \textbf{33.9 A } \angle \textbf{-75°}$$

c. $I_L = \sqrt{3} I_\phi = (1.73)(34 \text{ A}) = 58.82 \text{ A}$. Therefore,

$$I_{Aa} = I_{Bb} = I_{Cc} = \textbf{58.82 A}$$

EXAMPLE 18.8   For the Δ-Y system shown in Fig. 18.29:

a. Find the voltage across each phase of the load.
b. Find the magnitude of the line voltages.

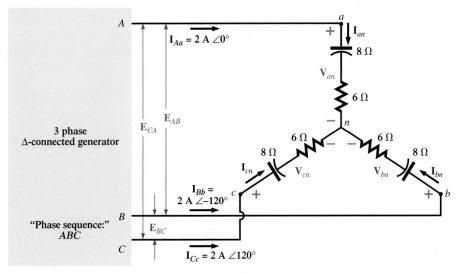

FIG. 18.29
Example 18.8: Δ-Y system.

Solutions:

a. $\mathbf{I}_{\phi L} = \mathbf{I}_L$. Therefore,

$$\mathbf{I}_{an} = \mathbf{I}_{Aa} = 2 \text{ A } \angle 0°$$
$$\mathbf{I}_{bn} = \mathbf{I}_{Bb} = 2 \text{ A } \angle -120°$$
$$\mathbf{I}_{cn} = \mathbf{I}_{Cc} = 2 \text{ A } \angle 120°$$

The phase voltages are

$$\mathbf{V}_{an} = \mathbf{I}_{an}\mathbf{Z}_{an} = (2 \text{ A } \angle 0°)(10 \ \Omega \ \angle -53.13°) = \textbf{20 V } \angle \textbf{-53.13°}$$
$$\mathbf{V}_{bn} = \mathbf{I}_{bn}\mathbf{Z}_{bn} = (2 \text{ A } \angle -120°)(10 \ \Omega \ \angle -53.13°) = \textbf{20 V } \angle \textbf{-173.13°}$$
$$\mathbf{V}_{cn} = \mathbf{I}_{cn}\mathbf{Z}_{cn} = (2 \text{ A } \angle 120°)(10 \ \Omega \ \angle -53.13°) = \textbf{20 V } \angle \textbf{66.87°}$$

b. $E_L = \sqrt{3}V_\phi = (1.73)(20\text{ V}) = 34.6\text{ V}$. Therefore,

$$E_{BA} = E_{CB} = E_{AC} = \mathbf{34.6\ V}$$

18.11 POWER

Y-Connected Balanced Load

Please refer to Fig. 18.30 for the following discussion.

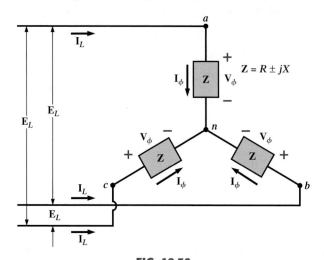

FIG. 18.30

Y-connected balanced load.

Average Power   The average power delivered to each phase can be determined by any one of Eqs. (18.18) through (18.20).

$$P_\phi = V_\phi I_\phi \cos\theta_{I_\phi}^{V_\phi} = I_\phi^2 R_\phi = \frac{V_R^2}{R_\phi} \qquad \text{(watts, W)} \qquad \textbf{(18.18)}$$

where $\theta_{I_\phi}^{V_\phi}$ indicates that θ is the phase angle between V_ϕ and I_ϕ. The total power to the balanced load is

$$P_T = 3P_\phi \qquad \text{(W)} \qquad \textbf{(18.19)}$$

or, since
$$V_\phi = \frac{E_L}{\sqrt{3}} \quad \text{and} \quad I_\phi = I_L$$

then
$$P_T = 3\frac{E_L}{\sqrt{3}}I_L \cos\theta_{I_\phi}^{V_\phi}$$

But
$$\left(\frac{3}{\sqrt{3}}\right)(1) = \left(\frac{3}{\sqrt{3}}\right)\left(\frac{\sqrt{3}}{\sqrt{3}}\right) = \frac{3\sqrt{3}}{3} = \sqrt{3}$$

Therefore,
$$P_T = \sqrt{3}E_L I_L \cos\theta_{I_\phi}^{V_\phi} = 3I_L^2 R_\phi \qquad \text{(W)} \qquad \textbf{(18.20)}$$

Reactive Power   The reactive power of each phase (in volt-ampere reactive) is

$$Q_\phi = V_\phi I_\phi \sin\theta_{I_\phi}^{V_\phi} = I_\phi^2 X_\phi = \frac{V_X^2}{X_\phi} \qquad \text{(VAR)} \qquad \textbf{(18.21)}$$

The total reactive power of the load is

$$Q_T = 3Q_\phi \qquad \text{(VAR)} \qquad \textbf{(18.22)}$$

or, proceeding in the same manner as above, we have

$$Q_T = \sqrt{3}E_L I_L \sin \theta_{I_\phi}^{V_\phi} = 3I_L^2 X_\phi \qquad \text{(VAR)} \qquad \textbf{(18.23)}$$

Apparent Power   The apparent power of each phase is

$$S_\phi = V_\phi I_\phi \qquad \text{(VA)} \qquad \textbf{(18.24)}$$

The total apparent power of the load is

$$S_T = 3S_\phi \qquad \text{(VA)} \qquad \textbf{(18.25)}$$

or, as before,

$$S_T = \sqrt{3}E_L I_L \qquad \text{(VA)} \qquad \textbf{(18.26)}$$

Power Factor   The power factor of the system is given by

$$F_p = \frac{P_T}{S_T} = \cos \theta_{I_\phi}^{V_\phi} \quad \text{(leading or lagging)} \qquad \textbf{(18.27)}$$

EXAMPLE 18.9   For the Y-connected load of Fig. 18.31:

a. Find the average power to each phase and the total load.
b. Determine the reactive power to each phase and the total reactive power.
c. Find the apparent power to each phase and the total apparent power.
d. Find the power factor of the load.

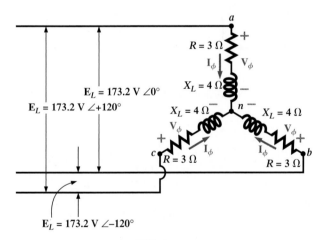

FIG. 18.31
Example 18.9.

Solutions:

a. The *average power* is

$$P_\phi = V_\phi I_\phi \cos \theta_{I_\phi}^{V_\phi} = (100\ \text{V})(20\ \text{A}) \cos 53.13° = (2000)(0.6) = \textbf{1200 W}$$
$$P_\phi = I_\phi^2 R_\phi = (20\ \text{A})^2 (3\ \Omega) = (400)(3) = \textbf{1200 W}$$
$$P_\phi = \frac{V_R^2}{R_\phi} = \frac{(60\ \text{V})^2}{3\ \Omega} = \frac{3600}{3} = \textbf{1200 W}$$
$$P_T = 3P_\phi = (3)(1200\ \text{W}) = \textbf{3600 W}$$

or $\quad P_T = \sqrt{3}E_L I_L \cos \theta_{I_\phi}^{V_\phi} = (1.732)(173.2\ \text{V})(20\ \text{A})(0.6) = \textbf{3600 W}$

b. The *reactive power* is

$$Q_\phi = V_\phi I_\phi \sin \theta_{I_\phi}^{V_\phi} = (100 \text{ V})(20 \text{ A}) \sin 53.13° = (2000)(0.8) = \textbf{1600 VAR}$$

or $\quad Q_\phi = I_\phi^2 X_\phi = (20 \text{ A})^2 (4 \text{ }\Omega) = (400)(4) = \textbf{1600 VAR}$

$$Q_T = 3Q_\phi = (3)(1600 \text{ VAR}) = \textbf{4800 VAR}$$

or $\quad Q_T = \sqrt{3}E_L I_L \sin \theta_{I_\phi}^{V_\phi} = (1.732)(173.2 \text{ V})(20 \text{ A})(0.8) = \textbf{4800 VAR}$

c. The *apparent power* is

$$S_\phi = V_\phi I_\phi = (100 \text{ V})(20 \text{ A}) = \textbf{2000 VA}$$

$$S_T = 3S_\phi = (3)(2000 \text{ VA}) = \textbf{6000 VA}$$

or $\quad\quad S_T = \sqrt{3}E_L I_L = (1.732)(173.2 \text{ V})(20 \text{ A}) = \textbf{6000 VA}$

d. The *power factor* is

$$F_p = \frac{P_T}{S_T} = \frac{3600 \text{ W}}{6000 \text{ VA}} = \textbf{0.6 lagging}$$

Δ-Connected Balanced Load

Please refer to Fig. 18.32 for the following discussion.

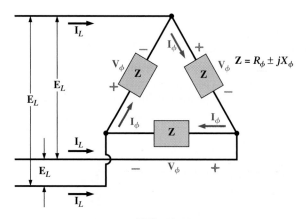

FIG. 18.32
Δ-connected balanced load.

Average Power

$$P_\phi = V_\phi I_\phi \cos \theta_{I_\phi}^{V_\phi} = I_\phi^2 R_\phi = \frac{V_R^2}{R_\phi} \quad \text{(W)} \tag{18.28}$$

$$P_T = 3P_\phi \quad \text{(W)} \tag{18.29}$$

Reactive Power

$$Q_\phi = V_\phi I_\phi \sin \theta_{I_\phi}^{V_\phi} = I_\phi^2 X_\phi = \frac{V_X^2}{X_\phi} \quad \text{(VAR)} \tag{18.30}$$

$$Q_T = 3Q_\phi \quad \text{(VAR)} \tag{18.31}$$

Apparent Power

$$S_\phi = V_\phi I_\phi \quad \text{(VA)} \tag{18.32}$$

$$S_T = 3S_\phi = \sqrt{3}E_L I_L \quad \text{(VA)} \qquad \textbf{(18.33)}$$

Power Factor

$$F_p = \frac{P_T}{S_T} \qquad \textbf{(18.34)}$$

EXAMPLE 18.10 For the Δ-Y connected load of Fig. 18.33, find the total average, reactive, and apparent power. In addition, find the power factor of the load.

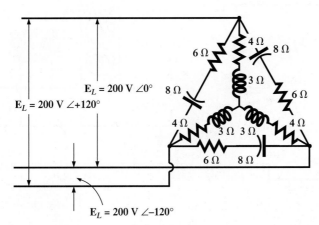

$\mathbf{E}_L = 200\text{ V }\angle 0°$
$\mathbf{E}_L = 200\text{ V }\angle +120°$
$\mathbf{E}_L = 200\text{ V }\angle -120°$

FIG. 18.33
Example 18.10.

Solution: Consider the Δ and Y separately.

For the Δ:

$$\mathbf{Z}_\Delta = 6\ \Omega - j\,8\ \Omega = 10\ \Omega\ \angle -53.13°$$

$$I_\phi = \frac{E_L}{Z_\Delta} = \frac{200\text{ V}}{10\ \Omega} = 20\text{ A}$$

$$P_{T_\Delta} = 3I_\phi^2 R_\phi = (3)(20\text{ A})^2(6\ \Omega) = \textbf{7200 W}$$
$$Q_{T_\Delta} = 3I_\phi^2 X_\phi = (3)(20\text{ A})^2(8\ \Omega) = \textbf{9600 VAR }(C)$$
$$S_{T_\Delta} = 3V_\phi I_\phi = (3)(200\text{ V})(20\text{ A}) = \textbf{12,000 VA}$$

For the Y:

$$\mathbf{Z}_Y = 4\ \Omega + j\,3\ \Omega = 5\ \Omega\ \angle 36.87°$$

$$I_\phi = \frac{E_L/\sqrt{3}}{Z_Y} = \frac{200\text{ V}/\sqrt{3}}{5\ \Omega} = \frac{116\text{ V}}{5\ \Omega} = 23.12\text{ A}$$

$$P_{T_Y} = 3I_\phi^2 R_\phi = (3)(23.12\text{ A})^2(4\ \Omega) = \textbf{6414.41 W}$$
$$Q_{T_Y} = 3I_\phi^2 X_\phi = (3)(23.12\text{ A})^2(3\ \Omega) = \textbf{4810.81 VAR}(L)$$
$$S_{T_Y} = 3V_\phi I_\phi = (3)(116\text{ V})(23.12\text{ A}) = \textbf{8045.76 VA}$$

For the total load:

$$P_T = P_{T_\Delta} + P_{T_Y} = 7200\text{ W} + 6414.41\text{ W} = \textbf{13,614.41 W}$$
$$Q_T = Q_{T_\Delta} - Q_{T_Y} = 9600\text{ VAR }(C) - 4810.81\text{ VAR }(I) = \textbf{4789.19 VAR }(C)$$
$$S_T = \sqrt{P_T^2 + Q_T^2} = \sqrt{(13,614.41\text{ W})^2 + (4789.19\text{ VAR})^2} = \textbf{14,432.2 VA}$$
$$F_p = \frac{P_T}{S_T} = \frac{13,614.41\text{ W}}{14,432.20\text{ VA}} = \textbf{0.943 leading}$$

EXAMPLE 18.11 For the residential supply appearing in Fig. 18.34, determine (assuming a totally resistive load) the following:

a. the value of R to ensure a balanced load
b. the magnitude of I_1 and I_2
c. the line voltage V_L
d. the total power delivered
e. the turns ratio $a = N_p/N_s$

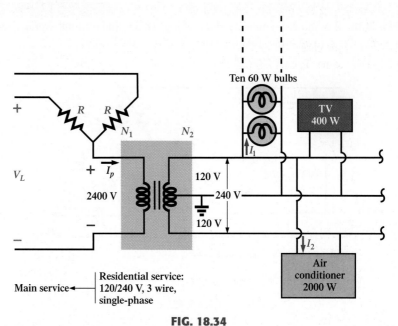

FIG. 18.34

Single-phase residential supply (Example 18.11).

Solutions:

a. $P_T = (10)(60 \text{ W}) + 400 \text{ W} + 2000 \text{ W} = 600 \text{ W} + 400 \text{ W} + 2000 \text{ W} = 3000 \text{ W}$

$P_{\text{in}} = P_{\text{out}}$

$V_p I_p = V_s I_s = 3000 \text{ W}$ (purely resistive load)

$(2400 \text{ V})I_p = 3000 \text{ W}$ and $I_p = 1.25 \text{ A}$

$R = \dfrac{V_\phi}{I_p} = \dfrac{2400 \text{ V}}{1.25 \text{ A}} = \mathbf{1920 \ \Omega}$

b. $P_1 = 600 \text{ W} = VI_1 = (120 \text{ V})I_1$

and $I_1 = \mathbf{5 \ A}$

$P_2 = 2000 \text{ W} = VI_2 = (240 \text{ V})I_2$

and $I_2 = \mathbf{8.33 \ A}$

c. $V_L = \sqrt{3}V_\phi = 1.73(2400 \text{ V}) = \mathbf{4152 \ V}$

d. $P_T = 3P_\phi = 3(3000 \text{ W}) = \mathbf{9 \ kW}$

e. $a = \dfrac{N_p}{N_s} = \dfrac{V_p}{V_s} = \dfrac{2400 \text{ V}}{240 \text{ V}} = \mathbf{10}$

18.12 POWER AND PHASE-SEQUENCE MEASUREMENTS

The Three-Wattmeter Method

The power delivered to a balanced or an unbalanced four-wire, Y-connected load can be found by the **three-wattmeter method,** that is, by using three wattmeters in the manner shown in Fig. 18.35. Each wattmeter measures the power delivered to each phase. The potential coil of each wattmeter is connected in parallel with the load, while the current coil is in series with the load. The total average power of the system can be found by summing the three wattmeter readings. That is,

$$P_{T_Y} = P_1 + P_2 + P_3 \qquad \textbf{(18.35)}$$

For the load (balanced or unbalanced), the wattmeters are connected as shown in Fig. 18.36. The total power is again the sum of the three wattmeter readings:

$$P_{T_\Delta} = P_1 + P_2 + P_3 \qquad \textbf{(18.36)}$$

If, in either of the cases just described, the load is balanced, the power delivered to each phase will be the same. The total power is then just three times any one wattmeter reading.

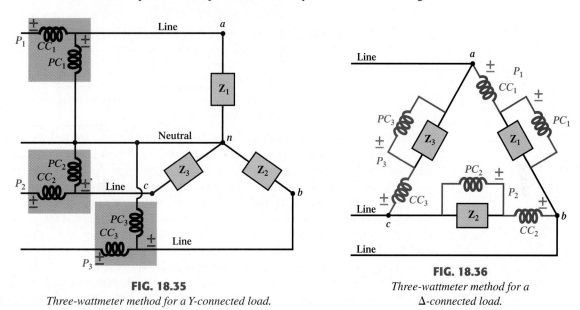

FIG. 18.35

Three-wattmeter method for a Y-connected load.

FIG. 18.36

*Three-wattmeter method for a
Δ-connected load.*

The Two-Wattmeter Method

The power delivered to a three-phase, three-wire, Δ- or Y-connected, balanced or unbalanced load can be found using only two wattmeters if the proper connection is employed and if the wattmeter readings are interpreted properly. The basic connections of this **two-wattmeter method** are shown in Fig. 18.37. One end of each potential coil is connected to the same line. The current coils are then placed in the remaining lines.

The connection shown in Fig. 18.38 will also satisfy the requirements. A third hookup is also possible, but this is left to the reader as an exercise.

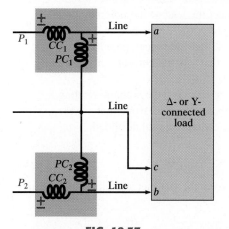

FIG. 18.37

*Two-wattmeter method for a Δ- or
a Y-connected load.*

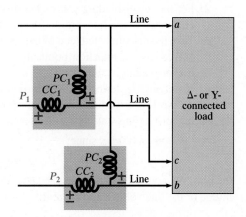

FIG. 18.38

*Alternative hookup for the
two-wattmeter method.*

The total power delivered to the load is the algebraic sum of the two wattmeter readings. For a *balanced* load, we will now consider two methods of determining whether the total power is the sum or the difference of the two wattmeter readings. The first method to be described requires that we know or be able to find the power factor (leading or lagging) of any one phase of the load. When this information has been obtained, it can be applied directly to the curve of Fig. 18.39.

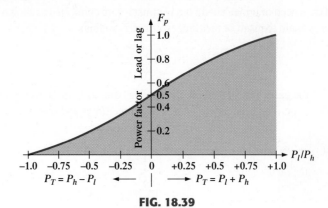

FIG. 18.39

Determining whether the readings obtained using the two-wattmeter method should be added or subtracted.

The curve in Fig. 18.39 is a plot of the power factor of the load (phase) versus the ratio P_l/P_h, where P_l and P_h are the magnitudes of the lower- and higher-reading wattmeters, respectively. Note that for a power factor (leading or lagging) greater than 0.5, the ratio has a positive value. This indicates that both wattmeters are reading positive, and the total power is the sum of the two wattmeter readings. That is, $P_T = P_l + P_h$. For a power factor less than 0.5 (leading or lagging), the ratio has a negative value. This indicates that the lower-reading wattmeter is reading negative, and the total power is the difference of the two wattmeter readings. That is, $P_T = P_h - P_l$.

A closer examination will reveal that when the power factor is 1 ($\cos 0° = 1$), corresponding to a purely resistive load, $P_l/P_h = 1$ or $P_l = P_h$, and both wattmeters will have the same wattage indication. At a power factor equal to 0 ($\cos 90° = 0$), corresponding to a purely reactive load, $P_l/P_h = -1$ or $P_l = -P_h$, and both wattmeters will again have the same wattage indication but with opposite signs. The transition from a negative to a positive ratio occurs when the power factor of the load is 0.5 or $\theta = \cos^{-1} 0.5 = 60°$. At this power factor, $P_l/P_h = 0$, so that $P_l = 0$, while P_h will read the total power delivered to the load.

The second method for determining whether the total power is the sum or the difference of the two wattmeter readings involves a simple laboratory test. For the test to be applied, both wattmeters must first have an up-scale deflection. If one of the wattmeters has a below-zero indication, an up-scale deflection can be obtained by simply reversing the leads of the current coil of the wattmeter. To perform the test:

1. Take notice of which line does not have a current coil sensing the line current.
2. For the lower-reading wattmeter, disconnect the lead of the potential coil connected to the line without the current coil.
3. Take the disconnected lead of the lower-reading wattmeter's potential coil, and touch a connection point on the line that has the current coil of the higher-reading wattmeter.
4. If the pointer deflects downward (below zero watts), the wattage reading of the lower-reading wattmeter should be subtracted from that of the higher-reading wattmeter. Otherwise, the readings should be added.

For a *balanced system*, since

$$P_T = P_h \pm P_l = \sqrt{3}E_L I_L \cos \theta_{I_\phi}^{V_\phi}$$

the power factor of the load (phase) can be found from the wattmeter readings and the magnitude of the line voltage and current:

$$\boxed{F_p = \cos \theta_{I_\phi}^{V_\phi} = \frac{P_h \pm P_l}{\sqrt{3}E_L I_L}} \qquad \textbf{(18.37)}$$

EXAMPLE 18.12 For the unbalanced Δ-connected load of Fig. 18.40 with two properly connected wattmeters:

a. Determine the magnitude and angle of the phase currents.
b. Calculate the magnitude and angle of the line currents.
c. Determine the power reading of each wattmeter.
d. Calculate the total power absorbed by the load.
e. Compare the result of part (d) with the total power calculated using the phase currents and the resistive elements.

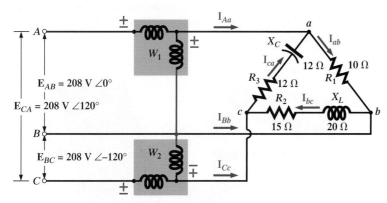

FIG. 18.40
Example 18.12.

Solutions:

a. $\mathbf{I}_{ab} = \dfrac{\mathbf{V}_{ab}}{\mathbf{Z}_{ab}} = \dfrac{\mathbf{E}_{AB}}{\mathbf{Z}_{ab}} = \dfrac{208 \text{ V } \angle 0°}{10 \text{ Ω } \angle 0°} = \mathbf{20.8 \text{ A } \angle 0°}$

$\mathbf{I}_{bc} = \dfrac{\mathbf{V}_{bc}}{\mathbf{Z}_{bc}} = \dfrac{\mathbf{E}_{BC}}{\mathbf{Z}_{bc}} = \dfrac{208 \text{ V } \angle -120°}{15 \text{ Ω } + j\,20 \text{ Ω}} = \dfrac{208 \text{ V } \angle -120°}{25 \text{ Ω } \angle 53.13°} = \mathbf{8.32 \text{ A } \angle -173.13°}$

$\mathbf{I}_{ca} = \dfrac{\mathbf{V}_{ca}}{\mathbf{Z}_{ca}} = \dfrac{\mathbf{E}_{CA}}{\mathbf{Z}_{ca}} = \dfrac{208 \text{ V } \angle +120°}{12 \text{ Ω } + j\,12 \text{ Ω}} = \dfrac{208 \text{ V } \angle +120°}{16.97 \text{ Ω } \angle -45°} = \mathbf{12.26 \text{ A } \angle 165°}$

b. $\mathbf{I}_{Aa} = \mathbf{I}_{ab} - \mathbf{I}_{ca} = 20.8 \text{ A } \angle 0° - 12.26 \text{ A } \angle 165°$

$\quad = 20.8 \text{ A } - (-11.84 \text{ A } + j\,3.17 \text{ A})$

$\quad = 20.8 \text{ A } + 11.84 \text{ A } - j\,3.17 \text{ A } = 32.64 \text{ A } - j\,3.17 \text{ A}$

$\quad = \mathbf{32.79 \text{ A } \angle -5.55°}$

$\mathbf{I}_{Bb} = \mathbf{I}_{bc} - \mathbf{I}_{ab} = 8.32 \text{ A } \angle -173.13° - 20.8 \text{ A } \angle 0°$

$\quad = (-8.26 \text{ A } - j\,1 \text{ A}) - 20.8 \text{ A}$

$\quad = -8.26 \text{ A } - 20.8 \text{ A } - j\,1 \text{ A } = -29.06 \text{ A } - j\,1 \text{ A}$

$\quad = \mathbf{29.08 \text{ A } \angle -178.03°}$

$\mathbf{I}_{Cc} = \mathbf{I}_{ca} - \mathbf{I}_{bc} = 12.26 \text{ A } \angle 165° - 8.32 \text{ A } \angle -173.13°$

$\quad = (-11.84 \text{ A } + j\,3.17 \text{ A}) - (-8.26 \text{ A } - j\,1 \text{ A})$

$\quad = -11.84 \text{ A } + 8.26 \text{ A } + j\,(3.17 \text{ A } + 1 \text{ A}) = -3.58 \text{ A } + j\,4.17 \text{ A}$

$\quad = \mathbf{5.5 \text{ A } \angle 130.65°}$

c. $P_1 = V_{ab} I_{Aa} \cos \theta^{\mathbf{V}_{ab}}_{\mathbf{I}_{Aa}} \qquad \mathbf{V}_{ab} = 208 \text{ V } \angle 0° \qquad \mathbf{I}_{Aa} = 32.79 \text{ A } \angle -5.55°$

$P_1 = (208 \text{ V})(32.79 \text{ A}) \cos 5.55° = \mathbf{6788.35 \text{ W}}$

$\mathbf{V}_{bc} = \mathbf{E}_{BC} = 208 \text{ V } \angle -120°$

However, $\qquad \mathbf{V}_{cb} = \mathbf{E}_{CB} = 208 \text{ V } \angle -120° + 180° = 208 \text{ V } \angle 60°$

with $\qquad\qquad\qquad \mathbf{I}_{Cc} = 5.5 \text{ A } \angle 130.65°$

Therefore, $\quad P_2 = V_{cb} I_{Cc} \cos \theta^{\mathbf{V}_{ab}}_{\mathbf{I}_{Cc}} = (208 \text{ V})(5.5 \text{ A}) \cos 70.65° = \mathbf{379.1 \text{ W}}$

d. $P_T = P_1 + P_2 = 6788.35 \text{ W } + 379.1 \text{ W } = \mathbf{7167.45 \text{ W}}$

e. $P_T = (I_{ab})^2 R_1 + (I_{bc})^2 R_2 + (I_{ca})^2 R_3$
$= (20.8 \text{ A})^2 10 \ \Omega + (8.32 \text{ A})^2 15 \ \Omega + (12.26 \text{ A})^2 12 \ \Omega$
$= 4326.4 \text{ W} + 1038.34 \text{ W} + 1803.69 \text{ W} = \mathbf{7168.43 \text{ W}}$

(The slight difference is due to the level of accuracy carried through the calculations.)

Phase-Sequence Indicator

The phase-sequence indicator of Fig. 18.41 will indicate the phase sequence when hooked up to the supply lines. The brightest bulb will indicate the phase sequence. The phase sequence can be quite important because, among other things, it will affect the direction of rotation of some ac motors.

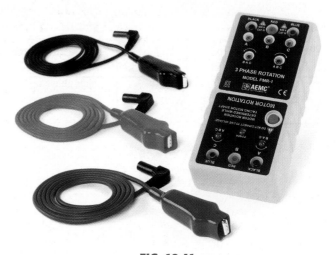

FIG. 18.41

Phase and motor rotation tester.
(Courtesy of AEMC® Instruments, Foxborough, MA.)

18.13 UNBALANCED, THREE-PHASE, FOUR-WIRE, Y-CONNECTED LOAD

For the three-phase, four-wire, Y-connected load of Fig. 18.42, conditions are such that *none* of the load impedances are equal—hence we have an **unbalanced polyphase load.** Since the neutral is a common point between the load and source, no matter what the

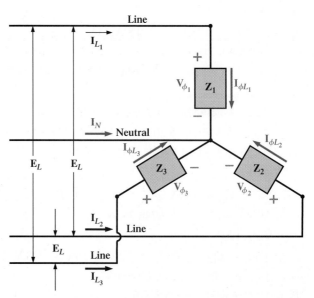

FIG. 18.42
Unbalanced Y-connected load.

impedance of each phase of the load and source, the voltage across each phase is the phase voltage of the generator:

$$\mathbf{V}_\phi = \mathbf{E}_\phi \qquad (18.38)$$

The phase currents can therefore be determined by Ohm's law:

$$\mathbf{I}_{\phi_1} = \frac{\mathbf{V}_{\phi_1}}{\mathbf{Z}_1} = \frac{\mathbf{E}_{\phi_1}}{\mathbf{Z}_1} \quad \text{and so on} \qquad (18.39)$$

The current in the neutral for any unbalanced system can then be found by applying Kirchhoff's current law at the common point n:

$$\mathbf{I}_N = \mathbf{I}_{\phi_1} + \mathbf{I}_{\phi_2} + \mathbf{I}_{\phi_3} = \mathbf{I}_{L_1} + \mathbf{I}_{L_2} + \mathbf{I}_{L_3} \qquad (18.40)$$

Because of the variety of equipment found in an industrial environment, both three-phase power and single-phase power are usually provided, with the single-phase obtained from the three-phase system. In addition, since the load on each phase is continually changing, a four-wire system (with a neutral) is normally employed to ensure steady voltage levels and to provide a path for the current resulting from an unbalanced load.

The system of Fig. 18.43 has a three-phase transformer dropping the line voltage from 13,800 V to 208 V. All the lower-power-demand loads, such as lighting, wall outlets, security, and so on, use the single-phase, 120 V line to neutral voltage. Higher power loads, such as air conditioners, electric ovens or dryers, and so on, use the single-phase, 208 V available from line to line. For larger motors and special high-demand equipment, the full three-phase power can be taken directly from the system, as shown in Fig. 18.43. In the design and construction of a commercial establishment, the National Electric Code requires that every effort be made to ensure that the expected loads, whether they be single- or multiphase, result in a total load that is as balanced as possible between the phases, thus ensuring the highest level of transmission efficiency.

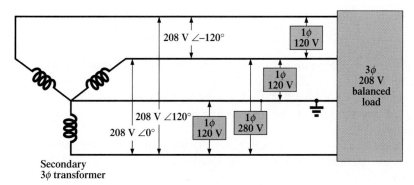

FIG. 18.43

Three-phase/one-phase, 208 V/120 V industrial supply.

18.14 APPLICATIONS

Low-Voltage Compensation

At times during the year, peak demands from the power company can result in a reduced voltage down the line. In mid-summer, for example, the line voltage may drop from 120 V to 100 V because of the heavy load often due primarily to air conditioners. However, air conditioners do not run as well under low-voltage conditions, so the following option using an autotransformer may be the solution.

In Fig. 18.44(a), an air conditioner drawing 10 A at 120 V is connected through an autotransformer to the available supply that has dropped to 100 V. Assuming 100% efficiency,

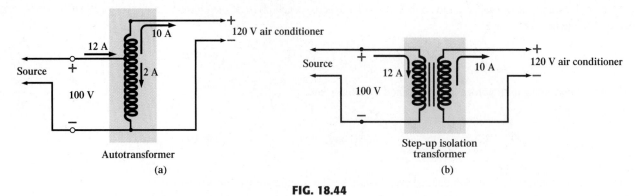

FIG. 18.44

Maintaining a 120 V supply for an air conditioner:
(a) using an autotransformer; (b) using a traditional step-up transformer.

the current drawn from the line would have to be 12 A to ensure that $P_i = P_o = 1200$ W. Using the analysis introduced in Section 18.2, we will find that the current in the primary winding is 2 A with 10 A in the secondary. The 12 A will exist only in the line connecting the source to the primary. If the voltage level were increased using the traditional step-up transformer shown in Fig. 18.44(b), the same currents would result at the source and load. However, note that the current through the primary is now 12 A which is 6 times that in the autotransformer. The result is that the winding in the autotransformer can be much thinner due to the significantly lower current level.

Let us now examine the turns ratio required and the number of turns involved for each setup (associating one turn with each volt of the primary and secondary).

For the autotransformer:

$$\frac{N_s}{N_p} = \frac{V_s}{V_p} = \frac{10 \text{ V}}{100 \text{ V}} \Rightarrow \frac{10 \text{ t}}{100 \text{ t}}$$

For the traditional transformer:

$$\frac{N_s}{N_p} = \frac{V_s}{V_p} = \frac{120 \text{ V}}{100 \text{ V}} \Rightarrow \frac{120 \text{ t}}{100 \text{ t}}$$

In total, therefore, the autotransformer has only 10 turns in the secondary, whereas the traditional has 120. For the autotransformer we need only 10 turns of heavy wire to handle the current of 10 A, not the full 120 required for the traditional transformer. In addition, the total number of turns for the autotransformer is 110 compared to 220 for the traditional transformer.

The net result of all the above is that even though the protection offered by the isolation feature is lost, the autotransformer can be much smaller in size and weight and, therefore, less costly.

Ballast Transformer

Until just recently, all fluorescent lights such as appearing in Fig. 18.45(a) had a ballast transformer as shown in Fig. 18.45(b). In many cases, the weight of the ballast alone is almost equal to that of the fixture itself. In recent years a solid-state equivalent transformer has been developed that in time may replace most of the ballast transformers. However, for now and the near future, because of the additional cost associated with the solid-state variety, the ballast transformer will appear in most fluorescent bulbs.

The basic connections for a single-bulb fluorescent light are provided in Fig. 18.46(a). Note that the transformer is connected as an autotransformer with the full applied 120 V across the primary. When the switch is closed, the applied voltage and the voltage across the secondary will add and establish a current through the filaments of the fluorescent bulb. The starter is initially a short circuit to establish the continuous path through the two filaments. In early fluorescent bulbs, the starter was a cylinder with two contacts, as shown in

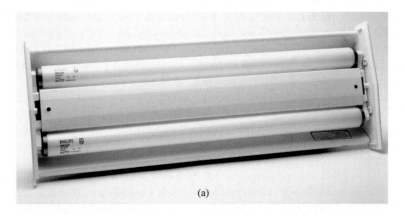

(a)

(b)

FIG. 18.45

Fluorescent lamp: (a) general appearance; (b) internal view with ballast.

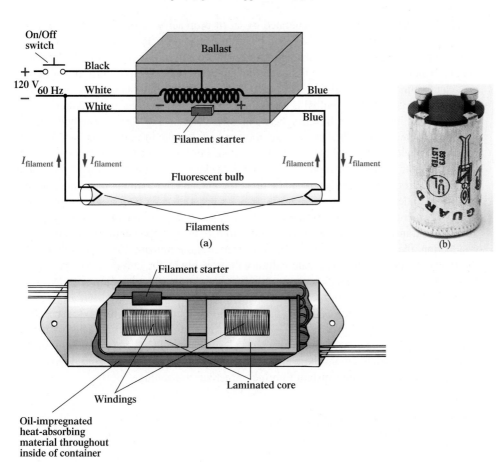

(a)

(b)

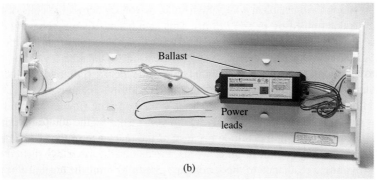

(c)

FIG. 18.46

(a) Schematic of single-bulb fluorescent lamp; (b) starter; (c) internal view of ballast transformer.

Fig. 18.46(b), which had to be replaced on occasion. It sat right under the fluorescent bulb near one of the bulb connections. Now, as shown by the sketch of the inside of a ballast transformer in Fig. 18.46(c), the starter is commonly built into the ballast and can no longer be replaced. The voltage established by the autotransformer action is sufficient to heat the filaments but not light the fluorescent bulb.

The fluorescent lamp is a long tube with a coating of fluorescent paint on the inside. It is filled with an inert gas and a small amount of liquid mercury. The distance between the electrodes at the ends of the lamp is too much for the applied autotransformer voltage to establish conduction. To overcome this problem, the filaments are first heated as described above to convert the mercury (a good conductor) from a liquid to a gas. Conduction can then be established by the application of a large potential across the electrodes. This potential is established when the starter (a thermal switch that opens when it reaches a particular temperature) opens and causes the inductor current to drop from its established level to zero amperes. This quick drop in current will establish a very high spike in voltage across the coils of the autotransformer as determined by $v_L = L(di_L/dt)$. This significant spike in voltage will also appear across the bulb and will establish current between the electrodes. Light will then be given off as the electrons hit the fluorescent surface on the inside of the tube.

It is the persistence of the coating that helps hide the oscillation in conduction level due to the low-frequency (60 Hz) power that could result in a flickering light. The starter will remain open until the next time the bulb is turned on. The flow of charge between electrodes will then be maintained solely by the voltage across the autotransformer. This current is relatively low in magnitude because of the reactance of the secondary winding in the resulting series circuit. In other words, the autotransformer has shifted to one that is now providing a reactance to the secondary circuit to limit the current through the bulb. Without this limiting factor, the current through the bulb would be too high, and the bulb would quickly burn out. This action of the coils of the transformer generating the required voltage and then acting as a coil to limit the current has resulted in the general terminology of *swinging choke*.

The fact that the light is not generated by an *IR* drop across a filament of a bulb is the reason fluorescent lights are so energy efficient. In fact, in an incandescent bulb, about 75% of the applied energy is lost in heat, with only 25% going to light emission. In a fluorescent bulb more than 70% goes to light emission and 30% to heat losses. As a rule of thumb, you can assume that the lighting from a 40 W fluorescent lamp [such as the unit of Fig. 18.45(a) with its two 20 W bulbs] is equivalent to that of a 100 W incandescent bulb.

One other interesting difference between incandescent and fluorescent bulbs is the method of determining whether they are good or bad. For the incandescent light, it is immediately obvious when it fails to give light at all. For the fluorescent bulb, however, assuming that the ballast is in good working order, the bulb will begin to dim as its life wears on. The electrodes will get coated and be less efficient, and the coating on the inner surface will begin to deteriorate.

Rapid-start fluorescent lamps are different in operation only in that the voltage generated by the transformer is sufficiently large to atomize the gas upon application and initiate conduction, thereby removing the need for a starter and eliminating the warm-up time of the filaments. In time the solid-state ballast will probably be the unit of choice because of its quick response, higher efficiency, and lighter weight, but the transition will take some time. The basic operation will remain the same, however.

Because of the fluorine gas (hence the name *fluorescent* bulb) and the mercury in fluorescent lamps, they must be discarded with care. Ask your local disposal facility where to take the bulbs. Breaking them for insertion in a plastic bag could be a very dangerous proposition. If you happen to break a bulb and get cut in the process, be sure to go right to a medical facility since you could sustain fluorine or mercury poisoning.

18.15 COMPUTER ANALYSIS

PSpice

Transformer (Controlled Sources)   The simple transformer configuration of Fig. 18.47 will now be investigated using controlled sources to mimic the behavior of the transformer as defined by its basic voltage and current relationships.

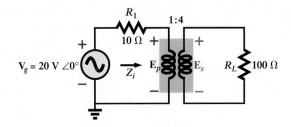

FIG. 18.47

Applying PSpice to a step-up transformer.

For comparison purposes, a theoretical solution of the network would yield the following:

$$Z_i = a^2 Z_L = \left(\frac{1}{4}\right)^2 100\ \Omega = 6.25\ \Omega$$

and

$$E_p = \frac{(6.25\ \Omega)(20\ \text{V})}{6.25\ \Omega + 10\ \Omega} = 7.692\ \text{V}$$

with

$$E_s = \frac{1}{a} E_p = \frac{1}{(1/4)}(7.692\ \text{V}) = 4(7.692\ \text{V}) = 30.77\ \text{V}$$

and

$$V_L = E_s = \mathbf{30.77\ V}$$

For the ideal transformer, the secondary voltage is defined by $E_s = N_s/N_p(E_p)$ which is $E_s = 4E_p$ for the network of Fig. 18.47. The fact that the magnitude of one voltage is controlled by another requires that we use the **Voltage-Controlled Voltage Source (VCVS)** source in the **ANALOG** library. It appears as **E** in the **Parts List** and has the format appearing in Fig. 18.48. The sensing voltage is **E1,** and the controlled voltage appears across the two terminals of the circular symbol for a voltage source. Double-clicking on the source symbol will permit setting the **GAIN** to 4 for this example. Note in Fig. 18.48 that the sensing voltage is the primary voltage of the circuit of Fig. 18.47, and the output voltage is connected directly to the load resistor **RL.** There is no real problem making the necessary connections because of the format of the **E** source.

The next step is to set up the current relationship for the transformer. Since the magnitude of one current will be controlled by the magnitude of another current in the same configuration, a **Current-Controlled Current Source (CCCS)** must be employed. It also

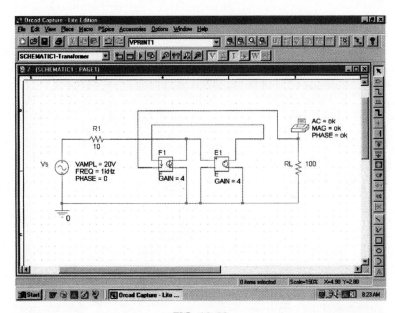

FIG. 18.48

*Using PSpice to determine the magnitude and phase angle for the
load voltage of the network of Fig. 18.47.*

appears in the **ANALOG** library under the **Part List** as **F** and has the format appearing in Fig. 18.48. Note that both currents have a direction associated with them. For the ideal transformer, $I_p = N_s/N_p(I_s)$ which is $I_p = 4I_s$ for the network of Fig. 18.47. The gain for the part can be set using the same procedure defined for the **E** source. Since the secondary current will be the controlling current, its level must be fed into the **F** source in the same direction as indicated in the controlled source. When making this connection, be sure to click the wire in location before crossing the wire of the primary circuit, and then click it again after crossing the wire. If you do this properly, a connection point indicated by a small red dot will not appear. The controlled current I_{R_1} can be connected as shown because the connection **E1** is only sensing a voltage level, essentially has infinite impedance, and can be looked upon as an open circuit. In other words, the current through **R1** will be the same as through the controlled source of **F.**

A simulation was set up with **AC Sweep** and 1 kHz for the **Start** and **End Frequencies.** One data point per decade was selected, and the simulation was initiated. After the **SCHEMATIC1** screen appeared, the window was exited, and **PSpice-View Output File** was selected to result in the **AC ANALYSIS** solution of Fig. 18.49. Note that the voltage is 30.77 V, which is an exact match with the theoretical solution.

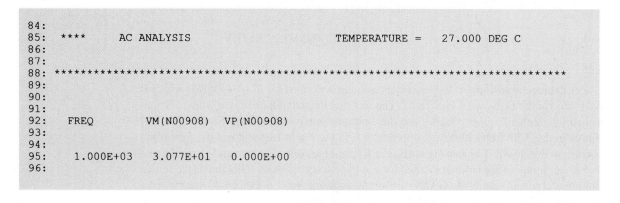

```
84:
85:   ****      AC ANALYSIS                        TEMPERATURE =   27.000 DEG C
86:
87:
88:   ******************************************************************************
89:
90:
91:
92:   FREQ          VM(N00908)   VP(N00908)
93:
94:
95:    1.000E+03    3.077E+01    0.000E+00
96:
```

FIG. 18.49
The output file for the analysis indicated in Fig. 18.48.

Transformer (Library)   The same network can be analyzed by choosing one of the transformers from the **EVAL** library as shown in Fig. 18.50. The transformer labeled **K3019PL_3C8** was chosen, and the proper attributes were placed in the **Property Editor** dialog box.

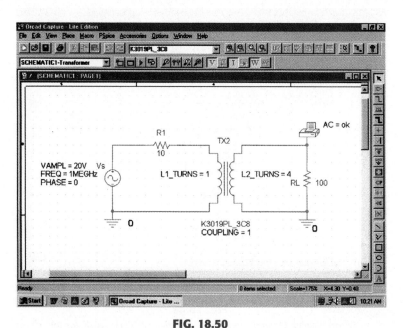

FIG. 18.50
*Using a transformer provided in the **EVAL** library to analyze the network of Fig. 18.47.*

The only three required attributes were **COUPLING** set at 1, **L1_TURNS** set at 1, and **L2_TURNS** set at 4. In the **Simulation Settings, AC Sweep** was chosen and **1MEGHz** used for both the **Start and End Frequency** because it was found that it acted as an almost ideal transformer at this frequency—a little bit of run and test. When the simulation was run, the results under **PSpice-View Output File** appeared as shown in Fig. 18.51—almost an exact match with the theoretical solution of 30.77 V.

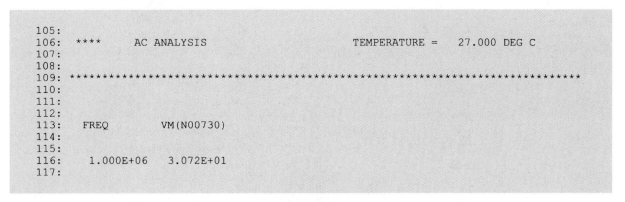

FIG. 18.51

The output file for the analysis indicated in Fig. 18.50.

CHAPTER SUMMARY
IRON-CORE TRANSFORMER

- The core of a transformer is made of laminated sheets of ferromagnetic material such as iron and steel.
- The voltages of an ideal transformer are related by the same ratio as the respective windings, whereas the currents are related by the inverse of the respective windings.
- The reflected impedance at the primary of an iron-core transformer is the transformation ratio squared times the applied load.
- For the ideal transformer, the applied power equals the power to the load.
- Frequency is an important part of the nameplate data of a transformer because it affects the line current drawn at the primary to establish the flux in the core.
- Using a two-circuit transformer as an autotransformer will result in a loss of isolation between the two circuits, but it will provide a higher kVA level.

THE THREE-PHASE SYSTEM

- The three phase voltages of a three-phase generator have the same peak value but are out of phase by 120°.
- For a Y-connected generator, the line current is the same as the phase current, but the line voltage is $\sqrt{3}$ times the phase voltage.
- For a Y-Y connected system, the phase and line currents are all the same, whereas all the line voltages are $\sqrt{3}$ times the phase voltages. The phase voltages of the load are the same as the phase voltages of the generator.
- For a Δ-connected load, the line and phase voltages are the same, but the line current is $\sqrt{3}$ times the phase current.
- The total power dissipated is three times the power dissipated in each phase of the load. The total reactive power is also three times the reactive power of each phase. The total apparent power is three times the apparent power of each phase.
- The total power to a three-phase system can be measured with two or three wattmeters. If only two are used, you must be aware of the power factor of the load to determine whether the two readings are added or subtracted.

Important Equations

Transformer:

$$\frac{\mathbf{E}_p}{\mathbf{E}_s} = \frac{N_p}{N_s} = a$$

$$\frac{\mathbf{I}_p}{\mathbf{I}_s} = \frac{N_s}{N_p} = \frac{1}{a}$$

$$\mathbf{Z}_p = a^2 \mathbf{Z}_s$$

$$P_{\text{in}} = P_{\text{out}} \qquad E_p I_p = E_s I_s$$

Three-phase systems:
Y-Y:

$$I_L = I_{\phi_g} = I_{\phi_L}$$

$$E_L = \sqrt{3} E_\phi = \sqrt{3} V_{\phi_L}$$

Δ-Δ:

$$I_L = \sqrt{3} I_{\phi_g} = \sqrt{3} I_{\phi_L}$$

$$E_L = V_{\phi_g} = V_{\phi_L}$$

Power:

Y-connected load:

$$P_T = 3P_\phi = \sqrt{3} E_L I_L \cos \theta_{I_\phi}^{V_\phi} = 3I_L^2 R_\phi$$

$$Q_T = 3Q_\phi = \sqrt{3} E_L I_L \sin \theta_{I_\phi}^{V_\phi} = 3I_L^2 X_\phi$$

$$S_T = 3S_\phi = \sqrt{3} E_L I_L$$

$$F_p = \frac{P_T}{S_T}$$

Δ-connected load:

$$P_T = 3P_\phi = 3I_\phi^2 R_\phi$$

$$Q_T = 3Q_\phi = 3I_\phi^2 X_\phi$$

$$S_T = 3S_\phi = \sqrt{3} E_L I_L$$

$$F_p = \frac{P_T}{S_T}$$

PROBLEMS

SECTION 18.2   Iron-Core Transformer

1. For the iron-core transformer ($k = 1$) of Fig. 18.52:
   a. Find the magnitude of the induced voltage E_s.
   b. Find the maximum flux Φ_m.

2. Repeat Problem 1 for $N_p = 240$ and $N_s = 30$.

3. Find the applied voltage of an iron-core transformer if the secondary voltage is 240 V and $N_p = 60$ with $N_s = 720$.

4. For the iron-core transformer of Fig. 18.53:
   a. Find the magnitude of the current I_L and the voltage V_L if $a = 1/5$, $I_p = 2$ A, and $Z_L = 2\ \Omega$ resistor.
   b. Find the input resistance for the data specified in part (a).

5. Find the input impedance for the iron-core transformer of Fig. 18.53 if $a = 2$, $I_p = 4$ A, and $V_g = 1600$ V.

6. Find the voltage V_g and the current I_p if the input impedance of the iron-core transformer of Fig. 18.53 is 4 Ω, $V_L = 1200$ V, and $a = 1/4$.

7. If $V_L = 240$ V, $Z_L = 20\ \Omega$ resistor, $I_p = 0.05$ A, and $N_s = 50$, find the number of turns in the primary circuit of the iron-core transformer of Fig. 18.53.

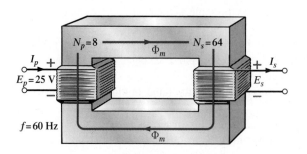

FIG. 18.52
Problems 1 and 2.

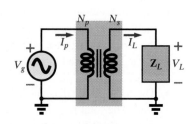

FIG. 18.53
Problems 4–8, 44, 47–49.

8. a. If $N_p = 400$, $N_s = 1200$, and $V_g = 100$ V, find the magnitude of I_p for the iron-core transformer of Fig. 18.53 if $Z_L = 9\,\Omega + j\,12\,\Omega$.

  b. Find the magnitude of the voltage V_L and the current I_L for the conditions of part (a).

9. For the circuit of Fig. 18.54:

  a. Find the transformation ratio required to deliver maximum power to the speaker.

  b. Find the maximum power delivered to the speaker.

10. An ideal transformer is rated 10 kVA, 2400/120 V, 60 Hz.

  a. Find the transformation ratio if the 120 V is the secondary voltage.

  b. Find the current rating of the secondary if the 120 V is the secondary voltage.

  c. Find the current rating of the primary if the 120 V is the secondary voltage.

  d. Repeat parts (a) through (c) if the 2400 V is the secondary voltage.

11. Determine the primary and secondary voltages and currents for the autotransformer of Fig. 18.55.

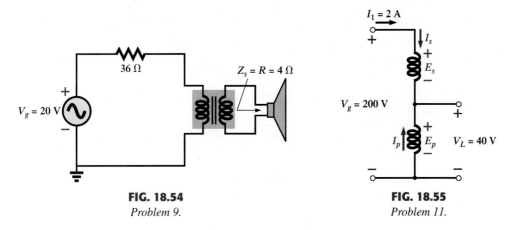

FIG. 18.54	**FIG. 18.55**
Problem 9.	*Problem 11.*

SECTION 18.7 The Y-Connected Generator with a Y-Connected Load

12. A balanced Y load having a 10 Ω resistance in each leg is connected to a three-phase, four-wire, Y-connected generator having a line voltage of 208 V. Calculate the magnitude of

  a. the phase voltage of the generator.

  b. the phase voltage of the load.

  c. the phase current of the load.

  d. the line current.

13. Repeat Problem 12 if each phase impedance is changed to a 12 Ω resistor in series with a 16 Ω capacitive reactance.

14. The phase sequence for the Y-Y system of Fig. 18.56 is *ABC*.

  a. Find angles θ_2 and θ_3 for the specified phase sequence.

  b. Find the voltage across each phase impedance in phasor form.

  c. Find the current through each phase impedance in phasor form.

  d. Draw the phasor diagram of the currents found in part (c), and show that their phasor sum is zero.

  e. Find the magnitude of the line currents.

  f. Find the magnitude of the line voltages.

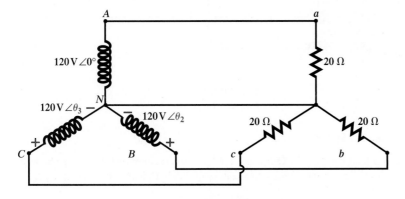

FIG. 18.56

Problems 14, 15, and 33.

15. Repeat Problem 14 if the phase impedances are changed to a 6 Ω resistance in parallel with an 8 Ω capacitive reactance.

16. For the system of Fig. 18.57, find the magnitude of the unknown voltages and currents.

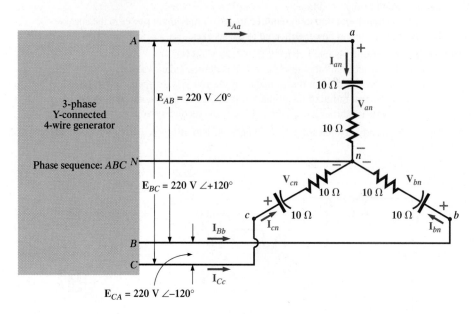

FIG. 18.57
Problems 16, 34, and 41.

17. For the Y-Y system of Fig. 18.58:
   a. Find the magnitude and angle associated with voltages E_{AN}, E_{BN}, and E_{CN}.
   b. Determine the magnitude and angle associated with each phase current of the load: I_{an}, I_{bn}, and I_{cn}.
   c. Find the magnitude and phase angle of each line current: I_{Aa}, I_{Bb}, and I_{Cc}.
   d. Determine the magnitude and phase angle of the voltage across each phase of the load: V_{an}, V_{bn}, and V_{cn}.

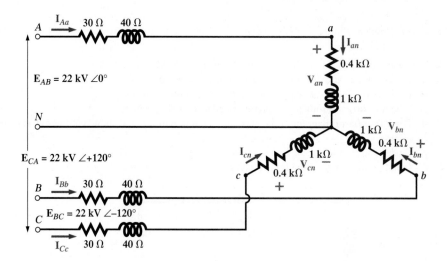

FIG. 18.58
Problem 17.

SECTION 18.8   The Y-Δ System

18. A balanced Δ load having a 20 Ω resistance in each leg is connected to a three-phase, three-wire, Y-connected generator having a line voltage of 208 V. Calculate the magnitude of
   a. the phase voltage of the generator.
   b. the phase voltage of the load.
   c. the phase current of the load.
   d. the line current.

19. Repeat Problem 18 if each phase impedance is changed to a 6.8 Ω resistor in series with a 14 Ω inductive reactance.

20. The phase sequence for the Y-Δ system of Fig. 18.59 is *ABC*.
    a. Find angles θ_2 and θ_3 for the specified phase sequence.
    b. Find the voltage across each phase impedance in phasor form.
    c. Draw the phasor diagram of the voltages found in part (b), and show that their sum is zero around the closed loop of the Δ load.
    d. Find the current through each phase impedance in phasor form.
    e. Find the magnitude of the line currents.
    f. Find the magnitude of the generator phase voltages.

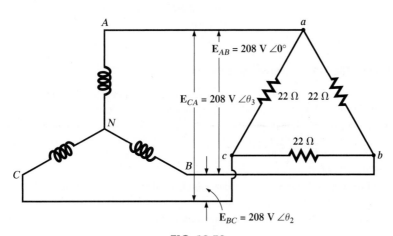

FIG. 18.59
Problems 20, 21, and 42.

21. Repeat Problem 20 if the phase impedances are changed to a 3 Ω resistor in parallel with an inductive reactance of 4 Ω.

22. For the system of Fig. 18.60, find the magnitude of the unknown voltages and currents.

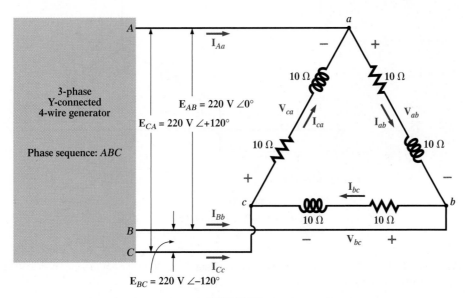

FIG. 18.60
Problems 22 and 35.

*23. For the Δ-connected load of Fig. 18.61:
    a. Find the magnitude and angle of each phase current $\mathbf{I}_{ab}$, $\mathbf{I}_{bc}$, and $\mathbf{I}_{ca}$.
    b. Calculate the magnitude and angle of each line current $\mathbf{I}_{Aa}$, $\mathbf{I}_{Bb}$, and $\mathbf{I}_{Cc}$.
    c. Determine the magnitude and angle of the voltages $\mathbf{E}_{AB}$, $\mathbf{E}_{BC}$, and $\mathbf{E}_{CA}$.

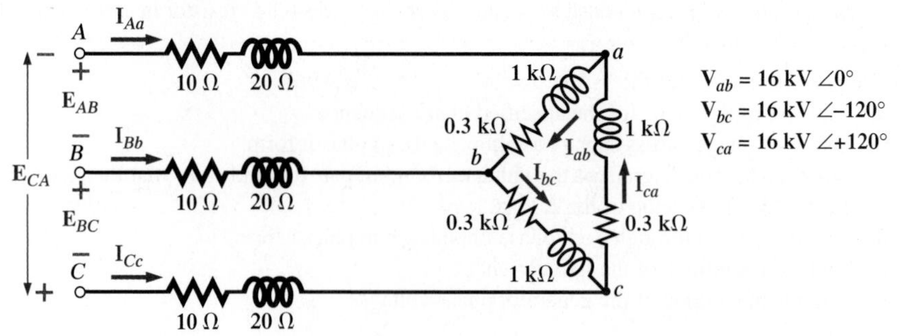

$V_{ab} = 16\ \text{kV}\ \angle 0°$
$V_{bc} = 16\ \text{kV}\ \angle -120°$
$V_{ca} = 16\ \text{kV}\ \angle +120°$

FIG. 18.61
Problem 23.

SECTION 18.10   The Δ-Δ, Δ-Y Three-Phase Systems

24. A balanced Y load having a 30 Ω resistance in each leg is connected to a three-phase, Δ-connected generator having a line voltage of 208 V. Calculate the magnitude of
   a. the phase voltage of the generator.
   b. the phase voltage of the load.
   c. the phase current of the load.
   d. the line current.

25. Repeat Problem 24 if each phase impedance is changed to a 15 Ω resistor in parallel with a 20 Ω capacitive reactance.

*26. For the system of Fig. 18.62, find the magnitude of the unknown voltages and currents.

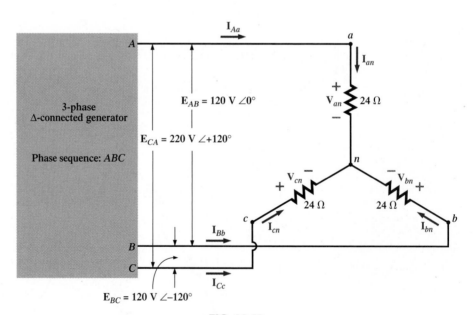

FIG. 18.62
Problems 26, 27, and 37.

27. Repeat Problem 26 if each phase impedance is changed to a 10 Ω resistor in series with a 20 Ω inductive reactance.

28. A balanced Δ load having a 220 Ω resistance in each leg is connected to a three-phase, Δ-connected generator having a line voltage of 440 V. Calculate the magnitude of
   a. the phase voltage of the generator.
   b. the phase voltage of the load.
   c. the phase current of the load.
   d. the line current.

29. Repeat Problem 28 if each phase impedance is changed to a 12 Ω resistor in series with a 9 Ω capacitive reactance.

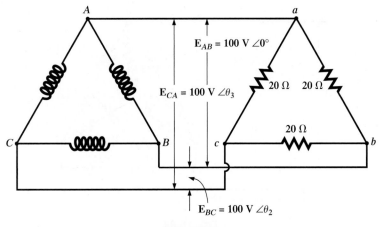

FIG. 18.63
Problems 30 and 31.

30. The phase sequence for the Δ-Δ system of Fig. 18.63 is *ABC*.
    a. Find angles θ_2 and θ_3 for the specified phase sequence.
    b. Find the voltage across each phase impedance in phasor form.
    c. Draw the phasor diagram of the voltages found in part (b), and show that their phasor sum is zero around the closed loop of the Δ load.
    d. Find the current through each phase impedance in phasor form.
    e. Find the magnitude of the line currents.

31. Repeat Problem 30 if each phase impedance is changed to a 20 Ω resistor in parallel with a 20 Ω capacitive reactance.

SECTION 18.11 Power

32. Find the total watts, volt-ampere reactive, volt-ampere, and F_p of the three-phase system of Problem 13.

33. Find the total watts, volt-ampere reactive, volt-ampere, and F_p of the three-phase system of Problem 14.

34. Find the total watts, volt-ampere reactive, volt-ampere, and F_p of the three-phase system of Problem 16.

35. Find the total watts, volt-ampere reactive, volt-ampere, and F_p of the three-phase system of Problem 22.

36. Find the total watts, volt-ampere reactive, volt-ampere, and F_p of the three-phase system of Problem 25.

37. Find the total watts, volt-ampere reactive, volt-ampere, and F_p of the three-phase system of Problem 27.

38. A balanced, three-phase, Δ-connected load has a line voltage of 200 and a total power consumption of 4800 W at a lagging power factor of 0.8. Find the impedance of each phase in rectangular coordinates.

39. A balanced, three-phase, Y-connected load has a line voltage of 208 and a total power consumption of 1200 W at a leading power factor of 0.6. Find the impedance of each phase in rectangular coordinates.

*40. Find the total watts, volt-ampere reactive, volt-ampere, and F_p of the system of Fig. 18.64.

SECTION 18.12 Power and Phase-Sequence Measurements

41. **a.** Sketch the connections required to measure the total watts delivered to the load of Fig. 18.57 using three wattmeters.
    b. Determine the total wattage dissipation and the reading of each wattmeter.

42. Repeat Problem 41 for the network of Fig. 18.59.

43. For the three-wire system of Fig. 18.65:
    a. Properly connect a second wattmeter so that the two will measure the total power delivered to the load.
    b. If one wattmeter has a reading of 200 W and the other a reading of 85 W, what is the total dissipation in watts if the total power factor is 0.8 leading?
    c. Repeat part (b) if the total power factor is 0.2 lagging and $P_l = 100$ W.

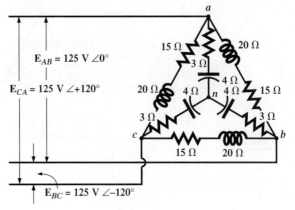

FIG. 18.64
Problem 40.

44. Sketch three different ways that two wattmeters can be connected to measure the total power delivered to the load of Problem 7.

*__**45.**__ For the Y-Δ system of Fig. 18.66:
    a. Determine the magnitude and angle of the phase currents.
    b. Find the magnitude and angle of the line currents.
    c. Determine the reading of each wattmeter.
    d. Find the total power delivered to the load.

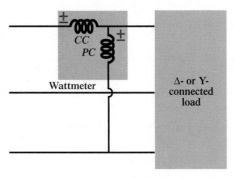

FIG. 18.65
Problem 43.

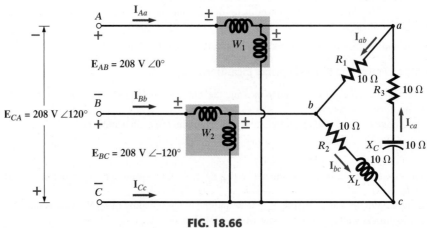

FIG. 18.66
Problem 45.

stem.

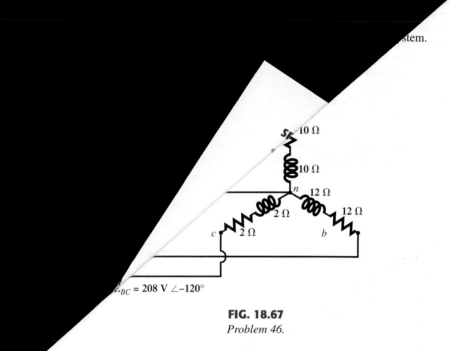

FIG. 18.67
Problem 46.

ECTION 18.15   Computer Analysis

*47. Using PSpice or EWB, generate the schematic for the transformer of Problem 6, and find voltage V_g and current I_p.

*48. Using PSpice or EWB, develop a technique to find the input impedance at the source for the network of Problem 5.

*49. Using PSpice or EWB, using a transformer from the library, find the load voltage for the network of Fig. 18.53 using the parameters of Problem 8.

GLOSSARY

Autotransformer   A transformer with one winding common to both the primary and the secondary circuits. A loss in isolation is balanced by the increase in its kilovolt-ampere rating.

Δ-connected ac generator   A three-phase generator having the three phases connected in the shape of the capital Greek letter delta (Δ).

Line current   The current that flows from the generator to the load of a single-phase or polyphase system.

Line voltage   The potential difference that exists between the lines of a single-phase or polyphase system.

Nameplate data   Information such as the kilovolt-ampere rating, voltage transformation ratio, and frequency of application that is of primary importance in choosing the proper transformer for a particular application.

Neutral connection   The connection between the generator and the load that, under balanced conditions, will have zero current associated with it.

Phase current   The current that flows through each phase of a single-phase or polyphase generator load.

Phase sequence   The order in which the generated sinusoidal voltages of a polyphase generator will affect the load to which they are applied.

Phase voltage   The voltage that appears between the line and neutral of a Y-connected generator and from line to line in a Δ-connected generator.

Polyphase ac generator   An electromechanical source of ac power that generates more than one sinusoidal voltage per rotation of the rotor. The frequency generated is determined by the speed of rotation and the number of poles of the rotor.

Primary   The coil or winding to which the source of electrical energy is normally applied.

Reflected impedance   The impedance appearing at the primary of a transformer due to a load connected to the secondary. Its magnitude is controlled directly by the transformation ratio.

Secondary   The coil or winding to which the load is normally applied.

'ire, Y-Connected Load

ach phase of the load.
phase of the load.
pere, and F_p of the sy

e of ac power that generates a single
ed of rotation and the number of poles

ge is less than its primary voltage.

reater than its primary voltage.

delivered to a three-phase

sformer.
o a Δ- or Y-connected
he load.
se.
he three phases

Appendixes

APPENDIX A

Conversion Factors

To Convert from	To	Multiply by
Btus	Calorie-grams	251.996
	Ergs	1.054×10^{10}
	Foot-pounds	777.649
	Hp-hours	0.000393
	Joules	1054.35
	Kilowatthours	0.000293
	Wattseconds	1054.35
Centimeters	Angstrom units	1×10^8
	Feet	0.0328
	Inches	0.3937
	Meters	0.01
	Miles (statute)	6.214×10^{-6}
	Millimeters	10
Circular mils	Square centimeters	5.067×10^{-6}
	Square inches	7.854×10^{-7}
Cubic inches	Cubic centimeters	16.387
	Gallons (U.S. liquid)	0.00433
Cubic meters	Cubic feet	35.315
Days	Hours	24
	Minutes	1440
	Seconds	86,400
Dynes	Gallons (U.S. liquid)	264.172
	Newtons	0.00001
	Pounds	2.248×10^{-6}
Electronvolts	Ergs	1.60209×10^{-12}
Ergs	Dyne-centimeters	1.0
	Electronvolts	6.242×10^{11}
	Foot-pounds	7.376×10^{-8}
	Joules	1×10^{-7}
	Kilowatthours	2.777×10^{-14}
Feet	Centimeters	30.48
	Meters	0.3048
Foot-candles	Lumens/square foot	1.0
	Lumens/square meter	10.764
Foot-pounds	Dyne-centimeters	1.3558×10^7
	Ergs	1.3558×10^7
	Horsepower-hours	5.050×10^{-7}
	Joules	1.3558
	Newton-meters	1.3558

To Convert from	To	Multiply by
Gallons (U.S. liquid)	Cubic inches	231
	Liters	3.785
	Ounces	128
	Pints	8
Gauss	Maxwells/square centimeter	1.0
	Lines/square centimeter	1.0
	Lines/square inch.	6.4516
Gilberts	Ampere-turns	0.7958
Grams	Dynes	980.665
	Ounces	0.0353
	Pounds	0.0022
Horsepower	Btus/hour	2547.16
	Ergs/second	7.46×10^9
	Foot-pounds/second	550.221
	Joules/second	746
	Watts	746
Hours	Seconds	3600
Inches	Angstrom units	2.54×10^8
	Centimeters	2.54
	Feet	0.0833
	Meters	0.0254
Joules	Btus	0.000948
	Ergs	1×10^7
	Foot-pounds	0.7376
	Horsepower-hours	3.725×10^{-7}
	Kilowatthours	2.777×10^{-7}
	Wattseconds	1.0
Kilograms	Dynes	980,665
	Ounces	35.2
	Pounds	2.2
Lines	Maxwells	1.0
Lines/square centimeter	Gauss	1.0
Lines/square inch	Gauss	0.1550
	Webers/square inch	1×10^{-8}
Liters	Cubic centimeters	1000.028
	Cubic inches	61.025
	Gallons (U.S. liquid)	0.2642
	Ounces (U.S. liquid)	33.815
	Quarts (U.S. liquid)	1.0567
Lumens	Candle power (spher.)	0.0796
Lumens/square centimeter	Lamberts	1.0
Lumens/square foot	Foot-candles	1.0
Maxwells	Lines	1.0
	Webers	1×10^{-8}
Meters	Angstrom units	1×10^{10}
	Centimeters	100
	Feet	3.2808
	Inches	39.370
	Miles (statute)	0.000621

(continued)

To Convert from	To	Multiply by
Miles (statute)	Feet	5280
	Kilometers	1.609
	Meters	1609.344
Miles/hour	Kilometers/hour	1.609344
Newton-meters	Dyne-centimeters	1×10^7
	Kilogram-meters	0.10197
Oersteds	Ampere-turns/inch	2.0212
	Ampere-turns/meter	79.577
	Gilberts/centimeter	1.0
Quarts (U.S. liquid)	Cubic centimeters	946.353
	Cubic inches	57.75
	Gallons (U.S. liquid)	0.25
	Liters	0.9463
	Pints (U.S. liquid)	2
	Ounces (U.S. liquid)	32
Radians	Degrees	57.2958
Slugs	Kilograms	14.5939
	Pounds	32.1740
Watts	Btus/hour	3.4144
	Ergs/second	1×10^7
	Horsepower	0.00134
	Joules/second	1.0
Webers	Lines	1×10^8
	Maxwells	1×10^8
Years	Days	365
	Hours	8760
	Minutes	525,600
	Seconds	3.1536×10^7

APPENDIX B

PSpice, Electronics Workbench, and Mathcad

B.1 PSpice

The PSpice software package employed throughout this text is derived from programs developed at the University of California at Berkeley during the early 1970s. SPICE is an acronym for Simulation Program with Integrated Circuit Emphasis. Although a number of companies have customized SPICE for their particular use, Cadence Design Systems offers both a commercial and a demo version of OrCAD. The commercial or professional versions employed by engineering companies can be quite expensive, so Cadence offers free distribution of the demo version to provide an introduction to the power of the simulation package. For this text, the OrCAD family release 9.2 Lite Edition was employed. Free copies can be obtained by contacting EMA Mid-Atlantic at 877-362-3321 or visiting the EMA website at **http://www.ema-eda.com.**

Minimum system requirements are the following:

Pentium 90MHz PC
32MB RAM
Hard disk space:
    Capture CIS 89MB
    Layout Plus 66MB
    PSpice A/D 46MB
800×600, 256 color VGA display
Microsoft Windows 95/98, or Windows NT 4.0 Service Pack 3
$4\times$ CD-ROM drive
16-bit audio (recommended)

B.2 Electronics Workbench

Multisim is a product of Electronics Workbench. For this text, the Multisim 2001 Education Version was employed. Copies can be obtained using one of the following three options:
  Phone: 800-282-0693
  Fax: 800-835-5327
  Website: **http://www.prenhall.com**

Minimum system requirements for the Student Version are as follows:

Pentium 90 or equivalent
Windows 95/98/NT/2000/XP
64MB RAM (128MB RAM recommended)
200MB hard disk space (minimum)
CD-ROM drive

B.3 Mathcad

Mathcad is a product of MathSoft Engineering & Education, Inc., located at 101 Main Street, Cambridge, MA 02142-1521. The Internet address is **http://www.mathsoft.com.** For this text, Mathcad 2000 was employed.

The current academic version of Mathcad is available online for purchase from **http://www.edu.com** and **http://www.journeyed.com.** Professors can purchase the current Mathcad 11 Academic Edition from the customer service department at 800-628-4223 or 617-444-8000 with the e-mail address of **sales-info@mathsoft.com.**

Minimum system requirements for Mathcad 11 are as follows:

PC with Pentium/Celeron 233MHz (300MHz or higher recommended)
Windows 98, XE, ME, NT with 4.0 SP6, 2000, SP2, XP, or higher
Minimum of 96MB RAM (256MB or higher recommended)
At least 150MB hard disk space for the single-user version
CD-ROM or DVD drive
SVGA or higher graphics card and monitor

Determinants

Determinants are used to find the mathematical solutions for variables in two or more simultaneous equations. Once you understand the procedure, you can obtain solutions with a minimum of time and effort and usually with fewer errors than when using other methods.

Consider the following equations, where x and y are the unknown variables and a_1, a_2, b_1, b_2, c_1, and c_2 are constants:

Col. 1		Col. 2		Col. 3	
$a_1 x$	$+$	$b_1 y$	$=$	c_1	**(C.1a)**
$a_2 x$	$+$	$b_2 y$	$=$	c_2	**(C.1b)**

It is certainly possible to solve for one variable in Eq. (C.1a) and substitute into Eq. (C.1b). That is, solving for x in Eq. (C.1a),

$$x = \frac{c_1 - b_1 y}{a_1}$$

and substituting the result in Eq. (C.1b),

$$a_2 \left(\frac{c_1 - b_1 y}{a_1} \right) + b_2 y = c_2$$

It is now possible to solve for y, because it is the only variable remaining, and then substitute into either equation for x. This is acceptable for two equations, but it becomes a very tedious and lengthy process for three or more simultaneous equations.

Using determinants to solve for x and y requires that the following formats be established for each variable:

$$x = \frac{\begin{vmatrix} c_1 & b_1 \\ c_2 & b_2 \end{vmatrix}}{\begin{vmatrix} a_1 & b_1 \\ a_2 & b_2 \end{vmatrix}} \qquad y = \frac{\begin{vmatrix} a_1 & c_1 \\ a_2 & c_2 \end{vmatrix}}{\begin{vmatrix} a_1 & b_1 \\ a_2 & b_2 \end{vmatrix}} \qquad \textbf{(C.2)}$$

First note that only constants appear within the vertical brackets and that the denominator of each is the same. In fact, the denominator is simply the coefficients of x and y in the same arrangement as in Eqs. (C.1a) and (C.1b). When solving for x, replace the coefficients of x in the numerator by the constants to the right of the equals sign in Eqs. (C.1a) and (C.1b), and simply repeat the coefficients of the y variable. When solving for y, replace the y coefficients in the numerator by the constants to the right of the equals sign, and repeat the coefficients of x.

Each configuration in the numerator and denominator of Eqs. (C.2) is referred to as a *determinant* (D), which can be evaluated numerically in the following manner:

$$\text{Determinant} = D = \begin{vmatrix} \overset{\text{Col.}}{\underset{1}{a_1}} & \overset{\text{Col.}}{\underset{2}{b_1}} \\ a_2 & b_2 \end{vmatrix} = a_1 b_2 - a_2 b_1 \qquad \textbf{(C.3)}$$

The expanded value is obtained by first multiplying the top left element by the bottom right and then subtracting the product of the lower left and upper right elements. This particular determinant is referred to as a *second-order* determinant, because it contains two rows and two columns.

When using determinants, it is important to remember that the columns of the equations, as indicated in Eqs. (C.1a) and (C.1b), must be placed in the same order within the determinant configuration. That is, since a_1 and a_2 are in column 1 of Eqs. (C.1a) and (C.1b), they must be in column 1 of the determinant. (The same is true for b_1 and b_2.)

Expanding the entire expression for x and y, we have the following:

$$x = \frac{\begin{vmatrix} c_1 & b_1 \\ c_2 & b_2 \end{vmatrix}}{\begin{vmatrix} a_1 & b_1 \\ a_2 & b_2 \end{vmatrix}} = \frac{c_1 b_2 - c_2 b_1}{a_1 b_2 - a_2 b_1} \qquad \textbf{(C.4a)}$$

$$y = \frac{\begin{vmatrix} a_1 & c_1 \\ a_2 & c_2 \end{vmatrix}}{\begin{vmatrix} a_1 & b_1 \\ a_2 & b_2 \end{vmatrix}} = \frac{a_1 c_2 - a_2 c_1}{a_1 b_2 - a_2 b_1} \qquad \textbf{(C.4b)}$$

EXAMPLE C.1 Evaluate the following determinants:

a. $\begin{vmatrix} 2 & 2 \\ 3 & 4 \end{vmatrix} = (2)(4) - (3)(2) = 8 - 6 = \textbf{2}$

b. $\begin{vmatrix} 4 & -1 \\ 6 & 2 \end{vmatrix} = (4)(2) - (6)(-1) = 8 + 6 = \textbf{14}$

c. $\begin{vmatrix} 0 & -2 \\ -2 & 4 \end{vmatrix} = (0)(4) - (-2)(-2) = 0 - 4 = \textbf{-4}$

d. $\begin{vmatrix} 0 & 0 \\ 3 & 10 \end{vmatrix} = (0)(10) - (3)(0) = \textbf{0}$

EXAMPLE C.2 Solve for x and y:

$$2x + y = 3$$
$$3x + 4y = 2$$

Solution: $x = \dfrac{\begin{vmatrix} 3 & 1 \\ 2 & 4 \end{vmatrix}}{\begin{vmatrix} 2 & 1 \\ 3 & 4 \end{vmatrix}} = \dfrac{(3)(4) - (2)(1)}{(2)(4) - (3)(1)} = \dfrac{12 - 2}{8 - 3} = \dfrac{10}{5} = \textbf{2}$

$y = \dfrac{\begin{vmatrix} 2 & 3 \\ 3 & 2 \end{vmatrix}}{5} = \dfrac{(2)(2) - (3)(3)}{5} = \dfrac{4 - 9}{5} = \dfrac{-5}{5} = \textbf{-1}$

$$2x + y = (2)(2) + (-1)$$
$$= 4 - 1 = 3 \quad \text{(checks)}$$
$$3x + 4y = (3)(2) + (4)(-1)$$
$$= 6 - 4 = 2 \quad \text{(checks)}$$

EXAMPLE C.3 Solve for x and y:

$$-x + 2y = 3$$
$$3x - 2y = -2$$

Solution: In this example, note the effect of the minus sign and the use of parentheses to ensure that the proper sign is obtained for each product:

$$x = \frac{\begin{vmatrix} 3 & 2 \\ -2 & -2 \end{vmatrix}}{\begin{vmatrix} -1 & 2 \\ 3 & -2 \end{vmatrix}} = \frac{(3)(-2) - (-2)(2)}{(-1)(-2) - (3)(2)} = \frac{-6 + 4}{2 - 6} = \frac{-2}{-4} = \frac{1}{2}$$

$$y = \frac{\begin{vmatrix} -1 & 3 \\ 3 & -2 \end{vmatrix}}{-4} = \frac{(-1)(-2) - (3)(3)}{-4} = \frac{2 - 9}{-4} = \frac{-7}{-4} = \frac{7}{4}$$

EXAMPLE C.4 Solve for x and y:

$$x = 3 - 4y$$
$$20y = -1 + 3x$$

Solution: In this case, the equations must first be placed in the format of Eqs. (C.1a) and (C.1b):

$$x + 4y = 3$$
$$-3x + 20y = -1$$

$$x = \frac{\begin{vmatrix} 3 & 4 \\ -1 & 20 \end{vmatrix}}{\begin{vmatrix} 1 & 4 \\ -3 & 20 \end{vmatrix}} = \frac{(3)(20) - (-1)(4)}{(1)(20) - (-3)(4)} = \frac{60 + 4}{20 + 12} = \frac{64}{32} = 2$$

$$y = \frac{\begin{vmatrix} 1 & 3 \\ -3 & -1 \end{vmatrix}}{32} = \frac{(1)(-1) - (-3)(3)}{32} = \frac{-1 + 9}{32} = \frac{8}{32} = \frac{1}{4}$$

The use of determinants is not limited to the solution of two simultaneous equations. Determinants can be applied to any number of simultaneous linear equations. First we will examine a shorthand method that is applicable to third-order determinants only, since most of the problems in the text are limited to this level of difficulty. We will then investigate the general procedure for solving any number of simultaneous equations.

Consider the three following simultaneous equations:

Col. 1		Col. 2		Col. 3		Col. 4
a_1x	$+$	b_1y	$+$	c_1z	$=$	d_1
a_2x	$+$	b_2y	$+$	c_2z	$=$	d_2
a_3x	$+$	b_3y	$+$	c_3z	$=$	d_3

in which x, y, and z are the variables, and $a_{1,2,3}$, $b_{1,2,3}$, $c_{1,2,3}$, and $d_{1,2,3}$, are constants.

The determinant configuration for x, y, and z can be found in a manner similar to that for two simultaneous equations. That is, to solve for x, find the determinant in the numerator by replacing column 1 with the elements to the right of the equals sign. The denominator is the determinant of the coefficients of the variables (the same applies to y and z). Again, the denominator is the same for each variable.

$$x = \frac{\begin{vmatrix} d_1 & b_1 & c_1 \\ d_2 & b_2 & c_2 \\ d_3 & b_3 & c_3 \end{vmatrix}}{D} \qquad y = \frac{\begin{vmatrix} a_1 & d_1 & c_1 \\ a_2 & d_2 & c_2 \\ a_3 & d_3 & c_3 \end{vmatrix}}{D} \qquad z = \frac{\begin{vmatrix} a_1 & b_1 & d_1 \\ a_2 & b_2 & d_2 \\ a_3 & b_3 & d_3 \end{vmatrix}}{D}$$

where

$$D = \begin{vmatrix} a_1 & b_1 & c_1 \\ a_2 & b_2 & c_2 \\ a_3 & b_3 & c_3 \end{vmatrix}$$

A shorthand method for evaluating the third-order determinant consists simply of repeating the first two columns of the determinant to the right of the determinant and then summing the products along specific diagonals as shown below:

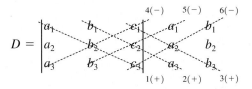

The products of the diagonals 1, 2, and 3 are positive and have the following magnitudes:

$$+a_1b_2c_3 + b_1c_2a_3 + c_1a_2b_3$$

The products of the diagonals 4, 5, and 6 are negative and have the following magnitudes:

$$-a_3b_2c_1 - b_3c_2a_1 - c_3a_2b_1$$

The total solution is the sum of the diagonals 1, 2, and 3 minus the sum of the diagonals 4, 5, and 6:

$$\boxed{+(a_1b_2c_3 + b_1c_2a_3 + c_1a_2b_3) - (a_3b_2c_1 + b_3c_2a_1 + c_3a_2b_1)} \qquad \textbf{(C.5)}$$

Warning: **This method of expansion is good only for third-order determinants!** It cannot be applied to fourth- and higher-order systems.

EXAMPLE C.5   Evaluate the following determinant:

$$\begin{vmatrix} 1 & 2 & 3 \\ -2 & 1 & 0 \\ 0 & 4 & 2 \end{vmatrix} \rightarrow \begin{matrix} \overset{(-)(-)(-)}{} \\ \begin{matrix} 1 & 2 & 3 & 1 & 2 \\ -2 & 1 & 0 & -2 & 1 \\ 0 & 4 & 2 & 0 & 4 \end{matrix} \\ \underset{(+)(+)(+)}{} \end{matrix}$$

Solution:

$$[(1)(1)(2) + (2)(0)(0) + (3)(-2)(4)] - [(0)(1)(3) + (4)(0)(1) + (2)(-2)(2)]$$
$$= (2 + 0 - 24) - (0 + 0 - 8) = (-22) - (-8)$$
$$= -22 + 8 = \mathbf{-14}$$

EXAMPLE C.6   Solve for x, y, and z:

$$1x + 0y - 2z = -1$$
$$0x + 3y + 1z = +2$$
$$1x + 2y + 3z = 0$$

Solution:

$$x = \frac{\begin{vmatrix} -1 & 0 & -2 \\ 2 & 3 & 1 \\ 0 & 2 & 3 \end{vmatrix}\begin{matrix} -1 & 0 \\ 2 & 3 \\ 0 & 2 \end{matrix}}{\begin{vmatrix} 1 & 0 & -2 \\ 0 & 3 & 1 \\ 1 & 2 & 3 \end{vmatrix}\begin{matrix} 1 & 0 \\ 0 & 3 \\ 1 & 2 \end{matrix}}$$

$$= \frac{[(-1)(3)(3) + (0)(1)(0) + (-2)(2)(2)] - [(0)(3)(-2) + (2)(1)(-1) + (3)(2)(0)]}{[(1)(3)(3) + (0)(1)(1) + (-2)(0)(2)] - [(1)(3)(-2) + (2)(1)(1) + (3)(0)(0)]}$$

$$= \frac{(-9 + 0 - 8) - (0 - 2 + 0)}{(9 + 0 + 0) - (-6 + 2 + 0)} = \frac{-17 + 2}{9 + 4} = -\frac{\mathbf{15}}{\mathbf{13}}$$

$$y = \frac{\begin{vmatrix} 1 & -1 & -2 \\ 0 & 2 & 1 \\ 1 & 0 & 3 \end{vmatrix}\begin{matrix} 1 & -1 \\ 0 & 2 \\ 1 & 0 \end{matrix}}{13}$$

$$= \frac{[(1)(2)(3) + (-1)(1)(1) + (-2)(0)(0)] - [(1)(2)(-2) + (0)(1)(1) + (3)(0)(-1)]}{13}$$

$$= \frac{(6 - 1 + 0) - (-4 + 0 + 0)}{13} = \frac{5 + 4}{13} = \frac{\mathbf{9}}{\mathbf{13}}$$

$$z = \frac{\begin{vmatrix} 1 & 0 & -1 \\ 0 & 3 & 2 \\ 1 & 2 & 0 \end{vmatrix}\begin{matrix} 1 & 0 \\ 0 & 3 \\ 1 & 2 \end{matrix}}{13}$$

$$= \frac{[(1)(3)(0) + (0)(2)(1) + (-1)(0)(2)] - [(1)(3)(-1) + (2)(2)(1) + (0)(0)(0)]}{13}$$

$$= \frac{(0 + 0 + 0) - (-3 + 4 + 0)}{13} = \frac{0 - 1}{13} = -\frac{\mathbf{1}}{\mathbf{13}}$$

or from $0x + 3y + 1z = +2$,

$$z = 2 - 3y = 2 - 3\left(\frac{9}{13}\right) = \frac{26}{13} - \frac{27}{13} = -\frac{\mathbf{1}}{\mathbf{13}}$$

Check:

$$\left.\begin{array}{r} 1x + 0y - 2z = -1 \\ 0x + 3y + 1z = +2 \\ 1x + 2y + 3z = 0 \end{array}\right\} \left.\begin{array}{r} -\dfrac{15}{13} + 0 + \dfrac{2}{13} = -1 \\ 0 + \dfrac{27}{13} + \dfrac{-1}{13} = +2 \\ -\dfrac{15}{13} + \dfrac{18}{13} + \dfrac{-3}{13} = 0 \end{array}\right\} \begin{array}{l} -\dfrac{13}{13} = -1 \checkmark \\ \dfrac{26}{13} = +2 \checkmark \\ -\dfrac{18}{13} + \dfrac{18}{13} = 0 \checkmark \end{array}$$

The general approach to third- or higher-order determinants requires that the determinant be expanded in the following form. There is more than one expansion that will generate the correct result, but this form is typically employed when the material is first introduced.

$$D = \begin{vmatrix} a_1 & b_1 & c_1 \\ a_2 & b_2 & c_2 \\ a_3 & b_3 & c_3 \end{vmatrix} = a_1 \left(+ \begin{vmatrix} b_2 & c_2 \\ b_3 & c_3 \end{vmatrix} \right) + b_1 \left(- \begin{vmatrix} a_2 & c_2 \\ a_3 & c_3 \end{vmatrix} \right) + c_1 \left(+ \begin{vmatrix} a_2 & b_2 \\ a_3 & b_3 \end{vmatrix} \right)$$

Minor   Cofactor   Multiplying factor

Minor   Cofactor   Multiplying factor

Minor   Cofactor   Multiplying factor

This expansion was obtained by multiplying the elements of the first row of D by their corresponding cofactors. It is not a requirement that the first row be used as the multiplying factors. In fact, any *row* or *column* (not diagonals) may be used to expand a third-order determinant.

The sign of each cofactor is dictated by the position of the multiplying factors (a_1, b_1, and c_1 in this case) as in the following standard format:

$$\begin{vmatrix} + & \rightarrow & - & + \\ \downarrow & & & \\ - & & + & - \\ + & & - & + \end{vmatrix}$$

Note that the proper sign for each element can be obtained by simply assigning the upper left element a positive sign and then changing sign as you move horizontally or vertically to the neighboring position.

For the determinant D, the elements would have the following signs:

$$\begin{vmatrix} a_1^{(+)} & b_1^{(-)} & c_1^{(+)} \\ a_2^{(-)} & b_2^{(+)} & c_2^{(-)} \\ a_3^{(+)} & b_3^{(-)} & c_3^{(+)} \end{vmatrix}$$

The minors associated with each multiplying factor are obtained by covering up the row and column in which the multiplying factor is located and writing a second-order determinant to include the remaining elements in the same relative positions that they have in the third-order determinant.

Consider the cofactors associated with a_1 and b_1 in the expansion of D. The sign is positive for a_1 and negative for b_1 as determined by the standard format. Following the procedure outlined above, we can find the minors of a_1 and b_1 as follows:

$$a_{1(minor)} = \begin{vmatrix} a_1 & b_1 & c_1 \\ a_2 & b_2 & c_2 \\ a_3 & b_3 & c_3 \end{vmatrix} = \begin{vmatrix} b_2 & c_2 \\ b_3 & c_3 \end{vmatrix} \qquad b_{1(minor)} = \begin{vmatrix} a_1 & b_1 & c_1 \\ a_2 & b_2 & c_2 \\ a_3 & b_3 & c_3 \end{vmatrix} = \begin{vmatrix} a_2 & c_2 \\ a_3 & c_3 \end{vmatrix}$$

It was pointed out that any row or column may be used to expand the third-order determinant, and the same result will still be obtained. Using the first column of D, we obtain the expansion

$$D = \begin{vmatrix} a_1 & b_1 & c_1 \\ a_2 & b_2 & c_2 \\ a_3 & b_3 & c_3 \end{vmatrix} = a_1 \left(+ \begin{vmatrix} b_2 & c_2 \\ b_3 & c_3 \end{vmatrix} \right) + a_2 \left(- \begin{vmatrix} b_1 & c_1 \\ b_3 & c_3 \end{vmatrix} \right) + a_3 \left(+ \begin{vmatrix} b_1 & c_1 \\ b_2 & c_2 \end{vmatrix} \right)$$

The proper choice of row or column can often effectively reduce the amount of work required to expand the third-order determinant. For example, in the following determinants, the first column and third row, respectively, would reduce the number of cofactors in the expansion:

$$D = \begin{vmatrix} 2 & 3 & -2 \\ 0 & 4 & 5 \\ 0 & 6 & 7 \end{vmatrix} = 2\left(+\begin{vmatrix} 4 & 5 \\ 6 & 7 \end{vmatrix}\right) + 0 + 0 = 2(28 - 30) = \mathbf{-4}$$

$$D = \begin{vmatrix} 1 & 4 & 7 \\ 2 & 6 & 8 \\ 2 & 0 & 3 \end{vmatrix} = 2\left(+\begin{vmatrix} 4 & 7 \\ 6 & 8 \end{vmatrix}\right) + 0 + 3\left(+\begin{vmatrix} 1 & 4 \\ 2 & 6 \end{vmatrix}\right)$$

$$= 2(32 - 42) + 3(6 - 8) = 2(-10) + 3(-2) = \mathbf{-26}$$

EXAMPLE C.7   Expand the following third-order determinants:

a. $D = \begin{vmatrix} 1 & 2 & 3 \\ 3 & 2 & 1 \\ 2 & 1 & 3 \end{vmatrix} = 1\left(+\begin{vmatrix} 2 & 1 \\ 1 & 3 \end{vmatrix}\right) + 3\left(-\begin{vmatrix} 2 & 3 \\ 1 & 3 \end{vmatrix}\right) + 2 + \left(\begin{vmatrix} 2 & 3 \\ 2 & 1 \end{vmatrix}\right)$

$\qquad = 1[6 - 1] + 3[-(6 - 3)] + 2[2 - 6] = 5 + 3(-3) + 2(-4)$

$\qquad = 5 - 9 - 8 = \mathbf{-12}$

b. $D = \begin{vmatrix} 0 & 4 & 6 \\ 2 & 0 & 5 \\ 8 & 4 & 0 \end{vmatrix} = 0 + 2\left(-\begin{vmatrix} 4 & 6 \\ 4 & 0 \end{vmatrix}\right) + 8\left(+\begin{vmatrix} 4 & 6 \\ 0 & 5 \end{vmatrix}\right)$

$\qquad = 0 + 2[-(0 - 24)] + 8[(20 - 0)] = 0 + 2(24) + 8(20)$

$\qquad = 48 + 160 = \mathbf{208}$

APPENDIX D

Color Coding of Molded Tubular Capacitors (Picofarads)

Color	Significant Figure	Decimal Multiplier	Tolerance ±%
Black	0	1	20
Brown	1	10	—
Red	2	100	—
Orange	3	1000	30
Yellow	4	10,000	40
Green	5	10^5	5
Blue	6	10^6	—
Violet	7	—	—
Gray	8	—	—
White	9	—	10

Note: Voltage rating is identified by a single-digit number for ratings up to 900 V and by a two-digit number above 900 V. Two zeros follow the voltage figure.

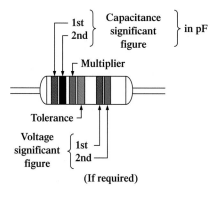

FIG. D.1

The Greek Alphabet

Letter	Capital	Lowercase	Used to Designate
Alpha	A	α	Area, angles, coefficients
Beta	B	β	Angles, coefficients, flux density
Gamma	Γ	γ	Specific gravity, conductivity
Delta	Δ	δ	Density, variation
Epsilon	E	ϵ	Base of natural logarithms
Zeta	Z	ζ	Coefficients, coordinates, impedance
Eta	H	η	Efficiency, hysteresis coefficient
Theta	θ	θ	Phase angle, temperature
Iota	I	ι	
Kappa	K	κ	Dielectric constant, susceptibility
Lambda	Λ	λ	Wavelength
Mu	M	μ	Amplification factor, micro, permeability
Nu	N	ν	Reluctivity
Xi	Ξ	ξ	
Omicron	O	o	
Pi	Π	π	3.1416
Rho	P	ρ	Resistivity
Sigma	Σ	σ	Summation
Tau	T	τ	Time constant
Upsilon	Υ	υ	
Phi	Φ	ϕ	Angles, magnetic flux
Chi	X	χ	
Psi	Ψ	ψ	Dielectric flux, phase difference
Omega	Ω	ω	Ohms, angular velocity

APPENDIX F

Magnetic Parameter Conversions

	SI (MKS)	CGS	English
Φ	webers (Wb)	maxwells	lines
	1 Wb	$= 10^8$ maxwells	$= 10^8$ lines
B	T	gauss	lines/in.^{2}
		(maxwells/cm^2)	
	1 T = 1 Wb/m^2	$= 10^4$ gauss	$= 6.452 \times 10^4$ lines/in.^{2}
A	1 m^2	$= 10^4$ cm^2	$= 1550$ in.^{2}
μ_o	$4\pi \times 10^{-7}$ Wb/Am	$= 1$ gauss/oersted	$= 3.20$ lines/Am
$\mathcal{F}$	NI (ampere-turns, At)	$0.4\pi NI$ (gilberts)	NI (At)
	1 At	$= 1.257$ gilberts	1 gilbert $= 0.7958$ At
H	NI/l (At/m)	$0.4\pi NI/l$ (oersteds)	NI/l (At/in.)
	1 At/m	$= 1.26 \times 10^{-2}$ oersted	$= 2.54 \times 10^{-2}$ At/in.
H_g	$7.97 \times 10^5 B_g$	B_g (oersteds)	$0.313 B_g$ (At/in.)
	(At/m)		

Maximum Power Transfer Conditions

The following discussion describes the derivation of maximum power transfer conditions for the situation where the resistive component of the load is adjustable but the load reactance is set in magnitude.*

For the circuit of Fig. G.1, the power delivered to the load is determined by

$$P = \frac{V_{R_L}^2}{R_L}$$

Applying the voltage divider rule:

$$\mathbf{V}_{R_L} = \frac{R_L \mathbf{E}_{Th}}{R_L + R_{Th} + X_{Th} \angle 90° + X_L \angle 90°}$$

The magnitude of $\mathbf{V}_{R_L}$ is determined by

$$V_{R_L} = \frac{R_L E_{Th}}{\sqrt{(R_L + R_{Th})^2 + (X_{Th} + X_L)^2}}$$

and

$$V_{R_L}^2 = \frac{R_L^2 E_{Th}^2}{(R_L + R_{Th})^2 + (X_{Th} + X_L)^2}$$

with

$$P = \frac{V_{R_L}^2}{R_L} = \frac{R_L E_{Th}^2}{(R_L + R_{Th})^2 + (X_{Th} + X_L)^2}$$

Using differentiation (calculus), maximum power will be transferred when $dP/dR_L = 0$. The result of the preceding operation is that

$$R_L = \sqrt{R_{Th}^2 + (X_{Th} + X_L)^2}$$

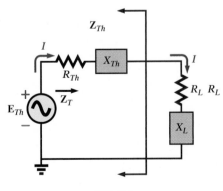

FIG G.1

---

*With sincerest thanks for the input of Professor Harry J. Franz of the Beaver Campus of Pennsylvania State University.

The magnitude of the total impedance of the circuit is

$$Z_T = \sqrt{(R_{Th} + R_L)^2 + (X_{Th} + X_L)^2}$$

Substituting this equation for R_L and applying a few algebraic maneuvers will result in

$$Z_T = 2R_L(R_L + R_{Th})$$

and the power to the load R_L will be

$$P = I^2 R_L = \frac{E_{Th}^2}{Z_T^2} R_L = \frac{E_{Th}^2 R_L}{2R_L(R_L + R_{Th})}$$

$$= \frac{E_{Th}^2}{4 \left(\dfrac{R_L + R_{Th}}{2} \right)}$$

$$= \frac{E_{Th}^2}{4R_{av}}$$

with

$$R_{av} = \frac{R_L + R_{Th}}{2}$$

Answers to Selected Odd-Numbered Problems

Chapter 1

5. 29.05 mph
11. 737.46 ft·lb
13. (a) 10^4 (b) 10^6 (c) 10^3 (d) 10^{-3} (e) 10 (f) 10^{-1}
15. (a) 52.2×10^3 (b) 450×10^3 (c) 440×10^{-6} (d) -4.75×10^3
17. (a) 15 (b) 4.4 (c) 229.6 (d) 8400
19. (a) 2.5×10^7 (b) 16.67×10^{-9} (c) 4.4 (d) 19.5×10^{24}
21. (a) 16×10^4 (b) 216×10^{-9} (c) 1440 (d) 110.59×10^9
23. (a) 300 (b) 2×10^5 (c) 9×10^{12} (d) 1.5×10^{-7} (e) 24×10^{12} (f) 8×10^{20}
    (g) 282.24×10^{-12}
25. (a) 50 ms (b) 2 ms (c) 40 μs (d) 0.0084 μs (e) 4000 mm (f) 0.26 km
27. (a) 10^5 pF (b) 500 m (c) 8 cm (d) 60×10^{-5} km (e) 11.52×10^6 ms (f) 16 μm
    (g) 60×10^{-4} m^2
29. 5280 ft, 1609.35 m, 1.61 km
31. 2.045 s
33. 73.31 days
35. $900
37. 345.6 m
39. 47.30 min/mi
41. (a) 4.74×10^{-3} Btu (b) 7.098×10^{-4} m^3 (c) 1.21×10^5 s (d) 2113.38 pints
43. 13
45. 0.643
47. 2.95
49. 1.2×10^{12}

Chapter 2

3. (a) 1.11 μN (b) 28.8 kN (c) 1138.34 kN
5. 10 mm
7. 25 V
9. 6 C
11. 13 A
13. 2400 C
15. 3 s
17. 1.873×10^{18} electrons
19. 22.43 mA
21. 6.67 V
23. 3.34 A
25. 60.8 Ah
27. W (60 Ah) : W (40 Ah) = 1.5 : 1
35. 600 C

Chapter 3

1. **(a)** 500 mils **(b)** 20 mils **(c)** 250 mils **(d)** 1000 mils **(e)** 240 mils **(f)** 39.37 mils
3. **(a)** 0.04 in. **(b)** 0.03 in. **(c)** 0.2 in. **(d)** 0.025 in. **(e)** 0.0025 in. **(f)** 0.011 in.
5. 92.81 Ω
7. 3.581 ft
9. **(a)** $R_{silver} > R_{copper} > R_{aluminum}$ **(b)** R (silver), 9.9 Ω; R (copper), 1.037 Ω; R (aluminum), 0.34 Ω
11. 21.71 $\mu\Omega$
13. 942.28 mΩ
15. **(a)** #8, 1.131 Ω; #18, 11.493 Ω **(b)** #18 : #8 $\cong$ 10 : 1 **(c)** #18 : #8 $\cong$ 1 : 10
17. **(a)** 1.087 mA/CM **(b)** 1.384 kA/in.^{2} **(c)** 3.709 in.^{2}
19. 0.028 Ω
21. 0.559 Ω
23. **(a)** 27.85°C **(b)** -186.8°C
25. **(b)** 83.61°C
27. 1.751 Ω
31. 33 kA $\gg$ 20 A
33. 7.2 kΩ
37. **(a)** red, red, brown, silver **(b)** orange, orange, red, silver **(c)** blue, gray, red, silver **(d)** red, black, green, silver
39. No
41. **(a)** 156.6 mS **(b)** 95.54 mS

Chapter 4

1. 1.915 A
3. 4 kΩ
5. 72 mV
7. 54.55 Ω
9. 25 Ω
11. 1.2 kΩ
13. **(a)** 12.632 Ω **(b)** 4.1×10^6 J
15. 16 s
17. 550 W
19. 96 W
21. 27 μW
23. 22.36 mA
25. 360 mW
27. 120 V, 32 Ω
29. 70.71 mA, 1.414 kV
31. **(a)** 4.095 W **(b)** 19.78 Ω **(c)** 88.45 kJ
33. **(a)** 864 J
35. 6.67 h
39. \$2.19
41. 82.89%
43. 84.77%
45. 14.03 A
47. 56.52 A
49. 38.4 J
51. 40% and 80%
53. **(a)** 1 Wh = 3600 J; 1 kWh = 3.6×10^6 J

Chapter 5

1. **(a)** E and R_1 **(b)** R_1 and R_2 **(c)** E and R_1 **(d)** E and R_2, R_3 and R_4
3. **(a)** 7.7 kΩ **(b)** 17.5 kΩ
5. **(a)** 62 Ω **(b)** 1.8 kΩ **(c)** 27 kΩ **(d)** R_1 = 8 kΩ; R_2 = 16 kΩ

 7. **(a)** 40 Ω **(b)** 3 A **(c)** $V_1 = 30$ V, $V_2 = 36$ V, $V_3 = 54$ V
 9. **(a)** 88 V **(b)** 20 V
11. **(a)** 8.18 mA, 18 V **(b)** 2.5 mA, 20 V **(c)** 9.935 μA, 99.35 V
13. **(a)** $R_T = 112\ \Omega$; $I_s = 214.29$ mA; $V_{R_1} = 4.714$ V; $V_{R_2} = 2.143$ V;
        $V_{R_3} = 10.072$ V; $V_{R_4} = 7.072$ V
    (b) $P_{R_1} = 1.01$ W; $P_{R_2} = 0.459$ W; $P_{R_3} = 2.158$ W; $P_{R_4} = 1.515$ W
    (c) $P_T = 5.142$ W **(d)** $P_s = 5.143$ W **(e)** the same **(f)** 47 Ω, the largest
    (g) dissipated **(h)** R_1, 2 W; R_2, 1/2 W; R_3, 5 W; R_4, 2 W
15. **(a)** 8/15 A **(b)** 8 W **(c)** 15 V **(d)** All go out!
17. **(a)** 0 V **(b)** 6 V **(c)** 14 V
19. **(a)** 10 V, 2 kΩ **(b)** 42 V, 1.5 kΩ
21. **(a)** 28 V, **(b)** 4 V
23. **(a)** $V_1 = 9$ V; $V_2 = 8$ V **(b)** $V_1 = 11$ V; $V_2 = 7$ V
25. **(a)** 8.2 kΩ **(b)** $V_3 : V_2 = 8.2 : 1$; $V_3 : V_2 = 82 : 1$ **(c)** 52.9 V **(d)** 59.36 V
27. **(a)** $V_1 = 60$ V; $V_2 = 40$ V; $E = 120$ V **(b)** $V_1 = 40$ V; $V_3 = 70$ V
    (c) $V_1 = 10$ V; $V_2 = 20$ V **(d)** $V_1 = 8$ V; $V_2 = 4$ V
29. **(a)** 1.6 kΩ **(b)** 1.5 Ω
31. **(a)** series resistor $= 80\ \Omega$ **(b)** 1/4 W
33. $V_{R_1} = 12$ V; $V_{R_2} = 42$ V; $V_{R_3} = 6$ V
35. **(a)** 2 Ω **(b)** 7.14%
37. **(a)** 1.2 mA **(b)** 1.171 mA **(c)** not for most applications

Chapter 6

 1. **(a)** R_2 and R_3 **(b)** E and R_3 **(c)** E and R_1 **(d)** R_2, R_3, and R_4 **(e)** E, R_1, R_2, R_3, and R_4
    (f) E, R_1, R_2, and R_3 **(g)** R_2 and R_3
 3. **(a)** 6.04 Ω **(b)** 545.55 Ω **(c)** 90.09 Ω **(d)** 5.99 kΩ **(e)** 2.62 Ω **(f)** 0.999 Ω
 5. **(a)** 8 Ω **(b)** 18 kΩ **(c)** 20 kΩ **(d)** 1.8 kΩ **(e)** $R_1 = R_2 = 6.4$ k Ω; $R_3 = 3.2$ kΩ
 7. **(a)** 1.6 Ω **(b)** $\infty\ \Omega$ **(c)** $\infty\ \Omega$ **(d)** 1.6 Ω
 9. **(a)** 2.119 Ω **(b)** 18 V **(c)** $I_s = 8.5$ A; $I_1 = 6$ A; $I_2 = 2$ A; $I_3 = 0.5$ A
11. **(a)** 1 kΩ **(b)** 1.003 kΩ **(c)** I_3, I_4, respectively **(d)** $I_1 = 4.4$ mA; $I_2 = 2$ mA; $I_3 = $
    36.67 mA; $I_4 = 0.786$ mA **(e)** 43.87 mA **(f)** equal
13. $I' = 12$ A; $I'' = 8$ A, 12 V
15. **(a)** $R_T = 867.86\ \Omega$; $I_1 = 100$ mA; $I_2 = 3.03$ mA; $I_3 = 12.195$ mA
    (b) $P_{R_1} = 10$ W; $P_{R_2} = 0.303$ W; $P_{R_3} = 1.22$ W **(c)** 11.52 W **(d)** the same **(e)** R_1
17. 1.26 kW
19. $I_1 = 4.1$ mA; $I_2 = 4.5$ mA
21. **(a)** $I_1 = 3$ mA; $I_2 = 1.5$ mA; $I_3 = 2$ mA; $I_4 = 5$ mA **(b)** $I_1 = 6$ μA; $I_2 = 4$ μA;
    $I_3 = 2$ μA; $I_4 = 5.5$ μA
23. **(a)** $R_1 = 5\ \Omega$; $R_2 = 10\ \Omega$; $I_2 = 1$ A **(b)** $I = 4.333$ A; $I_2 = 1.333$ A; $I_3 = 1$ A;
    $R_3 = 12\ \Omega$; $E = 12$ V **(c)** $I_1 = 64$ mA; $I_2 = 20$ mA; $I_3 = 16$ mA; $I = 36$ mA;
    $R = 3.2$ kΩ **(d)** $E = 30$ V; $I_1 = 1$ A; $I_3 = 0.5$ A; $R_2 = R_3 = 60\ \Omega$; $P_{R_2} = 15$ W
25. **(a)** $I_1 = 16$ mA; $I_2 = 4$ mA **(b)** $I_1 = 1.301$ mA; $I_2 = 2.385$ mA; $I_3 = 14.313$ mA;
    $I_4 = 18$ mA **(c)** $I_1 = 3.273$ A; $I_2 = 1.637$ A; $I_3 = 1.091$ A; $I_4 = 6$ A **(d)** $I_1 = I_2 = 6$ A;
    $I_3 = 3$ A; $I_4 = 9$ A
27. **(a)** $I_1 = 3$ A; $I_2 = 4$ A; $I = 4$ A **(b)** $I_1 = 2$ μA; $I_2 = 2$ μA; $I_3 = 6$ μA; $R = 9\ \Omega$
29. $R_1 = 2$ kΩ; $R_2 = 1$ kΩ; $R_3 = 0.5$ kΩ
31. $I_1 = 0.857$ A; $I_2 = I_3 = 1.714$ A
33. **(a)** 16.479 V **(b)** 16.474 V **(c)** 16.321 V **(d)** 13.333 V, 13.253 V, 11.428 V
35. not operating properly; 6 kΩ resistor not connected

Chapter 7

 1. **(a)** E and R_1 in series; R_2, R_3, and R_4 in parallel **(b)** E and R_1 in series; R_2, R_3, and R_4
    in parallel **(c)** R_1 and R_2 in series; E, R_3, and R_4 in parallel **(d)** E and R_1 in series;
    R_4 and R_5 in series; R_2 and R_3 in parallel **(e)** E and R_1 in series; R_2 and R_3 in parallel
    (f) E, R_1, and R_4 in parallel; R_6 and R_7 in series; R_2 and R_5 in parallel

3. **(a)** yes **(b)** 6 A **(c)** yes **(d)** 6 V **(e)** 3.733 Ω **(f)** 1 A **(g)** 20 W
5. **(a)** 4 Ω **(b)** $I_s = 9$ A; $I_1 = 6$ A; $I_2 = 3$ A **(c)** 6 V
7. **(a)** $I_s = 16$ mA; $I_2 = 2.333$ mA; $I_6 = 2$ mA **(b)** $V_1 = 28$ V; $V_5 = 7.2$ V
   (c) 261.33 mW
9. **(a)** 1.741 kΩ **(b)** 20.11 V **(c)** 11.89 V **(d)** 2.055 mA
11. **(a)** $I_2 = 1.667$ A; $I_6 = 1.111$ A; $I_8 = 0$ A **(b)** $V_4 = 10$ V; $V_8 = 0$ V
13. 16.667 Ω
15. **(a)** $I_1 = 10$ mA; $I_2 = 10.303$ mA **(b)** $V_a = 0$ V; $V_b = -8$ V; $V_{ab} = +8$ V
17. **(a)** $I_1 = 6$ A; $I_2 = 8$ A; $I_3 = 0.8$ A; $I_s = 14$ A **(b)** $V_a = -24$ V; $V_b = -8$ V; $V_{ab} = -16$ V
19. **(a)** 2.039 mA **(b)** 40 V
21. **(a)** 24 A **(b)** 8 A **(c)** $V_3 = 48$ V; $V_5 = 24$ V; $V_7 = 16$ V
23. 4.44 W
25. **(a)** 64 V **(b)** $R_{L_2} = 4$ kΩ; $R_{L_3} = 3$ kΩ **(c)** $R_1 = 0.5$ kΩ; $R_2 = 1.2$ kΩ; $R_3 = 2$ kΩ
27. **(a)** 200 Ω **(b)** 21.43 V **(c)** 23.72 V
29. **(a)** $V_{ab} = 32$ V; $V_{bc} = 8$ V **(b)** $V_{ab} = 31.51$; $V_{bc} = 8.49$ V **(c)** 16.015 W **(d)** 16 W
31. **(a)** 6.13 V **(b)** 9 V **(c)** 9 V
33. $I = 3$ A; $V = 12$ V
35. **(a)** 6.75 A **(b)** 32 V

Chapter 8

1. **(a)** $I_2 = I_3 = 10$ mA **(b)** 10 V **(c)** 37.6 V
3. 28 V
5. $V_3 = 1.6$ V; $I_2 = 0.1$ A
7. **(a)** $E = 4.5$ V; $R_s = 3$ Ω **(b)** $E = 28.2$ V; $R_s = 4.7$ kΩ
9. **(a)** $E = 8$ V; $R_s = 2$ Ω **(b)** 2.8 A **(c)** 2.8 A **(d)** -2.4 V
11. $V_2 = 9.6$ V; $I_1 = 2.4$ A
13. **(a)** $I_s = 5.455$ mA; $R_p = 2.2$ kΩ **(b)** 17.375 V **(c)** 5.375 V **(d)** 2.443 mA
15. $I_{R_1} (\downarrow) = 2.76$ A; $I_{R_2} (\uparrow) = 3.01$ A; $I_{R_3} (\downarrow) = 0.25$ A
17. $I_{R_1} (\rightarrow) = 2.032$ mA; $I_{R_2} (\leftarrow) = 0.8$ mA; $I_{R_3} (\uparrow) = 1.232$ mA
19. **(a)** $I_B = 63.02$ μA; $I_C = 4.416$ mA; $I_E = 4.479$ mA **(b)** $V_B = 2.985$ V;
    $V_E = 2.285$ V; $V_C = 10.285$ V **(c)** $\beta \cong 70.7$
21. $I_{R_1} (\rightarrow) = 2.032$ mA; $I_{R_2} (\uparrow) = 3.01$ A; $I_{R_3} (\downarrow) = 0.248$ A
23. $I_{R_1} (\rightarrow) = 2.034$ mA; $I_{R_2} (\leftarrow) = 0.802$ mA; $I_{R_3} (\uparrow) = 1.232$ mA
25. **(a)** All CW: $I_1 = 0.0321$ mA; $I_2 = -0.8838$ mA; $I_3 = -0.968$ mA; $I_4 = -0.639$ mA
    (b) All CW: $I_1 = -3.8$ A; $I_2 = -4.2$ A; $I_3 = 0.2$ A
27. $I_{R_1} (\uparrow) = -2.7$ A; $I_{R_2} (\uparrow) = 3.01$ A; $I_{R_3} (\downarrow) = 0.25$ A
29. $I_{R_1} (\rightarrow) = 2.032$ mA; $I_{R_2} (\leftarrow) = 0.8$ mA; $I_{R_3} (\uparrow) = 1.232$ mA
31. **(a)** All CW: $I_1 = 0.0321$ mA; $I_2 = -0.8838$ mA; $I_3 = -0.968$ mA; $I_4 = -0.639$ mA
    (b) All CW: $I_1 = -3.8$ A; $I_2 = -4.2$ A; $I_3 = 0.2$ A
33. $V_1 = -0.4$ V; $V_2 = -1.6$ V; $V_{R_1} = -0.4$ V; $V_{R_2} = 1.2$ V; $V_{R_3} = -1.6$ V
35. **(a)** $I_s = 3$ A; $R_p = 4$ Ω **(b)** $V_1 = -14.86$ V; $V_2 = -12.57$ V; $V_{R_1} = V_{R_4} = -14.86$ V;
    $V_{R_2} = -12.57$ V; $V_{R_3} = 9.71$ V
37. **(a)** $V_1 = -5.311$ V; $V_2 = -0.6219$ V; $V_3 = 3.751$ V **(b)** -5.311 V
39. $V_1 = 4.283$ V; $V_2 = -47.12$ V; $V_{R_1} = 4.283$ V; $V_{R_2} = 51.403$ V; $V_{R_3} = -47.12$ V
41. $V_1 = 4.8$ V; $V_2 = 6.4$ V; $V_{R_1} = 4.8$ V; $V_{R_2} = -1.6$ V; $V_{R_3} = V_{R_4} = 6.4$ V
43. $V_1 = -2.556$ V; $V_2 = 4.03$ V; $V_{R_1} = -2.556$ V; $V_{R_2} = V_{R_5} = 4.03$ V;
    $V_{R_4} = V_{R_3} = -6.586$ V
45. **(a)** $V_1 = -6.917$ V; $V_2 = 12$ V; $V_3 = 2.3$ V **(b)** 5 A: $V_s = -18.917$ V. 2 A: $V_s = 9.7$ V
47. $R_p = 100$ kΩ: $I_1 = 19.46$ mA (vs. 20 mA); $I_2 = 14.15$ mA (vs. 14.17 mA);
    $I_{R_1} (\downarrow) = 5.29$ mA (vs. 5.83 mA); $I_{R_2} (\downarrow) = 14.15$ mA (vs. 14.17 mA)
49. $R_p = 100$ Ω for both sources: $I_1 = 5.508$ A (vs. 5.53 A); $I_2 = 7.896$ A (vs. 8 A);
    $I_3 = 8.385$ A (vs. 8.53 A); $I_{4\Omega} (\uparrow) = 5.508$ A (vs. 5.53 A); $I_{6\Omega} (\leftarrow) = 2.388$ A
    (vs. 2.47 A); $I_{8\Omega} (\rightarrow) = 0.489$ A (vs. 0.53 A); $I_{1\Omega} (\downarrow) = 8.385$ A (vs. 8.53 A)
51. $R_s = 0.1$ Ω: $V_1 = -13.889$ V (vs. 14.015 V); $V_2 = 5.91$ V (vs. 6.006 V);
    $V_{4\Omega} = -13.889$ V (vs. 14.015 V); $V_{12\Omega} = 5.91$ V (vs. 6.006 V); $V_{10\Omega} = -19.799$ V
    (vs. 20.021 V)

53. $R_s = 1\ \Omega$: $V_1 = 46.38$ V (vs. 48 V); $V_2 = 67.76$ V (vs. 64 V); $V_{20\Omega} = 46.38$ V (vs. 48 V); $V_{40\Omega} = 67.76$ V (vs. 64 V)

RS TO SELECTED** **A-23**
ODD-NUMBERED
PROBLEMS

55. **(b)** 0.1967 V **(c)** no **(d)** no
57. 3.33 mA
59. 7.358 A
61. 2.33 A
63. **(b)** 5.714 mA

Chapter 9

1. 1.556 A
3. 3 A
5. 4.455 mA
7. **(a)** $R_{Th} = 12\ \Omega$; $E_{Th} = 6$ V **(b)** 0.75 A; 0.167 A; 0.057 A
9. **(a)** $R_{Th} = 10.667\ \Omega$; $E_{Th} = 13.333$ V **(b)** 4.162 W; 1.452 W
11. $R_{Th} = 2\ \Omega$; $E_{Th} = 60$ V
15. $R_{Th} = 10\ \Omega$; $E_{Th} = 2$ V
17. **(a)** $R_{Th} = 8.36$ kΩ; $E_{Th} = 3.28$ V **(b)** 4.444 mA **(c)** 42.82 μA **(d)** 10.223 V
19. **(a)** $R_{Th} = 3.2$ kΩ; $E_{Th} = 28$ V **(b)** 3.2 kΩ **(c)** 61.25 mW **(d)** 50% **(e)** 33.33%
21. 0 Ω
23. 500 Ω
25. **(a)** $R_N = 3$ kΩ; $I_N = 12$ mA **(b)** $E_{Th} = 36$ V; $R_{Th} = 3$ kΩ
27. $R_N = 1.58$ kΩ; $I_N = 0.727$ mA
29. **(a)** $R_N = 10\ \Omega$; $I_N = 0.2$ A **(b)** $E_{Th} = 2$ V; $R_{Th} = 10\ \Omega$

Chapter 10

1. 120 μF
3. 50 V/m
5. 8 kV/m
7. 375 pF
9. 3.54 μm
11. **(a)** 3.54 nF **(b)** 1 MV/m **(c)** 0.708 μC
13. 100 nF
15. 6.096 mm
17. 40 pF, 38 pF → 42 pF
19. 33,000 pF, 29,700 pF → 36,300 pF
21. **(a)** 5 s **(b)** $v_C = 20$ V $(1 - e^{-t/5s})$ **(c)** 1τ: 12.64 V. 3τ: 19 V. 5τ: 19.87 V
  (d) $i_C = 12\ \mu A e^{-5s}$; $v_R = 20$ V e^{-5s}
23. **(a)** 200 ms **(b)** 4 kΩ **(c)** 4.716 V **(d)** 11.999 V **(e)** 120 μC **(f)** 13.89 h
25. **(a)** 3 ms **(b)** $v_C = 30$ V $(1 - e^{-t/3ms})$; $v_R = 30$ V $e^{-t/3ms}$; $i_C = 1$ mA $e^{-t/3ms}$
27. **(a)** $v_C = 50$ V $(1 - e^{-t/10ms})$; $i_C = 10$ mA$e^{-t/10ms}$; $v_{R_1} = 30$ V $e^{-t/10ms}$
  (b) $v_C = 49.998$ V; $i_C = 0.454\ \mu$A; $v_{R_1} = 1.362$ mV **(c)** $v_C = 50$ V$e^{-t/4ms}$;
  $i_C = -25$mA $e^{-t/4ms}$; $v_{R_2} = -50$ V$e^{-t/4ms}$
29. **(a)** $v_C = 80$ V $(1 - e^{-t/1\ \mu s})$; $i_C = 0.8$ mA $e^{-t/1\mu s}$ **(b)** $v_C = 80$ V $e^{-t/4.9\mu s}$;
  $i_C = 163\ \mu$A $e^{-t/4.9\ \mu s}$
31. **(a)** $v_C = 50$ V $- 60$ V $e^{-t/20\ ms}$ **(b)** $i_C = 6$ mA$e^{-t/20ms}$
33. $v_C = -30$ V $+ 50$ V $e^{-t/3.44\mu s}$; $i_C = -58.14$ mA$e^{-t/3.44\mu s}$
35. **(a)** 4.722 V **(b)** 11.999 V **(c)** 13.86 μs **(d)** 127.94 μs
37. $R = 54.567$ kΩ
39. **(a)** 0.5 s: $v_C = 55.07$ V; $i_C = 4.93\ \mu$A; $v_{R_1} = 4.93$ V. 1 s: $v_C = 59.576$ V;
  $i_C = 0.404\ \mu$A; $v_{R_1} = 0.404$ V **(b)** $i_C = 8\ \mu$A: $t = 0.405$ s. $v_C = 10$ V: $t = 1.387$ s
41. **(a)** $v_C = 15$ V $(1 - e^{-t/0.15s})$; $i_C = 1.5$ mA$e^{-t/0.15s}$
43. **(a)** $v_C = 36$ V $- 40$ V $e^{-t/78ms}$; $i_C = 10.256$ mA$e^{-t/78\ ms}$
45. **(a)** $v_C = 20$ V $- 17$ V $e^{-t/265.2\ ms}$; $i_C = 2.5$ mA$e^{-t/265.2\ ms}$
47. $0 \to 4$ ms: 10 mA. $4 \to 6$ ms: 0 mA. $6 \to 7$ ms: 40 mA. $7 \to 9$ ms: 0 mA.
  $9 \to 11$ ms: -40 mA

49. $0 \to 2$ ms: 0 V. $2 \to 6$ ms: -16 V. $6 \to 16$ ms: $+20$ V. $16 \to 18$ ms: 0 V.
$18 \to 20$ ms: -12 V. $20 \to 25$ ms: 0 V
51. $4.364 \ \mu$F
53. $V_1 = 10$ V; $Q_1 = 60 \ \mu$C. $V_2 = 6.67$ V; $Q_2 = 40 \ \mu$C. $V_3 = 3.33$ V; $Q_3 = 40 \ \mu$C
55. $V_{C_1} = 48$ V; $Q_{C_1} = 1.92 \ \mu$C. $V_{C_2} = 32$ V; $Q_{C_2} = 2.56 \ \mu$C
57. 0.12 C
59. (a) 5 J (b) 0.1 C (c) 200 A (d) 10 kW (e) 10 s

Chapter 11

1. (a) 0.04 Wb/m^2 (b) 0.04 T (c) 88 At (d) 0.4×10^3 gauss
3. 15.65 μH
5. (a) 45 mH (b) 1.667 mH (c) 80 mH (d) 1875 mH
7. 4.25 V
9. 14 turns
11. 5 V
13. (a) 2.27 μs (b) $i_L = 5.45$ mA $(1 - e^{-t/2.27\mu s})$ (c) $v_L = 12$ V $e^{-t/2.27\mu s}$;
$v_R = 12$ V$(1 - e^{-t/2.27\mu s})$ (d) i_L: $1\tau = 3.45$ mA; $3\tau = 5.179$ mA; $5\tau = 5.413$ mA;
v_L: $1\tau = 4.415$ V; $3\tau = 0.598$ V; $5\tau = 0.081$ V
15. (a) $i_L = 9.23$ mA $- 1.23$ mA $e^{-t/30.77\mu s}$; $v_L = 4.8$ V $e^{-t/30.77\mu s}$
17. (a) $i_L = 1.765$ mA $- 4.765$ mA $e^{-t/588.2\mu s}$; $v_L = 16.2$ V $e^{-t/588.2\mu s}$
19. (a) $i_L = 6$ mA $(1 - e^{-t/0.5\mu s})$; $v_L = 12$ V $e^{-t/0.5\mu s}$ (b) $i_L = 5.188$ mA$e^{-t/83.3ns}$;
$v_L = -62.25$ V $e^{-t/83.3ns}$
21. (a) $i_L = 2$ mA $(1 - e^{-t/25\mu s})$; $v_L = 8$ V $e^{-t/25\mu s}$ (b) i_L: 1.264 mA. v_L: 2.943 V
23. (a) $i_L = 3.64$ mA $(1 - e^{-t/6.68\mu s})$; $v_L = 5.45$ V $e^{-t/6.68\mu s}$ (b) $i_L = 2.825$ mA;
$v_L = 1.219$ V (c) $i_L = 2.825$ mA $e^{-t/2.128\mu s}$; $v_L = -13.278$ V $e^{-t/2.128\mu s}$
25. (a) $i_L = 0.680$ mA $+ 1.320$ mA $e^{-t/484.9\mu s}$; $v_L = -5.44$ V $e^{-t/484.9\mu s}$
27. (a) 4.877 mA (b) 99.326 mA (c) 13.863 ms (d) 92.1 ms
29. 0–3 ms: 0 V. 3–8 ms: 1.6 V. 8–13 ms: -1.6 V. 13–14 ms: 0 V. 14–15 ms: 8 V.
15–16 ms: -8 V. 16–17 ms: 0 V
31. 0–5 μs: 4 mA. 5–10 μs: -12 mA. 10–12 μs: 12 mA. 12–16 μs: 4 mA.
16–24 μs: -4 mA
33. (a) $L_T = 16$ mH in series with $C_T = 18 \ \mu$F (b) $L_T = 25$ mH in series with $C_T = 18 \ \mu$F
35. (a) $i_L = 3.556$ mA $1 - e^{-t/8.333\mu s}$; $v_L = 12.8$ V $e^{-t/8.333\mu s}$
37. $I_1 = I_2 = 0$ A; $V_1 = V_2 = 60$ V
39. $I_1 = 2$ A; $I_2 = 1.333$ A; $V_1 = 10$ V; $V_2 = 4$ V

Chapter 12

1. (a) 20 mA (b) 15 ms: -20 mA. 20 ms: 0 mA. (c) 40 mA (d) 20 ms (e) 2.5 cycles
3. (a) 8 mV (b) 3 μs: -8 V. 9 μs: 0 mV (c) 16 V (d) 4.5 μs (e) 2.22 cycles
5. (a) 60 Hz (b) 100 Hz (c) 25 Hz (d) 40 kHz
7. 0.3 ms
9. (a) 150 mV (b) 40 μs (c) 25 kHz
11. (a) 45° (b) 30° (c) 18° (d) 108°
13. (a) 314.16 rad/s (b) 3769.91 rad/s (c) 12.56×10^3 rad/s (d) 25.12×10^3 rad/s
15. 2.08 ms
17. (a) 20, 60 Hz (b) 5, 120 Hz (c) 10^6, 1591.55 Hz (d) -6.4, 149.92 Hz
21. 0.476 A
23. 11.537°, 168.463°
27. (a) $v = 25 \sin(377t + 30°)$ (b) $i = 3 \times 10^{-3} \sin(6.28 \times 10^3 - 60°)$
29. v leads i by 10°.
31. i leads v by 80°.
33. i leads v by 190°.
35. 13.948 μs
37. 2 V
39. 3.866 mA
41. (a) 40 μs (b) 25 kHz (c) 17.125 mV

43. (a) $v = 14.14 \sin 377t$ (b) $i = 70.7 \times 10^{-3} \sin 377t$ (c) $v = 2.828 \times 10^3 \sin 377t$

45. 2.16 V

47. (a) 28.28 mV (b) 0.212 V

Chapter 13

1. (b) $v_T = 30 \sin \omega t$

3. (b) $v_T = 12.649 \sin(\omega t + 18.435°)$

7. (a) $5.196 + j\,3.0$ (b) $2.531 \times 10^3 + j\,6.954 \times 10^3$ (c) $396.107 \times 10^{-6} + j\,55.669 \times 10^{-6}$ (d) $8.561 \times 10^{-3} + j\,3.634 \times 10^{-3}$ (e) $-56.292 + j\,32.5$ (f) $5.177 \times 10^3 - j\,3.625 \times 10^3$ (g) $-4.313 - j\,6.160$ (h) $5.142 \times 10^{-3} - j\,6.128 \times 10^{-3}$

9. (a) $-12.0 + j\,34.0$ (b) $86.80 + j\,312.40$ (c) $8 \angle 82°$ (d) $49.68 \angle -64.0°$ (e) $-16; 740 \angle 160°$

11. (a) $10.0 - j\,5.0$ (b) $53.946 \times 10^{-3} + j\,15.028 \times 10^{-3}$ (c) $813.542 + j\,1.606 \times 10^3$ (d) $138.194 \times 10^{-3} + j\,5.058$ (e) $-144.372 + j\,400.79$

13. (a) $i = 56.569 \sin(377t + 20°)$ (b) $v = 169.68 \sin 377t$ (c) $i = 11.314 \times 10^{-3} \sin(377t + 120°)$ (d) $v = 6000 \sin(377t - 180°)$

15. $i_1 = 25.37 \times 10^{-6} \sin(\omega t + 96.79°)$

17. $i_s = 18 \times 10^{-3} \sin 377t$

19. (a) $v = 66 \sin 754t$ (b) $v = 4.4 \sin (400t - 120°)$

21. (a) $v = 19.8 \times 10^{-3} \sin(\omega t - 90°)$ (b) $v = 13.2 \sin(\omega t - 30°)$

23. (a) $i_R = 53.317 \times 10^{-3} \sin(\omega t + 80°)$

25. (a) 1.592 H (b) 2.654 H (c) 0.841 H

27. (a) $v = 100 \sin(\omega t + 90°)$ (b) $v = 8 \sin(\omega t + 150°)$

29. (a) $i = 1 \sin(\omega t - 90°)$ (b) $i = 0.6 \sin(\omega t - 70°)$

31. (a) $v = 96 \sin(\omega t - 120°)$ (b) $v = 48 \sin(\omega t + 190°)$

33. (a) $v = 0.5 \sin(1000t + 130°)$

35. (a) $\infty \ \Omega$ (b) 530.79 Ω (c) 17.693 Ω

37. (a) 4.66 Hz (b) 1.59 Hz

39. (a) $i = 6 \times 10^{-3} \sin(200t + 90°)$ (b) $i = 33.96 \times 10^{-3} \sin(377t + 90°)$

41. (a) $v = 1334 \sin(300t - 90°)$ (b) $v = 37.17 \sin(377t - 90°)$

43. (a) $L = 132.63$ mH (b) $C = 147.36 \ \mu F$ (c) $R = 7 \ \Omega$

45. 17.595 W

47. (a) 6.8 W (b) 6.8 mJ

49. (a) 0 W (b) 0.6 mVAR (c) 0 J

51. (a) 0 W (b) 0.48 mJ (c) 78.34 nJ

55. 318.47 mH

57. 5.067 nF

Chapter 14

1. (a) $6.8 \ \Omega + j\,6.8 \ \Omega = 9.167 \ \Omega \angle 45°$ (b) $10 \ \Omega - j\,6 \ \Omega = 11.66 \ \Omega \angle -30.96°$ (c) $5 \ k\Omega - j\,3 \ k\Omega = 5.83 \ k\Omega \angle -30.96°$

3. (a) $10 \ \Omega \angle 36.87°$ (c) $\mathbf{I} = 10$ A$\angle -36.87°$; $\mathbf{V}_R = 80$ V$\angle -36.87°$; $\mathbf{V}_L = 60$ V$\angle 53.13°$ (f) 800 W (g) $i = 14.14 \sin(377t - 36.87°)$; $v_R = 113.12 \sin(377t - 36.87°)$; $v_L = 84.84 \sin(377t + 53.13°)$

5. (a) $4.47 \ \Omega \angle -63.43°$ (c) 16 mH; 265 μF (d) $\mathbf{I} = 11.19$ A$\angle 63.43°$; $\mathbf{V}_R = 22.38$ V$\angle 63.43°$; $\mathbf{V}_L = 67.14$ V$\angle 153.43°$; $\mathbf{V}_C = 111.9$ V$\angle -26.57°$ (g) 250.43 W (h) $i = 15.82 \sin(377t + 63.43°)$; $e = 70.7 \sin 377t$; $v_R = 31.65 \sin(377t + 62.43°)$; $v_L = 94.94 \sin(377t + 153.43°)$; $v_C = 158.227 \sin(377t - 26.57°)$

7. (a) 40 mH (b) 220 Ω

9. (a) 2261.95 Ω (b) 2560.55 $\Omega \angle 62.05°$ (c) 23.432 mA $\angle -62.05°$ (d) 0.469 lagging (e) 659.38 mW (f) 1.242 VAR

11. (a) $\mathbf{V}_1 = 37.97$ V$\angle -51.57°$; $\mathbf{V}_2 = 113.92$ V$\angle 38.43°$ (b) $\mathbf{V}_1 = 55.80$ V$\angle 26.55°$; $\mathbf{V}_2 = 12.56$ V$\angle -63.45°$

13. (a) I = 39 mA $\angle 126.65°$; V_R = 1.17 V $\angle 126.65°$; V_C = 25.86 V $\angle 36.65°$
(b) 0.058 leading **(c)** 45.63 mW
(f) V_R = 1.17 V $\angle 126.65°$; V_C = 25.84 V $\angle 36.65°$
(g) $Z_T = R - jX_C = 30\ \Omega - j\,512.2\ \Omega$

15. R = 31.34 Ω; X_L = 22.10 Ω

19. (a) 5 Ω + j 5 Ω **(b)** $-j$ 240 Ω **(c)** 718.99 Ω $- j$ 449.45 Ω **(d)** 3.85 Ω + j 19.23 Ω
(e) 8.8 Ω $- j$ 4.4 Ω **(f)** 2.919 kΩ + j 486.44 Ω

21. (a) 1.859 Ω $\angle 21.8°$ **(c)** E = 3.71 V $\angle 21.8°$; I_R = 1.855 A $\angle 21.8°$;
I_L = 0.742 A $\angle -68.2°$ **(f)** 6.88 W **(g)** 0.928 lagging **(h)** e = 5.25 sin(377t + 21.8°);
i_R = 2.62 sin(377t + 21.8°); i_L = 1.049 sin(377t - 68.2°); i_s = 2.828 sin 377t

23. (a) 1.13 Ω $\angle 19.81°$ **(c)** 531 μF, 5.31 mH **(d)** E = 2.397 V $\angle 79.81°$;
I_R = 1.998 A $\angle 79.81°$; I_L = 1.199 A $\angle -10.19°$; I_C = 0.479 A $\angle 169.81°$
(f) 4.79 W **(g)** 0.941 lagging **(h)** e = 3.389 sin(377t + 79.81°);
i_R = 2.825 sin(377t + 79.81°); i_L = 1.695 sin(377t - 10.19°);
i_C = 0.677 sin(377t + 169.81°)

25. (a) I_1 = 18.09 A $\angle 65.241°$; I_2 = 8.528 A $\angle -24.759°$ **(b)** I_1 = 11.161 A $\angle 0.255°$;
I_2 = 6.656 A $\angle 153.69°$

31. (I): (a) v_1 leads v_2 by 72°. **(b)** $V_{1(p\text{-}p)}$ = 2.5 V; $V_{1(\text{rms})}$ = 0.884 V;
$V_{2(p\text{-}p)}$ = 1.2 V; $V_{2(\text{rms})}$ = 0.424 V **(c)** 1.25 kHz **(II): (a)** v_1 leads v_2 by 132°.
(b) $V_{1(p\text{-}p)}$ = 5.6 V; $V_{1(\text{rms})}$ = 1.98 V; $V_{2(p\text{-}p)}$ = 8 V; $V_{2(\text{rms})}$ = 2.828 V
(c) 16.67 kHz

33. (a) 5.154 mS $\angle -14.036°$ **(c)** I_R = 200 mA $\angle 0°$; I_2 = 100 mA $\angle -90°$;
I_C = 50 mA $\angle 90°$ **(d)** 206.16 mA $\angle -14.036°$ **(f)** 0.970 lagging **(g)** 8 W

Chapter 15

1. (a) 1.2 Ω $\angle 90°$ **(b)** 10 A $\angle -90°$ **(c)** 10 A $\angle -90°$ **(d)** I_2 = 6 A $\angle -90°$;
I_3 = 4 A $\angle -90°$ **(e)** 60 V $\angle 0°$

3. (a) Z_T = 3.87 Ω $\angle -11.817°$; Y_T = 0.258 S $\angle 11.817°$ **(b)** 15.504 A
(c) 3.985 A $\angle 82.826°$ **(d)** 47.809 V $\angle -7.174°$ **(e)** 910.71 W

5. (a) 24 A $\angle 6.87°$ **(b)** 260 V $\angle 60°$ **(c)** 224.32 V $\angle 74.81°$

7. (a) Z_T = 1.227 Ω $\angle 16.032°$; Y_T = 0.815 S $\angle -16.032°$
(b) I_1 = 26.834 A $\angle -26.565°$; I_2 = 20 A $\angle 0°$; I_3 = 3.354 A $\angle -26.565°$
(c) I_s = 48.9 A $\angle -16.032°$ **(d)** 0.961 (lagging)

9. 33.201 A $\angle 38.889°$

11. 139.71 mW

13. (a) R: 300 W. L: 0 W. C: 0 W **(b)** R: 0 VAR. L: 500 VAR. C: 900 VAR
(c) R: 300 VA. L: 500 VA. C: 900 VA **(d)** P_T = 300 W; Q_T = 400 VAR (L); S_T = 500 VA;
F_p = 0.6 lagging **(f)** W_R = 5 J **(g)** W_C = 1.327 J; W_L = 2.389 J

15. (a) R: 300 W. L: 0 W. C: 0 W **(b)** R: 0 VAR. L: 400 VAR. C: 250 VAR
(c) R: 300 VA. L: 400 VA. C: 250 VA
(d) P_T = 300 W; Q_T = 150 VAR (L); S_T = 335.41 VA; F_p = 0.894 (lagging)
(f) 6.71 A $\angle 33.38°$

17. (a) P_T = 900 W, Q_T = 0 VAR, S_T = 900 VA, F_p = 1 **(b)** 9 A $\angle 0°$ **(d)** 1: X_C = 20 Ω;
R = 0 Ω; X_L = 0 Ω. 2: R = 2.83 Ω; X_L = 0 Ω; X_C = 0 Ω. 3: R = 5.66 Ω;
X_L = 4.717 Ω; X_C = 0 Ω

19. (b) 438 μF **(c)** 14.43 A **(d)** 230 μF, 10.69 A

21. (b) 35.28 μF **(c)** 2.442 A

23. (a) 120 V, 5.5 A, 396 W **(b)** $Z_T = R + jX_L = 13.09\ \Omega + j\,17.46\ \Omega$

Chapter 16

1. (a) Z = 9.93 Ω $\angle 55.67°$; I = 2.014 A $\angle -35.67°$ **(b)** Z = 1.86 Ω $\angle 21.8°$;
I = 32.26 A $\angle 8.2°$

3. (a) 5.15 A $\angle -24.5°$ **(b)** 0.442 A $\angle 143.48°$

5. (a) 13.07 A $\angle -33.71°$ **(b)** 48.33 A $\angle -77.57°$

7. (a) V_1 = 14.68 V $\angle 68.89°$; V_2 = 12.97 V $\angle 155.88°$ **(b)** V_1 = 5.12 V $\angle -79.36°$;
V_2 = 2.71 V $\angle 39.96°$

9. $\mathbf{V}_1 = 11.74$ V $\angle -4.611°$; $\mathbf{V}_2 = 22.53$ V $\angle -36.48°$

11. $\mathbf{V}_1 = 5.839$ V $\angle 29.4°$; $\mathbf{V}_2 = 28.06$ V $\angle -89.15°$; $\mathbf{V}_3 = 31.96$ V $\angle -77.6°$

13. (a) $\mathbf{V}_1 = 5.74$ V $\angle 122.76°$; $\mathbf{V}_2 = 4.04$ V $\angle 145.03°$; $\mathbf{V}_3 = 25.94$ V $\angle 78.07°$
    (b) $\mathbf{V}_1 = 15.13$ V $\angle 1.29°$; $\mathbf{V}_2 = 17.24$ V $\angle 3.73°$; $\mathbf{V}_3 = 10.59$ V $\angle -0.11°$

15. (a) no (b) $\mathbf{I}_C = 1.76$ mA $\angle -71.54°$ (c) $\mathbf{V}_C = 7.03$ V $\angle -18.46°$

17. yes, balanced

19. (a) 12.02 A $\angle 38.16°$ (b) 7.02 A $\angle 20.56°$

21. (a) 6.095 A $\angle -32.115°$ (b) 3.77 A $\angle -93.8°$

23. $i = 0.5$ A $+ 1.581 \sin(\omega t - 26.565)$

25. 6.261 mA $\angle -63.43°$

27. (a) $\mathbf{Z}_{Th} = 21.312$ Ω $\angle 32.196°$; $\mathbf{E}_{Th} = 2.131$ V $\angle 32.196°$
    (b) $\mathbf{Z}_{Th} = 6.813$ Ω $\angle -54.228°$; $\mathbf{E}_{Th} = 57.954$ V $\angle 11.099°$

29. (a) $\mathbf{Z}_{Th} = 4$ Ω $\angle 90°$ (b) $i = 0.5$ A $+ 1.581 \sin(\omega t - 26.565°)$

31. (a) $\mathbf{Z}_{Th} = 4.472$ kΩ $\angle -26.565°$; $\mathbf{E}_{Th} = 31.31$ V $\angle -26.565°$
    (b) $\mathbf{I} = 6.26$ mA $\angle 63.435°$

33. (a) $\mathbf{Z}_N = 21.312$ Ω $\angle 32.196°$; $\mathbf{I}_N = 0.1$ A $\angle 0°$
    (b) $\mathbf{Z}_N = 6.813$ Ω $\angle -54.228°$; $\mathbf{I}_N = 8.506$ A $\angle 65.324°$

35. (a) $\mathbf{Z}_N = 9.66$ Ω $\angle 14.93°$; $\mathbf{I}_N = 2.15$ A $\angle -42.87°$
    (b) $\mathbf{Z}_N = 4.37$ Ω $\angle 55.67°$; $\mathbf{I}_N = 22.83$ A $\angle -34.65°$

37. (a) $\mathbf{Z}_N = 9$ Ω $\angle 0°$; $\mathbf{I}_N = 1.333$ A $+ 2.667$ A $\angle 0°$
    (b) $\mathbf{V}_C = 12$ V $+ 2.65$ V $\angle -83.66°$

39. (a) $\mathbf{Z}_L = 8.31$ Ω $- j\,0.462$ Ω; 1198.2 W (b) $\mathbf{Z}_L = 1.512$ Ω $- j\,0.39$ Ω; 1.614 W

41. (a) 1.414 kΩ (b) 0.518 W

Chapter 17

1. (a) $\omega_s = 250$ rad/s; $f_s = 39.79$ Hz (b) $\omega_s = 3535.53$ rad/s; $f_s = 562.7$ Hz
   (c) $\omega_s = 21,880$ rad/s; $f_s = 3482.31$ Hz

3. (a) 40 Ω (b) 10 mA (c) $V_R = 20$ mV; $V_L = 400$ mV; $V_C = 400$ mV (d) 20 (high Q)
   (e) $L = 1.27$ mH; $C = 0.796$ μF (f) 250 Hz (g) $f_1 = 4.875$ kHz; $f_2 = 5.125$ kHz

5. (a) 400 Hz (b) $f_1 = 5800$ Hz; $f_2 = 6200$ Hz (c) $X_L = X_C = 45$ Ω (d) 375 mW

7. (a) 10 (b) 20 Ω (c) $L = 1.59$ mH; $C = 3.98$ μF (d) $f_1 = 1900$ Hz; $f_2 = 2100$ Hz

9. $R = 10$ Ω; $L = 13.26$ mH; $C = 27.07$ nF; $f_1 = 8340$ Hz; $f_2 = 8460$ Hz

11. (a) 1 MHz (b) 160 kHz (c) $R = 720$ Ω; $L = 0.716$ mH; $C = 35.37$ pF (d) 56.25 Ω

13. (a) 159.155 kHz (b) 4 V (c) $I_L = I_C = 40$ mA (d) 20

15. (a) 104 Ω (b) 342.11 Ω (c) $\mathbf{I}_C = 16.45$ mA $\angle 90°$; $\mathbf{I}_L = 16.78$ mA $\angle -78.69°$
    (d) $L = 0.796$ mH; $C = 76.52$ nF (e) $Q_p = 3.29$; $BW = 6079.03$ Hz

17. (a) 102.69 kHz (b) $X_L = 51.62$ Ω; $X_C = 51.66$ Ω (c) 1.51 kΩ (d) $Q_p = 29.25$;
    $BW = 3.51$ kHz (e) $I_L = I_C = 292.14$ mA (f) 15.1 V

19. (a) 3558.81 Hz (b) 138.2 V (c) 691 mW (d) 575.86 Hz

21. $R_s = 3.244$ kΩ; $C = 31.66$ nF

23. (a) 251.65 kHz (b) 4.444 kΩ (c) 14.05 (d) 17.91 kHz (e) $f_p = 194.93$ kHz;
    $Z_{T_p} = 49.94$ Ω; $Q_p = 2.04$; $BW = 95.55$ kHz (f) $f_p = 251.65$ kHz; $Z_{T_p} = 13.33$ kΩ;
    $Q_p = 21.08$; $BW = 11.94$ kHz (g) Parts (a)–(d): $L/C = 100 \times 10^3$. Part (e): $L/C = 1 \times 10^3$. Part (f): $L/C = 400 \times 10^3$ (h) Yes, as L/C ratio increased, BW decreased.

25. (a) 5 (b) -4 (c) 8 (d) -6 (e) 1.301 (f) 3.937 (g) 4.748 (h) -0.498

27. (a) 11.513 (b) -9.21 (c) 2.996 (d) 9.065

29. -0.699 for each

31. 1.431 for each

33. 25.12 W

35. 20.792

37. 251.785 mV

39. $P_2 = 31.623 P_1$

41. (a) 6.318 mW (b) 0.317 mW

43. (a) 4.472 mV (b) 9.95 mV

45. (a) 67.726 kHz (b) $0.995 \cong 1$ (c) $0.0995 \cong 0.1$ (d) 6.77 MHz

47. (a) 0.707 (b) 0.970 (c) $0.995 \cong 1$ (d) 5 nW

49. (a) 79.577 kHz (b) $0.01 \cong 0$ (c) $0.995 \cong 1$ (d) 45.95 kHz

51. $R_1 = 1$ kΩ; $C_1 = 39.8$ nF; $R_2 = 20$ kΩ; $C_2 = 99.47$ pF
53. **(a)** $f_1 = 140.4$ kHz; $f_2 = 177.9$ kHz; $BW = 37.5$ kHz **(b)** $Q_p = 4.24$; $BW = 37.5$ kHz
55. **(a)** 2.78 **(c)** $Q_p = 2.78$; R_L little effect **(d)** Q_p increases; BW decreases
57. **(a)** $L_s = 12.68$ mH; $L_p = 128.19$ mH
59. **(a)** $f_c = 1989.44$ Hz **(e)** 0.841
61. **(a)** $f_c = 994.72$ Hz **(e)** 0.804

Chapter 18

1. **(a)** 200 V **(b)** 11.73 mWb
3. 20 V
5. 400 Ω
7. 12,000 turns
9. **(a)** 3 **(b)** 2.78 W
11. $E_p = 40$ V; $E_s = 160$ V; $I_p = 8$ A; $I_s = 2$ A
13. **(a)** $E_\phi = 120.1$ V **(b)** $V_\phi = 120.1$ V **(c)** $I_\phi = 6$ A **(d)** $I_L = 6$ A
15. **(a)** $\theta_2 = -120°$; $\theta_3 = 120°$ **(b)** $\mathbf{V}_{an} = 120$ V $\angle 0°$; $\mathbf{V}_{bn} = 120$ V $\angle -120°$;
$\mathbf{V}_{in} = 120$ V $\angle 120°$ **(c)** $\mathbf{I}_{an} = 25$ A $\angle 36.87°$; $\mathbf{I}_{bn} = 25$ A $\angle -83.13°$;
$\mathbf{I}_{cn} = 25$ A $\angle 156.87°$
17. **(a)** $\mathbf{E}_{AN} = 12.7$ kV $\angle -30°$; $\mathbf{E}_{BN} = 12.7$ kV $\angle -150°$; $\mathbf{E}_{CN} = 12.7$ kV $\angle 90°$
(b) and **(c)** $\mathbf{I}_{Aa} = \mathbf{I}_{an} = 11.285$ A $\angle -97.54°$; $\mathbf{I}_{Bb} = \mathbf{I}_{bn} = 11.285$ A $\angle -217.54°$;
$\mathbf{I}_{Cc} = \mathbf{I}_{cn} = 11.285$ A $\angle 22.46°$ **(d)** $\mathbf{V}_{an} = 12,154.28$ V $\angle -29.34°$;
$\mathbf{V}_{bn} = 12,154.28$ V $\angle -149.34°$; $\mathbf{V}_{cn} = 12,154.28$ V $\angle 90.66°$
19. **(a)** $E_\phi = 120.1$ V **(b)** $V_\phi = 208$ V **(c)** $I_\phi = 13.364$ A **(d)** $I_L = 23.15$ A
21. **(a)** $\theta_2 = -120°$; $\theta_3 = 120°$ **(b)** $\mathbf{V}_{ab} = 208$ V $\angle 0°$; $\mathbf{V}_{bc} = 208$ V $\angle -120°$
$\mathbf{V}_{ca} = 208$ V $\angle 120°$ **(d)** $\mathbf{I}_{ab} = 86.67$ A $\angle -36.87°$; $\mathbf{I}_{bc} = 86.67$ A $\angle -156.87°$;
$\mathbf{I}_{ca} = 86.67$ A $\angle 83.13°$ **(e)** $I_L = 150.11$ A **(f)** $E_\phi = 120.1$ V
23. **(a)** $\mathbf{I}_{ab} = 15.325$ A $\angle -73.30°$; $\mathbf{I}_{bc} = 15.325$ A $\angle -193.30°$; $\mathbf{I}_{ca} = 15.325$ A $\angle 46.7°$
(b) $\mathbf{I}_{Aa} = 26.54$ A $\angle -103.31°$; $\mathbf{I}_{Bb} = 26.54$ A $\angle 136.68°$; $\mathbf{I}_{Cc} = 26.54$ A $\angle 16.69°$
(c) $\mathbf{E}_{AB} = 17,013.6$ V $\angle -0.59°$; $\mathbf{E}_{BC} = 17,013.77$ V $\angle -120°59$;
$\mathbf{E}_{CA} = 17,013.87$ V $\angle 119.41°$
25. **(a)** $E_\phi = 208$ V **(b)** $V_\phi = 120.1$ V **(c)** $I_\phi \cong 10$ A **(d)** $I_L \cong 10$ A
27. $V_{an} = V_{bn} = V_{cn} = 69.28$ V; $I_{an} = I_{bn} = I_{cn} = 3.098$ A; $I_{Aa} = I_{Bb} = I_{Cc} = 3.098$ A
29. **(a)** $E_\phi = 440$ V **(b)** $V_\phi = 440$ V **(c)** $I_\phi = 29.33$ A **(d)** $I_L = 50.8$ A
31. **(a)** $\theta_2 = -120°$; $\theta_3 = 120°$ **(b)** $\mathbf{V}_{ab} = 100$ V $\angle 0°$; $\mathbf{V}_{bc} = 100$ V $\angle -120°$;
$\mathbf{V}_{ca} = 100$ V $\angle 120°$ **(d)** $\mathbf{I}_{ab} = 7.072$ A $\angle 45°$; $\mathbf{I}_{bc} = 7.072$ A $\angle -75°$;
$\mathbf{I}_{ca} = 7.072$ A $\angle 165°$ **(e)** $I_L = 12.27$ A
33. $P_T = 2160$ W; $Q_T = 0$ VAR; $S_T = 2160$ VA; $F_p = 1$
35. $P_T = 7.263$ kW; $Q_T = 7.263$ kVAR; $S_T = 10.272$ kVA; $F_p = 0.7071$ (lagging)
37. $P_T = 287.93$ W; $Q_T = 575.86$ VAR; $S_T = 643.83$ VA; $F_p = 0.4472$ (lagging)
39. $\mathbf{Z}_\phi = 12.98$ $\Omega - j\,17.31$ Ω
41. **(b)** $P_T = 2419.2$ W. Each wattmeter: 806.4 W
43. **(b)** $P_T = 285$ W **(c)** $P_T = 100$ W
45. **(a)** $\mathbf{I}_{ab} = 20.8$ A $\angle 0°$; $\mathbf{I}_{bc} = 14.708$ A $\angle -165°$; $\mathbf{I}_{ca} = 14.708$ A $\angle 165°$
(b) $\mathbf{I}_{Aa} = 35.213$ A $\angle -6.207°$; $\mathbf{I}_{Bb} = 35.213$ A $\angle -173.79°$; $\mathbf{I}_{Cc} = 7.614$ A $\angle 90°$
(c) $P_1 = 4336.81$ W; $P_2 = 4326.81$ W **(d)** $P_T = 8663.62$ W

INDEX